The chemistry of
carboxylic acids and esters

THE CHEMISTRY OF FUNCTIONAL GROUPS

A series of advanced treatises under the general editorship of
Professor Saul Patai

The chemistry of alkenes (published)
The chemistry of the carbonyl group (published)
The chemistry of the ether linkage (published)
The chemistry of the amino group (published)
The chemistry of the nitro and nitroso group (edited by Prosssor Henry Feuer)
(published)

—COOH(R)

The chemistry of
carboxylic acids and esters

Edited by

SAUL PATAI

The Hebrew University
Jerusalem, Israel

1969

INTERSCIENCE-PUBLISHERS

a division of John Wiley & Sons Ltd.

LONDON — NEW YORK — SYDNEY — TORONTO

First published by John Wiley & Sons Ltd. 1969

Library of Congress catalog card number 70–82547

SBN 471 66919 9

Printed in Hungary by Franklin Printing House

Contributing authors

L. D. Bergelson — Institute for Chemistry of Natural Products, Academy of Sciences of the U.S.S.R., Moscow, U.S.S.R.

Sergio Carrà — Institute of Chemical Physics, University of Milan, Italy.

Louis W. Clark — Western Carolina University, Cullowhee, North Carolina, U.S.A.

E. H. Cordes — Indiana University, Bloomington, Indiana, U.S.A.

Shawn Doonan — University College, London, England.

Lennart Eberson — University of Lund, Sweden.

Erkki K. Euranto — University of Turku, Finland.

Matthys J. Janssen — The University, Groningen, Netherlands.

Kenneth King — University of Delaware, Newark, Delaware, U.S.A.

G. Kohnstam — University of Durham, England.

Jouko Koskikallio — University of Oulu, Finland.

V. F. Kucherov — N. D. Zelinsky Institute of Organic Chemistry, Academy of Sciences of the U.S.S.R., Moscow, U.S.S.R.

Harold Kwart — University of Delaware, Newark, Delaware, U.S.A.

Sven-Olov Lawesson — University of Aarhus, Denmark.

T. S. Ma — City University of New York, U.S.A.

D. P. N. Satchell — King's College, London, England.

R. S. Satchell — Queen Elizabeth College, London, England.

Gustav Schroll — University of Aarhus, Denmark.

M. M. Shemyakin — Institute for Chemistry of Natural Products, Academy of Sciences of the U.S.S.R., Moscow, U.S.S.R.

Massimo Simonetta — Institute of Chemical Physics, University of Milan, Italy.

R. P. A. Sneeden — Monsanto Research S.A., Zürich, Switzerland.

D. L. H. Williams   University of Durham, England.

L. A. Yanovskaya   N. D. Zelinsky Institute of Organic Chemistry, Academy of Sciences of the U.S.S.R., Moscow, U.S.S.R.

Mieczyslaw Zielinski   University of Warsaw, Poland.

Foreword

The principles governing the plan and pattern of the present volume have been summarized in the Preface to the series 'The Chemistry of Functional Groups'.

Out of the originally planned contents, two chapters failed to materialize. These should have been chapters on 'Formation of COOH Groups' and 'Radiation and Photochemistry'.

SAUL PATAI

Jerusalem, September 1968

The Chemistry of the Functional Groups
Preface to the series

The series 'The Chemistry of the Functional Groups' is planned to cover in each volume all aspects of the chemistry of one of the important functional groups in organic chemistry. The emphasis is laid on the functional group treated and on the effects which it exerts on the chemical and physical properties, primarily in the immediate vicinity of the group in question, and secondarily on the behaviour of the whole molecule. For instance, the volume *The Chemistry of the Ether Linkage* deals with reactions in which the C—O—C group is involved, as well as with the effects of the C—O—C group on the reactions of alkyl or aryl groups connected to the ether oxygen. It is the purpose of the volume to give a complete coverage of all properties and reactions of ethers in as far as these depend on the presence of the ether group, but the primary subject matter is not the whole molecule, but the C—O—C functional group.

A further restriction in the treatment of the various functional groups in these volumes is that material included in easily and generally available secondary or tertiary sources, such as Chemical Reviews, Quarterly Reviews, Organic Reactions, various 'Advances' and 'Progress' series as well as textbooks (i.e. in books which are usually found in the chemical libraries of universities and research institutes) should not, as a rule, be repeated in detail, unless it is necessary for the balanced treatment of the subject. Therefore each of the authors is asked *not* to give an encyclopaedic coverage of his subject, but to concentrate on the most important recent developments and mainly on material that has not been adequately covered by reviews or other secondary sources by the time of writing of the chapter, and to adress himself to a reader who is assumed to be at a fairly advanced post-graduate level.

With these restrictions, it is realized that no plan can be devised for a volume that would give a *complete* coverage of the subject with *no* overlap between the chapters, while at the same time preserving the readability of the text. The Editor set himself the goal of attaining *reasonable* coverage with *moderate* overlap, with a minimum of cross-references between the chapters of each volume. In this manner, sufficient freedom is given to each author to produce readable quasimonographic chapters.

The general plan of each volume includes the following main sections:

(a) An introductory chapter dealing with the general and theoretical aspects of the group.

(b) One or more chapters dealing with the formation of the functional group in question, either from groups present in the molecule, or by introducing the new group directly or indirectly.

(c) Chapters describing the characterization and characteristics of the functional groups, i.e. a chapter dealing with qualitative and quantitative methods of determination including chemical and physical methods, ultraviolet, infrared, nuclear magnetic resonance, and mass spectra; a chapter dealing with activating and directive effects exerted by the group and/or a chapter on the basicity, acidity or complex-forming ability of the group (if applicable).

(d) Chapters on the reactions, tranformations and rearrangements which the functional group can undergo, either alone or in conjunction with other reagents.

(e) Special topics which do not fit any of the above sections, such as photochemistry, radiation chemistry, biochemical formations and reactions. Depending on the nature of each functional group treated, these special topics may include short monographs on related functional groups on which no separate volume is planned (e.g. a chapter on 'Thioketones' is included in the volume *The Chemistry of the Carbonyl Group*, and a chapter on 'Ketenes' is included in the volume *The Chemistry of Alkenes*). In other cases, certain compounds, though containing only the functional group of the title, may have special features so as to be best treated in a separate chapter as e.g. 'Polyethers' in *The Chemistry of the Ether Linkage*, or 'Tetraaminoethylenes' in *The Chemistry of the Amino Group*.

This plan entails that the breadth, depth and thought-provoking nature of each chapter will differ with the views and inclinations of the author and the presentation will necessarily be somewhat uneven. Moreover, a serious problem is caused by authors who deliver their manuscript late or not at all. In order to overcome this problem at least to some extent, it was decided to publish certain volumes in several parts, without giving consideration to the originally planned logical order of the chapters. If after the appearance of the originally planned parts of a volume, it is found that either owing to non-delivery of chapters, or to new developments in the subject, sufficient material has accumulated for publication of an additional part, this will be done as soon as possible.

It is hoped that future volumes in the series 'The Chemistry of the Functional Groups' will include the topics listed below:

The Chemistry of the Alkenes (*published*)
The Chemistry of the Carbonyl Group (*published*)
The Chemistry of the Ether Linkage (*published*)
The Chemistry of the Amino Group (*published*)
The Chemistry of the Nitro and Nitroso Group (*published*)
The Chemistry of Carboxylic Acids and Esters (*published*)
The Chemistry of the Carbon–Nitrogen Double Bond (*in press*)
The Chemistry of the Cyano Group (*in press*)
The Chemistry of the Amides (*in press*)
The Chemistry of the Carbon–Halogen Bond
The Chemistry of the Hydroxyl Group (*in preparation*)
The Chemistry of the Carbon–Carbon Triple Bond
The Chemistry of the Azido Group (*in preparation*)
The Chemistry of Imidoates and Amidines
The Chemistry if the Thiol Group
The Chemistry of the Hydrazo, Azo and Azoxy Groups
The Chemistry of Carbonyl Halides
The Chemistry of the SO, SO_2, —SO_2H and —SO_3H Groups
The Chemistry of the —OCN, —NCO and —SCN Groups
The Chemistry of the —PO_3H_2 and Related Groups

Advice or criticism regarding the plan and execution of this series will be welcomed by the Editor.

The publication of this series would never have started, let alone continued, without the support of many persons. First and foremost among these is Dr. Arnold Weissberger, whose reassurance and trust encouraged me to tackle this task, and who continues to help and advise me. The efficient and patient cooperation of several staff-members of the Publisher also rendered me invaluable aid (but unfortunately their code of ethics does not allow me to thank them by name). Many of my friends and colleagues in Jerusalem helped me in the solution of various major and minor matters and my thanks are due especially to Prof. Y. Liwschitz, Dr. Z. Rappoport

and Dr. J. Zabicky. Carrying out such a long-range project would be quite impossible without the non-professional but none the less essential participation and partnership of my wife.

The Hebrew University,   SAUL PATAI
Jerusalem, ISRAEL

Contents

CHAPTER 1

General and theoretical aspects of the COOH and COOR groups

MASSIMO SIMONETTA

and

SERGIO CARRÀ

Institute of Chemical Physics, University of Milan, Italy

I. INTRODUCTION

The carboxyl group is one of the most interesting entities met in organic molecules, especially from the physicochemical point of view. Some interesting features emerge even from the study of its geometry. There are two carbon–oxygen bonds, with different lengths. The value of the $O\overset{C}{}O$ angle and the distance to the carbon to which the group is directly bound have some significance. The group is planar and the hydrogen atom might be in the *cis* or the *trans* position with respect to the carbonyl bond. The presence of a carboxyl group makes a molecule an acid, and it is of interest to know how the nature of the fragment to which the carboxyl group is connected influences the acidic properties. The group can also participate in hydrogen bonding, acting as a hydrogen donor and/or acceptor.

Carboxylic acids show a tendency to dimerization through the formation of two hydrogen bonds. Cyclic dimers or polymolecular chains are often found in acid crystals. Dimerization equilibria in liquid, solution or gas phase have been the subject of thermodynamic investigations. The dimerization constants also show the influence of the radical to which the carboxyl group is bound, and relationships between acid dissociation and dimerization constants have been found.

The interactions of the group with electromagnetic fields show some very interesting features. In the range of ultraviolet radiation, light absorption by the group may occur through the $n-\pi^*$ or the $\pi-\pi^*$ mechanisms, and many examples of charge-transfer bands have been identified. When the group is bound to a residue able to undergo conjugation (*e.g.* phenyl), charge transfer can occur and show up in the appropriate band. Analysis of infrared spectra permits a study of the bonds present in the group, including hydrogen bonds, and the determination of the vibrational force constants.

By esterification, the carboxyl group can be converted to the —COOR group, and it is important to show how the different properties of the group are changed and how the nature of R influences these changes.

It seems to us useful to start the theoretical study of the carboxyl group with a review of some of the most important physicochemical properties. Instead of discussing many examples we prefer to discuss in some detail a few of the most fundamental ones. For this reason, in the experimental data quoted, only such examples are included which are interesting and necessary for the discussion.

At the end of the chapter, a theoretical discussion of the electronic properties, based on quantum mechanics, is given. In this field also the carboxyl group has many special characteristic features. It is a system for which the σ–π approximation may be appropriate, but the inclusion of the lone pair electrons does not involve insurmountable difficulties. The variation of the $\widehat{OCO}$ angles in different compounds can be related to the hybridization of the carbon atom. The problem of the presence of heteroatoms is encountered, with the two oxygens offering to the π-electron system one and two electrons respectively. It is worthwhile to notice how many of the problems of quantum chemistry show up in the discussion of such a simple system as the carboxyl group.

II. GEOMETRY, ENERGY AND POLARITY

A. Geometry and Structure

The geometry of the carboxyl group can be experimentally determined by means of x-rays, electron diffraction and microwave spectroscopy. The first method is used with acids or esters in the crystalline state, while the latter two methods are particularly useful for investigations in the gas or liquid phase. Interatomic distances and angles may be remarkably different in the various phases due to formation of dimers or polymers.

1. X-ray Methods

The crystal structures of a wide number of carboxylic acids and their esters have been investigated. For the determination of the geometry of the carboxyl group, the principal data are the $CO_{(1)}$, $CO_{(2)}$ distances and the $\widehat{O_{(1)}CO_{(2)}}$ angles (see 1).

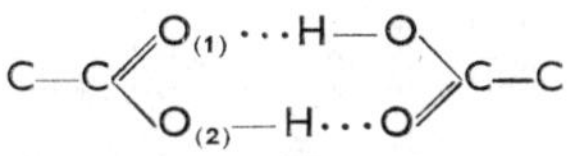

$$C-C\begin{matrix} O_{(1)} \cdots H-O \\ O_{(2)}-H \cdots O \end{matrix}C-C$$

(1)

The length of the adjacent carbon–carbon bond and the position of the hydroxylic hydrogen (when known) is also relevant.

A sample of values taken from the literature from the most recent sources is shown in Table 1. The range of the $C=O$ bond-length is $1\cdot187$–$1\cdot26$ Å; the average value is $1\cdot225$ over 27 measured groups. The corresponding values for the $C-O$ bond are $1\cdot257$–$1\cdot37$ and $1\cdot303$ Å as shown in Figure 1. In most cases one of the two carbon–oxygen bonds is considerably shorter. One remarkable exception is formic acid whose structure was determined

1*

TABLE 1. Bond-lengths and angles in carboxylic acids

Acid	C—O$_{(1)}$ (Å)	C—O$_{(2)}$ (Å)	OCO (degrees)	O—H...O (Å)	C—C (Å)	Reference
Formic	1·23	1·26	123	2·58	—	1
Acetic	1·24	1·29	122	2·61	1·54	2
Propionic	1·23	1·32	122	2·64	1·50	3
Butyric	1·22	1·35	123	2·62	1·54	4
Valeric	1·26	1·35	118	2·63	1·53	5
α-Oxalic	1·194	1·289	128·1	2·71	1·56	6
Oxalic dihydrate	1·187	1·285	125·48	—	1·529	7
Suberic	1·22	1·37	124·2	2·67	1·51	8
Adipic	1·231	1·291	122	2·64	1·500	9
3,4-Dimethyladipic	1·26	1·31	119	—	1·51	10
Sebacic	1·23	1·277	123	2·64	1·501	11
Dodecanedioic	1·244	1·294	122	2·65	1·497	12
Cyclooctane-1,2-trans-dicarboxylic	1·245	1·296	123·0	2·61	1·509	
	1·225	1·315	122·6	2·66	1·510	13
DL-3-bromo-octadecanoic	1·21	1·33	122	2·66	1·50	14
Cis-DL-8,9-methylene heptadecanoic	1·244	1·299	123·2	2·625	1·487	15
Acrylic	1·26	1·28	122	2·66	1·47	16
Maleic	1·20	1·275	125·5	2·46	1·44	17
α-Fumaric	1·224	1·293	124·3	2·684	1·465	18
β-Fumaric	1·228	1·289	124·4	2·673	1·49	19
Tartronic	1·208	1·305	126·2	2·66	1·531	
	1·212	1·303	125·9	2·88	1·539	20
Benzoic	1·24	1·29	122	2·64	1·48	21
Salicylic	1·234	1·307	121·2	2·620	1·457	22
Potassium acid phthalate	1·210	1·305	124·4	2·546	1·498	23
3-Thiophenic	1·235	1·332	123·3	2·66	1·474	24
3-Indolylacetic	1·223	1·298	123	2·665	1·495	25
N-acetylglycine	1·196	1·310	125·4	2·567	1·514	26
L-alanine	1·239	1·257	—	—	1·533	27

at $-50°$c: C—$O_{(1)}$ = $1·23\pm0·03$ Å and C—$O_{(2)}$ = $1·26\pm0·03$ Å The molecules are linked by hydrogen bonds at both ends to form infinite chains. The structure of liquid formic acid was also determined by means of x-rays[28]. The C—$O_{(1)}$ and C—$O_{(2)}$ lengths are similar, since the experimental electron radial distribution is consistent with the results obtained with a model in which both distances are taken as equal to $1·30$ Å. Liquid formic acid therefore consists of chain-like associates in which neighbouring molecules are connected by hydrogen bonds $2·7$ Å long.

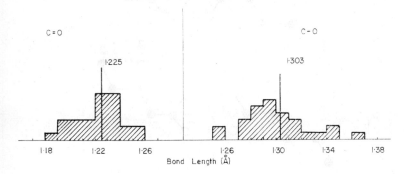

FIGURE 1. Occurrence of single and double CO bond lengths in carboxy groups (from Table 1).

Careful determinations of the crystal structure of normal fatty acids of low molecular weight (C_3 to C_5) have been recently accomplished. In all cases the molecules occur as hydrogen-bonded dimers, as shown in Figure 2.

For dicarboxylic acids, two crystalline forms may exist. In the β form the cyclic hydrogen-bonded system typical of monocarboxylic acids is present; the dicarboxylic acids are arranged in infinite chains. In the α form hydrogen bonds and carboxyl groups form an extended chain system in which the molecules are linked to form a puckered layer structure. Usually the β form is the most stable, but in oxalic acid α is the stable form[29]; its structure is shown schematically in Figure 3.

The crystal structure of oxalic acid dihydrate has also been determined as shown in Figure 4.

An interesting point arises in the examination of the structure of monoclinic and triclinic fumaric acid and of maleic acid. Single and double carbon–carbon bonds in maleic acid show very small differences as compared with corresponding ones in fumaric acid.

Maleic acid also shows a very short intermolecular hydrogen bond. It was suggested that the essential equivalence of single and double carbon–

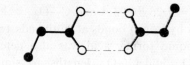

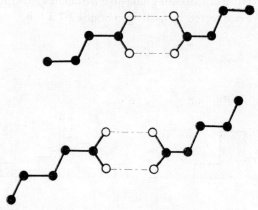

FIGURE 2. Dimers present in crystals of propionic, butyric and valeric acids;
○ : oxygen atom, ● : carbon atom.

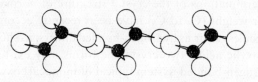

FIGURE 3. Crystal structure of α-oxalic acid; ○ : oxygen atom, ● : carbon atom.

TABLE 2. Comparison of single and double carbon–carbon bond-
lengths in fumaric and maleic acids

	Triclinic fumaric	Monoclinic fumaric	Maleic
C—C (Å)	1·490	1·460 1·462 1·473	1·44 1·465
C=C (Å)	1·315	1·361 1·334	1·43

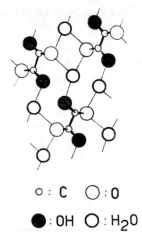

o : C　　◯ : O

⬤ : OH　◯ : H₂O

FIGURE 4. Crystal structure of oxalic acid dihydrate.

carbon bond-lengths in maleic acid might indicate the presence of aromatic character in this ring-like structure[19].

The crystal structure of benzoic acid has shown that the molecules occur in pairs connected by hydrogen bonds. The benzene rings are planar while the carboxylic carbon and singly bonded oxygen atoms are significantly out of molecular plane.

In ethyl stearate[30] the $C—O_{(1)}$ and $C—O_{(2)}$ distances are 1·36 Å and 1·15 Å respectively, a bit longer and shorter than the mean values of the same bonds in acids.

From the analysis of the values of the lengths of the carbon–carbon bond adjacent to the carboxyl group it seems that in non-conjugated molecules the measured value for such a distance is 1·516 Å, significantly shorter than the usual value for a carbon–carbon single bond. This may be due to the sp^2 hybridization of one of the carbon atoms involved in the bond. In conjugated molecules the mean value for the same distance is 1·474 Å, showing the adjunctive effect of conjugation. An interesting feature of the structural study of meso-3,4-dimethyladipic acid is the fact that the structure was determined theoretically by application of the close-packing principle[31].

2. Electron diffraction methods

The structure of the monomer of formic acid has been investigated[32] by electron diffraction. The equilibrium distances are $C—O_{(1)} = 1·23$ Å, $C—O_{(2)} = 1·36$ Å, $\widehat{O_{(1)}CO_{(2)}} = 122·4°$. No indication of the presence of a significant amount of dimer has been found under the conditions of this experiment.

The structures of methyl formate, acetate and chloroformate molecules in the gas phase were found to be as shown in Table 3[33]. All these molecules have an approximately planar skeleton, apart from the hydrogen atoms.

TABLE 3. Bond-lengths and angles in various methyl compounds

Compound	$C-O_{(1)}$ (Å)	$C-O_{(2)}$ (Å)	$\overset{\frown}{O_{(1)}CO_{(2)}}$ (degrees)
Methyl formate	1·22	1·37	123
Methyl acetate	1·22	1·36	124
Methylchloroformate	1·19	1·36	126

3. Microwave spectroscopy

The structure of formic acid has been the subject of extensive investigation by means of microwave spectroscopy[34, 35]. The results favour a planar molecule with the hydroxylic hydrogen *cis* to the doubly-bonded oxygen and the O—H bond undergoing a torsional oscillation around the carbon–oxygen bond. The *cis* isomer is probably stabilized by the formation of an internal hydrogen bond. The barrier height to this torsional oscillation has been estimated to be at least 17 kcal/mole. This suggests a large amount of double-bond character in the $C-O_{(2)}$ bond for which a length

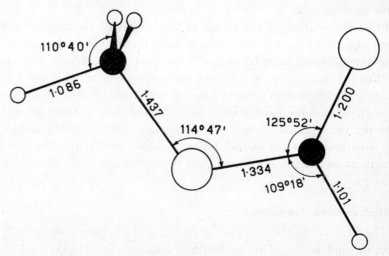

FIGURE 5. Bond lengths and bond angles of methyl formate determined by microwave spectroscopy, ●: carbon atom, ○: oxygen atom, ○: hydrogen atom.

of 1·343 Å was found. The values of the $C\!-\!O_{(1)}$ distance and the $O_{(1)}\overset{\frown}{C}O_{(2)}$ angle are 1·202 Å and 124°53′ respectively.

The structure of methyl formate has also been determined by the interpretation of its microwave spectrum[36]. The values of bond-lengths and angles are shown in Figure 5. The $C\!-\!O_{(1)}$ bond is found to have about 20% double-bond character but the $C\!-\!O_{(2)}$ bond shows no appreciable lengthening. The conformation of the methyl group was determined and it was found that the methyl hydrogens stagger the $C\!=\!O$ bond, that is, hydrogen appears to be repelled by oxygen. The barrier to internal rotation of the methyl group was estimated to be 1·19 kcal/mole. This is in good agreement with the more recent determination by far-infrared spectrum[37]. The question of why the *cis* conformer is more stable than the *trans* one has not yet been clearly settled.

B. Energy

1. Heats of formation and ionization potentials

From the measurement of the heats of combustion the heats of formation $(-\Delta H_f)$ of a number of monocarboxylic acids were calculated[38]. Starting from the fifth member of the series a linear relation was found between the heat of formation and the number of carbon atoms. The results fit the formula (1), where ΔH_f is in kcal/mole, n is the number of

$$-\Delta H_f = 109\cdot5 + 6\cdot1n - \Delta_n \tag{1}$$

CH_2 groups and Δ_n a correction factor for the first members of the series as shown in Table 4. Using the values calculated by equation (1) it is possible

TABLE 4. Values of the correction factor Δ_n for equation (1)

	$n = 1$	$n = 2$	$n = 3$	$n = 4$
Δ_n	7·8	−0·20	−0·40	0·20

to derive a mean value for the heat of formation of the carboxylic group: $-\Delta H_f[\text{COOH}] = 35\cdot71$ kcal/mole.

The heats of formation for a number of positive ions of carboxylic acids and their esters were determined by photoionization or electron impact experiments[39]. This method consists of the electron bombardment of the molecule X—Y of the gas under consideration, of measuring the *appearance potential* $A[X^+]$, that is, the minimum energy of the bombarding electrons at which the appropriate ions make their appearance, and of measuring the energy of these ions. The process is shown in equation (2):

$$X{-}Y+e \longrightarrow X^+ +Y+2e \qquad (2)$$

The heat of formation of the ion X^+ can be calculated by equation (3)

$$A[X^+] = \Delta H_f[X^+] + \Delta H_f[Y] - \Delta H_f[X-Y] + E_k \qquad (3)$$

where $\Delta H_f[\ \ ]$ is the enthalpy of formation of the appropriate species and E_k is the sum of the excitation energy and kinetic energy of fragmentation. Results obtained for ions produced from carboxylic acids and esters are collected in Table 5.

TABLE 5. Heats of formation of ions of carboxylic acids and esters

Parent molecule	$-\Delta H_f$ (eV)	Reference
HCOOH	−7·13	40
HCOOCH$_3$	−7·19	40
CH$_3$COOH	−5·80	40
CH$_2$=CHCOOH	−7·50	41
CH$_3$CH$_2$COOH	−5·49	40
HCOOC$_2$H$_5$	−6·76	40
CH$_3$COOCH$_3$	−6·39	40
CH$_3$CH$_2$CH$_2$COOH	−5·2	40
iso-C$_3$H$_7$COOH	−4·88	40
CH$_3$COOC$_2$H$_5$	−5·55	40
C$_2$H$_5$COOCH$_3$	−5·93	40
HCOOCH$_2$C$_2$H$_5$	−6·5	40

The heats of formation of acetate, propionate and butyrate radicals were determined by a different method based on the use of the heats of formation of the corresponding peroxides and the dissociation energies of the O—O bonds[42] (equation 4). The O—O bond dissociation energies (E_D)

$$2\Delta H_f[RC(O)O^\bullet] = E_D[RC(O)OOC(O)R] + \Delta H_f[RC(O)OOC(O)R] \qquad (4)$$

have been determined by the kinetic method, on the assumption that they are equal to the activation energies of the unimolecular fission of the peroxides into two acyl radicals. The following results were obtained for the radicals in gas phase:

$$\Delta H_f[CH_3COO^\bullet] = -45 \pm 2 \text{ kcal/mole}$$
$$\Delta H_f[C_2H_5COO^\bullet] = -54 \pm 2 \text{ kcal/mole}$$
$$\Delta H_f[C_3H_7COO^\bullet] = -60 \pm 2 \text{ kcal/mole}$$
$$\Delta H_f[PhCOO^\bullet] = -21 \quad \text{ kcal/mole}$$

From these data and from the known values of ΔH_f for the alkyl and phenyl radicals it is possible to calculate the heat involved in the decomposition of radicals according to equation (5).

$$RCOO^{\cdot} \longrightarrow R^{\cdot} + CO_2 \qquad (5)$$

It appears that reaction (5) is a strongly exothermic process ($\Delta H = -15$ kcal/mole) for the alkyl radicals, while it is almost thermally neutral for the phenyl derivative. This confirms the relative chemical stability of these radicals.

The heats of formation of the radicals can be used in conjunction with the heats of formation of the gaseous acids to calculate the dissociation energy of the acids:

$$RCOOH \text{ (gas)} \longrightarrow RCOO^{\cdot} \text{ (gas)} + H^{\cdot} \text{ (gas)} \qquad (6)$$

The results are:

$$E_D[R = CH_3] \quad = 111\cdot5 \ \text{kcal/mole}$$
$$E_D[R = C_2H_5] \quad = 109\cdot5 \ \text{kcal/mole}$$
$$E_D[R = C_3H_7) \quad = 102\cdot5 \ \ \text{kcal/mole}$$
$$E_D[R = Ph] \quad = 102 \quad \text{kcal/mole}$$

The interesting feature is that even in carboxylic acids the O—H bond dissociation energies are higher than the C—H bond dissociation energies. They are in fact of the same order as the O—H bond dissociation energy in alcohols (100 kcal/mole). The fact that in aqueous solution it is much easier to abstract a proton from carboxylic acids than from alcohols is due to the high value of the electron affinity and solvation energy of the carboxyl radical.

An alternative value for the heat of formation of the benzoyloxy radical was obtained from the difference between the activation energiesfor the decomposition reaction of this radical into the phenyl radical and CO_2, and the addition reaction of the benzoyloxy radical to the styrene molecule[43]. From these values a dissociation energy of 86 ± 3 kcal/mole was calculated for benzoic acid.

The kinetic method has also been used for the determination of the dissociation energy of the carbon–carbon bond adjacent to the carboxyl group in phenylacetic acid and diphenylacetic acid, yielding 55 kcal/mole and 52 kcal/mole respectively[44].

Standard enthalpies of formation in the gaseous state for a number of esters are available and reported in Table 6. From these data and the heat

TABLE 6. Heats of formation, $-\Delta H_f$ (kcal/mole), of esters

Ester	Formate[45]	Acetate[45]	Propionate[45]	Butyrate[45]	Benzoate[46]
Methyl	83·71	99·39	105·15	110·92	71·7
Ethyl	88·09	106·77	113·69	120·14	79·1
n-Propyl	—	112·99	118·53	125·91	—
n-Butyl	—	119·22	—	—	—
Isopropyl	—	115·99	—	128·21	—
Isobutyl	—	121·30	126·60	132·83	—
Isoamyl	—	127·52	—	—	—

of formation of acyl and alkyl radicals the enthalpy changes for reaction

$$R'COOR \longrightarrow R'CO^{\cdot} + RO^{\cdot} \qquad (7)$$

(7) have been calculated. The results give the C—O bond dissociation energy in organic esters. The most recent values are reported in Table 7.

TABLE 7. Values of dissociation energy of the C—O bond[46]

R'COOR	$E_D[R'COOR]$ (kcal/mole)
HCOOCH$_3$	94
CH$_3$COOCH$_3$	97
CH$_3$COOC$_2$H$_5$	96
CH$_3$COO-n-C$_3$H$_7$	98
CH$_3$COO-iso-C$_3$H$_7$	97
CH$_3$COO-n-C$_4$H$_9$	98
CH$_3$COO-iso-C$_4$H$_9$	99
PhCOOCH$_3$	90
PhCOOC$_2$H$_5$	88

In Table 8 the experimental values of ionization potential for a few acids and esters as obtained by electron impact, photoionization or spectroscopic experiments are collected.

From the available experimental data the thermodynamic properties of some acids were calculated. They are reported in Table 9.

TABLE 8. Ionization potentials for acids and esters

Compound	I (eV)	Method[a]	Reference
HCOOH	11·05	P	47
	11·33	S	48
	11·05	I	40
CH_3COOH	10·35	P	47
	10·37	I	40
	10·88	I	49
C_2H_5COOH	10·24	I	40
n-C_3H_7COOH	10·1	I	40
$CH_2{=}CHCOOH$	10·90	I	41
iso-C_3H_7COOH	9·98	I	40
PhCOOH	9·73	I	49
$HCOOCH_3$	11·14	I	50
	11·1	I	51
	10·815	P	40
$HCOOC_2H_5$	10·16	I	50
	10·50	I	51
	10·61	P	40
CH_3COOCH_3	10·58	I	50
	10·95	I	52
	10·51	I	53
	10·5	I	51
	10·27	P	40
$CH_3COOC_2H_5$	10·13	I	50
	9·97	I	41
	10·67	I	52
	10·11	P	40
	10·09	P	47
$NH_2C_6H_4COOCH_3$	8·08	I	49
$CH_3OC_6H_4COOCH_3$	8·43	I	49
$CH_3C_6H_4COOCH_3$	8·94	I	49
$PhCOOCH_3$	9·35	I	49
$NO_2C_6H_4COOCH_3$	10·20	I	49

[a] I = electron impact, P = photoionization, S = spectroscopy.

C. Dipole Moments

The dipole moments of a number of carboxylic acids and esters have been measured by the usual techniques, that is, from the dielectric constant in solution and by microwave spectroscopy. A sample of measured values taken from the literature is shown in Table 10. It is noticeable that the dipole moments of esters formed between saturated monohydroxylic alcohols and saturated monocarboxylic acids are approximately equal and lie in the range 1·7–2·0 Debye. This is explained by the fact that these

TABLE 9. Thermodynamic properties of some acids at 25°C[54]

Compound[a]	$-\Delta H_f^\circ$ (kcal/mole)	S° (cal/°K mole)	$-\Delta G_f^\circ$ (kcal/mole)
Formic (l)	101·52	30·82	86·39
Formic monomer (g)	90·49	59·45	83·89
Formic dimer (g)	195·12	82·89	171·19
Acetic (l)	115·7	38·2	93·1
Acetic monomer (g)	103·8	67·5	89·9
Acetic dimer (g)	223·0	96·7	183·7
Butyric (l)	127·2	54·1	89·9
Palmitic (s)	211·2	104·8	75·1
Lactic (s)	165·89	34·0	125·0
Lactic (l)	161·1	45·9	123·7
Benzoic (s)	91·812	40·04	59·11
o-Hydrobenzoic (s)	140·0	42·6	100·0
m-Hydrobenzoic (s)	141·1	42·3	101·0
p-Hydrobenzoic (s)	142·0	42·0	101·8
Oxalic (s)	196·7	28·7	165·9
Fumaric (s)	193·83	39·7	150·2
Maleic (s)	188·28	38·1	150·2
Succinic (s)	224·77	42·0	178·5
Phthalic (s)	186·88	49·7	61·30

[a] s = solid, l = liquid, g = gas.

esters exist in only one conformation *(cis)*, with a high energy barrier hindering rotation around the C—O bond[72].

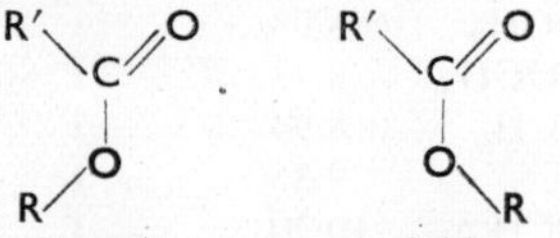

(trans)   (cis)
(2a)   (2b)

This fact was confirmed for methyl formate by the study of its microwave spectrum as discussed before.

D. Ionization and Dimerization Equilibria

1. Ionization

The equilibrium constants for the ionization of most carboxylic acids have been measured. The pK values at 25° for the reaction (8) of some

$$RCOOH + H_2O \rightleftharpoons RCOO^- + H_3O^+ \tag{8}$$

carboxylic acids are reported in Table 11.

TABLE 10. Electric dipole moments of carboxylic acids and esters

Compound	Dipole moment (Debye)	State or solvent	Temperature[a] (°c)	Reference
HCOOH	1·35	Gas	— (M)	55
	1·7	Gas	— (M)	56
	1·78	Benzene	30	54
	2·00	Dioxane	—	58
(HCOOH)$_2$	0	Gas	35–75	59
CH$_3$COOH	1·75	Gas	20	60
	1·92	Liquid	20	61
	2·17	Liquid	60	61
(CH$_3$COOH)$_2$	0·84	Benzene	20	62
	1·77	Dioxane	—	58
	1·64	Benzene	30	57
C$_6$H$_5$COOH	2·12	Benzene	15–45	63
	1·86	Dioxane	25	64
(C$_6$H$_5$COOH)$_2$	0·97	Benzene	15–45	63
HCOOCH$_3$	1·77	Gas	— (M)	36
CH$_3$CH$_2$COOH	1·76	Gas	—	60
	1·23	Liquid	25	61
HCOOC$_2$H$_5$	1·94	Gas	20–160	65
	1·96	Liquid	25	66
	1·96	Benzene	25–50	67
CH$_3$COOCH$_3$	1·706	Gas	34–110	68
	1·74	Liquid	40	69
	1·77	Benzene	25	70
CH$_3$CH$_2$CH$_2$COOH	1·23	Liquid	25	61
	1·9	Benzene	30	57
(CH$_3$)$_2$CHCOOH	1·09	Liquid	25	61
CH$_3$COOC$_2$H$_5$	1·78	Gas	30–195	65
	2·05	Liquid	20	71
	1·88	Benzene	50	66

[a] Except when indicated by M (microwave) the reported values were obtained from measurements of dielectric constants.

The higher acid strength of carboxylic acids as compared to that of alcohols may be understood, apart from solvation effects, in terms of π-electron delocalization. The resulting effect is the presence on the hydroxyl oxygen of a positive charge which repels the proton.

Acetic acid has a higher pK than formic acid, due to the electron-releasing character of the methyl group. This phenomenon is enhanced in trimethylacetic acid. Electron-withdrawing groups like halogens produce the opposite effect, that is, a lowering of pK. In the series of unbranched aliphatic carboxylic acids the pK values are almost constant.

TABLE 11. Thermodynamic data for dissociation of some
carboxylic acids

Acid	pK	$-\Delta H^a$ (kcal/mole)	Reference
Formic	3·752	−0·07	73
Acetic	4·756	0	74
Propionic	4·875	0·12	75
Butyric	4·818	0·61	76
Valeric	4·843	0·61	75
Isovaleric	4·781	1·11	75
Trimethylacetic	5·032	0·61	75
Iodoacetic	3·182	1·31	77
Bromoacetic	2·902	1·13	77
Chloroacetic	2·868	1·01	77
Fluoroacetic	2·586	1·28	77
Benzoic	4·213	−0·53	78
p-Bromobenzoic	4·002	−0·22	78
p-Chlorobenzoic	3·986	−0·34	78
m-Chlorobenzoic	3·827	0·07	78
m-Bromobenzoic	3·809	−0·05	78
p-Nitrobenzoic	3·442	−0·14	78

a Enthalpy variation for reaction (8).

The fact that benzoic is a stronger acid than acetic can be explained in terms of delocalization energy. The gain in delocalization energy on going from benzoic acid to the benzoate ion is larger than when going from acetic acid to the acetate ion. The effect of halosubstituents on benzoic acid is in the same direction as in acetic derivatives, but it is smaller in magnitude since it is counteracted by the delocalization of the π electrons of the halogen atoms.

The dissociation constants of carboxylic acids can be increased by intramolecular hydrogen bond formation. For this reason salicylic acid is a stronger acid than its *meta* and *para* analogues, and 2,6-dihydroxybenzoic acid is even stronger.

In dicarboxylic acids the second dissociation constant is smaller than the first one, but the effect decreases with increasing distance between the two carboxyl groups, as shown in Table 12. This effect can be easily explained on the basis of a simple electrostatic model since in the doubly-charged ion the two negative charges repel each other.

TABLE 12. The successive dissociation of dicarboxylic acids[79]

Acid	$n(CH_2)$	pK_1	pK_2	ΔpK
Oxalic	0	1·23	4·19	2·96
Malonic	1	2·83	5·69	2·86
Succinic	2	4·19	5·48	1·29
Adipic	4	4·42	5·41	0·99
Azelaic	7	4·55	5·41	0·89

2. Dimerization

The equilibrium constants of the dimerization reaction (9) have been

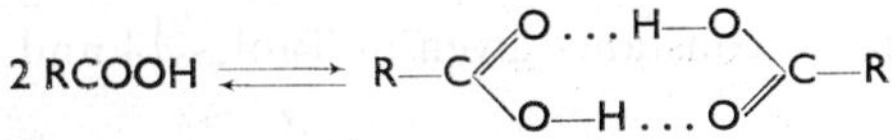

$$2\ RCOOH \rightleftharpoons R-C\overset{O\cdots H-O}{\underset{O-H\cdots O}{\Big\langle}}C-R \qquad (9)$$

determined in the pure acids and in their solutions.

The most widely used methods are the investigation of pressure–volume–temperature relations, density determinations and infrared spectroscopy for the gas phase. Cryoscopy, ebullioscopy, dielectric constant determinof tion, and infrared and ultraviolet spectra are the most useful methods fameasuring the dissociation constants in solution.

These values are of special interest for the derivation of the enthalpy or reaction which is a measure of the energy of the hydrogen bond. Values of equilibrium constants K and of $-\Delta H$ for a number of compounds are shown in Table 13. These data show that for the energy of the hydrogen

TABLE 13. Thermodynamic data for dimerization equilibria of some carboxylic acids

Acid	K	$-\Delta H$ (kcal/mole H bond)	Reference
Formic	260 (atm $^{-1}$, 300°K)	7·05	80
Acetic	539 (atm $^{-1}$, 300°K)	6·8	81
Propionic	1258 (atm $^{-1}$, 300°K)	7·6	80
Butyric	—	6·9	82
Trimethylacetic	1296 (atm $^{-1}$, 300°K)	7·0	83
Trifluoroacetic	177 (atm $^{-1}$, 300°K)	7·0	84
Acetic (liquid)	43·6 (m.f. $^{-1}$, 303°K)[a]	3·1	85
Propionic (liquid)	868 (m.f. $^{-1}$, 304°K)[a]	4·6	85
Stearic (liquid)	400 (m.f. $^{-1}$, 365°K)[a]	6·7	86
Benzoic (in benzene)	589 (m.f. $^{-1}$, 300°K)[a]	4·02	87

[a] m.f. = mole fraction.

bond in gaseous dimeric carboxylic acids 7 kcal/mole is the most useua value.

III. SPECTROSCOPIC PROPERTIES

A. Infrared

The internal force field of the carboxyl group has been investigated by infrared studies of carboxylic acids in monomeric and dimeric forms. The most detailed analyses have been carried out for formic and acetic acids. The infrared spectra of monomeric HCOOH, HCOOD, DCOOH and DCOOD have been studied in the gaseous phase[88] and in a solid nitrogen matrix[89].

The observed frequencies and the geometry as obtained by microwave spectroscopy[35] were used for a detailed analysis leading to the frequency assignments and force constants given in Tables 14 and 15. This calcula-

TABLE 14. Comparison of calculated and observed
frequencies of the formic acid monomer

Calculated[90] (cm^{-1})	Observed[88] (cm^{-1})	Assignment
3597	3570	ν(O—H)
2981	2943	ν(C—H)
1781	1770	ν(C=O)
1381	1387	δ(HC—O)
1254	1229	δ(COH)
1134	1105	ν(C—O)
624	625	δ(OCO)

TABLE 15. Force constants of formic acid[90]

Stretching (10^5 dyn/cm)	Bending (10^5 dyn/cm)	Repulsive (10^5 dyn/cm)
K (O—H) = 6·90	H (O—C=O) = 0·50	F (C...H) = 0·55
K (C—O) = 4·60	H (H—C—O) = 0·19	F (H...O$_2$) = 0·60
K (C=O) = 11·20	H (H—C=O) = 0·25	F (H...O$_1$) = 0·80
K (H—C) = 4·00	H (C—O—H) = 0·40	F (O$_1$...O$_2$) = 1·00

tion was performed for the seven in-plane vibrations, using symmetry coordinates related to the internal coordinates by the following equations:

$$S_1 = \Delta r, \quad \nu(O\text{—}H); \quad S_2 = \Delta d_1, \quad \nu(C\text{—}O)$$
$$S_3 = \Delta d_2, \quad \nu(C\text{=}O); \quad S_4 = \Delta D, \quad \nu(H\text{—}C)$$
$$S_5 = \Delta \delta, \quad \delta(C\text{—}O\text{—}H); S_6 = (1/\sqrt{6})(2\Delta\alpha - \Delta\beta - \Delta\gamma),$$
$$\delta(O\text{—}C\text{—}O) \tag{10}$$
$$S_7 = (1/\sqrt{2})(\Delta\beta - \Delta\gamma), \quad \delta(H\text{—}C\text{—}O)$$
$$S_8 = (1/\sqrt{3})(\Delta\alpha + \Delta\beta + \Delta\gamma) = 0 \quad \text{redundant}$$

Internal coordinates are shown in Figure 6.

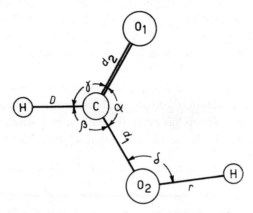

FIGURE 6. Internal coordinates for formic acid.

The normal coordinate analysis has been made in terms of the Urey–Bradley force field defined as follows:

$$V = \sum_i [K_i' r_{i0}(\Delta r_i) + 1/2 K_i(\Delta r_i)^2]$$
$$+ \sum_i [H_i' r_{i\alpha}^2(\Delta\alpha_i) + 1/2 H_i r_{i\alpha}^2(\Delta\alpha_i)^2]$$
$$+ \sum_i [F_i' q_{i0}(\Delta q_i) + 1/2 F_i(\Delta q_i)^2] \tag{11}$$

Δr_i, $\Delta\alpha_i$ and Δq_i are the changes of bond lengths, bond angles and all distances between non-bonded atoms. Symbols K_i, H_i and F_i represent the stretching, bending and repulsive force constants, respectively, and the prime indicates the interaction between non-bonded atoms. The values of the distances at equilibrium position, r_{i0}, $r_{i\alpha}$ and q_{i0} are introduced to make the force constants dimensionally consistent.

In Table 15 only the K_i, H_i and F_i force constants are reported because K_i' and H_i' can be expressed in terms of F_i' by the equilibrium conditions,

2*

while F_i' was assumed to be $(-F_i/10)$ as it follows from a repulsive energy between non-bonded atoms proportional to r^{-9}.

To get definitive assignments of low frequency bands, a calculation of the potential energy distribution[91] was necessary. It was shown that for the HCOOH molecule almost no vibrational coupling is present.

The observed frequencies for the two out-of-plane vibrations, that is, the C—H bending motion and the O—H torsional vibration, were 1041 cm^{-1} and 669 cm^{-1}, respectively[88]. A potential constant equal to 0.14×10^{-11} erg/rad^2 was obtained for the torsional vibration. For the experimental frequencies and the moments of inertia the standard entropy of HCOOH was evaluated ($S°_{298.16} = 59.43$ cal/°K mole). From thermal data a value of 59.44 was obtained[92]. A very satisfactory agreement between calculated and observed frequencies was also found for the isotopic species.

Accurate infrared spectra for acetic acid and its deuterated forms are available[93, 94]. A normal coordinate analysis similar to the previous one, in which the methyl group has been assumed to be a single atom, leads to the results collected in Table 16. The potential energy distribution calcu-

TABLE 16. Comparison of calculated and observed frequencies of acetic acid monomer

Calculated[90] (cm^{-1})	Observed[94] (cm^{-1})	Assignment
3597	3577	ν(OH)
1803	1799	ν(C=O)
1321	1279	ν(C—O) + ν(CH$_3$—C) + δ(C—O—H)
1175	1192	δ(C—O—H)
873	846	ν(CH$_3$—C)
667	654	δ(CH$_3$—C—O)
552	536	δ(O—C=O)

lation indicates that C—O stretching, CH$_3$—C stretching and C—O—H bending modes are coupled in the 1279 cm^{-1} band.

The infrared spectra of dimeric formic acid and its deutero-analogs have been measured by Millikan and Pitzer[95] and Miyazawa and Pitzer[96].

A normal coordinate analysis including all the atoms has been carried out for the 17 in-plane vibrations, using a modified Urey–Bradley force field[97]. The force constants reported in Table 17 were obtained. For the interaction between the two monomer units stretching–stretching and bending–bending force constants were considered. The origin of this vibra-

TABLE 17. Force constants of formic acid dimer[97] (10^5 dyn/cm)

Stretching	Bending	Repulsive
$K\,(O\!-\!H) = 4\cdot70$	$H\,(O\!-\!C\!\!=\!\!O) = 0\cdot45$	$F\,(C\ldots O_{(2)}\ldots H) = 0\cdot60$
$K\,(C\!-\!O) = 5\cdot50$	$H\,(H\!-\!C\!-\!O) = 0\cdot19$	$F\,(H\ldots C\ldots O_{(2)}) = 0\cdot60$
$K\,(C\!\!=\!\!O) = 10\cdot00$	$H\,(H\!-\!C\!\!=\!\!O) = 0\cdot25$	$F\,(H\ldots C\ldots O_{(1)}) = 0\cdot70$
$K\,(H\!-\!C) = 4\cdot00$	$H\,(C\!-\!O\!-\!H) = 0\cdot45$	$F\,(O_{(1)}\ldots C\ldots O_{(2)}) = 0\cdot80$
$K\,(O\ldots H) = 0\cdot36$	$H\,(C\!-\!O\ldots H) = 0\cdot01$	$F\,(C\ldots O_{(1)}\ldots H) = 0\cdot01$
	$H\,(O\!-\!H\ldots O) = 0\cdot015$	

tional interaction can be understood if we consider the tautomerism between the two structures as shown in (12). However, the frequency of this

$$H\!-\!C\!\!\underset{O\!-\!H\ldots O}{\overset{O\ldots H\!-\!O}{\diagup\!\!\diagdown}}\!\!C\!-\!H \;\;\rightleftharpoons\;\; H\!-\!C\!\!\underset{O\ldots H\!-\!O}{\overset{O\!-\!H\ldots O}{\diagup\!\!\diagdown}}\!\!C\!-\!H \tag{12}$$

tautomerism is such that the identities of C=O and C—O are maintained.

The comparison between the force constants of monomeric and dimeric formic acid shows that the C—H stretching force constant is practically unchanged. The C=O stretching force constant decreases upon formation of the hydrogen bond, indicating that the bond orders of C=O and C—O tend to average in the dimer. The O...H bond force constant is about 5% of the O—H stretching force constant of the monomer.

From the study of the out-of-plane vibrations[88] it was found that on dimerization there is an increase of 39% in the O—H torsional frequency, corresponding to an increase of 90% in the force constant. Since the value of the O—H torsional potential constant is $0\cdot21\times10^{-11}$ erg/rad^2, that is, only 50% larger than that of the monomer, the additional increase is due to the potential energy associated with hydrogen bonding.

The same kind of analysis has been carried out for acetic acid dimer[94]. The comparison of the force constants of monomer and dimer follows closely the trend found for formic acid. Whilst for formic acid dimer very little vibrational coupling was found, for acetic acid dimer strong vibrational coupling was found for at least eight vibrations.

The spectra of acetic, butyric, mono- and trichloroacetic and benzoic acids in the vapour phase were recorded and discussed[98]. The infrared spectra of sixty carboxylic acids in thin liquid or solid films, have been recorded[99], and the presence of five bands as tests to identify the carboxyl group has been suggested[99, 100]. The position of these bands is shown in Figure 7.

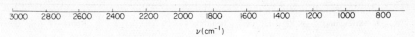

| 3000 | 2800 | 2600 | 2400 | 2200 | 2000 | 1800 | 1600 | 1400 | 1200 | 1000 | 800 |

$\nu(\text{cm}^{-1})$

FIGURE 7. Characteristic infrared frequencies of the carboxyl group.

For substituted benzoic acids, linear relationships have been established between the pK values and the O—H and C=O stretching frequencies[101].

B. Ultraviolet

In the ultraviolet spectra of saturated carboxylic acids and esters two bands have been clearly identified. The first band is due to a $n \rightarrow \pi^*$ transition, that is, a transition from an oxygen $2p$ lone-pair orbital to the carboxylic antibonding π^* orbital. It is found at about 210 mμ with a molar extinction coefficient $\varepsilon = 40$–60. The second band corresponds to a $\pi \rightarrow \pi^*$ transition; it is located at about 160 mμ with $\varepsilon = 2500$–4200.

The $n \rightarrow \pi^*$ transition in formic acid is blue-shifted with respect to the same transition in formaldehyde[102]. This effect is similar to the one observed in acetaldehyde and acetone but more pronounced. In both cases it is interpreted as due to a raising of the carbonyl π^* orbital owing to the interaction with the adjacent occupied π orbital.

The highest occupied methyl orbital is about 1 eV more stable than the non-bonding $2p$ orbital in oxygen, as shown by the difference of ionization potential for ethane (11·7 eV) and methanol (10·9 eV). This explains the fact that the shift is larger in formic acid. The same band is blue-shifted in the spectrum of the dimers of formic and acetic acids as compared with the corresponding monomers[102]. This can be explained through stabilization of the n orbital on hydrogen bond formation.

The effects of alkyl substitution and solvent polarity on the same electronic transition of carboxylate esters have been studied[103]. The effect of alkyl substitution in the acyl group on the spectra in different solvents is shown in Table 18.

TABLE 18. $n \rightarrow \pi^*$ Absorption maxima of methyl esters[103]

Ester	Isooctane		Ethanol	
	$\lambda_{\max}$ (mμ)	ε	$\lambda_{\max}$ (mμ)	ε
Formate	215·2	71	213·0	77
Acetate	209·7	57	207·2	57
Propionate	211·0	61	208·7	60
Isobutyrate	212·8	77	210·5	80
Pivalate	213·7	97	211·6	100

There is a considerable shift to shorter wavelengths on going from formate to acetate and the effect is probably due to hyperconjugation between the carbonyl groups and the alkyl groups. Further substitution on the methyl group shifts the transition energy back towards lower values, probably as a consequence of inductive effects, whilst the hyperconjugation effects remain roughly constant.

The absorption maxima of acetate and formate esters in the sequence methyl, ethyl, isopropyl, t-butyl, show a shift towards longer wavelengths. This may be due to steric effects.

The $n \rightarrow \pi^*$ transition energy E_T is sensitive to solvent polarity and a linear relationship has been established between this energy, and the empirical index of solvent polarity Z (equation 13)[104]. Values of the slope n are given in Table 19.

TABLE 19. Values of the slope m for
equation (13)

Ester	m
Methyl formate	4·25
Methyl acetate	4·79
Methyl propionate	4·37
Methyl isobutyrate	4·13
Methyl pivalate	3·27
Ethyl acetate	4·31
Isopropyl acetate	5·10
t-Butyl acetate	4·36

$$E_T = mZ + b \qquad (13)$$

The $\pi \rightarrow \pi^*$ bands for formic acid, acetic acid and ethyl acetate are shown in Figure 8[105].

These spectra were taken in the gas phase and the concentration of dimers was negligibly small. This band has been interpreted as the transition from the ground state to an excited state which results from a mixture of a locally-excited configuration and a charge-transfer configuration. The locally-excited configuration corresponds to the excited state of formaldehyde, whilst in the charge-transfer configuration an electron from the electron donor O—H group has been shifted to the electron acceptor C=O group.

As an example of conjugated acids we shall consider in some detail the u.v. spectra of benzoic acid and its derivatives. The absorption spectra of

benzoic acid and methyl benzoate in n-hexane solution are shown in
Figure 9[106]. Three bands appear at about 280, 230 and 200 mµ. The study
of the spectrum of benzoic acid is complicated by the fact that in non-
polar solvents it forms dimers. Evidence for this has been obtained by the
study of the effect of concentration and temperature on the band at circa
280 mµ[104, 108]. At relatively high concentrations $(10^{-3}-10^{-2}$ mole/l)
only dimeric molecules are present in solution, whose spectrum is shown
in Figure 9. On decreasing the concentration the band at 283·5 mµ is

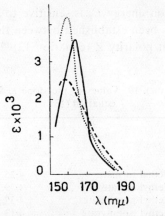

FIGURE 8. Vacuum ultraviolet absorption spectra of (1) (– – – – –), formic
acid (2) acetic acid (. . . . .), and (3) ethyl acetate (————).

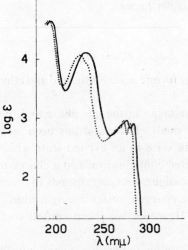

FIGURE 9. Ultraviolet absorption spectra in n-heptane of: (1) benzoic acid
(dimer) (————) and (2) methyl benzoate (. . . . .).

shifted towards shorter wavelengths. The same effect is observed when increasing the temperature. For instance with a concentration of 2×10^{-4} (mole/l) at $-180°$c the band of the dimer largely predominates, while at $90°$c only the monomer band appears, at $281 \cdot 5$ mµ. The absorption spectrum of methyl benzoate is not affected by changes in concentration.

When proton-accepting or -donating molecules are present in solution, spectral shifts are observed. For example at $-70°$c the addition of a small amount of ethyl ether weakens the monomer band, but the dimer bands are not affected. A new, blue-shifted band appears, which is ascribed to the species 3. The effect of a proton-donating molecule such as trichloroace-

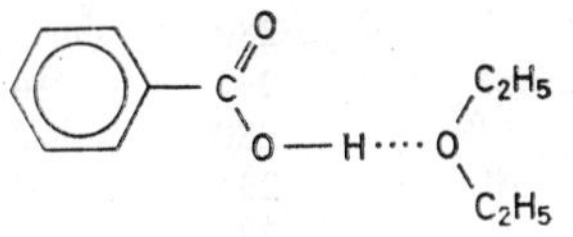

(3)

tic acid on methyl benzoate produces the appearance of new bands at longer wavelengths, probably due to the complex 4. A similar red shift is

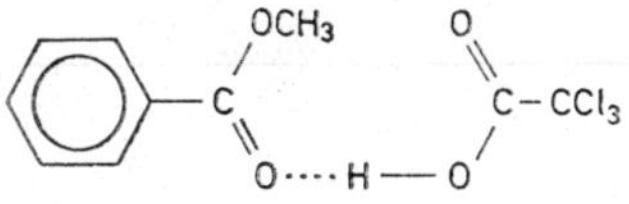

(4)

expected when benzoic acid is hydrogen-bonded at the oxygen atom of the carbonyl group by any proton-donating molecule. This effect also justifies the observed red shift on formation of the dimer.

The 230 mµ band of benzoic acid is also concentration-sensitive (Figure 10), and is also affected by variation of temperature or by addition of proton acceptors. This band has been interpreted as an intramolecular charge-transfer band, and is used for the spectroscopic determination of the monomer–dimer equilibrium[106].

The observed data for the first bands of some mono-substituted benzoic acids are collected in Table 20.

The charge-transfer band in mesitoic acid is very weak at 243 mµ, $\varepsilon = 4800$ in n-heptane (as compared to 12,800 for benzoic acid). This fact was attributed to steric hindrance, which decreases the conjugation between the benzene ring and the carboxyl group[111]. Comparing the spectra of the two acids in concentrated sulphuric acid with those in aqueous solution is interesting. The three bands in the spectrum of benzoic acid are shifted

TABLE 20. Ultraviolet absorption bands of some substituted benzoic acids

Substituent	Solvent	λ_{max} (mμ)	log ε	Reference
—	Ethanol	226	3·99	109
		272	2·93	
2-CH$_3$	Ethanol	228	3·81	109
		278	3·08	
3-CH$_3$	Ethanol	230	3·95	109
		280	3·06	
4-CH$_3$	Ethanol	235	4·15	109
		280	2·72	
4-C$_2$H$_5$	Ethanol	233	4·08	109
		280	2·76	
4-CH(CH$_3$)$_2$	Ethanol	235	4·11	109
		280	2·81	
4-C(CH$_3$)$_3$	Ethanol	235	4·11	109
		280	2·80	
2-NO$_2$	Water	270	3·72	110
3-NO$_2$	Water	214	4·38	110
		265	3·95	
4-NO$_2$	Water	278	3·92	110

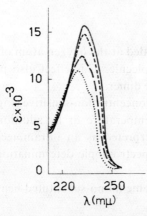
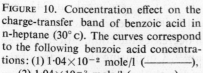

FIGURE 10. Concentration effect on the charge-transfer band of benzoic acid in n-heptane (30° c). The curves correspond to the following benzoic acid concentrations: (1) $1·04 \times 10^{-2}$ mole/l (————),
(2) $1·04 \times 10^{-3}$ mole/l (— — — —),
(3) $1·04 \times 10^{-4}$ mole/l (— — —) and
(4) $1·04 \times 10^{-5}$ mole/l (. . . .).

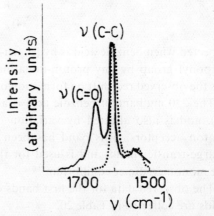

FIGURE 11. Raman spectra of benzoic acid, (1) in carbon tetrachloride solution (————) and (2) in 95% sulphuric acid solution (. . . .).

towards longer wavelengths. The greatest shift is for the charge-transfer band, from 228 mμ to 262 mμ, which has been shown to be the wavelength of the protonated benzoic acid. The structure of this cation has been investigated by measuring the Raman spectra of benzoic acid in carbon tetrachloride and in 95% sulphuric acid solution (Figure 11). The band observed at 1650 cm^{-1} in carbon tetrachloride, assigned to the C=O stretching vibration, is absent in sulphuric acid. This does not fit a protonated group like **5** and suggests for the protonated species the structure **6**. The

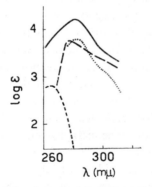

                    (5)   (6)

charge-transfer band of mesitoic acid appears very strongly at 282 mμ in concentrated sulphuric acid. To account for this, the presence of the acylium ion $(CH_3)_3C_6H_2CO^+$ has been suggested. The increased intensity is

FIGURE 12. Ultraviolet absorption spectra of (1) mesitoic acid in concentrated sulphuric acid (—), (2) mesitoic acid in trifluoroacetic anhydride (. . . .), (3) mesitoyl chloride in stannic chloride (— — —) and (4) mesitoic acid in aqueous solution (— — — —).

probably due to the coplanarity of the benzene ring and the substituent group, and the spectrum is similar to that obtained for the systems mesitoic acid–trifluoroacetic anhydride and mesitoyl chloride–stannic chloride (Figure 12). For these systems the following reactions have been demonstrated:

$$RCOOH + (CF_3CO)_2O \longrightarrow RCO^+ + CF_3CO_2^- + CF_3CO_2H$$
$$RCOCl + SnCl_4 \longrightarrow RCO^+ + SnCl_5^-$$

$$(14)$$

C. Nuclear Magnetic Resonance

Chemical shifts as measured by n.m.r. spectroscopy were found to be useful in the investigation of dissociation, chemical exchange and dimerization of carboxylic acids. In non-polar solvents, owing to the presence of stable dimers, the carboxylic proton frequencies are not concentration-dependent in concentrated solutions. The hydroxylic proton signal for aliphatic carboxylic acids in the pure liquid state appears at about $\tau = -2.2$[112, 113], and is unaffected by dilution with carbon tetrachloride down to 25 mole per cent. The proton magnetic resonance spectra of solutions of formic, acetic and benzoic acids in benzene have been studied in the temperature range 20–100°C. The observed shift of the carboxylic proton

$$\delta = (m/a)\,\delta_M + [(a-m)/a]\,\delta_D \tag{15}$$

is given by equation (15), where δ_M and δ_D are the monomer and dimer shift respectively, a is the total number of moles of acid and m the number of moles of the monomer form. From an analysis of the effect of concentration on the dimerization equilibrium, δ_M and δ_D were obtained[114]. The monomer shift is strongly temperature-dependent while the dimer shift is almost constant. This suggests a strong interaction of the monomer with the benzene solvent. In protic solvents the proton resonance is also concentration-dependent. Owing to the fast exchange of the proton among the solvent molecules, undissociated solute molecules and protonated ions, the proton resonance in such a solution is single. The position of the single line is an average of the resonance frequencies of the nuclei in the absence of chemical exchange. For acetic acid in aqueous solution there was, apart from the line for the proton in the methyl group, one concentration-dependent line due to the proton in the carboxylic acid group and the water[115]. Since dissociation of acetic acid was negligible in the experimental conditions the concentration dependence of the resonance will follow the law as shown in equation (16), where p_1 and p_2 are the proton fractions in the

$$\delta = p_1\delta_{HA} + p_2\delta_{H_2O} \tag{16}$$

acid and the water respectively. The linear dependence is clearly shown in Figure 13.

When using a polar aprotic solvent, n.m.r. measurements can be used to determine the acid dissociation constants. A correlation has been found between the chemical shifts in liquid sulphur dioxide and the corresponding dissociation constants in aqueous solutions[116].

The n.m.r. spectra of ^{17}O carboxylic acids and esters show the presence of a single absorption band[117, 118], suggesting that the two oxygen atoms are structurally equivalent, probably due to resonance between forms **7a**

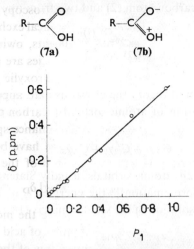

(7a)   (7b)

FIGURE 13. Concentration dependence of the chemical shift δ of proton magnetic resonance of acetic acid in aqueous solution; p_1 is the proton fraction in the acid.

and **7b**. Some characteristic ^{17}O chemical shifts are shown in Table 21.

TABLE 21. Some characteristic ^{17}O chemical shifts in carboxylic acids and esters

Range of chemical shifts (p.p.m. from water)	Compounds	Type of O atom
−130 to −150	Methyl esters	—O—
−160 to −180	Ethyl, propyl esters	—O—
−240 to −260	Acids	—O—
−350 to −370	Esters	O=

IV. THEORETICAL ASPECTS

A. Electronic Structure

1. Ground state

The electronic structure of the carboxyl group can be treated in a simplified manner by applying the $\sigma-\pi$ separation[119]. In this approximation the system of σ electrons is treated as a rigid non-polarizable core, building a field in which the motion of π electrons is described. For the carboxyl group, neglecting hyperconjugation effects, the problem is reduced to

a four electron and three orbital system. One electron comes from oxygen
atom (1), one from carbon atom (2) and two from oxygen atom (3).

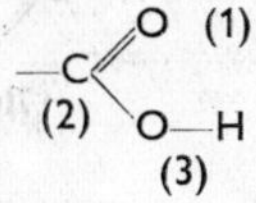

$$
\begin{array}{c}
\text{O (1)} \\
-\text{C} \\
\text{(2)} \quad \text{O—H} \\
\text{(3)}
\end{array}
\qquad (8)
$$

In the molecular orbital theory the electrons are assigned to orbitals built
as a linear combination of atomic orbitals (LCAO approximation) as in
equation (17)

$$\Psi_i = C_{i1}\chi_1 + C_{i2}\chi_2 + C_{i3}\chi_3 \qquad i = 1, 2 \tag{17}$$

where $\chi_{1,2,3}$ are the $2p_z$ atomic orbitals, usually Slater orbitals.

In the Hückel approximation[119] the energies ε_i of the molecular orbitals are obtained by solution of the secular matrix (18):

$$
\begin{vmatrix}
\alpha_1 - \varepsilon & \beta_{12} & 0 \\
\beta_{21} & \alpha_2 - \varepsilon & \beta_{23} \\
0 & \beta_{32} & \alpha_3 - \varepsilon
\end{vmatrix} = 0 \tag{18}
$$

where α_p are the Coulomb integrals:

$$\alpha_p = \int \chi_p \mathscr{H} \chi_p \, d\tau \tag{19}$$

and β_{pq} the resonance integrals:

$$\beta_{pq} = \int \chi_p \mathscr{H} \chi_q \, d\tau \tag{20}$$

$\mathscr{H}$ is an effective one-electron operator that contains some average of electron interaction terms and $d\tau$ is the element volume for one electron.

A precise evaluation of the values of the energies ε_i depends on the empirical choice of the parameters α_p and β_{pq}. Despite this difficulty it is
interesting to investigate the relative values of the molecular orbital
energy levels; a reasonable choice of empirical parameters is shown in
(21)[120].

$$\alpha_1 = \alpha_C + \beta$$

$$\alpha_3 = \alpha_C + 2\beta \tag{21}$$

$$\beta_{12} = \beta$$

$$\beta_{23} = 0.8\beta$$

The energy level diagram obtained is shown in Figure 14.

The four π electrons are placed in orbitals Ψ_1 and Ψ_2 with energies ε_1
and ε_2.

A more refined analysis can be performed applying the self-consistent molecular orbital method introduced by Roothaan[121], through which it is possible to take into account explicitly the electron interactions.

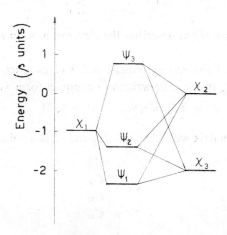

FIGURE 14. Hückel energy levels diagram of formic acid.

The wave-function for the ground state of the overall system of π electrons is constructed from normalized antisymmetrized product functions of the type (22) where

$$\Phi = 1/\sqrt{24} \begin{vmatrix} \Psi_1(1)\,\alpha(\omega_1) & \Psi_1(1)\,\beta(\omega_1) & \Psi_2(1)\,\alpha(\omega_1) & \Psi_2(1)\,\beta(\omega_1) \\ \Psi_1(2)\,\alpha(\omega_2) & \Psi_1(2)\,\beta(\omega_2) & \Psi_2(2)\,\alpha(\omega_2) & \Psi_2(2)\,\beta(\omega_2) \\ \Psi_1(3)\,\alpha(\omega_3) & \Psi_1(3)\,\beta(\omega_3) & \Psi_2(3)\,\alpha(\omega_3) & \Psi_2(3)\,\beta(\omega_3) \\ \Psi_1(4)\,\alpha(\omega_4) & \Psi_1(4)\,\beta(\omega_4) & \Psi_2(4)\,\alpha(\omega_4) & \Psi_2(4)\,\beta(\omega_4) \end{vmatrix} \quad (22)$$

the numbers 1, 2 . . . . . stand for the space coordinates of the electrons and ω is the spin coordinate. α and β are the usual spin functions and Ψ_t are the previously defined[88] molecular orbitals.

The coefficients C_{ip} are obtained by the variational procedure minimizing the energy:

$$E = \int \Phi^* \tilde{H} \Phi \, d\tau_1 \, d\tau_2 \, d\tau_3 \, d\tau_4 \tag{23}$$

subjected to the conditions:

$$\int \Psi_i^*(\mu)\,\Psi_j(\mu)\,d\tau_\mu = \sum_{pq} C_{ip} S_{pq} C_{jq} = \delta_{ij} \tag{24}$$

The integrals are extended over all the space of the electron coordinates; $d\tau_\mu$ is the volume element for electron μ. S_{pq} indicates the overlap integral:

$$S_{pq} = \int \chi_p^*(\mu)\,\chi_q(\mu)\,d\tau_\mu \tag{25}$$

and $\tilde{H}$ is the Hamiltonian operator:

$$\tilde{H} = \sum_{\mu=1}^{4} \tilde{H}_c(\mu) + 1/2 \sum_{\substack{\mu,\,\nu=1 \\ \mu \neq \nu}}^{4} e^2/r_{\mu\nu} \qquad (26)$$

where the operator $\tilde{H}_c(\mu)$ describes the electron motion in the core of the molecule.

Minimization of the energy (equation 23) gives a set of equations for the coefficients C_{ip} that can be written in compact form as follows:

$$\underline{F}\underline{C}_i = \varepsilon_i \underline{S}\underline{C}_i \qquad (i = 1, 2) \qquad (27)$$

$\underline{S}$ is the overlap matrix with elements S_{pq} and $\underline{C}_i$ is a column matrix:

$$\underline{C}_i = \begin{vmatrix} C_{1i} \\ C_{2i} \\ C_{3i} \\ C_{4i} \end{vmatrix} \qquad (28)$$

$\underline{F}$ is the Fock matrix with elements:

$$F_{pq} = H_{pq} + G_{pq} \qquad (29)$$

where

$$H_{pq} = \int \chi_p^*(\mu)[\tilde{T}(\mu) + U_c(\mu)]\chi_q(\mu)\,d\tau_\mu \qquad (30)$$

$$G_{pq} = \sum_{i=1}^{2} \sum_{r,\,t=1}^{3} C_{ri}^*[2(pq/rt) - (pt/rq)]C_{ti} \qquad (31)$$

In the equations (30) and (31) $\tilde{T}(\mu)$ is the kinetic energy operator, $U_c(\mu)$ is the core potential for electron μ and (pq/rt) are electron interaction integrals defined in equation (32):

$$(pq/rt) = \int \chi_p^*(\mu)\,\chi_q(\mu)\,e^2/r_{\mu\nu}\chi^*(\nu)\,\chi_t(\nu)\,d\tau_\mu\,d\tau_\nu \qquad (32)$$

They may be classified as follows:

Coulomb integrals if $p = q$ and $r = t$,

hybrid integrals if $p = q$ and $r \neq t$,

exchange integrals if $p \neq q$ and $r \neq t$.

ε_i are the diagonal elements of one hermitian diagonal matrix. These matrix elements are found by solution of the secular equation:

$$|F - \varepsilon S| = 0 \qquad (33)$$

Once the ε_i's are known they are introduced into equation (27) and the coefficients C_{ip} can be calculated. However the matrix elements F_{pq} contain the coefficients C_{ip}, so the problem must be solved by an iterative procedure.

One must start with arbitrary $\underline{C_i}$ matrices, subject to equation (24). The $\overline{F}$ matrix can then be calculated and used in the solution of equation (27). A new set of coefficients C_{ip} is obtained and used in the next cycle. Calculations are carried on until consistent results are obtained.

In an application of the Roothaan method to formic acid[122] the geometry of the molecule reported by Lerner, Dailey and Friend[34] was assumed and the two hydrogen atoms were neglected throughout the calculation.

Theoretical values were used for overlap integrals[123] and Coulomb integrals[124, 125]. Approximate values for hybrid and exchange integrals were obtained from overlap and Coulomb integrals through the Mulliken approximation (equation 34)[126].

$$\chi_p(\mu)\,\chi_q(\mu) \simeq (S_{pq}/2)\left[\chi_p^2(\mu) + \chi_q^2(\mu)\right] \tag{34}$$

The integrals H_{pq} express the potential energy of electrons in the *core* of charged nuclei, *i.e.* inner shell and σ electrons. According to Goeppert–Mayer and Sklar[127], they can be evaluated by means of the expression (35),

$$H_{pq} = S_{pq}W_p - \sum_{r\neq p} \int \chi_p^*(\mu)U_r^{z^+}(\mu)\,\chi_q(\mu)\,\mathrm{d}\tau_\mu \tag{35}$$

where $U_r^{z^+}$ is the potential energy of one electron in the field of atom r with charge $z^+ \cdot W_p$ is the ionization potential of atom p in its appropriate valence state.

The following values were used for ionization potentials:

$$\begin{array}{l}
C(tr^3, p_z, V_4) \longrightarrow C^+(tr^3, V_3) - 11\cdot54 \text{ eV}\\
O(s^2p_x^2p_yp_z, V_2) \longrightarrow O^+(s^2p_x^2p_y, V_1) - 17\cdot28 \text{ eV}\\
O^+(s^2p_xp_yp_z, V_3) \longrightarrow O^{2+}(s^2p_xp_y, V_2) - 34\cdot14 \text{ eV}
\end{array} \tag{36}$$

Penetration integrals appearing in equation (35) were calculated with formulae given in the literature[128]. The initial set of molecular orbitals was obtained through a Hückel calculation with inclusion of overlap integrals in the secular equation. The final SCF orbitals are:

$$\begin{array}{l}
\Psi_1 = 0\cdot355129\chi_1 + 0\cdot55282\chi_2 + 0\cdot61723\chi_3\\
\Psi_2 = 0\cdot67665\chi_1 + 0\cdot23897\chi_2 - 0\cdot69102\chi_3\\
\Psi_3 = 0\cdot68007\chi_1 - 0\cdot84146\chi_2 + 0\cdot4100\chi_3
\end{array} \tag{37}$$

and the corresponding energies are:

$$\varepsilon_1 = -15\cdot911 \qquad \varepsilon_2 = -9\cdot092 \qquad \varepsilon_3 = 8\cdot861$$

The ionization potential, taken as equal to the energy of the highest occupied orbital according to Koopman's theorem, is in reasonable agreement with the experimental value.

The charge density distribution in the molecule is, according to Mulliken[129], expressed by the gross population, given by:

$$q_p = v_i \sum_i C_{pi}^2 + 2v_i \sum_i \sum_r (C_{pi}C_{ri}S_{pr}) \tag{38}$$

where r indicates the atoms bonded to atom p and v_i is the number of electrons in the i-th molecular orbital. For formic acid the following values are obtained:

$$q_1 = 1\cdot304; \quad q_2 = 0\cdot926; \quad q_3 = 1\cdot770$$

The bond orders defined from the relation (39)[120]:

$$p_{pq} = \sum_i v_i C_{pi} C_{qi} \tag{39}$$

were found to be:

$$p_{C-O} = 0\cdot35; \quad p_{C=O} = 0\cdot71$$

Again, this electronic distribution can be interpreted as due to resonance between the structures **9a** and **9b**.

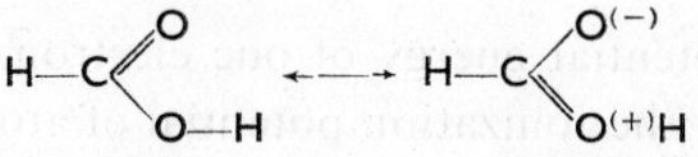

$$
\begin{array}{ccc}
\text{H—C} \displaystyle{\begin{matrix} \text{O} \\[-2pt] \diagdown \\[-2pt] \text{O—H} \end{matrix}} & \longleftrightarrow & \text{H—C} \displaystyle{\begin{matrix} \text{O}^{(-)} \\[-2pt] \diagdown \\[-2pt] \text{O}^{(+)}\text{H} \end{matrix}} \\
\textbf{(9a)} & & \textbf{(9b)}
\end{array}
\tag{40}
$$

Bond orders are in good agreement with values deduced from interatomic distances[34].

A modified Hückel treatment has been used to investigate the electronic structure of the —COOH group[131]. Bond orders were calculated and correlated with bond-lengths using order–length relations characterized by a σ-skeleton parameter $\gamma = \sqrt{S_A S_B}$ where S_A and S_B are the s characters of the two orbitals used in forming the σ bond and are obtained from experimental bond angles[132].

π-electron energies in HCO_2^+, HCO_2 and HCO_2^- were calculated[133] by different methods including molecular orbital, valence bond, configurations interaction and the so-called non-pairing spin orbital method, using different orbitals for different spins. The aim of this work was mainly to test the different approximations in their capability of reproducing the effect of electron correlation.

The modification of LCAO–MO theory known as the ω-technique[134] was used to calculate the ionization potential in many benzene derivatives including methyl benzoates[135]. The standard Hückel technique has been modified as follows.

The Coulomb integrals are given by:

$$\alpha_r^1 = \alpha_0 + h_r\beta_0 + 0.1 \sum_s h_s\beta_0 \qquad (41)$$

and the resonance integrals by:

$$\beta_{rs} = k_{rs}\beta_0 \qquad (42)$$

α_0 and β_0 are the Coulomb and resonance integrals for benzenic carbons ($\alpha_0 = -9.878$ eV, $\beta_0 = -2.11$ eV), h_r is a constant dependent on the nature of the valence state of the $r's$ atoms and the sum is taken on the next neighbours and k_{rs} is an empirical parameter. The values of this parameter are shown in Table 22. Taking $x = (\alpha_0 - \varepsilon)/\beta_0$ the secular matrix

TABLE 22. Comparison between the calculated and experimental values of the ionization potentials of benzoic ester derivatives

Substituent	q_r^0	k_{rs}	Calculated ionization potential[135] (eV)	Experimental ionization potential[49] (eV)
NH$_2$	2	0.85	8.31	8.08
OH	2	0.65	8.99	8.76
OCH$_3$	2	0.5	8.64	8.43
CH$_3$	2	0.75	8.97	8.94
H	0	1.0	9.09	9.35
Br	2	0.25	9.12	9.40
Cl	2	0.3	9.15	9.45
CN	2	0.91	9.62	9.24

can be written as follows:

$$\begin{vmatrix} \left(\dfrac{\alpha_1 - \alpha_0}{\beta_0} + x\right) & k_{12} & k_{13} \cdots\cdots k_{1n} \\ k_{21} & \left(\dfrac{\alpha_2 - \alpha_0}{\beta_0} + x\right) & k_{23} \cdots\cdots k_{2n} \\ \cdots\cdots\cdots\cdots & & \cdots\cdots\cdots\cdots \\ k_{n1} & k_{n2} \cdots\cdots\cdots & \left(\dfrac{\alpha_n - \alpha_0}{\beta_0} + x\right) \end{vmatrix} = 0 \quad (43)$$

After solution of the secular equation charge densities were calculated using the expression:

$$q_r^1 = \sum_i v_i(C_{ri}^1)^2 \qquad (44)$$

where the sum is over the occupied levels and v_i is the number of electrons in the i-th molecular orbital.

Then the Coulomb integrals are modified as follows:

$$\alpha_r^2 = \alpha_r^1 + \omega(q_r^1 - q_r^0)\beta_0 \tag{45}$$

where the value used for ω is 1·4. The number q_r^0 of electrons participating in the π system are also reported in Table 22. The calculations are then repeated in a new cycle, and so on, until consistent values for the charge densities q_r are reached. The total energy of the π system is given by:

$$E = \sum_i v_i \varepsilon_i = \sum_i n_i(\alpha_0 + x_i\beta_0) \tag{46}$$

where x_i are the roots of the secular matrix (43), at the last cycle.

The ionization potential I evaluated as the difference in energy between the molecules and their positive ions is then given by:

$$I = -[\alpha_0 + (E^+ - E)\beta_0] \tag{47}$$

where E^+ is the energy obtained in a similar calculation for the positive ion. The results are reported in Table 22.

Five molecular orbital calculations, all in the frame of the Hückel theory but using different assumptions about the choice of empirical parameters α and β were performed on the benzoate ion[136]. Bond orders p were calculated and converted into bond lengths l using the expression:

$$l = s - \frac{(s-d)}{1 + k(1-p)/p} \tag{48}$$

where s and d are single and double bond-lengths respectively, and k was assumed equal to 0·8095 to fit the carbon–carbon triple bond value.

The five calculations show good agreement and the average bond-lengths as shown in **10** are given in Table 23.

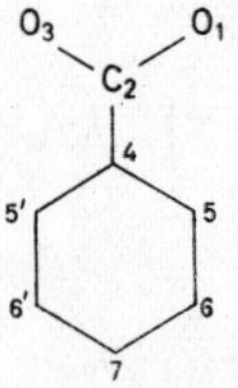

(10)

These distances can be compared with the ones obtained from the crystal structure study of potassium hydrogen dibenzoate[137]. The distances of this molecule are: C—C in the ring $= 1\cdot38$ Å, $C_{(2)}$—$C_{(4)} = 1\cdot53$ Å, $C_{(2)}$—$O_{(1)}$

TABLE 23. Bond-lengths for
structure **10**

Bond	Calculated distances (Å)
1–2	1·244
2–4	1·458
4–5	1·407
5–6	1·396
6–7	1·398

= 1·22 Å, $C_{(2)}$—$O_{(3)}$ = 1·24 Å. It must be noted however, that the two $O_{(3)}$ atoms in two neighbouring molecules are connected by a hydrogen bond. The main discrepancy is in the $C_{(2)}$—$C_{(4)}$ distance which, seems to be significantly larger than the predicted value.

2. Excited states

A theoretical treatment of the u.v. spectrum of formic acid was given by Nagakura, Kaya and Tsukomura[105], in which the molecules in molecules method[138] has been applied. Formic acid was separated into electron donor (OH) and electron acceptor (C=O) groups. Configuration interaction was assumed among the ground, locally excited (C=O) and charge-transfer configurations. The electron configurations used in this calculation are depicted in Figure 15. The energy zero was taken as the energy of the ground configuration. The energy of the locally excited configuration was taken from the observed excitation energy of the first band in formaldehyde, at 7·92 eV.

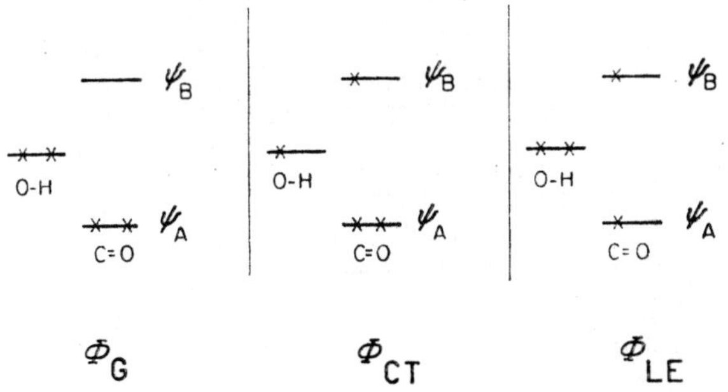

FIGURE 15. Electronic configurations of formic acid: Φ_G: ground state, Φ_{CT}: charge-transfer state, Φ_{LE}: locally-excited state.

To evaluate the energy of the charge-transfer configuration the molecular orbitals for the C=O group are needed. These were taken as the wavefunctions evaluated by Kon[139] for formaldehyde:

$$\psi_B = 0\cdot5472\chi_C + 0\cdot8370\chi_O$$
$$\psi_A = 0\cdot8370\chi_C - 0\cdot5472\chi_O$$

(49)

where ψ_B is the bonding and ψ_A the antibonding molecular orbital.

The energy for the charge-transfer configuration is then given by

$$E_{CT} = I_D - A_{C=O} - (0\cdot5472)^2\,(aa/bb) - (0\cdot8370)^2\,(aa/cc)$$

(50)

where (aa/bb) and (aa/cc) are the electron repulsion integrals defined in (32).

I_D was taken to be equal to the ionization potential of water which is 12·59 eV; $A_{C=O}$ was assumed to be $-1\cdot2$ eV, the mean value of electron affinity of the carbonyl group in a series of compounds. The electron repulsion integrals were evaluated according to Pariser and Parr[140]. A value of 6·63 eV was obtained for E_{CT}. The off-diagonal matrix elements were calculated according to Longuet-Higgins and Murrell[138]. The solution of the secular equation leads to the following eigenvalues and eigenvectors:

$$E_0 = -1\cdot156\ \text{eV}; \quad \Phi_0 = 0\cdot9300\Phi_G + 0\cdot3632\Phi_{CT} - 0\cdot0548\Phi_{LE}$$
$$E_1 = 6\cdot572\ \text{eV}; \quad \Phi_1 = 0\cdot3012\Phi_G - 0\cdot6688\Phi_{CT} + 0\cdot6797\Phi_{LE}$$
$$E_2 = 9\cdot132\ \text{eV}; \quad \Phi_2 = 0\cdot2013\Phi_G - 0\cdot6489\Phi_{CT} - 0\cdot7312\Phi_{LE}$$

(51)

The energy level diagram is shown in Figure 16. The calculated energy for the first transition is 7·73 eV (experimental 7·79 eV). The calculated oscillator strength for this transition is 0·29 (experimental 0·10). The contributions

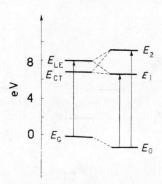

FIGURE 16. Energy level diagram of formic acid, after configuration interactions.

X of each configuration to the ground and excited states are given in Table 24. It is seen that in both excited states there is a substantial contribution

TABLE 24. Contributions X to the ground and excited states

	X_G (%)	X_{CT} (%)	X_{LE} (%)
E_0	86·49	13·19	0·30
E_1	9·07	44·73	46·20
E_2	4·05	42·11	53·47

of charge-transfer and locally excited configurations.

The molecules in molecules method has also been applied to substituted benzoic acids[141], where the two interacting groups are the substituted phenyl and the carboxyl group. Experimental ionization potentials of substituted benzenes and electronic absorption spectra of the two interacting groups were used in the calculations. Theoretical and experimental results are compared in Table 25. The agreement between theory and experiment

TABLE 25. Calculated and observed $\pi - \pi^*$ transitions in substituted benzoic acids

Substituent	ΔE (calculated) (eV)	ΔE (observed) (eV)	Reference
H	5·42	5·37	141
p-NH$_2$	4·54	4·46	141
p-OH	4·83	4·92	141
p-Cl	5·20	5·25	141
p-Br	5·18	5·15	141
p-NO$_2$	4·80	4·46	110
m-NO$_2$	4·94	4·67	110
o-NO$_2$	4·37	4·59	110

clearly indicates that the transition is of the charge-transfer type. Agreement is in fact less satisfactory in the nitro compounds, in which charge transfer from benzoic acid to the vacant orbital of the nitro group may occur.

B. Hydrogen bond

Hydrogen bond lengths and angles in the crystals of carboxylic acids have been the object of extensive investigations by means of x-ray, neutron and electron diffraction. These data have been collected by Donohue[142] and more recently by Fuller[143]. The histograms of Figure 17 from which the distribution of the hydrogen bond-lengths is apparent are taken from the second of these reviews.

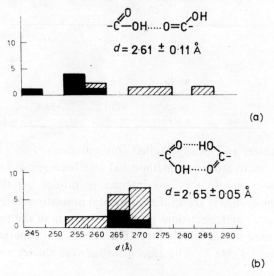

(a)

(b)

FIGURE 17. Histogram of the occurrence of the O...O distance in dimers of carboxylic acids. Black rectangles represent bond lengths with estimated limits of error of less than 0·05 Å.

The dependence of the O—H stretching frequencies on the O—H...O distance has also been investigated. The frequency shift as a function of the hydrogen bond-length is shown in Figure 18[144]. The dependence is linear in the range 2·45–2·75 Å; then the frequency shift becomes less sensitive to

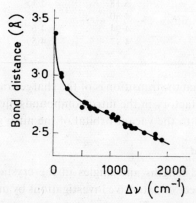

FIGURE 18. Frequency shift versus O—H...O bond distance in hydrogen bonds.

distance. From a similar analysis Pimentel and Sederholm[145] derived the following relationship:

$$\Delta\nu = 4\cdot43\times10^3(2\cdot84 - R_e) \tag{52}$$

where R_e is the O—H...O distance in Angstrom, and $\Delta\nu$ is in cm^{-1}.

The relationship between the O—H...O distance and the O—H distance[144] is shown in Figure 19.

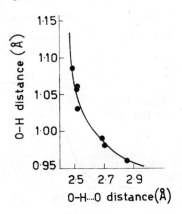

FIGURE 19. O—H...O distance versus O—H distance in hydrogen bonds.

The energy values of hydrogen bonds are reported in Table 13; typical values are in the range 6–7 kcal/mole/H bond. For the intramolecular hydrogen bond in salicylic acid ΔH was found to be of the order of −5 kcal/mole[146]; this value is in the normal range although the bond is far from being linear. The energy is about the same whether the acceptor group is —COO⁻, —COOH, or —COOC$_2$H$_5$.

Theoretical work on the hydrogen bond has been reviewed by Coulson[147]. Four energy contributions have been recognized as significant; these are the electrostatic energy, the delocalization energy, the repulsive energy, and the dispersion forces. The fact that hydrogen bonds occur only between electronegative elements suggests the importance of the electrostatic contribution. Calculation of the bond energy can be made by assuming, for the electrons in the bonds and in the lone pairs, point charges placed so as to give the correct dipole moments of the molecules[148]. Applying this procedure, Pople obtained for the water molecule an energy of 6·0 kcal/mole /H bond, in agreement with the experimental value of 5–6 kcal/mole[149]. The increase of the O—H distance can also be satisfactorily calculated with this model[150], which, however, is unable to account for the large increase in intensity of the O—H stretching band.

An alternative description may be given by a valence bond scheme in which the following structures are included:

(1) $O_{(1)}$—H $O_{(2)}$ covalent O—H bond

(2) $O_{(1)}^-$ $H^+ \cdot \cdot \cdot O_{(2)}$ ionic (no charge transfer)

(3) $O_{(1)}^-$ H —$O_{(2)}^+$ covalent H—O bond (with charge transfer)

(4) $O_{(1)}^+$ $H^- \cdot \cdot \cdot O_{(2)}$ ionic (no charge transfer)

(5) $O_{(1)}$ H^- $O_{(2)}^+$ covalent O—O bond (with charge transfer)

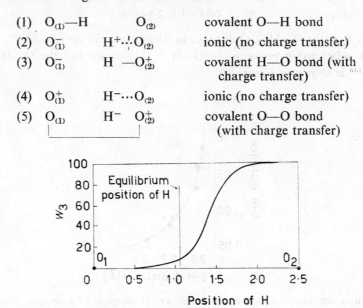

FIGURE 20. Weight of the structure (3) as a function of the position of the proton along the O...O line.

Calculations of this kind were carried out by Coulson and Danielsson[151] who considered only structures (1), (2) and (3) and obtained the weight of structure (3) as a function of the position of the proton along the O...O line for an O...O distance of 2·5 Å (Figure 20). At the equilibrium position of the hydrogen atom, the weight of structure (3) is about 11%. The calculation also showed that as the O...O distance decreases, the equilibrium position of the proton changes in such a way as to increase the O—H distance.

Improved calculations were performed by Tsubomura[149] who included all five structures. He obtained a value of 8·1 kcal/mole for the delocalization energy defined as the difference in energy when all the structures or only structures (1), (2) and (4) are included. Structures (3) and (5) appear to be equally important.

The repulsive energy term comes mainly from the H and $O_{(2)}$ atoms. It is very difficult to evaluate such a term since the two atoms are partially bonded. An estimate by Verwey[152] gives a value of 8·4 kcal/mole for ice. Verwey also evaluated the attractive dispersion forces and obtained 2·7 kcal/mole. Coulson[147] suggests 3 kcal/mole as a reliable value. Considering all energy terms, the sum obtained for the hydrogen bond in ice is: $6+8-8\cdot4+3 = 8\cdot6$ kcal/mole, which is of the order of the experimental value.

A semiempirical treatment of the hydrogen bond has been developed[153], based on the Lippincott potential function[154], originally formulated for diatomic molecules:

$$U(r) = D_0 \left\{ 1 - \exp\left[-\frac{n(r-r_0)^2}{2r} \right] \right\} \qquad (53)$$

where D_0 is the dissociation energy, r_0 the equilibrium distance and

$$n = k_0 r_0 / D_0 \qquad (54)$$

where k_0 is the force constant. The potential energy for the O—H...O system can be written as follows:

$$
\begin{aligned}
V(r, R_e) = D_0 &\left\{ 1 - \exp\left[-\frac{n(r-r_0)^2}{2r} \right] \right\} \\
&+ C(R_e)D_0 \left\{ 1 - \exp\left[-\frac{n(R_e-r-r_0)^2}{2(R_e-r)C} \right] \right\} + W(R_e)
\end{aligned} \qquad (55)
$$

R_e is the equilibrium O...O distance; the factor $C(R_e)$ takes care of the fact that even at equal distances the O—H bond is stronger than the H...O bond and is chosen to give the correct r_e value, since:

$$\left[\frac{dV(R_e, r)}{dr} \right]_{r=r_e} = 0 \qquad (56)$$

where r_e is the equilibrium O—H distance in the hydrogen-bonded molecules. $W(R_e)$ represents the non-bonding interaction energy between the oxygen atoms and has the form:

$$W(R) = -2595/R^6 + 4\cdot55 \times 10^6 e^{-4\cdot8R} \qquad (57)$$

where the parameters have been fixed in such a way as to reproduce the repulsive and electrostatic contributions to the hydrogen bond energy in ice and to give a correct value of the dissociation energy E for a shorter (i.e., less than 2·65 Å) hydrogen bond (equation 58). Curves corresponding

$$V(R_e, r_e) - U(r_0) = E \qquad (58)$$

to equation (55) can be drawn for a range of R_e values; they are reproduced in Figure 21(a). It is evident that there is a double minimum for R_e greater than 2·65 Å; this double minimum disappears for shorter hydrogen bonds where the curves show a flat bottom.

In Figure 21(b) the same surface is represented by means of potential contours. Figures 22(a) and 22(b) show the potential surface for the specific hydrogen bonds, one with $R_e = 2\cdot74$ Å, the other with $R_e = 2\cdot42$ Å.

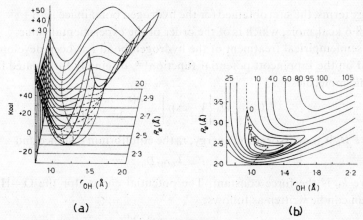

FIGURE 21. (a) Potential energy profiles and (b) contours for the hydrogen bond as a function of the distances R_e (O...O) and r_{OH}.

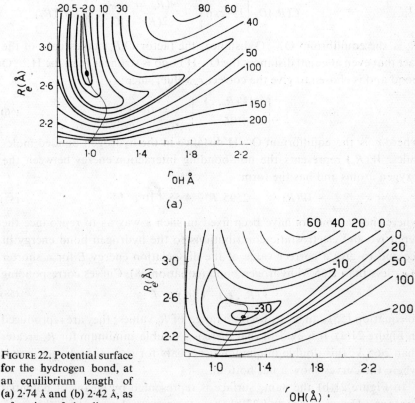

FIGURE 22. Potential surface for the hydrogen bond, at an equilibrium length of (a) 2·74 Å and (b) 2·42 Å, as a function of the distances R_e (O...O) and r_{OH}.

In calculating these energies allowance has been made for change of both the O—H and O...O distances; the same values of W have been used as in equation (57). By means of this diagram the infrared spectra due to O—H stretching can be explained. Broadening and intensity increase of the band corresponding to O—H stretching are observed in hydrogen bonds as compared to the same band in the free hydroxyl group. These effects are greater in hydrogen bonds of intermediate lengths and they are weak in bonds

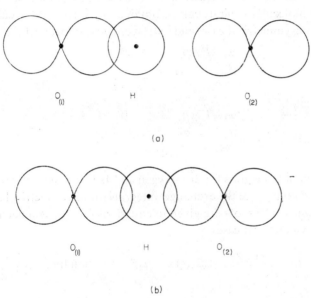

FIGURE 23. The atomic orbitals of (a) the unsymmetrical hydrogen bond and (b) the symmetrical hydrogen bond.

which are longer than 2·8 Å, or which are symmetrical. This can be justified by the fact that O—H bond-lengths change very slightly with O...O distance in such bonds.

From the potential surfaces for individual hydrogen bonds an estimation of the O...O stretching motion frequency is possible, assuming harmonicity. The results suggest that for $R_e > 2 \cdot 55$ Å the band should lie in the range of wave-numbers greater than 400 cm^{-1}.

An interesting point arises in connexion with the relative stability of symmetrical and unsymmetrical hydrogen bonds, using the naïve molecular orbital theory. The atomic orbitals of the system are shown in Figure 23.

The molecular orbitals have the form:

	unsymmetrical	symmetrical
Ψ_1 (bonding)	$(\chi_1 - c_1\chi_2) + c_2 s$	$(\chi_1 - \chi_2) + c_1 s$
Ψ_2 (non-bonding)	$c_3\chi_1 + \chi_2$	$\chi_1 + \chi_2$
Ψ_3 (antibonding)	$(\chi_1 - c_4\chi_2) - c_5 s$	$(\chi_1 - \chi_2) - c_2 s$

where χ_1 and χ_2 are the $2p\sigma$ oxygen orbitals, s the $1s$ hydrogen orbital and $c_1, c_2 \ldots\ldots$ are numerical coefficients[155].

Assuming the O...O interaction to be the same for both cases, the energies of the molecular orbitals can be obtained by solving the secular equations (59) (unsymmetrical case) and (60) (symmetrical case), where α_1 and α_2

$$\begin{vmatrix} \alpha_1-E & \beta_u & 0 \\ \beta_u & \alpha_2-E & 0 \\ 0 & 0 & \alpha_1-E \end{vmatrix} = 0 \tag{59}$$

$$\begin{vmatrix} \alpha_1-E & \beta_s & 0 \\ \beta_s & \alpha_2-E & \beta_s \\ 0 & \beta_s & \alpha_1-E \end{vmatrix} = 0 \tag{60}$$

are the Coulomb integrals for $2p\sigma$ oxygen and $1s$ hydrogen orbitals respectively, and β_u and β_s are the exchange integrals for unsymmetrical and symmetrical bonds. The roots are given in equations (61) (unsymmetrical case) and (62) (symmetrical case).

$$E_1 = \frac{\alpha_1+\alpha_2}{2} - 1/2\sqrt{(\alpha_1-\alpha_2)^2+4\beta_u^2} \quad \text{bonding} \tag{61a}$$

$$E_2 = \alpha_1 \quad \text{non-bonding} \tag{61b}$$

$$E_3 = \frac{\alpha_1+\alpha_2}{2} + 1/2\sqrt{(\alpha_1-\alpha_2)^2+4\beta_u^2} \quad \text{antibonding} \tag{61c}$$

$$E_1 = \frac{\alpha_1+\alpha_2}{2} - 1/2\sqrt{(\alpha_1-\alpha_2)^2+8\beta_s^2} \quad \text{bonding} \tag{62a}$$

$$E_2 = \alpha_1 \quad \text{non-bonding} \tag{62b}$$

$$E_3 = \frac{\alpha_1+\alpha_2}{2} + 1/2\sqrt{(\alpha_1-\alpha_2)^2+8\beta_s^2} \quad \text{antibonding} \tag{62c}$$

Since each system contains four electrons, the unsymmetrical situation has lower energy if $|\beta_u| > \sqrt{2}|\beta_s|$. In the isolated molecule β can be taken as equal to the binding energy and can be evaluated from the force constants for the O—H stretching motion in the isolated molecule. In the case of carboxylic acid dimers where $r_{OH} \simeq 1 \cdot 1$ Å, β_u is 101 kcal. Hence at the symmetrical–unsymmetrical transition point, $\beta_s = 71 \cdot 5$ which corresponds to a value of $r_e = 1 \cdot 22$ Å, that is $R_e = 2 \cdot 44$ Å. This is consistent with the fact that the proton is unsymmetrically placed $(R_e \simeq 2 \cdot 6 - 2 \cdot 7$ Å$)^{156}$.

In carboxylic acid dimers where two hydrogen bonds are present, when one hydrogen moves from one molecule to the other, the second hydrogen moves in the opposite direction. The potential energy for this double migration is represented in Figure 24. In chemical terms this again corresponds to the tautomerism shown in equation (12).

C. Reactivity

The theoretical approach to the interpretation of the chemical reactivity of organic compounds is one of the most difficult subjects in quantum chemistry. The transition state theory provides a framework in terms of which even complex chemical reactions can be discussed. Mechanistic studies are complementary to theoretical calculations in obtaining an understanding of reaction mechanisms, identifying the different stages of reactions and describing the nuclear configuration and electronic organization of transition states. Calculations are however confined to systems for which it is possible and justifiable to introduce wide simplifications[157–159].

The most successful theory has been the LCAO–MO method for conjugated systems, mainly aromatic compounds. The first theoretical discussion is perhaps the one given by Wheland and Pauling[160] for orientation in aromatic substitution; they used the Hückel theory to calculate charge densities at different sites, which, when modified by the polarizing effect to the

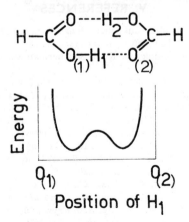

FIGURE 24. Potential energy for double migration of protons in formic acid dimers.

attacking group, determine the orientation. The calculations were carried out for the benzoic acid molecule, among others.

The counterpart of the charge density in nucleophilic or electrophilic reactions is, in the field of radical reactions, the free valence, defined as:

$$F_q = \sqrt{3} - \sum_r p_{qr} \qquad (64)$$

where p_{qr} is the bond order of the bond q—r. The sum is to be taken over all the bonds starting from atom q.

A different reactivity index is the localization energy[161] which is the difference between the π-electron energy in the initial molecule and the transition state in which the atom at the site of the reaction has been removed from the conjugated system. The number of π electrons in the transition state is equal to, or one less, or two less, than the number in the initial molecule for nucleophilic, radical or electrophilic reactions, respectively.

In his molecular orbital calculations on the benzoate ion Moser[136] evaluated charge densities, free valencies and localization energies for nucleophilic, radical and electrophilic reactions. No definite conclusions can be obtained from this calculation as to the chemical reactivity at different sites because the results depend on the assumptions involved.

The dissociation constants of benzoic acid and its derivatives have been taken by Hammett[162] as the basis for the definition of σ constants.

The σ values may be chosen as a sensitive gauge for the choice of the numerical values for the α and β parameters of the Hückel method for benzoic derivatives[163, 164].

V. REFERENCES

1. F. Haltzberg, B. Past and I. Fankuchen, *Acta Cryst.*, **6**, 127 (1953).
2. R. E. Jones and D. H. Templeton, *Acta Cryst.*, **11**, 484 (1958).
3. F. J. Strieter, D. H. Templeton, R. F. Schenerman and R. L. Sass, *Acta Cryst.*, 15,1233 (1962).
4. F. J. Strieter and D. H. Templeton, *Acta Cryst.*, **15**, 1240 (1962).
5. R. F. Schenerman and R. L. Sass, *Acta Cryst.*, **15**, 1244 (1962).
6. E. G. Cox, M. W. Dongill and G. A. Jeffrey, *J. Chem. Soc.*, 4854 (1952).
7. F. R. Ahmed and D. W. J. Cruickshank, *Acta Cryst.*, **6**, 385 (1953).
8. J. Housty and M. Hospital, *Acta Cryst.*, **17**, 1387 (1964).
9. J. Housty and M. Hospital, *Acta Cryst.*, **18**, 693 (1965).
10. P. Ganis, E. Martuscelli and G. Avitabile, *Riv. Sci.*, **36**, 689 (1966).
11. J. Housty and M. Hospital, *Acta Cryst.*, **20**, 325 (1966).
12. J. Housty and M. Hospital, *Acta Cryst.*, **21**, 553 (1966).
13. M. Dobler, J. D. Dunitz and A. Mugnoli, *Helv. Chim. Acta*, **49**, 2492 (1966).
14. S. Abrahamsson and M. M. Harding, *Acta Cryst.*, **20**, 377 (1966).
15. G. A. Jeffrey and M. Sax, *Acta Cryst.*, **16**, 1196 (1963).
16. M. A. Higgs and R. L. Sass, *Acta Cryst.*, **16**, 657 (1963).
17. M. Shahat, *Acta Cryst.*, **5**, 763 (1952).

18. C. J. Brown, *Acta Cryst.*, **21**, 1 (1966).
19. A. L. Bednowitz and B. Post, *Acta Cryst.*, **21**, 566 (1966).
20. B. P. Van Eijck, J. A. Kanters and J. Kroon, *Acta Cryst.*, **19**, 435 (1965).
21. G. A. Sim, J. M. Robertson and T. H. Goodwin, *Acta Cryst.*, **8**, 157 (1955).
22. M. Sundaralingom and L. H. Jensen, *Acta Cryst.*, **18**, 1053 (1965).
23. Y. Okaya, *Acta Cryst.*, **19**, 8791, (1965).
24. P. Hudson and J. H. Robertson, *Acta Cryst.*, **17**, 1497 (1964).
25. I. L. Karle, K. Britts and P. Gum, *Acta Cryst.*, **17**, 496 (1964).
26. J. Donohue and R. E. Marsh, *Acta Cryst.*, **15**, 941 (1962).
27. J. D. Dunitz and R. R. Ryan, *Acta Cryst.*, **21**, 617 (1966).
28. H. Geisenfelder and H. Zimmermann, *Ber. Bunsenges. Physik. Chem.* **67**, 480 (1963).
29. E. G. Cox, M. W. Dougill and G. A. Joffrey, *J. Chem. Soc.*, 4854 (1952).
30. Stig Aleby, *Acta Cryst.*, **15**, 1248 (1962).
31. G. Avitabile, P. Corradini, P. Ganis and E. Martuscielli, *Ric. Sci.*, **36**, 681 (1966).
32. I. L. Karle and J. Karle, *J. Chem. Phys.*, **22**, 43 (1954).
33. J. M. O'Gorman, W. Shand and V. Shomaker, *J. Am. Chem. Soc.*, **72**, 4222 (1950).
34. R. G. Lerner, B. P. Dailey and J. P. Friend, *J. Chem. Phys.*, **26**, 680 (1957).
35. G. H. Kwei, *J. Chem. Phys.*, **32**, 1592 (1960).
36. R. F. Curl, Jr., *J. Chem. Phys.*, **30**, 1529 (1959).
37. W. G. Fateley and F. A. Miller, *Spectrochim. Acta*, **17**, 857 (1961).
38. N. D. Lebedeva, *Russ. J. Phys. Chem.*, *(English Transl.)*, **38**, 1435 (1964).
39. R. R. Bernecker and F. A. Long, *J. Phys. Chem.*, **65**, 1565 (1961).
40. K. Watanabe, T. Nakayama and J. Mott, *Final Report on Ionization of Molecules by Photoionization Method*, University of Hawaii, 1959.
41. J. D. Morrison and A. I. Nicholson, *J. Chem. Phys.*, **20**, 1021 (1952).
42. L. Jaffe, E. J. Prosen and M. Szwarc, I. *Chem. Phys.*, **27**, 416 (1957).
43. Kh. S. Bagdasar'yan, *Russ. J. Phys. Chem. (English Transl.)*, **35**, 872 (1961).
44. M. H. Back and A. H. Sehon, *Can. J. Chem.*, **38**, 1261 (1960).
45. C. E. Brion and W. J. Dunning, *Trans. Faraday Soc.*, **59**, 647 (1963).
46. P. Gray and A. Williams, *Chem. Rev.*, **59**, 239 (1959).
47. K. Watanabe, *J. Chem. Phys.*, **26**, 542 (1957).
48. W. C. Price and W. M. Evans, *Proc. Roy. Soc. (London)*, **A 162**, 110 (1937).
49. A. Foffani, S. Pignataro, B. Cantone and F. Grasso, *Z. Physik. Chem. (Frankfurt)*, **42**, 221 (1964).
50. C. E. Brion and W. J. Dunning, *Trans. Faraday Soc.*, **59**, 648 (1963).
51. A. B. King and F. A. Long, *J. Chem. Phys.*, **29**, 374 (1958).
52. S. S. Friedland and R. E. Strakna, *J. Phys. Chem.*, **60**, 815 (1956).
53. K. Higasi, I. Omura and H. Baba, *Nature*, **178**, 652 (1956).
54. J. H. S. Green, *Quart. Rev. (London)*, **15**, 125 (1961).
55. G. Erlandsson and H. Selén, *Arkiv. Fysik.*, **14**, 61 (1958).
56. C. H. Townes and A. L. Schawlow, *Microwave Spectroscopy*, McGraw-Hill, New York, 1955.
57. H. A. Pohl, M. E. Hobbs and P. M. Gross, *Ann. N.Y. Acad. Sci.*, **40**, 389 (1940).
58. G. Potapenko and D. Wheeler, *Rev. Mod. Phys.*, **20**, 143 (1948).
59. I. E. Coop, N. R. Davidson and L. E. Sutton, *J. Chem. Phys.*, **6**, 905 (1938).
60. N. N. Stepanenko, B. A. Agranat and T. Novikova, *Acta Physicochem. USSR* **20**, 923 (1945).
61. R. P. Phadke, *J. Indian Inst. Sci.*, **34**, 189 (1952).
62. A. P. Kapustin, *Zh. Eksperim. Teor. Fiz.*, **17**, 30 (1947).
63. K. Palm and H. Dunken, *Z. Physik. Chem. (Leipzig)*, **217**, 248 (1961).
64. Cl. Bèguin and T. Gaumann, *Helv. Chim. Acta*, **41**, 1376 (1958).
65. C. T. Zahn, *Z. Physik*, **33**, 730 (1932).

66. I. Sakurada and M. Taniguchi, *Bull. Inst. Phys.-Chem. Res. (Tokyo)*, **12**, 224 (1933).
67. C. P. Smyth and W. S. Walls, *J. Am. Chem. Soc.*, **53**, 527 (1931).
68. S. Mizushima and M. Kubo, *Bull. Chem. Soc. Japan*, **13**, 174 (1938).
69. S. R. Phadke, S. D. Gokhale, N. L. Phalnikar and B. V. Bhide, *J. Indian Chem. Soc.*, **22**, 235 (1945).
70. H. Müller and H. Sack, *Z. Physik.*, **31**, 815 (1930).
71. T. Hanai, N. Koizumi and R. Gotoh, *Bull. Inst. Chem. Res., Kyoto Univ.*, **39**, 195 (1961).
72. G. W. Wheland, *Resonance in Organic Chemistry*, J. Wiley and Sons, New York, 1958.
73. H. S. Harned and N. D. Embree, *J. Am. Chem. Soc.*, **56**, 1042 (1934).
74. H. S. Harned and B. B. Owen, *Physical Chemistry of Electrolyte Solutions*, Reinhold Publishing Corporation, New York, 1943.
75. D. H. Everett, D. A. Landsman and B. R. W. Pinsent, *Proc. Roy. Soc. (London)*, A **215**, 403 (1952).
76. H. S. Harned and R. O. Sutherland, *J. Am. Chem. Soc.*, **56**, 2039 (1934).
77. D. J. G. Ines and J. H. Pryor, *J. Chem. Soc.*, 2104 (1955).
78. L. F. Nims, *J. Am. Chem. Soc.*, **58**, 987 (1936).
79. R. P. Bell, *The Proton in Chemistry*, Cornell University Press, New York, 1959.
80. M. D. Taylor and J. Bruton, *J. Am. Chem. Soc.*, **74**, 4151 (1952).
81. A. E. Potter, P. Bender and H. L. Ritter, *J. Phys. Chem.*, **59**, 250 (1955).
82. R. E. Lundin, F. E. Harris and L. K. Nash, *J. Am. Chem. Soc.*, **74**, 743 (1952).
83. E. W. Johnson and L. K. Nash, *J. Am. Chem. Soc.*, **72**, 547 (1950).
84. M. D. Taylor and M. B. Templeman, *J. Am. Chem. Soc.*, **78**, 2950 (1956).
85. E. Freedman, *J. Chem. Phys.*, **21**, 1784 (1953).
86. D. S. Sarkadi and J. H. De Boer, *Rec. Trav. Chim.*, **76**, 628 (1957).
87. G. Allen and E. F. Caldin, *Trans. Faraday Soc.*, **49**, 895 (1953).
88. R. C. Millikan and K. S. Pitzer, *J. Chem. Phys.*, **27**, 1305 (1957).
89. T. Miyazawa and K. S. Pitzer, *J. Chem. Phys.*, **30**, 1076 (1959).
90. K. Nakamoto and S. Kishida, *J. Chem. Phys.*, **41**, 1554 (1964).
91. Y. Morino and K. Kuchitsu, *J. Chem. Phys.*, **20**, 1809 (1952).
92. W. Waring, *Chem. Rev.*, **51**, 171 (1952).
93. W. Weltner, *J. Am. Chem. Soc.*, **77**, 3941 (1955).
94. J. K. Wilmshurt, *J. Chem. Phys.*, **25**, 1171 (1956).
95. R. C. Millikan and K. S. Pitzer, *J. Am. Chem. Soc.*, **80**, 3515 (1958).
96. T. Miyazawa and K. S. Pitzer, *J. Am. Chem. Soc.*, **81**, 74 (1959).
97. K. Nakamoto and S. Kishida, *J. Chem. Phys.*, **41**, 1558 (1964).
98. D. Hadzi and M. Pintar, *Spectrochim. Acta*, **12**, 162 (1958).
99. M. C. Flett, *J. Chem. Soc.*, 962 (1951).
100. K. Kakanishi, *Infrared Absorption Spectroscopy*, Holden-Day, Inc., San Francisco, 1962.
101. D. Peltier and A. Pichevin, *Bull. Soc. Chim. France*, 1141 (1960).
102. E. E. Barnes and W. T. Simpson, *J. Chem. Phys.*, **39**, 670 (1963).
103. W. D. Closson and P. Hang, *J. Am. Chem. Soc.*, **86**, 2384 (1964).
104. E. M. Kosower, *J. Am. Chem. Soc.*, **80**, 3253 (1958).
105. S. Nagakura, K. Kaya and H. Tsubomura, *J. Mol. Spectr.*, **13**, 1 (1964).
106. H. Hosoya, J. Tanaka and S. Nagakura, *J. Mol. Spectr.*, **8**, 257 (1962).
107. M. Ito, H. Tsukioka and S. Imanishi, *J. Am. Chem. Soc.*, **82**, 1559 (1960).
108. M. Ito, *J. Mol. Spectr.*, **4**, 144 (1960).
109. R. P. Ossorio, *An. Real Soc. Fis. Quim.*, B **56**, 379 (1960).
110. A. K. Chandra, *J. Phys. Chem.*, **66**, 562 (1962).
111. H. Hosoya and S. Nagakura, *Spectrochim. Acta*, **17**, 324 (1961).

112. L. W. Reevs, *Trans. Faraday Soc.*, **55**, 1684 (1959).
113. J. W. Emsley, J. Feeney and L. H. Stutcliffe, *High Resolution NMR Spectroscopy*, Vol. 2, Pergamon Press, Oxford, 1966, p. 816.
114. J. C. Davis and K. S. Pitzer, *J. Phys. Chem.*, **64**, 886 (1960).
115. H. S. Gutowsky and A. Saika, *J. Chem. Phys.*, **21**, 1688 (1953).
116. S. Brownstein and A. E. Stillman, *J. Phys. Chem.*, **63**, 2061 (1959).
117. S. S. Dharmatti, K. J. Sundara Rao and R. Vijayaraghavan, *Nuovo Cimento*, **11**, 656 (1959).
118. H. A. Christ, *Helv. Phys. Acta*, **33**, 572 (1960).
119. C. A. Coulson and E. T. Stewart, *The Chemistry of Alkenes*, (Ed. S. Patai), Interscience Publishers, London, 1954, pp. 106, 110.
120. A. Streitwieser, *Molecular Orbital Theory for Organic Chemists*, John Wiley and Sons, New York, 1961, p. 123.
121. C. C. J. Roothaan, *Rev. Mod. Phys.*, **23**, 69 (1961).
122. S. Carrà and M. Simonetta, *Gazz. Chim. Ital.*, **89**, 2456 (1959).
123. R. S. Mulliken, C. A. Rieke, D. Orloff and H. Orloff, *J. Chem. Phys.*, **17**, 1248 (1949).
124. R. G. Parr and B. L. Grawford, *J. Chem. Phys.*, **16**, 1049 (1948).
125. C. C. J. Roothaan, *J. Chem. Phys.*, **19**, 1445 (1951).
126. R. S. Mulliken, *J. Chim. Phys.*, **46**, 497 (1949).
127. M. Goeppert–Mayer and A. L. Sklar, *J. Chem. Phys.*, **6**, 645 (1938).
128. J. F. Mulligan, *J. Chem. Phys.*, **19**, 347 (1951).
129. R. S. Mulliken, *J. Chem. Phys.*, **23**, 1833, 1840 (1955).
130. C. A. Coulson, *Proc. Roy. Soc. (London)*, **A 169**, 413 (1939).
131. F. E. Morris and W. J. Orville Thomas, *J. Mol. Spectr.*, **6**, 572 (1961).
132. C. A. Coulson, *Victor Henry Memorial Volume*, Maison Desoer, Liège, 1948, p. 15.
133. W. H. Kirchoff, J. Farren and J. W Linnett, *J. Chem. Phys.*, **42**, 1410 (1965).
134. A Streitwieser, *Molecular Orbital Theory For Organic Chemists*, John Wiley and Sons, New York, 1961, p. 115.
135. B. Cantone, F. Grasso and S. Pignataro, *Mol. Phys.*, **11**, 221 (1966).
136. C. M. Moser, *J. Chem. Soc.*, 1073 (1953).
137. J. M. Skinner, G. M. D. Stewart and J. C. Speakman, *J. Chem. Soc.*, 180 (1954).
138. H. C. Longuet-Higgins and J. N. Murrell, *Proc. Phys. Soc.*, **A 68**, 601 (1955).
139. H. Kon, *Bull. Chem. Soc. Japan*, **28**, 275 (1955).
140. R. Pariser and R. G. Parr, *J. Chem. Phys.*, **21**, 466, 767 (1953).
141. J. Tanaka, S. Nagakura and M. Kobayashi, *J. Chem. Phys.*, **24**, 311 (1956).
142. J. Donohue, *J. Phys. Chem.*, **56**, 502 (1952).
143. W. Fuller, *J. Phys. Chem.*, **63**, 1705 (1959).
144. K. Nakamoto, M. Margoshes and R. E. Rundle, *J. Am. Chem. Soc.*, **77**, 6480 (1955).
145. G. C. Pimentel and C. H. Sederholm, *J. Chem. Phys.*, **24**, 639 (1956).
146. J. Hermans, S. J. Leach and H. A. Sheraga, *J. Am. Chem. Soc.*, **85**, 1390 (1963).
147. C. A. Coulson, *Research*, **10**, 149 (1957).
148. J. A. Pople, *Proc. Roy. Soc. (London)*, **A 205**, 163 (1951).
149. H. Tsubomura, *Bull. Chem. Soc. Japan*, **27**, 445 (1954).
150. N. D. Coggeshall, *J. Chem. Phys.*, **18**, 978 (1950).
151. C. A. Coulson and U. Danielsson, *Arkiv. Fysik*, **8**, 239, 245 (1954).
152. E. J. W. Verwey, *Rec. Trav. Chim.*, **60**, 887 (1941).
153. C. Reid, *J. Chem. Phys.*, **30**, 182 (1959).
154. E. R. Lippincott, *J. Chem. Phys.*, **23**, 603 (1955); **26**, 1678 (1957).
155. G. C. Pimentel and A. L. McClellan, *The Hydrogen Bond*, W. H. Freeman and Company, S. Francisco and London, 1960.

156. A. Sintes, J. Housty and M. Hospital, *Acta Cryst.*, **21**, 965 (1966).
157. R. D. Brown, in *Molecular Orbitals in Chemistry, Physics and Biology*, Academic Press, New York, 1964, p. 485.
158. K. Fukui, in *Molecular Orbitals in Chemistry, Physics and Biology*, Academic Press, New York, 1964, p. 531.
159. M. J. S. Dewar, in *Advances in Chemical Physics*, Vol. 8, Interscience Publishers, London, 1965, p. 65.
160. G. W. Wheland and L. Pauling, *J. Am. Chem. Soc.*, **57**, 2086 (1935).
161. G. W. Wheland, *J. Am. Chem. Soc.*, **64**, 900 (1942).
162. L. P. Hammett, *Chem. Rev.*, **17**, 125 (1935).
163. H. H. Jaffè, *J. Chem. Phys.*, **20**, 279 (1952).
164. M. Simonetta and A. Vaciago, *Nuovo Cimento*, **11**, 596 (1954).

CHAPTER 2

Electrochemical reactions of carboxylic acids and related processes

LENNART EBERSON

University of Lund, Sweden

I. INTRODUCTION

In principle, electrochemical methods would seem to be ideal for performing oxidations or reductions of organic compounds. An electrode acts as a sink (anodic oxidation) or a source (cathodic reduction) of electrons, and the 'electron activity' can be varied within wide limits by varying the potential between the electrode and the electrolyte. We can simply regard the electrode surface as an aggregate of orbitals, to or from which electrons can be transferred; the energy of these orbitals is reflected in the electrode potential. Overlap between an electrode orbital and a suitable molecular orbital in the organic molecule will result in an electron transfer in either direction if the corresponding energy levels are properly related to each other, *i.e.* if the electrode potential is positive (or negative) enough to allow removal of one or possibly two electrons from the highest occupied molecular orbital of the molecule (or supply of one or two electrons to the lowest empty molecular orbital). In an ideal case, the result of the electron transfer will be a cationic, radical, or anionic intermediate, which, after leaving the electrode surface, will undergo further *chemical* changes in the solution surrounding the electrode, or, in favorable cases, accumulate in the solution. A wide choice of solvents and added reagents would, again in principle, provide us with a variety of synthetic possibilities once we have been able to generate a particular intermediate by an electrode process.

Practical organic electrochemistry uses this unique possibility of being able to control both the activity of the electrode and the composition of the electrolyte in a number of ways. First of all, a number of simple and versatile synthetic methods have been developed[1-5]. Secondly, interest has been focussed on the investigation of the mechanisms of organic electrode processes[6, 7], in which field much progress has been made during recent years. A review by Perrin[8] summarizes much of this work, which aims at a description of electrode reactions in terms of transition states and intermediates in much the same way as for reactions in homogeneous media. Here electron spin resonance spectroscopy has contributed a powerful tool for the detection and identification of electrochemically generated paramagnetic species[9]. Thirdly, polarographic investigations in combination with molecular orbital theory have made possible some remarkably good corroborations of the simple molecular orbital calculations[10, 11].

This is all very well. But when did you last see an organic chemist perform an electrochemical synthesis? Most readers will probably answer 'never' to this question, and I would certainly believe them. Ignorance of electrochemical methods in courses and textbooks of organic chemistry, combined with the fact that these methods are *sometimes* tricky and do *sometimes* require equipment which is not accessible in most laboratories, has contributed heavily to this situation. Today, when reliable potentiostats are commercially available at less than one tenth of the cost of almost any other instrument used by organic chemists, one of the major obstacles to a widespread use of electrolytic syntheses has been eliminated. This chapter is intended to remove part of the inhibitions on the educational side.

The anodic oxidation of carboxylic acids, commonly denoted as the Kolbe electrosynthesis[12], is probably the most well-known and used electrolytic synthesis method. Not unexpectedly, it is a method which is very simple from the experimental point of view. In fact, a couple of platinum electrodes, an amperemeter, a slidewire resistance, and a d.c. source are the only prerequisites for the successful application of this method in the overwhelming majority of cases.

We shall not deal much here with the historical development of the Kolbe synthesis, nor with experimental techniques and details. These aspects have been adequately treated in monographs[1-5] and review articles[13-15], some of which cover the literature up to 1960. Earlier synthetic applications will not be covered except in cases where the results are of fundamental interest to the mechanism of the reaction. Instead, we shall concentrate on new developments of electrochemical reactions of carboxylic acids at the anode, reactions which may not sometimes be of immediate preparative interest but may be developed into synthetically useful methods in the future. However, it will be necessary to reexamine and reevaluate early experimental material with regard to the now well-proven suggestion that cationic intermediates can account for part of the product spectrum in the anodic oxidation of carboxylic acids. Earlier reviews do not treat the Kolbe reaction from this viewpoint, simply because experimental verifications of the formation of carbonium ions at the anode are of more recent date.

II. DEFINITIONS

The Kolbe electrosynthesis is not actually an oxidation of the carboxylic acid itself, but rather of its ion. Without implying anything about the potential-determining step (the mechanism will be discussed in section IV. B) for the moment, we shall write the reaction as an initial discharge of the carboxylate ion[16] by a one-electron transfer to the anode, followed by a decarboxylation step (equations 1 and 2). The radical R· is the first product-forming intermediate, giving coupling, disproportionation, or other products characteristic of radical reactions

$$RCOO^- \longrightarrow RCOO^· + e^- \tag{1}$$

$$RCOO^· \longrightarrow R^· + CO_2 \tag{2}$$

(equation 3), such as that formed by hydrogen abstraction from solvent or solute C—H bonds. The isolation of R_2 (R—R), the coupling product, has been the objective of most preparative applications hitherto, be it a symmetrical coupling between two identical radicals as indicated in equation (3)

$$
R^· \quad
\begin{cases}
\text{Coupling} & \longrightarrow R\!-\!R \\
\text{Disproportionation} & \longrightarrow RH + RH - H_2 \\
\text{Attack on C\!-\!H} & \longrightarrow RH \\
\text{Abstraction of H}^· & \longrightarrow RH - H_2
\end{cases}
\tag{3}
$$

or a mixed coupling between two different radicals produced by coelectrolysis of two acids[13-15].

The question of the possible intermediacy of the acyloxy radical in the Kolbe reaction will be discussed later (sections IV. B and V. A).

However, products are very seldom limited to the above-mentioned types. It was early observed[17] that electrolysis of $RCOO^-$ in aqueous solution often produced an alcohol ROH in addition to coupling and disproportionation products. This reaction is commonly referred to as the Hofer–Moest reaction. Analogously, a methyl ether, ROMe, or an acetate, ROAc, could be isolated if the electrolysis was carried out in methanol or acetic acid, respectively. Also, the ester formed between the starting material RCOOH and the alcohol ROH has sometimes been isolated in small amounts. Frequently, rearrangements of the hydrocarbon group were noticed in these products, and for a long time attempts were made to account for these phenomena within more or less extended radical mechanisms.

In 1957, following some intriguing studies by Muhs[18] on the electrolysis of some cycloalkaneacetic acids, Walling[19] suggested that these products originated from anodically generated carbonium ions, *i.e.* that $R^{\cdot}$ could be further oxidized to a carbonium ion R^{+} at the anode (equation 4),

$$R^{\cdot} \longrightarrow R^{+} + e^{-} \qquad (4)$$

followed by reaction of R^{+} or a rearranged carbonium ion R'^{+} with a nucleophile present in the electrolyte (equation 5), or by other carbonium ion reactions, such as proton elimination or cyclization. This idea was simultaneously and independently conceived in other laboratories[20, 21], the final experimental verification being provided by Corey and coworkers[20].

$$
\begin{array}{ccc}
& \xrightarrow{\text{Rearrangement}} R'^{+} & \xrightarrow{R''OH,\ RCOO^{-}} R'OR'' + RCOOR' \\
R^{+} & \xrightarrow{\qquad R''OH,\ RCOO^{-}\qquad} & ROR'' + RCOOR \\
& R'' = H,\ Me,\ Ac &
\end{array}
\qquad (5)
$$

As we shall see later, this suggestion proved to be a very fruitful one. Carbonium ion production by anodic oxidation of carboxylic acids is now almost as well established as the methods used in homogeneous systems, *i.e.* deoxidation, deamination, and solvolysis[22].

We shall here adopt a definition of the Kolbe reaction as the anodic formation of products through intervention of both radical and carbonium ion species. Only products which cannot be derived from these intermediates shall be referred to as by-products. This definition broadens the scope and utility of the Kolbe reaction considerably as compared to the earlier one which emphasized the formation of coupling products and more or less considered other products as undesirable. The extended definition covers both radical and carbonium ion aspects, and thus one of the important problems of the field becomes the question of which structural and experimental factors favor either kind of process. A second important task will be the systematic study of anodically generated radicals and carbonium ions to see how they differ from radicals and carbonium ions in homogeneous systems. Much of the work done in this field during the last 5–10 years has been conducted along these lines.

III. EXPERIMENTAL CONDITIONS

As in other fields of organic chemistry, the outcome of electrochemical reactions is determined by structural and experimental factors. In electrochemistry, the latter ones are sometimes both more critical and less easily defined, so that comparisons between different investigations are much more difficult to make. Apart from the normal variables, such as pH, temperature, solvent, influence of foreign electrolytes, *etc.*, an electrochemical reaction is governed by the nature of the electrode material and the electrode potential, which in its turn determines the current density. Only a short summary of the influence of experimental factors will be given here, since earlier reviews[13–15] cover the subject adequately.

The electrode potential is the potential set up between an electrode and the solution during electrolysis. It can easily be measured against a suitable reference electrode by means of an electronic voltmeter. By using an automatic control device, a potentiostat, the electrode potential can be kept constant during the electrolysis. The total applied voltage (*i.e.*, that applied across the anode and cathode) is of no interest in this connection, especially when the reaction is conducted in a high-resistance non-aqueous solvent, since it is mainly determined by the *i.r.* drop in the solution.

A. Anode Potential and Current Density

In aqueous solution, a plot of anode potential *versus* the logarithm of the current density for acetate electrolysis (and for the electrolysis of other carboxylates) has the typical appearance shown in Figure 1 (see also Table 2). In the low-potential region, oxygen evolution due to oxidation of water

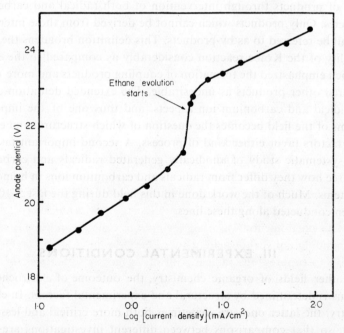

FIGURE 1. Anode potential (referred to the hydrogen electrode) versus the logarithm of the current density for the electrolysis of 0·5 M aqueous sodium acetate[45].

is observed up to about $+2 \cdot 1$ V, until a more or less pronounced transition to a high-potential region with a concurrent switch to anodic oxidation of acetate ion at about $+2 \cdot 3$ V (this is commonly referred to as the critical potential) occurs, as evidenced by the formation of almost quantitative amounts of ethane and carbon dioxide, accompanied by small amounts of oxygen. The same behavior is observed in anhydrous solvents, such as methanol and ethylene glycol, except that the oxygen evolution process is not possible in the low-potential region.

This is one of the most characteristic features of the anodic oxidation of carboxylate ions at platinum anodes: it takes place at a very high positive potential, above $+2 \cdot 3$ V (in actual preparative runs, probably in the region between $+2 \cdot 5$ and $+3 \cdot 0$ V), and can almost totally suppress other electrode processes which normally occur at much lower potentials, *e.g.* the oxidation of water in the region around $+1 \cdot 7$ V. This is probably the reason why the plug-into-the-wall variety of the Kolbe electrolysis is so successful. The experimenter's wish to see his reaction finished within a reasonable period of time will make him use as high currents as his d.c. source permits, thus automatically establishing an anode potential high enough for the process. The finding that high current densities increase the yield of coupled product is in part a reflection of the necessity of maintaining a high anode potential, and in part a result of the favored bimolecular reactions between radicals when the concentration of radicals in the immediate vicinity of the electrode is kept high.

B. Anode Material

In aqueous solution, only a smooth platinum or iridium anode will support the formation of radical and/or cationic intermediates from carboxylate ions. Gold, nickel, and platinized platinum electrodes give oxygen evolution only. Carbon electrodes do give some products formed via radicals, but the formation of carbonium ions seems to be the predominant reaction, as has been demonstrated by Koehl[23] for a number of aliphatic acids. This phenomenon was observed both in aqueous solutions and in the corresponding anhydrous acid. Thus, a carbon anode is preferable if it is desirable to suppress the radical pathway, which is sometimes the case in anodic acetoxylations (section V. A). In non-aqueous solvents, the choice of anode material is much less critical, although smooth platinum is preferred by most investigators. For large-scale runs the use of other materials should be contemplated, since disintegration of platinum electrodes at an appreciable rate has been observed in some cases[24]. In normal laboratory practice this phenomenon can be neglected.

At a smooth platinum anode, assuming that all other factors are kept constant, the distribution of products formed via either the radical or the cation pathway will be largely a function of the structure of the electrolyzed acid. In this review, the nature of the anode material will not be specified unless it differs from smooth platinum.

C. Effect of pH

The yield of products formed via radicals is optimal if the electrolyte is kept slightly acid during the run. Alkaline electrolytes such as pyridine–water tend to favor the carbonium ion-based products. The pH can be kept in the acid range during most of the run by using a large excess of the carboxylic acid as compared to the carboxylate ion. Since the cathodic process (discharge of sodium or potassium ions) will liberate hydroxide ions, the acid will be continuously transformed into its salt until all of it has been consumed at the anode. The reaction is discontinued when the electrolyte has turned slightly alkaline.

This procedure (the salt deficit method) will not work in aqueous solution if the acid is insoluble or only slightly soluble in water. However, this is not a very serious disadvantage, since non-aqueous solvents, especially methanol, are actually better solvents for the Kolbe reaction (see below). If for some reason it is necessary to use water as solvent, an alternative method employs a mercury cathode and the fully neutralized (to a pH of 7·5–8) carboxylic acid in aqueous solution as electrolyte[25]. The alkali metal formed at the cathode forms an amalgam with mercury and does not give hydroxide ions by reaction with water. This method is also advantageous in mixed electrolyses of carboxylic acids of widely differing pK, where both acids have to be completely converted into their salts. In the salt deficit method, the stronger acid will be preferentially electrolyzed and little or no mixed coupled product will be formed.

D. Effect of Solvent

Water was the preferred solvent in early applications of the anodic oxidation of carboxylic acids. The emphasis laid on maximizing the yield of coupling product soon led to experimentation with non-aqueous solvents, from which methanol emerged as the superior one for this purpose. A still better solvent for the coupling process was found in N,N-dimethylformamide (DMF)[26, 27], although it has later been observed that it has a slight disadvantage in being anodically oxidized itself (equation 6)[28–31]. This reaction will be discussed in connection with anodic acetoxylation (section V. A).

$$\text{HCON}\diagdown\begin{matrix}\text{CH}_3\\\text{CH}_3\end{matrix} \longrightarrow \text{HCON}\diagdown\begin{matrix}\text{CH}_2^+\\\text{CH}_3\end{matrix}+\text{H}^++2\text{e}^- \xrightarrow{\text{RCOO}^-} \text{HCON}\diagdown\begin{matrix}\text{CH}_2\text{OCOR}\\\text{CH}_3\end{matrix} \qquad (6)$$

Acetonitrile has been used in a few cases[20, 32, 33] and is of great potential interest, since anodically generated carbonium ions react with it to form nitrilium salts (1) which with water give N-substituted acetamides[33] (equation 7). This reaction is analogous to the Ritter reaction (synthesis of an

$$\text{RCOO}^- \xrightarrow[-\text{CO}_2]{-2\text{e}^-} \text{R}^+ \xrightarrow{\text{CH}_3\text{CN}} \underset{(1)}{\text{R}-\text{N}^+\!\!=\!\!\text{C}-\text{CH}_3} \xrightarrow{\text{H}_2\text{O}} \text{RNHCOCH}_3 \qquad (7)$$

N-alkylacetamide from an alkene or alcohol, acetonitrile, and concentrated sulfuric acid), but is much more lenient towards sensitive functional groups in the acid molecule.

Anhydrous carboxylic acids have often been used as solvents in the electrolysis of the corresponding salts, which for obvious reasons is a rather limited application. However, anhydrous acetic acid has also found some use in the electrolysis of other carboxylates[34-37], mainly for obtaining acetates via the cationic pathway. The predominant competing radical process is the formation of a mixed coupling product between acetic acid present in large excess and the acid in question, provided the pK difference between RCOOH and CH_3COOH is not too large. The reaction scheme is outlined in equation (8).

$$\begin{matrix} \text{RCOO}^- & & \text{R}\cdot & \xrightarrow{-\text{e}^-} & \text{R}^+ & \xrightarrow{\text{CH}_3\text{COOH}} & \text{ROCOCH}_3 \\ + & \xrightarrow[-\text{CO}_2]{-\text{e}^-} & + & & & & \\ \text{CH}_3\text{COO}^- & & \text{CH}_3^\cdot & & & & \\ & & \downarrow & & & & \\ & & \text{RCH}_3 & & & & \end{matrix} \qquad (8)$$

E. Effect of Temperature

An increase in temperature tends to decrease the yield of coupling product in those cases which have been systematically investigated. It is not known whether this decrease is accompanied by an increase in the yield of products formed via the carbonium ion path. Woolford[38] and Woolford, Arbic and Rosser[39] have described several cases where a small change in temperature causes a very marked change. As an example (equation 9), the electrolysis of 11-bromoundecanoic acid in methanol[38] at or below 50° produced a good yield of the coupling product (2), whereas at 65° methyl 11-bromoundecanoate (3) and methyl 11-methoxyundecanoate (4) were formed exclusively.

$$Br(CH_2)_{10}COOH \xrightarrow[MeOH]{65°} Br(CH_2)_{10}COOCH_3 + MeO(CH_2)_{10}COOCH_3$$
$$\text{(3)} \qquad\qquad\qquad \text{(4)}$$
$$50° \downarrow MeOH \qquad\qquad 71\% \qquad\qquad\qquad 2\cdot4\%$$
$$Br(CH_2)_{20}Br$$
$$\text{(2)}$$
$$64\%$$

$$(9)$$

The possible mode of formation of the ester of the solvent methanol at the anode will be discussed later (section VI.).

The case cited above is an extreme one. Generally, the reaction is fairly unsensitive towards temperature changes, especially in non-aqueous solution.

F. Effect of Added Foreign Electrolytes

Certain foreign anions inhibit the formation of radicals from the electrolysis of carboxylate ions and promote the formation of carbonium ions; such ions include bicarbonate, sulfate, perchlorate, dihydrogen phosphate, and fluoride ions. Metallic cations, such as Pb^{2+}, Mn^{2+}, Cu^{2+}, Fe^{2+}, and Co^{2+}, suppress the Kolbe reaction completely or almost so, presumably because the platinum anode is rapidly covered by a surface layer of the corresponding metal oxide, and this does not support the process. Sodium, potassium, calcium, and barium ions have no adverse effect.

G. Summary of the Influence of Experimental Factors

Table 1 is a summary of experimental factors and their influence on the mechanism of the Kolbe process. It should be stressed that experimental factors can be used to modify the outcome of the reaction to a limited

TABLE 1. Experimental factors and their influence on the radical and carbonium ion pathway in the Kolbe electrolysis of carboxylates, $RCOO^-$

Experimental factor	Formation of $R^{\cdot}$ favored by	Formation of R^+ favored by
Current density	High	Low (?)
Anode material	Smooth platinum	Carbon
pH of electrolyte	Neutral or slightly acid	Alkaline
Solvent	Methanol, DMF	Water, water–pyridine
Temperature	Low	Too little known
Added ions	—	ClO_4^-, SO_4^{2-}, HCO_3^-, $H_2PO_4^-$, F^-

extent only, especially with regard to the radical pathway. Structural features sometimes favor carbonium ion formation so strongly that no coupling can be achieved even under the most favorable conditions.

IV. MECHANISTIC CONSIDERATIONS

A. General

In organic electrochemistry, mechanistic studies are conducted along several lines, the choice of which is highly dependent on the individual investigator's educational background and main interests[6]. Three general approaches to the problem of organic electrode mechanisms are recognizable.

(1) Elucidation of the *chemical* reaction mechanism, with the aim of describing the process in terms of the usual language of physical organic chemistry[8]. Product analysis, polarography, controlled potential electrolysis, and electron spin resonance studies are important tools for this purpose. In most cases analogies from homogeneous solution chemistry provide important clues to the chemical mechanism, and the electrode is only considered to be a convenient source of reactive intermediates, be it radicals, anions, cations, or carbenes[40], but not otherwise of importance for the reaction.

(2) Study of the *electrochemical* reaction mechanism, in which the heterogeneous nature of electrode processes is properly acknowledged, *i.e.* the various steps occurring at the electrode surface (adsorption, electron transfer, desorption) are specified.

(3) Study of the *energetic* reaction mechanism, in which a detailed mathematical description of the electron transfer steps is the ultimate goal.

In this review we shall emphasize studies on the first, least sophisticated (at least from the electrochemical point of view) level, since after all, we as organic chemists are mostly interested in using anodic oxidation either as a probe into the chemistry of short-lived intermediates or as a synthetic tool. Consequently, we shall postulate that the possible intermediates behave in very much the same way as their counterparts in homogeneous solution, which means that specific electrode interactions (such as adsorbed species being the product-forming entities) shall not be invoked in order to explain a particular result. To exemplify, one would tend to regard anomalous (*i.e.* with regard to similar homogeneous reactions) stereochemical results as being due to adsorbed species which would be shielded from attack by molecules or ions in the solution on one side by the electrode surface. In fact, very few, if any, anodically formed products display such features as to make such an assumption necessary, and we shall therefore regard the postulate as a valid first approximation. However, before considering the chemical reaction mechanism, some features of the electrochemical mechanism must be discussed.

B. The Electrochemical Mechanism

In section II the Kolbe reaction was written as involving an initial discharge of the carboxylate ion onto the electrode surface (this theory is often denoted 'the discharged ion theory' and was proposed as early as 1891 by Brown and Walker[16]) to form an acyloxy radical which then undergoes further reaction. Although other theories have been proposed[13-15], the discharged ion theory or modifications thereof has remained the most satisfactory general theory of the Kolbe reaction.

Much interest has been devoted to the fact that there seems to exist a critical potential below which a platinum anode does not support the formation of Kolbe products (see section III. A and Figure 1). In the transition region the anode potential suddenly jumps by some tenths of a volt for a very slight change in current density, and this phenomenon is observed in aqueous as well as non-aqueous media. In an aqueous medium, the passage through the transition region is accompanied by a switch from the oxygen evolution process to the Kolbe reaction. In non-aqueous solution the formation of Kolbe product (C_2F_6) has also been observed in the low-potential region[41] in the system CF_3COO^-/CF_3COOH. It is not known whether this behavior is general in non-aqueous media; in aqueous solution, it has been shown that carboxylates are completely oxidized in a non-specific manner to carbon dioxide at low anode potentials[42, 43]. Propionate and butyrate do form olefins at low anode potentials in aqueous solution, but there is still a critical potential for the formation of coupled product[44].

Table 2 is a survey of critical potentials for the formation of *coupled products* from a number of carboxylic acids under different conditions. In most cases where aqueous solutions are used no hydrocarbons at all are formed below the critical potential. It is evident that the value of the critical potential is difficult to measure accurately and is not related in any easily discernible manner to the structure of the acid. The most notable feature is that it is high and fairly constant with respect to structural changes.

Originally it was thought that the critical potential represented the discharge potential of the carboxylate ion, but this assumption has later

TABLE 2. Critical potential for the formation of R_2 by electrolysis of $RCOO^-$ at a platinum anode

R in $RCOO^-$	Solvent	Critical potential versus the hydrogen electrode	Reference
CH_3	H_2O	2·5	46
CH_3	H_2O	2·7	47
CH_3	H_2O	2·1–2·2	48
CH_3	H_2O	2·2	49
CH_3	H_2O	2·1–2·2	45
CH_3	$HOCH_2CH_2OH$	2·1	50
C_2H_5	H_2O	2·7	51
C_2H_5	H_2O	2·8–3·0	47
$n\text{-}C_4H_9$	H_2O	2·1	44
$n\text{-}C_5H_{11}$	H_2O	2·2	44
$n\text{-}C_6H_{13}$	H_2O	2·2	44
$n\text{-}C_7H_{14}$	H_2O	2·1	44
$n\text{-}C_{11}H_{23}$	H_2O	2·7	52
$EtOCOCH_2$	H_2O	2·8	53
$EtOCOCH_2$	$HOCH_2CH_2OH$	3·4	54
CF_3	CF_3COOH	2·3	41

proved to be wrong. For acetate electrolyses in aqueous solution, Dickinson and Wynne–Jones[45] concluded that the oxygen evolution reaction was gradually suppressed with increasing anode potential and current density by an increased proportion of the discharge sites being occupied by adsorbed acetoxy radicals. With a sufficiently high current density all of the discharge sites will be occupied by adsorbed acetoxy radicals, and at even higher current densities the discharge of acetate ions onto already occupied sites must occur. It was postulated that ethane formation occurs only under these conditions. This modified discharged ion theory implies that an adsorbed acetoxy radical is a fairly stable species — as compared to its behavior in homogeneous solution (see below) — but that the presence of two acetoxy radicals at the same site will lead to instability and decarboxylation to form ethane and carbon dioxide, via methyl radicals. The high value of the critical potential was believed to be the result of the difficulty of discharging an acetate ion onto an occupied site. The formation of small amounts of methane was easily accounted for by assuming that methyl radicals abstract hydrogen atoms from acetate ions or acetic acid molecules. Formation of methanol was believed to occur via coupling of simultaneously formed methyl and hydroxyl radicals (at an anode potential intermediate between the low- and high-potential region).

Conway and Dzieciuch[55] studied the anodic oxidation of formate, acetate, and trifluoroacetate ions in the corresponding anhydrous acids or in aqueous solution and concluded that the transition region corresponds to filling of the electrode surface with adsorbed intermediates (acyloxy radicals) both in the anhydrous and aqueous cases. These radicals form a passive film on the platinum surface and this is a prerequisite for the Kolbe reaction (although this statement seems to be at variance with the observation that C_2F_6 is actually formed from CF_3COO^- in CF_3COOH in the low-potential region). Again, alcohol formation was ascribed to mixed coupling between $R^\cdot$ and $OH^\cdot$.

Using a repetitive potential pulse technique, *i.e.* non-steady state conditions, Fleischmann, Mansfield and Wynne–Jones[42, 43] have confirmed that the formation of ethane from aqueous acetate requires a build-up of certain critical conditions at the electrode surface, analogously to the mechanisms discussed above. However, the formation of an oxide layer rather than a film of adsorbed acetoxy radicals was considered to be the physical background of the transition region. The build-up of the acetate ion discharge-supporting oxide layer is completed within 10^{-3} sec, whereas electrolysis times of the order of 10^{-4} sec or less produced no ethane, no oxygen, but an amount of carbon dioxide corresponding to a complete, non-specific oxidation of acetate ion.

Thus, starting with an uncovered platinum anode, acetate ion is initially oxidized to carbon dioxide, presumably at 'pure' platinum sites. Within a period of 10^{-3} sec at a sufficiently high anode potential ($> 2\cdot2$ V) an oxide layer has been formed and discharge of acetate ions to form acetoxy radicals can take place. This step is the rate- and potential-determining step. Subsequent decarboxylation of the acetoxy radical gives ethane and other products. The same mechanism was also assumed to be valid in non-aqueous media, since under oxidizing conditions a layer of surface oxide may well be formed even in rigorously dried systems. Fleischmann and coworkers did not consider methyl cations to be likely as intermediates in the formation of methanol.

An alternative potential-determining step was considered by Eberson[56]. The calculated standard potential for the process $CH_3COO^- \rightarrow CH_3COO$ $+e^-$, $+2\cdot41$ V, would place the Kolbe reaction among the group of electrochemically reversible reactions, which is highly improbable in view of the extreme anode potential conditions required. Instead a concerted process, involving simultaneous one-electron transfer and decarboxylation (equation 10) might possibly be invoked (the standard potential of this process

$$RCOO^- \;\text{------}\; R^\cdot + CO_2 + e^- \tag{10}$$

was calculated to be $+1.55$ V). This would be attractive both from energetic and kinetic points of view, since the decarboxylation of the acetoxy radical is exothermic by $\simeq 17$ kcal/mole[57] and its half-life time is of the order of 10^{-10} sec[58]. It should be noted, however, that such calculations, and reasoning based on them, pertain to free, solvated species participating in the electrode process, whereas in the actual reaction adsorbed species may be involved, making the calculated standard potentials uncertain by an unknown term (although this may be expected to be fairly constant for a series of different RCOO. and R˙).

It has often been claimed that the acetoxylation of added aromatic substrates[37, 59, 60], e.g. naphthalene or anisole, during electrolysis of acetate in acetic acid is evidence for the intervention of acetoxy radicals. This view is now untenable, since anodic acetoxylation has been shown to be an initial oxidation of the aromatic substrate to form a cationic intermediate[31, 32, 61–63], which then reacts with acetate ion (section V.B). From this point of view, no objection can be raised against the concerted mechanism.

C. The Chemical Reaction Mechanism

In previous sections reference has often been made to radicals and carbonium ions as product-forming species in the anodic oxidation of carboxylate ions. We shall now examine the *chemical* evidence for the existence of such species and also try to ascertain whether our postulate that electrolytically generated species behave essentially as their counterparts in homogeneous solution is a reasonably good one. Special emphasis will be laid on the carbonium ion aspects, since these have not been discussed in previous reviews.

1. Radical aspects

a. Radical disproportionation and attack on solvent.   The intervention of radicals R˙ as product-forming intermediates from the anodic oxidation of RCOO⁻ is so well demonstrated by now[13–15] that only a selection of the most important arguments will be provided here.

By inference from the behavior of radicals in homogeneous solution, one would expect to find R—R, the coupling product, and RH and $RH - H_2$, the products resulting from disproportionation, hydrogen atom abstraction by R from another molecule, and/or hydrogen atom loss from the β carbon of R (equation 3), from anodically generated radicals also. Indeed these three types are found in most cases among the products isolated, although it must be stressed that *only the formation of* R—R *can be considered an unambiguous proof for the intermediacy of* R˙. In principle, an alkene $RH - H_2$ might be formed via R^+, whereas RH might be formed by a secondary *cathodic* reduction of the alkene (the majority of electrolytic oxidations of carboxylates are performed in undivided cells with platinum cathodes, so that anodic products may undergo secondary reactions at the cathode, at least in principle. Ethylenic double bonds *can* be reduced on platinized platinum[5], and since platinum cathodes gradually acquire the character of platinized platinum during the run even if they are smooth at the start, this possibility cannot be dismissed).

Deuterium labelling experiments have, however, convincingly demonstrated the occurrence of hydrogen atom abstraction by attack of $R^\cdot$ on other molecules. In aqueous trideuterioacetate electrolysis, the methane formed in small amounts with ethane was tetradeuterated[64], consistent with the known tendency of radicals to attack CH bonds in preference to OH bonds (equation 11).

$$CD_3COO^- \xrightarrow[-CO_2]{-e^-} CD_3^\cdot$$

$$CD_3^\cdot + CD_3COO^- \text{ (or } CD_3COOH) \longrightarrow CD_4 + {}^\cdot CD_2COO^- \text{ (or } {}^\cdot CD_2COOH) \quad (11)$$

Aqueous electrolysis of deuterium-labelled propionate furnished evidence[65] (even if not unambiguous) for the occurrence of disproportionation and hydrogen atom loss from $R^\cdot$ (equations 12 and 13). The deuterium content of the ethane portion was in agreement with the assumption that it had been formed via disproportionation of two radicals. However, the amount of ethane was only about 5% of that of ethylene, so this must be formed predominantly via other processes. Clusius and coworkers conclud-

$$CH_3CD_2COO^- \longrightarrow CH_3CD_2 \longrightarrow CH_2{=}CD_2 + CH_3CD_2H \quad (12)$$

$$CD_3CH_2COO^- \longrightarrow CD_3CH_{\cdot 2} \longrightarrow CD_2{=}CH_2 + CD_3CH_2D \quad (13)$$

ed that hydrogen atom loss from the β carbon of $R^\cdot$ was responsible for the major part of the olefin. Since their method of deuterium analysis was based on molecular weight determination of the gases formed, it did not yield any information about the distribution of deuterium in the ethylene. As we shall see in section IV. C. 2, this may be of importance in deciding whether the ethylene originates from $R^\cdot$ or R^+ (although in this particular case there seems to be little doubt that Clusius and coworkers were correct in their choice of mechanism, since the molecular weight in both cases was in excellent agreement with that of dideuterioethylene). Other electrode processes of the concerted type have been considered in the propionate case[56].

b. Additive dimerization. Another characteristic radical reaction which can be performed during Kolbe electrolysis is additive dimerization[66-68]. This type of reaction takes place in homogeneous solution between aliphatic free radicals and 1,3-dienes. The radical (R·) adds to the diene (M) to form a resonance-stabilized radical (5) which can dimerize to produce straight- or branched-chain products arising by 1,4- or 1,2-incorporation of the diene, or add a second radical R· (equations 14, 15, and 16).

$$R· + M \longrightarrow RM·$$ (14)

(5)

$$2\,RM· \longrightarrow RMMR$$ (15)

$$RM· + R· \longrightarrow RMR$$ (16)

Lindsey and Petersen[66] electrolyzed CH_3COOH/CH_3COO^- in methanolic solution in the presence of butadiene and obtained 11–26% yields of 3-hexene and 12–58% yields (the difference in the yields given are probably due to experimental difficulties) of a fraction consisting of C_{10} dienes, the major part of which was 3,7-decadiene (equations 17, 18, and 19).

$$CH_3· + CH_2=CHCH=CH_2 \longrightarrow [CH_3CH_2\overset{.}{C}HCH=CH_2 \longleftrightarrow CH_3CH_2CH=CHCH_2·]$$

(6) (17)

$$2\,CH_3CH_2CH=CHCH_2· \longrightarrow EtCH=CHCH_2CH_2CH=CHEt$$ (18)

$$CH_3CH_2CH=CHCH_2· + CH_3· \longrightarrow EtCH=CHEt$$ (19)

Small amounts of isomeric branched-chain C_{10} dienes were also obtained but not identified. Similarly, 1,1,1,10,10,10-hexafluoro-3,7-decadiene was identified among the products from the electrolysis of CF_3COOH/CF_3COO^- and butadiene in methanol. An interesting development of this reaction was the isolation of a 40% yield of diethyl 3,7-decadiene-1,10-dioate by electrolysis of a solution of potassium ethyl oxalate, monoethyl oxalate, and butadiene in methanol. Obviously, ethoxycarbonyl radicals are formed in the electrolysis of potassium ethyl oxalate (equation 20) and take the place of the methyl radical in equation (17).

$$EtOCO-COO^- \longrightarrow \overset{.}{C}OOEt + CO_2 + e^-$$ (20)

The same reaction was investigated at the same time by Smith and Gilde[67, 68]. By electrolysis of acetate and butadiene in methanol at pH 6–8 they obtained a complicated mixture of products, among which 1-pentene (and no 2-isomer), 3-methyl-1-pentene, *trans*-3-hexene (and no *cis*), 3,7-decadiene, and 3-ethyl-1,5-octadiene were identified, in fair agreement with the results obtained by Lindsey and Petersen. These products are easily accounted for by assuming that the radical 6 is an intermediate, the 1-pentene being formed by a hydrogen atom abstraction reaction.

In addition, a mixture of unidentified acetates was isolated. Hydrogenation and hydrolysis of this fraction gave a mixture of 1-butanol, 2-methyl-

1-butanol, and 1-pentanol. Smith and Gilde assumed that the acetates were formed via anodically generated acetoxy radicals. A discussion of this problem will be deferred to a later section (section V. A), since it is intimately related to other anodic acetoxylation processes which in all probability proceed via cationic intermediates.

The formation of the *trans* isomer only of 3-hexene led Smith and Gilde to extend their study to isoprene, which on coelectrolysis with acetate in methanol gave the *cis* isomer only of 3-methyl-3-hexene. Similarly, coelectrolysis of butadiene and propionate in methanol gave the *trans* isomer only of 4-octene. The cyclic diene, 1,3-cyclohexadiene, after being subjected to electrolysis together with acetate in methanol gave a product which after hydrogenation contained, among other products, methylcyclohexane, *cis*- and *trans*-1,2-dimethylcyclohexane, and *cis*- and *trans*-1,4-dimethylcyclohexane.

The results with the open-chain dienes indicate a high degree of stereospecificity in the addition process, and it was proposed that the diene is adsorbed on the electrode surface in its most stable conformation (the *s-trans* form for butadiene and *s-cis* form for isoprene) at the time when radical attack occurs. For the cyclic diene, the possibility that the reaction takes place in solution was not completely ruled out but was considered unlikely in view of the results with the acyclic dienes.

This hypothesis would evidently be in conflict with the postulate not to invoke adsorbed species in the chemical reaction mechanism. However, in the absence of knowledge about the stereochemistry of similar processes in homogeneous systems[69], opinion must be reserved as to whether it is actually necessary to make this assumption.

An additive dimerization was also observed in the electrolysis of acetate ion in anhydrous acetic acid in the presence of styrene[70] (equation 21). In addition, an acetate (7) was identified, the possible mode of formation of which will be discussed in connection with anodic acetoxylation (section V. A).

$$PhCH=CH_2 \xrightarrow{CH_3^{\cdot}} Ph\dot{C}HCH_2CH_3 \longrightarrow Ph-CH-CH-Ph \qquad (21)$$
$$\qquad\qquad\qquad\qquad\qquad\qquad\qquad\qquad | \quad\; |$$
$$\qquad\qquad\qquad\qquad\qquad\qquad\qquad\qquad Et \quad Et$$

meso and *dl*

$$CH_3COOCH_2CH(Ph)CH_3$$

(7)

An additive dimerization product was also isolated from a similar experiment in $C_2H_5COOH/C_2H_5COO^-$.

c. *Initiation of polymerization.* Goldschmidt and Stöckl[70] also isolated small amounts of polystyrene (molecular weight $\simeq 3000$) in the above-mentioned experiments, again confirming the intermediacy of radicals in the anodic oxidation of acetate and propionate (the reaction was performed in a divided cell and the polymer was formed in the anolyte only). Acrylonitrile could be polymerized in a similar manner. Smith and coworkers[71, 72] later extended these studies to vinyl acetate, methyl methacrylate, vinyl chloride, and acrylic acid, which all gave polymers upon rapid stirring together with an electrolyzing aqueous solution of potassium acetate. Again it was shown that the polymerization takes place in the anode compartment of a divided cell. Furthermore, if the electrolysis was run with aqueous potassium acetate-2-^{14}C as the electrolyte, the polymer formed from vinyl acetate was radioactive. Acetate exchange between the electrolyte and the polymer could not alone account for the uptake of ^{14}C, so it was concluded that acetoxy or methyl radicals generated at the anode had served as polymerization initiators.

The reactions discussed above, *viz.* dimerization, additive dimerization, and initiation of polymerization are some of the best criteria known for radical processes, and their occurrence during Kolbe electrolysis unequivocally proves the intervention of radicals. *Qualitatively*, these radicals undergo the same type of reactions as radicals produced in homogeneous systems, and we will therefore now have to turn our attention to the quantitative or at least semiquantitative aspects. Do we make an unsound assumption when we postulate that products are formed from radicals *in solution?*

d. *Comparison between anodically and chemically generated radicals.* To begin with, it must be emphasized that it is a difficult task to compare a heterogeneous and a homogeneous process, even if the intermediate(s) is (are) the same. In this particular case, neglecting the possible role of adsorbed radicals, the anode will produce a high concentration of radicals in the immediate vicinity of the electrode surface and there will be a more or less rapidly falling concentration gradient as the distance from the electrode increases. These conditions cannot possibly be simulated in homogeneous solution, where a uniform, very low concentration of radicals will be maintained. In reactions where two radicals are formed simultaneously from the same molecule, as is the case in diacyl peroxide and bis-azonitrile decomposition, cage reactions will be of importance, and this will further complicate the comparison.

Another, although less serious problem, is the necessity to distinguish between the primary products from the electrode process and secondary products formed by anodic oxidation of the primary ones. Any carboxylic acid formed in a primary reaction will of course be liable to electrolysis forming secondary products, *etc.* In view of present refinements in separation and identification techniques, this difficulty should be easily circumvented by running the electrolysis to an extent of 1% or less, thus ensuring that the primary reactant is present in large excess all the time.

One obvious prediction from the differences in radical concentration in the anodic and homogeneous cases, is that in the former case bimolecular reactions between radicals will be favored at the expense of the hydrogen atom abstraction reaction between radicals and solvent, since the solvent concentration around the anode will be lower. Apart from possible complications from cage reactions, the opposite behavior would be expected from radicals in homogeneous systems. However, if we observe such a difference, it is largely a matter of personal judgment and conviction to decide whether the difference is due to a concentration effect only or to the heterogeneous nature of the electrode process (or to any other effect).

Goldschmidt, Leicher, and Haas[73] compared the electrolysis of potassium propionate in anhydrous propionic acid at 100° and the decomposition of propionyl peroxide in the same solvent at 100° (there was no salt added in this experiment). The results are shown in Table 3. In agreement with the prediction above, the product (ethane) formed by reaction be-

TABLE 3. Products formed by decomposition of propionyl peroxide or by electrolysis of propionate in anhydrous propionic acid[73]

Product	Composition of product from decomposition of 1 mole of $(C_2H_5COO)_2$	Composition of product from electrolysis of 2 moles of $C_2H_5COO^-$
CO_2	61·6[a]	63·4[b]
C_2H_4	3·0[a]	5·6[b]
C_2H_6	28·1[a]	14·0[b]
C_4H_8	—	Very little
C_4H_{10}	6·9[a]	15·8[b]
$CO+O_2$	0·4[a]	1·0[b]
Ethyl propionate	0·13 mole	0·02–0·06 mole
sec-Butyl propionate	—	Traces
Diethyl ketone	0·001 mole	Traces
Propionic acid	0·28 mole	Not detectable
Methylethylacetic acid	0·038 mole	0·002–0·014 mole
Ethylpropylacetic acid	0·0006 mole	Traces
2,3-Dimethylsuccinic acid	0·106 mole	0·00015–0·002 mole

[a] Percentage of total gaseous products.
[b] Percentage of anode gases.

tween $C_2H_5^-$ and solvent is less favored in the electrolytic reaction. In addition, it is noticeable that the fate of the $CH_3\dot{C}HCOOH$ radical (8) formed is different in the two cases. In the neighborhood of the anode, ethyl radicals are in abundant supply, and methylethylacetic acid is the major product from (8) (equation 22). In peroxide decomposition, the concentration of $C_2H_5^-$ is very low, thus favoring the formation of 2,3-dimethylsuccinic acid (equation 23).

$$C_2H_5COO^- \xrightarrow[- CO_2]{- e^-} C_2H_5^- \xrightarrow{CH_3CH_2COOH} C_2H_6 + CH_3\dot{C}HCOOH \xrightarrow{C_2H_5^-} \begin{array}{c} CH_3 \\ \diagdown \\ \diagup \\ C_2H_5 \end{array}\!\!CH\!-\!COOH$$

$$\tag{22}$$

$$(C_2H_5COO)_2 \xrightarrow{\Delta} C_2H_5^- \xrightarrow{CH_3CH_2COOH} C_2H_6 + CH_3\dot{C}HCOOH \tag{23}$$

$$\textbf{(8)}$$

$$\downarrow CH_3\dot{C}HCOOH$$

$$CH_3\!-\!CH\!-\!COOH$$
$$CH_3\!-\!CH\!-\!COOH$$

The formation of ethyl propionate in the decomposition of propionyl peroxide is commonly assumed to occur via a cage reaction between $C_2H_5COO\cdot$ and $C_2H_5^-$. Ethyl propionate formed during electrolysis may be accounted for in the same way, but is most likely a carbonium ion mediated product (section IV. C. 2).

Alkylation of an added aromatic substrate is occasionally observed during electrolysis of carboxylates, and can also be accomplished by decomposing acyl peroxides in the appropriate solvent or solvent mixture. Goldschmidt and Minsinger[74] compared electrolytic and peroxide-mediated alkylation of pyridine under similar conditions and obtained mixtures of 2- and 4-alkylpyridines (which in addition contained small amounts of higher alkylated products). The results are shown in Table 4. Generally,

TABLE 4. Yields of alkylpyridines and isomer ratios in the decomposition of diacylperoxides and in the electrolysis of carboxylate ions, respectively, in pyridine[74]

Reaction	Total yield of 2- and 4-alkylpyridines (%)	Isomer ratio, 2-/4-alkylpyridine
Electrolysis of CH_3COO^- (100°)	3·5	2·8
Decomposition of $(CH_3COO)_2$ (100°)	86	7·6
Electrolysis of $C_2H_5COO^-$ (100°)	8·7	1·4
Decomposition of $(C_2H_5COO)_2$ (100°)	87	2·1
Electrolysis of n-$C_3H_7COO^-$ (100°)	4·4	5·1
Decomposition of $(n$-$C_3H_7COO)_2$ (100°)	84	2·4

the yield of alkylpyridines from electrolysis is much less than in the peroxide reaction, again indicating that reaction with solvent is less important
than reactions between radicals (main products were those shown in
Table 3) in the electrolytic reaction. Since the ratio between 2- and 4-alkyl-
pyridine was somewhat dependent on the ratio between the amounts of
pyridine and propionic acid used in the electrolytic experiments and the
medium was different in the two cases due to experimental difficulties, it
is difficult to attach any significance to these figures.

Hey and Bunyan[75] determined the ratio between 2-, 3-, and 4-phenyl-
pyridine formed during electrolysis of benzoic acid in pyridine, and compared the result with those obtained in homogeneous homolytic phenylation reactions (Table 5). The isomer distributions are almost identical in

TABLE 5. Ratios of isomers obtained in the phenylation of
pyridine by different methods[75]

Method	Isomers (%)		
	2-	3-	4-
Benzoyl peroxide	54	32	14
Lead tetrabenzoate	52	32·5	15·5
Phenyl iodosobenzoate	58	28	14
Electrolysis of PhCOO⁻	56	35	9

all cases and demonstrate the close similarity between anodically and chemically formed radicals, although it must be remembered that benzoic
acid is not a good model compound for carboxylic acids in the Kolbe reaction. The benzoyloxy radical is much more stable than aliphatic acyloxy
radicals (the half-life of PhCOO$\cdot$ being of the order of seconds[56], compared
to that of $CH_3COO\cdot$, 10^{-9}–10^{-10} sec) and can diffuse away from the
electrode and decarboxylate in the bulk of the solution (*i.e.* > 100 Å from
the electrode), thus creating a phenyl radical in essentially the same surroundings as one formed in a homogeneous process. Incidentally, decarboxylation is a minor pathway for the disappearance of the benzoyloxy radical, the predominant one being reaction with solvent to form benzoic acid.

Intramolecular anodic alkylations of aromatic rings have been observed
in a few cases, *e.g.* formation of tetralin in the electrolysis of δ-phenyl-
valeric acid[76].

Azobisisobutyronitrile (9) is thermally decomposed (equation 24) in
carbon tetrachloride to give initially a ketenimine (10) in an estimated
54% yield and tetramethylsuccinonitrile in 46% yield, corresponding to

carbon to nitrogen and carbon to carbon coupling of the intermediate cyanoisopropyl radical (11)[77].

$$
\underset{(9)}{(CH_3)_2\overset{\displaystyle CN}{\underset{|}{C}}-N=N-\overset{\displaystyle CN}{\underset{|}{C}}(CH_3)_2} \longrightarrow N_2 + [(CH_3)_2\overset{\displaystyle \cdot}{C}-C\equiv N \longleftrightarrow \underset{(11)}{(CH_3)_2C=C=N^\cdot}]
$$

$$
\xrightarrow[\text{coupling}]{\text{C—N or C—C}} \underset{(10)}{(CH_3)_2C=C=N-\overset{\displaystyle CN}{\underset{|}{C}}(CH_3)_2} + \underset{(CH_3)_2\overset{|}{C}-CN}{(CH_3)_2\overset{|}{C}-CN} \qquad (24)
$$

Anodic generation of α-cyanoalkyl radicals from two α-cyanoacetic acids, $NCCH_2COOH$ and t-$BuCH(CN)COOH$, in methanolic solution[78, 79] gave a mixture of dinitrile and the nitrile amide 12 in the approximate proportions 2 : 3 and 1 : 1, respectively (equation 25, where R = H or t-Bu). The nitrile amide was evidently formed by addition of water to the

$$
\underset{CN}{\overset{RCHCOO^-}{\underset{|}{}}} \xrightarrow[-e^-]{-CO_2} \underset{CN}{\overset{RCH^\cdot}{\underset{|}{}}} \longrightarrow \underset{RCHCN}{\overset{RCHCN}{\underset{|}{}}} + RCH_2CONHCH(CN)R \qquad (25)
$$

$$
(12)
$$

initially formed ketenimine. Cyanoisopropyl radicals generated by electrolysis of dimethylcyanoacetic acid[79] gave a ratio between carbon to carbon and carbon to nitrogen coupling products of about 4 : 1. Thus, cyanoalkyl radicals generated by either electrolytic or chemical means behave qualitatively in the same manner.

   e. Stereochemistry of anodically generated radicals. Little is known about the stereochemistry of radicals formed in the Kolbe electrolysis. Kharasch, Kuderna, and Urry[80] have reported that optically active monoethyl methylethylmalonate gives inactive diethyl 2,3-dimethyl-2,3-diethyl-succinate, and Wallis and Adams[81] isolated inactive 3,4-dimethylhexane from the electrolysis of (+)-2-methylbutanoic acid. Likewise, an additive dimerization product formed by coelectrolyzing sodium (+)-2-methylbutanoate and butadiene in methanol was inactive[68]. However, none of these reactions gives a really satisfactory demonstration that the intermediate radical has lost its optical activity completely. All three products were presumably mixtures of *meso* and *dl* forms and would be anyway expected to possess small specific rotations, thus making it possible that some optical activity was retained but not experimentally detected. It appears safe to conclude, though, that anodically generated radicals lose their optical activity to a large extent, and it remains to be seen if racemization really is complete. Such stereochemical studies would be of great importance since reactions between adsorbed radicals (section IV. B) would be predicted to proceed with at least some retention of configuration.

In homogeneous solution, allylic radicals possess a defined stereoche-mistry, the configuration of the double bond in the starting material being largely retained in the product(s). Kolbe electrolysis[82] of *cis*- and *trans*-hex-3-enoic acid (equation 26) in methanol gave products **13, 14** and **15** in the proportions 42 : 45 : 13, and the configuration of the double bond in the acid was retained to an extent of $> 90\%$ in **13** and **14**.

$$EtCH\!=\!CHCH_2COO^-$$

$$\downarrow \begin{array}{l} -e^- \\ -CO_2 \end{array}$$

$$[EtCH\!=\!CH\!-\!CH_2^\cdot \longleftrightarrow Et\dot{C}H\!-\!CH\!=\!CH_2] \xrightarrow{coupling}$$

$$\underset{\textbf{(13)}}{\begin{array}{c} EtCH\!=\!CH\!-\!CH_2 \\ | \\ EtCH\!=\!CH\!-\!CH_2 \end{array}} + \underset{\textbf{(14)}}{\begin{array}{c} EtCH\!=\!CH\!-\!CH_2 \\ | \\ Et\dot{C}H\!-\!CH\!=\!CH_2 \end{array}} + \underset{\textbf{(15)}}{\begin{array}{c} EtCH\!-\!CH\!=\!CH_2 \\ | \\ EtCH\!-\!CH\!=\!CH_2 \end{array}} \quad (26)$$

On the other hand, generation of the same allylic radical by electrolysis of $RCH(COO^-)CH\!=\!CH_2$ gave, as expected, all geometrical isomers of **13** and **14**.

Summarizing, radicals formed by the anodic oxidation of carboxylates behave qualitatively and sometimes even quantitatively in the same way as radicals generated via ordinary reactions in solution. The small differenc-es observed are best explained by the widely differing concentration rang-es prevalent in the two cases. However, the search for possible conse-quences of the participation of adsorbed species in product formation should be continued. These will most likely be stereochemical in nature.

2. Carbonium ion aspects

a. Introductory discussion. As mentioned above, Walling[19] was the first to suggest that carbonium ions might be formed from anodically ge-nerated radicals by a second one-electron transfer to the anode (equa-tion 4). Similar ideas had been vaguely expressed earlier ('oxidation of radicals to form alcohols', 'partial carbonium ion character in the radical') but the ultimate consequence, *i.e.* formation of full-fledged carbonium ions, was not elaborated. Experimental verifications of this hypothesis soon followed[20], and it also became evident that most of the 'by-products' found in earlier investigations were easily accounted for by the carbonium ion mechanism. We shall now discuss anodically-formed carbonium ions in the same way as was done for radicals, starting with the problem of defining criteria for demonstrating the intermediacy of carbonium ions in the Kolbe reaction.

In section IV. B, it was mentioned that the formation of methanol from acetate ions was considered to occur by reaction between simultaneously generated methyl and hydroxyl radicals. By analogy, one might assume that alcohols, ethers, and acetates are formed via a similar reaction between R· (or a rearranged radical) and HO·, MeO·, and AcO· in water, methanol, and acetic acid, respectively. However, this approach immediately causes serious, if not insurmountable, difficulties. Most importantly, one finds that in all cases the alcohol, ether, or acetate has been found to contain a *rearranged* carbon skeleton, while the accompanying dimer (R_2) contains *unrearranged* R groups. Thus, one is forced to assume that anodic radicals behave in a dramatically different way in the self-coupling reaction as compared to coupling with HO·, MeO·, or AcO·. Even if this might in principle be explained by a suitable manipulation of kinetic parameters, this gives a highly unsatisfactory explanation because of the additional variables introduced.

In fact, no case of a *radical* rearrangement has been found in the Kolbe reaction, if the only unambiguous criterion for the occurrence of such a rearrangement is rigorously adapted, *i.e.* it must be demonstrated by analysis of the self-coupling product (R_2) or the mixed coupling product with a different hydrocarbon radical (R—R′). As already mentioned none of the other possible products can be unequivocally traced back to a radical reaction (section IV. C. 1).

The search for radical rearrangements during Kolbe electrolysis has been concentrated to neopentyl- and neophyl-type radicals, since such radicals, generated chemically, are known to undergo rearrangements. Thus, Breederveld and Kooyman[36] coelectrolyzed 3-phenylisovaleric acid with acetic acid in methanol and obtained *t*-amylbenzene (**16**) as the sole mixed coupling product and bineophyl (**17**) as the sole self-coupling product (equation 27). Eberson and Sandberg[83] electrolyzed a number of malonic half esters of the general formula $RC(CH_3)_2CH(COO^-)COOEt$

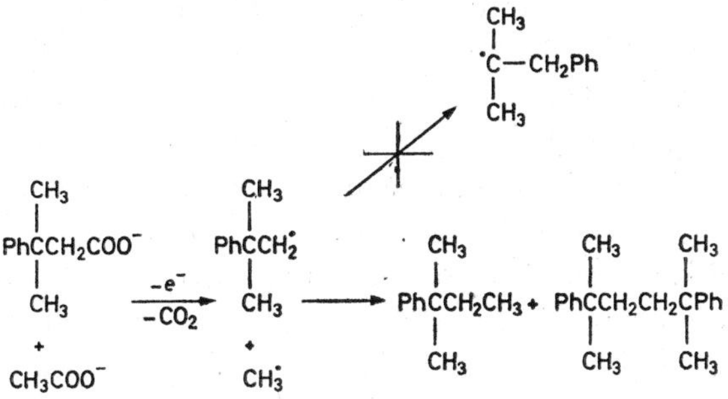

(27)

(R = Me, Et, Pr, iso-Pr, Ph) and obtained unrearranged coupling products only. Muhs[18] isolated bineopentyl as the only coupling product from the Kolbe electrolysis of t-butylacetic acid. Also, neopentyl-type radicals formed from alicyclic acids (*e.g.* equation 28) gave unrearranged dimers only[18].

$$\text{(18)} \qquad\qquad \xrightarrow[-CO_2]{-e^-} \qquad\qquad \longrightarrow \qquad\qquad \text{(28)}$$

Having established that a case of a radical rearrangement during Kolbe electrolysis still remains to be found, we can now use the formation of rearranged products as the best criterion for the intermediacy of carbonium ions. In equation (5), two possible rearranged products are shown (R'OR'' and R'OCOR), and we can also add the rearranged alkene R'H $-$ H$_2$ formed by proton loss from R'$^+$. The unrearranged products in equation (5), ROR'' and ROCOR, are in all probability formed from R$^+$, but it must be stressed that radical processes cannot *a priori* be excluded (*i.e.*, R$\cdot$ $+$ $\dot{\text{O}}$R'' $\rightarrow$ ROR''; R$\cdot$ $+$ RCOO$\cdot$ $\rightarrow$ ROCOR). Any unrearranged alkene, RH $-$ H$_2$, can in principle originate either from R$\cdot$ (equation 3, disproportionation) or from R$^+$ by proton loss. Some idea of the importance of these routes can be obtained by considering the amount of RH formed, since RH is formed in equimolar amounts in the disproportionation step. It must be emphasized, though, that this cannot be done quantitatively because of the other possible pathways leading to RH (section IV. C. 1). Again, work-up and analysis at a very early stage of the reaction should help considerably to minimize secondary processes.

We shall now discuss some examples of electrolytic carbonium ion formation and try to make comparisons with results obtained in ordinary chemical reactions. It is left to the reader to judge for himself if alternative explanations along the lines discussed above are to be preferred. In most cases to be discussed below, yields will be given as relative percentages of the total amount of products explicitly mentioned (*i.e.* products which are obviously of radical origin will be omitted) in order to facilitate comparison with other carbonium ion reactions.

b. Fatty acids. Some old observations of what must now be considered to be electrolytic carbonium ion formation from aliphatic acids have been summarized in Table 6, together with results obtained in the corresponding homogeneous carbonium ion processes. It must be remembered

TABLE 6. Carbonium ion mediated products (except the gaseous ones) in the electrolysis of fatty acids or in the corresponding deamination reaction

Reaction	Products, yields in mole % within parenththeses	Reference
Electrolysis of $CH_3CD_2COO^-$	Mixture of ethanols with approximately statistical distribution of D over the C atoms	84
Deamination of $CH_3^{14}CH_2NH_2$	CH^3 $^{14}CH_2OH$ (98), $^{14}CH_3CH_2OH$ (2)	85
Electrolysis of n-PrCOO$^-$	iso-PrOH (82), n-PrCOO-iso-Pr (13), n-PrCOO-n-Pr (5)	86
Deamination of n-PrNH$_2$	iso-PrOH (70), n-PrOH (30)	87

that the results of these early electrolytic experiments are uncertain and incomplete due to the experimental difficulties in separating and isolating closely similar products at the time they were done, so we can only conclude that electrolytic carbonium ion formation has taken place and that hydride and alkyl shifts occur to at least to the same extent as in deamination.

Recently, Skell and Maxwell[88] studied the electrolysis of 3-methylvaleric acid in detail and Table 7 shows a comparison of the electrolytic products with those obtained in the deoxidation of the corresponding alcohol. The product spectrum is very similar in the two cases, and it is especially interesting to note that cyclopropane formation is observed in electrolysis

TABLE 7. Comparison between gaseous products formed in the deoxidation of 2-methyl-1-butanol and electrolysis of 3-methylvaleric acid, respectively[88]

Product	Reaction	
	Deoxidation of 2-methyl-1-butanol, yields (%)	Electrolysis of 3-methylvaleric acid, yields (%)
2-Methyl-1-butene	48·2	36·4
2-Methyl-2-butene	11·3	9·3
trans-2-Pentene	13·5	8·5
cis-2-Pentene	7·9	6·6
1-Pentene	12·3	16·1
Ethylcyclopropane	2·1	3·0
1,2-Dimethylcyclopropane	2·0	3·0
3-Methyl-1-butene	1·2	1·9
Minor unidentified products	1·5	15·6

also, demonstrating the close similarity between cations produced by electrolysis and deoxidation.

Koehl[23] investigated the electrolysis of three isomeric C_5 acids at a carbon anode (which strongly promotes the formation of carbonium ion mediated products: see section III. B   ). Table 8 gives the results obtained

TABLE 8. Gaseous and liquid products formed in the electrolysis of C_5 acids[23]

Acid	Total yield of gaseous products	Composition of gaseous products (mole %)				
		Butenes				Methylcyclopropane
		1-	iso-	trans-2	cis-2	
Valeric	~35	58·3	3·0	19·2	9·5	8·4
2-Methylbutyric	~35	51·5	2·2	27·1	12·7	4·9
3-Methylbutyric	~35	30·7	19·5	24·9	11·9	9·3

Acid	Total yield of liquid products (%)	Composition of liquid products (%)					
		Butanols			Esters		
		t-	2-	1-	t-	2-	1-
Valeric	~10	—	15·0	—	—	55·9	11·0
2-Methylbutyric	~20	—[a]	40·0	—[a]	2·1	39·8	5·4
3-Methylbutyric	~20	5·4	19·5	—	12·9	50·6	—

[a] Compound formed in trace amounts.

in aqueous solution under slightly acidic conditions (pH $\simeq$ 6), and Table 9 shows a comparison of the electrolytic products with those obtained in the deoxidation[89] and deamination of the corresponding alcohol and amine, respectively. The comparison is based on the 2- to 1-butene and trans- to cis-2-butene ratios and the total yield of isobutene and methylcyclopropane obtained. Generally, all three reactions give similar results, although there are some distinct differences in the 2- to 1-butene and trans- to cis-2-butene ratios in some cases. Koehl interpreted these results in terms of a mechanism involving 'hot' carbonium ions formed by anodic oxidation of alkyl radicals (the possibility that cations were formed via the reaction $RCOO^- \rightarrow RCOO^+ + 2e^-$, followed by decarboxylation of $RCOO^+$, was dismissed). The differences in the above-mentioned ratios between the electrolytic and homogeneous reactions were thought to indicate that pro-

TABLE 9. Comparison between products formed from n-butyl, *sec*-butyl, and isobutyl cations produced by electrolysis, deamination, or deoxidation

Reaction	Butene ratios		Yield in % of		Reference
	2-/1-	*trans-/cis-*	isobutene	methyl-cyclo-propane	
Electrolysis of n-BuCOO⁻	0·5–0·7	1·5–2·0	3–4	8–13	23
Deamination of n-BuNH₂	0·41	2·2	0	—	90
	0·67	1·7	—	—	91
Deoxidation of n-BuOH	—	1·85	—	2	89
Electrolysis of *sec*-BuCOO⁻	0·6–0·8	1·9–2·2	2	3–5	23
Deamination of *sec*-BuNH₂	3·0	2·9	—	—	90
	3·55	2·1	—	—	91
Deoxidation of *sec*-BuOH	—	1·67	—	0·5	89
	1·37	1·66	—	Trace	23
Electrolysis of iso-BuCOO⁻	0·7–1·2	1·6–2·1	20–30	7–10	23
Deamination of iso-BuNH₂	2·67	1·40	34	—	92
Deoxidation of iso-BuOH	—	1·82	—ᵃ	4	89
	0·77	1·60	55	5	23

ᵃ Compound formed in trace amounts.

duct formation might occur via cations specifically associated with the anode. In the absence of any detailed studies of the electrochemical mechanism it is difficult to test this idea, neither is it clear whether deamination and deoxidation are good chemical analogues of the electrochemical process. These reactions produce carbonium ions by loss of neutral molecules or ions from cationic or carbenic precursors (equations 29 and 30), whereas

$$R—N_2^+ \longrightarrow R^+ + N_2 \qquad (29)$$

$$RO\ddot{C}Br \longrightarrow R^+ + CO + Br^- \qquad (30)$$

anodic carbonium ions are formed either by direct oxidation of alkyl radicals (equation 4) or possibly by loss of carbon dioxide from $RCOO^+$. In the former case the oxidation of alkyl radicals by copper(II) salts[93] should be a better chemical analogue to electrolytic carbonium ion oxidation. However, although the butene ratios found in the oxidation of *sec*-butyl radicals by simple cupric salts are in very good agreement with those obtained in the electrolysis of 2-methylbutyric acid, this analogy fails completely for n-butyl and isobutyl cations.

    c. Substituted aliphatic acids. We shall deal with α-substituted aliphatic acids only, since there is ample evidence that substituents which lead to anomalous results in the α position, exert only a minor influence when further removed from the carboxyl group[13-15], *i.e.*, such substituted acids behave similarly to the unsubstituted ones. It is in the α position that a substituent can influence the properties of an intermediate radical or carbonium ion, *e.g.* by conjugation with the reactive center, and thus the product distribution, most strongly.

Table 10 provides a number of examples of the effect of α substitution. The results are presented as total yields of products formed via the radical

TABLE 10. Yields of products formed via the radical or carbonium ion path in Kolbe electrolysis of substituted aliphatic acids in methanol (R′ = alkyl or phenyl)

R in RCOO⁻	Isolated yield of products formed via radicals (%)	Isolated yield of products formed via carbonium ions (%)	Reference
R′CONHCH₂	<15	60–80	34
R′(CH₃)₂CCH(COOEt)	20–85	3–16	83
(CH₃)₃CCH(CONH₂)	55	~20	94
(CH₃)₃CCH(CN)	61	<2	79
PhCH₂	55	0 ?	37
4-NO₂C₆H₄CH₂	33	16	95
2-NO₂C₆H₄CH₂	0	17	95
4-CH₃OC₆H₄CH₂	0	88	96
Ph₂CH	8	35	97
Ph₂CH	0	80	26
Ph₃C	0	60	37
PhCH(OCH₃)	0	62	98
Ph₂C(OCH₃)	0	74	98

and carbonium ion pathway, respectively. It is evident that certain substituents (CN, COOR, CONH₂) favor the radical step, whereas others (OR, NHCOR, Ph) favor the carbonium ion process. As will be discussed later (section IV. C. 2. h) a low ionization potential of the intermediate radical will favor the carbonium ion reaction.

d. Alicyclic and bicyclic acids. The chemistry of alicyclic and bicyclic compounds is abundantly rich in carbonium ion rearrangements, and it is therefore not surprising that a number of investigations have been devoted to the study of the anodic oxidation of these acids. In fact, the first convincing verifications of anodic carbonium ion formation were obtained in the alicyclic and bicyclic series[20].

Anodic oxidation of cyclobutanecarboxylic acid in aqueous solution[20] afforded a 30% yield of a mixture of cyclopropylcarbinol, cyclobutanol, and allylcarbinol identical in composition with that resulting from deamination of cyclobutylamine.

Electrolysis of 1 methylcyclohexaneacetic acid (18) in methanol[18] produced dimer (58%), 1-methylcycloheptene (11%), and methyl 1-methylcycloheptyl ether (13%), the last two products being formed by rearrangement of the initially formed primary cation (equation 31). Analogous results were obtained in the electrolysis of 1-methylcyclopentaneacetic acid.

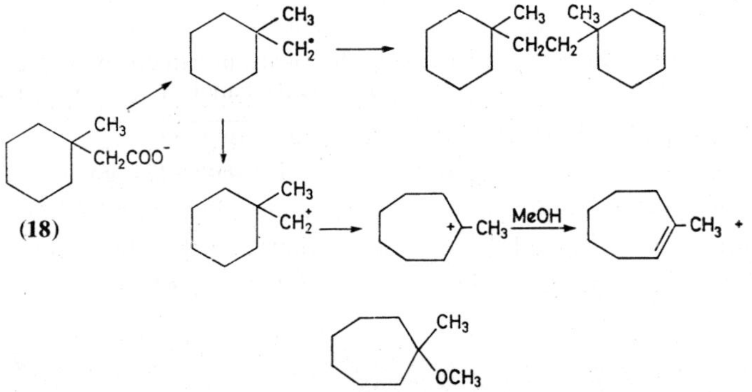

(31)

Anodic oxidation of *exo*- or *endo*-norbornane-2-carboxylic acids[20] (19 or 20) gave *exo*-norbornyl methyl ether (22) in 35–40% yield as the sole volatile product with no *endo* isomer detectable (equation 32).

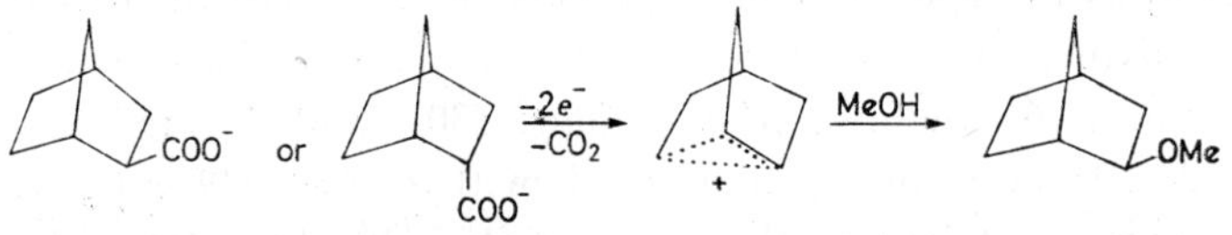

(32)

(19)   (20)   (21)   (22)

Moreover, the *exo* methyl ether 22 obtained from optically active 20 was racemic. Electrolysis of *exo*- or *endo*-5-norbornene-2-carboxylic acid[20] (23 or 24) gave 3-methoxynortricyclene (26) in 56% yield. The formation of ethers 22 and 26 in these cases correspond exactly to the products formed via the bridged ions 21 and 25 in solvolysis, whereas the corresponding deamination reactions indicate a higher degree of retention

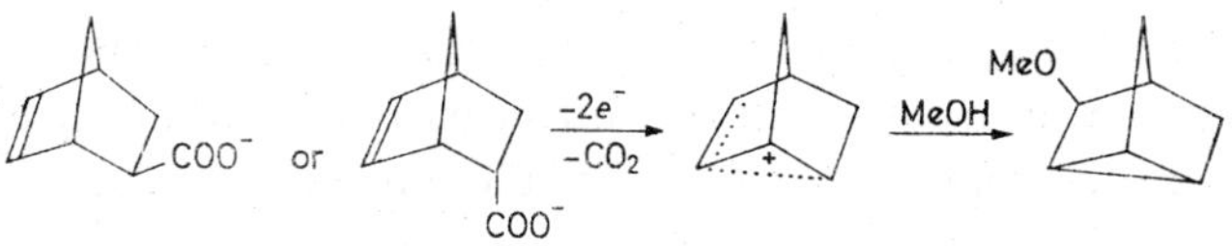

(33)

(23)   (24)   (25)   (26)

6*

of structural integrity (*e.g.*, optically active *endo*-norbornylamine on deamination in acetic acid[99] gives 5% of the *endo*-acetate and 95% of an *exo*
-acetate with 19% retention of optical activity, a total of 24% retention of
structural integrity). This difference further illustrates that the behavior of
anodic and deamination carbonium ions need not necessarily be similar
in all respects. In this particular case, participation by neighboring carbon
in the formation of electrolytic carbonium ions is indicated, although it
may be that the stereochemistry found is only a reflection of the properties of a poorly solvated norbornyl cation, which has a life-time long
enough to collapse completely into the bridged ion (**21**).

Electrolysis of apocamphane-1-carboxylic acid (**27**) gave 1,1′-biapo-
camphane (33%), 1-apocamphyl methyl ether (32%), and 1-apocamphyl
apocamphane-1-carboxylate[18], showing that even highly strained carbonium ions can be generated by anodic oxidation of carboxylic acids.

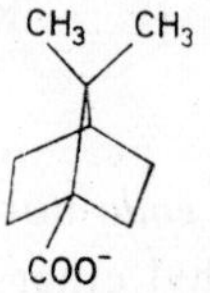

CH₃ CH₃

CH_3 CH_3

COO⁻

(27)

Electrolysis of Δ^5-cholestene-3β-carboxylic acid (**28**) in methanol[20, 100]
gave a mixture of 6β-methoxy-3,5-cyclocholestane (**29**, isocholesteryl
methyl ether), 6β-methoxy-Δ^4-cholestene (**30**), and 4β-methoxy-Δ^6-cho-
lestene (**31**). In addition to cholesteryl methyl ether, **29** is the characteristic ether product from methanolysis of cholesteryl tosylate, whereas the
two other products are obtained from the methanolysis of epicholesteryl

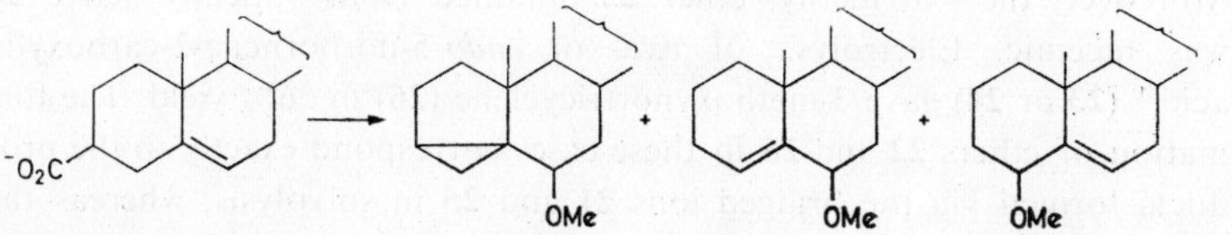

⁻O₂C

OMe   OMe   OMe
(28)   **(29)**   **(30)**   **(31)**   **(34)**

tosylate. These results indicate that the cholesteryl radical (**32**) can be
oxidized anodically with or without direct participation of the Δ^5 double
bond, *i.e.*, either from the β or α side of the molecule, oxidation from the
β side giving rise to the homoallylic carbonium ion **33** and from the α
side to a classical ion which rearranges to the allylic cation **34** via a

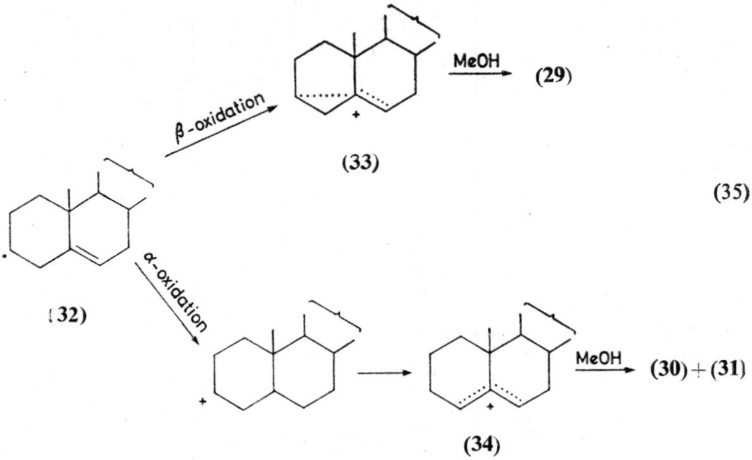

(29)

(33)

(35)

(32)

(34)

(30) + (31)

$4 \rightarrow$ 3-hydride shift. Oxidation from the β side is slightly predominant. No 3,5-diene was formed in the electrolysis, which was attributed to the fact that the anodic cation is generated without any basic leaving group and thus collapse with solvent occurs predominantly at $C_{(6)}$.

The electrolysis of cis- and trans-bicyclo[3.1.0]hexane-3-carboxylic acid (35 and 36) has recently been investigated[101]. This reaction is of particular interest since the 3-bicyclo[3.1.0]hexyl cation generated by solvolysis of cis-3-bicyclo[3.1.0]hexyl tosylate has been postulated to be a homoaromatic species (37)[102-104].

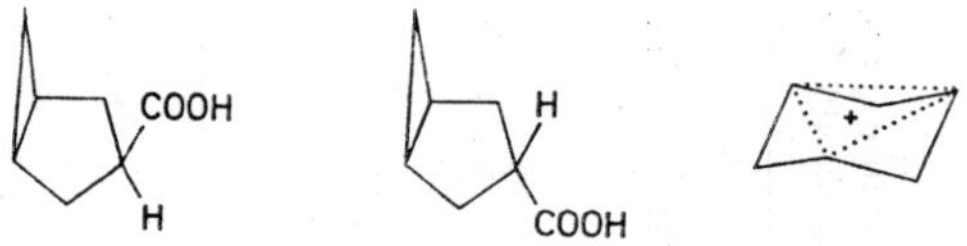

(35)   (36)   (37)

The electrolysis of 35 and 36 was performed in pyridine–water, a solvent which favors the formation of carbonium ion products. There is a slight disadvantage in using water as a nucleophile since the alcohols formed can be oxidized further to ketones at the anode. The possibility that this may occur stereospecifically to some extent is a source of uncertainty in the determination of the composition of the epimeric alcohol mixtures formed.

The compositions of oxygen-containing products from solvolysis, deamination, and electrolysis of the appropriate precursors are given in Table 11. It is immediately apparent that both deamination and electrolysis give drastically different results from those obtained in solvolysis and

TABLE 11. Composition of oxygen-containing products from tosylate acetolyses[104], amine deaminations[105], and electrolytic oxidations[100] in the 3-bicyclo-[3.1.0] hexyl system

Reaction	Composition of product (%)					
	cis-3-ROH	trans-3-ROH	cis-2-ROH	trans-2-ROH	3-ketone	2-ketone
Acetolysis of cis-3-ROTs[a]	99	1	<0.3	<0.3	—	—
Acetolysis of trans-3-ROTs[a,b]	54	0.3	7.5	12	—	—
Deamination of cis-3-RNH₂	5	30.5	27.5	37	—	—
Deamination of trans-3-RNH₂	41	14	9	36	—	—
Electrolysis of cis-3-RCOO⁻	4.3	4.6	37	30.8	10.7	12.6
Electrolysis of trans-3-RCOO⁻	4.3	2.0	34.5	41	7.1	11.1

[a] Products analyzed as acetates.
[b] Δ^3-Cyclohexenyl and Δ^2-cyclohexenyl acetate were also obtained (24 and 2·5%, respectively).

that the trishomocyclopropenyl cation cannot possibly be the product-forming species in the two former processes. The proportion of 3-substitution products is rather small in electrolysis as compared to deamination and solvolysis. The ratio of *cis*- to *trans*-3-alcohol indicates little steric control of the product ratio in the electrolytic reactions, although preferential consumption of either isomer might take place in the ketone-forming oxidation process. The composition of the mixture of 2-alcohols in either case indicates partial stereospecificity, the mixture from the *cis* acid having a slight excess of *cis*-2-alcohol and that from the *trans* acid a slight excess of *trans*-2-alcohol. This was thought to be due to the intermediacy of acyloxonium ions (the 3-bicyclo[3.1.0]hexyl radical and cation in homogeneous solution cannot differentiate whether they were formed from the *cis* or *trans* acid) **38** and **39** which would undergo a concerted hydrogen-bridging carbon dioxide loss (equations 36 and 37). Attack on $C_{(2)}$ by solvent from the side opposite to the hydrogen bridge would account for the stereospecificity observed. The possibility that the stereochemistry might be explained on the basis of cation association with the anode was

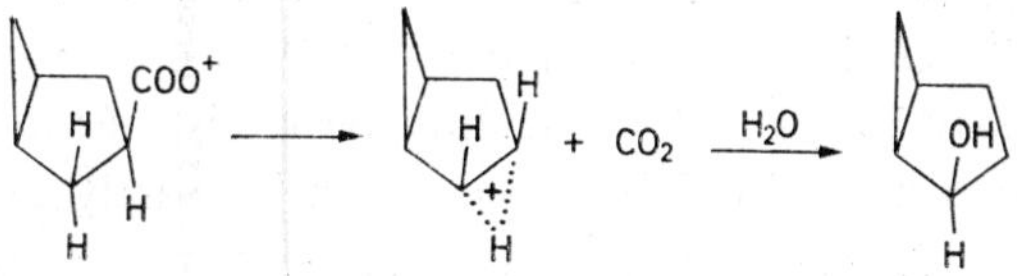

(38)

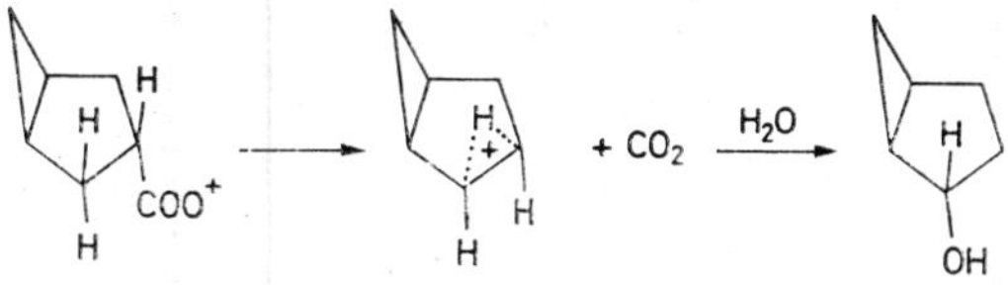

(39)

considered unlikely in view of the results obtained with the bicyclic systems discussed above (equations 32 and 33). No scrambling of a deuterium label in the 6-position of (**35**) was observed, showing that the 3-bicyclo[3.1.0]hexyl cation, be it formulated as a non-classical ion (**37**) or a rapidly equilibrating mixture of classical ions, does not play any role in the electrolytic decarboxylation of either epimer.

The electrolytic decarboxylation of 3β-acetoxybisnorallocholanic acid (**40**)[106] and isostevic acid (**41**)[107] in methanol has been shown to give 70–85% yields of carbonium ion mediated products.

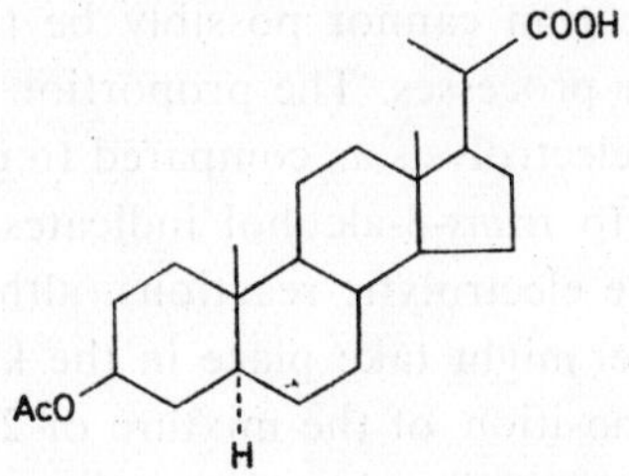

COOH

AcO

H

(40)

H
COOH

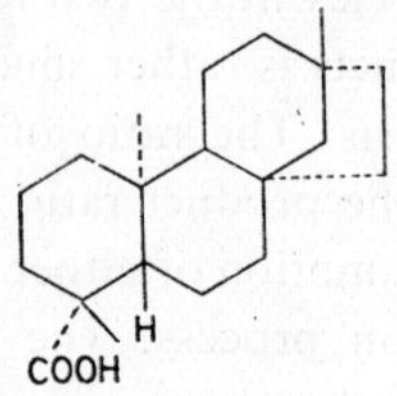

(41)

e. Dicarboxylic acids. In contrast to their half esters, dicarboxylic acids have received little interest as substrates for anodic oxidation, probably due to the early finding that cycloalkanes could not be prepared from diacids via an intramolecular coupling reaction[13-15]. However, some recent developments demonstrate that anodic oxidation of diacids to form cationic intermediates might have some synthetic value.

The oxidative bisdecarboxylation (Grob degradation) of 1,2-diacids by lead tetraacetate (see section VII.) could be duplicated anodically under certain conditions. Thus, electrolytic oxidation of *meso-* and *dl*-2,3-diphenylsuccinic acid in 90% pyridine–water[108] with an excess of triethylamine present gave 36% and 40% yield, respectively, of *trans*-stilbene with no *cis* isomer detectable in either case, which is exactly the same stereochemistry as in lead tetraacetate bisdecarboxylation of these acids. Since this result ruled out a concerted mechanism, a mechanism involving a labile monocarboxylic zwitterion was proposed (equation 38). A number

$$
\begin{array}{ccc}
\text{PhCHCOO}^- & & \overset{+}{\text{PhCH}} & & \text{Ph} \\
| & \xrightarrow[-\text{CO}_2]{-2e^-} & | & \longrightarrow & \diagdown \\
\text{PhCHCOO}^- & & \text{PhCHCOO}^- & & \quad\ \ \text{CH}{=}\text{CH} \\
& & & & \diagdown \\
& & & & \quad\text{Ph}
\end{array}
\quad +\text{CO}_2 \qquad (38)
$$

of other examples of anodic bisdecarboxylations are known[13-15, 109]. 1,2-Ethylenedicarboxylic acids give alkynes by this method[15].

Lactone formation has been observed in a number of anodic oxidations of 1,3- and 1,4-dicarboxylic acids. α-Truxillic acid **(42)** gave at least eight products on electrolysis in methanol[110], the major component being the lactone **(44)**. Formation of a carbonium ion intermediate **(43)** followed by rearrangement and internal attack of carboxyl at the cationic center is a probable mechanism (equation 39). Deamination of γ-truxillamic acid **(45)** gave the same lactone.

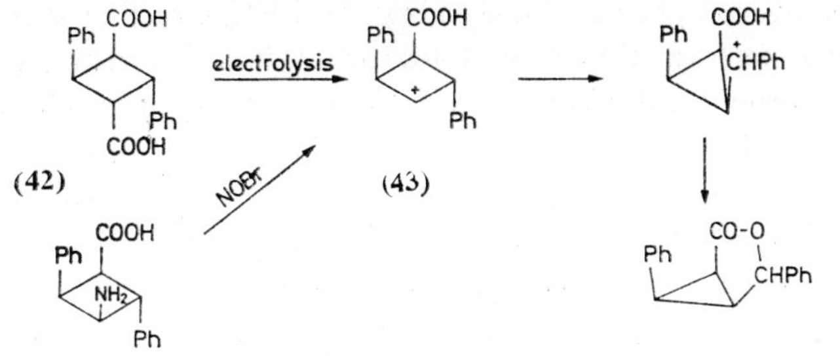

(39)

$$(42) \quad (43) \quad (44) \quad (45)$$

The only hitherto reported case of an internal coupling reaction was found in the electrolysis of *trans, trans, trans*-1,3-dicarboxyl-1,3-dicarbomethoxycyclobutane (46)[110]. The major product was the bicyclobutane derivative (47). The reaction is formally an intramolecular coupling process, but in view of the results in the 1,2-diacid series an ionic mechanism is also feasible (equation 40).

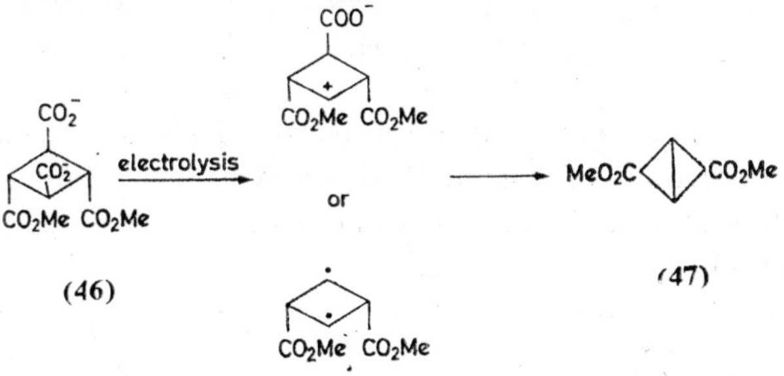

(40)

On electrolysis in methanol, poly(methacrylic acid), which has its carboxyls in 1,3-positions, is eventually converted into a polymer[111] containing 5–6% of its carboxyl groups as free COOH, 35–40% as γ-lactone functions, and 30% as ester groups. The rest was converted into partially unsaturated hydrocarbon groups. The formation of this product is best formulated as a carbonium ion reaction (equation 41).

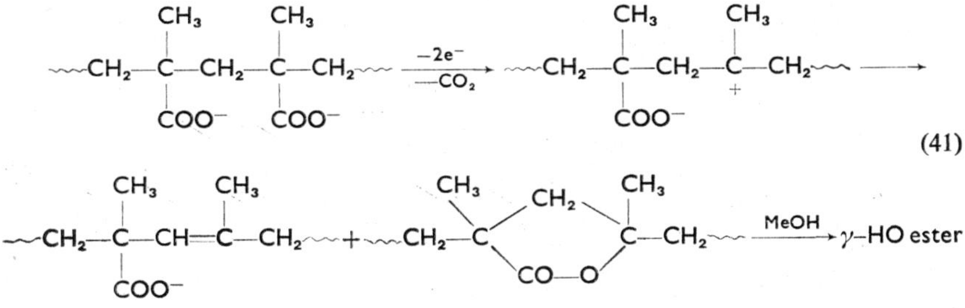

(41)

The perhydrodiphenic acids[21] also furnish interesting examples of anodic lactone formation from 1,4-dicarboxylic acids. Thus, *trans-anti-trans*-perhydrodiphenic acid on electrolysis in methanol produced 5% of the δ-lactone **48**, 45% of the γ-lactone **49**, and 20–25% of a polymeric material, probably according to the ionic mechanism outlined in equation (42). Overberger and Kabasakalian[21] did mention the possibility of this mechanism but preferred an alternative one involving radicals.

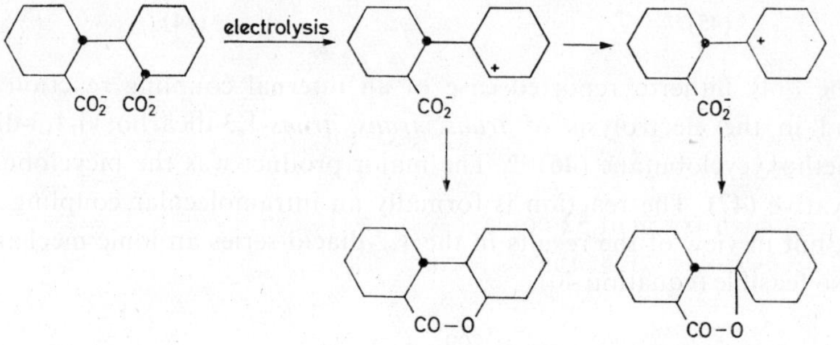

$$(42)$$

(48)   **(49)**

f. Hydroxy acids. Certain types of hydroxy acids undergo anodic carbonium ion reactions of potential preparative value. Thus, the electrolytic oxidation of β-hydroxy acids[20] can be used as a ring expansion method (equation 43):

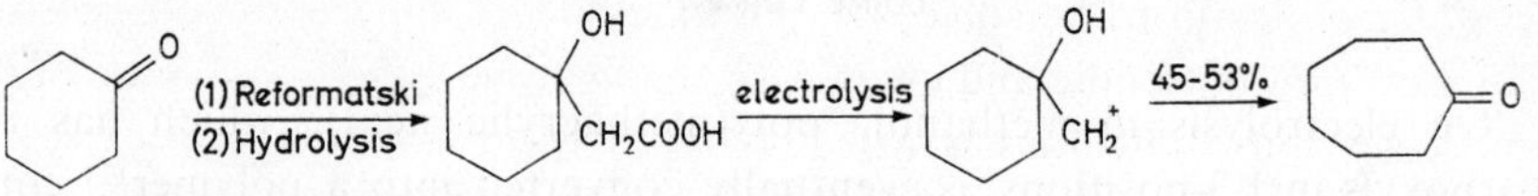

$$(43)$$

On electrolysis γ-hydroxy acids undergo an interesting C_β–C_γ cleavage reaction[112], first demonstrated for the acid **50** and later investigated in more detail with **51**, γ-benzyl-γ-hydroxyvaleric acid[20, 100] (equation 45,

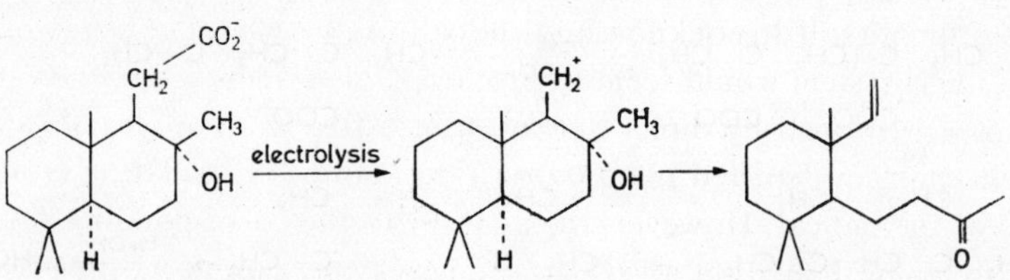

$$(44)$$

(50)

yields are based on unrecovered starting material). Deamination of the amine corresponding to **51** gave the two rearranged ketones but no cleavage product[100].

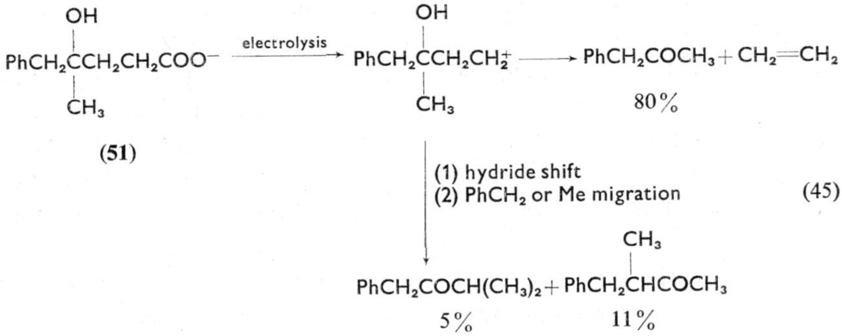

$$\underset{\underset{CH_3}{|}}{\overset{\overset{OH}{|}}{PhCH_2CCH_2CH_2COO^-}} \xrightarrow{\text{electrolysis}} \underset{\underset{CH_3}{|}}{\overset{\overset{OH}{|}}{PhCH_2CCH_2CH_2^+}} \longrightarrow PhCH_2COCH_3 + CH_2{=}CH_2$$

(51)   80%

(1) hydride shift
(2) PhCH₂ or Me migration   (45)

$$PhCH_2COCH(CH_3)_2 + \underset{\underset{CH_3}{|}}{PhCH_2CHCOCH_3}$$
5%   11%

The γ-hydroxy acid **52** could be converted into *trans*-5-cyclodecenone (**56**) by electrolysis in methanol (equation 46)[113].

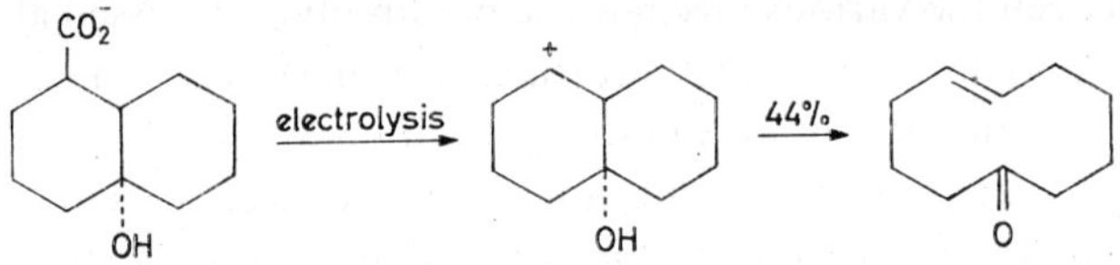

(52)   (53)   (46)

g. Comparison between anodic and non-anodic carbonium ions. Several problems remain to be solved before a detailed comparison between carbonium ions of anodic and chemical origin can be undertaken. One of the more obvious difficulties is to estimate effects due to the heterogeneous nature of the electrode reaction. The 2- to 1-butene and the *trans*- to *cis*-2-butene ratios shown in Table 9 as well as the partial stereospecificity observed in the bicyclo[3.1.0]hexyl system (Table 11) might be indicative of such effects, although it is not possible to draw definite conclusions on the basis of these rather small differences.

Even if it were possible to ascertain that product formation takes place in solution we still do not know how the cation is formed. The behavior of the cholesteryl system would seem to provide some evidence that the cation in this case is formed by direct oxidation of a free cholesteryl radical, since the mechanism depicted in equation (35) allows for oxidation from both sides of the radical. However, the analogy to the corresponding solvolytic reaction is probably not a very good one. As an example, no diene is formed in the electrolytic process which is probably due to the absence of a basic leaving group in the correct position. Furthermore, it is not known if there is a need for any special orientation of the carboxylate ion with regard to the anode for the electron transfer to occur.

The other possibility, that the cation is formed by decarboxylation of an acyloxonium ion $RCOO^+$, was discussed by Gassman and Zalar[100] in connection with their explanation of the partial stereospecificity observed in the bicyclo[3.1.0]hexyl system, and cannot be rejected. Thirdly, a concerted mechanism involving a two-electron transfer and simultaneous cleavage of the C—COO bond to form R^+ and CO_2 directly must be taken into account even if any observed stereospecificity then has to be explained by steric control exerted by the anode surface.

Apart from these difficulties, it is for the moment a safe generalization to state that anodic carbonium ions are of the exceptionally reactive, 'hot' type[114], *i.e.* their properties bear a close resemblance to those of carbonium ions produced in the deamination or deoxidation reaction. The differences observed are not very pronounced and do not lend themselves to a differentiation between the various mechanistic possibilities discussed above. As in the case of anodically generated radicals (section IV. C. 1), it is, however, a fair approximation to discuss anodic carbonium ions without any consideration of anode effects in the light of present evidence.

h. Radical versus carbonium ion path. The Kolbe reaction is unique among organic reactions with its blend of strict radical and carbonium ion mechanisms. For preparative purposes it is of great interest to have some means of predicting whether the radical or carbonium ion path will be the favored one for a particular acid. Assuming that experimental conditions are kept constant (*e.g.* platinum anode, methanolic solution) and that the simple mechanism outlined in equations 1, 2, and 4 is valid, it has been shown[56] that the ionization potential of the intermediate radical $R^{\cdot}$ (as measured by the electron impact or the photoionization method) is the dominant parameter in determining the ease of further oxidation of the radical. Radicals with high ionization potentials (*i.e.* > 8 eV) are fairly resistant towards oxidation to carbonium ions and thus the radical mechanism will be favored in these cases. Among these we find the primary aliphatic racidals $RCH_2^{\cdot}$ (methyl 9·95, ethyl 8·78, propyl 8·33, butyl 8·64, isobutyl 8·35 eV) and resonance-stabilized radicals of the type $R\dot{C}HX$ ($\dot{C}H_2CN$ 10·87, $Me_2\dot{C}CN$ 9·15 eV). Ionization potentials for radicals of the type $R_2\dot{C}COOEt$ and $R_2\dot{C}COHN_2$ are not known, but the good to moderate yields of coupling products in the electrolysis of substituted malonic half esters and malonamic acids is an indication that these radicals belong to this group.

Among the group of radicals with low ionization potentials (less than 8 eV) we find the cases where products formed via carbonium ions dominate, e.g., secondary and tertiary aliphatic radicals (isopropyl 7·90, *t*-butyl 7·42 eV), alicyclic radicals (cyclobutyl 7·88, cyclopentyl 7·80, cyclohexyl 7·60 eV) and phenyl-substituted radicals (benzyl 7·76, benzhydryl 7·32 eV). Radicals with ionization potentials around 8 eV constitute borderline cases.

V. ANODIC ACYLOXYLATION AND RELATED REACTIONS

A. Acyloxylation

The formation of acetoxylated products during Kolbe electrolysis of acetate ions in acetic acid in the presence of organic substrates, such as naphthalene or anisole, has frequently been cited as evidence[37, 59, 60] for the intermediacy of acetoxy radicals in the Kolbe reaction (equations 1 and 47). It has later been shown[31] that a number of organic compounds, among

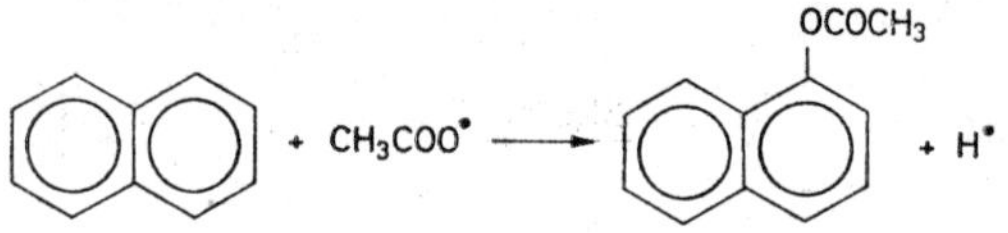

$$\text{+ } CH_3COO^{\bullet} \longrightarrow \text{+ } H^{\bullet} \qquad (47)$$

them napthalene and anisole, display polarographic waves in CH_3COO^-/CH_3COOH far below the critical potential (section III. A) for discharge of acetate ion. Table 12 is a compilation of anodic half-wave potentials obtain-

TABLE 12. Half-wave potentials for the oxidation of organic compounds in HOAc–0·5 M NaOAc at a rotating platinum electrode[31]

Compound	$E_{1/2}$(V) versus saturated calomel electrode	Compound	$E_{1/2}$(V) versus saturated calomel electrode
Mesitylene	1·90	Pyrene	1·20
Durene	1·62	Triphenylene	1·74
Pentamethylbenzene	1·62	Perylene	1·00
Hexamethylbenzene	1·52	Fluoranthene	1·64
Naphthalene	1·72	Anisole	1·67
1-Methylnaphthalene	1·53	Phenyl acetate	1·30
2-Methylnaphthalene	1·55	*trans*-Stilbene	1·51
Biphenyl	1·91	1,1-Diphenylethylene	1·52
Biphenylene	1·30	Biphenyl-2-carboxylic acid	1·71
Acenaphthene	1·36	Azulene	0·91
Acenaphthylene	1·53	Cyclooctatetraene	1·42
Fluorene	1·65	Furan	1·70
Anthracene	1·20	2,5-Dimethylfuran	1·20
Phenanthrene	1·68	DMF	1·90

ed in this medium. These data suggest that it is actually the organic substrate which is oxidized in the electrode process[61-63]. Unambiguous proof of this was obtained by controlled potential electrolysis[31], in which the anode potential was kept well below the critical potential for discharge of the acetate ion. In all these experiments, the characteristic acetoxylation products were isolated in low to fair yields. In the case of naphthalene the competition between the acetoxylation reaction and the Kolbe reaction (formation of methyl radicals, which give methylnaphthalene in analogy to the reactions discussed in section IV. C. 1) could be demonstrated by running the reaction at different anode potentials. At a low potential, the Kolbe reaction is almost completely suppressed.

These experiments verify that anodic acetoxylation occurs via a two-electron transfer from the organic substrate to the anode to form a dicationic intermediate (54), followed (or possibly assisted) by reaction with acetate ions (equation 48).

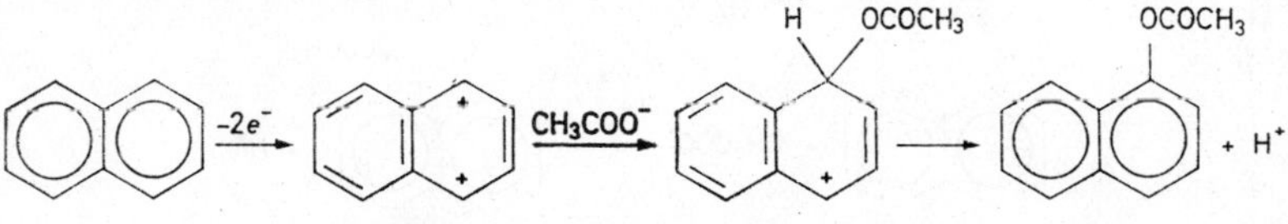

(54)   (55)   (49)

An alternative mechanism (equation 49) has been suggested by Perrin[8]. It is, however, very difficult to distinguish between the two mechanisms.

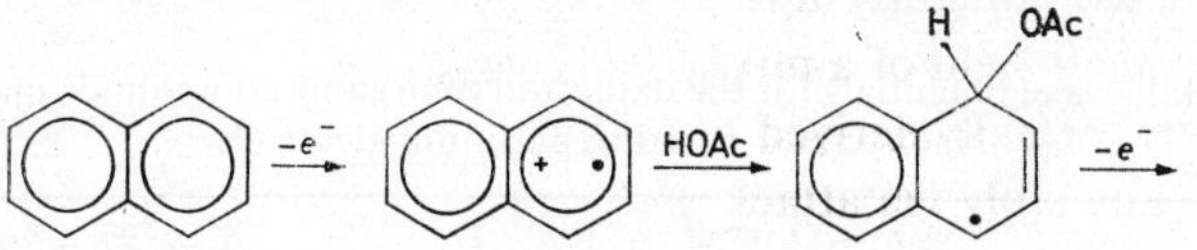

The most important feature of both is that the intermediate 55 is of the same type as in ordinary electrophilic substitution.

In addition to nuclear acetoxylation, side-chain acetoxylation (equation 50) is also observed for alkyl-substituted benzenes, presumably via the same kind of intermediate as in equation (48) which by loss of a proton would form a benzyl cation (for further evidence regarding the intermediacy of benzyl cations in these reactions, see reference 115).

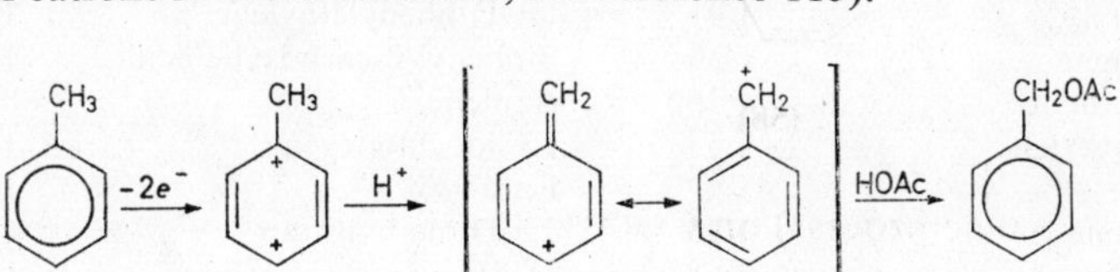

(50)

Mango and Bonner have investigated the electrolysis of *trans*-stilbene and 1,1-diphenylethylene in anhydrous and wet CH_3COO^-/CH_3COOH[116]. The stereochemistry of the products formed from *trans*-stilbene is in agreement with the assumption that a cyclic acetoxonium ion is the product-forming intermediate (equation 51). 1,4-Addition of two acetoxy groups is also observed in the anodic oxidation of furan in CH_3COO^-/CH_3COOH[117].

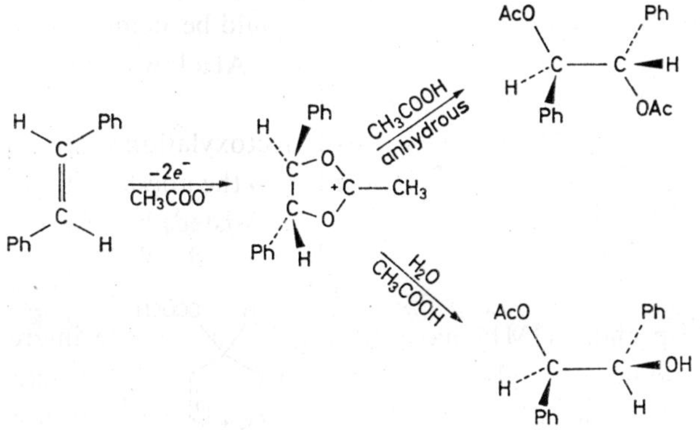

$$(51)$$

Cyclooctatetraene (COT) is of particular interest, since removal of two electrons might possibly give the planar six π-electron system COT^{2+} as an intermediate. On controlled potential oxidation in CH_3COO^-/CH_3COOH, COT gives predominantly a mixture of **56, 57,** and **58** (equation 52)[118]. In addition, a small yield of a mixture of two epimeric products **59** was isolated. These are formally derived by addition of the elements of CH_3COOCH_3 to COT, presumably via attack of a methyl radical

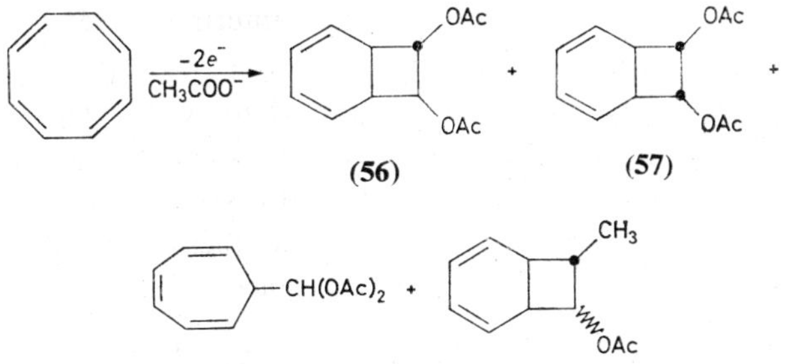

$$(52)$$

(56)   **(57)**

(58)   **(59)**

(from the Kolbe process) on COT, followed by a second electron transfer, ring contraction, and reaction with acetate ion. Accordingly, this reaction could be totally suppressed by running the electrolysis with a carbon anode, which is known to inhibit the formation of radicals (section III. B).

The formation of acetates from the electrolysis of butadiene[67], styrene[70], (section IV. C. 1) or alkenes[119] in CH_3COO^-/CH_3COOH presumably occurs via a similar ionic mechanism.

On anodic oxidation in CH_3COOH/CH_3COO^-, phenols are converted into a mixture of o- and p-quinol acetates[120]. This reaction has been shown to be ionic, and a phenoxonium ion has been postulated to be the intermediate (equation 53).

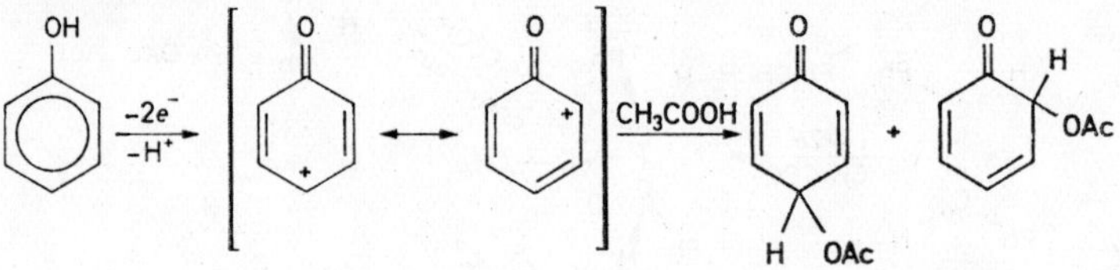

$$ \tag{53} $$

Amides, especially DMF, have attracted considerable interest as substrates for anodic acyloxylation[27-30]. On the basis of polarographic measurements and controlled potential electrolysis, Eberson and Nyberg[31] proposed that acyloxylation, except possibly formyloxylation, of DMF proceeds via the cationic intermediate **60** (equation 54). Anodic formyloxylation,

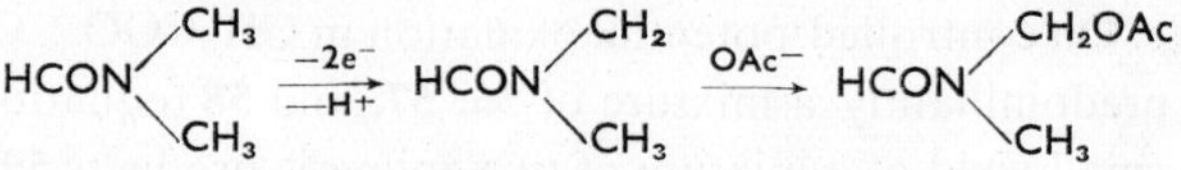

$$ \tag{54} $$

(60)

as well as acetoxylation in the presence of nitrate ion, presents special problems and it is difficult to decide between a radical and a polar mechanism[28-30].

Anodic aroyloxylation can be achieved by coelectrolysis of benzoate ion and a suitable organic compound (naphthalene [121], anisole[122]) in acetonitrile. In this case a homolytic mechanism involving benzoyloxy radicals cannot be excluded on the basis of present data, although the similarity with the corresponding acetoxylation processes makes an ionic mechanism more attractive.

Various cases of intramolecular anodic acyloxylations have been described, e.g., from biphenyl-2-carboxylic acid[31] (equation 55), β-(p-hydroxyphenyl)-propionic acid[123, 124] (equation 56) and 3,3-diphenylpropionic acid[35] (equation 57). All these reactions probably conform to a mechanism similar to that given in equations (48) or (49).

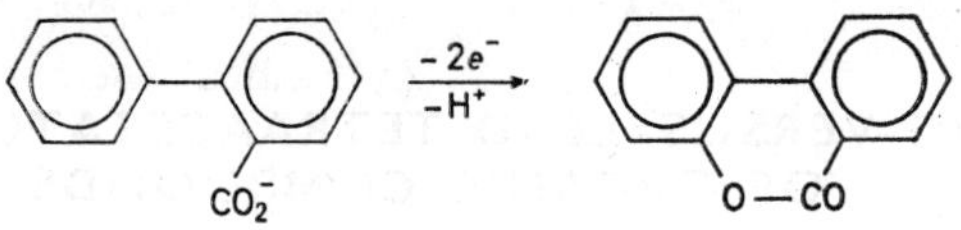

$$(55)$$

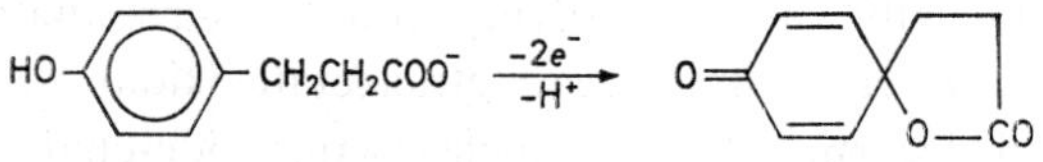

$$(56)$$

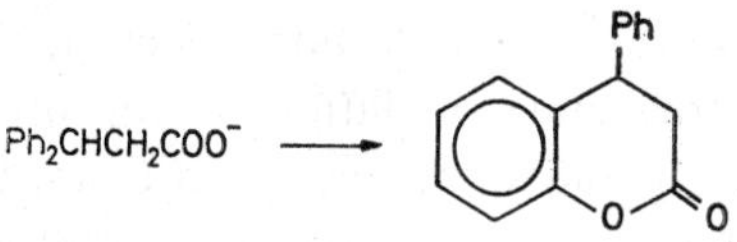

$$(57)$$

B. Anodic Cyanation

Nitriles, and consequently carboxylic acids, can be prepared by electrolytic cyanation of certain organic compounds in the presence of cyanide ion[125, 126], and again a mechanism involving electron transfer from the organic substrate rather than cyanide ion is probable[127] (equation 58). This method has great potential synthetic value.

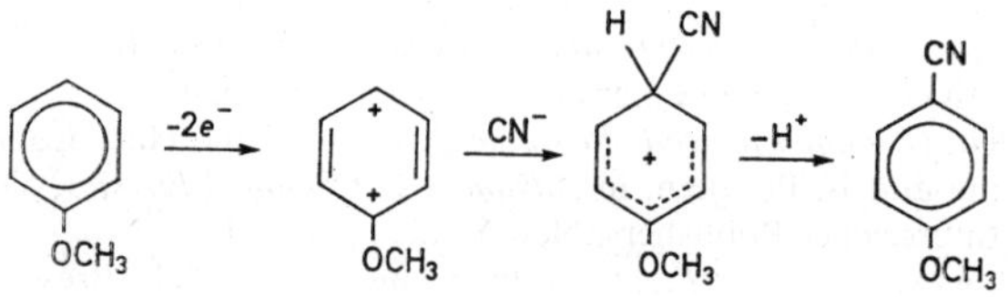

$$(58)$$

VI. ANODIC ESTERIFICATION, ALCOHOLYSIS AND HYDROLYSIS

It has sometimes been noted that esterification of the starting acid occurs during Kolbe electrolysis[13-15, 38, 39] or that alcoholysis of an ethyl ester takes place when methanol is used as solvent[128-129, 130]. If the reaction is run in a divided cell, esterification or alcoholysis occurs in the anode compartment only[130, 131]. In a suitable solvent, such as dioxane–water, it is also possible to hydrolyze an ester function anodically[130]. A probable mechanism for these reactions involving anodic acyl-oxygen cleavage is outlined in equation (59).

$$RCOOR' \longrightarrow RCO^+ + R'O^+ + 2e^- \xrightarrow{R''OH} RCOOR'' + H^+ \qquad (59)$$

7 C.C.A.E.

VII. ANODIC VERSUS LEAD TETRAACETATE OXIDATION OF ORGANIC COMPOUNDS

It is of interest to note that many of the electrochemical processes mentioned above, especially those involving ionic intermediates, have their counterparts in the chemistry of lead tetraacetate oxidations[132]. Thus decarboxylation, bisdecarboxylation, nuclear and side-chain acetoxylation, quinol acetate formation, addition of two acetoxy groups or one acetoxy and one methyl group to a double bond, lactone formation, *etc.* can be achieved by lead tetraacetate oxidation of appropriate compounds. The results are qualitatively and sometimes even quantitatively the same as in the electrolytic reactions. From the economic point of view, it would therefore seem to be advantageous to use the electrolytic reaction for large-scale preparations.

VIII. REFERENCES

1. C. J. Brockman, *Electro-Organic Chemistry*, John Wiley and Sons, New York, 1926.
2. F. Fichter, *Organische Electrochemie*, T. Steinkopff, Dresden and Leipzig, 1942.
3. S. Swann, Jr., in *Technique of Organic Chemistry*, Vol. 2, *Catalytic, Photochemical, Electrolytic Reactions*, 2nd ed., (Ed. A. Weissberger), Interscience Publishers, New York, 1956, p. 385.
4. F. Müller, in *Methoden der Organischen Chemie (Houben-Weyl)*, Vol. 4/2 (Ed. E. Müller), 4th ed., G. Thieme Verlag, Stuttgart, 1955, p. 461.
5. M. J. Allen, *Organic Electrode Processes*, Chapman and Hall, London, 1958.
6. P. J. Elving and B. Pullman, in *Advances in Chemical Physics*, Vol. I (Ed. I. Prigogine), Interscience Publishers, New York, 1961, p. 1.
7. B. E. Conway, *Theory of Electrode Processes*, The Ronald Press Co., New York, 1965.
8. C. L. Perrin, in *Progress in Physical Organic Chemistry*, Vol. 3 (Eds. S. G. Cohen, A. Streitwieser, Jr. and R. W. Taft), Interscience Publishers, New York, 1965, p. 165–316.
9. R. N. Adams, *J. Electroanal. Chem.*, **8**, 151 (1964).
10. A. Streitwieser, Jr., *Molecular Orbital Theory for Organic Chemists*, John Wiley and Sons, New York, 1961, p. 173–187, and references cited therein.
11. R. Zahradnik and C. Párkányi, *Talanta*, **12**, 1289 (1965), and references cited therein.
12. H. Kolbe, *Ann. Chem.* **69**, 257 (1849).
13. B. C. L. Weedon, *Quart. Rev. (London)*, **6**, 380 (1952).
14. B. C. L. Weedon, in *Advances in Organic Chemistry*, Vol. 1 (Eds. R. A. Raphael, E. C. Taylor and H. Wynberg), Interscience Publishers, New York, 1960, p. 1.
15. G. E. Svadkovskaya and S. A. Voitkevich, *Russian Chem. Rev. (English Transl.)*, **29**, 161 (1960).
16. A. C. Brown and J. Walker, *Ann. Chem.* **261**, 107 (1891); *Trans. Roy. Soc. Edinburgh*, **36**, 291 (1891).
17. H. Hofer and M. Moest, *Ann. Chem.* **323**, 285 (1902).
18. M. A. Muhs, *Ph. D. Thesis*, University of Washington, Seattle, 1954.

19. C. Walling, *Free Radicals in Solution*, John Wiley and Sons, New York, 1957, p. 581.
20. E. J. Corey, N. L. Bauld, R. T. LaLonde, J. Casanova, Jr. and E. T. Kaiser, *J. Am. Chem. Soc.*, **82**, 2645 (1960).
21. C. G. Overberger and P. Kabasakalian, *J. Am. Chem. Soc.*, **79**, 3182 (1957).
22. N. C. Deno, in *Progress in Physical Organic Chemistry*, Vol. 2 (Eds. S. G. Cohen, A. Streitwieser, Jr. and R. W. Taft), Interscience Publishers, New York, 1964, p. 129–193.
23. W. J. Koehl, Jr., *J. Am. Chem. Soc.*, **86**, 4686 (1964).
24. Yu. M. Tyurin, E. P. Kovsman and E. A. Karavaeva, *J. Appl. Chem. USSR, (English Transl.)*, **38**, 1776 (1965).
25. N. Dinh-Nguyen, *Acta Chem. Scand.*, **12**, 585 (1958).
26. M. Finkelstein and R. C. Petersen, *J. Org. Chem.*, **25**, 136 (1960).
27. L. Rand and A. F. Mohar, *J. Org. Chem.*, **30**, 3156, 3885 (1965).
28. S. D. Ross, M. Finkelstein and R. C. Petersen, *J. Am. Chem. Soc.*, **86**, 2745 (1964).
29. S. D. Ross, M. Finkelstein and R. C. Petersen, *J. Org. Chem.*, **31**, 128 (1966).
30. S. D. Ross, M. Finkelstein and R. C. Petersen, *J. Am. Chem. Soc.*, **88**, 4657 (1966).
31. L. Eberson and K. Nyberg, *J. Am. Chem. Soc.*, **88**, 1686 (1966); *Acta Chem. Scand.*, **18**, 1568 (1964).
32. C. D. Russell and F. C. Anson, *Anal. Chem.*, **33**, 1282 (1961).
33. L. Eberson and K. Nyberg, *Acta Chem. Scand.*, **18**, 1567 (1964).
34. R. P. Linstead, B. R. Shephard and B. C. L. Weedon, *J. Chem. Soc.*, 2854 (1951).
35. W. A. Bonner and F. D. Mango, *J. Org. Chem.*, **29**, 430 (1964).
36. H. Breederveld and E. C. Kooyman, *Rec. Trav. Chim.*, **76**, 297 (1957).
37. R. P. Linstead, B. R. Shephard and B. C. L. Weedon, *J. Chem. Soc.*, 3624 (1952).
38. R. G. Woolford, *Can. J. Chem.*, **40**, 1846 (1962).
39. R. G. Woolford, W. Arbic and A. Rosser, *Can. J. Chem.*, **42**, 1788 (1964).
40. A case of what appears to be an electrolytic carbene formation has been described: S. Wawzonek and R. C. Duty, *J. Electrochem. Soc.*, **108**, 1135 (1961).
41. B. E. Conway and M. Dzieciuch, *Proc. Chem. Soc.*, 121 (1962).
42. M. Fleischmann, J. R. Mansfield and W. F. K. Wynne-Jones, *J. Electroanal. Chem.*, **10**, 511 (1965).
43. M. Fleischmann, J. R. Mansfield and W. F. K. Wynne–Jones, *J. Electroanal. Chem.*, **10**, 522 (1965).
44. G. S. Pande and S. N. Shukla, *Electrochim. Acta*, **4**, 215 (1961).
45. T. Dickinson and W. F. K. Wynne-Jones, *Trans. Faraday Soc.*, **58**, 382, 388, 400 (1962).
46. G. Preuner, *Z. Physik. Chem. (Leipzig)*, **59**, 670 (1907).
47. D. A. Fairweather and O. J. Walker, *J. Chem. Soc.*, 3116 (1926).
48. S. N. Shukla and O. J. Walker, *Trans. Faraday Soc.*, **27**, 722 (1931).
49. A. Hickling and J. V. Westwood, *J. Chem. Soc.*, 1039 (1938).
50. S. Glasstone and A. Hickling, *J. Chem. Soc.*, 820 (1936).
51. G. Preuner and E. B. Ludlam, *Z. Physik. Chem. (Leipzig)*, **59**, 682 (1907).
52. R. Matsuda, T. Hisano, T. Terazawa and N. Shinohara, *Bull. Soc. Chem. Japan*, **35**, 1233 (1962).
53. A. J. Hickling and J. V. Westwood, *J. Chem. Soc.*, 1039 (1938).
54. A. Hickling and J. V. Westwood, *J. Chem. Soc.*, 1109 (1939).
55. B. E. Conway and M. Dzieciuch, *Can. J. Chem.*, **41**, 21, 38, 55 (1963).
56. L. Eberson, *Acta Chem. Scand.*, **17**, 2004 (1963).
57. L. Jaffe, E. J. Prosen and M. Szwarc, *J. Chem. Phys.*, **27**, 416 (1957).
58. W. Braun, L. Rajbenbach and F. R. Eirich, *J. Phys. Chem.*, **66**, 1591 (1962).

59. C. L. Wilson and W. T. Lippincott, *J. Am. Chem. Soc.*, **78**, 4291 (1956); K. E. Kolb and C. L. Wilson, *Chem. Comm.*, 271 (1966).
60. D. R. Harvey and R. O. C. Norman, *J. Chem. Soc.*, 4860 (1964).
61. M. Leung, J. Herz and H. W. Salzberg, *J. Org. Chem.*, **30**, 310 (1965).
62. H. W. Salzberg and M. Leung, *J. Org. Chem.*, **30**, 2873 (1965).
63. S. D. Ross, M. Finkelstein and R. C. Petersen, *J. Am. Chem. Soc.*, **86**, 4139 (1964).
64. K. Clusius and W. Schanzer, *Z. Physik. Chem. (Leipzig)*, **192 A**, 273 (1943).
65. P. Hölemann and K. Clusius, *Chem. Ber.*, **70**, 819 (1937).
66. R. V. Lindsey, Jr. and M. L. Peterson, *J. Am. Chem. Soc.*, **81**, 2073 (1959).
67. W. B. Smith and H. Gilde, *J. Am. Chem. Soc.*, **81**, 5325 (1959).
68. W. B. Smith and H. Gilde, *J. Am. Chem. Soc.*, **83**, 1355 (1961).
69. C. Walling and E. S. Huyser, *Org. Reactions*, **13**, 91 (1963).
70. S. Goldschmidt and E. Stöckl, *Chem. Ber.*, **85**, 630 (1952).
71. W. B. Smith and H. Gilde, *J. Am. Chem. Soc.*, **82**, 659 (1960).
72. W. B. Smith and D. T. Manning, *J. Polymer Sci.*, **59**, S45 (1962).
73. S. Goldschmidt, W. Leicher and H. Haas, *Ann. Chem.* **577**, 153 (1952).
74. S. Goldschmidt and M. Minsinger, *Chem. Ber.*, **87**, 956 (1954).
75. D. H. Hey and P. J. Bunyan, *J. Chem. Soc.*, 3787 (1960).
76. D. H. Hey and P. J. Bunyan, *J. Chem. Soc.*, 1360 (1962).
77. M. Talat-Erben and A. N. Isfendiyaroglu, *Can. J. Chem.*, **36**, 1156 (1958).
78. L. Eberson, *J. Org. Chem.*, **27**, 2329 (1962).
79. L. Eberson and S. Nilsson, unpublished results.
80. M. S. Kharasch, J. G. Kuderna and W. H. Urry, quoted from G. W. Wheland, *Advanced Organic Chemistry*, 2nd ed., John Wiley and Sons, New York, 1949, p. 714.
81. E. S. Wallis and F. H. Adams, *J. Am. Chem. Soc.*, **55**, 3838 (1933).
82. R. F. Garwood, C. J. Scott and B. C. L. Weedon, *Chem. Comm.*, 14 (1965).
83. L. Eberson and B. Sandberg, *Acta Chem. Scand.*, **20**, 739 (1966).
84. A. Kruis and W. Schanzer, *Z. Physik. Chem. (Leipzig)*, **191 A**, 301 (1942).
85. J. D. Roberts and J. A. Yancey, *J. Am. Chem. Soc.*, **74**, 5943 (1952).
86. F. Fichter and A. Bürgin, *Helv. Chim. Acta*, **14**, 90 (1931).
87. G. J. Karabatsos and C. E. Orzech, Jr., *J. Am. Chem. Soc.*, **84**, 2838 (1962).
88. P. S. Skell and R. J. Maxwell, quoted from N. C. Deno, *Chem. Eng. News*, **42**, 88 (1964).
89. P. S. Skell and I. Starer, *J. Am. Chem. Soc.*, **81**, 4117 (1959); **82**, 2971 (1960); **84**, 3962 (1962).
90. A. Streitwieser, Jr. and W. D. Schaeffer, *J. Am. Chem. Soc.*, **79**, 2888 (1957).
91. W. B. Smith and W. H. Watson, Jr., *J. Am. Chem. Soc.*, **84**, 3174 (1962).
92. L. G. Cannell and R. W. Taft, Jr., *J. Am. Chem. Soc.*, **78**, 5812 (1956).
93. J. K. Kochi, *J. Am. Chem. Soc.*, **85**, 1958 (1963).
94. L. Eberson, *Acta Chem. Scand.*, **17**, 1196 (1963).
95. B. Wladislaw and A. Giora, *J. Chem. Soc.*, 1037 (1964).
96. B. Wladislaw and H. Viertler, *Chem. Ind. (London)*, 39 (1965).
97. A. J. v. d. Hoek and W. T. Nauta, *Rec. Trav. Chim.*, **61**, 845 (1942).
98. B. Wladislaw and A. M. J. Ayres, *J. Org. Chem.*, **27**, 281 (1962).
99. J. A. Berson, in *Molecular Rearrangements*, Vol. 1 (Ed. P. de Mayo), Interscience Publishers, New York, 1963, p. 111–231.
100. N. L. Bauld, *Ph. D. Thesis*, University of Illinois, 1959.
101. P. G. Gassman and F. V. Zalar, *J. Am. Chem. Soc.*, **88**, 2252 (1966).
102. S. Winstein and J. Sonnenberg, *J. Am. Chem. Soc.*, **83**, 3235 (1961).
103. S. Winstein and J. Sonnenberg, *J. Am. Chem. Soc.*, **83**, 3244 (1961).
104. S. Winstein, E. C. Friedrich, R. Baker and Y. Lin, *Tetrahedron, Suppl.* **8** Part II, 621 (1966).

105. E. J. Corey and R. L. Dawson, *J. Am. Chem. Soc.*, **85**, 1782 (1963).
106. J. A. Waters, *J. Org. Chem.*, **29**, 428 (1964).
107. J. A. Waters, E. D. Becker and E. Mosettig, *J. Org. Chem.*, **29**, 4689 (1964).
108. E. J. Corey and J. Casanova, Jr., *J. Am. Chem. Soc.*, **85**, 165 (1963).
109. W. W. Paudler, R. E. Herbener and A. G. Zeiler, *Chem. Ind. (London)*, 1909 (1965).
110. A. F. Vellturo and G. W. Griffin, *J. Org. Chem.*, **31**, 2241 (1966).
111. G. Smets, X. van der Borght and G. van Haeren, *J. Polymer Sci.*, **2**, 5187 (1964).
112. E. J. Corey and R. R. Sauers, *J. Am. Chem. Soc.*, **81**, 1743 (1959).
113. P. S. Wharton, G. A. Hiegel and R. V. Coombs, *J. Org. Chem.*, **28**, 3217 (1963).
114. Ref. 99, p. 205.
115. L. Eberson and K. Nyberg, *Tetrahedron Letters*, 2389 (1966).
116. F. D. Mango and W. A. Bonner, *J. Org. Chem.*, **29**, 1367 (1964).
117. A. J. Baggaley and R. Brettle, *Chem. Comm.*, 106 (1966).
118. L. Eberson, K. Nyberg, M. Finkelstein, R. C. Petersen, S. D. Ross and J. J. Uebel, *J. Org. Chem.*, **31**, 16 (1966).
119. P. Courbis and A. Guillemonat, *Compt. Rend. Acad. Sci. Paris*, **262C**, 1435 (1966).
120. F. W. Steuber and K. Dimroth, *Chem. Ber.*, **99**, 258 (1966).
121. J. F. K. Wilshire, *Australian J. Chem.*, **16**, 432 (1963).
122. K. Koyama, K. Yoshida and S. Tsutsumi, *Bull. Chem. Soc. Japan*, **39**, 516 (1966).
123. H. Iwasaki, L. A. Cohen and B. Witkop, *J. Am. Chem. Soc.*, **85**, 3701 (1963).
124. A. I. Scott, P. A. Dodson, F. McCapra and M. B. Meyers, *J. Am. Chem. Soc.*, **85**, 3702 (1963).
125. K. Koyama, T. Susuki and S. Tsutsumi, *Tetrahedron Letters*, 627 (1965).
126. K. Yoshida and S. Tsutsumi, *Tetrahedron Letters*, 2501 (1966).
127. V. D. Parker and B. E. Burgert, *Tetrahedron Letters*, 4065 (1965).
128. S. F. Birch, V. E. Gripp, D. T. McAllan and W. S. Nathan, *J. Chem. Soc.*, 1363 (1962).
129. L. Eberson, *Acta Chem. Scand.*, **13**, 40 (1959); **14**, 641 (1960).
130. L. Eberson, unpublished observations.
131. R. G. Woolford, J. Soong and W. S. Lin, *Can. J. Chem.*, **45**, 1837 (1967).
132. R. Crigee, in *Oxidation in Organic Chemistry*, part A (Ed. K. B. Wiberg), Academic Press, New York, 1965, p. 277–366.

CHAPTER **3**

Alcoholysis, acidolysis and redistribution of esters

JOUKO KOSKIKALLIO

University of Oulu, Finland

I. INTRODUCTION

Although no reviews seem to have been published about alcoholysis, acidolysis and redistribution of esters in particular, these reactions have been discussed in reviews concerned mainly with ester hydrolysis[1-4]. Often the mechanisms of hydrolysis and alcoholysis arc similar.

An ester can react to produce another ester, either in a reaction with an alcohol in the alcoholysis reaction (equation 1), or with a carboxylic acid in the acidolysis reaction (equation 2), or with another ester in the redistribution reaction (equation 3).

$$R^1COOR^2 + R^3OH \quad = R^1COOR^3 + R^2OH \tag{1}$$

$$R^1COOR^2 + R^4COOH = R^4COOR^2 + R^1COOH \tag{2}$$

$$R^1COOR^2 + R^4COOR^3 = R^1COOR^3 + R^4COOR^2 \tag{3}$$

Alcoholysis may also occur between an ester and an alkoxide ion in a base-catalysed reaction or between a protonated ester and an alcohol in an acid-catalysed reaction. Acidolysis may involve a carboxylate ion or a protonated ester.

The mechanisms of hydrolysis and alcoholysis of esters may sometimes differ due to differences in the nucleophilic behaviour of water and alcohols or to the different solvents used. Both hydrolysis and alcoholysis can occur by acyl-oxygen fission (Ac) or alkyl-oxygen fission (Al). The products of these two mechanisms are identical in hydrolysis, but in alcoholysis acyl-

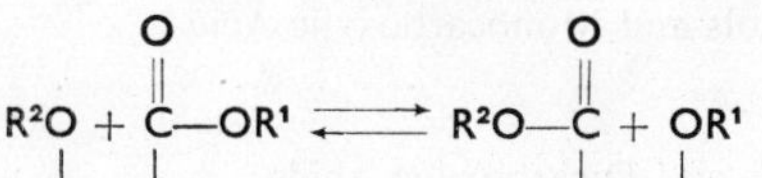

$$
\begin{array}{ccc}
& O & O \\
& \parallel & \parallel \\
R^2O + C-OR^1 & \rightleftharpoons R^2O-C + OR^1 \quad \text{acyl-oxygen fission} \\
| \quad | & | \quad | \\
H \quad R & R \quad H
\end{array}
\tag{4}
$$

$$R^2O + R^1O-\overset{\overset{\displaystyle O}{\|}}{\underset{\underset{\displaystyle R}{|}}{C}} \; \rightleftharpoons \; R^2O-R^1 + HO-\overset{\overset{\displaystyle O}{\|}}{\underset{\underset{\displaystyle R}{|}}{C}} \quad \text{alkyl-oxygen fission} \quad (5)$$

oxygen fission yields an alcohol and an ester, whereas with alkyl-oxygen fission an ether and a carboxylic acid are obtained (equations 4 and 5). Thus, for alcoholysis the two mechanisms are easily identified by product analysis, whilst for hydrolysis they can only be identified by using isotopically-labelled reactants.

In addition to uncatalysed alcoholysis, acid (A) base (B) or nucleophilic catalysis of the alcoholysis reaction has been observed. Each reaction may occur by a bimolecular or a unimolecular mechanism. Eight different mechanisms are possible, four involving alkyl-oxygen fission ($A_{Al}1$, $A_{Al}2$, $B_{Al}1$ and $B_{Al}2$) and four involving acyl-oxygen fission($A_{Ac}1$, $A_{Ac}2$, $B_{Ac}1$, $B_{Ac}2$).

The bimolecular acyl-oxygen fission alcoholysis occurs by a one-step direct substitution mechanism or by a two-step mechanism involving an initially-formed tetrahedral intermediate decomposing either to starting materials or to products. The experimental results[2, 5] indicate that the latter mechanism is more usual (equation 6).

$$\underset{\underset{\displaystyle H-OR^2}{+}}{R-\overset{\overset{\displaystyle O}{\|}}{C}-OR^1} \; \rightleftharpoons \; R-\overset{\overset{\displaystyle OH}{|}}{\underset{\underset{\displaystyle OR^2}{|}}{C}}-OR^1 \; \rightleftharpoons \; R-\overset{\overset{\displaystyle O}{\|}}{\underset{\underset{\displaystyle OR^2}{|}}{C}}+HOR^1 \quad (6)$$

Investigations of alcoholysis reactions were comparatively rare. Previous analytical difficulties in measuring alcoholysis rates are now overcome by the use of gas chromatography. The advantages in studying alcoholysis are the easy distinction of Ac and Al routes and the possibility of varying the nucleophile. In ester hydrolysis of esters, acyl-oxygen fission is usually much faster than alkyl-oxygen fission and it is therefore not usually possible to measure the rate of the latter. In alcoholysis, the rates of both reactions can be measured if R^1 is identical with R^2. Alcoholysis by acyl-oxygen fission (equation 4) yields products identical with the starting materials and no net reaction is observed; the rate of the simultaneous, much slower reaction by alkyl-oxygen fission can then be measured. The rate of alcoholysis by acyl-oxygen fission can be obtained using isotopically-labelled molecules.

Alcoholysis has been used as a synthetic method for preparing esters, especially in the fields of fats and polyesters, and is also useful in analysing fats or esters of high molecular weight.

There have been very few investigations of the mechanism of ester acidolysis, except in some studies of the catalysis of ester hydrolysis in water by carboxylate anions, involving a nucleophilic attack by the carboxylate ion on the ester. These are acidolysis reactions, although the anhydride produced is usually not stable in water and rapidly hydrolyses to a carboxylic acid. In a few cases acidolysis has been also used as a method of synthesizing esters.

A redistribution reaction between two esters yielding two different esters has not yet been proved. In the investigations of redistribution reaction published so far[6-9] the reaction mixture may contain small amounts of moisture, alcohols or carboxylic acids and hydrolysis, esterification, alcoholysis or acidolysis may occur yielding the same products as would be obtained in a redistribution reaction. A discussion of redistribution reactions has to be postponed until more experimental results are available. In the following section, however, a method for obtaining equilibrium constants of redistribution reactions will be discussed.

II. EQUILIBRIA IN ALCOHOLYSIS, ACIDOLYSIS AND REDISTRIBUTION REACTIONS

A. Equilibria in Alcoholysis

In the following, equilibria of alcoholysis reactions occurring by acyl-oxygen fission and involving two different alcohols and two esters are considered. Values of equilibrium constants of alcoholysis reactions occurring by alkyl-oxygen fission and producing an ether and carboxylic acid have not been published so far.

1. Effect of electrolytes

Equilibrium constants of alcoholysis reactions have usually been measured in mixtures containing added electrolytes, acids or bases as catalysts. The salt effects are usually small. For example a variation of the concentration of hydrochloric acid between 0·2 and 0·9 mole per cent did not alter the equilibrium constant of methanolysis of ethyl butyrate[10].

2. Effect solvent

No wide range of solvents has been studied Most of the alcoholysis equilibria have been measured in alcohol–ester mixtures and no result in dilute solutions of these substances in inert solvents are available except one investigation reported for an alcoholysis equilibrium in dimethylformamide[11]. In initially different alcohol–ester mixtures constant values of the equilibrium constant have been obtained[10, 12]. Solvent effects on alcoholysis equilibria must therefore be small.

3. Enthalpy values

The equilibrium constants of alcoholysis reactions are found to remain unaltered when the temperature is altered[10, 13]. The enthalpy values are then approximately zero.

4. Calculation of equilibrium constants from those of ester hydrolysis

The equilibrium constant K_7 of the alcoholysis reaction (7) can be calculated from the equilibrium constants K_8 and K_9 of the two hydrolysis reactions (8) and (9) of the esters involved because $K_7 = K_8/K_9$.

$$R^1COOR^2 + R^3OH = R^1COOR^3 + R^2OH \qquad (7)$$

$$R^1COOR^2 + H_2O = R^1COOH + R^2OH \qquad (8)$$

$$R^1COOR^3 + H_2O = R^1COOH + R^3OH \qquad (9)$$

The equation is only strictly valid if the three equilibrium constants are measured in the same solvent, which is usually not the case. Alcohol–water mixtures have been used for the hydrolysis, and alcohol–ester mixtures for the alcoholysis.

The solvent effects on alcoholysis equilibria have been found to be neglible in alcohol–ester mixtures, whereas the equilibrium constants for the hydrolysis of esters vary considerably when the composition of the alcohol–water mixtures is altered[14]. However the ratio K_{Me}/K_{Et} of the equilibrium constants for the hydrolysis of methyl and ethyl formates in methanol–water and ethanol–water mixtures containing the same concentration of water respectively remains approximately constant over a wide range of alcohol–water mixtures[14]. For example, K_{Me}/K_{Et} is equal to 0·67 and 0·63 in alcohol–water mixtures containing a mole fraction of water of 0·1 and 0·9 respectively. The solvent effects are thus largely cancelled in the ratio of the two equilibrium constants and the calculated value should be close to the true equilibrium constant of the alcoholysis. For instance the ratio $K_{Me}/K_{Et} = 0·67$ calculated using the equilibrium constants of the hydrolysis of methyl acetate, $K_{Me} = 0·173$, and of ethyl acetate, $K_{Et} = 0·257$[15], is close to the experimental value $K = 0·66$ of the relevant alcoholysis reaction[16].

5. Substituent effects

Both polar and steric substituent effects on alcoholysis equilibria have been observed. The equilibrium constants for ethanolysis of methyl esters of normal carboxylic acids are approximately constant, being 0·67, 0·66, 0·59 and 0·64 for methyl formate (calculated using values of equilibrium constants of hydrolysis reactions[14]), acetate[16] propionate[10] and butyrate[10] respectively. Somewhat larger differences are obtained in a series of equilibrium constants for the alcoholysis of methyl acetate in normal alcohols varying from a value of 1 in methanol, to about 0·6–0·7 for ethanol and long-chain normal alcohols (with one exception of 0·95 for n-pentanol)[16]. Decreasing K values of 1, 0·66 and 0·30 were obtained for the alcoholysis of methyl acetate in methanol, ethanol and isopropanol respectively[16].

Electron-withdrawing substituents decrease the equilibrium constants of alcoholysis reactions and the values 0·38 and 0·35 were obtained for the alcoholysis of methyl acetate in allyl and benzyl alcohols respectively[16]. A linear relationship has been established[17] in the hydrolysis of esters between the logarithm of the equilibrium constant and the acidity constant pK_a of the alcohol. Since the equilibrium constants of alcoholysis and hydrolysis reactions are closely related, similar substituent effects are to be expected for both. Substituents which increase the acidity of an alcohol decrease the stability of an ester derived from it towards hydrolysis and presumably also towards alcoholysis.

6. Equilibria of polyhydric alcohols

When an alcohol contains more than one hydroxyl group intramolecular alcoholysis equilibria are also established. For example[18], the equilibrium constant of the two different monoesters of glycerol and stearic acid at 100°C is $K = $ 1-monostearate/2-monostearate = 9. Two different diesters are also known[18] and the equilibrium constant at 165°C is $K = $ 1,2-distearate/1,3-distearate = 0·72[18]. Values of 8·9 and 0·72 have been reported[19] for the respective equilibrium constants for the two glycerides of linseed oil at 235°C.

The equilibrium constant of the alcoholysis of methyl myristate by sucrose to give sucrose monomyristate[11] in N, N-dimethylformamide at 80°C is $K = $ 1·73 and between sucrose and sucrose dimyristate is $K = $ 1·9.

7. Equilibria of lactones

Alcoholysis equilibria of γ-butyrolactone have been studied[20] in methanol ethanol and isopropanol at 25°C and the values of the equilibrium constants obtained are $K = $ 0·136, 0·117 and 0·091 M^{-1}, respectively.

B. Equilibria in Alcoholysis and Redistribution Reactions

No values relating to acidolysis or redistribution reactions seem to have been published so far. The values of the equilibrium constants of acidolysis reactions can be calculated from those of the hydrolysis reactions of the two esters involved, as for alcoholysis equilibria. Because of possible solvent effects, values obtained by this method have to be considered as only approximate.

Values of the equilibrium constants of redistribution reactions can also be calculated using values for two alcoholysis reactions or two acidolysis reactions. As already mentioned, the solvent effects observed in calculating equilibrium constants of alcoholysis reactions in alcohol–ester mixtures are negligible. The values calculated using those of alcoholysis equilibria are therefore expected to be good approximations not disturbed by solvent effects. The values for the enthanolysis of methyl esters of formic, acetic, propionic and butyric acids are approximately equal. Consequently, all the equilibrium constants of redistribution reactions calculated using these values are approximately equal to unity. A random distribution has been assumed to be valid for the redistribution reactions of esters[21], which is in accordance with the above result. However, the available experimental results are still too few for a general acceptance of the statement that all equilibrium constants of redistribution reactions are close to unity.

III. BASE-CATALYSED ALCOHOLYSIS BY B_{Ac} MECHANISM

Uncatalysed bimolecular alcoholysis of esters is very slow compared to base-catalysed alcoholysis, because the alkoxide ion is much more nucleophilic than the alcohol molecule. However, the uncatalysed unimolecular alcoholysis of some esters may occur at a rate comparable with the base-catalysed alcoholysis. The bimolecular reaction is favoured when steric effects are small, *i.e.* when the transition state of the bimolecular reaction is not crowded by bulky substituents, and when the reaction involves strong nucleophiles such as alkoxide ions. The nucleophiles usually attack the carbonyl carbon atom in acyl-oxygen fission alcoholysis which has been verified by use of ^{18}O-labelled esters[22]. The simultaneous alkyl-oxygen fission involving attack of the nucleophile at the alkoxyl carbon atom of the ester usually occurs at a much slower rate than the $B_{Ac}2$ alcoholysis. This $B_{Al}2$ alcoholysis has only been observed in a few cases.

Alcoholysis by $B_{Ac}2$ mechanism usually proceeds by formation of a tetrahedral intermediate 1[2, 5].

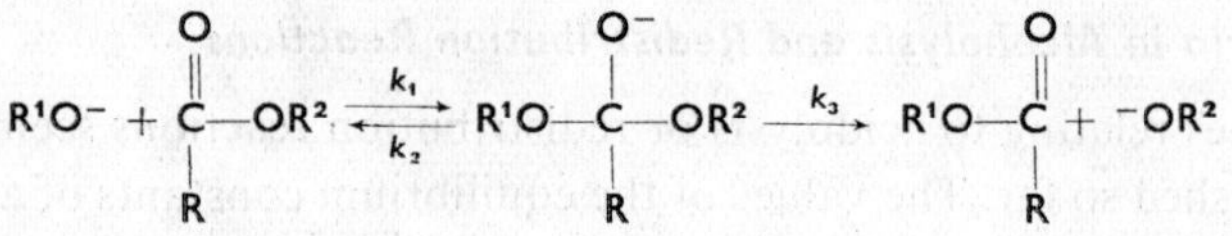

$$R^1O^- + \overset{\overset{\displaystyle O}{\|}}{\underset{\underset{\displaystyle R}{|}}{C}}\!-\!OR^2 \underset{k_2}{\overset{k_1}{\rightleftharpoons}} R^1O\!-\!\overset{\overset{\displaystyle O^-}{|}}{\underset{\underset{\displaystyle R}{|}}{C}}\!-\!OR^2 \overset{k_3}{\longrightarrow} R^1O\!-\!\overset{\overset{\displaystyle O}{\|}}{\underset{\underset{\displaystyle R}{|}}{C}} + {}^-OR^2 \qquad (10)$$

$$(1)$$

The experimental rate constant k_{exp} can be expressed using a steady state approximation:

$$k_{exp} = \frac{k_1}{1 + k_2 k_3} \qquad (11)$$

A. Substituent Effects

1. Substituents in the acyl group of the ester

Electron-withdrawing substituents in the acyl group of the ester accelerate its alcoholysis (Table 1). The ethanolysis of *para*- and *meta*-substituted 1-menthyl benzoates follows the Hammett equation (12)[23]:

$$\log k = \log k_0 + \varrho\sigma \qquad (12)$$

The value of the constant ϱ is 2·530 at 25°C. Similar values of ϱ have been obtained for hydroxyl ion catalysed hydrolysis of esters, *e.g.* $\varrho = 2·373$ for ethyl benzoates[24-26] and $\varrho = 2·305$ for methyl benzoates[24, 27] both in 56% acetone–water mixtures, and $\varrho = 2·498$ for ethyl benzoates in 87·8% ethanol–water mixtures[5, 24]. The substituent effects are almost identical for both alcoholysis and hydrolysis, indicating a similar mechanism for both reactions.

TABLE 1. Methanolysis of esters by $B_{Ac}2$ mechanism, catalysed by sodium methoxide in methanol

Ester	$T(°C)$	$10^4 k$ (1/mole sec)	ΔH^* (kcal/mole)	ΔS^* (cal/deg.)	Reference
Methyl *p*-methoxybenzoate	30·1	106	14·3	−20·3	28
Methyl benzoate	30·1	481	12·7	−22·8	28
Methyl *p*-nitrobenzoate	30·1	41 600	8·1	−29·5	28
1-Menthyl *p*-methylbenzoate	30	0·218	17·43	−22·4	29
1-Menthyl benzoate	30	0·519	16·91	−22·4	29
1-Menthyl *p*-nitrobenzoate	30	58·9	14·21	−21·9	29
1-Menthyl *o*-methylbenzoate	30	0·0383	17·56	−24·6	29
t-butyl 2,4,6-trimethyl-benzoate	65	0·01	—	—	29
Triphenylmethyl acetate	20	4·0	—	—	30

The effects of substituents on the experimental rate constant k_{exp} are due to changes in all three rate constants k_1, k_2 and k_3 in equation (10). Jones and Sloane[28] measured the rates of a symmetrical alcoholysis reactions, i.e. the methanolysis of para-substituted methyl benzoates in methanol catalysed by methoxide ions. As $R^1 = R^2$ in equation (10), $k_2 = k_3$ and $k_{exp} = k_1/2$. The substituent effects observed are then caused by changes in the rate constant k_1 only. The values of the constant in the Hammett equation are $\varrho = 2\cdot41$, $2\cdot32$ and $2\cdot18$ at $30\cdot1°$c, $41\cdot0°$c and $51\cdot7°$c respectively.

Electron-withdrawing substituents decrease the activation enthalpy and entropy, whereas in methanolysis of 1-methyl benzoates[23] the experimental activation entropy was approximately constant, probably because of an accidental compensation of substituent effects on the three rate constants.

Steric effects in ortho-substituted benzoates strongly retard alcoholysis reactions by $B_{Ac}2$ mechanism[23] (Table 1). No alcoholysis of t-butyl 2,4,6-trimethyl benzoate was observed in methanol containing sodium methoxide, although the methanolysis of t-butyl benzoate proceeds easily[29]. No base-catalysed alcoholysis of t-butyl 2,4,6-triphenyl benzoate was observed in methanol[31]. Methyl 2,4,6-triphenylbenzoate, however, reacts with methoxide ions in 90% methanol–water mixture by a $B_{Al}2$ mechanism, the $B_{Ac}2$ reaction being sterically hindered by two bulky ortho substituents[31].

2. Substituents in the alkoxyl group of the ester

No kinetic results are available regarding the effects of substituents in the alkoxyl group of the ester. One semiquantitative study[29] shows that methyl substituents in the alkoxyl part of the ester do not strongly retard the $B_{Ac}2$ alcoholysis, as they do closer to the reaction site in the acyl group. Similar results have also been obtained regarding steric effects in the hydrolysis of esters[32]. Electron-withdrawing substituents in the alkoxyl groups of the ester accelerate hydrolysis[33] and probably also alcoholysis.

B. Reactivity of Nucleophiles toward Esters

In water containing added nucleophiles such as alcohols, both hydrolysis and alcoholysis occur simultaneously. The rate constants of these two reactions can be calculated from product analysis or from the increase of the rate when the concentration of the nucleophile is increased. The equilibrium constants of reaction (13) have to be known in order to calcu-

$$ROH + {}^-OH \rightleftharpoons RO^- + H_2O \qquad (13)$$

TABLE 2. Relative rate constants of the reactions of p-nitrophenyl acetate (A), phenyl acetate (B) and acetyl-L-phenylalanine methyl or ethyl ester (C) with alcoholate and hydroxyl ion in water at 25°C. Values of the reactions with methoxide and ethoxide ions were recalculated using new values[33] of pK_{ROH}

Parent alcohol	pK_{ROH}[33, 34, 36]	Relative rate constants		
		A[36]	B[37]	C[37, 38]
Phenol	10·0	0·075	0·020	—
2,2,2-Trichloroethanol	12·24	2·25	—	—
2,2,2-Trifluoroethanol	12·37	4·3	—	—
2,2,3-Tetrafluoropropanol	12·79	4·1	—	—
2-Propyn-1-ol	13·55	28	—	—
Choline	13·9	13·9	—	—
2-Chloroethanol	14·31	7·3	—	—
2-Methoxyethanol	14·8	6·0	—	—
Methanol	15·09	16	—	1·95
(Water)	(15·75)	(1·00)	(1·00)	(1·00)
Ethanol	15·93	14·6	—	0·89

late the concentrations of hydroxide and alkoxide ions. Using values of these constants, the acidity constants of the alcohols can be calculated, some of which are shown in Table 2. Most uncertain are the values for the very weak acids ethanol and methanol. A recent review has been published by Murto[34].

The reactivity of alkoxide ions towards esters increases, though not linearly, with the basicity of the alkoxide ion (Table 2). The reactivity of nucleophiles is related not only to their basicity, but possibly also to other properties such as polarizability[35].

C. General Base and Nucleophilic Catalysis

The alcoholysis of an ester is usually catalysed by alkoxide ions only. When the ester contains strongly electron-withdrawing substituents in either the acyl or the alkoxyl group, catalysis by other bases has also been observed. Both general base catalysis and nucleophilic catalysis has been shown to occur. Nucleophilic catalysis is expected when the basicity of the alcoholic entity of the ester is at least equal to that of the attacking base[39].

1. Alcoholysis of esters containing strongly electron-withdrawing substituents in the acyl group

General base catalysis by pyridine and methyl-substituted pyridines has been observed in the ethanolysis of ethyl trifluoroacetate[5, 40]. Pyridine and 2,6-lutidine are approximately equally effective as catalysts. The negligible steric effect of the two o-methyl substituents indicates that the reaction occurs by general base catalysis and not by nucleophilic catalysis. In a discussion concerning possible mechanisms for the ethanolysis of ethyl trifluoroacetate Johnson[5] has shown that the number of possible mechanisms is reduced to three because of the symmetry of the reaction. Two of these mechanisms are ruled out because they involve unreasonably fast reactions in one of the steps. The only mechanism remaining involves a general base-catalysed formation of an anionic tetrahedral intermediate (equation 14). This is the first time that evidence of a tetrahedral interme-

$$
\begin{array}{c}
\quad\quad\quad O \quad\quad\quad\quad\quad\quad\quad O \\
\quad\quad\quad \| \quad\quad\quad\quad\quad\quad\quad : \| \\
C_2D_5OH + B + C-OC_2H_5 \longrightarrow C_2D_5O-C\cdots OC_2H_5 \rightleftharpoons \\
\quad\quad\quad | \quad\quad\quad\quad\quad\quad : \quad | \\
\quad\quad\quad CF_3 \quad\quad\quad\quad\quad B\cdots H \quad CF_3
\end{array}
$$

$$
\begin{array}{c}
\quad\quad\quad O^- \\
\quad\quad\quad | \\
\rightleftharpoons C_2D_5O-C-OC_2H_5 + BH^+ \longrightarrow \text{products} \quad\quad (14) \\
\quad\quad\quad | \\
\quad\quad\quad CF_3
\end{array}
$$

diate has been presented for reactions at the carbonyl carbon atom of an ester. As a termolecular mechanism is less likely, the reaction probably proceeds via a hydrogen-bonded alcohol–base complex ROH $\cdots$ B.

General base catalysis by pyridine was also observed in the methanolysis of ethyl trifluoroacetate but no alcoholysis was observed in t-butanol[5], probably due to its low values of acidity and dielectric constant compared with ethanol or methanol [41] and also to steric effects. The formation of opposite charges in the base–catalysed reaction is not favoured by solvents of low ionization power such as t-butanol.

2. Alcoholysis of esters containing electron-withdrawing substituents in the alkoxyl group

Methanolysis of esters of 2,4-dinitrophenol, 2,4,6-trinitrophenol and γ-phenyltetronic acid catalysed by pyridine and methyl-substituted pyridines occurs by acyl-oxygen fission. The catalytic rate constants are given in Table 3. The acidity constants (in water) for 2,4-dinitrophenol[42], 2,4,6-trinitrophenol[42] and γ-phenoltetronic acid enol[43] are $pK_a = 4·0$, 0·77 and 3·76 respectively. All are good leaving groups in alcoholysis.

The Brønsted equation is obeyed in the methanolysis of 2,4-dinitrophenyl acetate catalysed by the sterically unhindered pyridines and the value of the slope is $\varrho = 0·92$. The Brønsted equation is obeyed in general base-catalysed reactions, and also usually in nucleophile-catalysed reactions but the values of ϱ are generally about 0·7–0·8 for the latter[47, 48]. The low values obtained with 2-methyl- and 2,6-dimethylsubstituted

TABLE 3. Rate constants, $10^3 k_b$ (1/mole sec), of the methanolysis of 2,4-dinitrophenyl acetate (DNPAc), 2,4,6-trinitrophenyl acetate (TNPAc), 2,4,6-trinitrophenyl benzoate (TNPBz), γ-phenyltetronic acid enol acetate (PTAc) and γ-phenyltetronic acid enol benzoate (PTBz), catalysed by pyridine (Py), picolines (Pi), lutidine (Lu) and collidine (Col)

Ester	$T(^\circ C)$	Py	2–Pi	3–Pi	4–Pi	2,6–Lu	2,4,6–Col	Reference
DNPAc	45	5·60	0·269	15·0	31·3	0·207	2·73	42, 44
TNPAc	0	781	38·0	–	–	10·1	279	45
TNPBz	54	381	17·2	–	–	6·0	129	45
PTAc	55	44·5	7·5	–	–	20·8	39·7	46
PTBz	55	21·5	32·9	–	–	38·9	102	43

pyridines indicate steric hindrance and consequently nucleophilic catalysis, because steric effects in proton transfer reactions are small. The catalytic rate constant of 2,6-dimethylpyridine is not much smaller than that of 2-methylpyridine for the substituted phenyl esters, and an opposite order of rate constants was observed for γ-phenyltetronic acid esters. The small steric hindrance observed with a second methyl substituent near the reaction centre seems to indicate a general base-catalysed reaction occuring simultaneously with nucleophilic catalysis. The general base-catalysed reaction becomes dominant when the nucleophilic catalysis is sterically hindered by methyl substituents situated close to the reaction centre. The methanolysis of γ-phenyltetronic acid enol benzoate also occurs mainly by general base catalysis in the presence of unhindered pyridines.

IV. BASE-CATALYSED ALCOHOLYSIS BY B_{Al} MECHANISM

Alcoholysis of an ester by alkyl-oxygen fission yields an ether and a carboxylic acid (equation 5). Bimolecular base-catalysed alcoholysis by the $B_{Al}2$ mechanism is generally much slower than the $B_{Ac}2$ alcoholysis. For example, at 100°C methanolysis of methyl benzoate catalysed by sodium methoxide occurring by the $B_{Ac}2$ mechanism is $4·3 \times 10^5$ times faster than methanolysis by the $B_{Al}2$ mechanism[28, 49]. Alcoholysis by the $B_{Al}2$ mechanism is therefore only observed when the $B_{Ac}2$ reaction is sterically hindered[50] or produces the same ester initially present in the reaction mixture.

Electron-withdrawing substituents in the carboxylic part of the ester accelerate the $B_{Al}2$ alcoholysis and the effect is more pronounced than in reactions occurring by the $B_{Ac}2$ mechanism as can be estimated from the

TABLE 4. Methanolysis of esters by $B_{Al}2$ mechanism, catalysed by sodium methoxide in methanol

Ester	$T(°C)$	$10^6 k$ (1/mole sec)	ΔH^* (kcal/mole)	ΔS^* (cal/deg)	Reference
Methyl benzoate	100	9	—	—	48,49
Methyl p-nitrobenzoate	65	215	—	—	52
Methyl 2,4,6-trimethyl-benzoate	100	2	—	—	49
Methyl 2,4,6-tribrom-benzoate	100	50	—	—	49
Methyl 2,4,6-tri-t-butyl-benzoate (in 90% methanol–water)	95	120	24·5	−11·0	50
Methyl t-butylacetate	65	0·1	—	—	53

rate constants of methyl p-nitrobenzoate and methyl benzoate (Table 4). A rapid alkaline $B_{Al}2$ methanolysis of methyl trifluoracetate has also been observed[51].

Steric effects of subsituents near the carbonyl group have a very pronounced retarding effect on reactions occurring by the $B_{Ac}2$ mechanism, whilst in $B_{Al}2$ reactions they are hardly detectable because the substituents are further away from the reaction centre. The somewhat slower methanolysis of methyl 2,4,6-trimethylbenzoate compared to methyl benzoate (Table 4) is at least partly due to polar substituent effects of the three electron-donating methyl groups and the steric retardation of the methyl groups, if present, is quite small. The hydrolysis of methyl 2,4,6-tri-t-butylbenzoate by the $B_{Ac}2$ mechanism is completely inhibited by the bulky t-butyl substituents and only $B_{Al}2$ methanolysis has been observed[50].

The alcoholysis of β-propiolactone with phenolate ions occurs[54] by alkyl-oxygen fission, whereas methanolysis with methoxide ions occurs by acyl-oxygen fission. As it is a weaker nucleophile, the phenolate ion reacts with β-propiolactone by bimolecular attack at the alkoxyl carbon atom. Kinetic evidence for establishing the order of the alcoholysis reaction with respect to phenolate ions has, however, not yet been presented.

V. UNCATALYSED ALCOHOLYSIS BY B_{Ac} MECHANISM

Uncatalysed alcoholysis of esters is usually slow and masked by the faster acid- or base-catalysed reactions. When electron-withdrawing substituents are introduced into either the acyl or the alkoxyl group of the ester, the reactivity towards nucleophiles increases. As the acid-catalysed alcoholy-

8 *

sis remains almost unaffected by these substituents, conditions become more favourable for the detection of the uncatalysed reaction.

Electron-withdrawing substituents in the acyl group accelerate both the spontaneous and the base-catalysed reactions. With primary and secondary alcohols the rates of uncatalysed alcoholysis of the ester RCH_2CO_2Et decrease in the order of R: $C_6H_5CO \rangle CH_3CO \rangle EtO_2C \rangle EtO \rangle CN \rangle C_6H_5 \rangle$ Me[55, 56]. Uncatalysed alcoholysis reactions have also been observed with diethyl oxalate, malonate and fumarate. The reactivity of the esters is not related to the acidity constants of the respective carboxylic acids[55, 57].

Intramolecular catalysis by the hydrogen atom of the enol hydroxyl group as in the alcoholysis of ethyl acetoacetate was suggested to explain the reactivity of ethyl diethylacetoacetate[54], but is probably not present, since copper and aluminium chelates of the ester decrease rather than increase the rate of alcoholysis[58]. Alcoholysis of β-keto esters in t-butanol are slow and do not go to completion[59]. Too few kinetic results are available to decide whether these uncatalysed reactions occur by a bimolecular or unimolecular mechanism, although the $B_{Ac}2$ mechanism is more likely.

Electron-withdrawing substituents in the alkoxyl group also accelerate spontaneous alcoholysis reactions, which have been observed with esters of highly acidic alcohols such as phenols and γ-phenyltetronic acid enol, (Table 5). The uncatalysed methanolysis of picryl benzoate occurs simultaneously by a B_{Al} and a B_{Ac} mechanism[45]. The low value of the activation entropy $\Delta S^* = -30 \cdot 29$ cal/deg. obtained for the uncatalysed methanolysis of 2,4-dinitrophenyl acetate indicates a bimolecular $B_{Ac}2$ mechanism. In the methanolysis of picryl acetate, a much higher value, $\Delta S^* = -4 \cdot 21$ cal/deg., was obtained[45] and the reaction possibly occurs by a unimolecular $B_{Ac}1$ reaction.

TABLE 5. Rate constants of uncatalysed B_{Ac} methanolysis of esters in methanol

Ester	T (°c)	$10^6 \times k$ (1/sec)	ΔH^* (kcal/mole)	ΔS^* (cal/deg.)	Reference
2,4-Dinitrophenyl acetate	55·1	4·64	18·14	−30·29	42
Picryl acetate	54·1	5260	21·10	−4·21	45
Picryl benzoate	54·1	16·3	—	—	45
γ-Phenyltetronic acid enol acetate	55	146	—	—	46
γ-Phenyltetronic acid enol benzoate	55	22·5	—	—	43

VI. UNCATALYSED ALCOHOLYSIS BY B_{Al} MECHANISM

Alkyl-oxygen fission is usually observed in the uncatalysed alcoholysis of esters of tertiary and some secondary alcohols (see Table 6). A unimolecular $B_{Al}1$ mechanism is expected as for the hydrolysis reactions.

TABLE 6. Rate constants of uncatalysed B_{Al} methanolysis of esters in methanol

Ester	T (°C)	$10^6 \times k$ (1/sec)	$\overline{\Delta}H^*$ (kcal/mole	ΔS^* (cal/deg.)	Reference
Triphenylmethyl acetate	35·1	994	19·89	−8·50	30
Triphenylmethyl acetate[a]	25	560	—	—	60
Triphenylmethyl acetate[b]	25	47·3	—	—	60
Trimethylphenyl benzoate[c]	54·5	179	—	—	61
t-Butyl benzoate	64·6	<10	—	—	29
t-Butyl 2,4,6-trimethyl-benzoate	64·6	<1	—	—	29
Picryl benzoate	54·1	20	—	—	45
α-Methyl-γ-phenylallyl p-nitrobenzoate	65	15·1	—	—	62
α-Phenyl-γ-methylallyl p-nitrobenzoate	65	83·8	—	—	62
α-Methyl-γ-p-tolylallyl p-nitrobenzoate	65	127	—	—	62

[a] In 96% methanol–water.
[b] In ethanol.
[c] In 50% ethanol–methylethylketone.

A bimolecular attack by an alcohol molecule at the tertiary or secondary alkoxyl carbon atom of the ester is sterically hindered. In a unimolecular reaction the interaction between non-bonded atoms is partly released when the transition state is formed, resulting in steric acceleration. Only when strong nucleophiles such as alkoxide ions are present in the reaction mixture can a bimolecular $B_{Ac}2$ alcoholysis compete with the comparatively fast unimolecular $B_{Al}1$ reaction[63] (Table 1).

Alkyl-oxygen fission has also been observed in the uncatalysed alcoholysis of picryl benzoate[41]. Since a $B_{Al}2$ reaction would be sterically hindered by the two nitro groups in the *ortho* position, the uncatalysed alcoholysis probably occurs by a $B_{Al}1$ mechanism.

The reactivity of a carbon atom in small-ring compounds is well known. The uncatalysed methanolysis of β-propiolactone occurs by alkyl-oxygen fission[51, 54, 64]. In the uncatalysed hydrolysis of this compound, alkyl-oxygen fission was shown by $H_2^{18}O$ experiments, and a $B_{Al}2$ mechanism is in accordance with the negative activation volume ($\Delta V^* = -10\cdot 2\pm$ $0\cdot 5$ cm³/mole) and the negative activation entropy ($\Delta S^* = -17\cdot 2$ cal/deg.). The $B_{Al}2$ mechanism is also expected for the methanolysis reaction, but in alkaline methanol a $B_{Ac}2$ mechanism is found[51]. The reactivity of the carbonyl and alkoxyl carbon atoms is selective towards various nucleophiles. Methanol and phenoxide ions react preferably with the alkoxyl carbon atom, and methoxide ions with the carbonyl carbon atom[51, 64, 65].

The rate-determining step in a unimolecular alcoholysis is usually the heterolysis of the carbon–oxygen bond. The ion pair of the carbonium ion with the carboxylate ion reacts with solvent molecules to form first a solvent-separated ion pair and then the free ions or the products of the reaction (equation 15). If nucleophiles (N) other than solvent molecules (S)

$$R\text{—}X \rightleftharpoons R\cdots X \rightleftharpoons R^+X^- \overset{S}{\rightleftharpoons} R^+SX^- \rightleftharpoons R^+ + X^- \qquad (15)$$
$$\downarrow N \qquad\qquad \downarrow N \qquad\quad \downarrow N$$
$$RN^+ \qquad\qquad RN^+ \qquad\quad RN^+$$

are present in the reaction mixture, the ion pairs and free ions may also react with these, or with the carboxylate anion X^- to produce starting materials in a reverse reaction. Esters of optically active alcohols yield racemic products in the reactions of free carbonium ions or ion pairs with nucleophiles. Rearrangements of the carbonium ions may occur before they react with the nucleophiles[66, 63]. The two oxygen atoms of the ester become identical in the carboxylate anion, and products where they are exchanged are obtained[67]. In methanolysis of α-phenyl-γ-methylallyl p-nitrobenzoate[62, 68] (RX) in methanol, 24·9% of unrearranged methyl ether (RS), 47·7% of rearranged methyl ether (R′S) and 27·7% of rearranged ester (R′X) are obtained.

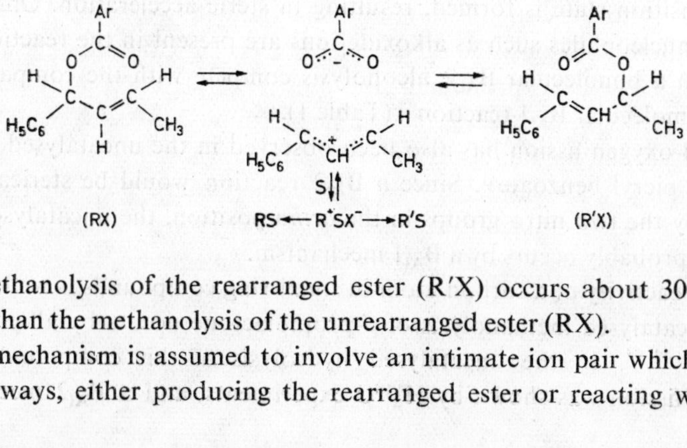

$$(16)$$

The methanolysis of the rearranged ester (R′X) occurs about 300 times slower than the methanolysis of the unrearranged ester (RX).

The mechanism is assumed to involve an intimate ion pair which reacts in two ways, either producing the rearranged ester or reacting with the solvent molecules to form solvent-separated ion pairs or free ions which produce the two ethers by reacting with methanol. When for example, azide ions are added to the reaction mixture, alkyl azides are produced. The amount of rearranged ester is however not altered, nor is the rate of disappearance of the unrearranged ester. Accordingly, the azide ion does not react with the carbonium ion in a stage which is rate-determining or in which the rearranged ester is produced. The rate-determining stage must be the formation of the intimate ion pair from the ester, and the azide ion reacts with the solvent-separated or free carbonium ion.

VII. ACID-CATALYSED ALCOHOLYSIS BY A_{Ac} MECHANISM

In acid-catalysed alcoholysis of esters a fast pre-equilibrium protonation of the ester is expected to occur. The protonated ester forms products in a consecutive bimolecular or a unimolecular reaction. The experimental rate constant is equal to the ratio of the rate constant k of the alcoholysis of the protonated ester and the acidity constant K_a of the ester, i.e. $k_{exp} = k/K_a$ in solutions where the ester is protonated to a small extent only, as is usually the case. The effect of substituents, solvents, electrolytes, temperature etc. on the rate of an acid-catalysed alcoholysis of esters in then twofold: changes in both k and K_a affect k_{exp}. As the acidity constant cannot usually be determined separately, values of the rate constants k of the protonated ester are not available.

Acid-catalysed alcoholysis reactions of esters of primary and secondary alcohols[69-71] and phenols[72] usually occur by acyl-oxygen fission (A_{Ac}). The low values found[69, 70, 72] for both activation enthalpy (12–13 kcal/mole) and activation entropy (about -20 to -30 cal/deg.) are of the magnitude usually obtained for bimolecular solvolysis reactions. The substituent effects are very similar in both acid-catalysed hydrolysis and alcoholysis. In both reactions, the difference between the values of the activation entropies of n-butyl propionate and n-butyl acrylate is almost zero[70], and the difference in the activation energies is about 1·7 kcal/mole. Obviously hydrolysis and methanolysis occur by similar mechanisms.

The acid-catalysed alcoholysis of β-naphtyl esters of carboxylic acids is retarded by α-methyl substituents in the acyl groups, and also by methyl substituents in the α position of the attacking alcohol[72].
Similar results were obtained in the acid-catalysed alcoholysis of methyl acetate[73] and menthyl formate[74, 75] in different alcohols.

The decrease of the reaction rate by substituents in the acyl group or in the nucleophile, is due to steric effects and is of the order usually observed in

TABLE 7. The effect of α-methyl substituents in the acyl groups and methyl substituents in the α position of the attacking alcohol on the acid-catalysed alcoholysis of β-naphthyl esters

	CH₃COOR	MeCH₂COOR	Me₂CHCOOR	Me₃CCOOR
Relative rates in MeOH	100	63	14·5	0·71

	in CH₃OH	CH₃CH₂OH	CH₃CH₂CH₂OH	(CH₃)₂CHOH
Relative rates of CH₃COOR	100	24	23	1·44

bimolecular solvolysis. Acyl-oxygen fission of esters in tertiary alcohols or phenols does not usually occur, and when it does only small amounts of products are obtained, because of unfavourable equilibria and side-reactions. Only one case has been reported, the alcoholysis of methyl benzoate with -naphtol catalysed by sulphuric acid yielding 2-naphtyl benzoate[76]. Methyl cetate and other esters yield only alkoxy 2-naphtyl ethers and these reacions possibly occur by intermediate formation of hydrogen alkyl sulphate.

The alcoholysis of ethyl acetate with 1-octanol h s been studied in toluene using m-xylenesulphonic acid as catalyst. The reaction was found to be approximately zero order with respect to 1-octanol over a wide range of concentrations. This result can be explained assuming a proton-transfer

$$\text{ROH}_2^+ + \text{CH}_3\text{COOR} \rightleftharpoons \text{ROH} + \text{CH}_3\text{COO(H}^+)\text{CH}_3 \qquad (17)$$

equilibrium (equation 14). The alcoholysis probably occurs between an unprotonated alcohol and a protonated ester. Increase of the alcohol concentration in the reaction mixture decreases the concentration of the protonated ester because of the proton-transfer equilibrium. The rate is proportional to the concentrations of alcohol and protonated ester, and remains almost unaffected by changes in the alcohol concentrations. Another kinetically undistinguishable reaction is that between a protonated alcohol and an unprotonated ester. This mechanism is less likely because on protonation the nucleophilicity of the alcohol decreases and the reactivity of the ester increases.

Acid-catalysed methanolysis of β-propiolactone occurs by acyl-oxygen fission[51, 54] although the uncatalysed reaction occurs by alkyl-oxygen fission[51, 54]. The acid-catalysed hydrolysis of β-butyrolactone occurs by the $A_{Ac}1$ mechanism[77] and a similar unimolecular mechanism can also be assumed for the acid-catalysed methanolysis. The acid-catalysed alcoholysis of β-propiolactone in phenol also occurs by acyl-oxygen fission[51, 54]. The alcoholysis of β-propiolactone is an interesting example of solvolysis reactions occurring by different mechanisms, $B_{Ac}2$, $B_{Al}2$ and $A_{Ac}1$, in alkaline, neutral and acid solutions, respectively.

VIII. ACID-CATALYSED ALCOHOLYSIS BY A_{Al} MECHANISM

Alcoholysis reactions occurring by the $B_{Al}1$ mechanism are often very sensitive to acid catalysis, which usually involves a unimolecular ($A_{Al}1$) mechanism. This reaction is of first order with respect to both the ester and acid catalyst and of zero order with respect to the nucleophile, the alcohol. As an example, the acid-catalysed methanolysis of triphenylmethyl benzoate may be mentioned. This was too fast to be measured by conventional methods in methanol or 50% benzene–methanol mixture[61].

The values of activation enthalpy, $\Delta H^* = 28 \cdot 7$ kcal/mole, and activation entropy, $\Delta S^* = +10 \cdot 7$ cal/deg., at 25°C for acid-catalysed methanolysis of t-butyl 2,4,6-triphenylbenzoate in 95% methanol–water mixture 31 are of the order usually obtained for unimolecular ester solvolysis and much larger than those obtained for $A_{Ac}2$ alcoholysis of esters (see section VII).

IX. ALCOLYSIS REACTIONS IN PREPARATIVE CHEMISTRY
A. Simple Esters

Alcoholysis has frequently been used to obtain esters with the alcohol group exchanged. Low yields due to unfavourable equilibria can be improved if one of the products, usually the alcohol, is more volatile than the others and can be continously removed by distillation; the reaction then proceeds to completion. Acid catalysts, such as sulphuric acid[78, 79] or p-toluenesulphonic acid[78, 80], acidic ion exchange resins[81–84] and basic catalysts, such as sodium alcoholate[9, 83–87] aluminium alcoholate[79] or potassium cyanide[89] have been used. In a few cases the reaction is fast even without a catalyst[55, 83, 89]. Strong basic catalysts have to be avoided when undesiderable sidereactions, such as elimination[88], C-alkylation[53] or polymerization[79] may result. When the alcoholysis occurs by alkyl-oxygen fission in neutral or acid solutions, an ether and a carboxylic acid are obtained instead of the desired esters. In the presence of strong nucleophiles, such as a sodium alkoxide, these reactions may be forced to occur by acyl-oxygen fission and to yield esters as products. For example, esters of *t*-butyl alcohol are obtained in high yields when large amounts of potassium *t*-butoxide are used[87, 90]. Large amounts of alkoxide are probably also needed in order to alter the unfavourable alcoholysis equilibrium by the additional alkoxide equilibrium (equation 18).

$$MeOH + t\text{-BuOK} \rightleftharpoons MeOK + t\text{-BuOH} \tag{18}$$

Because methanol is more acidic than *t*-butanol this reaction proceeds far to the right[41].

Alcoholysis has been used to prepare, for example, methyl myristate[78], esters of acrylic acid[79, 80], alkylbenzyl phthalates[84], alkyl-β-ketoesters[83, 88], dialkyl oxalates[59], alkyl *p*-methoxypropionates[89], dialkyl phthalates[86], esters of cholesterol[9] and esters of *t*-butyl alcohol[87, 90]. Phenyl esters[91] and esters of tertiary alcohols[92] are usually difficult to obtain by alcoholysis and special methods are needed[90, 93].

Alcoholysis has also been used for obtaining alcohols from esters, *e.g.*, cholesterol from cholesteryl acetate[88].

Unsymmetrical esters of dicarboxylic acids can be obtained if the exchange rates of the two alkoxyl groups are different. For example, alkyl benzyl phthalates can be prepared from butyl benzyl phthalate by alcoholysis in alkaline solutions because the exchange of the butoxy group proceeds much faster than the exchange of benzoxy group[84].

The acyl group of an ester can also be changed in two concurrent alcoholysis reactions occurring in the presence of two esters and sodium alcoxide as catalyst. If one of the esters produced in the reaction is more volatile than the other three in the reaction mixture, it can be removed continuously by distillation and the reaction is forced almost to completion. For example, cholesteryl palmitate is obtained from cholesteryl acetate and methyl palmitate by removing the methyl acetate by blowing dry nitrogen into the reaction mixture[9].

The alcoholysis of esters with large amounts of alkoxides (usually aluminium alkoxide), have been used to prepare esters which are difficult to prepare by usual methods (equation 19).

$$3\,RCOOR^1 + Al(OR^2)_3 \rightleftharpoons 3\,RCOOR^2 + Al(OR^1)_3 \tag{19}$$

If $RCOOR^2$ is more volatile than $RCOOR^1$, the yields can be improved by removing the former by continuous distillation[94, 95]. By this method, for example, *t*-butyl formate has been obtained from n-butyl formate and aluminium *t*-butoxide in about 45% yield[93]. Esters of isopropyl alcohol have been obtained in the same way[95]. A mixture of lithium and aluminium alkoxides formed from lithium aluminium hydride and an alcohol has also been used as catalyst in alcoholysis[96]. Alkoxides of titanium, zirconium and hafnium give facile alcoholysis reactions, whereas tetraalkoxy silanes do not react[97].

B. Esters of Polyhydric Alcohols and Monocarboxylic Acids

1. Glycerides

A large number of investigations on the alcoholysis of fats have been published. A complete discussion of all these results is outside the scope of this review, and only a few examples of general interest will be given.

Naudet has published a review of the alcoholysis of glycerides[98]. The rates of methanolysis of triglycerides are nearly independent of the chain length of the acid, but equilibria of the alcoholysis reactions have been reported to vary, depending on the length of the carbon chain of normal alcohols[99]. Esters of unsaturated fatty acids react more slowly in alcoholysis than do the esters of saturated fatty acids[100]. Intramolecular exchange of the alkoxy groups has been observed with triglycerides in the presence of sodium methoxide[101] or acids[102]. In ethanol–water mixtures, three reactions occur simultaneously, both hydrolysis and ethanolysis of the original ester as well as hydrolysis of the ethyl ester produced in the ethanolysis[103].

2. Esters of carbohydrates

Alcoholysis of alkyl esters of fatty acids by carbohydrates has been used to prepare esters of carbohydrates. In dimethylformamide as solvent and with potassium carbonate as catalyst at 100°C, a mixture of 81% monostearate, 15% distearate and 2% stearic acid was obtained from sucrose and alkyl stearates when the alcohol obtained in the reaction was removed by distillation[104]. A random distribution of mono-, di-, and tristearides of glucose has been reported[105]. A second-order reaction has been observed between sucrose and methyl stearate[106]. Tricarbonates of D-glucitol, galactitol and D-mannitol have been obtained by alcoholysis of diphenyl carbonate with the respective alcohols in dimethyl sulphoxide or N,N-diethylformamide as solvent[107] using alkaline catalysts. Three ester groups are easily introduced into pentaerythritol in alcoholysis reactions of esters of fatty acids, using sodium methoxide as catalyst, but the fourth hydroxyl group is esterified to an extent of about 50% only[108].

C. Esters of Dihydric Alcohols and Dicarboxylic Acids

Cyclic or long-chain polymeric esters are formed in alcoholysis reactions of dialkyl esters of dicarboxylic acids with dihydric alcohols. A large number of investigations on alcoholysis of polyesters have been published and they will be only briefly reviewed in the following.

1. Polyesters of phthalic acid and glycols

Polyethylene terephthalate has been prepared by alcoholysis of dimethyl terephthalate with ethylene glycol. The reaction proceeds stepwise (equations 20, 21 and 22).

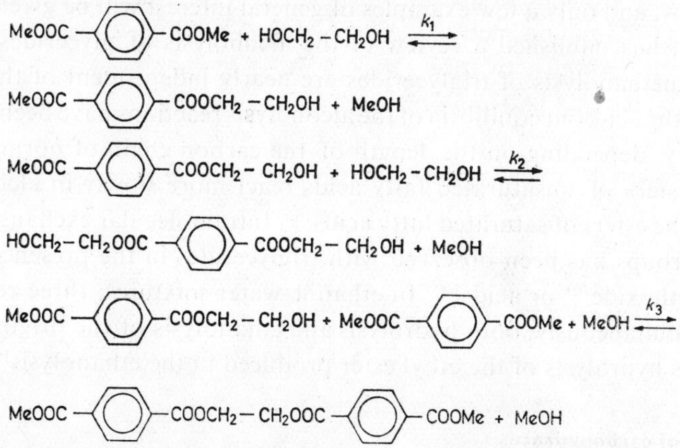

$$MeOOC - \bigcirc - COOMe + HOCH_2 - CH_2OH \underset{}{\overset{k_1}{\rightleftharpoons}}$$

$$MeOOC - \bigcirc - COOCH_2 - CH_2OH + MeOH \tag{20}$$

$$MeOOC - \bigcirc - COOCH_2 - CH_2OH + HOCH_2 - CH_2OH \overset{k_2}{\rightleftharpoons}$$

$$HOCH_2 - CH_2OOC - \bigcirc - COOCH_2 - CH_2OH + MeOH \tag{21}$$

$$MeOOC - \bigcirc - COOCH_2 - CH_2OH + MeOOC - \bigcirc - COOMe + MeOH \overset{k_3}{\rightleftharpoons} \tag{22}$$

$$MeOOC - \bigcirc - COOCH_2 - CH_2OOC - \bigcirc - COOMe + MeOH$$

If the methanol produced in the reaction is removed from the reaction mixture, high molecular weight polymers are obtained. In the absence of catalysts at 175°C, $k_1 = 7 \cdot 1 \times 11^{-5}$ and $k_2 = 23 \times 10^{-5}$ 1/mole sec[109].

Various catalysts have been used for this reaction, e.g. salts of copper, manganese (II), zinc, cadmium, lead and cobalt[110, 111], oxides such as PbO and Al_2O_3[112-114]. p-toluenesulphonic acid[115], lithium hydroxide[114] and iodine[116]. Salts of mercury (II) nickel, iron (III), antimony, germanium, copper or sodium do not catalyse the reaction[110]. In the presence of acetates of cadmium, zinc or lead as catalysts the reaction was found to be first order with respect to both the dimethyl terephthalate and the catalyst but of zero order with respect to ethylene glycol[117]. The activation energies are 9·5 and 10·6 kcal/mole when zinc acetate or cobalt acetate, respectively, are used as catalysts. When p-toluenesulphonic acid is used as catalyst the activation energy is 12 kcal/mole[115].

The rates of alcoholysis reactions of dimethyl esters of dicarboxylic acids with ethylene glycol decrease in the order: dimethyl terephthalate > dimethyl sebacate > dimethyl isophthalate > dimethyl phthalate[111, 115], but the opposite order has also been reported for dimethyl isophthalate and phthalate[117]. There are only small differences in the rates of alcoholysis of dimethyl terephthalate with alcohols such as ethylene glycol, 1,3-butanediol, 1,4-butanediol, benzyl alcohol or octyl alcohol[111, 117].

Polyethylene terephthalate has been prepared using the product of reaction (20), bis(2-hydroxyethyl)terephthalate, as starting material[118]. In addition to high polymer products, ethylene glycol is obtained. The rate constants decrease during the reaction.

Polyesters can be depolymerized by alcoholysis. Methanolysis of polyethylene terephthalate in methanol is catalysed by salts of zinc, manganese or lead[119, 120], sodium methoxide or sodium hydroxide[121]. The reaction is bimolecular and the activation energy is 32 kcal/mole in the absence of catalysts[122] and 13·18 kcal/mole with sodium hydroxide as catalyst[121]. Polyhexamethylene sebacate has been depolymerized by cetyl alcohol[123]. When two polyesters are mixed in liquid form, redistribution reactions occur producing new polyesters[8].

2. Polycarbonates

Polycarbonates are obtained in the alcoholysis of a dialkyl carbonate with a diol. In addition to long chain polyesters, a cyclic ester is obtained, and the equilibrium between these two products depends on the diol used[124]. Sodium methoxide[124] and copper acetate[125] have been used as catalysts. The reaction[126] between 2,2-bis(4-hydroxyphenyl)propane and diphenyl carbonate is of first order at 216°C and the activation energy is 12·80 kcal/mole.

The alcoholysis of polycarbonates[127] in methanol is about 3·7 times faster than in ethanol at 75°C. Hydrolysis of polycarbonates has only been observed in concentrated aqueous mixtures containing more than about 30% of potassium hydroxide[127].

3. High polymers containing ester groups

Alcoholysis of ester groups situated in chains of high polymer molecules has been studied. Methanolysis of polyvinyl acetate in methanol is catalysed by sodium methoxide ($E = 14·98$ kcal/mole)[128], sodium hydroxide ($E = 12·80$ kcal/mole)[128] or by anion exchange resins ($E = 14·59$ kcal/mole)[129]. In 40 : 60% methanol–water mixture[130] the rates of simultaneous alcoholysis and hydrolysis are related as 5·2 : 1. During the alcoholysis a decrease of the molecular weight of the polymer has been observed indicating the presence of ester linkages in the polymer chains[131]. Methanolysis of cellulose triacetate in chloroform has also been studied[132].

D. Alcoholysis of Esters Containing Additional Functional Gorups

When the ester contains a primary or secondary amine group, both intra-or intermolecular aminolysis reactions may occur simultaneously with the alcoholysis. Alcoholysis only was obtained (95% yield) in the reaction between the methyl ester of leucine and n-hexanol when sodium methoxide was used as catalyst and the methanol produced in the reaction was removed by distillation[133]. Similarly, no aminolysis was observed in the reaction between methyl p-aminobenzoate and diethylaminoethanol[134] and in some other reactions[135, 136]. However, in the alcoholysis of acrylic or methyl-acrylic esters with 2-alkylaminoethanol, aminolysis dominates unless the alkyl group is highly branched. With t-alkylaminoethanol only, products of alcoholysis are obtained when aluminium isopropoxide is used as catalyst[134]. Intramolecular catalysis of the alcoholysis by a tertiary amino group in the ester molecule was observed in the alcoholysis of diethylaminoethyl p-nitro-benzoate[137, 138].

When the ester molecule contains an additional acetal group, alcoholysis is observed in methanol solution with sodium methoxide as the catalyst, whereas in acid solution with potassium hydrogen sulphate as catalyst mainly an exchange of the alkoxy group of the acetal occurs[139, 140].

X. ANALYTICAL USE OF ALCOHOLYSIS REACTIONS

Alcoholysis has been used in the analysis of fats and high molecular weight polyesters to convert them to more soluble and volatile esters. Methyl esters are obtained in methanolysis and they can usually be conveniently analysed by a gas chromatograph. Both acid catalysts such as hydrochloric acid[141−143] and basic catalysts such as lithium methoxide[144] or sodium methoxide[142, 145, 146] have been used in addition to boron trifluoride[142, 147] for the methanolysis of fats[141−144], cholesteryl esters[147] or polyesters[145]. When acid catalysts are used, 2,2-dimethoxypropane can be added to the reaction mixture to obtain isopropylideneglycerol which can be used as the internal standard in the gas-chromatographic analysis[143]. The same method could obviously be used to identify esters of glycols and other diols.

XI. ACIDOLYSIS REACTIONS

Acidolysis of esters can occur by either alkyl-oxygen (Al) or acyl-oxygen (Ac) fission of the ester.

$$R^1\text{---}\overset{\displaystyle O}{\overset{\|}{C}}\text{---}OR + R^2\overset{\displaystyle O}{\overset{\|}{C}}\text{---}OH \;\rightleftharpoons\; R^1\text{---}\overset{\displaystyle O}{\overset{\|}{C}}\text{---}OH + R^2\text{---}\overset{\displaystyle O}{\overset{\|}{C}}\text{---}OR \quad \text{alkyl-oxygen fission} \quad (23)$$

$$R^1\!-\!\overset{\displaystyle O}{\overset{\|}{C}}\!-\!OR + R^2\!-\!\overset{\displaystyle O}{\overset{\|}{C}}\!-\!OH \;\rightleftharpoons\; \overset{\textstyle R^1\!-\!C=O}{\underset{\textstyle R^2\!-\!C=O}{\diagdown \!\!\! O \!\!\! \diagup}} + ROH \qquad (24)$$

$$\overset{\textstyle R^1\!-\!C=O}{\underset{\textstyle R^2\!-\!C=O}{\diagdown \!\!\! O \!\!\! \diagup}} + ROH \;\rightleftharpoons\; R^1\!-\!\overset{\displaystyle O}{\overset{\|}{C}}\!-\!OH + R^2\!-\!\overset{\displaystyle O}{\overset{\|}{C}}\!-\!OR \quad \text{acyl-oxygen fission} \qquad (25)$$

In alkyl-oxygen fission acidolysis the exchange of the carboxyl group can occur by a bimolecular, one-step substitution reaction or also by a uni-molecular, two-step reaction. If the acidolysis occurs by acyl-oxygen fission, an anhydride is first obtained as an intermediate, which then reacts with the alcohol, producing either the original ester in a reverse reaction, or a new ester, with the acyl group exchanged. Identical final products are obtained in both acyl-oxygen and alkyl-oxygen fission reactions. They can be distinguished by using isotopically-labelled compounds or by detecting the anhydride formed as an intermediate in one of the reactions. One method consists of detecting the racemic products obtained in the acidolysis by alkyl-oxygen fission of esters of optically active alcohols[148].

A. Acidolysis by Alkyl-Oxygen Fission

Rate constants of the acidolysis of esters occurring by alkyl-oxygen fission are given in Table 8. Acidolysis in formic acid is usually much faster than in acetic acid, due to the higher acidity and better ionizing properties of the former. Acetolysis of p-chlorobenzhydryl acetate was found to be catalysed by perchloric acid but the rate was almost unaffected by lithium acetate[149]. The small effect observed was of the order usually obtained for electrolytes in a salt effect.

TABLE 8. Acidolysis of esters by unimolecular alkyl-oxygen fission

Ester	Solvent	$T(°c)$	$\dfrac{10^6 k}{(1/\text{sec})}$	Reference
p-Chlorbenzhydryl acetate	CH_3COOH	75	1·00	149
Triphenylmethyl acetate	CH_3COOH	25	7890	60
Cyclohexylphenylmethyl acetate	CH_3COOH	100	0·8	150
Methyl-α-naphthylmethyl acetate	CH_3COOH	80	12	150
Cyclohexylphenylmethyl acetate	HCOOH	17	160	150
Methylphenylmethyl acetate	HCOOH	19	120	150
Methyl-α-naphthylmethyl acetate	HCOOH	15	1800	150
Cyclohexylphenylmethyl formate	HCOOH	21	200	150
Methylphenylmethyl formate	HCOOH	21	420	150

The unimolecular alkyl-oxygen fission acidolysis of esters yields as an unstable intermediate an ion pair, which can partly dissociate to free ions. Acidolysis of optically active and oxygen-labelled p-chlorobenzhydryl acetate[149] occurs by unimolecular alkyl-oxygen fission. The exchange of acetate ions of the intermediate ion pair occurs about 2·6 times faster than the recombination of the carbonium ion with the original acetate ion in the ion pair to give the starting materials.

β-Propiolactone reacts in water with acetate ions producing β-acetoxy-propionate ions[151]. This reaction is expected to occur by alkyl-oxygen fission, as do reactions of the lactone with other weak nucleophiles such as phenolate ions[62].

B. Acidolysis by Acyl-Oxygen Fission

1. Intermolecular acidolysis

Acyl-oxygen fission has been observed in a few cases of ester acidolysis[152-154]. Esters of primary alcohols are usually expected to react by this mechanism. The rates of acetolysis of acetyl-L-malic acid in acetic acid were measured by using a ^{14}C tracer method at 120°C and the results were interpreted in terms of acyl-oxygen fission[155]. Acyl-oxygen fission is also assumed to occur in the acidolysis of esters of acetoacetic acids[156]. Catalysis by sulphuric acid has been observed in the acidolysis of ethyl stearate with succinic or adipic acids[157]. Hydrolysis of esters is in some cases catalysed by carboxylate ions, due either to a general base catalysis or a nucleophilic catalysis. The latter involves a reaction between the ester and the carboxylate ion, producing an anhydride as an intermediate, which is rapidly hydrolysed to carboxylic acids in water. The reaction is then an acidolysis of the ester by a carboxylate ion and can be described as $B_{Ac}2$ acidolysis.

Using water enriched with $H_2^{18}O$ it has been shown that the hydrolysis of 2,4-dinitrophenyl benzoate, catalysed by acetate ions[152] and the hydrolysis of acetyl salicylate catalysed intramolecularly by the carboxylate ion in the *ortho* position[153] occur by an acidolysis mechanism involving intermediate anhydride formation (equation 26).

In nucleophilic catalysis both acids become labelled by ^{18}O, whereas in base-catalysed hydrolysis of esters only the carboxylic acid derived from the ester becomes labelled by ^{18}O.

Catalysis of ester hydrolysis by acetate ions has only been observed for esters having a good leaving group, *i.e.* when the alcohol produced in the hydrolysis is a comparatively strong acid. The hydrolyses of p-nitrophenyl acetate and of phenyl acetate catalysed by acetate ions, are expected to

$$
\begin{array}{c}
\underset{\substack{\| \\ O}}{R^1\!-\!C}\!-\!OR + \underset{\substack{\| \\ O}}{R^2\!-\!C}\!-\!O^- \longrightarrow
\underset{\substack{\| \\ C}}{\overset{\substack{O \\ \| \\ R^1\!-\!C \\ | \\ O}}{R^2\!-\!C}} + R\!-\!OH
\end{array}
$$

$$
\underset{\substack{\| \\ O}}{\overset{H_2{}^{18}O}{\nearrow\!\!\!\searrow}}
\begin{cases}
\underset{\substack{\| \\ O}}{R^1\!-\!C}\!-\!{}^{18}O\!-\!H + \underset{\substack{\| \\ O}}{R^2\!-\!C}\!-\!O\!-\!H \\[2mm]
R^1\!-\!C\!-\!O\!-\!H + R^2\!-\!C\!-\!{}^{18}O\!-\!H
\end{cases}
$$

(26)

occur though intermediate anhydride formation[152]. The activation enthalpies are $\Delta H^* = 16\cdot6$ kcal/mole and $15\cdot3$ kcal/mole, and the activation entropies are $\Delta S^* = -31\cdot2$ cal/deg. and $-28\cdot7$ cal/deg. for the two respective reactions. The small increase of the rate of hydrolysis of methyl hydrogen phthalate observed in the presence of acetate ions is probably not due to acetate ion catalysis, since it is of the order usually observed for a salt effect[153].

2. Intramolecular acidolysis

A suitably situated carboxyl group in an ester molecule may cause intramolecular acidolysis, observed as a rate-enchancement of the ester hydrolysis prominent in the pH region corresponding to the ionization of the carboxyl group[158, 159]. Intramolecular catalysis of a carboxyl group in either unionized or ionized form has been observed[160]. The hydrolyses of alkyl hydrogen phthalates with a good leaving group, as in trifluoroethyl hydrogen phthalate and phenyl hydrogen phthalate are catalysed by the carboxylate group[160]. Esters derived from weakly acid alcohols such as methyl hydrogen phthalate and chloroethyl hydrogen phthalate are catalysed by the unionized carboxyl group[160]. Simultaneous catalysis by carboxyl and carboxylate groups is observed for esters derived from alcohols of intermediate acidity such as propargyl hydrogen phthalate. It seems doubtful whether an earlier observation of the intramolecular catalysis in the hydrolysis of methyl hydrogen phthalate by carboxylate groups is correct[153]. Nucleophilic catalysis is expected for the intramolecular carboxylate ion catalysis with intermediate anhydride formation. For the intramolecular carboxylate-ion catalysed hydrolysis of phenyl hydrogen phthalate the observed rate constant, $k_2 = 0\cdot276$ min^{-1} at 25°C, is close to the rate of hydrolysis of phthalic anhydride, $k_2 = 0.739$ min^{-1} at 30°c, and the anhydride hydrolysis becomes almost rate-determining in the hydrolysis of this ester[160].

Hydrolysis of alkyl hydrogen phthalates, intramolecularly catalysed by the carboxyl group, does probably not involve a nucleophilic acidolysis reaction yielding anhydride as intermediate, since the nucleophilicity of an unionized carboxyl group is much smaller than that of the carboxylate[160]. The observed effect is possible due to acid catalysis by the carboxyl group. The increase of the acidity of the alcohol by electron-withdrawing substituents in alkyl hydrogen phthalates accelerates the hydrolysis catalysed by a carboxyl group. However, the reactions catalysed by the carboxylate group are much more sensitive towards changes in the acidity of the alcoholic part of the ester molecule. Such a difference between nucleophilic and acid catalysis is expected in reactions involving acid catalysis: the electron-withdrawing substituents of the alcohol part decrease the basicity of the ester and this effect partly cancels the rate-increasing effect of a better leaving group.

Hydrolysis of methyl hydrogen 3,6-dimethylphthalate yields 3,6-di-methylphthalic anhydride which is not completely hydrolysed in water to the corresponding carboxylic acid[161]. The appearance of a relatively stable anhydride as a product does not, however, prove by itself that the anhydride is formed directly from the ester in an intramolecular acidolysis, because the rate of the anhydride formation from the free carboxylic acid, $k_1 = 0.37$ min^{-1}, is rapid compared to the rate of disappearance of the ester, $k = 0.040$ min^{-1} (at 25°c).

The hydrolyses of phenyl hydrogen succinates and phenyl hydrogen glutarates are intramolecularly catalysed by a carboxylate group[162], and intermediate anhydride formation is expected to occur in these reactions too. Substituents at the phenyl group have a large effect on the rates, e.g. a 540 fold increase has been observed for a p-nitro substituent. The intermolecular acetate-catalysed hydrolysis of phenyl acetate is only accelerated about 15 times by introducing a p-nitro substituent into the phenyl group. This is similar to the alkaline hydrolysis in which p-nitrophenyl acetate reacts about 15 times faster than phenyl acetate. These intramolecular reactions are then expected to occur by a similar mechanism involving a formation of a tetrahedral intermediate from an ester and a hydroxyl ion or from an ester and a carboxylate ion. For the intramolecular carboxylate-catalysed hydrolysis of phenyl hydrogen succinates and glutarates a different mechanism has been proposed, involving a direct substitution of the phenoxy group by a carboxylate group[162].

$$\left[\begin{array}{c} CH_2-C \overset{\displaystyle O}{\diagdown} O-Ph \\ CH_2 \qquad O \\ CH_2-C=O \end{array} \right]$$

An increase of the negative charge at the oxygen atom of the phenoxy group in the transition state is supported by a similar large-substituent effect of a p-nitro group on the ionization of phenol (increase of K_a about 640–680 times) and on the rate of ester hydrolysis (increase of k about 540 times). The activation energy of the acetate ion-catalysed hydrolysis of p-nitrophenyl acetate is 15·7 kcal/mole and for the intramolecularly catalysed hydrolysis of p-nitrophenyl hydrogen glutarate 19·1 kcal/mole. The intramolecular reaction occurs by an energetically unfavourable path possibly because the energetically more favourable path is sterically hindered.

The rates of intramolecular and intermolecular reactions can best be compared by calculating the concentration of catalyst needed to give equal rates for the two reactions. Hydrolysis of phenyl acetate in a solution containing 8 mole/l of acetate ions would occur at a rate equal to the hydrolysis of acetyl salicylate involving intramolecular catalysis by the carboxylate ion[152]. Similarly, the hydrolysis of p-nitrophenyl acetate would require a solution containing 600 mole/l of acetate ions in order to occur at a rate equal to that of the hydrolysis of the mono-p-nitrophenyl glutarate ion. When intramolecular and intermolecular reactions occur by a similar mechanism, the difference in the rates is mainly due to entropy effects, the activation energies being approximately equal for both reactions[152].

Intramolecular catalysis of the hydrolysis of partly-esterified polymethylmetacrylic acid by carboxylate groups has been used to investigate the steric structure of the polymers[163, 164].

C. Acidolysis Reactions in Preparative Chemistry

Acidolysis of esters has been used as a synthetic method in organic chemistry. For example, monoesters of dicarboxylic acids have been prepared by heating the diester with a dicarboxylic acid, using sulphuric acid[165] or a cation-exchange resin in acid form as catalyst[166]. Boron trifluoride has been used as catalyst[167] in the acetolysis of n-butyl formate and lithium or sodium salts have been used in the redistribution reactions of two glycerides[8]. Acidolysis has to be used instead of the usual alcoholysis reactions in preparing vinyl esters because of the instability of vinyl alcohol. Mercury salts, sulphuric acid and bases have been used as catalysts for such reactions[168-171]. Acidolysis reactions can be also used to obtain carboxylic acid from the corresponding esters. Such reactions occur most conveniently in formic acid, which reacts much faster with esters than, for example, acetic acid[172].

XII. REFERENCES

1. A. G. Davies and J. Kenyon, *Quart. Rev. (London)*, **9**, 203 (1955).
2. M. L. Bender, *Chem. Rev.*, **60**, 53 (1960).
3. C. K. Ingold, *Structure and Mechanism*, Cornell University Press, Ithaca, 1953, p. 752.
4. A. Frost and G. Pearson, *Kinetics and Mechanism*, 2nd ed., John Wiley and Sons, New York, 1962, p. 316.
5. S. L. Johnson, *J. Am. Chem. Soc.*, **86**, 3819 (1964).
6. V. V. Korshak and S. V. Vinogradova, *Izv. Akad. Nauk. SSSR, Otd. Khim. Nauk*, 334 (1951); *Chem. Abstr.*, **46**, 1440d (1952).
7. P. Kresse, *Faserforsch. Textiltech.*, **11**, 353 (1960); *Chem. Abstr.*, **54**, 23340c (1960).
8. G. Y. Brokaw, *Chem. Abstr.*, **53**, P11864a (1959).
9. V. Mahadevan and V. O. Lundberg, *J. Am. Oil Chemists' Soc.*, **37**, 685 (1960); *Chem. Abstr.*, **55**, 4012b (1961).
10. R. S. Juvet and F. M. Wachi, *J. Am. Chem. Soc.*, **81**, 6110 (1959).
11. R. U. Lemieux and A. G. McInnes, *Can. J. Chem.*, **40**, 2376, 2394 (1962).
12. P. R. Fehlandt and H. Adkins, *J. Am. Chem., Soc.*, **57**, 193 (1935).
13. G. Challa, *Rec. Trav. Chim.*, **79**, 90 (1960).
14. R. F. Schultz, *J. Am. Chem. Soc.*, **61**, 1443 (1939).
15. St. I. Taussig and F. Petreanu, *Acad. Rep. Populare Romine, Studii Cercetari Stiint, Chim.*, **3**, 103 (1956), *Chem Abstr.*, **51**, 16063e (1957).
16. G. B. Hatch and H. Adkins, *J. Am. Chem. Soc.*, **59**, 1694 (1937).
17. W. P. Jencks and M. Gilchrist, *J. Am. Chem. Soc.*, **86**, 4651 (1964).
18. A. Crossley, I. P. Freeman, B. J. F. Hudson and J. H. Pierce, *J. Chem. Soc.*, 760 (1959).
19. S. M. Rybika, *Chem. Ind.*, 1947 (1962).
20. H. C. Brown and K. A. Keblys, *J. Org. Chem.*, **31**, 485 (1966).
21. G. Calingaert and H. A. Beatty, *J. Am. Chem. Soc.*, **61**, 2748 (1939).
22. R. V. Kudryavtsev and D. N. Kursanov, *Zh. Obshch. Khim.*, **27**, 1686 (1957); *Chem. Abstr.*, **52**, 3676g (1958).
23. R. W. Taft Jr., M. S. Newman and F. H. Verhoek, *J. Am. Chem. Soc.*, **72**, 4511 (1950).
24. L. P. Hammett, *Physical Organic Chemistry*, McGraw-Hill, New York, 1940, p. 191, ref. 3.
25. E. Tommila and C. N. Hinshelwood, *J. Chem. Soc.*, 1801 (1938).
26. E. Tommila, *Ann. Acad. Sci. Fennicae, Ser. A 57*, 13 (1941).
27. E. Tommila, L. Brehmer and H. Elo, *Ann. Acad. Sci. Fennicae, Ser. A II*, 16 (1945).
28. L. B. Jones and T. M. Sloane, *Tetrahedron Letters*, 831 (1966).
29. S. G. Cohen and A. Schneider, *J. Am. Chem. Soc.*, **63**, 3382 (1941).
30. C. A. Bunton and A. Konasiewicz, *J. Chem. Soc.*, 1354 (1955).
31. C. A. Bunton, A. E. Comyns, J. Craham and J. R. Quayle, *J. Chem. Soc.*, 3817 (1955).
32. L. P. Hammett, *Physical Organic Chemistry*, McGraw-Hill, New York, 1940, p. 213.

33. J. Murto, *Acta Chem. Scand.*, **28**, 1043 (1964).
34. J. Murto, *Ann. Acad. Sci. Fennicae, Ser. A II*, 117 (1962).
35. J. O. Edwards, *J. Am. Chem. Soc.*, **78**, 1819 (1956).
36. W. P. Jencks and M. Gilchrist, *J. Am. Chem. Soc.*, **84**, 2910 (1962).
37. M. L. Bender and W. A. Glasson, *J. Am. Chem. Soc.*, **81**, 1590 (1959).
38. M. L. Bender, G. E. Clement, E. R. Gunther and F. J. Kezdy, *J. Am. Chem. Soc.*, **86**, 3697 (1964).
39. J. F. Kirsch and W. P. Jencks, *J. Am. Chem. Soc.*, **86**, 837 (1964).
40. S. L. Johnson, *Tetrahedron Letters*, **23**, 1481 (1964).
41. J. Hine and M. Hine, *J. Am. Chem. Soc.*, **74**, 5266 (1952).
42. W. R. Ali, A. Kirkien-Konasiewicz and A. Maccoll, *J. Chem. Soc.*, 6409 (1965).
43. A. Kirkien-Konasiewicz and A. Maccoll, *J. Chem. Soc.*, 5421 (1961.)
44. W. R. Ali, A. Kirkien-Konasiewicz and A. Maccoll, *Chem. Ind.*, 809 (1964).
45. A. Kirkien-Konasiewicz and A. Maccoll, *J. Chem. Soc.*, 1267 (1964).
46. A. Kirkien-Konasiewicz, *J. Chem. Soc.*, 5430 (1961).
47. J. F. Bunnett, *Ann. Rev. Phys. Chem.*, **14**, 274 (1963).
48. R. Hakala and J. Koskikallio, *Suomen Kemistilehti* B **40**, 172 (1967).
49. J. F. Bunnett, M. M. Robison and F. C. Fennington, *J. Am. Chem. Soc.*, **72**, 2378 (1950).
50. L. R. C. Barclay, G. A. Cooke and N. O. Hall, *Chem. Ind.*, 346 (1961).
51. C. A. Bunton and T. Hadwick, *J. Chem. Soc.*, 943 (1961).
52. R. A. Sneen and A. M. Rosenberg, *J. Org. Chem.*, **26**, 2099 (1961).
53. J. G. Trayham and M. A. Battiste, *J. Org. Chem.*, **22**, 1551 (1957).
54. P. D. Bartlett and P. N. Rylander, *J. Am. Chem. Soc.*, **73**, 4273 (1951).
55. M. F. Carroll, *Proc. Intern. Congr. Pure Appl. Chem. 11th, London*, **II**, 39 (1947); *Chem. Abstr.*, **45**, 7015 (1951).
56. J. Hrivnek, Z. Vesala, E. Sohler and J. Dabek, *Chem. Prumysl*, **15**, 7 (1965); *Chem. Abstr.*, **63**, 8144e (1965).
57. A. R. Bader and H. A. Vogel, *J. Am. Chem. Soc.*, **74**, 3992 (1952).
58. J. Reeder and J. Schlabitz, *J. Org. Chem.*, **31**, 3415 (1966).
59. A. R. Bader, L. O. Cummings and H. A. Vogel, *J. Am. Chem. Soc.*, **73**, 4195 (1951).
60. C. G. Swain, T. E. C. Knee and A. MacLachlan, *J. Am. Chem. Soc.*, **82**, 6101 (1960).
61. G. S. Hammond and J. T. Rudesill, *J. Am. Chem. Soc.*, **72**, 2769 (1950).
62. R. A. Sneen and A. M. Rosenberg, *J. Am. Chem. Soc.*, **83**, 900 (1961).
63. W. E. Doering and H. H. Zeiss, *J. Am. Chem. Soc.*, **75**, 4733 (1953).
64. T. L. Gresham, J. E. Jansen, F. W. Shaver, J. T. Gregory and W. L. Bears, *J. Am. Chem. Soc.*, **70**, 1004 (1948).
65. T. L. Gresham, J. E. Jansen, F. W. Shaver, R. A. Bankert, W. L. Beears and M. G. Prendergast, *J. Am. Chem. Soc.*, **71**, 661 (1949).
66. H. Hart and J. M. Dandri, *J. Am. Chem. Soc.*, **81**, 320 (1959).
67. H. L. Goering and J. F. Levy, *J. Am. Chem. Soc.*, **84**, 3853 (1962).
68. R. A. Sneen and A. M. Rosenberg, *J. Am. Chem. Soc.*, **87**, 895 (1961).
69. L. Farkas, O. Schächter and B. H. Vromen, *J. Am. Chem. Soc.*, **71**, 1991 (1949).
70. D. Buess-Thiernagand and P. J. C. Fierens, *Bull. Soc. Chim. Belges*, **61**, 403 (1952).
71. M. F. Carroll, *J. Chem. Soc.*, 557; 2188; 2192 (1949).
72. H. Harfenist and R. Baltzly, *J. Am. Chem. Soc.*, **69**, 362 (1947).
73. A. J. Rao, *J. Indian Chem. Soc.*, **20**, 69 (1943).
74. J. Ducasse, *Bull. Soc. Chim. France*, **12**, 918 (1945).
75. J. Ducasse and G. Vavon, *Compt. Rend.*, **218**, 412 (1949).

76. S. Patai and M. Bentov, *J. Am. Chem. Soc.*, **74**, 6118 (1952).
77. R. J. Withey, J. E. McAlduff and E. Whalley, *The Physics and Chemistry of High Pressures*, Society of Chemical Industry, London, 1963, p. 196.
78. J. C. Sauer, B. E. Hain and P. W. Boutwell, *Organic Synthesis*, Coll. Vol. III, John Wiley and Sons, New York, 1955, p. 605.
79. C. E. Rehberg, *Organic Synthesis*, Coll. Vol. III, John Wiley and Sons, New York, 1955, p. 146.
80. J. K. Haken, *Tetrahedron Letters*, 551 (1964).
81. G. B. Ulvild, H. W. Tatum and W. C. Hopkins, *U.S. Pat.* 2,862,962 (1958), *Chem. Abstr.*, **53**, P7991a (1959).
82. F. Helfferich, *Angew. Chem.*, **66**, 241 (1954).
83. K. Thinius and W. Schwarz, *Plaste Kautschuk,* **6**, 383 (1959); *Chem. Abstr.*, **54**, 5152c (1960).
84. L. O. Raether and H. R. Gamrath, *J. Org. Chem.*, **24**, 1997 (1959).
85. J. Klosa, *Arch. Pharm.*, **285**, 364 (1952); *Chem. Abstr.*, **48**, 3911f (1954).
86. J. Klosa, *Arch. Pharm.*, **287**, 457 (1954); *Chem. Abstr.*, **51**, 15456f (1957).
87. J. Baltes, F. Weghorst and O. Wechman, *Fette, Seifen, Anstrichmittel*, **63**, 413 (1961); *Chem. Abstr.*, **55**, 18139i (1961).
88. Q. R. Petersen, *J. Am. Chem. Soc.*, **77**, 1743 (1955).
89. C. E. Rehberg, M. B. Dixon and C. H. Fischer, *J. Am. Chem. Soc.*, **68**, 544 (1946).
90. J. Baltes and O. Wechmann, *Fette, Seifen, Anstrichmittel*, **63**, 601 (1961); *Chem. Abstr.*, **55**, 24655i (1961).
91. E. S. Gyngell, *J. Chem. Soc.*, 2484 (1926).
92. K. W. Rosenmund, F. Zymalkowski and E. Gussnow, *Arch. Pharm.*, **286**, 324 (1953); *Chem. Abstr.*, **49**, 8220d (1955).
93. C. Barkenbus, M. B. Naff and K. E. Rapp, *J. Org. Chem.*, **19**, 1317 (1954).
94. E. Kaiser and E. P. Gunther, *J. Am. Chem. Soc.*, **78**, 3841 (1956).
95. R. H. Baker, *J. Am. Chem. Soc.*, **60**, 2673 (1938).
96. P. R. Stapp and N. Rabjohn, *J. Org. Chem.*, **24**, 1798 (1959).
97. R. C. Mehrotra, *J. Am. Chem. Soc.*, **76**, 2266 (1954).
98. M. Naudet, *Rev. Ferment. Ind. Aliment.*, **14**, 268 (1959); *Chem. Abstr.*, **54**, 19451e (1960).
99. K. Kagud and S. Nagasawa, *J. Agr. Chem. Soc. Japan*, **19**, 933 (1943); *Chem Abstr.*, **43**, 1715f (1949).
100. J. Poré, *Oleagineux*, **7**, 21 (1952); *Chem. Abstr.*, **46**, 5341i (1952).
101. M. Naudet and P. Desnuelle, *Bull. Soc. Chim. France*, 595 (1946); 323 (1947).
102. P. E. Verkade, *Rec. Trav. Chim.*, **85**, 426 (1966).
103. H. H. G. Jellinek and A. Gordon, *J. Appl. Chem.*, **1**, 185 (1951).
104. A. R. Menning, *Tr. Vses. Nauchn. Issled. Inst. Zhirov*, 110 (1961); *Chem. Abstr.*, **60**, 770a (1964).
105. K. Kunugi, *Chem. Pharm. Bull. (Tokyo)*, **11**, 478 (1963); *Chem. Abstr.*, **59**, 8848g (1963).
106. K. Kunugi, *Chem. Pharm. Bull. (Tokyo)*, **11**, 918 (1963); *Chem. Abstr.*, **59**, 12232a (1963).
107. L. Hough, J. E. Priddle and R. S. Theobald, *J. Chem. Soc.*, 1934 (1962).
108. V. G. Ostroverkhov and A. A. Kornienko, *Zh. Prikl. Khim.*, **38**, 405 (1965); *Chem. Abstr.*, **62**, 16044 (1965).
109. L. H. Peebles Jr. and W. S. Wagner, *J. Phys. Chem.*, **63**, 1206 (1959).
110. K. Yoda, K. Kimoto and T. Toda, *Kogyo Kagaku Zasshi*, **67**, 909, 919 (1964); *Chem. Abstr.*, **51**, 9372f, 10549h (1957).
111. W. Griehl and G. Schnock, *J. Polymer Sci.*, **30**, 413 (1958).
112. Goodyear Tire and Rubber Co, *Brit. Pat.* 765,609 (1957); *Chem. Abstr.*, **51**, P14812e (1957).

113. G. Torraca and R. Turriziani, *Chim. Ind. (Milan)*, **44**, 482 (1962); *Chem. Abstr.*, **57**, 5848i (1962).
114. V. V. Korshak, S. V. Vinogradova and V. M. Belyakov, *Izv. Akad. Nauk. SSSR Otd. Khim. Nauk*, 730, 737, 746 (1957); *Chem. Abstr.*, **52**, 2799i (1958).
115. M. Sumoto, *Kogyo Kagaku Zasshi*, **66**, 1867 (1963); *Chem. Abstr.* **61**, 3198g (1964).
116. M. Sumoto, *Kogyo Kagaku Zasshi*, **66**, 1663 (1963); *Chem. Abstr.*, **61**, 605f (1964).
117. W. Griehl and G. Schnok, *Faserforsch. Textiltech.*, **8**, 408 (1957); *Chem. Abstr.*, **52**, 11781c (1958).
118. G. Challa, *Makromol. Chem.*, **38**, 105, 123, 138 (1960); *Chem. Abstr.*, **54**, 2344a (1960).
119. A. Lupu, L. Dascalu and J. Cristescu, *Khim. Volokna*, **3**, 25 (1962); *Chem. Abstr.*, **57**, 12698i (1962).
120. Vereinigte Glanzstoff-Fabriken AG., *Chem. Abstr.*, **53**, P18547d (1959).
121. M. Kudra and V. Pavelcova, *Chem. Prumysl.*, **14**, 12 (1964); *Chem. Abstr.*, **60**, 7950e (1960).
122. M. Sumoto, A. Kito and R. Inoue, *Kobunshi Kagaku*, **15**, 664 (1958); *Chem. Abstr.*, **54**, 16001d (1960).
123. V. V. Korshak and S. V. Vinogradova, *Izv. Akad. Nauk SSSR Otd. Khim. Nauk*, 756 (1951); *Chem. Abstr.*, **46**, 7527d (1952).
124. S. Sarel, L. A. Pohoryles and R. Ben-Shoshan, *J. Org. Chem.*, **24**, 1873 (1959).
125. Y. Ueno and Y. Tachikawa, *Japan Pat.* 15,598 (1960), *Chem. Abstr.*, **55**, P10965a (1961).
126. I. P. Losev, O. V. Smirnova and E. V. S. Murova, *Vysokomolekul. Soedin*, **5**, 57 (1963); *Chem. Abstr.*, **59**, 2955b (1963).
127. M. Sumoto, *Kogyo Kagaku Zasshi*, **66**, 1870 (1963); *Chem. Abstr.*, **61**, 1961f (1964).
128. B. Takigawa, *J. Chem. Soc. Japan, Ind. Chem. Sect.*, **55**, 354 (1925); *Chem. Abstr.*, **48**, 1129h (1954).
129. I. P. Losev, O. Ya. Fedotova and G. N. Freidlin, *Izv. Akad. Nauk Arm. SSR, Ser. Khim. Nauk*, **10**, 403 (1957); *Chem. Abstr.*, **52**, 12529i (1958).
130. J. Sakurada, K. Ohashi and S. Morikawa, *J. Soc. Chem. Ind. Japan*, **45**, 450 (1942); *Chem. Abstr.*, **44**, 8161b (1950).
131. O. L. Wheeler, S. L. Ernest and R. N. Crozier, *J. Polymer Sci.*, **8**, 409 (1952).
132. K. Ward Jr., C–C. Tu and M. Lakstigala, *J. Am. Chem. Soc.*, **77**, 5679 (1955).
133. M. Brenner and W. Huber, *Helv. Chim. Acta*, **36**, 1109 (1953).
134. H. J. Sims, P. L. de Benneville and A. J. Kresge, *J. Org. Chem.*, **22**, 787 (1957).
135. M. Matter, *Swiss Pat.* 331,982; (1958) *Chem. Abstr.*, **53**, P5203d (1959).
136. L. Sardana and O. S. Partor, *Anales Real Soc. Espan. Fis. Quim. (Madrid)*, **52B**, 671 (1956); *Chem. Abstr.*, **54**, 4480i (1960).
137. Th. Eckert, *Arch. Pharm.*, **296**, 527 (1963); *Chem. Abstr.*, **59**, 11570b (1963).
138. J. B. Miller, D. L. Fields and D. D. Reynolds, *J. Org. Chem.*, **30**, 247 (1965).
139. E. H. Pryde, D. J. Moore, H. M. Teeter and J. C. Cowan, *J. Chem. Eng. Data*, **10**, 62 (1965).
140. E. H. Pryde, D. J. Moore, H. M. Teeter and J. C. Cowan, *J. Org. Chem.*, **29**, 2083 (1964).
141. M. L. Vorbeck, L. R. Mattinick, F. A. Lee and C. S. Pederson, *Anal. Chem.*, **33**, 1512 (1961).
142. G. R. Jamieson and E. H. Reid, *J. Chromatog.*, **17**, 230 (1965).
143. M. E. Mason and G. R. Waller, *Anal. Chem.*, **36**, 583 (1964).
144. G. G. Esposito and M. H. Swamm, *Anal. Chem.*, **34**, 1048 (1962).
145. D. F. Percival, *Anal. Chem.*, **35**, 236 (1963).

146. L. Hartman, *J. Am. Oil. Chemists' Soc.*, **33**, 129 (1956); *Chem. Abstr.*, **50**, 6813i (1956).
147. S. A. Hyrun, G. V. Vahoruny and C. R. Treadwell, *Anal. Biochem.*, **10**, 193 (1965); *Chem. Abstr.*, **62**, 14492c (1965).
148. M. P. Balfe, E. A. W. Downer, A. A. Evans, J. Kenyon, R. Poplett, C. E. Searle, A. L. Tarnoky and K. D. Nandi, *J. Chem. Soc.*, **797**, 803 (1946).
149. A. F. Diaz and S. Winstein, *J. Am. Chem. Soc.*, **86**, 4484 (1964).
150. M. P. Balfe, G. H. Beaven and J. Kenyon, *J. Chem. Soc.*, 376 (1951).
151. T. L. Gresham, J. E. Jansen, P. W. Shaver, J. T. Gregory and W. L. Beears, *J. Am. Chem. Soc.*, **70**, 1004 (1948).
152. M. L. Bender and M. C. Neveu, *J. Am. Chem. Soc.*, **80**, 5388 (1958).
153. M. L. Bender, F. Chloupek and M. C. Neveu, *J. Am. Chem. Soc.*, **80**, 5384 (1958).
154. M. L. Bender, E. J. Pollock and M. C. Neveu, *J. Am. Chem. Soc.*, **84**, 595 (1962).
155. E. Tascher, C. Wasilewski, G. Kupryszewski and T. Ulmiski, *Bull. Acad. Polon. Sci., Ser. Sci. Chim., Geol. Geograph.*, **7**, 873 (1959); *Chem. Abstr.*, **55**, 16600a (1961).
156. E. Cherbuliez and M. Fuld, *Helv. Chim. Acta*, **35**, 1280 (1952).
157. V. V. Korshak and S. V. Vinogradova, *Izv. Akad. Nauk SSSR Otd. Khim. Nauk*, 180 (1952); *Chem. Abstr.*, **47**, 1591a (1952).
158. E. R. Garrett, *J. Am. Chem. Soc.*, **79**, 3401 (1957).
159. H. Morawetz and I. Ousker, *J. Am. Chem. Soc.*, **80**, 2591 (1958).
160. J. W. Thanassi and T. C. Bruice, *J. Am. Chem. Soc.*, **88**, 747 (1966).
161. L. Eberson, *Acta Chem. Scand.*, **18**, 2015 (1964).
162. E. Gaetjens and H. Morawetz, *J. Am. Chem. Soc.*, **82**, 5328 (1960).
163. H. Morawetz and P. E. Zimmering, *J. Phys. Chem.*, **58**, 753 (1954).
164. H. Morawetz and E. Gaetjens, *J. Polymer Sci.*, **32**, 526 (1958).
165. J. Stanek and Z. Zekja, *Chem. Listy*, **46**, 565 (1952); *Chem. Abstr.*, **47**, 9921h (1953).
166. H. Henecka, *Methoden der organischen Chemie*, Vol. 3, 4th ed., (Ed. E. Müller), G. Thieme Verlag, Stuttgart, 1952, p. 526.
167. F. J. Sowa, *J. Am. Chem. Soc.*, **60**, 654 (1938).
168. R. L. Adelman, *J. Org. Chem.*, **14**, 1057 (1949).
169. D. Swerny and E. F. Jordan, *Organic Synthesis*, Coll. Vol. 4, John Wiley and Sons, New York, 1963, p. 977.
170. A. N. Kost and A. M. Yurkevich, *Zh. Obshch. Khim.*, **23**, 1738 (1953); *Chem. Abstr.*, **48**, 13622e (1954).
171. H. Akashi, *Kogyo Kagaku Zasshi*, **66**, 1909 (1963); *Chem. Abstr.*, **61**, 730a (1964).
172. C. E. Rehberg, *Organic Synthesis*, Coll. Vol. 3, John Wiley and Sons, New York, 1955, p. 33.

CHAPTER **4**

The formation of carboxylic acids and their derivatives from organometallic compounds

R. P. A. SNEEDEN

Monsanto Research S. A., Zürich, Switzerland

I. INTRODUCTION

The carbonation of organometallic compounds is one of the oldest known insertion reactions. However, despite the fact that the reaction has been used extensively for the preparation of carboxylic acids, and for the characterization of organometallic species, very little is known about the overall mechanism of the process. The present chapter treats of the use of the carbonation and related processes as a preparative route to carboxylic acids. Particular emphasis is placed on the general conditions, the scope and, when possible, the overall steric course of the reaction.

II. SATURATED ACIDS

A. Alkyl Alkali Metal Derivatives

The carbonation of alkyl and aralkyl alkali metal derivatives can, in certain instances, be an excellent synthetic route to saturated carboxylic acids. The yield of the *desired* saturated carboxylic acid depends upon a judicious choice of organometallic species and carbonation procedure. The average yields of acid, based on halide used, are with alkylmetal derivatives 30–60%, and with aralkyl metal derivatives 70–90%.

1. Reaction conditions

The preparation of alkyllithium, -sodium and -potassium compounds has been reviewed[1-4], and may be summarized as follows (the first three methods are best adapted for the preparation of alkyl metal compounds).

a. Halogen–metal exchange. ($RBr + 2 M \rightarrow RM + MBr$). The metal and the alkyl chloride or bromide are allowed to interact in an inert solvent (*e.g.* benzene, hexane[5-7]).

b. Metal–metal exchange. ($R_2Hg + M \rightarrow 2 RM + HgM$). In this reaction, the pure organomercury compound is allowed to interact with a large excess of alkali metal in an inert solvent[3]. This is one of the oldest methods of preparing organometallic compounds and is ideally adapted for the preparation of moderate quantities of salt-free alkali metal alkyls.

c. Metal–metal interconversion. ($R_nM + R'M' \rightleftharpoons R_{n-1}R'M + RM'$). This reversible reaction between two different organometallic compounds (*e.g.* an organolithium with an organotin or -mercury compound) is of limited application. It has, however, been used successfully in the preparation of some alkyllithium compounds[1].

d. Metalation. ($RM + R'H \rightleftharpoons R'M + RH$). In this process an organometallic compound is allowed to react with a hydrocarbon. The success of the preparative route depends on a reasonable difference in the acid strengths of the parent hydrocarbons, and is thus better adapted to the preparation of aralkyl and arylmetallic compounds (equation 1)[8, 9].

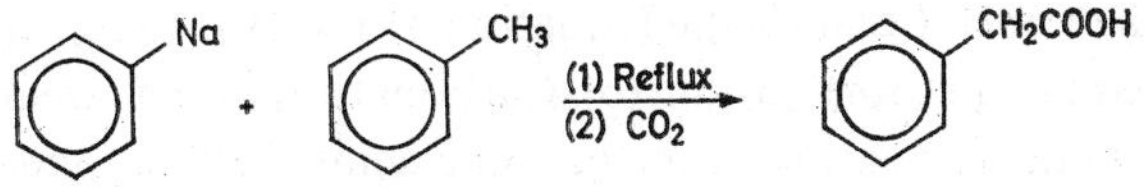

$$\text{(structures)} \quad \xrightarrow[\text{(2) CO}_2]{\text{(1) Reflux}} \quad \text{(structure)} \tag{1}$$

e. Halogen–metal interconversion. $(RM + R'X \rightleftharpoons R'M + RX)$. In certain instances organo alkali metal compounds react with organic halides, in an inert solvent, to give a new organo alkali metal compound. Again it is better adapted for the preparation of aralkyl and arylmetallic compounds. Since the metal atom always assumes the same position as the leaving halide this method is superior to the metalation procedure as a route to organometallic compounds. The reaction has been extensively used to prepare organolithium compounds[10] but has not been so widely used in the preparation of alkylsodium or potassium compounds.

f. Addition.   Organometallic compounds add to some activated double and triple bonds to give a new organometallic species (equation 2). This reaction is of limited application, since in some instances both addition and metalation can occur (equation 3)[11].

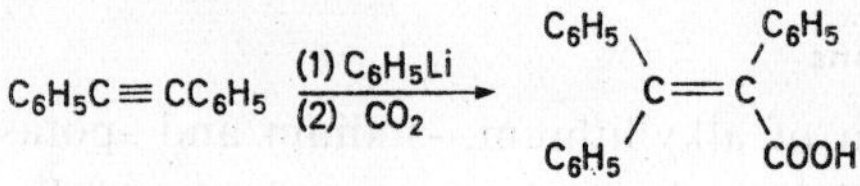

$$C_6H_5C \equiv CC_6H_5 \quad \xrightarrow[\text{(2) } CO_2]{\text{(1) } C_6H_5Li} \quad
\begin{array}{c}
C_6H_5 \\
\diagdown \\
C_6H_5
\end{array}
C = C
\begin{array}{c}
C_6H_5 \\
\diagup \\
COOH
\end{array}
\tag{2}$$

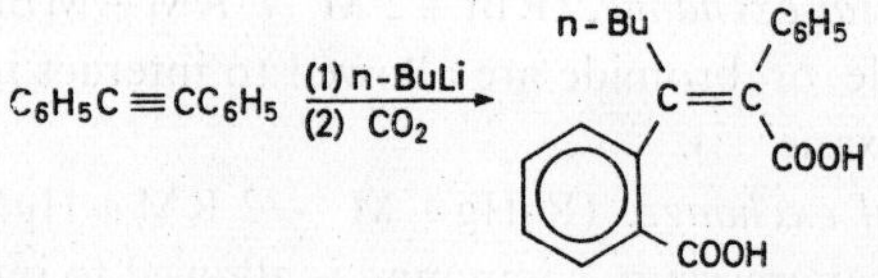

$$C_6H_5C \equiv CC_6H_5 \quad \xrightarrow[\text{(2) } CO_2]{\text{(1) } n\text{-BuLi}} \quad
\begin{array}{c}
n\text{-Bu} \\
\diagdown \\
\end{array}
C = C
\begin{array}{c}
C_6H_5 \\
\diagup \\
COOH
\end{array}
\tag{3}$$

The carbonation of the organometallic compound is best effected by spraying its solution onto an excess of dry, powdered, solid carbon dioxide.

2. Scope of the reaction

Whilst undoubtedly the carbonation of alkyl alkali metal compounds constitutes an excellent method of characterization, it is not always a good synthetic route to saturated carboxylic acids. The main disadvantages are (i) the methods of preparation of the alkyl alkali metal compounds (restricted to reactions *a*, *b* and *c* above) are more complicated than simple Grignard formation; (ii) the great reactivity of the alkyl alkali metal compounds occasions side-reactions which considerably reduce the overall yield of the desired saturated carboxylic acid. Thus in the carbonation of the alkyl alkali metal compound (equation 4) the two main side-reactions are illustrated in equations (5) and (6). Carbonation with a stream of carbon dioxide favors the formation of the malonic acids, ketones and carbinols[12].

$$RCH_2M \xrightarrow{CO_2} RCH_2COOM \tag{4}$$

$$RCH_2M \xrightarrow{CO_2} RCH_2M + RCH_2COOM \longrightarrow RCH_3 + R\overset{\displaystyle M}{\underset{\displaystyle COOM}{C}}H \xrightarrow{CO_2} R\overset{\displaystyle COOM}{\underset{\displaystyle COOM}{C}}H \tag{5}$$

$$RCH_2M \xrightarrow{CO_2} RCH_2M + RCH_2COOM \longrightarrow (RCH_2)_2CO + (RCH_2)_3COM \tag{6}$$

Spraying the solution of the organometallic compound onto a large excess of dry solid carbon dioxide (local high carbon dioxide concentration and low temperature) favors formation of the desired saturated acid[13, 14]. In general, however, the overall yield of carboxylic acid, based on alkyl halide used, is in the order of 30–60%[7, 9, 10], whilst with aralkyl halides it is in the order of 70–90%[8–10]. The steric course of the preparation and carbonation of alkyl- and cycloalkyllithium compounds has been investigated and reviewed[15–17]. This topic will be discussed later in the section dealing with alkyl- and cycloalkylmagnesium halides.

B. Alkyl Alkaline Earth Metal Derivatives

Alkylberyllium, -barium, -calcium and -strontium compounds have been prepared[18–22]. Like dimethylberyllium* and organocalcium halides[22] all these alkyl alkaline earth metal derivatives should interact with carbon dioxide to give alkyl carboxylic acids. The most practical and expedient route is however the carbonation of alkylmagnesium halides.

1. Reaction conditions

The conventional way of preparing organomagnesium halides, from an alkyl halide and magnesium in an inert solvent, has been reviewed[23, 24]. Recently, however, it has been shown that organomagnesium halides can be prepared by the interaction of a Grignard reagent and an olefin in the presence of titanium tetrachloride (equations 7–9)[†25, 26].

$$n\, RCH_2CH_2MgBr \xrightarrow{TiCl_4} (RCH_2CH_2)_nTiCl_{4-n} \longrightarrow RCH{=}CH_2 + H_nTiCl_{4-n} \tag{7}$$

$$H_nTiCl_{4-n} + R'CH{=}CH_2 \longrightarrow (R'CH_2CH_2)_nTiCl_{4-n} \tag{8}$$

$$(R'CH_2CH_2)_nTiCl_{4-n} \xrightarrow{RCH_2CH_2MgBr} R'CH_2CH_2MgBr + (RCH_2CH_2)_nTiCl_{4-n} \tag{9}$$

* It is stated that organoberyllium halides are inert to carbon dioxide[19].

† The volatility of the olefin $RCH{=}CH_2$, derived from the original organomagnesium compound favors the formation of the new organomagnesium compound, derived from the added olefin $R'CH{=}CH_2$.

Organomagnesium halides, which are thus accessible from either alkyl halides or terminal olefins, can be readily converted to carboxylic acids or their derivatives by any one of the following reactions (Scheme 1).

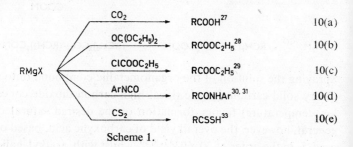

Scheme 1.

The factors influencing the yield of carboxylic acid obtained by the carbonation of organomagnesium halides have been reviewed[27, 32]. When a slow stream of carbon dioxide is passed through the reaction mixture the two competing reactions are again (i) ketone and carbinol formation[34, 35] and (ii) formation of malonic acids[36] (equations 4–6).

2. Scope of the reaction

Despite the fact that the reactions (10a)–(10e) have been used extensively for the preparation of carboxylic acids and their derivatives[27, 33, 37], little is known concerning the mechanism of the reactions. It has been suggested that the carbonation of organomagnesium compounds involves the rearrangement of a four-membered transition state complex (1) (equation 10f)[38].

$$RMgX \xrightarrow{CO_2} \begin{array}{c} R \underset{\displaystyle\,}{-} Mg - X \\ \Vert \quad \Vert \\ O = C = O \end{array} \longrightarrow \begin{array}{c} R \\ | \\ C - O - MgX \\ \overset{\displaystyle \parallel}{O} \end{array}$$

(10f)

(1)

The steric course of the formation and carbonation of alkyl- and cycloalkyllithium and -magnesium compounds has, however, been investigated and recently reviewed[17]. There are three discrete aspects of the reaction which have to be considered: (i) the formation of the organometallic compound, (ii) its configurational stability and (iii) its carbonation. No systematic study has been made concerning the steric course of the formation of alkyl and cycloalkyl metal compounds. The available information indicates that the halogen–metal exchange reaction when used in the preparation of alkyl magnesium halides is not stereospecific[31]. However, the metal–metal interconversion and halogen–metal interconversion reactions when used in the preparation of alkyllithium and alkylmagnesium halides proceed with some degree of stereospecificity[39-41].

Alkyllithium compounds and alkylmagnesium halides are not configurationally stable at room temperature. The rate of conversion and consequently the stability of these organometallic compounds vary with solvent and temperature[17, 40]. Configurational stability of the alkylmetallic compound is favored in hydrocarbon solvents and at low temperatures.

The available experimental data imply that the carbonation of alkyllithium and alkylmagnesium halides takes place stereospecifically.

With substituted cyclopropyl halides it is found that the halogen–metal exchange reaction is non-stereospecific (equation 11)[42] whereas the metal–metal interconversion (equation 12)[42] and the halogen–metal interconversion (equations 13 and 14)[15, 16] are stereospecific (100%).

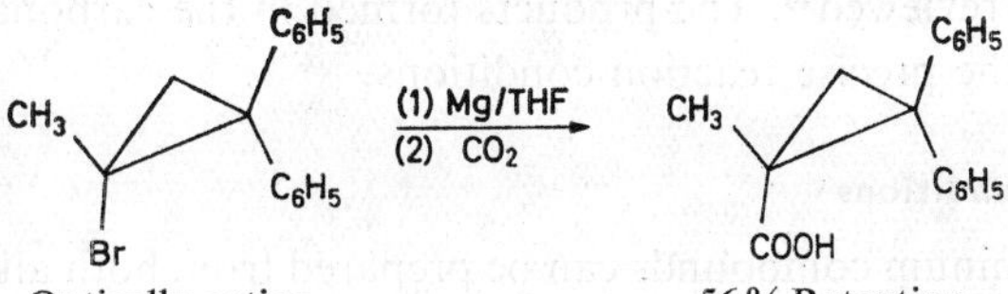

$$(11)$$

Optically active   56% Retention

$$(12)$$

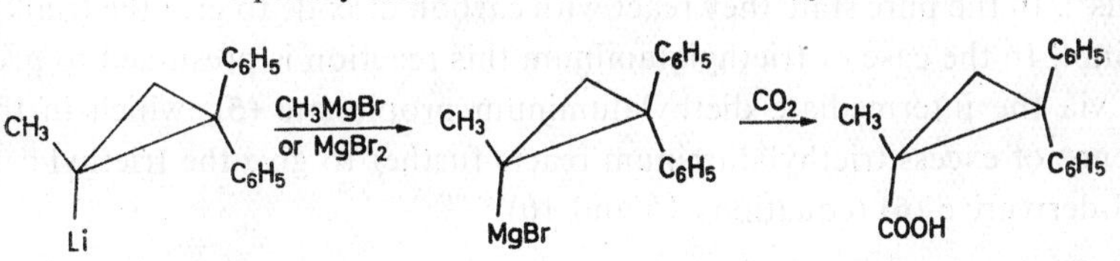

Optically active   **(2)**   100% Retention

$$(13)$$

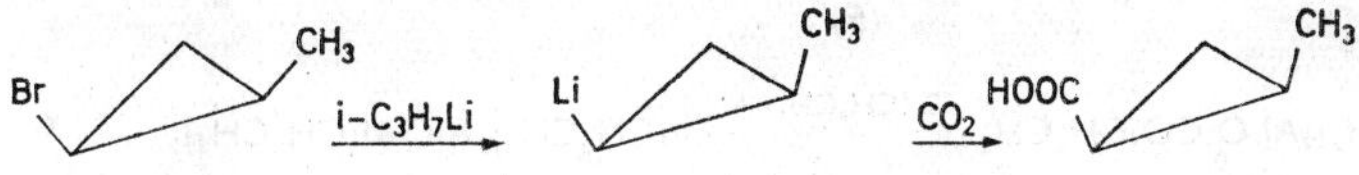

cis   **(3)**   *cis*

$$(14)$$

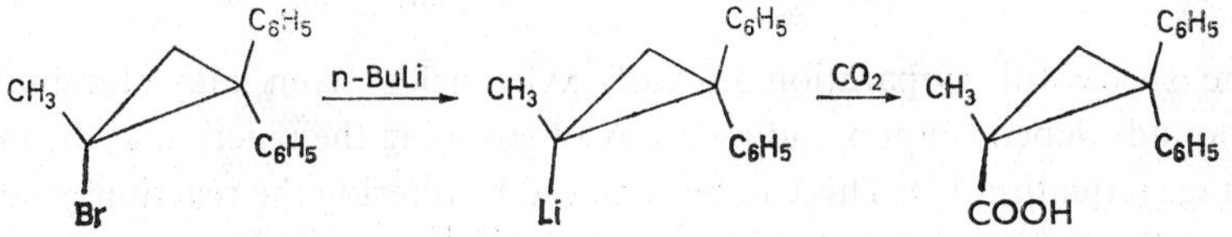

Optically active   **(4)**   100% Retention

The cyclopropylmetallic compounds **2**, **3** and **4**, which are intermediates in the above reactions, have been shown to possess a high degree of configurational stability. This stability is not adversely affected by changes in solvent and temperature[15-17].

The results summarized in equations (12), (13) and (14) also prove that the carbonation of cyclopropyllithium and -magnesium bromide proceeds stereospecifically. Thus in certain cases a cycloalkyl halide can be converted via an organometallic compound to the cycloalkyl carboxylic acid with overall retention of configuration. It is assumed that the individual steps in the reaction proceed with retention of configuration[15, 16, 42].

Alkylzinc and -cadmium compounds are stable to carbon dioxide under normal conditions of temperature and pressure. It is reported, however, that in pyridine solution methylzinc iodide reacts slowly with carbon dioxide to yield acetone[43].

C. Alkylaluminum Compounds

The preparation and carbonation of alkylaluminum compounds has been briefly reviewed[44]. The products formed in the carbonation stage depend upon the precise reaction conditions.

1. Reaction conditions :

Alkylaluminum compounds can be prepared from both alkyl halides and olefins[44]. In the pure state they react with carbon dioxide to give the trialkyl carbinol. In the case of triethylaluminum this reaction is presumed to proceed via the intermediate diethylaluminum propionate (5), which in the presence of excess triethylaluminum reacts further to give the triethyl carbinol derivative (6) (equations 15 and 16)[45, 46].

$$(CH_3CH_2)_3Al \xrightarrow[20°]{CO_2} (CH_3CH_2)_2AlOCOCH_2CH_3 \tag{15}$$

$$\textbf{(5)}$$

$$(CH_3CH_2)_2Al.O.CO.CH_2CH_3 \xrightarrow{2 (CH_3CH_2)_3Al} (CH_3CH_2)_2AlOAl(CH_2CH_3)_2$$
$$+$$
$$(CH_3CH_2)_2AlOC(CH_2CH_3)_3 \tag{16}$$
$$\textbf{(6)}$$

The successful preparation of carboxylic acids from alkylaluminum compounds depends upon finding a way of stopping the reaction at the initial stage (equation 15). This can be achieved by altering the reaction conditions or the nature of the aluminum complex[46, 47].

a. Use of solvent. The carbonation of a trialkylaluminum compound in a hydrocarbon solvent furnishes, after hydrolysis, the carboxylic acid and the hydrocarbon[46] (equation 17, R = iso-octyl, n-octyl, iso-butyl, n-butyl, l-vinylcyclohex-3-ene).

$$R_3Al \xrightarrow[\text{(2) } H_2O]{\text{(1) } CO_2/\text{octane}} RCOOH + 2 RH \tag{17}$$

b. Variation in aluminum complex. Alkylaluminum etherates, which are inert to carbon dioxide at low temperatures, are smoothly carbonated at high temperatures to give one mole of acid (equation 18)[46, 48]. It has also been claimed that lithium aluminum alkyls are carbonated to a mixture of acid and ketone (equation 19).[47] In the two former reactions (equations 17 and 18) only *one* of the alkyl groups in the trialkylaluminum is converted

$$R_3Al\ Et_2O \xrightarrow[(2)\ H_2O]{(1)\ CO_2/100\text{--}120°} RCOOH + 2\ RH \qquad (18)$$

$$LiAlR_4 \xrightarrow{CO_2} RCOOH + R_2CO \qquad (19)$$

to the carboxylic acid, the other two being lost as hydrocarbon. However, if the carbonation mixture is treated with oxygen, the acid and alcohol are the final reaction products[46, 49].

c. Reactions under pressure. The conflicting information indicates that the products formed by the carbonation of alkylaluminum compounds under pressure depend upon the solvent, the temperature and the pressure used for the reaction (equations 20 and 21)[46, 50–52].

$$R_3Al \xrightarrow[180\text{--}300\ Atm]{CO_2/220°} 2\ RCOOH\ (60\%) \qquad (20)$$

$$R_3Al \xrightarrow[330\ Atm]{CO_2/240°} RCOOH\ (15\%)\ R_2CO\ (65\%) \qquad (21)$$

d. Other reactions. The interaction of one mole a trialkylaluminum compound with ethylchloroformate gives one mole of the corresponding ester (equation 22)[53].

$$R_3Al \xrightarrow{ClCOOEt} RCOOEt \qquad (22)$$

2. Scope of the reaction

In view of the fact that usually only one (at the best two) of the alkyl groups attached to aluminum is converted to the carboxylic acid, and the highly flammable nature of these compounds, this reaction is of little practical use in the laboratory preparation of carboxylic acids. It does, however, represent a route whereby olefins (by interaction with aluminum hydrides) can be converted to the homologous acids.

D. Other σ-Bonded Organometallic Compounds

The scant information in the literature indicates that σ-bonded alkyl transition metal compounds are relatively stable to carbon dioxide. Thus in the preparation of alkyltitanium and -chromium compounds (equation 23, $M = Ti$, Cr^{III}, Cr^{II}; $n = 4, 3, 2$ respectively) the excess methyllithium is selectively destroyed by gaseous carbon dioxide[54]. A similar technique has been used to destroy the excess allylmagnesium chloride used in the preparation of triallylchromium (III)[55].

$$x\ LiCH_3 + MCl_n \longrightarrow M(CH_3)_n + x - n\ LiCH_3 \xrightarrow{CO_2} M(CH_3)_n + x - n\ LiOOCCH_3 \quad (23)$$

III. ARYLCARBOXYLIC ACIDS

Arylcarboxylic acids are conveniently prepared by the interaction of aryl-metallic compounds and carbon dioxide. There are four distinct synthetic routes involving essentially the carbonation of an 'organometallic species'.

(*i*) Carbonation of true σ-bonded organometallic compounds

$$\left(C_6H_5M \xrightarrow{CO_2} C_6H_5COOM\right)^{56}.$$

(*ii*) The Kolbe–Schmitt synthesis involving alkali phenolates

$$\left(C_6H_5ONa \xrightarrow{CO_2} C_6H_4\,(ONa)COONa\right)^{57}.$$

(*iii*) The Henkel reaction, involving the thermal disproportionation of the potassium salts of arylcarboxylates

$$\left(C_4H_5COOK \xrightarrow{heat} C_6H_6 + p\text{-}C_6H_4(COOK)_2\right)^{58}.$$

(*iv*) The Friedel–Crafts reaction in which the interaction of an aromatic hydrocarbon and carbon dioxide (or a derivative) is catalyzed by a Lewis acid[59, 60].

A. Carbonation of Arylmetallic Compounds

1. Aryl alkali metal and alkaline earth metal compounds

The carbonation of alkali metal and alkaline earth metal derivatives of aromatic compounds constitutes a convenient laboratory and industrial route to aromatic carboxylic acids[3, 61].

a. Reaction conditions. The preparation of arylmetallic compounds has been reviewed[2, 3, 9, 10, 23, 24]. The methods used are analogous to those employed in the preparation of alkylmetallic compounds (see section II. A.1) The choice of the intermediate organometallic species and the preparative route is determined largely by the starting material.

Aromatic hydrocarbons can be converted to arylmetallic compounds by transmetalation. The high reactivity of alkylsodium and -potassium compounds enables them to even react with pure aromatic hydrocarbons (equation 24)[2, 3]. Alkyllithium compounds however will only react with

$$C_6H_6 \xrightarrow[\text{(2) CO}_2]{\text{(1) i-C}_4H_9Na} C_6H_5COOH\ (78\%)^{62} \tag{24}$$

those hydrocarbons which contain 'active hydrogen' (equation 25)[2, 9].

$$\text{(structure)} \xrightarrow[\text{(2) } CO_2]{\text{(1) } n\text{-BuLi}} \text{(structure with COOH)} \quad (75\%)^{63} \tag{25}$$

The main disadvantage of these reaction sequences is that the precise site of metalation cannot always be predicted. Therefore the precise structure of the aromatic acid finally isolated depends not only upon the reaction conditions (see later) but also upon the reactivity of various sites of the starting material.

Aromatic halogen compounds are conveniently converted to the organometallic species by either halogen–metal interconversion, or halogen–metal exchange[2, 9, 10, 23, 24]. It is normal laboratory practice to use either the aryllithium compounds or the arylmagnesium halides. The former are readily prepared from either aryl bromides or chlorides, by halogen–metal interconversion[10], whereas the preparation of organomagnesium halides is restricted to halogen–metal exchange with the aryl bromides[23] the arylchlorides being too sluggish in their reaction with magnesium.

In the carbonation stage, the reaction conditions influence the overall yield of acid obtained. The main competing reaction is the formation of ketone and carbinol. Passing a stream of carbon dioxide through a solution of the organometallic species favors ketone formation[56], whereas spraying the solution of organometallic compound onto a large excess of solid carbon dioxide favors acid formation[10, 23].

The preparation and carbonation of arylcalcium halides has recently been discussed[22]. The overall yield of carboxylic acid depends critically on the reaction conditions used in making the organocalcium compound (ether cleavage and Wurtz coupling are major side-reactions). There is no advantage to be gained therefore in using these compounds in place of the more commonplace aryllithium and -magnesium compounds.

b. Scope of the reaction.   The carbonation of arylmetallic compounds has been used extensively both for characterization of the organometallic species and for the preparation of carboxylic acids, or their derivatives. The full scope of the reaction is described in the reviews already cited[2, 3, 9, 10, 23, 24]. By analogy with alkylmagnesium halides the arylmagnesium halides and presumably the other arylalkali metal and alkaline earth derivatives will react with ethyl chloroformate and ethyl carbonate and isocyanates to give derivatives of carboxylic acids (Scheme 1). It must always be borne in mind, however, that the reaction conditions often influence the yield and

10*

nature of product obtained. This is illustrated in the case of indene, where the reaction temperature determines the nature of the acid formed (equations 26 and 27)[64-66].

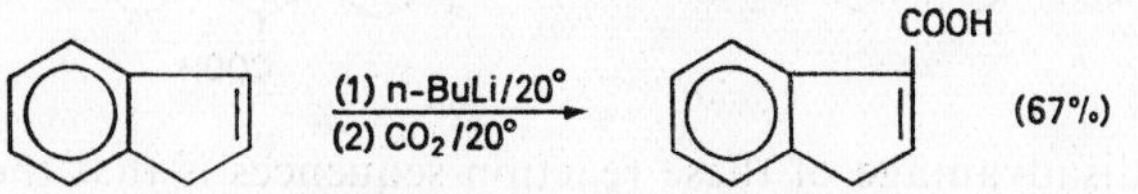

$$
\begin{array}{ccc}
\text{(indene)} & \xrightarrow[\text{(2) } CO_2/20^\circ]{\text{(1) } n\text{-BuLi}/20^\circ} & \text{(1-indenecarboxylic acid)} \quad (67\%) & (26)
\end{array}
$$

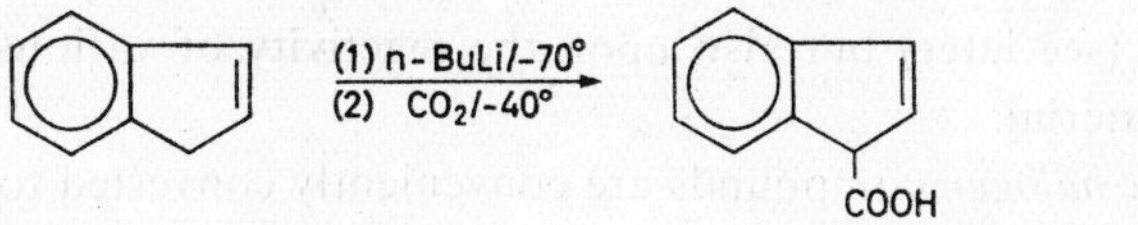

$$
\begin{array}{ccc}
\text{(indene)} & \xrightarrow[\text{(2) } CO_2/-40^\circ]{\text{(1) } n\text{-BuLi}/-70^\circ} & \text{(3-indenecarboxylic acid)} & (27)
\end{array}
$$

2. Aryl transition metal compounds

There are two main types of aryl transition metal compounds, the σ-bonded metal complexes and the bis-arene–metal-π-complexes. In the case of chromium the former may be converted to the latter by treatment with diethyl ether[67]. Pure σ-bonded triphenylchromium reacts slowly with carbon dioxide to give benzoic acid[68a]. Carbonation of the intermediate leading to the π-complex gives the π-benzoic acid-π-diphenylchromium cation[68b, 68c].

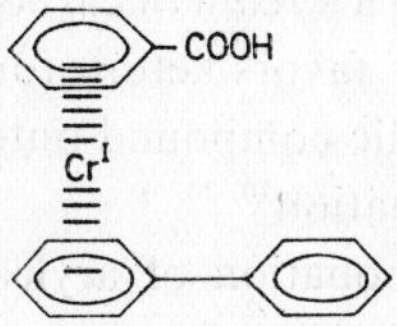

$$
\text{(benzoic acid-}\pi\text{-diphenylchromium cation, Cr}^{\text{I}}\text{)}
$$

The bis (arene) chromium π-complexes are stable to carbon dioxide and may be metalated with amylsodium. Subsequent treatment with carbon dioxide gives a mixture of arenecarboxylic acid-chromium π-complexes[69].

B. Kolbe–Schmitt Synthesis and Henkel Reaction

Both these reactions are high temperature and/or pressure reactions. The first involves the carbonation of an alkali phenolate (equation 28), the second, the thermal disproportionation of the alkali metal salt of a carboxylic acid (equation 29).

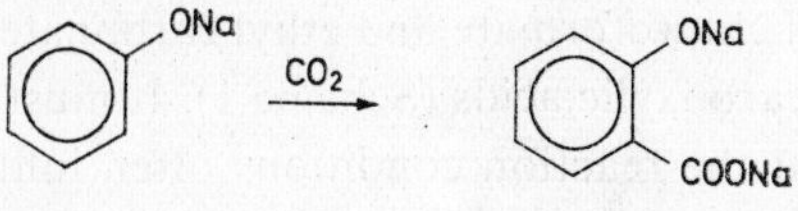

$$
\begin{array}{ccc}
\text{(sodium phenolate, ONa)} & \xrightarrow{CO_2} & \text{(sodium salicylate, ONa/COONa)} & (28)
\end{array}
$$

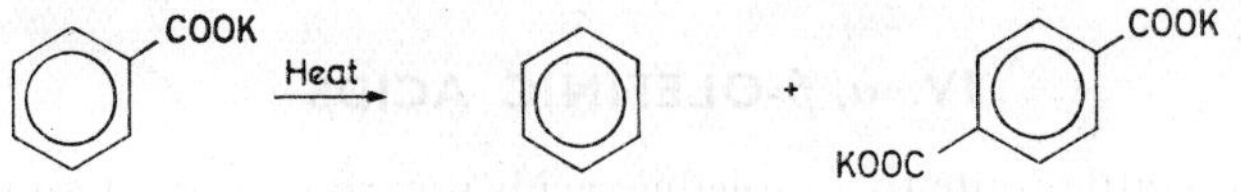

$$(29)$$

In the case of the Henkel reaction the intermediate formation of an organopotassium compound (equation 30) has been suggested[70]. Both reactions have been reviewed and will therefore not be discussed in detail[58, 71].

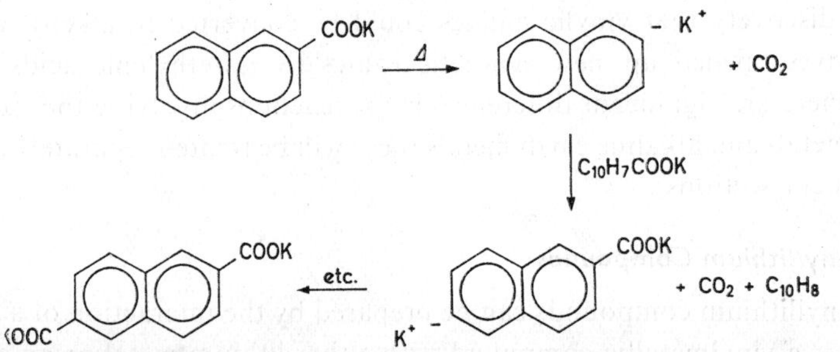

$$(30)$$

C. The Friedel-Crafts Reaction

Aromatic hydrocarbons and carbon dioxide interact, in the presence of aluminum chloride (or other Lewis acid) to give aromatic carboxylic acids*. The mechanism of this reaction is not clearly understood, however a σ-bonded arylaluminum compound ($C_6H_5Al_2Cl_5$) has been suggested as an intermediate[60]. The yields in the reaction are low, however excellent yields of acid derivative can be obtained when derivatives of carbon dioxide are used. These reactions have been reviewed in detail[60], and are summarized in Scheme 2; oxalyl chloride can be used in place of phosgene.

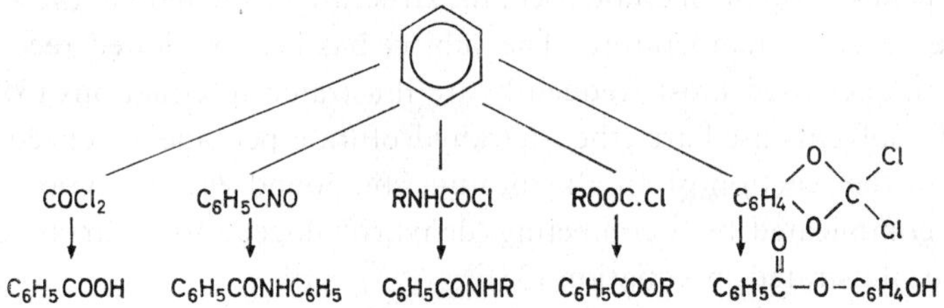

SCHEME 2.

The limiting feature of this synthetic route is that the hydrocarbon used must itself be stable towards aluminum chloride.

*It has also been reported that alkanes react with carbon dioxide, in the presence of aluminum chloride, at high temperatures and pressures to give mixtures of aliphatic carboxylic acids (1% based on aluminum chloride[72]).

IV. α, β-OLEFINIC ACIDS

The most versatile route to α,β-olefinic acids is carbonation of a metal acetylide and subsequent reduction of the resulting α,β-acetylenic acid. Catalytic reduction usually gives the *cis* isomer and chemical reduction the *trans*. The scope and applicability of this synthetic sequence has been reviewed[73].

The discovery that vinylic halides could be converted to alkenyl-metal derivatives opened up new synthetic routes to α,β-ethylenic acids[2, 3, 74]. Since there are significant differences in the reactions involving the various alkali metals and alkaline earth metals these will be treated separately in the subsequent sections.

A. Alkenyllithium Compounds

Alkenyllithium compounds can be prepared by the interaction of a vinyl halide or a vinylmetallic compound with either lithium metal or an alkyl- or aryllithium compound (equations 31 and 32). Subsequent carbonation of

$$RCH{=}CHBr \xrightarrow[\text{or } n-BuLi/Et_2O]{2\ Li/Et_2O} RCH{=}CHLi \xrightarrow{CO_2} RCH{=}CHCOOH \qquad (31)$$

$$(RCH{=}CH)_2Hg \xrightarrow[\text{or } n-BuLi/Et_2O]{2\ Li/Et_2O} RCH{=}CHLi \xrightarrow{CO_2} RCH{=}CHCOOH \qquad (32)$$

the organolithium compound usually results in moderate to good yields of the α, β-ethylenic acid.

1. Reaction conditions

The successful preparation of alkenyllithium compounds from vinyl halides depends upon the method used, the structure of the halide, the solvent and the reaction temperature. The subject has been reviewed recently[75]; the techniques used most frequently are illustrated in equations (31) and (32). The solvents used are ether, tetrahydrofuran, pentane or benzene. The course of the reaction of an alkyllithium compound with the vinyl halide can be complicated by a competing 'dehydrohalogenation rearrangement reaction' illustrated in equation (33)[76].

$$\begin{array}{c} Ph \\ {>}C{=}C{<} \\ Ph \end{array}\begin{array}{c} H \\ \\ Br \end{array} \xrightarrow[(2)\ CO_2]{(1)\ BuLi/Et_2O\ -35°} \begin{array}{c} Ph \\ {>}C{=}CHCOOH + PhC{\equiv}CPh \\ Ph \end{array} \qquad (33)$$

The carbonation step is best effected by spraying the solution of the alkenyl lithium compound onto a large excess of dry solid carbon dioxide. In this way the formation of side products (*e.g.* ketones and carbinols) is reduced to a minimum.

2. Scope of the reaction

Provided the vinyl halide is available, the present reaction sequence affords a good route to α, β-ethylenic acids. In particular, one advantage of the reaction is that the formation of alkylvinyllithium compounds and their subsequent carbonation proceed with overall retention of configuration[77, 78]. With arylvinyl halides the steric course of the reaction is determined by the configurational stability of the arylvinyllithium compound and depends upon both the solvent and the temperature of the reaction (equations 34a, b)[45]. It has been suggested that the difference in configurational

$$
\begin{array}{c}
\text{Ph}\diagdown \quad \diagup \text{Ph} \\
\text{C}{=}\text{C} \\
\text{Li}\diagup \quad \diagdown \text{H}
\end{array}
\quad
\begin{array}{l}
\xrightarrow[\text{CO}_2]{\text{ether/}-35°} \\[4mm]
\xrightarrow[\text{CO}_2]{\text{THF/}-45°}
\end{array}
\quad
\begin{array}{ll}
\begin{array}{c}\text{Ph}\diagdown\quad\diagup\text{Ph}\\ \text{C}{=}\text{C}\\ \text{HOOC}\diagup\quad\diagdown\text{H}\end{array} & (100\%) \qquad (34a) \\[6mm]
\begin{array}{c}\text{Ph}\diagdown\quad\diagup\text{H}\\ \text{C}{=}\text{C}\\ \text{HOOC}\diagup\quad\diagdown\text{Ph}\end{array} & (99\%) \qquad (34b)
\end{array}
$$

stability between alkyl- and aryl-substituted vinyllithium compounds can be attributed to the resonance stabilization of the arylvinyl carbanion (equation 35). Such stabilization is not possible in the case of alkyl-vinyllithium compounds[79]. Another advantage of the use of lithium al-

$$
\begin{array}{c}
\text{R}\\ \diagdown\\ \text{C}{=}\bar{\text{C}}{-}\bigcirc\\ \diagup\\ \text{R}
\end{array}
\quad\longleftrightarrow\quad
\begin{array}{c}
\text{R}\\ \diagdown\\ \text{C}{=}\text{C}{=}\bigcirc{-}\\ \diagup\\ \text{R}
\end{array}
\quad\longleftarrow\quad etc. \qquad (35)
$$

kenyls is that by careful control of reaction conditions (*i.e.* operating at low temperatures) it is now possible to prepare otherwise difficultly accessible compounds, (equation 36[80], *c.f.* equation 33).

$$
\begin{array}{c}
\text{Ar}\diagdown\quad\diagup\text{H}\\ \text{C}{=}\text{C}\\ \text{Ar}\diagup\quad\diagdown\text{Cl}
\end{array}
\quad
\xrightarrow[\text{(2) CO}_2]{\text{(1) BuLi/THF }-105°}
\quad
\begin{array}{c}
\text{Ar}\diagdown\quad\diagup\text{COOH}\\ \text{C}{=}\text{C}\\ \text{Ar}\diagup\quad\diagdown\text{Cl}
\end{array}
\quad (85\%) \qquad (36)
$$

It is important to note that the carbonation of more highly substituted lithium alkenyls, like the carbonation of lithium alkyls, often leads to mixtures of acids and ketones[78] (equation 37). This is due to the fact that the lithium alkenyls are sufficiently reactive to react with the lithium salt of the

$$
\text{R}_2\text{C}{=}\text{CHLi} \xrightarrow{\text{CO}_2} \text{R}_2\text{C}{=}\text{CHCOOLi} \xrightarrow{\text{R C}{=}\text{CHLi}}
\begin{array}{c}
\text{R}_2\text{C}{=}\text{CH}\\ |\\ \text{CO}\\ |\\ \text{R}_2\text{C}{=}\text{CH}
\end{array}
\qquad (37)
$$

carboxylic acid. Ketone formation can be kept to a minimum by using a large excess of carbon dioxide.

B. Alkenylsodium, -potassium, -rubidium and -caesium Compounds

Alkenylsodium and -potassium compounds may be prepared by (i) metalation of an alkene (CH_2=CH_2 + MC_5H_{11} ⟶ CH_2=CHM)[81, 82], (ii) reaction of vinylic halides with the metals (CH_2=$CHCl$ + M ⟶ CH_2=CHM)[82, 83] and (iii) metal–metal exchange reactions ([CH_2=$CH]_2Hg$ + + 2 M ⟶ $2CH_2$=CHM + Hg)[83]. Subsequent carbonation gives the α, β-ethylenic acids in good yields.

1. Reaction conditions

The preparation of alkenylsodium and -potassium compounds has recently been reviewed[2, 3, 75] and will not be discussed in detail. The organometallic compounds are usually prepared in a hydrocarbon or ether solvent; however, owing to their great reactivity they must be handled with appropriate care. It is again desirable to add the organometallic compound to a large excess of solid carbon dioxide in order to avoid side-reactions (e.g. formation of ketones and alcohols).

2. Scope of the reaction

The high reactivity of organosodium and -potassium compounds occasions side-reactions[84] which severely limit the scope of the present reaction. With the homologues of ethylene these are (i) double bond migration, (ii) allylic metalation and (iii) dimetalation, illustrated in equations (38) and (39)[81, 85].

$$CH_3CH{=}CH_2 \xrightarrow[\text{(2) } CO_2]{\text{(1) } NaC_5H_{11}} H_2C{=}CHCH_2COOH \text{ (69\%)} + \text{Dicarboxylic acids (15\%)}$$

$$(38)$$

$$CH_3(CH_2)_3CH{=}CH_2 \xrightarrow[\text{(2) } CO_2]{\text{(1) } NaC_5H_{11}} \begin{array}{ll} CH_3(CH_2)_3CH{=}CHCOOH & (6\%) \\ CH_3(CH_2)_2CH{=}CHCH_2COOH & (32\%) \\ CH_3(CH_2)_2{-}CH{-}CH{=}CH_2 & (51\%) \end{array}$$

$$\begin{array}{l} \quad\quad\quad\quad\quad\quad | \\ \quad\quad\quad\quad\quad COOH \end{array}$$

$$\text{Dicarboxylic acids} \quad\quad\quad (11\%)$$

$$(39)$$

Alkyl compounds of sodium, potassium, rubidium and caesium react with linear olefins to give both the isomerized olefins and the alkenyl metal compounds. Carbonation of the total reaction mixture gives isomerized olefins and isomeric β, γ-olefinic acids as products (equation 40)[86, 87].

$$CH_3(CH_2)_8CH{=}CHCH_3 \xrightarrow[\text{(2) } CO_2]{\text{(1) } MR} \begin{array}{l} CH_3(CH_3)_8CH{=}CHCH_3 \\ CH_3(CH_2)_9CH{=}CH_2 \\ CH_3(CH_2)_8CH{=}CHCH_2COOH \\ CH_3(CH_2)_8{-}CH{-}CH{=}CH_2 \end{array}$$

$$\begin{array}{l} \quad\quad\quad\quad\quad | \\ \quad\quad\quad\quad COOH \end{array}$$

$$(40)$$

An interesting reaction is the synthesis of dibasic acids by the coupling of butadiene[88, 89] or styrene[90] in the presence of sodium terphenyl and subsequent carbonation of the reaction mixture (equations 41 and 42).

$$CH_2{=}CH{-}CH{=}CH_2$$

$$
\begin{array}{c}
\text{(1) Na/}o\text{-terphenyl/(EtOCH}_2)_2{}^{88} \\
\text{or (1) Na/}p\text{-terphenyl/(EtOCH}_2)_2{}^{89} \\
\text{(2) CO}_2 \qquad\qquad\qquad \downarrow
\end{array}
\tag{41}
$$

3,7-decanedioic acid	(38%)	(63%)
2-vinyl-5-octenedioic acid	(51%)	(10%)
2,5-divinyladipic acid	(11%)	(24%)

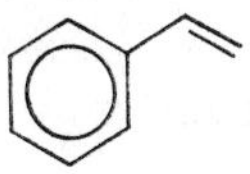

$$\xrightarrow[\text{(2) CO}_2]{\text{(1) Na/}o\text{-terphenyl in (EtOCH}_2)_2} \text{racemic and } meso \text{ diphenyladipic acid (80%)}^{90} \tag{42}$$

C. Alkaline Earth Metal Derivatives of Alkenes

To date the only known alkaline earth derivatives of alkenes are the alkenylmagnesium halides. These are readily prepared by the interaction of a vinyl halide and magnesium in an inert solvent: subsequent carbonation gives the carboxylic acid.

1. Reaction conditions

The preparation of alkenylmagnesium halides and their carbonation has recently been reviewed[74, 75]. The accumulated evidence indicates that the solvent used plays a determining role in the success of the reaction. In tetrahydrofuran or diethylene glycol diethyl ether the alkenylmagnesium halides can be obtained in high yields (70–80%).

2. Scope of the reaction

The formation and carbonation of alkenylmagnesium halides are free of the many side-reactions associated with the alkenylsodium compounds. The good overall yields (50–80%)[74] make this a potentially useful synthetic route to α, β-ethylenic acids. To date, however, it has not been extensively used for this purpose. From the existing data it would seem that in many instances the conversion of the alkenyl halide to α, β-ethylenic acid proceeds with overall retention of configuration. Thus the organomagnesium derivative from trans-2-bromobut-2-ene gives the trans carboxylic acid (angelic acid) on carbonation[91]. In some instances however (equations 43 and 44) the steric course of the reaction is not so clear-cut[92], implying that certain

$$PhCH{=}CHBr \xrightarrow[\text{(2) } CO_2]{\text{(1) } Mg/Et_2O} PhCH{=}CHCOOH + PhCH{=}CHCOOH \qquad (43)$$

cis   cis, 19%   trans, 9%
                                   (overall yield of acid 35%)

$$PhCH{=}CHBr \xrightarrow[\text{(2) } CO_2]{\text{(1) } Mg/Et_2O} PhCH{=}CHCOOH + PhCH{=}CHCOOH \qquad (44)$$

trans   cis, 20%   trans, 30%
                                   (overall yield of acids 62%)

substituents (*e.g.* aryl) reduce the configurational stability of the alkenyl-magnesium halides. This is in keeping with the lack of configurational stability observed in the arylvinyllithium compounds (see section IV. A. 2)

D. Other Alkenyl Metal Derivatives

Divinylzinc[93] and trivinylaluminum[94] have been prepared and should undergo reaction with carbon dioxide, under pressure, to give the α, β-ethylenic acids. Dialkylalkenylaluminum compounds are readily accessible[95, 96] and it would be of interest to determine which of the groups (*i.e.* the alkenyl or the alkyl group) attached to aluminum is converted to the carboxylic acid on carbonation.

V. NON-CONJUGATED OLEFINIC ACIDS

With the exception of β, γ-unsaturated acids, (equation 45, $n = 0$), non-conjugated olefinic acids can be prepared from the appropriate halides, like other alkanoic acids, by the carbonation of the derived organometallic species.

$$RCH{=}CH(CH_2)_nCH_2X \longrightarrow RCH{=}CH(CH_2)CH_2M_n \xrightarrow{CO_2} RCH{=}CH(CH_2)_nCH_2COOH$$
$$(45)$$

However, the high reactivity of the alkali metal compounds engenders side-reactions (*e.g.* cross metalation and addition to the double bond) leading to poor overall yields of the desired acids (equation 45, $n = 0$, M = Li, Na, K, Rb, Cs). The organomagnesium compounds on the other hand do not undergo these side-reactions and are therefore best suited for synthesis of these unsaturated acids.

A. Allyl Alkali Metal Compounds

Allyl derivatives of lithium, sodium and potassium cannot be prepared conveniently by the direct interaction of allyl halides and the metal. They can, however, be prepared by the interaction of phenyllithium with the appropriate allyltin compound (equation 46, R = H or CH_3)[97], or by the cleavage of allyl ethers with the metal (equation 47, R = allyl, M = Na or K; R = C_6H_5, M = Li)[98-100] or in some instances by the allylic metalation of olefins[85, 101].

$$(C_6H_5)_3SnCH_2CR{=}CH_2 \xrightarrow{C_6H_5Li} CH_2{=}C(R)CH_2Li \qquad (46)$$

$$CH_2{=}CHCH_2OR \xrightarrow{M} CH_2{=}CHCH_2M \qquad (47)$$

There is very little factual information available concerning the carbonation of allyl alkali metal compounds and their homologues; allyllithium[97] and allylsodium[99] are carbonated to give vinylacetic acid (36% and 17% respectively). Cinnamyllithium, -sodium and -potassium are reported as reacting with carbon dioxide to give a mixture of acids consisting mainly of β-benzalpropionic acid (PhCH$=$CHCH$_2$COOH)[102]. Similarly the potassium derivative of 1,1-diphenyl-2-methylethylene gives on carbonation 4,4-diphenylvinylacetic acid in 74% yield, (equation 48)[101]. In these cases the allylmetallic compound

$$Ph_2C{=}CHCH_2K \xrightarrow{CO_2} Ph_2C{=}CHCH_2COOH \qquad (48)$$

is carbonated *without rearrangement*. This is in sharp contrast to the reactions of allylmagnesium and -zinc compounds both of which undergo carbonation with rearrangement (see sections V.B. and C).

B. Allylmagnesium Compounds

β, γ-Unsaturated acids are readily accessible by the carbonation of allylmagnesium halides. The latter may be prepared by the interaction of an allylic halide and magnesium in a suitable solvent. However, in contrast to allyllithium compounds, the overall conversion of halide to acid may in some instances involve skeletal rearrangement (allylic rearrangement) (equation 49).

$$\left. \begin{array}{l} RCH{=}CH{-}CH_2Br \\ RCHBrCH{=}CH_2 \end{array} \right] \xrightarrow[(2)\ CO_2]{(1)\ Mg/Et_2O} \begin{array}{l} RCH{-}CH{=}CH_2 \\ \quad | \\ \quad COOH \end{array} \qquad (49)$$

1. Reaction conditions

Allyl halides react with magnesium in an inert solvent to give either the organomagnesium compound or the Wurtz coupling product (*i.e.* the biallyl). The reaction has been reviewed[103, 104], and it would appear that the factors which favor the formation of the organomagnesium compound are the dilution of the reaction mixture and the choice of solvent (*e.g.* diethyl-

or di-n-butyl-ether or tetrahydrofuran[103, 105]. The use of tetrahydropyran would seem to favor Wurtz coupling[104].

2. Scope of the reaction

The reaction sequence affords a practical route to β, γ-unsaturated acids, with the major disadvantage that the reaction may be accompanied by structural rearrangement (equation 49). The detailed mechanism of the reaction is not known, however spectral data indicate that both crotyl bromide and α-methallyl bromide react with magnesium to give the same 'butenylmagnesium halide'. The structure of this organomagnesium compound is best formulated as consisting essentially of either the linear form (7, $R = CH_3$) or a rapidly equilibrating mixture of the linear and branched forms, in which the former preponderates (equation 50, $R = CH_3$)[106, 107].

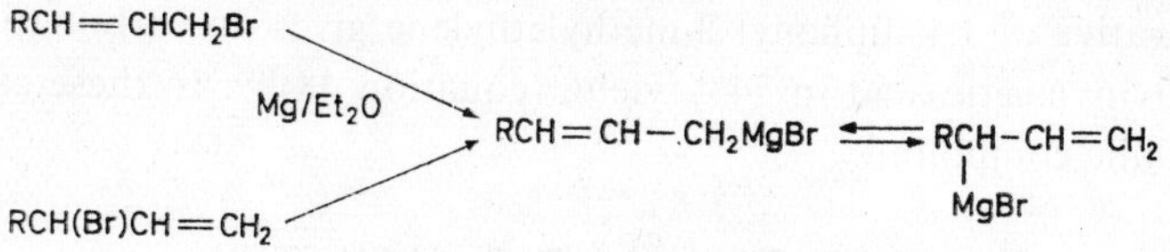

$$RCH = CHCH_2Br$$
$$\searrow$$
$$Mg/Et_2O$$
$$RCH = CH - CH_2MgBr \rightleftharpoons RCH - CH = CH_2 \quad (50)$$
$$\nearrow \qquad\qquad\qquad\qquad\qquad\qquad |$$
$$RCH(Br)CH = CH_2 \qquad\qquad\qquad\qquad\qquad MgBr$$

$$\qquad\qquad\qquad\qquad (7)\qquad\qquad\qquad (8)$$

Similarly spectral evidence supports structure 7, $R = C_6H_5$, for cinnamylmagnesium bromide[108]. Nevertheless, both the butenyl- and cinnamylmagnesium halides give the branched acids ($R—CH(COOH)—CH = CH_2$, $R = CH_3$ or C_6H_5) on carbonation. Of the many possible explanations for this, the most plausible would seem to be either that (i) interaction of the organomagnesium compound with the substrate favors the formation of a transition state leading to the branched form 8 or its derived anion; or (ii) subject to certain steric requirements, the substrate interacts with the linear form 7 to give an intermediate which subsequently rearranges by a concerted cyclic mechanism to the branched-chain acid (equation 51)[109-111].

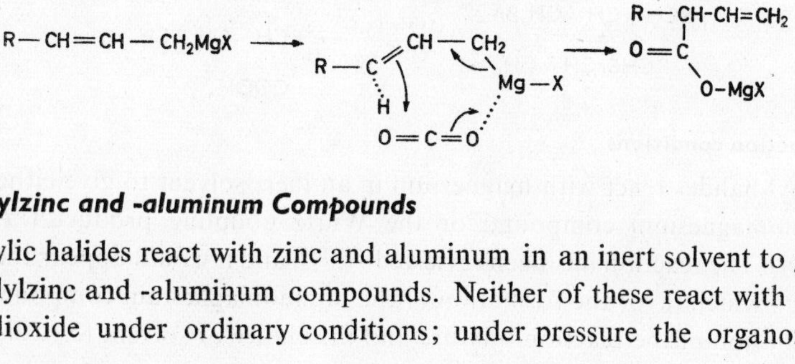

$$R—CH=CH — CH_2MgX \longrightarrow R—C \overset{CH — CH_2}{\underset{H}{\diagup}} \longrightarrow 0 = C \overset{R—CH-CH=CH_2}{\underset{O-MgX}{|}} \quad (51)$$
$$\qquad\qquad\qquad Mg—X$$
$$\qquad\qquad 0=C=0$$

C. Allylzinc and -aluminum Compounds

Allylic halides react with zinc and aluminum in an inert solvent to give the allylzinc and -aluminum compounds. Neither of these react with carbon dioxide under ordinary conditions; under pressure the organozinc compound gives good yields of acid (CH_2=CH—CH_2COOH) whereas the organoaluminum compound gives triallyl carbinol ([CH_2=CH——CH_2]$_3$COH). With diethyl carbonate, both compounds give triallyl carbinol, which is also obtained from the organoaluminum compound and ethyl chloroformate[107, 112, 113]. The scope of these reactions has not been investigated extensively; however, the ease of formation of the organometallic compounds make them potentially useful intermediates in the synthesis of unsaturated acids. Substituted allylic halides also react with zinc in tetrahydrofuran to give the organozinc compound. Carbonation of the latter, under pressure, furnishes moderate yields of β, γ-unsaturated acid. The overall reaction, from halide to acid is again accompanied by allylic rearrangement (equation 52, R = C_6H_5)[114].

$$RCH=CHCH_2Br \xrightarrow[\text{(2) } CO_2]{\text{(1) } Zn/THF} RCHCH=CH_2 \quad (45\%) \qquad (52)$$
$$\underset{\displaystyle COOH}{\mid}$$

VI. α, β-ACETYLENIC ACIDS

A. Alkali Metal Derivatives of Acetylenes

The alkali metal derivatives of acetylenes react with carbon dioxide to give directly the corresponding carboxylic acids, (equation 53, M = Li, Na, K, Rb, Cs)[4, 23, 24, 61, 73].

$$RC{\equiv}CM \xrightarrow{CO_2} RC{\equiv}CCOOM \qquad (53)$$

The good yields and ready accessibility of the alkali metal derivatives make this the best synthetic route to α, β-acetylenic acids. The reaction was first used by Glaser[115] in 1870 for the preparation of phenylpropiolic acid (equation 53, R = C_6H_5, M = Na). The reaction mechanism has not been elucidated.

1. Reaction conditions

The direct carbonation of solid sodium acetylides is a slow process. Traces of sodium hydroxide in the reaction mixture lead to extensive carbonization and consequently poor yields of carboxylic acids[116]. The reaction is therefore best carried out by either (i) first mixing the sodium salt with sand and carbonating the mixture for prolonged periods (4–7 weeks)[116] (ii) carbonating the sodium salt in the presence of a promoter (e.g. N, N-dimethylacetamide)[117] or (iii) carbonating the sodium salt under pressure (e.g. 800 p.s.i. for 8–40 h)[118, 119].

The more usual technique, however, is to treat a suspension of the sodium acetylide, in ether, benzene, toluene or some other inert solvent with either gaseous[115, 117] or solid carbon dioxide. The sodium acetylide can either be prepared directly in an inert solvent by the action of sodium or sodamide on the acetylene, or by the interaction of the acetylene and sodamide in liquid ammonia[120]. In the latter case the liquid ammonia must be replaced by an inert solvent (e.g. benzene or ether) before carbonation. The use of sodium acetylides in this reaction is probably a question of practical convenience. The potassium, rubidium and caesium derivatives of phenylacetylene, however, can also be prepared by the direct interaction of the metal and the acetylene in diethyl ether[121]. The lithium derivatives on the other hand are best prepared by the interaction of an alkyl- or aryllithium compound and the acetylene[121]. All these phenylethynyl alkali metal compounds react smoothly with solid carbon dioxide to give phenylpropiolic acid in about 60% yield[121].

2. Scope of the reaction

The scope of this reaction has been reviewed[73, 122, 123] and will be discussed later together with the carbonation of ethynylmagnesium halides.

Alkali metal acetylides react with carbon dioxide and ethyl chloroformate to give either the α, β-acetylenic carboxylic acids or the corresponding esters (equations 54–56)[124, 125]. With diethyl carbonate the reaction is more complex and the final products are mixtures containing the substituted ethoxyacrylates and diethoxypropionates (equation 57)[126].

$$\text{EtSC}\equiv\text{CH} \xrightarrow[\text{(2) CO}_2]{\text{(1) PhLi}} \text{EtSC}\equiv\text{CCOOH} \tag{54a}$$

$$\text{EtOC}\equiv\text{CH} \xrightarrow{\text{(2) CO}_2} \text{EtOC}\equiv\text{CCOOH} \tag{54b}$$

$$\text{RC}\equiv\text{CNa} \xrightarrow{\text{ClCOOEt}} \text{RC}\equiv\text{CCOOEt} \tag{55}$$

$$\text{EtSC}\equiv\text{CH} \xrightarrow[\text{(2) ClCOOEt}]{\text{(1) CH}_3\text{Li}} \text{EtSC}\equiv\text{CCOOEt} \tag{56a}$$

$$\text{EtOC}\equiv\text{CH} \xrightarrow{} \text{EtOC}\equiv\text{CCOOEt} \tag{56b}$$

$$\text{RC}\equiv\text{CNa} \xrightarrow{\text{OC(OEt)}_2} \underset{\overset{|}{\text{OEt}}}{\text{RC}}=\text{CHCOOEt} + \underset{\overset{|}{\text{OEt}}}{\text{RCCH}_2\text{COOEt}} \tag{57}$$

B. Alkaline Earth Metal Derivatives of Acetylenes

The alkaline earth metal derivatives of phenylacetylene (equation 58, M = Ba, Sr, Ca, Mg) in diethyl ether suspension, react with solid carbon dioxide to give phenylpropiolic acid[21]. The yields, which vary with the

$$(\text{PhC}\equiv\text{C})_2\text{M} \xrightarrow{\text{CO}_2} (\text{PhC}\equiv\text{CCOO})_2\text{M} \tag{58}$$

alkaline earth were found to be $M = Ba$ 74%, Sr 42%, Ca 36%, Mg 21%[21].
However, owing to the relative inaccessibility of the bis-alkynyl alkaline
earth compounds it is more usual and more convenient to carbonate the
'Grignard derivatives' of acetylenic compounds[4, 24, 61, 73].

1. Reaction conditions

The procedure most commonly employed for the preparation of alkynylmagnesium halides is the interaction of a terminal acetylene and an organomagnesium halide in an inert solvent (*e.g.* ether or benzene) (equation
59). The carbonation can be effected with either gaseous or solid carbon

$$RC{\equiv}CH \xrightarrow{\text{R'MgX/Et}_2\text{O}} R'H + RC{\equiv}CMgX \xrightarrow{\text{CO}_2} RC{\equiv}CCOOH \qquad (59)$$

dioxide in an autoclave[127, 128], by pouring the reaction mixture on to solid
carbon dioxide[129] or by passing a stream of gaseous carbon dioxide through
the reaction mixture[130]. In the latter case long reaction times are required
and the yield of acids tends to be low[130].

2. Scope of the reaction

The carbonation of alkynylsodium and -lithium compounds and of
alkynylmagnesium halides has been used extensively in the preparation of
α, β-acetylenic acids[73]. The great value of these synthetic routes lies in their
flexibility and also in the fact that the acetylenic acids may be transformed
into otherwise difficultly accessible compounds. The flexibility of the synthesis is illustrated in equations (60)–(67). Substituted terminal acetylenes
may be used provided the substituents are inert towards sodium metal,
sodamide, organolithium compounds or organomagnesium halides. The
synthetic utility of the α, β-acetylenic acids thus produced is illustrated by
the transformation of the methyl esters (**9, 10,** $R = CH_3$) to the lactones

$$NaC{\equiv}CH \xrightarrow{RX} RC{\equiv}CH \xrightarrow[\text{(2) CO}_2]{\text{(1) NaNH}_2} RC{\equiv}CCOOH^{120} \qquad (60)$$

$$NaC{\equiv}CH \xrightarrow{\text{Br(CH}_2)_n\text{Br}} HC{\equiv}C(CH_2)_nC{\equiv}CH \xrightarrow[\text{(2) CO}_2]{\text{(1) NaNH}_2} RC{\equiv}CCOOH^{131} \qquad (61)$$

$$NaC{\equiv}CH \xrightarrow{\underset{\text{CH}_3}{\text{CH}_2{=}\text{CCHO}}} \underset{\underset{\text{OH}}{|}}{\overset{\overset{\text{CH}_3}{|}}{CH_2{=}CCHC{\equiv}CH^{132}}} \xrightarrow[\text{(2) CO}_2]{\text{(1) NaNH}_2} \underset{\underset{\text{OH}}{|}}{\overset{\overset{\text{CH}_3}{|}}{CH_2{=}CCHC{\equiv}CCOOR^{133}}}$$

$$\qquad\qquad (9) \qquad\qquad (62)$$

$$NaC{\equiv}CH \xrightarrow{\text{PhCHO}} \underset{\underset{\text{OH}}{|}}{PhCHC{\equiv}CH^{134}} \xrightarrow[\text{(2) CO}_2]{\text{(1) NaNH}_2} \underset{\underset{\text{OH}}{|}}{PhCHC{\equiv}CCOOR^{135}} \qquad (63)$$

$$\qquad\qquad\qquad\qquad\qquad\qquad (10)$$

$$\underset{\displaystyle |\atop OH}{RCH}CH_2C{\equiv}CH \xrightarrow[(2)\ CO_2\ \text{pressure}]{(1)\ EtMgBr} \underset{\displaystyle |\atop OH}{RCH}CH_2C{\equiv}CCOOH^{127} \quad (64)$$

$$EtOC{=}\underset{\displaystyle |\atop R}{C}HC{\equiv}CH \xrightarrow[(2)\ CO_2\ \text{solid}]{(1)\ EtMgBr} EtOC{=}\underset{\displaystyle |\atop R}{C}HC{\equiv}CCOOH^{129} \quad (65)$$

R = Me, Et, n-Pr, i-Pr

$$CH_3(CH_2)_nC{\equiv}CC{\equiv}CH \xrightarrow[(2)\ CO_2\ \text{pressure}]{(1)\ EtMgBr} CH_3(CH_2)_nC{\equiv}CC{\equiv}CCOOH^{136,\ 137} \quad (66)$$

(11)

$n = 0, 2, 3$

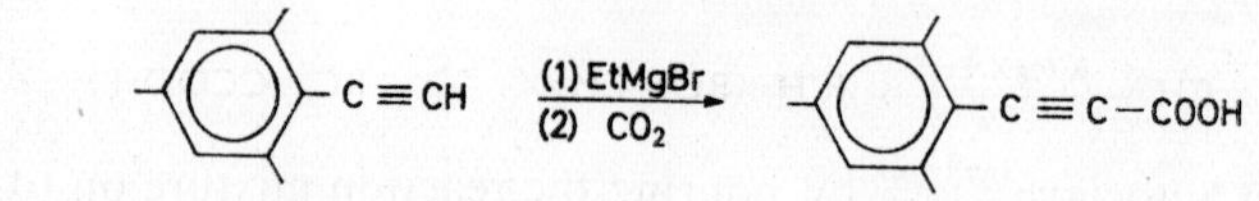
(67)

(12)

(13, 14) and of the methyl ester (10) to the keto ester (15). The α, β-acetylenic acids are valuable intermediates in the preparation of α, β-ethylenic

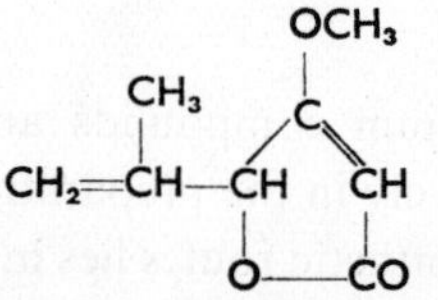

PhCOCH=CHCOOMe

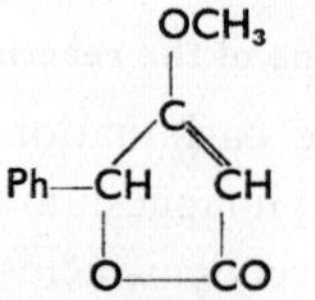

(13) (15) (14)

acids[73]. Thus catalytic reduction of the acids (11 and 12) gave high yields of the ethylenic acids (16 and 17)[128, 137]. Treatment of the acetylenic Gri-

HOOCCH = CHCH = CHCH$_3$

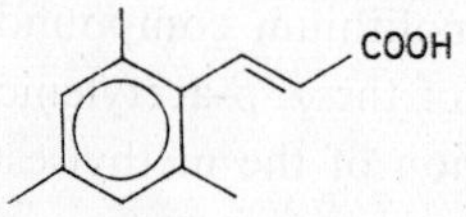

(16) (17)

gnard with ethyl chloroformate or ethyl carbonate yields directly the ethyl ester of the α, β-acetylenic acid (equations 68 and 69)[138, 139].

$$RC{\equiv}CMgX \xrightarrow{ClCOOEt} RC{\equiv}CCOOEt \quad (68)$$

$$CH_3OCH{=}CHC{\equiv}CMgBr \xrightarrow{OC(OEt)_2} CH_3OCH{=}CHC{\equiv}CCOOEt \quad (17\%) \quad (69)$$

An interesting reaction sequence for converting a terminal acetylene to a higher homologous acid is given in equation (70) ($R = CH_3CH_2$)[140].

$$RC{\equiv}CMgBr \xrightarrow{HCHO} RC{\equiv}CCH_2OH \xrightarrow[(3)\ NaOH]{(1)\ PBr_3\ (2)\ CuCN} RC{\equiv}CCH_2COOH \quad (70)$$

C. Aluminum Acetylides

The aluminum acetylides $Al(C \equiv CR)_3$, prepared by the interaction of the sodium acetylide and aluminum chloride, are reported as being unreactive towards carbon dioxide even up to temperatures of $100°$[141]. The complex sodium, lithium and potassium aluminoacetylides, on the other hand, react with carbon dioxide to give the α, β-acetylenic acids (equation 71)[142].

$$MAlH_4 + HC \equiv CR \longrightarrow MAl(C \equiv CR)_4 \xrightarrow{CO_2} RC \equiv CCOOH \qquad (71)$$

1. Reaction conditions

The reaction is carried out in ether, tetrahydrofuran or diglyme. The yield of acid is independent of the alkali metal (*i.e.* equation 71, M = Li, Na, K), but is very dependent upon the temperature of the reaction. Thus, at 60–70° only *one* of the acetylenic groups is converted to the acid (final yield of acid 25%), whereas at 120–160° the overall yield of acids is much higher (60–75%)[142].

2. Scope of the reaction

The terminal acetylenes which can be used in the reaction are necessarily restricted to those which are unreactive towards alkali metal aluminohydrides. This, together with the variable yields of acids obtained, severely limits the use of the reaction as a practical laboratory synthesis.

VII. NON-CONJUGATED ACETYLENIC ACIDS

With the exception of β, γ-acetylenic acids (equation 72, $n = 0$), non-conjugated acetylenic acids can be prepared from the appropriate halide, by converting the latter to an organometallic species and subsequent carbonation (equation 72). Since this sequence of reactions is analogous to

$$RC \equiv C(CH_2)_nCH_2X \xrightarrow{M} RC \equiv C(CH_2)_nCH_2M \xrightarrow{CO_2} RC \equiv C(CH_2)_{n+1}COOH \qquad (72)$$

that used in the preparation of alkanoic acids it will not be discussed separately here.

A. Propargyllithium, -sodium and -magnesium Compounds

The conversion of propargylic halides to carboxylic acids, by the carbonation of a derived organometallic intermediate (equation 72, $n = 0$), depends critically upon the structure of the starting material, the choice of organometallic intermediates and the reaction conditions. Thus, when propargylic halides containing an acetylenic hydrogen are treated with lithium methyl (at $-20°$ to $-50°$) and the reaction mixture carbonated, α, β-acetylenic acids are obtained, (equation 73, $R = R' = H$ and $R = H$, $R' = CH_3$)[143].

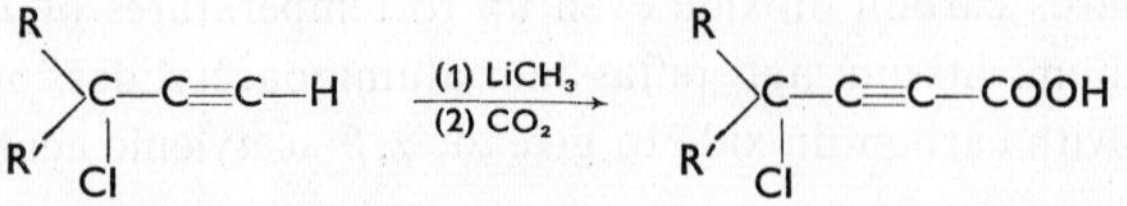

$$\begin{array}{c}R\\ {\scriptstyle\diagdown}\\ \overset{}{C}-C\equiv C-H\\ {\scriptstyle\diagup}\\ R'\overset{|}{Cl}\end{array} \xrightarrow[\text{(2) CO}_2]{\text{(1) LiCH}_3} \begin{array}{c}R\\ {\scriptstyle\diagdown}\\ \overset{}{C}-C\equiv C-COOH\\ {\scriptstyle\diagup}\\ R'\overset{|}{Cl}\end{array} \qquad (73)$$

On the other hand, when the propargyl halide is treated (under the appropriate reaction conditions) with magnesium and the reaction mixture carbonated, the products consist essentially of the β, γ-acetylenic acid and the allenic acid, admixed with a dimeric species and traces of an α, β-acetylenic acid (equation 74)[144, 145].

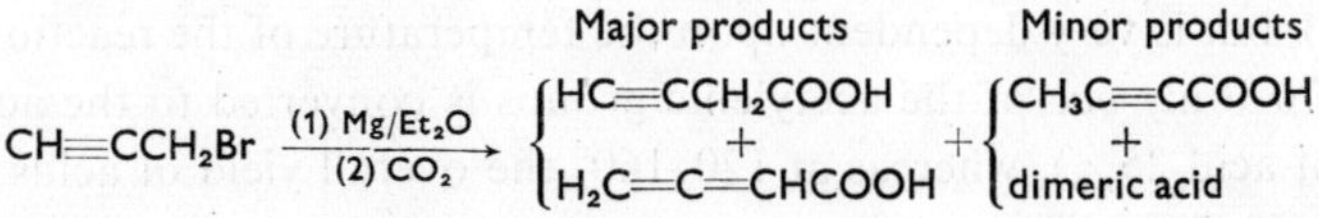

$$\text{CH}\equiv\text{CCH}_2\text{Br} \xrightarrow[\text{(2) CO}_2]{\text{(1) Mg/Et}_2\text{O}} \underbrace{\left\{\begin{array}{c}\text{HC}\equiv\text{CCH}_2\text{COOH}\\ +\\ \text{H}_2\text{C}=\text{C}=\text{CHCOOH}\end{array}\right.}_{\text{Major products}} + \underbrace{\left\{\begin{array}{c}\text{CH}_3\text{C}\equiv\text{CCOOH}\\ +\\ \text{dimeric acid}\end{array}\right.}_{\text{Minor products}} \qquad (74)$$

Similarly, the more highly substituted propargylic halides are converted to allenic acids by treatment with sodium amalgam and subsequent carbonation (equation 75)[146].

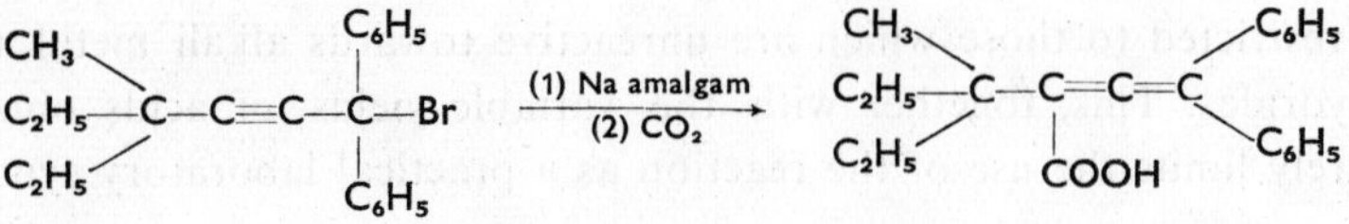

$$\begin{array}{c}\text{CH}_3\\ \text{C}_2\text{H}_5{-}\overset{}{C}-\text{C}\equiv\text{C}-\overset{C_6H_5}{\underset{C_6H_5}{\overset{|}{C}}}-\text{Br}\\ \text{C}_2\text{H}_5\end{array} \xrightarrow[\text{(2) CO}_2]{\text{(1) Na amalgam}} \begin{array}{c}\text{CH}_3\\ \text{C}_2\text{H}_5{-}\overset{}{C}-\text{C}=\text{C}=\overset{C_6H_5}{\underset{C_6H_5}{C}}\\ \text{C}_2\text{H}_5\overset{|}{COOH}\end{array} \qquad (75)$$

1. Reaction conditions

As is illustrated in Scheme 3, the products formed by the interaction of propargyl bromide and magnesium depend on the solvent and the tempe-

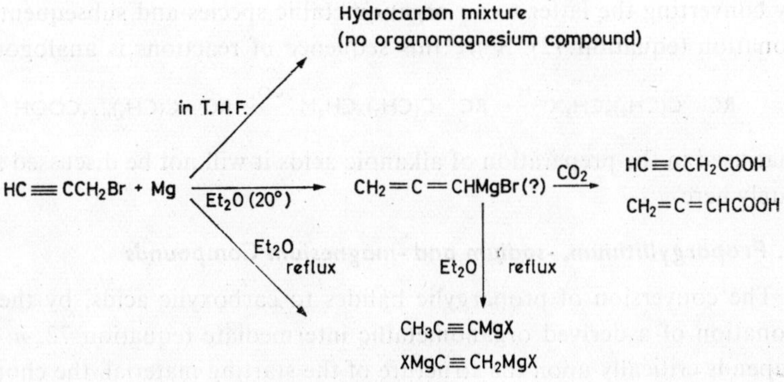

Hydrocarbon mixture
(no organomagnesium compound)

$$\text{HC}\equiv\text{CCH}_2\text{Br} + \text{Mg}$$

in T.H.F.

$\text{Et}_2\text{O (20°)}$ → $\text{CH}_2=\text{C}=\text{CHMgBr(?)}$ $\xrightarrow{\text{CO}_2}$ $\begin{array}{c}\text{HC}\equiv\text{CCH}_2\text{COOH}\\ \text{CH}_2=\text{C}=\text{CHCOOH}\end{array}$

Et_2O reflux Et_2O reflux

$$\text{CH}_3\text{C}\equiv\text{CMgX}$$
$$\text{XMgC}\equiv\text{CH}_2\text{MgX}$$

SCHEME 3.

rature of the reaction[112, 147]. Under the conditions usually employed for making Grignard reagents substituted propargylic halides are inert towards magnesium[148].

The introduction[149] of high dilution techniques to this field makes it now possible, however, to obtain the organomagnesium compound in high yields[150]. More recently it has been found that the tertiary propargylic halides $(CH \equiv C—CXR_2)$ react smoothly with magnesium either in diethyl ether (with irradiation)[151] or in tetrahydrofuran solution[152].

2. Scope of the reaction

The carbonation of metallic derivatives of propargylic halides gives mixtures of β, γ-acetylenic and allenic acids. The total and the relative yields of acids obtained both depend upon the structure of the acetylenic halide. Highly substituted propargylic halides can be converted, in good yield, by the carbonation of the organometallic derivative to the allenic acid (equation 76, M = Na or Mg), or by interaction of the organometallic compound with methyl chloroformate, to the ester, (equation 77)[146, 153]. Less highly substituted propargylic halides react to give organomagnesium

$$R_3CC \equiv CC(X)R_2 \xrightarrow[\text{(2) } CO_2]{\text{(1) } M/Et_2O} \underset{\underset{COOH}{|}}{R_3CC = C = CR_2} \qquad (76)$$

$$R_3CC \equiv CC(X)R_2 \xrightarrow[\text{(2) } ClCOOCH_3]{\text{(1) } Mg/Et_2O} \underset{\underset{COOCH_3}{|}}{R_3CC = C = CR_2} \qquad (77)$$

compounds which, upon carbonation give moderate yields (40–70%) of a variable mixture of β, γ-acetylenic acid (9–20%) and allenic acid (9–50%), (equation 78, (20 and 21) R = H)[154–157]. The overall mechanism for the

$$RC \equiv CCH_2Br \xrightarrow{Mg} \begin{cases} RC \equiv CCH_2MgBr \\ \quad \quad \updownarrow \quad (18) \\ RC = C = CH_2 \\ \quad \quad | \\ \quad \quad MgBr \\ \quad \quad (19) \end{cases} \xrightarrow{CO_2} \begin{cases} RC \equiv CCH_2COOH \\ \quad \quad (20) \\ RC = C = CH_2 \\ \quad \quad | \\ \quad \quad COOH \\ \quad \quad (21) \end{cases} \qquad (78)$$

conversion of a propargylic halide to the corresponding carboxylic acid, by the above routes, remains uncertain. Two main problems are involved, the first is the structure of the intermediate organometallic species, the second, the mechanism of the carbonation reaction. Spectroscopic evidence[111, 158, 159] reveals that both propargyl bromide and bromoallene react with magnesium to give the same allenylmagnesium bromide, (equa-

11*

tion 78, **19**, R = H). Phenylpropargyl bromide, on the other hand, reacts with magnesium to give a mixture of acetylenyl- and allenylmagnesium halides, (equation 78, (**18** and **19**) R = C_6H_5). In both instances carbonation of the organomagnesium halide gives mixtures of the acetylenic and the allenic acids (equation 78, (**20** and **21**) R = H and C_6H_5). The simplest explanation of the foregoing is that 'propargylmagnesium halides' are in effect equilibrium mixtures of the acetylenic and allenic forms. The relative proportions of the two forms will depend upon the nature of the substituent, R, and the reaction conditions. Subsequent reactions of the substrate (in this case carbon dioxide) with the two forms of the organomagnesium compound (or the derived anion) need not occur at the same rate. The difference in the rates of reaction will determine the relative proportions of the acetylenic and allenic acid found in the final products. In this mechanism the actual carbonation proceeds via a four-membered cyclic transition state (see equation 10f). Alternative mechanisms based upon a cyclic six-membered transition state, analogous to that postulated for the 'allylic rearrangement' (see equation 51), seem unlikely in view of the rigidity of the allene system.

In general the reaction can be considered as an effective route to allenic or β, γ-acetylenic acids. However, since the products are formed by the rearrangement of the starting propargylic halide, and the yields are variable, the reaction does not possess the flexibility normally associated with a good synthetic route.

It is of interest to note in passing that propargylic halides react with the sodio derivative of malonic ester to give, without rearrangement, the mono- or bis-propargyl malonic esters (equation 79)[160–162]. This is an excellent route to γ, δ-acetylenic acids and esters.

$$HC{\equiv}CCH_2X \xrightarrow{\text{NaCH (COOEt)}_2} HC{\equiv}CCH_2CH(COOEt)_2 \text{ or } (CH{\equiv}CCH_2)_2C(COOEt)_2$$

$$\text{(79)}$$

B. Propargyl-zinc and -aluminum Compounds

Propargyl halides react directly with metallic zinc and aluminum, in tetrahydrofuran, to give the corresponding organometallic compounds[104, 112, 113]. Spectroscopic evidence[112, 114] indicates that both the organozinc and -aluminum compounds consist essentially of the allenic form, (equation 80, **22**, R = H). Neither of these compounds reacts with carbon dioxide

$$HC{\equiv}CCHBrR \longrightarrow \begin{cases} (M)CH{=}C{=}CHR \\ \quad \text{(22)} \\ HC{\equiv}CCH(M)R \\ \quad \text{(23)} \end{cases} \xrightarrow{CO_2} \begin{cases} HOOCCH{=}C{=}CHR \\ \quad \text{(24)} \\ HC{\equiv}CCH(COOH)R \\ \quad \text{(25)} \end{cases}$$

$$\text{(80)}$$

under ordinary conditions. Under pressure (50 atm), the organozinc compounds react with carbon dioxide to give either the acetylenic acid (**25**, R=H) or the allenic acid (**24**, R=CH$_3$ and n-Pr) in moderate yield (about 40%)[112, 114]. The organoaluminum compounds on the other hand react with carbon dioxide (under pressure) and with ethyl chloroformate to give the carbinol (CH≡CCH$_2$)$_3$COH[112, 163].

VIII. OTHER REACTIONS, INCLUDING DIRECT CARBOXYLATION

There are many other reactions in which carbon dioxide or a derivative thereof is used to introduce the hydroxy- or ethoxycarbonyl group into an organic molecule. These reactions, however, involve molecules containing 'active hydrogen' or reactions in which the substrate is activated (*e.g.* irradiation) and therefore their discussion has been postponed to the present section.

A. Carboxylation and Carbethoxylation Reactions

The direct introduction of the hydroxycarbonyl or ethoxycarbony, group into molecules containing an activated hydrogen (*e.g.* ketones, esterl *etc.*) can be accomplished by the interaction of the derived sodio-derivar tive with carbon dioxide[164] or diethyl carbonate[165] (equation 81, R'=H or CH$_2$CH$_3$).

$$\text{RCOCH}_3 \xrightarrow[\text{or NaH}^{167}]{\substack{\text{NaOR}^{166} \\ \text{or NaNH}_2^{164}}} [\text{RCOCH}_2]^- \text{Na}^+ \xrightarrow[\text{or OC(OCH}_2\text{CH}_3)_2]{\text{CO}_2} \text{RCOCH}_2\text{COOR'} \quad (81)$$

These reactions have been used to prepare a variety of β-ketoesters and malonic esters in varying yields (25–70%)[164–167]. A recent modification of this synthesis involves the condensation of the ketone with ethyl diethoxyphosphonyl formate [EtOCOPO(OEt)$_2$] in the presence of sodium hydride[168]. The intermediate α-oxophosphonate (**26**) is not isolated but is cleaved by acid, in the presence of an alcohol, to give the β-keto ester in high yield (equation 82)[169]. Diethyl oxalate also condenses with ketones in the pre-

$$[\text{RCOCH}_2\text{COPO(OEt)}_2]^- \text{Na}^+ \xrightarrow[\text{H}^+]{\text{R'OH}} \text{RCOCH}_2\text{COOR'} \quad (82)$$
$$(\mathbf{26})$$

sence of sodamide or sodium hydride. The initially formed β-ketoethoxalyl compound (**27**) is thermally unstable, losing carbon monoxide on heating to give the β-keto ester (equation 83)[170].

$$\text{RCOCH}_3 \xrightarrow[\text{(COOEt)}_2]{\text{NaH}} \text{RCOCH}_2\text{COCOOEt} \xrightarrow{\text{heat}} \text{RCOCH}_2\text{COOEt} \quad (83)$$
$$(\mathbf{27})$$

The carbonation of the organozinc compounds from ethyl α-bromace-tate and methyl α-bromisobutyrate, prepared in tetrahydrofuran, has been reported to give malonic- and dimethylmalonic acid half-ester (16% and 50% respectively)[171].

B. Carboxylation with Methyl Magnesium Carbonate

Carbethoxylation, with diethylcarbonate, is essentially a reversible pro-cess and an excess of sodium amide (or hydride) is normally used to dis-place the equilibrium concentration in favor of the β-keto ester, (equation 84)[167].

$$
\begin{array}{c}
(RCOCH_2)Na \\
+ \\
CO(OEt)_2
\end{array}
\rightleftharpoons
\begin{array}{c}
RCOCH_2COOEt \\
+ \\
NaOEt
\end{array}
\xrightarrow{NaH}
(RCOCHCOOEt)^-Na^+ \qquad (84)
$$

However, this excess of sodium hydride (or amide) can, in some instances, promote secondary reactions (self-condensation of the ketone and amide formation) thereby reducing the overall yield of β-keto ester. These diffi-culties can be largely overcome by reacting the compound containing the active hydrogen (*e.g.* ketone or nitroparaffin) with methylmagnesium car-bonate (MMC) (equation 85)[172-176].

$$
\begin{array}{c}
R' \\
| \\
R-CH \\
| \\
NO_2
\end{array}
+ MMC \rightleftharpoons
\left[
\begin{array}{c}
R' \\
| \\
R-C-COOMg \\
| \\
NO_2
\end{array}
\right]
\xrightarrow{when\ R'=H}
\underset{(29)}{\text{chelate}}
\qquad (85)
$$

$$
\qquad\qquad\qquad (28) \qquad\qquad\qquad\qquad (29)
$$

In this reaction it is presumed that chelation, in the final product, favors the β-keto acid or α-nitro acid formation. This is substantiated by the obser-vation that whereas 1-nitro-methane, -ethane, -propane and -butane (28, R=alkyl, R'=H) all reacted smoothly with MMC, 2-nitropropane did not, *i.e.* the initial condensation product (28, R = R' = CH$_3$) contains no enolizable hydrogen and therefore cannot give the chelate (29).

The reaction conditions are very mild and consist of heating the ketone or nitroparaffin with methyl magnesium carbonate[177] (CH$_3$OMgOCO OCH$_3$ + x CO$_2$)[175] in dimethylformamide. The simplicity of reaction and the high yields obtained make this an excellent synthetic route to β-keto and α-nitro (α-amino) acids. The reaction has been used to carboxylate nitroparaffins (35–60%)[172], acetophenone (68%)[174], 1-indanone (91%)[174], cyclohexanone (48%)[174], 1-tetralone[174], and 5-methoxy-2-tetralone[176].

C. Carboxylation with Ethylsodium Carbonate

Ethylsodium carbonate, prepared by treating sodium ethoxide solutions with carbon dioxide[178], has been used to carboxylate phenol, resorcinol and benzoic acid[178 179]. However, in contrast to the reaction with methyl-magnesium carbonate, the only acidic product from the interaction of cyclohexanone and ethylsodium carbonate was adipic acid.

An interesting method for carboxylating molecules containing active methylene groups involves treating the ketone in dimethyl formamide with carbon dioxide in the presence of the potassium salt of a dialkyl hydroxythiazole (30, R = alkyl) or potassium phenoxide[180].

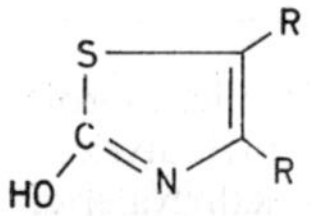

(30)

In this way acetophenone has been converted to potassium benzoylacetate and cyclohexanone to a mixture of dipotassium cyclohexanone-2,6-dicarboxylate and potassium cyclohexanone-2-carboxylate[180]. The high yields and the mildness of the reaction conditions make this a potentially valuable synthetic route to β-keto acids.

D. Carboxylation with Oxalyl Chloride

Oxalyl chloride reacts with olefinic and acetylenic hydrocarbons to give substituted acrylic acid chlorides and chloracrylic acid chlorides (equations 86 and 87)[181].

$$\underset{R'}{\overset{R}{\diagdown}}C{=}CH_2 \xrightarrow[\text{Reflux}]{(COCl)_2} \underset{R'}{\overset{R}{\diagdown}}C{=}CHCOCl \qquad (86)$$

$$PhC{\equiv}CH \xrightarrow[\text{Reflux}]{(COCl)_2} PhCCl{=}CHCOCl \qquad (87)$$

This reaction constitutes a general synthesis of diarylacrylic acids, (equation 86, R = R' = aryl); however, the yield of the aryl crotonic acids are poor (3–40%) (equation 86, R = aryl, R' = CH$_3$)[182].

Oxalyl chloride does not react with simple olefins, (equation 86, R = R' = alkyl and R' = alkyl, R = H) or alkanes under similar reaction conditions[181]. When, however, an alkane is treated with oxalyl chloride in the presence of benzoyl peroxide, or when the reaction mixture is irradiated[183], the alkane is converted by radical processes to the corresponding acyl chloride (equation 88)[183].

$$RH + (COCl)_2 \xrightarrow{h\nu} RCOCl \tag{88}$$

E. Direct Carboxylation of Hydrocarbons

The direct interaction of saturated and unsaturated hydrocarbons with carbon dioxide can be achieved by high energy radiations (*e.g.* x-, β- and γ-rays)[184–189]. In the reactions studied to date (Table 1), the products are

TABLE 1. Interaction of hydrocarbons and carbon dioxide induced by radiation

Hydrocarbon	Radiation	Products Acids	Other	Reference
Pentane	^{86}Kr	Formic, caproic, 2-ethylbutyric, 2-methylvaleric, valeric	Ketones, polymers	187
Pentane	x-Ray	Mixture as above	–	186
Cyclohexane	^{60}Co	Cyclohexane mono- and di-carboxylic acids	Cyclohexanol, cyclohexanone bicyclohexenyl resins	185
Methane	2 MeV electron beam	Acetic acid	–	184
Toluene	2 MeV electron beam	Phenylacetic acid	–	184
Ethanol	2 MeV electron beam	Lactic acid	–	184
Cyclohexane	2 MeV electron beam	Cyclohexyl-carboxylic acid	–	184
Ethylamine	2 MeV electron beam	Alanine	–	184
Ethylene	^{60}Co	Mixture of carboxylic acids	–	189

many and complex. They consist of (i) acids (in yields of 0·5–2%), (ii) alcohols and ketones and (iii) oligomers. The acids are formed by the radical processes[186, 189], involving the activated hydrocarbon species[RH]*, outlined in equation 89. The ketones, on the other hand, are presumed to be

$$RH \xrightarrow{\text{radiation}} [RH]^* \longrightarrow R \cdot \xrightarrow{CO_2} [RCOO^\cdot] + \xrightarrow{RH} RCOOH + R^\cdot \text{ etc.} \tag{89}$$

formed by the interaction of CO^+ (from the radiation of CO_2) with the activated hydrocarbon substrate[189].

At present the radiation reactions are not particularly well adapted for the laboratory preparation of a particular carboxylic acid, though they are claimed to be practicable for the preparation of labelled carboxylic acids (using $C^{14}O_2$)[187, 188].

IX. ACKNOWLEDGMENTS

I wish to thank Dr. H. H. Zeiss for his valuable comments and suggestions during the preparation of the manuscript.

X. REFERENCES

1. E. G. Rochow, D. T. Hurd and R. N. Lewis, *The Chemistry of Organometallic Compounds*, John Wiley and Sons, New York, 1957, p. 65.
2. M. Schlosser, *Angew. Chem., Intern. Ed. Engl.*, **3**, 287 (1964).
3. M. Schlosser, *Angew, Chem., Intern. Ed. Engl.*, **3**, 362 (1964).
4. E. Krause and A. von Grosse, *Die Chemie der Metallorganischen Verbindungen*, Verlag von Gebrüder Borntraeger, Berlin, 1937.
5. K. Ziegler and H. Colonius, *Ann. Chem.*, **479**, 135 (1930).
6. H. Gilman, E. A. Zoellner and W. M. Selby, *J. Am. Chem. Soc.*, **55**, 1252 (1933)
7. A. A. Morton, G. M. Richardson and A. T. Hallowell, *J. Am. Chem. Soc.*, **63**, 327 (1941).
8. J. F. Nobis and L. F. Moormeier, *Ind. Eng. Chem.*, **46**, 539 (1954).
9. H. Gilman and J. W. Morton, Jr., *Org. Reactions*, **8**, 258 (1954).
10. R. G. Jones and H. Gilman, *Org. Reactions*, **6**, 339 (1951).
11. J. E. Mulvaney, Z. G. Gardlund, S. L. Gardlund and D. J. Newton, *J. Am. Chem. Soc.*, **88**, 476 (1966).
12. H. Gilman and P. R. van Ess, *J. Am. Chem. Soc.*, **55**, 1258 (1933).
13. H. Gilman, H. A. Pacevitz and O. Baine, *J. Am. Chem. Soc.*, **62**, 1514 (1940).
14. A. A. Morton, F. Fallwell, Jr. and L. Palmer, *J. Am. Chem. Soc.*, **60**, 1426 (1938).
15. D. E. Applequist and A. H. Peterson, *J. Am. Chem. Soc.*, **83**, 862 (1961).
16. H. M. Walborsky, F. J. Impastato and A. E. Young, *J. Am. Chem. Soc.*, **86**, 3283 (1964).
17. D. J. Cram, *Fundamentals of Carbanion Chemistry*, Academic Press, New York, 1965, p. 116.
18. E. Krause and A. von Grosse, *Die Chemie der Metallorganischen Verbindungen*, Verlag von Gebrüder Borntraeger, Berlin, 1937, p. 110.
19. E. G. Rochow, D. T. Hurd and R. N. Lewis, *The Chemistry of Organometallic Compounds*, John Wiley and Sons, New York, 1957, p. 77.
20. H. Gilman and F. Schulze, *J. Chem. Soc.*, 2663 (1927).
21. H. Gilman and L. A. Woods, *J. Am. Chem. Soc.*, **67**, 520 (1945).
22. D. Bryce–Smith and A. C. Skinner, *J. Chem. Soc.*, 577 (1963).
23. M. S. Kharasch and O. Reinmuth, *Grignard Reactions of Nonmetallic Substances*, Prentice-Hall, New York, 1954, p. 5.
24. F. Runge, *Organometallverbindungen*, Vol. 1, *Organomagnesiumverbindungen*, Wissenschaftliche Verlagsgesellschaft MbH., Stuttgart, 1932, p. 8.
25. G. D. Cooper and H. L. Finkbeiner, *J. Org. Chem.*, **27**, 1493 (1962).
26. H. L. Finkbeiner and G. D. Cooper, *J. Org. Chem.*, **27**, 3395 (1962).
27. Ref. 23, p. 913.
28. A. E. Tschitschibabin, *Chem. Ber.*, **38**, 561 (1905).
29. J. Houben, *Chem. Ber.*, **36**, 3087 (1903).
30. E. E. Blaise, *Compt. Rend.*, **132**, 38 (1901).
31. A. M. Schwartz and J. R. Johnson, *J. Am. Chem. Soc.*, **53**, 1063 (1931).

32. H. Gilman and H. H. Parker, *J. Am. Chem. Soc.*, **46**, 2816 (1924).
33. M. S. Kharasch and O. Reinmuth, *Grignard* Reactions of Nonmetallic Substances, Prentice-Hall, New York, 1954, p. 1286.
34. V. Grignard, *Ann. Chim. (Paris)*, [7] **24**, 433 (1901).
35. F. Bodroux, *Bull. Soc. Chim. France*, [3] **31**, 24 (1904).
36. D. Ivanoff and A. Spassoff, *Bull. Soc. Chim. France*, **49**, 19 (1931).
37. F. Runge, *Organometallverbindungen*, Vol. 1, *Organomagnesiumverbindungen*, Wissenschaftliche Verlagsgesellschaft MbH., Stuttgart, 1932, p. 204.
38. G. Roberts and C. W. Shoppee, *J. Chem. Soc.*, 3418 (1954).
39. O. A. Reutov, *Bull. Soc. Chim. France*, 1383 (1963).
40. D. Y. Curtin and W. J. Koehl, Jr., *J. Am. Chem. Soc.*, **84**, 1967 (1962).
41. R. L. Letsinger, *J. Am. Chem. Soc.*, **72**, 4842 (1950).
42. H. M. Walborsky and A. E. Young, *J. Am. Chem. Soc.*, **86**, 3288 (1964).
43. A. P. Terent'ev and N. I. Shor, *J. Gen. Chem. USSR*, **17**, 2075 (1947); *Chem. Abstr.*, **43**, 8964 (1949).
44. K. Ziegler, *Organometallic Chemistry* (Ed. H. Zeiss), American Chemical Society Monograph No. 147, Reinhold Publishing Corporation, New York, 1960, p. 197.
45. K. Ziegler, *Angew. Chem.*, **68**, 721 (1956).
46. K. Ziegler, F. Krupp, K. Weyer and W. Larbig, *Ann. Chem.*, **629**, 251 (1960).
47. H. A. Walter, *U.S. Pat.* 2, 864, 842; *Chem. Abstr.*, **53**, 7014 (1959).
48. E. B. Baker and H. H. Sisler, *J. Am. Chem. Soc.*, **75**, 5193 (1953).
49. S. B. Mirviss and E. J. Inchalik, *U.S. Pat.* 2,827,458; *Chem. Abstr.*, **52**, 13778 (1958).
50. L. I. Zakharkin and V. V. Gavrilenko, *Dokl. Akad. Nauk SSSR*, **118**, 713 (1958); *Chem. Abstr.*, **52**, 11738 (1958).
51. L. I. Zakharkin and V. V. Gavrilenko, *USSR. Pat.* 112,349; *Chem. Abstr.*, **53**, 2093 (1959).
52. D. W. Marshall, *US. Pat.* 3, 168, 570; *Chem. Abstr.*, **62**, 16062 (1965).
53. D. W. Marshall, *US. Pat.* 3, 089, 884; *Chem. Abstr.*, **59**, 11269 (1963).
54. K. Clauss and C. Beermann, *Angew. Chem.*, **71**, 627 (1959).
55. P. Klimsch and E. Kurras, *East German Pat.* 45708; *Chem. Abstr.*, **65**, 8963 (1966).
56. A. Kekulé, *Ann. Chem.*, **137**, 178 (1866).
57. H. Kolbe, *Ann. Chem.*, **113**, 125 (1860).
58. B. Raecke, *Angew. Chem.*, **70**, 1 (1958).
59. G. A. Olah, *Friedel–Crafts and Related Reactions*, Vol. 1, (Ed. G. A. Olah), Interscience Publishers, New York, 1963, pp. 121–122.
60. G. A. Olah and J. A. Olah, *Friedel–Crafts and Related Reactions*, Vol. 3 (Ed. G. A. Olah), Interscience Publishers, New York, 1964, p. 1257.
61. H. Henecka, *Houben Weyl, Methoden der Organischen Chemie*, Vol. VIII, Sauerstoffverbindungen III, Georg Thieme Verlag, Stuttgart, 1952, p. 369.
62. A. A. Morton and I. Hechenbleikner, *J. Am. Chem. Soc.*, **58**, 2599 (1936).
63. R. R. Burtner and J. W. Cusic, *J. Am. Chem. Soc.*, **65**, 262 (1943).
64. N. H. Cromwell and D. B. Capps, *J. Am. Chem. Soc.*, **74**, 4448 (1952).
65. A. Malera, M. Claesen and H. Vanderhaeghe, *J. Org. Chem.*, **29**, 3705 (1964).
66. O. Meth-Cohn and S. Gronowitz, *Chem. Comm.*, 81, (1966).
67. H. H. Zeiss, ref. 44, p. 380.
68a. H. H. Zeiss and T. F. Burger, private communication.
68b. T. F. Burger and H. H. Zeiss, *Chem. Ind.* (London), 183 (1962).
68c. H. H. Zeiss and W. Herwig, *J. Am. Chem. Soc.*, **78**, 5959 (1956).
69. P. J. Wheatley, H. J. S. Winkler and H. H. Zeiss, *Benzenoid–Metal Complexes, Structural Determinations and Chemistry*, The Ronald Press Co., New York, 1966.

The formation of carboxylic acids

70. E. McNelis, *J. Org. Chem.*, **30**, 1209 (1965).
71. A. S. Lindsey and H. Jeskey, *Chem. Rev.*, **57**, 583 (1957).
72. H. Hopff and Th. Zimmermann, *Helv. Chim. Acta*, **47**, 1293 (1964).
73. A. W. Johnson, *Acetylenic Compounds*, Vol. II, *Acetylenic Acids*, E. Arnold and Co., London, 1950, p. 41.
74. H. Normant, *Advances in Organic Chemistry, Methods and Results*, Vol. II (Ed. R. A. Raphael, E. C. Taylor and H. Wynberg), Interscience Publishers, New York, 1960.
75. D. Seyferth, *Progr. Inorg. Chem.*, **3**, 129 (1962).
76. D. Y. Curtin and E. W. Flynn, *J. Am. Chem. Soc.*, **81**, 4714 (1959).
77. A. S. Dreiding and R. J. Pratt, *J. Am. Chem. Soc.*, **76**, 1902 (1954).
78. E. A. Braude, *Progr. in Org. Chemistry*, Vol. 3 (Ed. J. W. Cook), Butterworths, London, 1955, p. 172.
79. D. Y. Curtin and J. W. Crump, *J. Am. Chem. Soc.*, **80**, 1922 (1958).
80. G. Köbrich, *Angew. Chem.*, **79**, 15 (1967).
81. A. A. Morton, F. D. Marsh, R. D. Coombs, A. L. Lyons, S. E. Penner, H. E. Ramsden, V. B. Baker, E. L. Little and R. L. Letsinger, *J. Am. Chem. Soc.*, **72**, 3785 (1950).
82. D. J. Foster *US. Pat.* 2, 985, 691; *Chem. Abstr.*, **55**, 22134 (1961).
83. B. Bartocha, C. M. Douglas and M. Y. Gray, *Z. Naturforsch.*, **14b**, 809 (1959).
84. A. A. Morton, *Solid Organoalkali Metal Reagents*, Gordon and Breach, New York (1964).
85. A. A. Morton and M. E. T. Holden, *J. Am. Chem. Soc.*, **69**, 1675 (1947).
86. C. D. Broaddus, *French Pat.* 1, 352, 735; *Chem. Abstr.*, **60**, 15738 (1964).
87. C. D. Broaddus, *French Pat.* 1, 353, 392; *Chem. Abstr.*, **61**, 1760 (1964).
88. C. E. Frank and W. E. Foster, *J. Org. Chem.*, **26**, 303 (1961).
89. P. A. Moshkin, N. I. Kutsenko and L. K. Filippenko, *sintez i Svoistva Monomerov, Akad. Nauk SSSR, Inst. Neftekhim. sinteza, Sb. Rabot 12-oi (Dvenadtsatoi) Konf. po Vysokomolekul. Soedin.*, *1962*, 1964, pp. 212–215; *Chem. Abstr.*, **62**, 5184 (1965).
90. C. E. Frank, J. R. Leebrick, L. F. Moormeier, J. A. Scheben and O. Homberg, *J. Org Chem.*, **26**, 307 (1961).
91. H. Normant and P. Maitte, *Bull. Soc. Chim. France*, 1439 (1956).
92. G. F. Wright, *J. Org. Chem.*, **1**, 457 (1936).
93. B. Bartocha, C. M. Douglas and M. Y. Gray, *Z. Naturforsch.*, **14b**, 809 (1959).
94. B. Bartocha, A. J. Bilbo, D. E. Bublitz and M. Y. Gray, *Z. Naturforsch.*, **16b**, 357 (1961).
95. G. Wilke and H. Müller, *Ann. Chem.*, **629**, 222 (1960).
96. G. Wilke and H. Müller, *Ann. Chem.*, **618**, 267 (1958).
97. D. Seyferth and M. Weiner, *J. Org. Chem.*, **26**, 4797 (1961).
98. A. A. Morton, E. E. Magat and R. L. Letsinger, *J. Am. Chem. Soc.*, **69**, 950 (1947).
99. R. L. Letsinger and J. G. Traynham, *J. Am. Chem. Soc.*, **70**, 3342 (1948).
100. J. J. Eisch and A. M. Jacobs, *J. Org. Chem.*, **28**, 2145 (1963).
101. K. Ziegler, F. Crössmann, H. Kleiner and O. Schäfer, *Ann. Chem.*, **473**, 1 (1929).
102. W. G. Young, *Am. Chem. Soc.*, *Div. of Petroleum Chem.*, Reprints 4, No. 4 B. 45 (1959); *Chem. Abstr.*, **58**, 432 (1963).
103. M. S. Kharasch and O. Reinmuth, *Grignard Reactions of Nonmetallic Substances*, Prentice Hall, New York, 1954, Chapter II.
104. L. Miginiac-Groizeleau, *Bull. Soc. Chim. France*, 1449 (1963).
105. H. Gilman and E. A. Zuech, *J. Am. Chem. Soc.*, **81**, 5925 (1959).
106. J. E. Nordlander, W. G. Young and J. D. Roberts, *J. Am. Chem .Soc.*, **83**, 494 (1961).
107. M. Gaudemar, *Bull. Soc. Chim. France*, 1475 (1958).

108. R. H. DeWolfe, D. L. Hagmann and W. G. Young, *J. Am. Chem. Soc.*, **79**, 4795 (1957).
109. W. G. Young and J. D. Roberts, *J. Am. Chem. Soc.*, **68**, 649 (1946).
110. W. G. Young and J. D. Roberts, *J. Am. Chem. Soc.*, **68**, 1472 (1946).
111. M. Andrac, F. Gaudemar, M. Gaudemar, B. Gross, L. Miginiac, P. Miginiac and Ch. Prévost, *Bull. Soc. Chim. France*, 1391 (1963).
112. M. Gaudemar, *Ann. Chim. (Paris)*, [13], **1**, 161 (1956).
113. M. Gaudemar, *Bull. Soc. Chim. France*, 974 (1962).
114. M. Gaudemar and J. Pansard, private communication.
115. C. Glaser, *Ann. Chem.*, **154**, 137 (1870).
116. F. Straus and W. Voss, *Chem. Ber.*, **59**, 1681 (1926).
117. A. N. Kurtz, *Brit. Pat.* 986, 083; *Chem. Abstr.*, **62**, 14504 (1965).
118. W. J. Pope, *Chem. Ind. (London)*, **42**, 117 (1923).
119. A. D. Macallum, *U.S. Pat.* 2, 194, 362; *Chem. Abstr.*, **34**, 4745 (1940).
120. A. O. Zoss and G. F. Hennion, *J. Am. Chem. Soc.*, **63**, 1151 (1941).
121. H. Gilman and R. V. Young, *J. Org. Chem.*, **1**, 315 (1939).
122. W. Ziegenbein, *Aethinylierung und Alkinylierung*, Monograph zu 'Angewandte Chemie' und 'Chemie–Ingenieur–Technik', No. 79, Verlag Chemie GmbH, Weinheim/Bergstrasse, 1963.
123. R. A. Raphael, *Acetylenic Compounds in Organic Synthesis*, Butterworths, London, 1955, pp. 18, 199. (Experimental details).
124. J. F. Arens, H. C. Volger, T. Doornbos, J. Bonnema, J. Greidnaus and J. van den Hende, *Rec. Trav. Chim.*, **75**, 1459 (1956).
125. Ch. Moureu and R. Delange, *Bull. Soc. Chim. France*, [3], **29**, 648 (1903).
126. W. J. Croxall and M. F. Fegley, *J. Am. Chem. Soc.*, **71**, 1261 (1949).
127. L. J. Haynes and E. R. H. Jones, *J. Am. Chem. Soc.*, 503 (1946).
128. R. Adams and C. W. Theobald, *J. Am. Chem. Soc.*, **65**, 2208 (1943).
129. M. Bertrand and C. Rouvier, *Compt. Rend.*, **259**, 594 (1946).
130. R. Lespieau, *Ann. Chim. (Paris)*, **27**, 178 (1912).
131. J. H. Wotiz and B. F. Adams, *U.S. Pat.* 3, 153, 072, *Chem. Abstr.*, **62**, 449 (1965).
132. I. M. Heilbron, E. R. H. Jones, J. T. McCombie and B. C. L. Weedon, *J. Chem. Soc.*, 84 (1945).
133. R. A. Raphael, *J. Chem. Soc.*, 805 (1947).
134. E. R. H. Jones and J. T. McCombie, *J. Chem. Soc.*, 733 (1942).
135. A. W. Nineham and R. A. Raphael, *J. Chem. Soc.*, 118 (1949).
136. J. L. H. Allan, E. R. H. Jones and M. C. Whiting, *J. Chem. Soc.*, 1862 (1955).
137. E. R. H. Jones, J. M. Thompson and M. C. Whiting, *J. Chem. Soc.*, 2012 (1957)
138. G. Dupont, *Compt. Rend.*, **148**, 1522 (1909).
139. A. Dornow and F. Ische, *Chem. Ber.*, **89**, 876 (1956).
140. K. E. Schulte and W. Engelhardt, *Arch. Pharm.*, **287**, 495 (1954).
141. P. Chini, A. Baradel, E. Pauluzzi and M. De Malde, *Chim. Ind. (Milan)*, **44**, 1220 (1962); *Chem. Abstr.*, **58**, 13973 (1963).
142. L. I. Zakharkin, V. V. Gavrilenko and L. L. Ivanov, *Izv. Akad. Nauk. SSSR, Ser. Khim.*, **11**, 2066 (1964); *Bull. Acad. Sci. USSR, Div. Chem. Sci. (English Transl.)*, 1959 (1964).
143. J. P. Battioni and W. Chodkiewicz, *Compt. Rend.*, **263**, 761 (1966).
144. C. Prévost, M. Gaudemar and J. Honigberg, *Compt. Rend.*, **230**, 1186 (1950)
145. J. H. Wotiz, J. S. Matthews and J. A. Lieb, *J. Am. Chem. Soc.*, **73**, 5503 (1951).
146. J. Harmon and C. S. Marvel, *J. Am. Chem. Soc.*, **55**, 1716 (1933).
147. L. Miginiac-Groizeleau, *Ann. Chim. (Paris)*, [13], **6**, 1071 (1961).
148. G. Lappin, *J. Am. Chem. Soc.*, **71**, 3966 (1949).
149. D. C. Rowlands, *M. S. Thesis*, Ohio State University, 1948; see also E. Campaigne and O. E. Yokley, *J. Org. Chem.*, **28**, 914 (1963).

150. M. S. Newman and J. H. Wotiz, *J. Am. Chem. Soc.*, **71**, 1292 (1949).
151. Y. Pasternak, *Compt. Rend.*, **255**, 1750 (1962).
152. G. F. Hennion and C. V. DiGiovanna, *J. Org. Chem.*, **31**, 970 (1966).
153. J. H. Ford, C. D. Thompson and C. S. Marvel, *J. Am. Chem. Soc.*, **57**, 2619 (1935).
154. J. H. Wotiz, *J. Am. Chem. Soc.*, **72**, 1639 (1950).
155. J. H. Wotiz, J. S. Matthews and J. A. Lieb, *J. Am. Chem. Soc.*, **73**, 5503 (1951).
156. J. H. Wotiz and R. J. Palchak, *J. Am. Chem. Soc.*, **73**, 1971 (1951).
157. J. H. Wotiz and J. S. Matthews, *J. Org. Chem.*, **20**, 155 (1955).
158. Ch. Prévost, M. Gaudemar, L. Miginiac, F. Bardone–Gaudemar and M. Andrac, *Bull. Soc. Chim. (France)*, 679 (1959).
159. T. L. Jacobs and T. L. Moore, *Abstract of Papers, 141st Meeting of the American Chemical Society*, Washington, D. C., March 1962, p. 19.0.
160. O. K. Behrens, J. Corse, D. E. Huff, R. G. Jones, Q. F. Soper and C. W. Whitehead, *J. Biol. Chem.*, **175**, 788 (1948).
161. J. Colonge and R. Gelin, *Bull. Soc. Chim. (France)*, 797 (1954).
162. K. E. Schulte, I. Mleinek and K. H. Schar, *Arch. Pharm.*, **291**, 227 (1958).
163. M. Gaudemar, *Bull. Soc. Chim. (France)*, 1475 (1963).
164. R. Levine and C. R. Hauser, *J. Am. Chem. Soc.*, **66**, 1768 (1944).
165. C. R. Hauser and B. E. Hudson, Jr., *Org. Reactions*, **1**, 297 (1942).
166. V. H. Wallingford, A. H. Homeyer and D. M. Jones, *J. Am. Chem. Soc.*, **63**, 2252 (1941).
167. F. W. Swamer and C. R. Hauser, *J. Am. Chem. Soc.*, **72**, 1352 (1950).
168. P. Nylén, *Chem. Ber.*, **57**, 1023 (1924).
169. I. Shahak, *Tetrahedron Letters*, 2201 (1966).
170. H. Henecka, *Chemie der β-Dicarbonylverbindungen*, Springer Verlag, Berlin, 1950, p. 103.
171. G. Bottaccio and G. P. Chiusoli, *Chem. Ind. (London)*, 1457 (1966).
172. M. Stiles and H. L. Finkbeiner, *J. Am. Chem. Soc.*, **81**, 505 (1959).
173. M. Stiles, *J. Am. Chem. Soc.*, **81**, 2598 (1959).
174. M. Stiles, *Ann. N. Y. Acad. Sci.*, **88**, 332 (1960); *Chem. Abstr.*, **55**, 2475 (1961).
175. H. L. Finkbeiner and M. Stiles, *J. Am. Chem. Soc.*, **85**, 616 (1963).
176. S. W. Pelletier and P. C. Parthasarathy, *Tetrahedron Letters*, 103 (1964).
177. E. Szarvasy, *Chem. Ber.*, **30**, 1836 (1897).
178. J. I. Jones, *Chem. Ind. (London)*, 228 (1958).
179. R. F. Ruthruff, *U.S. Pat.* 3, 038, 006; *Chem. Abstr.*, **57**, 12389 (1962).
180. G. Bottaccio and G. P. Chiusoli, *Chem. Comm.*, 618 (1966).
181. M. S. Kharasch, S. S. Kane and H. C. Brown, *J. Am. Chem. Soc.*, **64**, 333 (1942).
182. F. Bergmann, M. Weizmann, E. Dimant, S. Patai and J. Szmuskowicz, *J. Am. Chem. Soc.*, **70**, 1612 (1948).
183. M. S. Kharasch and H. C. Brown, *J. Am. Chem. Soc.*, **64**, 329 (1942).
184. B. C. McKusick, *U.S. Pat.* 2, 940, 913; *Chem. Abstr.*, **54**, 24557 (1960).
185. D. Hummel and H. Barzynski, *Z. Elektrochem.*, **64**, 1015 (1960).
186. F. Cacace, A. Guarino and E. Possagno, *Gazz. Chim. Ital.*, **89**, 1837 (1959). *Chem. Abstr.*, **55**, 3421 (1961).
187. B. Aliprandi and F. Cacace, *Gazz. Chim. Ital.*, **89**, 2268 (1959). *Chem. Abstr.*, **55**, 6364 (1961).
188. E. P. Kalyazin and V. I. Makarov. *Neftekhimiya*, **3**, No. 2, 227 (1963); *Chem. Abstr.*, **59**, 6278 (1963).
189. V. E. Glushnev, Yu. A. Kolbanovskii, I. I. Patalakh, L. S. Polak, V. T. Popov and V. A. Shakhrai, *Kinetika i Kataliz*, **5** (1), 196 (1964); *Chem. Abstr.*, **60**, 13917 (1964).

CHAPTER 5

Synthesis of di- and polycarboxylic acids and esters

V. F. KUCHEROV

and

L. A. YANOVSKAYA

N.D. Zelinsky Institute of Organic Chemistry, Academy of Sciences of the U.S.S.R., Moscow, U.S.S.R.

I. INTRODUCTION

Di-and polycarboxylic acids and esters occur widely in nature (*e.g.* oxalic, malonic and tricarballylic acids and crocetin). Many of them have found considerable technical application (*e.g.* terephthalic acid) and they have been used variously in organic synthesis. It is not surprising therefore, that there are numerous methods for synthesizing these compounds. Most of them are based on principles which are usually applied in the preparation of monocarboxylic acids and esters but polyfunctional derivatives are used[1,2].

II. DIRECT INTRODUCTION OF THE CARBOXYL GROUP

A. Carboxylation by Carbon Monoxide (for recent reviews see refs. 3, 4)

Aliphatic dicarboxylic acids are formed from hydroxy and unsaturated acids[5,6], diols[5-7], cyclic ethers (*e.g.* dioxane, tetrahydrofuran)[6] and lactones (*e.g.* butyrolactone, valerolactone)[6] by reaction with carbon monoxide in the presence of boron fluoride hydrate, phosphoric acid[7,8] or nickel carbonyl[5,6] (the latter often in the presence of nickel iodide).

With catalysts such as boron fluoride hydrate or phosphoric acid the reaction is carried out at 75–300° and at a carbon monoxide pressure of 400–1000 atm[7,8]. A mechanism based on reaction of carbonium ions is proposed for these reactions[9].

In the presence of nickel carbonyl the reaction with olefins proceeds at 160–250° and at a carbon monoxide pressure of 50–250 atm under aqueous acidic conditions[5, 6]. In this case the reaction can be envisaged as proceeding via dissociation of nickel carbonyl to a coordinatively unsaturated species which then coordinates with the olefins. The nickel carbonyl complex reacts further:

$$\text{Ni(CO)}_{n-1}$$
$$\text{RCH}\cdots\text{CH}_2 \quad + \text{H}^+ \longrightarrow \text{RCH}_2\text{CH}_2\text{Ni(CO)}_{n-1} \rightleftharpoons \text{RCH}_2\text{CH}_2\text{CONi(CO)}_{n-2} \xrightarrow{\text{OH}^-}$$
$$\longrightarrow \text{RCH}_2\text{CH}_2\text{COOH} + \text{Ni(CO)}_{n-1}$$

The reaction with hydroxyacids, diols, lactones and cyclic ethers appears to proceed via formation of olefins.

The best results were obtained by the carboxylation of diols, which appear to give mainly straight-chain acids (yield about 70%).

The reaction of carbon monoxide with benzoic anhydrides or aryl halides may be used for the synthesis of phthalic anhydrides[3].

B. Carboxylation by Carbon Dioxide[10–14]

Adaptation of carboxylation by carbon dioxide to the preparation of dicarboxylic acids and esters requires the use of bis-organometallic compounds or organometallic reagents already containing a carboxyl function.

However, many of the double Grignard reagents (except reagents of the type $\text{BrMg(CH}_2)_n\text{MgBr}$, $n = 4-14$, and some aromatic reagents) and organodimetallic derivatives of alkali metals are difficult to obtain or inaccessible[14]. Moreover, many organometallic reagents have no synthetic value, due to complications and side-reactions occuring during their carboxylation. Some dilithium compounds, such as those obtained from arylated alkenes are formed rapidly and are soluble and stable in 1,2-dimethoxymethane or tetrahydrofuran. These have a limited use for the preparation of aryl-substituted dicarboxylic acids, formed in good yields by carboxylation of the dilithium adducts with carbon dioxide[15].

$$\text{PhCH}=\text{CHPh} \xrightarrow{\text{Li}} \text{PhCHLiCHLiPh} \xrightarrow{\text{CO}_2} \text{HOOCCH(Ph)CH(Ph)COOH}$$

The use of organometallic reagents containing carboxyl groups is also limited due to their poor accessibility and complications during their carboxylation.

Grignard reagents may be prepared from arylacetic acids by the following scheme[16]:

$$\text{ArCH}_2\text{COOH} + 2\,\text{RMgX} \longrightarrow \text{ArCH(MgX)COOMgX} + 2\,\text{RH}$$
$$\text{ArCH(MgX)COOMgX} + \text{CO}_2 \longrightarrow \text{ArCH(COOMgX)}_2$$

Thus, arylmalonic acids were obtained with yields of 56–65%.

Carboxyl-containing Grignard reagents have been obtained from ethyl magnesium bromide with α-allencarboxylic acids[17]. The carboxylation of these reagents yields highly branched unsaturated malonic acids (54–90%):

$$\text{HOOCC(n-Bu)}\!=\!\text{C}\!=\!\text{CH}_2 \xrightarrow{\text{RMgBr}} \text{BrMgOOCC(MgBr)(n-Bu)CR}\!=\!\text{CH}_2 \xrightarrow{\text{CO}_2}$$

$$\xrightarrow{\text{CO}_2} \text{MgOOCC(n-Bu)(CR}\!=\!\text{CH}_2)\text{COOMg}$$

Phenylmalonic acid is formed in 70% yield by slow carboxylation of benzylsodium[18]:

$$\text{PhCH}_2\text{Na} \xrightarrow{\text{CO}_2} \text{PhCH}_2\text{NaCOONa} \xrightarrow{\text{PhCH}_2\text{Na}} \text{PhCHNaCOONa} \xrightarrow{\text{CO}_2} \text{PhCH(COONa)}_2$$

C. Kolbe–Schmitt Reaction

The Kolbe–Schmitt reaction has not found wide application in the synthesis of hydroxy-substituted di- or polycarboxylic acids[19], though in principle it is possible to obtain such compound by this means. For instance, 2-hydroxyisophthalic acid was prepared by heating a mixture of salicylic acid and sodium at 400° in an atmosphere of carbon dioxide. Also, 2, 3, 4-trihydroxyisophthalic acid was obtained from 1,2,3-trihydroxybenzene by heating at 200° with excess of potassium carbonate under carbon dioxide pressure.

D. Rearrangement of Benzene Carboxylic Acids

Potassium benzoate and dipotassium phthalate or isophthalate rearrange to dipotassium terephthalate in high yield by heating them at 400–450° in the presence of catalysts (cadmium or zinc salts[20], or zinc, cadmium, iron or lead oxides[21–23]) at atmospheric or elevated pressure of carbon dioxide:

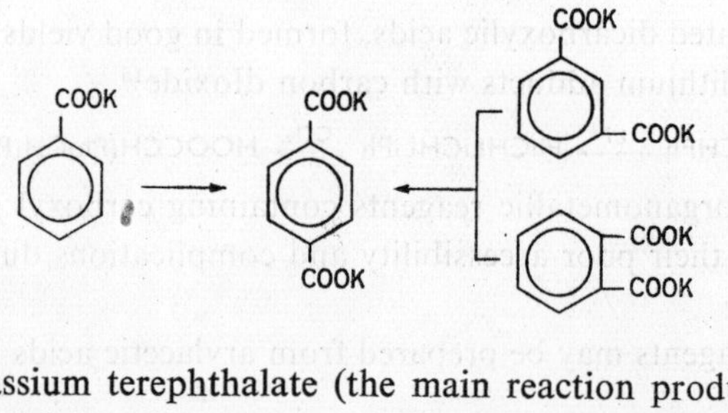

Besides dipotassium terephthalate (the main reaction product) and benzene, small quantities of potassium benzoate and potassium salts of benzenetri-, -tetra- and -pentacarboxylic acids were identified in the reaction mixture. 2 mole of potassium benzoate produced 1 mole of terephthalate and 1 mole of benzene.

In the case of potassium benzoate a disproportionation mechanism was suggested for the reaction:

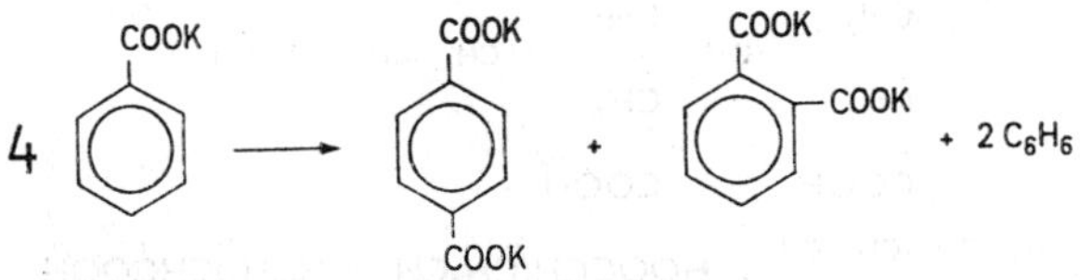

Experiments with ^{14}C-labelled cadmium carbonate confirmed this mechanism[20]. Both the recovered benzoic acid and the terephthalate produced by reaction of benzoate in the presence of labelled cadmium carbonate, proved to be inactive[20]. This conclusion was also supported by rate investigations[20-23].

In the cases of phthalate or isophthalate, the reaction involves an intramolecular rearrangement[21-23].

III. DIMERIZATION OF CARBOXYLIC ACIDS AND ESTERS

A. Radical Dimerization in Water Solution

The dimerization of carboxyalkyl or carbalkoxyalkyl radicals, and the additive dimerization of these radicals in the presence of dienes, is used for the synthesis of long-chain saturated and unsaturated acids and esters. The carboxyalkyl and carbalkoxyalkyl radicals may be generated by the action of hydroxyl (or other) radicals on acids or esters in aqueous media; e.g. hydroxyl radicals were generated by the reaction of hydrogen peroxide with ferrous salt[24]:

$$H_2O_2 + Fe^{2+} \longrightarrow HO\cdot + HO^- + Fe^{3+}$$
$$H(R)COOH + HO\cdot \longrightarrow \cdot RCOOH + H_2O$$
$$2(\cdot RCOOH) \longrightarrow HOOC(R)_2COOH$$

Thus, a mixture of straight- and branched-chain aliphatic acids was formed; e.g. propionic acid produced adipic and dimethylsuccinic acid, and n-butyric acid produced a mixture of isomeric C_8 dicarboxylic acids.

Cyclohexanone peroxide cleaves in the presence of ferrous salts, forming a free carboxyalkyl radical. This radical may react with butadiene in water solution to give a mixture of long-chain unsaturated dicarboxylic acids[25].

By action of hydroxyl radicals, monocarboxylic acids in the presence of butadiene undergo an additive dimerization to a mixture of unsaturated dicarboxylic acids. In this condition dicarboxylic acids produce a mixture of unsaturated tetracarboxylic acids[26]:

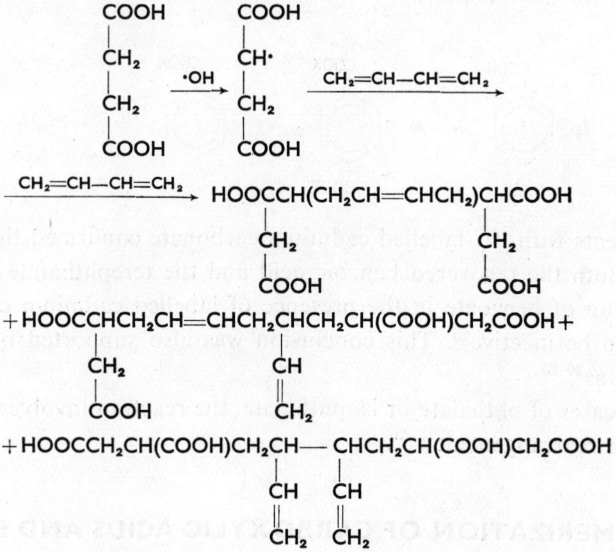

$$
\begin{array}{ccc}
\text{COOH} & \text{COOH} \\
| & | \\
\text{CH}_2 & \text{CH·} \\
| & \xrightarrow{\cdot\text{OH}} \quad | \quad \xrightarrow{\text{CH}_2=\text{CH}-\text{CH}=\text{CH}_2} \\
\text{CH}_2 & \text{CH}_2 \\
| & | \\
\text{COOH} & \text{COOH}
\end{array}
$$

$$\xrightarrow{\text{CH}_2=\text{CH}-\text{CH}=\text{CH}_2} \text{HOOCCH(CH}_2\text{CH}=\text{CHCH}_2)\text{CHCOOH}$$

$$
\begin{array}{cc}
| & | \\
\text{CH}_2 & \text{CH}_2 \\
| & | \\
\text{COOH} & \text{COOH}
\end{array}
$$

$$+ \text{HOOCCHCH}_2\text{CH}=\text{CHCH}_2\text{CHCH}_2\text{CH(COOH)CH}_2\text{COOH} +$$

$$
\begin{array}{cc}
| & | \\
\text{CH}_2 & \text{CH} \\
| & || \\
\text{COOH} & \text{CH}_2
\end{array}
$$

$$+ \text{HOOCCH}_2\text{CH(COOH)CH}_2\text{CH}-\text{CHCH}_2\text{CH(COOH)CH}_2\text{COOH}$$

$$
\begin{array}{cc}
| & | \\
\text{CH} & \text{CH} \\
|| & || \\
\text{CH}_2 & \text{CH}_2
\end{array}
$$

Additive dimerization of butadiene with radicals formed from the H_2O_2 adduct of an acyclic keto ester leads to the formation of long-chain diesters[27]:

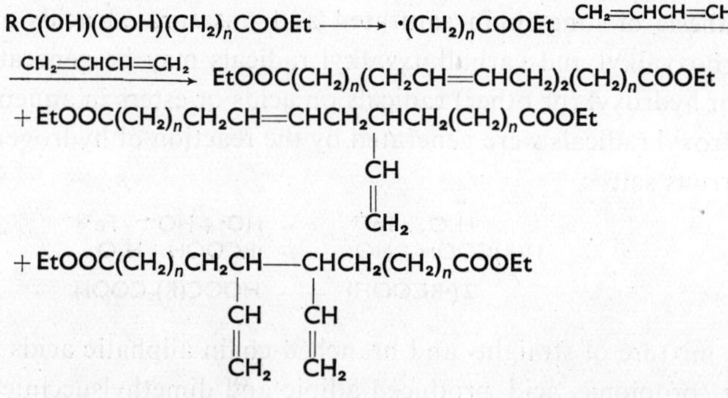

$$\text{RC(OH)(OOH)(CH}_2)_n\text{COOEt} \longrightarrow \cdot(\text{CH}_2)_n\text{COOEt} \xrightarrow{\text{CH}_2=\text{CHCH}=\text{CH}_2}$$

$$\xrightarrow{\text{CH}_2=\text{CHCH}=\text{CH}_2} \text{EtOOC(CH}_2)_n(\text{CH}_2\text{CH}=\text{CHCH}_2)_2(\text{CH}_2)_n\text{COOEt}$$

$$+ \text{EtOOC(CH}_2)_n\text{CH}_2\text{CH}=\text{CHCH}_2\text{CHCH}_2(\text{CH}_2)_n\text{COOEt}$$

$$
\begin{array}{c}
| \\
\text{CH} \\
|| \\
\text{CH}_2
\end{array}
$$

$$+ \text{EtOOC(CH}_2)_n\text{CH}_2\text{CH}-\text{CHCH}_2(\text{CH}_2)_n\text{COOEt}$$

$$
\begin{array}{cc}
| & | \\
\text{CH} & \text{CH} \\
|| & || \\
\text{CH}_2 & \text{CH}_2
\end{array}
$$

These methods are of limited use because the initial attack of the hydroxyl (or other) radical is not sufficiently selective, and mixtures of isomers are obtained. Moreover, the acids or esters to be dimerized have to be soluble in aqueous media and should not be easily oxidized.

Hydrodimerization of unsaturated acids or esters by electroreduction on a mercury cathode or by reduction with sodium or aluminium amalgam often involves many complications and seldom gives satisfactory results[28-31].

B. Oxidative Dimerization of Acetylenic Acids and Esters

Oxidative coupling of acetylenic monocarboxylic acids or esters in water–alcohol or pyridine solutions by air or oxygen in the presence of ammonium chloride and cuprous chloride is a very important method for the preparation of polyenynic dicarboxylic acids and esters. This method allows the doubling of the carbon chain and gives high yields.

Oxidative coupling is often used for the synthesis of naturally occurring and related diacids and diesters[32–35]. For instance, the first synthesis of corticrocin was based on the dimerization of hept-2-en-6-ynoic acid[34]:

$$HC{\equiv}CCH_2CH_2CH{=}CHCOOH \longrightarrow$$

$$\longrightarrow HOOCCH{=}CHCH_2CH_2(C{\equiv}C)_2CH_2CH_2CH{=}CHCOOH \xrightarrow{OH^-}$$

$$\xrightarrow{OH^-} HOOC(CH{=}CH)_6COOH$$

C. Miscellaneous Methods of Dimerization

The coupling of haloacids or haloesters by the action of metals in Wurtz-type reactions does not usually give satisfactory results and can be used only in rare cases; e.g. methyl 3-bromo-2-naphthoate can be converted to dimethyl 2,2′-dinaphthyl-3,3′-dicarboxylate by heating with copper bronze at 190–200°[36], and diethyl tetraphenylsuccinate was obtained by refluxing ethyl diphenylchloroacetate with molecular silver in benzene[37].

Dimerization via intermediate formation of organometallic compounds is much more successful. Thus, bis-(methyl ethyl veratroylacetate) was prepared from methyl ethyl veratroylacetate via the sodium derivative by action of iodine[38].

The dimerization of α-bromopropionate or α-bromocaprylate using sodium dialkylphosphite is believed to proceed via organometallic compounds[39].

Reductive dimerization of methyl pyruvate by trimethylphosphite results in a mixture of diastereomers of cyclic phosphoranes, the hydrolysis of which leads to diastereomeric 2,3-dimethyltartrates[40].

Ethyl crotonate dimerizes in the presence of potassium/benzyl potassium in toluene at 100° and yields a dimer, 81% of which consists of diethyl 2-ethylidene-3-methylglutarate with the carbethoxy group cis to the vinylic hydrogen[41].

The additive dimerization of butadiene, styrene or α-methylstyrene using sodium, followed by carbonation of the disodium dimers with carbon dioxide, was suggested as a method for the synthesis of aryl-substituted dicarboxylic acids[42, 43].

The alkyl-substituted aromatic acids are converted into dehydrodimers on heating with sulphur. The method is convenient for the preparation of stilbene-[44] or bibenzyl-dicarboxylic acids[45] in good yields.

Diazotized aromatic aminocarboxylic acids are converted into biaryl dicarboxylic acids by reduction with cupro-ammonia ion in dilute ammonium hydroxide[46, 47].

IV. CONDENSATION METHODS OF SYNTHESIS

Ester condensation, Knoevenagel condensation, malonic synthesis and oxalic ester condensation proceed in the presence of bases according to the following scheme:

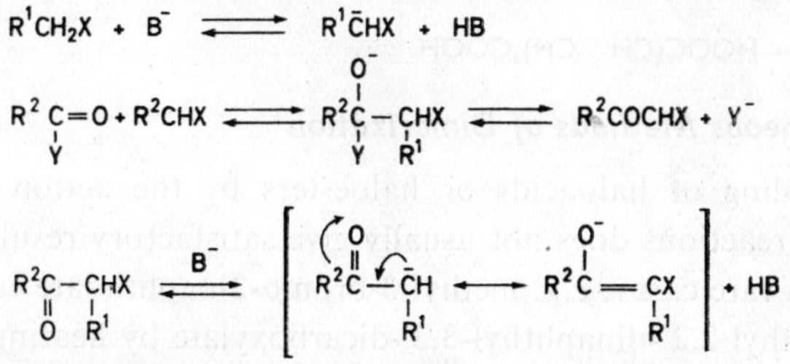

$$R^1CH_2X + B^- \rightleftharpoons R^1\bar{C}HX + HB$$

$$R^2C{=}O + R^2CHX \rightleftharpoons \underset{Y}{\underset{|}{R^2}}\overset{O^-}{\underset{|}{C}} - \underset{R^1}{\underset{|}{CHX}} \rightleftharpoons R^2COCHX + Y^-$$

$$R^2\underset{O}{\overset{||}{C}} - \underset{R^1}{\overset{|}{CHX}} \overset{B^-}{\rightleftharpoons} \left[R^2\overset{O}{\overset{||}{C}}{\overset{\curvearrowleft}{}} \underset{R^1}{\overset{|}{CH}} \longleftrightarrow R^2\overset{O^-}{\underset{R^1}{\overset{|}{C}}} {=} \underset{R^1}{\overset{|}{CX}} \right] + HB$$

With a variety of starting materials, these methods find wide application in the syntheses of various di- and polycarboxylic acids and esters.

A. Ester Condensation

The intramolecular condensation of aliphatic and alicyclic polycarboxylic esters is a synthetic method of great importance[48, 49] which is often used for the synthesis of polycyclic compounds, especially in the steroid series[50-53]:

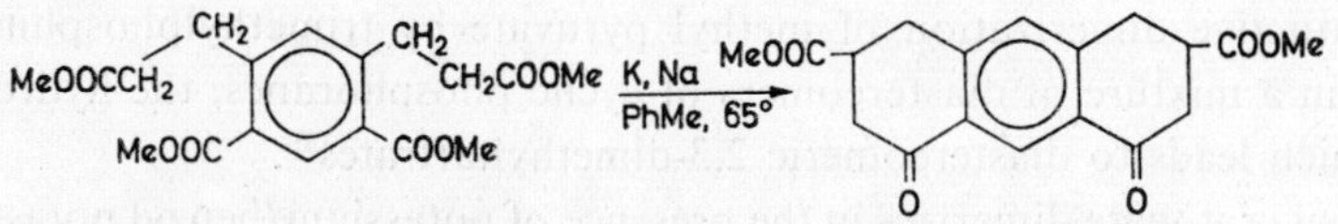

This cyclization has also found wide use in the preparation of hetero-cyclic di- or polyesters, e.g. derivatives of thiophen[54].

B. Knoevenagel Condensation

The condensation involving the reaction of aldehydo or keto esters with esters containing an acidic methylene group, affords a route to unsaturated di- and polyesters. Thus, t-butyl β-(carboxyphenyl)cinnamate is obtained by adding at 80° a mixture of t-butyl acetate and t-butanol to a mixture of t-butyl o-benzoylbenzoate and sodium hydride in benzene[55]. In a similar manner 2,2-dimethyl-3-ethoxycarbonylcyclopentanone condenses with ethyl cyanoacetate in acetic acid in the presence of ammonium acetate to form 2,2-dimethyl-3-carbethoxy-1-(α-cyano-α-carbethoxy)methylenecyclopentane[56]. Ethyl acetoacetate reacts with ethyl cyanoacetate and potassium t-butylate at room temperature to give diethyl α,β-dialkyl-γ-cyanoglutarate[57].

The condensation of one mole of aliphatic or aromatic aldehydes with two moles of ester in the presence of piperidine leads to bis-esters of the type: $RCH(COOEt)CHR'CH(COOEt)R$[58-62]. The reaction can also be carried out with semi-acetals instead of aldehydes[63].

The Knoevenagel condensation may also be used for the preparation of aromatic di- and polyesters; for example, ethyl sodium acetopyruvate treated with diethyl acetonedicarboxylate in alcohol gives triethyl isocochenillate[64].

C. Malonic Syntheses

The condensation of malonic acid or ester with aldehydes is used for the synthesis of α,β-unsaturated di- and polycarboxylic acids or esters. Alkylidenemalonic esters[65] or β-methylglutaconic ester[66] can also be used. The method is limited only by the accessibility of dialdehydes. Many polyenic diesters were prepared thus:

$$OHCC(Me)=CHCH=CHCH=C(Me)CHO + 2 MeCH=C(COOEt)_2 \longrightarrow$$
$$(EtOOC)_2C=CHCH=CHC(Me)=CHCH=CHCH=C(Me)CH=CHCH=C(COOEt)_2$$

Saturated polycarboxylic acids and esters can be prepared from one molecule of aldehyde with two molecules of malonic acid or ester. It is also possible to use two different malonic esters. Thus, formaldehyde condenses with dimethyl malonate and dimethyl acetaminomalonate in the presence of sodium hydride by refluxing in toluene to give tetramethyl 3-acetaminopropane-1,1,3,3-tetracarboxylate[67]. In certain cases Mannich bases may be used instead of the aldehydes[68, 69].

D. Oxalic Ester Condensation

Oxalic ester condensation involves the acylation of an ester containing an acidic methylene group by an oxalic ester. The method can be applied to the synthesis of saturated keto dicarboxylic esters from esters of the type RCH_2COOEt[70-72] or of unsaturated keto dicarboxylic acids from esters of the type $Me(CH=CH)_nCOOEt$[73]. The yields are poor when $n > 3$.

In many cases sodium hydride is a better catalyst then metal alkoxides *e.g.* for the introduction of an ethoxalyl group into ethyl caproate[72].

E. Organometallic Synthesis

The reactions between organometallic derivatives containing a carboxyl group (such as sodium malonic ester) and mono- or dihalocompounds are valuable for the synthesis of aliphatic and alicyclic di- and polycarboxylic acids and esters. The components of the reaction may be widely varied and the products are as a rule obtained in high yields.

t-Butyl sodium acetoacetate condenses with α-bromoesters to give aceto-succinates of the type *t*-BuOOCCH(COMe)CRR'COOEt[74]. The reaction between *t*-butyl acetate with dihaloalkanes in the presence of lithium amide in liquid ammonia is a convenient method for the synthesis of long-chain dicarboxylic acids[75].

The alkylation of esters containing acidic methylene groups may be carried out much more successfully by using sodium hydride[76], lithium hydride[77] or potassium *t*-butoxide[78] as condensing agents. Disodium malonate may also be used in the reaction[79].

An interesting new scheme for the synthesis of polyspirotetracarboxylates was recently proposed. The method was based on condensation of sodium malonic ester with pentaerythritol tetratosylate by refluxing in xylene[80].

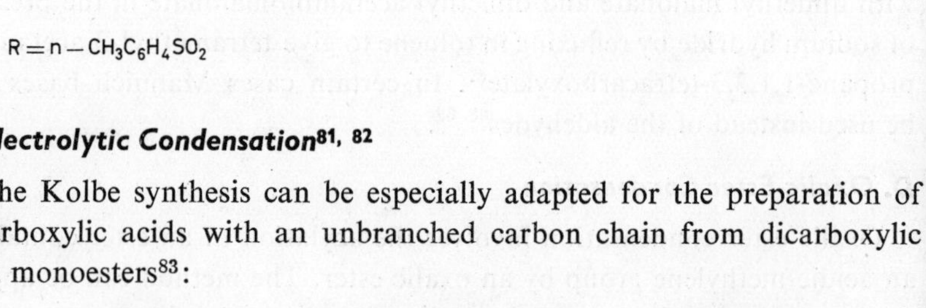

$R = n - CH_3C_6H_4SO_2$

F. Electrolytic Condensation[81, 82]

The Kolbe synthesis can be especially adapted for the preparation of dicarboxylic acids with an unbranched carbon chain from dicarboxylic acid monoesters[83]:

$$MeOOC(CH_2)_{16}COOH \xrightarrow[\text{aq. acetone}]{\text{Pt anode}} MeOOC(CH_2)_{32}COOMe$$

Electrolytic condensation is also suitable for the preparation of long-chain unsaturated diesters[84-86]. Thus, dicarboxylic acid monoesters react in the presence of butadiene according to the following scheme:

$$2\ ROOC(CH_2)_nCOOH + CH_2 =\!\!= CHCH =\!\!= CH_2 - 2e \longrightarrow$$
$$\longrightarrow ROOC(CH_2)_nCH_2CH =\!\!= CH(CH_2)_nCOOR$$

In contrast to radical dimerization (section III. A), electrolytic condensation usually leads to single products.

G. Reformatsky Reaction[87, 88]

The Reformatsky reaction can be applied to the preparation of dicarboxylic acids or esters, starting with dialdehydes, diketones, aldehydes or ketones containing a carboxyl group. Bromoacetic ester[89] or its vinylogues (γ-bromocrotonic ester or γ-bromotiglic ester)[89, 90] may be used.

This reaction was used for the synthesis of many naturally occurring esters (e.g. crocetin, bixin) using α-hydroxy derivatives. The reaction with polyunsaturated dialdehydes may only be carried out in dry tetrahydrofuran with carefully-activated zinc:

$$OHCC(Me) =\!\!= CHC \equiv CCH =\!\!= C(Me)CHO + 2\ BrCH_2CH =\!\!= C(Me)COOMe \longrightarrow$$
$$MeOOCC(Me) =\!\!= CHCH_2CH(OH)C(Me) =\!\!= CHC \equiv CCH =\!\!= C(Me)CH(OH)CH_2CH =\!\!= C(Me)COOMe$$

Various ketones (e.g. 2,2,5,5,-tetramethylhexanedione[91], 2,3-diketo-cis-dekaline[92], cycloxexanedione[93] and benzil[94]) react with α-halo esters in Reformatsky reaction conditions.

H. Wittig Reaction[95, 96]

At present the Wittig reaction is the most versatile and convenient method of synthesizing the α,β-unsaturated and polyenic dicarboxylic esters.

The dicarbonyl compounds react with carbalkoxymethylene triphenylphosphoranes in benzene, methylenechloride, alcohol etc., or with carbalkoxymethylphosphonic acids in dimethylformamide in the presence of metal alkoxides at room temperature or on heating. In both cases the yields of diesters are high, but the isolation of the products is more convenient in the latter case.

The Wittig reaction can be adapted for the preparation of many naturally occurring diesters[97-102], especially isoprenoid ones. Carbalkoxy-α methylmethylene triphenylphosphorane was used in this reaction for the introduction of side-chain methyl groups. The application of phosphoranes obtained from γ-bromocrotonate, 4-methyl-6-bromosorbate or 2,6-dimethyl-8-bromoocta-2,4,6-trienoate[99] is the most promising path for the synthesis of highly unsaturated diesters. The following examples[99, 100] illustrate the scope of the Wittig reaction:

$$\overset{+}{\text{OHC(CH=CH)}_n\text{CHO}+2\text{ PhP—}\overset{-}{\text{CHCOOEt}} \longrightarrow \text{EtOOC(CH=CH)}_{n+2}\text{COOEt}}$$

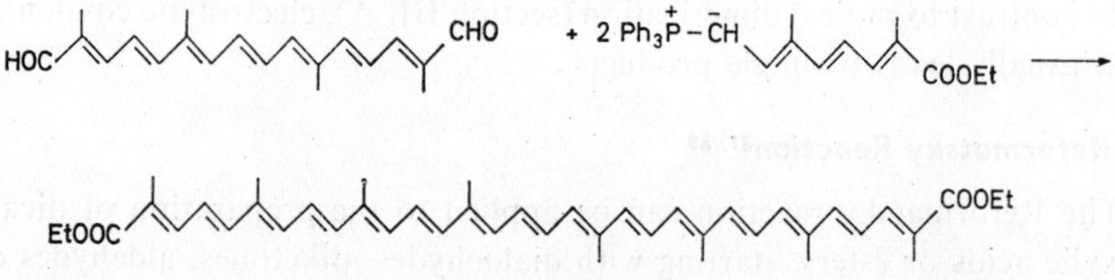

The reaction of ketones with carbalkoxymethylene triphenylphosphorane usually proceeds much less readily than with aldehydes. However, certain ketones, *e.g.* 3,4-benzocyclobutane-1,2-dione, react with two equivalents of carbomethoxymethylene triphenylphosphorane in methylene chloride at room temperature to yield 3,4-benzo-1,2-bis(carbomethoxymethylene)cyclobutane in high yields[103].

It is possible to prepare unsymmetrically substituted dicarboxylic esters by reaction of phosphoranes with monoaldehydes containing carbalkoxyl groups[101, 104]:

$$\text{EtOOCC(Me)=CH(CH=CH)}_n\text{CHO}+\text{Ph}_3\overset{+}{\text{P}}\text{—}\overset{-}{\text{CHCOOEt}}$$
$$\downarrow$$
$$\text{EtOOCC(Me)=CH(CH=CH)}_{n+1}\text{COOEt}$$
$$\downarrow$$
$$\text{EtOOC(CH=CH)}_n\text{CHO}+\text{Ph}_3\overset{+}{\text{P}}\text{—}\overset{-}{\text{C(Me)COOEt}}$$

The Wittig reaction with stable phosphoranes or phosphonates and aldehydes has a stereospecific course and gives mostly *trans* isomers[101, 102]. This is in line with the generally accepted reaction mechanism.

I. Diels-Alder Reaction[105]

Diene synthesis with maleic or other related anhydride is a general reaction path to cyclic 1,2-dicarboxylic acids. Cyclic dicarboxylic esters are also formed directly by condensation of dienes with unsaturated dicarboxylic esters. The ratio of the isomers formed depends essentially upon the structure of the initial dienes. For instance, the condensation of α-acetoxyvinyl-Δ^1-cyclohexene with maleic anhydride at 0° gives as the main product an isomer which corresponds to a reaction course against the rule of 'the preferential *endo* orientation of components (spatial orientation effect of the maximum accumulation of unsaturation)'[105]. In other cases this isomer forms in smaller quantities[106, 107]:

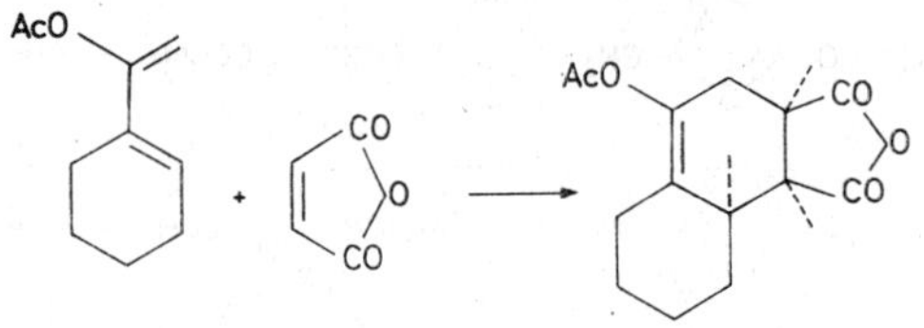

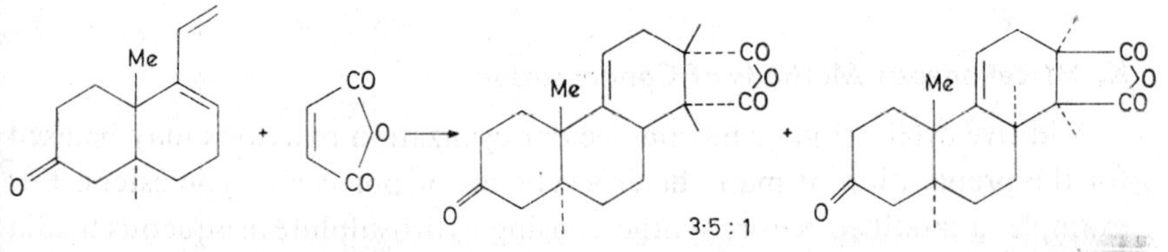

3·5 : 1

The elucidation of these regularities and steric features of the transformation of alicyclic anhydrides and dicarboxylic acids made possible stereoselective syntheses, not feasible by other methods, of many geometric isomers of mono-, di- and tricyclic dicarboxylic acids[107–113].

The adducts of vinylcyclene with formylacrylic Ψ-esters may also be used to obtain dicarboxylic acids[114].

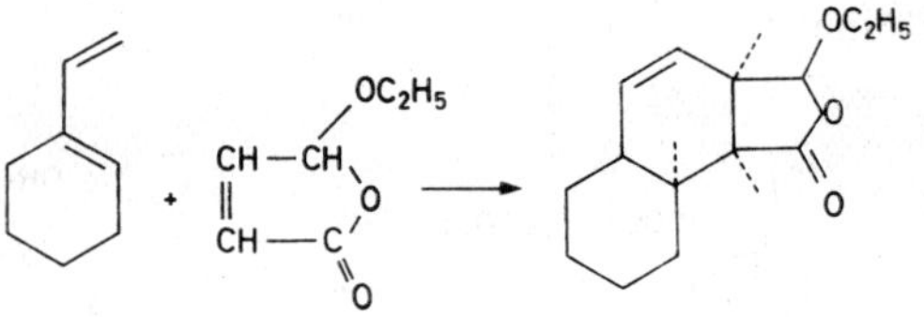

Condensation of dimethyl acetylenedicarboxylate with 6-carbomethoxy-α-pyrone at 210–215° gives trimethyl hemimellitate[115].

J. Friedel–Crafts Reaction[116]

The synthesis of the di-2-thienylmethanedicarboxylic acids by Friedel–Crafts reaction is of some interest[117]. The acylation of bis-2-thienylmethane with ester acid chlorides of dibasic acids proceesd with the best result in benzene at −14° to −18° in the presence of $SnCl_4$. The obtained diesters are reduced and subjected to hydrogenolysis at 90° in the presence of Raney nickel in an aqueous solution of sodium carbonate to give long-chain aliphatic dicarboxylic acids.

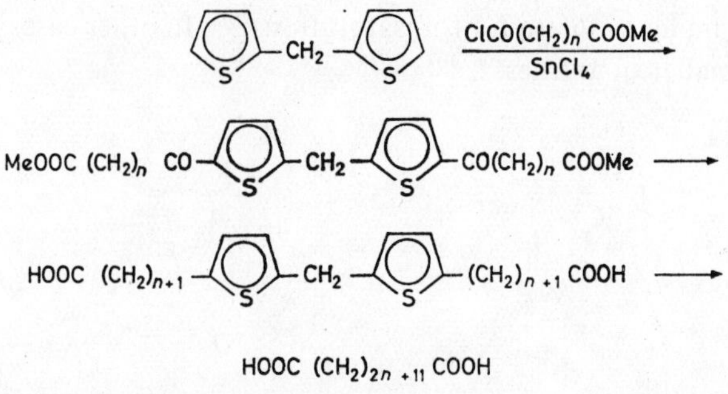

K. Miscellaneous Methods of Condensation

Additive cyclization or intramolecular cyclization reactions may be used for the preparation of many heterocyclic di- or polycarboxylic esters. For example, a modified Knorr synthesis using hydrosulphite in aqueous media for the reduction of isonitroso ketones was successfully used for the synthesis of pyrolledicarboxylic esters[118].

The condensation of ethyl β-aminocrotonate and similar enamines with ethyl ethoxymethylenoxaloacetate, ethyl ethoxymethyleneacetylpyruvate and related compounds proceeds with the formation of derivatives of pyridinedicarboxylate in excellent yields, without catalysts[119, 120].

An intramolecular cyclization of o-bis(β-dicarbethoxyvinylamino)-benzene by refluxing in diphenyl ether leads to 3,8-dicarbethoxy-4,7-dihydroxy-1,10-phenanthroline[121]:

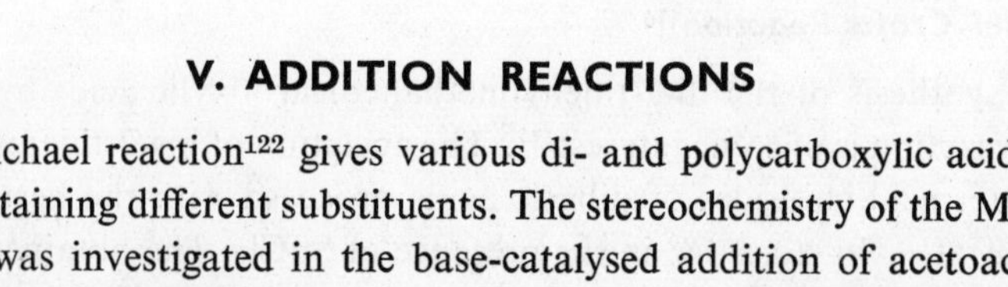

V. ADDITION REACTIONS

The Michael reaction[122] gives various di- and polycarboxylic acids and esters containing different substituents. The stereochemistry of the Michael addition was investigated in the base-catalysed addition of acetoacetate, malonate and methylmalonate to methyl bicyclo [1,2,2]hept-2,5-diene-carboxylate. The adducts proved to be exclusively of *trans* configuration, showing that 'exo addition' takes place[123].

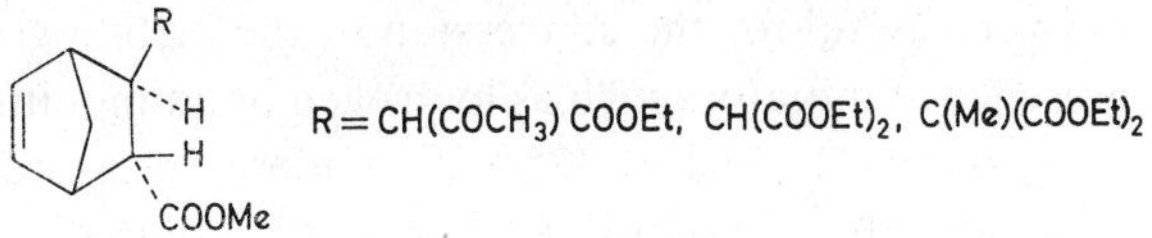

R = CH(COCH$_3$) COOEt, CH(COOEt)$_2$, C(Me)(COOEt)$_2$

The scope of the Michael addition is very wide. For example, ethyl n-butylmalonate adds to ethyl acrylate to give 3,3-di(carbethoxy)heptanoate[124]. Similarly, it is possible to obtain alicyclic compounds: diethyl 4-carbethoxy-4-methyl-3-oxocyclohexylmalonate is formed by reaction of ethyl 2-methyl-Δ^5-cyclohexenone-2-carboxylate with ethyl malonate[125].

Aromatic derivatives may be also prepared. Diethyl methylenemalonate reacts with ethyl 2-pyridylacetate to produce triethyl 3-(2-pyridyl)-1,1,3-propanetricarboxylate. The pyridine nitrogen acts here as a built-in basic catalyst[126]. Phenylacetic acid adds to ethyl cinnamate to form α,β-diphenylglutaric acid[127].

The Michael addition was also used successfully for the formation of small rings. Thus, the addition of ethyl 1,1,2,2-ethanetetracarboxylate to ethyl acetylenedicarboxylate produced ethyl 1,1,2,2,3,4-cyclobutanehexacarboxylate[128]. The condensation of α-halo esters with α,β-unsaturated esters gives 1,2-cyclopropanedicarboxylic acid diesters[129]. The mixture of stereoisomers obtained is rich in the less stable *cis* isomer.

The addition of acids and esters to multiple bonds is possible in radical conditions. A synthesis of adipic acid involving the addition of acetic acid to acetylene in the presence of butyl peroxide at 120° was recently described[130].

VI. OXIDATIVE AND HYDROLYTIC METHODS OF SYNTHESIS

.Oxidative Methods

Aliphatic, alicyclic, and aromatic di- and polycarboxylic acids may be prepared from the appropriate diols, dialdehydes, hydroxy aldehydes and hydroxy acids by oxidative methods. Such oxidations have played a great part in the synthesis of polyunsaturated dicarboxylic acids[131–133].

Many procedures were developed for the oxidation of the side chains in aromatic and heterocyclic compounds. Oxidative degradation of side chains may be carried out with common oxidants such as potassium permanganate (*e.g.*, the oxidation of 3-methyl-6-methoxyphthalic anhydride to 4-methoxybenzene-1,2,3-tricarboxylic acid[134]) or concentrated nitric acid (oxidation of hexamethylbenzene to mellitic acid[135]). Aqueous ammonium sulphate has been found very useful for the conversion of the methyl group to carboxyl in the presence of initiators such as hydrogen or ammonium sulphide. Thus, o-, m- or p-xylene heated at 325–350° under a pressure of 20 atm with ammonium sulphate containing sulphur, yielded phthalic, isophthalic or terephthalic acids[136].

Lead dioxide was found to oxidize 2-hydroxy-3-methylbenzoic acid to 2-hydroxysophthalic acid by heating at 200–240° in alkali solution[137]. 5-Ethyl-2-methylpyridine yields isocinchomeronic acid when heated with aqueous cupric nitrate under the pressure of an oxygen–nitrogen mixture[138].

Dilute aqueous hydrogen bromide catalyses the oxidation of water-soluble aromatic compounds such as p-toluic acid or γ-picoline to the corresponding dicarboxylic acids. The reaction requires high pressure of oxygen at 180–200°. The presence of vanadium compounds permits the oxidation of water-insoluble compounds (e.g. p-bromomethylbenzoic acid, p-xylene[139]).

Nitric acid oxidation of trans-bis-(iodomethyl)dioxane produced trans-2,3-bis-(dicarboxy)dioxane in good yield, without isomerization or ring cleavage[140].

The oxidative ring opening of cycloolefins leading to aliphatic, alicyclic or aromatic acids is of great importance. Thus, the oxidation of cyclohexene by heating it in alkaline solution with silver oxide produces adipic acid[141]. Glutaric acid is formed by heating dihydropyran with nitric acid[142].

The oxidation of norbornylene with aqueous potassium permanganate leads to cis-cyclopentane-1,3-dicarboxylic acid[143]. Methyl $\Delta^{2,3}$-allo-etiocholenate is oxidized in chloroform by ozone and after treatment with diazomethane produces trimethyl iso-allo-etiolithobilianate[144]. Indene reacts with ozone in emulsions of aqueous alkaline hydrogen peroxide to form homophthalic acid[145].

Oxidative ring opening may be applied to aromatic compounds. Thus, the oxidation of pyrocatechol in glacial acetic acid gives trans-muconic acid[146]. Under similar conditions β-naphthol yields o-carboxycinnamic acid[147].

Numerous procedures have been developed for the oxidative ring opening of alicyclic compounds containing oxygen functions. These methods yield aliphatic, especially branched, dicarboxylic acids. For instance, the oxidation of mathone with an alkaline solution of sodium hypochlorite gives β,β-dimethylglutaric acid[148]. α-Ethyl-α-butylglutaric acid is formed by treatment of γ-butyl-γ-ethyl-δ-valerolactone with potassium permanganate in alkaline solution at 50°[149]. Oxidative ring opening was widely used for the synthesis of steroid dicarboxylic acids. For the oxidative ring opening of the alicyclic part of compounds containing condensed aromatic and alicyclic rings, hydrogen peroxide is often used. Thus, 3-isopropyl-6-methylhomophthalic acid was obtained from the sodium salt of ethyl 4-methyl-7-isopropyl-α-indanonglyoxilate[152] and o-carboxycinnamic acid was obtained from 4,5-benzotropolone[153, 154].

Recently a new method of oxidative cleavage was proposed. It has been found that on treatment with aqueous hydrogen peroxide in t-butanol at 50–65°, 2-formylcycloalkanones undergo an unusual oxidative cleavage to aliphatic dicarboxylic acids containing the same number of carbon atoms as the starting compounds[155].

$$(CH_2)_n \begin{array}{c} CO \\ | \\ CHCHO \end{array} \xrightarrow{H_2O_2} HOOC(CH_2)_{n+1}COOH, \quad n = 3-16$$

The cleavage reaction of four-, five- and six-membered ketones is accompanied by ring contraction and formation of carboxylic acids with a smaller ring. 2-Acetylcyclohexanone reacts with migration of the methyl group:

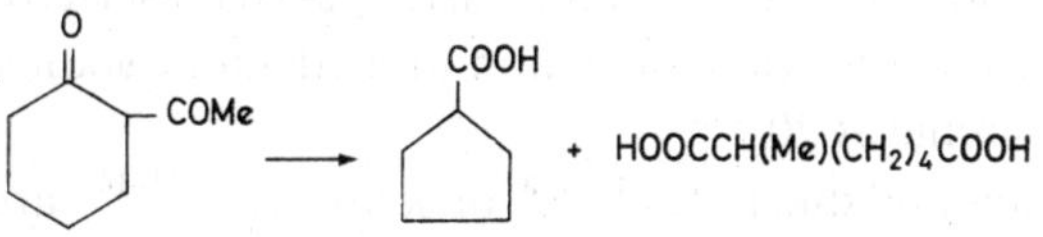

The following reaction mechanism was suggested:

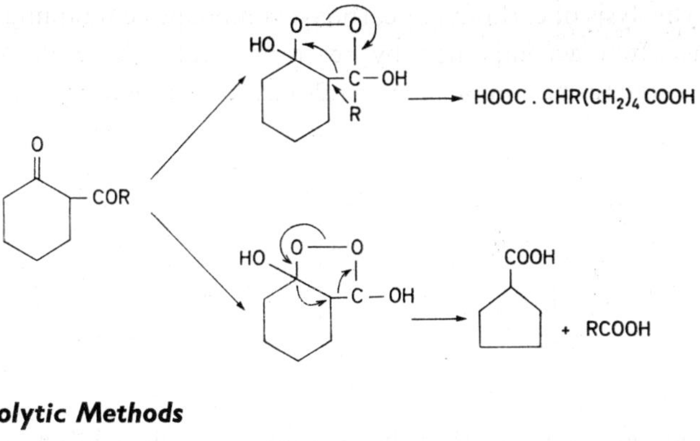

B. Hydrolytic Methods

The acid hydrolysis of 1,1,1-trichloroalkanes which are easily formed by telomerization of ethylene with carbon tetrachloride, achieved great industrial importance for the synthesis of aliphatic dicarboxylic acids (from succinic to azelaic)[156, 157].

$$n/2\ CH_2{=}CH_2 + CCl_4 \longrightarrow CCl_3(CH_2)_nCl \xrightarrow{NaCN} CCl_3(CH_2)_nCN$$

$$\downarrow \qquad\qquad\qquad \downarrow$$

$$HOOC(CH_2)_nCl \qquad HOOC(CH_2)_nCOOH$$

$$\downarrow OH^-$$

$$HOOC(CH_2)_nOH \xrightarrow{HNO_3} HOOC(CH_2)_{n-1}COOH$$

Concentrated sulphuric acid is usually used for the hydrolysis. Recently it was demonstrated that ferric chloride catalyses the hydrolysis of trichloro-methyl derivatives[158]. Thus, terephthalic acid was quantitatively obtained by refluxing 1,2-bis-(trichloromethyl)benzene in a water–chloroform mixture in the presence of ferric chloride.

The alcoholysis of α,ω-dichloroacetylenes has been developed for the production of diesters[159].

$$2\ NaC{\equiv}CCl + Br(CH_2)_6Br \longrightarrow ClC{\equiv}C(CH_2)_6C{\equiv}CCl \xrightarrow{ROH} ROOC(CH_2)_8COOR$$

Very interesting results were obtained during the study of the Favorsky rearrangement with 1,1-dibromoketones. It was found that methyl 1,1-dibromoundecanone-2-carboxylate rearranges stereoselectively to dimethyl cis-dodecen-2-dicarboxylate on action of methanolic sodium methoxide at 0° or triethylamine at 20°[160].

Cyclic mono- and diketones[161-163] or keto acids[164-167] are easily cleaved on treatment with bases, with the formation of aliphatic dicarboxylic acids. The synthesis of norcamphoric acid is an interesting example of the application of this method[168, 169].

The hydrolysis of carbonyl or carboxyl compounds containing α-halogen atoms is often accompanied by rearrangements. A recent interesting example is the formation of cubane-1,2-dicarboxylic acid[170] by the scheme:

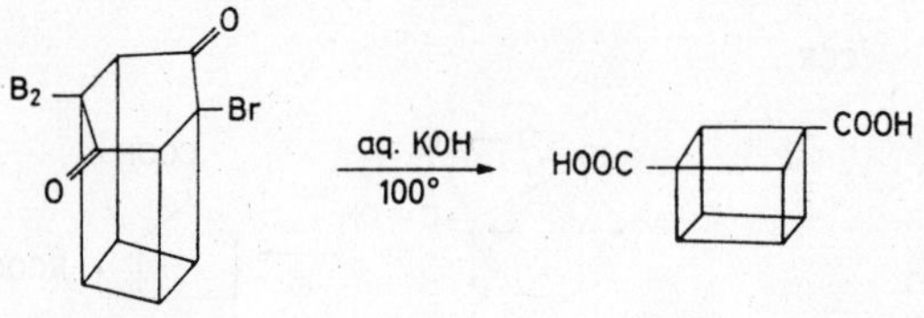

2-Chlorocyclohexene-1-carboxylic acid gives pimelic acid on boiling with 50% sodium hydroxide[171].

The alkali cleavage of β-dicarbonyl compounds may be carried out simultaneously with reduction[161, 162, 172] or alkylation with halo esters[173, 174].

The first method was used for the synthesis of long-chain saturated or unsaturated dicarboxylic acids from C-alkylated dihydroresorcines[161].

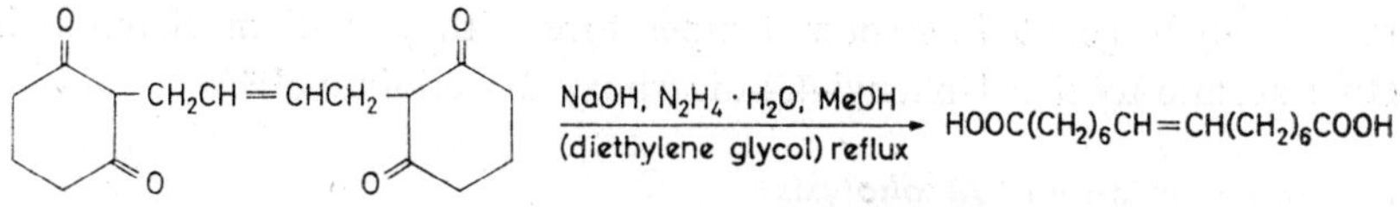

$$\text{(structure)} \xrightarrow[\text{(diethylene glycol) reflux}]{\text{NaOH, N}_2\text{H}_4\cdot\text{H}_2\text{O, MeOH}} \text{HOOC(CH}_2)_6\text{CH}=\text{CH(CH}_2)_6\text{COOH}$$

The second method was used in the preparation of 4-carboxyazelaic acid by reaction of the sodio derivative of 2-carbethoxycyclopentanone with esters of 4-bromobutyric acid in alcohol solution[174].

A radical cleavage of cyclohexanone in the presence of hydrogen peroxide and ferrous sulphate proceeds with simultaneous self-condensation and yields decane-1,2-dicarboxylic acid[175].

VII. SYNTHESES BASED ON ACID DERIVATIVES

A. Ester Hydrolysis

The hydrolysis of di- or polyesters is one of the most important methods of synthesizing di- and polycarboxylic acids. Sensitive compounds are usually hydrolysed under mild conditions. Thus tetraethyl butane-1,1,4,4-tetracarboxylate yields the tetracarboxylic acid on shaking with aqueous potassium hydroxide[176]. Tetraethyl dispiro[3,1,3,1]decane-2,2-8,8-tetracarboxylate is hydrolysed by alcoholic potassium hydroxide at room temperature[177].

It is possible to carry out selective hydrolysis of a compound containing several cleavable groups. Thus, diethyl α-acetamido-β-(3-indolyl)-α-carboxypropionate in aqueous sodium hydroxide gives quantitative amounts of α-acetamido-α-carboxy-β-(3-indolyl)propionic acid[178]. Selective cleavage of α,α'-diacetyldicarboxylic esters by boiling with alkali solution yields many dicarboxylic acids. This method was used for the synthesis of some β-arylglutaric acids[179].

Saponification of tetraesters with partial decarboxylation is often used for the synthesis of dicarboxylic acids. Thus, sodium tetraethyl hept-2,4,6-triene-1,1,8,8-tetracarboxylate gives nona-2,4,6-triene-1,9-dioic acid on careful heating with alcoholic sodium hydroxide[180]. Hydrolysis with partial decarboxylation, but with retention of the acetyl groups may also be carried out in acidic media[181].

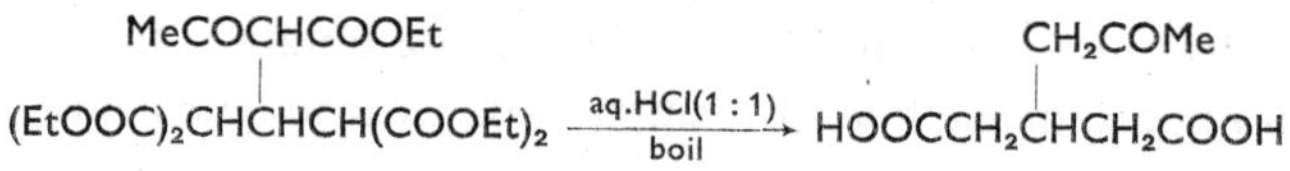

$$\begin{array}{c} \text{MeCOCHCOOEt} \\ | \\ \text{(EtOOC)}_2\text{CHCHCH(COOEt)}_2 \end{array} \xrightarrow[\text{boil}]{\text{aq.HCl(1 : 1)}} \begin{array}{c} \text{CH}_2\text{COMe} \\ | \\ \text{HOOCCH}_2\text{CHCH}_2\text{COOH} \end{array}$$

The hydrogenolysis of benzylic esters may be successfully used for the preparation of dicarboxylic acids from alkali- and acid- sensitive diesters[182, 183]. For example, 1-phenyl-4,4-dicarbobenzyloxy-2-azetidinone may be hydrogenated at room temperature with palladium charcoal in ethyl acetate to give 1-phenyl-4,4-dicarboxy-2-azetidinone[182].

B. Esterification and Alcoholysis

Esterification of acids with alcohols in the presence of catalysts is the most widely used method of ester synthesis. Esterification with diazomethane is usually employed for sensitive acids.

Other methods e.g., the boiling of a mixture of a dicarboxylic acid, acetone dimethyl ketal and methanol in the presence of n-toluene sulphonic acid[184], or the esterification of dicarboxylic acids with pyrocarbonates at 50–80° without any catalyst[185], yield dimethyl diesters.

Alcoholysis is useful especially for the preparation of high-boiling diesters from low-boiling diesters. Thus, dimethyl 4-methoxyisophthalate gives bis-(2-diethylaminoethyl)4-(2-diethylaminoethoxy)isophthalate by distillation with 2-diethylaminoethanol and toluene after treatment with sodium[186].

Partial or preferential trans-esterification may be carried out when the two ester groups in a diester are unequal. For instance, the trans-esterification of 1-n-carbobutoxy-2-carbobenzyloxybenzene by azeotropic distillation with isodecanol in the presence of sodium methoxide involves the ester group which contains the lower boiling alcohol, and yields 1-carboisodecyloxy-2-carbobenzyloxybenzene[187].

Diethyl 3-ketoadipate gives 1-benzyloxy-2-ethyl 3-ketoadipate on heating with benzyl alcohol at 140°, due to the retarding effect of the keto group on the velocity of the trans-esterification of the adjacent carbethoxy group[188].

C. Salt Alkylation

While ester formation by reaction of silver salts with alkyl halides is rather inconvenient from the preparative point of view, sodium or potassium salts of aliphatic saturated and unsaturated acids, as well as aromatic acids or arylaliphatic acids heated with alkyl halogenides in dimethylformamide[189] or in the presence of pyridine[190], give the corresponding diesters in high yields.

D. Anhydride Hydrolysis and Alcoholysis

The hydrolysis of anhydrides, especially those obtained in diene synthesis, is an important method of formation of dicarboxylic acids.

The opening of the anhydride in unsaturated cyclic compounds by water or in acidic media is usually accomplished without isomerization of the double bond[111]:

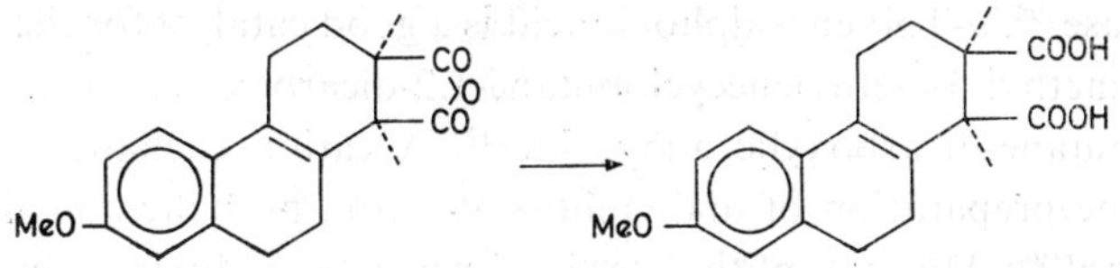

However, the cleavage of the anhydride in basic conditions is often accompanied by isomerization[112]:

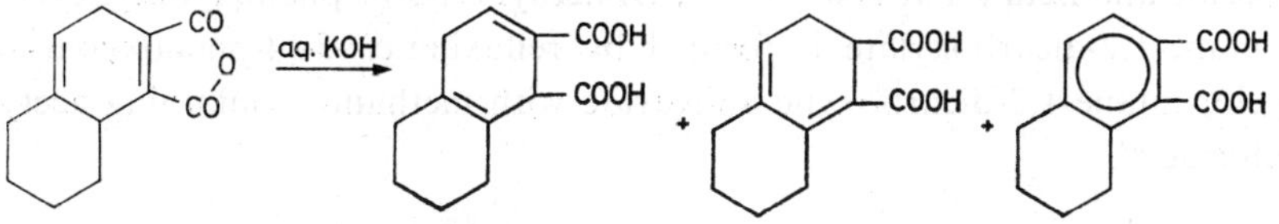

When heated at 200–240°, the anhydrides often undergo *cis–trans* isomerization; this procedure therefore proved to be very useful in the synthesis of geometric isomers of the cyclic 1,2-dicarboxylic acids[191]. Both *cis*- and *trans*-anhydrides as well as anhydrides with ring-substituents were found to undergo thermal reactions to give the more stable isomers[110, 192].

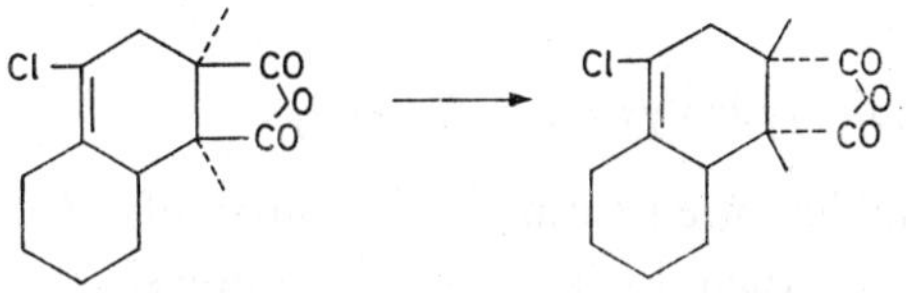

Boiling with 40% aqueous potassium hydroxide was used for the hydrolysis of sterically hindered anhydrides, such as tetraethylsuccinic anhydride[193]. The anhydrides obtained by the reaction between olefins and maleic anhydride are easily opened by heating with potash solution[194].

Some thio anhydrides are hydrolysed with simultaneous desulphurization by refluxing with concentrated acid[195]:

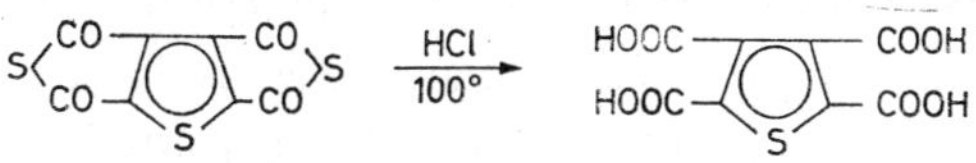

13*

The alcoholysis of anhydrides may be accomplished in the presence or absence of catalysts. For instance, chlorosulphonic acid is an excellent catalyst for the esterification of anhydrides with alcohols, particularly in the gas phase[196]. p-Toluenesulphonic acid is a good catalyst for the preparation of dimethyl 3-methylenecyclobutane-1,2-dicarboxylate from 3-methilenecyclobutane-dicarboxylic anhydride[197]. A cation exchange resin was used for the preparation of diallylphthalate from phthalic anhydride and allyl alcohol[198]. Dibutyl phthalate is formed by refluxing phthalic anhydride with butanol and boric acid[199] or ferric sulphate[200]. Aluminium powder catalysed the formation of dioctyl phthalate from phthalic anhydride and octanol at 180–195°[201]. Dimethyl cis-3-(4-phenoxybutyl)cyclopentane-1,2-dicarboxylate is formed by refluxing cis-3-(4-phenoxybutyl)-cyclopentane-1,2-dicarboxylic anhydride with methanol containing acetyl chloride[202].

E. Hydrolysis and Alcoholysis of Acyl Chlorides

Acyl chlorides react with oxalyl bromide at 100–110° to form malonic acid mixed diacyl halides, the hydrolysis or alcoholysis of which yields substituted malonic acids or esters[203]:

$$RCH_2COCl + (COBr)_2 \longrightarrow RCH(COCl)(COBr) \longrightarrow RCH(COOR')_2$$

Direct esterification of hindered carboxyl groups proceeds slowly and incompletely. Therefore, it was proposed to prepare such derivatives as dimethyl α-ethyl-α-butylglutarate from the readily accessible α-ethyl-α-butyl-α-carboxybutyryl chloride with methanol in the vapour phase at 170–180°[204].

F. Amide and Imide Hydrolysis and Alcoholysis

Amides and imides often form in the course of nitrile hydrolysis. The usual procedure for their hydrolysis or alcoholysis involves refluxing in aqueous or alcoholic media in the presence of bases or acids.

The scope of the method may be illustrated by a few examples: cis-cyclopentane-1,2-dipropionic acid diamide yields cis-cyclopentane-1,2-dipropionic acid by boiling with aqueous potassium hydroxide[204]. β-Methylglutaric acid is obtained by refluxing of β-methylglutaric diamide with concentrated hydrochloric acid[205].

Imides are hydrolysed by boiling with concentrated hydrochloric[206], hydrobromic[207] or sulphuric acid[208]. In some cases, basic hydrolysis is also used, as, for example, in one of the steps in the synthesis of cis-bergamotic acid[209]:

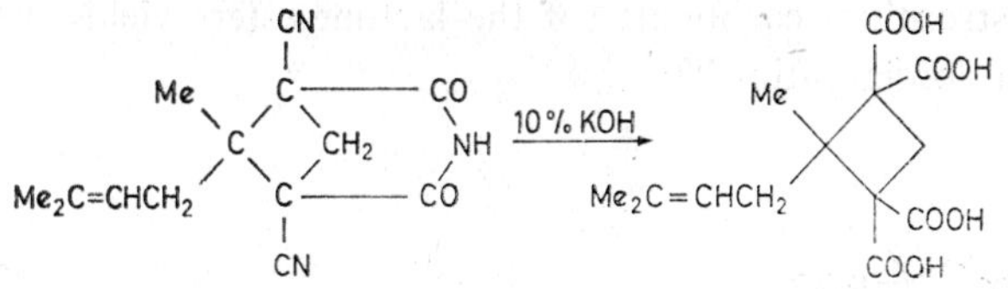

A new method of ester synthesis involves the heating of amides (*e.g.* adipic diamide) at 180–200° with primary alkyl halides in water–alcohol media[210]. The alkaline hydrolysis of accessible aromatic thioimides[211] gives aromatic, heterocyclic or arylaliphatic dicarboxylic acids[212]. In this way, phthalic, homophthalic, thiophene-2,3-dicarboxylic and other acids were obtained.

G. Lactone Hydrolysis and Alcoholysis

Lactone hydrolysis and alcoholysis is of some interest for the preparation of cyclic dicarboxylic acids and esters. For instance, the reduction of cyclic keto lactones with zinc in hydrochloric acid leads to dicarboxylic acids without change of configuration[108]:

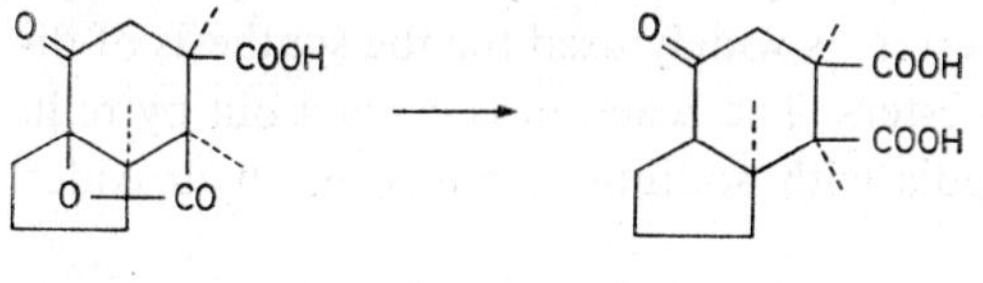

Acidic or alkaline cleavage of lactones may be accompanied by rearrangement of the double bonds in the system[213]:

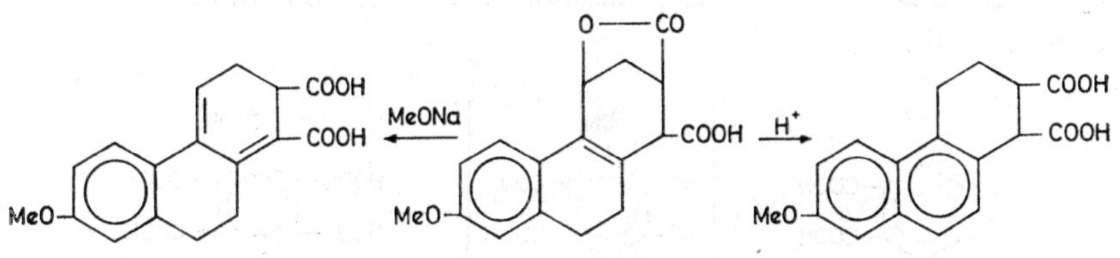

Alkaline hydrolysis of bromo-γ-lactones with *cis*-configuration of all substituents gives keto dicarboxylic acids[214]:

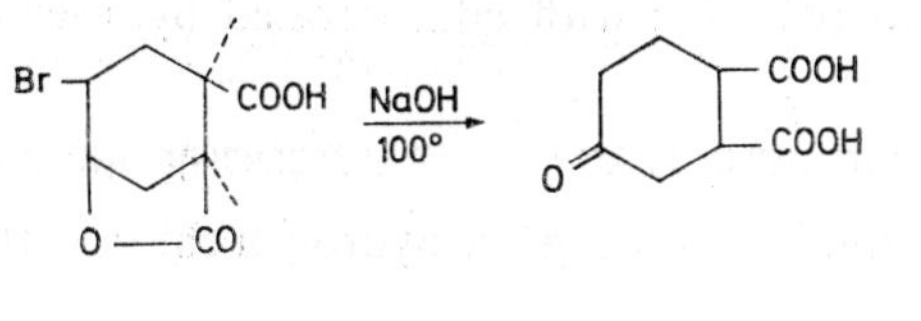

Alkaline hydrolysis in alcohol, or hydrogenolysis in the presence of palladium/strontium carbonate, of the lactone-esters yields diesters with retention of configuration[215]:

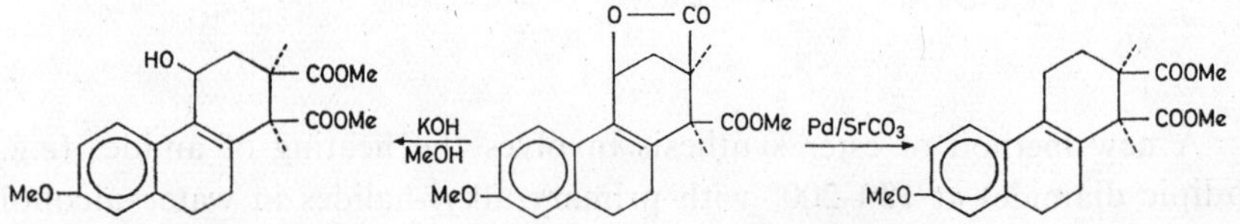

Hydrogenolysis of the di-γ-lactone of diphenacylfumaric acid with platinum oxide in ethyl acetate leads to α,α'-di(β-phenylethyl)succinic acid[216].

In some cases hydrolysis or alcoholysis of lactones may be carried out without catalysts. For instance, the δ-lactone of 5-hydroxy-2-phenyl-β,β-bis-trifluoromethyl-4-oxazolepropionic acid gives N-benzoyl-β,β-bis-trifluoromethyl-D,L-glutamic acid or ester by treatment with water or methanol[217].

H. Nitrile Hydrolysis and Alcoholysis

Hydrolysis or alcoholysis of easily accessible di- and polynitriles, cyano acids or cyano esters is widely used for the synthesis of di- and polycarboxylic acids and esters. The reaction is carried out by refluxing in aqueous or alcoholic media with sodium or potassium hydroxide[218-222], or barium hydroxide[223].

An interesting example is the formation, via tetracyanophthalic acid[224], of mellitic acid from tetraiodophthalic acid by heating with cuprous cyanide, potassium cyanide, potassium hydroxide and water in an autoclave at 180°.

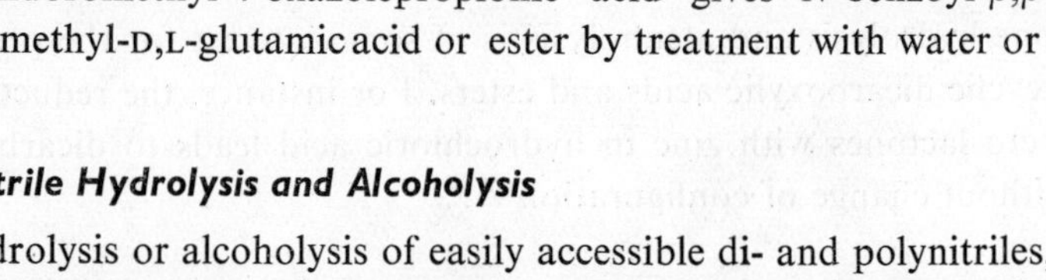

The acidic hydrolysis of dinitriles or cyano acids and esters is carried out by prolonged refluxing with concentrated hydrochloric[225] or hydrobromic acid[226].

$$RR'C(CN)CH(CN)R'' \longrightarrow RR'C(COOH)CH(R'')COOH$$

Diesters are formed by passing dry hydrogen chloride through a solution of a dinitrile in alcohol, followed by refluxing. In this manner diethyl 1,2,5,6-dibenzo-1,3,5,7-cyclooctatetraene-3,8-dicarboxylate was obtained from 3,8-dicyano-1,2,5,6-dibenzo-1,3,5,7-cyclooctatetraene[227].

VIII. MISCELLANEOUS METHODS OF SYNTHESIS

Keto diazoesters give dicarboxylic acids by the action of methanol and silver oxide, according to the Arndt–Eistert method[228]. Examples are the preparation of diethyl cyclopropylmalonic acid from ethyl cyclopropionyldiazoacetic acid[229] or dimethyl homoquinolinic acid from 2-carbomethoxy-3-diazocetopyridine[230].

Diethyl α-fluoro-α'-ketosuccinate yields triethyl $\alpha\alpha'$-difluoro-β-hydroxyglutarate by the action of aqueous potassium acetate[231].

IX. SYNTHESIS OF MONOESTERS OF DI- AND POLYCARBOXYLIC ACIDS

A. Partial Esterification

Partial esterification of dicarboxylic acids is possible if the esterification rate of the two carboxyl groups is different. This may depend on the steric factors, as is the case in the preferential esterification of the primary carboxyl in group trans-β-(2-carboxycyclohexyl)propionic acid by refluxing with ethanol and a little p-toluenesulphonic acid[232]. Electronic factors may also cause differences in the reactivity of the carboxyl groups. These factors (combined with steric effects) lead to partial esterification of α,α-diphenylsuccinic acid to β-methyl ester when dry hydrogen chloride is passed into a refluxing methanolic solution of the acid[233]. The pure electronic factors play a part in the partial esterification of oxalic, fumaric and some other carboxylic acids[2].

B. Partial Hydrolysis

For partial hydrolysis it is necessary that the hydrolysis rates of the two ester groups should be different. This difference may arise on account of steric or electronic factors. Partial hydrolysis is mostly carried out in mild conditions, with an equivalent quantity of alkali.

The partial hydrolysis of 1-acetyl-2,2-dicarbethoxybisnordeoxyeseroline may be explained by steric factors[234]:

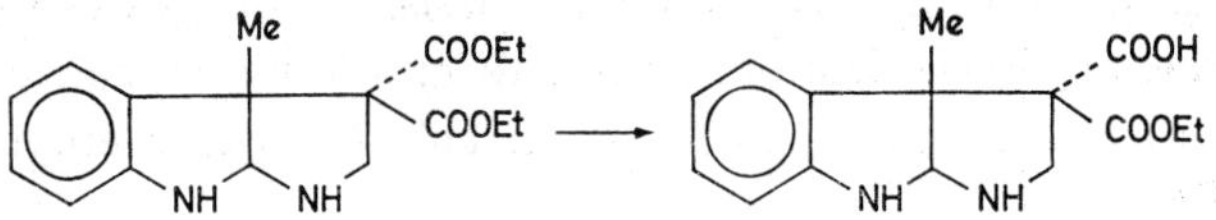

Electronic factors are involved in the partial hydrolysis of dibenzyl N-carbobenzoxy-L-aspartate by a calculated amount of potassium hydroxyde in benzyl alcohol at room temperature with formation of β-benzyl ester[235].

The formation of malonic acid monoesters from malonic acid diesters on mild hydrolysis at room temperature with an equivalent amount of alkali in ethanol[235–239] may be considered a result of electronic effects, as also may the partial hydrolysis of dimethyl 1,10-decanedicarboxylate with a calculated amount of barium hydroxide to monomethyl 1,10-decanedicarboxylate[240–242].

In some cases the difference in the direction of partial hydrolysis may be determined by the change in reaction mechanism under different conditions[243]:

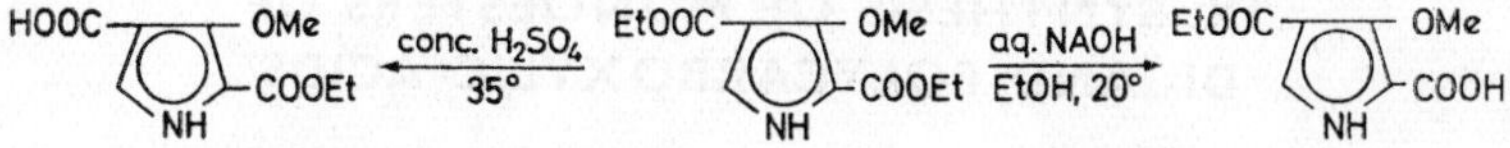

The most convenient method for the formation of half-esters from mixed benzyl alkyl di- or polyesters is hydrogenolysis. Thus, benzyl 5-carbethoxy-4-carbethoxymethyl-2-methylpyrrole-3-carboxylate gives 5-carbethoxy-4-carbethoxymethyl-2-methylpyrrole-3-carboxylic acid when hydrogenated with Raney nickel in absolute ethanol[244].

C. Reaction between Acids and Esters

The reaction between di- or poly-acids and di- or poly-esters is a general method of obtaining partial esters. The reaction is carried out without catalysts[245, 246] or in the presence of acids[247]. For example, if perfluoroadipic acid and ethyl perfluoroadipate are heated to 170° ethyl hydrogen perfluoroadipate is obtained[245]. ω-Carbobutoxylauric acid is formed by boiling 1,2-undecanedicarboxylic acid with its di-n-butylic ester in an alcohol–water medium in the presence of concentrated hydrochloric acid[247].

D. Alcoholysis of Anhydrides and Lactones

The action of alcohols on anhydrides without catalysts[248], or in the presence of sodium methoxide[249, 250], triphenylsodium[251], or pyridine[252, 253] is a general method for partial ester synthesis. In some cases the reaction is complicated by isomerization. Thus a warm solution of maleic anhydride in methanol treated with some sodium chloride yields ethyl hydrogen fumarate[254].

The study of the alcoholysis of cyclic anhydrides was connected with the elucidation of their structures and configurations and with the synthesis of the corresponding dicarboxylic acids, obtained also from steroids. The half-esters of dicarboxylic acids on a sterically hindered carboxyl group are formed by the action of sodium ethoxide on an alcoholic solution of the anhydride[255, 256]:

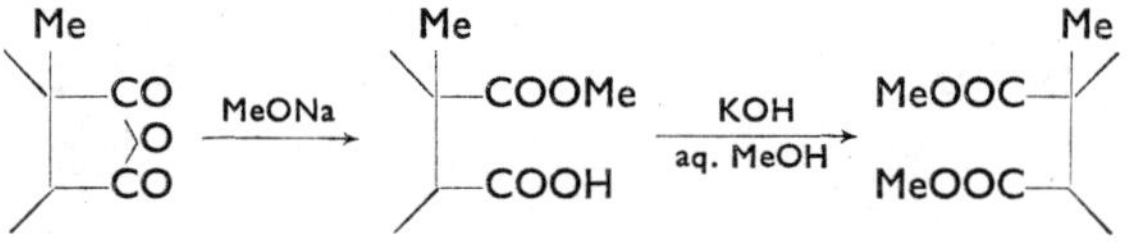

These half-esters are also obtained by partial hydrolysis of dicarboxylic esters with an equivalent quantity of alkali in aqueous methanol[257].

The half-esters on the sterically least hindered group are mainly formed by action of alcohols on anhydrides or by partial esterification of dicarboxylic acids[258]:

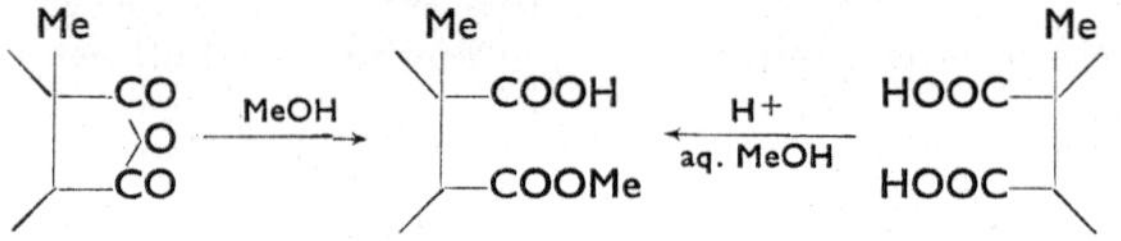

These regularities are often observed, but in some cases mixtures of the various half-esters are formed[111]. In the case of a tricyclic anhydride selective formation of a half-ester is observed only in the reaction with sodium methoxide[259]:

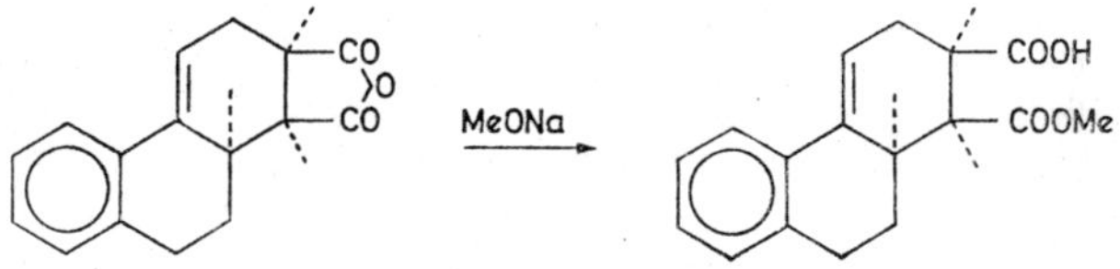

The isomeric half-esters are formed by crystallization of 1,3-dihydro-1-hydroxy-1-methoxy-3-oxofuro-[3,4-d]pyridine from methanol or methylene chloride. In the former case 2-carboxy-3-carbomethoxypyridine was obtained and in the latter case 2-carbomethoxy-3-carboxypyridine[260].

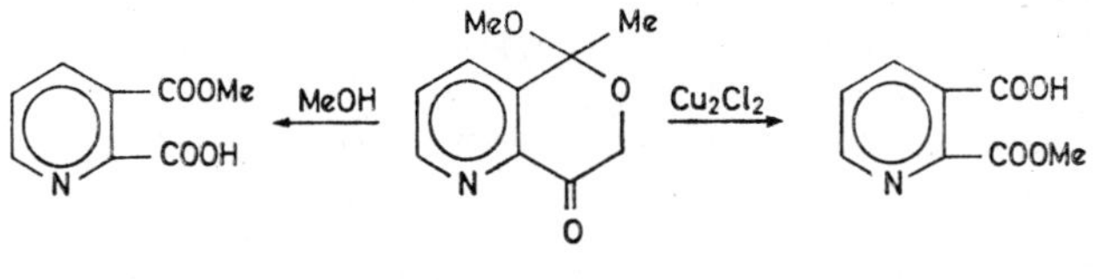

E. Miscellaneous Methods of Synthesis

Monoesters are formed by hydrolysis of anhydrides[261] or lactones[262-265] containing ester groups. In some cases it is possible to introduce a carboxyl group by oxidation[266]:

$$\text{MeOOC(CH}_2)_7\text{CH}-\text{CH(CH}_2)_5\text{CH}_2\text{OH} \xrightarrow{\text{KMnO}_4} \text{MeOOC(CH}_2)_7\text{CH}-\text{CH(CH}_2)_5\text{COOH}$$

$$\underset{\text{CH}_2}{\overset{O\qquad O}{\diagdown\diagup}} \qquad\qquad \underset{\text{CH}_2}{\overset{O\qquad O}{\diagdown\diagup}}$$

Monoesters containing acidic methylene groups may be carboxylated[267, 268]. Monomethyl succinate was formed from 2-formylacrylic acid by refluxing in methanol with potassium cyanide and sodium bicarbonate[269]. Monoesters can be also obtained by condensing malonic acid with aldehydoesters in the presence of basic agents[270]. The coupling of acetylenic esters with bromoacetylenic acids, by the reaction of oxygen in the presence of cuprous chloride is a new promising method for half-ester synthesis[271]:

$$\text{MeOOC(CH}_2)_2\text{C}\equiv\text{CC}\equiv\text{CH} + \text{BrC}\equiv\text{C(CH}_2)_2\text{COOH} \longrightarrow$$

$$\downarrow$$

$$\text{MeOOC(CH}_2)_2(\text{C}\equiv\text{C})_3(\text{CH}_2)_2\text{COOH}$$

X. REFERENCES

1. E. Müller (Ed.), *Methoden der organischen Chemie (Houben-Weil)*, Vol. 8, 4th ed., Georg Thieme Verlag, Stuttgart, 1952.
2. Rodd, *Chemistry of Carbon Compounds*, Vol. 1D, 2nd ed., 1965; Vol. 2B, 1956.
3. C. W. Bird, *Chem. Rev.*, **62**, 283 (1962).
4. Ya. T. Eidus and K. B. Pusitsky, *Usp. Khim.*, **33**, 991 (1964).
5. G. Du Pont, P. Piganiol and J. Vialle, *Bull. Soc. Chim. France*, 529 (1948).
6. W. Reppe, H. Kroper, N. Kutepov, H. J. Pistor and O. Weissbarth, *Ann. Chem.*, **582**, 38, 72, 87 (1953).
7. R. Sakurai, *Japan. Pat.* 73 (1953); *Chem. Abstr.*, **48**, 1065 (1954).
8. E. C. Kirkpatrick, *Brit. Pat.* 599868 (1945); *Chem. Abstr.*, **41**, 1699 (1947).
9. J. R. Roland, J. D. C. Wilson and W. E. Hanford, *J. Am. Chem. Soc.*, **72**, 2122 (1950).
10. K. Bláha, *Preparativni reakce v organické chemii*, VI, Nakladatelstvi československi akademie ved., Praha, 1961.
11. C. T. Ioffe and A. N. Nesmeyanow, *Methods of Organometallic Chemistry, Magnesium, Calcium, Strontium and Barium* (in Russian), Izdatelstvo, Akademii Nauk SSSR, Moscow, 1963.
12. M. S. Kharasch and O. Reinmuth, *Grignard Reactions of Non-metallic Substances*, Prentice-Hall, New York, 1957.
13. K. A. Kocheshkov and T. V. Talalaeva, *Synthetic Methods in the Field of Organometallic Compounds of Lithium, Sodium, Potassium, Rubidium and Cesium* (in Russian), Izdatelstvo Akademii Nauk SSSR, Moscow, 1949.
14. J. T. Millar and H. Heaney, *Quart. Rev.* **11**, 109 (1957).

15. A. G. Brook, H. L. Cohen and C. T. Wright, *J. Org. Chem.*, **18**, 447 (1953).
16. D. Ivanoff and A. Spassof, *Bull. Soc. Chem. France*, **49**, 19 (1931).
17. J. H. Wotiz and H. E. Merril, *J. Am. Chem. Soc.* **80**, 866 (1958).
18. J. T. Nobis and L. F. Moormeier, *Ind. Eng. Chem.*, **46**, 539 (1954).
19. A. S. Lindsey and H. Jeskey, *Chem. Rev.*, **57**, 583 (1957).
20. Y. Ogata, M. Tsuchida and A. Muramoto, *J. Am. Chem. Soc.*, **79**, 6005 (1957).
21. K. Chiba and T. Murakami, *J. Chem. Soc. Japan, Ind. Chem. Sec.*, **69**, 1285 (1966).
22. K. Chiba, *J. Chem. Soc. Japan, Ind. Chem. Sec.*, **69**, 1289, 1294, 1474 (1966).
23. K. Chiba, A. Yokemura and M. Kasai, *J. Chem. Soc. Japan, Ind. Chem. Sec.*, **69**, 1470 (1966).
24. D. D. Coffman, E. L. Jenner and R. D. Lipscomb, *J. Am. Chem. Soc.*, **80**, 2864 (1958).
25. M. S. Kharasch and W. Nudenberg, *J. Org. Chem.*, **19**, 1921 (1954).
26. D. D. Coffman and E. L. Jenner, *J. Am. Chem. Soc.*, **80**, 2872 (1958).
27. D. D. Coffman and H. N. Gripps, *J. Am. Chem. Soc.*, **80**, 2880 (1958).
28. I. L. Knunyantz and N. P. Gambaryan, *Usp. Khim.*, **23**, 781 (1954).
29. I. L. Knunyantz and K. S. Vyazankin, *Dokl. Akad. Nauk SSSR*, **113**, 1899 (1957).
30. U. R. Snyder and B. E. Purham, *J. Am. Chem. Soc.*, **76**, 1899 (1954).
31. L. Crombie, J. E. H. Hancock and R. P. Linstead, *J. Chem. Soc.*, 3496 (1953).
32. R. Ahmad and B. C. L. Weedon, *J. Chem. Soc.*, 3826 (1953).
33. E. R. H. Jones, B. L. Shaw and M. C. Whiting, *J. Chem. Soc.*, 3212 (1954).
34. B. L. Shaw and M. C. Whiting, *J. Chem. Soc.*, 3217 (1954).
35. B. G. Kovalev, L. A. Yanovskaya, V. F. Kucherov and G. A. Kogan, *Izv. Akad. Nauk SSSR, Otd. Khim. Nauk*, 145 (1963).
36. R. H. Martin, *J. Chem. Soc.*, 679 (1941).
37. B. Witten and F. Y. Wiselogle, *J. Org. Chem.*, **6**, 584 (1941).
38. S. B. Baker, T. H. Evans and H. Hibbert, *J. Am. Chem. Soc.*, **70**, 60 (1948).
39. V. Chavane and P. Rumpf, *Compt. Rend*, **225**, 1322 (1947).
40. F. Ramirez, N. B. Desai and N. Ramanathan, *Tetrahedron Letters*, 323 (1963).
41. J. Shabtai and H. Pines, *J. Org. Chem.*, **30**, 3854 (1965).
42. C. E. Frank and W. E. Föster, *J. Org. Chem.*, **26**, 303 (1961).
43. C. E. Frank, J. R. Leebrick, L. F. Moormeier, J. A. Scheben and O. Homberg, *J. Org. Chem.*, **26**, 397 (1961).
44. E. J. Brutschy, *J. Am. Chem. Soc.*, **75**, 2263 (1953).
45. W. Toland, J. B. Wilkins and E. J. Brutschy, *J. Am. Chem. Soc.*, **76**, 307 (1954).
46. E. R. Atkinson, *J. Am. Chem. Soc.*, **67**, 1513 (1945).
47. L. Chardonnens and A. Würmli, *Helv. Chim. Acta*, **29**, 922 (1946).
48. Ch. R. Hauser and B. E. Hudson, *Organic Reactions* (Ed. R. Adams), Vol. 1, John Wiley and Sons, New York, 1942, p. 266.
49. D. Wolf and K. Volkers, *Organic Reactions*, (Ed. R. Adams), Vol. 6, John Wiey and Sons, New York, 1951, p. 410.
50. P. Karrer, R. Keller and E. Ustei, *Helv. Chim. Acta*, **27**, 237 (1944).
51. P. Rüggli and P. Bücher, *Helv. Chim. Acta*, **30**, 2048 (1947).
52. H. T. Openshaw and R. Robinson, *J. Chem. Soc.*, 912 (1946).
53. W. Johnson, R. Christiansen and R. Ireland, *J. Am. Chem. Soc.*, **79**, 1995 (1957).
54. P. Karrer and F. Kehrer, *Helv. Chim. Acta*, **27**, 142 (1944).
55. W. S. Johnson, A. L. McCloskey, D. A. Dunningan, *J. Am. Chem. Soc.*, **72**, 514 (1950).
56. P. B. Talukdar and P. Bagchi, *J. Org. Chem.*, **20**, 21 (1955).
57. A. S. Bailey and J. S. Brunskill, *J. Chem. Soc.*, 2554 (1959).
58. A. H. Cook, I. M. Heilbron and L. Steger, *J. Chem. Soc.*, 413 (1943).
59. F. Šorm and B. Keil, *Collection Czech. Chem. Commun*, **12**, 655 (1947).

60. E. C. Horning and R. E. Field, *J. Am. Chem. Soc.*, **68**, 384 (1946).
61. W. T. Smith and R. W. Shelton, *J. Am. Chem. Soc.*, **76**, 2731 (1954).
62. R. Anliker, A. S. Lindsey, D. E. Nettelton and R. B. Turner, *J. Am. Chem. Soc.*, **79**, 220 (1957).
63. J. S. Moffat, G. Nowbery and W. Webster, *J. Chem. Soc.*, 451 (1946).
64. H. Muhlemann, *Helv. Pharm. Acta*, **24**, 351, 356 (1949).
65. R. Ahmad and B. C. L. Weedon, *J. Chem. Soc.*, 3299 (1959).
66. B. C. L. Weedon, *J. Chem. Soc.*, 4168 (1954).
67. H. Hellmann and R. Lingens, *Angew. Chem.*, **66**, 201 (1954).
68. H. Hegedus, *Helv. Chim. Acta*, **29**, 1499 (1946).
69. E. E. Howe, A. G. Zambito, H. R. Snyder and M. Tishler, *J. Am. Chem. Soc.*, **67**, 38 (1945).
70. S. B. Soloway and F. B. La Forge, *J. Am. Chem. Soc.*, **69**, 2677 (1947).
71. J. Schreiber, *Compt. Rend.*, **232**, 980 (1951).
72. R. E. Miller and F. F. Nord, *J. Org. Chem.*, **16**, 1380 (1951).
73. R. Kuhn and Ch. Grundmann, *Ber.*, **69**, 1757, 1978 (1936); **70**, 1318 (1937).
74. S. O. Lawesson, M. Dahlen and C. Frisell, *Acta Chem. Scand.*, **16**, 1191 (1962).
75. K. Sisido, Y. Kazawa, H. Kodama and H. Nozaki, *J. Am. Chem. Soc.*, **81**, 5817 (1959).
76. W. Green and F. B. La Forge, *J. Am. Chem. Soc.*, **70**, 2287 (1948).
77. N. A. Babiyan, G. A. Mednikyan, A. A. Gamburyan, J. A. Shakaryan and O. L. Mndjoyan, *Army. Khim. Zh.*, **19**, 434 (1966).
78. J. Cason, G. Sumrell and R. S. Mitchell, *J. Org. Chem.*, **15**, 850 (1950).
79. E. Hardegger, P. A. Plattner and F. Blank, *Helv. Chim. Acta*, **27**, 793 (1944).
80. E. Buchta and W. Merk., *Ann. Chem.*, **694**, 1 (1966).
81. B. C. L. Weedon, *Quart. Rev.*, **6**, 380 (1952).
82. B. C. L. Weedon in *Advances in Organic Chemistry* (Ed. R. A. Raphael, E. C. Taylor, H. Wynberg), Vol. 1, Interscience Publishers, New York, 1960, p. 1.
83. F. Beck, *Chem. Ing. Tech.*, **37**, 607 (1965).
84. R. Lindsey and M. Peterson, *J. Am. Chem. Soc.*, **81**, 2073 (1959).
85. M. Ya. Fioshin, A. I. Kameneva, L. A. Mirkind, L. A. Calmin and A. G. Kornienko, *Neftekhimiya*, **2**, 557 (1962).
86. M. Ya. Fioshin, L. A. Mirkind, L. A. Calmin, A. G. Kornienko, Zh. Vses. Khim. Obshch. (Ed. D. J. Mendelleva) **10**, 238, 594 (1965).
87. N. I. Sheverdina and K. A. Kocheshkov, *Methods of Elementoorganic Chemistry*, Zinc, Cadmium (in Russian) ed., Izdatelstvo 'Nauka', Moscow, 1965.
88. R. Schreiner, in *Organic Reactions* (Ed. R. Adams), Vol. 1, John Wiley and Sons, New York, 1942 p. 1.
89. H. H. Inhoffen and G. Raspe, *Ann. Chem.*, **592**, 214 (1955).
90. H. H. Inhoffen, O. Isler, G. van der Bey, G. Raspe, P. Zeller and R. Ahrens, *Ann. Chem.*, **580**, 7 (1953).
91. M. S. Neumann and G. P. Kahle, *J. Org. Chem.*, **23**, 669 (1958).
92. E. V. Shevlyagina, E. I. Gushtina and V. N. Belov, *Tr. Vses. Inst. Synthetitcheskykh i Naturalnykh Dushistykh Vestchestv* (USSR), **44**, 4 (1958).
93. A. Siegel and K. H. Bröll-Keckeis, *Monatsh. Chem.*, **88**, 910 (1957).
94. H. W. Bost and P. S. Bailey, *J. Org. Chem.*, **21**, 803 (1956).
95. S. Trippett, in *Advances in Organic Chemistry* (Ed. R. A. Raphael, E. C. Taylor, H. Wynberg), Vol. 1, Interscience Publishers, New York, 1960, p. 83.
96. A. Maerker, in *Organic Reactions* (Ed. R. Adams), Vol. 14, John Wiley and Sons, New York, 1965, p. 270.
97. O. Isler, H. Gutmann, M. Montavon, R. Rüegg, G. Ryser and P. Zeller, *Helv. Chim. Acta*, **40**, 1242 (1957).
98. E. Buchta and F. Andree, *Chem. Ber.*, **92**, 3111 (1959); **93**, 1349 (1960).

99. F. Hoffmann–La Roche & Co., *Brit. Pat.* 838,926 (1960); *Chem. Abstr.*, **54**, 24408 (1960).
100. L. A. Yanovskaya and V. E. Kucherov, *Izv. Akad. Nauk SSSR, Ser. Khim.* 1341 (1964).
101. L. A. Yanovskaya, R. N. Stepanova, G. A. Kogan and V. F. Kucherov, *Izv. Akad. Nauk SSSR, Ser. Khim.*, 857 (1963).
102. V. F. Kutcherov, B. G. Kovalev, G. A. Kogan and L. A. Yanovskaya, *Dokl. Akad. Nauk SSSR*, **138**, 1115 (1961).
103. M. P. Cava and R. J. Pohl, *J. Am. Chem. Soc.*, **82**, 5242 (1960).
104. E. Truscheit and K. Eiter, *Ann. Chem.*, **658**, 65 (1962).
105. A. S. Onishchenko, *Diene Synthesis* (in Russian) (Ed. V. F. Kucherov), Izdatelstvo Akademii Nauk SSSR, Moscow, 1963; Diene Synthesis (English Translation), Israel Programme for Scientific Translations, Jerusalem, 1964.
106. V. F. Kutcherov, I. M. Milshtein and I. A. Gurvitsh, *Zh. Obshch. Khim.*, **31**, 2832 (1961).
107. I. N. Nazarov, V. F. Kucherov, V. I. Andreev, G. M. Segal *Croat. Chim. Acta*, **29**, 360 (1957).
108. V. F. Kucherov, E. P. Serebryakov and A. V. Usova, *Izv. Akad. Nauk SSSR, Otd. Khim. Nauk*, 106 (1962).
109. V. F. Kucherov, V. M. Andreev, L. K. Lysantchuk, *Izv. Akad. Nauk SSSR, Otd. Khim. Nauk*, 1797 (1960).
110. V. F. Kucherov, V. M. Andreev and N. Y. Grigorieva, *Bull. Soc. Chim. France*, 1406 (1960).
111. V. F. Kucherov and E. P. Serebryakov, *Zh. Obshch. Khim.*, **32**, 426 (1962).
112. V. F. Kucherov and N. Y. Grigorieva, *Zh. Obshch. Khim.*, **31**, 447, 457, 2894 (1961).
113. I. A. Gurvitch and V. F. Kucherov, *Izv. Akad. Nauk SSSR, Ser. Khim. Nauk*, 1456, 1507 (1964).
114. V. M. Andreev, S. A. Kasaryan and V. F. Kucherov, *Izv. Akad. Nauk SSSR, Otd. Khim. Nauk*, 1996, 2003 (1963).
115. E. Wenkert, D. B. R. Johnston and K. G. Dave, *J. Org. Chem.*, **29**, 2534 (1964).
116. G. A. Olah (Ed.), *Friedel–Crafts and Related Reactions*, Vol. 1–4, Interscience Publishers, New York and London, 1963–1965.
117. Y. Li. Goldfarb and M. L. Kirmalova, *Izv. Akad. Nauk SSSR, Otd. Khim. Nauk*, 479 (1954).
118. A. Treibs, R. Schmidt and R. Zinsmeister, *Chem. Ber.*, **90**, 79 (1957).
119. E. M. Bottorff, R. G. Jones, E. C. Kornfeld and M. J. Mann, *J. Am. Chem. Soc.*, **73**, 4380 (1951).
120. A. G. Jones, *J. Am. Chem. Soc.*, **73**, 5610 (1951).
121. H. R. Snyder and H. E. Freier, *J. Am. Chem. Soc.*, **68**, 1320 (1946).
122. E. Bergmann, D. Ginsburg and R. Pappo in *Organic Reactions* (Ed. R. Adams), Vol. 10, John Wiley and Sons, New York, 1959, p. 149.
123. K. Alder, H. Wirtz and H. Hoppelberg, *Ann. Chem.*, **601**, 138 (1951).
124. D. E. Floyd and S. E. Miller, *J. Org. Chem.*, **16**, 882 (1951).
125. S. M. Mukherjee, *J. Ind. Chem. Soc.*, **25**, 155 (1948).
126. E. E. van Tamelen and J. S. Daran, *J. Am. Chem. Soc.*, **80**, 4659 (1958).
127. C. R. Hauser and M. T. Tetenbaum, *J. Org. Chem.*, **23**, 1146 (1958).
128. E. B. Reid and M. Sack, *J. Am. Chem. Soc.*, **73**, 1985 (1951).
129. L. L. McCoy, *J. Am. Chem. Soc.*, **80**, 6568 (1953).
130. J. Di Pietro and W. J. Roberts, *Angew. Chem.*, **78**, 388 (1966).
131. P. Mildner and B. C. L. Weedon, *J. Chem. Soc.*, 3294 (1953).
132. G. A. Lapitsky, S. M. Makin and G. M. Dymshakova, *Zh. Obshch. Khim.*, **34**, 2564 (1964).

133. A. C. Cope and H. E. Johnson, *J. Am. Chem. Soc.*, **80**, 1504 (1958).
134. E. Sherman and A. Dunlop, *J. Org. Chem.*, **25**, 1309 (1960).
135. M. Chaigneau, *Compt. Rend.*, **233**, 657 (1951).
136. W. C. Toland, *J. Am. Chem. Soc.*, **82**, 1911 (1960).
137. D. Todd and A. E. Martell., *Org. Syn.*, **40**, 48 (1960).
138. T. Kato, *Bull. Chem. Soc. Japan*, **34**, 636 (1961).
139. J. E. McIntre and D. A. S. Ravens, *J. Chem. Soc.*, 4082 (1961).
140. R. K. Summerbell and G. J. Lestina, *J. Am. Chem. Soc.*, **79**, 3878 (1957).
141. H. Wilms, *Ann. Chem.*, **567**, 96 (1950).
142. J. English and J. E. Dayan, *Org. Syn.*, **30**, 48 (1950).
143. S. F. Birch, W. J. Oldham and E. A. Johnson, *J. Chem. Soc.*, 818 (1947).
144. P. A. Plattner and A. Fürst, *Helv. Chim. Acta*, **28**, 173 (1945).
145. M. J. Fermery and E. K. Fields, *J. Org. Chem.*, **28**, 2537 (1963).
146. A. Waček and R. Fiedler, *Monatsh. Chem.*, **80**, 170 (1949).
147. F. Greenspan, *Ind. Eng. Chem.*, **39**, 847 (1947).
148. W. T. Smith and G. L. McLeod, *Org. Syn.*, **31**, 40 (1951).
149. J. Cason, K. W. Kraus and W. D. McLeod, *J. Org. Chem.*, **24**, 392 (1959).
150. N. G. Brink and E. S. Wallis, *J. Biol. Chem.*, **162**, 667 (1946).
151. J. Heer and K. Miescher, *Helv. Chim. Acta*, **28**, 156 (1945).
152. J. Eicheberger, *Helv. Chim. Acta*, **31**, 1663 (1948).
153. H. Fernholz, E. Hartwig and J. C. Salfeld, *Ann. Chem.*, **576**, 131 (1952).
154. V. Sykora, *Collection Czech. Chem. Commun.*, **23**, 2207 (1958).
155. S. I. Zavialov, L. P. Vinogradova and G. V. Kondratieva, *Tetrahedron*, **20**, 2745 (1964).
156. A. N. Nesmeyanow, R. Kh. Freidlina and L. U. Zakharkin, *Quart. Rev.*, **10**, 330 (1956).
157. A. N. Nesmeyanow, A. A. Strepichejew, R. Kh. Freidlina, L. I. Sacharkin, E. I. Wasiljewa, G. B. Owakimjan, A. A. Beer, R. G. Petrowa, S. A. Karapetjan, V. M. Toptschibaschewa, T. I. Schein and M. A. Besproswanni, *Chem. Tech.*, **9**, 139 (1957).
158. M. E. Hill, *J. Org. Chem.*, **25**, 1115 (1960).
159. H. G. Viehe, *Chem. Ber.*, **92**, 1270 (1959).
160. J. Kennedy, N. Y. McCorkindal, R. A. Raphael, W. T. Scott and W. Zwanenburg, *Proc. Chem. Soc.*, 148 (1964).
161. H. Stetter and W. Dilrichs, *Chem. Ber.*, **85**, 290 (1952).
162. H. Stetter and M. Coehen, *Chem. Ber.*, **87**, 869 (1954).
163. H. Lettre and A. Jahn, *Chem. Ber.*, **85**, 346 (1952).
164. V. Prelog and S. Szpilfogel, *Helv. Chim. Acta*, **28**, 173 (1945).
165. V. Prelog and U. Geyer, *Helv. Chim. Acta*, **28**, 576 (1945).
166. R. Lukeš and J. Poleček, *Collection Czech. Chem. Commun.*, **20**, 1253 (1955).
167. A. Horeau and J. Jacques, *Bull. Soc. Chim. France*, 1001 (1945).
168. H. Gault and L. Daltriff, *Compt. Rend.*, **209**, 997 (1939).
169. L. Daltroff, *Ann. Chim. (Paris)*, **14**, 207 (1940).
170. J. C. Barborak, L. Watts and R. Pettit, *J. Am. Chem. Soc.*, **88**, 1328 (1966).
171. W. Ziegenbein and W. Lang, *Chem. Ber.*, **93**, 2743 (1960).
172. H. Stetter, H. Kessler and H. Meisel, *Chem. Ber.*, **87**, 1617 (1954).
173. V. Prelog, P. Barman and M. Zimmerman, *Helv. Chim. Acta*, **32**, 1284 (1949).
174. J. E. Tinker, *J. Am. Chem. Soc.*, **73**, 4493 (1951).
175. W. Cooper and W. T. H. Davies, *J. Chem. Soc.*, 1180 (1952).
176. L. Crombie, J. E. H. Hancock and R. P. Linstead, *J. Chem. Soc.*, 3496 (1953).
177. C. M. Sharts and A. H. McLeod, *J. Org. Chem.*, **30**, 3308 (1965).
178. J. Koo, S. Avakian and G. J. Martin, *J. Org. Chem.*, **24**, 179 (1959).
179. W. T. Smoth and P. G. Kort, *J. Am. Chem. Soc.*, **72**, 1877 (1950).

180. R. Grewe and W. von Bonin, *Chem. Ber.*, **94**, 234 (1961).
181. R. Lukeš and J. Paleček, *Collection Czech. Chem. Commun.*, **29**, 1073 (1964).
182. J. C. Sheehan and A. K. Bose, *J. Am. Chem. Soc.*, **72**, 5158 (1950).
183. C. Heidelberger and R. B. Hurlber, *J. Am. Chem. Soc.*, **72**, 4704 (1950).
184. N. B. Lorette and J. H. Brown, *J. Org. Chem.*, **24**, 261 (1959).
185. W. Thoma and H. Rinks, *Ann. Chem.*, **624**, 30 (1959).
186. J. M. Z. Gladych and E. P. Taylor, *J. Chem. Soc.*, 2720 (1960).
187. L. O. Raether and U. R. Gamrath, *J. Org. Chem.*, **24**, 1997 (1949).
188. W. G. Laver, A. Neuberger and J. J. Scott, *J. Chem. Soc.*, 1474 (1959).
189. S. Yoneda, Z. Yoshda and K. Fukui, *J. Chem. Soc. Japan, Ind. Chem. Sec.*, **69**, 641 (1966).
190. V. V. Dovlatyan, T. O. Chakryan, *Izv. Akad. Nauk Arm. SSR, Ser. Khim. Nauk*, **17**, 651 (1964).
191. V. F. Kucherov, N. Y. Grigorieva, I. N. Nazarov, *Zh. Obshch. Khim.*, **29**, 792, 848 (1959).
192. G. M. Segal, L. P. Rybkina, V. F. Kucherov, *Izv. Akad. Nauk SSSR, Otd. Khim. Nauk*, 1424 (1962).
193. L. Eberson, *Acta Chem. Scand.*, **13**, 40 (1959).
194. C. Aganti, M. Andrac-Taussig, C. Justin and Ch. Prevost, *Bull. Soc. Chim. France*, 1195 (1966).
195. O. Scherer and F. Klüge, *Chem. Ber.*, **99**, 1973 (1966).
196. J. Erdos, *J. Am. Chem. Soc.*, **73**, 4976 (1951).
197. H. B. Stevenson, H. N. Cripps and J. K. Williams, *Org. Syn.*, **43**, 27 (1963).
198. J. Lindeman, W. Trochimczuk, *Chem. Techn. (Berlin)*, **11**, 32 (1959).
199. V. K. Kuskov and V. A. Zhukova, *Izv. Akad. Nauk SSSR, Otd. Khim. Nauk*, 733 (1956).
200. S. Niraz, *Roczniki Chem.*, **31**, 1047 (1957).
201. T. P. Li, *J. Chinese Chem. Soc.*, Ser. II, **1**, 147 (1954); *Chem. Abstr.*, **50**, 2490 (1956).
202. M. N. Donin, S. L. Burson, J. H. Müller, C. Chen, W. E. Behnke and K. Hofmann, *J. Am. Chem. Soc.*, **73**, 4286 (1951).
203. W. Treibs and H. Ortmann, *Chem. Ber.*, **96**, 297 (1958).
204. J. Cason and K. W. Kraus, *J. Org. Chem.*, **26**, 2624 (1961).
205. J. M. Osboud, J. D. Fulton and D. F. Spooner, *J. Chem. Soc.*, 4785 (1952).
206. P. Linstead and M. Whaley, *J. Chem. Soc.*, 3722 (1954).
207. H. O. House, V. Paragamian and D. J. Wluka, *J. Am. Chem. Soc.*, **83**, 2714 (1961).
208. H. K. Farmer and W. Rabjohn, *Org. Syn.*, **36**, 28 (1956).
209. C. S. Narayanan, S. S. Welankiwar, S. W. Kulkarni and S. S. Bhattacharyya, *Tetrahedron Letters*, 985 (1965).
210. L. V. Kaabak, A. P. Tomilov, S. A. Varshavski, *Problemy Organicheskogo Synteza*, Izdatelstvo Akad. Nauk. SSSR, 1965, p. 32.
211. P. A. S. Smith and R. O. Kan, *J. Am. Chem. Soc.*, **82**, 4753. (1960).
212. P. A. S. Smith and R. O. Kan, *Org. Syn.*, **44**, 62, 91 (1964).
213. V. M. Andreev, L. K. Lysanchuk and V. F. Kucherov, *Izv. Akad. Nauk SSSR, Otd. Khim. Nauk*, 96 (1962).
214. V. F. Kucherov, A. L. Shabanov and A. S. Onitschenko, *Izv. Akad. Nauk SSSR Otd. Khim. Nauk*, 844 (1963).
215. V. F. Kucherov, L. K. Lysantchuk and V. M. Andreev, *Izv. Akad. Nauk SSSR, Otd. Khim. Nauk*, 96 (1962).
216. Chen Shang Fang and W. Bergmann, *J. Org. Chem.*, **16**, 1231 (1951).
217. E. M. Rochlin, N. P. Gambaryan and I. L. Knunyantz, *Izv. Akad. Nauk SSSR, Otd. Khim. Nauk*, 1952 (1963).

218. E. E. van Tamelen, T. A. Spencer, D. S. Allen and R. L. Ovris, *Tetrahedron*, **14**, 8 (1961).
219. M. S. Newman and D. K. Phillips, *J. Am. Chem. Soc.*, **81**, 3667 (1959).
220. L. Pichat, C. Baret and M. Audinot, *Bull. Soc. Chim. France*, 88 (1954).
221. A. Morrison and T. C. P. Mulholland, *J. Chem. Soc.*, 2537 (1958).
222. B. R. Baker, M. V. Querry and A. F. Kadish, *J. Org. Chem.*, **13**, 123 (1948).
223. G. B. Brown, *Org. Syn.*, **26**, 54 (1946).
224. H. Brusset and A. Uny, *Bull. Soc. Chim. France*, 565 (1951).
225. K. Sen, P. Bagchi, *J. Org. Chem.*, **20**, 854 (1955).
226. R. A. Barnes and R. Muller, *J. Am. Chem. Soc.*, **82**, 4960 (1960).
227. A. C. Cope and S. W. Fenton, *J. Am. Chem. Soc.*, **73**, 1668 (1951).
228. W. Bachman, W. Struve, in *Organic Reactions* (Ed. R. Adams), Vol. 1, John Wiley and Sons, New York, 1942, p. 38.
229. L. J. Smith and S. McKenzie, *J. Org. Chem.*, **15**, 74 (1950).
230. K. Mischer and H. Kagi, *Helv. Chim. Acta*, **24**, 1471 (1941).
231. D. Rouge and H. Gault, *Compt. Rend.*, **251**, 94 (1960).
232. L. A. Paquette and N. A. Nelson, *J. Org. Chem.*, **27**, 2272 (1962).
233. B. H. Chase and D. H. Hey, *J. Chem. Soc.*, 553 (1952).
234. B. Witkop and R. H. Hill, *J. Am. Chem. Soc.*, **77**, 6592 (1955).
235. M. Frankel and A. Berger, *J. Org. Chem.*, **16**, 1513 (1951).
236. H. Hellmann, K. Teichmann and E. Lingens, *Chem. Ber.*, **91**, 2487 (1958).
237. G. Stork and G. Singh, *J. Am. Chem. Soc.*, **73**, 4742 (1951).
238. C. Uhle, *J. Am. Chem. Soc.*, **71**, 761 (1949).
239. S. Widequist, *Arkiv. Kemi*, **26A**, 16 (1948).
240. R. Singer and P. Sprecher, *Helv. Chim. Acta*, **30**, 1001 (1947).
241. J. Cason, P. B. Taylor and D. E. Williams, *J. Am. Chem. Soc.*, **73**, 1187 (1951).
242. J. A. Elvidge, R. P. Linstead and P. Sims, *J. Chem. Soc.*, 3386 (1951).
243. H. Rapoport and K. G. Holden, *J. Am. Chem. Soc.*, **84**, 635 (1962).
244. S. F. MacDonald, *J. Chem. Soc.*, 4176 (1952).
245. W. H. Rauscher and H. Tucker, *J. Am. Chem. Soc.*, **76**, 3599 (1954).
246. F. L. M. Pattison, W. C. Howell, A. J. McNamara, J. C. Schneider and J. F. Walker, *J. Org. Chem.*, **21**, 739 (1956).
247. R. G. Jones, *J. Am. Chem. Soc.*, **69**, 2350 (1947).
248. J. Cason, *Org. Syn.*, **25**, 19 (1945).
249. V. du Vigneaud and G. L. Miller, *Biochem. Prep.* **2**, 79 (1952).
250. N. N. Saha, B. K. Gamguly and P. C. Dutta, *J. Am. Chem. Soc.*, **81**, 3670 (1959).
251. J. M. Prokipcak and D. C. Fung, *J. Org. Chem.*, **28**, 582 (1963).
252. J. Kenyon and R. Poplett, *J. Chem. Soc.*, 273 (1945).
253. J. Mathieu, *Ann. Chim. (Paris)*, **20**, 215 (1945).
254. U. Eisner, J. A. Elvidge and R. P. Linstead, *J. Chem. Soc.*, 1501 (1951).
255. J. Heer and K. Miescher, *Helv. Chim. Acta*, **31**, 229 (1948).
256. W. Bachmann and J. Confronlis, *J. Am. Chem. Soc.*, **73**, 2636 (1951).
257. W. Bachmann and S. Kushner, *J. Am. Chem. Soc.*, **65**, 1905 (1943).
258. W. Bachmann and W. Struve, *J. Am. Chem. Soc.*, **63**, 1262 (1941).
259. V. M. Andreev, L. K. Lysanchuk, V. F. Kucherov, *Izv. Akad. Nauk SSSR, Otd. Khim. Nauk*, 1804 (1960).
260. J. L. Norula, *J. Proc. Inst. Chemists (India)*, **35**, 51 (1963).
261. E. Wenkert, D. B. R. Johnston and K. G. Dave, *J. Org. Chem.* **29**, 2534 (1964).
262. W. E. Bachmann and G. D. Johnson, *J. Am. Chem. Soc.*, **71**, 3463 (1949).
263. D. Kostermans, *Rec. Trav. Chim.*, **70**, 79 (1951).
264. G. R. Clemo, *J. Chem. Soc.*, 263 (1951).
265. J. A. Elvidge, R. P. Linstead and P. Sims, *J. Chem. Soc.*, 3386 (1951).
266. S. D. Sabnis, H. H. Mathur and S. C. Bhattacharyya, *J. Chem. Soc.*, 2477 (1963).

267. C. R. Hauser, R. Levine and R. F. Kibler, *J. Am. Chem. Soc.*, **68**, 26 (1946).
268. R. A. Raphael, *J. Chem. Soc.*, 805 (1947).
269. V. Franzen and L. Fikentserer, *Ann. Chem.*, **623**, 60 (1959).
270. I. N. Nazarov, V. M. Andreev and I. V. Torgov, *Zh. Obshch. Khim.*, **29**, 775 (1959).
271. A. P. Derzjinski, M. V. Mavrov and V. Г. Kucherov, *Izv. Akad. Nauk SSSR, Ser. Khim. Nauk*, 544 (1965).

CHAPTER **6**

Acidity and hydrogen bonding of carboxyl groups

LENNART EBERSON

University of Lund, Sweden

I. INTRODUCTION

One of the most important aims of organic chemistry is to relate energy differences between molecules with changes in molecular structure in theoretical terms. Besides thermochemical, kinetic, and spectral data, equilibrium
constants of acid–base equilibria constitute the main body of experimental
quantities suitable for such correlations. One obvious reason for this is the
relative ease and speed with which reasonably accurate ionization constant
measurements can be carried out for large series of compounds without
resort to expensive and complicated equipment. The direct connection between the logarithm of the ionization constant and the free energy change of
the process then provides a set of data which can be used for testing theories of substituent effects.

If the experimental part of the problem is fairly simple, the theoretical
treatment of equilibrium data is the more difficult. A qualitative discussion
of the relative magnitudes of a series of constants can use any combination
of polar, resonance, steric, and solvation effects, working parallel or opposite to each other, and will always furnish an 'explanation' for the observed regularities or anomalies. Quantitative treatments, having available
something around a dozen substituent parameter sets and a corresponding
number of two- or multiparameter equations, are not much better off in
this respect. If we then add some other important factors, such as the influence of solvent composition, temperature, intramolecular hydrogen bonding, and conformational equilibrium, and also some trivial ones, such as
contributions to the entropy of ionization due to differences in entropy of
mixing and/or symmetry number between acid and ion, it is obvious that
the organic chemist faces a formidable problem when he tries to interpret,
let alone predict, equilibrium data on the basis of such a number of more
or less adjustable parameters.

The indisputable success of some quantitative treatments is, however, an
indication that differences in the free energy of ionization between different
acids in aqueous solution around room temperature actually parallel differences in potential energy, *i.e.* those energy differences between molecules
which are due to electronic effects. Furthermore, the general trends for the
effect of structural change on molecular properties obtained by the most
diverse experimental techniques are closely parallel to those obtained from
studies of acid–base equilibria, enabling us to use our theories with some
confidence. It must nevertheless be stressed that utmost care must be exer-
cized in attempts to explain small energy differences. Unfortunately, small
energy differences between molecules are more common than not, and it is

not surprising that a number of apparently conflicting interpretations can exist side by side within the theory of organic chemistry. Our theoretical tools are simply too crude to handle the complex situation in any chemical system of moderate size.

It is the purpose of this chapter to examine the ionization equilibrium of carboxylic acids with respect to the above mentioned factors, especially those related to inter- and intramolecular hydrogen bond formation. There will be no attempt to review the experimental methods for the determination of ionization constants, since excellent modern treatises on this topic are available[1-4]. For a more general discussion of all types of acid–base equilibria the reader is referred to recent monographs by King[1] and Bell[5], and the classical review article by Brown, McDaniel, and Häfliger[6].

II. DEFINITIONS AND STANDARDIZATION PROCEDURES

A. Definition of Ionization Constants

The ionization scheme for a monocarboxylic acid in aqueous solution is generally written as in equation (1) where K_a denotes the thermodynamic ionization constant, defined by equation (2).

$$RCOOH + H_2O \overset{K_a}{\rightleftharpoons} RCOO^- + H_3O^+ \tag{1}$$

$$K_a = \frac{a_{RCOO^-} a_{H^+}}{a_{RCOOH}} = \frac{c_{RCOO^-} c_{H^+}}{c_{RCOOH}} \cdot \frac{f_{RCOO^-} f_{H^+}}{f_{RCOOH}} = K_c \frac{f_{RCOO^-} f_{H^+}}{f_{RCOOH}} \tag{2}$$

Here a, c, and f represent activities, concentrations, and activity coefficients respectively, while K_c is the classical dissociation constant, expressed in terms of concentrations. The water activity or concentration is considered to be constant and as usual set equal to the activity or concentration of pure water and incorporated into K_a or K_c. Solvation of acid and ions is not explicitly accounted for in equation (1) but it is obvious that the measured ionization constant is actually referring to a number of different solvated species. This problem has been discussed in detail by Ives and Marsden[7].

In practice it is however customary to measure and report a 'mixed' ionization constant, K_m, defined as in equation (3) since the majority of

$$K_m = \frac{c_{RCOO^-}}{c_{RCOOH}} a_{H^+} \tag{3}$$

measurements have been made by potentiometric titration using a glass electrode, which measures a_{H^+}. This means that f_{RCOO^-} and f_{RCOOH} are set equal to unity, which in most cases is a valid approximation at the low concentrations used. For doubly or higher charged anions the omittance of activity coefficients is more serious, but as long as comparisons are made between constants measured by a single author using a highly standardized method, the use of K_m instead of K_a will not affect relative acidities or discussions based on these. Since it is common practice to measure ionization constants for a series of compounds, including one or several reference acids of known acidity, it is usually also possible to compare results obtained by different authors, although it is necessary to make certain that differences in experimental technique and calculation method do not invalidate the comparison. In later sections it will frequently be necessary to compare ionization constants from different sources, especially in discussions of the influence of structure on acidity. Ionization constants K_a, K_c, and K_m will then be treated as being equal and denoted by K, if it is not obvious that the small errors introduced by this convention will affect the discussion. In other cases, the use of K_a, K_c, and K_m will be mentioned explicitly. Throughout this chapter the more convenient symbol $pK = -\log K$ will be used instead of K in discussions of acidities, because pK is directly related to the free energy of ionization through equation (4) where zero as a superscript

$$\Delta G^0 = 2 \cdot 3RT\, pK \qquad (4)$$

indicates that ΔG is referred to a hypothetical 1 M ideal solution as standard state.

B. Comparison with Other Ionization Processes

Ionization constants of carboxylic acids range over more than ten powers of ten in aqueous solution, from the completely ionized trifluoroacetic acid to the very weak second ionization step of rac-2,3-di-(t-butyl)-succinic acid with $pK = 10 \cdot 3$. For comparison with some other ionization processes representative pK values for COOH, NH_3^+, OH, and SH ionizations have been assembled in Table 1, in which the acids are arranged pairwise in the order of functional groups given above. For each type, the commonly used reference processes for ionization of aliphatic and aromatic acids are given. It can be noted that only carboxylic acids have ionization constants conveniently accessible by simple pH-meter technique both in the aliphatic and aromatic series. Qualitatively, all four types are influenced in the same way by structural changes.

A large number of ionization constants in aqueous solution have been tabulated by Perrin[8] and Kortüm, Vogel, and Andrussow[9].

TABLE 1. Representative pK values of organic ionization processes at 25°

Process	pK	Reference
$CH_3COOH + H_2O \rightleftharpoons CH_3COO^- + H_3O^+$	4·76	9
$PhCOOH + H_2O \rightleftharpoons PhCOO^- + H_3O^+$	4·20	9
$CH_3NH_3^+ + H_2O \rightleftharpoons CH_3NH_2 + H_3O^+$	10·66	8
$PhNH_3^+ + H_2O \rightleftharpoons PhNH_2 + H_3O^+$	4·60	8
$CH_3OH + H_2O \rightleftharpoons CH_3O^- + H_3O^+$	15·5	10
$PhOH + H_2O \rightleftharpoons PhO^- + H_3O^+$	10·00	9
$EtSH + H_2O \rightleftharpoons EtS^- + H_3O^+$	10·54	11
$PhSH + H_2O \rightleftharpoons PhS^- + H_3O^+$	6·52	12

C. Ionization Constants in Mixed Aqueous Organic Solvents

A more serious difficulty is encountered when ionization constants of water-insoluble acids are determined in aqueous organic solvents. This is a frequently occurring situation and has unfortunately led to a proliferation of solvent mixtures, making comparisons between differents sets of data difficult and risky. Common solvent mixtures are mixtures of water with methanol, ethanol, 2-alkoxyethanols, acetone or dioxane in different proportions and it sometimes seems as if every single author has his own favorite solvent composition. In the light of recent developments in the definition of a useful pH scale in certain aqueous solvents, to be discussed below, this seems rather unnecessary and it is only to be hoped that a reasonable degree of standardization will be introduced in this field.

The usual method of obtaining the pH of a solution in a mixed aqueous solvent is to standardize the conventional pH-meter set-up (glass electrode/saturated KCl(aq) bridge/calomel electrode) against an aqueous buffer of known acidity and then to use the same cell for measurement of the pH of the unknown solution. Normally, the pH values so obtained are used directly for calculation of an *apparent* ionization constant, which will be denoted by K' in this chapter. This treatment implies that the liquid junction potential, although it may be of a considerable magnitude in solvents low in water, is sufficiently reproducible not to affect relative acidities. Fortunately, this appears to be the case, but nevertheless it is of considerable importance to introduce a meaningful pH scale in aqueous solvents, *i.e.* the value of a_{H+} used for calculation of ionization constants should have a clear interpretation in terms of chemical equilibrium.

If the unknown solution is denoted by X and the aqueous standard buffer by S, the operational (measured) pH is defined by equation (5) where E_X

and E_S represent the e.m.f. of the cell.

$$\mathrm{pH}(X) = \mathrm{pH}(S) + \frac{E_X - E_S}{(RT\ln 10)/F} \qquad (5)$$

The electrode reversible to hydrogen ions may be the hydrogen, glass, or quinhydrone electrode, or any other electrode having this property. It has been shown by Bates and coworkers[13-15] that in alcoholic solvents, such as methanol–water and ethanol–water, it is possible to obtain pH^* (which is related to $a_{\mathrm{H}+}^*$, the activity referred to the standard state *in the alcoholic solvent*, in the same way as pH is related to $a_{\mathrm{H}+}$ in aqueous solution) by subtracting an empirical correction term δ from the pH measured by the procedure described above and defined by equation (5). Values of δ for a number of methanol–water mixtures up to 70% by weight methanol[15] and ethanol–water mixtures up to 100% ethanol[13] have been tabulated and thus make possible the determination of pH^* by simple pH-meter technique using the glass electrode in a cell with a liquid junction. Table 2 gives values of δ for a number of water–methanol mixtures.

TABLE 2. Recommended δ values[15] in aqueous methanol mixtures

Weight % MeOH	0	10	20	30	40	50	60	70
δ	0	0·01	0·02	0·04	0·07	0·11	0·15	0·13

It is also possible to use selected standard buffer solutions of known pH^* in the solvent to be used for the determination of unknown ionization constants and to standardize the pH-meter with this buffer. Then the pH-meter readings can be directly equated with pH^*. Such reference buffers are now available for the solvent 50% methanol–water by weight[14] and no doubt other solvent systems will be investigated systematically in the future.

Another attempt, of wide scope and utility, to standardize measurements of ionization constants is the $\mathrm{p}K_{\mathrm{MCS}}^*$ scale introduced by Simon[16]. In practice, this method measures the apparent pH value of a half-neutralized acid in the system methyl cellosolve–water (80 : 20 by weight) and this value is equated with $\mathrm{p}K_{\mathrm{MCS}}^*$. The cell consists of a glass electrode/calomel electrode combination, which is calibrated in aqueous standard buffers. A high degree of standardization with respect to temperature, titration procedure and concentration leads to excellently reproducible $\mathrm{p}K_{\mathrm{MCS}}^*$ values, which correlate reasonably well with the corresponding $\mathrm{p}K$ values. A large number of $\mathrm{p}K_{\mathrm{MCS}}^*$ values have been tabulated[17-19] and from this values list it is apparent that this solvent system has an excellent dissolving power and should be a preferable choice in many cases.

It must be emphasized that consideration of the above factors does not affect reasoning based on relative acidity data but merely means that a series of pK' values obtained under standardized conditions should be reduced by the same constant term δ, characteristic of the particular solvent mixture. As long as aqueous organic solvents of relatively high dielectric constants are used, pK' values correlate well with pK values with few exceptions. In investigations aimed at the detailed studies of medium effects, measurements are almost invariably made in cells without liquid junctions, so that pK values have a direct relation to chemical equilibria. Also, some very important developments have been made in the study of acid–base equilibria in non-aqueous aprotic solvents, ranging from benzene and chloroform, where it is actually the association equilibrium between a series of acids and a common base which is measured, to acetonitrile, dimethyl sulphoxide, and dimethyl formamide, where an ionization equilibrium can be defined and studied but where ion-pair formation and formation of higher aggregates have to be taken into account. Such studies are usually made by spectroscopic techniques but it has been possible to use pH-meter techniques in some cases. These investigations, as well as those dealing with medium effects will be discussed in later sections.

For investigations of substituent effects by pK measurements of water-insoluble acids in aqueous organic solvents, *it is, however, strongly recommended that the solvent systems referred to above, particularly 50% by weight methanol–water and 80% by weight methyl cellosolve–water, be used,* since these appear to satisfy even strong demands for dissolving power. This would introduce a certain amount of standardization within this field and allow direct comparisons between data obtained by different authors. It would probably also save a lot of duplicate work, which is now needed for the purpose of calibration.

Some authors[2, 5] have expressed grave doubts regarding the use of mixed aqueous organic solvents for the determination of ionization constants. The presence of two solvent species introduces complications in several ways. There will be a number of different acidic and basic species present, *e.g.* H_2O, MeOH, H_3O^+, $MeOH_2^+$, OH^-, and MeO^- in aqueous methanol. Also, the possibility of preferential solvation of the neutral acid and especially the ions by either solvent species will cause problems, since then the macroscopic properties of the solvent will be less relevant than they are in pure solvents. However, when faced with the enormous number of ionization constants measured in mixed aqueous solvents, one can hardly dismiss them on these grounds but must adopt a more pragmatic view. In cases where comparison is possible, pK' values do correlate well with pK values, and this is at least an empirically valid reason for accepting them as meaningful quantities.

D. Relative Acidity Constants

It is sometimes advantageous to compare acidity constants in terms of a relative acidity constant K_r, *i.e.* two ionization equilibria (6 and 7) are combined to give equation (8).

$$R^0COOH + H_2O \xrightleftharpoons{K_{R^0COOH}} R^0COO^- + H_3O^+ \qquad (6)$$

$$R^xCOOH + H_2O \xrightleftharpoons{K_{R^xCOOH}} R^xCOO^- + H_3O^+ \qquad (7)$$

$$R^0COO^- + R^xCOOH \xrightleftharpoons{K_r} R^0COOH + R^xCOO^- \qquad (8)$$

K_r ('r' denoting that the value is a relative one) is then a measure of how easily a proton is transferred from a given acid R^xCOOH to the ion R^0COO^- of the reference acid, and is related to K_{R^xCOOH} and K_{R^0COOH} through equations (9) and (10).

$$K_r = K_{R^xCOOH}/K_{R^0COOH} \qquad (9)$$

$$\log K_r = pK_{R^0COOH} - pK_{R^xCOOH} \qquad (10)$$

The use of relative acidity constants has the advantage that K_r is dimensionless and the corresponding thermodynamic functions $\Delta G_r^0 = -2\cdot3\,RT \log K_r$, ΔH_r^0, and ΔS_r^0 are thus independent of the concentration unit employed. Also, if the reference acid is chosen with some care, kinetic energy factors (contributions due to the fact that the molecules and ions are not in their ground state rotational and vibrational levels) probably cancel to a large extent. Although the solvent is not explicitly included in equation (8) it must be remembered that *solvated* species are involved; therefore contributions due to differences in solvation of molecules and ions cannot be expected to cancel.

Although not always explicitly mentioned, equation (8) represents the basis of all comparisons between acidities and thus underlies the quantitative treatments of substituent effects. The use of pK instead of $\log K_r$ in this connection merely means that the unit scale employed is shifted by a constant factor, equal to the pK of the reference acid.

III. SYMMETRY FACTORS

Before considering other factors, it is necessary to deal with the effect of difference in symmetry between the species participating in an equilibrium[20]. These effects arise from the contribution of the rotational partition function to the entropy of ionization. A well-known case is the so-called statistical factor of 4, which is always introduced in discussions of the ionization behaviour of symmetrical dicarboxylic acids. According to simple reasoning, and with the assumption that there are no interactions whatsoever between the acid centers, the diacid has two ionizable groups and should dissociate twice as fast as the monoanion, making $K_1/K_2 = 2$; by the same reasoning the dianion with its two equivalent carboxylate

$$HOCO(CH_2)_nCOOH + H_2O \xrightarrow{K_1} HOCO(CH_2)_nCOO^- + H_3O^+ \qquad (11)$$

$$HOCO(CH_2)_nCOO^- + H_2O \xrightarrow{K_2} {}^-OCO(CH_2)_nCOO^- + H_3O^+ \qquad (12)$$

groups should pick up protons from the solvent twice as fast as the mono-anion, again giving $K_1/K_2 = 2$ (equations 11 and 12). The total effect on K_1/K_2 will then be $2 \times 2 = 4$, which is the theoretical ratio expected for a diacid with the carboxyl groups at an infinite distance from each other. The same result is obtained by considering, in a way which will be obvious from the following discussion[21], the symmetry numbers of the species involved.

The experimentally accessible equilibrium constant of a reaction involving the species A, B, C, and D, e.g. equation (13):

$$A + B \xrightarrow{K_{eq}} C + D \qquad (13)$$

can be written as

$$K_{eq} = \frac{Q_C Q_D}{Q_A Q_B} \cdot \Delta E_0 \qquad (14)$$

where Q_i is the total partition function of species i and ΔE_0 is the difference in electronic energy between $C + D$ and $A + B$. As usual, Q_i is considered to be separable into translational, rotational, and vibrational partition functions (the partition function for electronic excitation is neglected and set equal to 1):

$$Q_i = (Q_i)_{tr}(Q_i)_{rot}(Q_i)_{vib} \qquad (15)$$

The rotational partition function for any species can be written as:

$$(Q_i')_{rot} = (Q_i)_{rot}/\sigma_i \qquad (16)$$

where σ_i is the overall symmetry number of the i-th species. Then K_{eq} can be rewritten as

$$K_{eq} = \frac{\sigma_A \sigma_B}{\sigma_C \sigma_D} \frac{Q'_C Q'_D}{Q'_A Q'_B} e^{-\Delta E_0 RT} = \frac{\sigma_A \sigma_B}{\sigma_C \sigma_D} K_{chem} \qquad (17)$$

so that the symmetry numbers have been factorized from Q_i, and K_{eq} is given as the product between the quotient of symmetry numbers and a 'chemical' equilibrium constant K_{chem}, from which the effect of symmetry has been eliminated. Thus K_{chem} is the equilibrium constant to be used in the comparison of the influence of substituent effects.

The symmetry number σ_i is a product of an 'external' symmetry number $(\sigma_i)_{ext}$ and an internal symmetry number $(\sigma_i)_{int}$. The external symmetry number is simply the number of ways in which the molecule as a whole may be rotated into an indistinguishable configuration. As an example, methane has four three-fold axes of symmetry and thus there are twelve ways in which the hydrogens may be interchanged, and the external symmetry number is twelve. Similarly, for the benzoate ion the symmetry number is 2, since the ion has a two-fold axis of symmetry, whereas the symmetry number of benzoic acid is 1, since it has no axis of symmetry.

The internal symmetry number is a symmetry number for rotation around bonds in the molecule. It corresponds to the number of equivalent positions reproduced in the course of 360° rotation of the group considered. Thus, the internal symmetry number for ethane is 3, corresponding to the number of reproduced (e.g. staggered or eclipsed) conformations, for the acetate ion 6, and for acetic acid 3. It should be noticed that the consideration of internal symmetry numbers as described here is strictly valid only for unhindered or nearly unhindered rotations, with a potential barrier to rotation of less than 0·5 kcal/mole. Acetic acid, having a barrier to internal rotation around the C—C bond of about 0·5 kcal/mole, may be treated in this way. The barrier in the acetate ion is not known, but if the situation is assumed to be similar to that of the isoelectronic nitromethane molecule, it should be almost completely free. For higher barriers no explicit formula covering the influence of $(\sigma_i)_{int}$ can be given and it is therefore not possible to evaluate the effect directly. For simplicity, it is best to omit the consideration of the internal symmetry number when the barrier is known to be, or may be suspected to be, higher than 0·5 kcal/mole.

Some examples will demonstrate the application and importance of symmetry considerations in the interpretation of ionization equilibria. o-Toluic acid is stronger than benzoic acid by a factor very close to 2·0, and this is usually explained as a result of steric inhibition of resonance.

If only external symmetry numbers are considered for the equilibrium (18)

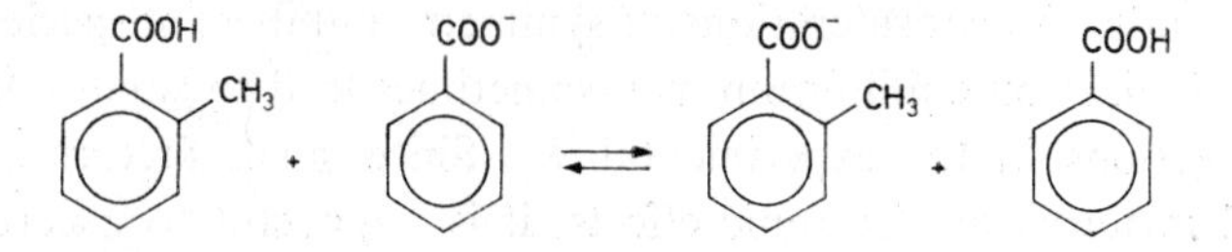

$$(\sigma_i)_{ext} \quad 1 \qquad\qquad 2 \qquad\qquad 1 \qquad\qquad 1 \tag{18}$$

K_r will be a product of the quotient between the symmetry numbers and K_{chem} in accordance with equation (17):

$$K_r = \frac{2 \times 1}{1 \times 1} K_{chem}$$

Experimentally[9], K_r is found to be $1 \cdot 235 \times 10^{-4}/6 \cdot 312 \times 10^{-5} = 1 \cdot 96$, so that K_{chem} will be equal to 1 within the limits of experimental error. Thus, the 'chemical' influence of the methyl group on the ionization of o-toluic acid as compared to benzoic acid is negligible. To get unambiguous confirmation for the postulate of steric inhibition of resonance, one has to compare benzoic acid with an acid possessing the same symmetry properties, e.g. 2,6-dimethylbenzoic acid. It is also evident that the same factor must be taken into account when comparing the acidity of benzoic acid with any *meta*- or *ortho*-substituted benzoic acid. Fortunately, the symmetry factor usually cancels in Hammett-type treatments.

Another example is the above-mentioned case of ionization of a symmetrical dibasic acid. Again assuming completely independent acid centers, equations (11) and (12) can be converted into (19) by subtraction:

	$HOCO(CH_2)_nCOOH + {}^-OCO(CH_2)_nCOO^-$	$\xrightarrow{K_1/K_2}$	$2\,HOCO(CH_2)_nCOO^-$	(19)
$(\sigma_i)_{ext}$	2	2	1	
$(\sigma_i)_{int}$	1	4	2	
σ_i	2	8	2	

Then
$$K_1/K_2 = \frac{2 \times 8}{2 \times 2} = 4$$

Other examples have been discussed by Benson[20] and Welvart[22]. The above treatment involving rotational partition functions is equivalent to ascribing to a species with symmetry number σ_i an entropy of symmetry of $R \ln \sigma_i$.

Another contribution to the free energy of ionization is the entropy of mixing due to a particular species being a *dl* form. This amounts to $-R\ln 2$

entropy units, and Welvart[22] has demonstrated a case where this factor is of importance.

To conclude, the considerations of symmetry numbers of species involved in an ionization equilibrium may sometimes lead to small but significant corrections in the experimental K_r. Since small factors are often discussed in terms of electronic effects, it is important to ascertain that they are not of this origin.

IV. CONFORMATIONAL ASPECTS

A. Rotation Around the C—COOH Bond

It has long been recognized that a characteristic difference in acidity exists between axial and equatorial carboxyl groups in cyclohexane systems, the axial conformer being the weaker acid (see section VIII. A). Only recently has it become possible to demonstrate that the conformational situation around the C—C bond connecting the carboxyl group to the rest of the molecule, in this particular case the cyclohexane ring, is also of importance in connection with acidity[23-25]. It is therefore appropriate to review the unfortunately somewhat meagre data concerning this problem.

From its microwave spectrum, acetic acid was shown to possess a threefold barrier to internal rotation around the C—C bond of 0·48 kcal/mole[26]. An earlier value of $2·5 \pm 0·7$ kcal/mole determined by heat capacity measurements[26], was dismissed since the result was probably influenced by association phenomena. The structure of the stable conformer was not given, but it is highly probable that it is analogous to the situation in acetaldehyde and acetyl fluoride, chloride, bromide, and cyanide, in which the stable conformer has one C—H bond and the C=O bond eclipsed (1)[28].

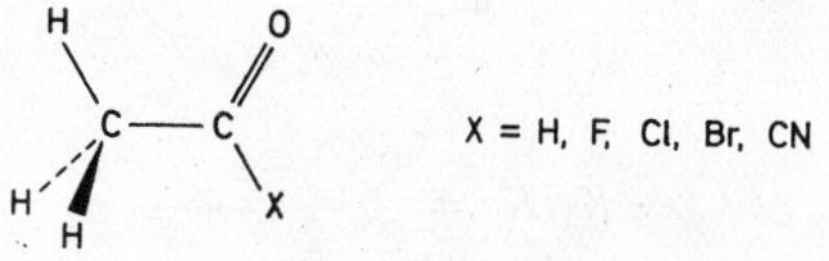

X = H, F, Cl, Br, CN

(1)

Microwave studies on the mixed acetic–trifluoroacetic acid dimer indicate that in both acid parts the rotation around the C—C bonds is almost completely free[29]. For the acetate ion itself nothing appears to be known in this respect, but some data on substituted acetic acids and their ions are available. On the basis of Raman spectra in aqueous solution, Spinner[30] concluded that mono- and dichloroacetic acid and dibromoacetic acid, as well as the corresponding ions, exist solely in a conformation (2 and 3) with eclipsing C—X and CO bonds (C=O in the acid). In mono-bromoacetic acid and its ion two conformations in roughly equal amounts seem to coexist (4, 5, 6, and 7), whereas in iodoacetic acid and its ion the

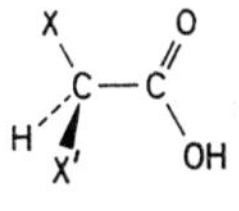

X = Cl, X' = H
X = X' = Cl
X = X' = Br

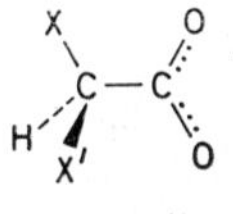

(2)   (3)

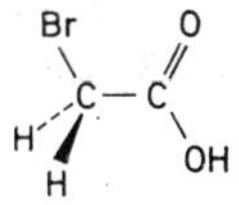

(4) 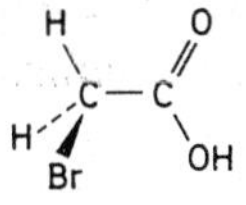   (5) 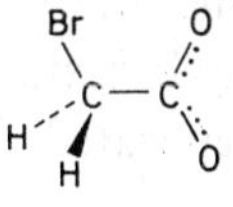   (6) 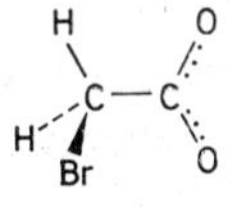   (7)

gauche form predominates. A similar trend was observed in substituted acetophenones[31]: the conformation with eclipsed C—X and CO bonds is increasingly favored in the series iodo < bromo < chloro ≈ dichloro ≈ ≈ dibromo derivatives. A similar trend is observed for 2-halocyclohexanones[32]. In the series 8, X = F, Cl, and Br, the equatorial conformer becomes less and less favored, the equilibrium mixture in benzene solution

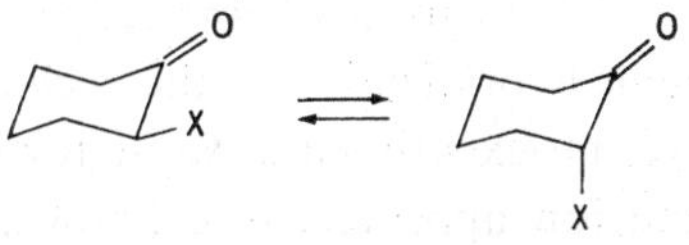

(8)

containing 77%, 44%, and 24%, respectively, of the equatorial form. Chloroacetyl chloride, bromoacetyl chloride, and bromoacetyl bromide have stable conformations analogous to 2, and for the latter two compounds the energy difference between this conformation and the less stable one (probably the one with eclipsed C—H and C=O bonds) was found to be 1·0 and 1·9 kcal/mole, respectively[33].

Unfortunately, nothing appears to be known about the conformational situation in propionic acid and the higher fatty acids or their ions. A comparison with some aldehyde data may be helpful: in propionic[34] and iso-butyric aldehyde[35] the favored conformation has a methyl group eclipsed with the carbonyl oxygen (9 and 11). In the latter compound, the free energy difference between 11 and the less stable conformation 12

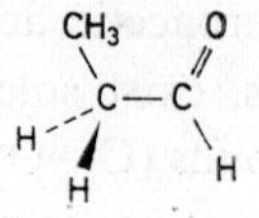

(9) 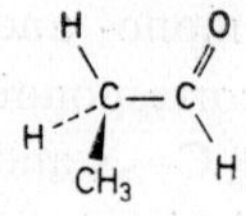   **(10)**

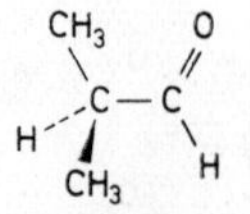

(11) 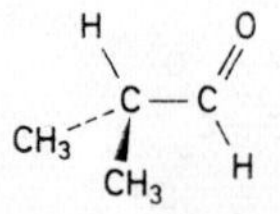   **(12)**

is about 1·3 kcal/mole, and the barrier to rotation is 2 kcal/mole. In cyclopropane carboxaldehyde[36] the two conformers **13** and **14** are of almost equal energy and separated

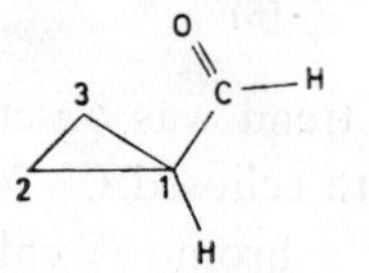

(13) 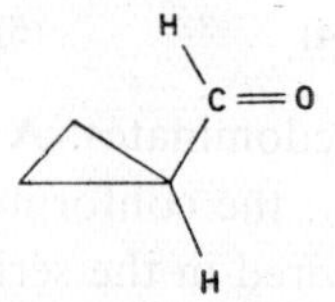   **(14)**

by a barrier to rotation of 2·5 kcal/mole. Note that in **13** and **14** the plane of the aldehyde group *bisects* the angle $C_{(2)}$—$C_{(1)}$—$C_{(3)}$, rather than the carbonyl oxygen and a CH_2 hydrogen being eclipsed. This was considered to be due to the different hybridization in the cyclopropane ring.

As mentioned above, no experimental value for the internal barrier to rotation in the acetate ion appears to be known. On the basis of the similarity to the isoelectronic molecule nitromethane, which has a six-fold barrier of 0·006 kcal/mole, it has been concluded that rotation should be essentially free[29]. Thus, it appears to be a reasonable assumption that the preferred conformation of carboxylic acids (and their ions) of the type RCH_2COOH and $R_2CHCOOH$, where R is a group of small or moderate size (*e.g.* F, Cl, Br, CH_3, or —CH_2), has an R group and the carbonyl oxygen eclipsed. If R has greater steric demand (*e.g.* I) the conformation with eclipsing hydrogen and carbonyl oxygen will dominate. A crude estimate using Spinner's results shows that for chloro-, dichloro-, and dibromoacetic acids and their ions in aqueous solution the former conformation is favored by an enthalpy difference of at least 2 kcal/mole and for bromoacetic acid and its ion by 0·4 kcal/mole. For iodoacetic acid and its ion the *gauche* form appears to be more stable than that with eclipsing iodine and oxygen by at least 2 kcal/mole.

It would be highly desirable to have an estimate of the enthalpy differences between the conformers of propionic acid corresponding to **9** and **10**, and of isobutyric acid corresponding to **11** and **12**. These enthalpy differences might be expected to reflect the behavior of acids in the fatty and alicyclic acid series, since in such acids the conformational situation around the C—COOH bond is similar to that in either propionic or isobutyric acid. The importance of conformational factors has recently been emphasized by ionization constant studies of certain cyclohexane carboxylic acids[23-25].

B. Rotation Around the CO—OH Bond

In the carboxyl group itself rotation around the OH bond produces a *cis* and a *trans* conformation (**15** and **16**). The *cis*

$$R-C\begin{smallmatrix}\nearrow O \\ \searrow O-H\end{smallmatrix} \qquad\qquad R-C\begin{smallmatrix}\nearrow O \\ \searrow O \\ H\end{smallmatrix}$$

(15)   (16)

conformer is more stable than the *trans* by 2·0 kcal/mole and the barrier to rotation is about 11 kcal/mole for the *cis–trans* conversion (known from studies[37] on the microwave spectrum of formic acid). Thus, the situation in the carboxyl group is similar to that in amides and esters. This conformational change is of some importance in connection with intramolecular hydrogen bonding in dicarboxylic acids and their monoanions (section VIII. C. 1).

The above discussion has been concerned with the conformational situation within the carboxyl group and around the bond connecting the carboxyl group with the rest of the molecule. Of course, the conformation of the rest of the molecule is of no less importance for acidity, but a treatment of these aspects will be deferred to later sections (VIII. A and VIII. C). It is, however, appropriate to stress that any experimental ionization constant is a weighted mean of those of all possible conformations, and consequently interpretations of data from conformationally flexible acids, such as the aliphatic ones, must be undertaken with great care.

V. THERMODYNAMIC FUNCTIONS

A. Calculation of Thermodynamic Functions of Ionization Equilibria

The ionization constant K is related to the standard free energy of ionization ΔG^0 through the familiar equation (20). Additional, although not

$$\Delta G^0 = -RT \ln K = 2 \cdot 3\, RT\, \mathrm{p}K \qquad (20)$$

always easily interpretable, information can be obtained by accurate measurements of K over a range of temperatures[38]. As a first approximation, the standard enthalpy and entropy of ionization (ΔH^0 and ΔS^0, respectively) are related to K according to the simple van't Hoff equation (21),

$$\ln K = A + B/T \tag{21}$$

where $A = \Delta S^0/R$ and $B = -\Delta H^0/R$. Equation (21) assumes constancy of ΔH^0 and ΔS^0, but since these conditions cannot be expected to be fulfilled except in very narrow temperature intervals, other empirical equations have been developed. Normally, one finds that ΔH^0 is a function of temperature (equation 22), *i.e.* the

$$\frac{d\Delta H^0}{dT} = \Delta C_p^0 \tag{22}$$

ionization process is accompanied by a change in heat capacity (ΔC_p^0). Assuming that ΔC_p^0 is constant in the temperature range investigated, $\ln K$ can be expressed in the form (23),

$$\ln K = -A/RT - B \ln T/R + D \tag{23}$$

from which ΔH^0 and ΔC_p^0 can be obtained as $A - BT$ and $-B$, respectively[39]. As a next higher approximation one attempts to account for the dependence of ΔC_p^0 on temperature, *e.g.* by the equation (24)

$$\log K = -A^*/T - C^*T + D^* \tag{24}$$

used by Harned and Robinson[40] for the systematic computation of thermodynamic functions of ionization equilibria. From A^*, C^*, and D^* the thermodynamic functions are obtained as $\Delta G^0 = A' - D'T + C'T^2$, $\Delta H^0 = A' - C'T^2$, $\Delta C_p^0 = -2C'T$, and $\Delta S^0 = D' - C'T$, where A', C', and D' equal $2 \cdot 3T$ times A^*, C^*, and D^*.

Ives and Pryor[41] used the simple quadratic equation (25) to

$$\ln K = A + BT + CT^2 \tag{25}$$

represent their early results on the ionization of haloacetic acids. Equations (23), (24), and (25) are relatively easy to fit to the experimental results using a least squares treatment and have been used for a large number of early computations. With the widespread availability of digital computers during recent years, more complicated equations including cubic or higher terms in T can be easily handled, and it is also possible to get a more rigorous statistical treatment of the errors involved. Thus, Clarke and Glew[42] have demonstrated that equation (26) is the best one to use for obtaining un-

$$\ln K = A + B/T + C \ln T + DT + ET^2 + FT^3 \tag{26}$$

biased thermodynamic functions for ionization equilibria.

Since ΔH^0 and ΔC_p^0 are computed by successive differentiations of $\ln K$ with respect to temperature, it is important to note that these functions, especially ΔC_p^0, are less precisely known than $\ln K$[38]. For equation (24), using pK values at $0°, 10°, 20°, 30°, 40°$, and $50°$ with an accuracy of -0.02pK units, typical of measurements with a pH-meter, the errors in the corresponding values of $\Delta H^0, \Delta S^0$, and ΔC_p^0 are ± 0.2 kcal/mole, ± 0.65 cal/mole/degree, and ± 27 cal/mole/degree, respectively. With a precision of ± 0.001 in pK, attainable only in very careful work, the corresponding errors will be ± 0.01 kcal/mole, ± 0.003 cal/mole/degree, and 1.3 cal/mole/degree. For a narrower temperature interval the errors will be appreciably larger. It is evident that any meaningful calculation of ΔC_p^0 must rely on very accurate pK data. Ives and Marsden[7] have measured pK values with an accuracy of ± 0.00017, which is sufficient not only for the accurate calculation of ΔC_p^0 but also to demonstrate that ΔC_p^0 actually varies with temperature.

With recent developments in reaction calorimetry it has become possible to measure enthalpies of ionization directly[43]. Improvements in experimental technique will probably also make accurate measurements of ΔC_p^0 feasible in the near future[44].

B. Discussion of Thermodynamic Functions of Ionization Equilibria

Table 3 is a compilation of thermodynamic functions for carboxylic acid equilibria at $25°$. Care has been taken to include values of high accuracy since any discussion of such data relies heavily on the significance of differ-

TABLE 3. Thermodynamic functions for carboxylic acids relative to those of acetic acid at $25°$ in water

Acid	$\log K_r$	ΔG_r^0 (kcal/mole)	ΔH_r^0 (kcal/mole)	ΔS_r^0 (e.u.)	Reference
Acetic	0	0	0	0	45
Formic	0.99	-1.35	-0.06	5	45
Propionic	-0.12	0.16	-0.01	-1	45
Butyric	-0.06	0.08	-0.67	-3	45
Isobutyric	-0.10	0.14	-0.94	-4	45
Valeric	-0.08	0.11	-0.65	-3	46
Isovaleric	-0.02	0.03	-1.15	-4	46
Hexanoic	-0.10	0.14	-0.63	-3	46
Isohexanoic	-0.09	0.12	-0.65	-3	46
Trimethylacetic	-0.27	0.37	-0.65	-3	46
Diethylacetic	0.02	-0.03	-1.96	-6	46
Fluoroacetic	2.17	-2.97	-1.32	6	41
Difluoroacetic	3.45	-4.71	0.0	16	47
Trifluoroacetic	4.54	-6.20	0.0	21	47

(Table 3: continued)

Acid	log K_r	ΔG_r° (kcal/mole)	ΔH_r° (kcal/mole)	ΔS_r° (e.u.)	Reference
Chloroacetic	1·89	−2·58	−1·05	5	41
Dichloroacetic	3·52	−4·80	0·2	16	47
Trichloroacetic	4·12	−5·52	1·57	24	47
2-Chloropropionic	1·88	−2·57	−1·43	4	43
3-Chloropropionic	0·77	−1·05	−0·25	5	43
Bromoacetic	1·86	−2·50	−1·17	5	41
Dibromoacetic	3·36	−4·58	0·57	14	47
Tribromoacetic	4·90	−6·70	0·87	20	47
2-Bromopropionic	1·79	−2·44	−1·24	4	43
3-Bromopropionic	0·77	−1·05	−0·25	4	43
Cyanoacetic	2·29	−3·13	−0·83	8	48
Diisopropylcyanoacetic	2·20	−3·00	−3·33	−1	7
Glycolic	0·93	−1·28	0·18	5	43
Methoxyacetic	1·19	−1·63	−0·89	2	49
Lactic	0·90	−1·23	−0·10	4	50
Trimethylammonioacetic	2·94	−4·01	−0·01	13	51
Glycine	2·41	−3·29	1·00	14	52
α-Alanine	2·41	−3·29	0·69	13	53
β-Alanine	1·21	−1·65	1·25	10	54
Oxalic (1)	3·49	−4·77	−0·95	13	43
Oxalic (2)	−1·06	1·45	−1·43	−3	43
Malonic (1)	1·93	−2·64	0·36	10	43
Malonic (2)	−0·94	1·28	−0·99	−7	43
Succinic (1)	0·55	−0·75	0·87	5	43
Succinic (2)	−0·88	1·20	0·13	−4	43
Glutaric (1)	0·42	−0·56	−0·05	2	43
Glutaric (2)	−0·66	0·90	−0·51	−5	43
Diethylmalonic (1)	2·55	−3·48	−1·18	8	43
Diethylmalonic (2)	−2·53	3·45	−0·75	−14	43
Ethyl(isoamyl)malonic (1)	2·26	−3·08	−1·24	8	43
Ethyl(isoamyl)malonic (2)	−2·55	3·48	−0·29	−13	43
Fumaric (1)	1·66	−2·26	0·18	8	43
Fumaric (2)	0·16	−0·22	−0·61	1	43
Maleic (1)	2·85	−3·89	0·15	13	43
Maleic (2)	−1·57	2·14	−0·76	−10	43
Tartaric (1)	1·72	−2·35	0·81	11	55
Tartaric (2)	0·45	−0·61	0·31	3	55
Citric (1)	1·63	−2·22	1·07	11	56
Citric (2)	0·00	0·00	0·65	2	56
Citric (3)	−1·64	2·24	−0·73	−11	56
Benzoic	0·56	−0·76	0·17	3	43
p-Bromobenzoic	0·76	−1·04	0·22	4	57
p-Nitrobenzoic	1·32	−1·80	0·14	7	57
Salicylic	1·76	−2·40	0·80	11	58
2-Furoic	1·49	−2·04	−1·19	3	59
Phthalic (1)	1·81	−2·47	−0·57	6	60
Phthalic (2)	−0·66	0·90	−0·43	−4	60

ences between small numbers. Thermodynamic functions are given relative to acetic acid (equation 8, $R^0 = CH_3$) and are denoted ΔG_r^0, ΔH_r^0, and ΔS_r^0. For acetic acid, the following values were used: $pK = 4.76$, $\Delta H^\circ = -0.07$ kcal/mole, $\Delta G^0 = 6.49$ kcal/mole, and $\Delta S^0 = -22$ cal/mole/degree.

Inspection of Table 3 reveals some important factors which must be considered in any discussion of acid strength in terms of molecular structure. First of all, small differences in pK should be interpreted with great

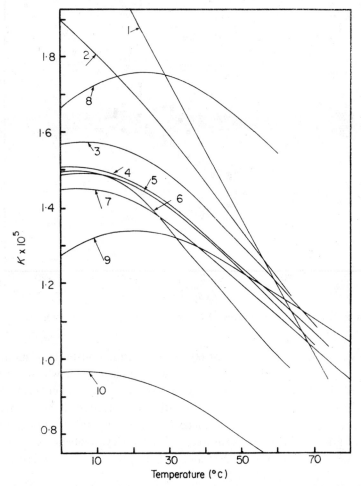

FIGURE 1. Plot of K versus temperature for alkanecarboxylic acids. (1) Diethylacetic acid, (2) isovaleric acid, (3) butyric acid, (4) valeric acid, (5) isocaproic acid, (6) isobutyric acid, (7) caproic acid, (8) acetic acid, (9) propionic acid, and (10) trimethylacetic acid. Data were taken from reference 9.

care, since it is often possible to invert the order of acid strength by a relatively small change in temperature. Figure 1 shows plots of K versus temperature for ten alkanecarboxylic acids in the pK range 4·75–5·03, and it is
easy to see that a large number of acid pairs invert their relative acid strength
at some temperature. Also, if pK values are compared at 60° instead of the
convenient but arbitrarily chosen temperature of 25°, a slightly different
picture is obtained. At 25°, the ionization constants are fairly well spread

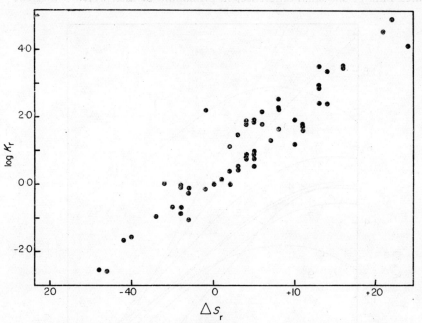

FIGURE 2. Plot of log K_r against ΔS_r^o for the acids tabulated in Table 3.

over the whole interval, whereas at 60° the ionization constants of mono-
and dialkylacetic acids fall within a very narrow range, outside which acetic
and trimethylacetic acids emerge as slightly stronger and weaker, respectively. The appearance of inversions in relative acid strength is not limited
to the fatty acid series but can also be found among substituted acetic
acids. To take an extreme example, diisopropylcyanoacetic acid and
β-alanine with pK values at 25° 2·55 and 3·55, respectively, will invert their
relative acid strength at about 110°.

For a symmetrical reaction like (8) one would expect that the entropy term
would be approximately zero. As is seen from Table 3 and Figure 2, which
shows a plot of log K_r against ΔS_r^0, this is certainly not the case. Instead,

the log K_r value is almost entirely governed by the entropy term, and there is no correlation whatsoever between log K_r and ΔH_r^0, nor can any enthalpy–entropy relationship be found. There is every reason to believe that this entropy dependence is caused by differences in solvent–solute interactions.

From the point of view of current electronic theories about substituent effects, the lack of correlation between log K_r and ΔH_r^0 is somewhat disturbing, since it is thought that differences in potential energy, i.e. inductive and resonance effects, should be largely reflected in this term[61]. Thus, according to the simplest electronic view, the inductive effect from a dipole is transmitted by successively weaker polarizations of the σ bond framework, thereby either strengthening (17) or loosening (18) the O—H bond. It is

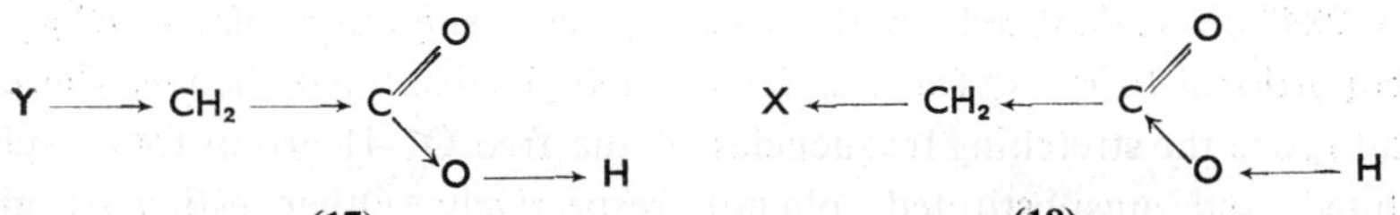

(17)   (18)

evident that any such effect does not occur in the ΔH_r^0 term, nor can this theory account for inversions in relative acid strength on temperature changes. The extended electronic view that an electron-attracting substituent increases acid strength by decreasing solvent orientation around the anion through better dispersion of its negative charge, leads to the prediction that ΔH_r^0 and ΔS_r^0 must always have opposite sign and hence this theory also does not provide for the occurrence of inversions in relative acid strength.

Entropy-controlled ionization equilibria are not unique to the carboxylic acid series. It has been shown that the ionization of both phenols[62] and thiols[11] are dependent on entropy changes. Conversely, ΔH^0 is the major contribution to the variation in pK of cyanocarbon acids[63] over a pK range of $-5\cdot8$ to $11\cdot2$.

Hepler[64a], stressing that any theory of the thermodynamics of ionization in solution must be concerned with the properties of the solvent used and with solvent–solute interactions, tried to solve the problem by dividing enthalpy and entropy contributions to the free energy change accompanying reaction (8) into external and internal ones. External contributions are associated with solvent–solute interactions, whereas internal contributions affect enthalpy and entropy differences within the molecules and ions. Resonance and inductive effects obviously belong to the latter type and it is therefore important to try to separate these from the external contributions.

Hepler accordingly wrote ΔH_r^0 and ΔS_r^0 as a sum of external and internal contributions (equations 27 and 28). $(\Delta S_r^0)_{int}$ must be very close to zero

$$\Delta H_r^0 = (\Delta H_r^0)_{int} + (\Delta H_r^0)_{ext} \tag{27}$$

$$\Delta S_r^0 = (\Delta S_r^0)_{int} + (\Delta S_r^0)_{ext} \tag{28}$$

for the symmetrical reaction (8), as can be shown by Pitzer's method[64b], and it is further assumed that $(\Delta H_r^0)_{ext}$ is proportional to $(\Delta S_r^0)_{ext}$ (proportionality constant β), which follows from the assumption that Born's equation[64c] is valid. It is then easy to derive the expression (29) for $(\Delta H_r^0)_{int}$, and it only remains to assign a reasonable value to β. A value of

$$(\Delta H_r^0)_{int} = \Delta H_r^0 - \beta \, \Delta S_r^0 \tag{29}$$

$\beta = 284°$ K was deduced on the basis of the assumption that $(\Delta H_r^0)_{int}$ is proportional to $\nu_s^2 - \nu_u^2$ for a series of methyl-substituted phenols where ν_s and ν_u are the stretching frequencies of the free O—H group for a substituted and unsubstituted phenol, respectively. Other estimates give similar values of β, and Hepler chose $\beta = 280°$ for his calculations and discussions, values of $(\Delta H_r^0)_{int}$ being fortunately not very sensitive to the choice of β. Substitution of equation (29) into $\Delta H_r^0 = \Delta G_r^0 - T \Delta S_r^0$ gives equation (30), where the second right-hand term is much smaller

$$\frac{\Delta G_r^0}{(\Delta H_r^0)_{int}} = 1 + \frac{(\beta - T) \, \Delta S_r^0}{(\Delta H_r^0)_{int}} \tag{30}$$

than unity for β values close to T. Accordingly $\Delta G_r^0 \cong (\Delta H_r^0)_{int}$, demonstrating that the free energy change of an ionziation process is actually a valid approximation for changes in potential energy within the species participating in the equilibrium.

Using a more elaborate treatment based on the consideration of hydrational equilibria, Ives and Marsden[7] have arrived at the same conclusion, with the difference that β in equation (29) is actually equated with T, the absolute temperature at which the measurements are made.

Thus, it follows that ΔG^0 is the thermodynamic function best suited for discussion in relation to molecular models and electronic effects operating within these, since it is much less sensitive than ΔH^0 or ΔS^0 to complications introduced by the solvent employed in a particular investigation. It must be emphasized, though, that no thermodynamic function can completely eliminate the effects of solvent–solute interactions; therefore there will always be an inherent uncertainty attached to any substituent parameter derived from thermodynamic data.

VI. KINETICS OF CARBOXYLIC ACID IONIZATION

During the last decade it has become possible to measure the kinetics of very fast proton transfer reactions[65-67], enabling an analysis to be made of ionization equilibria in terms of a dissociation and a recombination reaction with experimental rate constants k_d and k_r, respectively. Using the high level faradaic rectification method, Nürnberg and Dürbeck[68] have determined k_d for a number of carboxylic acids (equation 31), and k_r is then obtained from k_d and K_c,

$$\text{RCOOH (aq)} + H_2O \; \underset{k_r}{\overset{k_d}{\rightleftharpoons}} \; \text{RCOO}^- \text{(aq)} + H_3O^+ \text{(aq)} \qquad (31)$$

the equilibrium constant under the conditions employed, through the expression $k_d/k_r = K_c$. Their kinetic data, together with the corresponding pK values, are given in Table 4, while Figure 3 shows plots of log k_d and

TABLE 4. Rate constants[68] for the ionization (k_d) and recombination (k_r) steps of carboxylic acid equilibria in 1 M LiCl at 20°

Acid	$10^{-6}k_d$ sec^{-1}	$10^{-10}k_r$ l/mole/sec	pK[9]
Formic	18·0	4·8	3·75
Acetic	1·39	3·8	4·76
Propionic	0·82	3·1	4·87
Butyric[a]	0·67	2·2	4·81
Valeric	0·87	2·9	4·83
Lactic	3·85	0·86	3·86
Phenylacetic	4·48	4·4	4·31
Mandelic	40·9	4·5	3·40
Benzoic	4·0	2·8	4·20
Salicylic	9·3	0·4	2·99
m-Hydroxybenzoic	7·7	4·2	4·08
p-Hydroxybenzoic	2·27	4·0	4·59

[a] At 16·5°.

log k_r against pK. Except for some acids capable of giving intramolecular hydrogen bonds in the anion (lactic, salicylic), log k_d is linearly correlated with pK, whereas log k_r is constant within the limits of error ($\pm 30\%$ in k_r) for the range of acids investigated. These acids are said to exhibit 'normal' behavior.

The numerical average value of $k_r = 3\cdot7 \times 10^{10}$ l/mole/sec for the normal acids is in accordance with the rate-determining step for recombination being the diffusion-controlled approach of a hydrated hydrogen ion and a hydrated anion to a critical distance of 7–8 A, for which a theoretical value of $k_r = 4 \cdot 9 \times 10^{10}$ 1/mole/sec can be estimated. At this distance the two hydrated species can form a 'latent ion pair' (19), in

$$H^+ (aq) + RCOO^- (aq) \xrightarrow[k_D]{k_D'} \begin{array}{c} H \\ O-H\cdots O \\ H \end{array} \begin{array}{c} H \\ O \\ \vdots H \cdots O \end{array} \begin{array}{c} O \\ C-R \\ O \end{array} \xleftarrow[k_u]{k_u'}$$

$$(19)$$

$$\xrightarrow[k_u]{k_u'} \begin{array}{c} H \\ O\cdots H-O \\ H \end{array} \begin{array}{c} H \\ \vdots \\ H \end{array} \begin{array}{c} O \\ C-R \\ O \end{array} \qquad (32)$$

which the proton and the anion are separated by a well-defined hydrogen-bonded system (equation 32). Once the latent ion pair has been formed valence bond reorganization of the system can occur at a much higher rate than the latent ion pairs are formed by diffusion-controlled encounters of the hydrated species. This type of ultra-fast proton transfer has been de-

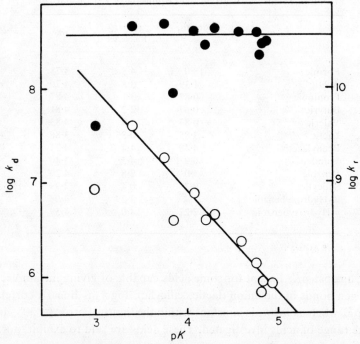

FIGURE 3. Plot of log k_d (open circles) and log k_r (solid circles) versus pK for the acids given in Table 4. The two pairs of points deviating from the linear relationships correspond to lactic and salicylic acid, respectively.

monstrated to occur in ice, where the hydrogen bond system in the crystal structure allows the proton-jump mechanism over distances limited only by the number of disruptions of the crystal lattice. The same mechanism can also occur in the liquid state, although the ice-like order in water is likely to be of short range, such as is the case in the latent ion pair **19**. The rate constant of the proton-jump reaction is of the order of 10^{13}/sec.

Equation (32) defines rate constants for the individual steps in the formation of the latent ion pair, the proton-jump reaction, and the reversal of these reactions. The experimentally accessible rate constants k_d and k_r are related to k_D, k_D', k_u, and k_u' for the case of a diffusion-controlled reaction between $RCOO^-$(aq) and H_3O^+(aq) $(k_D \ll k_u')$ through equations (33) and (34).

$$k_d = k_D \frac{k_u}{k_u'} = k_D K_u \qquad (33)$$

$$k_r = k_D' \qquad (34)$$

The dissociation step consists of a transformation of RCOOH (aq) into the latent ion pair, for which either a retransformation into RCOOH (aq) or a diffusive separation into $RCOO^-$ (aq) and H_3O^+ (aq) is possible. For the latter process, the rate constant k_D for a normal acid (*i.e.* free from complications due to intramolecular hydrogen bonding) should be independent of the structure of the acid just as k_D' is. The theoretical value of k_D at zero ionic strength is calculated to be about 2×10^{10}/sec. Thus, k_D being essentially constant for normal acids, the experimental rate constant k_d for the dissociation process is determined by the ratio $k_u/k_u' = K_u$, the equilibrium constant for the dissociation of RCOOH (aq) into the latent ion pair. It is K_u which is sensitive to changes in the structure of RCOOH, and since k_r is also approximately constant, variations in the equilibrium constants of carboxylic acids originate from variations in the rate constants of dissociation k_d (Figure 3).

For acids capable of forming an intramolecular hydrogen bond in the anion (*e.g.* salicylic acid) there will be an equilibrium between anions having an internal hydrogen bridge and anions which have been opened by an internal rotation process. Opening and closing of the intramolecular hydrogen bond occurs at a very high rate. This means that only the fraction of the anions which have just been opened is accessible for intermolecular hydrogen bond formation to form latent ion pairs, and thus k_r will be correspondingly smaller. Table 4 shows that this is the case for salicylic and lactic acid.

For an extended discussion of the kinetics of proton transfer processes the reader is referred to Eigen's excellent reviews[66, 67].

VII. SOLVENT EFFECTS

A. Classification of Solvents

For the purpose of discussing medium effects on carboxyl group acidity, solvents are conveniently divided into four groups.

(1) Protic solvents of high dielectric constant, such as water ($D = 78.5$), ethylene glycol ($D = 37.7$), methanol ($D = 31.5$), ethanol ($D = 24.2$), and formamide ($D = 110$). These solvents can function in hydrogen bonding both as donors and acceptors, and their dielectric constants are high enough to exclude ion pair formation, which causes complications in protic solvents with dielectric constants lower than about 25.

(2) Protic solvents of low dielectric constant, such as n-butanol ($D = 17.4$) and m-cresol ($D = 11.8$), which possess good hydrogen-bonding properties both as donors and acceptors but which will tend to favor the formation of ion pairs or larger aggregates.

(3) Dipolar aprotic solvents[69] of a relatively high dielectric constant, such as N,N-dimethylformamide (DMF, $D = 36.7$), dimethyl sulfoxide (DMSO, $D = 46.7$), and acetonitrile ($D = 37$), which can function only as hydrogen bond acceptors, acetonitrile being considerably weaker as such than the two others. In these solvents, care has to be taken to exclude effects of ion pairing and the formation of conjugated species, such as RCOOH....OCOR, which is especially pronounced in acetonitrile because of its low hydrogen bond acceptor strength.

(4) Inert or nearly inert solvents of low dielectric constant, such as benzene ($D = 2.3$) or chloroform ($D = 4.8$). Here one actually measures association constants between a series of acids and a common added base, such as triethylamine, and measurements are complicated by the formation of conjugated species and by the existence of the carboxylic acid monomer–dimer equilibrium.

B. Electrostatic Theories of Solvent Effects

Since excellent reviews on medium effects upon acidity from the electrostatic point of view are available[1, 5], only a short summary of the most important conclusions will be given here. Instead, emphasis will be laid on specific solvent effects, such as hydrogen-bonding and dispersion effects, in which field considerable progress has been made during recent years. The discussion will also be limited to pure solvents because of the objections which can be raised against the use of mixed solvents (section II. C). For a discussion of ionization equilibria in water–organic solvents the reader is referred to the book by King[1].

Electrostatic theories of solvent effects upon ionization equilibria are based on the simple Born model of an ion as a conducting sphere of radius r immersed in a homogeneous isotropic linear medium of dielectric constant D. To eliminate the problem of the medium effect of the proton, comparisons between relative acidity constants (K_r in equation 8) in different solvents are preferred to the considerably more difficult task of deriving absolute values of acidity constants. Using the Born model, Wynne–Jones[70] made an electrostatic calculation of the effect of a change in dielectric constant on K_r, resulting in equation (35) for matched pairs of carboxylic acids, where K_r^∞ is the value of K_r in a hypothetical medium of

$$\ln K_r = \ln K_r^\infty - \frac{e^2}{2kTD}\left(\frac{1}{r_0} + \frac{1}{r}\right) \tag{35}$$

infinite dielectric constant, e is the electronic charge, k is the Boltzmann constant, r_0 is the common radius of the reference acid and this ion, and r is the common radius of the acid of interest and its ion. Thus, equation (35) predicts a linear correlation between $\log K_r$ and $1/D$ with different slopes for different acids, depending on r_0 and r.

Equation (35) is approximately valid for carboxylic acids in a series of solvents which are chemically very closely related, such as water, ethylene glycol, methanol, and ethanol, whereas it does not hold if a solvent of high dielectric constant but of different type, such as formamide, is included[71]. Although the dielectric constant of formamide is larger than that of water, $\log K_r$ values are nevertheless *larger* in formamide than water. This effect has been ascribed to the formation of a cyclic hydrogen-bonded complex between formamide and the undissociated carboxylic acid[72]. A similar effect is observed for ionization equilibria in N-methylacetamide ($D = 165$), where the pK of acetic acid is 7·14. No other acids have been studied in this medium, but it is known that trichloroacetic acid gives a solid 1 : 1 complex with N-methylacetamide[73]. Of course even greater deviations are to be expected for comparisons between solvents chosen from the different groups defined above, since the simple electrostatic picture does not account for specific solvent–solute interactions.

Also, any theory which predicts a linear correlation between $\log K_r$ and $1/D$ will lead to the prediction that ΔG_r^0 should be linearly correlated with ΔS_r^0 for a single solvent[38, 74], as is indeed found for a wide range of carboxylic acids (Table 3 and Figure 2).

Electrostatic theories of substituent effects in acids containing a charged or a dipolar substituent usually predict an inverse variation with the macroscopic dielectric constant of the solvent (the Bjerrum equation 53) or the effective dielectric constant (the Kirkwood–Westheimer equation 54). These equations will be dealt with in section VIII. B. 2 in connection with substituent effects. Like Wynne–Jones' treatment, they do not include terms for specific solvation effects, and it is difficult to test them with respect to medium effects because of the lack of suitable data. In the few cases where testing is possible, King[75] has demonstrated that the variation in log K_r with dielectric constant is best represented by the Kirkwood–Westheimer treatment.

C. Specific Solvent–Solute Interactions

In view of the substantial differences in solvent properties between the solvent groups (1)–(4) referred to above, there is no *a priori* reason to expect that log K_r values (K_r taken in the wide sense, *i.e.* including relative association constants in solvents of low dielectric constant) should correlate well with those measured in aqueous solution. In his survey on the effects of the medium upon acidity, Bell[5] nevertheless concluded that the relative strengths of acids of *the same charge and chemical type* are independent of solvent. Later investigations in other solvents do not invalidate this statement as far as gross effects are concerned, but it must be remembered that inversions in relative strength between acid pairs of similar acidity do sometimes occur, just as they do in a single solvent when the temperature is changed slightly.

The electrostatic contribution to the solvent effect, as defined by equation (35), depends on only one structural parameter of the species participating in an ionic equilibrium, namely size. This is of no particular chemical interest. The most important and chemically interesting solvent effect originates from interactions between solvent and solute, such as hydrogen-bonding, dipole–dipole, and dispersion interactions, which are closely related to the molecular structure of the acid and its ion. Hydrophobic bonding may also be included to complete the list of intermolecular forces, although it is a type of bonding unique to compounds having nonpolar hydrocarbon groups in aqueous solution.

A well-known manifestation of the strong hydrogen-bonding ability of carboxylic acids[76] is the formation of cyclic dimers (20) in the gas phase or

$$2\ RCOOH \ \rightleftharpoons \ R-C\overset{O\cdots H-O}{\underset{O-H\cdots O}{\diagdown}}C-R$$

(20)

in nonpolar solvents, such as benzene or carbon tetrachloride[77]. In protic and dipolar aprotic solvents at low acid concentrations the dimers are broken up to form 1 : 1 or 1 : 2 hydrogen-bonded complexes with the solvent. This has been demonstrated by a variety of techniques in water[76], alcohols[78], amides[72, 73, 79], sulfoxides[80, 81], acetonitrile[78, 82], and acetone[78, 82]. The predominant formation of heteroassociated species in such solvents is sometimes due to a simple mass action effect, as for example for acetic acid[80] in DMSO, where $K_{selfassn.}$ is 1730 and $K_{heteroassn.}$ is 380, and sometimes to an inherent instability of the dimer, as for example for acetic acid in water[83], where $K_{selfassn.}$ is about 0·1. Even in solvents like carbon tetrachloride and benzene, there are indications of hydrogen-bonding between RCOOH and chlorine or RCOOH and the π-electron system, resulting in lower dimerization constants than in completely nonpolar solvents[76].

Hydrogen bonding of the type RCOOH...B, where B is a nitrogen base, has been studied extensively both in the solid and liquid phase[84, 85], and it is possible to demonstrate the transformation from RCOOH...B into a hydrogen-bonded ion pair RCOO$^-$....HB$^+$ at a critical pK difference between the acid and the nitrogen base. Some of these complexes have hydrogen bonds of the single minimum type, and it appears that the type of bond is the same whether the complex is in the solid phase or in an acetonitrile solution.

The ability of RCOO$^-$ to form hydrogen bonds with proton donors is also well documented. Apart from the above-mentioned case of hydrogen-bonded ion pairing, carboxylate groups can form hydrogen-bonds to a phenol[86] (21a) or to a molecule of the corresponding acid (21b). Species (21b) has been denoted as a 'homoconjugated' anion by Kolthoff and

$$
\begin{array}{cc}
\overset{\displaystyle O \cdots HOAr}{\underset{\displaystyle O}{RC \!\!\! < \; -}} & \overset{\displaystyle O \cdots HO}{\underset{\displaystyle O \quad\;\; O}{RC \!\!\! < \; - \quad C-R}} \\[2mm]
\textbf{(21a)} & \textbf{(21b)}
\end{array}
$$

Chantooni[87]. Homoconjugated anions have attracted considerable interest during recent years, since it has been shown by X-ray[88–92], neutron diffraction[93], i.r.[94–96], and n.m.r.[94] measurements that some of them are held together by a symmetrical single minimum hydrogen bond in the solid phase. In solution, homoconjugated species tend to form in solvents of low hydrogen-bonding acceptor ability, e.g. acetonitrile, in which a number of association constants have been determined[97–99] (Table 5). Figure 4 shows that the stability of the homoconjugated anion decreases linearly with the pK of the corresponding acid in aqueous solution. The difference of about one power of ten between the two sets of data hitherto

TABLE 5. Stability constants of some homoconjugated anions RCOOH...OCOR⁻ in acetonitrile

Acid	log $K_{HA_2^-}$	
	Kolthoff and Chantooni[97]	Gordon[98]
Benzoic	3·6	2·63
p-Methylbenzoic	—	2·61
p-Hydroxybenzoic	3·1	—
Salicylic	3·3[a]	—
p-Methoxybenzoic	—	2·58
m-Chlorobenzoic	—	2·72
m-Bromobenzoic	3·8	—
p-Nitrobenzoic	3·8	2·78
3,5-Dinitrobenzoic	4·0	—

[a] Another study[99] gave a value of 3·0–3·2.

reported is probably due to some systematic error in one or both of the widely different methods used for their determination. Generally, hydrogen-bonding equilibrium constants tend to be linearly correlated with pK[100].

It is interesting that the formation of carboxylic acid dimers and homo-

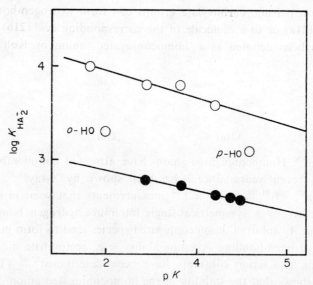

FIGURE 4. Plot of log K_{HA_2} versus pK for the benzoic acids given in Table 5. Open circles correspond to Kolthoff and Chantooni's values[97], and filled the ones to Gordon's values[98].

conjugated anions cannot always be neglected even in aqueous solution at high concentrations of acid and salt. Thus, Farrer and Rossotti[83] were unable to described the hydrolytic equilibrium in sodium acetate solutions by means of a single parameter, and suggested that formation of $(CH_3COOH)_2$ and $(CH_3COO)_2H^-$ might account for the experimental results. Log $K_{assn.}$ for $(CH_3COOH)_2$ and $(CH_3COO)_2H^-$ were then found to be -0.96 and -1.0, respectively, whereas log K for equilibrium (36) was found to be 5.06, i.e. the pK for ionization of the dimer, the reverse of

$$(CH_3COO)_2H^+ + H^+ \xrightarrow{\;\;K\;\;} (CH_3COOH)_2 \qquad (36)$$

equation (36), is not very much different from the pK of monomeric acetic acid.

Schrier, Pottle, and Scheraga[101] extended the study of the dimer equilibrium in water to a few other n-alkanecarboxylic acids, and found that dimer formation was favored by increasing chain length. From these results they concluded that the carboxylic acid dimer in water is not cyclic (20) but open (22), and that the increase in stability with increasing length of the nonpolar part of the acid is due to hydrophobic bonding[102] between the hydrocarbon portions. Support for this view is provided by studies on association equilibria in dilute aqeous solutions of carboxylic acid soaps[103].

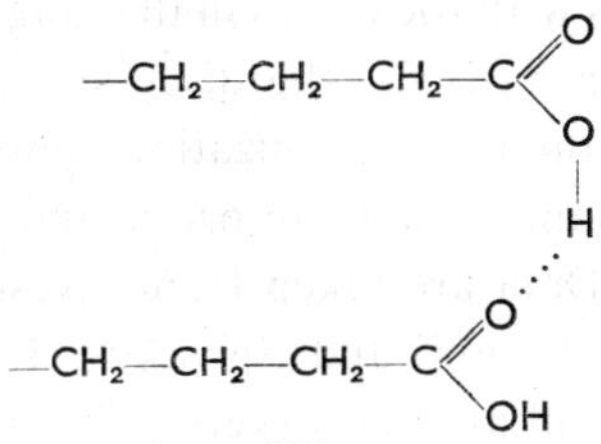

(22)

Apart from the hydrolytic equilibrium (equation 37) which predicts the

$$RCOO^- + H_2O \rightleftharpoons RCOOH + OH^- \qquad (37)$$

hydrolytic behavior of carboxylic acid soaps only at concentrations below 10^{-4} M (C_{12} soap) or 10^{-6} M (C_{14} and C_{16} soaps), Eagland and Franks considered the formation of homoconjugated species (equation 38) with the equilibrium constant K_D and its ionization (equation 39) with an ioni-

$$RCOOH + RCOO^- \xrightarrow{\;\;K_D\;\;} (RCOO)_2H^- \qquad (38)$$

$$(RCOO)_2H^- + H_2O \xrightarrow{\;\;K_a'\;\;} (RCOO)_2^{2-} + H_3O^+ \qquad (39)$$

TABLE 6. pK values associated with equilibria (38) and
(39) at 25° in aqueous solution[103]

	C_{12}	C_{14}	C_{16}
$-pK_D$	4·5	6·8	7·6
pK'_a	10·5	10·0	9·5

zation constant K'_a. Values of pK_D and pK'_a are shown for C_{12}, C_{14}, and C_{16} soaps in Table 6. The association constant K_D for the formation of the homoconjugated species increases appreciably with increasing chain length, just as is observed for dimer formation in water, and this can best be explained in terms of hydrophobic bonding, since there is no reason to expect any great change in hydrogen-bonding donor or acceptor ability in this series of closely related acids. The same factor must be responsible for the aggregation of ions of long-chain fatty acids into dimeric ions, as has been shown by Mukerjee[104]. The stability of the dimeric ion makes possible the determination of its ionization constant K'_a (Table 6), which is formally analogous to that of an intramolecularly hydrogen-bonded monoanion of a dicarboxylic acid (section VIII. C. 1). Such monoanions have abnormally low ionization constants with pK values in the range 6·5–10.

In order to provide a basis for the discussion of solvent–solute interactions and their effect on acidity, ionization constant data for carboxylic acids and, for comparison, phenols, in protic and dipolar aprotic solvents are shown in Table 7. Data are taken from investigations carried out by Verhoek[105], Izmailov, Chernyi, and Spivak[106], Konovalov[107], Clare and coworkers[108], and Kolthoff and coworkers[97, 109], and were chosen to cover a representative number of acids and solvents. All values of log K_r are given relative to benzoic acid, i.e. in the case of phenols ArOH will take the place of R^xCOOH in equation (8). Due to the inhomogeneity of the material, log K_r values are probably not known with better precision than $\pm 0·2$ log units, except for those in water.

Figure 5 shows plots of log $(K_r)_{solv}$ against log $(K_r)_{H_2O}$ for carboxylic acids. Considering the magnitude of possible errors, the lines for methanol, ethanol, formamide, and DMF are hardly distinguishable from each other, whereas the lines for acetone and acetonitrile differ appreciably from the others. Data for carboxylic acids in DMSO are unfortunately very few, but the existing ones fit reasonably well with the regression line for acetonitrile. Regression lines for the correlations shown in Figure 5 were

TABLE 7. Log K values for carboxylic acids and phenols in different protic and dipolar aprotic solvents[97, 105-108]

Acid	log K_r							
	H_2O	HCONH$_2$	MeOH	EtOH	DMF	DMSO	MeCN	Acetone
Trichloroacetic acid	3·50	—	4·48	4·43	—	—	—	—
Dichloroacetic acid	2·91	3·48	2·7	2·99	3·0	—	—	—
Chloroacetic acid	1·33	1·63	1·4	1·62	1·2	—	—	2·14
Acetic acid	-0·56	-0·55	-0·5	-0·27	-0·9	-1·4	—	-0·61
Propionic acid	-0·68	-0·84	—	—	—	—	—	—
Benzoic acid	0·00	0·00	0·00	0·00	0·00	0·00	0·00	0·00
4-Nitrobenzoic acid	0·78	—	1·02	1·26	—	—	2·0	1·35
3-Nitrobenzoic acid	0·72	0·86	1·05	1·13	—	—	—	1·28
3,5-Dinitrobenzoic acid	1·38	—	—	—	—	—	3·8	—
3-Bromobenzoic acid	0·34	—	0·55	0·66	—	—	1·2	—
Salicylic acid	1·22	1·83	1·53	1·53	—	3·1	4·0	2·72
Phenol	-5·78	—	-5·1	—	< -5	—	—	—
4-Nitrophenol	-2·95	-2·25	-2·1	-0·87	-0·7	0·1	—	—
2,6-Dinitrophenol	0·62	2·10	1·74	—	—	—	—	3·19
2,5-Dinitrophenol	-0·92	0·27	1·95	—	—	—	—	1·65
2,4-Dinitrophenol	0·10	1·77	1·2	1·92	4·2	4·8	—	3·19
Picric acid	3·82	—	5·3	6·20	9·0	11·9	9·6	8·78
2,4,6-Trichlorophenol	-2·21	-1·01	—	—	—	—	—	—

16*

calculated using the method of least squares and were found to be:

$$\log (K_r)_{MeOH} = 0 \cdot 1 + 1 \cdot 1 \log (K_r)_{H_2O} \qquad (40)$$

$$\log (K_r)_{EtOH} = 0 \cdot 2 + 1 \cdot 1 \log (K_r)_{H_2O} \qquad (41)$$

$$\log (K_r)_{HCONH_2} = 0 \cdot 1 + 1 \cdot 2 \log (K_r)_{H_2O} \qquad (42)$$

$$\log (K_r)_{DMF} = -0 \cdot 2 + 1 \cdot 1 \log (K_r)_{H_2O} \qquad (43)$$

$$\log (K_r)_{acetone} = 0 \cdot 2 + 1 \cdot 7 \log (K_r)_{H_2O} \qquad (44)$$

$$\log (K_r)_{MeCN, DMSO} = 0 \cdot 1 + 2 \cdot 7 \log (K_r)_{H_2O} \qquad (45)$$

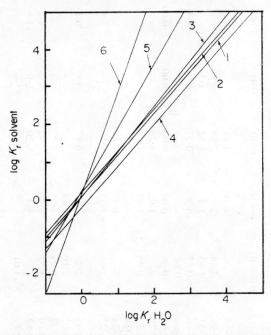

FIGURE 5. Plot of log (K_r) solvent versus log $(K_r)H_2O$ for carboxylic acids (data from Table 7). The lines shown correspond to equations (40—(45). For clarity, individual points have not been plotted. (1) Methanol, (2) ethanol, (3) form-amide, (4) dimethylformamide, (5) acetone, and (6) acetonitrile and dimethyl-sulfoxide.

The existence of different lines for different solvents or groups of solvents indicates a differentiating effect of the particular solvent or solvent group compared to water, whereas any deviation of the slope of the regression line from unity indicates a differentiating action of a solvent within a series of acids of given type.

A similar behavior is exhibited by phenols (Figure 6), the order of differentiating power being approximately the same as for carboxylic acids (except for DMF).

Next, attention is turned to log K_r values determined in solvents of low dielectric constant. Here log K_r values are derived from association constants for the equilibria between a series of acids and a common reference base, *e.g.* triethylamine[110] or 1,3-diphenylguanidine[77, 111]. Some represen-

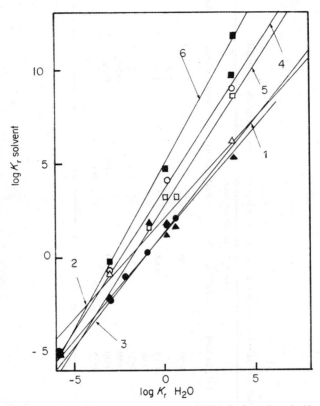

FIGURE 6. Plot of log (K_r) solvent against log $(K_r)H_2O$ for phenols (data from Table 7). (1) Methanol (filled triangles), (2) ethanol (open triangles), (3) formamide (filled circles), (4) dimethylformamide (open circles), (5) acetone (open squares), and (6) acetonitrile and dimethylsulfoxide (filled squares).

tative data covering different reference bases and solvents have been collected in Table 8, while Figure 7 shows that log $(K_r)_{solv}$ is again linearly correlated with log $(K_r)_{H_2O}$. As before, data from only a limited number of investigations are included in order to have a consistent set of data for every base and solvent employed. For other references to this work in this field the reader is referred to a recent paper by Davis and Paabo[77].

As can be seen from Figure 7, different solvent–base combinations differ slightly in their ability to differentiate within the carboxylic acid

TABLE 8. Log K_r values for selected carboxylic acids in solvents of low dielectric constant

Acid	Log K_r in solvent/reference base					
	$(CH_2Cl)_2/Et_3N$[110]	CCl_4/Et_3N[110]	$PhCl/Et_3N$[110]	PhH/Et_3N[110]	$CHCl_3/BuNH_2$[112]	$PhCl/i\text{-}BuNH_2$[113]
Trichloroacetic	2·20	1·90	1·90	2·35	4·15	—
Dichloroacetic	—	—	—	—	2·85	3·10
Chloroacetic	0·85	1·15	0·65	0·75	1·85	1·33
Formic	0·35	0·60	—	—	—	—
Acetic	−0·40	−0·25	−0·30	−0·40	−0·2	−0·42
Propionic	−0·75	−0·50	—	—	−0·50	−0·68
Benzoic	0·00	0·00	0·00	0·00	0·00	0·00
Salicylic	1·10	1·35	—	0·85	1·7	1·32

series. The effect is, however, not very pronounced, and the general conclusion from the data in Table 8 is that within a group of solvents of the same nature the differentiating action is small or absent, as long as the same reference base is used.

In order to analyze the solvent effect on the individual species participating in an ionization process, Izmailov and Izmailova[114] related K_{H_2O}

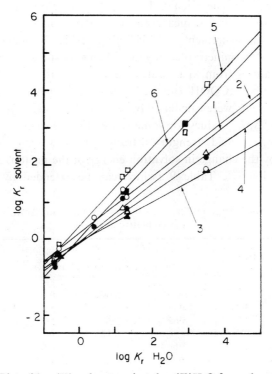

FIGURE 7. Plot of log (K_r) solvent against log $(K_r)H_2O$ for carboxylic acids in low dielectric constant solvents (data from Table 8). (1) 1,2-dichloroethane/triethylamine, (2) carbon tetrachloride/triethylamine, (3) chlorobenzene /triethylamine, (4) benzene/triethylamine, (5) chloroform/butylamine, and (6) chlorobenzene/isobutylamine.

to K_M, the ionization constant, in a solvent of interest, through equation (46), where γ_{H+}, γ_{RCOO-}, and γ_{RCOOH} are the *medium activity coeffi-*

$$K_{H_2O} = K_M \frac{\gamma_{H+}\gamma_{RCOO-}}{\gamma_{RCOOH}} \qquad (46)$$

cients (not to be confused with the Debye concentration activity coefficients, here denoted by *f)* for the species H^+, $RCOO^-$, and $RCOOH$, defining the change in free energy on transfer from the solvent M to water by the expression $\log \gamma = \Delta G / 2 \cdot 3RT$. Taking logarithms of equation (46) and rearranging the terms, one obtains equation (47) which expresses the

$$\log K_{H_2O} - \log K_M = \Delta pK_M = \log \gamma_{H^+} \log \gamma_{RCOO^-} - \log \gamma_{RCOOH} \quad (47)$$

change in pK, ΔpK_M, in transferring the process from water to M, in terms of medium activity coefficients. Log γ_{RCOOH} and ($\log \gamma_{H^+} + \log \gamma_{RCOO^-}$) can be obtained by e.m.f. measurements, either alone or in combination with solubility measurements, and if log γ_{H^+}, which is independent of the acid type and determined solely by the change in solvation energy of the proton on transfer from M to water, can be estimated, the solvent effect on RCOO⁻ and RCOOH can be evaluated. Values of log γ_{H^+} for a number of solvents were calculated by Izmailov[115], and are given in Table 9 together with ΔpK_M values for some representative carboxylic acids.

If the change in acid strength on transfer from water to M was solely determined by the change in solvation energy of the proton, ΔpK_M would be equal to $\log \gamma_{H^+}$. Table 9 (which can be extended by many more

TABLE 9. Values of $\log \gamma_{H^+}$ in different solvents together with ΔpK_M values for some representative acids

Solvent	$\log \gamma_{H^+}$	ΔpK_M			
		CH₃COOH	ClCH₂COOH	PhCOOH	p-NO₂C₆H₄COOH
H₂O	0·0	0·00	0·00	0·00	0·00
MeOH	3·2	4·95	4·98	5·02	4·99
EtOH	4·2	5·65	5·65	5·93	5·46
PrOH	4·2	—	6·12	—	—
BuOH	4·7	5·60	5·44	6·04	5·69
Acetone	3·9	—	—	7·75	7·18

ΔpK_M values) shows that it is indeed true that a considerable proportion of ΔpK_M is determined by log γ_{H^+} and that there is a very rough parallelism between these parameters for closely related solvents, such as the lower alcohols; but it also shows that changes in solvation of RCOOH and RCOO⁻ are important. For the dipolar aprotic solvent acetone, the parallelism does not hold, however, $\log \gamma_{RCOO^-} - \log \gamma_{RCOOH}$ being the dominant contribution to ΔpK_M.

Table 10 is taken from Konovalov's work[107], in which $\log \gamma_{RCOOH}$ values were determined by an e.m.f. method instead of the less accurate solubility measurement method, and gives ΔpK_M, $\log \gamma_{RCOOH}$ and

TABLE 10. ΔpK_M, $\log \gamma_{RCOOH}$, and $\log \gamma_{RCOO-}$ for selected carboxylic acids on transfer from water to methanol, ethanol, and acetone, respectively[107]

Acid	MeOH			EtOH			Acetone		
	ΔpK_M	$-\log \gamma_{RCOOH}$	$\log \gamma_{RCOO-}$	ΔpK_M	$-\log \gamma_{RCOOH}$	$\log \gamma_{RCOO-}$	ΔpK_M	$-\log \gamma_{RCOOH}$	$\log \gamma_{RCOO-}$
Acetic	4.95	1.38	0.37	5.65	1.57	-0.12	7.80	0.71	3.19
Chloroacetic	4.95	1.37	0.38	5.66	1.61	-0.15	6.95	0.41	2.64
Benzoic	5.20	1.62	0.38	5.93	1.96	-0.23	7.75	1.54	2.31
Salicylic	4.89	2.04	-0.35	5.59	2.45	-1.06	6.25	2.02	0.33
2-Nitrobenzoic	5.43	2.20	0.03	6.28	2.52	-0.44	7.58	2.13	1.55
3-Nitrobenzoic	4.81	2.40	-0.79	5.48	2.75	-1.54	7.17	2.56	0.71
4-Nitrobenzoic	4.99	2.43	-0.64	5.46	2.83	-1.57	7.18	2.60	0.68

log $\gamma_{\text{RCOO}}-$ for a number of carboxylic acids on transfer from water to methanol, ethanol, and acetone, respectively. The most important trend (Figure 8) is that acid molecules are more stabilized in these solvents than in water (log $\gamma_{\text{RCOOH}} < 0$). Anions are not much affected in methanol, log γ_{RCOOH} ranging from 0·4 to −0·7, whereas in ethanol stabilization occurs in all cases. Conversely, anions are destabilized in acetone, which

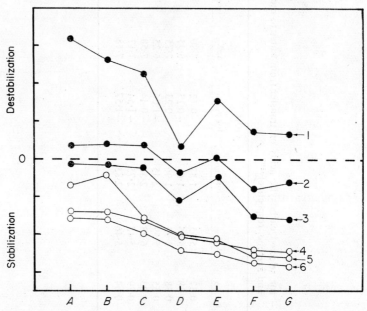

FIGURE 8. The variation of log γ_{RCOOH} and log $\gamma_{\text{RCOO}}-$ with changes in acid structure. (A) Acetic acid, (B) chloroacetic acid, (C) benzoic acid, (D) salicylic acid, (E) 2-nitrobenzoic acid, (F) 3-nitrobenzoic acid, and (G) 4-nitrobenzoic acid. (1) log $\gamma_{\text{RCOO}}-$ in acetone, (2) log $\gamma_{\text{RCOO}}-$ in methanol, (3) log $\gamma_{\text{RCOO}}-$ in ethanol, (4) log γ_{RCOOH} in methanol, (5) log γ_{RCOOH} in acetone, and (6) log γ_{RCOOH} in ethanol.

can be accounted for qualitatively because of its inability to function as a hydrogen bond donor. Generally, molecules and anions of aromatic acids seem to be more stabilized than those of aliphatic ones, although more data are clearly needed to establish this firmly. It may be noted that the salicylate anion is least affected by solvent interactions in the series of acids in Table 10, probably due to its ability to stabilize itself by 'internal solvation' through intramolecular hydrogen bonding (section VIII. C. 2).

Thus, the hydrogen-bonding properties of the solvent appear to be of primary importance in determining $\Delta p K_M$ on transfer from water to M, especially if M is an aprotic solvent. Both anions and molecules are affected by solvent, the stabilizing effect being roughly parallel for the two kinds of species (Figure 8).

Clare and coworkers[108] have attempted to assess the importance of hydrogen-bonding on the relative acidities of a wide variety of acids, including some carboxylic ones, on transfer from a dipolar aprotic solvent (DMF) to a protic one (MeOH). For equation (48), where B^- is a suitable reference base and HA represents a series of protonic acids, they expres-

$$HA + B^- \overset{K_r}{\rightleftarrows} HB + A^- \tag{48}$$

sed $(K_r)_{DMF}$ as a function of $(K_r)_{MeOH}$ and the medium activity coefficients, analogously to equation (46). These were factorized into γ_i^H, related only to the change in hydrogen-bonding interactions on transfer from DMF to MeOH, and γ_i^*, which accomodates all other interactions (equation 49).

$$(K_r)_{DMF} = (K_r)_{MeOH} \frac{\gamma_{HB}^* \gamma_{A^-}^* - \gamma_{HB}^H \gamma_{A^-}^H}{\gamma_{HA}^* \gamma_{B^-}^* - \gamma_{HA}^H \gamma_{B^-}^H} \tag{49}$$

On taking logarithms, equation (49) can be rearranged to equation (50). If the differences in $\log (K_r)_{DMF} - \log (K_r)_{MeOH}$ for a series of anions A^-

$$\log (K_r)_{DMF} = \log (K_r)_{MeOH} + \log \frac{\gamma_{A^-}^H}{\gamma_{B^-}^H} + \log \frac{\gamma_{HB}^* \gamma_{A^-}^*}{\gamma_{HA}^* \gamma_{B^-}^*} + \log \frac{\gamma_{HB}^H}{\gamma_{HA}^H} \tag{50}$$

are due entirely to differences in hydrogen-bonding solvation of A^- relative to B^-, then the final two terms in equation (50) must be constant. This means that a plot of $[\log (K_r)_{MeOH} + \log \gamma_{A^-}^H]$ against $\log (K_r)_{DMF}$ should be linear and have unit slope. Values of $\log \gamma_{A^-}^H$ were calculated as the quotient between the second-order rate constants for the reaction $A^- + CH_3I \rightarrow ACH_3 + I^-$ in DMF and methanol, respectively; this could be shown to be a justifiable procedure. The plot thus obtained was indeed linear and of unit slope, demonstrating the validity of the approach. The conclusions of this work may be stated as follows: 'If A^- and B^- differ markedly in their hydrogen bond acceptor properties, large changes in K_r occur on transfer from DMF to MeOH. If A^- and B^- are equivalent acceptors, the change, if any, will be small.'

Applied to carboxylic acid ionizations, the latter statement implies that anions of different carboxylic acids should be approximately equivalent acceptors of hydrogen bonds from methanol, since $\log K_r$ values on transfer from DMF to methanol are largely unchanged (Figure 5, Table 7, and equations 40 and 43).

However, changes in $\log K_r$ do occur on transfer from methanol to acetone, acetonitrile, or DMSO (Figure 5 and equations 44 and 45), indicating that the statements cited above cannot be extended to other dipolar aprotic solvents. Obviously, the ability of the solvent to accept hydrogen bonds from acid molecules cannot always be neglected ($\log \gamma_{HB}^H/\gamma_{HA}^H$ in equation 50 is not a constant).

D. Dispersion effects

Dispersion forces are weakly attractive forces between molecules, resulting from interactions between electronic oscillators, *i.e.* instantaneous dipoles which are created by the motion of electrons. In the previous discussion these forces were included in γ_i^*. Grunwald and Price[116] have considered the effect of dispersion forces on the equilibrium constant of equation (48) on transfer from water to methanol or ethanol, B^- being the acetate ion and the acids HA being picric and trichloroacetic acid, respectively. Now the picrate ion is a *delocalized* electronic oscillator, being capable of stronger dispersion interactions with solvent than *localized* ones, to which category the acetate and trichloroacetate ions, and undissociated picric, acetic, and trichloroacetic acid belong. If in equation (48) B^- is the acetate ion and A^- the picrate ion, one can qualitatively predict that K_r will increase when the solvent molecules become stronger centers of dispersion, *i.e.*, when A^-, being the only delocalized electronic oscillator of the equilibrium, can have its chemical potential lowered relative to the three other localized oscillators of the system. Such a change in K_r is expected on transfer from water to methanol or ethanol, since the molecules of these solvents are stronger centers of dispersion than water molecules. Table 11 gives K_r values for picric and trichloroacetic acid relative to acetic acid in water, methanol, and ethanol, showing that K_r actually unchanged for trichloroacetic acid. Grunwald and Price were also able to demonstrate by theoretical calculations that the effect on K_r is of the correct magnitude for a dispersion effect.

TABLE 11. K_r values for picric and trichloroacetic acids in water, methanol, and ethanol (acetic acid as reference)[116]

Solvent	$10^{-4}K_r$	
	Picric acid	Trichloroacetic acid
H_2O	2·6	~5
MeOH	78	6·0
EtOH	210	7·3

E. Isotope Effects

A special case of solvent effects is the effect on ionization constants of the transfer of the ionization process from ordinary water to D_2O. Since the general principles of this subject have been thoroughly discussed by Bell[5] and King[1], only recent developments will be summarized here.

A pD scale has been defined by Gary, Bates, and Robinson[117], and two reference buffer systems are available for calibration purposes.

Earlier data on the solvent isotope effect[118] suggested that the difference $(pK_{DA}-pK_{HA})$ was linearly correlated with pK_{HA}, as in equation (51) where subscript HA refers to any protic acid in H_2O, and DA to the cor-

$$pK_{DA}-pK_{HA} = a+b \, pK_{HA} \qquad (51)$$

responding deuterio acid in D_2O, and a and b are constants, provided that acids capable of giving intramolecular hydrogen bonds were excluded. Such acids were thought to give anomalously high or low $(pK_{DA}-pK_{HA})$ values. The value of $b \cong 0.02$ was of the correct order of magnitude to be expected if the origin of the isotope effect was a difference in zero-point energies for the H- and D-acids and if allowance was made for changes in hydrogen bonding in the dissociation process.

As more data have become available, equation (51) has been shown to be of doubtful validity, at least for carboxylic acids[118, 119]. However, phenols and alcohols follow this relationship[118], and for carboxylic acids forming strong intramolecular hydrogen bonds, $(pK_{DA}-pK_{HA})$ values do not always show anomalous behavior[119].

Hepler[120] extended his studies of substituent effects on acidity to a comparison between water and D_2O. Substituent effects were considered in terms of relations of the usual symmetrical type, as in equations (52) and (53) where HR and DR represent reference acids. Starting from equation

$$HA+R^- \xrightarrow{\text{H}_2\text{O}} A^-+RH \qquad (52)$$

$$DA+R^- \xrightarrow{\text{D}_2\text{O}} A^-+RD \qquad (53)$$

(51), he derived equation (54), which predicts that substituent effects should

$$\frac{pK_{DA}-pK_{DR}}{pK_{HA}-pK_{HR}} = 1+b \qquad (54)$$

be uniformly greater for D-acids in D_2O than for H-acids in H_2O. Although available data were mostly in agreement with this qualitative prediction (except for intramolecularly hydrogen-bonded acids, where the quotient above is generally < 1), Hepler concluded that $(pK_{DA}-pK_{DR})/(pK_{HA}-$

pK_{HR}) was not a constant as demanded by equation (54); the reason for this was considered to be the failure of equation (51) for carboxylic acids. However, in the author's opinion, the disagreement is not very serious if experimental errors in the measured pK values are taken into account. Symptomatically, the largest deviations occur for acids where pK_{HA} and pK_{HR} are very close to each other, making $(pK_{DA} - pK_{DR})/(pK_{HA} - pK_{HR})$ very sensitive to experimental errors. This is evident from Figure 9, which shows a plot of $(pK_{DA} - pK_{DR})/(pK_{HA} - pK_{HR})$ versus pK_{HA};

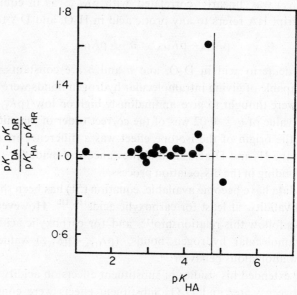

FIGURE 9. Plot of $(pK_{DA} - pK_{DR})/(pK_{HA} - pK_{HR})$ versus pK_{HA} (for sources of data, see reference 120). The solid vertical line indicates the position of pK_{HR} (acetic acid).

acids having intramolecular hydrogen bonds have been omitted. The dotted line corresponds to $b = 0.02$, and the solid vertical line indicates the position of pK_{HR} (acetic acid, $pK = 4.76$). Excluding the point at $pK_{HA} = 4.60$ (second pK of fumaric acid), a b value of 0.036 would seem to fit the data fairly well, considering the possible errors involved.

Hepler also analyzed the solvent isotope effect in terms of internal and external contributions to ΔH_r^0 and ΔS_r^0 of the equilibria (52) and (53), and concluded that *internal* energy effects, *i.e.* isotope effects on the O—H or O—D bonds, rather than external effects, such as isotope effects on dielectric constants of solvents or hydrogen bonding between solvent and solute, were important in determining the solvent isotope effect.

Bellamy, Osborn, and Pace[121] have measured the monomeric O—H and O—D stretching frequencies of some aliphatic carboxylic acids in a number of solvents. The relative frequency shifts $(\Delta v/v)$ for any of the acids in the range of solvents were plotted against the corresponding values for acetic H-acid as reference, giving a series of good linear correlations. The slopes of these lines should be a function of the proton donating power of the acids relative to that of the reference acid. Plotting the slopes for the H-acids and the D-acids gave two linear plots, one for each kind of acid, indicating that the proton donating power of any particular acid as defined above is directly related to its acid strength in water. It could also be shown that H-acids are stronger than D-acids in non-aqueous solvents.

In the vapor phase, v_{OH} and v_{OD} were found to be constant for the acids investigated, which covered a range from the very strong trichloroacetic acid to the weak trimethylacetic acid. Thus, in the free gaseous state the acids all have the same OH(D) force constant, unaffected by the presence of strongly polar substituents. However, when the acid is transferred to a medium where association between solvent and acid becomes possible, the influence of the substituent is important in determining the strength of the hydrogen bond formed. This kind of interaction is detectable even in carbon tetrachloride, since values of v_{OH} in this solvent reflect differences in aqueous pK values, and thus the difference between carbon tetrachloride and other more polar solvents as proton acceptors is one of magnitude.

Isaacs[122] determined the extent of protonation of weakly basic indicators in CH_3COOH and CH_3COOD and concluded that protonation occurs more readily ($\sim 10\%$ more) in the protic acid.

VIII. STRUCTURAL EFFECTS

A. Conformational Effects

The influence of conformational equilibria on acidity constants was first studied in the cyclohexane series, using the general principles set out by Eliel and Lukach[123] and Winstein and Holness[124]. A relationship between ionization constants and conformational equilibria in the cyclohexane carboxylic acid system (see scheme below) was derived and used for the calculation of ΔG^0_{COOH} and $\Delta G^0_{COO^-}$, defined as $-RT \ln K_{COOH}$ and $-RT \ln K_{COO^-}$, respectively. Stolow[125] combined the ionization constant data in water for cis- and trans-4-methyl cyclohexane and cyclohexanecarboxylic acid, and, using the known value of $\Delta G^0_{CH_2}$,

$$\text{COOH-cyclohexane} + H_2O \quad \underset{K_{COOH}}{\rightleftharpoons} \quad \text{cyclohexane-COOH} + H_2O$$

$$\big\updownarrow K_{axial} \qquad\qquad\qquad\qquad \big\updownarrow K_{equatorial}$$

$$\text{COO}^- \text{-cyclohexane} + H_3O^+ \quad \underset{K_{COO^-}}{\rightleftharpoons} \quad \text{cyclohexane-COO}^- + H_3O^+$$

obtained $\Delta G^0_{COOH} = -1.7 \pm 0.2$ kcal/mole and $\Delta G^0_{COO^-} = -2.4 \pm 0.4$ kcal/mole. He also chose the ionization constants of *cis*- and *trans*-4-*t*-butylcyclohexanecarboxylic acid (in 66% DMF–water) to represent acidity constants of pure conformations with an axial and equatorial carboxyl group, respectively, and arrived at $-\Delta G^0_{COOH} = 1.2$–2.0 kcal/mole. Using the same approach, Tichy, Jonas and Sicher[126] calculated $\Delta G^0_{COOH} = -1.6 \pm 0.3$ kcal/mole and $\Delta G^0_{COO^-} = -2.2 \pm 0.3$ kcal/mole from data in 80% methyl cellosolve–water, whereas van Bekkum, Verkade, and Wepster[25] obtained $-\Delta G^0_{COOH} = 1.5$–1.6 kcal/mole from data in 50% ethanol. Recently, Sicher, Tichy and Sipos[24] have redetermined ΔG^0_{COOH} and $\Delta G^0_{COO^-}$ by ionization constant measurements on some other systems and have arrived at slightly different values, -1.2 kcal/mole and -1.8 kcal/mole, respectively. Table 12 gives the data used for the abovementioned calculations.

Inspection of Table 12 reveals that an acid with an axial carboxyl group is distinctly weaker than one with an equatorial carboxyl, as measured by the pK difference, 0.4–0.5 pK units, between *cis*- and *trans*-4-*t*-butylcyclohexanecarboxylic acid. The acid-weakening effect has been attributed to steric inhibition of solvation of the carboxylate anion in the axial conformation[125, 126], where the bulk of the molecule should prevent solvation from one side. This effect is normally found in carboxylic acids containing groups with great steric demands (section VIII. B. 1), and it appears reasonable to explain the difference between an axial carboxyl and a relatively unhindered equatorial one in this way.

The use of ionization constant data for evaluating ΔG^0_{COOH} and $\Delta G^0_{COO^-}$ is not entirely satisfactory from several points of view. Firstly, the method of calculation involves differences between ionization constants which are fairly close to each other and is thus very sensitive to experimental errors. Secondly, it is not known how the temperature influences the ionization constants employed in the calculations. In section V. B it has been

TABLE 12. Ionization constant data for cyclohexanecarboxylic acids

Acid	pK in			
	H_2O^{125}	DMF–H_2O 66%[125]	Methyl cellosolve– H_2O 80%[24, 126]	EtOH–H_2O 50%[25]
Cyclohexanecarboxylic	4·89	7·81	7·43	6·30
trans-4-Methylcyclohexane- carboxylic	4·90	—	7·41	—
cis-4-Methylcyclohexane- carboxylic	5·04	—	7·67	—
trans-4-Isopropylcyclohexane- carboxylic	—	—	7·39	—
cis-4-Isopropylcyclohexane- carboxylic	—	—	7·75	—
trans-4-t-Butylcyclohexane- carboxylic	—	7·79	7·39	6·28
cis-4-t-Butylcyclohexane- carboxylic	—	8·23	7·84	6·78

shown that ionization constants change with temperature in a seemingly erratic manner (probably due to changes in hydrational equilibria), and since ionization constants of a supposedly sterically hindered acid (axial COOH) and a relatively unhindered one (equatorial COOH) are used in the calculation of the conformational equilibria, it is possible that a slight change in temperature may change the conformational free energies without necessarily being due to changes in the equilibrium positions. In other words, the choice of 25°c as a standard temperature gives one set of values, whereas measurements at, say, 40°c, might well give considerably different values as a result of solvation changes and not because of changes in the position of the conformational equilibrium. To exemplify, the Winstein–Holness treatment might as well be applied to the conformational situation around the C—COOH bond in propionic acid, taking acetic and trimethylacetic acids to represent pure conformations; but as is evident from Figure 1, such a calculation would be of rather doubtful value because of the lack of knowledge as to which factors really make the ionization constants differ from each other. Thirdly, the use of aqueous organic solvents may influence pK values in the manner described in section II. C, and an error of unknown magnitude may thus be introduced.

It is therefore gratifying to note that conformational free energies from ionization constant data are in reasonable agreement with values obtained by equilibration of cis- and trans-4-t-butylcyclohexanecarboxylic acids, a method which is more accurate and in addition allows the determination

17 C.C.A.E.

TABLE 13. Thermodynamic functions for the conformational equilibria between axial and equatorial carboxyl, carboxylate, and carbonyl chloride groups[127]

Group	Solvent	$-\Delta G^0_{25}$ (kcal/mole)	$-\Delta H^\circ$ (kcal/mole)	ΔS° (e.u.)
COOH	Dodecane	1.38 ± 0.10	1.63 ± 0.02	-0.83 ± 0.05
	Aq. diglyme, 10 mole % H_2O	1.36 ± 0.22	1.56 ± 0.05	-0.68 ± 0.09
	Aq. diglyme, 50 mole % H_2O	1.20	1.12^a	$+0.28$
COO⁻		2.0 ± 0.5	2.12 ± 0.06	-0.54 ± 0.12
COCl	Dodecane	1.3 ± 0.5	1.38 ± 0.06	-1.3 ± 0.5

^{a} Calculated from two points only.

of both conformational enthalpies and entropies. Table 13 gives data obtained by Reese and Eliel[127] for the carboxyl group in three different solvents and for the carboxyl chloride group in dodecane. Within the limits of error ΔG^0_{COOH} and $\Delta G^0_{COO^-}$ determined by the equilibration method are in good agreement with the values calculated from ionization constant data.

Some other important conclusions can be drawn from the data in Table 13. From dodecane, a completely nonpolar solvent, to the polar solvent 10 mole % water–diglyme, the change in thermodynamic functions for the conformational equilibrium of the carboxyl group is very small. Changes are somewhat greater on transfer to 50 mole % water–diglyme (data are less precisely known in this solvent, however), especially in the entropy term, which is slightly positive, whereas it is negative for the two former solvents. The negative entropy contribution in dodecane is not likely to originate from differences between dimerization constants of the acids, as indicated by the very similar values of ΔG^0, ΔH^0, and ΔS^0 for the non-hydrogen-bonding COCl group in dodecane. Actually, an axial carboxyl, being weaker as an acid than an equatorial one, would be expected to give the higher dimerization constant[100], so that any correction for differences in dimerization constants would tend to make the entropy change more negative.

A more attractive explanation can be found in terms of preferred conformations around the C—COOH bond in the axial and equatorial conformations (see section IV. A). The axial carboxyl group must exist in two enantiomeric conformations with eclipsed carbonyl and C—C bonds (23 and 24), whereas 25 with eclipsed carbonyl and C—H bonds is very

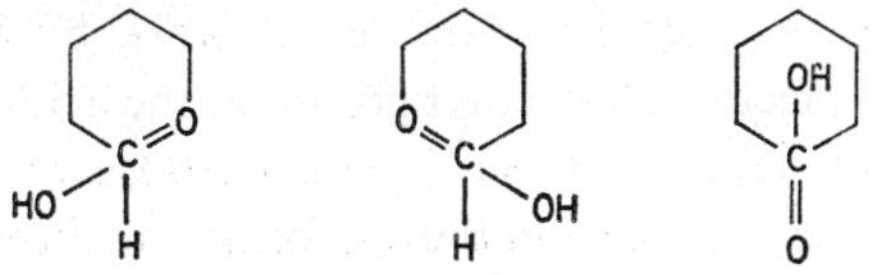

(23)   (24)   (25)

unfavorable due to repulsion between the OH group and the *syn* axial hydrogens. The situation in the equatorial carboxyl conformation is less obvious. Two enantiomeric conformations (26 and 27) with eclipsed carbonyl and C—C bonds are possible in addition to a third one with eclipsed carbonyl and C—H bonds (28), and since nothing is known about the

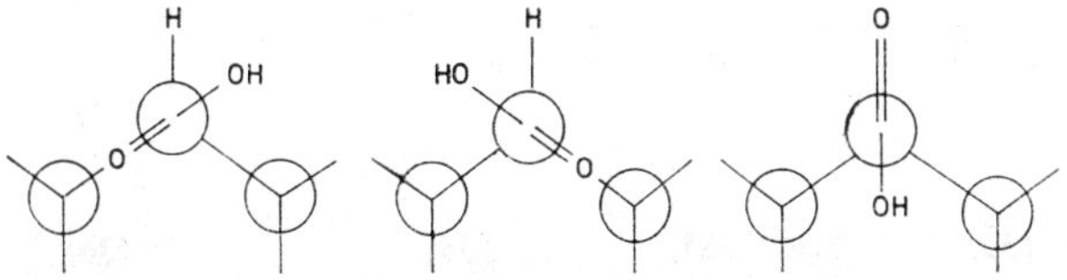

(26)   (27)   (28)

conformational preference of the carboxyl group in this type of system, it is necessary to rely on assumptions. It is therefore postulated that an equatorial carboxyl group prefers conformation 28 so that this becomes the predominant one. With this assumption the axial conformer will be favored by an entropy of mixing factor R ln 2, since it exists in two enantiomeric conformations (23 and 24), but the equatorial conformer will prefer one conformation (28) to a large extent. This picture is in good agreement with the values of ΔS^0_{COOH} and ΔS^0_{COCl} in dodecane given in Table 13. In the hydrogen-bonding solvent 10 mole % water–diglyme, dimerization does not occur in competition with solvent–solute interactions (section VII. C). Again, the value of ΔS^0_{COOH} is consistent with the explanation discussed above. The value reported for 50 mole % water-diglyme is obtained from only two measurements and is less accurate than the others. The change to a slightly positive ΔS^0_{COOH} might indicate a change in conformational preference of the carboxyl group in the equatorial conformation, but may also be caused by solvation effects.

Irrespective of the method used for determination, $\Delta G^0_{COO^-}$ comes out as more negative than ΔG^0_{COOH}, indicating a larger effective size of the carboxylate ion. This has been ascribed to solvation effects on the ion[126].

Additional, if not altogether conclusive, evidence for the above assumptions about the conformational preference of the carboxyl group in axial

17*

or equatorial positions comes from recent work by Sicher, Tichy and Sipos[23, 24] on the effect of a 2-methyl group on the acidity of cyclohexane-carboxylic acid. Values of pK in 80% methyl cellosolve–water (pK_{MCS}^*) for acids **29–36** were determined and the conformational free energy, ΔG_{COOH}^0

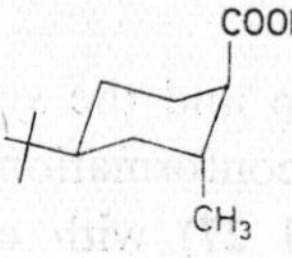

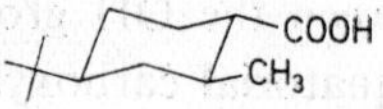

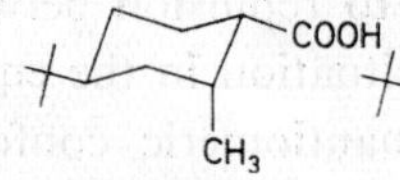

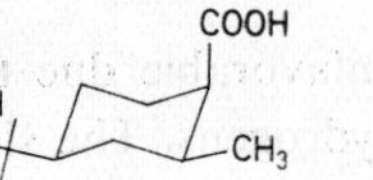

	(29)	**(30)**	**(31)**	**(32)**
pK_{MC}^*	7·80	7·42	7·60	8·27

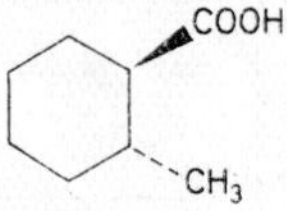

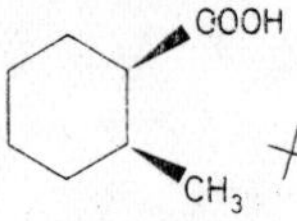

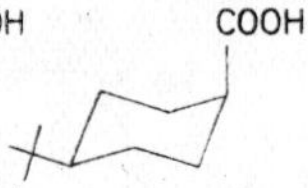

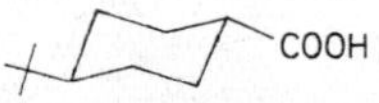

	(33)	**(34)**	**(35)**	**(36)**
pK_{MCS}^*	7·42	7·70	7·85	7·39

for cis-2-methylcyclohexanecarboxylic acid (equation 52) was calculated

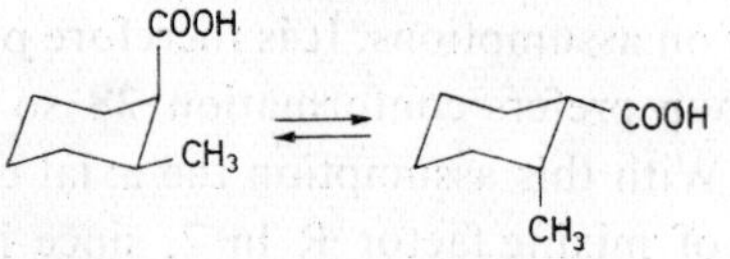

$$(52)$$

to be $-0\cdot6\pm0\cdot2$ kcal/mole, and ΔG_{COO-}^0 to be $-1\cdot5$ kcal/mole. If additivity of the individual conformational free energies is assumed, ΔG_{COOH}^0 for equation (52) should actually be $-1\cdot15-(-1\cdot7) = +0\cdot6$ kcal/mole, a difference of more than 1 kcal/mole. For the ionized form the deviation from additivity is even larger, about 1·5 kcal/mole. It was concluded that the axial conformer must be subject to a powerful destabilizing effect, which could be further substantiated by comparison of the conformationally rigid 4-t-butyl-2-methyl-substituted acids **29–32** with their parent acids (**35 and 36**). The difference in pK between the methyl-substituted and the appropriate parent acid was denoted ΔpK_{Me} (see below). These

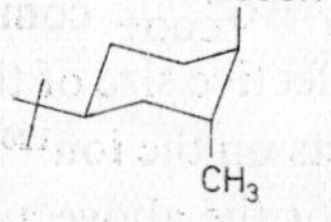

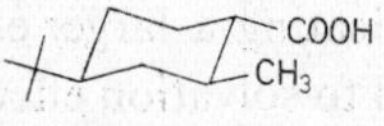

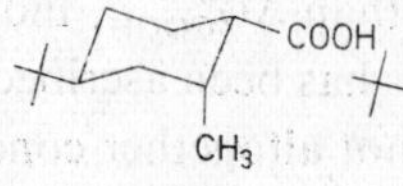

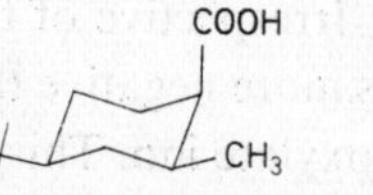

ΔpK_{Me}	$-0\cdot05$	$0\cdot03$	$0\cdot21$	$0\cdot42$

ΔpK_{Me} values were rationalized by assuming that conformations **23** and **28** were the preferred ones for an axial and equatorial carboxyl group, respectively (and for the corresponding ions). A 2-methyl group tends to force the carboxyl group into an unfavorable conformation and this effect would be especially pronounced in acids **31** and **32**. Since the effective size of the carboxylate ion is larger than that of the carboxyl group, this effect should be stronger in the ionized form and hence an acid-weakening effect is observed.

Van Bekkum, Verkade, and Wepster[25] have reached similar conclusions in the course of their investigations on substituted cyclohexanecarboxylic acids. Both groups have commented on the validity of Simon's rules[128], which were intended to correlate structure and acidity in certain cyclohexanecarboxylic acids in terms of one increment assigned to any γ-*syn* hydrogen and one assigned to 1-methyl substitution or ring junction in the 1-position. These rules do not allow for changes in the rotational orientation of the carboxyl group, and van Bekkum and coworkers revised them accordingly to provide for this complication.

B. Steric, Polar, and Resonance Effects

As mentioned in the introduction, ionization and rate constants constitute an important part of the experimental basis for testing theories of substituent effects and their propagation in organic molecules. The general subject of structure–reactivity relationships has been critically examined recently, both from the theoretical[129] and empirical[130] point of view, and treatises on the effect of structural changes on acidity can be found in any modern textbook in physical organic chemistry[74], in special monographs[1, 5], or in review articles[6, 131], and needs no repetition here. The following discussion will be concerned with some selected problems in the field of substituent effects, mainly with the intention of demonstrating the limitations of current concepts and theories. For convenience, the usual separation into steric, polar, and resonance effects will be retained.

1. Steric effects

Steric effects on carboxyl group acidity can be classified as *primary*, being caused by a direct steric interaction between a substituent or other part of the molecule and the acidic center, and *secondary*, being caused by an overall change in the molecule by one or several bulky alkyl groups[132]. To exemplify, the forcing of the carboxyl group into an axial conformation by a *cis*-4-*t*-butyl group in cyclohexanecarboxylic acid would be a secondary steric effect, whereas the steric inhibition of solvation inflicted by the *syn* hydrogens in this conformation would be a primary steric effect. In this section primary steric effects will be discussed, with one exception. Intramolecular hydrogen bonding, which can be classified as a particular primary steric effect, heavily dependent on secondary ones, will be reserved for a separate discussion (section VIII. C).

Some effects of steric origin have already been encountered in connection with conformational equilibria (section VIII. A). Thus, the fact that an axial carboxyl is less acidic than an equatorial one has been ascribed to steric inhibition of solvation in the anion, as is generally postulated for explaining the lower acidity of di- and trialkylacetic acids as compared to acetic acid. Table 14 gives some examples of this effect. This assumption implies that solvation of the acid molecule itself is more or less unchanged by increasing the number of, and/or steric demands of, alkyl groups attached to the α carbon. Since a protic solvent will form stronger hydrogen

TABLE 14. Ionization constant data for sterically hindered aliphatic acids in 50% (v/v) methanol at 40°C

R^2 R^2—C—COOH R^3			$-\log K_r$	Reference
R^1	R^2	R^3		
H	H	H	0	—
t-Bu	H	H	0·55	133
i-Pr	i-Pr	H	0·85, 0·79	134, 133
t-Bu	Me	H	0·70	134
t-Bu	Et	H	0·76, 0·81	134, 133
t-Bu	i-Pr	H	1·07	134
t-Bu	*t*-Bu	H	1·35	133
neopentyl	Me	H	0·49	133
neopentyl	neopentyl	H	0·96	134
Et	Et	Me	0·72	135
Et	Et	Et	0·86, 0·89, 0·96	135, 134, 133
Et	Et	Pr	0·89	135
Et	Et	i-Pr	1·09	135
Et	Et	i-Bu	1·01	135
Et	Et	neopentyl	1·05	135
Me	Et	i-Pr	0·93	135
Me	i-Pr	i-Pr	1·28	133
Et	i-Pr	i-Pr	1·54	133
Me	*t*-Bu	neopentyl	1·33, 1·62	134, 133
Me	Me	neopentyl	0·94	135
Me	Me	*t*-Bu	1·19	134
i-Pr	i-Pr	*i*-Pr	1·67	133

bonds to an ion than to an acid molecule due to the partly electrostatic character of the hydrogen bond, *e.g.*, compare the enthalpy of formation of the HF....HF bond (about -7 kcal/mole[76]) to that of the F...H...F$^-$ hydrogen bond (-37 kcal/mole[136]), this picture is in at least qualitative agreement with facts. An exception is, however, the case of cyanoacetic versus diisopropylcyanoacetic acids[7], which are of equal strength in water (actually, there is a reversal in acid strength at 285°K, the diisopropyl acid being the stronger one at lower temperatures). Ives and Marsden[7], starting from thermodynamic data for the ionization of fatty acids, attempted to rationalize this result as being due to a *solvent exclusion* effect. Every carbon atom in the hydrocarbon group except the α carbon was assigned an increment in terms of $-\Delta H_r^0$ and $-T \Delta S_r^0$ (relative to acetic acid), and it was then found that two of the β carbons had an appreciable effect, and that the γ carbons all contributed with large increments, whereas δ, ε and

$$\begin{array}{c} \beta \\ | \\ \varepsilon\!-\!\delta\!-\!\gamma\!-\!\beta\!-\!\alpha\!-\!\text{COOH} \\ |\quad |\quad | \\ \delta\quad \gamma\quad \beta \\ | \\ \gamma \end{array}$$

carbons further away had negligible effects. On rotation about the α-carbon–carboxylate bond, the hydrocarbon group sweeps out a certain volume of low dielectric constant in the surrounding solvent, β and γ carbons being most effective. Thus, the ion is divided into two regions, one where solvent exclusion has led to a reduction in dielectric constant and redistribution of the ionic field, and one 'watery' region at the carboxylate end which is much the same for all types of carboxylate ions and which is the center of primary hydration by hydrogen bonding to water. By the field redistribution the field strength will increase in the carboxylate region and this will increase the extent and stability of hydration.

Although consistent with some data for fatty acids and the cyano acid observations, this model hardly accounts for the data from highly alkyl-substituted acetic acids. Qualitatively, the model by itself would predict an acid-strengthening effect of the accumulation of alkyl groups, which is contrary to what is observed.

Another model related to ideas about hydrophobic bonding[137] can be briefly described in the following way. The primary hydration shell around the carboxylate end of the ion consists of a highly ordered arrangement of water molecules, outside which a zone more disordered than unperturbed water prevails[138]. Large alkyl groups will penetrate into this zone and increase the order by 'hydrophobic hydration' (hydration due to hydrophobic bonding), thus decreasing the entropy of the ion relative to acetate ion. Very few thermodynamic data on the ionization of di- and trialkylacetic acids are available, but the existing ones are in qualitative agreement with this model (Table 3).

Clearly, whichever of these models one prefers, the mode of operation classifies this acid-weakening effect as an *external* one in Hepler's treatment (section V.B), *i.e.* $(\Delta H_r^0)_{int}$ relative to acetic acid (equation 29) should be close to zero. However, to meet this requirement for a difference of 0·5 pK units, it must be assumed that ΔS_r^0 is about -40 e.u. (for a β value of 280°K), which appears quite unreasonable in view of the data in Table 3. Alternatively, one would have to admit an electron-repelling inductive effect of alkyl groups which for some reason does not operate in the cyano acid case and is anyway strongly dependent on temperature (Figure 1). It is evident that any interpretation of fatty acid ionizations in terms of the inductive effect of alkyl groups will be virtually meaningless. Accurate data on the thermodynamics of ionization for highly alkyl-substituted acetic acids in water are urgently needed to resolve these difficulties.

A second type of primary steric effect[23-25] has been defined in the axial (the carboxyl group) conformer of 2-methylcyclohexanecarboxylic acid (section VIII. A), where the equatorial methyl group is assumed to force the carboxyl group into the unfavorable conformation **25**. This effect should be more pronounced in the ionic form because of the larger effective size of the carboxylate ion, leading to a larger energy difference between acid and ion and consequently a larger pK. An alternative explanation based on the concept of steric inhibition of solvation in the anion cannot be excluded, however, as long as no independent experimental evidence for a purely conformational effect is available.

Another steric effect related to the above is the acid-strengthening effect due to the presence of ortho alkyl groups in aromatic acids. The alkyl groups will force the ionizing group out of the plane of the phenyl group thereby decreasing the conjugative interaction between the groups. Because of its negative charge, the decrease will be less pronounced in the ion than in the acid molecule, and hence an increase in acid strength will occur. Some representative data on this effect have been collected in Table 15. As already mentioned in section III, symmetry factors must be considered when benzoic acid is the reference acid, making log K_r for o-alkylbenzoic acids 0·3 pK units higher. Thus, o-toluic acid is of almost exactly the same strength as benzoic acid after applying this correction (section III). In benzene with 1,3-diphenylguanidine as the reference base[111],

TABLE 15. Log K_r values for some sterically hindered benzoic acids

Substituents	log K_r in H_2O	log K_r in 50% MeOH	Reference
None	0	0	—
2-Methyl	−0·01[a]	—	9
2-Ethyl	0·11[a]	—	9
2-Isopropyl	0·27[a]	—	9
2-t-Butyl	0·37[a]	—	9
2,3-Dimethyl	0·17[a]	—	9
2,6-Dimethyl	0·96	—	9
2,6-Diisopropyl	1·01[b]	—	139
2,5-Di-t-butyl	—	0·14[a]	140
2,4,6-Tri-t-butyl	—	−1·04	140

[a] Corrected for symmetry differences.
[b] Extrapolated value.

o-toluic acid is actually weaker than benzoic acid, which was accounted for by assuming that steric inhibition of resonance is augmented in water because of the larger size of the carboxyl group in this medium (due to hydration).

However, turning to other o-alkylbenzoic acids, the acid-strengthening effect is observed in the 2-ethyl, -isopropyl, and -t-butyl derivatives (see Table 15) the trend being that expected for a steric effect. Introduction of two o-methyl groups has a still larger effect in the same direction, as is probably also true for the 2,6-diisopropyl acid[139] (the value of log K_r is somewhat uncertain since it is obtained by extrapolation from data in ethanol–water mixtures). 2,6-t-Butyl groups, on the other hand, lower the acidity by about a power of ten as compared to benzoic acid, and again steric inhibition of solvation has been invoked[140].

Steric inhibition of resonance has also been proposed to explain the higher acidity of cis-3-alkylacrylic acids compared to the corresponding trans acids, but this has recently been questioned since one can find just as many data contradicting as supporting this view[141].

2. Polar effects

The effect of a dipolar or charged substituent on acidity is commonly assumed to be electrical in origin, but two conflicting views exist as to its mode of operation and propagation[129]. According to one view, it is thought that the charge separation at the site of the substituent, be it a dipole or a point charge, is propagated through the bonds connecting it with the acidic center only by successively weaker polarizations, thereby influencing the strength of the ionizing bond (section V.B). Alternatively, the polarization induced in the bond connecting the acidic group to the carbon chain is assumed to change the ability of the ion to disperse its charge, an electron-withdrawing substituent resulting in a greater dispersion of charge in a carboxylate ion than an electron-repelling one. In section V.B it was mentioned that none of these views of the inductive effect (the term 'inductive' will be used here to denote a through-the-bonds propagated effect) was in accordance with thermodynamic data on carboxylic acid ionizations, or, for that matter, phenol and thiol ionizations, except possibly in Hepler's modification. Neither temperature nor solvent effects are accomodated within the concept of the inductive effect, which would be expected to be invariant with respect to these variables.

According to the second view, the polar effect is an electrostatic effect (a field effect) from the dipole of charge on the dissociating proton, propagated through the surrounding solvent (the original Bjerrum treatment) or through both solvent and molecule (the Kirkwood–Westheimer treatment[142]). The work of transfer of the proton from the ionizing group to a point at an infinite distance away in the solvent is dicided into an electrostatic part, w_{el}, a non-electrostatic part, w_{non-el}, and a statistical part, $RT \ln \sigma$, where σ is the symmetry number defined in section III.

$$\Delta G^0 = w_{el} + w_{non-el} + RT \ln \sigma.$$

The non-electrostatic part includes bond dissociation energies, changes in translational, rotational, and vibrational partition function, specific solvation effects, and intramolecular hydrogen bonding effects. By comparing two matched acids as in equation (8) it is hoped that the non-electrostatic parts will cancel, so that the $\log K_r$ value will be a true reflection of the difference in electrostatic effect from the substituents in the two acids, allowance being made for possible differences in symmetry properties. $RT \ln \sigma$ is generally incorporated into $\log K_r = pK$ (reference acid) $- pK$ (given acid) $- \log \sigma$.

For a charged substituent (e.g. COO^- in a dicarboxylic acid) $\log K_r$ is calculated as the work required to move the proton from the position of the ionizing group (in a carboxylic acid, the proton is generally placed on the line bisecting the O—C—O angle at a distance of 1.45 Å away from the carbon atom) to infinity under the influence of the field of the charge, and is given by equation (53), where z is the charge number, e is the electronic charge, D is the dielectric constant of the solvent, r is the distance between

$$\log K_r = ze^2/2.3kTDr \qquad (53)$$

the charge and the proton, and k and T have their usual meanings. For a dipolar substituent the relation between μ, the dipole moment of the X—C bond, θ, the angle between the dipole moment vector and the line connecting its midpoint with the proton, and the quantities defined about, is given by equation (54).

$$\log K_r = e\mu \cos \theta / 2 \cdot 3kTDr^2 \qquad (54)$$

These equations were later modified by Kirkwood and Westheimer, who exchanged D for D_E, the effective dielectric constant, which has a value between that of the solvent and that of the low dielectric constant cavity defined by the molecule. The value of the internal dielectric constant is usually set equal to 2·00, the dielectric constant of a liquid n-alkane. D_E is a complicated function of the shape and size of the molecule, and the locations of the charge or dipole and the acidic center within it. Due to mathematical difficulties, molecules have to be treated as being either spherical or ellipsoidal.

The electrostatic model is conceptually more attractive than the inductive one, since it is directly related to ΔG^0, the maximum useful work obtainable from the process. It does provide for the solvent effect connected with a change in the dielectric constant (section VII. B) but not for temperature effects. It also accounts for differences in pK between stereoisomeric molecules (cis and trans forms, different conformations), although admittedly in a crude way. The assumption that specific solvent effects cancel in equation (8) is a questionable one, and it has been shown that intramolecular hydrogen-bonding effects do not cancel (section VIII. C. 1). A certain arbitrariness in the choice of the molecular parameters necessary for the computation of D_E is another weakness, as is also the assumption that the solvent can be treated as an isotropic, linear, and homogeneous medium.

Tanford[143] modified the Kirkwood–Westheimer treatment by considering the molecular cavities to be ellipsoids of revolution and the depth of the charge or dipole in the cavity to be the important parameter. Dewar and Grisdale[144] introduced a field effect term in their treatment of substituent effects, but used an r^{-1} term to account for the electrostatic interaction between an effective monopole, located at the carbon atom to which the substituent is attached, and the acidic center.

Necessary requirements for the application of the Kirkwood–Westheimer method or the Tanford modification of it are that the molecules of interest must be free from conjugative, steric, and intramolecular hydrogen-bonding effects, be approximately spherical or ellipsoidal in shape, and be sterically rigid in order to allow fixation of structural parameters and avoid complications due to conformational equilibria. Only a few series of carboxylic acids satisfying these requirements have been used for testing the electrostatic models, 1,4-derivatives of bicyclo[2.2.2]octane being the favorite choice after the pioneering study by Roberts and Moreland[145]. Holtz and Stock[146] calculated log K_r values for a number of 4-substituted (dipolar substituents) bicyclo[2.2.2]octanecarboxylic acids, using both the original Kirkwood–Westheimer and the Tanford approach. The former model underestimates log K_r by a factor of two or more, whereas the latter one is in considerably better agreement with experiments. A plot of μ/r^2 versus pK for all acids except the H- and alkylsubstituted ones gave an approximately linear correlation, from the slope of which a value of $D_E = 5 \cdot 6$ could be calculated. This value was in good agreement with those predicted by the Tanford model for the various acids but not with the Kirkwood–Westheimer theory, which predicts a fairly wide range of D_E values. Hydrophobic hydration was considered a possible cause of the anomalous behaviour of the H- and alkyl-substituted acids.

Wilcox and McIntyre[147] investigated charged substituents in the 4-position of the same system and treated the results using the Tanford model. A reverse treatment of the problem was adopted, the depth of charge in the cavity being calculated from the observed log K_r values. The theoretical log K_r values for different values of the solvent dielectric constant were then calculated in order to uncover the factor responsible for the strong solvent dependence of σ values for charged substituents. The great sensitivity of the theoretical log K_r values for charged substituents towards changes in solvent dielectric constant, which contrasts strongly with the calculated behavior of dipolar substituents, led Wilcox and McIntyre to conclude that the observed variation in σ constants of charged substituents is actually a bulk dielectric constant effect and not a specific solvent effect.

Wilcox and McIntyre considered the COO^- substituent to possess two opposing effects; (1) a negative potential field created by a point charge centered between the oxygens, and (2) a positive potential field originating from the dipole of the anion. Again using the Tanford model in the reverse manner, the dipole moment was calculated to be 3·2 Å, which appears to be a reasonable value.

Bell and Wright[148], using ionization constant data on sulfocarboxylic acids of the type $HOCO(CH_2)_nSO_3^-$ in combination with electrostatic theory, were similarly forced to conclude that the SO_3^- group acts as a superposition of a charge and a dipole, thus accounting for the fact that it is electron-withdrawing at short distances and electron-repelling at longer distances.

Ritchie and Lewis[149] studied the ionization behavior of 4-substituted bicyclo[2.2.2]octane-1-carboxylic acids in a range of solvents, including methanol, ethanol, DMSO, and acetone, as well as mixtures of these with water, and concluded that no single electrostatic theory could correlate all of the data obtained. Only for the H- and $(CH_3)_3N^+$-substituted acids were linear correlations between $\log K_r$ values and $1/D_E$ observed, and it was concluded that specific interactions between solvent and solute might alter the effect of the substituents, which is a reasonable assumption in view of the discussion in section VII.C.

Ionization constant data for *trans*-4-substituted cyclohexanecarboxylic acids have also been found to agree reasonably well with predictions based on the electrostatic model, especially in Tanford's modification[150].

K_1/K_2 ratios for saturated dicarboxylic acids[151] have been extensively used in connection with the Kirkwood–Westheimer theory, but in most cases suffer from complications due to the uncertainty as to which conformation is the preferred one. An attempt to estimate the barrier to internal rotation around the central C—C bond in succinic acid from the K_1/K_2 ratio and the Kirkwood–Westheimer theory[152] gives a reasonable value for the barrier, though this is probably fortuitous. The theory would generally not be expected to differentiate between such small effects.

Unsaturated systems, such as 3-substituted acrylic acids and substituted benzoic acids, are less suitable for comparison with electrostatic models because of possible conjugative interactions between the substituent and the acidic center. For references to earlier work in this field and critical discussions, the reader is referred to papers published by Bowden[153] and Roberts and Jaffé[154]. The general conclusion appears to be that both the inductive and field effects are important, although a strict separation is difficult to make on the basis of the experimental material available. In *ortho*-substituted benzoic and *cis*-3-substituted acrylic acids the field affect is a major constituent of the total electrical effect[155, 156].

In a recent paper, Exner[157] has presented evidence that the polar effect is transmitted more effectively to the acidic center from the *para* than the *meta* position, *i.e.* λ in equation (55) is larger than unity. This cannot be

$$\log K_r^p = \lambda \log K_r^m \tag{55}$$

rationalized on the basis of an inductive effect, and Exner also claimed that the field effect, as calculated by the Kirkwood–Westheimer theory, was not adequate either. However, the calculations[153] referred to by Exner

were the result of a simplified treatment, which equated the D_E for *meta*
and *para* derivatives, *i.e.*, only the $\mu \cos \theta / r^2$ factor was considered. Even
so a λ slightly larger than unity is obtained for dipolar substituents if the
structural parameters originally used by Kirkwood and Westheimer are
used (the proton located on the line bisecting the O—C—O angle 1·45 Å
from the carbon atom, and the dipole located at the mid- point of the dipolar bond). A correction for differing D_E values for *meta* and *para* derivatives would lead to a still larger λ, since a larger D_E is to be expected for
a *meta* compound. Of course, this merely demonstrates the shortcomings
of the electrostatic theories and by no means implies that the π-inductive
effect introduced to rationalize a value of $\lambda > 1$, does not operate in aromatic systems.

3. Conjugation effects

To assess the importance of conjugation effects on carboxylic acid equilibria one is faced with the problem of separating conjugation effects from
polar and steric ones. For a carboxylic acid it would be expected that
conjugative interactions are more effective in the acid molecule than in the
ion, resulting in a stabilization of the acid molecule and a decrease in
acid strength. The small change in pK between benzoic and acetic acids
(4·20 and 4·76, respectively) is accordingly the result of a superposition of
two opposing effects, an electron-withdrawing polar effect of the phenyl
group and a resonance interaction between the benzene ring and the carboxyl group, the two effects being approximately equal in magnitude.
Some idea of the magnitude of the conjugation effect can be obtained from
ionization constant data on 2,6-dimethyl- and 2,6-diisopropylbenzoic acids,
in which the carboxyl group is forced out of the plane of the benzene ring
(Table 15) and the conjugation is accordingly weakened or absent. These
acids are stronger than benzoic by roughly a factor of 10, but since it is not
known to what extent steric inhibition of solvation (compare with the
2,4,6-tri-*t*-butyl acid) may influence the acidity the conjugation effect may
be even greater, perhaps a factor of 50. This would make the polar effect
of a phenyl group, as compared to methyl, connected directly to the carboxyl group correspond to a pK change of about 2·3. This estimate is in
reasonable agreement with the polar effect of the phenyl group on anilinium ion acidity, experimentally established to correspond to 2·7 pK
units[158].

Considering aromatic acids without added complications due to substituents, it does not appear possible to interpret the observed differences in
strength of the various positional isomers in polynuclear systems in terms

of resonance alone[6]. The differences are generally small and may depend on solvent, e.g. the pK difference between 2- and 1-naphthoic acid[159], is 0·46 in water, 0·36 in 20% dioxane–water, 0·06 in 50% butyl cellosolve–water, and 0·05 in 78% ethanol–water. It is therefore doubtful if any consistent and meaningful interpretation of these data is possible at present.

The ionization constants of 1-, 5-, and 6-azulenecarboxylic acids have been measured (6·67, 5·91, and 5·75, respectively, in 50% ethanol–water, where benzoic acid has pK 5·46) and are found to correlate well with the electron density of the carbon to which the carboxyl is attached[160]. The agreement may be fortuitous, since hydrolysis rates for the corresponding esters did not fit in at all with the electron densities.

The conjugative interaction between substituents and the carboxyl group, transmitted by conjugation through the unsaturated hydrocarbon residue, has been the subject of much recent work[130, 157]. Generally an empirical separation in terms of substituent parameters for the conjugation and polar effects, σ_R and σ_I, is possible, and it happens that resonance interactions are generally small, except in the case of substituents with conjugative character opposite to that of the carboxyl group (e.g. the OH and NH_2 group). Steric inhibition of resonance between the substituent and the benzene ring has been studied in 3,5-dimethyl-4-substituted benzoic acids[161], and for sterically demanding substituents like the nitro, carbomethoxy, and dimethylamino groups the observed pK changes were in close agreement with predictions based on σ_R values.

Summarizing, resonance effects on carboxyl group acidity are generally small, except when mutual conjugation between the substituent and the acidic center is possible. In many cases involving small pK changes, judgment regarding the relative importance of the polar and resonance effects should be reserved until studies have been performed at other temperatures and in other solvents.

The same general conclusion might well be applied to all qualitative and quantitative treatments of substituent effects on acidity. Often small, and sometimes even large, alterations in pK may be due to solvent or temperature, and any interpretation in terms of electronic effects will therefore have to be modified to fit in with the results, a far from satisfactory state of affairs. 'Anomalies' are often ascribed to solvation effects, which is probably closer to the truth than anything else, inasmuch as the solvent must be of fundamental importance in connection with ionization equilibria. It actually seems as if the acidity of a given acid is more a reflection of the solvent properties than of those of the solute, and future investigations must aim at solving this important problem. A beginning realization of this aspect is reflected in recent articles on dispersion effects[116], hydrogen-bonding effects[108, 162], and general solvation effects[163, 164].

C. Intramolecular Hydrogen Bonding and Acidity

Following a proposal by Jones and Soper[165] to ascribe the anomalously high K_1/K_2 ratio of cis-caronic acid (3,3-dimethyl-cis-cyclopropane-1,2-dicarboxylic acid) to intramolecular hydrogen bonding, McDaniel and Brown[166] suggested in 1953 that this concept might be extended to a number of other dicarboxylic acids, e.g. dialkylmalonic acids, maleic acid, and tetramethylsuccinic acid. In their opinion, the field effect, as calculated by the Kirkwood–Westheimer treatment, could not alone be responsible for the high ratios observed. Intramolecular hydrogen bonding in the acid ion would facilitate the first ionization step and make the second one more difficult, thereby increasing the K_1/K_2 ratio. The crucial role of steric effects, such as minimization of rotation of carboxyl groups by alkyl substitution or relief of steric strain in the cyclic hydrogen-bonded structure as compared to the open one, was emphasized. Some years later Westheimer and Benfey[167] partly corrected this picture by showing that the K_1/K_E ratio, where K_E is the ionization constant of the monomethyl ester of the diacid, forces an upper limit to the factor by which intramolecular hydrogen bonding can contribute. Thus, for diethylmalonic, maleic, and tetramethylsuccinic acids internal hydrogen bonding makes an appreciable contribution to the K_1/K_2 ratio, but is not a dominant factor. For phthalic acid, K_1/K_E in water is 2, which is equal to the statistical factor; this criterion would therefore seem to indicate that the acid phthalate ion does not exist in an internally hydrogen-bonded form.

A large number of investigations related to this subject have been published since the time of the above-mentioned reviews, and the following discussion will attempt to summarize the results and to obtain a consistent picture of the role of internal hydrogen bonding in determining acidity.

1. Dicarboxylic acids

As already mentioned in section VII.C, a COOH group can act both as a hydrogen bond donor and acceptor, whereas a COO⁻ group can function only as an acceptor. On the basis of the simple electrostatic model of the hydrogen bond, anions would be expected to form stronger hydrogen bonds than neutral molecules, and available data support this view for the carboxylate ion as compared to the unionized carboxyl group. In a number of homoconjugated carboxylate ions (section VII. C) of the type RCOOH...OCOR⁻, infrared, n.m.r. and x-ray diffraction results indicate that the intermolecular hydrogen bond is of the single minimum, symmetrical, very strong type in the solid state, whereas intermolecular hydrogen bonds between —COOH and neutral molecules belong to the asymmetrical, double minimum type. The strength of the HF..HF bond compared to the F..H..F$^-$ bond is also indicative of this large difference (section VIII.B.1).

For intramolecular hydrogen bonds the situation is analogous. In the solid state, maleic acid possesses an asymmetrical hydrogen bond[168], whereas in the acid ion the internal hydrogen bond appears from n.m.r.[94], infrared[169], x-ray diffraction[170], and neutron diffraction[171] data to be a symmetrical one. The symmetrical hydrogen bond persists in the asymmetrically substituted acid chloromaleate ion[172]. Conversely, phthalic acid[173] and the acid phthalate ion[174] in the solid state do not possess intramolecular hydrogen bonds. In the acid the carboxyl group planes are forced out of the plane of the benzene ring by about 30°, whereas in the monoanion the carboxyl and carboxylate group planes make an angle with the benzene ring of 30 and 75°, respectively.

An x-ray investigation of the crystal structure of dipotassium ethylenetetracarboxylate[175] showed that the ion is not planar and that no intramolecular hydrogen bonds exist in it.

The existence of a particular hydrogen-bonded structure in the solid state does not necessarily prove that the same structure will remain stable in solution. The intramolecular hydrogen bond may be broken by competing external hydrogen bonding to solvent molecules or the molecule may adopt a different conformation in solution. Borderline cases involving the simultaneous presence of internally and externally hydrogen-bonded conformations are to be expected and the experimental ionization constants K_1 and K_2 will then be the weighted means of those of the individual conformations.

Before considering this problem, the pertinent experimental material will be examined. Table 16 is a compilation of pK_1, pK_2, log K_1/K_2 and log K_1/K_E values for alkyl-substituted malonic, succinic, and glutaric acids, open-chain and cyclic ethylene-1,2-dicarboxylic acids, cycloalkanedicarboxylic acids, and bicycloalkanedicarboxylic acids, selected to illustrate the different factors involved in determining the K_1/K_2 ratio of a given dicarboxylic acid. Additional data can be found in the references of Table 16, as well as in other papers treating diacids of unsymmetrical[176] and more unusual structures[177–180, 183]. Only diacids without polar substituents have been included in order to avoid the uncertainty involved in estimating the magnitude of the polar effect.

TABLE 16. Ionization constant data for dicarboxylic acids

Acid	pK_1	pK_2	$\log K_1/K_2^{a}$	$\log K_1/K_E^{a}$	Solvent[b]	Reference
Malonic	2·83	5·69	2·86	0·55	W	6
Methylmalonic	3·05	5·76	2·71	0·34	W	6
Ethylmalonic	2·99	5·83	2·84	0·50	W	6
i-Propylmalonic	2·94	5·88	2·94	—	W	6
t-Butylmalonic	2·92	7·04	4·12	—	W	181
Dimethylmalonic	3·17	6·06	2·89	0·40	W	6
Diethylmalonic	2·21	7·29	5·08	1·50	W	182
Ethylbutylmalonic	2·15	7·25	5·10	—	W	182
Ethylisopropylmalonic	2·07	8·10	6·07	—	W	182
Diisopropylmalonic	2·18	8·60	6·42	—	W	182
Succinic	4·19	5·48	1·29	0·33	W	6
rac-2,3-Dimethylsuccinic	5·44	7·35	1·91	0·44	E	184
rac-2,3-Diethylsuccinic	3·94	6·20	2·26	0·94	W	6
rac-2,3-Diisopropylsuccinic	5·04	8·17	3·13	—	E	184
rac-2,3-Di-t-butylsuccinic	3·51	6·60	3·09	1·25	W	6
rac-2,3-Diisopropylsuccinic	4·76	9·22	4·46	2·97	E	184
rac-2,3-Di-t-butylsuccinic	3·66	11·4	7·7	3·37	E	184
rac-2,3-Di-t-butylsuccinic	2·20	10·3	8·1	3·86	W	119
meso-2,3-Dimethylsuccinic	3·58	13·1	9·5	—	E	184
meso-2,3-Diethylsuccinic	3·77	5·94	2·17	0·50	W	6
meso-2,3-Diisopropylsuccinic	4·97	7·58	2·61	0·95	E	184
meso-2,3-Di-t-butylsuccinic	3·63	6·46	2·83	—	W	6
meso-2,3-Diisopropylsuccinic	5·37	7·43	2·06	0·67	E	184
meso-2,3-Di-t-butylsuccinic	5·98	8·10	2·12	0·41	E	184
meso-2,3-Diisopropylsuccinic	6·43	8·29	1·86	0·40	E	184
meso-2,3-Di-t-butylsuccinic	3·56	7·41	3·85	1·42	W	119
Tetramethylsuccinic	4·84	10·1	5·3	1·96	E	184

Acid	pK_1	pK_2	ΔpK			Ref.
rac-2,3-Diethyl-2,3-dimethylsuccinic[e]	4·45	10·4	6·0	—	E	185
meso-2,3-Diethyl-2,3-dimethylsuccinic	5·73	9·66	3·93	1·95	E	185
Tetraethylsuccinic	3·39	8·06	4·47	2·57	W	119
	4·78	11·4	6·6	—	E	184
Glutaric	4·42	5·44	1·02	—	W	186
3-Methylglutaric	4·35	5·44	1·09	—	W	186
3,3-Dimethylglutaric	3·85	6·45	2·60	—	W	186
3,3-Diethylglutaric	3·67	7·42	3·75	—	W	186
3,3-Diisopropylglutaric	3·63	7·68	4·05	—	W	186
3-Ethyl-3-methylglutaric	3·78	6·92	3·14	—	W	186
3-t-Butyl-3-methylglutaric	3·61	7·49	3·88	—	W	186
Fumaric	3·02	4·38	1·36	0·30	W	6
Maleic	1·92	6·34	4·42	1·02	W	187
Dimethylmaleic	3·1	6·1	3·0	—	W	188
Diethylmaleic	3·2	6·1	2·9	—	W	188
Cyclohex-1-ene-1,2-dicarboxylic	3·3	6·2	2·9	0·34	W	188
Phthalic	2·95	5·41	2·46	1·30	W	6
Cyclopent-1-ene-1,2-dicarboxylic	1·64	7·27	5·63	1·70	W	187
Bicyclo[2.2.1]hepta-2,5-diene-2,3-dicarboxylic	1·32	7·77	6·45	1·87	W	187
Bicyclo[2.2.1]hept-2-ene-2,3-dicarboxylic	1·32	8·00	6·68	2·31	W	187
Furan-3,4-dicarboxylic	1·44	7·84	6·40	2·20	W	187
Cyclobut-1-ene-1,2-dicarboxylic	1·12	7·63	6·51	—	W	187
trans-Cyclopropane-1,2-dicarboxylic	3·80	5·08	1·28	—	W	141
cis-Cyclopropane-1,2-dicarboxylic	3·56	6·65	3·09	—	W	141
trans-1,2-Dimethylcyclopropane-1,2-dicarboxylic	3·70	5·24	1·54	—	W	141
cis-1,2-Dimethylcyclopropane-1,2-dicarboxylic	4·06	6·56	2·56	—	W	141
trans-1,2-Diethylcyclopropane-1,2-dicarboxylic	3·53	3·03	1·50	—	W	141
cis-1,2-Diethylcyclopropane-1,2-dicarboxylic	4·16	6·58	2·42	—	W	141
trans-Caronic	3·82	5·32	1·50	—	W	165

(Table 16. continued)

Acid	pK_1	pK_2	$\log K_1/K^a$	$\log K_1/K_E$	Solvent[b]	Reference
cis-Caronic	2·34	8·31	5·97	—	W	165
trans-Cyclobutane-1,2-dicarboxylic	3·94	5·55	1·39	—	W	189
cis-Cyclobutane-1,2-dicarboxylic	4·16	6·23	2·07	—	W	189
trans-Cyclobutane-1,3-dicarboxylic	4·11	5·13	1·02	—	W	189
cis-Cyclobutane-1,3-dicarboxylic	4·08	5·12	1·04	—	W	189
trans-Cyclopentane-1,2-dicarboxylic	3·89	5·91	2·02	—	W	6
cis-Cyclopentane-1,2-dicarboxylic	4·37	6·51	2·14	—	W	6
trans-Cyclohexane-1,2-dicarboxylic	4·18	5·93	1·75	—	W	6
cis-Cyclohexane-1,2-dicarboxylic	4·34	6·76	2·42	—	W	6
Bicyclo[2.2.1]heptane-trans-2,3-dicarboxylic	4·07	5·66	1·59	—	W	9
Bicyclo[2.2.1]heptane-endo-cis-2,3-dicarboxylic	4·09	7·38	3·29	—	W	9
Bicyclo[2.2.1]heptane-exo-cis-2,3-dicarboxylic	4·52	6·72	2·20	—	W	9
Bicyclo[2.2.2]octane-trans-2,3-dicarboxylic	3·89	6·17	2·28	—	W	190
Bicyclo[2.2.2]octane-cis-2,3-dicarboxylic	4·63	6·90	2·27	—	W	190

[a] Not corrected for the statistical factor, since this will be the same for all acids except in a few cases.

[b] W = water, E = 50% (w/w) ethanol–water.

[c] This acid has been shown by resolution to be the racemic form[191].

Inspection of Table 16 reveals two opposite effects of alkyl substitution. In mono- and dialkylmalonic, racemic 2,3-dialkyl- and tetraalkylsuccinic and 3-mono- and 3,3-dialkylglutaric acids, increasing size and number of alkyl groups strongly increases log K_1/K_2, whereas in *meso*-2,3-dialkyl-succinic, maleic and dialkylmaleic acids, and cyclopropane-1,2-dialkyl-1,2-dicarboxylic acid, the trend is opposite and not so pronounced. Some of the cyclic and bicyclic unsaturated acids exhibit extremely high ratios, whereas the cycloalkanedicarboxylic acids appear to have normal log K_1/K_2 values. Generally, high K_1/K_2 ratios are accompanied by high K_1/K_E ratios, as demanded by the criterion for intramolecular hydrogen bonding in the acid ion developed by Westheimer and Benfey[167].

It should be pointed out that there is a fundamental difference (at least from the experimental point of view) between inter- and intramolecular hydrogen bonds[192]. Since an intermolecular hydrogen bond is formed between two molecules, the entropy term for the equilibrium A—H+ +B ⇌ A—H...B will be negative. Only relatively strong hydrogen bonds will be experimentally detectable, since an appreciable $(-\Delta H)$ term is required to balance the unfavorable $T\Delta S$ term, which in solution was estimated to be about 3 kcal/mole. Conversely, the formation of an intra-molecular hydrogen bond within a single molecule was not considered to give rise to any large entropy changes, since the contributions to the entropy of the molecule from the various bond-stretching vibrations would be small and only moderate changes in the associated force constants were to be expected upon hydrogen bond formation. Contributions to the entropy change from changes in solvation were also assumed to be negligible, and accordingly the total entropy change was approximately set to zero. In this case even weak hydrogen bonds will be reflected in experimentally accessible quantities, such as ionization constants and spectral changes.

Eberson[193] compared infrared spectra of alkylated succinic acids, their monoacid salts, and solutions of these in methanol and D_2O. The salts of acids with very high K_1/K_2 ratios displayed some very characteristic features, which were present in both the solid and liquid state. In D_2O, the asymmetric stretching frequency of the COO^- group was uniformly about 35 cm^{-1} higher than expected for a non-hydrogen-bonded structure and the carbonyl stretching frequency of the COOH group was about 30 cm^{-1} lower than that of a normal COOH group. Also, the carbonyl band was of low intensity, whereas a very strong and broad absorption occurred in the region 1600–1400 cm^{-1}. These spectra were quite different from those of the salts of diacids having low K_1/K_2 ratios, which were essentially superpositions of the simple COOH and COO^- group spectra.

Similarly, Dodd, Miller, and Wynne-Jones[194] studied solutions of maleic acid and potassium maleate in water, D_2O, methanol, and dioxane. They concluded that the intramolecular hydrogen bond present in maleic acid is not strong enough to withstand competition from external hydrogen bonding to these solvents, whereas the hydrogen bond in the acid maleate ion appeared to remain stable in water, D_2O, and methanol, but not in dioxane, a very good hydrogen bond acceptor. In D_2O, both internally and externally hydrogen-bonded species appeared to coexist in comparable equilibrium concentrations, indicating that the intramolecular hydrogen bond is actually a fairly weak one.

Chapman, Lloyd, and Prince[195] reached similar conclusions regarding the intramolecular hydrogen bond in the acid maleate ion, the acid di-n-propyl malonate ion, and the acid phthalate ion, which all appear to exist in the internally bonded form in D_2O. Likewise, Hanrahan[196] demonstrated intramolecular hydrogen bonding by infrared methods in some selected substituted acid malonate ions in D_2O. On the other hand, an infrared study[197] of the hydrogen malonate ion itself demonstrated that the ion is not internally hydrogen bonded, as was proposed on the basis of thermodynamic data[198]. The same conclusion was reached from an investigation of the carbon isotope effect in the formation of the hydrogen malonate ion[199].

Forsén[200] has shown that sodium hydrogen maleate and potassium hydrogen phthalate in DMSO show n.m.r. signals at very low field (about 15 p.p.m. downfield from an external water standard) and ascribed these to the internal hydrogen bond, which is retained in this solvent. Similar shifts[201] were observed in DMSO for potassium hydrogen rac-2,3- diisopropyl, rac-2,3-di-(t-butyl), rac-2,3-dicyclohexyl, and tetraethyl succinate, which all correspond to diacids with extremely high K_1/K_2 ratios. The meso forms of the above acid salts did not show any signal attributable to an internal hydrogen bond. If the magnitude of the chemical shift is an approximate measure of hydrogen bond strength, these acid ions must indeed possess very strong hydrogen bonds, even if opinion must be somewhat reserved because of the possible influence of the negative charge upon the chemical shift[151].

Another approach was used by Silver and coworkers[202], who measured the proton chemical shifts of the non carboxylic hydrogens of a number of dicarboxylic acids as a function of the degree of neutralization in water and methanol–water mixtures. For aliphatic saturated acids (except formic) neutralization was accompanied by an upfield chemical shift, whereas in phthalic, pyromellitic, and maleic acids, a downfield shift was observed on addition of the first equivalent of base, the second equivalent again causing an upfield shift. This result was interpreted in terms of a hydrogen-bonded, approximately planar acid ion, in which a ring current in the chelate ring would shift the protons in the aromatic ring towards lower fields. The solvent effect was in agreement with this interpretation.

Although spectral data also seem to support the intramolecularly hydrogen-bonded structure in solutions of the acid salts discussed above, some cases may still be doubtful. Potassium maleate has a strong, symmetrical hydrogen bond in the solid state, but the almost planar ion is considerably strained as indicated by the increase of the $C{=}C{-}C$ angles by about $10°$. Darlow and Cochran[170] estimated the strain to be about $7 \cdot 5$ kcal/mole, and it is quite possible that intermolecular forces in the crystal lattice may help to maintain the planar structure. This is not possible in aqueous solution, where in addition intermolecular hydrogen bonding may interfere with the internal hydrogen bond. The results obtained by Dodd, Miller, and Wynne-Jones[194] indicate that cyclic and open ions exist in comparable concentrations in D_2O, so that the measured log K_1/K_2 value may be a blend of those of several species. A Kirkwood–Westheimer calculation demonstrated that the field effect might actually be solely responsible for the K_1/K_2 ratio observed[194]. Similar calculations for highly alkylated succinic acids[184] were more conclusive as regards the influence of intramolecular hydrogen bonding. For rac-2,3-(t-butyl)-succinic acid, the calculated log K_1/K_2 ratio (assuming COOH groups to be gauche) is $3 \cdot 6$, which is far below the observed value, $9 \cdot 5$. Even if the shortcomings of the theoretical model are taken into account, no possible explanation can be given for this large difference except in terms of intramolecular hydrogen bonding. An estimate of this factor from the K_1/K_E ratio in combination with the calculated field effect was in good agreement with experiment[151, 184].

The data on phthalic acid are even more confusing. There is no intramolecular hydrogen bond in solid potassium hydrogen phthalate, whereas both i.r. and n.m.r. data in solution point to its existence. The K_1/K_E value is close to the statistical factor 2 and is a strong indication against it.

McCoy[187, 203] has provided an elegant demonstration of the very important effect of slight changes in geometry on the K_1/K_2 ratio. Assuming that the intramolecular hydrogen bond would have its largest effect on the K_1/K_2 ratio in a diacid where an O.....O distance of about $2 \cdot 4$ Å could be attained without serious steric distortions, he calculated this distance in a number of sterically rigid acids in their (hypothetical) undistorted planar structure. The results of these calculations are shown in Table 17. Again returning to maleic and phthalic acids, one can see that the O.....O dis-

TABLE 17. Log K_1/K_2 values and calculated O····O distances in the O····H····O grouping for some selected dicarboxylic acids[187]

Acid	log K_1/K_2	O..H..O (Å)
Cyclohex-1-ene-1,2-dicarboxylic	2·9	1·44
Phthalic	2·46	1·60
Maleic	4·42	1·68
Cyclopent-1-ene-1,2-dicarboxylic	5·63	1·92
Bicyclo[2.2.1]hepta-2,5-diene-2,3-dicarboxylic	6·45	2·07
Bicyclo[2.2.1]hept-1-ene-2,3-dicarboxylic	6·68	2·11
Furan-3,4-dicarboxylic	6·40	2·23
Cyclobut-1-ene-1,2-dicarboxylic	6·51	2·61

tance is much smaller than that required in a strain-free, planar, hydrogen-bonded ion. By small changes in the bond angles, which can be accomplished by incorporating the ethylene-1,2-dicarboxylic acid structure in cyclic systems, the O.....O distance increases until one arrives at acids with extremely high K_1/K_2 ratios, such as 1-cyclobutene-1,2-dicarboxylic acid. The increase in K_1/K_2 is paralleled by a strong increase in K_1/K_E.

This treatment demonstrates the importance of internal hydrogen bonding in determining the K_1/K_2 ratio in a most convincing way. By increasing the distance between the carboxyl groups the field effect will decrease and yet an enormous increase in log K_1/K_2 is observed. The crucial importance of the O....O distance is emphasized, and it appears highly probable that ions which cannot attain a planar structure having an O....O distance of about 2·4 Å without inducing severe steric strain, do not exist in solution as internally hydrogen-bonded forms. The acid phthalate ion almost certainly belongs to this class, and the acid maleate ion is probably a borderline case.

The decrease in K_1/K_2 ratio in the dialkylmaleic acids[185] and the 1,2-dialkylcyclopropane-1,2-dicarboxylic acids[141] as compared to the parent compounds becomes intelligible on the basis of McCoy's treatment. The steric interference between the alkyl groups and the carboxyl groups will increase the steric strain in a hypothetical planar ion, and the planar ion will thus be still more disfavored.

Another question is posed by the above treatment. What is the K_1/K_2 ratio in a saturated 1,2-dicarboxylic acid, where the carboxyl groups are so close as to preclude intramolecular hydrogen bond formation? Such an acid would have the carboxyl group planes almost parallel to each other and the field effect would be at its maximum. A consideration of molecular models suggests that bicyclo[2.2.1]heptane-*exo-cis*-2,3-dicarboxylic acid (37) with a log K_1/K_2 value of 2·20 in water might be the best model (the *endo* isomer probably has too strong interactions between the *endo*-5,6

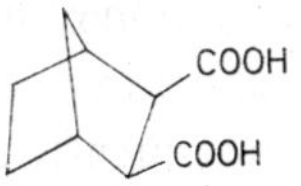

(37) 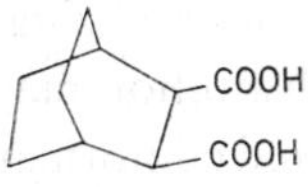   (38)

hydrogens and the carboxyl groups to be a suitable choice). *Cis*-bicyclo [2.2.2]octane-2,3-dicarboxylic acid (38) (log $K_1/K_2 = 2·27$) may also represent this type of acid. Since small changes in the conformation around the C—COOH bond may cause small changes in K_1/K_2, it is obviously futile to try to find better examples. As a crude estimate it is enough to know that the field effect between two *cis*-situated carboxyl groups on adjacent carbon atoms corresponds to a log K_1/K_2 ratio of 2–2·5, and that log K_1/K_2 ratios in excess of the latter value indicate that intramolecular hydrogen bonding may influence the K_1/K_2 ratio.

Having recognized that the attainment of an O...O distance of about 2·4 Å in an unstrained or nearly unstrained structure is the crucial factor, the data in Table 16 are now amenable to a more detailed discussion. In the malonic acid series the K_1/K_2 ratio of the parent compound is not influenced by intramolecular hydrogen bonding, since the field effect is enough to account for it. Alkyl and especially dialkyl substitution strongly increases the K_1/K_2 ratio, much more than the Kirkwood–Westheimer treatment can accomodate. From the geometry of the malonic acid molecule it is easily seen that an O....O distance of 2·5–2·6 Å is attainable without steric distortions (39), but that the hydrogen bond formed is not linear.

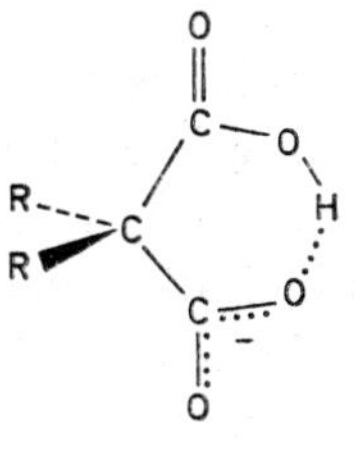

(39)

Nonlinear hydrogen bonds are considered to be less favorable than linear ones[76], and this probably means that any internal hydrogen bond in the malonate system is fairly weak. Another factor to be considered is the conformational situation around the C—COOH bonds. As pointed out in section IV, conformations with eclipsing CO and CH or eclipsing CO and CC are favored. To form the internal hydrogen bond, one of the carboxyl groups has to be somewhat removed from its favored position, thereby introducing a second destabilizing factor. Inspection of models reveals that alkyl and dialkyl substitution may possibly assist in pushing the carboxyl group out of its favored conformation and consequently ease the formation of the hydrogen-bonded form. A second conformational factor not hitherto recognized is the rotation around the C—OH bond within the carboxyl group itself. The stable *cis* conformation cannot possibly form an internal hydrogen bond, so that about 2 kcal/mole have to be spent before the *trans* conformation can be attained, as illustrated below. The same reasoning

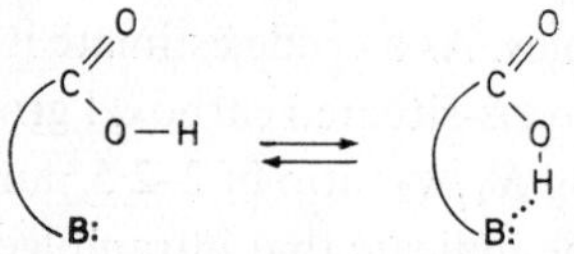

applies to almost all situations where the COOH group acts as a proton donor in an intramolecular hydrogen bond.

As they have at least two factors opposing intramolecular hydrogen bond formation, one may ask why it occurs at all in malonic acids. The answer probably lies in the explanation given by McDaniel and Brown[166], namely that steric strain in the open molecule is relieved in the cyclic form. This is commonly assumed to be the reason why alkyl substitution in a carbon chain leads to an increase in cyclization tendency (denoted as the gem-dialkyl effect, although the alkyl groups need not always to be geminal). Thus, in spite of the unfavorable conformational situation, intramolecular hydrogen bonding is still favored because of the decrease in steric strain possible in the cyclic arrangement.

This becomes still more evident in the succinic acid series. Succinic acid itself in its *gauche* conformation needs very little readjustment in the dihedral angle to form an almost linear hydrogen bond of the desired length (40). Yet there is no evidence from ionization constant data of intramolecular hydrogen bonding in succinic acid (in fact, dipole moment studies[204] show that the *trans* conformation is favored in dioxane solution), and this must be ascribed to the conformational factors mentioned above. A tetraalkyl-

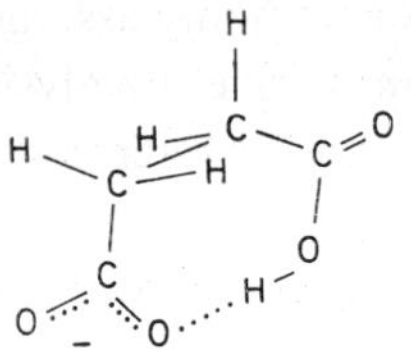

(40)

succinic acid, on the other hand, is severely strained and some of the strain can be relieved in a cyclic internally hydrogen-bonded conformation in spite of the increase in conformational strain.

The importance of secondary steric effects is also evident from the results in the *meso*- and *rac*-2,3-dialkylsuccinic acid series. Increasing the size of the alkyl groups leads to a complete dominance of a conformation with *gauche* carboxyls in the racemic series, and one with *trans* carboxyls in the *meso* series (**41** and **42**). The result is a very strong increase in log K_1/K_2 in the racemic series, whereas the opposite trend is observed in the *meso* series. The extreme example of this effect is 2,3-di-(*t*-butyl)-succinic acid[184], whose *meso* and racemic forms have log K_1/K_2 values of 1·86 and 9·5 (8·0 in water[119]), respectively, in 50% ethanol–water. Evidently, the bulky alkyl groups cause a severe steric strain in this molecule, and this is effectively relieved in the internally hydrogen-bonded structure.

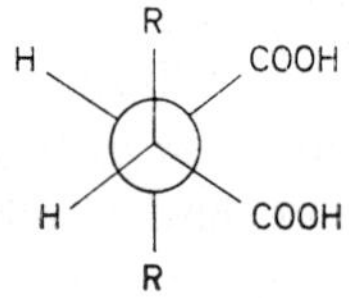

(41) 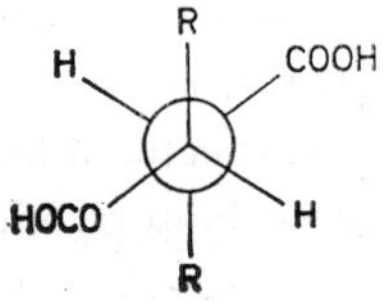   (42)

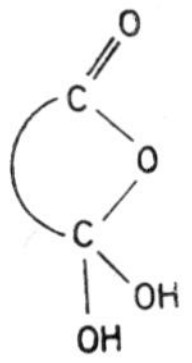

(43)

Recently, King[205] proposed that the old Anschütz structure (**43**) might be invoked to explain very high K_1/K_2 ratios. This possibility was discussed some years ago by the author[206] but dismissed on the basis of all the structural work known at that time. From the theoretical point of view, it is not very attractive, because **43** is commonly assumed to be the very short-lived tetrahedral intermediate in anhydride solvolysis, and is anyway completely ruled out today by the n.m.r. studies referred to above.

Glutaric acids are considerably more complicated. Even if it is evident that 3-alkyl substitution will favor conformations with carboxyl groups close to each other, the flexibility of the system nevertheless precludes any detailed discussion of the K_1/K_2 ratios[186]. Hydrogen-bonded and non-hydrogen-bonded conformations may exist in comparable equilibrium concentrations and there is no means of assessing the importance of hydrogen bonding from ionization constant data alone.

Acid ions of ethylene-1,2-dicarboxylic acids are theoretically best suited for intramolecular hydrogen bond formation. If an O.....O distance of 2·4–2·5 Å can be attained without steric distortion of the rest of the molecule, it will be almost linear. The conjugation between the double bond and the carboxyl groups will favor the planar arrangement, and the only destabilizing factor to be overcome is the *cis–trans* conformational change in the COOH group (**15 → 16**). It is also here that extreme log K_1/K_2 values are found and the best evidence for internal hydrogen bonding is obtained.

Additional evidence for intramolecular hydrogen bonding has been provided by Eyring and coworkers[182, 207–209] by a kinetic method. The reaction $HA^- + OH^- \xrightarrow{k} A^{2-} + H_2O$ can be studied by the temperature-jump method, and Table 18 gives a compilation of their data on several types of dicarboxylic acids. Figure 10 reveals that a high K_1/K_2 ratio of a diacid is accompanied by a low rate constant for the neutralization reaction. This is exactly the effect to be expected if an intramolecular hydrogen bond has to be broken in the process. Isotope effects in D_2O have also been studied but are not conclusive as regards the role of internal hydrogen bonding[209].

Except for cyclopropane-1,2-dicarboxylic acid and *cis*-caronic acid, intramolecular hydrogen bonding does not seem to be important in the ionization of cycloalkanedicarboxylic acids[211]. Due to the intricate conformational relationships in these molecules (compare section VIII. A) the small differences in log K_1/K_2 between *cis* and *trans* isomers are difficult to interpret and a discussion must be deferred until more data are available.

2. Hydroxy carboxylic acids

The *cis–trans* conformational change in the COOH group (**15 → 16**) required to form an internal hydrogen bond would seem to prevent the formation of weak intramolecular hydrogen bonds between the carboxyl group and acceptor groups other than the carboxylate ion. Thus, one would not expect that hydrogen bonds of the type COOH...Hal, COOH....S,

TABLE 18. Rate constants for the reaction $HA^- + OH^- \rightarrow A^{2-} + H_2O$ at 25° in aqueous solution

Acid	$\log K_1/K_2$	$10^{-7}k$ l/mole/sec	Reference
Diethylmalonic	4·90	28	182
Ethylbutylmalonic	5·10	16	182
Ethylisoamylmalonic	5·16	16	182
Dipropylmalonic	5·19	13	182
Ethylphenylmalonic	5·22	14	182
Ethylisopropylmalonic	6·07	5·5	182
Diisopropylmalonic	6·42	4·5	182
cis-Caronic	5·97	6·3	208
3,3-Diphenylcyclopropane-1,2-dicarboxylic	6·90	0·44	208
Tetramethylsuccinic	3·85	25	208
Tetraethylsuccinic	4·67	5·3	208
rac-2,3-Di-t-butylsuccinic	8·05	0·23	208
Maleic	4·42	9[a]	210

[a] Estimated from a value measured at 12°.

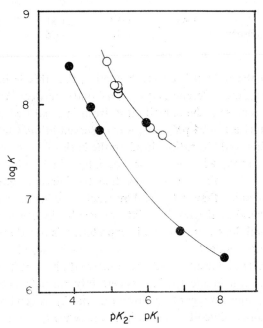

FIGURE 10. Plot of $\log k$ for reaction $HA^- + OH^- \rightarrow A^{2-} + H_2O$ against $pK_2 - pK_1$ for intramolecularly hydrogen-bonded acid ions (data from Table 18). Open circles correspond to 1,1-dicarboxylic acids, closed ones to 1,2-dicarboxylic acids.

COOH...OS, or COOH...O$_2$S would have any appreciable influence on the ionization constant of carboxylic acids containing halogen, sulfide, sulfoxide, or sulfone groups in suitable positions. Available data[9, 212] are in agreement with this prediction. This leaves us with cases where the carboxyl group or the carboxylate ion acts as a hydrogen bond acceptor, *e.g.* in hydroxy carboxylic acids, where no conformational obstacle to formation of OH....OCOH bonds exists.

Sicher and coworkers[213] have determined the ionization constants for the different axial (a) or equatorial (e) isomers of 3- and 4-*t*-butyl-2-hydroxy-cyclohexanecarboxylic acids and noticed some interesting regularities. Subtraction of the pK value of the 'parent' acid from that of the appropriate hydroxy acid gives the ΔpK_{OH} value, shown in Table 19. In the COOH^{a}OH^a conformation the difference is 0·45 pK units, considered to be

TABLE 19. Values of ΔpK_{OH} and ΔpK_{OH}(H-bond) for 4-*t*-butyl-2-hydroxycyclohexanecarboxylic acids[213]

Conformation	COOH^{a}OH^a	COOH^{e}OH^e	COOH^{e}OH^a	COOH^{a}OH^e
ΔpK_{OH}	0·45	0·57	0·81	1·39
ΔpK_{OH}(H-bond)	—	0·12	0·36	0·94

due solely to the inductive effect. Assuming that this is independent of stereochemistry, the difference, ΔpK_{OH}(H-bond), between ΔpK_{OH} and 0·45 was taken to represent the contribution from internal hydrogen bonding. The strongest effect, 0·94 pK units, was observed in the COOH^{a}OH^e conformation. This could be rationalized on the basis of a preferred conformation of the axial group which would be especially suited for internal hydrogen bond formation. Again, one assumes that the internal hydrogen bond is stronger in the anion than in the acid molecule, and hence an acid-strengthening effect would be the result of intramolecular hydrogen bonding. The effects are small, however, and the interpretation is based on the inductive effect being independent of stereochemistry (as it should be as a first approximation). The situation would be somewhat different if the polar effect were treated as a field effect, which is certainly not independent of stereochemistry. A judgment regarding intramolecular hydrogen bonding in these acids is obviously difficult to make in the absence of supporting physical measurements of other kinds.

A better case for the intervention of intramolecular hydrogen bonding can be made for salicylic acid[6, 214]. Ionization constant data for salicylic

acid and some reference compounds are given in Table 20. The first ionization step of salicylic acid is considerably stronger than that of p-hydroxy-
benzoic acid, whereas the OH ionization step is much weaker. An internally hydrogen-bonded acid ion is again in agreement with these data. The

TABLE 20. Ionization constant data for salicylic acid and some
reference compounds[214, 215] in water at 25°

Acid	pK of COOH group	pK of OH group
Benzoic acid	4·22	—
Phenol	—	9·86
p-Hydroxybenzoic acid	4·48	9·09
Salicylic acid	2·83	12·62
Ethyl p-hydroxybenzoate	—	8·34
Ethyl salicylate	—	9·92
3-Methylsalicylic acid	2·95	14·6

standard enthalpy and entropy of formation of the intramolecular hydrogen bond in the acid ion was calculated to be $-5\cdot0$ kcal/mole and $-4\cdot0$
e.u. Steric interference between a 3-methyl group and the hydroxyl group
in salicylic acid produces even more dramatic effects on the pK_1 and especially the pK_2 value, being 2·95 and 14·6, respectively, for 3-methylsalicylic
acid[215].

Intramolecular hydrogen bonding has also been invoked[6] to explain the
large difference in acidic behaviour between o-aminobenzoic and o-(N,N-
dimethylamino) benzoic acids (44 and 45). The decrease in pK_1 and increase
in pK_2 for the dimethylamino acid indicates an internally hydrogen-bonded
structure for its zwitterion (46). Temperature-jump measurements have
shown that the reaction between the zwitterion and hydroxide ion is slowed
down by at least three orders of magnitude from the diffusion-controlled
value[216], which is similar to what is observed for internally hydrogen-bonded acid ions of dicarboxylic acids (Table 18 and Figure 10).

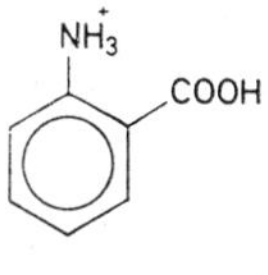

	(44)	(45)	(46)
pK_1	2·04	1·4	
pK_2	4·98	8·42	

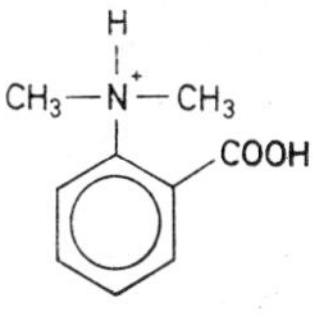

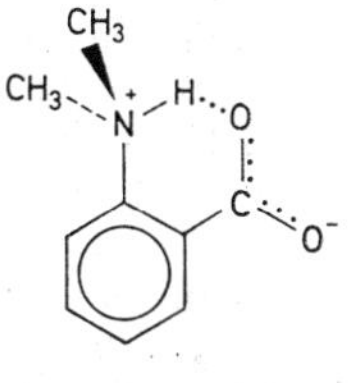

Intramolecular hydrogen bonding has also been studied in ions of EDTA and related compounds[217].

IX. REFERENCES

1. E. J. King, *Acid–Base Equilibria*, Pergamon Press, London, 1965.
2. A. Albert and E. P. Serjeant, *Ionization Constants of Acids and Bases*, Methuen, London, 1962.
3. C. Tanford and S. Wawzonek in *Physical Methods of Organic Chemistry*, Part IV (Ed. A. Weissberger), Interscience Publishers, New York, 1960, pp. 2915–3009.
4. T. Shedlovsky in *Physical Methods of Organic Chemistry*, Part IV (Ed. A. Weissberger), Interscience Publishers, New York, 1960, pp. 3011–3048.
5. R. P. Bell, *The Proton in Chemistry*, Methuen, London, 1959.
6. H. C. Brown, D. H. McDaniel and O. Häfliger in *Determination of Organic Structures by Physical Methods*, Vol. 1 (Eds. E. A. Braude and F. C. Nachod), Academic Press, New York, 1955, pp. 567–662.
7. D. J. G. Ives and P. D. Marsden, *J. Chem. Soc.*, 649 (1965).
8. D. D. Perrin, *Dissociation Constants of Organic Bases in Aqueous Solution*, Butterworths, London, 1965.
9. G. Kortüm, W. Vogel, and K. Andrussow, *Dissociation Constants of Organic Acids in Aqueous Solution*, Butterworths, London, 1961.
10. P. Ballinger and F. A. Long, *J. Am. Chem. Soc.*, **82**, 795 (1960).
11. R. J. Irving, L. Nelander, and I. Wadsö, *Acta Chem. Scand.*, **18**, 769 (1964).
12. M. M. Kreevoy, B. E. Eichinger, F. E. Stary, E. A. Katz, and J. H. Sellstedt, *J. Org. Chem.*, **29**, 1641 (1964).
13. R. G. Bates, M. Paabo, and R. A. Robinson, *J. Phys. Chem.*, **67**, 1833 (1963).
14. M. Paabo, R. A. Robinson and R. G. Bates, *J. Am. Chem. Soc.*, **87**, 415 (1965).
15. K. C. Ong, R. A. Robinson and R. G. Bates, *Anal. Chem.*, **36**, 1971 (1964).
16. For a review, see W. Simon, *Angew. Chemie Intern. Ed. Engl.*, **3**, 661 (1964).
17. W. Simon, G. H. Lyssy, A. Mörikofer and E. Heilbronner, *Zusammenstellung von scheinbaren Dissoziationskonstanten im Lösungsmittelsystem Methylcellosolve–Wasser*, Vol. I, Juris-Verlag, Zürich, 1959.
18. P. F. Sommer and W. Simon, *Zusammenstellung von scheinbaren Dissoziationskonstanten im Lösungsmittelsystem Methylcellosolve–Wasser*, Vol. 2, Juris–Verlag, Zürich, 1961.
19. W. Simon and P. F. Sommer, *Zusammenstellung von scheinbaren Dissoziationskonstanten im Lösungsmittelsystem Methylcellosolve–Wasser*, Vol. 3, Juris–Verlag, Zürich, 1963.
20. S. W. Benson, *J. Am. Chem. Soc.*, **80**, 5151 (1958).
21. A useful general introduction to this subject can be found in K. S. Pitzer, *Quantum Chemistry*, Prentice-Hall, New York, 1953, pp. 205–278.
22. Z. Welwart, *Bull. Soc. Chim. France*, 2203 (1964).
23. J. Sicher, M. Tichy and F. Sipos, *Tetrahedron Letters*, 1393 (1966).
24. J. Sicher, M. Tichy and F. Sipos, *Coll. Czech. Chem. Commun*, **31**, 2238 (1966).
25. H. van Bekkum, P. E. Verkade and B. M. Wepster, *Tetrahedron Letters*, 1401 (1966).
26. W. J. Tabor, *J. Chem. Phys.*, **27**, 974 (1957).
27. W. Weltner, Jr., *J. Am. Chem. Soc.*, **77**, 3941 (1955).
28. K. M. Sinnott, *J. Chem. Phys.*, **34**, 851 (1961).
29. C. C. Costain and G. P. Srivastava, *J. Chem. Phys.*, **41**, 1620 (1964).
30. E. Spinner, *J. Chem. Soc.*, 4217 (1964).
31. R. N. Jones and E. Spinner, *Can. J. Chem.*, **36**, 1020 (1958).

32. E. L. Eliel, N. L. Allinger, S. J. Angyal and G. A. Morrison, *Conformational Analysis*, Interscience Publishers, New York, 1965, p. 465.
33. I. Nagakawa, I. Ichisima, K. Kuratani, T. Miyazawa, T. Shimanouchi and S. Mizushima, *J. Chem. Phys.*, **20**, 1720 (1952).
34. R. J. Abraham and J. A. Pople, *Mol. Phys.*, **3**, 609 (1960).
35. J. P. Guillory and L. S. Bartell, *J. Chem. Phys.*, **43**, 654 (1965); *cf.* also W. D. Cotterill and M. J. T. Robinson, *Tetrahedron*, **20**, 777 (1964).
36. L. S. Bartell and J. P. Guillory, *J. Chem. Phys.*, **43**, 647 (1965).
37. T. Miyazawa and K. S. Pitzer, *J. Chem. Phys.*, **30**, 1076 (1959).
38. E. J. King, *Acid–Base Equilibria*, Pergamon Press, London, 1965, Chapter 8.
39. D. H. Everett and W. F. K. Wynne-Jones, *Trans. Faraday Soc.*, **35**, 1380 (1939).
40. H. S. Harned and R. A. Robinson, *Trans. Faraday Soc.*, **36**, 973 (1940).
41. D. J. G. Ives and J. H. Pryor, *J. Chem. Soc.*, 2104 (1955).
42. E. C. W. Clarke and D. N. Glew, *Trans. Faraday Soc.*, **66**, 539 (1966).
43. J. J. Christensen, R. M. Izatt, and L. D. Hansen, *J. Am. Chem. Soc.*, **89**, 213 (1967).
44. G. Öjelund and I. Wadsö, *Acta Chem. Scand.*, **21**, 1838 (1967).
45. W. J. Canady, H. M. Papée and K. J. Laidler, *Trans. Faraday Soc.*, **54**, 502 (1958).
46. D. H. Everett, D. A. Landsman and B. R. Pinsent, *Proc. Roy. Soc.*, **215A**, 403 (1952).
47. C. T. Mortimer, *Reaction Heats and Bond Strengths*, Pergamon Press, London, 1962, p. 180.
48. F. S. Feates and D. J. G. Ives, *J. Chem. Soc.*, 2802 (1956).
49. E. J. King, *J. Am. Chem. Soc.*, **82**, 3575 (1960).
50. L. F. Nims and P. K. Smith, *J. Biol. Chem.*, **113**, 145 (1936).
51. J. C. Ahluwalia, F. J. Millero, R. N. Goldberg and L. G. Hepler, *J. Phys. Chem.*, **70**, 319 (1966).
52. J. M. Sturtevant, *J. Am. Chem. Soc.*, **63**, 88 (1941).
53. J. M. Sturtevant, *J. Am. Chem. Soc.*, **64**, 762 (1942).
54. E. J. King, *J. Am. Chem. Soc.*, **73**, 406 (1951).
55. R. G. Bates and R. G. Canham, *J. Res. Natl. Bur. Std.*, **47**, 343 (1951).
56. R. G. Bates and G. D. Pinching, *J. Am. Chem. Soc.*, **71**, 1274 (1949).
57. G. Briegleb and A. Bieber, *Z. Elektrochem.*, **55**, 250 (1951).
58. Z. L. Ernst, R. J. Irving and J. Menashi, *Trans. Faraday Soc.*, **60**, 56 (1964).
59. D. S. Klett and B. F. Freasier, *J. Chem. Soc.*, 4741 (1962).
60. W. F. Hamer, G. D. Pinching and S. F. Acree, *J. Res. Natl. Bur. Std.*, **35**, 539 (1945); W. J. Hamer and S. F. Acree, *J. Res. Natl. Bur. Std.*, **35**, 381 (1945).
61. C. R. Allen and P. G. Wright, *J. Chem. Educ.*, **41**, 251 (1964); Ref. 47, Chap. 9.
62. D. T. Y. Chen and K. J. Laidler, *Trans. Faraday Soc.*, **58**, 480 (1962).
63. R. H. Boyd and C.-H. Wang, *J. Am. Chem. Soc.*, **87**, 430 (1965).
64a. L. G. Hepler, *J. Am. Chem. Soc.*, **85**, 3089 (1965).
64b. K. S. Pitzer, *J. Am. Chem. Soc.*, **59**, 2365 (1937).
64c. M. Born, *Z. Physik.*, **1**, 45 (1920).
65. E. Grunwald in *Progress in Physical Organic Chemistry*, Vol. 3 (Eds. S. G. Cohen, A. Streitwieser, and R. W. Taft), Interscience, New York, 1965, pp. 317–358.
66. M. Eigen, W. Kruse, G. Maass and L. De Maeyer in *Progress in Reaction Kinetics*, Vol. 2 (Ed. G. Porter), Pergamon Press, London, 1964, pp. 286–318.
67. M. Eigen, *Angew. Chemie, Intern. Ed. Engl.*, **3**, 1 (1964).
68. H. W. Nürnberg and H. W. Dürbeck, *Z. Anal. Chem.*, **205**, 217 (1964); H. W. Nürnberg and G. C. Barker, *Naturwissenschaften*, **51**, 191 (1964).
69. For reviews on the properties and use of dipolar aprotic solvents, see A. J. Parker in *Advances in Organic Chemistry*, Vol. 5 (Eds. R. A. Raphael, E. C. Taylor and

H. Wynberg), Interscience Publishers, New York, 1965, pp. 1–46; *Quart. Rev. (London)*, **16**, 163 (1962).

70. W. F. K. Wynne-Jones, *Proc. Roy. Soc.*, **140A**, 440 (1933).

71. R. P. Bell, *The Proton in Chemistry*, Methuen, London, 1959, Chapter 4. For recent discussions of the electrostatic solvent theory, see G. Aksnes, *Acta Chem. Scand.*, **16**, 1967 (1962); J. F. J. Dippy, S. R. C. Hughes and B. C. Kitchner, *J. Chem. Soc.*, 1275 (1964).

72. M. Mandel and P. Decroly, *Nature*, **201**, 290 (1964).

73. L. R. Dawson, J. W. Vaughan, M. E. Pruitt and H. C. Eckstrom, *J. Phys. Chem.*, **66**, 2684 (1962).

74. K. B. Wiberg, *Physical Organic Chemistry*, John Wiley and Sons, New York, 1964, p. 259.

75. E. J. King, *Acid–Base Equilibria*, Pergamon Press, London, 1965, Chap. 10.

76. G. C. Pimentel and A. L. Mc Clellan, *The Hydrogen Bond*, W. H. Freeman, San Francisco and London, 1960.

77. For a critical discussion, see M. M. Davis and M. Paabo, *J. Org. Chem.*, **31**, 1804 (1966).

78. N. A. Izmailov, V. A. Kremer, L. M. Kutsyna, and E. V. Titov, *Chem. Abstr.*, **54**, 23629f (1960); A. S. Naumova and V. S. Smorodinov, *Chem. Abstr.*, **61**, 12693e (1964).

79. L. B. Reeves, *Can. J. Chem.*, **39**, 1711 (1961).

80. J. J. Lindberg and C. Majani, *Suomen Kemistilehti*, **38**, 21 (1965).

81. D. Hadzi and N. Kopilarov, *J. Chem. Soc. (A)*, 439 (1966).

82. L. B. Reeves, *Trans. Faraday Soc.*, **55**, 1684 (1959).

83. H. N. Farrer and F. J. C. Rossotti, *Acta Chem. Scand.*, **17**, 1824 (1963).

84. S. L. Johnson and K. A. Rumon, *J. Phys. Chem.*, **69**, 74 (1965).

85. K. Bauge and J. W. Smith, *J. Chem. Soc. (A)*, 616 (1966).

86. D. Hadzi, A. Novak, and J. E. Gordon, *J. Phys. Chem.*, **67**, 1118 (1963).

87. I. M. Kolthoff and M. K. Chantooni, *J. Phys. Chem.*, **66**, 1675 (1962).

88. J. C. Speakman, *J. Chem. Soc.*, 3357 (1949).

89. J. M. Skinner and J. C. Speakman, *J. Chem. Soc.*, 185 (1951).

90. J. M. Skinner, G. M. D. Stewart, and J. C. Speakman, *J. Chem. Soc.*, 180 (1954).

91. J. C. Speakman, *Proc. Chem. Soc.*, 316 (1959).

92. J. C. Speakman and H. H. Mills, *J. Chem. Soc.*, 1164 (1961); L. Golic and J. C. Speakman, *J. Chem. Soc.*, 2530 (1965).

93. G. E. Bacon and N. A. Curry, *Acta Cryst.*, **10**, 524 (1957).

94. R. Blinc and D. Hadzi, *Spectrochim. Acta*, **16**, 852 (1960).

95. R. Blinc, D. Hadzi, and A. Novak, *Z. Elektrochem.*, **64**, 567 (1960).

96. D. Hadzi and A. Novak, *Infrared Spectra of, and Hydrogen Bonding in, Some Acid Salts of Carboxylic Acids*, University of Ljubljana, 1960.

97. I. M. Kolthoff and M. K. Chantooni, Jr., *J. Phys. Chem.*, **70**, 856 (1966); *J. Am. Chem. Soc.*, **87**, 4428 (1965).

98. J. E. Gordon, *J. Phys. Chem.*, **67**, 19 (1963).

99. J. F. Coetzee and G. P. Cunningham, *J. Am. Chem. Soc.*, **87**, 2534 (1965).

100. J. E. Gordon, *J. Org. Chem.*, **26**, 738 (1961).

101. E. E. Schrier, M. Pottle, and H. A. Scheraga, *J. Am. Chem. Soc.*, **86**, 3444 (1964).

102. W. Kauzmann, *Advan. Protein Chem.* **14**, 1 (1959); G. Némethny and H. A. Scheraga, *J. Phys. Chem.*, **66**, 1773 (1962); E. S. Hand and T. Cohen, *J. Am. Chem. Soc.*, **87**, 133 (1965).

103. D. Eagland and F. Franks, *Trans. Faraday Soc.*, **61**, 2468 (1965).

104. P. Mukerjee, *J. Phys. Chem.*, **69**, 2821 (1965).

105. F. H. Verhoek, *J. Am. Chem. Soc.*, **58**, 2577 (1936).

106. N. A. Izmailov, V. S. Chernyi, and L. L. Spivak, *Russ. J. Phys. Chem. (Engl. Transl.)*, **37**, 427 (1963).
107. O. M. Konovalov, *Russ. J. Phys. Chem. (Engl. Transl.)*, **39**, 364 (1965).
108. B. W. Clare, D. Cook, E. C. F. Ko, Y. C. Mac, and A. J. Parker, *J. Am. Chem. Soc.*, **88**, 1911 (1966).
109. I. M. Kolthoff and T. B. Reddy, *Inorg. Chem.*, **1**, 189 (1962).
110. N. A. Izmailov and L. L. Spivak, *Russ. J. Phys. Chem. (Engl. Transl.)*, **36**, 395 (1962).
111. M. M. Davis and H. B. Hetzer, *J. Res. Natl. Bur. Std.*, **60**, 569 (1958).
112. M. Rumeau and B. Trémillon, *Bull. Soc. Chim. France*, 1049 (1964).
113. D. C. Griffiths, *J. Chem. Soc.*, 815 (1938).
114. N. A. Izmailov and V. N. Izmailova, *Zh. Fiz. Khim.*, **29**, 1050 (1955).
115. N. A. Izmailov, *Russ. J. Phys. Chem. (Engl. Transl.)*, **34**, 1142 (1960); *cf.* J. E. Leffler and E. Grunwald, *Rates and Equilibria of Organic Reactions*, John Wiley and Sons, New York, 1963, p. 297.
116. E. Grunwald and E. Price, *J. Am. Chem. Soc.*, **86**, 4517 (1964).
117. R. Gary, R. G. Bates, and R. A. Robinson, *J. Phys. Chem.*, **69**, 2750 (1965).
118. R. P. Bell and A. T. Kuhn, *Trans. Faraday Soc.*, **59**, 1789 (1963), and references cited therein.
119. P. K. Glasoe and L. Eberson, *J. Phys. Chem.*, **68**, 1560 (1964).
120. L. G. Hepler, *J. Phys. Chem.*, **68**, 2645 (1964).
121. L. J. Bellamy, A. R. Osborn, and R. J. Pace, *J. Chem. Soc.*, 3749 (1963).
122. N. S. Isaacs, *Tetrahedron Letters*, 4553 (1965).
123. E. L. Eliel and C. A. Lukach, *J. Am. Chem. Soc.*, **79**, 5986 (1957).
124. S. Winstein and N. J. Holness, *J. Am. Chem. Soc.*, **77**, 5562 (1955).
125. R. D. Stolow, *J. Am. Chem. Soc.*, **81**, 5806 (1959).
126. M. Tichy, J. Jonas, and J. Sicher, *Coll. Czech. Chem. Comm.*, **24**, 3434 (1959).
127. M. C. Reese and E. L. Eliel, *J. Am. Chem. Soc.*, **90**, 1560 (1968).
128. P. F. Sommer, C. Pascual, V. P. Arya, and W. Simon, *Helv. Chim. Acta*, **46**, 1734 (1963).
129. S. Ehrenson in *Progress in Physical Organic Chemistry*, Vol. 2 (Eds. S. G. Cohen, A. Streitwieser, and R. W. Taft), Interscience Publishers, New York, 1964, pp. 195–251.
130. C. D. Ritchie and W. F. Sager in *Progress in Physical Organic Chemistry*, Vol. 2 (Eds. S. G. Cohen, A. Streitwieser, and R. W. Taft), Interscience Publishers, New York, 1964, pp. 323–400.
131. J. Clark and D. D. Perrin, *Quart. Rev. (London)*, **18**, 295 (1964); G. B. Barlin and D. D. Perrin, *Quart. Rev. (London)*, **20**, 75 (1966).
132. For a review, see G. S. Hammond in *Steric Effects in Organic Chemistry*, (Ed. M. S. Newman), John Wiley and Sons, New York, 1956, pp. 425–470.
133. M. S. Newman and T. Fukunaga, *J. Am. Chem. Soc.*, **85**, 1176 (1963).
134. G. S. Hammond and D. H. Hogle, *J. Am. Chem. Soc.*, **77**, 338 (1955).
135. L. Eberson, *J. Org. Chem.*, **27**, 3706 (1962).
136. S. A. Harrell and D. H. McDaniel, *J. Am. Chem. Soc.*, **86**, 4497 (1964).
137. H. S. Frank and M. Evans, *J. Chem. Phys.*, **13**, 507 (1945).
138. For a review about theories on water structure and water–solute interactions, see J. L. Kavanau, *Water and Solute–Water Interactions*, Holden-Day, San Francisco, 1964.
139. M. Crawford and M. Woodhead, *Tetrahedron Letters*, 1189 (1962).
140. E. E. Betts and L. R. C. Barclay, *Can. J. Chem.*, **33**, 1768 (1955).
141. L. L. McCoy and G. W. Nachtigall, *J. Am. Chem. Soc.*, **85**, 1321 (1963).
142. For a detailed escriptiodn of the Kirkwood–Westheimer theory, see E. J. King, *Acid–Base Equilibria*, Pergamon Press, London, 1965, Chap. 7.

143. C. Tanford, *J. Am. Chem. Soc.*, **79**, 5348 (1957).
144. M. J. S. Dewar and P. J. Grisdale, *J. Am. Chem. Soc.*, **84**, 3539, 3548 (1962); *cf.* W. Adcock and P. R. Wells, *Australian J. Chem.*, **18**, 1351 (1965).
145. J. D. Roberts and W. T. Moreland, *J. Am. Chem. Soc.*, **75**, 2167 (1953).
146. H. D. Holtz and L. M. Stock, *J. Am. Chem. Soc.*, **86**, 5188 (1964).
147. C. F. Wilcox, Jr. and J. S. McIntyre, *J. Org. Chem.*, **30**, 777 (1965).
148. R. P. Bell and G. A. Wright, *Trans. Faraday Soc.*, **57**, 1377 (1961).
149. C. D. Ritchie and E. S. Lewis, *J. Am. Chem. Soc.*, **84**, 591 (1962).
150. S. Siegel and J. M. Komarmy, *J. Am. Chem. Soc.*, **82**, 2547 (1960).
151. For references to this subject, see L. Eberson and I. Wadsö, *Acta Chem. Scand*, **17**, 1552 (1963).
152. H. M. Peek and T. L. Hill, *J. Am. Chem. Soc.*, **73**, 5304 (1951).
153. K. Bowden, *Can. J. Chem.*, **41**, 2781 (1963).
154. J. L. Roberts and H. H. Jaffé, *Tetrahedron (Suppl. 2)*, **19**, 455 (1963).
155. K. Bowden, *Can. J. Chem.*, **43**, 3354 (1965).
156. M. Charton, *J. Org. Chem.*, **30**, 974 (1965).
157. O. Exner, *Coll. Czech. Chem. Comm.*, **31**, 65 (1966).
158. B. M. Wepster, *Rec. Trav. Chim.*, **71**, 1159, 1171 (1952).
159. A. Fischer, W. J. Mitchell, J. Packer, R. D. Topsom, and J. Vaughan, *J. Chem. Soc.*, 2892 (1963).
160. P. A. Leermakers and W. A. Bowman, *J. Org. Chem.*, **29**, 3708 (1964).
161. J. P. Schaefer and T. J. Miraglia, *J. Am. Chem. Soc.*, **86**, 64 (1964).
162. P. Huyskens and T. Zeegers–Huyskens, *J. Chim. Phys.*, **61**, 81 (1964).
163. F. E. Condon, *J. Am. Chem. Soc.*, **87**, 4481, 4485, 4491, 4494 (1965).
164. D. J. Glover, *J. Am. Chem. Soc.*, **87**, 5275, 5279 (1965).
165. J. Jones and F. G. Soper, *J. Chem. Soc.*, 133 (1936).
166. D. H. McDaniel and H. C. Brown, *Science*, **118**, 370 (1953).
167. F. H. Westheimer and O. T. Benfey, *J. Am. Chem. Soc.*, **78**, 5309 (1956).
168. M. Shahat, *Acta Cryst.*, **5**, 763 (1952).
169. K. Nakamoto, Y. A. Sarma, and G. T. Behnke, *J. Chem. Phys.*, **42**, 1662 (1965).
170. S. Darlow and W. Cochran, *Acta Cryst.*, **14**, 1250 (1961); S. Darlow, *Acta Cryst.*, **14**, 1257 (1961).
171. S. W. Petersen and H. A. Levy, *J. Chem. Phys.*, **29**, 948 (1958).
172. R. D. Ellison and H. A. Levy, *Acta Cryst.*, **19**, 260 (1965).
173. G. Ferguson and G. A. Sim, *Acta Cryst.*, **14**, 1262 (1961).
174. Y. Okaya, *Acta Cryst.*, **19**, 879 (1965).
175. S. K. Kumra and S. F. Darlow, *Acta Cryst.*, **18**, 98 (1965).
176. F. Salmon-Legagneur and Y. Olivier, *Bull. Soc. Chim. France*, 1392 (1965); R. Robin, *Bull. Soc. Chim. France*, 2296 (1965).
177. R. C. Cookson and M. E. Trevett, *J. Chem. Soc.*, 3864 (1965).
178. J. H. Wotiz and H. E. Merrill, *J. Am. Chem. Soc.*, **80**, 866 (1958).
179. R. Darms, T. Threlfall, M. Pesaro, and A. Eschenmoser, *Helv. Chim. Acta*, **46**, 2893 (1963).
180. C. F. Huebner, E. Donoghue, L. Dorfman, E. Wenkert, W. E. Streth, and S. W. Donely, *Chem. Comm.*, 419 (1966).
181. H. F. van Woerden, *Rec. Trav. Chim.*, **83**, 920 (1963).
182. M. H. Miles, E. M. Eyring, W. W. Epstein, and R. E. Ostlund, *J. Phys. Chem.*, **69**, 467 (1965).
183. H. W. Ashton and J. A. Partington, *Trans. Faraday Soc.*, **30**, 598 (1934).
184. L. Eberson, *Acta Chem. Scand.*, **13**, 211 (1959).
185. L. Eberson, *Acta Chem. Scand.*, **14**, 641 (1960).
186. T. C. Bruice and W. C. Bradbury, *J. Am. Chem. Soc.*, **87**, 4851 (1965).
187. L. L. McCoy, *J. Am. Chem. Soc.*, **89**, 1673 (1967); *cf.* also S. Oae, N. Furukawa,

T. Watanabe, Y. Otsuji, and M. Hamada, *Bull. Chem. Soc. Japan*, **38**, 1247 (1965).

188. L. Eberson, *Acta Chem. Scand.*, **18**, 1276 (1964).
189. J. J. Bloomfield, D. van der Helm, and R. Fuchs, private communication.
190. H. Christol and M. Gaignon, *J. Chim. Phys.*, **57**, 707 (1960).
191. L. Eberson and H. Welinder, unpublished results.
192. H. H. Jaffé, *J. Am. Chem. Soc.*, **79**, 2373 (1957).
193. L. Eberson, *Acta Chem. Scand.*, **13**, 224 (1959).
194. R. E. Dodd, R. E. Miller, and W. F. K. Wynne-Jones, *J. Chem. Soc.*, 2790 (1961).
195. D. Chapman, D. R. Lloyd, and R. H. Prince, *J. Chem. Soc.*, 550 (1964).
196. E. S. Hanrahan, *Spectrochim. Acta*, **22**, 1243 (1966).
197. D. R. Lloyd and R. H. Prince, *Proc. Chem. Soc.*, 464 (1961).
198. S. N. Das and D. J. G. Ives, *Proc. Chem. Soc.*, 373 (1961).
199. W. E. Buddenbaum, W. G. Koch, and P. E. Yankwich, *J. Phys. Chem.*, **70**, 673 (1966).
200. S. Forsén, *J. Chem. Phys.*, **31**, 852 (1959).
201. L. Eberson and S. Forsén, *J. Phys. Chem.*, **64**, 767 (1960).
202. B. L. Silver, Z. Luz, S. Peller, and J. Reuben, *J. Phys. Chem.*, **70**, 1434 (1966).
203. L. L. McCoy, *J. Org. Chem.*, **30**, 3762 (1965).
204. H. B. Thompson, L. Eberson, and J. V. Dahlen, *J. Phys. Chem.*, **66**, 1634 (1962).
205. J. F. King in *Elucidation of Structures by Physical and Chemical Methods*, Part 1 (Ed. A. Weissberger), Interscience Publishers, New York, 1963, pp. 317–412.
206. L. Eberson, *Svensk Kem. Tidskr.*, **71**, 153 (1959).
207. J. L. Haslam, E. M. Eyring, W. W. Epstein, G. A. Christiansen, and M. H. Miles, *J. Am. Chem. Soc.*, **87**, 1 (1965).
208. J. L. Haslam, E. M. Eyring, W. W. Epstein, R. P. Jensen, and C. W. Jaget, *J. Am. Chem. Soc.*, **87**, 4247 (1965).
209. E. M. Eyring and J. L. Haslam, *J. Phys. Chem.*, **70**, 293 (1966); *cf.* alsc P P. Jensen, E. M. Eyring and W. M. Walsh, *J. Phys. Chem.*, **70**, 2264 (1966).
210. M. Eigen, W. Kruse, G. Maass and L. De Maeyer in *Progress in Reaction Kinetics*, Vol. 2 (Ed. G. Porter), Pergamon Press, London, 1964, p. 312.
211. J. Sicher, F. Sipos, and J. Jonas, *Coll. Czech. Chem. Comm.*, **26**, 262 (1961).
212. H. Hogeveen and F. Montanari, *J. Chem. Soc.*, 4864 (1963); *cf.* alsc D. J. Pasto and R. Kent, *J. Org. Chem.*, **30**, 2684 (1965).
213. J. Sicher, M. Tichy, F. Sipos, M. Svoboda, and J. Jonas, *Coll. Czech. Chem. Comm.*, **29**, 1561 (1964).
214. J. Hermans, Jr., S. J. Leach, and H. A. Scheraga, *J. Am. Chem. Soc.*, **85**, 1390 (1963).
215. Z. L. Ernst and J. Menashi, *Trans. Faraday Soc.*, **59**, 1803 (1963).
216. M. Eigen and E. M. Eyring, *J. Am. Chem. Soc.*, **84**, 3254 (1955).
217. D. Chapman, D. R. Lloyd, and R. H. Prince, *J. Chem. Soc.*, 3645 (1963).

CHAPTER 7

Introduction of COOH groups by carbonyl olefination

L. D. BERGELSON and M. M. SHEMYAKIN

Institute for Chemistry of Natural Products, Academy of Sciences of U.S.S.R., Moscow, U.S.S.R.

I. THE CARBONYL OLEFINATION REACTION

A. Introduction

Carbonyl olefination, one of the most general methods for chain elongation, has been widely used for the synthesis of carboxylic acids and their derivatives. The underlying principle of the method has been briefly discussed in the first volume of this series[1]. The present chapter treats the method in more detail with particular attention to such of its aspects that have a bearing on the stereospecific synthesis of various types of unsaturated carboxylic acids.

B. Method of Carbonyl Olefination

Under the general name of carbonyl olefination are classified a group of reactions in which phosphorus ylides (1) or their PO-activated analogs (2) react with carbonyl compounds to form olefins.

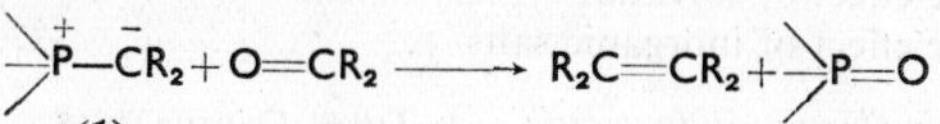

$$\overset{+}{P}-\overset{-}{C}R_2 + O{=}CR_2 \longrightarrow R_2C{=}CR_2 + \overset{}{P}{=}O$$
(1)

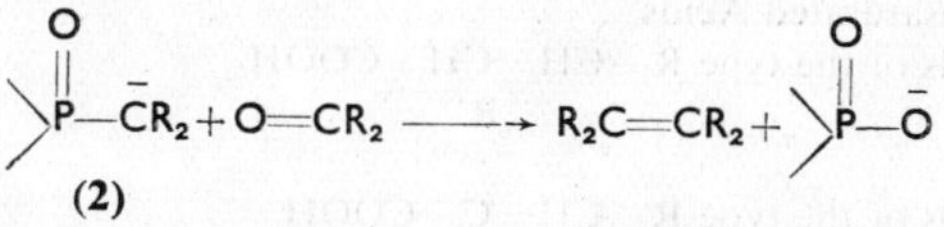

$$\overset{O}{\overset{\|}{P}}-\overset{-}{C}R_2 + O{=}CR_2 \longrightarrow R_2C{=}CR_2 + \overset{O}{\overset{\|}{P}}-\overset{-}{O}$$
(2)

The chief merits of this method, distinguishing it from other methods of building up the carbon chain, are that it occurs under mild conditions, that it does not induce isomerization or rearrangement and that it is frequently stereospecific.

Triphenylalkylenephosphoranes (3) (Wittig reagents) are most often used as olefinating agents[2-8]; trialkyl analogs (4)[9-15], or compounds of type 3 containing various substituents at the P-phenyls[16-20] have found much rarer application. Lately the PO-activated analogs of the phosphorus ylides, namely carbanions of phosphonic and phosphinic acids (5, 6) and diphenylphosphine oxides (7) are being used more and more extensively[21].

$$Ph_3\overset{+}{P}-\overset{-}{C}HR \qquad\qquad R_3\overset{+}{P}-\overset{-}{C}HR$$
(3)   (4)

$$
\underset{\textbf{(5)}}{RO_2\overset{\overset{\textstyle O}{\|}}{P}-\overset{-}{C}HR}
\qquad
\underset{\textbf{(6)}}{Ph(RO)\overset{\overset{\textstyle O}{\|}}{P}-\overset{-}{C}HR}
\qquad
\underset{\textbf{(7)}}{Ph_2\overset{\overset{\textstyle O}{\|}}{P}-\overset{-}{C}HR}
$$

In this article we shall discuss only the practically most important olefinating agents—the triphenylalkylenephosphoranes (3) and their PO-activated analogs (5) and (7).

Triphenylalkylenephosphoranes (3) are usually prepared by treating triphenylphosphonium halides (8) with bases[4, 8].

$$
\underset{\textbf{(8)}}{[Ph_3P-CH_2R]^+X^-} \underset{+HX}{\overset{-HX}{\rightleftarrows}} \underset{\textbf{(3)}}{\overset{+}{P}h_3P-\overset{-}{C}HR}
$$

Since the phosphorus ylides (3) are in turn capable of adding HX with the formation of phosphonium salts (8), the latter can be regarded as Brønsted acids, and the ylides (3) as their corresponding bases. The strength of the base deprotonating the salt (8) depends on the acidity of the latter, which is determined by the nature of the substituent R. Electron-accepting R substituents augment the acidity of the salt and lower the basicity of the phosphorus ylide. Because of this, when $R = CO_2Me$, the phosphonium salt, being a strong acid, is ylidized even by dilute soda solution[9], whereas when $R =$ alkyl, strong bases such as hydrides, amides and alcoholates of alkali metals or organolithium compounds are required for this purpose. Recently the methylsulfinyl carbanion, formed by the action of sodium hydride on dimethylsulfoxide, has been recommended as base[22].

The second method of practical importance involves the transylidization principle[23]. The interaction of the phosphorus ylides (3) with the phosphonium salt (8a) leads to the equilibrium:

$$
\underset{\textbf{(3)}}{\overset{+}{P}h_3P-\overset{-}{C}HR} + \underset{\textbf{(8a)}}{[Ph_3P-CH_2R']^+X^-} \rightleftarrows \underset{\textbf{(8)}}{[Ph_3P-CH_2R]^+X^-} + \underset{\textbf{(3a)}}{\overset{+}{P}h_3P-\overset{-}{C}HR'}
$$

Its position depends upon the relative acidity and basicity of the salt and ylide, i.e. on the nature of the substituents R and R′. If 3 and 3a differ strongly in basicity, the equilibrium is shifted in the direction of the less basic ylide and the less acid phosphonium salt. The ylides (3) can, therefore, be transylidized by the phosphonium salts (8a) if in the series COPh, CO_2Me, Ph, Alk, the substituent R occupies a position closer to the right than does R′.

Other methods of preparing triphenylphosphoranes are only of limited applicability. Among these is the reaction of triphenylphosphine with aliphatic diazo compounds in the presence of copper salts[24-26].

$$Ph_3P + N_2CHR \xrightarrow{Cu^+} Ph_3\overset{+}{P}—\overset{-}{C}HR + N_2$$

α-Halo-substituted triphenylphosphorus ylides can be obtained by the addition of carbenes to triphenylphosphine[27-34].

$$Ph_3P + CH_2X_2 \xrightarrow{base} Ph_3\overset{+}{P}—\overset{-}{C}HX$$

$$Ph_3P + CHX_3 \xrightarrow{base} Ph_3\overset{+}{P}—\overset{-}{C}X_2$$

$$2\,Ph_3P + CX_4 \xrightarrow{\Delta} Ph_3\overset{+}{P}—\overset{-}{C}X_2 + Ph_3PX_2$$

Ylides are also formed in the interaction of triphenylphosphine with derivatives of acrylic acid (9, X = OR or NH₂)[35],

$$Ph_3P + H_2C{=}CHCOX \longrightarrow Ph_3\overset{+}{P}—\overset{-}{C}HCH_2COX$$
$$(9)$$

or on thermal decomposition of α-methoxycarbonylalkylphosphonium salts (10)[4].

$$[Ph_3PCH(R)CO_2Me]^+X^- \xrightarrow{\Delta} Ph_3\overset{+}{P}—\overset{-}{C}HR + CO_2 + MeX$$
$$(10)$$

Finally, phosphorus ylides (13) with two electron-accepting α-substituents can be synthesized from triphenylphosphonium dichloride (11) and derivatives of malonic acid (12, X, Y = CO₂R or CN)[36].

$$Ph_3PCl_2 + CHXY \xrightarrow{Et_3N} Ph_3\overset{+}{P}—\overset{-}{C}\!\!\diagup^{X}_{\diagdown Y}$$
$$\quad (11) \qquad (12) \qquad\qquad (13)$$

PO-activated analogs of phosphorus ylides (5 and 7) are prepared by metalating the corresponding phosphinates (R′O)₂P(O)CH₂R or diphenylphosphine oxides Ph₂P(O)CH₂R. Hydrides, amides or the alcoholates of alkali metals, as well as butyllithium and phenyllithium are used as metalating agents[21].

The stability of the metalated compound depends upon the nature of the alkali metal. Lithium derivatives are the most stable and in some cases can be even isolated[37]. One of the advantages of the PO-activated ylides is that the alkylphosphorous acid salts formed in the carbonyl olefination reaction are water-soluble, so that in contrast to the insoluble triphenylphosphine oxide they can be readily separated from the olefins. Moreover, PO-activated ylides react more easily with ketones than do Wittig reagents. However, the set of PO-activated olefinating reagents is limited because the reaction proceeds satisfactorily only with ylides stabilized by electron-acceptor substituents on the ylide carbon (**5–7**, $R = CO_2Et$, Ph, *etc.*).

Usually an attempt is made to carry out the reaction in a single step, *i.e.* without isolating the olefinating agent. Hence when the latter is triphenylalkylenephosphorane (**3**), obtained from the corresponding phosphonium salts, the reaction mixture as a rule also contains alkali metal halides. Since the latter can have an important bearing on the yield of olefins and on the stereoselectivity of the reaction[38–41a], special methods were devised for obtaining salt-free triphenylphosphorus ylide solutions in polar and non-polar solvents[4, 41a].

In order to avoid side-reactions it is often necessary to carry out the carbonyl olefination in an absolutely dry, inert gas. The reaction is usually completed by treatment with water, so as to dissolve the inorganic salts (metal halides, alkylphosphites) and, in some cases, to decompose intermediates (see below) which rapidly yield the products in the presence of water.

C. Structure and Properties of Phosphorus Ylides*

Since a phosphorus atom is capable of enlarging its electron octet to a decet, phosphorus ylides can be regarded as ylide–ylene resonance hybrids

$$\overset{\diagup}{\underset{\diagdown}{P}}\!\!\overset{+}{}\!\!-\!\!\overset{-}{C}\!\!\overset{R}{\underset{R}{\diagdown}} \quad\longleftrightarrow\quad \overset{\diagup}{\underset{\diagdown}{P}}\!\!=\!\!C\!\!\overset{R}{\underset{R}{\diagdown}}$$

The degree of double bonding and consequently the reactivity of a given ylide depends on the nature of the substituents on the phosphorus and on the ylide carbon.

The effect of the substituents on the phosphorus atom is determined by their ability to decrease or increase the d orbital resonance, since a decrease in the latter lowers the degree of double bonding. Therefore, electron-accepting substituents on phosphorus decrease, and electron-donating ones increase, the reactivity (nucleophilicity) of the phosphorus ylide. It is for this reason that replacement of phenyl substituents on the ylide phosphorus by alkyl substituents considerably augments the reactivity of the phosphorus ylides[9, 11, 12]. Negatively-charged oxygen on the phosphorus

---

* The aim of this section is to give the general background necessary for the understanding of the mechanism and the stereochemistry of carbonyl olefination. A more thorough discussion can be found in the monograph of A. W. Johnson[7].

atom is even more effective. Phosphonate ions of the type 14 are much more nucleophilic than the corresponding triphenylphosphoranes because of the greater weight of the carbonyl structure in the latter[21].

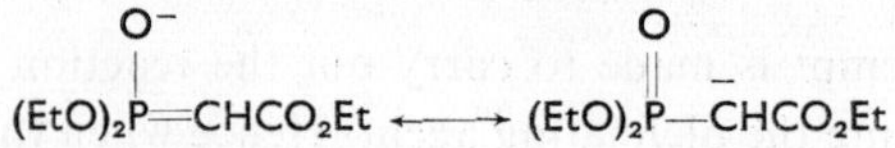

$$(EtO)_2P\overset{\overset{\displaystyle O^-}{|}}{=}CHCO_2Et \longleftrightarrow (EtO)_2P\overset{\overset{\displaystyle O}{\|}}{—}\overset{\displaystyle -}{C}HCO_2Et$$

(14a)   (14b)

With reference to the substituents on the ylide carbon it is to be noted that electron acceptors which diminish the negative charge, decrease (electron donors increase) the reactivity of the phosphorus ylides. Depending upon the nature of the substituent on the ylide carbon, triphenylphosphorus ylides (Wittig reagents) can be subdivided into three groups.

1. Non-stabilized ylides.

To this type belong triphenylalkylenephosphoranes in which the ylide carbon substituents have little effect upon the carbanionic character of the molecule. Non-stabilized phosphoranes display quite definite nucleophilic properties, being readily decomposed by water, oxidized by atmospheric oxygen and usually reacting vigorously with carbonyl compounds even at room temperature.

2. Stabilized ylides.

To this group belong triphenylphosphoranes with strong electron acceptor substituents on the ylide carbon. They have a less marked nucleophilicity than the ylides of the first group and are less reactive, not being hydrolyzed by water or oxidized by atmospheric oxygen.

3. Partly stabilized ylides

In ylides of this type the substituent on the ylide carbon does not possess marked electron acceptor properties. Their reactivity is intermediate between the ylides of the first and second groups.

Characteristic representatives of the above three groups are the alkylidene- (15), the alkoxylcarbonylmethylene-(16) and the benzylidene-(17) or alkenylidenetriphenylphosphoranes (18).

$Ph_3P\!=\!CH(CH_2)_nCH_3$ $Ph_3P\!=\!CHCO_2R$ $Ph_3P\!=\!CHPh$ $Ph_3P\!=\!CHCH\!=\!CHR$

(15)   (16)   (17)   (18)

Of considerable importance for understanding the mechanism and steric course of the carbonyl olefination reaction is the stereochemistry of the initial phosphorus ylides. At present little information is available on the geometry of the ylide carbon and phosphorus atoms. The results of an estimation of the valence angles of a stabilized phosphorus ylide[41b] are not inconsistent with tetrahedral hybridization of the ylide phosphorus (overlap of the phosphorus d orbital with a p orbital of the ylide carbon).

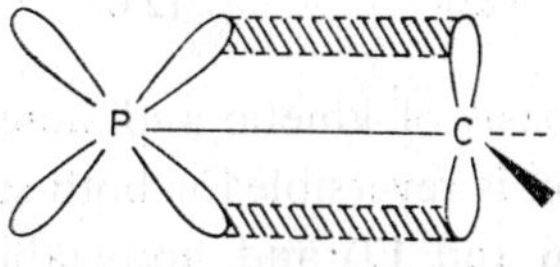

In the case of the stabilized ylides of type **16** the substituent, by lowering the electron density of the ylide carbon atom, itself acquires an excess negative charge. Because of this, such ylides have a predominantly cisoid structure (**19**)[42]. Apparently the negative charge on the ylide carbon is

$$\overset{+}{Ph_3P}\diagdown \underset{H}{\overset{}{C}}\text{---}\overset{\overset{-}{\cdot}}{C}\diagup \overset{O}{\underset{OR}{}} \qquad \overset{+}{Ph_3P}\diagdown \underset{H}{\overset{}{C}}\text{---}C\diagup \underset{\overset{}{O^-}}{\overset{OR}{}}$$

(19)   **(20)**

also delocalized in the partly stabilized phosphorus ylides (**17**):

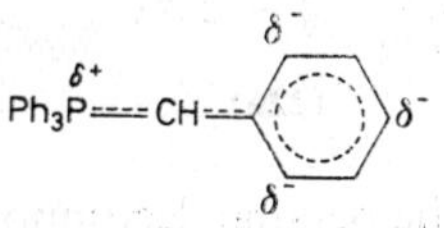

$$\overset{\delta^-}{}\\ \overset{\delta^+}{Ph_3P}\text{===}CH\text{===}\langle\bigcirc\rangle\delta^- \\ \underset{\delta^-}{}$$

On the other hand the phosphorus atom should exert a strong effect on the phenyl groups directly bound to it, lowering their electron density. The most prefered conformation in this case should therefore be **21**, wherein the C-phenyl is situated between two P-phenyls and is transoid to the third P-phenyl.

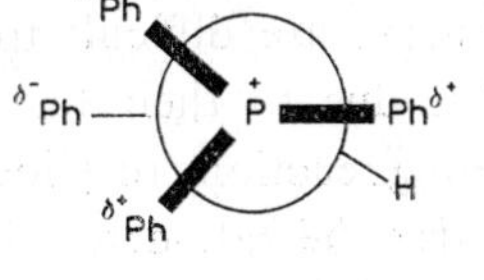

(21)

D. Mechanism of Carbonyl Olefination

It is well established that carbonyl olefination proceeds through the formation of the intermediate betaine (**22**), whose structure has been confirmed by a number of indirect data[43, 44, 51].

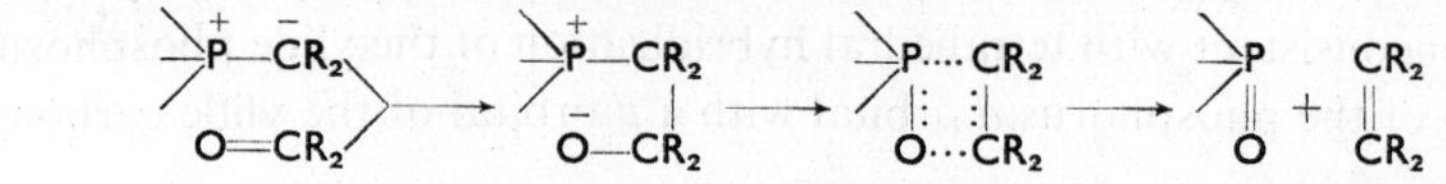

(22)   (23)   (24)

It was shown on the basis of kinetic and stereochemical data that the stage of betaine formation is reversible for both stabilized or semistabiliz-ed phosphorus ylides (16 and 17) and non-stabilized ylides (15)[50–54, 56].

As a rule the betaines themselves cannot be isolated, but betaine-type intermediates have been obtained in certain special cases[45–47]. It was shown in the example of optically active phosphorus ylides that the last stage of carbonyl olefination—decomposition of the betaine (22a) into phosphine oxide (26) and olefin—takes place with retention of the con-figuration at the phosphorus atom[48, 49].

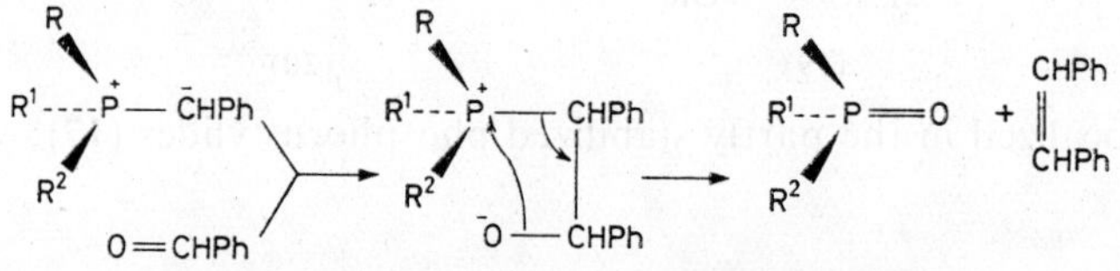

(25)   (22a)   (26)

From this it follows that the betaine decomposes via the cyclic transition state (23) by *cis* elimination. Hence the configuration of the olefin (24) is predetermined by the configuration of its betaine precursor (22).

Regarding the rate-determining stage of carbonyl olefination, it was unequivocally shown by kinetic studies[14, 50, 52] that for the relatively slug-gish, stabilized ylides of type (16) in non-polar media the rate of decom-position of the betaine into olefin and phosphine oxide exceeds the rate of its formation.

Direct kinetic measurements are difficult to carry out in the case of the partly stabilized ylides, due to their fast reaction rates. However, investigation of the competing reactions of ylide 17 with various aldehydes showed that in this case also the rate-determining stage is formation of the betaines[41a, 56].

An important feature of the non-stabilized ylides (15) is that they form betaines so fast that the rates of both stages become commensurate. How-ever at least under certain conditions (non-polar solvent, absence of foreign salts) betaine formation remains the rate-determining stage[56]. Another important characteristic of non-stabilized ylides (15) is that in the reaction with aldehydes the relative rates of the separate stages may depend upon the presence of inorganic salts forming complexes with the betaine

$$\left[Ph_3\overset{+}{P}-CHRCHR'\overset{-}{O}\right]\cdot LiX$$

In such complexes the charges on the phosphorus and oxygen are partly neutralized and, moreover, these atoms become sterically less accessible than in the initial betaine. Especially stable are the complexes with lithium salts, the equilibrium between the free betaine and the lithium complex being strongly shifted towards the latter. This slows down the second stage of the reaction (decomposition of the betaine into the endproducts) to a point where it becomes rate-determining[41a, 55, 56] (see section I.E).

The mechanism of the PO-activated olefination has been investigated in much less detail than the mechanism of the Wittig reaction and is still largely hypothetical. It has been suggested by analogy with the Wittig reaction that PO-activated carbonyl olefination proceeds through formation of the β-hydroxyphosphorus derivative (27) which decomposes via the transition state (28) into metal alkylphosphonate and olefin. In some

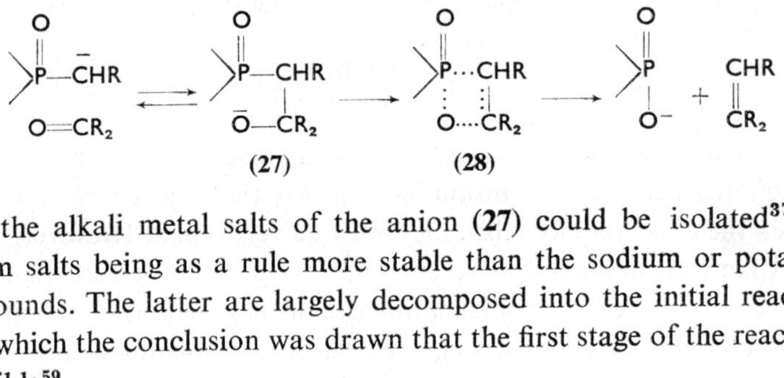

(27)   (28)

cases the alkali metal salts of the anion (27) could be isolated[37, 58, 59], lithium salts being as a rule more stable than the sodium or potassium compounds. The latter are largely decomposed into the initial reactants, from which the conclusion was drawn that the first stage of the reaction is reversible[59].

The question of the rate-determining stage of the PO-activated carbonyl olefination still remains unresolved. However, competition reactions of benzaldehyde with the ylide 17 and the carbanion (7, R = Ph) showed the latter to be more reactive[60]. From this, one may conclude that in PO-activated carbonyl olefination also, the betaine formation stage is rate-determining; otherwise the end-products should form more readily from the ylide 17, phosphorus being more electrophilic in the latter.

E. Stereochemistry of Carbonyl Olefination

Since olefination of aldehydes leads to a mixture of *cis* and *trans* olefins, the intermediate betaine should be in the form of both *erythro* and *threo* isomers (**29** and **30**; see Scheme 1)*.

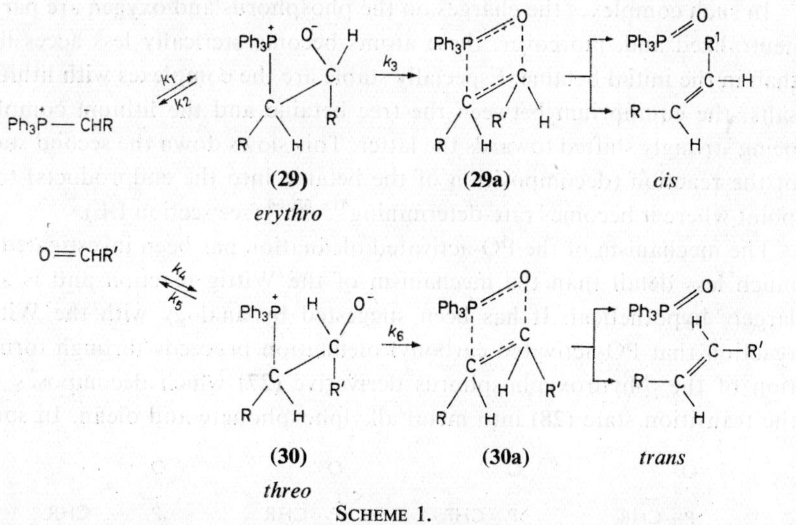

(29)   **(29a)**   *cis*
erythro

(30)   **(30a)**   *trans*
threo

SCHEME 1.

Early investigations showed that the Wittig reaction is often non-stereo-specific[6], whence the conclusion was drawn that the energy difference between the diastereomeric betaines must be very small. However, experimental facts accumulated in recent years have demonstrated that under certain conditions carbonyl olefination can display striking selectivity, with the steric course of the reaction much more dependent on the structure of the ylide than on that of the aldehyde. Thus, non-stabilized phosphorus ylides of type **15**, in non-polar solvents and in the absence of halides, give with various aliphatic and aromatic aldehydes high yields of *cis* olefins, whereas stabilized ylides of type (**16**) form the *trans* isomer (Table 1). Partly stabilized ylides (**17** and **18**) also give predominantly *trans* olefins with aldehydes, but the reaction is less specific.

On passing over from the stabilized or partly-stabilized triphenylphosphorus ylides to the trialkyl analogs the tendency to undergo *trans*-carbo-

---

* The term *erythro* refers to the betaine with both hydrogens and substituents R and R′ cisoid in one of the eclipsed conformations, whereas *threo* refers to the second isomer.

TABLE 1. Stereochemistry of the reaction
$EtCHO + Ph_3P=CHR \rightarrow EtCH=CHR + Ph_3PO$
(in non-polar solvents).

R	Type of ylide	cis:*trans* Ratio	Reference
Et	15	92 : 8	41a
CO$_2$Me	16	1 : 99	61
Ph	17	18 : 82	41a
CH=CH$_2$	18	6 : 94	62

nyl olefination markedly increases (Table 2). *Trans* olefins also prevail in PO-activated carbonyl olefination, the steric course of the reaction being practically independent of the nature of the substituent on the ylide carbon (Table 3)[57, 59, 63-66].

TABLE 2. Effect of the replacement of phenyl by alkyl groups at the ylide phosphorus atom upon the stereochemistry of the reaction of stabilized ylides with benzaldehyde (in ethanol).

Ylide	cis: *trans* Ratio	Reference
Ph$_3$P=CHCO$_2$Me	16 : 84	14
Bu$_3$P=CHCO$_2$Me	5 : 95	14
Ph$_3$P=CHPh	55 : 45	15
Ph$_2$MeP=CHPh	32 : 68	15
PhMe$_2$P=CHPh	12 : 88	15
(C$_6$H$_{11}$)$_3$P=CHPh	8 : 92	15

TABLE 3. Stereochemistry of the reaction of phosphonate and phosphine oxide carbanions with benzaldehyde.

Ylide	cis: *trans* Ratio	Reference
(EtO)$_2$P(O)$-\overset{-}{C}$HCO$_2$Et[a]	12 : 88	57, 63, 64
(EtO)$_2$P(O)$-\overset{-}{C}$HPh[b]	6 : 94	65
Ph$_2$P(O)$-\overset{-}{C}$HPh[b]	5 : 95	65
Ph$_2$P(O)$-\overset{-}{C}$HEt[c]	4 : 96	65

[a] In benzene.
[b] In cyclohexane.
[c] In dimethylformamide.

House and Rasmusson[67] attempted to explain the selective *trans* reaction of stabilized ylides (16) by differences in the rates of decomposition of the intermediate betaines due to overlap of the π electrons of the carbonyl group and the newly formed ethylene bond.

According to House and Rasmusson the coplanarity required for maximum overlap of the π electrons can be realized more easily in the case of the *threo* betaine with transoid arrangement of the substituents (see 30a), whence it follows that this isomer should decompose at a faster rate then the other. If it be assumed that the stereoisomeric betaines form at about the same rates and come to equilibrium with sufficient rapidity, the effect of such coplanarity should be predominant formation of the *trans* olefin. This hypothesis was used time and time again by various authors to explain the selective *trans* olefination by stabilized and partly stabilized phosphorus ylides[9, 14, 15, 28, 52, 67, 68]. However, it had to be rejected after it was shown that the rate-determining stage of carbonyl olefination by triphenylphosphorus ylides (15, 16 and 17) is the formation of the betaine (see section I.D). Indeed, if this stage is reversible and at the same time rate-determining, the reactions could be treated by the steady-state concentration method, which yields equation (1), expressing the *cis : trans* ratio of olefins as a function of the rate constants of six elementary steps (see Scheme 1)[69].

$$cis:trans = \frac{k_1\left(\dfrac{k_5}{k_6}+1\right)}{k_4\left(\dfrac{k_2}{k_3}+1\right)} \tag{1}$$

Since under normal conditions the experimental values for k_2/k_3 and k_5/k_6 differ little from each other[52], equation (1) shows that marked selectivity of the reaction is possible only in the case of large differences in the rates of formation (k_1 and k_4) of the *erythro* and *threo* betaines. A high stereospecificity of carbonyl olefination therefore bears evidence of considerable energy differences between the diastereomeric betaines (or between the transition states leading to the betaines)[*].

It might have been assumed that the source of such energy differences could have been differences in the steric interaction between the R and R' groups in the eclipsed conformations 31 and 32.

---

[*] On the grounds that with non-stabilized ylides in the presence of salts betaine formation step is the fast stage[53], Johnson[7] concluded that for such ylides steric influences are not pronounced in the transition state leading to betaine formation. This conclusion must be considered as not valid for salt-free reactions.

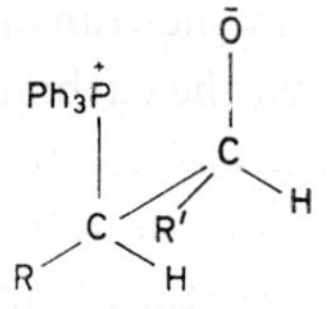

(31)   (32)
erythro   *threo*

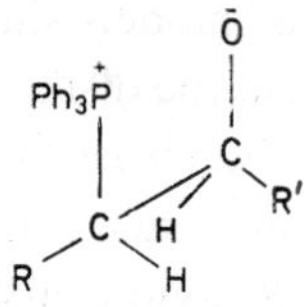

However, an examination of the models shows that owing to the bulkiness of the phosphorus atom and its three pendant phenyl groups, the eclipsed conformation cannot be realized without serious distortion of the atomic parameters[68]. The betaines must therefore form in the staggered conformers **29** and **30** (Scheme 1) with skewed phosphorus and oxygen and transoidal arrangement of the large triphenylphosphine and R′ groupings. In this case the *erythro* betaine (**29**) differs from the *threo* isomer (**30**) by an unfavorable interaction between the substituent R and the oxygen atom. Since, however the latter is relatively small, this effect could result only in some excess of *trans* over *cis* olefin, but it is without doubt too small to be responsible for the high *trans* selectivity of carbonyl olefination by stabilized and partly stabilized ylides of type **16** and **17**.

Since these ylides contain an electron acceptor group at the ylide carbon, interaction of these groups with other charged centers should also be taken into account. It is easy to see that in the case of the ylides **16** and **17** the mutual repulsion of the negative groups hinders the formation of the *erythro* betaine, leading to more rapid formation of the *threo* isomer, that is (providing the first stage is the rate-determining one) to *trans*-carbonyl olefination (*cf.* **33** and **34**, which are transition states in the formation of diastereomeric betaines by stabilized phosphorus ylides).

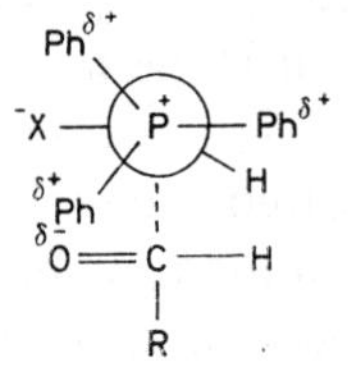

(33)   (34)
erythro   *threo*

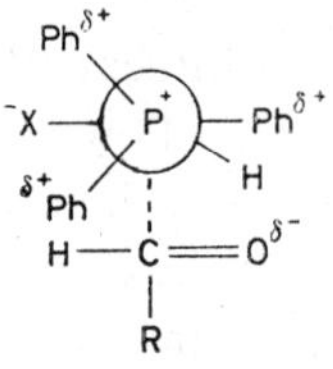

When the substituent on the ylide carbon is electrically neutral (nonstabilized ylides of type **15**), the decisive effect on the arrangement of the betaine-forming reactants is due to the interaction of the carbonyl-oxygen with the electrophilic phenyl groups of the ylide.

20*

A consideration of models shows that in the transition state **35** leading to the *erythro* betaine, the distance between the carbonyl-oxygen and nearest phenyl group is much less than in the case of the *threo* betaine (**36**). The *erythro* compound therefore forms more rapidly than its *threo* counterpart, which, under conditions of the first stage being rate-determining, leads to predominant *cis* olefination*.

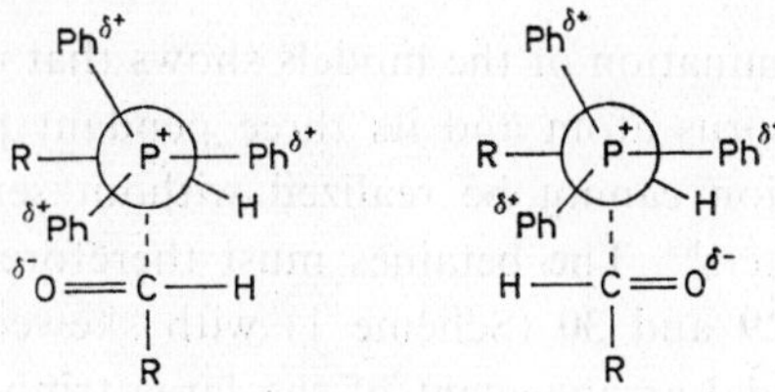

(35)
erythro

(36)
threo

This approach also makes clear a number of other observations on the effect of the ylide structure upon the stereochemistry of carbonyl olefination.

Thus the lesser *trans* selectivity of carbonyl olefination by partly stabilized ylides (**17**) than by stabilized ylides (**16**) (see Table 1) can be explained by the weaker electron-accepting properties of the phenyl group of **17** compared to the carboxylic ester group of **16**, with the corresponding lower tendency of the former ylide to form the transition state **34**.

In the reaction of diphenylphosphine oxide carbanions (**7**) with aldehydes, repulsion between two negatively charged oxygen atoms leads to the most stable conformations becoming **37** and **38** (see Scheme 2)[65]. In such a conformation the *erythro* betaine **37** differs from the *threo* isomer (**38**) by a highly unfavorable interaction between the phosphorus-containing grouping and the substituent R', so that it is the *threo* compound which forms predominantly[59]. Possibly it is for this reason that PO-activated carbonyl olefination is so highly *trans* selective (see Table 3).

The steric course of the carbonyl olefination reaction naturally depends upon the reaction conditions as well as upon the structure of the reactants. At present only the effect of solvation and of certain inorganic salts has been investigated in detail.

---

* Another possible cause for the *cis*-carbonyl olefination by non-stabilized ylides (**15**) in non-polar medium could be the formation of intermolecular associates between the ylides and betaines. However, cryoscopic measurements of the molecular weight of the ylides **15** and **17** showed that in benzene they are practically unassociated[56].

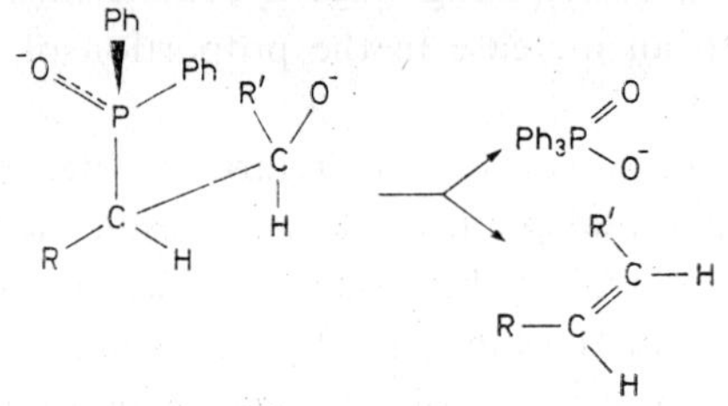

(37)
erythro   *cis*

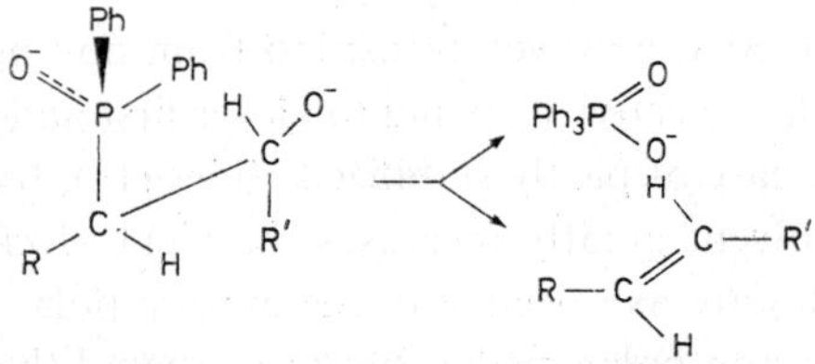

(38)
threo   *trans*

SCHEME 2.

I. The steric effect of solvation

Solvation of the initial reactants and of the betaines may have a dual effect on the stereochemistry of carbonyl olefination. The solvation of the oxygen and phosphorus atoms diminishes their mutual attraction, so that the preferred betaine conformation becomes that with transoidal arrangement of the solvated groups. In this conformation the *erythro* betaine (**39**) is energetically more favorable than the *threo* form (**40**) (see

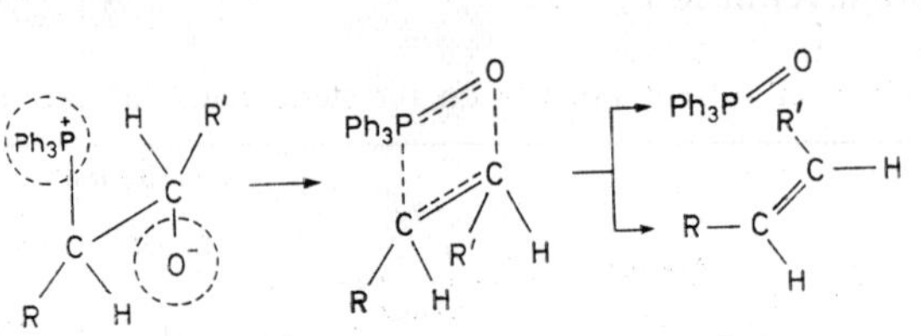

(39)
erythro   *cis*

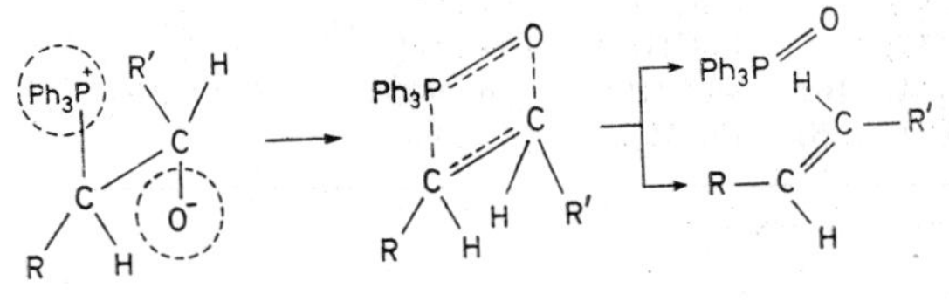

(40)
threo   *trans*

SCHEME 3.

Scheme 3). If the rate-determining stage is betaine formation, the above effect should lead to an increase in the proportion of *cis* olefin among the reaction products.

On the other hand, solvation may retard the decomposition of the betaine into olefins and phosphine oxide. As a result the reaction becomes more subject to the control of thermodynamic factors, which should lead to a relative increase in the amount of *trans* olefin.

The strength of each of these solvation effects depends upon the structure of the reactants and the nature of the solvent (Table 4).

Non-stabilized ylides (15) give predominantly *cis* olefins in both non-polar and polar media; however, transition from non-polar to polar solvent diminishes the selectivity, owing to closer first and second stage reaction rates. In the case of partly stabilized ylides (17), transition from non-polar to polar solvent greatly increases the yield of *cis* isomer.

Stabilized ylides are as selective in the aprotic polar solvent, dimethylformamide, as in non-polar media, but in methanol the reaction becomes non-stereospecific. The different effects of protic and aprotic solvents can be explained by the different ways in which they solvate the phosphorus and oxygen atoms in the betaine. The nucleophilic dimethylformamide solvates chiefly phosphorus. It therefore manifests a considerable *cis* effect in the case of partly stabilized ylides (17), but has little or no influence on the stereochemistry of the reaction of the CO-stabilized ylide (16), which is governed chiefly by mutual repulsion of the ylide and aldehyde carbonyls (see above). At the same time the *cis* effect of protic solvents is considerable, since by solvating the carbonyl groups, they diminish their mutual repulsion.

TABLE 4. The effect of solvent polarity on the steric course of carbonyl olefination.

Ylide	Aldehyde	In non-polar media[a]	cis: trans Ratio — In polar media		Reference
			ROH	DMF	
Ph_3P=CHEt (15)	PhCHO	91 : 9[b]	—	79 : 21	41a
Ph_3P=CHPh (17)	PhCHO	34 : 66[b]	55 : 45[c]	60 : 40	41a
Ph_3P=CHCO$_2$Me (16)	MeCHO	6 : 94[d]	38 : 62[f]	3 : 97	70a
$(EtO)_2P(O)$—CHPh (5)	EtCHO	3 : 96[e]	—	9 : 91	65

[a] In the absence of inorganic salts.
[b] In benzene.
[c] In ethanol.
[d] In methylene chloride.
[e] In cyclohexane.
[f] In methanol.

2. The steric effect of inorganic salts

It has already been mentioned that carbonyl olefination by non-stabilized and partly stabilized ylides **15** and **17** is most often carried out in the presence of alkali halides formed during generation of the ylides from the corresponding phosphonium salts. It was found that lithium halides can significantly diminish the stereospecificity of the reaction of such ylides, the steric salt effect becoming especially marked in non-polar solvents (Table 5), but not in polar (DMF) solvents[41a]. The steric effect depends upon the nature of the halogen, increasing in the order $Cl^- < Br^- < I^-$.

TABLE 5. The effect of lithium iodide on the stereospecificity of reactions of non-stabilized and partly stabilized ylides (in benzene)[41a].

Ylide	Aldehyde	cis: trans Ratio	
		In the absence of salts	In the presence of LiI
$Ph_3P{=}CHEt$ (**15**)	EtCHO	92 : 8	77 : 23
	PhCHO	91 : 9	35 : 65
$Ph_3P{=}CHPh$ (**17**)	EtCHO	18 : 82	41 : 59
	PhCHO	34 : 66	45 : 55

One might assume that the cause of the steric effect of lithium halides is complex formation with the betaines. In the case of non-stabilized ylides (**15**) lithium halide – betaine interaction hinders the second stage of the reaction (k_3 and k_6 in Scheme 1) so that it becomes the rate-determining one*, and the reaction is then controlled more by thermodynamic than by kinetic factors. The reaction therefore proceeds preferably via the *threo* betaine, which decomposes into the end-products more readily and into the initial reactants less readily than does the *erythro* isomer.

In the case of the partly stabilized ylides (**17**) lithium halides also coordinate with the betaines, without however hindering their decomposition sufficiently for this step to become the rate-determining one. The stereochemistry of the reaction therefore remains under kinetic control, and coordination of LiX with the betaine, lowering the attractive force between phosphorus and the carbonyl-oxygen, facilitates formation of the *erythro* isomer and hence of the *cis* olefin.

* In some cases, retardation of the second stage is great enough to permit isolation of the lithium halide–betaine complexes[16, 53, 55, 71, 72].

With CO-stabilized ylides (16), the effect of inorganic halides becomes noticeable only when the reaction is carried out in polar, aprotic solvents[70a] (Table 6). It was suggested that in this case the alkali metal halides act as Lewis acids, coordinating with the negative carbonyl-oxygen[70a]. By diminishing the electrostatic repulsion of the carbonyl groups such coordination should facilitate the formation of the *erythro* betaine and, providing the first stage of the reaction is rate-determining, the reaction should become less stereospecific.

TABLE 6. Effect of lithium bromide on the
stereoselectivity of the reaction
$$MeCHO + Ph_3P = CHCOOMe \longrightarrow$$
$$MeCH = CHCOOMe + Ph_3PO^{[70a]}.$$

Solvent	Fraction of *cis* isomer	
	In the absence of salts	In the presence of LiBr
CH_2Cl_2	3	3
DMF	18–22	31–34

Since in the presence of lithium halides the second stage of the reaction of non-stabilized ylides (15) is rate-determining (in non-polar solvents), its stereochemistry becomes dependent upon the reaction time and reactant ratio.

It was found that under the above conditions, the stereospecificity of the reaction is at first very high, the *cis* olefin being formed almost exclusively, but then the *cis: trans* ratio quickly falls until it finally reaches a constant value[41a]. Owing to the high energy level of the ylide 15, the k_2 and k_5 values are much smaller than those of k_1 and k_4 (see Scheme 1). Hence equilibrium between the diastereomeric betaines 29 and 30 is established relatively slowly. In the initial period, when the amount of *erythro* betaine is relatively large, the reaction is much more stereospecific, the specificity falling only with time.

With regard to the reactant ratios, an excess of one of the reactants under second stage rate-determining conditions sharply increases the relative yield of *cis* isomer[54]. This stereochemical effect is ascribed to the considerable fall in the k_2 and k_5 values in the presence of an excess of one of the reactants (see Scheme 1). In this case, therefore, the establishing of equilibrium between the diastereomeric betaines, *i.e.* transition of the initially formed *erythro* isomer into the *threo* form, is hindered. Naturally, in all other cases, when the rate of carbonyl olefination is determined by the betaine-forming stage, neither the duration of the reaction, nor the reactant ratio has any effect on the stereospecificity.

The above-mentioned correlations between the stereochemistry of carbonyl olefination and the structure of the reactants and environmental conditions reveal the existence of certain limitations with respect to the stereoselectivity of the reaction. Salt-free reactions in non-polar media permit the *cis* olefination of aldehydes by non-stabilized ylides (15) and their *trans* olefination by partly stabilized ylides (17 and 18) but in the presence of lithium salts the reaction becomes non-stereospecific. If the reaction of these ylides is to take place in the presence of lithium halides, it should be carried out in dimethylformamide when one desires to increase the yield of *cis* olefin. However, if the reaction with non-stabilized ylides (15) must be carried out in a non-polar solvent in the presence of lithium iodide, then in order to increase the yield of *cis* isomer excess ylide should be used*.

With the stabilized ylides (16), for maximum yield of *trans* olefin, the reaction should be carried out in a non-polar medium, independent of the presence or absence of inorganic salts.

With these limitations in mind, carbonyl olefination can be used for the stereoselective synthesis of 1,4-substituted 1,3-butadienes RCH=CHCH=CHR. For instance, condensation of saturated aldehydes with the partly stabilized *trans*-alkenylidenetriphenylphosphoranes (41)gives *trans-trans* dienes if the reaction is carried out in a non-polar medium, while the *cis* ylides (42) give selectively *trans-cis* dienes[62]. Another method of obtaining *trans-cis* dienes is by the *cis* olefination of *trans-α, β*-unsaturated aldehydes by non-stabilized ylides (15) in DMF or in non-polar media in the absence of salts[73].

$$\begin{array}{c} & trans \\ \text{RCHO} + \text{Ph}_3\text{P=CH—CH=CHR} \end{array} \longrightarrow \begin{array}{c} trans \qquad trans \\ \text{RCH=CH—CH=CHR} \end{array}$$
(41)

$$\begin{array}{c} & cis \\ \text{RCHO} + \text{Ph}_3\text{P=CH—CH=CHR} \end{array} \longrightarrow \begin{array}{c} cis \qquad trans \\ \text{RCH=CH—CH=CHR} \end{array}$$
(42)

$$\begin{array}{c} trans \\ \text{RCH=CHCHO} + \text{Ph}_3\text{P=CHR} \end{array} \longrightarrow \begin{array}{c} trans \qquad cis \\ \text{RCH=CH—CH=CHR} \end{array}$$
(15)

---

* *Trans* olefination of aldehydes with non-stabilized ylides may be achieved by treatment of the betaine–lithium halide complex with a second equivalent of butyllithium[70b]. However, this method seems not to be suitable for the synthesis of carboxylic acid derivatives.

Cis-cis dienes can be obtained by *cis* olefination of *cis*-α, β-unsaturated aldehydes; however, owing to the tendency of the latter to isomerize, it is more feasible to start with substituted propiolic aldehydes.

$$RC\equiv CCHO + Ph_3P=CHR \longrightarrow RC\equiv CCH\overset{cis}{=}CHR \xrightarrow{H_2/Pd} R\overset{cis}{CH}=CH-CH\overset{cis}{=}CHR$$

II. SYNTHESIS OF CARBOXYLIC ACIDS AND THEIR DERIVATIVES

Carbonyl olefination may be used to synthesize carboxylic acids and their derivatives by the following two general routes:

(i) The carbonyl compounds are olefinated by ylides with a potential carboxyl function (ester, amide or nitrile group).

(ii) Aldehydo or keto carboxylic acid derivatives are olefinated by phosphorus ylides of various structure.

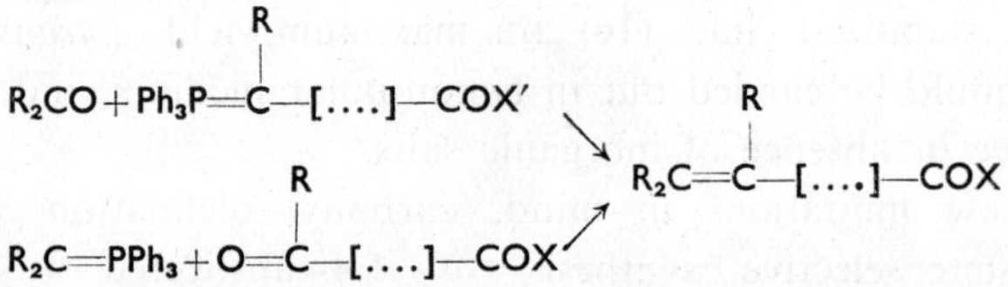

$$R_2CO + Ph_3P\overset{R}{-}C-[....]-COX'$$

$$R_2C=PPh_3 + O=\overset{R}{C}-[....]-COX'$$

$$R_2C=\overset{R}{C}-[....]-COX$$

When CO-stabilized phosphoranes of type **16** are used, the reaction usually proceeds without side-effects, because ylides of this type do not react with ester and similar groupings, although they still attack the aldehyde function. In the case of partly or non-stabilized phosphorus ylides (**17** and **15**) the reaction can become complicated by intra- or intermolecular acylation of the ylide carbon*[74-76]. However, in the presence of both ester and aldehyde groups the latter react first and undesirable side-reactions can be suppressed by selection of suitable conditions.

* Apparently this process is accompanied by transylidation and proceeds according to the following scheme:

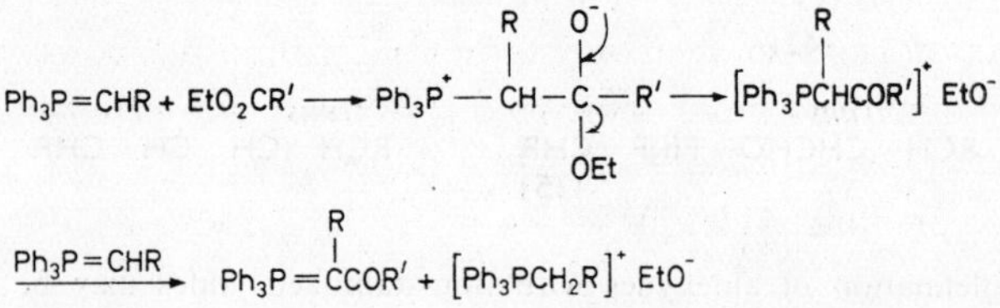

$$Ph_3P=CHR + EtO_2CR' \longrightarrow Ph_3\overset{+}{P}-\overset{R}{CH}-\overset{O^-}{\underset{OEt}{C}}-R' \longrightarrow [Ph_3\overset{R}{PCHCOR'}]^+ EtO^-$$

$$\xrightarrow{Ph_3P=CHR} Ph_3P=\overset{R}{CCOR'} + [Ph_3PCH_2R]^+ EtO^-$$

In the presence of lithium bromide, which facilitates the reaction, the phosphonium alcoholate formed is transformed into the corresponding bromide.

A. α,β-Unsaturated Acids

1. Acids of the type RCH═CH─COOH

Such acids can be prepared by olefination of glyoxylic ester[78] or its vinylog, β-alkoxycarbonylacrylic aldehyde[79].

$$RCH═PPh_3 + OCHCO_2R' \longrightarrow RCH═CHCO_2R'*$$
$$RCH═PPh_3 + OCHCH═CHCO_2R' \longrightarrow RCH═CHCH═CHCO_2R'$$

In the case of non-stabilized or partly stabilized phosphorus ylides these reactions are not stereoselective and are rarely used for the synthesis of non-substituted α, β-unsaturated acids, which are more frequently made by condensation of aldehydes with alkoxycarbonylmethylenetriphenyl-phosphoranes (16).

$$RCHO + Ph_3P═CHCO_2R' \longrightarrow RCH═CHCO_2R'$$
$$\text{(16)}$$

An elegant modification of this reaction is the reduction of the thiol esters of carboxylic acids by Raney nickel in the presence of ylide 16[80].

$$RCO_2H \xrightarrow[\text{(2) EtSNa}]{\text{(1) SOCl}_2} RCOSEt \xrightarrow[\text{Raney Ni}]{Ph_3P═CHCO_2Me} RCH═CHCO_2Me$$

The ylide 16 can be easily prepared by the addition of triphenylphosphine to bromoacetic ester, followed by treatment with alkali[81]. It is stable in air and is convenient to work with; at the same time its reaction with aldehydes is less complicated and more rapid than the corresponding Reformatsky reaction.

As has been pointed out in section I, reactions of the stabilized ylide (16) with aldehydes always lead to predominantly *trans* olefin. In non-polar solvents, the steric course of the reaction is practically independent of inorganic salts, while in polar aprotic solvents their presence lowers the stereoselectivity. Significant amounts of *cis* isomers can also form if the reaction is carried out in polar protic solvents[70a].

The reaction has been relatively little used with saturated aldehydes[61, 83]. It has been employed much more frequently with α,β-unsaturated aldehydes in the preparation of conjugated polyenic acids[78, 81, 83–87], especially with the isoprenoid skeleton[63, 85–87].

---

* Here and further on the triphenylphosphine oxide formed in the reaction is not shown in the schemes.

$$R(CH{=}CH)_n\,CHO + Ph_3P{=}CHCO_2R' \longrightarrow R(CH{=}CH)_{n+1}COOR'$$

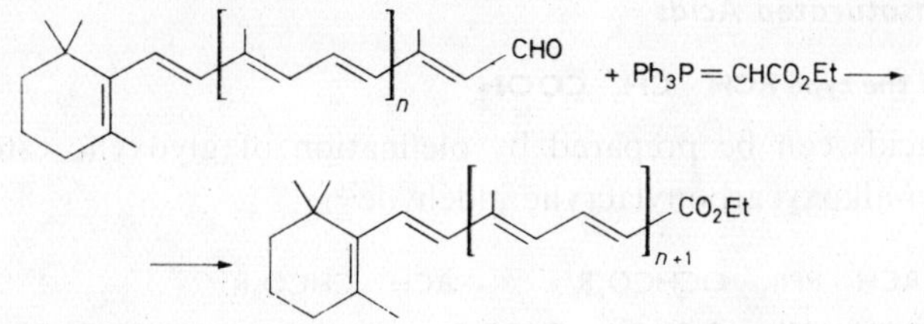

Polyenic aldehydes with triple bonds can also be used[81, 83, 85–87].

$$RC{\equiv}C\left[CH{=}\overset{CH_3}{\overset{|}{C}}CH{=}CH\right]_n\overset{CH_3}{\overset{|}{C}}{=}CHCHO + Ph_3P{=}CHCO_2R' \longrightarrow$$

$$\longrightarrow RC{\equiv}C\left[CH{=}\overset{CH_3}{\overset{|}{C}}CH{=}CH\right]_{n+1}CO_2R'$$

$R = CH_2{=}\overset{CH_3}{\overset{|}{C}}{-}$ or

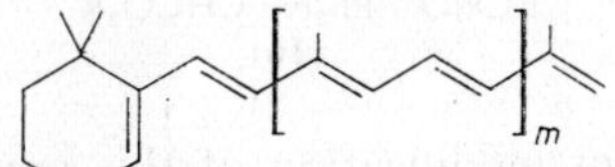

Unsaturated dialdehydes give with two moles of phosphorane all-*trans* polyenic dicarboxylic acids[81, 83, 88–92].

$$\underset{}{OCH\overset{trans}{(CH{=}CH)_n}CHO} + 2\,Ph_3P{=}CHCO_2Me \longrightarrow MeO_2C\overset{trans}{(CH{=}CH)_{n+2}}CO_2Me$$

By this method, the synthesis of methylbixin (**43**) was accomplished by a 'double' Wittig reaction according to the scheme:

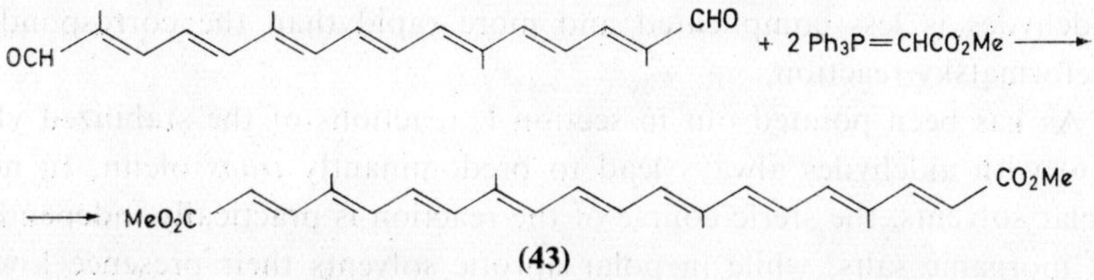

(43)

ω-Functional α,β-unsaturated acids can be synthesized by condensation of alkoxycarbonylmethylenetriphenylphosphorane (**16**) with the corresponding ω-substituted aldehydes. Thus, *trans*-7-hydroxyheptenoic ester (**45**) has been obtained from 5-hydroxyvaleraldehyde (the tautomer of 2-hydroxytetrahydropyrane) (**44**)[93].

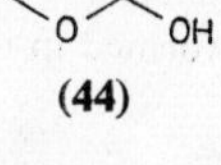

$$+ Ph_3P{=}CHCO_2R \longrightarrow HO(CH_2)_4CH{=}CHCO_2R$$

(44) **(16)** **(45)**

By carrying out the reaction of dialdehydes with one mole of phosphorane one can obtain α,β-unsaturated ω-aldehydoacids[94, 95].

$$\text{OCH(CH}_2)_n\text{CHO} + \text{Ph}_3\text{P}{=}\text{CHCO}_2\text{R} \longrightarrow \overset{trans}{\text{OCH(CH}_2)_n\text{CH}{=}\text{CHCO}_2\text{R}}$$

The reaction with ω-aldehydoesters leads to α,β-unsaturated dicarboxylic acids[96, 97].

$$\text{RO}_2\text{C(CH}_2)_n\text{CHO} + \text{Ph}_3\text{P}{=}\text{CHCO}_2\text{R} \longrightarrow \underset{95\%}{\overset{trans}{\text{RO}_2\text{C(CH}_2)_n\text{CH}{=}\text{CHCO}_2\text{R}}}$$

The first of these reactions has been used in the synthesis of *trans*-9-ketodec-2-enoic acid (46) a component of the royal jelly of the honey bee[94, 95].

$$\text{OCH(CH}_2)_3\text{CHO} + \text{Ph}_3\text{P}{=}\text{CHCO}_2\text{Me} \longrightarrow \overset{trans}{\text{OCH(CH}_2)_3\text{CH}{=}\text{CHCO}_2\text{Me}} \xrightarrow[12-15\%]{\text{4 stages}}$$

$$\underset{\textbf{(46)}}{\overset{trans}{\text{CH}_3\text{CO(CH}_2)_5\text{CH}{=}\text{CHCO}_2\text{H}}}$$

The second reaction was utilized to prepare bombycol (47), the attractant principle of the female silkworm *(Bombyx mori L.)*[96, 97].

$$\text{MeO}_2\text{C(CH}_2)_8\text{CHO} + \text{Ph}_3\text{P}{=}\text{CHCO}_2\text{Me} \longrightarrow \underset{67\%}{\overset{trans}{\text{MeO}_2\text{C(CH}_2)_8\text{CH}{=}\text{CHCOOMe}}}$$

$$\xrightarrow[\substack{(1)\ \text{SOCl}_2 \\ (2)\ \text{Rosenmund reduction} \\ (3)\ \text{LiAlH}_4 \\ (4)\ \text{Ac}_2\text{O}}]{} \overset{trans}{\text{AcO(CH}_2)_9\text{CH}{=}\text{CHCHO}} \xrightarrow[(2)\ \text{OH}^-]{(1)\ \text{Ph}_3\text{P}{=}\text{CH(CH}_2)_3\text{CH}_3}$$

$$\longrightarrow \underset{\textbf{(47)}}{\overset{trans\quad\quad cis}{\text{HO(CH}_2)_9\text{CH}{=}\text{CHHC}{=}\text{CH(CH}_2)_3\text{CH}_3}}$$

The reaction of the phosphorane (16) with aromatic aldehydes gives derivatives of *trans*-cinnamic acid in good yields[38-40, 61, 98-100].

$$\text{X}\langle\bigcirc\rangle\text{-CHO} + \text{Ph}_3\text{P}{=}\text{CHCN} \longrightarrow \text{X}\langle\bigcirc\rangle\text{-CH}{=}\text{CHCN}$$

Heterocyclic α,β-unsaturated acids can be prepared by condensation of the ylide 16 with various heterocyclic aldehydes such as furfural[61] and its derivatives[101], pyridinealdehydes[99] etc.

For the synthesis of *trans-trans*dienic acids of the type $\text{RCH}{=}\text{CHCH}{=}\text{CHCO}_2\text{H}$, instead of alkoxycarbonylmethylenetriphenylphosphorane (16) one can use its vinylog (48), prepared from γ-bromocrotonic ester[92]:

$$\text{BrCH}_2\text{CH}{=}\text{CHCO}_2\text{Me} + \text{Ph}_3\text{P} \longrightarrow [\text{Ph}_3\text{PCH}_2\text{CH}{=}\text{CHCO}_2\text{Me}]^+ \text{Br}^- \longrightarrow$$

$$\xrightarrow{\text{base}} \underset{\textbf{(48)}}{\text{Ph}_3\text{P}{=}\text{CHCH}{=}\text{CHCO}_2\text{Me}}$$

Another method of preparing the ylide **48** is by carbomethoxylation of allylenetriphenylphosphorane by methyl chloroformate; the phosphonium salt forming in the reaction being deprotonized by excess ylide[102].

$$Ph_3P=CHCH=CH_2 + ClCO_2Me \longrightarrow [Ph_3PCH=CHCH_2CO_2Me]^+ \, Cl^- \longrightarrow$$
$$\longrightarrow Ph_3P=CHCH=CHCO_2Me$$
$$\textbf{(48)}$$

The reaction of aldehydes with the ylide **48** is highly stereospecific (all-*trans* compounds being formed) and as a rule gives better results than the Reformatsky reaction. The condensation of the ylide **48** with aromatic aldehydes[92, 103] and also with hydroxytetrahydropyrane **(44)**[93] has been described.

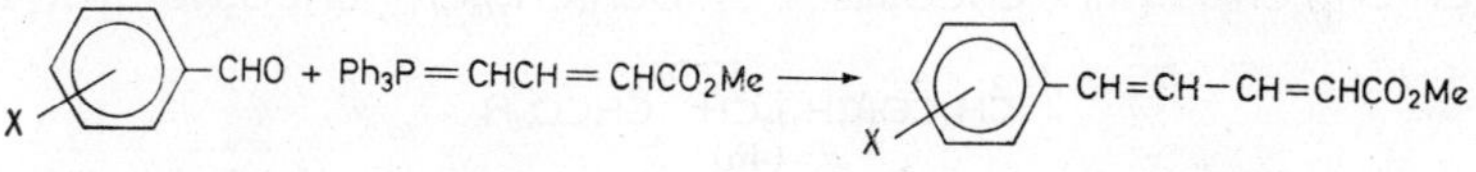

$$\textbf{(48)}$$

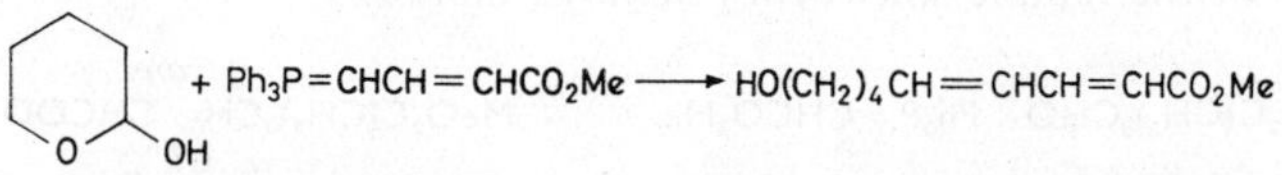

$$\textbf{(44)}$$

Another method of preparing α,β-unsaturated acids involves olefination of aldehydes by the phosphonate carbanion **(50)**[57, 63, 64, 104].

$$RCHO + (EtO)_2\overset{\overset{\displaystyle O}{\|}}{P}-\overset{-}{C}HCO_2Et \longrightarrow RCH=CHCO_2Et + (EtO)_2\overset{\overset{\displaystyle O}{\|}}{P}O$$
$$\textbf{(50)}$$

The carbanion **(50)** is formed on alkaline treatment of the phosphonic ester **(51)**, readily prepared from triethyl phosphite by the Arbuzov reaction.

$$(EtO)_3P + XCH_2CO_2Et \longrightarrow (EtO)_2\overset{\overset{\displaystyle O}{\|}}{P}CH_2CO_2Et + EtX$$
$$\textbf{(51)}$$

A complication in the reactions of aldehydes with the carbanion **50** is that due to the instability of the latter it must be prepared in the presence of the carbonyl component, which in alkaline medium undergoes polymerization and condensation. Hence the phosphonate ion **(50)** usually gives poorer yields with aldehydes than does the Wittig reaction. On the other hand it reacts quite smoothly with ketenes to form the esters of allenic acids[105]:

$$RCH=CO + (EtO)_2P(O)-\overline{C}HCO_2Et \longrightarrow RCH=C=CHCO_2Et$$

2. Acids of the type RCH=C—COOH with R¹

Esters of α-substituted α,β-unsaturated acids can be synthesized either by olefination of keto esters (non-stereospecific)[67] or of aldehydes by phosphorus ylides of the type **52**[67, 83, 87, 106, 107].

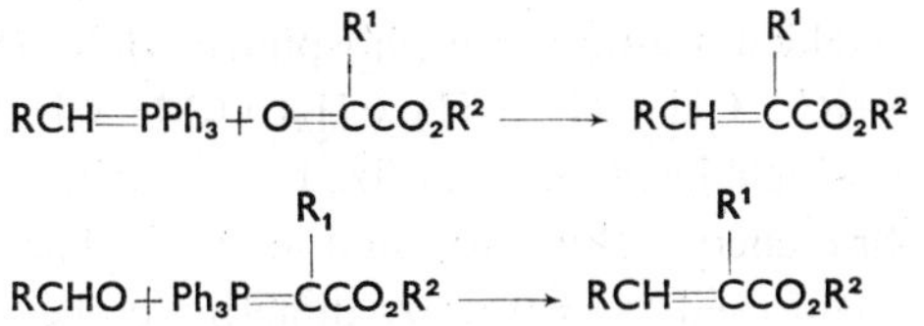

$$RCH=PPh_3 + O=\overset{R^1}{\underset{|}{C}}CO_2R^2 \longrightarrow RCH=\overset{R^1}{\underset{|}{C}}CO_2R^2$$

$$RCHO + Ph_3P=\overset{R_1}{\underset{|}{C}}CO_2R^2 \longrightarrow RCH=\overset{R^1}{\underset{|}{C}}CO_2R^2$$
$$\textbf{(52)}$$

The vinylogs (**53**) of these phosphorus ylides give α-substituted dienic acids (**54**)[89, 92].

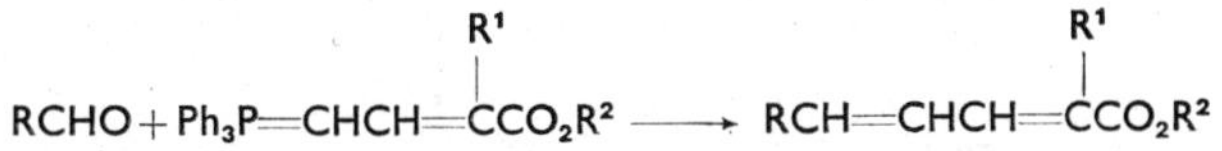

$$RCHO + Ph_3P=CHCH=\overset{R^1}{\underset{|}{C}}CO_2R^2 \longrightarrow RCH=CHCH=\overset{R^1}{\underset{|}{C}}CO_2R^2$$
$$\textbf{(53)} \qquad\qquad\qquad\qquad \textbf{(54)}$$

One can sometimes prepare branched ylides (**52**) directly from the esters of the corresponding α-bromocarboxylic acids, for example, from α-bromopropionate[83, 85–87] by reacting the latter with triphenylphosphine, followed by treatment of the phosphonium salt with sodium alcoholate. A more general method is the alkoxycarbonylation of triphenylphosphorus ylides by chloroformate[107, 108].

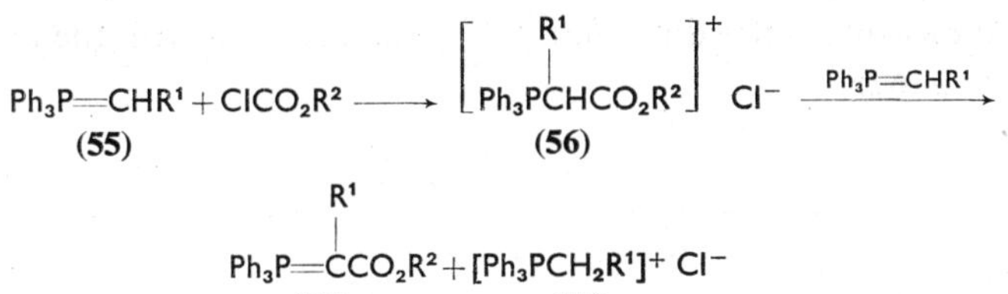

$$Ph_3P=CHR^1 + ClCO_2R^2 \longrightarrow \left[Ph_3P\overset{R^1}{\underset{|}{C}}HCO_2R^2\right]^+ Cl^- \xrightarrow{Ph_3P=CHR^1}$$
$$\textbf{(55)} \qquad\qquad\qquad\qquad \textbf{(56)}$$

$$Ph_3P=\overset{R^1}{\underset{|}{C}}CO_2R^2 + [Ph_3PCH_2R^1]^+ Cl^-$$
$$\textbf{(52)} \qquad\qquad \textbf{(57)}$$

The reaction is accompanied by transylidation (see section I. B) and the phosphonium salt (**57**) resulting therefrom can be reused for preparing the ylide **55**. By means of this method high yields of α-substituted alkoxycarbonylmethylenetriphenylphosphoranes (**52**, R¹ = Me, Et, Pr, PhCH₂, cyclohexyl) have been obtained[107, 108].

When the substituent R^1 is sufficiently electrophilic, ylides of the type **52** can be obtained by alkylation of unsaturated alkoxymethylenetriphenylphosphoranes (**16**).

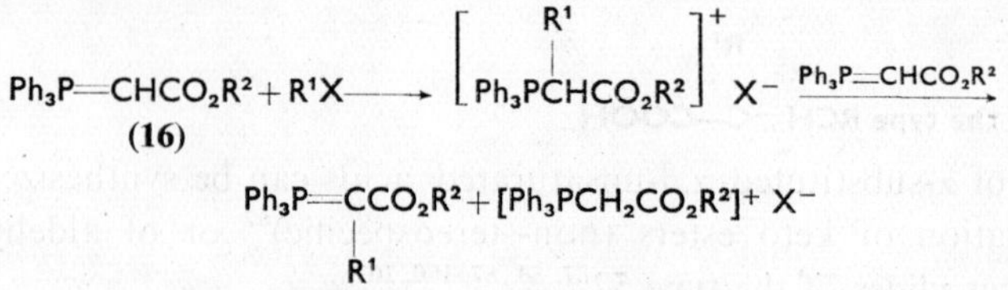
Ph₃P═CHCO₂R² + R¹X
R¹
Ph₃PCHCO₂R²
X⁻
Ph₃P═CHCO₂R²
(16)
Ph₃P═CCO₂R² + [Ph₃PCH₂CO₂R²]⁺ X⁻
R¹

(52)

This method has yielded a number of phosphorus ylides (**52**) with electrophilic substituents ($R^1 = CH_2CH{=}CH_2$, CH_2CO_2Me, CH_2Ph, CH_2CN and $CH_2CH{=}CHPh$) which have been utilized for further synthesis of the corresponding α-branched, α,β-unsaturated acids[106]. The ylides **52** (R^1= Me) and **60** were also used to synthesize dimethyl crocetinate (**61**)[89, 92].

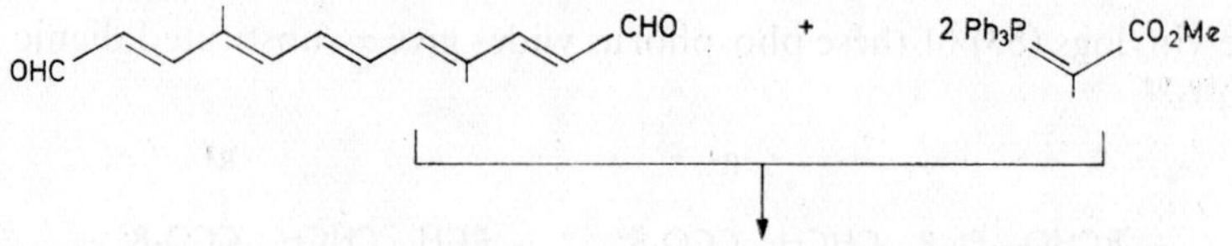
OHC
CHO
+
2 Ph₃P
CO₂Me

(58) **(52)**

(61)

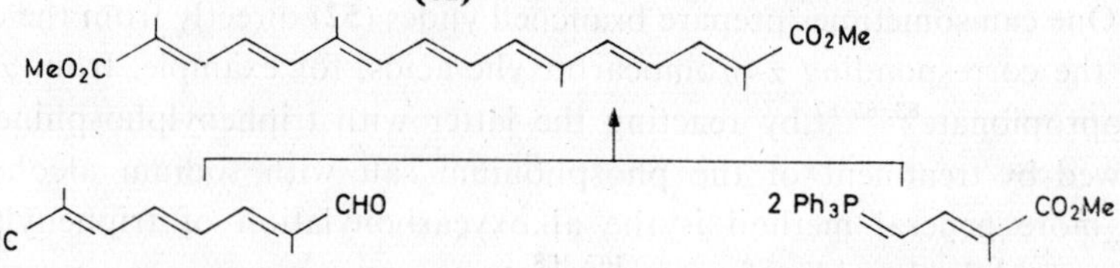
MeO₂C
CO₂Me
OHC
CHO
2 Ph₃P
CO₂Me

(59) **(60)**

The synthesis of torularidine (**63**)[81, 85–87] was achieved with the aid of the trienic ylide (**62**).

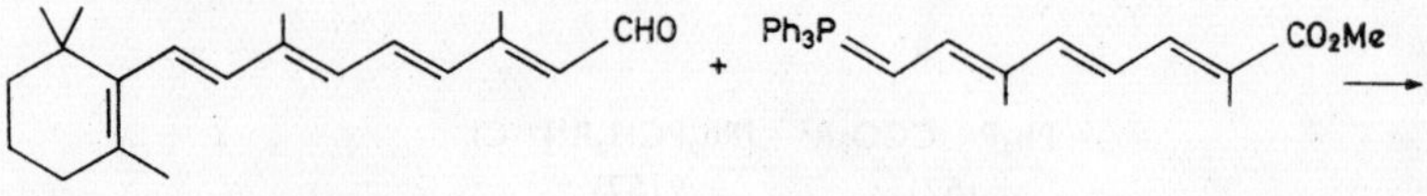
CHO
+
Ph₃P
CO₂Me

(62)

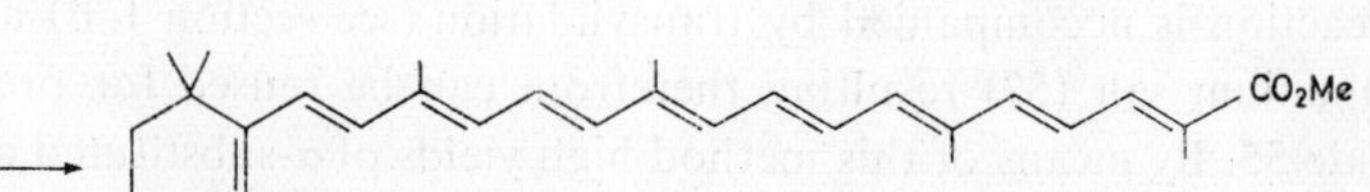
CO₂Me

(63)

α-Halogen-substituted α,β-unsaturated acids (65) (X = Cl, Br, I) were synthesized using halomethoxycarbonylmethylenetriphenylphosphoranes (64)[50, 109–113].

$$\underset{\textbf{(64)}}{RCHO + Ph_3P\!\!=\!\!\overset{\displaystyle X}{\overset{|}{C}}CO_2Me} \longrightarrow \underset{\textbf{(65)}}{RCH\!\!=\!\!\overset{\displaystyle X}{\overset{|}{C}}CO_2Me}$$

The ylide 64 is formed on halogenation of methoxycarbonylmethylenetriphenylphosphorane (16) with subsequent elimination of hydrohalic acids[109–111].

$$\underset{\textbf{(16)}}{Ph_3P\!\!=\!\!CHCO_2Me + X_2} \longrightarrow \underset{\textbf{(66)}}{\left[Ph_3P\overset{\displaystyle X}{\overset{|}{C}}HCO_2Me\right]^+ X^-} \overset{-HX}{\longrightarrow} \textbf{(64)}$$

With the α-chloroylides (64, X = Cl) optimum results were obtained by using ICl_3[109–111] or t-butylhypochlorite[110] as halogenating agent. Sodium hydroxide, Et_3N, pyridine or excess ylide (16) (transylidation) were used for cleavage of HX from the salt (66).

It has been shown that the reaction of the haloylides (64, X = Cl, Br, I) with aromatic aldehydes in chloroform is stereoselective, *trans* cinnamic acids being predominantly formed[112, 113].

α-Substituted α,β-unsaturated acids can also be synthesized by way of PO-activated carbonyl olefination[57]. In order to prepare the olefinating agent, phosphoneacetic ester is metalated with sodium hydride and the sodium derivative (67) is then treated with alkyl bromide. The resultant branched phosphonium ester (68) is again converted into a carbanion and condensed with the carbonyl compounds.

$$\underset{\textbf{(67)}}{(EtO)_2P(O)CH_2CO_2Et \overset{NaH}{\longrightarrow} (EtO)_2P(O)CHNaCO_2Et} \overset{R'Br}{\longrightarrow}$$

$$\underset{\textbf{(68)}}{(EtO)_2P(O)\overset{\displaystyle R'}{\overset{|}{C}}HCO_2Et} \overset{(1)\,NaH}{\underset{(2)\,RCHO}{\longrightarrow}} RCH\!\!=\!\!\overset{\displaystyle R'}{\overset{|}{C}}CO_2Et + (EtO)_2P\overset{\displaystyle O}{\underset{O^-}{\diagup\!\!\!\diagdown}}$$

$$\underset{\textbf{(69)}}{(EtO)_2P(O)CH\!\!=\!\!\overset{\displaystyle CH_3}{\overset{|}{C}}CO_2Me}$$

Condensation of the dialdehydes 58 and 59 with the phosphonates 68 (R = Me) and 69 was utilized in the synthesis of dimethyl crocetinate (61) using a scheme similar to that described above[114].

For the synthesis of α, β-unsaturated α-bromoacids, the sodium derivative (67) is brominated and the bromide (70) is converted into the phosphonate bromoanion (71), used in the carbonyl olefination reaction[57].

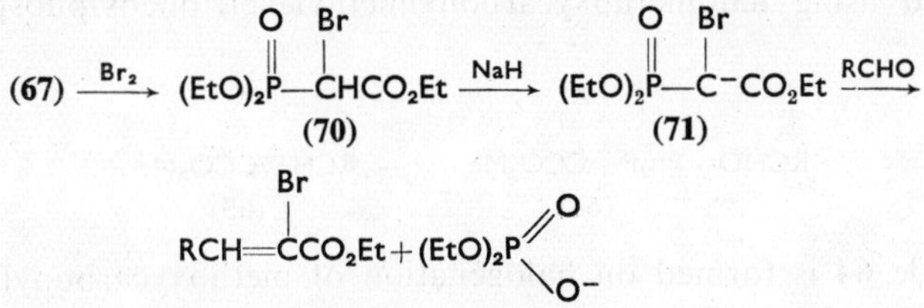

$$(67) \xrightarrow{\text{Br}_2} (EtO)_2\overset{\displaystyle O}{\overset{\|}{P}}-\overset{\displaystyle Br}{\underset{|}{C}}HCO_2Et \xrightarrow{\text{NaH}} (EtO)_2\overset{\displaystyle O}{\overset{\|}{P}}-\overset{\displaystyle Br}{\underset{|}{C}}{}^{-}-CO_2Et \xrightarrow{\text{RCHO}}$$

$$\text{(70)} \qquad\qquad \text{(71)}$$

$$RCH{=}\overset{\displaystyle Br}{\underset{|}{C}}CO_2Et + (EtO)_2P\overset{\displaystyle \diagup O}{\underset{\diagdown O^-}{}}$$

When iodine and excess sodium hydride are used instead of bromine, esters of the corresponding acetylenic acids, rather than of the ethylenic haloesters, are formed[57].

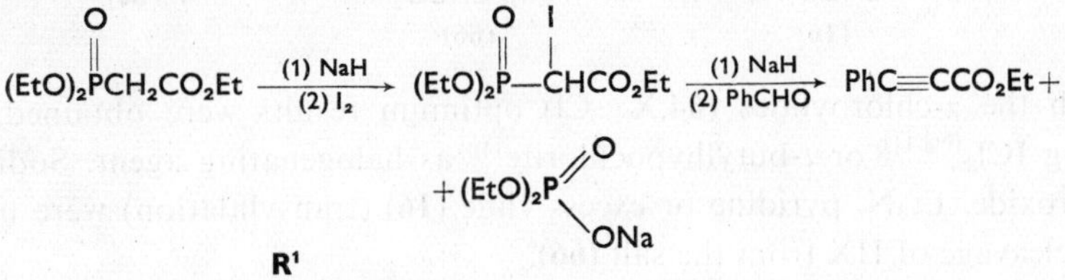

$$(EtO)_2\overset{\displaystyle O}{\overset{\|}{P}}CH_2CO_2Et \xrightarrow[\text{(2) I}_2]{\text{(1) NaH}} (EtO)_2\overset{\displaystyle O}{\overset{\|}{P}}-\overset{\displaystyle I}{\underset{|}{C}}HCO_2Et \xrightarrow[\text{(2) PhCHO}]{\text{(1) NaH}} PhC{\equiv}CCO_2Et +$$

$$+ (EtO)_2P\overset{\displaystyle \diagup O}{\underset{\diagdown ONa}{}}$$

3. Acids of the type R—C(R¹)=CH—COOH

The carbonyl olefination reaction makes it possible to synthesize β-substituted α,β-unsaturated acids through olefination of ketones by means of alkoxylcarbonylmethylenetriphenylphosphoranes (16) or olefination of glyoxylic esters by means of α,α-disubstituted phosphorylides (72).

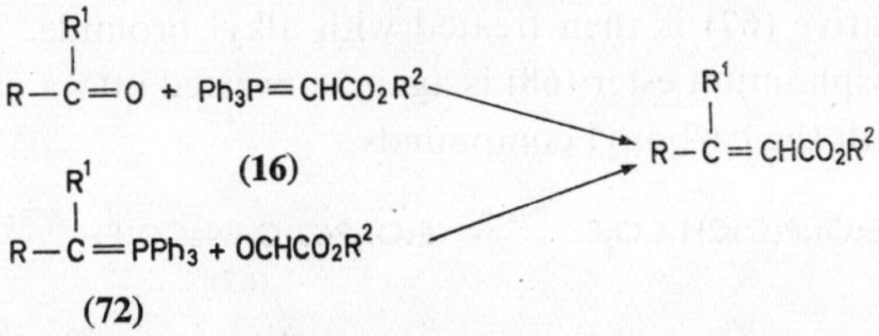

$$\overset{\displaystyle R^1}{\underset{|}{R-C}}{=}O + Ph_3P{=}CHCO_2R^2 \searrow$$

$$\text{(16)} \qquad\qquad\qquad \overset{\displaystyle R^1}{\underset{|}{R-C}}{=}CHCO_2R^2$$

$$\overset{\displaystyle R^1}{\underset{|}{R-C}}{=}PPh_3 + OCHCO_2R^2 \nearrow$$

$$\text{(72)}$$

As a rule olefination of ketones by the relatively less reactive ylide 16 requires drastic conditions and gives unsatisfactory yields of the unsaturated esters[99, 104, 115–117]. For instance, 2-methylcyclohexanone gives only a 30% yield of ethyl 2-methylcyclohexylideneacetate even if the reaction is carried out without solvents for several hours at 150°[115].

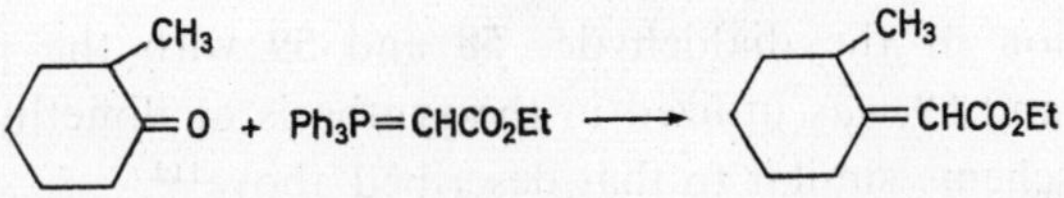

The ketones with an activated carbonyl group, such as the vigorously reacting fluoroacetone, are an exception to this rule[118].

$$\underset{\underset{\displaystyle FCH_2CO}{|}}{\overset{\displaystyle CH_3}{|}} + Ph_3P{=}CHCO_2Et \longrightarrow \underset{\underset{\displaystyle FCH_2C}{|}}{\overset{\displaystyle CH_3}{|}}{=}CHCO_2Et \quad (83\%)$$

The olefination of some ketones by stabilized ylides is catalyzed by benzoic acid[119].

The carbanion of phosphoneacetic acid (50) is much more suitable for the olefination of ketones. It reacts quite readily even with unreactive ketones and usually gives a high yield of the unsaturated ester[57, 63, 64, 105, 117]. For instance, quinuclidone is olefinated almost quantitatively at room temperature[120].

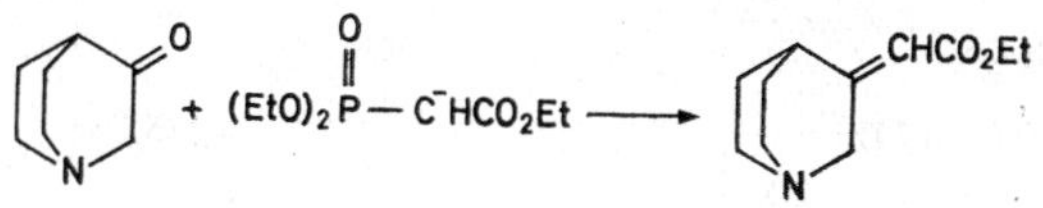

(50)

Olefination of ketones by means of the phosphonate carbanion (50) has been used for the synthesis of carotenoid acids[114, 121] and for the conversion of 3-ketosteroids into methoxycarbonylmethylene derivatives[125].

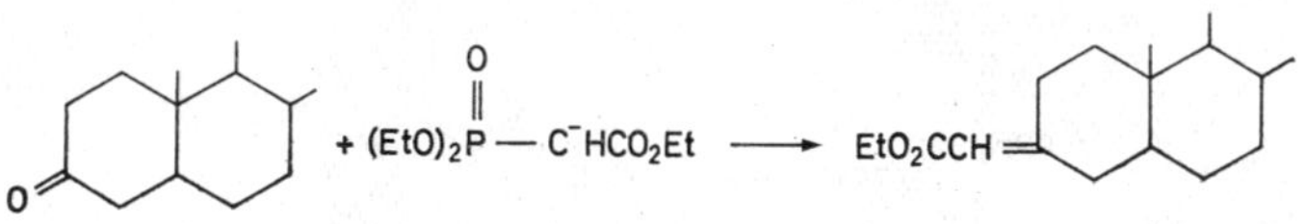

(50)

When the phosphoneacetic esters of optically active alcohols are used in the reaction, it proceeds with asymmetric induction. Thus, olefination of *p*-substituted cyclohexanones by diethyl-*p*-menthyloxycarbonylmethyl phosphonate yielded the laevorotatory isomers of substituted cyclohexylideneacetic acids[126].

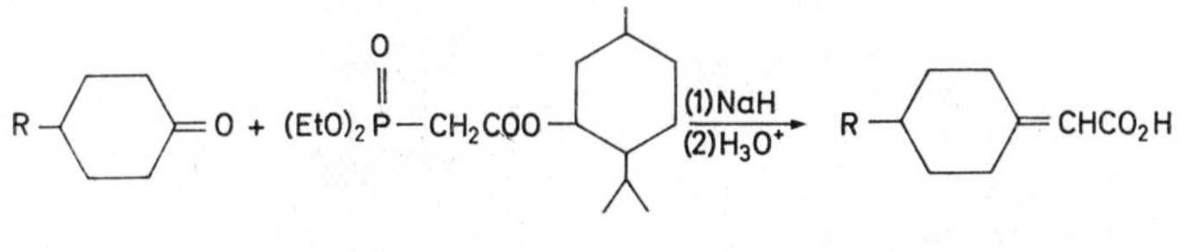

R = Me,   *t*-Bu

For the synthesis of β-substituted α,β-unsaturated acids by the second route (olefination of glyoxylic esters and their analogs) it is recommended that the reagents be mixed in 'reverse', *i.e.* that the ylide be added to

21*

the aldehyde. Such a procedure was used in the synthesis of ethyl β-ionyli-deneacetate[78].

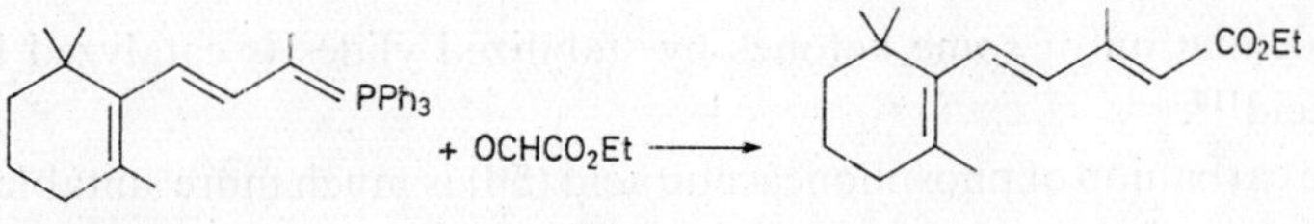

(73)

Methylated vinylogs of glyoxylic ester (74—77) were used in numerous syntheses in the vitamin A and carotenoid series[127–129].

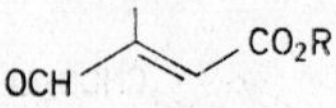

(74) 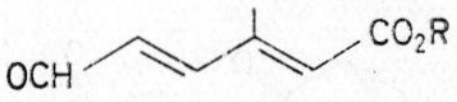   (75)

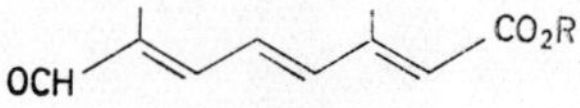

(76) 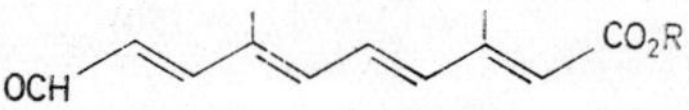   (77)

Thus, the ethyl ester of vitamin A acid was obtained in 65% yield from β-cyclogeranylidenetriphenylphosphorane (78) and the aldehydoester (76).

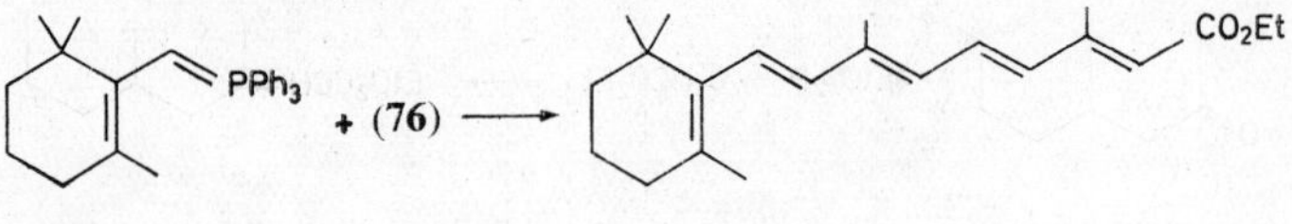

(78)

In the same way the carotenoid acids (79) were synthesized by olefination of the aldehydo esters 75 and 77 by the ylide 73[78, 130–134].

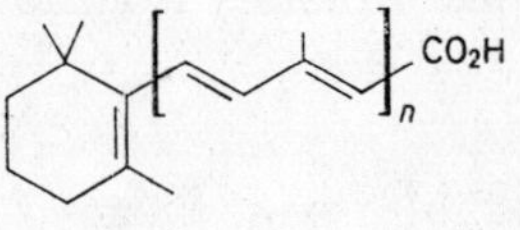

(79, $n = 2,3$)

B. β,γ-Unsaturated Acids

Acids of this type were obtained by olefination of ketones with carboxy-ethylidenetriphenylphosphorane (80) in dimethylsulfoxide[135].

$$RR'CO + Ph_3P{=}CHCH_2CO_2^- \longrightarrow RR'C{=}CHCH_2CO_2^-$$

(80)

The β-carboxy ylide **80** is very unstable and spontaneously decomposes into triphenylphosphine and acrylic acid. It is therefore prepared from the corresponding phosphonium chloride, by reacting the latter with a base (methylsulfinylsodium) in the presence of an equimolar amount of ketone. Under such conditions the ylide **80** cannot be used for olefinating aldehydes. It has been shown that in the interaction of the ylide **80** with m-methoxyacetophenone the main product is the '*cis*' isomer (**81**)[135].

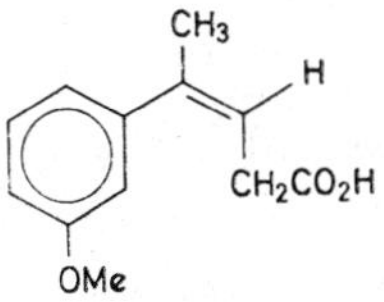

(**81**)

C. Acids with Remote Double Bonds

In the past years several hundred unsaturated higher acids have been isolated from the lipids of microbes, plants and animals. The chemical synthesis of many of these acids has not yet been achieved, since the classical synthetic methods are as a rule highly involved, requiring starting materials which are not readily available and only giving low yields of the desired product[136]. Moreover, these methods (with the exception of the ones starting with acetylene derivatives) are non-stereospecific. Carbonyl olefination opens up new vistas in this field, making possible the stereoselective synthesis of all the basic types of naturally occurring fatty acids[137].

1. Monoenic acids

The starting materials for the synthesis of *cis* alkenoic acids of the type $CH_3(CH_2)_mCH{=}CH(CH_2)_nCOOH$ by carbonyl olefination are aliphatic aldehydes and ω-haloalkanoic acids $XCH_2(CH_2)_nCOOH$. The latter may be prepared from dicarboxylic acids by the Hunsdiecker reaction[138] or by the recently developed selective reduction of the monoesters of dicarboxylic acids to hydroxy acids[139] followed by their halogenation. ω-Chlorocarboxylic acids with an odd number of carbon atoms are available in commercial quantities, being obtained by hydrolysis of carbontetrachloride–ethylene telomerization products[140].

The conversion of ω-haloalkanoic acids into higher alkenoic acids is carried out according to Scheme 4:

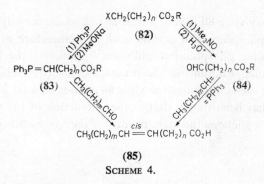

(85)
Scheme 4.

One of the methods represented in the scheme involves conversion of esters of chloroalkanoic acids into ω-alkoxycarbonylphosphorus ylides **(83)**, which are then utilized for the olefination of aliphatic aldehydes. In order to transform the chloroesters **(82, X = Cl)** into phosphonium salts, they should be first converted into the iodine derivatives which require much milder conditions for reaction with triphenylphosphine. Also ω-alkoxycarbonyltriphenylphosphonium chlorides have a greater tendency to decompose with formation of triphenylphosphine than the iodides. Although the latter also undergo such decomposition, they give satisfactory yields of the unsaturated esters providing one uses a large excess of the phosphonium salt. The ylides **83** cannot be obtained in non-polar media because of their reaction with the ester groupings (see above). In polar media when the activity of the ylide is lowered by solvation, the alkoxycarbonyl group is not affected. Of all the polar solvents tested, the ylides **83** are best formed and their condensation with aldehydes is best carried out in dimethylformamide. In this solvent the reaction is quite stereospecific. Thus oleic acid obtained by carbonyl olefination contains not more than 6% elaidic acid[141]. The method was used for preparing various naturally occurring *cis* alkenoic acids **(85)**, with $n = 6$, 7 or 9[76, 141, 142]. However, it proved to be unsuitable for preparing acids with $n = 3$ or 4, since the corresponding ω-alkoxycarbonylphosphorus ylides **(83, $n = 3$, 4)** behave anomalously undergoing intramolecular acylation with the formation of 2-oxocyclopentylidene- or 2-oxocyclohexylidenetriphenylphosphoranes[77, 143].

$$Ph_3P=CH(CH_2)_n CO_2R \longrightarrow Ph_3P=C(CH_2)_nCO$$

$$n = 3,4$$

In these cases one can make use of the second route of Scheme 4, namely condensation of alkylidenetriphenylphosphoranes with ω-alkoxycarbonylaldehydes **(84)**. The latter can be made from the acyl chlorides of the monoesters of dicarboxylic acids by the Rosenmund reduction[144] or by treatment of ω-iodoesters (82, X = I) with trimethylamine N-oxide followed by mild acid hydrolysis[143, 145].

2. Conjugated polyenic acids

The principles of the steric control of carbonyl olefination discussed in the first part of this paper can be applied to the stereoselective synthesis of trienic acids $CH_3(CH_2)_m(CH{=}CH)_3(CH_2)_n COOH$. Acids of this type with various combinations of *cis* and *trans* ethylenic bonds are constituent parts of the lipids of a number of higher plants[146]. The most widespread of these, acids with a *cis-trans-trans* trienic system (87), can be synthesized by *cis* olefination of *trans-trans* dienic aldehydes (86) with ω-alkoxycarbonylalkylidenetriphenylphosphoranes (83).

$$\overset{trans}{CH_3(CH_2)_m\,CH}{=}CH{-}CH{=}\overset{trans}{CH}CHO + Ph_3P{=}CH(CH_2)_n\,CO_2R \xrightarrow{\;DMF\;}$$
$$\qquad\qquad (86) \qquad\qquad\qquad\qquad (83)$$

$$\xrightarrow{} \overset{trans}{CH_3(CH_2)_m\,CH}{=}CH{-}CH{=}\overset{trans}{CH}{-}CH{=}\overset{cis}{CH}(CH_2)_n CO_2R$$
$$\qquad\qquad (87)$$

It was in this way that α-eleosteric acid (87, $m = 3$, $n = 7$), the fatty acid of tung oil and its isomers (87, $m = 0$, $n = 10$, and 87, $m = 4$, $n = 6$) were synthesized [147, 148].

The same scheme was used to carry out the total synthesis in about 50% yield of α-camlolenic acid (88), isolated from *Éuphorbia* seed oil[149].

$$\overset{trans}{AcO(CH_2)_4 CH}{=}CH{-}CH{=}\overset{trans}{CH}CHO + Ph_3P{=}CH(CH_2)_7 CO_2Et \xrightarrow{\;DMF\;}$$

$$\xrightarrow{} \overset{trans}{HO(CH_2)_4 CH}{=}CH{-}CH{=}\overset{trans}{CH}{-}CH{=}\overset{cis}{(CH_2)}_7 CO_2H$$
$$\qquad\qquad (88)$$

For synthesizing *trans-trans-cis* trienic acids (90), *trans* olefination of ω-alkoxycarbonylaldehydes (84) with the partly stabilized *cis-trans* dienic phosphorus ylides (89) was used[150].

$$\overset{cis}{CH_3(CH_2)_m CH}{=}CH{-}CH{=}\overset{trans}{CH}{-}CH{=}PPh_3 + OCH(CH_2)_n CO_2R \xrightarrow{\;hexane,\,Cl^-\;}$$
$$\qquad (89) \qquad\qquad\qquad\qquad (84)$$

$$\overset{cis}{CH_3(CH_2)_m CH}{=}CH{-}\overset{trans}{CH}{=}CH{-}\overset{trans}{CH}{=}CH(CH_2)_n CO_2H$$
$$\qquad\qquad (90)$$

In this way the total synthesis of catalpic acid (90, $m = 3$, $n = 7$), isolated from the *Catalpa ovata* tree, and calendic acid (90, $m = 4$, $n = 6$), found in *Calendula officinalis*, was possible[150].

Trienic acids occurring in plants include C_{18} acids with *cis-trans-cis* double bonds, *e.g.* punicic acid (92, $m = 3$, $n = 7$) contained in pomegranate *(Punica granatum)* seeds and jacarandic acid (**92**, $m = 4$, $n = 6$) isolated from the seed oil of *Jacaranda mimosifolia*. Acids of this type can be synthesized by *cis* olefination of *trans-cis* dienic aldehydes (**91**) with the aid of ω-alkoxycarbonylalkylidenetriphenylphosphoranes (**73**).

$$\overset{cis}{}\qquad\overset{trans}{}$$
$$CH_3(CH_2)_mCH\!=\!CH\!-\!CH\!=\!CHCHO + Ph_3P\!=\!CH(CH_2)_n CO_2R \xrightarrow{\ DMF\ }$$
$$\text{(91)}\qquad\qquad\qquad\text{(73)}$$

$$\overset{cis}{}\quad\overset{trans}{}\quad\overset{cis}{}$$
$$CH_3(CH_2)_mCH\!=\!CH\!-\!CH\!=\!CH\!-\!CH\!=\!CH(CH_2)_n CO_2H$$
$$\text{(92)}$$

This scheme was used for the total synthesis of jacarandic acid[151a].

3. Divinylmethanic acids

Acids containing the divinylmethane pattern comprise a large group of biologically important substances isolated from animal, plant and microbial lipids. The most common of these acids is the all-*cis* type, but acids with *cis-trans* divinylmethane groupings have also been found, Synthesis of *cis-cis* divinylmethane acids was achieved via acetylenic derivatives[151b]. They can also be obtained by *cis* olefination of *cis-β,γ*-unsaturated aldehydes (**93**) using ω-alkoxycarbonylalkylidenetriphenylphosphoranes (**83**)[152].

$$\overset{cis}{}$$
$$CH_3(CH_2)_mCH\!=\!CHCH_2CHO + Ph_3P\!=\!CH(CH_2)_nCO_2R \xrightarrow{\ DMF\ }$$
$$\text{(93)}\qquad\qquad\qquad\text{(83)}$$

$$\overset{cis}{}\quad\overset{cis}{}$$
$$\longrightarrow CH_3(CH_2)_mCH\!=\!CHCH_2CH\!=\!CH(CH_2)_nCO_2R$$

A detailed study of this reaction showed, however, that it leads to complex mixtures of products, because the aldehydes (**93**) are partly transformed into the α, β-unsaturated isomers under the mild alkaline conditions of carbonyl olefination[153]. In addition, the *cis* aldehydes (**93**) are not easily purified and often contain the *trans* isomers as impurities.

Another type of synthesis of these acids involves olefination of ω-alkoxycarbonylaldehydes (**84**) by *cis*-alkylidenetriphenylphosphoranes (**94**) which can be comparatively easily prepared from alkylacetylenes and ethylene oxide, with subsequent reduction of the triple bond.

$$CH_3(CH_2)_mC\!\equiv\!CH \longrightarrow CH_3(CH_2)_mC\!\equiv\!CCH_2CH_2OH \longrightarrow$$

$$\overset{cis}{}$$
$$\longrightarrow CH_3(CH_2)_mCH\!=\!CHCH_2CH_2OH \longrightarrow$$

$$\overset{cis}{}$$
$$\longrightarrow CH_3(CH_2)_mCH\!=\!CHCH_2CH\!=\!PPh_3 \xrightarrow{\ OCH(CH_2)_nCO_2R\ \text{(84)}\ }$$
$$\text{(94)}$$

$$\overset{cis}{}\quad\overset{cis\ or\ trans}{}$$
$$\longrightarrow CH_3(CH_2)_mCH\!=\!CHCH_2CH\!=\!CH(CH_2)_nCO_2R$$

Phosphoranes of type (94) react non-stereospecifically, forming a mixture of *cis-cis* and *trans-cis* divinylmethanic acids whose proportions differ, depending upon the conditions[153]. The stereoisomers can be separated chromatographically on a silver nitrate-impregnated silica-gel column[154]. An undesirable side-reaction in this method is the transformation of the γ,δ-alkenylidenetriphenylphosphoranes (94) into their β,γ-isomers, $CH_3(CH_2)_mCH{=}CHCH{=}PPh_3$. Conjugated dienic esters formed as the result of such isomerizations can be separated from the divinylmethanic compounds with alumina chromatography. In general, carbonyl olefination is less suitable for the synthesis of *cis-cis* divinylmethanic acids than the procedures based on the use of acetylenic derivatives. However, the method can be used for the synthesis of acids with *trans-cis* dienic systems[153].

4. Divinylethanic acids

Butter and the fat of marine animals contain polyenic acids which yield succinic acid on degradative oxidation, whence it has been concluded that they are acids of the divinylethane ($-CH{=}CHCH_2CH_2CH{=}CH-$) pattern. Synthesis of the *cis-cis* divinylethanic acids (99) has been carried out according to Scheme 5[82].

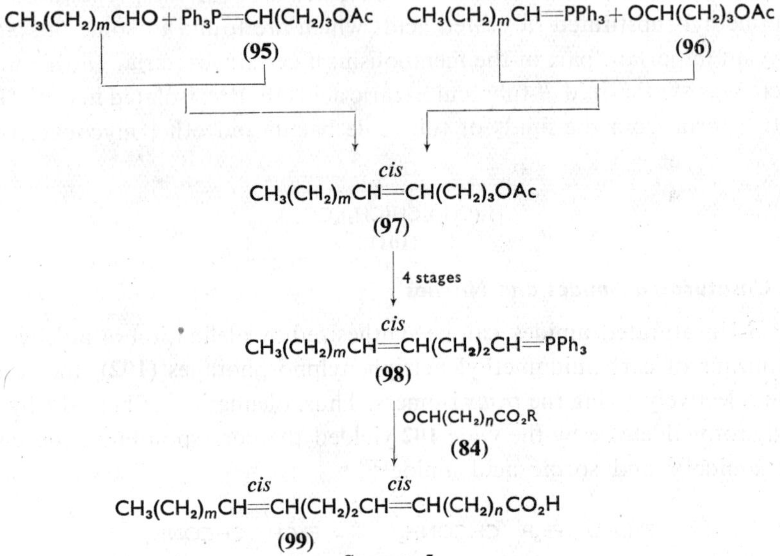

$$CH_3(CH_2)_mCHO + Ph_3P{=}CH(CH_2)_3OAc \qquad CH_3(CH_2)_mCH{=}PPh_3 + OCH(CH_2)_3OAc$$
$$(95) \qquad\qquad\qquad (96)$$

$$cis$$
$$CH_3(CH_2)_mCH{=}CH(CH_2)_3OAc$$
$$(97)$$

| 4 stages

$$cis$$
$$CH_3(CH_2)_mCH{=}CH(CH_2)_2CH{=}PPh_3$$
$$(98)$$

| $OCH(CH_2)_nCO_2R$
| (84)

$$cis \qquad\qquad cis$$
$$CH_3(CH_2)_mCH{=}CH(CH_2)_2CH{=}CH(CH_2)_nCO_2H$$
$$(99)$$

SCHEME 5.

The starting materials for these syntheses, the *cis* acetates (97), can be produced by *cis* olefination of aldehydes with ω-acetoxybutylidenetriphenylphosphorane (95) or by condensation of non-stabilized ylides with 4-acetoxybutanal (96). The *cis* acetates are converted through the corresponding bromides into the non-stabilized *cis* Δ^4-alkylidenetriphenylphosphoranes (98), which are subsequently used for *cis* olefination of ω-alkoxycarbonylaldehydes (84), leading to the desired acids (99).

5. Branched-chain acids

Monoenic fatty acids with a methyl group at the double bond (100) were prepared by olefinating methyl ketones with ω-alkoxycarbonylalkylidenetriphenylphosphoranes (83)[155].

$$CH_3(CH_2)_mCOCH_3 + Ph_3P{=}CH(CH_2)_nCO_2R \longrightarrow$$
$$(83)$$

$$\overset{\displaystyle CH_3}{\underset{\displaystyle |}{}}$$
$$\longrightarrow \quad CH_3(CH_2)_mC{=}CH(CH_2)_nCO_2H$$
$$(100)$$

Non-stabilized phosphorus ylides (83) react with ketones much more readily than CO-stabilized alkoxycarbonylmethylenetriphenylphosphoranes (16) and esters of the branched-chain acids (100) are formed in satisfactory yields on slightly heating excess ylide with methyl ketones in DMF.

The acids (100) are of interest as intermediates for the preparation of the methyl-substituted saturated acids which are found in some fats and play an important part in the metabolism of certain bacteria. In this way there was synthesized *dl*-tuberculostearic acid (101)[155], isolated in optically active form from the lipids of tubercule bacilli and other mycobacteria.

$$\overset{\displaystyle CH_3}{\underset{\displaystyle |}{}}$$
$$CH_3(CH_2)_8CH(CH_2)_8CO_2H$$
$$(101)$$

D. Unsaturated Amides and Nitriles

α,β-Unsaturated amides can be synthesized by olefination of aldehydes by means of carbamidomethylenetriphenylphosphoranes (102), the reaction selectively giving the *trans* isomers. Thus, olefination of benzaldehyde and crotonaldehyde by the ylide 102 yielded the corresponding *trans* cinnamamide[156] and sorbic acid amide[157].

$$PhCHO + Ph_3P{=}CHCONH_2 \longrightarrow \overset{\textit{trans}}{PhCH{=}CHCONH_2}$$
$$(102)$$

$$CH_3CH{=}CHCHO + Ph_3P{=}CHCONH_2 \longrightarrow \overset{\textit{trans}}{CH_3CH{=}CH}{-}\overset{\textit{trans}}{CH{=}CHCONH_2}$$
$$(102)$$

α,β-Unsaturated amides can also be obtained by olefination of glyoxylamides[158].

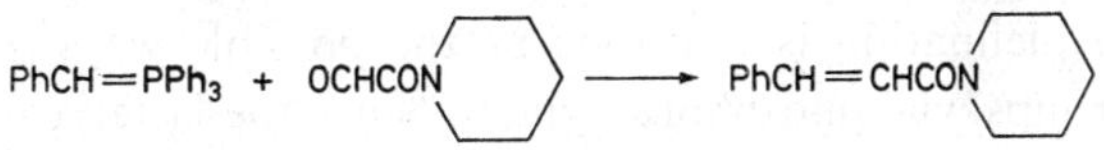
PhCH=PPh3 + OCHCON
PhCH=CHCON

α, β-Unsaturated nitriles were obtained by olefination of aldehydes with cyanomethylenetriphenylphosphorane (**103**). With benzaldehyde and its derivatives this ylide forms the corresponding cinnamonitriles in high yield[156, 159, 160].

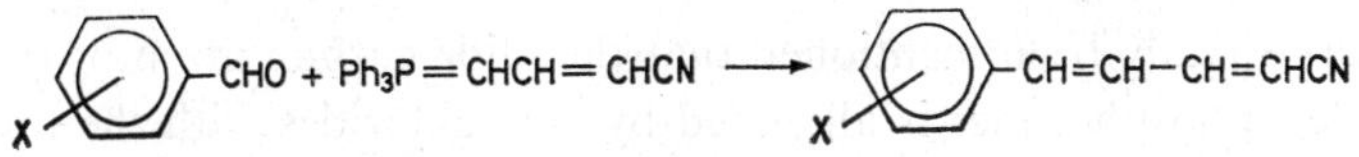
CHO + Ph3P=CHCH=CHCN
CH=CH−CH=CHCN
X
X

(103)

For the synthesis of vitamin A_2 nitrile (**104**) which was obtained in 86% yield, use was made of the methyl-substituted vinylog of the ylide **103**[161].

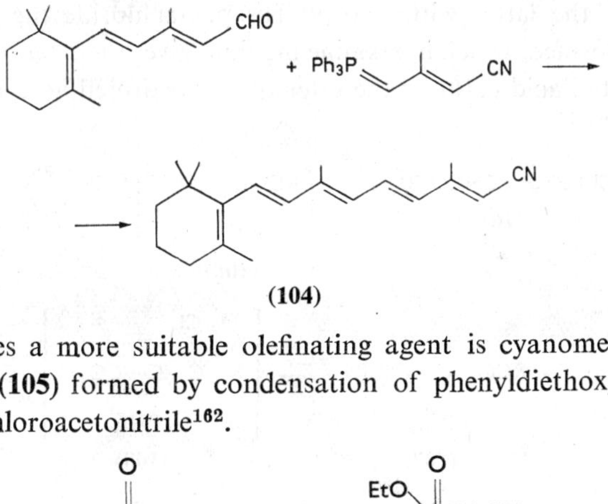
CHO
+ Ph3P
CN
CN

(104)

For ketones a more suitable olefinating agent is cyanomethylphenylphosphinate (**105**) formed by condensation of phenyldiethoxyphosphine oxide with chloroacetonitrile[162].

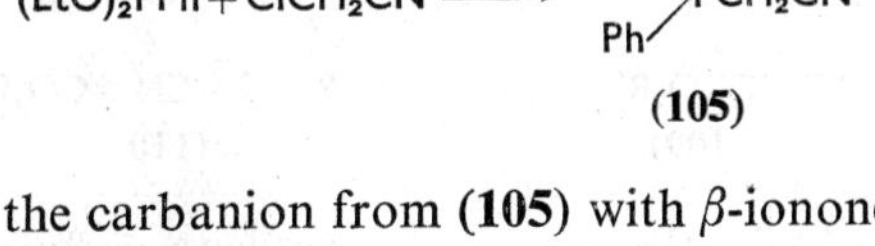
O
(EtO)2PPh + ClCH2CN
EtO
O
PCH2CN
Ph

(105)

Condensation of the carbanion from (**105**) with β-ionone gave high yields of β-ionylideneacetonitrile[162].

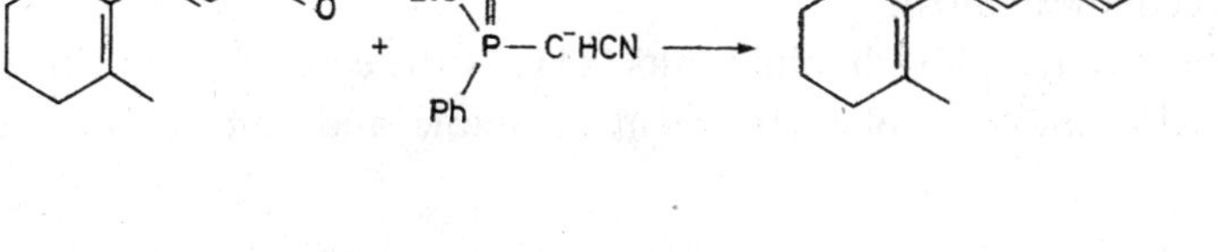
O
EtO
O
+
P−C⁻HCN
Ph
CN

E. Other Methods of Introducing the Carboxyl Group by Means of Phosphorus Ylides

Carbonyl olefination is only one of the possible ways of introducing carboxyl groups via phosphorus ylides, since the latter can also attack many other types of organic substances. Among such reactions we shall discuss only those that are useful for the synthesis of carboxylic acids and their derivatives.

1. Reaction of α-alkoxycarbonylalkylidenetriphenylphosphoranes with acid chlorides

It has already been mentioned that the ylide carbon atom of triphenyl-alkylidenephosphoranes is alkylated by acyl chlorides with the formation of phosphonium salts (see section II. A). The acid–base nature and reactivity of these salts depends upon the structure of the initial ylide. In the case of unsubstituted alkoxycarbonylmethylenetriphenylphosphoranes (16) the phosphonium salts (106) immediately undergo transylidation with a second molecule of phosphorus ylide to form the acylated ylide 107. In treating the latter with phosphorus pentachloride the phosphonium salt 108 is formed, which in alkaline medium gives the ester of a β-chloro-α, β-unsaturated acid (110) or the ester of a substituted acetylenecarboxylic acid (109)[162, 163].

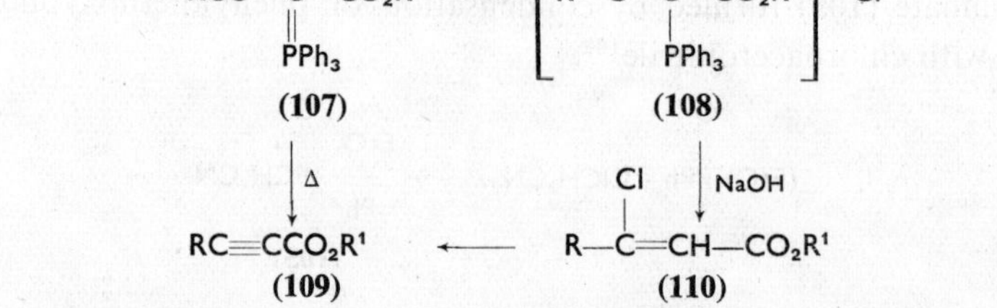

Once can also obtain the esters of acetylene carboxylic acid in high yields by pyrolysis of acyl ylides (107)[163, 164]. Bis-acetylenecarboxylic acids have also been synthesized by this procedure[165]. The ester group can be replaced by a nitrile group[166].

The acylated phosphonium salts (111) in the case of α-substituted ylides (60, R^1 = alkyl or phenyl) are more stable and can be isolated if the reaction is carried out at low temperatures. On electrolysis these salts eliminate triphenylphosphine, giving α-substituted β-keto esters[167].

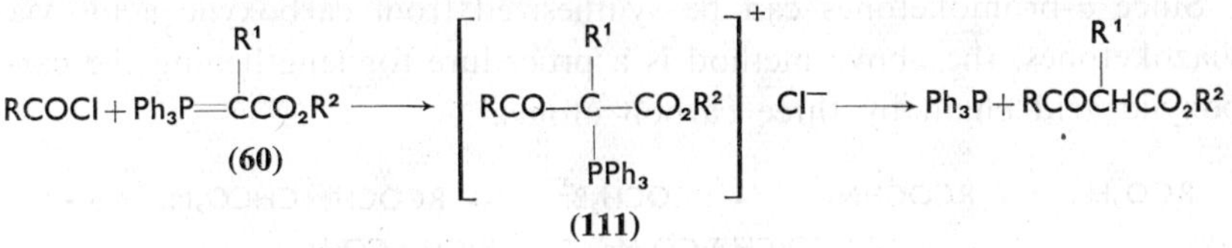

$$
RCOCl + Ph_3P\!\!=\!\!CCO_2R^2 \longrightarrow \left[\begin{array}{c} R^1 \\ | \\ RCO\!-\!C\!-\!CO_2R^2 \\ | \\ PPh_3 \end{array} \right]^{+} Cl^- \longrightarrow Ph_3P + RCOCHCO_2R^2
$$

$$\text{(60)} \hspace{3cm} \text{(111)}$$

In the case of α-unsubstituted acid chlorides RCH_2COCl, the phosphonium salts (112) first formed can react with a second molecule of the ylide[168].

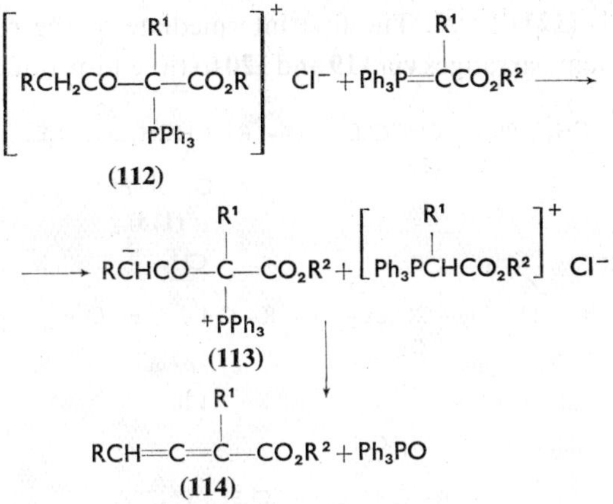

$$
\left[\begin{array}{c} R^1 \\ | \\ RCH_2CO\!-\!C\!-\!CO_2R \\ | \\ PPh_3 \end{array} \right]^{+} Cl^- + Ph_3P\!\!=\!\!CCO_2R^2 \longrightarrow
$$

$$\text{(112)}$$

$$
\longrightarrow R\overset{.}{C}HCO\!-\!\overset{\overset{\displaystyle R^1}{|}}{C}\!-\!CO_2R^2 + \left[Ph_3P\overset{\overset{\displaystyle R^1}{|}}{C}HCO_2R^2 \right]^{+} Cl^-
$$

$$+PPh_3$$
$$\text{(113)}$$

$$
RCH\!\!=\!\!\overset{\overset{\displaystyle R^1}{|}}{C}\!-\!CO_2R^2 + Ph_3PO
$$

$$\text{(114)}$$

The betaine (113) formed by γ-elimination then decomposes by the usual carbonyl olefination pathway to triphenylphosphine oxide and allenic ester (114)[168].

2. Reaction of alkoxycarbonylmethylenetriphenylphosphoranes with α-bromoketones

In the reaction of α-bromoketones with the ylide 16 the phosphonium salt 115 is first formed and this then undergoes transylidation with a second molecule of the ylide[169].

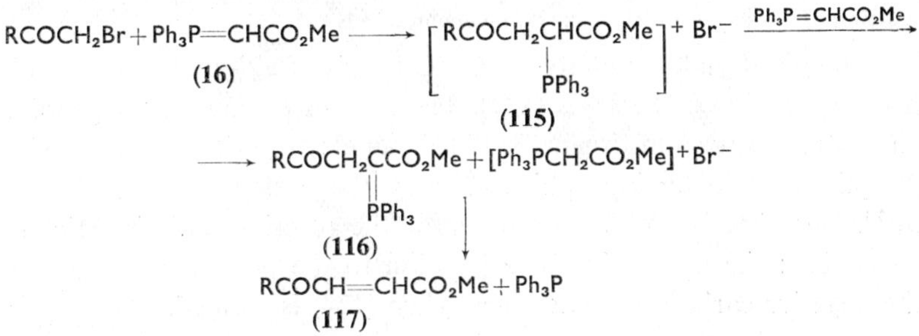

$$
RCOCH_2Br + Ph_3P\!\!=\!\!CHCO_2Me \longrightarrow \left[\begin{array}{c} RCOCH_2CHCO_2Me \\ | \\ PPh_3 \end{array} \right]^{+} Br^- \xrightarrow{\;Ph_3P=CHCO_2Me\;}
$$

$$\text{(16)} \hspace{4cm} \text{(115)}$$

$$
\longrightarrow RCOCH_2CCO_2Me + [Ph_3PCH_2CO_2Me]^+ Br^-
$$

$$\overset{||}{\underset{PPh_3}{}}$$
$$\text{(116)}$$

$$
RCOCH\!\!=\!\!CHCO_2Me + Ph_3P
$$

$$\text{(117)}$$

The resultant ylide (116) spontaneously gives off triphenylphosphine to form the acylacrylic ester (117)[170].

Since α-bromoketones can be synthesized from carboxylic acids via diazoketones, the above method is a procedure for lengthening the carboxylic acid chain by three carbon atoms.

$$RCO_2H \longrightarrow RCOCHN_2 \longrightarrow RCOCH_2Br \longrightarrow RCOCH{=}CHCO_2Me \longrightarrow$$
$$\longrightarrow RCO(CH_2)_2CO_2Me \longrightarrow R(CH_2)_3CO_2H$$

3. Reaction of alkoxycarbonylmethylenetriphenylphosphoranes with epoxides

The ylide 16 reacts with epoxides to form the esters of cyclopropane carboxylic acids (121)[171, 172]. The first intermediate of the reaction is the betaine 118, which rearranges via 119 and 120 to the end-product (121)[173-175].

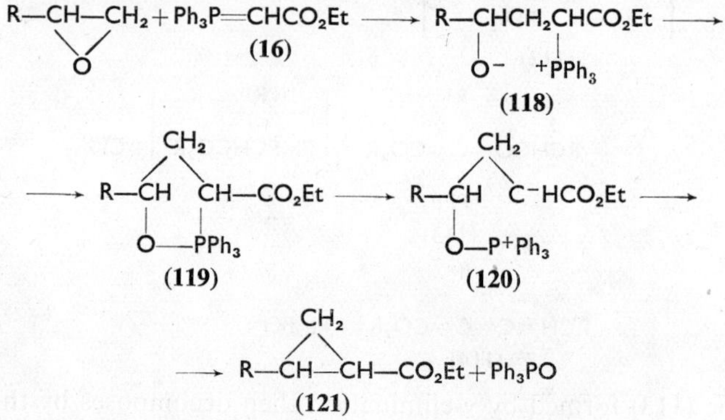

$$R{-}CH{-}CH_2 + Ph_3P{=}CHCO_2Et \longrightarrow R{-}CHCH_2CHCO_2Et \longrightarrow$$

(16)

(118)

(119)

(120)

(121)

By this means norcaranecarboxylic acid has been synthesized in 63% yield from cyclohexane epoxide.

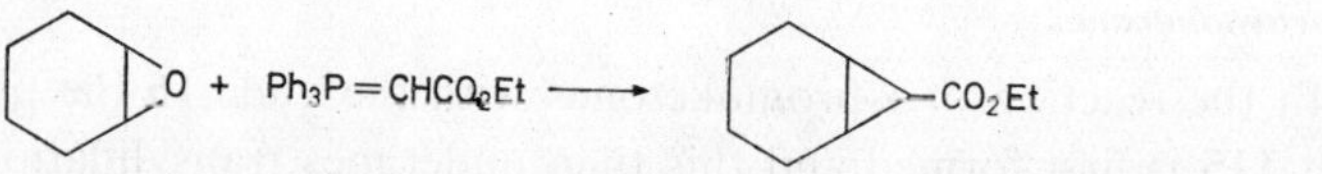

4. Addition of ylides to multiple carbon–carbon bonds

Non-stabilized ylides add to the C=C bonds of α,β-unsaturated acids with the formation of betaines (122). The latter are stabilized depending on the nature of the substituent R. If R is an electron donor, the betaine (122) splits off triphenylphosphine forming the cyclopropane derivative (123)[176]. But if R is an electron acceptor the reaction ends in Michael condensation: the proton in the α position to the phosphorus migrates to the negative carbon, so that a new ylide (124) is formed.

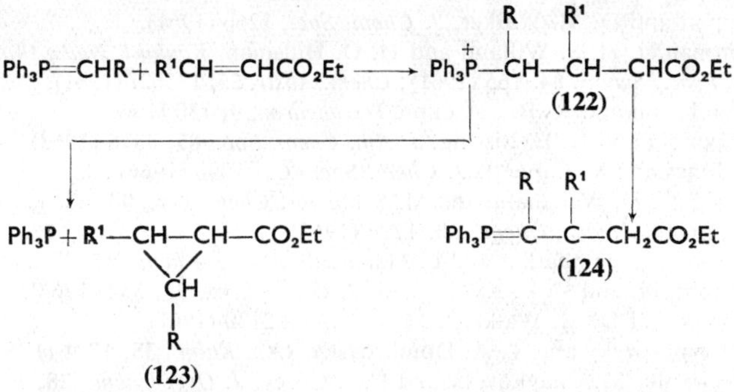

$$Ph_3P{=}CHR + R^1CH{=}CHCO_2Et \longrightarrow Ph_3\overset{+}{P}{-}\overset{\overset{R}{|}}{CH}{-}\overset{\overset{R^1}{|}}{CH}{-}\overset{-}{CH}CO_2Et$$

$$(122)$$

$$Ph_3P + R^1{-}CH{-}CH{-}CO_2Et \qquad Ph_3P{=}\overset{\overset{R}{|}}{C}{-}\overset{\overset{R^1}{|}}{C}{-}CH_2CO_2Et$$
$$\underset{\underset{\underset{R}{|}}{CH}}{\diagdown\diagup} \qquad\qquad\qquad (124)$$

$$(123)$$

Proton migration does not occur when the ylides add to an acetylenedi-carboxylic ester, the betaine **125** first formed being converted into the new ylide **(127)**[177–180] via the hypothetical four-membered compound **(126)**.

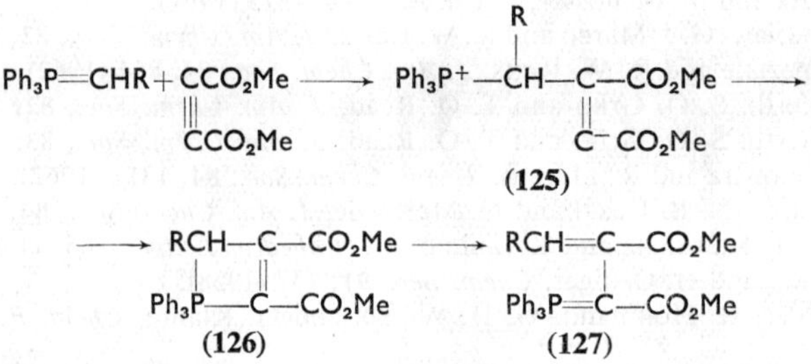

$$Ph_3P{=}CHR + \underset{\underset{CCO_2Me}{|||}}{CCO_2Me} \longrightarrow Ph_3\overset{+}{P}{-}\overset{\overset{R}{|}}{CH}{-}\underset{\underset{C^-CO_2Me}{||}}{C}{-}CO_2Me \longrightarrow$$

$$(125)$$

$$\longrightarrow \underset{\underset{Ph_3P{-}C{-}CO_2Me}{|}}{RCH{-}\overset{||}{C}{-}CO_2Me} \longrightarrow \underset{Ph_3P{=}C{-}CO_2Me}{RCH{=}C{-}CO_2Me}$$
$$(126) \qquad\qquad\qquad (127)$$

In one case the product of a Michael condensation **(128)** has been isolated as well as the ylide **127**[180]:

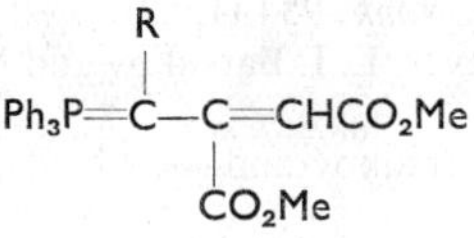

$$Ph_3P{=}\overset{\overset{R}{|}}{C}{-}\underset{\underset{CO_2Me}{|}}{C}{=}CHCO_2Me$$

$$(128)$$

III. REFERENCES

1. Th. I. Crowell, in *The Chemistry of Alkenes* (Ed. S. Patai), Interscience, London 1964, p. 241.
2. J. Levisalles, *Bull. Soc. Chim. France*, 1021 (1958)
3. U. Schöllkopf, *Angew. Chem.*, **71**, 260 (1959).
4. H. J. Bestmann, *Angew. Chem.*, **77**, 609 (1965).
5. L. A. Yanovskaya, *Usp. Khim.*, **30**, 813 (1961).
6. S. Trippett, *Advan. Org. Chem.*, **1**, 83 (1960); *Quart. Rev.*, **17**, 406 (1963).
7. A. W. Johnson, *Ylide Chemistry*, Academic Press, New York, 1966.
8. A. Maercker, *Org. Reactions*, **14**, 270 (1965).
9. H. J. Bestmann and O. Kratzer, *Chem. Ber.*, **95**, 1894 (1962).
10. S. Trippett and D. M. Walker, *Chem. Ind.*, 933 (1960).

11. S. Trippett and D. M. Walker, *J. Chem. Soc.*, 1266 (1961).
12. H. Heitman, J. H. S. Willand and H. O. Huisman, *Koninkl. Ned. Akad. Wetenschap., Proc., Ser. B.* **64**, 165 (1961); *Chem. Abstr.* **55**, 17562f (1961).
13. A. W. Johnson and R. B. LaCount, *Tetrahedron*, **9**, 130 (1960).
14. A. J. Speziale and D. E. Bissing, *J. Am. Chem. Soc.*, **85**, 3878 (1963).
15. M. E. Jones and S. Trippett, *J. Chem. Soc. (C)*, 1090 (1966).
16. G. Wittig, H. D. Weigmann and M. Schlosser, *Chem. Ber.*, **94**, 676 (1961).
17. D. E. Bissing, *J. Org. Chem.*, **30**, 1296 (1965).
18. R. Ketcham, D. Jambatkar and L. Martinelli, *J. Org. Chem.*, **27**, 4666 (1962).
19. A. W. Johnson and V. L. Kyllingstad, *J. Org. Chem.*, **31**, 334 (1966).
20. S. Trippett and D. M. Walker, *J. Chem. Soc.*, 2130 (1961).
21. A. V. Dombrovsky and V. A. Dombrovsky, *Usp. Khim.*, **35**, 1771 (1966).
22. R. Greenwald, M. Chaykovsky and E. J. Corey, *J. Org. Chem.*, **28**, 1128 (1963).
23. H. J. Bestmann, *Chem. Ber.*, **95**, 58 (1962).
24. G. Wittig and K. Schwarzenbach, *Angew. Chem.*, **71**, 652 (1959).
25. G. Wittig and K. Schwarzenbach, *Ann. Chem.*, **650**, 1 (1961).
26. G. Wittig and M. Schlosser, *Tetrahedron*, **18**, 1023 (1962).
27. G. Wittig and M. Schlosser, *Angew. Chem.*, **72**, 324 (1960).
28. G. Wittig and M. Schlosser, *Chem. Ber.*, **94**, 1373 (1961).
29. A. J. Speziale, G. J. Marco and K. W. Ratts, *J. Am. Chem. Soc.*, **82**, 1260 (1960).
30. A. J. Speziale and K. W. Ratts, *J. Am. Chem. Soc.*, **84**, 854 (1962).
31. D. Seyferth, S. O. Grim and T. O. Read, *J. Am. Chem. Soc.*, **82**, 1510 (1960).
32. D. Seyferth, S. O. Grim and T. O. Read, *J. Am. Chem. Soc.*, **83**, 1617 (1961).
33. R. Rabinowitz and R. Marcus, *J. Am. Chem. Soc.*, **84**, 1312 (1962).
34. F. Ramirez, N. B. Desai and N. McKelvie, *J. Am. Chem. Soc.*, **84**, 1745 (1962).
35. R. Oda, T. Kawabata and S. Tanimoto, *Tetrahedron Letters*, 1653 (1964).
36. L. Horner and H. Oediger, *Chem. Ber.*, **91**, 437 (1958).
37. L. Horner, H. Hoffmann, G. H. Wippel and G. Klahre, *Chem. Ber.*, **92**, 2499 (1959).
38. L. D. Bergelson, V. A. Vaver, L. I. Barsukov and M. M. Shemyakin, *Dokl. Akad. Nauk SSSR*, **143**, 111 (1962).
39. L. D. Bergelson, V. A. Vaver, L. I. Barsukov and M. M. Shemyakin, *Izv. Akad. Nauk SSSR, Otdel. Khim. Nauk*, 954 (1963).
40. L. D. Bergelson, V. A. Vaver, L. I. Barsukov and M. M. Shemyakin, *Tetrahedron*, **19**, 149 (1963).
41a. L. D. Bergelson, L. I. Barsukov and M. M. Shemyakin, *Tetrahedron*, **23**, 2709 (1967).
41b. T. C. W. Mark and J. Trotter, *Acta Cryst.* **18**, 81 (1965).
42. H. J. Bestmann, G. Joachim and J. Lengyel, *Tetrahedron Letters*, 3555 (1966).
43. H. Goetz, F. Nerdel and H. Michaelis, *Naturwissenschaften*, **50**, 496 (1963).
44. A. J. Speziale and K. W. Ratts, *J. Org. Chem.*, **28**, 465 (1963).
45. G. Wittig and U. Schöllkopf, *Chem. Ber.*, **87**, 1318 (1954).
46. G. Wittig and A. Haag, *Chem. Ber.*, **96**, 1535 (1963).
47. M. Schlosser and K. F. Christmann, *Angew. Chem.*, **77**, 682 (1965).
48. A. Bladé-Font, C. A. Van der Werf and W. E. McEwen, *J. Am. Chem. Soc.*, **82**, 2396 (1960).
49. L. Horner and H. Winkler, *Tetrahedron Letters*, 3265 (1964).
50. A. J. Speziale and K. W. Ratts, *J. Am. Chem. Soc.*, **85**, 2790 (1963).
51. A. J. Speziale, D. E. Bissing, *J. Am. Chem. Soc.*, **85**, 1888 (1963).
52. D. E. Bissing and A. J. Speziale, *J. Am. Chem. Soc.*, **87**, 2682 (1965).
53. S. Fliszar, R. F. Hudson and G. Salvadori, *Helv. Chim. Acta*, **46**, 1580 (1963).
54. L. D. Bergelson, V. A. Vaver, L. I. Barsukov and M. M. Shemyakin, *Tetrahedron Letters*, 2669 (1964).

55. M. Schlosser, G. Müller and K. F. Christmann, *Angew. Chem.* **78**, 677 (1966).
56. L. D. Bergelson, L. I. Barsukov and M. M. Shemyakin, *Zh. Obshch. Khim.*, **38**, 846 (1968).
57. W. S. Wadsworth and W. Emmons, *J. Am. Chem. Soc.*, **83**, 1733 (1961).
58. L. Horner, H. Hoffman and V. G. Toscano, *Chem. Ber.*, **95**, 536 (1962).
59. L. Horner and W. Klink, *Tetrahedron Letters*, 2467 (1964).
60. L. Horner, W. Klink and H. Hoffmann, *Chem. Ber.* **96**, 3133 (1963).
61. V. F. Kucherov, B. G. Kovalev, I. I. Nazarova and L. A. Yanovskaya, *Izv. Akad. Nauk SSSR, Otd. Khim. Nauk*, 1512 (1960).
62. L. D. Bergelson, V. D. Solodovnik and M. M. Shemyakin, *Izv. Akad. Nauk SSSR, Ser. Khim.*, 499 (1966).
63. L. A. Yanovskaya and V. F. Kucherov, *Izv. Akad. Nauk SSSR, Otd. Khim. Nauk*, 1341 (1964).
64. A. F. Tolochko, and A. V. Dombrovski, *Ukr. Khim. J.*, **31**, 220 (1965).
65. L. D. Bergelson, V. A. Vaver, L. I. Barsukov and M. M. Shemyakin, *Izv. Akad. Nauk SSSR, Ser. Khim.*, 506 (1966).
66. D. H. Wadsworth, O. E. Schupp, E. J. Sens and J. A. Ford, Jr., *J. Org. Chem.*, **30**, 681 (1965).
67. H. O. House and G. H. Rasmusson, *J. Org. Chem.*, **24**, 4278 (1961).
68. L. D. Bergelson and M. M. Shemyakin, *Pure Appl. Chem.*, **9**, 271 (1964).
69. S. Trippett, *Pure Appl. Chem.*, **9**, 255 (1964).
70a. H. O. House, V. K. Jones and G. A. Frank, *J. Org. Chem.* **29**, 3327 (1964).
70b. M. Schlosser and K. F. Christmann, *Angew. Chem.*, **78**, 115 (1966).
71. G. Wittig and M. Rieber, *Ann. Chem.*, **562**, 177 (1949).
72. M. Schlosser and K. F. Christmann, *Angew. Chem.*, **76**, 683 (1966).
73. L. D. Bergelson, V. D. Solodovnik, E. V. Dyatlovitskaya and M. M. Shemyakin, *Izv. Akad. Nauk SSSR, Otdel. Khim. Nauk*, 683 (1963).
74. F. Sondheimer and R. Mechoulam, *J. Am. Chem. Soc.*, **80**, 3087 (1958).
75. F. Sondheimer and R. Mechoulam, *J. Am. Chem. Soc.*, **79**, 5029 (1957).
76. L. D. Bergelson, V. A. Vaver, L. I. Barsukov and M. M. Shemyakin, *Izv. Akad. Nauk SSSR, Otd. Khim. Nauk*, 1417 (1963).
77. H. O. House and H. Babad, *J. Org. Chem.*, **28**, 90 (1963).
78. H. Pommer and W. Sarnecki, *Ger. Pat.* 1,068,706 (1959); *Chem. Abstr.* **56**, 512 (1962).
79. F. Bohlmann and H. H. Inhoffen, *Chem. Ber.*, **89**, 1276 (1956).
80. H. J. Bestmann, H. Schulz, R. Kunstmann and K. Rostok, *Chem. Ber.*, **99**, 1906 (1966).
81. O. Isler, G. Gutmann, M. Montavon, R. Rüegg, G. Ryser and P. Zeller, *Helv. Chim. Acta*, **40**, 1242 (1957).
82. L. D. Bergelson, V. A. Vaver, A. A. Bezzubov and M. M. Shemyakin, *Izv. Akad. Nauk SSSR, Otdel. Khim. Nauk*, 1453 (1964).
83. L. A. Yanovskaya, V. F. Kucherov and B. G. Kovalev, *Izv. Akad. Nauk SSSR, Otd. Khim. Nauk*, 674 (1962).
84. V. F. Kucherov, B. G. Kovalev, L. Kogan and L. A. Yanovskaya, *Dokl. Akad. Nauk SSSR*, **138**, 1115 (1961).
85. O. Isler, W. Guex, R. Rüegg, G. Ryser, G. Saucy, U. Schwieter, M. Walter and A. Winterstein, *Helv. Chim. Acta*, **42**, 864 (1959).
86. R. Rüegg, W. Guex, M. Montavon, U. Schwieter, G. Saucy and O. Isler, *Angew. Chem.*, **71**, 80 (1959).
87. W. Guex, R. Rüegg, O. Isler and G. Ryser, *Ger. Pat.* 1,088,951; *Chem. Abstr.* **55**, 17 541 (1961).
88. E. Buchta and F. Andree, *Ann. Chem.* **640**, 29 (1961).
89. E. Buchta and F. Andree, *Naturwissenschaften*, **46**, 74 (1959).

90. E. Buchta and F. Andree, *Naturwissenschaften*, **46**, 75 (1959).
91. E. Buchta and F. Andree, *Chem. Ber.*, **92**, 3111 (1959).
92. E. Buchta and F. Andree, *Chem. Ber.*, **93**, 1349 (1960).
93. L. D. Bergelson, E. V. Dyatlovitskaya and M. M. Shemyakin, *Izv. Akad. Nauk SSSR, Otd. Khim. Nauk*, 388 (1963).
94. K. Eiter, *Angew. Chem.*, **73**, 619 (1961).
95. K. Eiter, *Ann. Chem.*, **658**, 91 (1962).
96. E. Truscheit and K. Eiter, *Ann. Chem.*, **658**, 65 (1962).
97. E. Truscheit, K. Eiter, A. Butenandt and E. Hecker, *Ger. Pat.* 1,138,037 (1962); *Chem. Abstr.* **58**, 6694 (1963).
98. S. S. Novikov and G. A. Shvekhgeimer, *Izv. Akad. Nauk SSSR, Otd. Khim. Nauk*, 673 (1960).
99. S. Sugasawa and H. Matsuo, *Chem. Pharm. Bull. (Tokyo)*, **8**, 819 (1960); *Chem. Abstr.* **55**, 20 901 (1967).
100. H. Bredereck and B. Föhlisch, *Chem. Ber.*, **95**, 414 (1962).
101. K. A. Venter, *Dokl. Akad. Nauk SSSR*, **140**, 1073 (1961).
102. H. J. Bestmann, *Angew. Chem.*, **77**, 662 (1965).
103. F. Bohlmann, *Chem. Ber.*, **90**, 1519 (1957).
104. H. Takahashi, K. Fujwara and M. Ohta, *Bull. Chem. Soc. Japan*, **35**, 1408 (1962).
105. W. S. Wadsworth and W. Emmons, *J. Am. Chem. Soc.*, **83**, 1733 (1961).
106. H. J. Bestmann and H. Schulz, *Chem. Ber.*, **95**, 2921 (1962).
107. H. J. Bestmann and H. Schulz, *Angew. Chem.*, **73**, 27 (1961).
108. H. J. Bestmann and H. Schulz, *Ann. Chem.* **674**, 11 (1964).
109. G. Märkl, *Chem. Ber.*, **95**, 3003 (1962).
110. A. B. Denney and S. T. Ross., *J. Org. Chem.*, **27**, 998 (1962).
111. G. Märkl, *Chem. Ber.*, **94**, 2996 (1961).
112. A. J. Speziale and K. W. Ratts, *J. Org. Chem.*, **27**, 998 (1962).
113. A. J. Speziale and K. W. Ratts, *J. Org. Chem.*, **28**, 1353 (1963).
114. H. Pommer, *Angew, Chem.*, **72**, 911 (1960).
115. G. Fodor and Y. Tömösközi, *Tetrahedron Letters*, 579 (1961).
116. H. T. Openshaw and N. Whitaker, *Proc. Chem. Soc.*, 454 (1961).
117. S. Trippett and D. M. Walker, *Chem. Ind.*, 990 (1961).
118. H. Machleidt, V. Hartmann, R. Wessendorf and W. Grell, *Angew. Chem.*, **74**, 505 (1962).
119. C. Rüchardt, S. Eichler and P. Panse, *Angew. Chem.*, **75**, 858 (1963).
120. L. N. Yachontov, L. I. Mastafanov and M. V. Rubcov, *Zh. Obshch. Khim.*, **33**, 3211 (1963).
121. B. G. Kovalev, L. A. Yanovskaya and V. F. Kucherov, *Izv. Akad. Nauk SSSR, Otd. Khim. Nauk*, 1876 (1962).
122. K. Fujiwara, *J. Chem. Soc. Japan, Pure Chem. Sect.*, **84**, 656 (1963).
123. S. T. Young, I. R. Turner and D. S. Tarbell., *J. Org. Chem.* **28**, 928 (1963).
124. H. Machleidt, V. Hartmann and H. Bunger, *Ann. Chem.*, **667**, 35 (1963).
125. A. K. Bose, R. T. Hadill and N. J. Noboken, *Angew. Chem.*, **76**, 796 (1964).
126. Y. Tömösközi, and G. Janzso, *Chem. Ind.*, 2085 (1962).
127. G. Wittig and H. Pommer, *Ger. Pat.* 971,986 (1959); *Chem. Zentr.*, 16 097 (1959).
128. H. Pommer, W. Sarnecki and G. Wittig, *Ger. Pat.* 1,025,869 (1958); *Chem. Abstr.* **54**, 22 713 (1960).
129. G. Wittig, H. Pommer and E. Harting, *Ger. Pat.* 957,942 (1957); *Chem. Abstr.* **53**, 7232 (1959).
130. H. Pommer and W. Sarnecki, *Ger. Pat.* 1,068,703 (1959); *Chem. Abstr.* **55**, 13473 (1961).
131. H. Pommer and W. Sarnecki, *Ger. Pat.* 1,068,705 (1959); *Chem. Abstr.* **56**, 1487 (1962).

132. H. Pommer, and W. Sarnecki, *Ger. Pat.* 1,068,710 (1959); *Chem. Abstr.* **55**, 12446 (1961).
133. H. Pommer, and W. Sarnecki, *Ger. Pat.* 1,070,173 (1959); *Chem. Abstr.* **55**, 11332 (1961).
134. H. Pommer and G. Wittig, *Ger. Pat.* 951,212 (1957); *Chem. Zentr.* 437 (1959).
135. H. S. Corey, Jr., I. R. D. McCormick and W. E. Swenson, *J. Am. Chem. Soc.*, **86**, 1884 (1964).
136. W. J. Gensler, *Chem. Rev.*, **57**, 191 (1957).
137. L. D. Bergelson and M. M. Shemyakin, *Angew. Chem. Internat. Ed. Engl.* **3**, 250 (1964).
138. H. Hunsdiecker, *Chem. Ber.*, **75**, 291 (1942).
139. S. G. Batrakov and L. D. Bergelson, *Izv. Akad. Nauk SSSR, Ser. Khim.*, 369 (1965).
140. R. Kh. Freidlina and E. I. Vasilieva, *Khim. Nauka i Prom.* **2**, 2 (1957).
141. L. D. Bergelson, V. A. Vaver, V. Yu. Kovtun, L. B. Senyavina and M. M. Shemyakin, *Zh. Obsh. Khim.*, **32**, 1802 (1962).
142. V. A. Vaver, V. V. Dorogov and L. D. Bergelson, *Izv. Akad. Nauk SSSR, Ser. Khim.*, 2241 (1966).
143. L. D. Bergelson, V. A. Vaver, L. I. Barsukov and M. M. Shemyakin, *Izv. Akad. Nauk SSSR, Otd. Khim. Nauk*, 1135 (1963).
144. E. Mosettig and R. Mozingo, *Org. Reactions*, **4**, 362 (1948).
145. V. Franzen and S. Otto, *Chem. Ber.*, **94**, 1360 (1961).
146. V. D. Solodovnik, *Usp. Khim.*, **36**, 636 (1967).
147. L. D. Bergelson, V. D. Solodovnik, E. V. Dyatlovitskaya and M. M. Shemyakin, *Izv. Akad. Nauk SSSR, Otd. Khim. Nauk*, 683 (1963).
148. N. Petragoni and G. Schill, *Chem. Ber.*, **97**, 3293 (1964).
149. L. D. Bergelson, E. V. Dyatlovitskaya and M. M. Shemyakin, *Izv. Akad. Nauk SSSR, Ser. Khim.*, 2003 (1964).
150. L. D. Bergelson, V. D. Solodovnik and M. M. Shemyakin, *Izv. Akad. Nauk SSSR, Otd. Khim. Nauk*, 843 (1967).
151a. L. D. Bergelson, V. D. Solodovnik and M. M. Shemyakin, *Izv. Akad. Nauk. SSSR, Ser. Khim.*, 843 (1967).
151b. J. M. Osbond, *Progr. Chem. Fats Lipids*, **9**, 121 (1966).
152. L. D. Bergelson, V. A. Vaver and M. M. Shemyakin, *Izv. Akad. Nauk SSSR, Otd. Khim. Nauk*, 1894 (1962).
153. L. D. Bergelson, V. A. Vaver, A. A. Bezzubov and M. M. Shemyakin, *Zh. Obsch. Khim.* in press.
154. L. J. Morris, *J. Lipid Res.*, **7**, 717 (1966).
155. L. D. Bergelson, V. A. Vaver, B. A. Bezzubov and M. M. Shemyakin, *Zh. Obshch. Khim.*, **32**, 1807 (1962).
156. S. Trippett, and D. M. Walker, *J. Chem. Soc.*, 1266 (1961).
157. G. Wittig and H. Pommer, *Ger. Pat.* 943.648; *Chem. Abstr.*, **52.** 16292. (1958.)
158. H. Gross and J. Gloede, *Angew. Chem.*, **78**, 823 (1966).
159. S. S. Novikov, and G. A. Shvekhgeimer, *Izv. Akad. Nauk SSSR, Otd. Khim. Nauk*, 2061 (1960).
160. G. P. Schiemenz and H. Engelhard, *Chem. Ber.*, **94**, 578 (1961).
161. K. Eiter, H. Oediger and E. Tzuscheit, *Ger. Pat.* 1,110,633; *Chem. Abstr.*, **56**, 3522 (1962).
162. G. Märkl, *Angew. Chem., Internat. Ed. Engl.*, **1**, 160 (1962).
163. G. Märkl, *Chem. Ber.*, **94**, 3005 (1961).
164. S. T. D. Gough and S. Trippett, *J. Chem. Soc.*, 2333 (1962).
165. S. T. D. Gough and S. Trippett, *J. Chem. Soc.*, 543 (1964).
166. S. Trippett and D. M. Walker, *J. Chem. Soc.*, 3874 (1959).

2*

167. H. J. Bestmann, *Angew Chem.*, **77**, 663 (1965).
168. H. J. Bestmann and H. Hartung, *Angew. Chem.*, **75**, 297 (1963).
169. H. J. Bestmann, H. Häberlein and I. Pils, *Tetrahedron*, **20**, 2079 (1964).
170. H. J. Bestmann, F. Seng and H. Schulz, *Chem. Ber.*, **96**, 465 (1963).
171. D. B. Denney and M. J. Boskin, *J. Am. Chem. Soc.*, **81**, 6330 (1959).
172. D. B. Denney, J. J. Vill and M. J. Boskin, *J. Am. Chem. Soc.*, **84**, 3944 (1962).
173. W. E. McEwen, A. Bladé-Font and C. A. Van der Werf, *J. Am. Chem. Soc.*, **84**, 677 (1962).
174. W. E. McEwen and A. P. Wolf, *J. Am. Chem. Soc.*, **84**, 676 (1962).
175. S. Trippett, *Quart. Rev.*, **17**, 406 (1963).
176. H. J. Bestmann and F. Seng, *Angew. Chem.*, **74**, 154 (1962).
177. H. J. Bestmann and O. Rothe, *Angew. Chem.*, **76**, 569 (1964).
178. S. Trippett, *J. Chem. Soc.*, 4733 (1962).
179. G. W. Brown, R. C. Cookson, I. D. R. Stevens, T. C. Mak and J. Trotter, *Proc. Chem. Soc.*, **80** (1964).
180. J. B. Hendrickson, R. Rees and J. F. Templeton, *J. Am. Chem. Soc.*, **86**, 107 (1964).

CHAPTER 8

Rearrangement and cyclization reactions of carboxylic acids and esters

HAROLD KWART and KENNETH KING

University of Delaware, Newark, Delaware, U. S. A.

I. REARRANGEMENTS AND CYCLIZATIONS FORMING AND DESTROYING THE CARBOXYLIC ACID AND ESTER FUNCTIONS

With the exception of decarboxylation reactions induced by thermolytic or electrolytic means (which will not concern us here), the carboxyl group is possessed of a high degree of both thermodynamic and non-reactivity stability and is thus not readily destroyed. The alkoxycarbonyl (ester) group is also quite stable in the same sense though not so much as the corresponding carboxylic acid. Thus, while rearrangements destroying these groups do not occur very generally, rearrangements forming them can often be effected under fairly mild conditions. For example, the benzilic acid rearrangement of α-diketones to α-hydroxy acids and the Favorsky rearrangement of α-haloketones to esters occur in many cases in a fairly dilute basic solution. These two rearrangements have been recently reviewed[1] and so need not be discussed here.

It should be noted, however, that there has recently been a good deal of important work on the Favorsky reaction which might be cited to bring the subject up to date. Most notable is the work of Turro and Hammond[2], who have succeeded in preparing cyclopropanone and derivatives thereof, the proposed intermediates in the Favorsky reaction. Thus, tetramethylcyclopropanone, upon treatment with sodium methoxide in methanol or dimethoxyethane at room temperature gives the Favorsky product, methyl 2,2,3-trimethylbutanoate, in 97% yield. Of considerable significance, also, is the work of House and Frank[3] on the stereospecific Favorsky rearrangement, which occurs with inversion of configuration at the carbon originally bonded to halogen. Stereospecificity is favored by heterogeneous conditions, a strong base in a non-polar, aprotic solvent. A common method is to use dry, powdered sodium hydroxide in refluxing xylene, the so-called 'quasi-Favorsky' conditions[4]. The indications here are that the haloketone forms a cyclopropanone by a concerted S_N2 intramolecular displacement. Recently Rappe has obtained good yields of the *cis* unsaturated acid from the corresponding 1,3-dibromo-2-ones by treatment with aqueous sodium bicarbonate[5]. Rappe believes that here the structure-determining intermediate is the carbanion (1), formed from the cyclopropanone, which survives long enough to assume the most sterically favored conformation (equation 1).

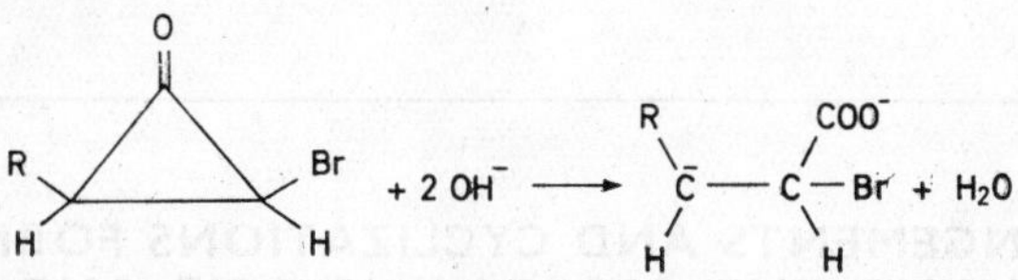

(1)

(1)

A. Dieckmann and Acyloin Condensations

While base converts ketones to esters in the Favorsky rearrangement, esters may also be converted to ketones by strong base. Thus, diesters may undergo an intramolecular Claisen condensation to give the cyclic β-keto-ester. The usual reaction condition is alkoxide in alcohol, but sodium metal in aromatic hydrocarbon solvent has also been used. Like all Claisen condensations, the reaction is reversible, although the equilibrium may be pushed to the right by reaction of the keto ester product with a second equivalent of base to form a stable (under the reaction conditions) carbanion.

Carrick and Fry have shown that the rate-determining step is —C—C— bond formation rather than ionization of the diester to a carbanion[6]. Thus cyclization of (2), labeled with ^{14}C in both the carbonyl and α carbons, shows a kinetic isotope effect in both cases (equation 2).

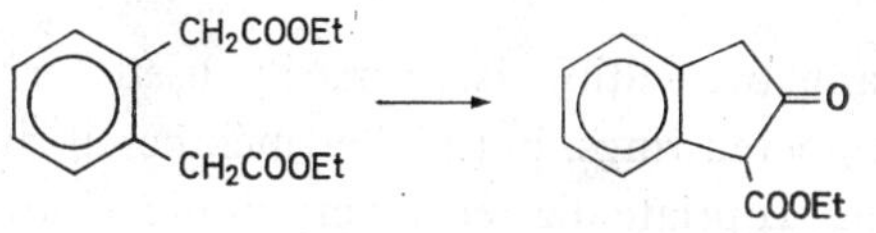

$$(2)$$

Dieckmann cyclization usually leads to five- and six-membered rings. The kinetics of formation of 2-ethoxycarbonylcyclopentanone from adipic ester and of 2-ethoxycarbonylcyclohexanone from pimelic ester have been studied[7]. As might be expected, cyclohexanone formation has a slightly lower enthalpy of activation but a much more negative entropy of activation. In fact, the results of Reed and Thornley[8] indicate that cyclopentanone formation has an entropy of activation greater than zero. Thus, in any case where both five- and six-membered rings may form, use of long reaction times and equilibrium conditions should favor the six-membered ring while shorter reaction times should give more of the kinetically-favored five-membered ring.

Bearing this in mind the cyclization of 3-carboethoxylpimelic ester(3) is particularly interesting[7]. The three possible products are 4, 5, and 6. Using ethoxide in ethanol, an 85% yield of 6 is obtained, while with sodium in benzene as cyclizing medium, the chief product is 4 in 81% yield. In neither case is any 5 found. In an aprotic solvent, such as benzene, formation of the carbanion is irreversible; thus the chief product will be that arising from the most stable carbanion, which is 7. What is most surprising is the high yield of the five-membered ring product (6) under equilibrium conditions. This could be due to axial interaction destabilizing the six-membered ring of 4 or to intramolecular solvation of the transition state or intermediate in the reaction of 6, resembling the structure shown in 8.

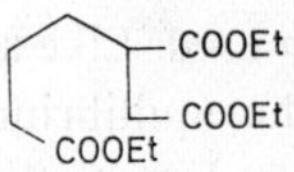

(3) 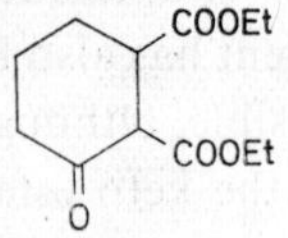   (4) 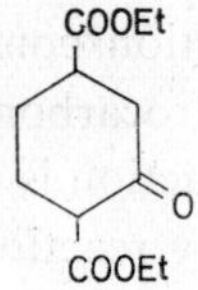   (5) 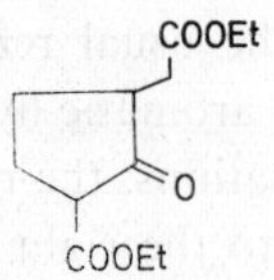   (6)

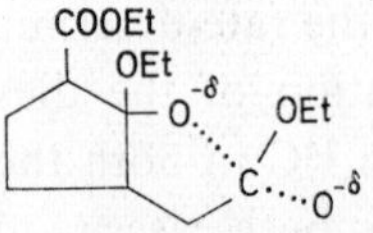

(8) 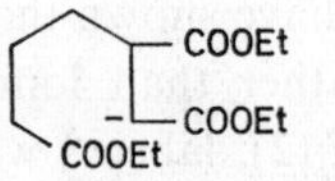   (7)

The Dieckmann cyclization is primarily used for the preparation of five- and six-membered rings, but higher-membered rings can be obtained by reacting the appropriate diester in very dilute solution to prevent inter-molecular condensation. Using this technique a 48% yield of cyclohexade-canone could be obtained from heptadecanedioic ester, with a suspension of potassium t-butoxide in xylene as the cyclizing agent[9]. However, even with high dilution, the medium-sized rings, with eight to twelve members, are obtained in very poor yield or not at all due to destabilization by trans-annular interactions.

The best, and often the only, method for the preparation of medium-sized ring ketones is the acyloin condensation. This very important synthetic tool has been most recently reviewed in 1964[10]. In this cyclization the two ester groups of a diester are reduced to an hydroxy ketone (acyloin) by a dispersion of molten sodium droplets in xylene. The actual reaction product is the enediol, which gives the acyloin upon acidification. Since the enediol is very sensitive to oxygen, the cyclization must be carried out under nitrogen atmosphere. By this method 2-hydroxycyclodecanone (sebacoin) can be obtained from dimethyl sebacate in 66% yield (equation 3)[11].

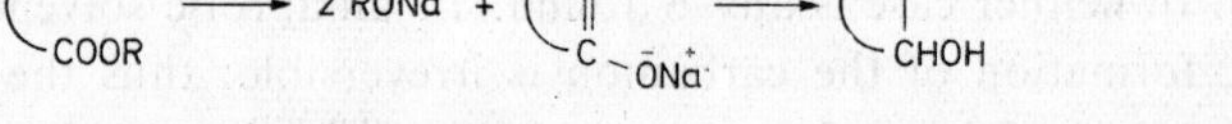

(3)

There has been very little mechanistic work done on the acyloin cyclization. However, the related formation of acyclic acyloins by intermolecular condensation of monoesters is known to involve free radical ions[12]. A recent communication[13] reports a systematic study of yields of cyclic acyloin and α-diketone from acyclic diester. On this basis, as well as from the earlier results (referred to above), it is possible to surmise the following plausible sequence of steps producing acyloin.

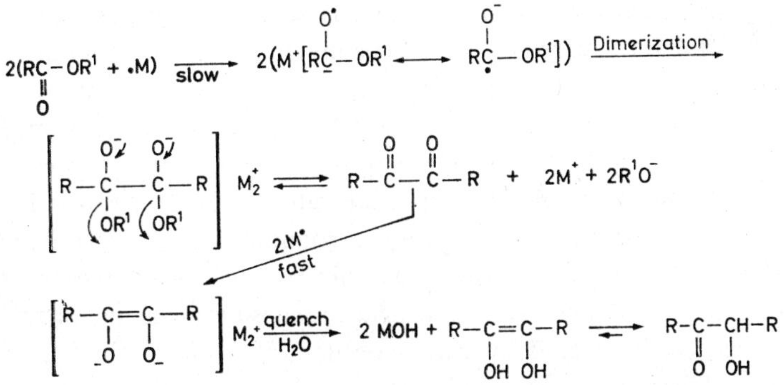

$$2(RC-OR^1 + \cdot M) \xrightarrow[\text{slow}]{} 2(M^+[R\overset{O\cdot}{\underset{\cdot}{C}}-OR^1 \longleftarrow R\overset{O^-}{\underset{\cdot}{C}}-OR^1]) \xrightarrow{\text{Dimerization}}$$

The Dieckmann cyclization is often carried out under conditions similar to those of the acyloin condensation. The chief difference is that the acyloin reduction requires four moles of sodium per mole of ester, while the Dieckmann reaction needs only one equivalent of sodium. Thus, it is not too surprising to find the two reactions occasionally competing. For example, while pimelic ester undergoes the acyloin condensation readily, **8a** and **9** give only the Dieckmann products (**10** and **11**), respectively. Gardner, Hayes, and Brandon[14] ascribe this to an interaction such as **12**, which reduces the electrophilicity of the carbonyl carbon. This hinders both the Dieckmann and acyloin condensations, but the latter is more

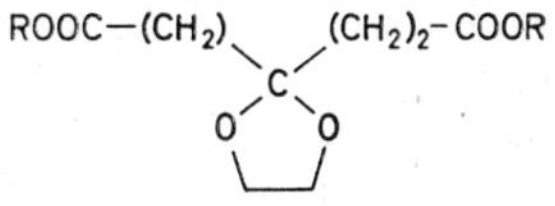

(8a) 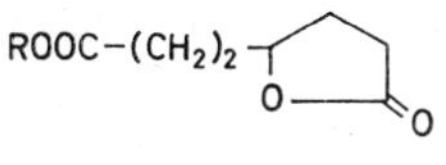　　　　(9)

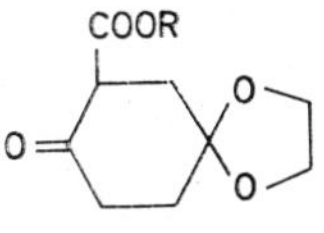

(10) 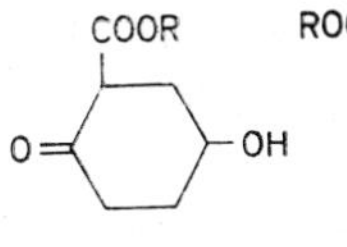　　(11) 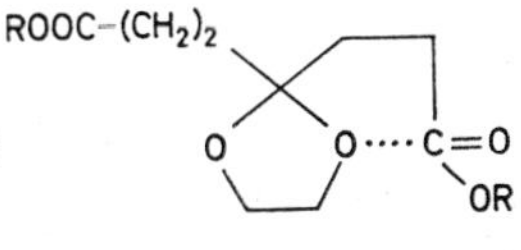　　(12)

easily blocked because it requires heterogeneous conditions and is thus far more sensitive to environment.

B. Cyclodehydration Reactions

Just as an ester may be cyclized to a ketone by a strong base, so unsaturated acids may be converted to cyclopentenones or cyclohexenones by strongly acidic dehydrating agents. This is actually an intramolecular acylation, in which the carboxyl group is converted to an acylium ion, which then attacks a benzene ring or olefinic double bond suitably positioned elsewhere in the molecule. The subject has been covered in a recent review[15].

Acid catalysis may also convert unsaturated acids to lactones. Here the double bond is protonated to form a carbonium ion, which then attacks the oxygen of a carboxyl group. Lactonization is favored by strong protonating agents such as hydrohalic acids, arenesulfonic acids, or trifluoroacetic acid, while cyclodehydration occurs with dehydrating agents like polyphosphoric acid, acetic anhydride, or Lewis acids such as $ZnCl_2$ or $SnCl_4$. Sulfuric acid usually leads to lactonization, but some unsaturated ketone may also be formed. However 13, upon treatment with sulfuric acid, gives equal amounts of ketone 14 and lactone 15[15].

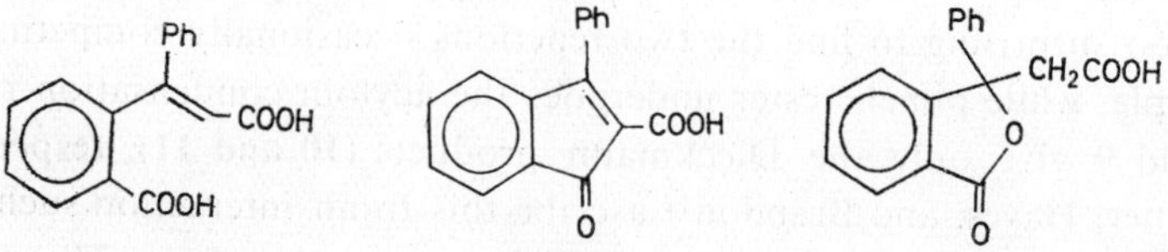

                    (13)   (14)   (15)

An interesting cyclodehydration is that of geranic acid, (16)[16]. Upon treatment with acetic anhydride and sodium acetate, the expected piperitenone 17 is obtained, but the bicyclic ketone 18 is also found. Another interesting cyclodehydration is that of the ether 19 which gives lactone 20 rather than ether 21[17]. This rearrangement involving migration (illustrated in 22) of the ether oxygen to the acylium ion center, occurs under a variety of conditions: polyphosphoric acid at 60° for 5 hours, p-toluenesulfonic acid at 90° for 5 hours, or boron trifluoride at 20° for 36 hours.

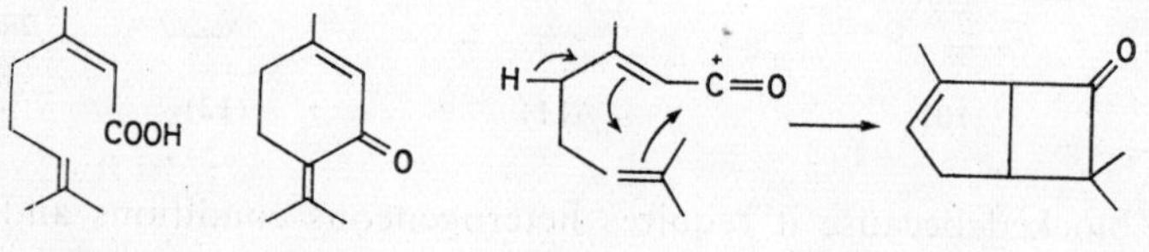

       (16)   (17)   (18)

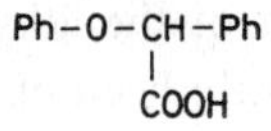

Ph-O-CH-Ph
|
COOH

(19)

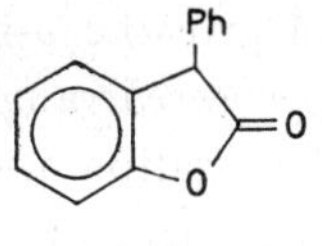

(20)

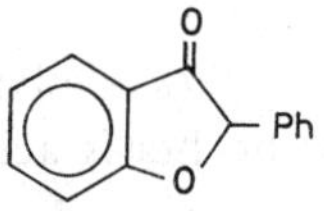

(21)

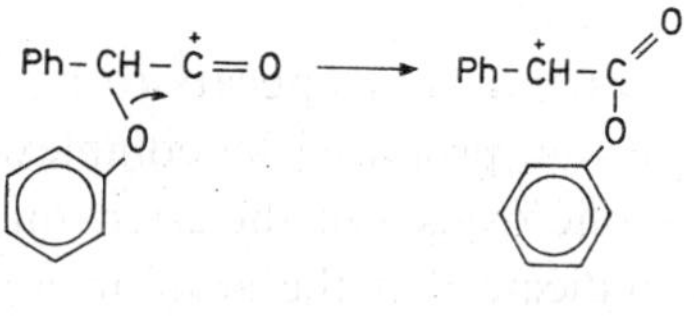

Ph-CH-C=O → Ph-CH-C

(22)

C. Fries Reaction and Related Rearrangements and Extensions

1. Monoesters

This reaction (which was reviewed last in 1942)[18] consists of the conversion of phenolic esters to *ortho* and *para* hydroxy ketones under the influence of Lewis acid catalysts, particularly aluminum chloride. The reaction is useful synthetically, for even though the product hydroxy ketones may be prepared in one step by Friedel–Crafts acylation of phenols, use of the two-step Fries rearrangement (preparing and reacting ester) gener ally leads to higher yields[18]. As in the Friedel–Crafts reaction, catalyst and substrate are used in equimolar quantities, or sometimes even larger quantities of catalyst are required due to complexation of catalyst and product.

Use of low temperature generally leads to the *p*-hydroxy ketone while high temperature (150–250°) and short reaction time favors the *ortho* product. The *ortho*/*para* ratio also depends on the catalyst used; boron trifluoride favors the *para* while titanium and tin tetrachlorides favor the *ortho* product[19]. The reaction is generally conducted in an inert solvent such as nitrobenzene, chlorobenzene, or dichloroethane. Nitrobenzene in particular is a useful solvent in the preparation of *p*-hydroxy ketones.

The range of acyl groups which may be used is wide. However, esters of aromatic acids, such as benzoates, generally give poor results. It is the phenolic group, however, which exerts the greatest influence on the course of the reaction. For ·example, 2-methoxyphenyl acetate gives 4-, 5-, and 6-acetyl-2-methoxyphenol· The occurrence of a *meta*hydroxy ketone as the product of a Fries rearrangement is very rare[18]. This must be due to the activating effect of the methoxy group, since when the acetyl group is in the 5 position, it is *para* to the methoxyl group. Similarly esters of *o*-cresol give the *o*-hydroxy ketone. With esters of poly-alkylphenols there may be rearrangement of the alkyl groups during the Fries reaction[20]. Alkyl migration has recently been observed with a mono-alkylphenol[21]. Thus, treatment of *p*-ethylphenyl chloroacetate with aluminum chloride gave 2-(chloroacetyl)-5-ethylphenol instead of the expected 4-ethyl compound. However, no migration occurred with *o*-ethylphenyl chloroacetate.

The Fries reaction now appears to be intermolecular, involving attack of an acylium ion, produced by complexation of the Lewis acid catalyst with the phenolic oxygen of the ester, on the benzene ring. Two crossing experiments indicate that the acylium ion is comparatively free and not bound up in a tight ion pair throughout the course of the reaction. In the first of these[22], it was found that phenyl benzoate, upon treatment with aluminum chloride in dichloroethane at 84°, gave *p*-benzoylphenyl benzoate in 17% yield, in addition to 37% of the expected *p*-benzoylphenol. At 60°, only the ester-ketone was found as the product of rearrangement, again in 17% yield. If the ester-ketone were formed from a *p*-benzoyl-phenol precursor then the yield of the latter should decrease with increasing temperature. Exactly the opposite result was found. Further proof was found in the reaction of phenyl acetate with benzoyl chloride and aluminum chloride at 50°. The products were *p*-acetylphenyl benzoate and *p*-benzoylphenyl benzoate in 23% and 17% yields respectively. This indicates a preliminary exchange between phenyl acetate and benzoyl chloride to produce phenyl benzoate and acetyl chloride. Thus, it is seen that the low-temperature, *para*-producing reaction is intermolecular. Similar proof for the high temperature, *ortho*-producing reaction is found in treating equimolar amounts of phenyl proprionate and *m*-cresol with aluminum chloride at 165° for one hour without solvent or at 200° for 10 minutes. In either case almost equal amounts of *o*-proprionylphenol and 2-proprionyl-5-methylphenol were found, with combined yields of the two products in excess of 90%[19].

The concept of a free acylium ion may, however, be an oversimplification. The generally high yields of the Fries reaction argue against any long lifetime for free acylium ions, which are rather unstable. In particular, treatment of α-naphthyl pivalate (trimethylacetate) with $SnCl_4$ gave the Fries product, 2-pivaloyl-1-naphthol, in good yield[23]. This acylium ion, $(Me_3CCO)^+$, would be expected to decarbonylate immediately. However, since the tin tetrachloride catalyst in this reaction is a weaker Lewis acid than aluminum chloride, it may be possible to regard this particular reaction as quasi-intramolecular, that is, involving a tightly bound ion pair.

The weaker Lewis acid would complex less strongly with the phenolic oxygen, thus decreasing the tendency of the acyl group to wander off by itself.

2. Diesters

Compounds of the type $p\text{-}CH_3C_6H_4OCO(CH_2)_nCOOC_6H_4CH_3$ undergo a double Fries rearrangement to give diketones of the type **23**. This occurs provided $n \geqq 2$[24]. Di-p-cresyl oxalate failed to give any diketone. With malonate, the only product was 4-hydroxy-6-methyl-coumarin (**24**). Esters of isophthalic and terephthalic acids also rearranged satisfactorily, but di-p-cresyl phthalate gave only **25** in 14%' yield, evidently due to

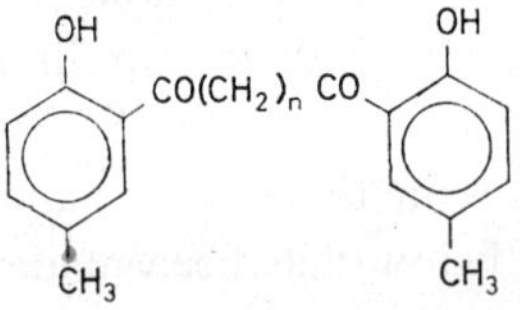

(23) 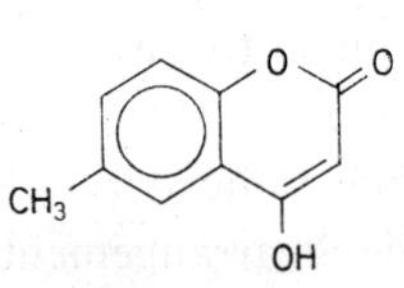   (24) 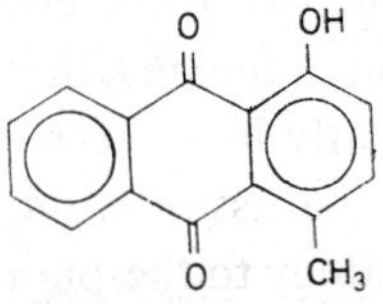   (25)

steric hindrance favoring the cyclodehydration of the intermediate mono-ketone.

3. Acid-catalyzed rearrangements of enol esters

Enol esters can undergo a Fries rearrangement to give β-diketones[25]. Reaction of a methyl ketone, RCH_2COCH_3, with ketene gives the two possible enol acetates, $AcOCMe{=}CHR$ and $CH_2{=}C(OAc)CH_2R$. When the enol acetate mixture is treated with boron trifluoride at 0–20°, the two possible β-diketones are obtained, Ac_2CHR and $AcCH_2COCH_2R$, in amounts which indicate that there has been no interchange between the terminal and internal double bonds of the enol acetates. Like the Fries reaction, this is intermolecular, as indicated by the occurrence of cross-products.

4. Thermic-Fries rearrangement of phenol and enol esters

Enol esters can also give β-diketones by an intramolecular pyrolysis. The reaction goes smoothly at 500° and yields of 75–80% can be obtained by recycling. In this case the amounts of the two possible diketones indicate that there has been interchange between the two possible enol acetates, with the product from the terminal enol acetate being slightly favored.

As might be expected, however, there are other pyrolysis products. The results of Ritchie and coworkers indicate that vinyl benzoate may decompose by three paths[26] The first of these is alkyl-oxygen cleavage to give benzoic acid and acetylene; this is analogous to the well-known pyrolysis of saturated esters to olefins. The second is decarboxylation to give styrene while the third and major path is rearrangement to benzoylacetaldehyde. Vinyl esters of aliphatic acids containing α hydrogen can also decompose by another fairly important path, consisting of acyl-oxygen scission to give acetaldehyde and a ketene $R_2C{=}C{=}O$. Vinyl acetate gives acetaldehyde but no ketene, which is known to be unstable at 500°. It should be noted that the β-keto aldehydes resulting from rearrangement of vinyl esters are unstable under these conditions and decarbonylate. Thus, the isolated product from rearrangement of vinyl benzoate is acetophenone. On the other hand, for those enol esters which can rearrange to the more stable β-diketones, the reaction is found to be reversible, although the equilibrium greatly favors the diketone.

The assumption of free radical intermediates in this process and the analogy to the photo-Fries rearrangement (see below) has been made[27-29]. Other authors have taken note of major differences between the products of the thermic- and photoreactions of vinyl esters, though this by no means rules out a free radical mechanism for the thermic-Fries reaction. However, in view of the tendency of thermolysis reactions, (particularly those of esters and related carbonyl derivatives), to take place via a cyclic mechanism, it is possible to conceive of the thermic-Fries reaction as passing through the four-center transition state shown as **26** (equation 4). This representation can certainly account for the principal products observed (see also reference 45). Nevertheless, there is no evidence, as might have been anticipated on

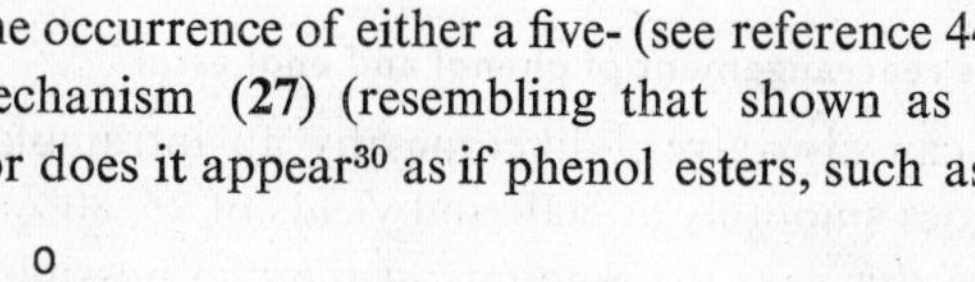

$$(4)$$

(26)

this basis, for the occurrence of either a five- (see reference 44) or six-membered cyclic mechanism (**27**) (resembling that shown as inoperative in equation 5). Nor does it appear[30] as if phenol esters, such as shown as **28**,

$$(5)$$

(27)

undergo any thermic rearrangement through the cyclic six-membered activated complex characteristic of the Claisen rearrangement of allyl phenyl ethers, (equation 6).

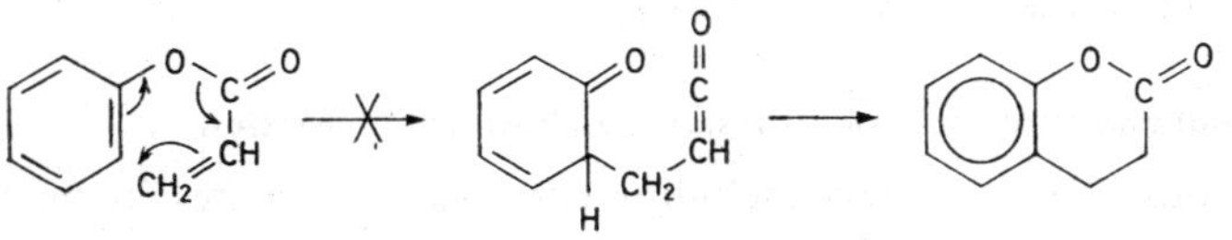

(28)

Furthermore, Louw and Kooyman[31] (see also references 32 and 33) have established a purely radical course of thermolysis of the homologous allyl esters. In the case of allyl acetate the radical chain is initiated by cleavage into carbon dioxide and the allyl and methyl radicals. However, decomposition into acrolein, methane, and carbon monoxide may also occur. This acyl-oxygen cleavage, whether it occurs during initiation or propagation, is the only one possible for the vinyl and alkenyl esters. Some acrolein is found among the decomposition products of allyl acetate, but the major products are 1-butene, carbon monoxide, carbon dioxide, and methane.

On the other hand the products from the thermolysis of vinyl chlorobenzoate all contained the chlorine in the same position on the benzene ring as in the starting material[26]. Thus, the decarboxylation product from vinyl *m*-chlorobenzoate was almost entirely *m*-chlorostyrene. The small amounts of *otho* and *para* isomers found are explained as the result of a heterolytic side-reaction occurring on the walls of the reactor. Ritchie and coworkers[26] express the decarboxylation as a concerted reaction involving a four-membered cyclic transition state similar to that illustrated in **35** for the photolytic decarboxylation. This and the reversibility of the rearrangement to diketones as mentioned above argue strongly against a free radical mechanism, but do not disprove it.

Similarly the pyrolysis[34] of allyl cyanoacetate (**29**) does not involve free radicals but is instead a bimolecular enolization followed by a rapid rearrangement to form 2-cyano-4-pentenoic acid as illustrated in equation (7)[35, 36]. The enolization may involve two molecules of ester or one molecule of ester and one of acid or base catalyst. Evidently this reaction occurs because the electron-withdrawing cyano group stabilizes the second double bond in the enol.

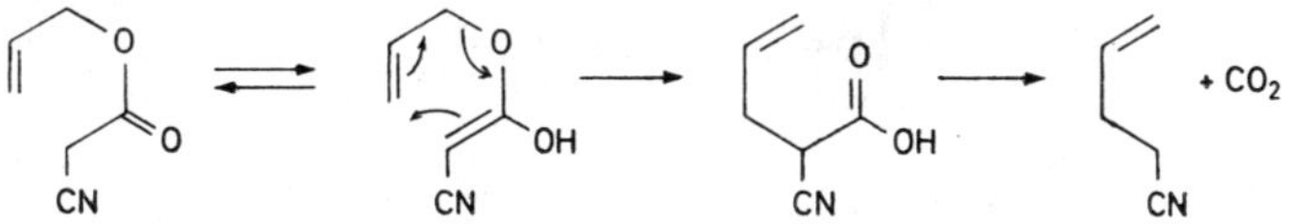

(29)

Further clarification of the driving forces which compel a cyclic transition state in such thermolytic processes of rearrangement in unsaturated esters, but which are notably absent in others (as discussed above), must await the results of further study.

5. Phenol and enol ester photolysis: the photo-Fries reaction

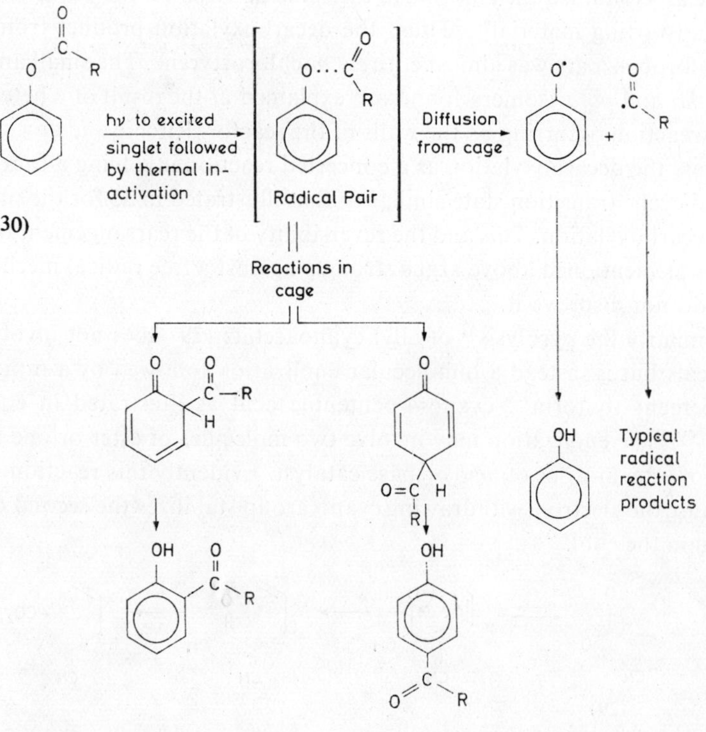

Various phenolic esters including carbonates, oxalates, formates[37] and fluorene-9-carboxylates[38] have been identified as phenoxyl free radical sources under photolytic conditions in a variety of solvents. However, the carboxylic esters afford substantial proportions of hydroxyketone (Fries arrangement) accompanying the cleavage products. This rearrangement reaction is clearly intramolecular and leads generally to the *ortho* isomer. The difference between this reaction and the normal Fries are dramatically illustrated by the work of Kobsa[39]; (see also references 40–42). Treatment of *p-t*-butylphenyl benzoate with aluminum chloride at 140° gives only *p*-benzoylphenol, but photolysis gives 2-benzoyl-4-*t*-butylphenol. In the normal Fries reaction alkyl groups which can form stable carbonium ions may cleave; this never occurs in the photolytic reaction. On the other hand photolysis of 2-chloro-4-*t*-butylphenyl benzoate gave a mixture of 2-chloro-4-*t*-butyl-6-benzoylphenol and 2-benzoyl-4-*t*-butylphenol. Halogen is never cleaved in the Lewis acid catalyzed reaction. Kobsa proposed that the ester is photolytically cleaved into phenoxyl and acyl free radicals, which recombine before they can diffuse out of the solvent cage. The recombination would give a cyclohexadienone, which would then rearrange to the observed phenolic ketone. Since the photolysis of *p*-*t*-butylphenyl benzoate always leads to some *p*-*t*-butylphenol, it is clear that some of the radicals must diffuse out of the solvent cage. The phenoxyl radicals would then abstract hydrogen from the solvent. These events may be most simply described[39] by equation (8), depicting the course of photorearrangement leading from the ester **30**.

Anderson and Reese[43] found that **31** failed to undergo a photolytic Fries reaction. The only possible rearrangement product of this compound is **32**, the *m*-hydroxy ketone. If the photolytic reaction involves a cyclohexadienone intermediate, one would not expect the *meta* product to occur very often. That it can occur under extreme conditions is demonstrated by the work of Finnegan and Mattice[41a], who irradiated **33** for 119 hours and obtained **34** in 10% yield.

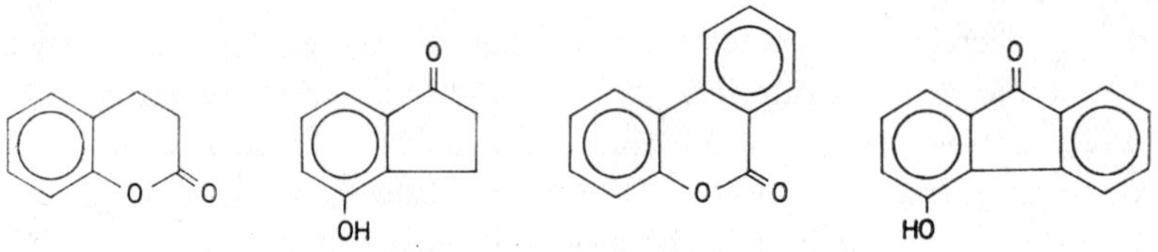

| (31) | (32) | (33) | (34) |

The results of Finnegan and Knutson[41b] emphasize the variety of reactions which may occur between the radical pair partners in the cage. However, a side-reaction leading to decarboxylation and formation of hydrocarbon product has been expressed as a concerted (cyclic) expulsion of carbon dioxide as in **35**.

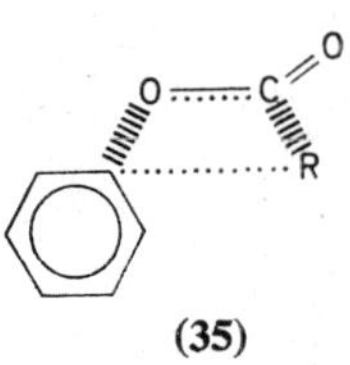

(35)

Recent work indicates some polar character in the transition state. Coppinger and Bell[44] measured rates and quantum yields for the rearrangement of a number of *para*-substituted phenyl esters of 3,5-di-*t*-butyl-4-hydroxybenzoic acid. The reaction was carried out in sheets of pressed polyethylene powder and followed spectroscopically. Under these conditions the reaction was clean; the *o*-hydroxy ketone (36a) was the only product formed. Thus 36b failed to react. It was found that all substituents on the phenolic ring decrease the rate of reaction, but electron-withdrawing substituents do so much more than electron-donating ones. The results were explained in

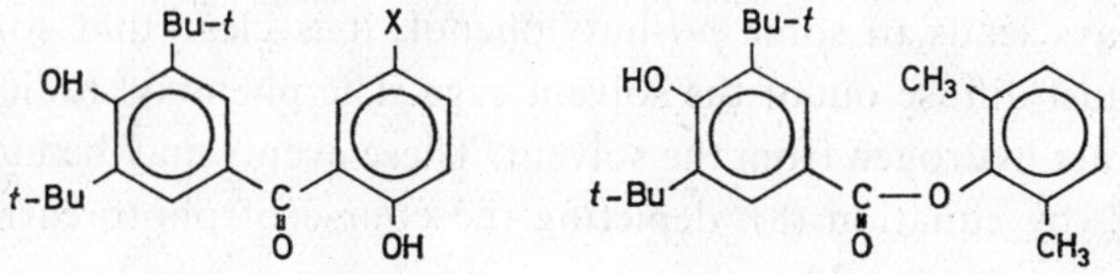

(36a)   (36b)

semiquantitative fashion on the basis of two rules. The first of these is based on a general rule applying in photochemical reactions, which states that, at any time, the concentration of molecules in quantum states leading to reaction (the so-called activated complex) is proportional to the fraction of all possible quantum states which can lead to reaction. Increasing complexity of the molecule increases the number of possible quantum states and thus decreases this fraction. This is turn decreases the quantum yield and thus the reaction rate. The second is the familiar Hammett equation, to take care of resonance and inductive effects. Electron-withdrawing substituents hinder reaction by lowering the electron density at the *ortho* position of the phenolic ring. The mechanism proposed is that excitation to a π^* excited state is followed by decay to a vibrationally-excited ground state, which then dissociates to a pair of radicals held in a charge-transfer complex rather than a solvent cage.

Enol esters can also undergo the photolytic Fries reaction. An interesting example of this is the photolysis of 37a which leads to 37b[29]. However, if the cyclohexene ring is part of a steroid system it does not open[45]. Irradiation

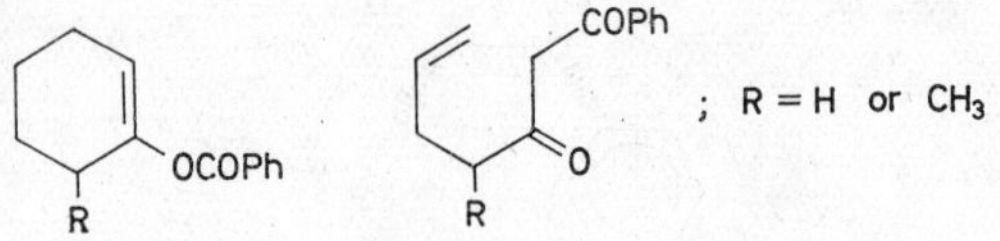

(37a)   (37b)

of dienol acetates in the steroid series gives rise to entirely analogous data which support a caged-radicals mechanism with stereoelectronic control of the overall photo-rearrangement results[46].

II. REARRANGEMENTS OF CARBOXYLATE SALTS

A. *Rearrangement of Aromatic Carboxylic Acid Functions (The Raecke Process)*

The most common high-temperature reaction of carboxylic acids and metal carboxylates is decarboxylation. This subject will not be treated here. However, there are some reactions of metal carboxylates which can be classified unhesitatingly as rearrangements. One of the most important of these is the isomerization of potassium phthalate to terephthalate at 400° with cadmium iodide catalyst. Because of the importance of terephthalic acid in plastics, this process is of some value in industry[47].

Raecke, the inventor of the process, has discussed the conditions under which it occurs[47]. The reaction is exothermic and goes best at 400–420°. At 350°, it is very slow. Above 450° decomposition sets in. The patents suggest conducting the reaction under a high pressure of carbon dioxide, but this is not essential. All that matters is that oxygen be excluded and that some pressure be maintained: the reaction fails in vacuum. In the laboratory, sealed tubes or autoclaves are used. It is also essential that no materials containing active hydrogen (such as the free acid) be present, since these materials facilitate decarboxylation. Potassium, rubidium, and cesium phthalates give good results, but the sodium salt does not isomerize below 450°, the decomposition temperature. The reaction fails completely with lithium phthalate. Cadmium iodide is not the only possible catalyst, but it is by far the best[48]. It should be noted that under the reaction conditions cadmium iodide and potassium phthalate are miscible liquids, while potassium terephthalate is solid and precipitates out. Thus, under equilibrium conditions high yields of potassium terephthalate will result even if its formation is not thermodynamically favorable.

The reaction is also applicable to other aromatic acids. For example, naphthalene-1,8-dicarboxylate gives naphthalene-2,6-dicarboxylate, furan-3,4-dicarboxylate gives furan-2,5-dicarboxylate, and pyridine-2,3-dicarboxylate gives pyridine-3,6-dicarboxylate. Alkali salicylates may be converted to the corresponding *p*-hydroxybenzoates at 200°[49]. This reaction will be readily recognized as typical of the Kolbe–Schmitt synthesis[50].

At first glance this isomerization would appear to be merely decarboxylation and carboxylation, with benzoate as intermediate. Indications of a greater degree of complexity are gleaned from the fact that while potassium benzoate, or benzoic anhydride and potassium carbonate, give potassium terephthalate, they do so at a much slower rate than the potassium phtha-

23*

late itself. In addition benzene is found as a product, suggesting that the benzoate has suffered disproportionation[48].

Tracer experiments have given somewhat confusing results. Upon reacting potassium phthalate with $Cd^{14}CO_3$ at 450°, Ogata and coworkers found most of the original radioactivity in the terephthalate product. The recovered phthalate was slightly radioactive[48]. When potassium benzoate was reacted with $Cd^{14}CO_3$ at 470° for 2·5 hours, the terephthalate obtained had 24% of the original activity. However, at 450°, the terephthalate obtained after 4 hours was found to be non-radioactive. Ratusky and Šorm[51] reacted phthalic and benzoic anhydrides with $K_2^{14}CO_3$ at 405° with 10% zinc powder as catalyst and found that the terephthalate recovered from the phthalic anhydride had 50% of the original activity while that from the benzoic anhydride had 65%. The isomerization of alkali salicylates at 200° was found to occur with about 5% exchange with $K_2^{14}CO_3$[49]. It is thus clear that exchange with an outside source of carbon dioxide can occur but that it need not be complete. The amount of exchange may depend to a large extent on physical factors such as the completeness of mixing[49].

Kinetic experiments reported by Ogata and Sakamoto[52] are illuminating. The isomerization of isophthalate to terephthalate with cadmium iodide catalyst in potassium cyanate solvent at 365° is cleanly third-order, first-order in catalyst and second-order in isophthalate. It thus appears that the rate-determining step is formation of a 2:1 complex, perhaps a sandwich-type complex, with the π electrons of the benzene rings. The isomerization of phthalate is much more complicated[53]. The products are benzoate, trimellitate, isophthalate, and terephthalate. The yield of benzoate reaches a maximum at about 0·5 hour reaction time and thereafter decreases slowly. At two hours benzoate is still the major product, but after four or five hours, almost everything has been converted to terephthalate. Evidently carboxyl groups exchange freely within the complex.

B. Rearrangement of α-Hydroxy Acids to α-Diketones and Related Reactions

It would not be proper to omit mention of a very unusual rearrangement undergone by lead salts of certain α-hydroxy acids. In 1898 Wallach found that on heating at atmospheric pressure the dry lead salt of fenchocarboxylic acid (**38**) gave the diketone **39**[54]. There are two forms of this acid, probably endo–exo isomers. The lead salts of both undergo this rearrangement but the free acids differ in their behavior upon destructive distillation. The more stable of the two isomers undergoes partial dehydration and partial rearrangement while the less stable suffers dehydration and decarboxylation

A similar rearrangement of **40** gives carbocamphenilone (**41**)[55]. Hückel[56] has called this a 'reverse benzilic acid rearrangement', but this would seem to be misleading nomenclature. Since the conditions under which this reaction occurs are so different from those of the benzilic acid rearrangement, it is hardly likely that the two have the same mechanism, although the bonds made and the bonds broken may be exactly reversed. In addition, since the driving force for the benzilic acid rearrangement is the formation of the resonance-stabilized carboxylate anion, it is most unlikely that the reverse reaction, which would destroy the carboxyl group, would occur at the low temperatures of the benzilic acid rearrangement[1].

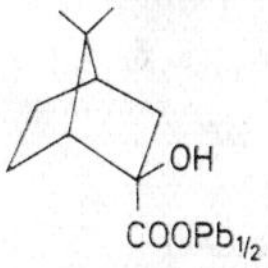

COOPb$_{1/2}$ O   COOPb$_{1/2}$ O

(38) 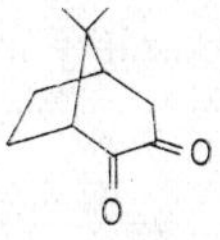   (39) 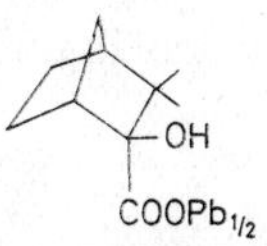   (40) 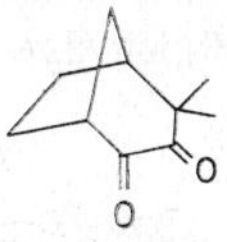   (41)

Kenner and Morton investigated the pyrolysis of other lead α-hydroxy carboxylates but found no further examples of this reaction[57]. Thus, pyrolysis of lead benzilate gave benzophenone in 50% yield and 25% each of diphenylacetic acid and tetraphenylethane, the last two products probably arising by a free radical mechanism. No benzil was found. Pyrolysis of **42** gave only fluorenone (**43**). On the other hand **44** failed to give cyclohexanone but underwent instead dehydration to cyclohexene-1-carboxylate (**45**).

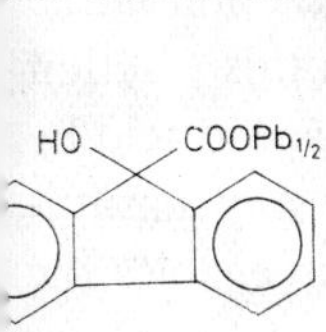

(42) 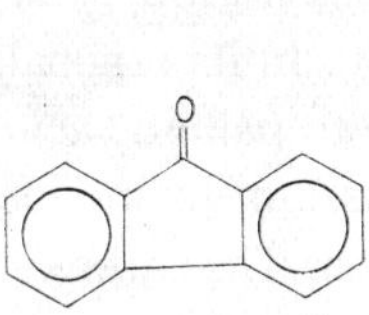   (43) 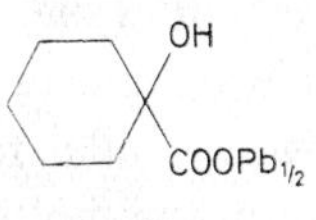   (44) 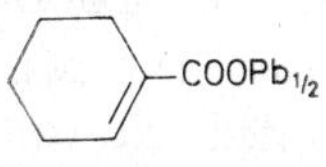   (45)

Decarboxylation is a very common pyrolytic reaction of all acids, including α-hydroxy acids, but here dehydration to the α,β-unsaturated acid may also occur. The use of metal salts, particularly lead or manganese, promotes the conversion to aldehydes or ketones over the competing dehydration. A cyclic mechanism **46** involving complexation of the metal atom with both the hydroxyl and carboxyl oxygens is almost certainly involved here. It should be noted that here the carboxyl group is lost as carbon monoxide rather than the dioxide. A similar complex **47** could also be involved in the rearrangement of **38** and **40**.

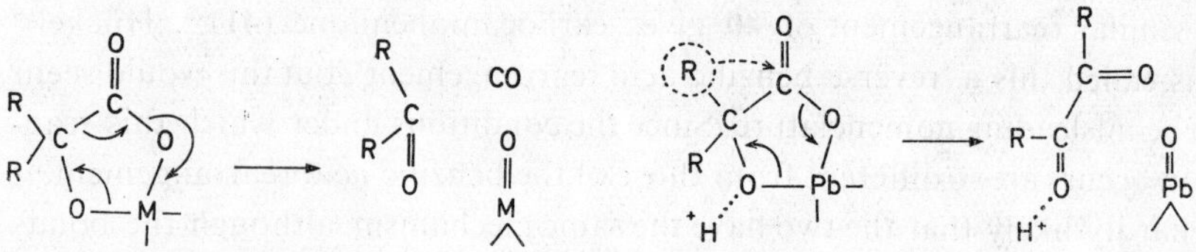

$$(46) \qquad\qquad\qquad (47) \qquad\qquad\qquad (9)$$

Oxidation with lead tetraacetate will also convert α-hydroxy acids to aldehydes or ketones[61]. Since this is a true oxidation, with the carboxyl being lost as carbon dioxide, it is clear that this is actually quite different from the thermolytic decarbonylation shown in **46** although a cyclic mechanism (equation 10) similar to the one given by Criegee for the lead tetraacetate cleavage of glycols is quite possibly involved[59, 60].

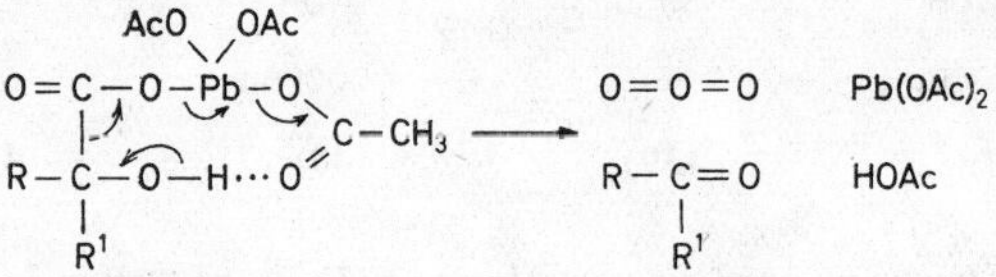

$$(48) \tag{10}$$

III. VARRENTRAPP AND RELATED REACTIONS

Compounds containing olefinic double bonds undergo a great many rearrangements. In the case of olefinic acids many of these rearrangements are due entirely to the presence of the carboxyl group. Such a reaction is the Varrentrapp reaction, discovered in 1840. In its original form this reaction consisted of the conversion of oleic acid to palmitic acid and acetic acid by treatment with molten sodium or potassium hydroxide, but any unsaturated acid will undergo a similar reaction. The carboxyl group enables the compound to dissolve in the molten alkali.

The most likely conception of the reaction would have it in two parts, first a migration of the double bond to the α, β position, and secondly a cleavage of the α, β bond. Apparently the cleavage is faster than the migration and involves a β-hydroxy acid as intermediate[61]. The cleavage would then occur as in equation (11).

$$RCH{=}CHCOOH \xrightarrow{H_2O} RCH(OH)CH_2COOH \xrightarrow{OH^-} RCHO^-{-}CH_2COO^- \longrightarrow$$
$$RCHO + {}^-CH_2COO^- \xrightarrow{H_2O} CH_3COO^- + OH^- \tag{11}$$

Aldehydes are known to undergo oxidation to the corresponding acid under these conditions.

The double bond migration has been amply proven. Weedon and cowork-ers treated oleic acid (9-octadecenoic acid) with molten alkali under ni-trogen and obtained palmitic acid in 50% yield[64]. By using shorter reaction times all the octadecenoic acids from 4- to 12- were found. The position of the double bond was checked by ozonolysis. Since the double bond of oleic

$$CH_3(CH_2)_mCH{=}CH(CH_2)_nCOOH \xrightarrow{[O]} CH_3(CH_2)_mCOOH + HOOC(CH_2)_nCOOH \quad (12)$$

acid is too far away from the carboxyl group to be influenced by it, the migration would be expected to be random. However the 2- and 3-octadecenoic acids were not found at all and the 4-, 5-, and 6- acids were found only in very small amounts. Evidently the double bond migrates faster as it approaches the carboxyl group. and the 2- and 3- acids are removed rapidly by the subsequent hydration and cleavage steps. Similarly, treatment of 10-undecenoic acid with molten KOD gave a nonanoic acid that was 70% deuterated. Oleic acid gave a palmitic acid with 29 of its 31 hydrogens replaced by deuterium[62]. A surprising result was that treatment of 2-octadecenoic acid with KOD gave a palmitic acid which had more deuterium in it than that obtained by treatment of palmitic acid with KOD. Also, when 2-octadecenoic acid was treated with alkali at 280° for three minutes, the only isolable materials were 70% of the starting material and 20% of 3-octadecenoic acid[61]. Evidently the cleavage is not fast enough to prevent migration of the double bond to the 3-position.

It is obvious that when the double bond cannot migrate to the α, β posi-tion, the Varrentrapp reaction will not occur. Thus, Pistor and Plieninger obtained pimelic acid by treatment of 4-cyclohexenecarboxylate with an equimolar mixture of potassium hydroxide and water in 2·6-fold excess at 320°, but 1-methyl-4-cyclohexenecarboxylate failed to cleave under these conditions[63]. However, under more extreme conditions cleavage of a dif-ferent sort will occur. Lukes and Hofman[64] have found that treatment of 3,3-dimethyl-8-nonenoic acid with potassium hydroxide in 24-fold excess at 370° for four hours gave acetic acid and 2-methyl-1,3-octadiene (49). This, of course, does not involve the double bond, as shown by the fact that 4,4-dimethyldecanoic acids could be cleaved to monoolefins under these conditions. (equation 13).

$$CH_3(CH_2)_3-CH{=}CH-\underset{\underset{CH_3}{|}}{C}{=}CH_2 \quad + \quad CH_3COOH$$

(49)

A concerted mechanism illustrated in equation (13) is admittedly only one of several possible ways of expressing the lack of rate influence of the iso-merizable double bond, as indicated by the fact that the saturated and un-saturated acids cleave at the same temperature. Normally this would mobi-lize the bond-making and bond-breaking steps through stabilization of the incipient double bond in dienic resonance in a non-cyclic transition state resembling that of the usual base-catalyzed $E2$ elimination. It is possible by means of the cyclic mechanism in equation (13) to understand the failure of the double bond to affect the scission, as illustrated in **50** (equation 14).

$$\text{(14)}$$

$$\text{(50)}$$

Double bond migration can also occur under milder conditions. When Lüttringhaus and Reif treated 10-undecenoic acid with 6N aqueous potas-sium hydroxide at 220° in an autoclave, they obtained all the undecenoic acid isomers from 6- to 10-. However, a plot of the yield of 9-undecenoic acid against time was asymptotic, indicating that the double bond would never reach the 2-position under these conditions[68]. This is perhaps not surprising since there is no particular thermodynamic advantage in leaving the 9-position, although there is free energy decrease in going from the ter-minal 10-position to the internal 9-position.

Under even milder conditions isolated double bonds fail to react, but polyunsaturated acids do react. An important analytical method for the determination of naturally occurring polyunsaturated fatty acids such as linoleic (*cis-cis*-9,-12-octadecadienoic) acid involves conjugation of the double bonds with potassium hydroxide in ethylene glycol at 180° followed by ultraviolet spectroscopy to determine the concentration of conjugated acid. As an analytical method this is complicated by the fact that geometric isomers react at different rates to give products with different absorption coefficients, although the absorption maxima are usually quite close. Con-jugation of naturally occurring linoleic acid gives maximum absorption at 233 mμ after treatment with potassium hydroxide–ethylene glycol for about half an hour. On the other hand the *trans-trans* isomer reaches an absorp-tion maximum at 231 mμ only after six hours of reaction[66]. Naturally-occurring polyunsaturated fatty acids are generally *cis*, but bringing into conjugation a *cis* double bond through an allylic carbanion leads primarily to the *trans* isomer[67]. Thus, the major products from this treatment of linoleic acid are mostly *trans-trans* 9,11- and 10,12-octadecadienoic acids.

As might be expected, acetylenic acids isomerize under extremely mild conditions. Reaction of 3-butynoic acid with 18% aqueous potassium carbonate at 40° for three hours gave the allene 2,3-butadienoic acid in 92% yield. At 90° the 2-butynoic acid was obtained in 60% yield[6s]. The ease with which the allene forms is in all likelihood a reflection of the stability of the intermediate carbanion $HC\equiv CCH\overline{COO}^-$. Through a similar carbanion the conjugated diene, 2,4-pentadienoic acid, could be obtained in 80% yield from 3,4-pentadienoic acid by treatment with 10% potassium carbonate at 60°. Esters react under even milder conditions than the corresponding carboxylate anion, since the latter group is less electron-withdrawing and provides less stabilization for the intermediate carbanion. Thus, treatment with aqueous potassium carbonate for 45 minutes at 20° under heterogeneous conditions gave the corresponding allene from esters of 3-butynoic acid.

Since acetylenes and conjugated dienes are interconvertible in base, it is generally found that acetylenic and diolefinic acids react similarly under Varrentrapp conditions. For example, linoleic and stearolic acids both gave myristic (tetradecanoic) acid in about 60% yield with about 5% of palmitic acid[69]. The intermediate 2,4-octadecadienoic acid loses two moles of acetic acid in a stepwise reaction. The formation of palmitic acid is explained as oxidation of the intermediate aldehyde $C_{13}H_{27}CHO$ by the 2,4-dienoic acid instead of by water. In either case the oxidizing agent acts as a hydride acceptor.

Other functional groups also react under Varrentrapp conditions. Hydroxy acids other than β-hydroxy acids dehydrogenate to keto acids which then cleave either to two monocarboxylic acids or to a saturated hydrocarbon and a dicarboxylic acid[70]. One interesting exception to this is 6-keto-stearic acid which undergoes an intramolecular aldol condensation to give 2-dodecyl-1-cyclopentenecarboxylic acid. Isolation of an α,β-unsaturated acid such as this under Varrentrapp conditions is particularly noteworthy: it occurs because cleavage of this acid can only give the original 6-keto-stearic acid. Under more extreme conditions, however, the double bond can migrate to give **51** which then cleaves to **52** (equation 15). At 230° acids containing *vic*-diol functions dehydrogenate to α-diketones which then undergo the benzilic acid rearrangement. The product[74] from 9,10-dihydroxystearic acid is thus **53**, in 52% yield. The α-hydroxy acid from the benzilic acid rearrangement is stable at this temperature but at higher

temperatures it loses the elements of formic acid to give a keto a
which then cleaves.

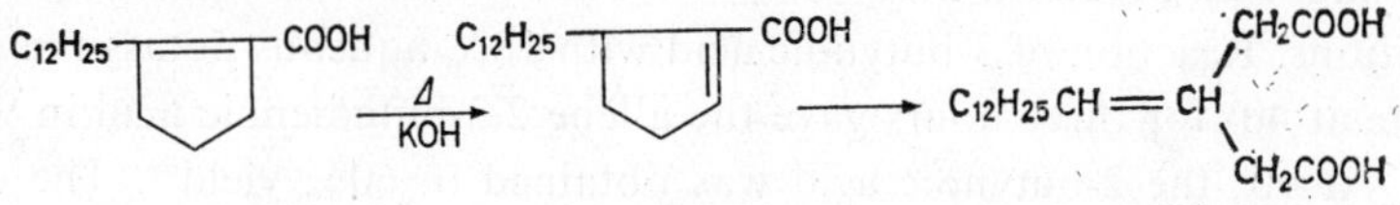

$$(51) \qquad\qquad (52) \qquad\qquad (15)$$

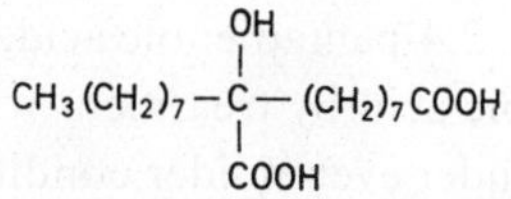

$$(53)$$

IV. CARBONYL-FORMING DECARBOXYLATION REACTIONS

A. Reaction Conditions

The most common reaction of carboxylic acids and their salts is decarboxylation. In most cases this does not involve rearrangements, being
simply of the type shown by equation (11). This is particularly the case

$$RCOOH \xrightarrow{\Delta} RH + CO_2 \qquad (16)$$

where R contains electron-withdrawing groups that can stabilize a carbanion, but it can be made to occur with almost any type of acid. There is,
however, another type of decarboxylation which does involve rearrangement, exemplified in the pyrolysis of metal carboxylates to give ketones,
according to equation (17). The dry distillation of calcium acetate to give
acetone

$$(RCOO)_2M \xrightarrow{\Delta} R_2CO + CO_2 + MO \quad \text{or} \quad MCO_3 \qquad (17)$$

is one of the oldest reactions in orgranic chemistry and for a long time constituted the chief commercial source of acetone. Salts of other aliphatic
acids also give ketones but the yields are frequently low and the products
often difficult to purify[72]. Using a mixture of two salts, the three possible
ketones are obtained, two symmetrical and one unsymmetrical. Thus, with
excess formate it is sometimes possible to obtain good yields of aldehydes.
Even calcium benzoate can be made to give benzophenone, but with such
aromatic acids the yields are more often quite poor and the reaction of
equation (16) usually predominates.

The carbonyl-froming decarboxylation reaction is not limited to calcium salts, other alkaline earth metals performing just as well. Thorium (IV) and manganese can also be used, but these metals are mostly used in the form of their oxides in the catalytic method. Lead (II) salts have also been reported to give good results[57]. In this case the evolution of carbon dioxide at temperatures from 280° to 310° leaves a gummy residue from which the desired ketone is isolated by vacuum distillation or by treatment with formic acid. Yields of 78% acetone from lead acetate, 80% stearone (18-pentatriacontanone) from lead stearate, and a 92% yield from 10-undecylenic acid are claimed. An interesting preparation is that of $MeOOC(CH_2)_6$—CO—$(CH_2)_6$—$COOMe$ from the monomethyl ester of suberic acid in 70% yield.

Not all metals are satisfactory. Finely divided copper and nickel metals tend to facilitate the decarboxylation reaction of equation (16)[73]. A preferred procedure for the preparation of styrene and stilbenes involves heating the corresponding cinnamic acid in quinoline in the presence of catalytic amounts of copper, copper chromite, or the copper cinnamate[74]. Shemakin and coworkers[75] claim the isolation of free acid and anhydride from the pyrolysis of copper and silver acetates and benzoates at 200–250°. The free acid is supposed to arise from the anhydride due to the presence of small amounts of water in the hygroscopic salts. Benzoic anhydride reacts further under these conditions to give phenyl benzoate, phenol, and salicylic acid. Sodium salts also give poor results, although for a different reason. Thus, while Nakai, Sugii, and Nakao[76] were able to obtain a 57% yield of 2-butanone by vacuum distillation of a mixture of sodium acetate and proprionate at 350°, Grün and Wirth found that sodium stearate on vacuum distillation at 550° gave mostly olefins, arising from cracking of the initially formed stearone[77]. It should also be noted that heating sodium or calcium acetate in fused sodium hydroxide gives methane by direct decarboxylation (equation 16)[74].

For synthetic purposes it is generally simpler to pass the vaporized acid over a catalyst, which is usually a metal oxide, although with the alkaline earth metals the carbonates are used. Many materials have been tried as catalysts but most, such as calcium carbonate, zinc oxide, and cadmium oxide, give good results only with acetic acid and the lower aliphatic acids[73]. Alumina and aluminum silicates are rather poor catalysts. The passage of butyric acid over activated clay at 200–300° for one hour gave only 30–38% 4-heptanone[78]. All of the catalysts decompose benzoic acid only above 500° and lead primarily to benzene. The best yield of benzophenone is obtained with lithium carbonate at 550°, but even here more benzene than ketone is formed[76]. With formic acid the primary result is dehydration to carbon monoxide although some formaldehyde may form. Alumina and silica give only carbon monoxide from formic acid.

The best catalysts are thorium and manganese oxides because these retain their activity longest. As might be expected, the method of preparation is crucial. A good example is the aerogel thoria of Swann, Kistler and Appel[79]. The thoria was precipitated from aqueous thorium nitrate with aniline, transferred to methanol, and heated under pressure until the critical point was reached (250°). Other forms of thoria prepared by directly drying the hydrated thorium oxide precipitate or by ignition of thorium oxalate give products of higher density which require higher ketonization temperatures and give lower yields than does the aerogel. A quantitative yield of 8-pentadecanone from caprylic acid is reported.

The vapor phase catalytic method is of course limited to those acids which have a reasonably high vapor pressure. However, the method can be extended to the higher fatty acids by use of the more volatile ethyl esters. For example, Swann reports an 82·5% yield of laurone from ethyl laurate over his aerogel thoria catalyst at 360°[79]. Sabatier[73] explains that the first step in this reaction is the cleavage of the ester into metal carboxylate and alcoholate. The carboxylate then ketonizes, and the alcoholate dehydrates to olefin. If the activity of the catalyst toward carboxylate is higher than its activity towards alcoholate, some free alcohol also forms; if its activity is higher towards alcoholate, free acid results. Since methoxide cannot dehydrate to olefin, the methyl esters require higher temperatures and give dimethyl ether as the side product.

Isotopic tracer experiments indicate that in all cases the first step of the catalytic ketonization of acids is formation of the metal carboxylates. Thus, pyrolysis of a mixture of phenylacetic and propionic acids, in which the propionic acid had been enriched with ^{13}C at the carbonyl carbon, over thoria, manganese oxide, and calcium carbonate gave the same degree of isotopic enrichment of the product $C_6H_5CH_2COEt$ as did pyrolysis of the mixed barium salts[80]. A similar result was found in the pyrolysis of labeled acetic acid and one of four other acids over thoria and as the thorium salt[81]. An even better indication of salt formation in catalytic ketonization is found with the alkaline earth carbonates, for the rate of carbon dioxide evolution in the ketonization of acetic acid is the same over two polymorphic forms of calcium carbonate (aragonite and calcite) and three forms of barium carbonate[82]. This indicates that the rate-determining step is decomposition of previously formed calcium (or barium) acetate, and does not involve adsorption of acetic acid onto the catalyst.

Higher fatty acids may also be ketonized in the liquid phase by use of the free metal. Kino[83] was able to obtain ketones from saponified vegetable oils using about 5% by weight of magnesium powder at 330°. The actual reagent is magnesium soap, formed by release of free hydrogen from the fatty acid. The gas evolution causes frothing, which Kino inhibited by use of a small amount of calcium or chromium soap. Better results are obtained with iron. Heating stearic acid in an iron vessel for four hours at 300° is reported to give a 91% yield of stearone[84]. With lauric acid a temperature of 270° was satisfactory.

Ketonization by iron has been throughly investigated by Schultz and Davis[85]. Aliphatic acids react with iron powder to release hydrogen and form the green ferrous carboxylate at temperatures of 80–124°. Most of these soaps decompose below 300° to give volatile ketone and a residue of black, non-magnetic ferrous oxide, which can still catalyze the ketonization of more acid. However, ferrous acetate and propionate are stable above 400°, thereby limiting this method to the higher acids. For example, a 96% yield of ketone is obtained from decanoic acid but only 70% from butanoic, 48% from propionic, and 25% from acetic acid. The ferrous carboxylates — and ferrous oxide — are very easily oxidized by air to the corresponding ferric salts, which are completely inactive as regards ketonization. Good yields of alkanophenones are obtained from a mixture of ferrous benzoate and the corresponding alkanoate; for example, a 78% yield of $PhCOC_{15}H_{31}$ results from a 1:1 mixture of palmitate and benzoate. Any substituent on the benzene ring lowers the yield of alkanophenone, but electron-withdrawing substituents do so more than electron-donating ones. Substituents *ortho* to the carboxyl group completely prevent ketone formation[86]. On the other hand, ferrous benzoate alone decomposes above 350° to give 13% benzene, 17% benzophenone, a trace of benzaldehyde, and the rest tar.

As might be expected these methods can also be used to prepare cycloalkanones from straight-chain dibasic acids, for instance, cyclohexanone from pimelic acid, a 40% yield of cycloheptanone from suberic acid, and a 20% yield of cyclooctanone from azelaic acid[88]. Iron powder also gives good results in these cases[87]. Higher ring ketones, however, are formed in very low yield, in all cases under 5%[74, 89]. For the preparation of macrocyclic ring ketones the acyloin condensation[11] is much to be preferred.

B. Results Contributing to the Development of a Reaction Mechanism

There has been a great deal of work on the mechanism of ketonic decarboxylation, but in spite of all this work only a relatively few definite conclusions can be reached. In fact, some of the data is even contradictory. The first mechanistic proposal was probably that of Bamberger, who suggested the anhydride as an intermediate[90]. This seems quite likely in the case of cyclic ketones since cyclopentanone and cyclohexanone may be prepared in 50% yield by merely distilling adipic and pimelic acids at atmospheric pressure with excess acetic anhydride[91]. The intermediate here is probably the polymeric anhydride, which decarboxylates at about 250°[74]. Vacuum distillation of the polymer gives the monomeric 7-membered cyclic adipic anhydride. With the exception of acetic and 2-methyl-butyric anhydrides, aliphatic anhydrides have also been found to decarboxylate upon atmospheric distillation[92]. Where metal catalysis was used it was found that passage of aliphatic acid vapors over manganese oxide gave both anhydrides and ketones, anhydride formation being favored by lower temperatures and higher flow rates[92]. Shemyakin[93] found that passage of acetic anhydride over calcium oxide gave acetone at 350° while calcium acetate did not decarboxylate until 400°.

Koch and Leibnitz[92] interpret the decarboxylation of anhydrides in terms of the β-keto acid intermediate first proposed by Neunhoeffer and Paschke[94]. The former authors note that while carboxylate salts should not lose α hydrogen readily (since this would form a doubly-charged anion), loss of a proton from a carbon α to the electron-withdrawing anhydride group should be fairly easy. In fact, however, the mechanism should not involve a carbanion but dissociation of the anhydride into acylium and carboxylate ions, followed by attack of the acylium ion on the α carbon of the carboxylate with transfer of a proton and formation of the β-keto acid, as illustrated in equation (21).

$$\begin{array}{c} RCO \\ {>}O \\ RCO \end{array} \longrightarrow \begin{array}{c} RCOO^- \\ ; \\ R\overset{+}{C}{=}O \end{array} \quad \underset{\underset{O}{\|}}{\overset{R}{\underset{|}{C^+}}}\cdots\overset{R'}{\underset{|}{C}}HCOO^- \longrightarrow RCO\overset{R'}{\underset{|}{C}}H{-}COOH \quad (21)$$

It should be noted that the acetylium ion, $CH_3C^+{=}O$, plays the same role in acetic anhydride solvent chemistry that the proton plays in water[95]. As a further example, the well-known acetylation reaction by means of pyridine in acetic anhydride may be cited.

Neunhoeffer and Paschke proposed the intermediacy of a β-keto acid on the basis of the fact that only those acids which possess an α hydrogen undergo ketonic decarboxylation readily. Branched-chain aliphatic acids give poorer yields than straight-chain acids, and aromatic acids undergo primarily simple decarboxylation (equation 16). For example, 2,2,5,5-tetramethyladipic acid failed to give the corresponding cyclopentanone, while 3,3,4,4-tetramethyladipic acid cyclized readily[96]. The electron-withdrawing keto group would cause the keto acid, once formed, to undergo rapid decarboxylation via the route of equation (16) and thus prevent its isolation. However, the data of Neunhoeffer and Paschke would exclude an anhydride precursor, since it appears that water or free acid is necessary for the success of the reaction. Thus, barium adipate, which usually decomposes to cyclopentanone at 430°, did so at 260° in the presence of free adipic acid. In the presence of barium oxide it decomposed at 325° to give 41% cyclopentene, 47% carbon dioxide, and no cyclopentanone. The authors explain that the β-keto acid can form an enolic complex (54) with the metal ion which would be stable in the absence of water or free acid.

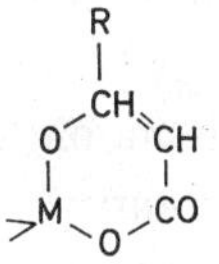

(54)   (55)   (19)

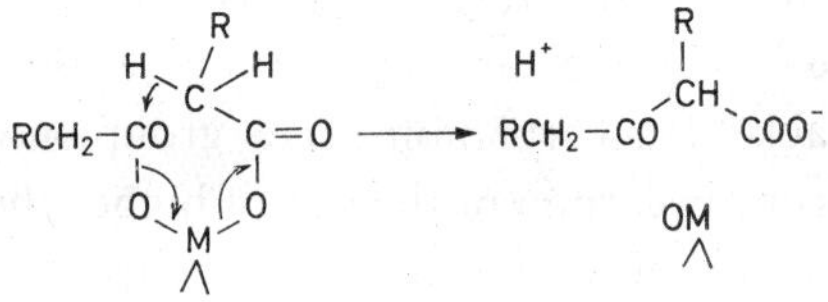

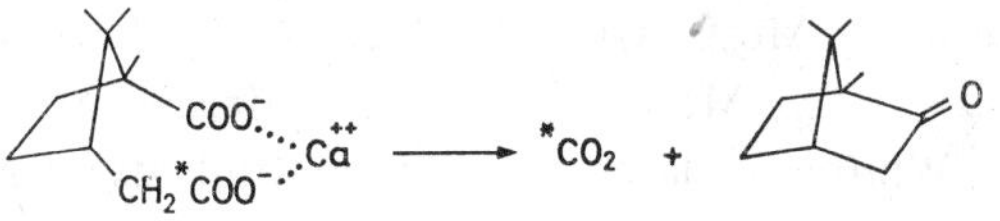

(20)

(56)

Isotopic tracer experiments generally support this mechanism. A case in point is the pyrolysis of the dicarboxylic acid salt (56), in which the starred carbon was labeled with ^{14}C. All of the radioactivity was found in the carbon dioxide and none in the camphor[97]. Similarly, when mixtures of labeled acetate with formate, benzoate, or p-toluate are pyrolyzed, all of the label is found in the carbon dioxide[98-100]. The results of pyrolysis of mixtures in which both salts contain α hydrogen can be readily explained by considering the formation of the β-keto acid.

All of these results seem to point to a concerted mechanism involving a cyclic six-membered transition state (55). Such a transition state could have any degree of ionic character. If the degree of ionic character is low, the results of labeling experiments should be statistical (in the following sense). Thus, the pyrolysis of thorium acetate with ^{14}C-labeled isobutyrate gave 2-methyl-3-butanone containing 80% of the original radioactivity[100]. On a statistical basis this product should have had 75% of the label, since isobutyrate has one hydrogen while acetate has three. The difference must reflect the ionic character of the transition state, which does not favor the formation of a carbanion from isobutyrate.

Similarly the decomposition of calcium phenylacetate in the presence of labeled calcium acetate gave phenylpropanone with one third of the original label; on a statistical basis 40% would be expected[99]. However, when the thorium salts were used, better than 90% of the carbon dioxide was derived from the phenylacetate[100]. Evidently this system entails a large degree of ionic character, reflecting the high stability of the carbanion derived by loss of a proton from phenylacetate. In line with this, Goto and Okubo[100] note that the decomposition temperatures of the thorium salts of phenylacetic (216°), formic (220–260°), acetic, butyric, isobutyric, and benzoic (300°) acids parallel the relative rates of saponification of the corresponding esters. Just as the first step in ester hydrolysis is attack by hydroxide ion on the carbonyl carbon, so this reaction is presumed to involve attack of a carbanion (or a group having carbanionic character) on the carbonyl carbon of the salt of higher thermal stability.

Such a mechanism may also be extended to acids having no α hydrogen. Whitmore, Miller, and Cook[101] studied the decarboxylation of trimethylacetic acid over aerogel thoria at 490°. The products were as follows: $(Me_3C)_2CO$—none, $Me_3CCOCH_2CHMe_2$—15%, Me_3CCOMe—9%, CO_2—20%, CO—2·7%, Me_3CCHO—3%, hydrocarbons—8%, starting material—21%. Whitmore claims that this multiplicity of products arises from decomposition of $Me_3CCOCH_2CHMe_2$, which, in turn, results from decarboxylation of a δ-keto acid formed by a concerted mechanism having a cyclic seven-membered transition state (57). Branched-chain acids having α hydrogen may form both β- and δ-keto acids. For example, the decomposition of sodium isobutyrate[105] at 380° gives 13·2% $(Me_2CH)_2CO$, $EtCH_2COCHMe_2$ in 43·7% yield and 43·1% of $(EtCH_2)_2CO$. Reed[102], however, considers this as evidence for a free radical mechanism in which a branched-chain alkyl radical rearranges to a straight-chain one.

(57)   (58)

Aromatic acids cannot form keto acids, but their ketonization can still be interpreted in terms of a concerted mechanism in which the ionic character arises by loss of carbon dioxide rather than by loss of α hydrogen, as in **58**. Thus, the benzaldehyde obtained from pyrolysis of a mixture of calcium formate and ^{14}C-labeled benzoate contained almost all of the original radioactivity[99]. The same result was obtained with the barium salts[102]. This may be considered as reflecting the higher stability of the benzoyl ion relative to the formyl ion, which would immediately decompose to carbon monoxide and hydrogen, or the lower thermal stability of the formate salt.

However, when the thorium salts are used, the benzaldehyde obtained possessed only 2·5% of the original radioactivity. This puzzling result is not in line with the thermal stabilities of thorium formate and benzoate, nor does it reflect the stabilities of the anionic species transferred in accompaniment to the loss of carbon dioxide; hydride would be expected to migrate in preference to phenide, $C_6H_5^-$. It may, in fact, correspond to an altered conformational preference introduced by changing the central metal atom organizing the cyclic structure of the transition state. The net result may be that in the thorium-centered cycle, phenyl is in a conformationally accomodated position for migration to the carbonyl compared to hydrogen.

It is obvious that a mechanism such as **58** could apply to any acid substrate and that there is no need to postulate a keto acid intermediate. If so, the far greater ease of reaction of acids containing α hydrogen would be merely indicative of the greater stability of primary over secondary and tertiary alkyl carbanions. Evidence for this conclusion is provided by measurements of the effect of reaction rate on the ^{13}C content of carbon dioxide formed by decarboxylation of barium adipate, as measured by mass spectroscopy, compared to the ^{13}C content of naturally occurring carbon dioxide. No isotope effect was found; the product carbon dioxide had the same ^{13}C content as the natural material[103]. In order for this to happen, the number and kinds of bonds broken and formed in the rate-determining step must be identical. If there is a keto acid intermediate, then a C—C bond must be formed in one step and broken in the next, whereas a concerted mechanism such as **58** has a —C—C— and a —C—O— both broken and formed in the same step. Similarly, Wiberg found no deuterium isotope effect in the decarboxylation of $EtCD_2COOM$, indicating that —C—COOM bond cleavage is rate-determining[104]. However, when the pyrolysis was conducted in the presence of unlabeled butyrate, deuterium exchange occurred. While this could be due to a side-reaction, further study appears to be in order.

We have yet to consider the effect of the metal ion. This is a fault of the mechanism proposed by Koch and Leibnitz[92], for it does not require metal ion catalysis and ascribes no role to this catalyst in the transition state. Okubo and Goto[100] discuss the depolarizability of the metal, while Kenner and Morton[57] consider it in terms of the basicity of the metal oxide. Expressed either way, the matter reduces to consideration of the affinity for oxide or other anion and the activity of the oxide as a base toward water or other proton sources. That is to say, metals with a higher affinity towards a given anion form salts of greater covalent character with that anion. This could explain the differences between the calcium and thorium acetate–phenylacetate systems[99, 100]. The presence of a cyclic transition state imposes another obvious requirement; the central metal atom must have an even valence. This would explain why ferrous ion catalyzes ketonic decarboxylation while ferric does not. It must be noted to the contrary, however, that Ruzicka and coworkers[88] claim a 45% yield of cycloheptanone by pyrolysis of cerium(III) sebacate.

V. REFERENCES

1. C. J. Collins and J. F. Eastham, in *Chemistry of the Carbonyl Group* (Ed. S. Patai), Interscience, London and New York, 1966, Chap. 15.
2. N. J. Turro, W. B. Hammond and P. A. Leermakers, *J. Am. Chem. Soc.*, **87**, 2776 (1965); N. J. Turro and W. B. Hammond, *J. Am. Chem. Soc.*, **87**, 3258 (1965); **88**, 2880, 3672 (1966).
3. H. O. House and G. A. Frank, *J. Org. Chem.*, **30**, 2948 (1965).
4. G. Hite and H. Patel, *J. Org. Chem.*, **30**, 4336 (1965).
5. C. Rappe and K. Anderson, *Arkiv Kemi*, **24**, 315 (1965); *Chem. Abstr.*, **63**, 4156c (1965).
6. W. L. Carrick and A. Fry, *J. Am. Chem. Soc.*, **77**, 4381 (1955).
7. M. J. D'Errico, *Thesis*, Columbia University, 1960; *Diss. Abstr.*, **21**, 52 (1960).
8. R. I. Reed and M. B. Thornley, *J. Chem. Soc.*, 2148 (1954).
9. N. J. Leonard and C. W. Schimelpfenig, *J. Org. Chem.*, **23**, 1708 (1958).
10. K. T. Finley, *Chem. Rev.*, **64**, 573 (1964).
11. R. C. Fuson, *Reactions of Organic Compounds*, John Wiley and Sons, New York, 1962, p. 560.
12. E. Van Heyningen, *J. Am. Chem. Soc.*, **77**, 4016 (1955).
13. K. T. Finley and N. A. Sasaki, *J. Am. Chem. Soc.*, **88**, 4267 (1966).
14. P. D. Gardner, G. R. Hayes and R. L. Brandon, *J. Org. Chem.*, **22**, 1206 (1957).
15. M. F. Ansell and M. H. Palmer, *Quart. Rev.*, **18**, 211 (1964).
16. J. Beereboom, *J. Org. Chem.*, **30**, 4230 (1965).
17. M. C. Khoska and N. Anand, *Indian J. Chem.*, **3**, 232 (1965).
18. A. H. Blatt, *Org. Reactions*, **1**, 342 (1942).
19. F. Krausz and R. Martin, *Bull. Soc. Chim. France*, 2192 (1965).
20. K. von Auwers, H. Bundesmann and F. Wieners, *Ann. Chem.*, **447**, 162 (1926).
21. J. N. Chatterjea, S. N. P. Gupta, and V. N. Mehrota, *J. Indian Chem. Soc.*, **42**, 205 (1965).
22. C. R. Hauser and E. H. Man, *J. Org. Chem.*, **17**, 390 (1952).
23. H. Zimmer and R. E. Eibeck, *Naturwissenschaften*, **11**, 263 (1958); *Chem. Abstr.*, **53**, 1266d (1958).

24. F. D. Thomas, M. Shamma and W. C. Fernelius, *J. Am. Chem. Soc.*, **80**, 5864 (1958).
25. F. G. Young, F. C. Frostick, J. J. Sanderson and C. R. Hauser, *J. Am. Chem. Soc.*, **72**, 3635 (1950).
26. R. J. P. Allan, R. L. Forman and P. D. Ritchie, *J. Chem. Soc.*, 2717 (1955); R. J. P. Allan, J. McGee and P. D. Ritchie, *J. Chem. Soc.*, 4701 (1957); P. E. Reimimger and P. D. Ritchie, *J. Chem. Soc.*, 2678 (1963); W. M. Muir and P. D. Ritchie, *J. Chem. Soc.*, 2692 (1963).
27. R. B. Rashbrook and G. W. Taylor, *Chem. Ind. (London)*, 215 (1962).
28. R. A. Finnegan and A. W. Hagen, *Tetrahedron Letters*, 365 (1963).
29. M. F. Gorodetsky and Y. Mazur, *Tetrahedron Letters*, 369 (1963).
30. Unpublished results of work currently in progress in these laboratories.
31. R. Louw and E. C. Kooyman, *Rec. Trav. Chim.*, **84**, 1511 (1965).
32. W. J. Bailey and R. Barclay Jr., *J. Org. Chem.*, **21**, 328 (1956).
33. R. J. P. Allan, E. Jones and P. D. Ritchie, *J. Chem. Soc.*, 524 (1957).
34. E. C. Kooyman, R. Louw and W. A. M. De Tonkelaar, *Proc. Chem. Soc.*, 66 (1963).
35. F. W. Schuler and G. W. Murphy, *J. Am. Chem. Soc.*, **72**, 3155 (1950).
36. Y. Pocker, *Proc. Chem. Soc.*, 141 (1961).
37. W. M. Horspool and P. L. Pauzon, *J. Chem. Soc.*, 5162 (1965).
38. D. H. R. Barton, Y. L. Chow, A. Cox and G. W. Kirky, *Tetrahedron Letters*, 1055 (1962).
39. H. Kobsa, *J. Org. Chem.*, **27**, 2293 (1962).
40. J. L. Stratenus, *Thesis*, University of Leiden, 1966.
41a. R. A. Finnegan and J. J. Mattice, *Tetrahedron*, **21**, 1015 (1965).
41b. R. A. Finnegan and D. Knutson, *Chem. Ind. (London)*, 1837 (1965).
42. J. L. Stratenus and E. Havinga, *Rec. Trav. Chim.*, **85**, 434 (1966).
43. J. C. Anderson and C. B. Reese, *Proc. Chem. Soc.*, 27 (1960); *J. Chem. Soc.*, 1781 (1963).
44. G. M. Coppinger and E. R. Bell, *J. Phys. Chem.*, **70**, 3479 (1966).
45. A. Yogev, M. Gorodetsky and Y. Mazur, *J. Am. Chem. Soc.*, **86**, 5208 (1964).
46. M. Gorodetsky and Y. Mazur, *J. Am. Chem. Soc.*, **86**, 5213 (1964).
47. B. Raecke, *Angew. Chem.*, **70**, 1 (1958).
48. Y. Ogata, M. Tsuchida and A. Muramoto, *J. Am. Chem. Soc.*, **79**, 6005 (1957).
49. J. I. Jones, A. S. Lindsey and H. S. Turner, *Chem. Ind. (London)*, 659 (1958).
50. A. S. Lindsey and H. Jeskey, *Chem. Rev.*, **57**, 583 (1957).
51. J. Ratusky and F. Sörm, *Coll. Czech. Chem. Comm.*, **24**, 2553 (1958).
52. Y. Ogata and K. Sakamoto, *Chem. Ind. (London)*, 749 (1964).
53. Y. Ogata and K. Nakajima, *Tetrahedron*, **21**, 2393 (1965).
54. O. Wallach, *Ann. Chem.*, **300**, 297 (1898).
55. S. V. Hintikka, *Chem. Ber.*, **47**, 512 (1914).
56. W. Huckel, *Theoretical Principles of Organic Chemistry*, Vol. 1, Elsevier, New York, 1955, p. 463.
57. J. Kenner and F. Morton, *Chem. Ber.*, **72**, 452 (1939).
58. A. Berka and J. Zyka, *Chem. Listy*, **52**, 930 (1958).
59. R. Criegee, E. Höger, G. Huber, P. Kruck, F. Marktscheffel and H. Schellenberger, *Ann. Chem.*, **599**, 81 (1956).
60. N. Sonntag, in *Fatty Acids* (Ed. K. Markley), Vol. 2, Interscience, New York, 1961, Chap. 11.
61. R. G. Ackman, P. Linstead, B. J. Wakefield and B. C. L. Weedon, *Tetrahedron*, **8**, 221 (1960).
62. B. C. L. Weedon, M. F. Ansell, A. N. Radziwill, D. J. Redshaw, I. S. Sheperd and D. Wallace, *J. Agr. Food. Chem.*, **13**, 399 (1965).

63. H. Pistor and H. Plieninger, *Ann. Chem.*, **562**, 239 (1947).
64. R. Lukes and J. Hofman, *Chem. Listy*, **52**, 1747 (1958).
65. A. Luttringhaus and R. Reif, *Ann. Chem.*, **618**, 221 (1958).
66. R. F. Paschke, J. E. Jackson, W. Tolberg, H. M. Boyd and D. H. Wheeler, *J Am. Oil Chem. Soc.*, **29**, 229 (1952).
67. P. L. Nichols, S. F. Herb and R. W. Riemenschneider, *J. Am. Chem. Soc.*, **77**, 247 (1951).
68. E. R. H. Jones, M. C. Whiting, G. Eglinton and G. H. Mansfield, *J. Chem. Soc.*, 3197 (1954); E. R. H. Jones, M. C. Whiting and G. H. Whitman, *J. Chem. Soc.*, 3201 (1954).
69. R. G. Ackman, R. A. Dytham, B. J. Wakefield and B. C. L. Weedon, *Tetrahedron*, **8**, 239 (1960).
70. R. A. Dytham and B. C. L. Weedon, *Tetrahedron*, **8**, 246 (1960).
71. R. A. Dytham and B. C. L. Weedon, *Tetrahedron*, **9**, 246 (1960).
72. H. Schultz and J. Sickels, *J. Chem. Educ.*, **38**, 300 (1961).
73. P. Sabatier and E. E. Reid, 'Catalysis in Organic Chemistry', *Catalysis Then and Now*, Part. II, Franklin Publishing Co., Engelwood, N. J., 1965, pp. 825–869 (original pagination).
74. L. F. Fieser and M. Fieser, *Advanced Organic Chemistry*, Reinhold Publishing Corp., New York, 1961, pp. 119, 567, 572, 579.
75. M. M. Shemyakin and S. J. Kanevskaya, *Chem. Ber.*, **69**, 2152 (1936); M. M. Shemyakin and E. M. Bamdas, *Zh. Obshch. Khim.*, **18**, 324 (1948); *Chem. Abstr.*, **43**, 125a (1948).
76. R. Nakai, M. Sugii and H. Nakao, *J. Am. Chem. Soc.*, **81**, 1003 (1959).
77. A. Grün and T. Wirth, *Chem. Ber.*, **53**, 1301 (1920).
78. A. V. Frost and A. V. Ochkin, *Vestn. Mosk. Univ.*, Which Series, No. **5**, 73, (1947); *Chem. Abstr.*, **42**, 2577b (1947).
79. S. Swann. S. S. Kistler and E. G. Appel, *Ind. Eng. Chem.*, **26**, 388, 1014 (1934).
80. R. I. Reed, *J. Chem. Soc.*, 4423 (1955).
81. V. D. Nefedov, M. A. Totopova and I. A. Skulskii, *Zh. Fiz. Khim.*, **29**, 2236 (1955).
82. A. M. Rubenshtein and N. A. Pribytkova, *Dokl. Akad. Nauk. SSSR*, **78**, 917 (1951); *Chem. Abstr.*, **46**, 33b (1951).
83. K. Kino, *J. Chem. Soc. Japan, Ind. Chem. Sect.*, **40**, Suppl. Binding, 311, 437 (1934); **41**, Suppl. Binding 91 (1938); *Chem. Abstr.*, **32**, 489, 1961, 1962, 6085 (1938).
84. A. Grün, E. Ulbrich and F. Krczil, *Angew. Chem.*, **39**, 421 (1926).
85. H. P. Schultz and R. Davis, *J. Org. Chem.*, **27**, 854 (1962).
86. H. P. Schultz and C. Granito, *J. Org. Chem.*, **28**, 879 (1963).
87. A. I. Vogel, *J. Chem. Soc.*, 2032 (1928); 721 (1929).
88. L. Ruzicka, W. Brugger, M. Pfeiffer, H. Schinz and M. Stoll, *Helv. Chim. Acta*, **9**, 499 (1926).
89. L. Ruzicka, M. Stoll and H. Schinz, *Helv. Chim. Acta*, **11**, 670 (1928).
90. E. Bamberger, *Chem. Ber.*, **43**, 3517 (1910).
91. G. Blanc, *Bull. Soc. Chim. (Belges)*, 4 Serie **3**, 778 (1908).
92. H. Koch and E. Leibnitz, *Periodica Polytech.*, **5**, 139 (1961).
93. M. M. Shemyakin and E. M. Bambas, *Zh. Obshch. Khim.*, **18**, 629 (1948); *Chem. Abstr.*, **43**, 1318a (1948).
94. O. Neunhoeffer and O. Paschke, *Chem. Ber.*, **72**, 919 (1939).
95. G. Jander, E. Rüsberg and H. Schmidt, *Z. Anorg. Chem.*, **255**, 238 (1948).
96. E. H. Farmer and J. Kracovski, *J. Chem. Soc.*, 680 (1927).
97. L. Ötvös and L. Noszko, *Tetrahedron Letters*, **2**, 19 (1960).
98. J. Bell and R. I. Reed, *J. Chem. Soc.*, 1383 (1952).

99. C. C. Lee and W. T. Spinks, *J. Org. Chem.*, **18**, 1079 (1953).
100. M. Okubo and R. Goto, *Nippon Kagaku Zasshi*, **81**, 1132, **82**, 620 (1961); *Chem. Abstr.*, **56**, 11439 (1961).
101. F. C. Whitmore, A. L. Miller and N. C. Cook, *J. Am. Chem. Soc.*, **72**, 2732 (1950).
102. R. I. Reed, *J. Chem. Phys.*, **21**, 377 (1953).
103. J. Bigeleisen, A. A. Bothnerby and L. Friedman, *J. Am. Chem. Soc.*, **75**, 2908 (1953).
104. K. B. Wiberg, *J. Am. Chem. Soc.*, **74**, 4381 (1952).

CHAPTER 9

Substitution in the groups COOH and COOR

D.P.N. Satchell

King's College, London, England

and

R.S. Satchell

Queen Elizabeth College, London, England

I. INTRODUCTION

A. Scope of the Chapter

Attention will be confined mainly to the type of substitution illustrated in general terms (omitting charges) by equation (1), in which R^1 and R^2

$$R^1COOR^2 + X \longrightarrow R^1COX + OR^2 \tag{1a}$$

$$R^1COOH + X \longrightarrow R^1COX + OH \tag{1b}$$

represent alkyl or aryl groups. We shall deal less thoroughly with the sometimes related, but less widespread, substitutions (2) and shall entirely

$$R^1COOR^2 + X \longrightarrow R^1COOX + R^2 \tag{2a}$$

$$R^1COOH + X \longrightarrow R^1COOX + H \tag{2b}$$

omit processes which lead to replacement, or permanent modification of the carbonyl group, for example (3). Of course, substitutions like (1)

$$RCOOH + 2 H_2 \longrightarrow RCH_2OH + H_2O \qquad (3)$$

may often proceed via intermediates which involve a *temporary* modification of the carbonyl group, and we shall deal with all varieties of process (1) whatever their exact mechanism. It must be remembered throughout that all processes — including (1) and (2) — are, in principle, equilibria and that any equilibrium position depends upon prevailing conditions, as well as upon chemical constitution.

B. Mechanism

With very few exceptions, all the known reactions which can be represented by equation (1) are considered to involve bond heterolysis. They comprise an important fraction of the general class of reactions known as acylation[1] (equation 4). In these processes RCOY is termed the acylating

$$RCOY + X \longrightarrow RCOX + Y \qquad (4)$$

agent and X is the substrate which undergoes acylation, *i.e.* it undergoes attachment of an acyl group, RCO. It is found that heterolysis of the $C-Y$ bond always occurs in the sense $RCO^+ + Y^-$, irrespective of the nature of Y. That atom of the substrate which becomes attached to the carbonyl carbon atom therefore always acts as a nucleophile, so that most processes like (1) may be formally described as nucleophilic substitutions at a carbonyl carbon atom. It happens that carboxylic esters and acids are comparatively poor acylating agents, a rough sequence of reactivity being

$$R^1COR^2 < R^1CONR_2^2 < R^1CO_2R^2 < R^1COOCOR^2 < RCOHal$$
$$< RCO_2SO_3H < RCOOClO_3 \simeq RCO^+BF_4^-$$

where R^2 represents H, alkyl or aryl. Spontaneous acylation by esters and acids is only common with the more powerfully nucleophilic substrates, and catalysis is necessary, or especially beneficial, in other cases.

Catalysis of heterolytic reactions arises from the presence of additional electrophiles, or nucleophiles, which facilitate the overall process by modifying the reactants[*]. In promoting acylation, catalysts function either by increasing the electrophilicity of the carbonyl carbon atom of the acylating agent (acidic catalysis) or by increasing the nucleophilic character of the substrate (basic catalysis).

---

[*] In this chapter limitations of space confine us to non-enzymic catalysis. Much current interest centres on the application of knowledge of non-enzymic catalysis to enzyme systems[2].

1. Acidic catalysis

Acids may be divided into two types: Brønsted acids and Lewis acids. Among the latter, those of the covalent metal halide type MX_n are the most important in the present contexts. Anhydrous conditions are normally required for the successful use of these Lewis acids. Acids may increase the electrophilicity of an acylating agent in the following general ways:

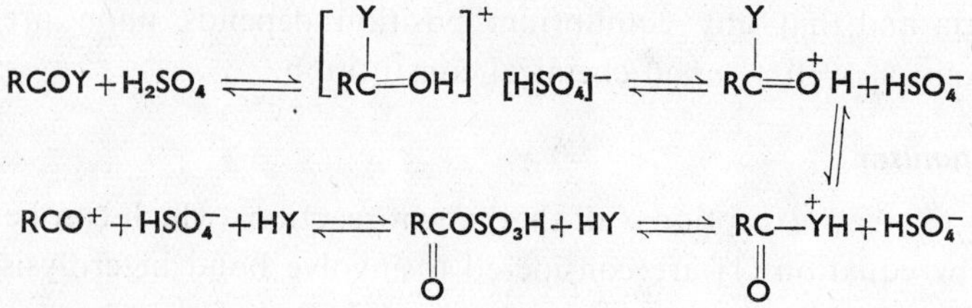

$$RCOY + H_2SO_4 \rightleftharpoons \left[\underset{\underset{\displaystyle RC=OH}{|}}{\overset{\overset{\displaystyle Y}{|}}{}} \right]^+ [HSO_4]^- \rightleftharpoons \underset{\underset{\displaystyle RC=\overset{+}{O}H}{|}}{\overset{\overset{\displaystyle Y}{|}}{}} + HSO_4^-$$

$$\quad (5)$$

$$RCO^+ + HSO_4^- + HY \rightleftharpoons \underset{O}{\overset{||}{RCOSO_3H}} + HY \rightleftharpoons \underset{O}{\overset{||}{RC-\overset{+}{Y}H}} + HSO_4^-$$

$$RCOY + MX_n \rightleftharpoons \underset{Y}{\overset{|}{RC=O:MX_n}} \rightleftharpoons \underset{O}{\overset{||}{RC-Y:MX_n}} \rightleftharpoons [RCO]^+ [MX_nY]^-$$

$$RCO^+ + MX_nY^-$$

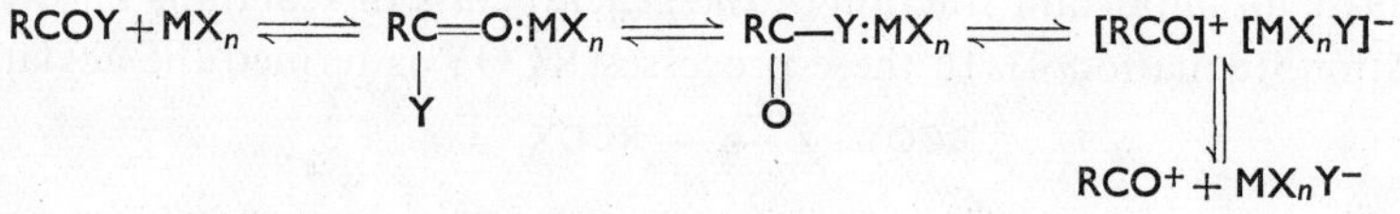

$$\quad (6)$$

Typical examples[3] relevant to carboxylic esters and acids are (7) and (8):

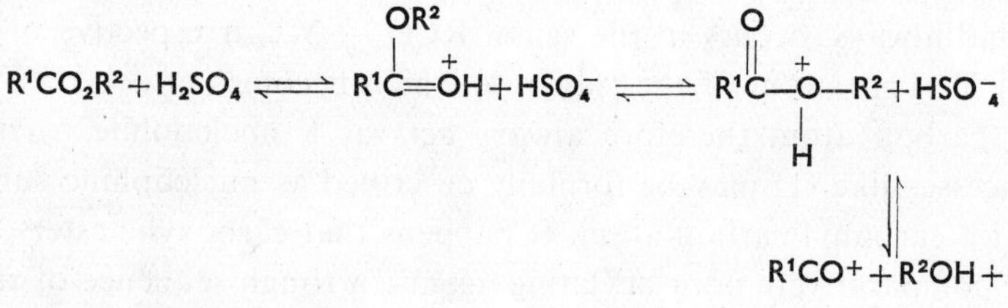

$$R^1CO_2R^2 + H_2SO_4 \rightleftharpoons \underset{\underset{\displaystyle R^1C=\overset{+}{O}H}{|}}{\overset{\overset{\displaystyle OR^2}{|}}{}} + HSO_4^- \rightleftharpoons \underset{\underset{\displaystyle H}{|}}{\overset{\overset{\displaystyle O}{||}}{R^1C-\overset{+}{O}-R^2}} + HSO_4^-$$

$$\quad (7)$$

$$R^1CO^+ + R^2OH + HSO_4^-$$

$$2 RCO_2H + SnCl_4 \rightleftharpoons$$

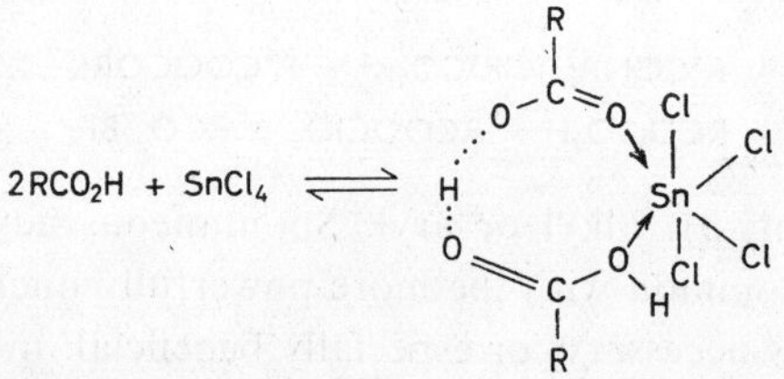

$$\quad (8)$$

It will be apparent that the actual effect of the acid catalyst is the production of a new acylating agent further along the reactivity series than the original reagent RCOY. Normally the new species is more effective because it possesses a better leaving group than the original leaving group Y. In equations (5)–(8) the relative positions of the various equilibria will depend upon RCOY, upon the catalyst, and upon the dielectric constant and solvating power of the medium; but it is clear that systems which favour the extreme of acylium ion (RCO^+) formation will, other factors being equal, provide the most powerful acylating agents. In general, therefore, an acid will coordinate with the acylating agent in a rapid equilibrium prior to the acylation step, and will then assist the departure of the leaving group either prior to (equations 9–11) or during (equations 12–13) the actual acylation.

$$R^1COY + HX \rightleftharpoons R^1CO^+ + HY + X^- \qquad\qquad \text{fast} \qquad (9)$$

$$R^1CO^+ + R^2OH \longrightarrow \overset{H}{\underset{+}{R^1COOR^2}} \qquad\qquad \text{slow} \qquad (10)$$

$$\overset{H}{\underset{+}{R^1COOR^2}} + X^- \longrightarrow R^1COOR^2 + HX \qquad \text{fast} \qquad (11)$$

$$R^1COY + MX_n \rightleftharpoons R^1COYMX_n \qquad\qquad \text{fast} \qquad (12)$$

$$R^1COYMX_n + R^2OH \longrightarrow R^1CO_2R^2 + HY + MX_n \qquad \text{slow} \qquad (13)$$

2. Basic catalysis

The base serves to increase the nucleophilicity of the substrate. Substrates which benefit from basic catalysis normally contain a hydrogen atom attached to the atom which undergoes acylation and which is replaced in the acylation process (14). It is the function of the catalyst to aid the

$$R^1COY + R^2NH_2 \longrightarrow R^1CONHR^2 + HY \qquad\qquad (14)$$

removal of this hydrogen atom either prior to (equations 15–17) or, during (equations 18–19) the acylation step, for, in general, anions are more

$$R^2NH_2 + OH^- \rightleftharpoons R^2NH^- + H_2O \qquad\qquad \text{fast} \qquad (15)$$

$$R^1COY + R^2NH^- \longrightarrow R^1CONHR^2 + Y^- \qquad\qquad \text{slow} \qquad (16)$$

$$H_2O + Y^- \rightleftharpoons HY + OH^- \qquad\qquad\qquad \text{fast} \qquad (17)$$

$$R^1COY + 2\ R^2NH_2 \longrightarrow R^1CONHR^2 + R^2NH_3 + Y^- \qquad \text{slow} \qquad (18)$$

$$R^2NH_3^+ + Y^- \rightleftharpoons R^2NH_2 + HY \qquad\qquad\qquad \text{fast} \qquad (19)$$

susceptible to acylation than are neutral substrates. The type of base most uitable will depend upon the nature of the substrate and upon the reaction medium. This point will become evident as different classes of substrate are dealt with in the text. In other cases the base temporarily takes the place of the substrate, becomes acylated itself, and in so doing provides a more reactive acylating agent for attacking the substrate (equations 20–22).

$$R^1COY + B^- \rightleftharpoons R^1COB + Y^- \qquad\qquad \text{fast} \qquad (20)$$

$$R^1COB + R^2NH_2 \longrightarrow R^1CONHR^2 + BH \qquad \text{slow} \qquad (21)$$

$$BH + Y^- \rightleftharpoons HY + B^- \qquad\qquad\qquad \text{fast} \qquad (22)$$

This latter type of effect is sometimes called nucleophilic catalysis.

Nucleophilic catalysis[4] is a useful and generally meaningful term whose meaning is also easy to grasp. Schemes like (9)–(11) and (15)–(17) and

(12)–(13) and (18)–(19), correspond respectively, for suitable reactants and media, to what is often termed specific and general (acid or base) catalysis. These terms are confusing—especially to the uninitiated. Moreover they are based only on the possibility (or impossibility) of representing the differential rate equation in a particular mathematical form, and the distinction they represent becomes meaningless in Lewis acid systems and in aprotic solvents. Their demise is therefore overdue. A more suitable term, if terms are needed, for catalytic schemes (9)–(11) and (15)–(17) would be 'precursory' catalysis and for schemes (12)–(13) and (18)–(19) 'synchronous' catalysis.

3. Catalytic efficiency

In the widest (Lewis) sense, all heterolytic reactions are acid–base in type. Acidic and basic catalysts ultimately owe their efficacy to the fact that they too *are* acids and bases, and can therefore sometimes facilitate the overall heterolysis involved (as described in sections I.B.1 and 2 above). Complicating factors (*e.g.* steric hindrance and varying degrees of solvation) apart, bases which coordinate strongly to hydrogen ions in aqueous solution are also likely to be strongly attracted to the various other sorts of electrophilic centre with which they engage in their catalytic roles. There is likely therefore to be a correlation between their association constant towards hydrogen ions in aqueous solution $(1/K_a)$, their equilibrium constants with more typical Lewis acids, and their catalytic effectiveness as measured by the relevant catalytic rate constant (k_c). The same type of argument also applies for acids. It is not surprising therefore that for all varieties of catalysed heterolysis (including nucleophilic substitution in esters) correlations between k_c and K_a are often found. These take the form $k_c = G_B(1/K_a)^\beta$ (*i.e.* $\log k_c = -\beta \log K_a +$ constant) or $k_c = G_A(K_a)^\alpha$ (i.e. $\log k_c = \alpha \log K_a +$ constant) of the well-known Brønsted equations for basic and acidic catalysis respectively[5]. The value of β(or α) reflects the sensitivity of the reaction rate to the basic (or acidic) strength of the catalyst as measured in a standard equilibrium. Values greater than unity may imply that more than one molecule of catalyst is involved. Since different catalytic mechanisms will involve different degrees of sensitivity to catalyst strength, β (or α) values can sometimes be used to distinguish between mechanisms. The origin of the foregoing correlations is simply the fact that a comparison is being made between electrically similar processes. All other correlation equations are ultimately successful for the same reason and fail when extra steric and/or electronic effects are present in one half of the comparison. For this reason they are not applicable to enzymic catalysis. A correlation equation which has been organized to express substituent effects is mentioned in section I.B.4.b.

4. Substituent effects

a. R^2 in $R^1CO_2R^2$. Examination of the rough reactivity series for acylating agents suggests the generalization[1] that the stronger the acid HY the more powerful will be the acylating agent RCOY. It follows that in esters, groups R^2 which withdraw electrons, and thus make the corresponding alcohol R^2OH more acidic, will increase the reactivity of the ester in acylation. This appears to be broadly true; for example, phenyl and cyanomethyl esters are more reactive than the corresponding methyl derivatives. However, this effect can be partially offset or even overridden under the influence of acid catalysts, when other factors enter, in particular the ability of the leaving group OR^2 to coordinate with the catalyst. Thus in the acid-catalysed hydrolysis of esters the sensitivity of the reaction rate to the polar effects of R^2 is rather small, whereas for the spontaneous or base-catalysed hydrolysis the dependence of the reaction rate on R^2 is appreciably greater.

In discussing the effects of changes in R^2 (or in Y of RCOY), play is often made with the varying possibilities for resonance structures involving interaction between Y and the carbonyl group as in **1**. Such considerations usually lead to complicated and indefinite conclusions. Whatever

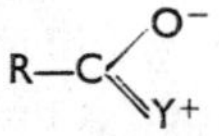

$$R-C\overset{\textstyle O^-}{\underset{\textstyle Y^+}{\big\langle}}$$

(1)

the exact effect such resonance has in different cases, it appears in practice rarely to invalidate the simple generalization about the reactivity of RCOY that we have given above.

A final matter which concerns R^2 is the preference of esters for the *trans* configuration, which minimizes the interaction between the ethereal oxygen lone pairs and the π electrons of the carbonyl double bond. It is in-

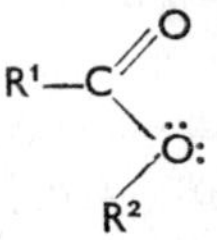

cis	*trans*
(2)	**(3)**

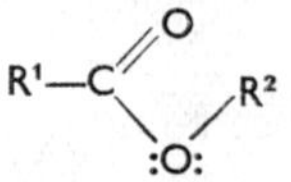

teresting in this context that small ring lactones (ring size < 8) must take a *cis* configuration. Their inherent dislike of this structure is illustrated in their observed rates of hydrolysis, which are much greater than those of larger ring lactones whose size permits them to assume the *trans* form[6].

   b. R^1 *in* $R^1CO_2R^2$ *and* R *in* RCO_2H. The effects on reactivity of changes in R^1 are not as straightforward as for variations in R^2. Substituents which, by provision of electrons, favour the departure of OR^2 (or OH) and stabilize the acylium ion, have the effect of reducing the positive charge on the carbonyl carbon atom of both the acylium ion and of the polarized reagent

Electron-withdrawing substituents increase this charge, but hinder the departure of the leaving group. A balance of factors is therefore involved and the effect of changes in R^1 will depend on the reaction involved. In particular it will depend upon whether the bond-forming or bond-breaking process is kinetically dominant in the overall acylation (4). Again, as for R^2, the effect of changes in R^1 may be modified under conditions of catalysis. Thus the sensitivity of the rate of the acid-catalysed hydrolysis of esters to changes in the polar effects of R^1 (changes which affect the ability of the leaving group to accept a proton and the vulnerability of the carbonyl carbon atom to nucleophilic attack in opposite senses) is much lower than the sensitivity of the rates of the corresponding spontaneous or base-catalysed hydrolyses, in which the effect of R^1 on the behaviour of the leaving group is not so important. This general finding of small polar effects in acid-catalysed ester hydrolysis has been utilized by Taft[7] to separate the polar and steric effects of groups, and to calculate numerical parameters (σ^* and E_s) which represent them. Once calculated, σ^* and E_s can be used to assess the sensitivity of other reactions to polar and steric effects via equation (23), in which k/k_0 represents the ratio of the rate of reaction of the compound carrying a particular substituent

$$\log (k/k_0) = \delta E_s + \varrho^*\sigma^* \tag{23}$$

to that of a reference compound, usually the corresponding species with hydrogen as the substituent. Large values of ϱ^*($|\varrho^*| \gtrsim 1$) indicate marked dependence on polar effects whereas smaller absolute values point to a lower sensitivity to such effects. While $\varrho\sigma$ relationships are often useful, in that they facilitate the combination of data for the prediction of the reactivity of compounds yet unstudied in particular reactions, nevertheless, so far as the *meaning* of the ϱ^* values goes — *i.e.* the degree of the reaction's dependence on polar effects — no more information is conveyed by them than is obtainable by inspecting the *available* experimental data. For this reason they are not emphasized in this chapter.

c. Substituent effects on substrate reactivity. In uncatalysed reactions and also under acidic catalysis (the substrate then not being directly involved with the catalyst) substituent effects are simple: electron-donating groups will normally favour reaction. In basic catalysis, however, electron release will hinder anion formation, although it will activate this ion once formed. A balance of factors is therefore involved, similar to that obtaining for the structure of the acylating agent under acidic catalysis (see section I.B.4.b.).

5. Mechanistic detail

As we have seen, when a substrate X reacts with an acylating agent RCOY it may react with either the ionized or the polarized reagent. On this basis mechanisms of acylation may be broadly divided into two classes[1], which are represented quite generally in equations (24) and (25) without indicating the relative rates of the steps, which will depend on the system concerned. When step (*a*) alone is rate-determining in equation (24), and especially when other molecules of X, apart from that actually

$$\text{RCOY} \underset{}{\overset{(a)}{\rightleftharpoons}} \text{RCO}^+ + \text{Y}^- \text{ (or RCO}^+\text{Y}^-) \xrightarrow[(b)]{\text{X}} \text{RCOX} + \text{Y} \qquad (24)$$

$$\text{RCOY} + \text{X} \longrightarrow \text{RCOX} + \text{Y} \qquad (25)$$

acylated, may assist the ionization, then a sharp distinction between (24) and (25) can only be made on the basis of particular definitions of molecularity and bonding change. 'Borderline' controversies — prominent in connexion with nucleophilic substitution[8] — are really concerned only with definitions. That route (25) may sometimes exhibit substituent effects more or less characteristic of route (24), depending on the presence or absence of catalysis and/or on the relative importance of bond-breaking and making in (25), has also led to some confusion of thought.

It is very probable that many examples of route (25) occur via addition of the substrate to the carbonyl group (equation 26) rather than via a synchronous displacement of Y by X (equation 27). Indeed some workers dismiss

$$\text{R}^1\text{CO}_2\text{R}^2 + \text{H}_2\text{O} \rightleftharpoons \overset{\text{OH}}{\underset{\text{OH}}{\text{R}^1\text{C}-\text{OR}^2}} \longrightarrow \text{R}^1\text{CO}_2\text{H} + \text{R}^2\text{OH} \qquad (26)$$

$$R^1CO_2R^2 + H_2O \longrightarrow \begin{matrix} H \\ \backslash \\ O \\ / \\ H \end{matrix} \cdots \overset{O^{\delta-}}{\underset{R^1}{\overset{\|}{C^{\delta+}}}} \cdots OR^2 \longrightarrow R^1CO\overset{+}{O}H_2 + R^2O^- \qquad (27)$$

synchronous processes altogether. In our view the assumption of such an extreme position is premature and reflects an uncritical consideration of the available data. Until very recently the bulk of the evidence for schemes like (26) consisted of the observation of oxygen exchange concurrent with acylation[9]. Such evidence is only circumstantial[10]. Lately, however, some independent kinetic evidence pointing to addition intermediates has been found[11]. It is likely that, in general, intermediates will be most important for the less reactive acylating agents, i.e. those with poor leaving groups. Carboxylic esters and acids fall in this category and detailed mechanisms involving them are therefore often based on schemes like (26), the individual steps of which can be elaborated to fit a variety of contexts.

The points outlined above and various appropriate, and detailed, mechanistic schemes are examined in the sections which follow, where we shall deal principally with nucleophilic substitution in esters and acids by substrates whose attacking atoms are either halogen, oxygen, sulphur, nitrogen, or carbon. This list covers most of the field[1]. Phosphorus and silicon nucleophiles have not yet proved important, and are unlikely to do so[1]. While the usual product is an alcohol, the controlled reduction of (appropriate) esters to aldehydes is possible[12] (e.g. using a metal hydride or partially deactivated Raney nickel) but is not particularly widely used preparatively. To conserve space we have omitted substitution by hydrogen from the chapter.

The major aspects of substitution by oxygen nucleophiles are dealt with at length in other chapters and our treatment of esterification, hydrolysis and related matters will therefore be restricted to the drawing of parallel between these reactions and the other substitutions with which we are here concerned. Some of these have been the subjects of much kinetic study; others, of great preparative importance, are still known only at that level. Our emphasis will be on principles and generalizations rather than upon facts. In such a treatment reactions known only as successful preparative recipes must inevitably take a minor place no matter how often they are used in practice. Our aim is to reflect light onto such mechanistically unexplored processes by concentrating it on formally related reactions which are better understood.

Substitutions like (2) nearly always involve (apparent) *electrophilic* attack by X on an oxygen atom of the ester or acid (equation 28). For esters they involve the relatively rare alkyl- (as opposed to acyl-) oxygen fission. These

$$R^1CO_2H + \overset{\delta+}{R^2} \overset{\delta-}{Cl} \longrightarrow R^1CO_2R^2 + HCl \tag{28a}$$

$$R^1CO_2R^2 + R^3OH \longrightarrow R^1CO_2H + R^2OR^3 \tag{28b}$$

processes will only be dealt with when they are relevant to the corresponding substitutions at carbon.

Esters and acids will be treated together when discussing different types of substrate. This arrangement emphasizes how, throughout the field, superficial differences in behaviour are all underlaid by a fundamental unity.

II. SUBSTITUTION BY VARIOUSLY BOUND HALOGEN ATOMS

A. Reaction with Hydrogen Halides

1. Carboxylic acids

The equilibrium corresponding to (1) will be (29).

$$RCO_2H + HHal \rightleftharpoons RCOHal + H_2O \tag{29}$$

As is well known, acyl halides react very readily with water and such equilibria normally appear to lie well to the left. Little work concerning the pure liquid hydrogen halides exists, but with hydrogen fluoride it is known[13] that carboxylic acids can be recovered in good yield after dissolution for long periods, and it seems unlikely that much acyl fluoride is formed. Nevertheless, anhydrous hydrogen fluoride is very powerfully acidic and has a high dielectric constant. Suitably substituted acids would be expected to undergo some proton-stimulated heterolysis (equation 30), as in anhydrous sulphuric acid (equation 5). This effect will be important in

$$RCO_2H + 2\,(HF)_n \rightleftharpoons RCO_2H_2^+ + F(HF)_{n-1}^- + (HF)_n \rightleftharpoons RCO^+ + H_3O^+ + 2\,F(HF)_{n-1}^- \tag{30}$$

hydrogen fluoride catalysed acylation by carboxylic acids (see section V. B. 1).

In a variety of solvents, dissolved hydrogen halides will protonate carboxylic acids (to some extent) on oxygen, and in suitable solvents (e.g. EtOH), the acid is thereafter solvolysed, but kinetic analyses of such catalysed solvolyses do *not* require even a low concentration of the acyl halide as an intermediate[14]. It is clear that hydrogen halides behave towards carboxylic acids simply as electrophilic reagents. They engage readily therefore in equilibria like (2) (equation 31).

$$RCOOH + DCl \rightleftharpoons RCOOD + HCl \tag{31}$$

2. Carboxylic esters

The structure of saturated esters suggests that they will, if anything, be less likely than the corresponding acids to acylate halogen in hydrogen halides (equation 32). The lack of reference to such reactions in the

$$R^1CO_2R^2 + HHal \rightleftharpoons R^1COHal + R^2OH \tag{32}$$

literature may therefore have a certain negative significance. In the special case when R^2OH is an unstable enol which rapidly rearranges to a ketone, then, as would be expected, equilibrium (32) is pulled well to the right. Thus, alone, or in inert solvents, enol esters and hydrogen halides give the ketone and the acyl halide[15], e.g. equation (33). It is

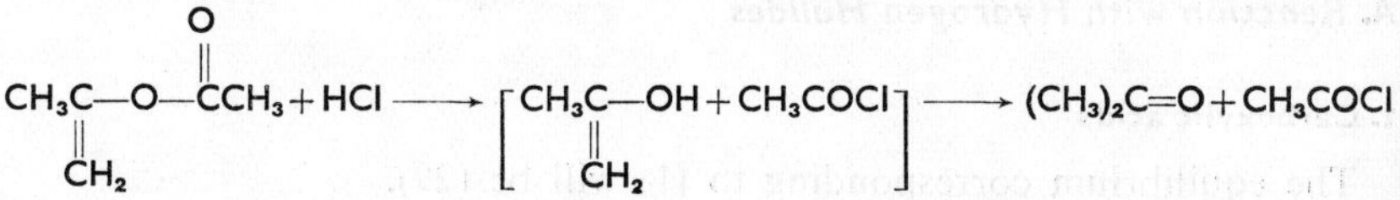

$$\tag{33}$$

probable, however, that the reactions do not proceed via an attack of the carbonyl carbon atom of the ester (or protonated ester) on halogen (equation 34), but via the initial addition of hydrogen halide to the ethylenic rather than to the carbonyl double bond, this step being followed by intramolecular acylation of halogen[16] (equation 35). It has been suggested[17] that in chlorobenzene, where the reaction has been shown to be first-

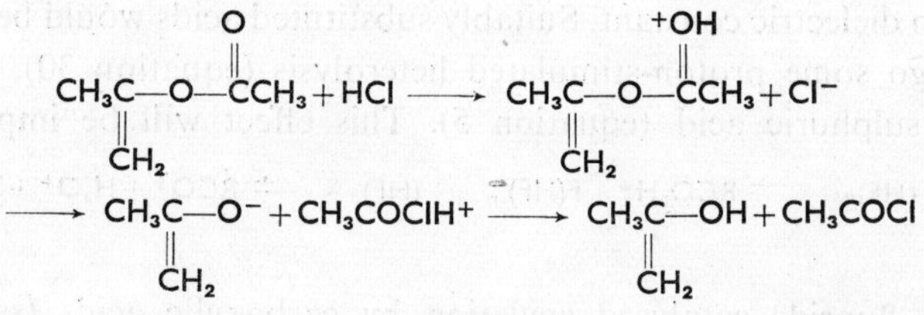

$$\tag{34}$$

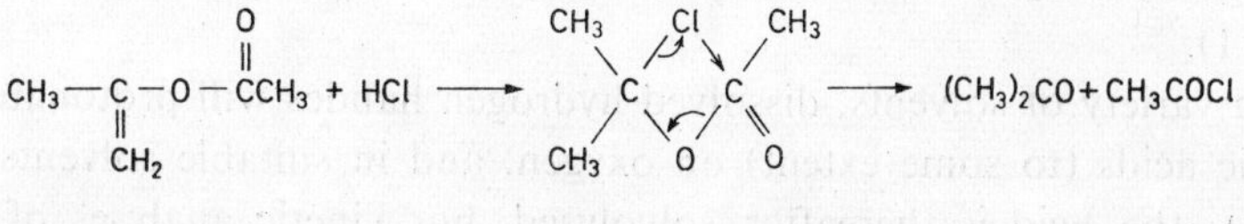

$$\tag{35}$$

order in both ester and hydrogen halide, the detailed mechanism may be as shown in equation (36), which is really only a variation of (35). A very

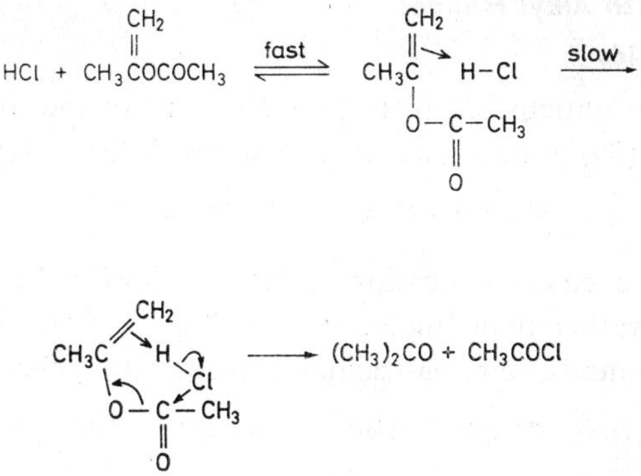

$$HCl + CH_3COCOCH_3 \underset{fast}{\rightleftharpoons} \begin{matrix} CH_2 \\ \| \\ CH_3\overset{\nearrow}{C} \quad H-Cl \\ | \\ O-C-CH_3 \\ \| \\ O \end{matrix} \xrightarrow{slow}$$

(36)

$$\begin{matrix} CH_2 \\ CH_3C\overset{\diagup}{} H \curvearrowright \\ \diagdown \overset{Cl}{} \\ O-C\overset{\curvearrowleft}{-}CH_3 \\ \| \\ O \end{matrix} \longrightarrow (CH_3)_2CO + CH_3COCl$$

similar scheme has recently been suggested[18] for the acid-catalysed hydrolysis of enol esters.

An interesting parallel is provided by the reactions of carboxylic acids with appropriate olefinic halides, *e.g.* α-chlorovinyl ethers. These processes result in acylation of halogen by acid, and cyclic transition states, similar to that in equation (35), have been proposed[19] (equation 37).

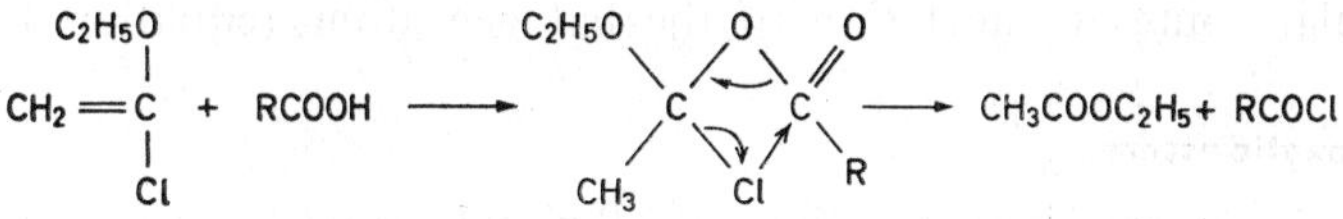

$$\begin{matrix} C_2H_5O \\ | \\ CH_2 = C \\ | \\ Cl \end{matrix} + RCOOH \longrightarrow \begin{matrix} C_2H_5O \quad O \quad O \\ \diagdown \diagup \diagdown \diagup \diagup \\ C \qquad C \\ \diagup \diagdown \diagdown \diagup \diagdown \\ CH_3 \quad Cl \quad R \end{matrix} \longrightarrow CH_3COOC_2H_5 + RCOCl \quad (37)$$

Because of these reactions (35) of enol esters, hydrogen halide catalysed acylation of other substrates with them, in solvents which permit acyl halide formation, commonly proceeds via these latter species as inter-

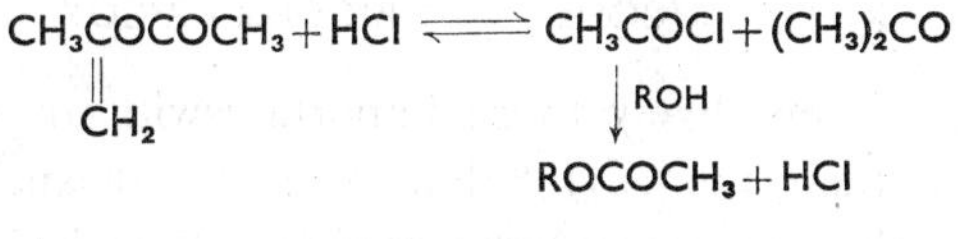

$$\begin{matrix} CH_3COCOCH_3 + HCl \rightleftharpoons CH_3COCl + (CH_3)_2CO \\ \| \qquad\qquad\qquad\qquad\qquad \downarrow ROH \\ CH_2 \\ \qquad\qquad\qquad ROCOCH_3 + HCl \end{matrix}$$

(38)

mediates[20] (equation 38). Analogous schemes obtain with other Brønsted acids[21], in particular with sulphuric acid (section III. G), a catalyst often used in preparative acylations employing enol esters[15]. However, in hydroxylic solvents, in common with the behaviour of the structurally rather similar anhydrides, it is unlikely that the acyl halide is formed as an intermediate in the hydrogen halide catalysed solvolysis[16]. Here the excess of solvent takes the halide's place as the nucleophile. Whether the mechanism is the analogue of (34) or of (35) is not known with certainty.

5*

B. Reaction with Alkyl Halides

1. Carboxylic acids

As would be anticipated from the behaviour of the hydrogen halides, processes like (39) appear rare. In keeping with (31), alkyl halides which

$$R^1CO_2H + R^2Hal \rightleftharpoons R^1COHal + R^2OH \tag{39}$$

ionize readily exchange a cation (equation 2) with the carboxylic acid (equation 40) rather than undergoing acylation of the halogen atom[22]. Certain compounds, *e.g.* ω-trichlorotoluene, do effect the conversion

$$R^1CO_2H + R^2Hal \rightleftharpoons R^1CO_2R^2 + HHal \tag{40}$$

of carboxylic acids to the acyl halide[23]. It appears likely to us that this in fact occurs via reaction (40) followed by an intramolecular step (equa-

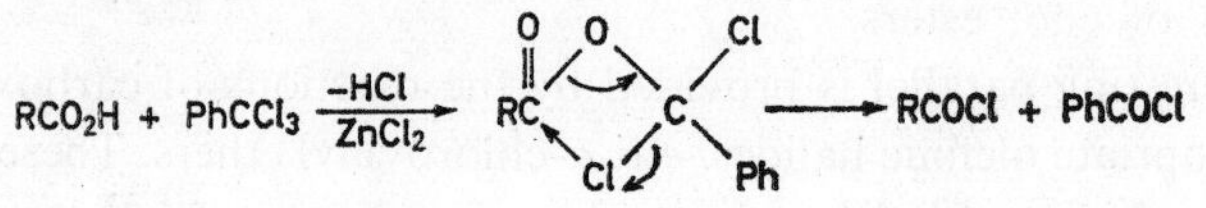

$$\tag{41}$$

tion 41). This scheme correctly accounts for the observed products, and the parallelism with (35) is obvious. Vinyl halides also provide intramolecular routes to acylation of the halogen atoms (equation 37).

2. Carboxylic esters

Little is known about reactions of alkyl halides with esters. Such processes are unlikely to be common.

C. Reaction with Acyl Halides

Acylation of halogen in an acyl halide involves acyl exchange (equation

$$R^1COY + R^2COHal \rightleftharpoons R^1COHal + R^2COY \tag{42}$$

42). This process is not likely to be important with an acylating agent, R^1COY, weaker than the acyl halide itself, for in such a system the former species will tend to undergo preferential acylation[1]. This generalization is borne out by what is known of the reactions of carboxylic acids and esters with acyl halides.

1. Carboxylic acids

Equation (43) portrays a fairly well-known reaction. The dissolution of a deficit of an aromatic, or aliphatic, acyl halide in a carboxylic acid eventually results, provided no halide is lost from solution, in the exchange of

$$R^1CO_2H + R^2COHal \rightleftharpoons R^1COHal + R^2CO_2H \tag{43}$$

acyl groups[24]. Reactive acyl halides appear to exchange fastest. Except perhaps when R^1 is strongly electron-attracting, the direct acylation (equation 43) is very unlikely for, as explained above, the preferred path should involve nucleophilic attack by the carboxylic oxygen atom, a process which leads to the anhydride (equation 44). And it is indeed known[25] that the reaction may be used to prepare anhydrides pro-

$$R^1CO_2H + R^2COHal \longrightarrow [R^2C(OH)(Hal)OCOR^1] \longrightarrow R^1COOCOR^2 + HHal \quad (44)$$

vided that the hydrogen halide is efficiently removed (compare section III. F). If it is not, since the equilibrium position for reactions like (44) lies far to the left, the following processes will occur:

$$R^1COOCOR^2 + HHal \longrightarrow R^1COHal + R^2CO_2H \quad (45)$$

$$R^1COOCOR^2 + HHal \longrightarrow R^2COHal + R^1CO_2H \quad (46)$$

The relative rates of (45) and (46) will depend on the natures of R^1 and R^2, but with R^1CO_2H in great excess any R^1COHal formed will be effectively trapped, because reaction of it with the medium will lead predominantly to $(R^1CO)_2O$, and thus back again to R^1COHal. In this way R^2COHal will be converted eventually to R^1COHal by R^1CO_2H. The result may be facilitated, when R^1COHal is low-boiling compared to the other components, by distilling it out of the mixture as it forms. Little hydrogen halide will be lost in this process over short periods because of its low equilibrium concentration. The relatively high-boiling benzoyl and phthalyl chlorides have been used in this way for the preparation of other acyl chlorides directly from their parent acids[26]. Acyl fluorides have also been made this way[27].

Analogous preparative methods[28] make use of acyl halides which lead to an unstable acid R^2COOH in (45), the acid decomposing to very volatile products so that isolation of R^1COHal is again facilitated. Oxalyl chloride and phosgene probably owe their efficacy to this type of effect (equations 47 and 48).

$$RCOOH + ClCOCl \longrightarrow RCOOCOCl + HCl \longrightarrow RCOCl + [ClCOOH] \longrightarrow$$
$$RCOCl + HCl + CO_2 \quad (47)$$
$$RCOOH + ClCOCOCl \longrightarrow RCOOCOCOCl + HCl \longrightarrow$$
$$RCOCl + [ClCOCOOH] \longrightarrow RCOCl + HCl + CO + CO_2 \quad (48)$$

It is emphasized that, as in all the previous examples given of effective acylation of halogen, we have very probably a mechanism which, in fact, involves an initial reaction like (2), here (44), the products of this process leading to acylation of the halogen atom.

2. Carboxylic esters

Acyl exchange between esters and acyl halides is represented in

$$R^1CO_2R^2 + R^3COHal \rightleftharpoons R^1COHal + R^3CO_2R^2 \qquad (49)$$

equation (49). The alternative process (corresponding to equation 2) is

$$R^1CO_2R^2 + R^3COHal \rightleftharpoons R^1CO_2COR^3 + R^2Hal \qquad (50)$$

equation (50). These processes have been little studied. Benzoyl chloride and phenyl acetate, with catalysis by zinc chloride, do yield[29] phenyl benzoate and acetyl chloride (equation 51). However, the nature of the

$$PhCOCl + CH_3COOPh \xrightarrow{ZnCl_2} PhCOOPh + CH_3COCl \qquad (51)$$

catalysis and the fact that acylation by R^3COHal will follow path (49) rather than (50) when R^2 cannot easily depart as a positive ion, suggest that here again the primary attack is on oxygen rather than on halogen.

D. Reaction with Inorganic Halogen Compounds

The most commonly used inorganic halides are thionyl chloride, phosphorus pentachloride, and phosphorus trichloride. We shall discuss them together.

1. Carboxylic acids

One of the oldest and most used methods[30] of preparing acyl chlorides* is to heat the free carboxylic acid or its alkali metal salt with thionyl chloride, phosphorus pentachloride, or phosphorus trichloride. The choice of inorganic component is usually influenced by the boiling point of the organic product, for with phosphorus pentachloride phosphorus oxychloride (b.p. 110°) is also produced, whereas with the other two inorganic compounds the by-products do not lead to potential complications of this kind.

Little is known definitely about the mechanisms of these reactions, but what is known concerning the reactions between thionyl chloride and alcohols is probably of relevance. Optically active aliphatic alcohols can react with thionyl chloride via a process which is first-order in both alcohol and thionyl chloride, but which does not lead to inversion[31]. A likely

---

* Acyl fluorides, bromides and iodides are most usually made in ways which do not employ acids or esters[1, 30].

scheme is given in equation (52). If the

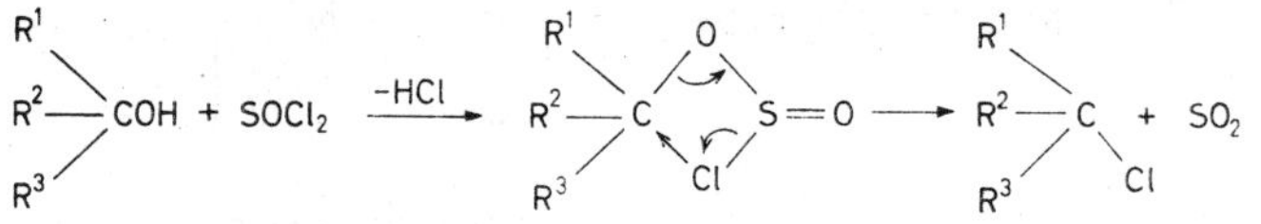

$$R^2 \overset{R^1}{\underset{R^3}{\diagdown}} COH + SOCl_2 \xrightarrow{-HCl} R^2 \overset{R^1}{\underset{R^3}{\diagdown}} C \overset{O}{\underset{Cl}{\diagup}} S=O \longrightarrow R^2 \overset{R^1}{\underset{R^3}{\diagdown}} C \overset{}{\underset{Cl}{\diagdown}} + SO_2 \quad (52)$$

reaction with acids takes a similar course it would correspond to equation (53). This type of process is an internal nucleophilic substitution by

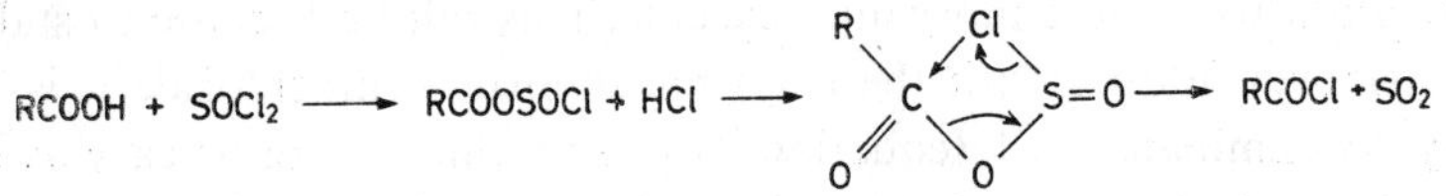

$$RCOOH + SOCl_2 \longrightarrow RCOOSOCl + HCl \longrightarrow \overset{R}{\underset{O}{\diagdown}} C \overset{Cl}{\underset{O}{\diagup}} S=O \longrightarrow RCOCl + SO_2 \quad (53)$$

halogen at the carbonyl carbon atom, and is very similar to schemes (35) and (41). Analogous equations can be written for the phosphorus penta-chloride and phosphorus trichloride reactions, *e.g.* equation (54). Like (41)

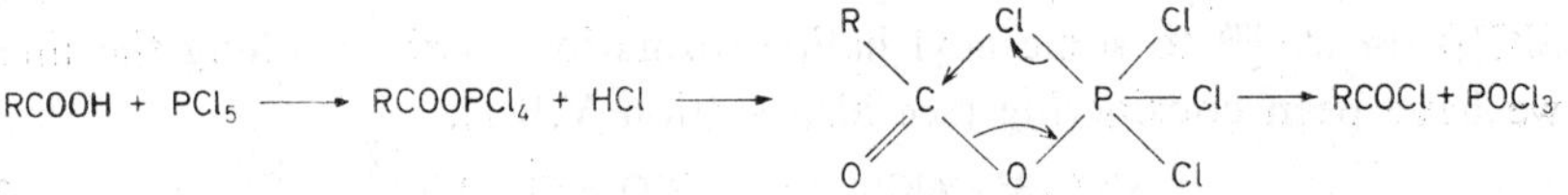

$$RCOOH + PCl_5 \longrightarrow RCOOPCl_4 + HCl \longrightarrow \overset{R}{\underset{O}{\diagdown}} C \overset{Cl}{\underset{O}{\diagup}} P \overset{Cl}{\underset{Cl}{\diagdown}} Cl \longrightarrow RCOCl + POCl_3$$

$$(54)$$

processes (53) and (54) involve an initial reaction of the carboxylic acid via scheme (2), *i.e.* as an oxygen nucleophile. This produces an ester containing a halogen atom so situated that a (kinetically favoured) intra-molecular attack on the carbonyl carbon atom can take place. Intramolec-ular processes are favoured, compared with the corresponding inter-molecular reactions, owing to the artificially enhanced collision numbers (sometimes called the proximity effect) and often the good steric position-ing (orientation effect) they involve[32]. It will have become clear from the various examples given above that unless a halogen atom can achieve enhanced nucleophilicity in some such way then it will not—whatever its origin—normally attack the carbonyl carbon atom of acids and esters successfully, and reaction will follow course (2) only, rather than (1). The postulation of equations (53) and (54) for these carboxylic acid halo-genations—which involve an initial nucleophilic role for the acid—is supported by the facts that the acid salt is often used rather than the free acid (the free anion is then available), and that phosphorus halides are known to behave as (weak) Lewis acids, so that the details of the first step could be akin to equation (55).

$$RCOO^- + PCl_5 + Na^+ \rightleftharpoons \left[\begin{array}{c} RC-O \rightarrow P\begin{array}{c} Cl \\ \diagdown Cl \\ \diagup Cl \\ \diagdown Cl \end{array} \\ \parallel \quad\quad | \\ O \quad\quad Cl \end{array} \right]^- + Na^+$$

$$\downarrow$$

2. Carboxylic esters

$$RCOOPCl_4 + NaCl \qquad (55)$$

Esters, as would be anticipated if the processes described in section II.D.1 above have been correctly represented, do not normally undergo halogenation with these inorganic reagents. This will be because to stabilize the (loose) adduct with the (say) phosphorus pentachloride it is necessary to eliminate RCl (equation 56), and this is not as easy as the elimination of HCl (or NaCl) with acids or their salts (equation 55).

$$\begin{array}{c} R^2 \\ | \\ R^1C-O : PCl_5 \quad \diagup\!\!\!\!\diagup\!\!\!\!\rightarrow \quad R^1C-O-PCl_4 + R^2Cl \\ \parallel \qquad\qquad\qquad\qquad \parallel \\ O \qquad\qquad\qquad\qquad O \end{array} \qquad (56)$$

However, powerful and/or easily solvolysed Lewis acids (*e.g.* $AlCl_3$, $TiCl_4$, $SiCl_4$) do lead[33] to some acyl halide formation, perhaps along the lines we have been considering (see also section V.B.1).

$$R^1CO_2R^2 + AlCl_3 \longrightarrow R^1CO_2AlCl_2 + R^2Cl \qquad (57)$$

$$\begin{array}{c} R^1 \qquad Cl \\ \diagdown \quad \diagup \\ C \qquad Al-Cl \quad \longrightarrow \quad R^1COCl + AlOCl \\ \diagup\!\!\parallel \quad \diagdown\!\!\diagup \\ O \qquad O \end{array} \qquad (58)$$

It is clear that the general reluctance of carboxylic esters to yield acyl halides is primarily due to their normal disinclination, compared to carboxylic acids, to undergo reactions of type (2), which as we have seen, can, in appropriate circumstances, lead to products which can undergo halogenation either by an intramolecular route or in other ways (*e.g.* equations 45 and 46).

Certain activated esters (*e.g.* diethyl oxalate and ethyl acetoacetate) can react with phosphorus pentachloride as in equation (59). The extent to which processes like (59) could, or do, operate with carboxylic acids is not clear[34].

$$\begin{array}{c} COOEt \\ | \quad\quad + PCl_5 \longrightarrow \\ COOEt \end{array} \quad \begin{array}{c} COOEt \\ | \quad\quad + POCl_3 \\ C(Cl)_2OEt \end{array}$$

$$\downarrow$$

$$\begin{array}{c} COOEt \\ | \quad\quad + EtCl \\ COCl \end{array} \qquad (59)$$

III. SUBSTITUTION BY VARIOUSLY BOUND OXYGEN AND SULPHUR ATOMS

A. Introduction

As we have seen, halogen atoms are normally not sufficiently nucleophilic to attack the carbonyl carbon atom of acids and esters directly. If they can gain a location in a suitable adjacent group, then the kinetically easier intramolecular substitution by them may be possible (section II.D.1 above). It seems unlikely that schemes like (35), (41), (53), or (54) will involve much disturbance of the carbonyl group of the acid or ester; in particular it does not seem necessary to invoke any temporary opening of the carbonyl group along lines shown in equation (26). Oxygen (and sulphur) atoms are in all contexts a good deal more nucleophilic than halogen atoms and there is, of course, much evidence that they can engage in successful, direct intermolecular attack on the carbonyl carbon atom of esters and acids—as in hydrolysis or alcoholysis. It is also to oxygen nucleophiles that the major part of the evidence for carbonyl addition intermediates (equation 26) in acylation pertains[9-11]. In discussing substitution by oxygen we are dealing with one of the most thoroughly studied aspects of the reactions of species RCOY and we shall find that the mechanisms involved serve as a pattern for most other nucleophiles apart from halogen[1]. Indeed the two fundamental mechanisms, given in outline in equations (24) and (25), sprang originally from the accumulated work in this field, so extensive because of the past concentration of interest on hydroxylic, and especially aqueous, media as solvents. For esters and acids the vast majority of the data referring to oxygen nucleophiles therefore concerns hydrolysis (equations 60 and 61), alcoholysis (equation 62), esterification (equation 63), and closely related processes, such as the intramolecular counterparts of equations (60) and (63) — the hydrolysis and

$$R^1CO_2R^2 + H_2O \longrightarrow R^1COOH + R^2OH \tag{60}$$

$$RCOOH + H_2^{18}O \longrightarrow RCO^{18}OH + H_2O \tag{61}$$

$$R^1CO_2R^2 + R^3OH \longrightarrow R^1COOR^3 + R^2OH \tag{62}$$

$$R^1COOH + R^2OH \longrightarrow R^1COOR^2 + H_2O \tag{63}$$

formation of lactones. For carboxylic acids the hydrolysis (61) is usually termed oxygen exchange because it may be followed by isotopic labelling. The equilibrium positions of reactions (60)–(63) — and this is true of the reactions of esters and acids with most nucleophiles — are not far to one side or to the other, but depend in general upon the nature of the substituents R^1, R^2, and R^3, and upon the relative concentrations of the reactants. Since processes (60)–(63) are dealt with separately in other chapters, we shall content ourselves with outlining the general features of these reactions in so far as these are relevant to similar substitutions. In particular the effects of catalysts are well exemplified in these reactions of oxygen nucleophiles.

In water, a medium in which ion pairs are largely dissociated, step (24b) will usually be fast, so that hydrolysis either involves a slow ionization (24a) or a rate-influencing attack by the substrate oxygen atom (equation 25). These possibilities exist for both spontaneous and catalysed reactions. The ionisation route is favoured by substituents R^1 which help to stabilize the acylium ion (*e.g.* *p*-methoxyphenyl) or hinder path (25) by blocking access to the carbonyl carbon atom (*e.g.* 2,6-dimethylphenyl). Path (25) is usually favoured by substituents which withdraw electrons from the carbonyl carbon atom[35] (compare section I.B.4). If the solvent is an alcohol, or some inert medium of low dielectric constant, this will tend to favour the occurrence of (25) rather than (24).

So far as the spontaneous reactions of esters and acids are concerned it is considered[31, 35-37] that most react via path (25). The evidence for this is not always unequivocal owing to the fact that when the substrate (*e.g.* water or alcohol) also serves as the solvent it is not possible to determine the molecularity in this reactant from the kinetic data. As shown in some of the examples given below, other lines of evidence can, however, sometimes be brought to bear on the problem. It is, in fact, probable that even under acid catalysis, which can aid route (24) via protonation as in equation (64) (*cf.* equation 5), only rather specially constituted acids and

$$RCO_2H + H^+ \longrightarrow RCO_2H_2^+ \longrightarrow RCO^+ + H_2O \tag{64}$$

esters for which route (25) is inhibited[37] (*e.g.* esters of 2,6-dimethyl- or 2,6-di-*t*-butylbenzoic acid) follow path (24). The usual route for Brønsted acid catalysis is probably[38] a modification of equation (25), given in its most elaborate form in equations (65) and (66). Whether or not the

$$\underset{\substack{\text{O}\\ \parallel}}{R^1C}-OR^2 + H^+ \rightleftharpoons \underset{\substack{+OH\\ \parallel}}{R^1C}-OR^2 \tag{65}$$

$$\underset{\substack{+OH\\ \parallel}}{R^1C}-OR^2 + H_2O \rightleftharpoons \underset{\substack{+OH_2\\ \mid \\ OH}}{R^1C}-OR^2 \rightleftharpoons \underset{\substack{OH\\ \mid \\ OH}}{R^1C}-OR^2 + H^+ \rightleftharpoons \underset{\substack{OH\ H\\ \mid \\ OH}}{R^1C}-\overset{+}{O}-R^2 \longrightarrow$$

$$R^1CO_2H + R^2OH + H^+ \tag{66}$$

successive protonation of both the carbonyl group and of the leaving group are always involved (as in equations 65–66) as essential features is not certain[38]. However, as we have long argued[1, 39], there is much evidence to suggest that the successful engagement of the leaving group is the crucial factor for the operation of acid catalysts, while it is by no means clear that the occurrence of carbonyl addition under such conditions plays a part in the acylation (see below).

Basic catalysts (*cf.* section I.B.3) either (i) temporarily take the place of the nucleophile as in equations (67)–(69), or (ii) increase its nucleophilicity as in equation (70). Both these schemes are, of course, variations on the general route (25) and can easily be elaborated to include a carbonyl addition intermediate where this is justified by the evidence[4]. Base-catalysed

$$R^1CO_2R^2 + R^3CO_2^- \rightleftharpoons R^1COOCOR^3 + R^2O^- \tag{67}$$

$$R^1COOCOR^3 + H_2O \longrightarrow R^1COOH + R^3COOH \tag{68}$$

$$R^3COOH + R^2O^- \rightleftharpoons R^2OH + R^3CO_2^- \tag{69}$$

$$R^1CO_2R^2 + R^3CO_2^- + H_2O \longrightarrow \underset{\substack{| \\ H}}{\overset{\substack{O \\ \| \\ R^1C-OR^2 \\ \uparrow}}{H-O}} \underset{\substack{\| \\ O}}{\overset{\substack{\frown \\ \overset{-}{O}CR^3}}{}} \longrightarrow R^1CO_2H + R^2OH + R^3CO_2^- \tag{70}$$

examples of hydrolysis via equation (24) are rare. This is because the base will need to be very weak (*e.g.* Cl$^-$) if the intermediate it produces is to have a chance of ionizing as in (24). This very weakness precludes the formation of the intermediate.

In appropriate cases both acidic and basic catalysis may occur together[40]. And in suitable reactants acidic and basic groups neighbouring on the reaction centre may provide intramolecular catalysis[4].

Many, indeed most, 'spontaneous' hydrolyses (and alcoholyses) will involve catalysis by other molecules of water (or alcohol) acting as acids or bases[41]. This is the probable reason for the greater reactivity of associated, compared to monomeric, alcohol species in dilute solution in aprotic solvents[42].

Below we illustrate some of these general principles in a discussion of those reactions between oxygen nucleophiles and carboxylic acids which are not dealt with in, or are only of marginal interest to, the chapters which are concerned mainly with esterification and ester solvolysis. The discussion also exemplifies some of the extra problems involved in quantitative studies of reactions of carboxylic acids, compared with those of esters.

B. Reaction with water

Interaction between carboxylic acids and water leads—apart from the rapid exchange of hydroxylic hydrogen atoms—to oxygen exchange (equations 61 and 71), the analogue of the hydrolysis undergone by esters.

$$RCO^{18}OH + H_2O \rightleftharpoons RCOOH + H_2^{18}O \tag{71}$$

The corresponding reaction for thiol acids is (72).

$$RCOSH + H_2O \rightleftharpoons RCOOH + H_2S \tag{72}$$

A little kinetic information in available about processes like (71) and (72). The majority of the data refer to oxygen exchange[10], but comparisons of the mechanisms and of the reactivity for corresponding sulphur and oxygen compounds can be made in two cases[43]. (The only other similar comparisons available concern the hydrolysis[44] and aminolysis[45] of oxygen and thiol esters—see section IV.A.2). Virtually all the data[10, 43] refer to acetic and benzoic acids and their derivatives. Like most acetates and benzoates, the free acids react rather slowly with water at 25° in the absence of catalysts, and processes (71) and (72) are more conveniently studied at 70° or above. All the available kinetic work refers to reactions with a large excess of water (acting as solvent) so that effectively only the forward steps of (71) and (72) were involved, and the molecularity in water in these steps was not discovered. One difficulty of working with carboxylic acids is that, except at high or low pH, the acid exists in solution as a mixture of the undissociated and anionic forms. These species will have different reactivities towards hydrolysis. If therefore the pH changes during the reaction—as it does in (72), the products leading to a lower pH than the reactants—the ratio $[RCOSH]/[RCOS^-]$ will change, and a simple first-order dependence on unreacted acid will not be found. The use of buffer solutions helps to overcome this difficulty, and also to demonstrate[43] the presence or absence of self-catalysis by either undissociated acid or anion in both reactions (71) and (72). In the event it is found (i) that the anion ($RCOS^-$ or RCO_2^-) is generally considerably less readily hydrolysed than the corresponding undissociated acid, (ii) that the reaction of the undissociated acid is significantly catalysed by hydroxonium ion, by hydroxyl ion, and by anions like RCO_2^-, (iii) that (except for mesitoic acid) the acid-catalysed rates parallel the hydroxonium ion concentration more closely than H_0, and exhibit very similar Arrhenius parameters for both the oxygen exchange and the thiol acid reactions, (iv) that benzoic acids in general react less rapidly than acetic acids, and that electron-withdrawing substituents in both favour reaction in water and in alkaline solution, (v) that the sulphur compounds exhibit a greater rate of hydrolysis than their oxygen analogues, (vi) that the sulphur analogues are less susceptible to acid catalysis, and (vii) that the acid-catalysed process exhibits a positive salt effect for the sulphur compounds but a negative salt effect for the oxygen compounds. These results may all be used to illustrate our generalized expectations (sections I and III.A above).

Firstly, a level of reactivity similar to that for structurally related esters is sensible. Secondly, the lower reactivity of benzoic compared to acetic acid argues against an ionization route (24) for the spontaneous reaction, and therefore for a scheme based on (25). This conclusion is supported by the effects of electron-withdrawing substituents. Indeed the effects of substituents[10] on the 'spontaneous' exchange rate of benzoic acid appear entirely compatible with such a scheme, but it is unfortunate that no proper kinetic analysis was undertaken to discover to what extent self-catalysis (different for each acid derivative) is involved in these 'spontaneous' rates. Their meaning is therefore somewhat ambiguous. All the data determined at a single concentration of mineral acid catalyst appear completely ambiguous, for in this case it is not known what proportion of the rate refers to the spontaneous, and what to the acid-catalysed path, and the effects of substituents on these two processes may be different (section I.B.4). Thirdly, the lower reactivity of the anion (compared with the undissociated acid) is sensible because the former species will be resistant to nucleophilic attack. This feature of the reactions of carboxylic acids is important because basic catalysts will tend to lead to the anion and thereby to negate their catalytic effect by simultaneously reducing both their own concentration and that of the more susceptible undissociated acid. It follows (i) that basic catalysis is not often of practical value in the reactions of carboxylic acids with nucleophiles and (ii) that for strong nucleophiles reaction is much easier (and less complicated kinetically) with the structurally related esters (see section IV). Fourthly, the close dependence of the acid-catalysed rate on the hydroxonium ion concentration and the actual values of the Arrhenius parameters for this route (similar for both exchange and thiol acid hydrolysis) indicate, in the light of much other related data on ester hydrolysis and similar reactions[1, 14, 46], an acid-catalysed path based on equation (25)—perhaps equations (73)–(75). There appears no need to invoke the significant participation of

$$RCOSH + H_3O^+ \rightleftharpoons RCOSH_2^+ + H_2O \qquad \text{fast} \qquad (73)$$

$$RCOSH_2^+ + H_2O \longrightarrow RCO_2H_2^+ + H_2S \qquad \text{slow} \qquad (74)$$

$$RCO_2H_2^+ + H_2O \rightleftharpoons RCO_2H + H_3O^+ \qquad \text{fast} \qquad (75)$$

carbonyl addition even for the oxygen exchange (the carbonyl and hydro-xylic oxygen become equivalent because of the acid–anion equilibrium) but it seems likely, and cannot be ruled out. Fifthly, the lower susceptibility of the thiol acids to acid catalysis reflects the lower basicity of their leaving group compared to that of the oxygen analogues. It is clear that proto-nation by catalyst at the leaving group is important (*cf.* section A above). The same effect is found for acid-catalysed thiol ester hydrolysis[44]. Sixthly, the greater reactivity of the thiol compounds in the spontaneous reactions reflects again the lower basicity of SH^- compared with OH^-. Seventhly, the fact that the salt effects on the acid-catalysed paths are in opposite senses for the oxygen and sulphur compounds indicates that the finer details of scheme (73)–(75) cannot be identical for the two sorts of com-pound. This may be understood in the following way[43]. Added salt in-creases the tendency of the medium to protonate either substrate ($-H_0$ increases) but renders molecular water less available to hydrogen bond to the leaving group of, and for nucleophilic attack on, the protonated sub-strate[47]. Once protonated (on sulphur) the thiol acid possesses a very good leaving group and the second (slow) step may be rather insensitive to the availability of additional hydrogen bonding. This is unlikely to be so true for the oxygen analogues where the effect of salt on the second step may outweigh its effect on the protonation equilibrium. Thus salt could aid the reaction of the sulphur compound but hinder that of the oxygen com-pound, as found. This type of mechanistic fine structure illustrates the catalytic role of molecules of water additional to the one undergoing acylation. The details of the spontaneous reactions which may be written formally as in equation (76) or (77) are undoubtedly much more com-

$$RCOSH + H_2O \longrightarrow H_2O \overset{\delta+}{\cdots} \overset{\overset{O}{\parallel}}{\underset{\underset{R}{\mid}}{C}} \overset{\delta-}{\cdots} SH \longrightarrow RCO_2H_2^+ + SH^- \longrightarrow RCO_2H + H_2S \quad (76)$$

$$RCOSH + H_2O \rightleftharpoons \underset{\underset{\overset{+}{OH_2}}{\mid}}{\overset{\overset{SH}{\mid}}{RC}} - O^- \rightleftharpoons \underset{\underset{OH}{\mid}}{\overset{\overset{\overset{+}{SH_2}}{\mid}}{RC}} - O^- \rightleftharpoons RCO_2H + H_2S \quad (77)$$

(4)

plicated in reality: it is possible that in some cases cyclic transition states are involved[48] (*e.g.* **4**). Finally, although polysubstituted thiol benzoic acids have not been studied, it is likely that both thiol acid hydrolysis and oxygen exchange for the 2,6-dialkyl- and 2,4,6-trialkyl-benzoic derivatives will exhibit an ionization type of mechanism (equation 24), at least under acidic catalysis. This follows from the well-known ester synthesis for hindered benzoic acids[37]. Such acids, resistant in their normal state to nucleophilic attack by alcohols, react when sufficient concentrated acid—usually sulphuric acid—is present to convert them to the corresponding acylium ions, ions which are both more reactive and less sterically encumbered. In a similar way mesityl esters undergo hydrolysis via an ionization route in strong acid[14]. The oxygen exchange data available for mesitoic acid bear a considerable resemblance to those for mesityl ester hydrolysis (dependence of rate on H_0 and not on $[H_3O^+]$, and a positive entropy of activation) and this acid doubtless[49] reacts with water via a route based on equation (24)—perhaps via equations (78) and (79).

$$\text{Me}\!-\!\underset{\underset{\text{Me}}{|}}{\overset{\overset{\text{Me}}{|}}{C_6H_2}}\!-\!CO^{18}OH + H_3O^+ \underset{}{\overset{fast}{\rightleftharpoons}} \text{Me}\!-\!\underset{\underset{\text{Me}}{|}}{\overset{\overset{\text{Me}}{|}}{C_6H_2}}\!-\!\overset{+}{C}O^{18}OH_2 + H_2O \overset{slow}{\longrightarrow} \text{Me}\!-\!\underset{\underset{\text{Me}}{|}}{\overset{\overset{\text{Me}}{|}}{C_6H_2}}\!-\!CO^+ + H_2{}^{18}O + H_2O$$

(78)

$$\text{Me}\!-\!\underset{\underset{\text{Me}}{|}}{\overset{\overset{\text{Me}}{|}}{C_6H_2}}\!-\!CO^+ + 2\,H_2O \overset{fast}{\longrightarrow} \text{Me}\!-\!\underset{\underset{\text{Me}}{|}}{\overset{\overset{\text{Me}}{|}}{C_6H_2}}\!-\!COOH + H_3O^+$$

(79)

C. Reaction with Hydrogen Peroxide

Interaction between hydrogen peroxide and a carboxylic acid leads to the peroxy acid (equation 80). A stronger acid, a weaker base[50], and

$$RCO_2H + H_2O_2 \underset{k_{-1}}{\overset{k_1}{\rightleftharpoons}} RCO_3H + H_2O; \qquad K_1 = \frac{k_1}{k_{-1}} \tag{80}$$

therefore probably a weaker nucleophile than water, unionized hydrogen peroxide may be correspondingly more difficult to acylate. Thus, in view of the slowness of their spontaneous reactions with water (equation 71, section B above) most carboxylic acids will be expected to react very sluggishly with hydrogen peroxide in the absence of catalysts. This may well be the case[51]. With acetic and the higher aliphatic acids the equilibrium constant, K_1, for equation (80) is not very dependent on structure[34]. Various values, lying between *ca.* 2 and *ca.* 5 at 25°, have been reported. The actual value observed depends upon the concentration of mineral acid catalyst used, for this tends to preferentially engage the free water and so to pull the equilibrium somewhat to the right[34, 51]. The equilibrium position is considerably more favourable with formic acid and here provides a practicable preparative route to the peroxy derivative[34]. Excess of hydrogen peroxide is used, at a temperature below 5°, with a little sulphuric acid as catalyst. More powerful acylating agents than the free carboxylic acid (*e.g.* the acyl halide) are normally employed with hydrogen peroxide to make the higher aliphatic and aromatic peracids[30, 52].

Comparisons of reactivity concerning hydrogen peroxide systems are potentially complicated owing to the unexpectedly great nucleophilicity of HOO^- compared with OH^-. The former species is the weaker base but reacts much the more easily in various nucleophilic substitutions[53]. One explanation[54] of this effect is that hydrogen bonding (possibly intramolecular and catalytic*) is of greater importance for the peroxide ion than for the hydroxyl ion (*cf*. section 1V.C.2). It may be that hydrogen peroxide itself enjoys a similar type of enhanced reactivity compared with water. The rate of the reaction between *p*-nitrophenyl acetate and hydrogen peroxide in aqueous solution probably greatly exceeds the rate of the hydrolysis of this ester[2]. However, few details about this comparison are available, and whether reaction involves the hydrogen peroxide molecule and/or its anion is uncertain. In general, esters appear to have been very little studied with hydrogen peroxide[55].

Quite recently the reaction with carboxylic acids (equation 80) has been examined, from both the equilibrium and kinetic viewpoints, using dilute solutions in dioxan[51]. Measureable rates were observed at room temperatures, using $\geqslant 0.2 \, M$ sulphuric acid as catalyst. As in solventless systems, K_1 (at 25°, $0.2 \, M \, H_2SO_4$) varies little in the series $R = Me$, Et, Pr ($K_1 \simeq 0.8$) falling slightly for *t*-Bu ($K_1 = 0.6$). When $R = CH_2Cl$ or CH_2OMe, $K_1 \simeq 0.2$.

The rate of establishment of the equilibrium increases with increasing catalyst concentration; but at a given catalyst concentration, the observed kinetics correspond closely to those expected for the two opposing second-order processes depicted in the stoicheiometric equation (80). It is very probable that, under the conditions used, the peroxide nucleophile is H_2O_2 and not HOO^-. Over the series Me, Et, Pr, *t*-Bu the second-order rate constant k_1 falls by a factor of *ca*.4. Since K_1 is roughly constant it follows that k_{-1} must fall by about the same factor. The Arrhenius parameters

---

* Catalytic effects will not, of course, be reflected in measures of (equilibrium) basicity.

closely resemble those common in the ordinary esterification of acids, and it therefore comes as no surprise (section I.B.4.b) that the substituent effects noted above for the acid-catalysed rate k_1, correlate well with Taft's steric parameter E_s. Polar effects seem quite unimportant, and this is true for k_1 even when $R = CH_2Cl$ or CH_2OMe. The small value of K_1 in these latter cases is due to enhanced k_{-1} values, possibly attributable to the existence of intramolecular hydrogen-bonded structures (5) for the relevant peroxy acids—structures not so readily formed with the ordinary acid—which facilitate nucleophilic attack.

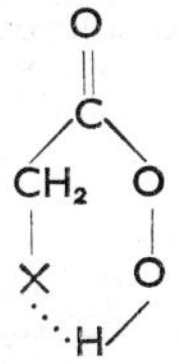

(5)

As for the reactions of H_2O and H_2S with carboxylic acids discussed in section B above, it is likely that the essential mechanism of these acid-catalysed H_2O_2 reactions closely resembles that of acid-catalysed ester hydrolysis based on equation (25)—cf. equations (65)–(66) and (73)–(75). A kinetic study of base-catalysed peroxy acid formation would be interesting.

D. Reaction with other Carboxylic Acids

Processes like (81), which lead to an anhydride, appear rare. The equilibrium positions will be well to the left but could be drawn to the right by effective removal of water. Phosphorus pentoxide can be used for

$$R^1CO_2H + R^2CO_2H \rightleftharpoons R^1COOCOR^2 + H_2O \qquad (81)$$

this purpose but the total extent of its role is uncertain[34, 56]. A free acid can also be converted to the corresponding anhydride via reaction (82) with an alkoxyacetylene[57]. It is probable that an intermediate 2:1 adduct is formed, via successive addition, which splits out the anhydride in an

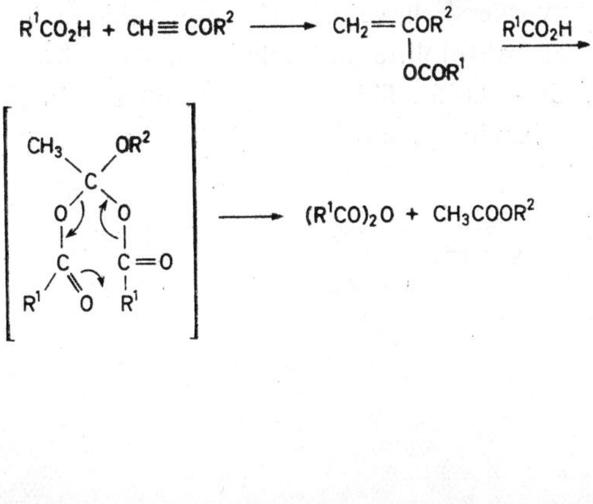

(82)

intramolecular decomposition[16, 58] (*cf.* equations 35, 37 and 83). This is a somewhat roundabout, but effective, route to substitution of OH^- by RCO_2^-. The first stage of addition leads to the enol ester, which is definitely an intermediate in the overall scheme. The actual acylation step in equation (82) involves attack of one ester group on another and is most easily written as the attack of one keto group on the other, this being facilitated in the cyclic transition state. (Normally ketones are not acylated at oxygen unless the enolate ion is formed; *cf.* section V.D.2.b). A similar cyclic transition state permits the facile re-arrangement[59] (83). As mentioned in section I.D. above (and as re-em-

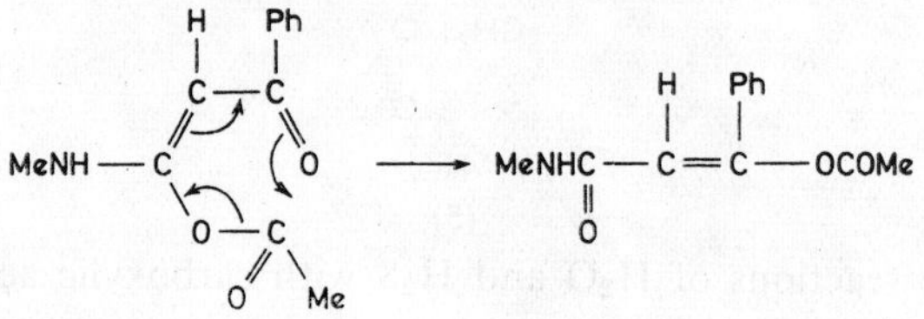

$$
\text{(83)}
$$

phasized in the next section) the best chance of observing inherently un-favourable substitutions is in an intramolecular context. The success of schemes (82) and (83) is in line with this. It is especially interesting that there is evidence[60] that in aqueous solution the aminolysis of appropriate dicarboxylic acids can proceed via a very small concentration of the anhydride as intermediate (equation 84). For succinic acid, the concentration of anhydride intermediate is approximately 3 parts/10^5

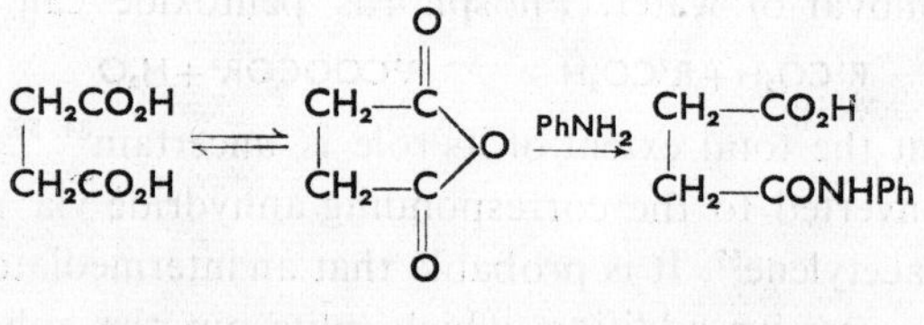

$$
\text{(84)}
$$

of acid.

Higuchi and coworkers[61] have recently shown that such anhydride formation is catalysed by sulphite and sulphonium salt formation, perhaps as in equations (85) and (86). This is very reasonable in view of the improvements in the leaving group involved.

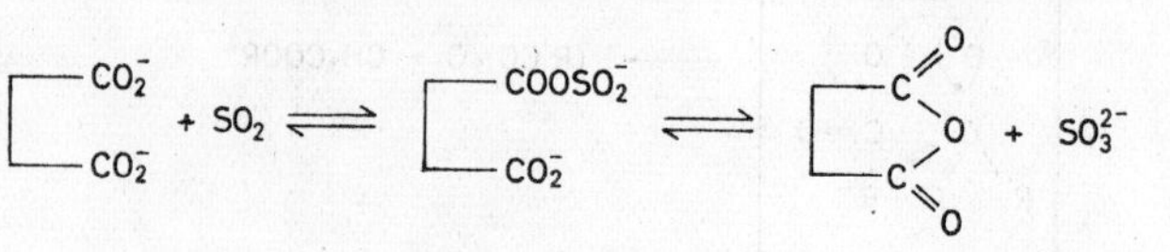

$$
\text{(85)}
$$

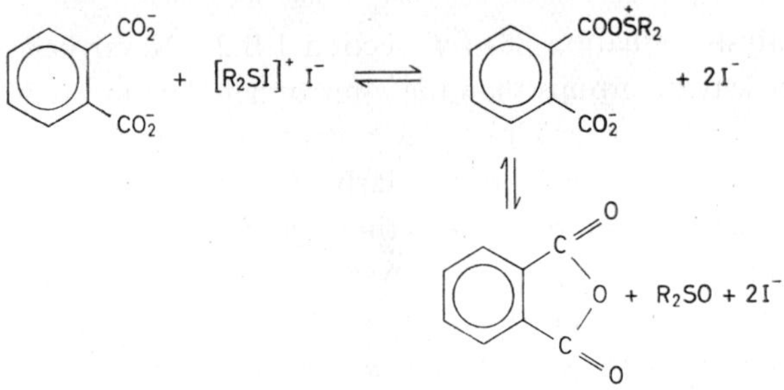

$$
\begin{array}{c}
\text{(structure)} + [R_2SI]^+\ I^- \rightleftharpoons \text{(structure)} + 2I^-
\end{array}
$$

$$
\Updownarrow \qquad\qquad (86)
$$

(structure) + R$_2$SO + 2I$^-$

The acylation of carboxylic acids by esters (equation 87) is, like reaction (81), not very common. It is clear, however, that the process is

$$R^1COOR^2 + R^3COOH \rightleftharpoons R^1COOCOR^3 + R^2OH \qquad (87)$$

likely to be greatly influenced by the nature of R^2: if R^{2}O$^-$ is a good leaving group, if that is, R^{2}OH is a reasonably strong Brønsted acid, then we can expect more success than when R^{2}OH is of a somewhat similar acidic nature to H$_2$O, as in common esters or in carboxylic acids. Moreover reaction (87) is more susceptible to basic catalysis than is (81), because whereas *both* carboxylic acids will tend to undergo conversion to the anion in (81)—and thus any increased nucleophilicity for one reagent will be cancelled out by the loss in electrophilicity of the other—in equation (87) bases will provide some R^{3}CO$_2^-$ (and also R^{2}O$^-$) and so facilitate the establishment of equilibrium. Evidence for the occurrence of (87) comes mostly from kinetic studies of the catalysis of the hydrolysis of appropriate esters by carboxylate ions. Thus it was once considered[62] that the observed catalysis of the hydrolysis of *p*-nitrophenyl acetate by acetate ions proceeds via acetic anhydride (equations 88 and 89). However, experiments[63] designed to trap the supposed intermediate anhydride failed

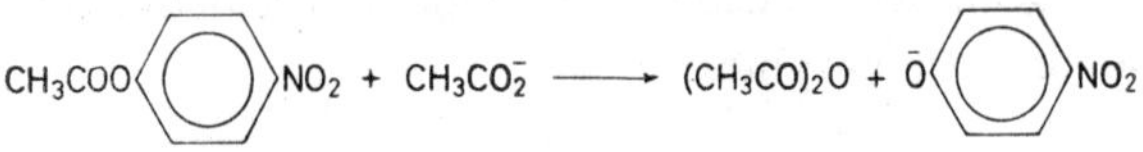

$$CH_3COO\text{(ring)}NO_2 + CH_3CO_2^- \longrightarrow (CH_3CO)_2O + \bar{O}\text{(ring)}NO_2 \qquad (88)$$

$$(CH_3CO)_2O + H_2O \longrightarrow 2\,CH_3CO_2H \qquad (89)$$

to detect any significant quantity of it, and this reaction probably involves synchronous catalysis by acetate ions (equation 90) and not the nucleo-

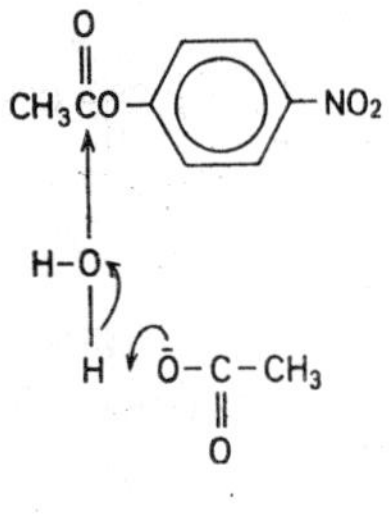

$$
\begin{array}{c}
\overset{O}{\underset{\parallel}{CH_3C}}O-\text{(ring)}-NO_2 \\[4pt]
\uparrow \\
H-O \\
| \\
H\quad O-C-CH_3 \\
\qquad\quad \parallel \\
\qquad\quad O
\end{array}
\qquad (90)
$$

philic catalysis (equation 88) (*cf.* section I.B.2). Nevertheless esters with even better leaving groups than the *p*-nitrophenolate ion (*e.g.* 2,4-dinitrophenyl esters) almost certainly[62] do undergo such nucleophilic catalysis*. This example illustrates a particularly important generalization about basic catalysis in acylation[66]: acylating agents with good leaving groups tend to undergo nucleophilic catalysis (rather than synchronous catalysis) because in the transition state appreciable bond breaking will have occurred and virtually any available nucleophile will be able to attack the carbonyl carbon atom, whereas for acylating agents with poor leaving groups bond breaking will have progressed much less in the transition state and only sufficiently powerful nucleophiles can successfully attack the carbonyl carbon atom. It will be necessary therefore to produce such nucleophiles by the method of synchronous (general base) catalysis.

For reasons already outlined (section II.D.1) all acylations are normally facilitated when the reactant groups occupy suitable relative positions in the same molecule, *i.e.* under intramolecular conditions. Good examples of intramolecular versions of equation (88) are the hydrolyses of 3,5-dinitro-aspirin (equation 91) and thioaspirin[67, 68]. Indeed species like

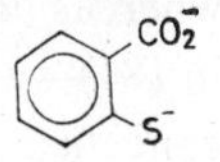

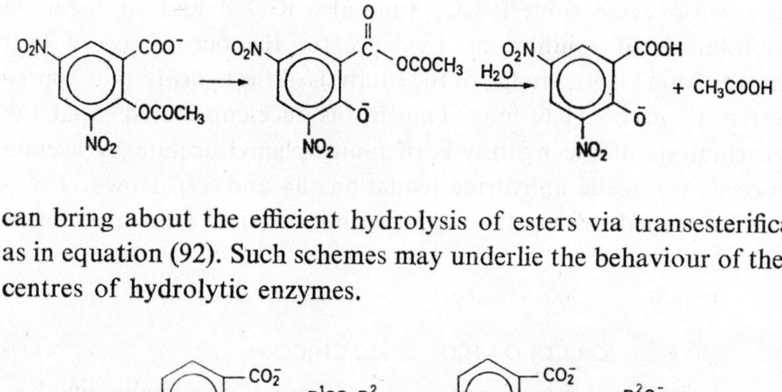

$$(91)$$

can bring about the efficient hydrolysis of esters via transesterification[68] as in equation (92). Such schemes may underlie the behaviour of the active centres of hydrolytic enzymes.

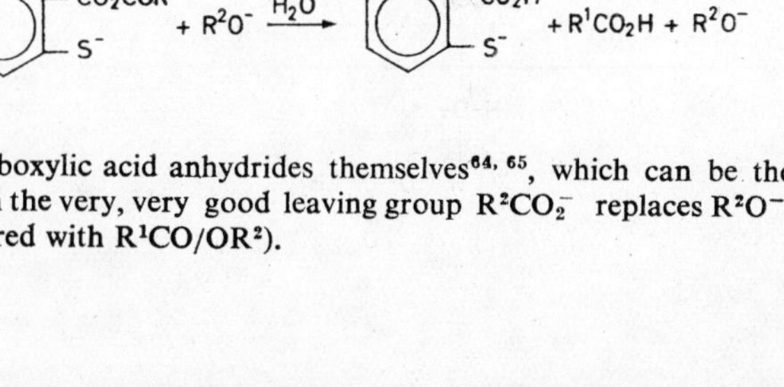

$$(92)$$

---

* As do carboxylic acid anhydrides themselves[64, 65], which can be thought of as 'esters' in which the very, very good leaving group $R^2CO_2^-$ replaces R^2O^- (*i.e.* $R^1CO/OCOR^2$ compared with R^1CO/OR^2).

Intramolecular opportunities can bring about reaction (87) even for quite poor leaving groups as in reaction (93), where $R = Me$, $-CH_2CH_2Cl$, $-CH_2C\equiv CH$, $-CH_2CF_3$, or Ph. As would be expected from our

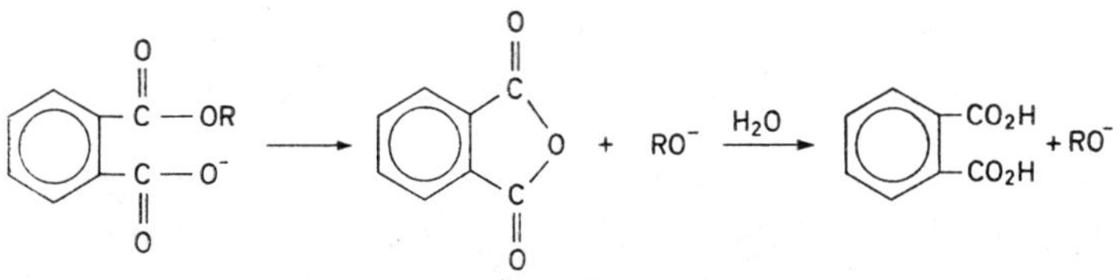

foregoing generalizations, the rate of hydrolysis increases from left to right[69] (*i.e.* it is a function of the acid strength of the conjugate acid of the leaving group, section I.B). Interestingly, acid catalysis by the adjacent carboxyl group (equation 94)—rather than nucleophilic attack as in equation (93)—is the best route for the very poor leaving groups —OMe and —OCH$_2$CH$_2$Cl. It is clear that with poor leaving groups acid-catalysed assistance of leaving group departure carries more kinetic weight than an improvement in the nucleophile.

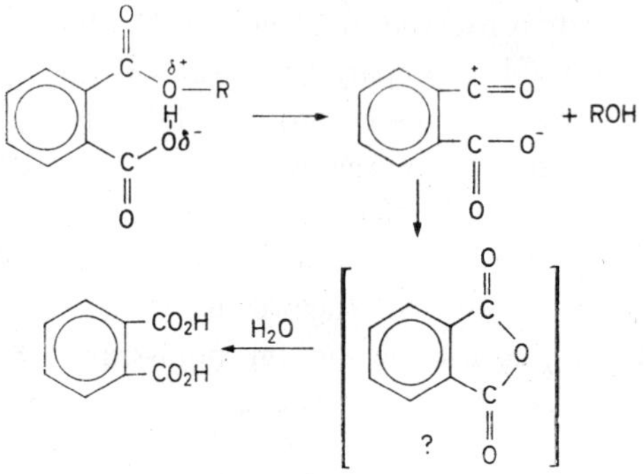

(94)

E. Reaction with Carboxylic Anhydrides

This reaction is often used to prepare uncommon anhydrides from the free acid and the readily available acetic anhydride[30, 70]. The overall process observed, after distillation of the mixture, is shown in equation (95). The equilibrium position depends on the reagents and their propor-

$$2\,RCO_2H + (CH_3CO)_2O \rightleftharpoons (RCO)_2O + 2\,CH_3CO_2H \qquad (95)$$

tions. In practice the relatively volatile acetic acid can often be distilled away as it forms. Since it is unlikely that the weaker acylating agent (RCO_2H) will lead to a dominant electrophilic attack on the stronger [($CH_3CO)_2O$] it is predictable that the overall reaction will involve a series of steps like equation (2), in this case involving nucleophilic attack of the carboxylic oxygen of RCOOH on any acyl anhydride present in the system (equations 96 and 97). What is known[71] about the rates

$$RCO_2H + (CH_3CO)_2O \rightleftharpoons RCOOCOCH_3 + CH_3CO_2H \tag{96}$$

$$RCO_2H + RCOOCOCH_3 \rightleftharpoons (RCO)_2O + CH_3CO_2H \tag{97}$$

of the individual processes making up reactions (96) and (97) for variously substituted acids, supports the concept of the free acid acting as the nucleophile. In reactions (96) and (97) the intermediate, unsymmetrical anhydrides often disproportionate at usual distillation temperatures, thus aiding the formation of a single product[72].

F. Reaction with Acyl Halides

Reaction (98) has been mentioned already (see equation 36), the point

$$R^1CO_2H + R^2COHal \rightleftharpoons R^1COOCOR^2 + HHal \tag{98}$$

there being that acylation of (equation 98), rather than by (equation 35), R^1CO_2H is primarily observed (cf. section E above on anhydrides). Scheme (98) can be used to prepare anhydrides if the hydrogen halide is effectively removed by the presence of a suitable base. To this end the reaction is often conducted in the presence of an excess of pyridine, which also serves to convert the free acid into the more nucleophilic anion[73]. A common alternative procedure, which also combines the mechanistic device of basic catalysis with the preparative device of engagement of one of the products, is to use the sodium, potassium, or silver salt of the acid. This has been perhaps the most used method for the preparation of anhydrides[30].

Since equilibria like (98) are established much more quickly than those like (96), hydrogen halides, which with anhydrides lead rapidly to acyl halides, will catalyse reaction (96), as will acids stronger than hydrogen halides. This fact is only occasionally exploited in preparations[34] of anhydrides via reaction (96).

It is emphasized that in reactions (96) and (98) the essential behaviour of the carboxylic acid is nucleophilic and that its reaction pattern is therefore based on equation (2) rather than (1), i.e. these processes really involve only substitution of H at oxygen, and not of OH at carbon. For the use of processes like (98) when hydrogen halide is not artificially removed see section II.C.1.

G. Reaction with Concentrated Sulphuric Acid

In the presence of sulphuric acid, carboxylic acids (and also esters) may react to some extent as in equation (99). If the dielectric constant is high enough (*e.g.* if excess of sulphuric acid is used) all the species on the right-hand side may be both ionized and dissociated*, HSO_4^- being a

$$RCO_2H + 2 H_2SO_4 \rightleftharpoons RCO_2SO_3H + H_3O^+ HSO_4^- \tag{99}$$

good leaving group. The equilibrium position will, of course, also depend on R, and in many systems[3] reaction will not proceed beyond some protonation of RCOOH (*cf.* equation 5). The formation of acylium ions (RCO^+) in concentrated sulphuric acid is favoured by substituents which stabilize such ions. The use of sulphuric acid induced acylium ion formation in certain esterifications and hydrolyses has been mentioned in section III. B. It is occasionally useful in the acylation of carbon compounds with carboxylic acids (section V). Other inorganic oxyacids (*e.g.* H_3PO_4) can sometimes replace sulphuric acid in schemes probably similar to equation (99), but the details in such cases appear less well understood[75].

Acyl hydrogen sulphates, $RCOOSO_3H$, (which are more easily made from the carboxylic anhydrides[76]), are powerful acylating agents even in an undissociated form (*i.e.* in media of low dielectric constant). They tend, however, to rearrange[77] to sulphocarboxylic acids (equation 100) and it is probable that a sample of free acyl hydrogen sulphate has never been

$$CH_3CO_2SO_3H \longrightarrow HSO_3CH_2CO_2H \tag{100}$$

isolated, although the corresponding metal salts are known[78]. Saturated esters[3] behave analogously to acids (equation 101) but enol esters[21] undergo sulphonation at the double bond (equation 102), this reaction therefore proceeding somewhat differently from that with hydrogen chloride (section II.A.2).

$$R^1CO_2R^2 + 2 H_2SO_4 \rightleftharpoons R^1CO_2SO_3H + R^2OH_2^+ + HSO_4^- \tag{101}$$

$$CH_2{=}C{-}OCOCH_3 + H_2SO_4 \longrightarrow \left[\begin{array}{c} HSO_3CH_2 \\ \diagdown C \diagup \\ CH_3 \diagdown OCOCH_3 \end{array} \begin{array}{c} OH \end{array} \right]$$
$$| \\ CH_3$$

$$\longrightarrow HSO_3CH_2COCH_3 + CH_3CO_2H \tag{102}$$

H. Conclusion

The reactions of carboxylic acids and esters with oxygen nucleophiles other than water or alcohols have, with certain notable exceptions, been somewhat fragmentarily studied. The data available are, however, compatible with predictions based on the more thoroughly studied topics.

---

* The sulphuric acid anions then present may actually be more complex[74].

IV. SUBSTITUTION BY VARIOUSLY BOUND NITROGEN ATOMS

The comparatively high basicity of organic nitrogen compounds[79] is well known, and in general acylation at nitrogen is comparatively easy to achieve. In many cases the acylations proceed readily even in the absence of catalysts and this is true even for such comparatively feeble reagents as carboxylic esters and acids[30, 34]. The reactions with carboxylic acids tend to be more limited in application (and have been less studied) than those with esters. This is because the acid inevitably protonates the nitrogen atom to be acylated to a greater or lesser extent. It thereby renders the attacking reagent less nucleophilic and at the same time, by producing the carboxylate ion, renders itself less electrophilic. The deactivation of the nitrogen compound in the presence of acids is also the reason why acidic catalysis is not usually employed in acylation at nitrogen, even when catalysis is necessary. Appropriate base catalysis is normally much more successful.

In the absence of strongly acidic catalysts carboxylic acids and esters do not lead to acylium ions so that acylation by them at nitrogen will almost invariably proceed via a route based on equation (25). This will become evident in the following discussion.

A. Reaction with Ammonia and Primary Amines

The overall process for both carboxylic acids and esters is amide formation (equations 103 and 104); such equilibria will be expected to

$$R^1CO_2H + R^2NH_2 \rightleftharpoons R^1CONHR^2 + H_2O \qquad (103)$$
$$R^1CO_2R^2 + R^3NH_2 \rightleftharpoons R^1CONHR^3 + R^2OH \qquad (104)$$

lie well to the right, but for acids the situation is complicated by the protonation equilibrium (equation 105) noted above, which tends to hold equation (103) on the left-hand side.

$$R^1CO_2H + R^2NH_2 \rightleftharpoons R^1CO_2^- \overset{+}{N}H_3R^2 \qquad (105)$$

1. Carboxylic acids

In spite of the occurrence of salt formation, carboxylic acids are often used to prepare amides[30, 34, 80]. The experimental procedure may simply amount to the dry distillation of the ammonium salt as, for example, in the well-known preparation of acetamide from ammonium acetate. Yields are often poor, but are improved by using an excess of the acid component which, under conditions of continuous distillation, helps to carry away the water produced. The factors ultimately controlling the yields in such processes do not appear to be clearly understood. A wide variety of acids and primary amines may be used, and thiol acids seem somewhat more satisfactory than their oxygen analogues[81]. This general method of amide synthesis is one of the oldest known.

Amide synthesis from amines and free acid is an important biochemical step in the formation of amino acids, and under physiological conditions is catalysed by enzymes (by mechanisms as yet unknown). Recent kinetic studies[82] of the reactions between carboxylic acids and ammonia, or primary amines, in dilute aqueous solution, in the pH range 10–11, have shown that the rate constant, k, for amide formation is increased by substituents which increase the strength of the acid or of the amine, and that the differential rate equation has the form of equation (106) where R^2 is H or Me, and R^1 is an aliphatic group. This equation is formally

$$d[\text{amide}]/dt = k[R^1CO_2^-][R^2NH_2] \qquad (106)$$

equivalent to (107) where K is the appropriate equilibrium constant. As noted, carboxylate ions are not likely to be very electrophilic, so that R^1CO_2H and R^2NH^- are decidedly the more promising pair of reactants,

$$d[\text{amide}]/dt = kK[R^1CO_2H][R^2NH^-] \qquad (107)$$

the substituent effects probably being compatible with either pair. The overall scheme may therefore be as shown in equations (108)–(110).

$$R^1CO_2H + OH^- \rightleftharpoons R^1CO_2^- + H_2O \qquad \text{fast} \quad (108)$$

$$R^2NH_2 + OH^- \rightleftharpoons R^2NH^- + H_2O \qquad \text{fast} \quad (109)$$

$$R^1CO_2H + R^2NH^- \longrightarrow R^2NH\cdots\overset{\overset{O}{\|}}{\underset{R^1}{C}}\cdots OH \longrightarrow R^1CONHR^2 + OH^- \qquad \text{slow} \quad (110)$$

Succinic acid[60] exhibits the above behaviour in alkaline solution (and also exhibits no significant rate enhancement due to intramolecular catalysis by the second carboxyl group), but at lower pH (3–5), when many fewer RNH^- or carboxylate ions—and little free amine—will exist, undergoes aminolysis by aniline via a pre-equilibrium formation of a *low* concentration of the anhydride (equation 111, section III.D above). The

$$(111)$$

reason that this process does not intrude at high pH is presumably the result of a less favourable pre-equilibrium position owing to the conversion of much of the succinic acid to its anion or, what is more likely, to the absence of pre-equilibrium conditions in the presence of much free amine, when the rate of the direct attack by amine, or its anion, on the acid (equations 108–110) may be faster than the rate of cyclization of the acid (k_1 in equation 111).

It will be apparent from the foregoing account that the rate of amide formation from amines and carboxylic acids in aqueous solution will be greatly affected by the pH of the system. This fact clearly has important implications for biological systems.

When forming amides of feebly basic amines it is sometimes feasible to use acidic catalysts (to activate the carboxylic acid) without totally inhibiting the acylation owing to protonation of the amine. Naturally this method tends to produce its best results with the least basic amines[83, 84]. Anhydrous phosphoric acid and solutions of boron fluoride in the carboxylic acid component have been used in this context. A kinetic and equilibrium study[84] of the acetylation of nitroanilines, at ordinary temperatures, in boron fluoride–acetic acid mixtures (20–50 mole % BF_3) shows, as expected, (i) that only the free aniline is acylated, the anilinium ion being inert, and (ii) that the less basic the free aniline the smaller its rate of acylation, but the more of it remains available for acylation. Mono- and dinitroanilines are effectively completely acylated under these conditions but 2,4,6-trinitroaniline and its N-acetyl derivative coexist at equilibrium. Boron fluoride and acetic acid form powerful Brønsted acids, e.g. BF_3AcOH and $BF_3(AcOH)_2$, which are responsible for the protonations in the system. It is suggested that the acylation is an example of route (24) being brought about by (solvated) acylium ions formed in equilibrium (112) (cf. equation 184).

$$AcOH_2^+(AcOH) + BF_3AcOH \rightleftharpoons Ac^+(AcOH)_2 + BF_3H_2O \qquad (112)$$

2. Carboxylic esters

The reaction of esters with aqueous, alcoholic, or pure liquid ammonia is a well recognized preparative route[30] to the corresponding amide (equation 113). In aqueous ammonia solution (normally used at room temperature) some concurrent base catalysed hydrolysis is inevitable and the

$$R^1CO_2R^2 + NH_3 \longrightarrow R^1CONH_2 + R^2OH \qquad (113)$$

feasibility of the procedure indicates the powerful nucleophilic character of ammonia. An alcohol is the preferred solvent for the less reactive esters, since any concurrent solvolysis leads only to another ester molecule. Very resistant esters are usually heated with alcoholic ammonia in sealed tubes, or some sodium alkoxide[85] is used as an additional* basic catalyst. Alkyl and dialkyl malonic esters are particularly resistant to aminolysis, presumably owing to steric hindrance of approach to their carbonyl carbon atoms.

In general primary amines behave similarly to ammonia, and exhibit a roughly comparable reactivity. Arylamines, particularly if containing electron-withdrawing groups (cf. section 1 above), are less reactive and often require catalysis, even with the more electrophilic esters. With very weakly basic amines concentrated mineral acid catalysts may be used (to activate the ester). Another procedure[86] is to preform the magnesium halide as in reaction (114) when the whole process may be conducted in ether. This method has recently been effectively used[87] to synthesize a wide range of thioamides from thionesters, R^1CSOR^2. The route is a form of basic catalysis for the amino-magnesium halide will be more nucleophilic than the original amine. Dialkyl malonic esters are again resistant.

$$ArNH_2 + MeMgI \longrightarrow CH_4 + ArNHMgI \qquad (114)$$

$$ArNHMgI + R^1CO_2R^2 \longrightarrow R^1CONHAr + R^2OMgI \qquad (115)$$

Of the various routes to amides, that using an ester as acylating agent is both relatively effective (compared with those based on the free carboxylic acid or another amide as acylating agent) and relatively mild (compared with those employing the anhydride or acyl halide). For these and other reasons it has proved very appropriate for the synthesis of peptides and proteins from amino acids[88]. For this purpose a number of special types of ester have been selected and developed, being so chosen as to lead to adequate reaction rates but, at the same time, to the minimum of complicating side-reactions, such as racemization of the optically active centres so important to peptide synthesis. Particularly suitable esters are those with such good leaving groups as $-SC_6H_4R$, $-OC_6H_4NO_2$, $-ON\langle\bigcirc\rangle$, $-OP(OR)_2$, $-OCO_2Et$, $-OC(=CH_2)OEt$, and other enol esters. At the present time scarcely a month passes without the report of a new type of ester suitable for peptide synthesis[89]. It is interesting that many of these esters are conveniently made from the carboxylic acid, not by a normal acylation of the corresponding hydrogen derivative of the leaving group, but via a substitution like reaction (2), e.g. reactions (116)–(118).

$$RCO_2H + ClCO_2Et \longrightarrow RCOOCO_2Et + HCl \qquad (116)$$

* The ammonia will tend to catalyse its own reaction (see below).

$$R^1CO_2H + ClP(OR^2)_2 \longrightarrow R^1COOP(OR^2)_2 + HCl \qquad (117)$$

$$RCO_2H + CH\!\equiv\!COEt \longrightarrow RCOOC(\!=\!CH_2)OEt \qquad (118)$$

While aminolysis of ordinary esters, and especially of substituted phenyl and thiol esters, has been extensively studied from a kinetic (*i.e.* mechanistic) viewpoint, such detailed knowledge about the aminolysis of esters with other special types of leaving group is still lacking.

Before turning to a discussion of the kinetics of ester aminolysis there are some further special routes to amides which deserve mention. These too have been largely developed in connexion with peptide synthesis[88]. They really involve the free acid, rather than the ester, and find an analogy[90] in process (118). The first, at present one of the most used in peptide synthesis, concerns the 1 : 1 adducts of carboxylic acids with carbodiimides, particularly dicyclohexylcarbodiimide[91]. These adducts, the evidence for which is largely circumstantial, are structurally similar to enol esters and even more reactive. They provide both intra-(B) and inter-(A) molecular acylation at nitrogen (equation 119). The intramolecular rearrangement (B) to the *N*-acylurea is prominent at ordinary temperatures, but this complicating feature can be minimized by appropriate choice of solvent (benzene, carbon tetrachloride and ethyl acetate are common) and by working at temperatures below 0°. Other acylatable species (*e.g.* alcohols and carboxylic acids) may replace R^4NH_2 in reaction (119). Hence

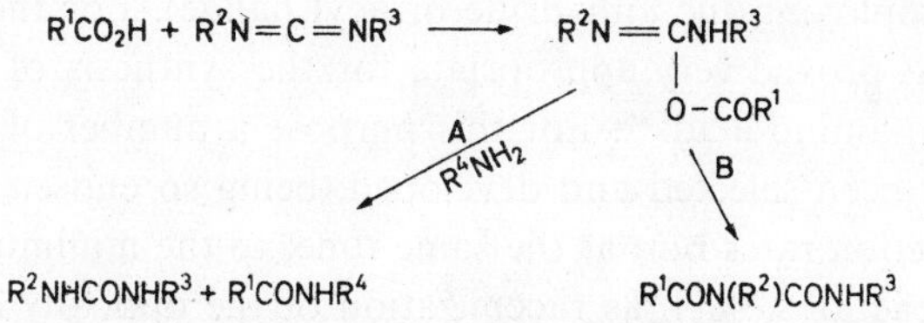

the use of excess acid leads (equation 120) to the symmetrical anhydride. The parallelism with reaction (82) suggests that (120) proceeds via species

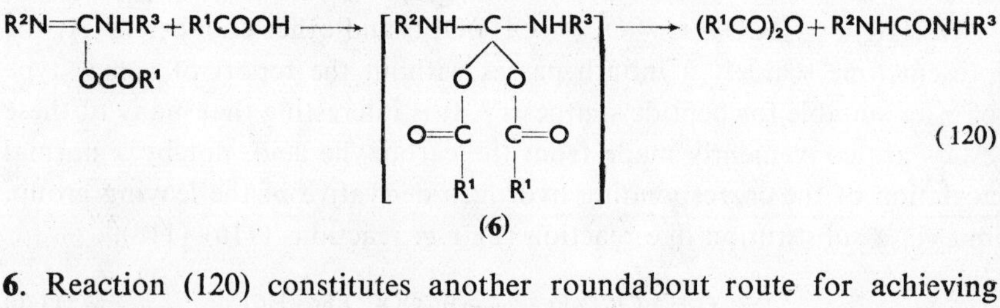

6. Reaction (120) constitutes another roundabout route for achieving the substitution of OH in RCO_2H by RCO_2 (*cf.* section II.D). Excess of R^1CO_2H is often used in aminolysis (equation 119), in order to repress

route B, and it is possible that under such conditions the acylation of the amine is, in fact, due to the anhydride. This point is still debated. A recent kinetic study[92] shows that in carbon tetrachloride or acetonitrile the reactions with acids (equation 121) is of the first-order in carbodiimide but of somewhat greater than first-order in acid. It is thought therefore that

$$R^1CO_2H + R^2N=C=NR^3 \xrightarrow{k_1} \begin{bmatrix} R^2NHC=NR^3 \\ | \\ OCOR^1 \end{bmatrix} \xrightarrow{k_3} \begin{matrix} R^2NHCONR^3 \\ | \\ COR^1 \end{matrix}$$

$$\text{(7)}$$

$$R^1CO_2H \downarrow k_2$$

$$R^2NHCONHR^3 + (R^1CO)_2O \tag{121}$$

dimeric R^1CO_2H may be involved to some extent, so facilitating both the formation of **7** and its decomposition to the anhydride and urea[*]. This notion is supported by the reduced proportion of acylurea obtained when the solvent is changed from acetonitrile to carbon tetrachloride, for in the latter solvent more of the acid will be dimeric. This result, and others, also eliminates the possibility that the acylurea is produced by acylation of the urea by the anhydride or by **7**.

When an amine, R^4NH_2, is also present (equation 119) the rate of loss of diimide is reduced, owing to deactivation of the acid on forming various ion-paired, salt-like species, e.g. $R^1CO_2^- \overset{+}{N}H_3R^4$. Alone the amine does not attack the diimide and its acylation will not occur in the absence of acid. Whether or not the amine is acylated by **7** or by the anhydride is not clearly settled by this kinetic study. The balance between paths A and B depends upon the amine but here also the observed phenomena have still to be rationalized.

Methods[94] closely related to the carbodiimide method employ ketene-imines (**8**), allenes (**9**), and cyanamides (**10**).

Another route[88, 95] related to the foregoing examples, concerns acyl

$$\begin{matrix} R^1 \\ \diagdown \\ C=C=N-R^3 \\ \diagup \\ R^2 \end{matrix} \qquad RCH=C=CH_2 \qquad \begin{matrix} R^1 \\ \diagdown \\ N-C\equiv N \\ \diagup \\ R^2 \end{matrix}$$

$$\text{(8)} \qquad\qquad \text{(9)} \qquad\qquad \text{(10)}$$

carbamates, which are the 1 : 1 adducts of carboxylic acids with isocyanic esters. They also lead to intra-(B) and inter-(A) molecular acylation (equation 122). Acyl azides also react with carboxylic acids and lead

---

* It is interesting that the dimer is also the reactive species in the addition of carboxylic acids to ketenes under similar conditions[93].

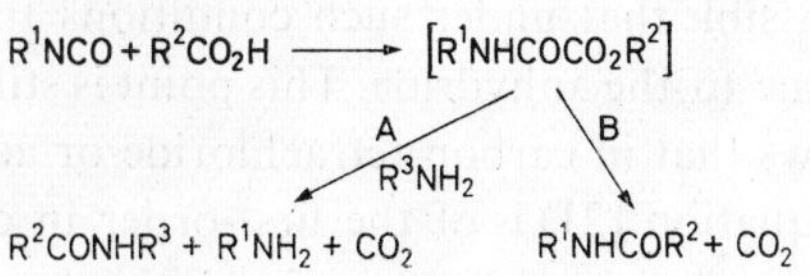

$$R^1NCO + R^2CO_2H \longrightarrow [R^1NHCOCO_2R^2]$$

$$\underset{R^2CONHR^3 + R^1NH_2 + CO_2}{\overset{A}{\swarrow}} \underset{R^3NH_2}{\overset{}{}} \overset{B}{\searrow} \atop R^1NHCOR^2 + CO_2$$

(122)

to the corresponding amide (equation 123). No doubt a preliminary Curtius rearrangement of the azide to the isocyanate is involved here[34]. (In the presence of free amines acyl azides provide acylation of the amine[96].)

$$R^1CO_2H + R^2CON_3 \longrightarrow R^1CONHR^2 + N_2 + CO_2 \tag{123}$$

Processes (119A) and (122A) constitute the indirect substitution of OH in R^1CO_2H by R^2NH; indirect because they occur via path (2), followed by an intermolecular acylation by an active ester of the acid. So far only the outline mechanisms, of these, and of the related reactions (118) and (82), are known. They all await more detailed study. It will be interesting to learn if a family resemblance exists between the mechanisms of such acylations and that of uncatalysed acylation by straightforward enol esters like isopropenyl acetate. The preparative aspects of most of the above processes are dealt with in detail in standard texts on peptide synthesis[88] and cannot be further elaborated here.

Detailed kinetic studies of aminolysis of esters have been made using both aqueous and alcoholic solutions. Studies with alcohol as solvent came first historically[97] and established that the reaction order with respect to ester was unity, but that with respect to amine the order could be greater than unity and usually appears to be 1·5. These studies culminated in the establishment, by Bunnett and Davis[98], of catalysis of the aminolysis by the solvent anion, and by other molecules of the amine. They suggested reactions (124)–(126). In their system R^1COOR^2 was ethyl formate, R^3NH_2 being n-butylamine. The solvent was ethyl alcohol (i.e. R^2OH)

$$R^1COOR^2 + R^3NH_2 \underset{k_{-1}}{\overset{k_1}{\rightleftharpoons}} \overset{O^-}{\underset{\overset{|}{NH_2R^3}}{\overset{|}{R^1C-OR^2}}} \tag{124}$$

$$\overset{O^-}{\underset{\overset{|}{\overset{NH_2R^3}{+}}}{\overset{|}{R^1C-OR^2}}} \overset{k_2}{\longrightarrow} R^1CONHR^3 + R^2OH \tag{125}$$

$$\overset{O^-}{\underset{\overset{|}{\overset{NH_2R^3}{\underset{-}{}}}}{\overset{|}{R^1C-OR^2}}} + B \overset{k_3}{\longrightarrow} R^1CONHR^3 + R^2OH + B \tag{126}$$

so that any concurrent solvolysis simply led back to ethyl formate. B represents bases able to catalyse step (125). Application of the steady-state approximation to equations (124)–(126) gives equation (127). Letting k_2 include the rate of equation (125) as written, and also the catalysis

$$-d[R^1COOR^2]/dt = \left\{ \frac{k_1k_2 + k_1k_3\,[B]}{k_{-1} + k_2 + k_3\,[B]} \right\} [R^1COOR_2]\,[R^3NH_2] \quad (127)$$

of equation (125) by solvent (equation 128), and letting B represent R^2O^- or R^3NH_2,

$$
\begin{array}{c}
O^- \\
| \\
R^1C\!-\!OR^2 + R^2OH \longrightarrow R^1CONHR^3 + 2\ R^2OH \qquad (128)\\
| \\
NH_2R^3 \\
+
\end{array}
$$

then, if $k_2 \ll k_{-1} \gg k_3[B]$,

$$-d[R^1COOR^2]/dt =$$

$$\left\{ \frac{k_1}{k_{-1}} (k_2 + k_{R^2O^-}[R^2O^-] + k_{RNH_2}[RNH_2]) \right\} [R^1COOR^2]\,[R^3NH_2] \quad (129)$$

but

$$R^3NH_2 + R^2OH \xrightleftharpoons{K_b} R^3NH_3^+ + R^2O^- \qquad (130)$$

$$\therefore\ -d[R^1COOR^2]/dt = \frac{k_1}{k_{-1}} \{ k_2 + k_{R^2O^-} K_b^{1/2}[R^3NH_2]^{1/2}$$

$$+ k_{R^3NH_2}[R^3NH_2] \} [R^1COOR^2]\,[R^3NH_2] \qquad (131)$$

If scheme (124)–(126) is correct, then, using an excess of amine, the observed pseudo first-order rate constant (k_{obs}) for loss of ester will be

$$k_{obs} = \frac{k_1}{k_{-1}} \{ k_2[R^3NH_2] + k_{R^2O^-} K_b^{1/2}[R^3NH_2]^{3/2} + k_{R^3NH_2}[R^3NH_2]^2 \} \quad (132)$$

Bunnett and Davis found that a plot of $k_{obs}/[R^3NH_2]^{3/2}$ versus $[R^3NH_2]^{1/2}$ gave a good straight line. This suggests that the scheme is satisfactory but that k_2 is negligible, i.e. there is no uncatalysed, or solvent-catalysed, path for the system studied, only those paths (126) involving self-catalysis by R^3NH_2 or catalysis by the solvent anion R^2O^-. It should be noted that the same form of rate equation will therefore result whether the mechanism is written as in equations (124)–(126) or without tetrahedral intermediates as in equation (133).

$$R^1COOR^2 + R^3NH_2 + B \longrightarrow R^1CONHR^3 + R^2OH + B \qquad (133)$$

Studies in aqueous solution[99, 100] using both ammonia and various amines as the amino component, and substituted phenyl acetates as the ester component, combine in indicating that with such systems the spontaneous (or solvent-catalysed; equation 125) aminolysis is significant, as well as the catalysed routes analogous to those in equation (126). Thus, under conditions of constant amine concentration, k_{obs} (corrected for loss of ester via hydrolysis) takes the form of equation (134). The relative importance of the different terms in equation (134)—not only among themselves

$$k_{obs} = \{k + k_{OH^-}[OH^-] + k_{R^3NH_2}[R^3NH_2]\}\,[R^3NH_2] \qquad (134)$$

but compared to the hydrolysis rate—depends both upon the amine and upon the ester. So far as the ester is concerned the situation is relatively simple: for esters with good leaving groups the catalytic term in $[R^3NH_2]$ (and probably that in $[OH^-]$) is relatively less important compared with the spontaneous aminolysis (k) than for esters with poor leaving groups. This situation may also be expressed by stating that k is more sensitive to the polar effects of substituents in the leaving group than is $k_{R^3NH_2}$ (for a series of such substituents $\varrho \approx 2$ for k but $\varrho \approx 0.5$ for $k_{R^3NH_2}$; see section I.B.4.b). This result is easily rationalized in terms of a synchronous mechanism. With good leaving groups much bond breaking will have occurred in the transition state and almost any nucleophile will be effective, whereas with poor leaving groups bond formation will be more important and synchronous catalytic routes which provide powerful nucleophiles locally (as in equation 135) will therefore be favoured (cf. section III.D). At the same time the shift of emphasis to bond forming for catalytic routes will decrease their dependence on the ability of the leaving group to depart. By the same arguments, if any concurrent hydrolysis

$$\begin{array}{c} O \\ \parallel \\ R^1C-O-R^2 \\ \uparrow \\ R^3-N-H \\ | \\ H \\ \uparrow \\ R^3-N-H \\ | \\ H \end{array} \qquad (135)$$

is predominantly due to attack by OH^- (a powerful nucleophile) then hydrolysis will be expected to be much more prominent for esters with poor leaving groups than for those with good leaving groups. In keeping with this expectation, oxygen esters exhibit relatively more hydrolysis than do thiol esters under such conditions[101]. It is quite unnecessary to explain these facts in terms of partitioning of tetrahedral intermediates.

The relatively greater significance of the spontaneous aminolysis with esters with good leaving groups is just another manifestation of the general prediction that esters (or acids) containing substituents which enhance the positive charge on the carbonyl carbon atom (either by possessing a good leaving group—one with electron-withdrawing substituents — or an electron-withdrawing group R^1) will generally react more easily (*i.e.* will need less catalysis) than those which do not possess such substituents (*cf.* sections I.B.4.a and b). Complications can arise if electron withdrawal by R^1, although favouring attack on the carbonyl carbon atom, hinders a sufficient departure of the leaving group. This will not normally occur because with attack on the carbonyl carbon atom dominant, very little departure will need to have taken place in the transition state and solvent assistance will usually be sufficient to achieve this. In non-coordinating, non-hydroxylic solvents, however, the requirement of even a very small amount of bond breaking can prove influential[102], process (135); becoming cyclic in order to assist leaving group departure. Even in aqueous media the opposing effects of electron withdrawal by R^1 are reflected in the tendency with such esters for anionic nucleophiles to be favoured relative to uncharged species of the same basicity[103], for in cases where departure has some importance a negatively charged transition state will possess advantages over a formally neutral transition state. The postulation of special secondary valency forces to explain the facts appears unnecessary. The finding that water as a nucleophile, *i.e.* the spontaneous hydrolysis, exhibits 'anionic' behaviour is convincing proof that such hydrolysis really involves synchronous catalysis by other water molecules[66] (**11**, *cf.* section III.B) resulting in local hydroxide ion formation.

$$
\begin{array}{c}
\overset{\displaystyle O}{\underset{\displaystyle \|}{}} \\
R^1 - C - OR^2 \\
\uparrow \\
H - O \\
\end{array}
\qquad (11)
$$

So far as the amine is concerned the effect of its structure on the relative importance of the different terms in equation (134) is not well understood[2]. This is true for both oxygen and thiol esters[2]. Secondary amines, especially bulky ones, seem particularly susceptible to hydroxyl ion catalysis in preference to that by a second amine molecule. This may reflect a steric

27 C.C.A.E.

difficulty in forming the transition state like (135) for secondary amines. For primary amines a consistent pattern for the relative importance of k, k_{OH-}, and $k_{R \cdot NH_2}$ has not yet been discerned, although this may be partly due to experimental errors[2, 100]. It may be significant that diamines like $NH_2(CH_2)_n NH_2$ exhibit a spontaneous term k and a term in k_{OH-} but no self-catalysis, $k_{R \cdot NH_2}$. Here again steric reasons may underlie the result. Such diamines (with $n > 2$) react via their uncatalysed path (k) about ten times more rapidly than monoamines of comparable basicity. Intramolecular self-catalysis may therefore be involved[104]. Powerful intramolecular catalysis is undoubtedly possible[102b].

As already noted, the kinetic orders found for ester aminolysis are predicted by an outline mechanism which can be written in two different ways; either including an addition intermediate present in low concentration (equation 124) or as a synchronous displacement, for example reaction (135). We have chosen to rationalize the effects that substituents have on the relative importance of the different paths—catalysed and uncatalysed—in terms of the synchronous picture. We do this because we think such rationalizations simpler (and more convincing) than those necessary on the assumption of an intermediate. It is a recognized scientific principle to postulate the simplest scheme compatible with the data and a scheme without intermediates is already simpler than one containing one or more such species. Intermediates are very frequently postulated in mechanisms of acylation, but most often they are not, in fact, demanded by the evidence. Their automatic postulation is therefore unjustified but depressingly common. It is encouraging that Samuel and Silver[10] in their discussion of oxygen exchange between solvent species and various acylating agents (the phenomenon upon which the postulation of addition intermediates in acylation has, until very recently, largely rested*) do not fall into this error. We do not say that intermediates do not often occur; the widespread occurrence of carbonyl addition in other contexts suggests they may well do so. What does not follow is that they *always* occur. We emphasize that in the majority of cases of acylation even by esters (which on the whole have comparatively poor leaving groups and are therefore suited to intermediate formation[105]) the actual evidence for intermediates is usually non-existent.

---

* It is not generally realised that, although the *existence* of oxygen exchange for species RCOY implies that tetrahedral intermediates must form, nevertheless the details of the exchange data suggest, if anything, that exchange and acylation (replacement of Y) usually proceed by *different* routes[10b]. This point is not normally stressed by advocates of intermediates in acylation. See however M. L. Bender and H. d'A. Heck, *J. Am. Chem. Soc.* **89**, 1211 (1967).

While schemes like (124)–(126) have disadvantages in terms of formal and interpretative complexity compared with those like (135), nevertheless there is no doubt that many, although not all[102a], of the data can be rationalized on their basis. Jencks and coworkers[66, 99] have been most active in this area. With an intermediate there are two transition states; one for formation of the intermediate, and one for its breakdown to products (equation 136). It is interesting that rationalization of substituent effects

$$
\text{>NH} + R^1COOR^2 \underset{k_{-1}}{\overset{k_1}{\rightleftharpoons}} \underset{\underset{+NH}{\vert}}{\overset{\overset{O^-}{\vert}}{R^1C-OR^2}} \overset{k_2}{\longrightarrow} R^1CON< + R^2OH \tag{136}
$$

T.S.1   T.S.2

on this basis leads, in spite of Jencks' assurances to the contrary[66], to exactly the opposite conclusions about the importance of the different phases of the overall process to those reached on the basis of a synchronous scheme (p. 416). Thus Jencks, in fact, concludes that with a good leaving group the bond forming process (k_1) dominates the rate, whereas with a poor leaving group bond breaking (k_2) controls it. Although under steady-state conditions k_2 can never wholly determine the rate because $k_{obs} = k_1 k_2/(k_{-1}+k_2)$ so that at best $k_{obs} = k_1 k_2/k_{-1}$, nevertheless one can understand what is meant. However, if with good leaving groups (i.e. $k_2 \gg k_{-1}$) k_1 is dominant it is difficult to understand why catalytic paths are relatively so unimportant under such conditions, for catalysis of attack of the amine will make it more nucleophilic and therefore increase k_1. The same dilemma is not incurred with a synchronous route for then the transition state is reached with very little bond forming and much bond breaking. Here the nature of the attacking reagent is of little importance and therefore catalysis is not particularly effective (cf. p. 416).

Considering, however, that an intermediate is involved, there are at least two possible variations of scheme (124)–(126) formally compatible with the observed unit order in base catalyst (equations 129 and 134). We prefer the variation proposed by Jencks and Carriuolo[99] (equation 137) to that of Bunnett and Davis[98]. One compelling reason is that equa-

$$
\begin{array}{c}
R^1COOR^2 + R^3NH_2 \\
+ B
\end{array}
\rightleftharpoons
\left[
\begin{array}{c}
\overset{\delta-}{O} \\
\| \\
R^1-C-OR^2 \\
\uparrow \\
H-N-R^3 \\
\vert \\
H \\
\uparrow \\
\ddot{B}^{\delta+}
\end{array}
\right]
\rightleftharpoons
\begin{array}{c}
\overset{O^-}{\vert} \\
R^1C-OR^2 + BH^* \\
\vert \\
NHR^3 \\
\downarrow \\
Products
\end{array}
\tag{137}
$$

Transition state

27*

tion (137) is entirely in line with what is believed about similar phenomena throughout the whole field of acylation[1,105a].

B. Reactions with Secondary and Tertiary Amines

I. Introduction

The majority of secondary amines behave in principle as do primary amines, except that steric effects—in particular steric hindrance of their approach to the carbonyl carbon atom—are likely to be of more importance. Relatively few secondary amines have received kinetic study and therefore we shall not discuss this group of nucleophiles in detail. Tertiary amines possess no replaceable hydrogen on their nitrogen atoms so that any acyl derivatives will inevitably involve quadrivalent nitrogen, as in (12), and be ionic both in the solid state and in solution. In the solid the ions (12a and 12b) will be uniquely associated with an appropriate

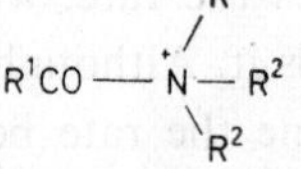

$$R^1CO \overset{R^2}{\underset{R^2}{\overset{|}{\overset{+}{N}}}} R^2 \qquad\qquad RCO \overset{+}{-} N \overset{}{\langle}$$

(12a) 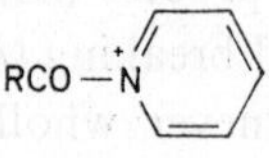   (12b)

anion (e.g. Hal⁻). Compounds of this type are known, but not unnaturally they are very susceptible to further reaction, especially hydrolysis[106]. Their ease of formation and their stability increase with the basicity of the amine and the stability of the acylium ion RCO⁺. They are not normally prepared from esters* but from acyl halides or acid anhydrides (equations 138 and 139). They doubtless occur as reactive intermediates in many tertiary amine catalysed acylations of other compounds

$$R_3N + (R^2CO_2)O \rightleftharpoons [R_3^1NCOR^2]^+ [R^2CO_2]^- \qquad\qquad (138)$$

$$C_5H_5N + RCOCl \rightleftharpoons [C_5H_5NCOR]^+Cl^- \qquad\qquad (139)$$

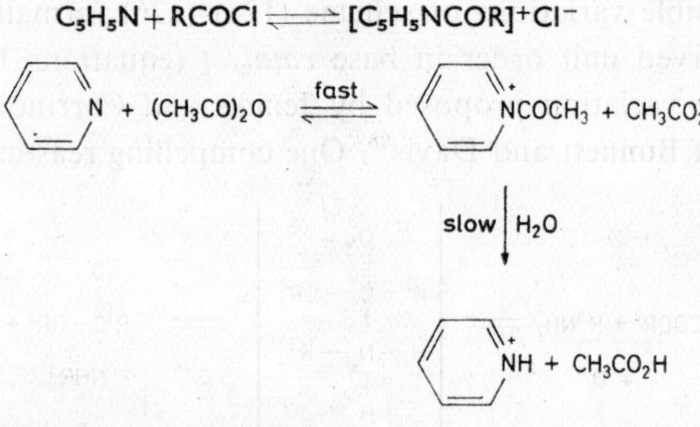

$$\langle N + (CH_3CO)_2O \overset{fast}{\rightleftharpoons} \langle \overset{+}{N}COCH_3 + CH_3CO_2^-$$

$$slow \downarrow H_2O \qquad\qquad\qquad (140)$$

$$\langle \overset{+}{N}H + CH_3CO_2H$$

* Carboxylic acids will be expected to give only the salt with tertiary amines: $R^1COOH + NR_3^2 \rightleftharpoons R^1CO_2^- {}^+NHR_3^2$.

by anhydrides and acyl halides (equation 140)[106, 107]. Such catalysis is of the nucleophilic type. Esters with good leaving groups (*e.g.* phenyl acetates) can, in fact, lead to quaternary acyl compounds, which are considered to participate in the tertiary amine catalysed solvolysis of such esters in a way analogous to reaction (140) above[108]. The solvolysis of esters with poor leaving groups (*e.g.* ethyl acetate, ethyl dichloroacetate) are also catalysed by tertiary amines but here the mechanism of catalysis is not nucleophilic but synchronous (section I.B.2) and therefore does not involve any substitution by the base in the ester group.

In this section we shall limit discussion to the reactions of one particular class of base—the imidazoles—which happen to be simultaneously both secondary and tertiary bases (**13** and **14**). Imidazoles therefore pro-

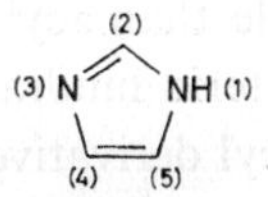

Imidazole   Benzimidazole
(**13**)   (**14**)

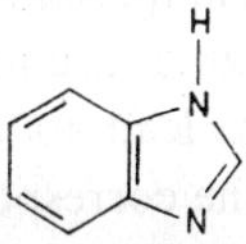

vide a useful and instructive bridge between substitution of OR in a CO_2R group by tri- and tetravalent nitrogen. For instance there is an appropriate gradation in reactivity: N-acylpyridines $>$ N-acylimidazoles $\gg$ amides. The influence of the acyl group on stability, noted above, is illustrated by the fact that N-acetylbenzimidazole is much more easily hydrolysed than is N-benzoyl benzimidazole.

2. Ester solvolysis catalysed by imidazoles

Catalysis by imidazoles has received a good deal of attention[2] because the imidazole groups of the histidine residues found in a variety of hydrolytic enzymes are probably involved in the enzymic catalysis[109].

The details of the behaviour of an imidazole in its reaction with an ester will depend upon three main factors: (i) the substituents carried by the ester, (ii) the substituents carried by the imidazole, and (iii) the solvent. Taking these factors in reverse order, the solvent will affect reaction because it will obviously be much easier to isolate any amide or quarternary acyl salt (imidazoles can form both) when the solvent contains no acylatable species for these compounds to attack. Thus in aqueous or alcoholic solvents the product normally isolated (unless special precautions are taken) is the solvolysed ester. It is in this context that catalysis of solvolysis has been studied.

It is found that the substituents in the imidazole affect reaction[110] (i) because 1-substituents force the formation of quarternary compounds (16) by preventing amide formation, which otherwise occurs preferentially (15), however, possibly via attack at the tertiary nitrogen atom, followed by stabilization of the cationic product by loss of a proton, (ii) because substituents in the 4-and/or 5-positions lead appparently to negligible steric hindrance, and fusion with another ring system at these sites leads to

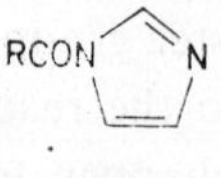

RCON⟨N⟩   RCON⟨+NMe⟩

(15) 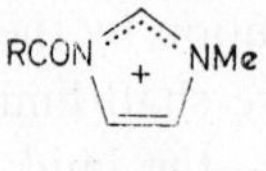   (16)

considerable increase in the stability of the corresponding acyl derivatives (thus acyl benzimidazoles are more stable than acyl imidazoles), and (iii) because 2-substituents lead to notable steric hindrance and therefore to smaller amounts of the corresponding acyl derivatives at equilibrium.

Substituents in the ester affect reaction in ways which have been hinted at above and which may be largely deduced from what has already been said concerning the details of reactions of esters with primary amines (section IV.A). Thus, in any ester hydrolysis catalysed by imidazole one would predict that, in general, there would be at least five terms in the differential rate equation (141). In equation (141), in which the concentration of the solvent has been omitted, IM represents the imidazole, k represents the rate constant for hydrolysis via solvent species only, k_n

$$-d[R^1COOR^2]/dt = \{k + k_n[IM] + k_{H_2O}[IM] + k_{IM}[IM]^2$$
$$+ k_{IM^-}[IM^-]\}[R^1CO_2R^2] \quad (141)$$

that for hydrolysis via nucleophilic catalysis (equations 142 and 143), k_{H_2O} that for hydrolysis via a synchronous catalysis of attack of a water molecule by imidazole (equation 144), k_{IM} that for synchronous catalysis of attack of imidazole by another imidazole molecule and therefore for hydrolysis via equations (145 and 146), and k_{IM^-} represents the rate constant for hydrolysis via nucleophilic catalysis by the imidazole anion (equations 147 and 148). It is the relative importance of these various

$$R^1COOR^2 + HN⟨N⟩ \underset{slow}{\overset{k_n}{\rightleftharpoons}} R^1CON⟨N⟩ + R^2OH \quad (142)$$

$$R^1CON⟨N⟩ + H_2O \xrightarrow{fast} R^1CO_2H + HN⟨N⟩ \quad (143)$$

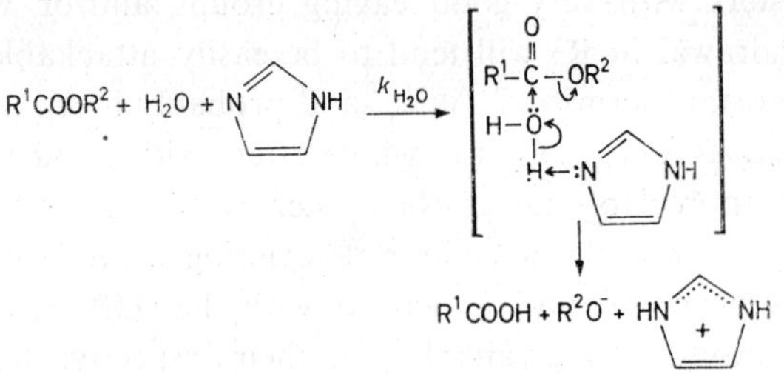

$$R^1COOR^2 + H_2O + N\overset{\frown}{\quad}NH \xrightarrow{k_{H_2O}} \left[\begin{array}{c} O \\ \| \\ R^1-C-OR^2 \\ \uparrow \\ H-\overset{..}{O} \\ | \\ H \leftarrow :N\overset{\frown}{\quad}NH \end{array} \right]$$

$$R^1COOH + R^2O^- + HN\overset{\frown}{\underset{+}{\quad}}NH \qquad (144)$$

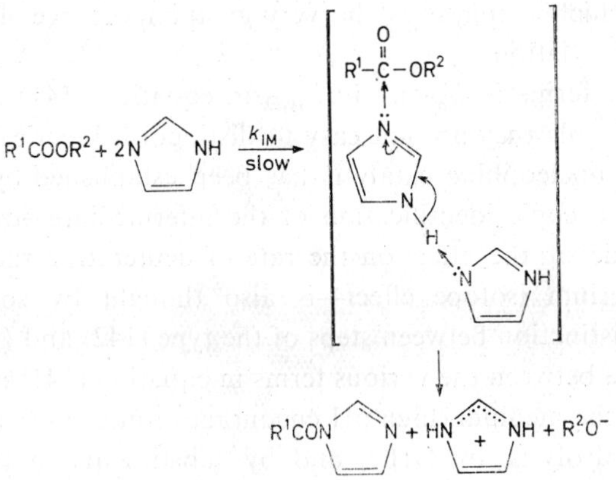

$$R^1COOR^2 + 2N\overset{\frown}{\quad}NH \xrightarrow[\text{slow}]{k_{IM}} \left[\begin{array}{c} O \\ \| \\ R^1-C-OR^2 \\ | \\ N \\ \\ N \\ | \\ H \\ \\ :N\overset{\frown}{\quad}NH \end{array} \right]$$

$$R^1CON\overset{\frown}{\quad}N + HN\overset{\frown}{\underset{+}{\quad}}NH + R^2O^- \qquad (145)$$

$$R^1CON\overset{\frown}{\quad}N + H_2O \xrightarrow{fast} R^1COOH + HN\overset{\frown}{\quad}N \qquad (146)$$

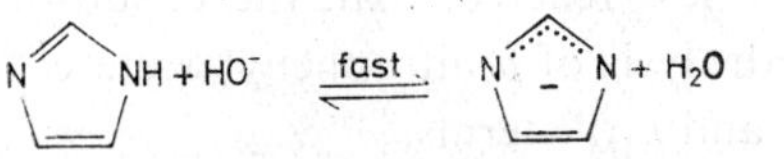

$$N\overset{\frown}{\quad}NH + HO^- \underset{fast}{\rightleftharpoons} N\overset{\frown}{\underset{-}{\quad}}N + H_2O \qquad (147)$$

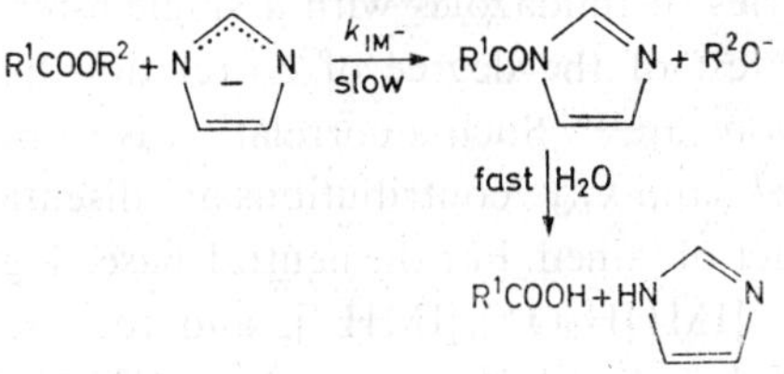

$$R^1COOR^2 + N\overset{\frown}{\underset{-}{\quad}}N \xrightarrow[\text{slow}]{k_{IM^-}} R^1CON\overset{\frown}{\quad}N + R^2O^-$$

$$fast \Big| H_2O$$

$$R^1COOH + HN\overset{\frown}{\quad}N \qquad (148)$$

paths that will be affected by the changes in the ester structure. Thus esters for which leaving group departure is not very easy (so that bond breaking has not much advanced in the transition state and bond forming is therefore of dominant importance) good attacking species will be necessary and paths (144), and especially (145)–(146), and (147)–(148) will be favoured. This is true for esters like ethyl chloroacetate, p-cresyl acetate, and phenyl benzoate compared with say p-chloro- or p-nitrophenyl acetate[108]. Esters with very good leaving groups and/or with powerful electron withdrawal in R^1 will tend to be easily attackable and exhibit relatively important terms in k, in k_n, and probably in k_{H_2O}, with respect to those in k_{IM} and k_{IM-}. On the whole esters with good leaving groups tend to react appreciably faster than esters with poor leaving groups[99]. This is in keeping with the behaviour of acylating agents generally (section I.C.1). It is a phenomenon—associated with the shifting balance of the different acylation routes available and their respective free energies of activation—which emphasizes the very great importance of leaving group departure in acylation.

Because the terms in k_n and in k_{H_2O} in equation (141) are both first-order in imidazole they are not easy to distinguish kinetically. In certain instances the nucleophilic catalysis has been established by actual isolation or spectroscopic identification of the intermediate acylimidazole[111]. The magnitude of the effect on the rate of deuterating the solvent—the solvent deuterium isotope effect—is also thought by some authors[112] to permit a distinction between steps of the type (142) and (144).

The balance between the various terms in equation (141) is also affected by the pH of the medium (high pH encourages route (147)–(148) and also unassisted hydrolysis by OH^-) and by substituents in the imidazole, particularly 1-substituents which prevent routes (145) and (147). N-Alkyl imidazoles can nevertheless still exhibit the nucleophilic route (142)–(143) the intermediate taking the form 16 rather than 15. In practice, N-methylimidazole is a little less reactive than the unsubstituted compound as a catalyst for the hydrolysis of p-nitrophenyl acetate[108, 111] which is an ester with dominant k_n and k_{IM-} terms.

The use of a series of imidazoles with a single ester like p-nitrophenyl acetate permits a test of the degree of correlation between nucleophile basicity and reaction rate[110]. Such a correlation is to be expected (section I.B.3) and once the k_n and k_{IM-} contributions are disentangled, good Brønsted plots are in fact obtained. For the neutral base, $\log k_n = -0.8 \log K_a -4.3$, where $K_a = [IM][H_3O^+]/[IMH^+]$, and for the imidazole anion, $\log k_{IM-} \approx -0.15 \log K_a' + 1.35$, where $K_a' = [IM^-][H_3O^+]/[IM]$. The lower dependence of k_{IM-} on the dissociation constant is sensible because highly reactive species are usually found to be associated with lower selectivity. Also, just as with the aminolysis of esters by primary amines (for the k_n and k_{IM-} terms represent rates of aminolysis; that is our justification for discussing imidazoles at all) we will expect Hammett–Taft type correlations (section I.B.4) between the ester structure and the nucleophilic catalysis rates[108, 112]. As always the simplest effects are obtained from variation of the leaving group. For imidazole itself, with a series of substituted phenyl acetates, $\varrho \approx 1 \cdot 8$ for k_n and $\varrho \approx 0 \cdot 5$ for k_{IM-}, values indicating that the sensitivity of nucleophilic attack to leaving group structure in imidazole reactions is very similar to that found in aminolysis of the same esters by ammonia, and has the same dependence on the nucleophilicity of the attacking group (cf. section IV.A.).

C. Reaction with Hydrazines

1. Introduction

The reactions between hydrazines and carboxylic esters and acids are in principle very similar to those of amines but contain one or two special features owing to the presence of the second amino group. The substitutions analogous to those considered for amines are reactions (149) and (150) and lead to hydrazides[30]. As with amines, (150) is infrequently of

$$R^1COOR^2 + NH_2NH_2 \longrightarrow R^1CONHNH_2 + R^2OH] \qquad (149)$$

$$RCOOH + NH_2NH_2 \longrightarrow RCONHNH_2 + H_2O \qquad (150)$$

any practical synthetic value owing to the concurrent process (151) which will often lie well to the right. Since two amino groups are present, dia-

$$RCO_2H + NH_2NH_2 \rightleftharpoons [RCO_2]^- [NH_2NH_3]^+ \qquad (151)$$

cylation (equation 152) is comparatively easy*.
These products, termed dihydrazides, predominate when active acylating agents (e.g. acyl halides or acid anhydrides) are used. It happens there-

$$2 RCOY + NH_2NH_2 \longrightarrow RCONHNHCOR + 2 HY \qquad (152)$$

fore that acylation by esters provides the best synthetic route to (mono) hydrazides[30, 34]. Diacylation can be minimized by using excess of the hydrazine. For unsubstituted hydrazides the starting materials are normally hydrazine hydrate and the appropriate ester. Warming the reactants together in the absence of additional catalysts (since hydrazines, like amines, will catalyse their own reactions) is usually sufficient to effect the substitution. So far as is known substituent effects parallel those in amide synthesis and in all other acylations by esters.

Aliphatic ω,ω'-dicarboxylic esters lead predominantly to substitution by hydrazine at both ester groups (equation 153) and to prepare the monohydrazide use of the monoester is necessary. The latter leads to a monohydrazide salt (equation 154). In the aliphatic series little cyclization is

---

* Diacylation at a single amino group is possible, although difficult. The products N,N-diacylamines[113] (or N-acylamides) are relatively unstable and are powerful acylating agents. Diacylamines are not normally formed from esters and are therefore omitted from this chapter.

usually apparent, but cyclic dihydrazides can be the main product with the aromatic dicarboxylic monoesters (equation 155).

$$(CH_2)_n(COOR)_2 + 2\,NH_2NH_2 \longrightarrow (CH_2)_n(CONHNH_2)_2 + 2\,ROH \qquad (153)$$

$$HOOC(CH_2)_nCOOR + 2\,NH_2NH_2 \longrightarrow [NH_2NHCO(CH_2)_nCOO]^- [NH_2NH_3]^+ \quad (154)$$

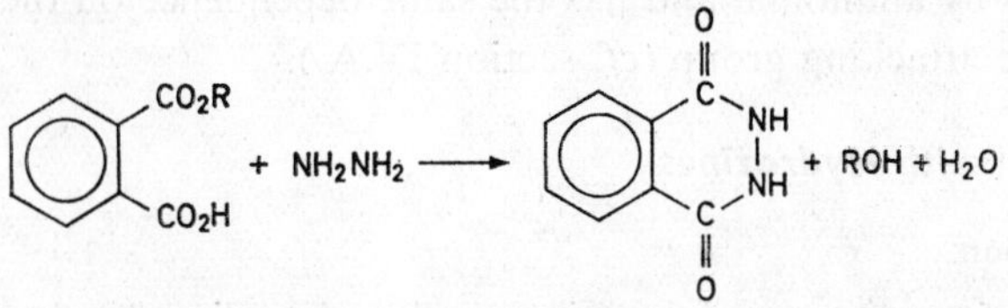

$$(155)$$

Reaction (155) presumably involves intramolecular attack by the amino group of the intermediate monohydrazide on the carboxyl group, the latter being at least partly in the carboxylate form. Such an unfavourable process appropriately only comes to the fore under intramolecular conditions.

Hydrazines will attack amides as well as esters but (as expected) less rapidly and with overall equilibrium positions generally less favourable than reaction (149) to hydrazide formation. The lower reactivity of amides compared with esters is important in peptide synthesis, when although intermediate products frequently contain both types of group, selective hydrazinolysis is possible[88].

2. Kinetic studies of hydrazinolysis of esters

Kinetic experiments[112a] using aqueous solutions have shown that reaction between hydrazines and various phenyl acetates* involve differential rate equations like equation (156). This equation indicates an uncatalysed (or solvent-catalysed) hydrazinolysis (k) as well as two catalytic paths

$$-\,d[R^1COOR^2]/dt = \{k + k_{NH_2NH_2}[NH_2NH_2]$$
$$+ k_{NH_2NH_3^+}[NH_2NH_3^+]\}\,[NH_2NH_2]\,[R^1COOR^2] \qquad (156)$$

($k_{NH_2NH_2}$ and $k_{NH_2NH_3^+}$) in which an additional molecule of hydrazine or of hydrazonium ion aids the substitution. Both these paths could be of the synchronous, base catalytic type, for example (157), although the very

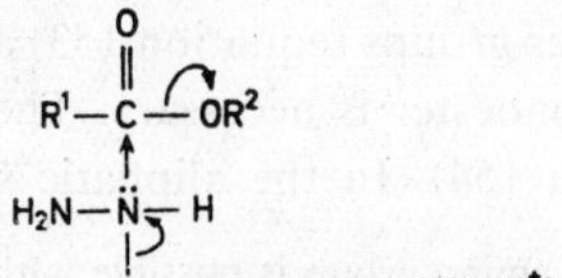

$$(157)$$

_____

* Phenyl esters owe their ubiquity in kinetic studies to the ease of the spectroscopic estimation of the phenolic products to which they lead.

different solvent isotope effects found for k_{NH,NH_2} and k_{NH,NH_3^+} perhaps suggest a different transition state for the latter path, for example (158).

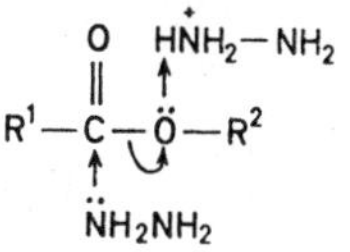

$$
\begin{array}{c}
\overset{\displaystyle O}{\underset{\displaystyle \underset{NH_2NH_2}{\uparrow}}{\overset{\parallel}{R^1-C-\underset{\uparrow}{O}-R^2}}}\quad \overset{H\dot{N}H_2-NH_2}{\uparrow}
\end{array}
\tag{158}
$$

The absence of a direct (uncatalysed) attack by $NH_2NH_3^+$ is not, perhaps, surprising. With substituted hydrazines, $RNHNH_2$ or R_2NNH_2, the path involving the cation is insignificant. As with the aminolysis of esters, substituents in the leaving group affect the value of k (direct attack) more than the value of k_{NH,NH_2} (catalysed attack), doubtless for the same reasons ($\varrho \approx 2 \cdot 9$ and $0 \cdot 5$ respectively; cf. section IV.B). Intramolecular base catalysis, superficially feasible for compounds of structure $R_2N(CH_2)_nNHNH_2$, ($n = 2$ or 3), is not marked, if present at all[100, 114]. The most notable, unique feature of ester hydrazinolysis is that hydrazine itself exhibits in its 'uncatalysed' rate (k) an anomalous reactivity compared with other bases of comparable strength; it falls off the Brønsted plot for the attack of different bases on phenyl acetate, being about 100-fold too reactive. This result is an example of what has been termed[53] the 'α effect'. Hydrazine and certain other nucleophiles, e.g. NH_2OH, NH_2OMe, $HOOH$ and HOO^-, all having a lone pair of electrons on an atom adjacent (α) to that undergoing electrophilic attack, exhibit anomalously great nucleophilicity in various reactions. The proper and complete explanation of the α effect is not yet clear. In the present context it may be significant that monosubstituted hydrazines show a decreased α effect, and disubstituted compounds R_2NNH_2 show no such effect, their rate constants fitting the usual Brønsted plot. In our view the reduced opportunities for hydrogen bonding to the solvent in the substituted compound (159) may represent an important reduction in base catalysis by the solvent, for those hydrogen atoms on the nitrogen atom actually undergoing electrophilic attack may not be so accessible, for geometrical reasons, to hydrogen bonding as those on the α-nitrogen atom. However, the reduction in

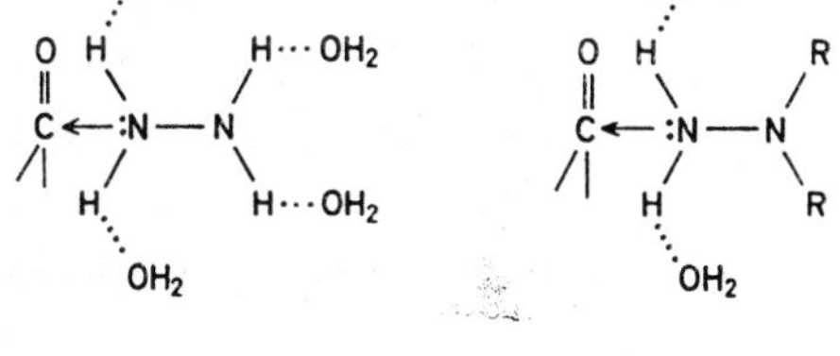

$$
\tag{159}
$$

intramolecular catalysis via hydrogen bonding (as in 160) could also explain the effect (*cf.* section III.C for hydrogen peroxide). We return to the α effect below, in our discussion of hydroxylamine.

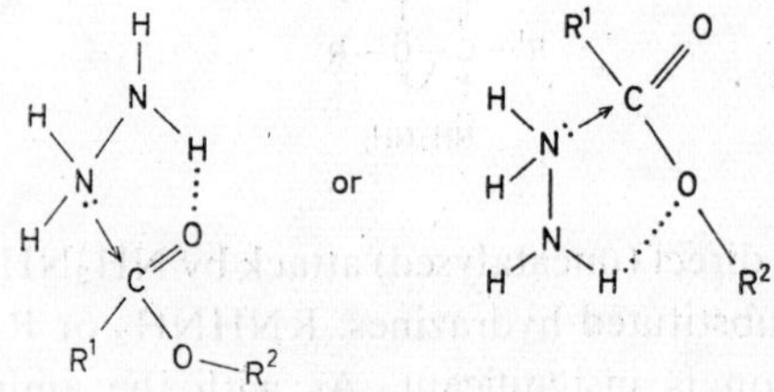

$$(160)$$

D. Reaction with Hydroxylamines

The main principles outlined for amines and hydrazines apply also to hydroxylamines. In this case substitution by nitrogen leads with esters to hydroxamic acids (equation 161). Equilibria like (161) normally lie

$$R^1COOR^2 + NH_2OH \rightleftharpoons R^1CONHOH + R^2OH \qquad (161)$$

well to the right and, as for hydrazides, this reaction with esters is a good method for preparation of the free hydroxamic acid because more powerful acylating agents tend, under preparative conditions[30, 34], to lead to some diacylation (equation 162). Commonsense suggests that

$$2 (RCO)_2O + NH_2OH \longrightarrow RCONHOCOR + 2 RCOOH \qquad (162)$$

while attack at nitrogen and oxygen is inevitably concurrent, the predominant, or initial, attack will be at the more basic nitrogen site. With excess of hydroxylamine, *i.e.* under conditions favouring monoacylation, path (161) would be expected to predominate over (163) no matter what the

$$R^1COOR^2 + NH_2OH \longrightarrow R^1COONH_2 + R^2OH \qquad (163)$$

acylating agent. However, kinetic studies and product determinations[115] using aqueous solutions (at pH ≈ 6–8) reveal the interesting fact that the balance between (161) and (163) depends critically upon the acylating agent. The less powerful this reagent, *i.e.* the poorer its leaving group, the greater is the relative importance of route (163) leading to the *O*-acylhydroxylamine. Thus for the series benzoyl chloride, acetic anhydride, and *p*-nitrophenyl acetate the percentages of *O*-acyl derivative are 4%, 51% and 75% respectively. In the presence of excess of hydroxylamine the *O*-acyl derivatives react further to give the *N*-acyl compound (equation 164) as the final product, but with phenyl acetates (at least) a very large

$$RCO_2NH_2 + NH_2OH \longrightarrow RCONHOH + NH_2OH \qquad (164)$$

percentage of the *initial* reaction involves substitution by oxygen. Moreover this reaction is very fast compared to the rate of hydrolysis of the
ester under the same conditions, *i.e.* compared to attack by the oxygen
atom of a water molecule, a more basic entity. These two reactions—
hydrolysis and *O*-acylation of hydroxylamine—will be expected to involve at least local formation of the anion of the nucleophile as a result
of solvent base catalysis (165). This is likely because it very probably

$$
\begin{array}{cc}
\underset{\substack{\displaystyle \text{H}-\overset{}{\underset{\displaystyle |}{\text{O}}} \\ \displaystyle \underset{\displaystyle \text{H} \longleftarrow :\text{OH}_2}{|}}}{\overset{\displaystyle \overset{\text{O}}{\underset{|}{\parallel}}}{\text{R}^1\text{C}-\text{OR}^2}}
&
\underset{\substack{\displaystyle \text{H}_2\text{NO} \\ \displaystyle \underset{\displaystyle \text{H} \longleftarrow :\text{OH}_2}{|}}}{\overset{\displaystyle \overset{\text{O}}{\underset{|}{\parallel}}}{\text{R}^1\text{C}-\text{OR}^2}}
\end{array}
\qquad (165)
$$

occurs even in the hydrolysis of acetic anhydride[116], which has an even
better leaving group than *p*-nitrophenyl acetate and is therefore in less
need of a good nucleophile (see also **11, p. 417**). Such a scheme, given
that the hydroxylic proton of hydroxylamine is of comparable or greater
lability than that of water, can, however, explain neither the difference in
rates of the respective *O*-acylations (otherwise phenol would be acylated
more readily than water) nor the occasional preference for *O*-rather than
N-acylation. The explanation of both these facts could be that hydroxylamine exists in solution to a small, but significant, extent as the zwitterion
$\overset{+}{\text{N}}\text{H}_3\text{O}^-$ which, because of its particular structure, is better able than
hydroxylamine itself to benefit from either (or both) of those types of
catalytic, hydrogen-bonding scheme suggested (section C above) to account
for the unusual reactivity of hydrazine (159, 160). This type of route would

$$
\begin{array}{ccc}
\underset{\substack{\displaystyle \text{H}-\underset{\displaystyle |}{\overset{\displaystyle |}{\text{N}}}-\text{H} \\ \displaystyle \text{H}}}{\overset{\displaystyle \overset{\text{O}}{\underset{}{\parallel}}}{\text{R}^1-\text{C}\underset{\overset{}{\text{O}}}{\overset{}{\text{O}}}\text{R}^2}}
& \text{or} &
\underset{\substack{\displaystyle \text{H}_2\text{O}\cdots\text{H}-\overset{+}{\underset{\displaystyle |}{\text{N}}}\text{H}\cdots\text{OH}_2 \\ \displaystyle \underset{\displaystyle \text{OH}_2}{\overset{}{\text{H}}}}}{\overset{\displaystyle \overset{\text{O}}{\underset{\displaystyle \overset{|}{\text{O}^-}}{\parallel}}}{\text{R}^1-\text{C}-\text{OR}^2}}
\end{array}
\qquad (166)
$$

be expected to be important when bond forming was at a premium (esters
with poor leaving groups). With acylating agents with good leaving groups
almost any nucleophile will be effective and the more readily available
uncharged hydroxylamine, which is most easily attacked at nitrogen, will
take a dominant part, as found. Hydroxylamine probably presents a com-

plicated manifestation of the α effect and a simple, single explanation of this phenomenon appears unlikely. More work is needed in this area[2, 53].

Kinetic studies[117] of the hydroxaminolysis of lactones surprisingly reveal a more conventional type of behaviour, with O-acylation not in evidence, and attack by nitrogen requiring base catalysis, just as in many standard aminolyses of esters. Thus the differential rate equation is equation (167).

$$-d[\text{lactone}]/dt =$$

$$\{k_{NH_2OH}[NH_2OH] + k_{NH_2O^-}[NH_2O^-]\}\,[NH_2OH]\,[\text{lactone}] \qquad (167)$$

E. Intramolecular Effects with Nitrogen Nucleophiles

Amino esters can exhibit one of two forms of intramolecular substitution: lactam formation (equation 168) or an $O \rightarrow N$ (or $S \rightarrow N$ for thiol esters) rearrangement (equation 169). Suitably positioned amino groups

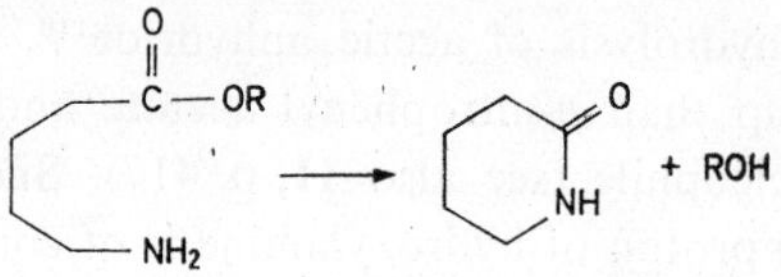

$$(168)$$

$$(169)$$

may also catalyse the hydrolysis of esters by any of the various routes available for this process (see sections IV.A,B). An intramolecular nucleophilic catalysis is shown in equation (170). Examples of all these effects are

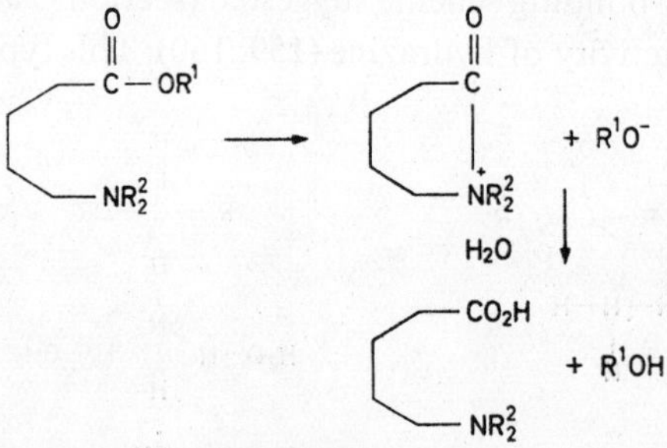

$$(170)$$

well known[2, 9] and have provided some of the more concrete evidence for the existence of carbonyl addition intermediates in substitution at the CO_2R group, as well as various quantitative estimates of the rate enhancement possible for an intra-, compared with the structurally analogous intermolecular process (see sections II.D and III.D). Particularly interesting in the latter context is the comparison[118] between the intermolecular, nucleophilic catalysis of the hydrolysis of substituted phenyl acetates by trimethylamine (equation 171) and the intramolecular nucleophilic catalysis of the hydrolysis of substituted phenyl esters of γ-(N,N-dimethylamino)butyric acid (equation 172). The intermolecular process leads to a second-order rate constant, the intramolecular mechanism to a first-order constant. The two rates are most meaningfully compared therefore at a concentration of trimethylamine of 1M. Such a comparison reveals rate enhancements of between 10^3 and 5×10^3 for the intramolecular route.

$$\underset{\substack{\| \\ O}}{Me-C}-OC_6H_4R + Me_3N \xrightarrow{k_2} \underset{\substack{\| \\ O}}{Me\overset{+}{C}}-NMe_3 + RC_6H_4O^- \xrightarrow{H_2O} MeCO_2H$$
$$+ NMe_3 + RC_6H_4OH \quad (171)$$

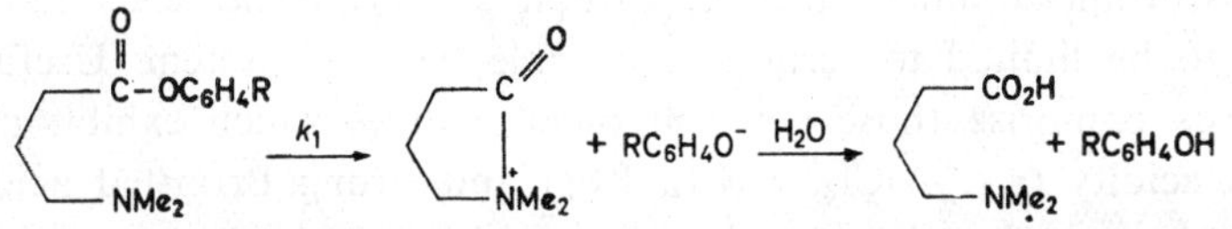

(172)

It is considered that much larger rate enhancements are possible in appropriate systems[2]. This type of effect is of considerable interest in connexion with mechanisms of enzymic catalysis.

The kinetic dependence of the S → N transfer in S-acetyl-β-mercaptoethylamine on pH, together with the concomitant thiazoline formation, strongly suggests a scheme like (173) involving a carbonyl addition intermediate[119]. A similar scheme may operate for the O-analogue[119, 120].

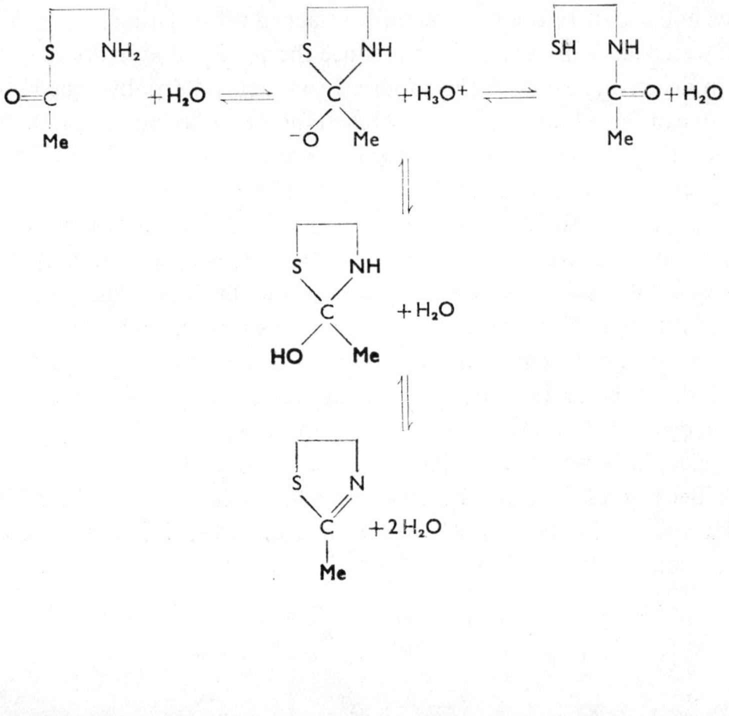

(173)

V. SUBSTITUTION BY VARIOUSLY BOUND CARBON ATOMS

A. Introduction

Nucleophilic substitution in RCOY by carbon nearly always leads to a ketone (equation 174). The saturated, and even the unsaturated, carbon

$$RCOY + {>}C\!-\!H \longrightarrow {>}C\!-\!COR + HY \qquad (174)$$

atom is a much less nucleophilic entity than are the corresponding nitrogen or oxygen atoms and powerful catalysis is usually necessary to achieve acylation at carbon even using reagents like acyl halides[121]. It follows that successful nucleophilic attack by carbon on carboxylic acids and esters tends to be limited to certain favourable types of system. Useful acid catalysts comprise those covalent metal halides which exhibit marked Lewis acidity (e.g. $AlCl_3$, $SnCl_4$, BF_3) and strong Brønsted acids like sulphuric, polyphosphoric and anhydrous (polymeric) hydrogen fluoride. The use of sulphuric acid is limited to systems which it does not attack in other ways. Among useful base catalysts are sodium hydride, and sodium alkoxides. The use of preformed organometallic derivatives of the carbon compound (e.g. RLi and R_2Cd) is really an exploitation of basic catalysis[1].

Various factors combine in restricting the majority of effective C-acylations to non-aqueous, indeed to non-hydroxylic, solvents. Hydroxylic oxygen atoms are, as a class, much more readily acylated than carbon atoms and are therefore preferentially attacked by acylating agents. Moreover, water-like substances tend to reduce the acidity of anhydrous systems and to deactivate, or even decompose, Lewis acids. Partially aqueous conditions can be tolerated in certain reactions of carboxylic acids (because hydrolysis leads only to the acid again) but the normal restriction to an aromatic, an ethereal, or a similar type of solvent, or to the use of an excess of one of the reactants, has meant that rather little kinetic study of these systems has been undertaken. Indeed many of the preparative recipes involve heterogeneous mixtures, a circumstance which acts as a powerful inhibitor of kinetic investigation. As a result only the bare outlines of reaction mechanisms are known with any certainty, and analysis in the depth possible in the preceeding sections concerning substitution by nitrogen must await further study. In short, work on most aspects of nucleophilic attack by carbon has yet to rise above the preparative level. Because of this, and because these topics have been covered in an earlier volume in the series[121], we shall deal rather briefly with carbon nucleophiles.

B. Reaction with Aromatic Compounds

1. Acid catalysis

Acid-catalysed nucleophilic substitution by aromatic carbon in carboxylic acids (equation 175) or esters (equation 176) comprises a rather small part of the Friedel–Crafts ketone synthesis[121]; to acylate an aromatic

$$RCOOH + ArH \xrightarrow{H_2SO_4} RCOAr + H_2O \qquad (175)$$

$$R^1COOR^2 + ArH \xrightarrow{AlCl_3} R^1COAr + R^2OH \qquad (176)$$

compound it is normal to choose a more powerful reagent (an acyl halide or acid anhydride) than an acid or ester. Nevertheless, both these reagents can be used with success in appropriate circumstances. The carboxylic acid is, in fact, very often chosen when intramolecular acylation is in-

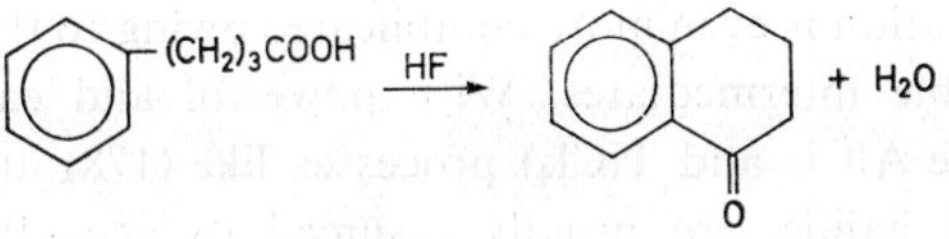

$$(177)$$

volved (equation 177). As well as being nucleophilic substitutions at the carboxylate group, these reactions are, of course, examples of electrophilic aromatic substitution. Not a great deal is known of the mechanism of the Friedel–Crafts ketone synthesis, and what is known has, in the main, been uncovered using acyl halides and anhydrides as reagents[121]. However, the general outlines will apply also to acylation by acids and esters. In keeping with what we have learnt about oxygen, sulphur, and nitrogen nucleophiles, the more basic the aromatic compound the more readily does it attack the acylating agent. Thus electron-repelling substituents favour, and electron-withdrawing substituents disfavour, reaction. (Hence the use of nitro- and halosubstituted benzenes as solvents for acylation of other aromatics.) In most cases of intermolecular reaction (for example reactions 175 or 176) it is very probable that even with reactive aromatic nucleophiles like anisole, only a carboxylic acid derivative with a very good leaving group will be successfully substituted; normally a species with significant acylium ion character will be required. It is the function of the catalyst to produce this species and this is why powerful catalysts are necessary in the present contexts. Some ways in which Brønsted and Lewis acid catalysts can interact with carboxylic acids and esters have been outlined (section I.B.1). Appreciable amounts of acylium ion will be likely only in media of high dielectric constant and in most other media coordinated species will predominate, coexisting with perhaps small quantities of ion pairs. The preference of ionic species for media of high dielectric constant is one reason for the spontaneous formation of heterogeneous mixtures in many Friedel–Crafts systems. Even when the acylating agent is certainly an acyl halide (RCOHal) interaction with the Lewis acid ($MHal_n$)—the type of catalyst normally used in such cases—can take place at two sites, **17** and **18**. Whether both **17** and **18** are generally capable of sustaining attack by aromatic carbon, whether the reaction of one

$$RCOCl \cdots MCl_n$$
$$\text{or} \quad RCO^+MCl_{n+1}^-$$
$$\text{or} \quad RCO^+ + MCl_{n+1}^-$$
$$(\mathbf{17})$$

$$RC\!\!=\!\!O : MCl_n$$
$$|$$
$$Cl$$
$$(\mathbf{18})$$

species normally predominates, or whether only acylium ions are effective, are topics which are still debated. With carboxylic acids and esters as reagents the situation is even more complicated owing to the greater multiplicity of possible intermediates. With powerful and easily solvolysed Lewis acids (like $AlCl_3$ and $TiCl_4$) processes like (178) and (179), which provide an acyl halide, are usually assumed to occur[33], although the detailed reaction scheme has not been rigidly established in any case so far

$$RCOOH + 2\,AlCl_3 \longrightarrow RCOClAlCl_3 + AlOCl + HCl \qquad (178)$$
$$R^1COOR^2 + 2\,AlCl_3 \longrightarrow R^1COClAlCl_3 + R^2Cl + AlOCl \qquad (179)$$

(*cf.* section II.D.2). Here the 'catalyst' is not regenerated, even in principle, so that strictly speaking the essential reaction involves neither acid nor ester, but acyl halide. Genuinely catalysed nucleophilic attack on acids or esters is probably confined to examples involving the weaker Lewis acids (*e.g.* BF_3, $SnCl_4$, $ZnCl_2$) where the evidence points to the absence of such processes as (178) and (179) and to the formation of coordination complexes (*e.g.* **19** and **20**). Species like HBF_3OCOR and $H_2SnCl_4(OCOR)_2$

$$R^1C\!\!=\!\!O : SnCl_4$$
$$|$$
$$OR^2$$
$$(\mathbf{19})$$

$$RC\!\!=\!\!O : BF_3$$
$$|$$
$$OH$$
$$(\mathbf{20})$$

are very powerful Brønsted acids[122] and may in turn protonate, and perhaps induce ionization of, other molecules of carboxylic acid rather as do concentrated mineral acids[3] (equations 180–183). With boron acids

$$RCOOH + H_2SO_4 \xrightleftharpoons{} RCOOH_2^+ + HSO_4^- \xrightleftharpoons{H_2SO_4} RCO^+ + H_3O^+ + 2\,HSO_4^- \qquad (180)$$
$$RCOOH + 2\,(HF)_n^* \xrightleftharpoons{} RCO^+ + H_3O^+ + 2\,F(HF)_{n-1}^- \qquad (181)$$

---

* Anhydrous hydrogen fluoride is highly polymerized, the stability of the polymeric anions accounting for its great acidity towards bases.

$$R^1COOR^2 + 2\,H_2SO_4 \;\rightleftharpoons\; R^1CO^+ + R^2OH_2^+ + 2\,HSO_4^- \tag{182}$$

$$RCOOH + HBF_3OCOR \;\rightleftharpoons\; RCOOH_2^+ + BF_3OCOR^- \xrightarrow{\;HBF_3OCOR]\;} RCO^+ + H_3O^+ + \\ + 2\,BF_3OCOR^- \tag{183}$$

$$RCOOH + BF_3 \;\rightleftharpoons\; RCO^+ + BF_3OH^- \tag{184}$$

(in particular) interaction like that shown in equation (184) is also conceivable[123]. In general, catalysis of aromatic acylation by strong Brønsted acids is presumably much along the lines of their catalysis of acylation in hydroxylic media[1] (equation 185; see section III), although the positions

$$\begin{array}{c}
\qquad\qquad\qquad\qquad\quad \overset{H}{|} \\
R^1COOR^2 + HX \;\rightleftharpoons\; R^1CO\!-\!OR^2 + X^- \;\rightleftharpoons\; R^1CO^+ + R^2OH + X^- \qquad\text{fast} \\
\qquad\qquad\qquad\qquad\quad\;\; \overset{+}{\downarrow}\text{ nucleophile} \qquad\qquad\qquad\quad\downarrow\text{ nucleophile} \qquad\text{slow} \\
\qquad\qquad\qquad\qquad\quad\;\; \text{Products} \qquad\qquad\qquad\qquad\qquad \text{Products} \qquad\quad (185)
\end{array}$$

of the equilibria involved will obviously depend upon the dielectric constant and solvating properties of the medium. A further point to remember concerning the Lewis acid catalysis of acylation by carboxylic acids and esters, is that one product will be either water or an alcohol (equations 186

$$RCOOH + ArH \longrightarrow RCOAr + H_2O \tag{186}$$

$$R^1COOR^2 + ArH \longrightarrow R^1COAr + R^2OH \tag{187}$$

and 187), and this will coordinate to, or decompose, some of the catalyst, possibly either activating or deactivating it. It may be, therefore, that however such acylations are initiated, the nature of the catalysis changes significantly during the course of the reaction. It will be apparent that while the general nature of the active acylating agent, and the role played by the catalysts, are clear enough, the exact description of these matters is very far from settled in Friedel–Crafts' acylation as a whole, and that this is particularly true when the carboxylic ester or acid is used. No doubt the details vary somewhat from reaction to reaction.

Because of the general nature of the active species, the slow steps of these aromatic acylations are not usually formulated in terms of carbonyl addition, but either as examples of scheme (24) (equation 188) or as examples of (25) with, or without, coordination to the carbonyl group as in equations (189) and (190). The loss of the aromatic proton can

$$\tag{188}$$

$$
\begin{array}{c}
\mathrm{RC}\!=\!\mathrm{O}\!:\!\mathrm{MX}_n + \bigcirc \longrightarrow \left[\begin{array}{c} \mathrm{H}\!\!\!\diagdown \!\!\!\diagup \overset{Y\cdots}{\underset{\delta^+}{\overset{\mathrm{R}}{\mathrm{C}}}}\!\!\diagdown \mathrm{O}\!:\!\mathrm{MX}_n \end{array}\right] \xrightarrow{\mathrm{MX}_n}
\end{array}
$$

$$
\left[\begin{array}{c} \mathrm{H}\!\!\!\diagdown \!\!\!\diagup \overset{\mathrm{R}}{\underset{+}{\overset{}{\mathrm{C}}}}\!\!\diagdown \mathrm{O}\!:\!\mathrm{MX}_n \end{array}\right] \ \mathrm{MX}_n\mathrm{Y}^- \longrightarrow \ \underset{\overset{\|}{\mathrm{O}}}{\mathrm{PhC}\!-\!\mathrm{R}} \ \ \ddot{\mathrm{M}}\mathrm{X}_n + \mathrm{HY} + \mathrm{MX}_n \tag{189}
$$

$$
\mathrm{RC}\!\!\underset{\overset{\|}{\mathrm{O}}}{-}\!\!\mathrm{Y}\!:\!\mathrm{MX}_n + \bigcirc \longrightarrow \overset{\overset{\mathrm{R}}{\underset{\mathrm{H}}{\diagdown}}\overset{\delta^-}{\mathrm{YMX}_n}}{\underset{\delta^+}{\mathrm{C}\!-\!\mathrm{O}}} \longrightarrow \left[\begin{array}{c} \overset{\mathrm{R}}{\underset{+}{\mathrm{H}\ \mathrm{C}\!-\!\mathrm{O}}} \end{array}\right] \mathrm{MX}_n\mathrm{Y}^- \longrightarrow \underset{\overset{\|}{\mathrm{O}}\ \ddot{\mathrm{M}}\mathrm{X}_n}{\mathrm{PhC}\!-\!\mathrm{R}} + \mathrm{HY} \tag{190}
$$

sometimes affect the rate. Slow steps like (188)–(190) are compatible with the observed effects (noted above) of substituents in the nucleophile. The effect of altering R (or Y) is, as in acid-catalysed acylation generally, difficult to predict in a given case: it depends on whether the extent of formation of the active intermediate (RCO^+ or $RYC\!=\!O : MX_n$) in say equation (181) or (8), or the reactivity of this intermediate in equations (188)–(190) is the dominant factor in controlling the overall rate constant. Little is known about this matter so far as acids and esters are concerned[121]. Indeed, it turns out that esters have rarely been used in Friedel–Crafts ketone syntheses, and very rarely with catalysis other than by aluminium chloride[124]—conditions which probably involve acylation via the acyl chloride rather than via the ester itself (equation 179). The complexity of reactions like (179) may account for the concurrent alkylation often observed in these systems[125]. Carboxylic acids have found their greatest application in intramolecular examples[124], but it has been demonstrated recently that very good yields of ketone can also be obtained in intermolecular reactions[126].

2. Intramolecular reactions

Aromatic compounds containing an acyl group suitably located in a side-chain can undergo ring closure[127] (equation 191). Five- and six-membered rings form most easily. Esters rarely appear in this context, but

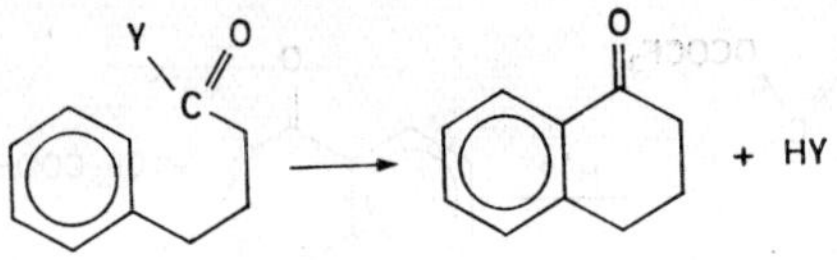

$$ \text{(191)} $$

carboxylic acids probably find as much application as acyl halides or anhydrides. Often the carboxylic acid is used with concentrated sulphuric acid, and especially with anhydrous hydrogen fluoride or polyphosphoric acid (PPA), as catalyst and solvent[128]. Under these conditions appreciable protonation, and significant acylium ion formation, will often be possible; and there is no reason to suppose that the overall mechanism in these intramolecular reactions differs in principle from that outlined above for intermolecular cases. Although there are few quantitative data on the point in the present context, it seems that the intramolecular reaction is easier to achieve than the equivalent intermolecular process. Thus *o*-benzoylbenzoic acids may be readily cyclized (equation 192), whereas benzophenone itself is resistant to further acylation. On several occasions in

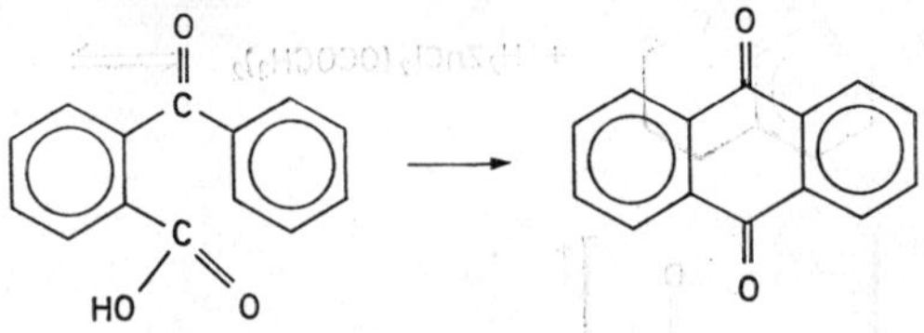

$$ \text{(192)} $$

earlier sections we have noted the comparative ease of intramolecular acylation (*e.g.* in sections II.D.1 and IV.E) and the present examples no doubt have the same underlying explanation. The positional and collisional advantages of the intramolecular reactions may be the reason that the carboxylic acid is entirely satisfactory in these contexts, the greater concentration of active intermediate which, in general, will be expected from catalysed processes involving the anhydride or acyl halide not being necessary. Indeed it is possible that as feeble a reagent as the free anhydride alone is adequate in some cases, provided that it is a relatively reactive anhydride (*e.g.* if it has a halogen-substituted leaving group). Thus one preparative method[129] simply involves adding trifluoroacetic anhydride to the carboxylic acid to be cyclized, thus forming the unsymmetrical anhydride, which then leads to the product (equations 193 and 194). It may

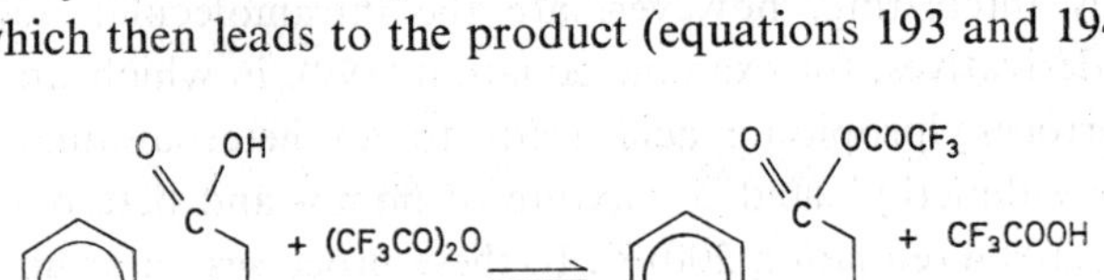

$$ + (CF_3CO)_2O \rightleftharpoons \quad + CF_3COOH \qquad \text{(193)} $$

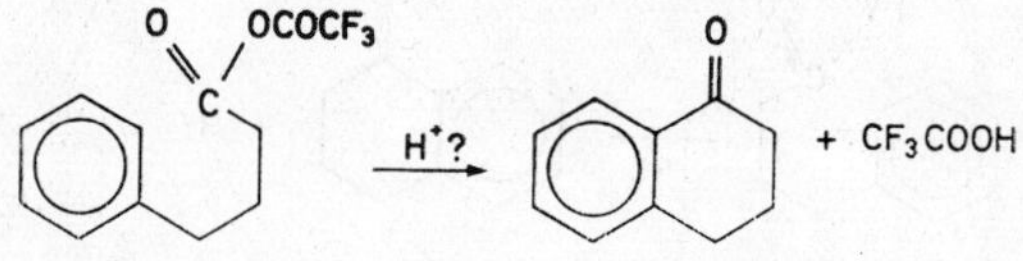

$$\text{(structure)} \xrightarrow{\text{H}^+?} \text{(structure)} + CF_3COOH \qquad (194)$$

be, however, that the trifluoroacetic acid formed also acts as a catalyst. Another preparative recipe which may owe its efficacy to a similar series of reactions (equations 195–198) is the treatment of a reactant with $Ac_2O/AcOH/ZnCl_2$ mixtures[128].

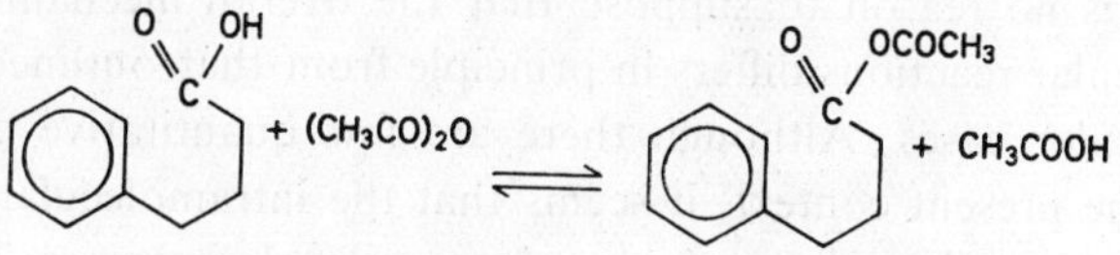

$$\text{(structure)} + (CH_3CO)_2O \rightleftharpoons \text{(structure)} + CH_3COOH \qquad (195)$$

$$ZnCl_2 + 2\,CH_3COOH \rightleftharpoons H_2ZnCl_2(OCOCH_3)_2 \qquad (196)$$

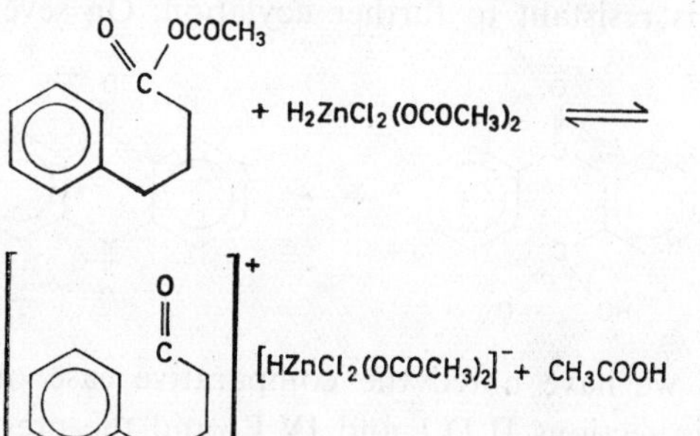

$$\text{(structure)} + H_2ZnCl_2(OCOCH_3)_2 \rightleftharpoons$$

$$\text{(structure)} \quad [HZnCl_2(OCOCH_3)_2]^- + CH_3COOH \qquad (197)$$

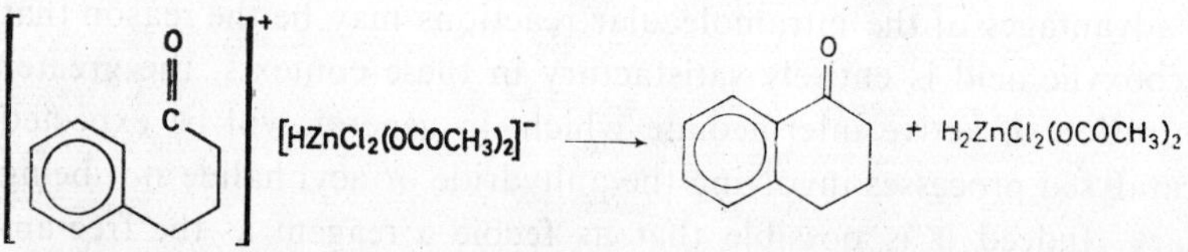

$$\text{(structure)} \quad [HZnCl_2(OCOCH_3)_2]^- \longrightarrow \text{(structure)} + H_2ZnCl_2(OCOCH_3)_2 \qquad (198)$$

Intramolecular acylation has many preparative ramifications and is widely used in syntheses of complex, multi-ring systems[127]. No new principles are involved in such reactions and they cannot be dealt with here. Particularly interesting, however, are the intramolecular acylations of ferrocene derivatives, for example equation (199), in which an α,β-substituted β-ferrocenylpropionic acid leads to an heteroannular system[130]. With the α,α-dimethyl acid a mixture of homo- and heteroannular products is obtained (equation 200)[131]. In these processes an aromatic carbon atom of a cyclopentenyl system is the nucleophile.

(199)

(200)

(201)

Finally, an instructive recent example[132], which illustrates several of the points made in the foregoing discussion, is shown in equation (201). The same product, 5,6-dimethoxy-1-indanone, is obtained whether the free acid or its methyl ester is used as starting material, but different routes are followed. These show (i) that the ester is less reactive than the free acid under comparable conditions as an intermolecular acylating agent and (ii) that esters *can* be used successfully in intramolecular cyclizations with polyphosphoric acid.

Basic catalysis is discussed in section V.D.2 below.

C. Reaction with Olefins and Acetylenes

1. Acidic catalysis

The acid-catalysed acylation of olefins exhibits many of the features shown by the analogous reactions involving aromatic compounds and displays some extra complications, one of which is the possibility of addition (equation 202) as well as of substitution (equation 203). As in

$$R^1COY + R^2CH{=}CHR^2 \longrightarrow R^2CHYCHR^2COR^1 \qquad (202)$$

$$R^1COY + R^2CH{=}CHR^2 \longrightarrow R^2CH{=}CR^2COR^1 + HY \qquad (203)$$

the aromatic ketone synthesis, carboxylic acids and esters are not normally used[121]—although they can be effective on occassion—the common reagents being $RCOHal/MX_n$ or $(RCO)_2O/ZnCl_2$. Alone, carboxylic acids tend to add to olefins forming esters (equation 204). If the acid is converted to a

$$R^1COOH + R^2CH{=}CHR^2 \longrightarrow \begin{array}{c} R^2CH{-}CH_2R^2 \\ | \\ OCOR^1 \end{array} \qquad (204)$$

suitable anhydride, for example, as in preparative recipes[128, 129] which involve $(CF_3CO)_2O$ or $(CH_3CO)_2O/ZnCl_2$, then acylation occurs (equations 205 and 206), but this hardly comprises nucleophilic substitution in a carboxylic acid. The addition compound **21** can often be isolated when

$$R^1COOH + (CF_3CO)_2O \;\rightleftharpoons\; R^1COOCOCF_3 + CF_3COOH \qquad (205)$$

$$R^1COOCOCF_3 + R^2CH{=}CHR^2 \xrightarrow{H^+?} \begin{array}{c} R^2CH{-}CHR^2COR^1 \\ | \\ OCOCF_3 \\ (21) \end{array} \longrightarrow R^2CH{=}CR^2COR^1$$
$$+ CF_3COOH$$
$$(206)$$

the trifluoroacetic anhydride method is employed, and with acetylenes comprises a relatively stable β-keto enol trifluoroacetate (**22**) which can be hydrolysed to a β-diketone.

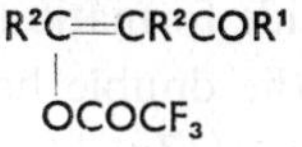
R²C=CR²COR¹
OCOCF₃

(22)

The use of the carboxylic acid in these contexts is, as for the aromatic systems, most common in intramolecular substitutions. However, unlike the aromatic systems, addition (lactonization) is likely to predominate over ketone formation unless suitable catalysts are used[133]. In the general case one will obtain a mixture of products (equation 207), both lactones and ketones, and both five- and six-membered rings, the double bond often being able to migrate under the experimental conditions. The balance between lactonization and acylation appears to depend primarily

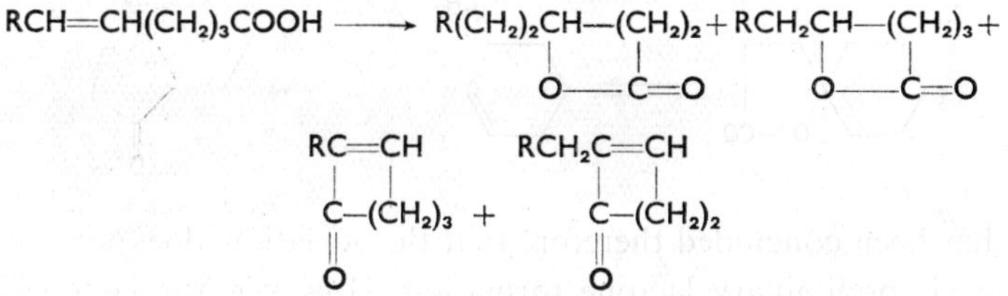
RCH=CH(CH₂)₃COOH
R(CH₂)₂CH—(CH₂)₂ +
O—C=O
RCH₂CH—(CH₂)₃ +
O—C=O
RC=CH
C—(CH₂)₃ +
O
RCH₂C=CH
C—(CH₂)₂
O

(207)

upon the catalyst and upon the degree of aromaticity of the olefin. Thus concentrated sulphuric acid, which leads mainly to lactones with open-chain olefinic acids, but to good yields of ketones with aromatic acids, provides with **23** approximately equal amounts of lactone and ketone.

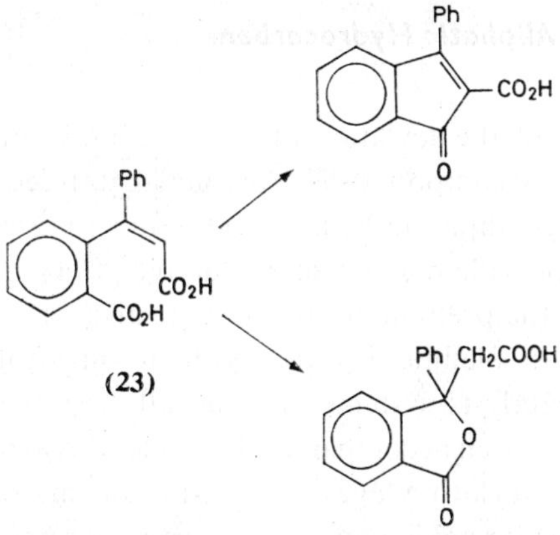
Ph
CO₂H
O
Ph
CO₂H
CO₂H
Ph CH₂COOH
O
O

(23)

(208)

It is clear that it is the susceptibility of the carbon–carbon double bond to protonation which determines the susceptibility to lactonization. With olefinic acids therefore the most desirable Brønsted acid catalysts are those which promote mixed anhydride formation at the carboxyl group with the minimum of protonation of the double bond. It appears that polyphosphoric acid, and the $Ac_2O/AcOH/ZnCl_2$ or $(CF_3CO)_2O/CF_3COOH$ combinations, do this best[133]. Polyphosphoric acid is a very effective catalyst in many of these cyclizations but the origin of its efficacy is not yet clear[128]. As with the substitution by aromatic carbon, one can only speculate about the finer details of the reaction mechanisms, which, in the present case, are complicated by the possibility of addition.

Whereas the acylations are essentially irreversible, lactonizations are not, and suitable γ- or δ-lactones can on occasion be used as the starting materials rather than the free acid (equation 209). The yields are poorer,

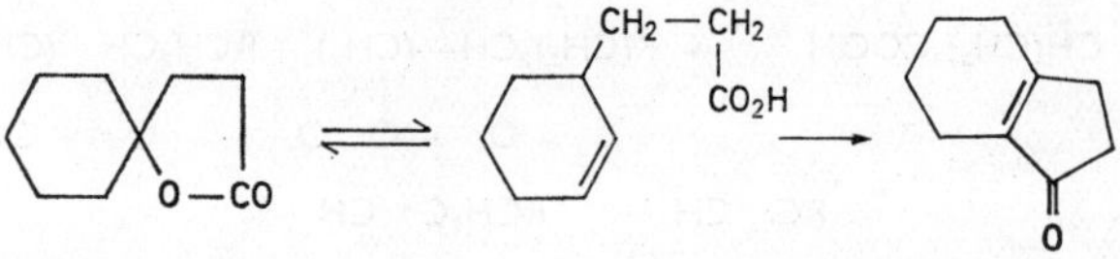

$$\text{(209)}$$

and it has been concluded therefore that the acylation does not normally proceed via preliminary lactone formation. However, the lactone/ketone product ratio should, in any particular case, depend upon the contact time. The yields in all these olefin reactions are often rather low owing to the variety of possible side-reactions which can be induced by the acid catalysts.

Basic catalysis is discussed in section V.D.2 below.

D. Reaction with Aliphatic Hydrocarbons

1. Acidic catalysis

Acidic catalysis of the acylation of hydrocarbons only appears successful with unsaturated compounds[121]. Any acid-catalysed acylation of saturated, unsubstituted aliphatic hydrocarbons, or saturated compounds substituted so as to provide one or more 'active' CH_2 groups, always, in fact, proceeds via the preliminary formation (also under the influence of the catalyst) of an olefinic centre. With unsubstituted hydrocarbons rather drastic catalyst systems are usually necessary to effect this oxidation, which is often accompanied by skeletal rearrangements. Correspondingly active acylating agents are also normally employed and it seems that carboxylic acids and esters have very rarely been used in any of these contexts[121, 134]. However, ketones are acylated by acids in the presence of polyphosphoric acid (equation 210). The exact mechanism can

$$R^1CO_2H + R^2CH_2COCH_2R^2 \xrightarrow{PPA} R^1COCHCOCH_2R^2 + H_2O \qquad (210)$$
$$\underset{R^2}{|}$$

only be conjectured[135], but very probably proceeds via the enol or enol ester, being similar in outline to the analogous reaction using the carboxylic anhydride with boron fluoride as catalyst[1, 121]. In polyphosphoric acid further condensation leads to pyranone derivatives[135].

2. Basic catalysis

It will be recalled that basic catalysts function by increasing the nucleophilicity of the substrate. When the substrate atom is carbon, basic catalysis consists essentially in the provision of a species approximating to the carbanion. This may be achieved either by preforming an organometallic derivative of the hydrocarbon (and this applies for aliphatic, olefinic or aromatic compounds) or, for substituted aliphatic compounds containing an 'active' CH group (*e.g.* esters, ketones, or nitriles), by interaction *in situ* with appropriate strong bases.

   a. *Reaction with organometallic compounds.* Typical reactions[134], which are usually conducted in inert media of low dielectric constant, such as

$$R^1ZnBr + R^2COBr \longrightarrow R^1COR^2 + ZnBr_2 \qquad (211)$$

$$R^1MgBr + R^2COOEt \longrightarrow R^2COR^1 + MgBrOEt \qquad (212)$$

$$R_2^1Cd + 2\ R^2COCl \longrightarrow 2\ R^1COR^2 + CdCl_2 \qquad (213)$$

ether or benzene, are given above. Most organometallic compounds tend to react further than shown in equations (211)–(213): they add to the carbonyl group of the ketone. Thus the usual product of the reaction between Grignard reagents and acyl derivatives is the tertiary alcohol. By conducting the reaction at low temperatures, and in the presence of excess of acyl derivative, the ketone may, however, be isolated[136]. In general the most useful reagents have proved to be those of cadmium, although there has been much confusion—and there remains some uncertainty—as to the reason for this[137]. From the restricted viewpoint of the present chapter these preparatively important processes are, somewhat paradoxically, of mainly theoretical interest. This is because (i) carboxylic acids, like other hydroxylic molecules, decompose organometallics to hydrocarbons (and are therefore usually avoided), and (ii) esters are rarely used in schemes like (212), while organocadmium derivatives have been considered unable to attack esters at significant speeds.

   One useful and promising route to ketones uses organolithium derivatives[138], with the lithium salt of a carboxylic acid, or even with the free acid (equations 214 and 215). The ketone is obtained in good yield upon hydrolysis. The carbonyl addition intermediate suggested in equation (215),

$$R^1Li + R^2COOH \longrightarrow R^2COOLi + R^1H \tag{214}$$

$$R^1Li + R^2C{-}OLi \longrightarrow \left[\begin{matrix} R^1 & OLi \\ & C \\ R^2 & OLi \end{matrix} \right] \xrightarrow{H_2O} R^1COR^2 + 2\,LiOH \tag{215}$$

and those usually written for the more conventional Grignard syntheses, correspond to the essential pattern of scheme (26). For further details we must await kinetic study. It appears probable that other syntheses similar to equation (215) will be developed.

   b. *Reactions with compounds containing 'active' C–H bonds.* The, as yet, most widely studied aliphatic compounds possessing 'active' hydrogen atoms are ketones, esters, and nitriles[12, 139]. The electron-withdrawing substituents in these hydrocarbons render the α-hydrogen atoms relatively acidic and thus lead to significant carbanion formation on contact with strong bases. The commonly used bases are sodium alkoxides, sodamide and sodium hydride; less often the very powerful triphenylmethylsodium is employed. In this context an ester is the commonest acylating agent, although others are used in certain circumstances. A very wide range of substrates has been studied, especially for ketones and esters[12, 139]. Typical reactions are shown in equations (216), (217), and (218). These reactions

$$2\,CH_3CO_2C_2H_5 \xrightleftharpoons{NaOC_2H_5} CH_3COCH_2CO_2C_2H_5 + C_2H_5OH \tag{216}$$

$$CH_3CO_2C_2H_5 + (CH_3)_2CO \xrightleftharpoons{NaOC_2H_5} CH_3COCH_2COCH_3 + C_2H_5OH \tag{217}$$

$$CH_3CO_2C_2H_5 + RCH_2CN \xrightleftharpoons{NaOC_2H_5} RCH(CN)COCH_3 + C_2H_5OH \tag{218}$$

are generally termed Claisen condensations[140]; reaction (216) is, of course, the acetoacetic ester condensation. The products are β-diketones, β-keto esters, β-keto nitriles, and related compounds. Reaction is often conducted in an excess of one reactant as solvent or in an inert medium like benzene or ether. The three essential components (substrate, acylating agent, and catalyst) are mixed at ordinary temperatures and the product, a stronger acid than the original substrate, usually precipitates as its sodium salt. Were this not to occur these equilibria would often lie largely on the reactant side. When ethyl esters are used as acylating agents the forward process is often further aided by distilling out the ethanol as it is formed. Because catalyst is removed by the product, molecular proportions are necessary, and a two-fold excess is common. It is clear that side-reactions will frequently limit the yields. These are, however, often good. Self-condensation of the acylating agent, the substrate or the product, polysubstitution, and O- rather than, or as well as, C-acylation are all possibilities among others. The last two complications seem particularly important in variations which employ the acid chloride or anhydride with the pre-formed (*i.e.* free from alkoxide, *etc.*) sodium derivative of the substrate, *e.g.* equations (219) and (220). For these reasons the reaction with acyl

$$2\,R^1COCl + NaCHR^2COR^3 \longrightarrow R^1COCR^2{=}CR^3OCOR^1 + HCl + NaCl \qquad (219)$$

$$2\,R^1COCl + NaCH_2COR^2 \longrightarrow (R^1CO)_2CHCOR^2 + HCl + NaCl \qquad (220)$$

halides and anhydrides has proved of less preparative value than that with esters, except in special contexts. Work on the factors which control the balance between O- and C-acylation is only just beginning[12, 141].

With suitable keto esters, intramolecular acylations may be effected. These produce normally five- or six-membered rings, the second keto group appearing endo- or exocyclic depending upon the original structure (equations 221 and 222). When the substrate is a diester the cyclization is known as the Dieckmann reaction[142].

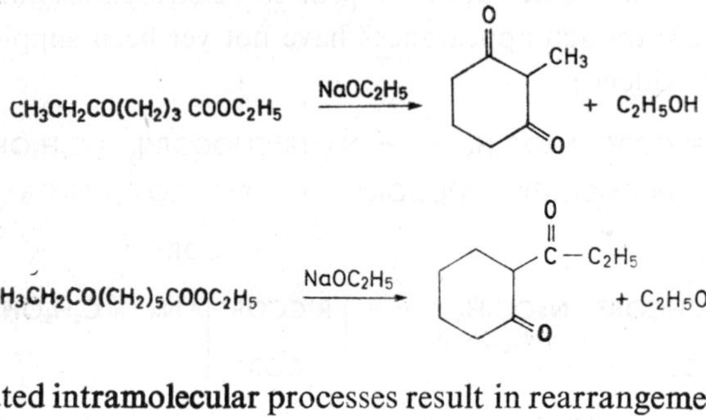

$$CH_3CH_2CO(CH_2)_3\,COOC_2H_5 \xrightarrow{\;NaOC_2H_5\;} \text{(cyclohexanone with CH}_3\text{)} + C_2H_5OH \qquad (221)$$

$$CH_3CH_2CO(CH_2)_5COOC_2H_5 \xrightarrow{\;NaOC_2H_5\;} \text{(cyclohexanone with } C{-}C_2H_5\text{)} + C_2H_5OH \qquad (222)$$

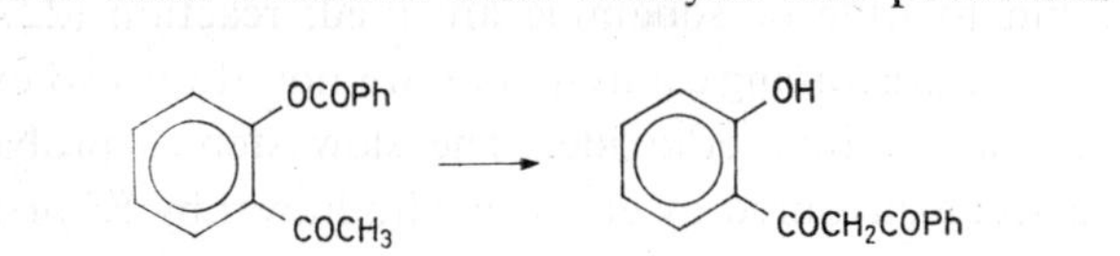

Other related **intramolecular** processes result in rearrangements[143] (equation 223). For these reactions mild catalysts like potassium carbonate

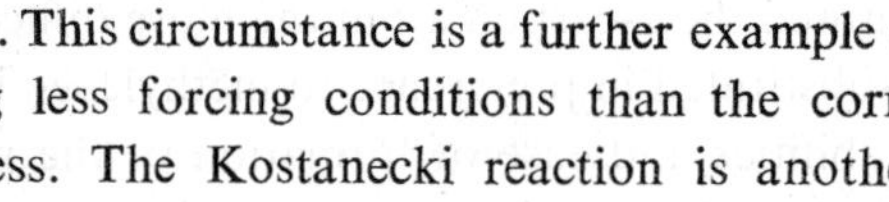

$$\text{(aryl OCOPh / COCH}_3\text{)} \longrightarrow \text{(aryl OH / COCH}_2\text{COPh)} \qquad (223)$$

appear adequate. This circumstance is a further example of intramolecular effects requiring less forcing conditions than the corresponding inter-molecular process. The Kostanecki reaction is another intramolecular variation[144].

The use of a formic ester as the acylating agent leads to an alde-hyde[12, 139, 145]. With ketones the product, a β-keto aldehyde, often exists mainly in its enolic form (equations 224 and 225).

$$(CH_3)_2CO + HCOOC_2H_5 \longrightarrow CH_3COCH{=}CHOH + C_2H_5OH \qquad (224)$$

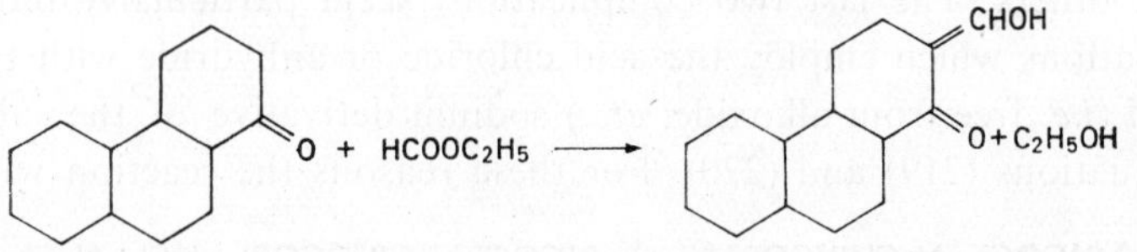

$$(225)$$

Very recently aliphatic sulphoxides, sulphones and related compounds have been subjected to Claisen-like reactions[146]. Their preformed carbanions react readily with esters. The acyl sulphoxides formed on hydrolysis can, if desired, be smoothly reduced to a sulphur-free ketone (equation 226).

$$R^1COOR^2 + 2\,CH_3SOCH_2^- Li^+ \longrightarrow [R^1COCHSOCH_3]^- Li^+ + Me_2SO + LiOR^2 \xrightarrow{H_3O^+}$$
$$R^1COCH_2SOCH_3 \xrightarrow[H_2O]{Al/Ag} R^1COCH_3 \qquad (226)$$

These base-catalysed acylations appear to be straightforward examples of equation (25) though appearances have not yet been supplemented by much kinetic evidence.

$$R^1CH_2COR^2 + NaOC_2H_5 \rightleftharpoons Na^+[R^1CHOCOR^2]^- + C_2H_5OH \qquad (227)$$

$$Na^+[R^1CHCOR^2]^- + R^3COOR^4 \rightleftharpoons R^1CHCOR^2 + NaOR^4 \atop \qquad\qquad\qquad\quad COR^3 \qquad (228)$$

$$R^1CHCOR^2 + NaOC_2H_5 \rightleftharpoons \left[R^1CCOR^2 \atop \;\; COR^3 \right]^- Na^+ + C_2H_5OH \atop COR^3 \quad (or\ Na[OR^4]) \qquad\qquad (229)$$

When sodium hydride or sodamide are used, reaction (227) lies well to the right. These are forcing catalysts but are not always so experimentally convenient as sodium ethoxide. The slow step is probably (228), since reaction seems favoured by electron withdrawal by R^3 or R^4: phenyl esters rather than ethyl esters are often advantageous. In the substrate the more highly substituted the α-carbon atom the more difficult the reaction: methyl is usually substituted in preference to methylene. This effect parallels the relative stabilities and ease of formation of the respective carbanions and implies that the relative amounts of these species available are more important than their relative reactivities. Steric factors may also intrude.

E. Reaction with Hydrogen Cyanide

The reaction of esters with hydrogen cyanide (equation 230) comprises a somewhat special case of substitution by carbon, the product being

$$R^1CO_2R^2 + HCN \longrightarrow R^1COCN + R^2OH \qquad (230)$$

an acyl cyanide, not a conventional ketone. Rather little is known about this type of process and most of the available data refer, in fact, to thiol esters[147]. The reaction can take place in aqueous solution and very probably proceeds via a nucleophilic attack on the ester by CN^-. Since acyl cyanides are powerful acylating agents[148] and react rapidly with water (equation

$$RCOCN + H_2O \longrightarrow RCO_2H + HCN \qquad (231)$$

231), the overall effect of cyanolysis in aqueous solution is the hydrolysis of the ester, the cyanide acting as nucleophilic catalyst.

There is no doubt that cyanide will react with acylating agents other than esters, but little is yet known of any such processes involving carboxylic acids.

VI. CONCLUSION

While there is still much to be learnt about the mechanisms of substitution in carboxylic esters and acids by oxygen, sulphur, and nitrogen nucleophiles, these topics are relatively very well understood compared with the analogous substitutions involving nucleophilic halogen or carbon. Further kinetic studies of these latter processes are therefore awaited with particular interest. We postulate that intramolecular mechanisms will prove very widespread in substitutions by halogen. Indeed one feature highlighted by this survey is the importance throughout the field of intramolecular substitutions. It is probable that in many more cases than are currently recognized, the ease of reaction not only of intra-, but also of formally intermolecular acylations, is influenced by the possibilities available for the formation of a cyclic transition state.

For oxygen, sulphur, and nitrogen nucleophiles current attention is rapidly shifting towards the problem of unravelling their involvement in biological systems, and in formulating models for enzymic catalysis in the light of knowledge, like that detailed in this chapter, for non-enzymic processes.

REFERENCES

1. D. P. N. Satchell, *Quart. Rev. (London)*, **17**, 160 (1963).
2. T. C. Bruice and S. Benkovic, *Bioorganic Mechanisms*, Vol. 1 and 2, Benjamin, New York, 1966.
3. M. H. Palmer, *Chem. Ind. (London)*, 589 (1963); A. Casadevall, G. Cauquil and R. Corriu, *Bull. Soc. Chim. France*, 187 (1964); G. A. Olah, M. E. Moffatt, S. J.

Kuhn and B. A. Hardie, *J. Am. Chem. Soc.*, **86**, 2198 (1964); D. P. N. Satchell and J. L. Wardell, *Trans. Faraday Soc.*, **61**, 1133 (1965).

4. M. L. Bender, *Chem. Rev.*, **60**, 53 (1960).

5. J. Hine, *Physical Organic Chemistry*, 2nd ed., McGraw-Hill, New York, 1962, Chap. 4; H. H. Jaffe, *Chem. Rev.*, **53**, 191 (1953).

6. R. Huisgen and H. Olt, *Tetrahedron*, **6**, 253 (1959); R. J. B. Marsden and L. E. Sutton, *J. Chem. Soc.* 3183 (1936).

7. R. W. Taft Jr., in *Steric Effects in Organic Chemistry* (Ed. M. S. Newman), John Wiley and Sons, New York, 1956, Chap. 13.

8. V. Gold, *J. Chem. Soc.*, 4633 (1956).

9. M. L. Bender, *J. Am. Chem. Soc.*, **73**, 1626 (1951).

10a. D. Samuel and B. L. Silver, in *Advances in Physical Organic Chemistry* (Ed. V. Gold), Vol. 3, Academic Press, London, 1965.

10b. D. P. N. Satchell, *J. Chem. Soc.*, 555 (1963).

11. L. R. Fedor and T. C. Bruice, *J. Am. Chem. Soc.*, **86**, 5697 (1964); G. E. Lienhard and W. P. Jencks, *J. Am. Chem. Soc.*, **87**, 3855 (1965); F. Hibbert and D. P. N. Satchell, *Chem. Commun.*, 516 (1966).

12. H. O. House, *Modern Synthetic Reactions*, Benjamin, New York, 1965; E. Mosetig, *Organic Reactions*, **8**, (1954), Chapter 5.

13. J. H. Simons, *Fluorine Chemistry*, Academic Press, London, 1954.

14. F. A. Long and M. A. Paul, *Chem. Rev.*, **57**, 935 (1957).

15. H. J. Hagemayer and D. C. Hull, *Ind. Eng. Chem.*, **41**, 2920 (1949).

16. H. H. Wasserman and P. S. Wharton, *J. Am. Chem. Soc.*, **82**, 661, 1411 (1960); E. Euranto and T. Kujanpaa, *Acta Chem. Scand.*, **15**, 1209 (1961).

17. E. A. Jeffery and D. P. N. Satchell, *J. Chem. Soc.*, 2889 (1963).

18. J. A. Landgrebe, *J. Org. Chem.*, **30**, 2997 (1965).

19. L. Heslinga and J. F. Arens, *Rec. Trav. Chim.*, **76**, 982 (1957).

20. E. A. Jeffery and D. P. N. Satchell, *J. Chem. Soc.*, 3002 (1963).

21. E. A. Jeffery and D. P. N. Satchell, *J. Chem. Soc.*, 1876, (1962).

22. A. H. Fainberg and S. Winstein, *J. Am. Chem. Soc.*, **78**, 2770 (1956).

23. E. Gryszkiewicz–Trochimowski, A. Sporzynski and J. Wunk, *Rec. Trav. Chim.*, **66**, 419 (1947).

24. D. P. N. Satchell, *J. Chem. Soc.*, 1752 (1960).

25. C. F. H. Allen, C. J. Kibler, D. M. McLachlin and C. V. Wilson, *Org. Syn.*, **26**, 1 (1946).

26. H. C. Brown, *J. Am. Chem. Soc.*, **60**, 1325 (1938); see also L. P. Kyrides, *J. Am. Chem. Soc.*, **59**, 207 (1937).

27. A. J. Meschentsev, *Chem. Tiel.*, **16**, 203 (1946); *Chem. Abstr.*, **41**, 706 (1947).

28. R. Adams and L. H. Ulrich, *J. Am. Chem. Soc.*, **42**, 599 (1920); J. Prat and A. Etienne, *Bull. Soc. Chim. France.*, **11**, 30 (1944).

29. A. Dobner, *Ann. Chem.*, **210**, 246 (1881).

30. W. J. Hickinbottom, *Reactions of Organic Compounds*, Longmans, London, 1959.

31. E. S. Gould, *Mechanism and Structure in Organic Chemistry*, Holt, Rinehart and Winston, New York, 1959, Chap. 8.

32. T. C. Bruice and S. Benkovic, *Biorganic Mechanisms*, Vol. 1, Benjamin, New York, 1966, p. 119.

33. H. Schmidt, C. Blohm and G. Jander, *Angew. Chem.*, **A59**, 233 (1947); P. H. Groggins, R. H. Nagel and A. I. Stirton, *Ind. Eng. Chem.*, **26**, 1313 (1934); S. H. Dandegaonker, *J. Karnatak Univ.*, **7**, 95 (1962); *Chem. Abstr.*, **62**, 2700 (1965).

34. Houben–Weyl, *Methoden der Organischen Chemie*, Vol. 8, Verlag, Stuttgart, 1952, Chap. 4 and 5.

35. C. K. Ingold, *Structure and Mechanism in Organic Chemistry*, Bell and Sons, London, 1953.

36. J. N. E. Day and C. K. Ingold, *Trans. Faraday Soc.*, **37**, 686 (1941).
37. H. P. Treffers and L. P. Hammett, *J. Am. Chem. Soc.*, **59**, 1708 (1937); M. S. Newman, *J. Am. Chem. Soc.*, **63**, 2431 (1941).
38. J. F. Bunnett, *J. Am. Chem. Soc.*, **83**, 4978 (1961); R. L. Hansen, and W. O. Ney, *J. Org. Chem.*, **27**, 2059 (1962); G. Aksnes and J. E. Prue, *J. Chem. Soc.*, 103 (1959); A. R. Osborn and E. Whalley, *Trans. Faraday Soc.*, **58**, 2144 (1962); R. B. Martin, *J. Am. Chem. Soc.*, **84**, 4130 (1962).
39. D. P. N. Satchell, *J. Chem. Soc.*, 555 (1963); J. A. Duffy and J. A. Leisten, *J. Chem. Soc.*, 853 (1960).
40. K. G. Wyness, *J. Chem. Soc.*, 2934 (1958); see also S. M. Kupchan, S. P. Eriksen and Y.-T. S. Liang, *J. Am. Chem. Soc.*, **88**, 347 (1966).
41. W. P. Jencks and J. Carriuolo, *J. Am. Chem. Soc.*, **83**, 1743 (1961); E. Berliner and L. H. Altschul, *J. Am. Chem. Soc.*, **74**, 4110 (1952); E. A. Moelwyn-Hughes, *J. Chem. Soc.*, 4301 (1962).
42. H. B. Friedman and G. V. Elmore, *J. Am. Chem. Soc.*, **63**, 864 (1941); R. F. Hudson and I. Stelzer, *Trans. Faraday Soc.*, **54**, 213 (1958); P. J. Lillford and D. P. N. Satchell, *J. Chem. Soc.*, (B), 889, (1968).
43. J. Hipkin and D. P. N. Satchell, *Tetrahedron*, **21**, 835 (1965).
44. P. N. Rylander and D. S. Tarbell, *J. Am. Chem. Soc.*, **72**, 3021 (1950); B. K. Morse and D. S. Tarbell, *J. Am. Chem. Soc.*, **74**, 416 (1952).
45. T. C. Bruice and S. Benkovic, *Biorganic Mechanisms*, Vol. 1, Benjamin, New York, 1966, Chap. 1 and 3.
46. See the relevant chapters in this volume.
47. C. A. Bunton, N. A. Fuller, S. G. Perry and I. H. Pitman, *J. Chem. Soc.*, 4478 (1962).
48. Ya. K. Syrkin and I. I. Moiseev, *Usp. Khim.*, **27**, 717 (1958); K. J. Laidler and P. A. Landskroener, *Trans. Faraday Soc.*, **52**, 200 (1956); C. A. Lane, *J. Am. Chem. Soc.*, **86**, 2521 (1964).
49. C. A. Bunton, D. H. James and J. B. Senior, *J. Chem. Soc.*, 3364 (1960).
50. A. J. Everett and G. J. Minkoff, *Trans. Faraday Soc.*, **49**, 410 (1953); A. G. Mitchell and W. F. K. Wynne-Jones, *Trans. Faraday Soc.*, **52**, 824 (1956).
51. Y. Ogata and Y. Sawaki, *Tetrahedron*, **21**, 3381 (1965).
52. A. G. Davies, *Organic Peroxides*, Butterworths, London, 1961.
53. R. G. Pearson and D. N. Edgington, *J. Am. Chem. Soc.*, **84**, 4607 (1962); J. O. Edwards and R. G. Pearson, *J. Am. Chem. Soc.*, **84**, 16 (1962); see also T. C. Bruice, A. Donzel, R. W. Huffman and A. R. Butler, *J. Am. Chem. Soc.*, **89**, 2106 (1967).
54. J. Epstein, M. M. Demek and D. H. Rosenblatt, *J. Org. Chem.*, **21**, 796 (1956).
55. W. P. Jencks and J. Carriuolo, *J. Am. Chem. Soc.*, **82**, 1778 (1960).
56. R. F. Clark and J. H. Simons, *J. Am. Chem. Soc.*, **75**, 6305 (1953).
57. J. F. Arens and H. C. Volger, *Rec. Trav. Chim.*, **77**, 1170 (1958) and earlier papers.
58. G. Eglinton, E. R. H. Jones, B. L. Shaw and M. C. Whiting, *J. Chem. Soc.*, 1860 (1954); J. C. Arnell, J. R. Dacey and C. C. Coffin, *Can. J. Research*, **18B**, 410 (1940) and earlier papers.
59. R. B. Woodward and R. A. Olofson, *J. Am. Chem. Soc.*, **83**, 1007 (1961).
60. T. Higuchi, T. Miki, A. C. Shah and A. K. Herd. *J. Am. Chem. Soc.*, **85**, 3655 (1963).
61. T. Higuchi and K-H. Gensch, *J. Am. Chem. Soc.*, **88**, 3874 (1966); T. Higuchi, J. D. McRae and A. C. Shah, *J. Am. Chem. Soc.*, **88**, 4015 (1966).
62. M. L. Bender and M. C. Neveu, *J. Am. Chem. Soc.*, **80**, 5380 (1958).
63. A. R. Butler and V. Gold, *J. Chem. Soc.*, 1334 (1962).
64. M. Kilpatrick and M. L. Kilpatrick, *J. Am. Chem. Soc.*, **52**, 1418 (1930); M. Kilpatrick, *J. Am. Chem. Soc.*, **50**, 2891 (1928).

65. For a review see S. L. Johnson in *Advances in Physical Organic Chemistry* (Ed. V. Gold), Vol. 5, Academic Press, London, 1967.
66. J. Hipkin and D. P. N. Satchell, *J. Chem. Soc.*, 1057 (1965); see also J. F. Kirsch and W. P. Jencks, *J. Am. Chem. Soc.*, **86**, 833, 837 (1964).
67. A. R. Fersht and A. J. Kirby, *J. Am. Chem. Soc.*, **89**, 5960 (1967).
68. G. R. Schonbaum and M. L. Bender, *J. Am. Chem. Soc.*, **82**, 1900 (1960).
69. J. Thanassi and T. C. Bruice, *J. Am. Chem. Soc.*, **88**, 747 (1966).
70. H. T. Clark and E. J. Rahrs, *Org. Syn.*, Coll. Vol. 1, 1941, p. 91.
71. T. G. Bonner, E. G. Gabb, and P. McNamara, *Rec. Trav. Chim.*, **84**, 1253 (1965); T. G. Bonner, E. G. Gabb, P. McNamara and B. Smethurst, *Tetrahedron*, **21**, 463 (1965) and earlier papers.
72. A. R. Emery and V. Gold, *J. Chem. Soc.*, 1443 (1950).
73. R. Kuhn and I. Low, *Chem. Ber.*, **77**, 211 (1944).
74. R. J. Gillespie and J. A. Leisten, *Quart Rev. (London)*, **8**, 40 (1954); J. A. Leisten, *J. Chem. Soc.*, 298 (1955).
75. F. D. Popp and W. D. McEwen, *Chem. Rev.*, **58**, 321 (1958).
76. E. A. Jeffery and D. P. N. Satchell, *J. Chem. Soc.*, 1887 (1962).
77. E. A. Jeffery and D. P. N. Satchell, *J. Chem. Soc.*, 1913 (1962).
78. A. van Peski, *Rec. Trav. Chim.*, **40**, 103 (1921); G. W. Kenner, *Chem. Ind. (London)* 15 (1951).
79. D. D. Perrin, *The Dissociation Constants of Organic Bases in Aqueous Solution*, Butterworths, London, 1965.
80. G. H. Coleman and A. M. Alvarado, *Org. Synth.*, Coll. Vol. 1, 3 (1941).
81. E. E. Reid, *Organic Chemistry of Bivalent Sulphur*, Vol. 4, Chemical Publishing Co., New York, 1962; P. J. Hawkins, D. S. Tarbell and P. Noble, *J. Am. Chem. Soc.*, **75**, 4462 (1953).
82. H. Morawetz and P. S. Otaki, *J. Am. Chem. Soc.*, **85**, 463 (1963).
83. H. R. Snyder and C. T. Elston, *J. Am. Chem. Soc.*, **76**, 3039 (1954).
84. V. E. Bel'skii, M. L. Ivankova and M. I. Vinnik, *Zh. Fiz. Khim.*, **39**, 1426,1624 (1965).
85. P. B. Russell, *J. Am. Chem. Soc.*, **72**, 1853 (1950).
86. H. C. Bassett and C. R. Thomas, *J. Chem. Soc.*, 1188 (1954).
87. P. Reynaud, R. C. Moreau and J.-P. Samara, *Bull. Soc. Chim. France*, 3623, 3628 (1965).
88. M. Bodanszky and M. A. Ondetti, *Peptide Synthesis*, Interscience Publishers, New York (1966); N. F. Albertson, *Org. Reactions*, **12**, 157 (1962).
89. L. A. Paquette, *J. Am. Chem. Soc.*, **87**, 5186 (1965); R. B. Woodward, R. A. Olofson and H. Mayer, *J. Am. Chem. Soc.*, **83**, 1010 (1961).
90. G. Tadema, E. Harryvan, H. J. Panneman and H. J. Arens, *Rec. Trav. Chim.*, **83**, 345 (1964).
91. J. C. Sheehan and G. B. Hess, *J. Am. Chem. Soc.*, **77**, 1067 (1955); H. G. Khorana, *Chem. Ind. (London)* 1087 (1955).
92. D. F. DeTar and R. Silverstein, *J. Am. Chem. Soc.*, **88**, 1013, 1020 (1966).
93. J. M. Briody and D. P. N. Satchell, *Chem. Ind.*, *(London)* 1427 (1965).
94. C. L. Stevens and M. E. Munk, *J. Am. Chem. Soc.*, **80**, 4069 (1958); D. G. Appleyard and C. J. M. Stirling, *J. Chem. Soc.* (C), 2686 (1967); G. Losse and H. Weddige, *Angew. Chem.*, **72**, 323 (1960).
95. C. Naegeli and A. Tyabji, *Helv. Chim. Acta*, **18**, 142 (1935); K. D. Kopple and R. A. Thursack, *J. Chem. Soc.*, 2065 (1962).
96. Th. Curtius, *J. Prakt. Chem.*, **95**, 327 (1917); J. S. Fruton, *Advan. Protein Chem.* **5**, 1 (1954).
97. R. L. Betts and L. P. Hammett, *J. Am. Chem. Soc.*, **59**, 1568 (1937); W. H. Watanabe and L. R. DeFonso, *J. Am. Chem. Soc.*, **78**, 4542 (1956).

98. J. F. Bunnett and G. T. Davis, *J. Am. Chem. Soc.*, **82**, 665 (1960).
99. W. P. Jencks and J. Carriuolo, *J. Am. Chem. Soc.*, **82**, 675 (1960); T. C. Bruice and M. F. Mayahi, *J. Am. Chem. Soc.*, **82**, 3067 (1960).
100. W. P. Jencks and M. Gilchrist, *J. Am. Chem. Soc.*, **88**, 104 (1966).
101. K. A. Connors and M. L. Bender, *J. Org. Chem.* **26**, 2498 (1961).
102a. J. M. Briody and D. P. N. Satchell, *J. Chem. Soc.*, 168 (1965); A.S.A.S. Shawali and S. S. Biechler, *J. Am. Chem. Soc.*, **89**, 3020 (1967).
102b. F. M. Menger, *J. Am. Chem. Soc.*, **88**, 3081 (1966); P. J. Lillford and D. P. N. Satchell, *J. Chem. Soc.*, (B), 54, 897 (1968).
103. K. Koehler, R. Skora and E. H. Cordes, *J. Am. Chem. Soc.*, **88**, 3577 (1966).
104. T. C. Bruice and R. G. Willis, *J. Am. Chem. Soc.*, **87**, 531 (1965).
105. See especially references 4 and 87 for some examples concerning esters.
105a. See, however, R. L. Schowen, H. Jayaraman and L. Kershner, *J. Am. Chem. Soc.*, **88**, 3373 (1966).
106. K. Freudenberg and D. Peters, *Chem. Ber.*, **52**, 1463 (1919); H. Adkins and Q. E. Thompson, *J. Am. Chem. Soc.*, **71**, 2242 (1949); H. K. Hall, *J. Am. Chem. Soc.*, **78**, 2717 (1956).
107. V. Gold and E. G. Jefferson, *J. Chem. Soc.*, 1416 (1953); A. R. Butler and V. Gold, *J. Chem. Soc.*, 4362 (1961); 589 (1962).
108. T. C. Bruice and G. L. Schmir, *J. Am. Chem. Soc.*, **79**, 1663 (1957).
109. E. A. Barnard and W. D. Stein, *Adv. Enzymol*, **20**, 51 (1958).
110. T. C. Bruice and G. L. Schmir, *J. Am. Chem. Soc.*, **80**, 148 (1958).
111. M. L. Bender and B. W. Turnquest, *J. Am. Chem. Soc.*, **79**, 1656 (1957); W. Langenbeck and R. Mahrwald, *Chem. Ber.*, **90**, 2423 (1957).
112. T. C. Bruice and S. Benkovic, *J. Am. Chem. Soc.*, **86**, 418 (1964); M. Caplow and W. P. Jencks, *Biochemistry*, **1**, 773 (1962); T. C. Bruice, T. H. Fife, J. J. Bruno and S. Benkovic, *J. Am. Chem. Soc.*, **84**, 3012 (1962).
113. J. F. J. Dippy and V. Moss, *J. Chem. Soc.*, 2205 (1952).
114. T. C. Bruice and R. G. Willis, *J. Am. Chem. Soc.*, **87**, 531 (1965).
115. W. P. Jencks, *J. Am. Chem. Soc.*, **80**, 4581, 4585 (1958).
116. J. Hipkin and D. P. N. Satchell, *J. Chem. Soc. (B)*, 345 (1966).
117. T. C. Bruice and J. J. Bruno, *J. Am. Chem. Soc.*, **83**, 3494 (1961).
118. T. C. Bruice and S. Benkovic, *J. Am. Chem. Soc.*, **85**, 1 (1963).
119. R. B. Martin and A. Parcell, *J. Am. Chem. Soc.*, **83**, 4830, 4835 (1961).
120. R. Greenhalgh, R. M. Heggie and M. A. Weinberger, *Can. J. Chem.*, **41**, 1662 (1963).
121. D. P. N. Satchell and R. S. Satchell in *The Chemistry of the Carbonyl Group* (Ed. S. Patai), Interscience Publishers, London, 1966, Chap. 5.
122. D. P. N. Satchell and J. L. Wardell, *Trans. Faraday Soc.*, **61**, 1127 (1965).
123. V. Gold and T. Riley, *J. Chem. Soc.*, 1676 (1961); see also ref. 84.
124. P. H. Gore in *Friedel–Crafts and Related Reactions* (Ed. G. A. Olah), Vol. 3, Interscience Publishers, London, 1964, Chap. 31.
125. J. M. Pepper and B. P. Robinson, *Can. J. Chem.* **44**, 1809 (1966).
126. A. M. Glatz and A. C. Razus, *Rev. Chim. Acad. Rep. Populaire Roumaine*, **11**, 551 (1966).
127. S. Sethna in *Friedel–Crafts and Related Reactions* (Ed. G. A. Olah), Vol. 3, Interscience Publishers, London, 1964, Chap. 35.
128. F. D. Popp and W. D. McEwen, *Chem. Rev.* **58**, 321 (1958); W. S. Johnson, *Org. Reactions*, **2** (1944), chap. 4.
129. J. M. Tedder, *Chem. Rev.*, **55**, 787 (1955).
130. J. W. Huffman and R. C. Asbury, *J. Org. Chem.*, **30**, 3941 (1965).
131. M. Rosenblum, A. K. Banerjee, N. Danieli, R. W. Fish and V. Schlatter, *J. Am. Chem. Soc.*, **85**, 316 (1963).

132. F. H. Marquardt, *Helv. Chim. Acta*, **48**, 1486 (1965).
133. M. F. Ansell and M. H. Palmer, *Quart. Rev. (London)*, **15**, 211 (1961).
134. C. D. Nenitzescu and A. T. Balaban, *Friedel–Crafts and Related Reactions* (Ed. G. A. Olah), Vol. 3, Interscience Publishers, London, 1964, Chap. 37.
135. R. L. Letsinger and O. Kolewe, *J. Org. Chem.*, **26**, 2993 (1961).
136. D. A. Shirley, *Org. Reactions*, **8**, (1954), Chap. 2.
137. J. Kollonitsch, *J. Chem. Soc. (A)*, 453 (1966).
138. H. Gilman and P. R. Van Ess, *J. Am. Chem. Soc.*, **55**, 1258 (1933); C. Tegner, *Acta Chem. Scand.*, **6**, 782 (1952).
139. C. R. Hauser and B. E. Hudson, *Org. Reactions*, **1**, (1942), Chap. 9; C. R. Hauser, F. W. Swamer and J. T. Adams, *Org. Reactions*, **8** (1954), Chap. 3.
140. L. Claisen and O. Lowman, *Chem. Ber.*, **20**, 651 (1887).
141. A. Brandstrom, *Arkiv Kemi*, **6**, 155 (1953): H. D. Murdoch and D. C. Nonhebel, *J. Chem. Soc.*, 2153 (1962); J. P. Ferris, C. E. Sullivan and B. E. Wright, *J. Org. Chem.*, **29**, 87 (1964).
142. W. Dieckmann, *Chem. Ber.*, **27**, 102 (1894).
143. W. Baker, *J. Chem. Soc.*, 1381 (1933).
144. S. M. Sethna and N. M. Shah, *Chem. Rev.*, **36**, 1 (1945).
145. G. A. Olah and S. J. Kuhn, *Friedel–Crafts and Related Reactions* (Ed. G. A. Olah), Vol. 3, Interscience Publishers, London, 1964, Chap. 38.
146. E. J. Corey and M. Chaykovsky, *J. Am. Chem. Soc.*, **86**, 1639 (1964).
147. H. Eggerer, E. R. Stadtman and J. M. Poston, *Arkiv. Biochem.*, **98**, 432 (1962); F. Hibbert and D. P. N. Satchell, *J. Chem. Soc. (B)*, 565 (1968).
148. F. Hibbert and D. P. N. Satchell, *J. Chem. Soc. (B)*, 653 (1967).

CHAPTER **10**

Syntheses and uses of isotopically labelled carboxylic acids

MIECZYSLAW ZIELINSKI

University of Warsaw, Poland

I. INTRODUCTION

Two carbon isotopes are commonly used in tracer and isotope effect studies: the stable isotope ^{13}C and the radioactive isotope ^{14}C, which has a half-life of 5570 years[1-3]. A natural isotopic mixture contains about 1·1% [13]C. The radioactive isotope ^{14}C is produced during neutron bombardment of nitrogen-containing molecules in atomic reactors:

$$^{14}N + n \longrightarrow {}^{14}C + p$$

The target material at present used for ^{14}C production is beryllium nitride Be_3N_2 and aluminium nitride AlN. In earlier production of ^{14}C other nitrogen-containing compounds such as ammonium nitrate, calcium nitrate *etc.*, were also used[4]. Depending on the nature of the material bombarded, various chemicals containing the 'hot' carbon atom are produced in the (n, p) nuclear transformation[5]. For instance, solution of the neutron-irradiated beryllium nitride in sulphuric acid gives $^{14}CH_4$, $^{14}CO_2$, ^{14}CO, $H^{14}CN$ and $^{14}C_2N_2$. Their specific activity is very high; but for the synthesis of ^{14}C-labelled carboxylic acids only two are useful, $^{14}CO_2$ and $H^{14}CN$. The others are usually oxidized over copper oxide to $^{14}CO_2$ and precipitated in the form of barium carbonate.

II. SYNTHESES OF LABELLED CARBOXYLIC ACIDS

A. Introduction

Radioactive barium or sodium carbonate and potassium or sodium cyanide are the usual starting materials used for further chemical transformations. The use of ^{14}C of high specific activity requires the use of vacuum systems to avoid any losses. Fortunately ^{14}C emits soft β radiation with a maximum energy of about 0·15 MeV and no additional shielding is required, since the glass walls of the vacuum apparatus are sufficient protection. Nevertheless, to avoid hazards due to breakage or explosion, all vacuum lines should be built inside protecting boxes with good ventilation.

B. Carbonation of Organometallic Compounds with $^{14}CO_2$ or $^{13}CO_2$

Preparation of the carboxyl-labelled acids is the first step in the synthesis of many ^{14}C-labelled compounds. The most convenient way is the carbonation of a Grignard reagent:

$$RMgX + {}^{14}CO_2 \longrightarrow R^{14}COOMgX \tag{1}$$

$$R^{14}COOMgX + H^+(H_2O) \longrightarrow R^{14}COOH \tag{2}$$

The reaction is carried out in the region of $-20°C$ to minimize carbinol and ketone formation.

The synthesis of carboxyl-labelled acetic acid is carried out accordingly[6]:

$$CH_3I + Mg \xrightarrow{\text{dry ether}} CH_3MgI \tag{3}$$

$$CH_3MgI + {}^{14}CO_2 \xrightarrow{-20°C} CH_3{}^{14}COOMgI \tag{4}$$

$$CH_3{}^{14}COOMgI + H^+(H_2O) \longrightarrow CH_3{}^{14}COOH \tag{5}$$

Carboxyl-labelled benzoic acid is obtained similarly[7]:

$$C_6H_5MgBr + {}^{14}CO_2 \longrightarrow C_6H_5{}^{14}COOMgBr \tag{6}$$

$$C_6H_5{}^{14}COOMgBr + H^+(H_2O) \qquad C_6H_5{}^{14}COOH \tag{7}$$

Radiocarbonation of the Grignard reagent has also been used in the labelling of naturally occurring fatty acids[8-10]. Howton and coworkers have developed the degradation and reconstitution technique which consists of removing the original unlabelled carboxyl group of the fatty acid and replacing it by $^{14}CO_2$[8, 10]. Double bonds can be protected during this process by bromination, e.g. in the preparation of oleic and linoleic-1-^{14}C acids[10]. The following series of reactions have been used for the preparation of ^{14}C-labelled linoleic acid[8]:

$$Am(CHBr)_2CH_2(CHBr)_2(CH_2)_7COONH_4 \tag{8}$$
$$\downarrow \text{AgNO}_3/\text{CH}_3\text{OH}$$

$$Am(CHBr)_2CH_2(CHBr)_2(CH_2)_7COOAg \tag{9}$$
$$\downarrow \text{Br}_2/\text{CCl}_4$$

$$Am(CHBr)_2CH_2(CHBr)_2(CH_2)_7Br \tag{10}$$
$$\downarrow \text{Zn}$$

$$AmCH=CHCH_2CH=CH(CH_2)_7Br \tag{11}$$

$$\downarrow Mg$$

$$AmCH=CHCH_2CH=CH(CH_2)_7MgBr$$

$$\downarrow {}^{14}CO_2 \tag{12}$$

linoleic-1-^{14}C acid

In some cases, especially in the synthesis of acids labelled with the stable isotope ^{13}C, when dilution of the labelled material must be avoided, carbonation of the organolithium compounds is a commonly applied procedure:

$$RLi + {}^*CO_2 \longrightarrow R^*COOLi \tag{13}$$

When R = alkyl, the ^{14}C-labelled lithium salts are obtained directly[11], as *e.g.* in the low-temperature carbonation of ethyllithium with $^{14}CO_2$:

$$C_2H_5Li + {}^{14}CO_2 \xrightarrow[-70°C]{\text{n-hexane}} C_2H_5{}^{14}COOLi \ (97-98\%) \tag{14}$$

Several ^{14}C-labelled aromatic carboxylic acids have been prepared by carbonation of aryllithium[12], which was prepared by the reaction of n-butyllithium with aryl halides.

Synthesis of the carboxyl-labelled nicotinic acid through a lithium aryl is illustrated in equations (15) and (16)[12].

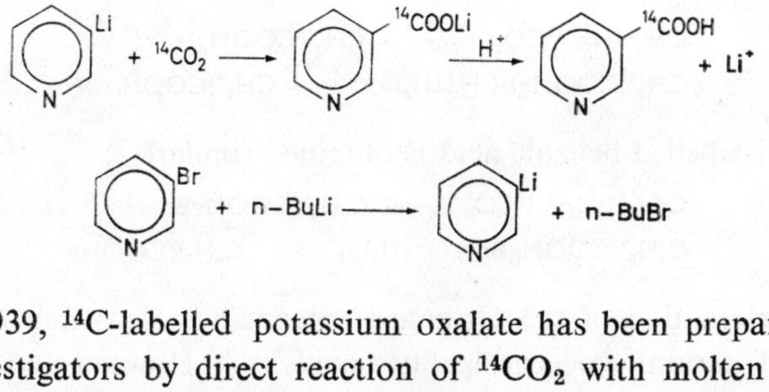

$$\tag{15}$$

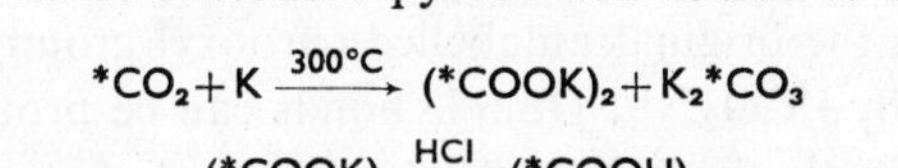

$$\tag{16}$$

Since 1939, ^{14}C-labelled potassium oxalate has been prepared[13-15] by many investigators by direct reaction of $^{14}CO_2$ with molten potassium mixed with sand in an evacuated pyrex flask heated to 270°–360°C.

$$^*CO_2 + K \xrightarrow{300°C} (^*COOK)_2 + K_2{}^*CO_3 \tag{17}$$

$$(^*COOK)_2 \xrightarrow{HCl} (^*COOH)_2 \tag{18}$$

Oxalic acid labelled in one carboxyl group has been prepared by exchange reaction between radioactive potassium oxalate and normal oxalic acid[15].

C. Syntheses Based on the Metal Cyanides

Cyanides are also readily available and are useful starting materials. A radioactive nitrile is obtained by reaction between an organic halide and a labelled metal cyanide, and is hydrolysed to the carboxyl-labelled acid. Malonic acid ^{14}C-labelled in one carboxyl group is usually prepared by this method[1]. Applying the degradation and reconstitution technique Bergström, Pääbo and Rottenberg[16] have prepared oleic-1-^{14}C acid by a nitrile synthesis protecting the *cis* double bond by hydroxylation.

Addition of isotopically labelled potassium cyanide or hydrogen cyanide to carbonyl compounds, followed by hydrolysis of the reaction products gives a variety of labelled hydroxy acids. Carboxyl-labelled lactic acid[6], for instance, has been prepared by the reaction of the acetaldehyde with potassium cyanide:

$$H^{14}CN + CH_3CHO \longrightarrow CH_3CHOH^{14}CN \qquad (19)$$

$$CH_3CHOH^{14}CN + 2\,H_2O + HCl \longrightarrow CH_3CHOH^{14}COOH + NH_4Cl \qquad (20)$$

In the similar way several α-amino acids labelled with isotopic carbon have been prepared[17].

In the synthesis of ^{14}C carboxyl-labelled picolinic acid[18], the following sequence of reactions have been used (equations 21–23):

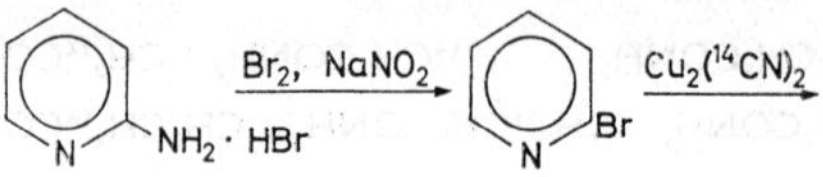

$$(21)$$

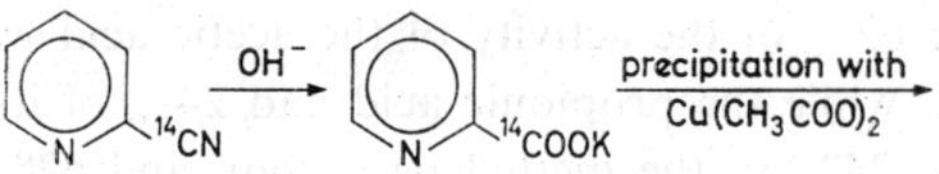

$$(22)$$

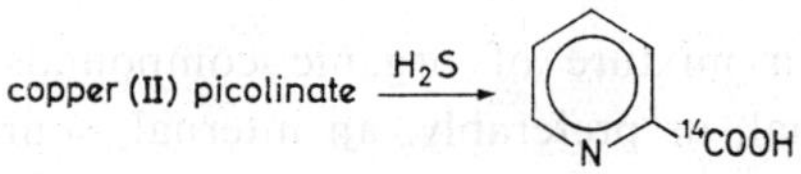

$$(23)$$

Carboxyl-labelled quinaldic acid (equation 24) has been obtained similarly[19].

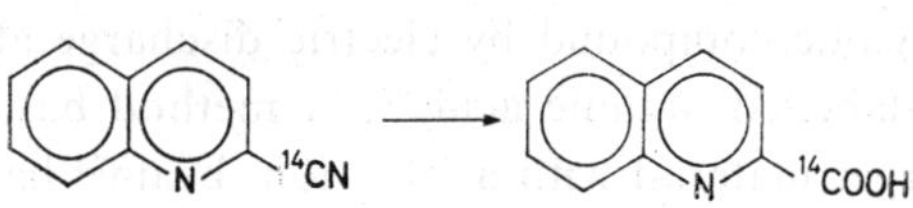

$$(24)$$

D. Nuclear Recoil and Other Labelling Methods

In the (n, p) nuclear reaction ^{14}C receives a recoil energy of about 40,000 eV. This leads to fission of chemical bonds and fast diffusion of the ^{14}C ion into the medium for a distance varying from 10 to 3000 Å[20]. The ^{14}C loses the greater part of its energy by Coulomb scattering, collisions with other atoms and other processes leading to the partial fragmentation and destruction of the surrounding molecules. In this period the transformations of the majority of the ^{14}C recoil energy are worthless. When the energy of ^{14}C has been slowed down to several electronvolts, it is able to react with its environment, substituting a carbon or a heteroatom either in the parent compound or in one produced in the process of its destruction. This process has been used for the preparation of a variety of labelled compounds[5, 21]. When the substance to be labelled does not contain nitrogen it is mixed with a radiation-stable nitrogen source having a high nitrogen to carbon ratio. The optimum irradiation times are based on two opposite requirements, namely on achieving a minimum of radiation damage and a maximum in the production of the chemically pure labelled material.

Application of the nuclear recoil method to the synthesis of variously ^{14}C-labelled acetic and propionic acids by the irradiation of acetamide in a nuclear reactor is illustrated in equations (25) and (26)[20].

$$CH_3CONH_2 \xrightarrow{(n,\ p)} \text{recoil } ^{14}C + \text{molecular fragments} \tag{25}$$

$$\begin{aligned}
\text{recoil } ^{14}C + n\ CH_3CONH_2 &\longrightarrow\ ^{14}CH_3CONH_2 + CH_3{}^{14}CONH_2 + \\
&+\ ^{14}CH_3CH_2CONH_2 + CH_3{}^{14}CH_2CONH_2 + CH_3CH_2{}^{14}CONH_2
\end{aligned} \tag{26}$$

After hydrolysis, 62% of the activity of the acetic acid was found in the carbonyl carbon, while the propionic acid had 24% of its activity in the carbonyl carbon, 24% in the methylene carbon and 52% in the methyl carbon.

The irradiation of a mixture of organic compounds together with $^{14}CO_2$, using an external, or preferably, an internal, source of radiation also leads to the formation of ^{14}C-labelled acids. Thus Turton[22] produced benzoic acid labelled in the carboxyl group by irradiation of a mixture of benzene and 1·3 mCi of $^{14}CO_2$ with 2Ci of ^{85}Kr. Activation of a mixture of $^{14}CO_2$ with an organic compound by electric discharge also leads to the formation of ^{14}C-labelled organic acids[23]. A method based on the bombardment of a target material with a $^{14}C^+$ ion beam[24] has been used for synthesis of ^{14}C-labelled benzoic acid.

Undesired radiation damages inherently present in nuclear methods of labelling are partially avoided when the material to be treated, spread on the cathode, is bombarded by ^{14}C-labelled carbon dioxide ions accelerated by submitting them to a voltage of 500 V. Under such conditions ^{14}C is able to react with organic compounds or radicals[25]. Acetamide, citric acid, succinic acid, benzoic acid and benzoates have been labelled by this method.

E. Final Remarks

The advantage of chemical syntheses in preference to other methods of labelling is that the compound can usually be labelled in a single, known position. The reaction schemes are chosen so as to avoid molecular rearrangements and exchange reactions. This cannot be achieved by radiochemical and other (*e.g.* biological) methods of synthesis of carboxyl-labelled substances.

III. TRACER APPLICATIONS OF LABELLED CARBOXYLIC ACIDS

A. Physical and Analytical Determinations

Carboxylic acids can be easily separated from complicated organic mixtures. This advantage has been used in the determination of the molecular weight of many compounds (complicated aryl and alkyl groups, polysaccharides, *etc.*) into which ^{14}C-labelled carboxyl groups can be easily introduced[26]. The same principle was used for the determination of functional groups in the molecule[27]. For instance, previous methods of estimation of acetylable hydrogen in the molecule have been replaced by treatment with ^{14}C acetic anhydride of known specific activity, followed by determination of the specific activity of the acetylated product[28].

Labelled reagents are very useful in the microdeterminations of biochemical materials. Reagents labelled with ^{14}C and ^{3}H are especially useful because of their long half-lifes. In this way the main disadvantage of applying reagents labelled with short-lived radioactive isotopes, namely, the frequent preparation of the labelled compound, is eliminated.

^{3}H- and ^{14}C-labelled reagents have found wide application in the quantitative analysis of mixtures of organic compounds by the isotope dilution method[29, 30]. ^{14}C- and ^{3}H-labelled acetic anhydrides were applied by Whitehead[31] for the determination of amino acids in protein hydrolysates. The technique consisted of the acetylation of the mixture of amino acids with tritiated acetic anhydride, subsequent addition of a standard solution of ^{14}C-labelled acetyl derivates of the amino acids being analysed to the acetylated sample and separation of the mixture of N-acetylamino acids by either two-dimensional paper chromatography or paper chromatography in one dimension followed by paper ionophoresis in the second dimension[32]. The areas of paper containing the labelled N-acetyl derivates of the amino acids being assayed are cut out from the developed chromatograms and oxidized in a stream of oxygen to carbon dioxide and water. Radioactivity of the carbon dioxide is then directly measured in the gaseous phase by the standard method in a Geiger–Müller counter or ionization chamber. Tritiated water is decomposed quantitatively over zinc at 400°c and the mixture of H_2 and HT obtained is transferred into a Geiger–Müller tube or an ionization chamber for radioassay.

The sensitivity of the isotope dilution technique is limited by the specific activity of the labelled reagent. Amounts of 0·001–0·002 µg of amino acids could be detected[31]. Quantitative analysis by the isotope dilution method does not demand full recovery of the labelled species to be assayed, but the separated compound must be chemically and isotopically pure. The purity of the labelled derivative can be tested by measuring the ratio $^{14}C : ^3H$ for each amino acid assayed. The accuracy obtained in quantitative determinations of the amino acids by the isotope dilution method was about 5%. This is a good result, considering that only as little as 2 µg of protein is required in the majority of experiments.

B. Ligand Exchange Studies in Organometallic Complexes

Metallic ions form complexes with many organic acids[33, 34]. Availability of acids labelled with ^{14}C or tritium in non-exchangeable position permits the study of the exchange of the ligand between the free, uncomplexed ligand ion and the corresponding ligand group in various organometallic complexes[35–39], for example between free oxalate-^{14}C ions and oxalate-ligand groups in $[Me(C_2O_4)_3]^{3-}$ complexes[35–37]. It was found that the exchange between $[^{14}C_2O_4]^{2-}$ and $[Fe(C_2O_4)_3]^{3-}$ or $[Al(C_2O_4)_3]^{3-}$ complexes is a very fast process, while the exchange with corresponding complexes of Co(III) and Cr(III) is slow. Ion exchange between ferrous picolinate and free ^{14}C-labelled picolinate ions is also very fast.

Exchange studies (of both ligand and central metal atoms) gave general correlations between the electronic configuration of the central metal ion and the rate of exchange. According to Taube[40] all complexes of the 'outer orbital' type give fast exchange of their ligands and central ion with free uncomplexed ligand or aquocomplex of free metal ion. Complexes of the 'inner orbital' type are very stable and exchange very slowly or not at all with free uncomplexed groups.

C. Reversibility of Organic Reactions

Rate constants of forward and reverse reactions can be determined by isotopic tracer studies. As an example let us consider the hydrolysis of the dipeptide, glycylglycine, into two glycine molecules[41]:

$$H_2NCH_2CONHCH_2COOH + H_2O \rightleftharpoons 2 H_2NCH_2COOH \qquad (27)$$

Addition of labelled glycine to the reaction mixture and subsequent radio-assay of the undecomposed dipeptide confirmed the reversibility postulated by equation (27).

Reversal of the carboxylation reaction was demonstrated by following the decarboxylation of carboxyl-labelled methylmalonyl coenzyme A in the presence of adenosine diphosphate and inorganic phosphate Pi[42]:

$$\text{Enzyme} + \text{ATP} + {}^{14}\text{CO}_2 \rightleftharpoons \text{Enzyme-}{}^{14}\text{CO}_2 + \text{ADP} + \text{Pi} \qquad (28)$$

$$\text{Enzyme-}{}^{14}\text{CO}_2 + \text{propionyl-CoA} \rightleftharpoons \text{methylmalonyl-CoA} + \text{Enzyme} \qquad (29)$$

In the biosynthesis of purine[43, 44], 5-aminoimidazole reacts with $^*\text{CO}_2$ (or HCO_3^-) yielding 5-amino-4-imidazolecarboxylic (*C) acid ribonucleotide (1). The reversibility of this reaction in the presence of enzyme was demonstrated using ^{14}C as a tracer.

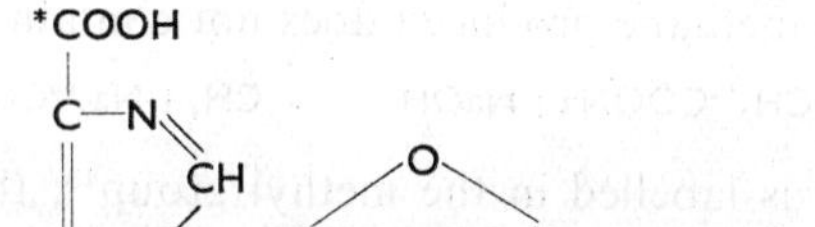

(1)

D. Pyrolytic Reactions

Alkaline-earth metal acetates and lithium acetate heated to 400°C and above decompose into acetone and the corresponding carbonate. It has been suggested that the pyrolytic decomposition proceeds by a chain mechanism.

The pyrolysis of mixtures of calcium acetate and calcium formate yields acetaldehyde as one of the products of decomposition. Bell and Reed[45] studied the distribution of ^{13}C amongst the decomposition products of mixtures of barium acetate enriched with ^{13}C in the carboxyl group and normal barium formate. The products consisted of a mixture of acetaldehyde and acetone, a residue of barium carbonate and a small quantity of tar. The ^{13}C content of the carbonyl carbon atom of the acetaldehyde was approximately normal, while the barium carbonate residue was ^{13}C-enriched. Such a distribution requires fission of the carbon–carbon bond in the acetate during the pyrolysis, as suggested schematically by equation (30):

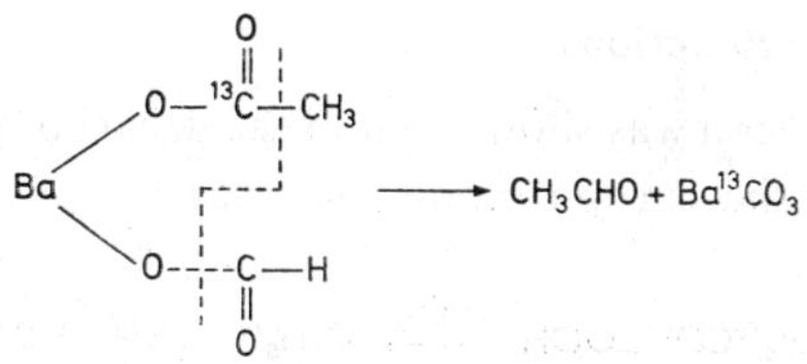

$$\longrightarrow \text{CH}_3\text{CHO} + \text{Ba}^{13}\text{CO}_3 \qquad (30)$$

By analogy with the pyrolysis of alkaline-earth metal acetates, it was
supposed that the reaction is initiated by the free methyl radical, and the
following probable sequence of reactions has been postulated (ba = 1/2Ba)

Chain initiation: $CH_3CO_2ba \longrightarrow \dot{C}H_3 + \dot{C}O_2ba$ (31)

Aldehyde formation: $\dot{C}H_3 + HCO_2ba \longrightarrow CH_3CHO + \dot{O}ba$ (32)

Acetone formation: $\dot{C}H_3 + CH_3CO_2ba \longrightarrow CH_3COCH_3 + \dot{O}ba$ (33)

Chain continuation: $CH_3CO_2ba + \dot{O}ba \longrightarrow \dot{C}H_3 + BaCO_3$ (34)

Chain termination: $2\ \dot{C}H_3 \longrightarrow C_2H_6$ or $CH_4 + :CH_2$ etc.   (35)

                        $2\ \dot{O}ba \longrightarrow (Oba)_2$ (36)

                        $\dot{C}O_2ba + \dot{O}ba \longrightarrow BaCO_3$ (37)

In the pyrolysis of sodium acetate mixed with an excess of sodium hydroxide, methane is formed[46]. Using acetate labelled with ^{14}C in the carboxyl group[47] the methane produced does not contain ^{14}C:

$$CH_3{}^{14}COONa + NaOH \longrightarrow CH_4 + Na_2{}^{14}CO_3 \qquad (38)$$

When the acetate is labelled in the methyl group[48], the carbonate is not
enriched with ^{14}C:

$$^{14}CH_3COONa + NaOH \longrightarrow {}^{14}CH_4 + Na_2CO_3 \qquad (39)$$

The pyrolysis of sodium acetate was also performed in the presence of
sodium tritoxide[49], when the methane produced was tritium-labelled:

$$CH_3COONa + NaOT \xrightarrow[\text{pyrolysis}]{340°C} CH_3T + Na_2CO_3 \qquad (40)$$

Reaction (40) has been used for the synthesis of tritium-labelled methane
formed by abstraction of tritium from NaOT and exchange of methyl
hydrogens with tritium. The absence of by-products suggests that the
mechanism of decomposition is heterolytic:

$$(41)$$

Before final conclusions concerning the mechanism of methane formation
can be reached, the reversible exchange according to equation (42), in the
fused state should be studied.

$$CH_3COONa + NaOT \rightleftharpoons CH_2TCOONa + NaOH \qquad (42)$$

. Decarbonylation Reactions

Using ^{14}C as a tracer it was shown that in the decarbonylation of benzoyl-
formic-1-^{14}C acid, the carbon monoxide released originates in the carboxyl group only[50].

$$C_6H_5{}^{12}CO{}^{14}COOH \xrightarrow{H_2SO_4} C_6H_5{}^{12}COOH + {}^{14}CO \qquad (43)$$

The same conclusion was reached by the use of benzoylformic-2-^{14}C acid which produced non-labelled carbon monoxide[51]:

$$C_6H_5{}^{14}CO^{12}COOH \xrightarrow{H_2SO_4} C_6H_5{}^{14}COOH + {}^{12}CO \qquad (44)$$

F. Reactions of Isocyanates with Carboxylic Acids

Isocyanates react with carboxylic acids by two different paths[52]:

$$R^1NCO + R^2COOH \longrightarrow CO_2 + R^2CONHR^1 \qquad (45)$$

$$2 R^1NCO + 2 R^2COOH \longrightarrow CO_2 + (R^2CO)_2O + (R^1NH)_2CO \qquad (46)$$

In order to establish the source of the carbon dioxide evolved in reactions (45) and (46), two different ^{14}C carboxyl-labelled acids were allowed to react with each of two different isocyanates and the distribution of radiocarbon among the resulting products studied. The N-substituted amide obtained in the reaction has the same specific activity as the acid used, while the carbon dioxide is inactive, and obviously originates from the isocyanate:

$$R^1NCO + R^2\,{}^{14}COOH \longrightarrow CO_2 + R^2\,{}^{14}CONHR^1 \qquad (47)$$

G. Oxidation of Aliphatic Acids

Propionic acid is oxidized by a dilute aqueous solution of chromic acid to carbon dioxide and acetic acid. Oxidation of the carboxyl-labelled propionic acid showed that the reaction proceeds according to equation (48)[53].

$$CH_3CH_2{}^{14}COOH \xrightarrow{H_2Cr_2O_7} CH_3COOH + {}^{14}CO_2 \qquad (48)$$

In the alkaline permanganate oxidation of the ^{14}C-labelled propionate ion, both products of partial oxidation, (oxalic acid and carbon dioxide) contain radioactive carbon[54].

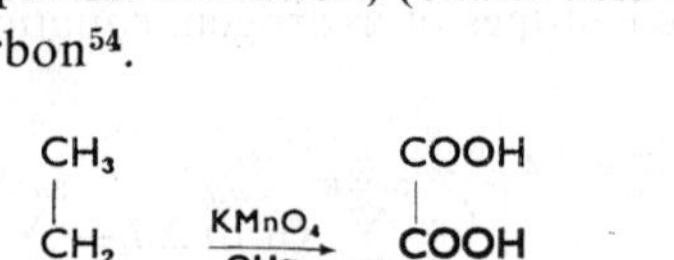

$$\begin{array}{c} CH_3 \\ | \\ CH_2 \\ | \\ {}^{14}COO^- \end{array} \xrightarrow[OH^-]{KMnO_4} \begin{array}{c} COOH \\ | \\ COOH \\ \\ {}^{14}CO_2 \ (30\%) \end{array} \qquad (49)$$

$$\begin{array}{c} CH_3 \\ | \\ CH_2 \\ | \\ {}^{14}COO^- \end{array} \xrightarrow[OH^-]{KMnO_4} \begin{array}{c} CO_2 \\ \\ COOH \\ | \ (70\%) \\ {}^{14}COOH \end{array} \qquad (50)$$

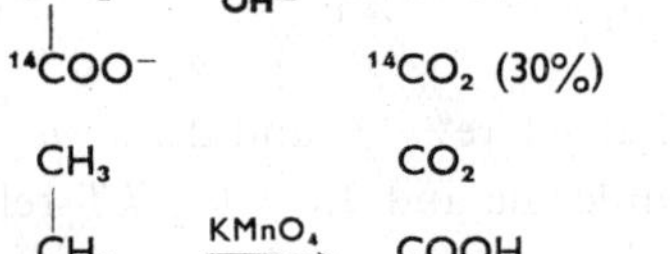

The distribution of the radioactivity between the products depends on the alkalinity of the oxidizing medium. In a similar way the mechanism of oxidation of many other carboxylic acids has been studied[1].

IV. ISOTOPE EFFECT STUDIES WITH CARBOXYLIC ACIDS

A. Introduction

Two chemically identical molecules with different isotopic compositions generally react with different rates. The ratio of rate constants k_1/k_2 of a chemical change obtained with two isotopic compounds is called the kinetic isotope effect.

The technique for calculation of *a priori* differences in isotopic reaction rates is based on the transition state theory. The assumption of the existence of the thermodynamic equilibrium between transition state complex (TS) and the reactants (R)[55-57] is used to establish the relation between the ratio k_1/k_2 and the geometry and vibration frequencies of the reactants and the transition state complex (equation 51),

$$\frac{k_1}{k_2} = \frac{x_1}{x_2} \left(\frac{m_2^{\ddagger}}{m_1^{\ddagger}}\right)^{1/2} \frac{\prod_{i=1}^{3n-y} \left(\frac{u_{2i}}{u_{1i}}\right) \frac{\sinh(u_{1i}/2)}{\sinh(u_{2i}/2)} \left(\frac{s_1}{s_2}\right)}{\prod_{i=1}^{3n-(y+1)} \left(\frac{u_{2i}^{\ddagger}}{u_{1i}^{\ddagger}}\right) \frac{\sinh(u_{1i}^{\ddagger}/2)}{\sinh(u_{2i}^{\ddagger}/2)} \left(\frac{s_1}{s_2}\right)^{\ddagger}} \tag{51}$$

where $\ddagger$ denotes the transition state complex, subscripts 1 and 2 denote the two isotopic species, $m^{\ddagger}$ is the effective mass along the reaction coordinate, $u_i = h\nu_i/kT$, ν_i is the fundamental frequency, h is Planck's constant, k is the Boltzmann constant, T is the absolute temperature, x is the transmission coefficient and sinh denotes a hyperbolic sine[58-60].

For heavy isotopes and small difference of masses, that is for practically all isotopes except the isotopes of hydrogen, equation (51) is approximated to equation (52)[58],

$$\frac{k_1 s_2 s_1^{\ddagger}}{k_2 s_1 s_2^{\ddagger}} = \left(\frac{m_2^{\ddagger}}{m_1^{\ddagger}}\right)^{1/2} \left[1 + \sum_{i=1}^{3n-6} G(u_i)\,\Delta u_i - \sum_{i=1}^{3n-7} G(u_i^{\ddagger})\,\Delta u_i^{\ddagger}\right] \tag{52}$$

where $G(u_i) = 1/2 - 1/u_i + 1/(e^{u_i} - 1)$ and $\Delta u_i = u_{1i} - u_{2i}$, if the subscript 1 refers to the lighter molecule and $u_{2i} = h\nu_{2i}/kT$ refers to the frequency of the heavier molecule.

Usually the exact form of the potential function is unknown and the *a priori* calculations are limited to approximate predictions of upper and lower boundary values. Exact experimental determinations of the kinetic isotope effect are used for establishing the details of the structure of the transition state complex of the reacting isotopic molecule.

Molecules having natural abundance of isotopes or labelled with radio-active isotopes have usually one isotope at the tracer concentration level. The most common method of experimental determination of the kinetic isotope effect consists of measuring the enrichment of the product of the reaction in one of the isotopes at a small percentage of chemical conversion or measuring the isotopic composition of the reactants at large (more than 50%) percentages of reaction and comparing it with the isotopic composition of the reactants at zero time.

In the first case equation (53) is used to evaluate the ratio of rate constants in terms of isotopic composition R and fraction reacted f:

$$\frac{k_1}{k_2} = \frac{\ln(1-f)}{\ln(1-fR_{tp}/R_0)} \tag{53}$$

and in the second case:

$$\frac{k_1}{k_2} = \frac{\ln(1-f)}{\ln(1-f)+\ln(R_{ts}/R_0)} \tag{54}$$

where f is the fraction reacted, R_0 is the initial specific activity (isotopic ratio) of the substrate, R_{ts} is the specific activity (isotopic ratio) of the substrate at time t and R_{tp} is the specific activity (isotopic ratio) of the product at time t. All the methods used for the determination of experimental kinetic isotope effects are discussed in detail in a review by Bigeleisen and Wolfsberg[58].

B. ^{14}C Isotope Effects

Since the construction of very precise mass spectrometers for isotope ratio measurements, the number of studies concerning the application of the 'tagging' technique to isotope effects has been small. Even using compounds with a very high degree of radiochemical and chemical purity the precision of the measurement of the specific radioactivity is at least one order of magnitude worse than that achieved by isotope ratio determinations. Nevertheless, in certain cases the 'tagging' technique, in spite of being time-consuming and inconvenient, has some advantages. For instance, it established the dependence of the kinetic carbon isotope effect in decarboxylation reactions on the structure of the labelled acid[61], and predicted correctly the effects of phenol and its derivatives on the kinetic carbon isotope effect in the decarboxylation of picolinic acid[64]. Research with carboxylic acids having a natural abundance of carbon isotopes creates some difficulties caused by the non-uniform distribution of the ^{13}C among all carbon positions in the molecule. This has been demonstrated in the decarboxylation of malonic acid[58, 62]. Determination of the kinetic isotope effects requires knowledge of the exact distribution of ^{13}C among all carbon atoms in the starting material. Synthesis of an acid labelled in a known position by ^{14}C facilitates the work. The time taken for the synthesis is compensated by avoiding the multistep degradation procedure necessary for the determination of the departure of the distribution of ^{13}C from the statistical one within the molecule.

1. Decarboxylation of organic acids

For the decarboxylation of 2-benzoylpropionic acid (equation 55)[63] label-

$$C_6H_5COCHMeCO_2H \longrightarrow C_6H_5COCH_2CH_3 \qquad (55)$$

led in one of the positions 1, 2 and the ketonic carbonyl group, the isotope effects listed in Table 1 were obtained. Application of the tagging technique

TABLE 1. Experimental isotope effects in the decarboxylation of 2-benzoylpropionic acid

Position of ^{14}C	Temperature (°C)	Isotope effect k/k^*	95% confidence limits
1	58	1·077	0·014
1	78	1·074	0·011
2	78	1·051	0·009
2[a]	78	1·053	0·013
carbonyl	78	1·000	0·004

[a] In a solution of 1,1-diphenylethane.

to the determination of the three k_1/k_2 ratios only required the measurement of the specific activity of the acid at zero time and of the undecomposed acid isolated at time t from the reaction mixture. Studies of the same problem with ^{13}C at natural abundance level of isotopes necessitates the determination of the $^{13}C/^{12}C$ ratios for all significant carbon positions.

Measurements of ^{14}C showed that the ^{14}C isotope effect in the decarboxylation of molten picolinic acid is about 1·046 in the temperature region 156–185°C and changes very negligibly with temperature[61]. This primary kinetic isotope effect should be compared with the value of 1·074 in the decarboxylation of 2-benzoylpropionic (1-^{14}C) acid (after correction for its temperature dependence) and with that for malonic acid (1·054), although subsequent ^{13}C measurements have shown that the actual difference between the effects with malonic and picolinic acids is much larger (see section IV. C.).

Calculations of k_1/k_2 ratios for the decarboxylation of ordinary and ^{14}C-labelled molten picolinic acid gave values ranging from 1·076 (at 156°C) to 0·997, depending on the assumed structure of the transition complex. At 195°C the maximum and minimum values of the calculated isotope effect were 1·068 and 0·997[64]. Calculations have been made for the sym-

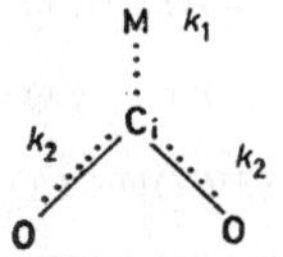

(2)

metrical complex **2**. Experimentally measured k_{12C}/k_{14C} ratios did not contradict the theoretical expectations based on this model. Nevertheless the temperature dependence of the calculated k_1/k_2 ratios was larger than that experimentally measured. It should be noted that the upper limit permitted by the symmetrical model of the transition complex is very close to the experimental ^{14}C isotope effect in the decarboxylation of malonic and 2-benzoylpropionic acids.

2. Decarbonylation reactions

Some organic acids release carbon monoxide in concentrated sulphuric acid[51, 65−68]. A mechanism of decarbonylation was proposed by Hammett (equations 56–60), but without deciding which is the rate-determining

$$RCOOH + H_2SO_4 \rightleftharpoons RC\overset{O}{\underset{\overset{+}{OH_2}}{}} + HSO_4^- \qquad (56)$$

$$RC\overset{O}{\underset{\overset{+}{OH_2}}{}} \longrightarrow RCO^+ + H_2O \qquad (57)$$

$$RCO^+ + H_2O \longrightarrow RC\overset{O}{\underset{\overset{+}{OH_2}}{}} \qquad (58)$$

$$RCO^+ + H_2SO_4 \longrightarrow (R \cdot H_2SO_4)^+ + CO \qquad (59)$$

$$(R \cdot H_2SO_4)^+ + 2 H_2O \xrightarrow{\text{quenching}} ROH + H_3O^+ + H_2SO_4 \qquad (60)$$

step. The ^{14}C isotope effect data collected in Table 2 were a great help in solving this problem.

Comparison of the large ^{14}C isotope effects observed in the decarbonylation of formic-^{14}C and benzoylformic-1-^{14}C acids (at 25°C) with the value of 11·9% estimated theoretically by Eyring and Cagle[69] for [14]C—[16]O (versus [12]C—[16]O) bond rupture revealed that [14]C—[16]O bond cleavage in

30*

TABLE 2. The ^{14}C isotope effects in the decarbonylation of monobasic acids

Reaction	$k_{^{12}C}/k_{^{14}C}$	Reference
$H^{14}COOH \xrightarrow{H_2SO_4} H_2O + {}^{14}CO$	1·09	51, 67
$C_6H_5CO^{14}COOH \xrightarrow{H_2SO_4} C_6H_5COOH + {}^{14}CO$	1·10	68
$C_6H_5{}^{14}COCOOH \xrightarrow{H_2SO_4} C_6H_5{}^{14}COOH + CO$	1·039	67
$(C_6H_5)_3{}^{14}CCOOH \xrightarrow{H_2SO_4} (C_6H_5)_3{}^{14}COH + CO$	1·01	67

the second step (equation 57) controls the rate of decarbonylation. Isotope effects in the decarbonylation of oxalic acid[70] also show that carbon–carbon bond cleavage is not rate-determining.

Simple one-bond model calculations do not explain the ^{14}C isotope effect in the decarbonylation of benzoyl-2-^{14}C acid. Ropp[51] suggests that this might be an equilibrium ^{14}C isotope effect in the reversible protonation of the α-keto group, as well as in the protonation of the carboxyl group:

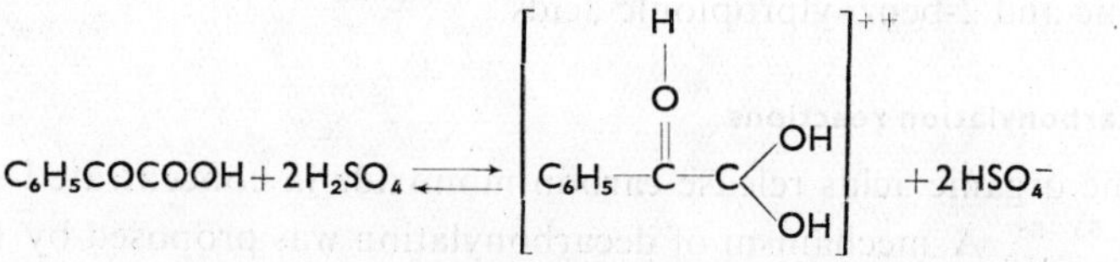

$$C_6H_5COCOOH + 2H_2SO_4 \rightleftharpoons \left[C_6H_5 - \overset{\overset{\displaystyle H}{\underset{\displaystyle O}{|}}}{\underset{}{C}} - C \overset{OH}{\underset{OH}{<}} \right]^{++} + 2HSO_4^- \tag{61}$$

The very small ^{14}C isotope effect in the decarbonylation of triphenylacetic-2-^{14}C acid is in disagreement with both extreme one-bond model calculations, for $^{14}C—^{16}O$ and for $^{14}C—^{12}C$ bond rupture. Moreover, 24% enrichment of the $C^{18}O$ was observed during the triphenylacetic acid decarbonylation in ^{18}O-enriched sulphuric acid. These two observations confirm the suggestion that the decarbonylation mechanisms of formic acid and triphenylacetic acid are different[71] but the structure of the transition state for the latter reaction is still unclear and additional isotope effect studies are required.

C. ^{13}C and ^{18}O Isotope Effects in Decarboxylation

1. Introduction

^{13}C Isotope effects in the decarboxylation of organic acids are the subject of continuous and precise mass spectrometric studies. Most of them are concerned with the decarboxylation of malonic acid and have been performed in the ^{13}C laboratories at the University of Illinois[72]. Long experience in this field and improvements of the Nier mass spectrometer have given very precise results in the determination of k_{12C}/k_{13C} of the two malonic acid species[58, 73, 74]. Measured ^{13}C kinetic isotope effects were found to be consistent with the calculations based on the one-bond model.

However, the temperature dependences do not fit, and it is necessary to use a more general model to find a better correlation between calculated and experimental isotope effects. In the case of decarboxylation of salts of organic acids, the symmetrical configuration of the COO^- group should facilitate the theoretical interpretation of the experimental results. To provide more material for comparative studies, quinaldic[75] and picolinic[76] acids have been investigated, since earlier work[64] revealed quite large differences between the ^{14}C kinetic isotope effects in the decarboxylation of picolinic acid and malonic acid. Because the measurements of ^{13}C isotope effects are intrinsically more accurate the ^{14}C measurements had to be confirmed by investigations of the decarboxylation of quinaldic and picolinic acids containing ^{13}C at the natural abundance level.

The carbon and oxygen isotope ratios reported in this section have been measured on a Nier-type mass spectrometer modified as described by McKinney and coworkers[77]. The method of measurement of small differences in the isotopic composition of the two samples of carbon dioxide consisted of the determination of the 'δ_m', that is in measuring the isotopic ratio enrichment (per ml) in a sample relative to the working standard:

$$\delta_{m(carbon)} = (R_{45(sample)}/R_{45(std)} - 1) \times 1000$$
$$\delta_{m(oxygen)} = (R_{46(sample)}/R_{46(std)} - 1) \times 1000$$

and calculating on their basis 'δ_c' i.e. the true enrichment of the sample relative to the standard, corrected for the contribution of the ^{17}O to the mass ratio R_{45}, and contribution of the ^{17}O and ^{13}C to the mass ratio R_{46}[78].

$$\delta_{c(carbon)} = (R_{13(sample)}/R_{13(std)} - 1) \times 1000$$
$$\delta_{c(oxygen)} = (R_{18(sample)}/R_{18(std)} - 1) \times 1000$$

The ratio-denoting subscripts are as follows:

$$R_{45} = \frac{\text{the mass-45 ion beam}}{\text{the mass-44 ion beam}}, \text{ that is the ratio } \frac{^{13}C^{16}O^{16}O + ^{12}C^{16}O^{17}O}{^{12}C^{16}O^{16}O}$$

$$R_{46} = \frac{\text{the mass-46 ion beam}}{\text{the mass-44 + mass-45 ion beam}}, \text{ that is }$$

$$\frac{^{12}C^{16}O^{18}O + ^{13}C^{16}O^{17}O + ^{12}C^{17}O_2}{^{12}C^{16}O_2 + ^{13}C^{16}O_2 + ^{12}C^{16}O^{17}O}$$

$$R_{13} = \frac{^{13}C^{16}O_2}{^{12}C^{16}O_2},$$

$$R_{18} = \frac{^{12}C^{16}O^{18}O + ^{13}C^{16}O^{18}O}{^{12}C^{16}O_2 + ^{13}C^{16}O_2} = \frac{^{12}C^{16}O^{18}O}{^{12}C^{16}O_2}$$

In naturally occurring mixtures the amounts of the species $^{13}C^{16}O^{18}O$ and $^{13}C^{16}O_2$ are by two orders of magnitude lower than of the $^{12}C^{16}O^{18}O$ and $^{12}C^{16}O_2$ molecules, respectively.

Earlier ^{13}C measurements with malonic acid have also been repeated on the same McKinney-type mass spectrometer.

2. Decarboxylation of malonic acid

Neglecting the ^{17}O secondary isotope effects there are seven isotopic reactions in the process of decarboxylation of naturally occurring malonic acid:

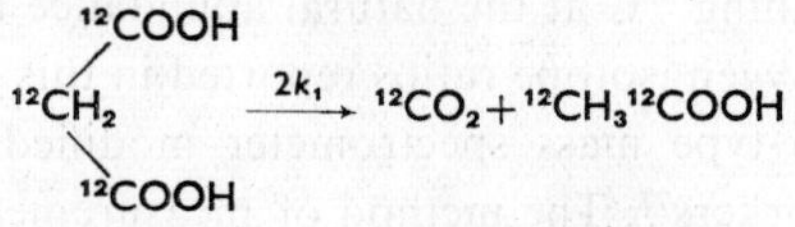

$$\begin{array}{c} ^{12}COOH \\ ^{12}CH_2 \\ ^{12}COOH \end{array} \xrightarrow{2k_1} {}^{12}CO_2 + {}^{12}CH_3{}^{12}COOH \tag{62}$$

$$\begin{array}{c} ^{12}COOH \\ ^{13}CH_2 \\ ^{12}COOH \end{array} \xrightarrow{2k_2} {}^{12}CO_2 + {}^{13}CH_3{}^{12}COOH \tag{63}$$

$$\begin{array}{c} ^{13}COOH \\ ^{12}CH_2 \end{array} \xrightarrow{k_3} {}^{13}CO_2 + {}^{12}CH_3{}^{12}COOH \tag{64}$$

$$^{12}COOH \xrightarrow{k_4} {}^{12}CO_2 + {}^{12}CH_3{}^{13}COOH \tag{65}$$

$$\begin{array}{c} C^{16}O^{16}O \\ CH_2 \\ C^{16}O^{16}O \end{array} \xrightarrow{2k_5} C^{16}O_2 + CH_3C^{16}O^{16}OH \tag{66}$$

$$\begin{array}{c} C^{16}O^{18}O \\ CH_2 \end{array} \xrightarrow{k_6} C^{16}O^{18}O + CH_3C^{16}O^{16}OH \tag{67}$$

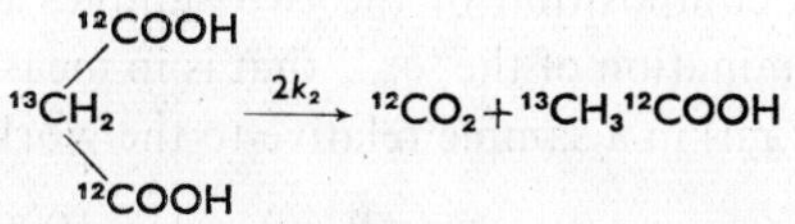

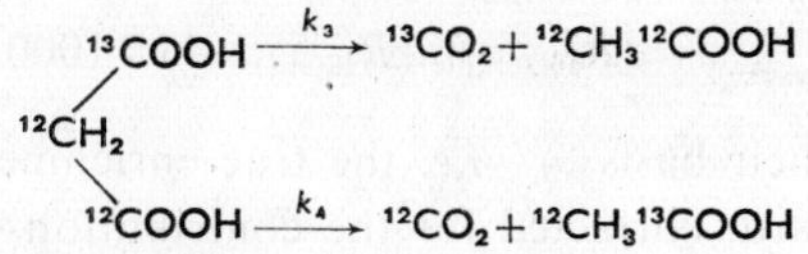

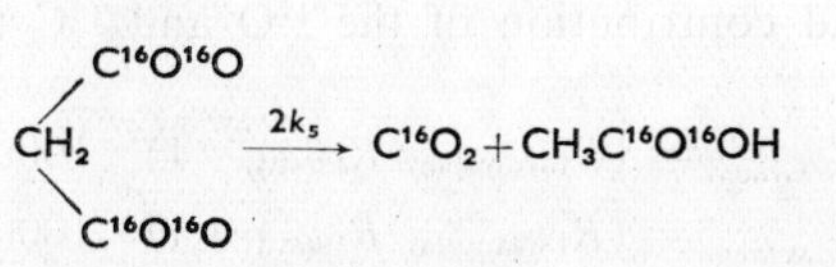

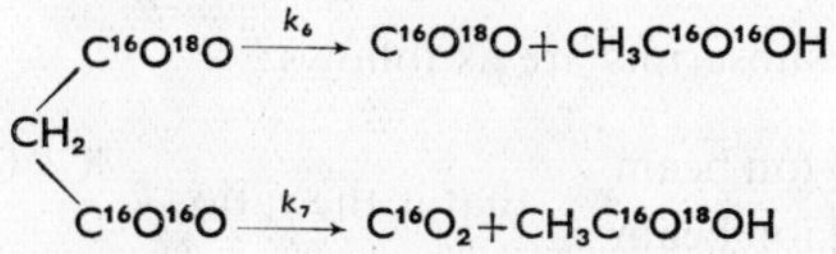

$$C^{16}O^{16}O \xrightarrow{k_7} C^{16}O_2 + CH_3C^{16}O^{18}OH \tag{68}$$

From equations (66)–(68) follows the relation:

$$\frac{[C^{16}O^{18}O]_t}{[C^{16}O^{16}O]_t} = \frac{k_6 M_6^0[1 - e^{-(k_6+k_7)\,t}]}{(k_6+k_7)\,M_5^0[1 - e^{-2k_5 t}]} \tag{69}$$

here M_5^0 and M_6^0 denote the initial concentrations of the isotopic molecules reacting in the reactions (66)–(68). At complete decomposition of
the malonic acid into carbon dioxide and acetic acid:

$$\frac{[C^{16}O^{18}O]_{t=\infty}}{[C^{16}O^{16}O]_{t=\infty}} = \frac{k_6}{k_6+k_7}\left(\frac{M_6^0}{M_5^0}\right) = R_{I\,18}^0 \tag{70}$$

t complete decomposition of the malonic acid into two carboxyl groups
and one methylene group:

$$M_{C^{16}O_2}^0 = 2M_5^0 + M_6^0 \tag{71}$$

$$M_{C^{16}O^{18}O}^0 = M_6^0 \tag{72}$$

t the natural isotopic abundance level:

$$\frac{M_6^0}{M_5^0} = 2R_{18}^0 \tag{73}$$

rom equations (70)–(73) one obtains:

$$\frac{k_7}{k_6} = \left(\frac{2R_{18}^0}{R_I^0} - 1\right) \tag{74}$$

r taking into account that:

$$R_{18}^0 = 1/2(R_I^0 + R_{II}^0)_{18} \tag{75}$$

here $(R_{II}^0)_{18}$ denotes the ^{18}O isotopic ratio in the carboxyl group of the
etic acid obtained at complete decomposition of the malonic acid, we
ave:

$$k_7/k_6 = R_{II\,18}/R_{I\,18}^0 \tag{76}$$

quations (74) and (76) are used for determinations of the intramolecular
condary isotope effects of ^{18}O in the decarboxylation of malonic acid.
  Let us introduce the notations:

$$f = (1-e^{-2k_5 t}) \quad \text{and} \quad R_{t\,18} = \frac{[C^{16}O^{18}O]_t}{[C^{16}O^{16}O]_t}$$

d rewrite equation (69) in the logarithmic form:

$$-(k_6+k_7)t = \ln\left[1-f\left(\frac{k_6+k_7}{k_6}\right)R_{t\,18}\left(\frac{M_5^0}{M_6^0}\right)\right] \tag{77}$$

viding equation (77) by $-2k_5 t = \ln(1-f)$, gives, after simple rearrange-
ents, the expression for the determination of the intermolecular ^{18}O

secondary isotope effect:

$$\frac{k_5}{k_6} = \frac{1}{2}\left(1+\frac{k_7}{k_6}\right)\frac{\ln(1-f)}{\ln\left(1-fR_{It\,18}/R^0_{I\,18}\right)} \tag{78}$$

The relations for the determination of the intermolecular and intra-molecular ^{13}C isotope effects can be derived in a similar manner:

$$\frac{k_4}{k_3} = \frac{R^0_{II\,13}}{R^0_{I\,13}} = \frac{1}{(1+a_t)}\left[3\frac{R^0_{13}}{R^0_{I\,13}}-1\right] \tag{79}$$

and

$$\frac{k_1}{k_3} = \frac{1}{2}\left(1+\frac{k_4}{k_3}\right)\frac{\ln(1-f)}{\ln\left[1-f\left(\dfrac{k_3+k_4}{k_3}\right)\left(\dfrac{R_{It\,13}}{3R^0_{13}-R^0_{III\,13}}\right)\right]} \tag{80}$$

where

$$a_t = \frac{R^0_{III\,13}}{R^0_{II\,13}} = \frac{R^0_{III\,13}}{3R^0_{13}-\left(R^0_{III\,13}+R^0_{I\,13}\right)}$$

or

$$a_t^i = \frac{R^0_{III\,13}}{2R^0_{AA}-R^0_{III\,13}}$$

R^0_{13} is the ^{13}C/^{12}C ratio in the starting material, obtained by measuring the carbon dioxide after complete combustion of the malonic acid, $R^0_{I\,13}$ i the ^{13}C/^{12}C ratio of the carbon dioxide collected at full decarboxylation of the malonic acid, $R^0_{III\,13}$ is the ^{13}C/^{12}C ratio of the methylene carbon in the malonic acid; R^0_{II} is the ^{13}C/^{12}C ratio of the carboxyl group of the acetic acid obtained at complete decarboxylation of the malonic acid, $R_{It\,13}$ is the ^{13}C/^{12}C ratio of carbon dioxide obtained at partial decarboxy-lation of malonic acid; R^0_{AA} is the ^{13}C/^{12}C ratio in the carbon dioxid from combustion of the acetic acid obtained at complete decarboxylatio of the malonic acid, f is the fraction of decarboxylation of malonic aci and k_i are the rate constants for the isotopic species appearing in equation (62)–(68).

In the case of equal distribution of ^{13}C among all carbon atoms in th malonic acid: $R^0_{II} = 2R^0_{13}-R^0_{I\,13}$, and correspondingly:

$$\frac{k_4}{k_3} = \left(2R^0_{13}/R^0_{I\,13}\right)-1 \tag{81}$$

Experimental kinetic isotope effects of ^{13}C and ^{18}O in the decarboxyla-tion of malonic acid are given in Tables 3–5. These values confirm th temperature dependence of intramolecular kinetic isotope effects of ^{13}C i

TABLE 3. Intramolecular ^{13}C isotope effects in the decarboxylation of malonic acid

Temperature (°C)	$k_4/k_3{}^a$	$k_4/k_3{}^b$
121·5	1·0335	1·0354
136·0	1·0316	1·0335
140·0^c	1·0330	1·0348
148·0	1·0301	1·0320
181·0	1·0275	1·0294

^{a} Calculated from equation (79).
^{b} Calculated from equation (81).
^{c} This experiment was performed in a gold tube at a pressure of 15,100 p.s.i.

TABLE 4. Intermolecular ^{13}C isotope effects in the decarboxylation of malonic acid

Temperature (°C)	k_4/k_3 used for the calculation of k_1/k_3	$k_1/k_3{}^a$	k_1/k_4
140·0^b	1·0330	1·0365	1·0038
136·5	1·0316	1·0363	1·0045
112·5	1·0349	1·0388	1·0038
112·0	1·0349	1·0410	1·0058

^{a} Calculated from equation (80).
^{b} This experiment was done in a gold tube at a pressure of 15,100 p.s.i.

TABLE 5. Secondary ^{18}O isotope effects in the decarboxylation of malonic acid

Temperature (°C)	$2k_5/k_6+k_7$	$k_5/k_6{}^b$
140·0^a	1·00077	1·00155
136·5	1·00109	1·00218
135·5	1·00104	1·00208
112·5	1·00238	1·00477
112·0	1·00250	1·00501

^{a} Experiment performed in a gold vessel at a pressure of 15,100 p.s.i.
^{b} Calculated under the condition that $k_5 = k_7$.

the decarboxylation of malonic acid, and show that secondary kinetic isotope effects of ^{13}C (see Table 4, column 4) are by one order of magnitude smaller than primary intramolecular and intermolecular ^{13}C isotope effects. Primary ^{13}C intermolecular isotope effects are larger than primary intramolecular isotope effects. Secondary ^{18}O isotope effects are also smaller than primary ^{18}O isotope effects by one order of magnitude.

These results are in satisfactory agreement with the theoretical calculations. According to equation (52) the ratio k_4/k_3 (equations 64 and 65) can be *a priori* estimated by equation (82).

$$\frac{k_4}{k_3} = \left(\frac{m_3^\ddagger}{m_4^\ddagger}\right)^{1/2}\left[1 + \sum_{i=1}^{3n-7} G(u_i^\ddagger)\,(u_{3i}^\ddagger - u_{4i}^\ddagger)\right] \tag{82}$$

As a first approach in the model calculations one can assume that one C—C bond in the transition complex is disrupted, the transition state being represented as consisting of two fragments (3):

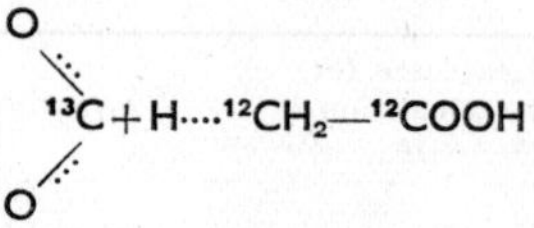

$$\begin{array}{c} O \\ \diagdown \\ ^{13}\text{C} + \text{H}\cdots ^{12}\text{CH}_2 -\!^{12}\text{COOH} \\ \diagup \\ O \end{array}$$

(3)

In this model the C—C bond in the acetic acid skeleton is contributing mainly to the temperature-dependent part of the intramolecular isotope effect, and similarily the C—C bond in the starting malonic acid contributes mainly to the temperature-dependent part of the intermolecular isotope effect ratio k_1/k_3:

$$k_1/k_3 = \left(\frac{m_3^\ddagger}{m_1^\ddagger}\right)^{1/2}[1 + G(u_{^{13}\text{C}-^{12}\text{C}})\,(u_{^{12}\text{C}-^{12}\text{C}} - u_{^{12}\text{C}-^{13}\text{C}})] \tag{83}$$

Within the validity of the first, one-bond approximation both the k_1/k_3 and k_4/k_3 ratios should be of the same order. The values presented in Tables 3–5 confirm this prediction. The different temperature dependences of the intermolecular and intramolecular ^{13}C isotope effects require the use of a more general model, reflecting the real structure of the malonic acid and of the transition state complex.

3. Decarboxylation of quinaldic and picolinic acid[75, 76]

Measured ^{13}C and ^{18}O isotope effects in the decarboxylation of quinaldic and picolinic acids are collected in Table 6, where the k_1/k_2 ratios for oxygen and carbon correspond to the following three reactions:

$$R—^{12}C^{16}O_2H \xrightarrow{\;k_1\;} RH + {}^{12}C^{16}O_2 \tag{84}$$

$$R—^{12}C^{16}O^{18}OH \xrightarrow{\;k_2(\text{oxygen})\;} RH + {}^{12}C^{16}O^{18}O \tag{85}$$

$$R—^{13}C^{16}O_2H \xrightarrow{\;k_2(\text{carbon})\;} RH + {}^{13}C^{16}O_2 \tag{86}$$

The k_{12C}/k_{13C} ratios for these acids are much lower than those published for malonic-type acids. ^{13}C isotope effects in the decarboxylation of picolinic acid above its melting point are slightly higher than the ^{13}C isotope effect in the decarboxylation of quinaldic acid. At 180°c they differ by 0·05%.

The temperature dependence of the ^{13}C isotope effects in both processes is small and positive above the melting points of the acids. Secondary ^{18}O isotope effects are by an order of magnitude smaller than primary ^{13}C kinetic isotope effects. Their temperature dependence is negative in the whole range of temperatures studied.

TABLE 6. ^{13}C and ^{18}O isotope effects in the decarboxylation of quinaldic and picolinic acid

Acid	Temperature °C	k_1/k_2	
		Oxygen	Carbon
Quinaldic acid	180·5	1·0023 ± 0·0003	1·0109 ± 0·0002
	171·5	1·0022 ± 0·0003	1·0111 ± 0·0005
	161·0	1·0019 ± 0·0004	1·0113 ± 0·0006
	151·0	1·0015 ± 0·0005	1·0109 ± 0·0003
	141·5	1·0003 ± 0·0003	1·0104 ± 0·0004
Picolinic acid	182·5	1·0013 ± 0·0002	1·0114 ± 0·0001
	149·5	0·9975 ± 0·0002	1·0122 ± 0·0001
	138·5	0·9972 ± 0·0002	1·0126 ± 0·0001

In the case of decarboxylation of malonic acid the rate ratios of decarboxylation of light and heavy molecules only have been measured. The ratio k_5/k_6 has been determined under the assumption that $k_5 = k_7$. This approximation does not contradict the assertion that in the decarboxylation of malonic acid also secondary ^{18}O isotope effects are by an order of magnitude smaller than primary ^{13}C isotope effects.

4. Interpretation of the experimental results

The results presented in this section reveal great differences between ^{13}C isotope effects and their temperature dependences in the decarboxylation of malonic acid on the one hand, and in that of quinaldic and picolinic acids on the other. The structural dependence of the primary kinetic isotope effect of ^{13}C becomes obvious, and the limits of the application of the one-bond treatment of the process have been shown.

In the general theory of the kinetic isotope effects the structure and characteristics of the whole molecule and of the transition complex are taken into account. The two series of decarboxylation reactions discussed above involve different substrates; different transition states might also be postulated. Let us consider the subject in detail.

Malonic acid and β-keto acids have only weak intramolecular bonds between the functional groups in the initial state, but a strongly bound cyclic transition state is generally accepted in the mechanism of their decarboxylation:

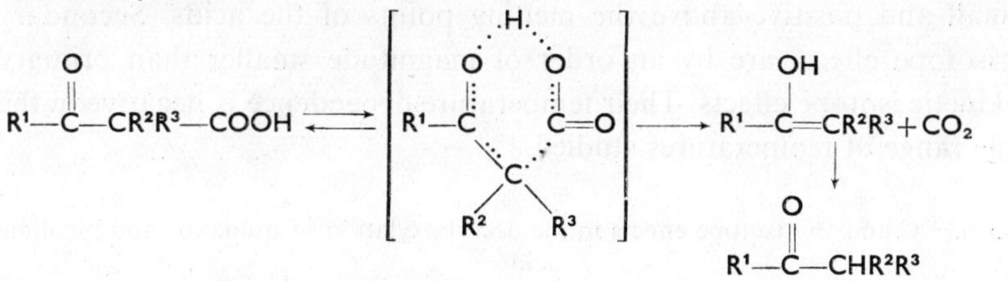

$$R^1-\overset{O}{\overset{\|}{C}}-CR^2R^3-COOH \rightleftharpoons \left[R^1-C \cdots \overset{H}{\cdots} \cdots C=O \right] \longrightarrow R^1-\overset{OH}{\overset{|}{C}}=CR^2R^3 + CO_2 \quad (87)$$

$$\downarrow$$

$$R^1-\overset{O}{\overset{\|}{C}}-CHR^2R^3$$

or

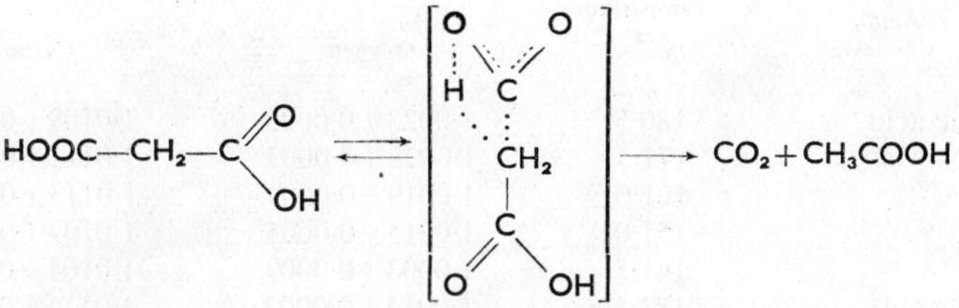

$$HOOC-CH_2-C\overset{\nearrow O}{\underset{\searrow OH}{}} \rightleftharpoons \left[\begin{array}{c} O \quad\quad O \\ H \quad C \\ \vdots \quad\quad \\ CH_2 \\ C \\ O \quad OH \end{array} \right] \longrightarrow CO_2 + CH_3COOH \quad (88)$$

In contrast, quinaldic and picolinic acids are, in the initial state, strongly bridged internally, and a zwitterion is formed (4). In this case, the struc-

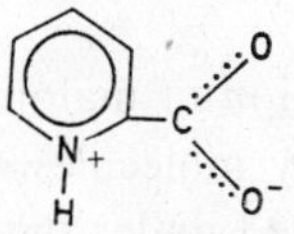

$$\text{(4)}$$

ture of the carboxyl group can be approximated by the symmetrical model (5)[61]. Hence, in the decarboxylation of quinaldic and picolinic acid,

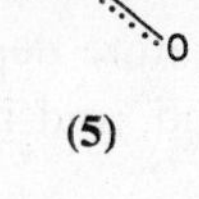

$$\text{(5)}$$

the transition state is less restricted in respect to rotation along the C—C bond. The reaction path can be represented by equation (89). Calculation

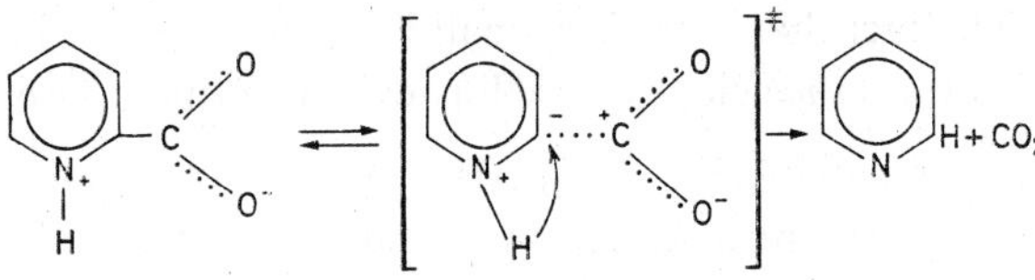

$$\tag{89}$$

based on the symmetrical model gives a maximum value of 1·038 for the ratio k_{12C}/k_{13C} at 156°C. The minimum value calculated under the assumption of the transition complex resembling the carbon dioxide molecule, is 0·998. The measured ^{13}C kinetic isotope effect (see Table 6) shows that the structure of the carboxyl group in the transition state of the decarboxylation of quinaldic acid is much closer to the structure of carbon dioxide than to the substrate zwitterion form of the acid (see Note 1, p. 503).

5. The role of deuterium isotope effects

In relation to the ^{13}C isotope effect studies it is worthwhile to mention measurements of the deuterium isotope effects (k_{RCOOH}/k_{RCOOD}) in the decarboxylation of substituted benzoylacetic acids[80]. At 50° the ratio k_H/k_D changes from 2·8 for m-nitrobenzoylacetic acid to 0·85 for p-methylbenzoylacetic acid. This large variation with substituents was interpreted by Swain and coworkers as favouring the cyclic proton transfer mechanism (6) and as being inconsistent with a cyclic hydride transfer mechanism (7).

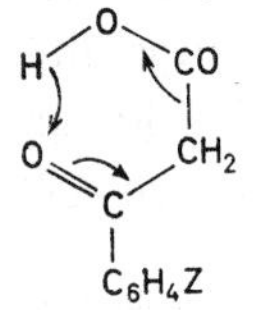

(6) 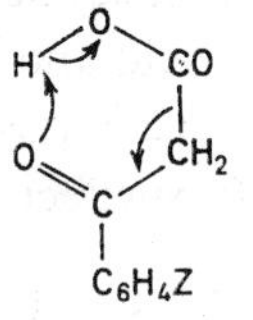   (7)

Similar measurements have not yet been taken for the decarboxylation of deuterated picolinic or quinaldic acid; neither has the influence of deuterium substitution on the ^{13}C kinetic isotope effect in their decarboxylation been studied. Similar studies should also be carried out with anthranilic acid, which decarboxylates when heated above its melting point to give aniline and carbon dioxide:

$$NH_2C_6H_4{}^{13}COOH \longrightarrow C_6H_5NH_2 + {}^{13}CO_2 \tag{90}$$

Decarboxylation also takes place in boiling water and in dilute aqueous solutions of sulphuric acid. Earlier mass spectrometric measurements[81] showed that the maximum ^{13}C isotope effect in reaction (90) is less than 0·5%, if any. The fact that there is a small ^{13}C isotope effect in the decarboxylation of anthranilic acid, was used as evidence that proton attack on the α carbon of the zwitterion $\overset{+}{N}H_3C_6H_4COO^-$ is the rate-determining step and that the rupture of the C—C bond takes place in a kinetically unimportant subsequent step (see Note 2, p. 503).

Finally the interesting observations made by Halevi[82] on the effect of deuterosubstitution on the ionization constants of deuterosubstituted acids should be mentioned. The ionization constants of $C_6H_5CD_2COOH$, CD_3COOH and $C_6H_5CD_2NH_3^+$ are by 10–13% smaller than those of the corresponding protium analogues. These effects are said to be due to the anharmonicity of the C—D oscillator. The real potential well of the C—D bond is not parabolic, but skewed. As a result of anharmonicity the mean C—D bond-length is shorter than the C—H bond-length. This causes a slightly greater electron density near the D-bonded carbon atom compared to that carrying only H. This means that deuterium should have an electron-donating inductive effect as compared to H, and in the case of α-deuterosubstituted acids this effect is revealed by their lower acidity.

6. Decarboxylation of picolinic acid hydrochloride

Picolinic acid hydrochloride decarboxylates at 180–210°c to give carbon dioxide and pyridine hydrochloride. The kinetic isotope effects of ^{13}C and ^{18}O are given in Table 7. Comparison of the data presented in Tables 6 and 7 shows large differences between the ^{13}C isotope effects in the decarboxylation of picolinic acid and its hydrochloride. These differences reflect the complexity of the decomposition of the hydrochloride and the dependence of the kinetic isotope effect on the mechanism of the decomposition.

TABLE 7. ^{13}C and ^{18}O isotope effects in the decarboxylation of picolinic acid hydrochloride[76]. (Decarboxylations have been done as previously in break-off tubes sealed under vacuum)

Temperature (°C)	Amount of decarboxylation	k_1/k_2	
		Oxygen	Carbon
181·0	0·2234	0·9987	1·0084
181·0	0·2856	0·9981	1·0081
195·3	0·1546	0·9983	1·0080
210·5	0·3429	0·9991	1·0072

Two routes may be postulated for the decarboxylation of picolinic acid hydrochloride. The first is a one-step decarboxylation (equation 91) the

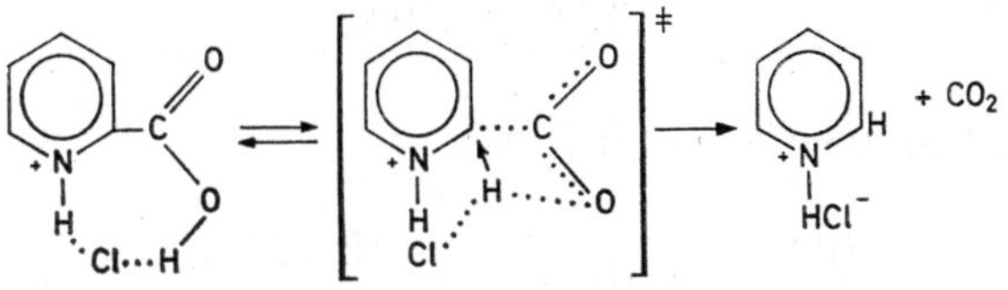

$$(91)$$

second a decarboxylation through a stable intermediate, (equation 92)

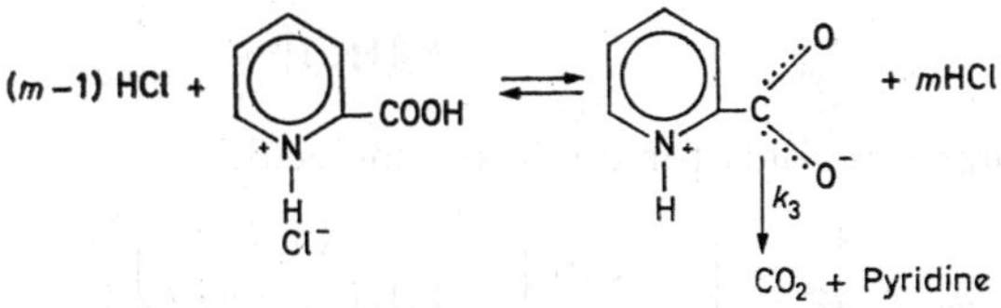

$$(92)$$

where picolinic acid hydrochloride dissociates first into hydrochloric acid and picolinic acid, which in turn decarboxylates rapidly to give carbon dioxide and pyridine.

In the first case, neglecting weak solvation effects, the rate should be independent of the presence of hydrochloric acid in the system and should obey equation (93).

$$\frac{d[CO_2]}{dt} = k[PA \cdot HCl] \qquad (93)$$

In the second case, i.e. in decarboxylation through a stable intermediate, picolinic acid (PA), we shall introduce the following symbols in order to obtain the dependence of the rate of decomposition on the concentration of the hydrochloride:

$$[^{12}CO_2]_t = X_t \qquad\qquad [^{13}CO_2] = X_t^*$$

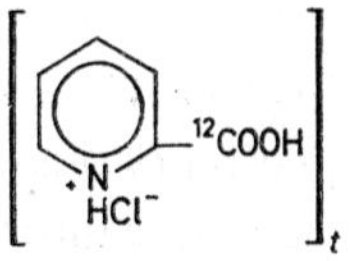

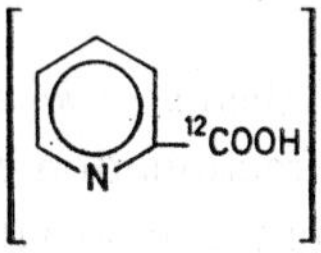

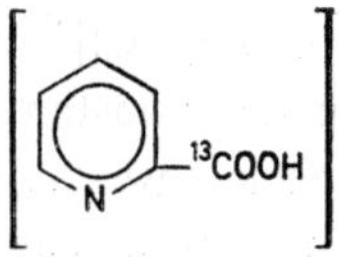

According to equation (92), the rate of formation of carbon dioxide is proportional to the concentration of free picolinic acid:

$$\frac{d[X]}{dt} = k_3[PA] \tag{94}$$

Steady state treatment gives for [PA]:

$$[PA] = \frac{k_1}{k_2[HCl]^m + k_3} [PA \cdot HCl] \tag{95}$$

Substitution of (95) into (94) and integration leads to equation (96).

$$\frac{X_t}{X_0} = \left[1 - \exp\left(-\frac{k_1 k_3 t}{k_2[HCl]^m + k_3} \right) \right] \tag{96}$$

A similar expression holds for the heavy molecule:

$$\frac{X_t^*}{X_0^*} = \left[1 - \exp\left(-\frac{k_1^* k_3^* t}{k_2^*[HCl]^m + k_3^*} \right) \right] \tag{97}$$

Equations (96) and (97) lead directly to (98):

$$\frac{k}{k^*} = \frac{\ln(1-f)}{\ln(1 - R_t/R_0)} \tag{98}$$

where $R_t = X_t^*/X_t$, $R_0 = X_0^*/X_0$, $f = 1 - e^{-kt}$, $k = k_1 k_3/(k_2[HCl]^n + k_3)$ and $k^* = k_1^* k_3^*/(k_2^*[HCl]^n + k_3^*)$. For short times or small percentages of decarboxylation equation (98) can be replaced by the approximate relation (99).

$$R_\theta/R_t = \frac{k_1 k_3 \{k_2^*[HCl]^m + k_3^*\}}{k_1^* k_3^* \{k_2[HCl]^m + k_3\}} \tag{99}$$

From equations (94)–(99) it follows that the rate constant is a function of the concentration of the hydrochloric acid in the medium in which the picolinic acid hydrochloride decomposes:

$$k = \frac{k_1 k_3}{k_2[HCl]^m + k_3} \tag{100}$$

If $k_3 \gg k_2[HCl]^m$, then $k \approx k_1$ and correspondingly $k/k^* = k_1/k_1^*$. In this case the isotopic fractionation estimated by equation (99) corresponds to the isotope effect in the dissociation of the hydrochloride into hydrochloric acid and free picolinic acid.

If $k_2[HCl] \gg k_3$ then $k \approx k_1 k_3/k_2[HCl]^m$ and $k/k^* = k_3 K/k_3^*$, where $K = (k_1/k_2)/(k_1^*/k_2^*)$. Expressing the constant K by the equilibrium concentrations of the isotopic species we obtain: $K = [HCl \cdot PA^*][PA]/[HCl \cdot PA][PA^*]$. The measured isotope effects favour the dissociation mechanism of decarboxylation, but further studies are required, particularly regarding the dependence of the carbon isotope fractionation on the pressure of the hydrochloric acid in the system. Direct measurements of the rate of decarboxylation as a function of the concentration of hydrochloric acid[76], and the fast exchange of chloride between picolinic acid hydrochloride and the gaseous hydrogen chloride fit the predissociation mechanism of decarboxylation[83]. We are now studying the tritium exchange between gaseous hydrogen chloride and the hydrochlorides of pyridine or picolinic acid. These studies will furnish additional data concerning the strength of the hydrogen bonding the hydrochlorides of the compounds studied.

V. SYNTHESES OF LABELLED CARBOXYLIC ESTERS

A. Chemical Methods

Synthesis of tritium-, ^{14}C-, ^{13}C- and ^{18}O-labelled esters can be achieved either by direct esterification or by alcoholysis of acyl chlorides or acid anhydrides using labelled reagents[2]. For instance ^{18}O-labelled ethyl propionate has been obtained by reaction (101)[84]:

$$C_2H_5C\overset{O}{\underset{Cl}{\diagdown}} + C_2H_5{}^{18}OH \longrightarrow C_2H_5\overset{O}{\overset{\|}{C}}{}-{}^{18}O{-}C_2H_5 \tag{101}$$

Tritium-labelled ethyl and methyl formates and methyl-^{14}C formate have been obtained by esterification in the presence of anhydrous calcium chloride[85, 86]:

$$TCOOH + CH_3OH \xrightarrow{\text{CaCl}_2} TCO_2CH_3 \tag{102}$$

$$HCOOH + {}^{14}CH_3OH \xrightarrow{\text{CaCl}_2} HCO_2{}^{14}CH_3 \tag{103}$$

The esterification will also proceed in the presence of concentrated sulphuric acid[87]:

$$CH_2TC_6H_4CO_2H + CH_3OH \xrightarrow{\text{H}_2\text{SO}_4} CH_2TC_6H_4CO_2CH_3 \tag{104}$$

Another useful synthesis of labelled esters is the reaction between the silver salt of the acid and an alkyl halide[1-3, 154]:

$$R^1COOAg + XR^2 \longrightarrow R^1COOR^2 + AgX \tag{105}$$

B. Radiochemical and Other Methods

Tritiation takes place when a saturated aliphatic ester is exposed to tritium gas[88, 89]. However it was found that methyl oleate, $C_8H_{17}CH{=}CH(CH_2)_7CO_2CH_3$, exposed to tritium yields radioactive methyl stearate

$C_{17}H_{34}TCO_2CH_3$ due to addition of the tritium to the double bond[90]. Methyl stearolate, $CH_3(CH_2)_7C{\equiv}C(CH_2)_7CO_2CH_3$, yields *cis-* and *trans*-9,10-tritiooctadecenoate, $C_{17}H_{34}TCO_2CH_3$, as the major radiochemical product on exposure to tritium gas[91]. The total amount of tritium incorporated into the ester upon irradiation of a 1 g sample of methyl stearolate with 2·6 Ci of tritium gas for 20 days was found to be 210 mCi. By analogy with other compounds the exchange should occur between the hydrogen of the esters and hot tritium atoms[92a] generated by $^3He(n, p)^3H$ and $^6(n, \alpha)T$ nuclear reactions[49, 92b, 92c]. Carboxylic esters can also be labelled with tritium by reaction with tritium atoms generated from molecular HT at a tungsten filament maintained at 1750°c and at pressures less than 10 μ[93]. With unsaturated esters both addition and substitution should take place. The degradation of the irradiated sample should be less in this method than in either the gas exposure or recoil-labelling techniques.

VI. TRACER STUDIES WITH LABELLED CARBOXYLIC ESTERS

A. Studies of the Alkaline, Acid and Enzymic Hydrolysis of Carboxylic Esters by Means of ^{18}O

1. Alkaline and acid hydrolysis

Since the time of the availability of ^{18}O-enriched samples[94, 95] many tracer studies have dealt with the mechanism of ester hydrolysis and the problem of acyl-oxygen or alkyl-oxygen fission.

$$R^1{-}\overset{\overset{\displaystyle O}{\|}}{C}{-}O{-}R^2 + H_2O \longrightarrow R^1{-}\overset{\overset{\displaystyle O}{\|}}{C}{-}OH + R^2OH \qquad (106)$$

Polanyi and Szabo[94] showed that hydrolysis of n-amyl acetate in $H_2^{18}O$ gives amyl alcohol with normal isotopic composition:

$$CH_3COOC_5H_{11} + H_2^{18}O \longrightarrow CH_3CO^{18}OH + C_5H_{11}OH \qquad (107)$$

Roberts and Urey[95] found that esterification of benzoic acid with $CH_3^{18}OH$ gives water of ordinary composition:

$$C_6H_5CO{\vdots}OH + CH_3^{18}O{\vdots}H \longrightarrow C_6H_5CO{-}^{18}OCH_3 + H_2O \qquad (108)$$

At the same time Datta, Day and Ingold[96] demonstrated that a similar mechanism holds for the acid hydrolysis of methyl hydrogen succinate in $H_2^{18}O$

$$COOH(CH_2)_2CO{\vdots}OCH_3 + H_2^{18}O \longrightarrow COOH(CH_2)_2CO{-}^{18}OH + CH_3OH \quad (109)$$

The experiments of Polanyi and Szabo have been confirmed by Kursanov and Kudriavcev[84]. ^{18}O-labelled ethyl propionate was used as a tracer:

$$C_2H_5CO\text{---}^{18}O\text{---}C_2H_5 + H_2O \xrightarrow{\text{NaOH}} C_2H_5CO\text{---}^{16}OH + C_2H_5{}^{18}OH \qquad (110)$$

It has also been demonstrated that in the reversible reaction of acetic anhydride with $C_2H_5{}^{18}OH$, ^{18}O-labelled ethyl acetate is formed[97]:

$$CH_3OC\text{---}OCOCH_3 + H\text{---}^{18}OC_2H_5 \rightleftharpoons CH_3OC\text{---}^{18}OC_2H_5 + CH_3COOH \quad (111)$$

Long and Friedman[98] found that γ-butyrolactone is hydrolysed by acyl-oxygen fission:

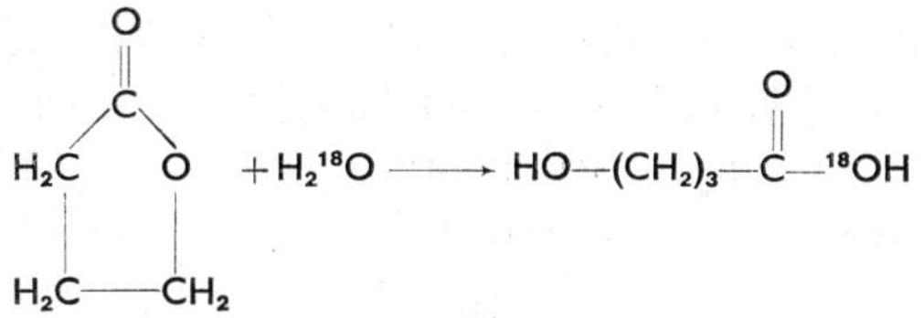

$$(112)$$

Ethyl benzoate and methyl 2,4,6-trimethylbenzoate also undergo acyl-oxygen fission[99], but in the methanolysis of methyl mesitoate a very slow alkyl-oxygen cleavage with formation of dimethyl ether is observed[100].

Fission of the alkyl-oxygen bond is of major importance in the solvolysis of esters of tertiary alcohols (equations 113 and 114)[101], although in alkaline

$$CH_3COO\text{---}CPh_3 + ROH \longrightarrow CH_3CO_2H + Ph_3COR \qquad (113)$$
$$(C_6H_5)_3C_6H_2COO\text{---}C(CH_3)_3 + H_2{}^{18}O \longrightarrow (C_6H_5)_3C_6H_2COOH + (CH_3)_3C\text{---}^{18}OH \quad (114)$$

solution some acyl-oxygen fission also occurs. The departure from the general acyl-oxygen fission mechanism is explained by the formation of relatively stable tertiary carbonium ions in the rate-determining step (equations 115 and 116), followed by the fast reaction (117).

$$CH_3COOCPh_3 + H^+ \xrightleftharpoons{\text{fast}} [CH_3CO_2HCPh_3]^+ \qquad (115)$$
$$[CH_3CO_2HCPh_3]^+ \xrightleftharpoons{\text{slow}} CH_3CO_2H + Ph_3C^+ \qquad (116)$$
$$Ph_3C^+ + H_2O \xrightarrow{\text{fast}} Ph_3COH + H^+ \qquad (117)$$

The nature of the intermediates formed in the process of hydrolysis or esterification has been the subject of many studies[102-110]. Studying the OH$^-$-catalysed ^{18}O exchange between the solvent and ^{18}O carbonyl-labelled ethyl benzoate, isopropyl benzoate and t-butyl benzoate during their hydrolysis, it has been concluded[106] that an unstable intermediate is formed. It has been suggested that for both acidic and basic hydrolysis, one of the intermediates must have the structure of the unionized hydrate of the

31*

ester, $R^1C(OH)_2OR^2$, with the symmetrical structure capable of exchange. The following scheme has been proposed for the base-catalysed exchange process:

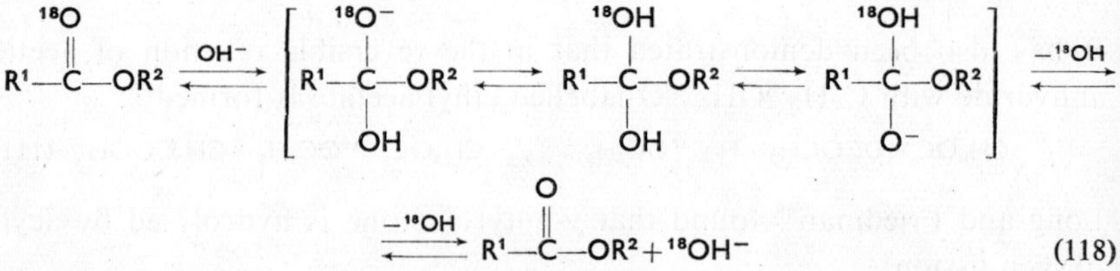

$$
\begin{array}{c}
^{18}O \\
\parallel \\
R^1{-}C{-}OR^2
\end{array}
\underset{\longrightarrow}{\overset{OH^-}{\longleftarrow}}
\left[
\begin{array}{c}
^{18}O^- \\
| \\
R^1{-}C{-}OR^2 \\
| \\
OH
\end{array}
\rightleftarrows
\begin{array}{c}
^{18}OH \\
| \\
R^1{-}C{-}OR^2 \\
| \\
OH
\end{array}
\rightleftarrows
\begin{array}{c}
^{18}OH \\
| \\
R^1{-}C{-}OR^2 \\
| \\
O^-
\end{array}
\right]
\underset{\longrightarrow}{\overset{-^{18}OH}{\longleftarrow}}
$$

$$
\underset{\longleftarrow}{\overset{-^{18}OH}{\longrightarrow}}
\begin{array}{c}
O \\
\parallel \\
R^1{-}C{-}OR^2 + {}^{18}OH^-
\end{array}
\tag{118}
$$

Equation (118) has been further extended for the ^{18}O exchange reaction between substituted benzoic acids and water[107–110]. It has also been concluded that the exchange reaction proceeds by means of addition of a molecule of water to the carboxylic acid to form an unstable orthoacid:

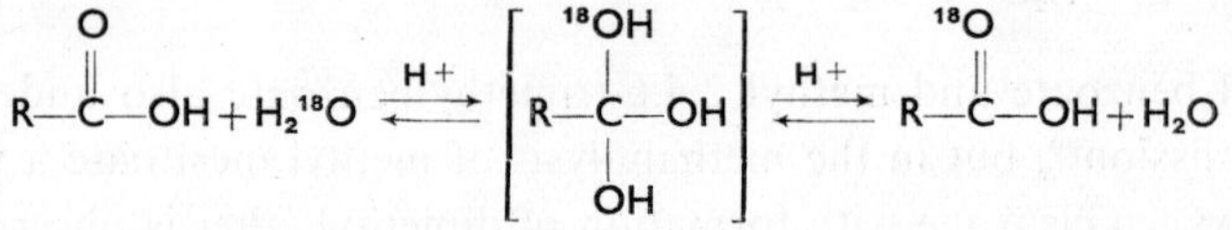

$$
\begin{array}{c}
O \\
\parallel \\
R{-}C{-}OH + H_2{}^{18}O
\end{array}
\underset{\longleftarrow}{\overset{H^+}{\longrightarrow}}
\left[
\begin{array}{c}
^{18}OH \\
| \\
R{-}C{-}OH \\
| \\
OH
\end{array}
\right]
\underset{\longleftarrow}{\overset{H^+}{\longrightarrow}}
\begin{array}{c}
^{18}O \\
\parallel \\
R{-}C{-}OH + H_2O
\end{array}
\tag{119}
$$

Finally it has been postulated that both in the hydrolysis of the ester and in the exchange of ^{18}O between carboxylic acids and water the addition of the water molecule to the carbonyl group of either the acid or the ester is a slow process while subsequent breakdown of the orthoacid or hydrate of the ester is rapid. Some evidence against this suggestion was given by Talvik and Pal'm[111]. They found that ethyl acetate hydrolyses by the unimolecular mechanism in concentrated sulphuric acid solution:

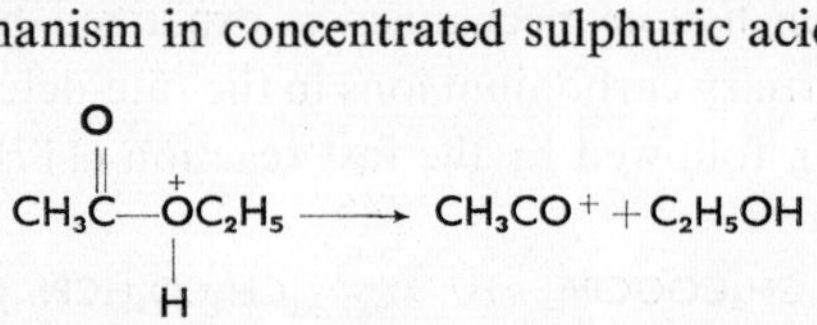

$$
\begin{array}{c}
O \\
\parallel \\
CH_3C{-}\overset{+}{\underset{|}{O}}C_2H_5 \\
H
\end{array}
\longrightarrow
CH_3CO^+ + C_2H_5OH
\tag{120}
$$

The results have been explained by assuming complete protonation of ethyl acetate by the sulphuric acid. Later it was found that in the acid hydrolysis of ethyl benzoate the carbonyl oxygen of the unhydrolysed ester exchanges with the oxygen of the solvent water, but at a slower rate than the rate of hydrolysis[109]. This suggests that the water molecule first adds in a reversible step to $[C_6H_5CO_2(C_2H_5)H]^+$ forming $[C_6H_5CO_2(C_2H_5)HH_2O]^+$ but it is possible that exchange and hydrolysis proceed by unrelated mechanisms. Recently it has been suggested, using volumes of activation measurements as a method of determining the reaction mechanism[112] that the decomposition of the adduct protonated on the ether oxygen might also be postulated as the slow step in the acid-catalysed hydrolysis of methyl and ethyl acetate:

$$
H_3O^+ + CH_3\overset{\overset{\displaystyle OCH_3}{|}}{\underset{\underset{\displaystyle OH}{|}}{C}} = O \xrightleftharpoons{\text{fast}} CH_3\overset{\overset{\displaystyle H\overset{+}{O}CH_3}{|}}{\underset{\underset{\displaystyle OH}{|}}{C}} - OH \xrightarrow{\text{slow}} CH_3\overset{+}{C}(OH)_2 + CH_3OH \xrightarrow[H_2O]{\text{fast}}
$$

$$
\xrightarrow[H_2O]{\text{fast}} CH_3COOH + CH_3OH + H_3\overset{+}{O} \tag{121}
$$

The reversible scheme (118) is not applicable to the alkaline hydrolysis of benzyl benzoate and phenyl benzoate[109, 113]. When phenyl benzoate is hydrolysed by $H_2^{18}O$ in the presence of sodium hydroxide in aqueous dioxane, the unhydrolysed ester has the natural $^{18}O/^{16}O$ ratio and therefore no appreciable oxygen exchange takes place between the carbonyl oxygen and the oxygen of the water. Thus if the intermediate in reaction (122) is formed at all there can be no hydrogen migration from one oxygen atom to another during the lifetime of the intermediate and it loses the phenoxide ion far more easily than the hydroxide ion ($k_3 \gg k_2$).

$$
C_6H_5CO_2C_6H_5 + {}^{18}OH^- \underset{k_2}{\overset{k_1}{\rightleftharpoons}} \left[C_6H_5\overset{\overset{\displaystyle O^-}{|}}{\underset{\underset{\displaystyle {}^{18}OH}{|}}{C}} - OC_6H_5 \right] \xrightarrow{k_3} C_6H_5CO^{18}O^- + C_6H_5OH \tag{122}
$$

2. Acetate ion catalysed hydrolysis

The acetate ion catalyses the hydrolysis of 2,4-dinitrophenyl benzoate[114]. Use of ^{18}O-labelled acetate provided indirect evidence for the formation of a mixed acetic benzoic anhydride as an intermediate in the process:

$$
CH_3\overset{\overset{\displaystyle {}^{18}O}{\|}}{C} - {}^{18}O^- + \overset{\overset{\displaystyle C_6H_5}{|}}{\underset{\underset{\displaystyle C_6H_3(NO_2)_2}{|}}{\underset{\displaystyle O}{C}}} = O \longrightarrow CH_3\overset{\overset{\displaystyle {}^{18}O}{\|}}{C} - {}^{18}O - \overset{\overset{\displaystyle O}{\|}}{C}C_6H_5 \longrightarrow
$$

$$
CH_3\overset{\overset{\displaystyle {}^{18}O}{\|}}{C} - O^- + C_6H_5\overset{\overset{\displaystyle O}{\|}}{C} - {}^{18}O^- \tag{123}
$$

The benzoic acid formed in the hydrolysis was found to contain about 75% of the ^{18}O derived from one of the oxygen atoms of the acetate $-^{18}O$ ion.

Formation of the mixed acetic salicylic anhydride in the hydrolysis of the aspirin anion was demonstrated[115] using ^{18}O-enriched water:

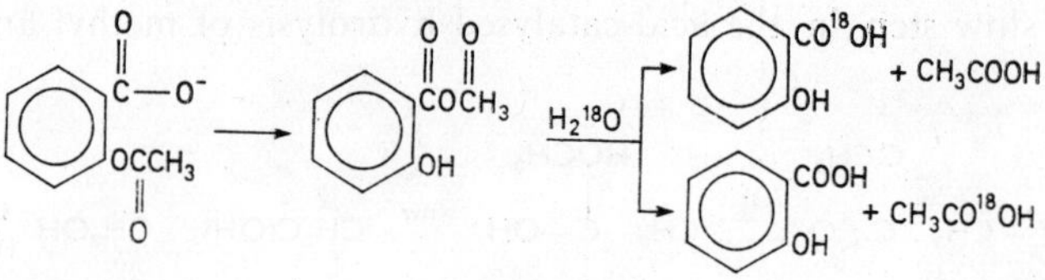

$$\text{(124)}$$

3. Enzymic hydrolysis

A number of enzymic hydrolyses have been investigated[116-122] and is has been shown that these proceed by means of acyl-oxygen fission. However, in contrast to non-enzymic alkaline hydrolysis, there is no exchange of the carbonyl oxygen of the ester during the α-chymotrypsin-catalysed hydrolysis[120]. Acetylcholine-sterase-catalysed hydrolysis is not accompanied by exchange of the unhydrolysed ester[117], although the same enzyme does catalyse the exchange reaction between the oxygen atoms of fatty acids and water[116]. The significance of isotopic exchange as a criterion for enzyme mechanism has been reviewed by Koshland[121]. He put forward the idea that the fission point of compounds described by the general formula R—O—Q is directly related to the specificity of the enzyme. Thus if the enzyme shows high specificity for R and low specificity for Q, then R—O cleavage occurs in the enzyme reaction. The lack of carbonyl-oxygen exchange can be explained by the formation of an acyl-enzyme intermediate and by the non-equivalence of the carbonyl-oxygen atom with other oxygen atoms, owing to the interaction of the former with the enzyme surface at the active site[109, 123, 124]:

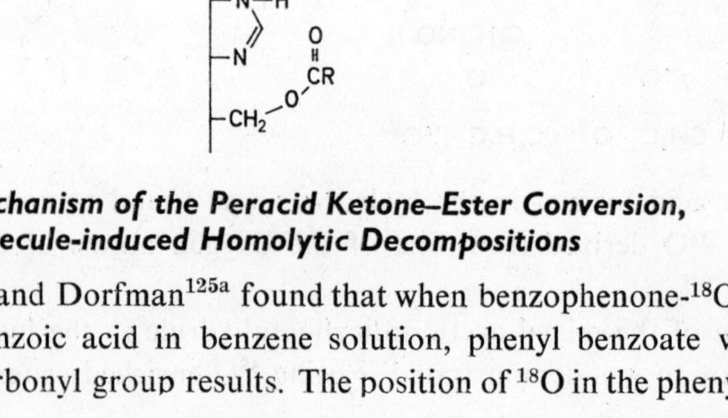

$$\text{(125)}$$

B. The Mechanism of the Peracid Ketone–Ester Conversion, and Molecule-induced Homolytic Decompositions

Doering and Dorfman[125a] found that when benzophenone-^{18}O is treated with perbenzoic acid in benzene solution, phenyl benzoate with a ^{18}O-labelled carbonyl group results. The position of ^{18}O in the phenyl benzoate was determined by reduction of the ester to phenol and benzyl alcohol:

$$C_6H_5\overset{\overset{\displaystyle ^{18}O}{\|}}{C}OC_6H_5 \xrightarrow{\text{LiAlH}_4} C_6H_5CH_2{}^{18}OH + C_6H_5OH \qquad (126)$$

It was thus established that the ketone–ester conversion must proceed via the 'Criegee mechanism'[125b] (equation 127), the other possible mechanisms being rejected as inconsistent with the ^{18}O distribution in the products.

$$C_6H_5\overset{\overset{\displaystyle ^{18}O}{\|}}{C}C_6H_5 + C_6H_5COOH \longrightarrow \underset{\underset{\|}{\underset{O}{OOCC_6H_5}}}{C_6H_5\overset{\overset{\displaystyle ^{18}OH}{|}}{C}C_6H_5} \longrightarrow C_6H_5\overset{\overset{\displaystyle ^{18}O}{\|}}{O}CC_6H_5 \qquad (127)$$

The mechanism of formation of cyclohexyl acetate has been established by decomposing, in boiling cyclohexene, acetyl peroxide labelled with ^{18}O in the carbonyl group[125c]. The resulting cyclohexyl acetate contains 58% of the label (^{18}O excess over natural abundance) in the carbonyl oxygen and 42% in the alkoxy oxygen. Such a distribution of ^{18}O rules out (within the limit of the uncertainty introduced by neglecting the ^{18}O isotope effect) the possibility that the cyclohexyl acetate results exclusively by the adition of free acetoxy radicals to cyclohexene and supports the simultaneous operation of a competing mechanism which involves attack of cyclohexene on the saturated peroxide oxygen to yield the intermediate 2-acetoxycyclohexyl radical:

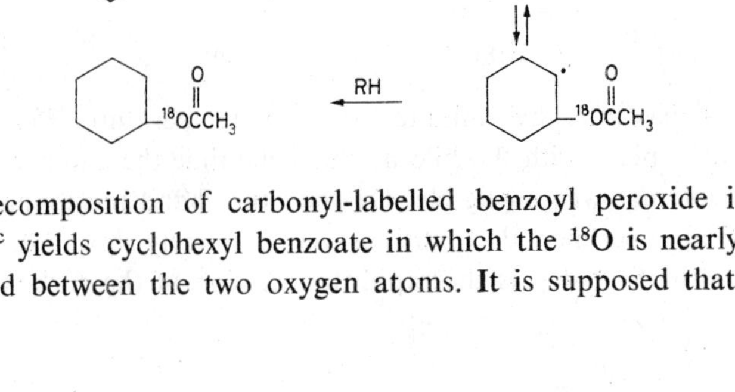

The decomposition of carbonyl-labelled benzoyl peroxide in cyclohexene[125c] yields cyclohexyl benzoate in which the ^{18}O is nearly equally distributed between the two oxygen atoms. It is supposed that the full equilibration of the ^{18}O proceeds through the intermediate radical stabilized by the aromatic ring:

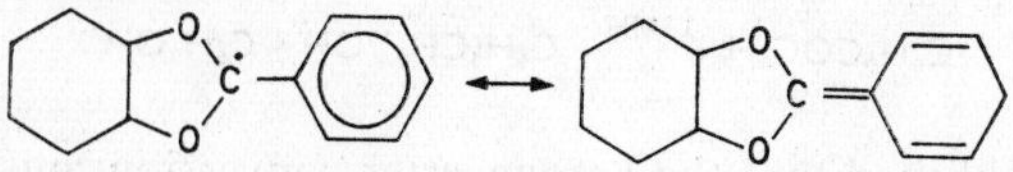

$$\tag{129}$$

However, studying the reaction of the triphenylmethyl radical with carbonyl-labelled benzoyl peroxide Doering, Akamoto and Krauch have found that the resulting triphenylmethyl benzoate contains most of the label in the carbonyl group[125d]. In that way they demonstrated that the new carbon–oxygen bond in the ester is formed from the ether-oxygen of the benzoyl peroxide and that the initiating attack of the triphenylmethyl radical on benzoyl peroxide occurs entirely at the peroxide oxygen atom (equation 130).

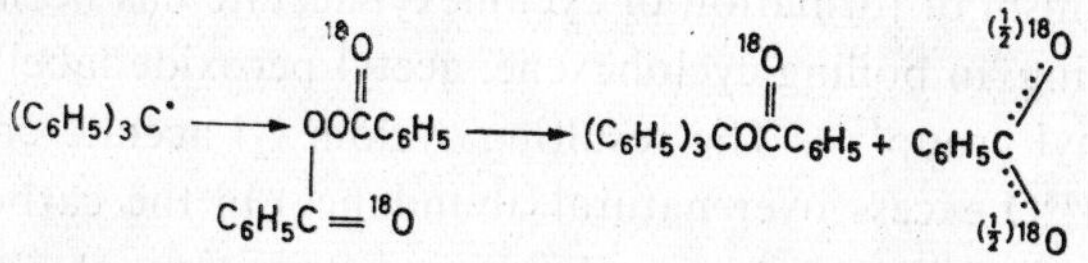

$$\tag{130}$$

C. Acetoxy Exchange Studies in Saccharide Esters of Carboxylic Acids

Application of ^{14}C-labelled reagents to the chemistry of the saccharide esters of carboxylic acids clarified to some extent the problem of anomerization of acetylated derivates of monosaccharides. Lemieux and Brice[126–128] studied the anomerization of 1,2,3,4,6-penta-O-acetyl-α-D-glucopyranose (8) and 1,2,3,4,6-penta-O-acetyl-β-D-glucopyranose in chloroform solution containing stannic chloride as the catalyst and stannic trichloride acetate [$SnCl_3OAc^*$] labelled with ^{14}C.

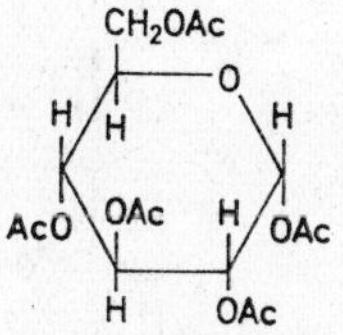

(8) 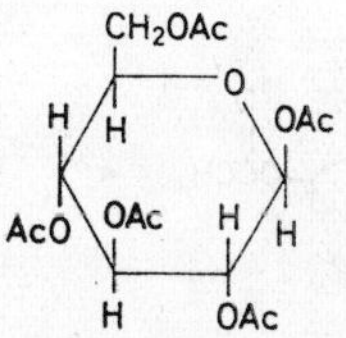   (9)

They found that after five minutes at room temperature, 75% acetoxy exchange took place with 9 while at the same time the anomerization of the β to α anomer was negligible. On the other hand the exchange with 8 in identical experimental conditions was extremely slow. According to the authors the first step is the rapid dissociation of the carboxyl group at $C_{(1)}$ with formation of the 1,2-α-D-cyclic carbonium ion (10). Recombination to the anomer 9 readily takes place by substitution (with acetate ion) and inversion at $C_{(1)}$ of 10, while for the formation of the anomer 8 the previous formation of ionic forms such as 11 and 12 is necessary and the anomerization process is therefore slow. To reveal the influence

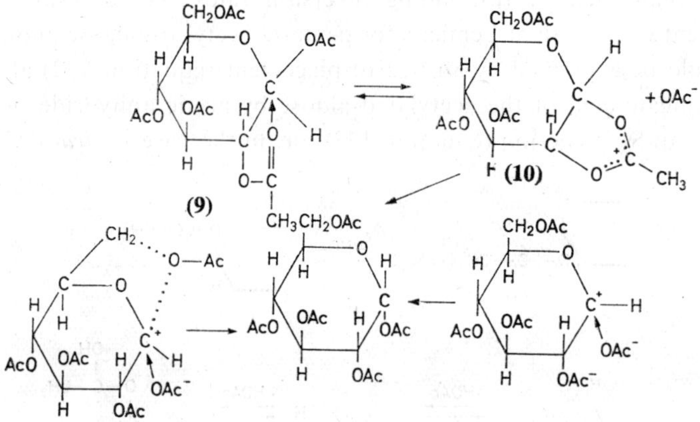

of the substituent at $C_{(2)}$ on the acetoxy exchange rate, the authors have measured the exchange with two anomeric pairs, 8 and 9 on the one hand,

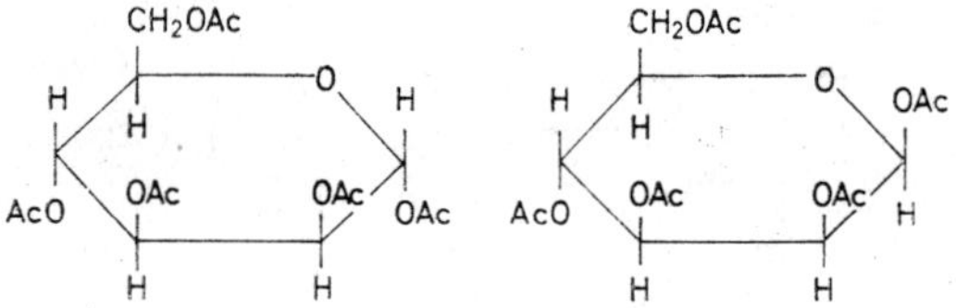

and 1,2,3,4,6-penta-O-acetyl-α-D-mannopyranose (13) and 1,2,3,4,6-penta-O-acetyl-β-D-mannopyranose (14), on the other.

It has been found that saccharides having 1,2-*trans* substituents (9 and 13), exchange the acetoxy group faster than saccharides with 1,2-*cis* substituents (8 and 14). Further studies revealed that the configuration of the substituents at $C_{(3)}$ also influences the rate of the acetoxy exchange[128].

Sulphuric acid catalysed $C_{(1)}$ acetoxy exchanges during the anomerization of $C_{(1)}$ acetoxy-labelled acetylated aldopyranoses in 1 : 1 acetic anhydride–acetic acid solvent have been studied by Bonner[129–131]. In the case of acetylated aldopyranose anomers with *cis* acetoxy groups at $C_{(1)}$ and

$C_{(2)}$, the polarimetrically determined $C_{(1)}$ inversion rate and the $C_{(1)}$ acetoxy exchange rates determined by the radioactive tracer method are identical within experimental error. Acetylated aldopyranose anomers having *trans* $C_{(1)}$–$C_{(2)}$ acetoxy groups show $C_{(1)}$ acetoxy exchange rates three (for acetylated pentoses) to fourteen (for acetylated hexoses) times greater than their corresponding inversion rates. These results are in agreement with those of Lemieux for penta-O-acetyl-D-glucose anomers[126], and could be explained by an S_N2 displacement (equation 132) at $C_{(1)}$ of the conjugate acid of the acetylated aldose by acetic anhydride or acetic acid, by an S_N1 reaction (equation 133), or, in the case of *trans* $C_{(1)}$–$C_{(2)}$

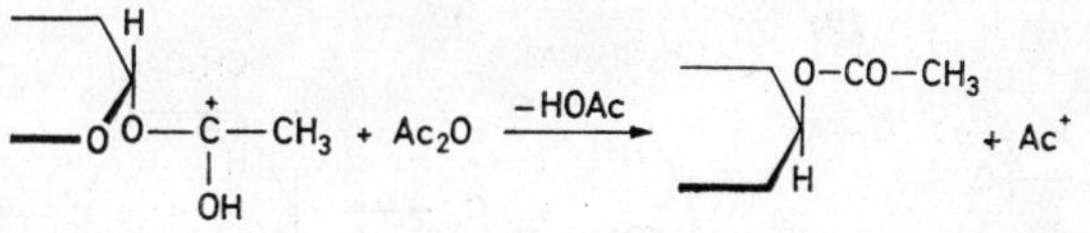

$$(132)$$

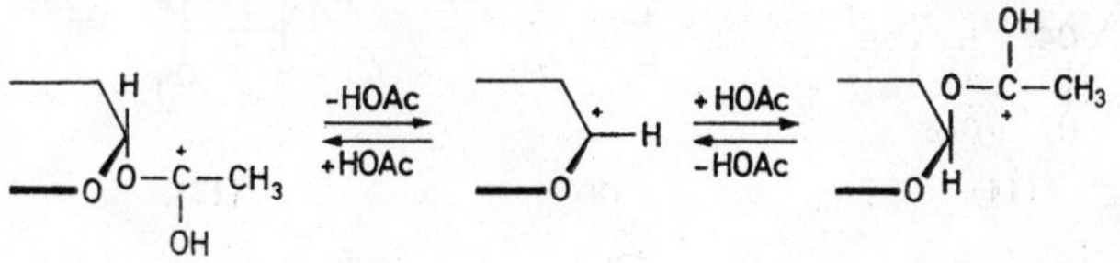

$$(133)$$

anomers by a $C_{(2)}$ acetoxy participation process (equation 134).

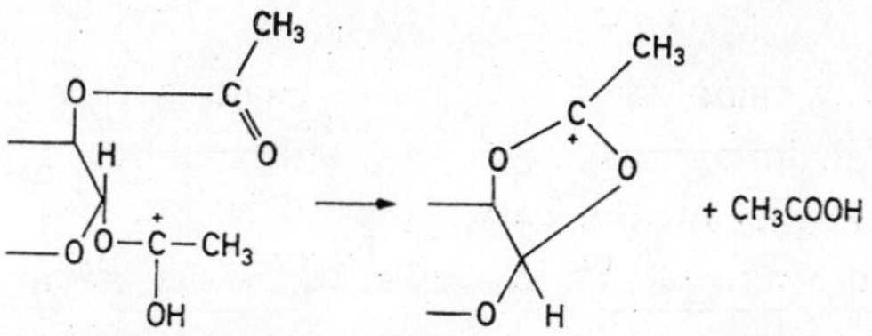

$$+ CH_3COOH \qquad (134)$$

To eliminate the ambiguity caused by possible participation of the $C_{(2)}$ acetoxy group in the anomerization of the 1,2-*trans* anomers, the anomers of tetra-O-acetyl-2-deoxy-D-glucopyranose, labelled in the $C_{(1)}$ acetoxy group with ^{14}C, were studied. With both anomers, the $C_{(1)}$ acetoxy exchange rate exceeded the inversion rate by a factor of 1·8–3·7. These observations are in agreement with the S_N1 mechanism (equation 133).

The S_N1 mechanism was further supported by deuterium isotope effect studies during acid-catalysed anomerization and $C_{(1)}$ acetoxy exchange reactions of acetylated D-aldopyranoses. It was found that in an AcOD–Ac$_2$O–D$_2$SO$_4$ medium the rates of both inversion and $C_{(1)}$ acetoxy exchange were approximately 1·7 times greater than the rates observed in AcOH–Ac$_2$O–H$_2$SO$_4$. The equilibrium (135) is shifted to the right when

D_2SO_4 is substituted for H_2SO_4, increasing the concentration of the conjugate acid of the acetylated aldose, thus increasing the rate of reactions (133) and (134) to comparable extents. The latter steps, involving no O—H bond fission, should not be subject to a deuterium isotope effect.

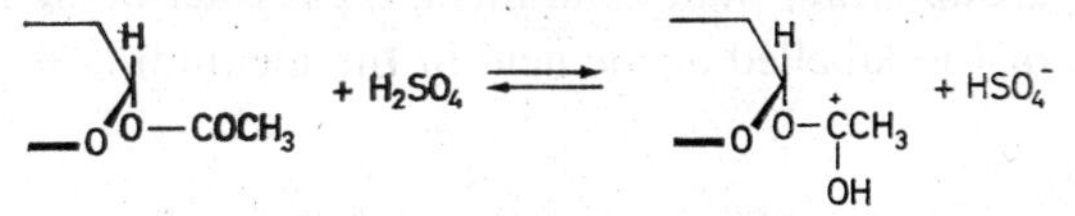

$$\text{(135)}$$

A similar S_N1 mechanism has been proposed by Bonner and Collins[132] to explain the acid-catalysed radiochemical isomerization and acetoxyl exchange of the 1,2,2-triphenylethyl acetates $Ph_2CHCH(OAc)Ph$ labelled discretely in the chain, phenyl and acetate portions of the molecule. The authors have explained the statistical redistribution of the radioactive labels in terms of open carbonium ion intermediates which are sufficiently long-lived to permit radiochemical isomerization (equation 136), before reacting with the solvent acetic acid.

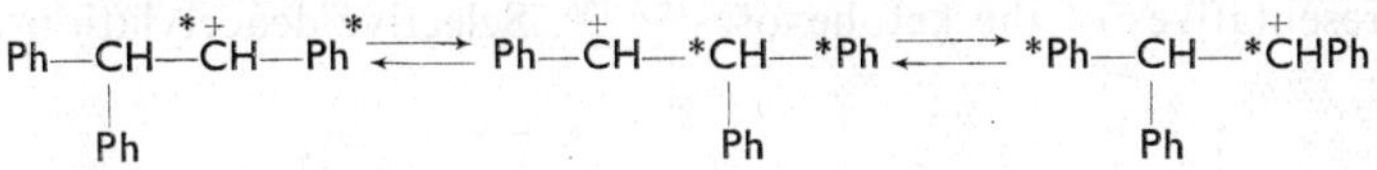

$$\text{(136)}$$

Swiderski and coworkers studied the acetoxyl exchange between anhydrous acetic acid labelled with ^{14}C in the carboxyl group and peracetylated monosaccharides without additional catalysts[133–138]. Acetoxyl exchange has been found in the case of per-O-acetylated derivates of pentoses[138], aldohexoses[133, 136] and ketohexoses[135, 137]. Applying selective deacetylation reactions, the position of the exchanging acetoxyl group has been determined. The exchange occurs mainly with the acetoxy group attached to the acetal-carbon. For instance, in the case of 1,2,3,4-tetra-O-acetyl-β-L-arabinopyranose (15) only the acetoxy group at $C_{(1)}$ participates in the exchange, but in the case of 9 some exchange takes place at $C_{(2)}$ (7·7%)

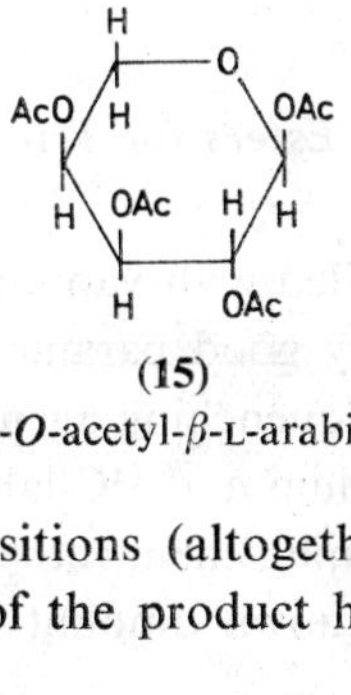

(15)

1,2,3,4-tetra-O-acetyl-β-L-arabinopyranose

but is negligible at other positions (altogether 1·3%). In the case of 8, 97·8% of the total activity of the product has been located at $C_{(1)}$ and

2·2% at $C_{(2)}$, and none at the other acetoxy groups. It should be noted that with these compounds no anomerization accompanies the exchange.

The acetoxy exchange rate depends on the configuration of the acetoxy groups and on the conformation of the compounds. β Anomers exchange their acetoxy group easier than α anomers, the former being more exposed to the attack of the labelled acetic acid in the medium.

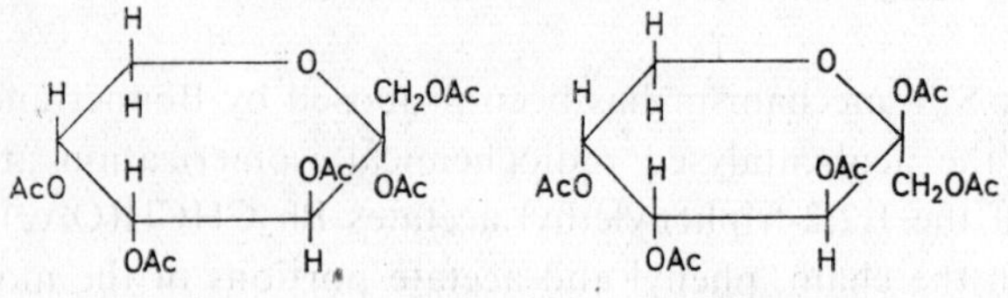

(16)   (17)

1,2,3,4,5-penta-O-acetyl-α-D-fructopyranose (16), 1,2,3,4,5-penta-O-acetyl-β-D-fructopyranose (17), 1,2,3,4,5-penta-O-acetyl-α-L-sorbopyranose (18) and 1,2,3,4,5-penta-O-acetyl-β-L-sorbopyranose (19) have been chosen as representatives of the ketohexoses[135, 136]. Selective deacetylation of the

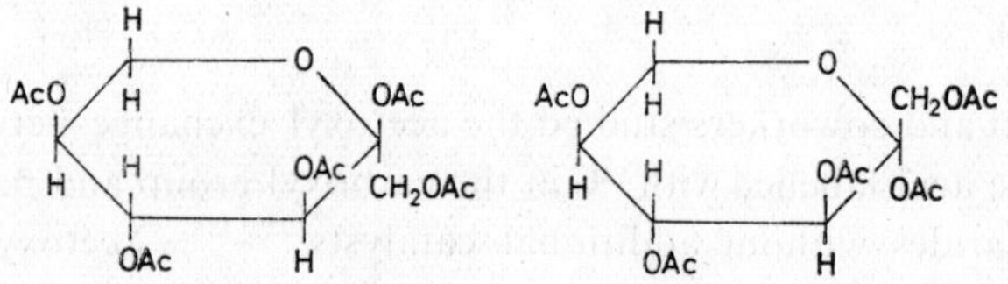

(18)   (19)

exchange products showed that about 98% of the total activity of the compound is located at $C_{(2)}$; 2·2% (for the α anomer) and 1·5% (for the β anomer) was found at $C_{(1)}$ and $C_{(3)}$. Again the exchange reaction is not accompanied by anomerization, but the rate of exchange is strongly dependent on steric effects caused by the presence of large groups near the anomeric carbon[135].

D. Application of Carboxylic Esters for Tritium and ¹⁴C Radioactivity Measurements

Geiger–Müller counters filled with vapours of carboxylic esters have a broad plateau and show very good parameters[85]. Methyl and ethyl formates are specially useful as quenching agents in the counters for tritium and ¹⁴C determinations. Tritium and ¹⁴C-labelled acids and alcohols can be radioassayed by transforming them into volatile esters (generally formates) and introducing a known amount of the labelled ester into a

Geiger–Müller counter for a specific activity determination. The repro-
ducibility of such measurements is very good and the adsorption of the
labelled compound inside the counter is small.

E. Mechanism of the Gas Phase Decomposition of Labelled Esters

Methyl formate undergoes thermal decomposition in two possible
ways[139–144]:

$$HCO_2CH_3 \begin{cases} \longrightarrow CH_4 + CO_2 & \text{(137a)} \\ \longrightarrow CH_3OH + CO & \text{(137b)} \end{cases}$$

Equation (137b) is the rapid and principal mode of decomposition, and is
followed by a slow secondary process of decomposition of methanol into
carbon monoxide and hydrogen[145]:

$$CH_3OH \longrightarrow [CH_2O + H_2] \longrightarrow CO + 2H_2 \qquad (138)$$

Danoczy studied the mechanism of deterioration of Geiger–Müller counters filled with methyl and ethyl formates and repeated earlier thermal
decomposition experiments[146] using a different analytical method[147] for the
determination of the composition of gases which do not condense in a
liquid air trap. Her results were similar to the data published by Steacie[139]
and Bairstow and Hinshelwood[140]. By making further studies of the decomposition of tritium-labelled formate TCO_2CH_3 and comparing the specific
activities of the components of the product mixture obtained at partial
decompositions with the specific activity of the starting material, she
concluded that the formyl hydrogen takes part in the rate-determining
step of the thermal decomposition of methyl formate. This confirmed an
old idea of Sabatier and Mailhe[148] that the specificity of the decomposition
of formates (equation 139) is caused by the special mobility of the formyl

$$HCO_2C_nH_{2n+1} = CO + C_nH_{2n+1}OH \qquad (139)$$

hydrogen atom leading to facile formation of the alcohol and elimination
of the carbon monoxide. However, the conclusion would require the support of further studies, e.g. of the pressure and temperature dependence of
the tritium isotope effect in reaction (137), including study of the tritium
isotope effect in the decomposition of HCO_2CH_2T.

According to equation (137b) methyl formate decomposes into CH_3OH,
which in turn splits at a much slower rate into carbon monoxide and
hydrogen. Measurements of the specific activity of the carbon monoxide
obtained at partial decompositions of $HCO_2{}^{14}CH_3$, combined with kinetic
studies of the $HCO_2{}^{14}CH_3$ decomposition for longer times, and studies of

the decomposition of pure $^{14}CH_3OH$ and HCO_2CH_3 mixed with known amounts of $^{14}CH_3OH$ permitted the evaluation in the gaseous product mixture of the amount of carbon monoxide originating from the methoxy group at various reaction times, and of the negligible contribution of the competitive free radical process (140d) to the total amount of the radioactive carbon monoxide formed[86].

$$HCO_2{}^{14}CH_3 \longrightarrow H\dot{C}{=}O + {}^{\cdot}O^{14}CH_3 \tag{140a}$$

$${}^{\cdot}O^{14}CH_3 \xrightarrow{H-R} {}^{14}CH_3OH \tag{140b}$$

$${}^{\cdot}O^{14}CH_3 \longrightarrow H_2{}^{14}\dot{C}OH \tag{140c}$$

$${}^{\cdot}O^{14}CH_3 \longrightarrow {}^{14}CO + 3/2\,H_2 \tag{140d}$$

VII. ISOTOPE EFFECTS IN THE REACTIONS OF CARBOXYLIC ESTERS

A. ^{14}C Isotope Effects in Hydrolysis

Comprehensive studies of the ^{14}C isotope effect in the hydrolysis of substituted ethyl benzoates, including changes in ring substituents, reaction temperature, solvent and concentration have been carried out by Ropp and Raaen[149–151]. Table 8 contains data showing the effect of varying the

TABLE 8. Isotope effects in the base-catalysed hydrolysis of carboxyl-labelled substituted ethyl benzoates

Nuclear substituent	p-OCH$_3$	p-Cl	p-CH$_3$	None	m-Cl	m-NO$_2$
k_{12C}/k_{14C}	1·0917	1·0810	1·0775	1·0764	1·0718	1·0672

group R on the isotope effect in the basic hydrolysis of ethyl benzoates (equation 141) labelled with ^{14}C in the carboxyl group.

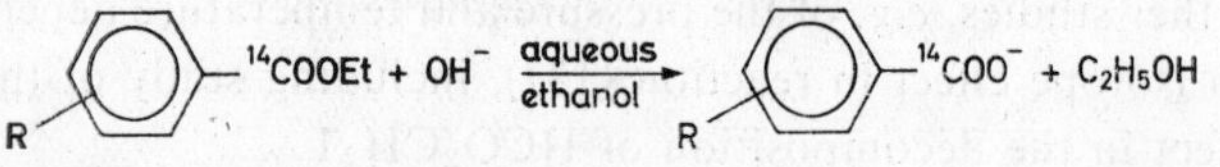

$$R{-}\langle\bigcirc\rangle{-}^{14}COOEt + OH^- \xrightarrow[\text{ethanol}]{\text{aqueous}} R{-}\langle\bigcirc\rangle{-}^{14}COO^- + C_2H_5OH \tag{141}$$

Data presented in Table 9 illustrate the dependence of the kinetic isotope effect in reaction (141) on the ester alkyl group and the temperature. The difference between the effects in acetone and in ethanol was within the experimental error and may be neglected. The precision of the measurements was good enough to reveal the variation of the k_{12C}/k_{14C} ratios

TABLE 9. The effect of varying the temperature and the ester-alkyl group on the isotope effect in the hydrolysis of carboxyl-labelled benzoates

Ester	Temperature (°c)	Solvent	k_{12C}/k_{14C}
$C_6H_5CO_2Et$	0	90% ethanol	1·0893
$C_6H_5CO_2Et$	25	90% ethanol	1·0764
$C_6H_5CO_2Et$	25	56% acetone	1·0775
$C_6H_5CO_2Et$	78·5	90% ethanol	1·0649
$C_6H_5CO_2Bu$-t	25	56% acetone	1·0649

with the type of substituents in the benzene ring. Table 8 shows smaller sotope effects with *meta*- than with *para*-substituted esters, with the largest effect observed in the hydrolysis of ethyl *p*-methoxybenzoate. The relatively large temperature dependence of the factors k_{12C}/k_{14C}, as well as previous kinetic studies of the hydrolysis[103-105] suggest that zero-point energy levels of the isotopic ester molecules depend largely on the nature of the substituents. For instance, in the case of the isotopic ethyl *p*-methoxy-benzoate molecules the zero-point energy levels are more widely separated than in the case of *meta*-substitued esters. This also means that the carbon atom in the labelled position has tighter bonding in the *para*-methoxy than for example in the *meta*-nitro ester. Larger separation of the zero-point energy levels, $E_{14} - E_{12}$, in the *para*-methoxy esters is determined by the large contributions of resonance forms of the type **20** to the ground

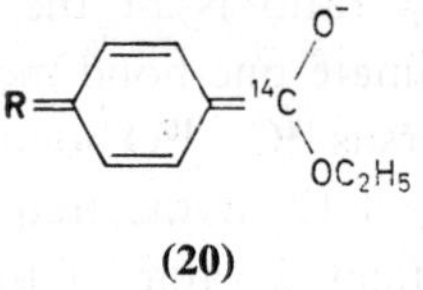

(20)

state of the esters, while in the case of *meta*-substituted esters large contributions are made by structures such as **21** and **22**. Additionally, it

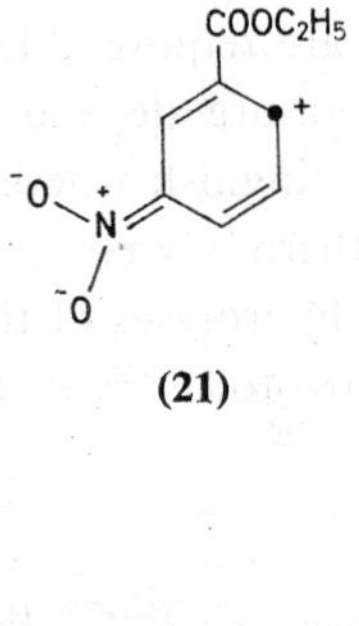

(21)

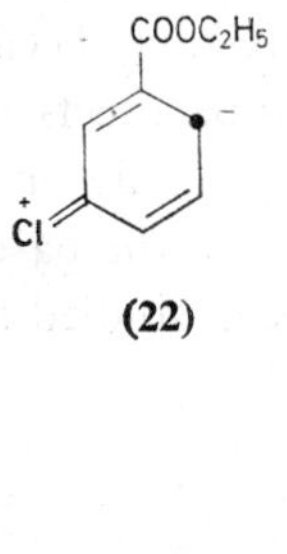

(22)

had been necessary to assume that in the transition complex of the reaction the difference $(E_{14}-E_{12})$ largely disappears and that the activated state can be represented by structure (23).

$$R-\text{C}_6\text{H}_4-\overset{14}{\underset{\text{OH}}{\overset{\text{O}^-}{\text{C}}}}-\text{OC}_2\text{H}_5$$

(23)

Bigeleisen[152] expressed the view, that in terms of the generally accepted scheme for the base-catalysed hydrolysis of esters (equation 142) (see section VI.A.), the ^{14}C isotope effect is the product of two factors: (1) a

$$\text{C}_6\text{H}_5-\overset{\text{O}}{\overset{\|}{\text{C}}}-\text{OEt}+\text{OH}^- \underset{\longleftarrow}{\overset{K}{\longrightarrow}} \left[\text{C}_6\text{H}_5-\overset{\text{O}}{\underset{\text{OH}}{\overset{|}{\underset{|}{\text{C}}}}}-\text{OEt}\right]^- \overset{k}{\longrightarrow} \text{C}_6\text{H}_5\text{COOH}+\text{EtO}^- \quad (142)$$

thermodynamic isotope effect in the fast equilibrium preceding the rate-determining step, and (2) a pure kinetic isotope effect in the subsequent rate-determining step. Therefore the ratio of the observed rate constants is expressed

$$k_{^{12}\text{C obs}}/k_{^{14}\text{C obs}} = K_{^{12}\text{C}}k_{^{12}\text{C}}/K_{^{14}\text{C}}k_{^{14}\text{C}} \quad (143)$$

in equation (143) where K is the equilibrium constant and k is the rate of decomposition of the intermediate, and subscripts 12 and 14 refer to isotopic species having ^{12}C and ^{14}C in the carboxyl group. Bigeleisen estimated that the $K_{^{12}\text{C}}/K_{^{14}\text{C}}$ ratio is of the order of $1/1 \cdot 03 - 1/1 \cdot 05$ at room temperature. Approximate one-bond model calculations of the isotope effect in the ^{12}C—^{16}O versus ^{14}C—^{16}O bond rupture (see section IV.A.) give the value of $k_{^{12}\text{C}}/k_{^{14}\text{C}} = 1 \cdot 12$, if the frequency ω (^{12}C—^{16}O) is taken as 1093 cm^{-1}. Thus the estimated value of the $k_{^{12}\text{C obs}}/k_{^{14}\text{C obs}}$ will be of the order $1 \cdot 09 - 1 \cdot 07$, in agreement with Ropp and Raaen's experimental findings. Nevertheless, this agreement does not give decisive proof of the reaction scheme used for the calculations by Bigeleisen. Strict calculations using more than one bond, as well as more precise measurements of the temperature dependence of the isotope effects are required. The problem is quite interesting, because the *exact* temperature dependence of the observed isotope effects would enable us to distinguish whether they are of kinetic origin only or of both kinetic and thermodynamic nature. ^{14}C Isotope effects in the base- and acid-catalysed hydrolyses of the aliphatic esters have been studied by Korshunov and Novotorov[153], in 1 : 1 water–

alcohol solutions. The experimental error of the determination was about 2%. The results are listed in Table 10.

TABLE 10. ^{14}C kinetic isotope effects in the base- and acid-catalysed hydrolysis of carboxyl-labelled aliphatic esters

Ester	Base-catalysed hydrolysis		Acid-catalysed hydrolysis	
	Temperature (°c)	k_{12C}/k_{14C}	Temperature (°C)	k_{12C}/k_{14C}
Me $^{14}CO_2Et$	57	1·16	69	1·11
Me $^{14}CO_2Et$	37	1·15	38	1·12
Me $^{14}CO_2Et$	0	1·16	13	1·12
Et $^{14}CO_2Et$	55	1·13	70	1·09
Et $^{14}CO_2Et$	37	1·13	37	1·09
Et $^{14}CO_2Et$	0	1·14	12	1·10
Pr $^{14}CO_2Et$	52	1·11	70	1·10
Pr $^{14}CO_2Et$	32	1·12	37	1·09
Pr $^{14}CO_2Et$	0	1·11	12	1·12
Me $^{14}CH_2CO_2Et$	50	1·03	—	—
Me $^{14}CH_2CO_2Et$	25	1·03	—	—

It should be noted that the ^{14}C isotope effects reported in Table 10 are very large. Calculation of the isotope effect in the rupture of the *C—O bond gives the value $k_{12C}/k_{14C} = 1·12$, if the frequency of the C—O bond is taken as 1093 cm^{-1}. This value is quite close to the primary isotope effects reported for acid-catalysed hydrolysis of ^{14}C-labelled esters. Unfortunately the large experimental error inherently present in ^{14}C isotope effect determinations did not allow the study of the temperature dependence and, consequently, the determination of the structure of the transition complex. A dependence of the isotope effect on the length of aliphatic chain is observed, and, as one would expect, the secondary ^{14}C isotope effect is smaller than the primary one.

B. Secondary Deuterium Isotope Effects in the Hydrolysis of Carboxylic Esters

Secondary deuterium kinetic isotope effects in basic hydrolysis have been studied by Bender and coworkers[154, 155]. At 25·0°C, CD_3COOEt reacts faster than CH_3COOEt, that is, a 'reverse' kinetic isotope effect is observed ($k_H/k_D = 0·90 \pm 0·01$). In the majority of known reactions the normal molecule reacts faster than the isotopic species containing deuterium and tritium atoms. The origin of this 'reverse' deuterium isotope effect is in the view of the authors[154] that the slow step of the reaction involves the addition of hydroxide ion to the trigonal ground carbon to produce a transition state approximating to a tetrahedral carbon atom. Moreover the carbonyl carbon atom to which the methyl group is attached becomes less electron-deficient in the transition state than in the ground state. This causes a difference in hyperconjugative stabilization between the ground state and the transition state, the former being stabilized to a greater degree by hyperconjugation.

The effect may also be explained by assuming that non-bonded repulsions are in operation[156-161]. In the present author's view it is more reasonable to assume that deuterium substitution in the methyl group of the acetic acid causes some changes in the frequencies of the skeletal vibrations of the ester, including those of the C—O bond being broken, so as to favour the hydrolysis of the deuterated species. Only if the calculations of the secondary deuterium isotope effects based on the effect of the mass of deuterium on the vibrational frequencies fail to explain the observed reverse isotope effect, should one add to the pure mechanistic effects further electronic effects, which are thoroughly discussed by Halevi and Pauncz[162].

Quite different kinetic isotope effects are observed in the hydrolysis of methyl p-(tritiomethyl) benzoate either in basic water–alcohol solution or in 99–100% sulphuric acid[87]:

$$CH_2TC_6H_4CO_2CH_3 \begin{array}{l} \xrightarrow{OH^-} CH_2TC_6H_4CO_2{}^- + CH_3OH \qquad (144a) \\ \xrightarrow{H^+} CH_2TC_6H_4CO_2H + CH_3OH \qquad (144b) \end{array}$$

The tritium-labelled ester reacts at a slower rate than the ordinary ester in alkaline solution (at 25°C, $k_T/k_H = 0.953 \pm 0.004$), but at practically the same rate in 99.9% sulphuric acid (at 20°C, $k_T/k_H = 0.999 \pm 0.006$). Hodnett and coworkers[87] assume that the rate-determining step in the basic hydrolysis is the attack of a hydroxyl ion on the carboxylate carbon (equation 145). The rate-determining step in the hydrolysis of esters in

$$\underset{\displaystyle \|}{\overset{\displaystyle O}{R^1-C-OR^2}} + OH^- \longrightarrow \underset{\displaystyle OH}{\overset{\displaystyle O^-}{R^1-C-OR^2}} \qquad (145)$$

99–100% sulphuric acid is thought to be the formation of an acylium ion (equation 146).

$$\underset{\displaystyle H}{\overset{\displaystyle O}{R^1-C-O-R^2}} \longrightarrow R^1-C^+ + R^2OH \qquad (146)$$

By analogy with the observations that electron-releasing groups attached to the phenyl ring of benzoates retard the rate of basic hydrolysis, it is supposed that a tritium atom in the p-methyl group reduces the reaction rate due to its greater ability for inductive electron release. The difference between the effects of heavy hydrogen isotope substitution on the rate of the hydrolysis of CD_3COOEt on one hand and of $CH_2TC_6H_4COOCH_3$ on the other, is explained by the simultaneous action of two opposing factors: (1) greater electron release of the tritium atom through the inductive effect, and (2) smaller electron release by the tritium atom through the electromeric effects. This explanation should be confirmed by isotope effect studies of the hydrolysis of p-$CD_3C_6H_4COOCH_3$.

VIII. ACKNOWLEDGEMENTS

I wish to thank Prof. J. Swiderski and Dr. A. Temeriusz for stimulating the writing of section VI. C and commenting on it, Prof. R. F. Nystrom and many other scientists for providing me with reprints of their papers, Mrs. E. Danoczy for making her data on the decomposition of tritium-labelled methyl formate available to me, Dr. E. Nalborczyk and Dr. H. Rybicka for helpful suggestions and Prof. M. Taube, head of the Department of Radiochemistry, for his goodwill during the time that I was writing the manuscript.

IX. REFERENCES

1. M. Calvin, C. Heidelberg, J. C. Reid, B. M. Tolbert and P. E. Yankwich, *Isotopic Carbon*, John Wiley and Sons, New York, 1949.
2. A. Murray and D. L. Williams, *Organic Synthesis with Isotopes*, Interscience Publishers, New York, 1958.
3. G. P. Miklukhin, *Isotopes in Organic Chemistry*, Ukrainian Academy of Science, Kiev, 1961.
4. P. E. Yankwich, *J. Chem. Phys.*, **15**, 374 (1947).
5. A. P. Wolf, *Ann. Rev. Nucl. Sci.*, **10**, 259 (1960).
6. W. Sakami, W. E. Evans and S. Gurin, *J. Am. Chem. Soc.*, **69**, 1110 (1947).
7. W. G. Dauben, J. C. Reid and P. E. Yankwich, *Anal. Chem.*, **19**, 828 (1947).
8. D. R. Howton, R. H. Davis and J. C. Novenzel, *J. Am. Chem. Soc.*, **74**, 1109 (1952); **76**, 4970 (1954).
9. R. F. Nystrom, Y. H. Loo and J. C. Leak, *J. Am. Chem. Soc.*, **74**, 3434 (1952).
10. J. C. Nevenzel and D. R. Howton, *J. Org. Chem.*, **22**, 319 (1957); 76, 4970 (1954).
11. N. F. Novotorov and I. A. Korshunov, *Zh. Obshch. Khim.*, **26**, 1959 (1956).
12. A. Murray, W. W. Foreman and W. Langham, *J. Am. Chem. Soc.*, **70**, 1037 (1948).
13. E. A. Long, *J. Am. Chem. Soc.*, **61**, 570 (1939).
14. K. Bernhard, G. Brubacher and H. Jaquet, *Helv. Chim. Acta.*, **36**, 1968 (1953).
15. M. Zieliński and H. Wincel, *Roczniki Chem.*, **32**, 1189 (1958).

16. S. Bergström, K. Pääbo and M. Rottenberg, *Acta Chem. Scand.*, **6**, 1127 (1952).
17. R. B. Lotfield, *Nucleonics*, **1**, 54 (1947).
18. I. Złotowski, M. Zieliński and P. Pańta, *Nukleonika*, **7**, 311 (1962).
19. E. Prokopowicz and M. Zieliński, unpublished results.
20. A. P. Wolf, C. S. Redvanly and C. Anderson, *J. Am. Chem. Soc.*, **79**, 3717 (1957).
21. A. P. Wolf in *Advances in Physical Organic Chemistry* (Ed. V. Gold), Vol. II, Academic Press, New York, 1964, p. 202.
22. C. N. Turton, *Proc. Intern. Conf. Peaceful Uses Atomic Energy, 2nd, Geneva, 1958*, **20**, 91, Paper 284 (1959).
23. F. Cacace, G. Ciranni, G. Giacomello and M. Zifferero, *Ric. Sci.*, **28**, 2131 (1958).
24. B. Aliprandi, F. Cacace and G. Giacomello, *Ric. Sci.*, **26**, 3029 (1956).
25. M. Guillaume, *Nature*, **182**, 1592 (1958).
26. H. S. Isbell, *Science*, **113**, 532 (1951).
27. V. P. Hollander and J. Vinecour, *Anal. Chem.*, **30**, 1492 (1958).
28. R. H. Benson, R. B. Turner, *Anal Chem.*, **32**, 1464 (1960).
29. F. Sanger, *Biochem. J.*, **53**, 353 (1953).
30. A. P. Ryle, F. Sanger, L. F. Smith and R. Kitai, Biochem. J., **60**, 541 (1955).
31. J. K. Whitehead, *Biochem. J.*, **68**, 662 (1958).
32. J. K. Whitehead, *Biochem. J.*, **68**, 653 (1958).
33. A. E. Martel and M. Calvin, *Chemistry of the Metal Chelate Compounds*, Prentice-Hall, New York, 1953.
34. A. A. Grinberg, *Introduction into the Chemistry of the Complex Compounds* (English Transl.), State Edition of the Chem. Lit., Leningrad–Moskow, 1951.
35. F. S. Dainton, G. S. Laurence, W. Schneider, D. R. Stranks and M. S. Vaidya, *International Conference on Radioisotopes in Scientific Research*, UNESCO/NS/RIC/211.
36. F. D. Graziano and G. M. Harris, *J. Phys. Chem.*, **63**, 330 (1959).
37. M. Zielinski, H. Wincel, R. Krupa, E. Prokopowicz, unpublished results.
38. G. Lapidus and G. M. Harris, *J. Am. Chem. Soc.*, **85**, 1223 (1963).
39. R. Krupa and M. Zieliński, unpublished results.
40. H. Taube, *Chem. Rev.*, **50**, 69 (1952).
41. K. B. Zaborenko, B. Z. Iofa, V. G. Lukianov and I. O. Bogatyriev, *Radioactive Tracer Method in Chemistry* (English Transl.) Highe School Moscow, 1964, p. 244.
42. M. D. Lane, D. R. Halenz, D. P. Kosov and C. S. Hegre, *J. Biol. Chem.*, **235**, 3082 (1960).
43. L. N. Lukens and J. M. Buchanan, *J. Am. Chem. Soc.*, **79**, 1511 (1957).
44. J. M. Buchanan in *The Nucleic Acids* (Ed. E. Chargaff and J. N. Davidson), Vol. 3, Academic Press, New York, 1960, p. 303.
45. J. Bell and R. I. Read, *J. Chem. Soc.*, 1383 (1952).
46. J. Packer and J. Vaughan, *Modern Approach to Organic Chemistry*, Clarendon Press, Oxford, 1958, p. 261.
47. P. Blicharski and J. Swiderski, *Roczniki Chem.*, **31**, 1317 (1957).
48. I. Złotowski and M. Zieliński, *Nucleonika*, **5**, 27 (1960).
49. M. Zieliński, *Nucleonika*, **7**, 789 (1962).
50. K. Banholzer and H. Schmid, *Helv. Chim. Acta*, **39**, 548 (1956).
51. G. A. Ropp, *J. Am. Chem. Soc.*, **82**, 842 (1960).
52. A. Fry, *J. Am. Chem. Soc.*, **75**, 2686 (1953).
53. P. Nahinsky and S. Ruben, *J. Am. Chem. Soc.*, **63**, 2275 (1941).
54. P. Nahinsky, C. N. Rice, S. Ruben and M. D. Kamen, *J. Am. Chem. Soc.*, **64**, 2299 (1942).
55. S. Glasstone, G. Laidler and H. Eyring, *Theory of Rate Process*, McGraw-Hill, New York, 1941.

56. G. Careri, in *Advances in Chemical Physics* (Ed. I. Prigogine), Vol. 1, Interscience Publishers, London, 1958, p. 119.
57. O. K. Rice, *J. Phys. Chem.*, **65**, 1972 (1961); **66**, 1058 (1962); **67**, 6 (1963).
58. J. Bigeleisen and M. Wolfsberg, in *Advances in Chemical Physics*, (Ed. I. Prigogine) Vol. 1, Interscience Publishers, London, 1958, p. 15.
59. H. S. Johnston, W. A. Bonner and D. J. Wilson, *J. Chem. Phys.*, **26**, 1002 (1957).
60. I. Złotowski and M. Zieliński, *Nukleonika*, **4**, 599 (1959).
61. I. Złotowski and M. Zlieliński, *Nukleonika*, **6**, 511 (1961).
62. M. Zieliński, *Nukleonika*, **10**, 337 (1965).
63. E. M. Hodnett and R. L. Bowton, *Radioisotopes in the Physical Sciences and Industry* International Atomic Energy Agency, Vienna, 1962, p. 225.
64. M. Zieliński, *Kernenergie*, **5**, 351 (1962).
65. L. P. Hammett, *Physical Organic Chemistry*, McGraw-Hill, New York, 1940, p. 283.
66. W. W. Eliot and D. L. Hammick, *J. Chem. Soc.*, 3402 (1951).
67. G. A. Ropp, A. J. Weinberg and O. K. Neville, *J. Am. Chem. Soc.*, **73**, 5573 (1951).
68. B. Fingerman and R. M. Lemmon, *Bio-Organic Chemistry Quarterly Report*, Dec., Jan. and Febr., 1957–1958, U.C.R.L. 8204.
69. H. Eyring and F. W. Cagle, *J. Phys. Chem.*, **56**, 889 (1952).
70. I. Fry and M. Calvin, *J. Phys. Chem.*, **56**, 897 (1952).
71. N. C. Deno and R. W. Tatt, Jr., *J. Am. Chem. Soc.*, **76**, 248 (1954).
72. P. E. Yankwich, A. L. Promislow and R. F. Nystrom, *J. Am. Chem. Soc.*, **76**, 5893 (1954).
73. P. E. Yankwich and H. S. Weber, *J. Am. Chem. Soc.*, **77**, 4513 (1955).
74. P. E. Yankwich and R. M. Ikeda, *J. Am. Chem. Soc.*, **81**, 1532 (1959); **82**, 1891 (1960).
75. M. Zieliński, *J. Chem. Phys.*, **41**, 3646 (1964).
76. M. Zieliński, *Habilitation Thesis*, University of Warsaw, 1966.
77. C. R. Mc Kinney, Y. M. Mc Crea, S. Epstein, K. A. Allen and H. C. Urey, *Rev. Sci. Instr.*, **21**, 724 (1950).
78. H. Craig, *Geochimica et Cosmochimica Acta*, **12**, 133 (1957).
79. M. Zieliński, *Nukleonika*, **10**, 337 (1965).
80. C. G. Swain, R. F. W. Bader, R. M. Esteve, Jr. and R. N. Griffin., *J. Am. Chem. Soc.*, **83**, 1951 (1961).
81. W. H. Stevens, J. M. Pepper and M. Lounsbury, *Can. J. Chem.*, **30**, 529 (1952).
82. E. A. Halevi, *Tetrahedron*, **1**, 174 (1957).
83. J. Szydłowski, T. Więch and M. Zieliński, unpublished results.
84. D. N. Kursanov and R. V. Kudriavcev, *Zh. Obshch. Khimii*, **26**, 1040 (1956).
85. M. Zieliński, I. Złotowski, *Isotopen Techn.*, **2**, 281, (1962).
86. M. Zieliński, M. Kamińska, unpublished results.
87. E. M. Hodnett, R. D. Taylor, J. V. Tormo and R. E. Lewis, *J. Am. Chem. Soc.*, **81**, 4528 (1959).
88. K. E. Wilzbach, *J. Am. Chem. Soc.*, **79**, 1013 (1957).
89. R. F. Nystrom, L. H. Mason, E. P. Jones and H. J. Dutton, *J. Am. Oil Chemists Soc.*, **36**, 212 (1959).
90. E. P. Jones, L. H. Mason, H. J. Dutton and R. F. Nystrom, *J. Org. Chem.*, **25**, 1413 (1960).
91. H. J. Dutton, E. P. Jones, V. L. Davison and R. F. Nystrom, *J. Org. Chem.*, **27**, 2648 (1962).
92a. J. K. Lee, B. Musgrave and F. S. Rowland, *J. Am. Chem. Soc.*, **81**, 3803 (1959).
92b. A. M. Elatrash, R. H. Johnsen and R. Wolfgang, *J. Phys. Chem.*, **64**, 785 (1960).

92c. A. M. Elatrash and R. H. Johnsen, *Chemical effects of Nuclear Transformations*, International Atomic Energy Agency, Vienna, 1961, p. 123.

93. R. D. Shores and H. C. Moser, *J. Phys. Chem.*, **65**, 570 (1961).

94. M. Polanyi and A. L. Szabo, *Trans. Faraday Soc.*, **30**, 508 (1934).

95. I. Roberts and H. C. Urey, *J. Am. Chem. Soc.*, **60**, 2391 (1938).

96. S. C. Datta, J. N. E. Day and C. K. Ingold, *J. Chem. Soc.*, **838** (1939).

97. N. J. Dedusenko and A. J. Brodzki, *Zh. Obshch. Khimii*, **12**, 361 (1942).

98. F. A. Long and L. Friedman, *J. Am. Chem. Soc.*, **72**, 3692 (1950).

99. M. L. Bender and R. S. Dewey, *J. Am. Chem. Soc.*, **78**, 317 (1956).

100. J. F. Bunnett, M. M. Robinson and F. C. Pennington, *J. Am. Chem. Soc.*, **72**, 2378 (1950).

101. C. A. Bunton and A. Konasiewicz, *J. Chem. Soc.*, 1354 (1955).

102. J. N. E. Day and C. K. Ingold, *Trans. Faraday Soc.*, **37**, 686 (1941).

103. C. K. Ingold and W. S. Nathan, *J. Chem. Soc.*, 222 (1936).

104. C. K. Ingold, *Structure and Mechanism in Organic Chemistry*, Cornell University Press, Ithaca, 1953, p. 754.

105. L. P. Hammett, *Physical Organic Chemistry*, McGraw-Hill, New York, 1940, p. 354–359.

106. M. L. Bender, *J. Am. Chem. Soc.*, **73**, 1626 (1951).

107. M. L. Bender, R. R. Stone and R. S. Dewey, *J. Am. Chem. Soc.*, **78**, 319 (1956).

108. M. L. Bender, R. D. Ginger and J. P. Unik, *J. Am. Chem. Soc.*, **80**, 1044 (1958).

109. M. L. Bender, *Chem. Rev.*, **60**, 53 (1960).

110. Ia. Syrkin and I. I. Moissev, *Usp. Khim.*, **27**, 717 (1958).

111. A. I. Talvik and V. A. Pal'm, *Zh. Fiz. Khim.*, **33**, 1214 (1959).

112. E. Whalley, in *Advances in Physical Organic Chemistry* (Ed. V. Gold), Vol. 2, Academic Press, New York, 1964, p. 142.

113. A. Bunton and D. N. Spachter, *J. Chem. Soc.*, 1079 (1956).

114. M. L. Bender and M. C. Neven, *J. Am. Chem. Soc.*, **80**, 5388 (1958).

115. M. L. Bender, F. Chloupek and M. C. Neven, *J. Am. Chem. Soc.*, **80**, 5384 (1958).

116. R. Bentley and D. Rittenberg, *J. Am. Chem. Soc.*, **76**, 4883 (1954).

117. S. S. Stein and D. E. Koshland, Jr., *Arch. Biochem. Biophys.*, **45**, 467 (1953).

118. D. B. Sprinson and D. Rittenberg, *Nature*, **167**, 484 (1951).

119. D. G. Doherty and F. Valsov, *J. Am. Chem. Soc.*, **74**, 931 (1952).

120. M. L. Bender and K. C. Kemp, *J. Am. Chem. Soc.*, **79**, 111 (1957).

121. D. E. Koshland Jr., *Discussions Faraday Soc.*, **20**, 142 (1956).

122. D. E. Koshland Jr. and S. S. Stein, *J. Biol. Chem.*, **208**, 139 (1954).

123. L. L. Ingraham, *Biochemical Mechanisms*, John Wiley and Sons, New York, 1962, p. 35. Russian translation (Ed. Ia. M. Varsavskij), Izdatielstvo "Mir", Moskow. 1964, pp. 59–100.

124. A. E. Braunstein, M. Ia. Karpieiskii and R. M. Khomutov, in *Enzymes* (Ed. A. E. Braunstein) (English Transl.) Edition 'Science', Moscow, 1964, pp. 237–268.

125a. W. E. Doering and E. Dorfman, *J. Am. Chem. Soc.*, **75**, 5595 (1953).

125b. R. Criegee, *Ann. chem.*, **560**, 127 (1948).

125c. J. C. Martin and E. H. Drew, *J. Am. Chem. Soc.*, **83**, 1232 (1961).

125d. W E. Doering, K. Akamoto and H. Krauch, *J. Am. Chem. Soc.*, **82**, 3579 (1960).

126. R. U. Lemieux and C. Brice, *Can. J. Chem.*, **30**, 259 (1952).

127. R. U. Lemieux and C. Brice, *Can. J. Chem.*, **33**, 109 (1955).

128. R. U. Lemieux and C. Brice, *Can. J. Chem.*, **34**, 1006 (1956).

129. W. A. Bonner, *J. Am. Chem. Soc.*, **81**, 5171 (1959).

130. W. A. Bonner, *J. Am. Chem. Soc.*, **83**, 962 (1961).

131. W. A. Bonner, *J. Am. Chem. Soc.*, **83**, 2661 (1961).
132. W. A. Bonner and C. J. Collins, *J. Am. Chem. Soc.*, **77**, 99 (1955).
133. J. Świderski and P. Blicharski, *Roczniki Chem.*, **32**, 1121 (1958).
134. J. Świderski, Z. Pawlak and P. Blicharski, *Roczniki Chem.*, **33**, 739 (1959).
135. Z. Pawlak, *Roczniki Chem.*, **37**, 457 (1963).
136. J. Świderski and J. Struciński, *Roczniki Chem.*, **36**, 115 (1962).
137. Z. Pawlak, J. Świderski and A. Temeriusz, *Roczniki Chem.*, **37**, 443 (1963).
138. J. Świderski, K. Ostalska, *Roczniki Chem.*, **39**, 621 (1965).
139. E. W. R. Steacje, *Proc. Roy. Soc.* (London), **127**, 314 (1930).
140. S. Bairstow and C. N. Hinshelwood, *J. Chem. Soc.*, 1148 (1933).
141. J. W. Mitchell and C. N. Hinshelwood, *Proc. Roy. Soc.* (London), **159**, 42 (1937).
142. L. A. K. Staveley and C. N. Hinshelwood, *J. Chem. Soc.*, 1570 (1937).
143. R. F. Faull and G. K. Rollefson, *J. Am. Chem. Soc.*, **58**, 1755 (1936).
144. G. M. Schwab and H. Knozinger, *Z. Phys. Chem. (Neue Folge)*, **37**, 230 (1963).
145. W. A. Bone and H. Davies, *J. Chem. Soc.*, **105**, 1691 (1914).
146. E. Danoczy, unpublished results.
147. M. Zieliński, *Nuclear Applications*, **2**, 51 (1966).
148. P. Sebatier and A. Mailhe, *Compt. Rend.*, **154**, 50 (1912).
149. G. A. Ropp and V. F. Raaen, *J. Chem. Phys.*, **20**, 1823 (1952).
150. G. A. Ropp and V. F. Raaen, *J. Chem. Phys.*, **22**, 1223 (1954).
151. V. F. Raaen and G. S. Ropp, *Anal. Chem.*, **25**, 174 (1953).
152. J. Bigeleisen, *J. Phys. Chem.*, **56**, 823 (1952).
153. I. A. Korshunov and I. F. Novotorov, *Res. Chem. and Chem. Eng.*, **2**, 221 (1958).
154. M. L. Bender and M. S. Feng, *J. Am. Chem. Soc.*, **82**, 6318 (1960).
155. J. M. Jones and M. L. Bender, *J. Am. Chem. Soc.*, **82**, 6322 (1960).
156. L. S. Bartell, *Tetrahedron Letters*, **6**, 13 (1960).
157. K. T. Leffek, J. A. Llewellyn and R. E. Robertson, *Chem. Ind. (London)*, 588 (1960).
158. M. M. Kreevoy and H. Eyring, *J. Am. Chem. Soc.*, **79**, 5121 (1957).
159. V. J. Shiner, *J. Am. Chem. Soc.*, **82**, 2655 (1960).
160. L. S. Bartell, *J. Am. Chem. Soc.*, **83**, 3567 (1961).
161. L. S. Bartell, *Iowa State J. Sci.*, **36**, 137 (1961).
162. E. A. Halevi and R. Pauncz, *J. Chem. Soc.*, 1974 (1959).

Notes Added in Proof

Note 1. The reader of section IVC. 4 is referred for further experimental results on this subject and their dicussion to the papers by M. Zieliński published recently in the Polish journals *Nucleonika* and *Roczniki Chemii*.

Note 2. Recent detailed investigations of the carbon-13 isotope effects in the decarboxylation of anthranilic acid in solid state and in solutions (see paper by M. Zieliński published recently in *Roczniki Chemii*) have shown that in water medium and in the temperature region 102–142° c, the carbon-13 isotope effect is independent of temperature and has the value $1\cdot0040 \pm 0\cdot0001$. This corresponds exactly to the ratio of the effective $^{12}C-^{12}C$ and $^{12}C-^{13}C$ bond distances (broken in the course of decarboxylation) corrected for their amplitudes of vibrations.

CHAPTER **11**

Esterification and ester hydrolysis

ERKKI K. EURANTO

University of Turku, Finland

I. INTRODUCTION

The hydrolysis of carboxylic esters (equation 1) and the reverse reaction, the esterification, have been the subject of intensive kinetic and mechanis-

$$R^1COOR^2 + H_2O \rightleftharpoons R^1COOH + R^2OH \qquad (1)$$

tic studies during the last century since Berthelot and Péan de Saint-

Gilles[1] investigated the esterification of acetic acid and the hydrolysis of ethyl acetate; their results were used by Guldberg and Waage[2] in the derivation of the Law of Mass Action. Thousands of papers dealing with the kinetics of these reactions have appeared and several excellent reviews have been published[3-6].

The ester hydrolysis is catalysed by hydrogen and hydroxide ions, but a base-catalysed esterification is not known. Other chemical species may also

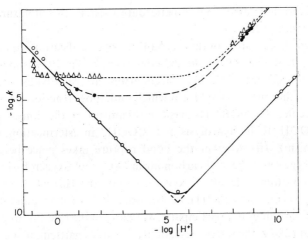

FIGURE 1. The logarithm of the experimental rate coefficient in water at 25°C for the hydrolysis of ethyl acetate[8-11] (○), methyl dichloroacetate[11, 12] (●), and chloromethyl chloroacetate[13, 14] (△) as a function of the logarithm of the hydrogen ion concentration

act as catalysts. In their absence the rate law for the hydrolysis of an ester E is given by equation (2), as first proposed by Wegscheider[7] in 1902.

$$-\frac{d[E]}{dt} = (k_0 + k_H[H^+] + k_{OH}[OH^-])[E] \qquad (2)$$

In equation (2), k_H, k_{OH}, and k_0 are the rate coefficients of the hydrogen-ion and hydroxide-ion catalysed reactions, and of the neutral or water hydrolysis, respectively. Only one of the three terms in equation (2) is usually significant at a definite pH range; therefore a plot of log $k_{experimental}$ versus log $[H^+]$ is formed of three linear parts with slopes 1, 0, and −1 (Figure 1). The minimum rate is observed at the acidic side because $k_{OH} > k_H$; the hydrogen ion concentration at that point is given[15] by equation (3), in which K_W is the ion product of water. A neutral hydrolysis is observable

$$[H^+] = (k_{OH}K_W/k_H)^{1/2} \qquad (3)$$

only if v (equation 4) is at least of the order of unity[8]. The rate coefficients

$$v = \frac{1}{2} k_0/(k_H k_{OH} K_W)^{1/2} \tag{4}$$

k_H and k_{OH} are usually easily measured by a suitable choice of pH; a high rate of neutral hydrolysis may mask acid hydrolysis and sometimes alkaline hydrolysis is inconveniently rapid. If buffer solutions are used in adjusting pH, a possible catalysis by buffer components has to be taken into consideration; erroneous kinetic data exist in the literature arising from ignorance of this.

A general classification of the mechanisms of esterification and ester hydrolysis was first given by Ingold and coworkers[16, 17] in 1939; small changes were introduced later[6, 18]. The mechanisms are subdivided depending upon (i) whether the carboxyl form which undergoes reaction is the neutral ester molecule, R^1COOR^2, (B, basic mechanism) or the ionic conjugate acid, $R^1COOH^+R^2$ in hydrolysis or $R^1COOH_2^+$ in esterification, (A, acidic mechanism), (ii) whether the bond rupture takes place between the ethereal oxygen and the acyl carbon atom (AC, acyl-oxygen fission) or the alkyl carbon atom (AL, alkyl-oxygen fission), and (iii) whether the rate-limiting step is unimolecular (1), *i.e.* the water in hydrolysis and the alcohol in esterification does not participate covalently in the rate-limiting step, or bimolecular (2), *i.e.* the water, base, or alcohol does participate. Of the eight possible mechanisms obtained by all combinations of these three factors, two, $B_{AC}1$ and $A_{AL}2$, have not been observed with certainty. The bimolecular hydrolyses with acyl-oxygen fission, $B_{AC}2$ and $A_{AC}2$, are by far the most common mechanisms.

In addition to the Ingold mechanisms in which the rate-limiting step is a reaction in the carboxyl group, some special esters hydrolyse by mechanisms in which the carboxyl group is not directly involved in the rate-limiting stage. For example, α-haloalkyl esters, $R^1COOCXR^2R^3$ (X = Cl, Br, or I), and chloroformates, ClCOOR, are found in some cases to solvolyse by mechanisms in which the halogen–carbon bond is ruptured in the rate-limiting step.

II. ALKALINE HYDROLYSIS BY ACYL-OXYGEN FISSION ($B_{AC}2$)

The alkaline hydrolysis or saponification of esters is often called hydroxide-ion catalysed hydrolysis, although there is no catalysis in the strict sense because the reacting hydroxide ion is consumed in the reaction (5).

$$R^1C(=O){-}OR^2 + OH^- \longrightarrow R^1COO^- + R^2OH \qquad (5)$$

The reaction is in principle reversible, but in practice it is driven completely to the right because of the stability of the carboxylate ion, R^1COO^-.

A. Reaction Mechanism

1. Evidence for acyl-oxygen fission

Holmberg found in 1912[19] that in the hydrolysis of acetoxysuccinic acid, $CH_3COO\overset{*}{C}H(COOH)CH_2COOH$, with an asymmetric α-carbon atom in the alkyl component, the configuration was fully retained during alkaline hydrolysis. This was taken as an indication that the alkyl-oxygen bond was not ruptured; the proof, however, is not complete because later work has shown occasions with retention of configuration in carbonium ions[6]. E. H. and C. K. Ingold[20] used two esters, 1-methylallyl and 3-methylallyl (crotyl) acetates, which would yield the same mesomeric carbonium ion. Because only the corresponding alcohols were produced[21], the hydrolysis had to occur through acyl-oxygen fission. A method which is in principle similar was applied by Quayle and Norton[22] to the neopentyl esters of acetic and the three chloroacetic acids, when they showed that the hydrolysis yielded neopentyl rather than rearranged t-amyl alcohol.

The most general method for demonstration of the place of fission is to hydrolyse the ester in water enriched in ^{18}O, when OH^- will also become labelled, and to investigate whether labelled alcohol or acid is formed. Acyl-oxygen fission was proved in this way in the alkaline hydrolysis of amyl acetate[23] and of γ-butyrolactone[24]. Kursanov and Kudryavtsev[25] obtained the same result by a modified method: ethyl propionate labelled with ^{18}O in the ethereal oxygen was hydrolysed in normal water and the resulting ethanol was found to have an excess of ^{18}O. Bunton and coworkers have shown by using ^{18}O that acyl-oxygen fission takes place in the alkaline hydrolysis of the following esters with large variations in structure: methyl 2,4,6-triphenylbenzoate[26], diphenylmethyl and 9-fluorenyl acetate[27], methyl trifluoroacetate[28], phenyl and diphenylmethyl trifluoroacetates[29], and bornyl and isobornyl acetates[30].

2. Evidence for the addition–elimination mechanism

The alkaline hydrolysis of esters (5) can be presented either as a one-stage substitution reaction (6) without any formal participation of the carbonyl

$$R^1C(=O){-}OR^2 + OH^- \xrightarrow{\text{slow}} R^1C(=O){-}OH + R^2O^- \xrightarrow{\text{fast}} R^1COO^- + R^2OH \qquad (6)$$

group or as an addition-elimination mechanism (7) with a tetrahedral

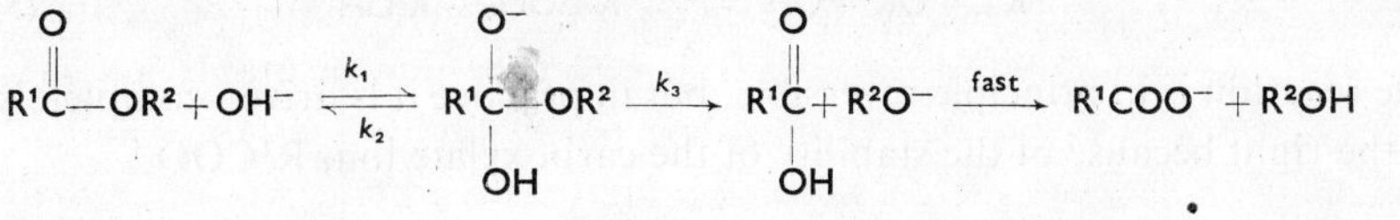

$$R^1C\overset{\text{O}}{\underset{}{\|}}{-}OR^2 + OH^- \underset{k_2}{\overset{k_1}{\rightleftharpoons}} R^1C\underset{OH}{\overset{O^-}{\underset{|}{-}}}OR^2 \overset{k_3}{\longrightarrow} R^1C\underset{OH}{\overset{O}{\underset{|}{\|}}} + R^2O^- \overset{fast}{\longrightarrow} R^1COO^- + R^2OH \qquad (7)$$

intermediate. The possible elimination–addition mechanism is discussed in section II. A. 5.

The direct displacement reaction (6), actually an S_N2 reaction at the carbonyl carbon, does not explain why esters react much more rapidly than ethers[3]. There is no direct indication that the hydroxide ion can add to the carbonyl group of an ester, but addition of sodium methoxide to the carbonyl groups of ethyl fluoroacetates has been detected by the aid of infrared absorption spectra[31]. The main evidence for the formation of a tetrahedral intermediate in ester hydrolysis comes from Bender's finding of concurrent hydrolysis and carbonyl-oxygen exchange with the solvent[32]. If there is a real intermediate in an energy minimum of the reaction path, rather than a transition state in an energy maximum as in mechanism (6), the carbonyl-oxygen and the added hydroxyl-oxygen may become equivalent and an exchange reaction will be possible (equation 8).

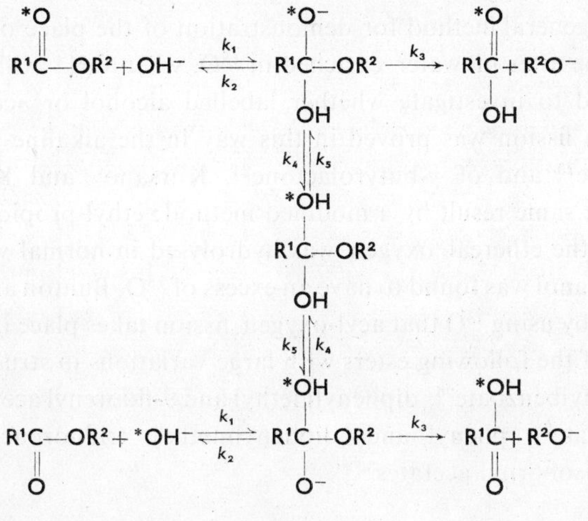

$$ (8) $$

By analysing the recovered ester for [18]O (marked above as *O) after incomplete hydrolysis, Bender[32] showed that a carbonyl-oxygen exchange really accompanies the ester hydrolysis. The rate of exchange was found to be about one fifth of the alkaline hydrolysis of ethyl, isopropyl, and t-butyl benzoates in water. The ratio of the rates of exchange and hydrolysis was found to depend somewhat on the temperature[33], the solvent composition[34], and the structure of the acyl component of p-substituted methyl benzoates[35] or the alkyl component of esters[36]. It can even be unmeasurably low, as, for example, in the case of phenyl benzoate[37] and p-methoxy- and p-chlorobenzyl benzoates[36]. An especially high ratio, about two, was found by Bunton, Khaleeluddin and Whittaker[30] in the hydrolysis of isobornyl acetate.

One may argue that exchange takes place as a side-reaction and that the tetrahedral intermediate does not lie on the reaction path. In the case of the analogous general base-catalysed ethanolysis of ethyl trifluoroacetate (*cf.* section III. B), Johnson[38] has given evidence for a mechanism where an intermediate lies on the reaction path in the solvolysis of a carboxylic ester. As stated by Burwell and Pearson[39], the principle of microscopic reversibility was applied incorrectly by Johnson, but this does not seem to invalidate the result. Fedor and Bruice[40a] have presented kinetic evidence, based on the retarding effect of acids at pH values below 2·5 and solvent isotope effects, for the formation of a tetrahedral intermediate in the general base-catalysed hydrolysis of ethyl trifluorothiolacetate. Bender and Heck[40b] have recently found that oxygen exchange is in agreement with the kinetic results, indicating that the same intermediate is being observed by both methods and that the intermediate observed by carbonyl-oxygen exchange, which is necessarily tetracovalent, therefore, at least in this reaction, lies on the reaction path of ester hydrolysis. Smith and O'Leary[41] found that ethyl benzoate (2) was formed in the hydrolysis of O-ethyl thiobenzoate (1); the relative rate coefficients for the production of ethyl benzoate (2), thiobenzoic acid (3), and benzoic acid (4) under alkaline conditions in 40% acetone–water at 25°c were found to be 0·3, 1·0, and 1, respectively. The results were rationalized by the addition–elimination mechanism (9).

$$
\begin{array}{c}
\underset{(1)}{\overset{\overset{\displaystyle S}{\|}}{R^1C-OR^2}} + H_2O \; \rightleftharpoons \; \underset{(1)}{\overset{\overset{\displaystyle SH}{|}}{R^1C-OR^2}} \\[-0.5em]
\hspace{2cm} |\\[-0.5em]
\hspace{2cm} OH
\end{array}
$$

$$
\begin{array}{c}
R^1C \!\!\! \overset{SH}{\underset{O}{<}} \; + R^2OH \quad (3) \\[1em]
R^1C \!\!\! \overset{OR^2}{\underset{O}{<}} \; + H_2S \longrightarrow R^1COOH + R^2OH \quad (4)\\
(2)
\end{array} \tag{9}
$$

$$R^1 = C_6H_5$$
$$R^2 = Et$$

The structures of the intermediates in the above mechanisms (7) and (8) are written without water molecules but formulae can be given with water participating covalently; *e.g.*, in the Lowry mechanism[42] the intermediate

is written as

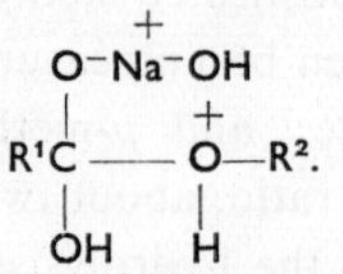

$$
\begin{array}{c}
\quad\; \overset{+}{\text{O}^-\text{Na}^-\text{OH}} \\[2pt]
\text{R}^1\text{C} \longrightarrow \overset{+}{\text{O}} \text{—R}^2. \\[2pt]
\quad\; \text{OH} \quad\; \text{H}
\end{array}
$$

Water certainly solvates the different species but its role does not seem to be clarified unequivocally.

3. Kinetic form

The simple reaction equation (5) calls for second-order kinetics, and as early as 1881 Warder[43] showed that the alkaline hydrolysis of ethyl acetate was first order in respect of both the ester and the hydroxide ion. If mechanism (7) is followed, it can be shown by a steady-state treatment, that the observed second-order rate coefficient k_{OH} is given by equation (10). If the

$$
k_{OH} = \frac{k_1}{1+(k_2/k_3)} \tag{10}
$$

proton transfer reactions (k_4 and k_5) in the reaction scheme (8) are much faster than the other rates, the partitioning ratio k_2/k_3 of the intermediate is given by $2\,k_{ex}/k_{OH}$ where k_{ex} is the observed rate coefficient for the carbonyl-oxygen exchange reaction. If however the proton transfer reactions must be taken into account, which seems to be the true situation[35, 40b], the expression for k_{OH}/k_{ex} is given by equation (11) and k_2/k_3 and k_1 can no longer be calculated from the experimental rate coefficients.

$$
\frac{k_{OH}}{k_{ex}} = 2\,\frac{k_3}{k_2} + 2\,\frac{k_3}{k_4}\left(1+\frac{k_3}{k_2}\right) \tag{11}
$$

Second-order kinetics is still followed.

Recently Tsujikawa and Inoue[44] have reported that the second-order rate coefficient for the alkaline hydrolysis of ethyl acetate, obtained by a continuous measurement of the electric conductivity, decreases as the reaction proceeds. They claimed that steady-state treatment could not be applied to reaction (7). In the light of the unobservable low concentration of the intermediate[31] and the unlikely result that k_2/k_3 should be strongly temperature-dependent (the calculated values ranging from about 7 to 0·3 at the temperature range 50–20°C, cf. ref. 33), the kinetic analysis seems to be wrong and may depend on some error in the rate measurements. Usually the experimental data are found to fit the second-order rate equation well (see, for example, the recent conductimetric data of Saadi and Lee[45] for the hydrolysis of methyl acetate).

Sacher and Laidler[46] found that in the alkaline hydrolysis of p-nitrophenyl acetate in 9·56 vol. % dioxane–water the rate coefficient falls off at high hydroxide ion concentrations and they attributed this to the establishment of an equilibrium between p-nitrophenyl acetate $+OH^-$ and the anion

$$\left\{ \begin{array}{ccc} \underset{\parallel}{\overset{O}{}} & & \underset{\mid}{\overset{O^-}{}} \\ -CH_2COC_6H_4NO_2 & \longleftrightarrow & CH_2{=}COC_6H_4NO_2 \end{array} \right\},$$

which is stabilized by resonance, $+H_2O$. This explanation is, however, suspected by Bruice and Benkovic[139] because the enolate anion should easily collapse to nitrophenoxide and ketene. On the other hand, Papoff and Zambonin[380] recently found by a quasi-adiabatic enthalpimetric method that k_{OH} for ethyl acetate in aqueous solution increases with increasing sodium hydroxide concentration but decreases with increasing sodium chloride content.

A pH dependence of the rate of alkaline hydrolysis differing from that in Figure 1 is observed if the ester has an acidic form E_a which reacts at a rate different from that of the basic form E_b. The experimental second-order rate coefficient k_{OH} at a definite hydroxide ion concentration is then given by equation (12), in which k_a and k_b are the rate coefficients of E_a and

$$k_{OH} = \frac{k_b[OH^-]+k_aK_b}{[OH^-]+K_b} \tag{12}$$

E_b, respectively, and K_b is the base constant of E_b. If now $k_aK_b \gg k_b[OH^-]$ (k_a is probably $\gg k_b$, see section II. B. 3) and $[OH^-] \gg K_b$, then k_{OH} is proportional to $1/[OH^-]$ which means that the rate is independent of the pH. This is found by Ågren, Hedsten and Jonsson[47] to be the case in the alkaline hydrolysis of several 2-diethylaminoethyl benzoates in limited pH ranges (cf. section III. E. 4).

4. Temperature and pressure dependence of rate coefficients

The rate coefficients for the alkaline hydrolysis of carboxylic esters are generally found to obey the Arrhenius equation if there is no change in mechanism with temperature. Typical for the reaction are low values of activation energy E or enthalpy $\Delta H^{\ddagger}$ and relatively low values of the frequency factor A or activation entropy $\Delta S^{\ddagger}$. This is illustrated by the data for ethyl acetate[48] in water at 25°C: $k = 0·111$ M^{-1} sec^{-1}, $E = 11·37$ kcal/mole, $\log A = 7·38$ (A in $M^{-1}sec^{-1}$), $\Delta H^{\ddagger} = 10·78$ kcal/mole, $\Delta S^{\ddagger} = -26·8$ E.U.

33 C.C.A.E.

More examples will be given when structural and solvent effects are considered.

Increased pressure was found in 1915 by Cohen and Kaiser[49] to increase the rate of the alkaline hydrolysis of ethyl acetate. More recent measurements of Laidler and Chen[50] gave the values -9.9 and -8.8 cm^3/mole for the activation volume $\Delta V^{\ddagger}$ of the alkaline hydrolysis of methyl and ethyl acetates, respectively, in water at 25°C. These values are in accordance with a bimolecular reaction and correlate well with the negative activation entropy.

5. Elimination–addition mechanism

Although the addition–elimination mechanism (7) is well documented for most $B_{AC}2$ hydrolyses, Bender and Homer[51] have presented strong evidence that the alkaline hydrolysis of aryl N-alkylcarbamates (5) takes place by an elimination–addition mechanism (13) first proposed by Dittert[52] and Christenson[53].

$$R^1NHC\overset{\displaystyle O}{\overset{\|}{{}}}OR^2 \underset{H_2O}{\overset{OH^-}{\rightleftharpoons}} R^1N\overset{\displaystyle O}{\overset{-\|}{{}}}C OR^2 \xrightarrow{slow} R^1N{=}C{=}O + OR^2$$

$$(5) \hspace{9cm} (13)$$

$$R^1N{=}C{=}O + H_2O \longrightarrow R^1NHCOOH \longrightarrow CO_2 + RNH_2$$

The reaction takes place at a rate 10^5–10^6 times higher than that of the corresponding N,N-disubstituted esters and has a positive activation entropy[53]. It is not general base catalysed, the nitrogen-bonded hydrogen exchanges rapidly with deuterium oxide, and there is a large kinetic solvent isotope effect[51] ($k_{D_2O}/k_{H_2O} = 1.8$, cf. section II. C. 1).

B. Structural Effects

1. Polar effects: meta- and para-substituted benzene derivatives, the Hammett equation

Since a negative hydroxide ion is added to the carbonyl group in the rate-limiting step according to the mechanism (7), electron-withdrawing groups would be expected to facilitate the reaction. This is seen most clearly in the hydrolysis of esters with substituents in the *meta* or *para* position of a benzene ring when steric and resonance effects are nearly constant. The kinetic data for some *meta*-substituted methyl benzoates (Table 1) illustrate this and also show that the rate variation is almost exclusively due to the differences in activation energy E, the frequency factor A being almost constant (see, however, the move recent data of Tommila and co-workers[132]).

TABLE 1. Kinetic data for the alkaline hydrolysis of *meta*-substituted methyl benzoates in 56 wt.% acetone–water at 25°C (Tommila, Brehmer, and Elo[54])

Substituent	σ^{55}	$10^3 k_{OH}$ $(M^{-1} sec^{-1})$	E (kcal/mole)	$\log A$
NO$_2$	0·710	347·0	12·79	8·95
Br	0·391	47·0	13·66	8·73
I	0·352	39·8	13·74	8·71
CH$_3$O	0·115	11·0	14·03	8·36
H	0	9·02	14·35	8·52
NH$_2$	−0·161	5·40	14·83	8·64

The data for reactions of *meta*- and *para*-substituted benzene derivatives can be fitted to the Hammett equation[4]:

$$\log (k/k°) = \varrho\sigma \tag{14}$$

This and other similar linear free energy structure–reactivity relationships are reviewed and their theory interpreted by Jaffé[55], Wells[56], Leffler and Grunwald[57], Ehrenson[58], and Ritchie and Sager[59].

The kinetic data for the alkaline hydrolysis of *meta*- and *para*-substituted alkyl benzoates usually give an excellent fit[55] to the Hammett equation (14). The reaction constant ϱ is usually a little over two in aqueous organic solvents at room temperature, indicating a high susceptibility of the reaction to substitution in the expected direction. The value of ϱ increases almost linearly with $1/T$, an indication of the isoentropic behaviour of the reaction. As expected, ϱ decreases when the substituted phenyl group Ar is moved further off the reaction centre as indicated by the ϱ values in 88% ethanol–water at 30 or 25°C for the following ethyl esters: ArCOOEt 2·43, *p*-ArC$_6$H$_4$COOEt 0·61, ArCH$_2$COOEt 0·82, ArCH$_2$CH$_2$COOEt 0·49, and ArCH=CHCOOEt 1·33. When the variable substituent is in the alkyl component of phenyl or benzyl acetates, the fit is only satisfactory[55]. The reaction constant ϱ in 60 vol. % acetone–water is 1·76 for the first-mentioned series[60] at 1°C and 0·74 for the last-mentioned series[56] at room temperature. The smaller values when compared with the data for substituted alkyl benzoates are caused by the greater distance from the reaction centre.

If there are several substituents in the benzene ring, the resultant effect is close to the sum of their separate effects provided that the groups do not interact[61].

Several extensions of the Hammett equation are possible[55, 56]. Wells and Adcock[62] have shown, that the Hammett equation applies to the hydrolysis of 6- and 7-substituted methyl 2-naphthoates, with high precision, but some of the reaction parameters are different for the 6β and 7β series as well as for the benzene and naphthalene systems.

It is to be remembered that the observed rate coefficient k_{OH} of the alkaline ester hydrolysis is a complex constant (equation 10). Theoretically the Hammett equation can only be applied to the rate coefficient of an elementary reaction, for example, to k_1. That the observed rate coefficients conform to a simple relationship, may depend on the fact that one of two assumptions is valid[35]: either (i) k_2/k_3 is independent of the substituent (as it may *a priori* assumed to be[35]) or (ii) $k_2/k_3 \ll 1$ when $k_{OH} \approx k_1$.

2. Resonance effects

The p-amino group has a σ value[55] of -0.66, differing much more than usual from the value, -0.16, for the m-amino group. This is due to the ground-state stabilization by the resonance (15) which is impossible in the transition state of ester hydrolysis according to equation (7) and leads to a

$$(15)$$

lower rate of hydrolysis of the *para* compound (the values of k_{OH} for ethyl p- and m-aminobenzoates in 56 wt.% acetone–water at 24.9°C are[63] 8.40 and 164.7×10^{-5} M^{-1} sec^{-1}, respectively). That this is a resonance effect is strongly supported by the results of Westheimer and Metcalf[64], who showed that ethyl 3,5-dimethyl-4-dimethylaminobenzoate, in which the two o-methyl groups prevent the dimethylamino group from lying in the plane of the benzene ring and hence inhibit the resonance, hydrolyses about 24 times as fast as the unhindered ester.

Different σ values are needed for a substituent depending upon whether it resonates with the reacting group or not. To avoid the duality or multiplicity of σ values, the resonance effect ψ has to be taken into account as in the extended Hammett equation (16) by Taft[65], and σ values have to

$$\log (k/k^\circ) = \varrho\sigma + \psi \qquad (16)$$

be based on reaction series without direct conjugative interactions between substituent and reaction centre[66, 67]. Norman and coworkers[68] and more recently Yukawa, Tsuno and Sawada[69] suggested that normal substituent constants, $\sigma°$, are best obtained from the rates of alkaline hydrolysis of *meta*- and *para*-substituted ethyl phenylacetates when there is no conjugative interaction between substituent and reaction site and the substituent effects are comparatively large.

A still more pronounced influence is to be expected if a saturated acyl component of an ester is replaced by a group with bonds conjugated with the carbonyl group; since the resonance energy stabilizes the ground state, an increased activation energy and a lower rate of hydrolysis is to be expected. The unsaturated groups, however, are electron-withdrawing and should inductively enhance the rate. The experimental data in Table 2 show that

TABLE 2. Kinetic data for the alkaline hydrolysis of ethyl esters RCOOEt in water at 25°C

R	k_{OH} $(M^{-1} sec^{-1})$	E (kcal/mole)	log A	Reference
CH_3CH_2	0·0870	10·72	6·78	70
$C_6H_5CH_2$	0·207	—	—	71
C_6H_5	0·0293	12·68	7·76	72
$CH_2{=}CH$	0·0779	11·95	7·64	73
$CH{\equiv}C$	4·68	12·26	9·64	74

the mesomeric influence often overweighs the inductive effect. The higher rate of ethyl propiolate compared with that of ethyl acrylate and propionate must be due to the inductive effect of the acetylenic group.

In esters with an unsaturated group in the alkyl component a conjugation with the carbonyl group is impossible and stabilization of the ground state does not take place.

3. Polar effects: aliphatic compounds

In the case of aliphatic and *ortho*-substituted aromatic compounds steric effects become important and may mask the polar effects. If, however, the polarities of the substituents differ considerably, the polar effect dominates, as can be seen from the examples in Tables 3 and 4, in which the substituents are arranged in the order of increasing polarity[65]. The rates are usually seen to follow the order of polarity, activation enthalpy being often the rate-determining factor. Exceptions are found, *e.g.*, in the case of alkyl monohaloacetates, where both the enthalpy and entropy of activation decrease with increasing atomic weight of the halogen causing an abnormal rate order with a maximum at the bromoacetate[11]. When structural varia-

TABLE 3. Kinetic data for the alkaline hydrolysis of ethyl esters RCOOEt in water at 25°C

R	k_{OH} (M^{-1} sec^{-1})	$\Delta H^{\ddagger}$ (kcal/mole)	$\Delta S^{\ddagger}$ (E.U.)	Reference
$^-OCOCH_2$	0·0135	11·2	−29·7	75
CH_3	0·111	10·8	−26·8	48
CH_2I	16·2	6·2	−32·2	11
CH_2Br	49·7	6·5	−29·1	11
CH_2Cl	33·2	8·3	−23·7	11
$CH_3SO_2CH_2$	12·8	—	—	76
CH_3CO	800	—	—	77
$CHBr_2$	202	6·6	−25·7	11
$CHCl_2$	680	5·9	−25·9	11
C_2H_5OCO	3960^a	7·1^a	−20·3^a	78
CHF_2	4500	—	—	79
CCl_3	2570	4·8	−26·9	11
CF_3	150,000^b	—	—	11

^{a} Without statistical corrections.
^{b} This reaction took place at 15°C.

TABLE 4. Kinetic data for the alkaline hydrolysis of esters of acetic acid CH_3COOR in water at 25°C

R	k_{OH} (M^{-1} sec^{-1})	$\Delta H^{\ddagger}$ (kcal/mole)	$\Delta S^{\ddagger}$ (E.U.)	Reference
t-Bu	0·00180	—	—	80
i-Pr	0·0247	10·5	−30·6	81
Et	0·111	10·8	−26·8	48
Me	0·184	11·56	−23·1	82
$C_6H_5CH_2$	0·197	—	—	83
$ClCH_2CH_2$	0·330	12·03	−20·4	84
CH_3OCH_2	1·13	—	—	85
$CH_2=CH$	4·9	12·8	−12·7	86
C_6H_5	1·37	—	—	83
CH_2Br	9·19	9·6	−22	14
CH_2Cl	6·66	10·6	−20	14
$Me_3N^+CH_2$	60	—	—	87

tions take place in the alkyl component of the ester (Table 4), the activation enthalpy and entropy sometimes exert their influence in a quite confused manner, but this depends at least partly on steric effects. A compensatory effect of activation enthalpy and entropy has been found in several basic hydrolysis reactions of esters[88, 89]. When such an isokinetic relationship is prevailing, the order of rate coefficients, *i.e.* the sign of the difference

TABLE 5. Comparison of kinetic data for the alkaline hydrolysis of esters of formic and acetic acids R^1COOR^2 at 25°C

Solvent	R^2	R^1	k_{OH} (M^{-1} sec^{-1})	$\Delta H^{\ddagger}$ (kcal/mole)	$\Delta S^{\ddagger}$ (E.U.)	Reference
Water	Me	H	38.4[a]	9.81±0.22	−18.41±0.77	90
			36.7	8.96	−21.29	91
			41.3	9.36±0.2	−19.8	92
		CH₃	0.152	9.62	−30.1	93, 90
			0.184	11.56	−23.1	82
			0.186	11.1 ±0.1	−24.6	45
Water	Et	H	26	8.35	−24.1	91
				9.09	−21.6	94
			26.4	9.16±0.2	−21.1	92
		CH₃	0.111	10.78	−26.8	48
Water	i-Pr	H	10.9	7.72	−27.9	91
			10.4	8.5 ±0.3	−25.3	92
		CH₃	0.0247	10.5	−30.6	81
36 wt.% EtOH–H₂O	Et	H		11.43	−14.9	94
		CH₃		11.2[a]	−28.6[a]	48
85% EtOH–H₂O	Et	H	2.82[a]	13.86±0.14	−9.83±0.45	90
		CH₃	0.00703[a]	14.1	−21.1	95, 90
			0.0070[a]	15.5[a]	−18[a]	48

[a] An interpolated value.

between the free energies of activation, is temperature-dependent and conclusions based on a single temperature may be erroneous.

The important case of formic and acetic esters deserves a little more detailed consideration (Table 5). The rates of hydrolysis for alkyl formates are a few hundred times higher than those of the corresponding acetates. This is the expected polar influence but was stated by Humphreys and Hammett[90] to be a pure entropy effect (cf. also the consideration of Schaleger and Long[96]) rather than an expected enthalpy effect. An inspection of Table 5 shows that the higher rates of the formates in water solution are due to both lower activation enthalpies and less negative entropies, the former being, however, more important, and that the above-mentioned statement is based on erroneous kinetic data, which is especially the case in the literature value[93] for methyl acetate. The situation changes, however, when organic solvents are added to water and, for example, in ethanol–water mixtures the activation energy of ethyl formate is in a definite solvent range

(x_{EtOH} between 0·075 and 0·3) higher than that for ethyl acetate, the rate difference being due to the activation entropy alone[94] (see also section II. C. 3).

4. Steric effects

The Hammett equation generally fails in the case of aliphatic esters and *ortho*-substituted benzene derivatives because of steric effects. The appreciable retarding influence of increasingly more branched aliphatic groups is illustrated by the data in Table 6, which also show that the substitution of a

TABLE 6. Kinetic data for the alkaline hydrolysis of alkyl acetates CH_3COOR in 70 vol.% acetone–water at 24·7°C (Jones and Thomas[97])

R	Me	Et	n-Pr	i-Pr
$10^3 k_{OH}$ (M^{-1} sec^{-1})	108	46·6	27·0	7·06
E (kcal/mole)	8·8	10·1	11·1	11·8
log A	5·5	6·1	6·6	6·6

R	i-Bu	n-Bu	s-Bu	t-Bu
$10^3 k_{OH}$ (M^{-1} sec^{-1})	18·2	23·0	3·27	0·265
E (kcal/mole)	11·3	11·3	12·8	13·5
log A	6·5	6·7	6·3	6·3

hydrogen atom by a methyl group produces an almost constant increase in the activation energy but that the frequency factor also varies. (Compare also the data in Tables 3 and 4, *e.g.*, regarding the different alkyl substituents). The steric nature of these effects is reflected in the fact that the rates and activation parameters vary much less in the alkaline hydrolysis of substituted γ-lactones than in that of the corresponding open-chain esters[98]; the more rigid structure of the former esters prevents the substituents from screening the reaction site with equal efficiency.

The exceptional behaviour of *ortho*-substituted benzene derivatives in ester hydrolysis has been known for a long time[99]. The main features of this *ortho* effect are seen from Table 7: *ortho* substituents, with the exception of fluorine, retard the rate more or accelerate it less than corresponding *para* substituents or even retard the reaction when *para* substituents enhance the rate (the resonance effect of a *p*-amino-substituent is considered in section

TABLE 7. Comparison of kinetic data for the alkaline hydrolysis of *ortho*- and *para*-substituted ethyl benzoates XC_6H_4COOEt in 60 vol.% acetone–water at 25°c (Tommila and coworkers[100])

X	ortho			para		
	k_X/k_H	ΔE (cal/mole)	$\Delta \log A$	k_X/k_H	ΔE (cal/mole)	$\Delta \log A$
NH_2	0·0882	1170	−0·197	0·0298	2140	0·050
CH_3	0·106	460	−0·635	0·394	480	−0·038
H	1	0	0	1	0	0
I	0·57	−700	−0·72	4·24	−700	0·15
Cl	1·25	−1240	−0·810	3·90	−770	0·022
F	4·03	−1700	−0·638	2·03	−300	0·09

II. B. 2). Whereas in the *para* series the frequency factors are almost equal, they are always seen to be smaller for *ortho*-substituted compounds than for the unsubstituted benzoate. The activation energies vary in quite a similar manner in the *ortho* and *para* series.

Attempts to separate polar and steric effects are considered in section IV. B. 4.

5. Effect of configuration and conformation

The configuration or conformation of the ester has an important influence especially on the steric effect as can be seen from the following examples. Huisgen and Ott[101] investigated the rates of the alkaline hydrolysis of ω-lactones (Table 8). The higher rates of lactones with 5- to 9-atom rings were attributed to the *cis* configuration of these lactones whose free energy in the ground state is higher than the free energy in the prevailing *trans* conforma-

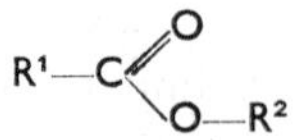

$$R^1—C{\overset{\displaystyle O}{\underset{\displaystyle O—R^2}{\diagup}}}$$

tion of the open-chain esters[3] and lactones that have 10 or more atoms in the ring. No large free energy differences in the tetrahedral transition state are to be expected.

Because of steric effects there are differences in the rates of hydrolysis of alicyclic esters depending on whether the carboxyl group is in the equatorial or in the more hindered axial position (for references, see Eliel[102]). Thus Chapman and coworkers[103] have shown that in the case of methyl 4-*t*-butylcyclohexanecarboxylates in 50 vol.% dioxane–water at 30°c an ester with the equatorial methoxycarbonyl group hydrolyses at least 17 times as

TABLE 8. Kinetic data for the alkaline hydrolysis of ω-lactones

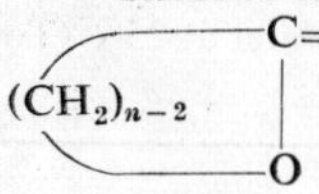

$$(CH_2)_{n-2} \underset{O}{\overset{C=O}{\diagup}} \quad \text{in 60 vol.\% dioxane–water at 0°c (Huisgen and Ott[101])}$$

n	5	6	7	8
$10^4 k_{OH}$ (M^{-1} sec^{-1})	1480	55,000	2550	3530
E (kcal/mole)	—	—	—	—
log A	—	—	—	—

n	9	10	11	12	16
$10^4 k_{OH}$ (M^{-1} sec^{-1})	116	0·22	0·55	3·3	6·5
E (kcal/mole)	11·7	15·2	13·7	11·7	10·8
log A	7·43	7·50	6·72	5·89	5·47

fast as an ester with the axial group, the difference being due to the activation enthalpy. For cyclohexyl esters the differences are smaller because the ring is one atom further from the reaction site. These kinetic differences can be used to derive conformations of esters.

The eleven times higher rate of hydrolysis of diethyl fumarate compared with diethyl maleate[73] and the five times higher rate of diethyl *trans*-1,2-cyclobutanedicarboxylate compared with the *cis* isomer[104] may be due to a steric hindrance in the *cis* form. The greater differences in the case of the corresponding ester ions (experimental rate ratios are 32 and 37, respectively) are possibly due to direct electrostatic influences of the negative carboxylate group of the *cis* isomers and are caused by higher activation energies of the last-mentioned esters. A similar, but opposite, effect is proposed[105] as an explanation for the unexpectedly high rate of hydrolysis of ethyl 3-(*m*-trimethylammoniophenyl)propionate, where the transition state (6) may be stabilized by a short-range electrostatic interaction.

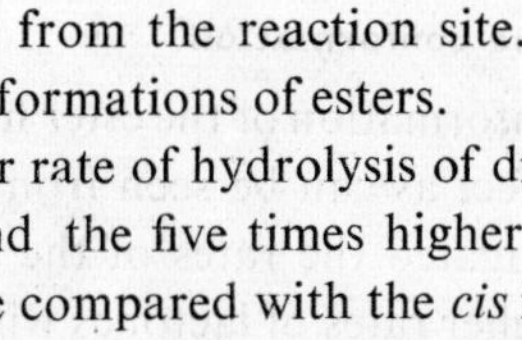

(6)

6. The Taft equation

A linear free energy equation (17), formally similar to the Hammett

$$\log (k/k°) = fA \qquad (17)$$

equation (14), was proposed by Taft[106] for ester hydrolysis. In equation (17), $k/k°$ is the relative rate coefficient of the ester in question as compared with that of the standard ester: acetate for changes in the acyl component, methyl ester for changes in the alkyl component, or *o*-toluate for *ortho* substituents in the benzoate series; f, the reaction constant, is a proportionality factor giving the susceptibility of the series to structural changes, and A, the substituent constant is obtained from $A \equiv \log (k/k°)$ when $f \equiv 1$ Although the usefulness of equation (17) is limited by the fact that different values for A and f are needed for acid and basic hydrolyses and different positions of the substituent, it can be employed, for example, for estimating rate coefficients in cases when they are not known, say in the determination of σ^* values from the Ingold–Taft equation (see section IV. B. 4). Changes in reaction mechanism may be discovered from large differences between observed rate coefficients and values calculated from equation (17).

7. Isotope effects

Ropp and Raaen[107] determined the carbonyl-carbon kinetic isotope effect in the alkaline hydrolysis of ethyl benzoates and found that $k(^{14}C)/k(^{12}C)$ is 0·916, 0·929, and 0·937 for ethyl *p*-methoxybenzoate, benzoate, and *m*-nitrobenzoate, respectively, in 90% ethanol–water at 25°C. The values for ethyl benzoate increase with increasing temperature giving an activation energy difference of 46 cal/mole. Solvent has little effect.

The secondary hydrogen isotope effect, determined by Bender and Feng[108] and Halevi and Margolin[109], in the alkaline hydrolysis of ethyl acetate and trideuteroacetate is strongly temperature-dependent; k_H/k_D has a minimum, 0·90±0·01, at 25°C, and is 1·00±0·01 at 0°C and 1·15±0·09 at 65°C.

C. Medium Effects

1. Solvent isotope effects

The alkaline hydrolysis of ethyl acetate was one of the first reactions to have its rate studied in deuterium oxide (Wynne–Jones[110] in 1935). The rate was found to be higher in heavy water, the ratio k_{H_2O}/k_{D_2O} being 0·75. Values for other esters varying from 0·2 to 0·5 have been found later[111] and it was shown that kinetic isotope ratios varying from 0·2 to 0·7 may be expected for a reaction involving nucleophilic attack of OH $^-$ and OD $^-$ on the ester carbonyl group[111].

2. Salt effects

As early as 1887 Arrhenius[112] investigated the influence of added salts on the rate of alkaline hydrolysis of ethyl acetate in a work often referred to as an example of the small effect of salts on the rate of reactions between an ion and a dipolar molecule. Most neutral salts, like alkali halides and nitrates, were found by Arrhenius[112] and subsequent investigators[113] to slow down the hydrolysis. The Brønsted–Bjerrum equation for medium effects on reaction rate may be written in the form of equation (18), in which k and $k°$ are the rate coefficients in the actual solution and at infi-

$$k/k° = f_E f_{OH}/f^{\ddagger} \tag{18}$$

nite dilution, and f_E, f_{OH} and $f^{\ddagger}$ are the activity coefficients of the ester, hydroxide ion, and transition state, respectively. The Debye–Hückel theory predicts that f_{OH} and $f^{\ddagger}$ vary in dilute solutions in equal manner. The observed small decrease in rate in spite of a small increase in f_E (see Long and McDevit[114]), could be caused by $f^{\ddagger}$ being higher than f_{OH} even in dilute solutions. The observed increase in rate in sulphate solutions[112] obviously depends on the higher values of f_E in alkali sulphate as compared with halide solutions. The approximate constancy of $f_{OH}/f^{\ddagger}$ explains why the rate of the alkaline hydrolysis of simple esters is proportional to the concentration rather than the activity of the hydroxide ion.

Connors and Bender[115] found that potassium chloride increases the rate of hydrolysis of ethyl p-nitrobenzoate (and p-nitrothiolbenzoate). Specific salt effects were found by Duynstee and Grunwald[116] to be important in the hydrolysis of methyl 1-naphthoate in dioxane–water where alkali halides produce small retardation, salts with organic anions large retardation, and salts with organic cations either small retardation or marked acceleration. It was assumed that the ester forms van der Waals complexes with the added organic ion. Becker and Hoffmann[117] studied the hydrolysis of 2-methoxyethyl acetate in moderately concentrated salt solutions (Table 9) and found that solvation of both the initial and the transition state is affected by the salt, and that specific influences of cations, anions, and hydrogen bond donors and acceptors are to be taken into account.

If the ester is charged, a greater salt effect is to be expected on the basis of equation (18) and the Debye–Hückel theory. According to them equation (19) should be valid in dilute solutions. In this equation α is the

$$\log (k/k°) = 2\alpha z_E z_{OH} \sqrt{\mu} \tag{19}$$

TABLE 9. Thermodynamic and kinetic data for the alkaline hydrolysis of 2-methoxyethyl acetate in 3 M aqueous salt solutions at 25°C. $\Delta G\ddagger$ is the free energy of activation and ΔH the enthalpy of solution (Becker and Hoffmann[117])

Salt	None	LiCl	KCl	CsCl
k_{OH} (M^{-1} sec^{-1})	0·186	0·146	0·181	0·202
$\Delta\Delta G\ddagger$ (kcal/mole)	0	+0·139	+0·016	−0·049
$\Sigma\Delta\Delta H\nu$ (kcal/mole)	0	−0·58	−0·69	−0·80

Salt	KF	KCl	KBr	KI
k_{OH} (M^{-1} sec^{-1})	0·225	0·181	0·150	0·122
$\Delta\Delta G\ddagger$ (kcal/mole)	−0·113	+0·016	+0·123	+0·250
$\Sigma\Delta\Delta H\nu$ (kcal/mole)	+0·48	−0·69	−1·21	−1·11

Debye–Hückel coefficient ($\approx$ 0·5 in water at 25°C), z_E and z_{OH} ($= -1$) the charges of the ester and the hydroxide ion, and μ the ionic strength. In the alkaline hydrolysis of half-esters of dicarboxylic acids a positive salt effect is to be expected and is also observed[118–122]. In the case of potassium ethyl oxalate and malonate[120, 122] cations give results which fit equation (19) but the rate is not affected by the valence of the anion and therefore not by the ionic strength, whereas in the case of adipates and sebacates[121] the results show a satisfactory agreement with the theoretical equation.

In the alkaline hydrolysis of $Et_3N^+CH_2COOEt$, Nielsen[123] found that equation (19) is followed in small salt concentrations and Bell and Lindars[124] observed that the rate was proportional to $[OH^-]f_Ef_{OH}$ rather than to $[OH^-]$ which is in accordance with equation (18) because the transition state is neutral. Aksnes and Prue[125] found in the alkaline hydrolysis of esters $CH_3COO(CH_2)_nN^+Me_3$ ($n = 2, 3$) and $Me_3N^+(CH_2)_3COOMe$, in addition to the negative effect according to equation (19), a positive specific salt effect, which increased as the distance between the positively charged quaternary nitrogen atom and the reaction centre increased. This is in accordance with the specific salt effects for half-esters of dicarboxylic acids observed by Hoppé and Prue[119].

3. Solvent effects

Solvent effects in the alkaline hydrolysis of esters in aqueous solutions of organic solvents have been widely studied, especially by Tommila and coworkers. Since the factors causing the observed solvent effects are imperfectly known (for a recent review, see Hudson[126]), only a few examples are considered in the following, and theoretical equations giving the dependence of rate on dielectric constant and other factors are not presented (a recent review has been given by Amis[127]).

Kinetic data for ethyl acetate as a function of solvent composition are given in Figures 2-4. In some solvents (dioxane; t-butyl alcohol, not shown), the rate has a maximum at $x_w \approx 0.98$ at 25°C, but in other solvents it continuously decreases with decreasing water content. This is in

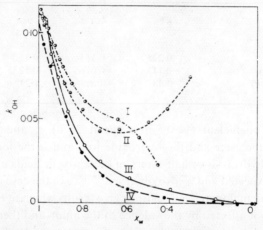

FIGURE 2. Variation of the rate coefficient k_{OH} for the alkaline hydrolysis of ethyl acetate at 25°C as a function of the mole fraction x_w of water (Tommila and coworkers[48, 137]). I: in dioxane–water mixtures, II: in acetone–water mixtures, III: in ethanol–water mixtures, IV: in methanol–water mixtures

contrast with what is expected from the Ingold–Hughes theory about solvent effects. This states that a transition state less polar than the ground state is stabilized less than the latter by increasing the polarity of the solvent. The explanation may be that the activity of the ester is higher in water than in aqueous solutions, in other words, the organic molecules stabilize the ground state by solvation more than the transition state (*cf.* equation 18). This is seen from the recent data of Villermaux, Villermaux and Gibert[128] who found that the rate coefficients for the alkaline hydrolysis of several alkyl acetates in alcohol–water mixtures are approximately proportional to the product of the mole fractional activity coefficient of the ester and the mole volume of the solution. In alcoholic solutions the hydroxide-alkoxide equilibrium (20)

$$OH^- + ROH \rightleftharpoons RO^- + H_2O \tag{20}$$

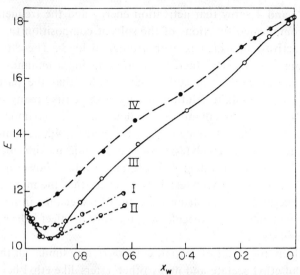

FIGURE 3. Variation of the activation energy E for the alkaline hydrolysis of ethyl acetate as a function of the mole fraction x_w of water (Tommila and coworkers[48]). I: in dioxane–water mixtures, II: in acetone–water mixtures, III: in ethanol–water mixtures, IV: in methanol–water mixtures

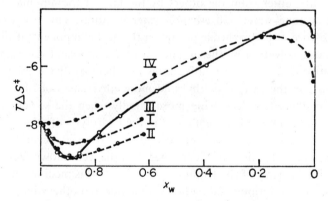

FIGURE 4. Variation of the entropy of activation $\Delta S^{\ddagger}$ for the alkaline hydrolysis of ethyl acetate at 25°C as a function of the mole fraction x_w of water (Tommila and coworkers[48]). I: in dioxane–water mixtures, II: in acetone–water mixtures, III: in ethanol–water mixtures, IV: in methanol–water mixtures

affects the rate, sometimes decisively[129], by diminishing the concentration of the hydroxide ion. In solutions of alcohols other than that produced from the ester, the alkoxide ion, which is more nucleophilic than the hydroxide ion[129], reacts with the ester and makes the situation still more complicated.

Figures 3 and 4 show that activation energy and the frequency factor are more complicated functions of the solvent composition than the free energy of activation, which is proportional to $\log k$. These two factors strongly compensate each other, both often having a minimum at about $x_w = 0.9$. Arnett and coworkers[130] have shown that the partial molal heat of solution at infinite dilution of ethyl acetate (like many other non-electrolytes) in aqueous ethanol has an endothermic maximum at about $x_w = 0.85$, the maximum value of ΔH_s being about 4.5 kcal/mole higher than in water. Also electrolytes have endothermic maxima between 0.8 and 0.9 mole fraction of water; their maximum values, however, are smaller than those of nonelectrolytes. This indicates that the minimum of the activation energy is to a great measure due to the change in heat of solution of the reacting ester molecule, *i.e.*, the solvent effect is largely caused by the ground state.

Qualitatively the effect of solvent composition is similar in the alkaline hydrolysis of ethyl acetate and many other esters like ethyl benzoate and benzyl acetate[72], saturated and unsaturated diesters of dicarboxylic acids[73], and even half-esters of the last-mentioned acids[73, 131] with the exception of the ethyl oxalate ion. The hydrolyses of ethyl formate[94] and of the ethyl oxalate ion[131] differ from the others by the lack of the minima in E and $\log A$ in acetone–water and ethanol–water mixtures. This was explained by the absence of hydrophobic groups in the acyl component of the ester, which makes solvation by the organic solvent component less important. Quantitative differences in relative rates in different solvent mixtures are found, *e.g.*, in the case of diethyl and monoethyl esters of fumaric acid, the rate of the latter decreasing more rapidly when the solvent polarity decreases[73]. Also the substituent effect in the hydrolysis of ethyl benzoates is considerably dependent on the solvent, and changes occur[100, 132] both in ΔE and $\Delta \log A$; this is in accordance with the well-known fact that the reaction parameter ϱ of the Hammett equation (14) is medium-dependent.

The aqueous solutions of dimethyl sulphoxide and other dipolar aprotic solvents are in a special position. The hydroxide ion is poorly solvated in these and thus very active[133]. It has been found that the rate of the alkaline hydrolysis of ethyl acetate[134, 135] and benzoic esters[132, 136] increases with increasing dimethyl sulphoxide concentration, especially when the concentration is high. In the case of ethyl benzoates[137] the rate initially decreases. In acetone–water mixtures also (acetone may be considered as a dipolar aprotic solvent[133]) the rate begins to increase when the water content becomes sufficiently low[137] (Figure 2). According to the data by Tommila and Murto[134] the variation of E and $\log A$ in the hydrolysis of ethyl acetate with the solvent composition is quite similar in aqueous dimethyl sulphoxide and acetone mixtures with minima at $x_w \approx 0.9$; the higher rates in the former are due to higher A values. Roberts[135, 136] has found a minimum in the activation enthalpy at a higher dimethyl sulphoxide concentration ($x_w \approx 0.55$), and his activation enthalpy values are lower than Tommila's. He also denies the importance of the anion desolvation as the major cause for the enhanced rate in dimethyl sulphoxide solutions. Probably the solvation of the ester is more important in solvents of high water content and the desolvation of the hydroxide ion becomes decisive when the water content is low. This view is supported by the results according to which the reaction in anhydrous dimethyl sulphoxide between various esters and suspended sodium hydroxide is 10^4–10^5 times as high as in hydroxylic solvents. The reaction takes place by $B_{AC}2$ mechanism and the reactive species is the small proportion of sodium hydroxide in solution[138].

III. CATALYSIS IN ESTER HYDROLYSIS BY ACYL-OXYGEN FISSION (B_{AC})

In this section various kinds of catalyses are discussed and the so-called 'uncatalysed' or 'spontaneous' hydrolysis is also included because it in fact takes place as general base catalysis (section III. B) and is better called neutral or water hydrolysis. There 'neutral' does not mean that the reaction would take place only in neutral solution but that the reacting species are uncharged ester and water molecules. Catalysis in ester hydrolysis was excellently reviewed by Bender[3] in 1960, and more recently by Bruice and Benkovic and by Johnson[139].

A. Neutral Ester Hydrolysis

1. Scope and kinetics of the reaction

The neutral hydrolyses of some lactones, dioxolones and aliphatic acylals, and esters of tertiary and secondary alcohols take place by alkyl-oxygen fission (section V. B). A hydrolysis independent of pH may also be an indication of intramolecular catalysis (section III. E) or a reaction of the positively charged conjugate acid of an ester with the hydroxide ion (section II. A. 3). The present section deals only with neutral ester hydrolysis by general base catalysis, which is found to take place in the case of esters with electronegative substituents. Probably all simple esters undergo this reaction which, however, is often masked by alkaline and acid hydrolyses (see equation 4 and Figure 1).

Kinetic data for neutral ester hydrolysis are collected in Table 10. This reaction has especially small values of activation enthalpy and entropy. Euranto and Cleve[12, 13] have shown that the reaction does not obey the Arrhenius equation accurately; but the activation enthalpy is temperature-dependent and the heat capacity of activation, $\Delta C_p^{\ddagger}$, has been found to be negative. The values of $\Delta C_p^{\ddagger}$ are approximately equal in this reaction and in the solvolyses of esters of strong acids such as alkyl halides[146] and methyl benzenesulphonates[147]. They are obviously connected with a highly polar transition state.

Bunton and Hadwick have shown by the ^{18}O method that the neutral hydrolyses of methyl[28] and phenyl[29] trifluoroacetates take place by acyl-oxygen fission. Carbonyl-oxygen exchange, comparable to that in the alkaline and acid hydrolyses of alkyl benzoates, has been found to accompany the neutral hydrolyses of methyl[148] and ethyl[40b] trifluoroacetates. The mechanism of this reaction, which must be of the B_{AC} type, is discussed in section III. B. 4.

2. Structural effects

The examples in Table 10 show that structural effects are similar in neutral and alkaline hydrolyses: electronegative substituents increase, electropositive substituents decrease the rate. Moffat and Hunt[149] found that equation (21), where a and b are constants, k the rate coefficient, and

$$\ln k = a/\mu + b, \quad \mu = M_1 M_2/(M_1 + M_2) \tag{21}$$

M_1 and M_2 the molecular weights of the acyl and alkyl component of the ester, respectively, can be applied to the data for both the neutral hydrolysis of alkyl trifluoroacetates and the alkaline hydrolysis of a number of esters. Equation (21) is closely related[149] to the Taft equation (17), and in fact neutral hydrolysis fits the Taft equation in all the cases where sufficient experimental data are available, and the substituent constants A have the same values as in alkaline ester hydrolysis[150]. The values of the reaction constant f were found to be greater than unity, which indicates that neutral hydrolysis is more susceptible to structural changes than alkaline hydrolysis[150]. The same can be observed by plotting $\log k_0$ versus $\log k_{OH}$ when a straight line of slope 1·4 is obtained for different methyl and ethyl esters[151]. Esters with varying alkyl components fall on a separate line with slope 2·1 at 25°C.

TABLE 10. Kinetic data for the neutral hydrolysis of carboxylic esters at 25°C (a = acetone, d = dioxane, w = water)

Ester	Solvent	$10^5 k_0$ (sec^{-1})	$\Delta H^{\ddagger}$ (k cal/mole)	$\Delta S^{\ddagger}$ (E.U.)	$\Delta C_p^{\ddagger}$ (cal deg^{-1} mole^{-1})	Reference
CH₃COOEt	w	0·000025	—	—	—	8
CH₃COOCH=CH₂	w	0·0113	—	—	—	140
CH₃COOCHClCCl₃	50% a–w	0·022	14·1	-42	—	141
HCOOCH₂Cl	w	8·5	11·4	-39	-40	13
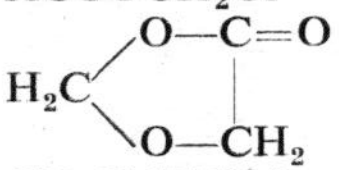 H₂C–C=O / O–CH₂	w	0·302	15·4	-32	—	142
CH₂ClCOOMe	w	0·0205	—	—	—	143
CH₂ClCOOCH₂Cl	w	10·82	11·8	-37	-40	13
CH₂ClCOOCH₂Cl	50% a–w	0·39	—	—	—	13
CHCl₂COOMe	w	1·56	11·4	-42	-43	12
CHCl₂COOCH₂Cl	50% a–w	28·3	10·9	-38	-53	12
CHCl₂COOCH₂CCl₃	50% d–w	3·12	8·1	-52	—	144
CCl₃COOMe	w	79·5	9·0	-42	—	145
CCl₃COOMe	50% d–w	2·67	8·6	-51	—	145
CF₃COOMe	60% d–w	24·3	9·6	-43	—	28
CF₃COOMe	70% d–w	10·0	9·9	-44	—	28

34*

3. Medium effects

The solvent isotope effect for neutral ester hydrolysis differs from that for alkaline hydrolysis (section II. C. 1), as the former reaction is considerably slower in deuterium oxide than in ordinary water; k_{H_2O}/k_{D_2O} has been found to be 2·1 for ethyl difluoroacetate, about 5 for ethyl dichloroacetate[79], and 3·4 for chloromethyl chloroacetate[152]. In the last-mentioned case the solvent isotope effect was also measured in $H_2O - D_2O$ mixtures and, by applying the method of calculation proposed by Salomaa, Schaleger, and Long[153], the results were found to be in good agreement with a transition state with four exchangeable protons[152].

Neutral salts retard the rate of neutral ester hydrolysis much more than that of alkaline hydrolysis[79, 13]. Thus the effect of added sodium perchlorate on the neutral hydrolysis of chloromethyl chloroacetate[13] is almost linear up to 0·5 M solution, obeying the equation $10^4 k_0 = 1·095 - 0·602$ [NaClO₄] at 25 °C.

Palomaa, Salmi, and Korte[145] observed that dioxane lowers the rate of the neutral hydrolysis of methyl trichloroacetate (Table 10) much more than the rates of alkaline and acid hydrolyses. An analogous effect is found in the neutral hydrolysis of ethyl difluoroacetate in ethanol–water[79] and chloromethyl chloroacetate in acetone–water mixtures[13]. A plot of log k_0 versus log $[H_2O]$ for the last-mentioned reaction is in solutions of high water content approximately linear with slope 5 (see Figure 10).

B. General Base Catalysis

The term general base catalysis is used here in the classical sense for reactions involving the attack of a general base on the substrate removing a proton in the rate-limiting step. It is to be distinguished from nucleophilic catalysis where a nucleophile attacks upon a substrate leading to the formation of an unstable intermediate which breaks down to give the products and regenerates the catalyst[3].

1. Scope of the reaction

Jencks and coworkers demonstrated that the hydrolysis of N,O-diacetylserinamide catalysed by imidazole[154], and the hydrolyses of ethyl haloacetates and other esters having electronegative substituents in the acyl component catalysed by several nitrogen and oxygen bases such as aniline, imidazole, and carboxylate and phosphate anions[79], take place by general base catalysis. Also the imidazole-catalysed hydrolysis of dimethyl oxalate, interpreted by Brouwer, Vlugt and Havinga[155] as an example of nucleophilic catalysis, was supposed to take place by general base catalysis[156]. The hydrogen phosphate catalysed hydrolyses of ethyl and methyl acetates, which Holland and Miller[157] detected and proposed to take place by nucleophilic attack by the phosphate dianion on the carbonyl carbon, may in the light of the low acidity of ethanol and methanol (see below) and of the kinetic data for the alkaline and neutral hydrolyses of ethyl acetate (Table 10) really take place by general base catalysis. Imidazole-catalysed hydrolysis of ethyl acetate, which is of the same magnitude as the above-mentioned hydrogen phosphate catalysed reaction, is shown to take place by general base catalysis[151].

Later general base catalysis was found in the hydrolyses of chloromethyl chloroacetate by carboxylate anions[158] and of alkyl acetates CH_3COOR in which pK_a of ROH is less than pK_w by imidazole[111]. In the acetate-ion catalysed hydrolysis of substituted phenyl acetates, Gold and coworkers[159, 160] found a change from nucleophilic to general base catalysis in the case of esters of phenols having pK_a higher than 8. The neutral hydrolysis of diketen

$$H_2C=C-CH_2$$
$$\;\;\;\;\;\;|\;\;\;\;\;\;|$$
$$\;\;\;\;\;O-C=O$$

studied by Briody and Satchell[161], seems to take place by acyl-oxygen fission, in contrast with other β-lactones, and to be subject to general base catalysis by carboxylate anions.

2. Characterization of general base catalysis

The rate of a general base catalysed reaction at constant pH is proportional to the concentration of the catalysing base, as are also the rates of reactions taking place by nucleophilic and general acid-specific hydroxide ion catalyses which are kinetically indistinguishable from general base catalysis.

The following facts are characteristic of general base rather than nucleophilic catalysis: (i) relatively strong deuterium oxide solvent isotope effect, k_{H_2O}/k_{D_2O} generally from 1·9 to 4 (cf. the interpretation by Bender, Pollock, and Neveu[162]); (ii) basicity rather than nucleophilicity determines the effectiveness of the catalyst, e.g. in the hydrolysis of ethyl dichloroacetate, imidazole and the hydrogen phosphate ion, which are of almost equal basicity, are almost equally effective catalysts although imidazole is 4000 times more reactive in nucleophilic catalysis[79]; (iii) no intermediate is formed from the ester and the base in contrast to nucleophilic catalysis[79, 159, 160]; (iv) the entropy of activation is more negative, by about 20 E.U., in the general base than in the nucleophilic catalysis[160]; (v) the Brønsted coefficient β (see below) is of normal magnitude (around 0·5), whereas in nucleophilic catalysis β may be unusually high; (vi) steric effects are of moderate size, whereas in nucleophilic catalysis they are large, *e.g.*, 2,4-lutidine is nearly as active in general base catalysis as predicted from its pK but is practically without effect as a nucleophilic catalyst[163].

3. The Brønsted relation

The Brønsted equation (22) gives the relationship between the catalytic

$$k_{\mathrm{B}} = G_{\mathrm{B}}K_{\mathrm{B}}^{\beta}, \quad \log k_{\mathrm{B}} = \log G_{\mathrm{B}} - \beta \mathrm{p}K_{\mathrm{B}} \tag{22}$$

rate coefficient k_{B} of a base B and its base strength K_{B}, β and G_{B} are constants depending on the nature of the reaction, solvent and temperature[164].

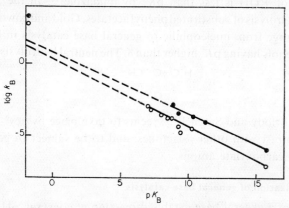

FIGURE 5. The Brønsted plot of log k_{B} in the general base catalysed hydrolysis of ethyl dichloroacetate[79] (○) and chloromethyl chloroacetate[158] (●) versus pK_{B} of the catalysing bases B (in water at 25°C). The bases employed include (from left to right): hydroxide ion, imidazole, hydrogen phosphate ion, 4-picoline, succinate ion, pyridine, acetate ion, aniline, formate ion, and water for ethyl dichloroacetate and hydroxide ion, pyridine, acetate ion, formate ion, chloroacetate ion, and water for chloromethyl chloroacetate

It has been found that the Brønsted equation is satisfactorily obeyed in the general base catalysed hydrolysis of ethyl dichloroacetate[79] with slope $\beta = 0.47$, and in that of chloromethyl chloroacetate[158] with slope $\beta = 0.42$ (Figure 5). In both cases the point for the neutral ester hydrolysis (k_0 being divided by the molar water concentration) falls on the same line with other bases. This is a direct indication that the neutral ester hydrolysis is, in fact, a water-catalysed reaction of water with ester. The point for the alkaline hydrolysis (k_{OH}), on the other hand, lies appreciably above the Brønsted line indicating that the hydroxide ion reacts as a nucleophile rather than only promoting the addition of water.

4. Reaction mechanism

All the data for the general base catalysed hydrolysis of carboxylic esters, including neutral ester hydrolysis, are in accordance with the reaction mechanism (23)[13, 111, 165], which is an extension of the accepted mechanism

$$
R^1C-OR^2 \underset{k_2}{\overset{k_1}{\rightleftharpoons}} \left[R^1C-OR^2 \right] \rightleftharpoons R^1C-OR^2 \underset{}{\overset{k_3}{\rightleftharpoons}} \left[R^1C\cdots OR^2 \right] \underset{}{\overset{}{\rightleftharpoons}} R^1C+:OR^2 \tag{23}
$$

(7) for the alkaline $B_{AC}2$ hydrolysis. The addition of the water molecule to the carbonyl group is according to (23) catalysed by a base B, which is a second water molecule in the neutral hydrolysis. The expulsion of the leaving group is catalysed by the general acid ^+HB. In the case of the neutral hydrolysis of ethyl trifluoroacetate Bender and Heck[40b] found that the ratio of carbonyl–oxygen exchange and hydrolysis is independent of hydrogen ion concentration; hence the steps 2 and 3 in equation (23) are similarly catalysed. Bunton and coworkers[148] have stated that it may be difficult to distinguish between mechanisms in which a specific hydrogen atom is transferred and those in which there are only strong hydrogen-bond interactions.

C. Nucleophilic Catalysis

1. Reaction mechanism and kinetics

The mechanism for the nucleophilic catalysis in ester hydrolysis may according to Bender and Turnquest[166] be written as equation (24), where N is the nucleophile. If N is OH^- mechanism (24) is identical with mecha-

$$
R^1C-OR^2 \underset{k_2}{\overset{k_1}{\rightleftharpoons}} R^1C-OR^2 \underset{k_4}{\overset{k_3}{\rightleftharpoons}} R^1C+{}^-OR^2 \underset{H_2O}{\overset{k_5}{\rightleftharpoons}} R^1C-OH+R^2OH \tag{24}
$$

$$
\underset{(7)}{+N} \qquad \underset{(8)}{N^+} \qquad N^+ \qquad +N
$$

nism (7) of alkaline ester hydrolysis, except that step 5 is different. If **8** is stable, no hydrolysis takes place, and if step 5 is rate-limiting, N may act as an inhibitor to the hydrolysis. In true nucleophilic catalysis, **8** must be an intermediate reacting faster than the ester.

The kinetic characteristics which distinguish nucleophilic catalysis from general base catalysis were considered in section III. B. 2. As there are no slow proton transfer reactions in mechanism (24), there should be no primary solvent isotope effect but secondary effects may be relatively large[162, 165];

when N is imidazole or some other nitrogen base, there is only a small isotope effect, but when N is a carboxylate ion, k_{H_2O}/k_{D_2O} is between 1 and 2.

The Brønsted relation (22) is often followed[166, 167], but different lines are obtained for different types of nucleophiles[168], and greater steric requirements may lead to large negative deviations[166]. Negatively charged bases are less effective than neutral bases in the hydrolyses of p-nitrophenyl and 2,4-dinitrophenyl acetates[168, 169], whereas the difference is smaller or inversed in the case of p-nitrophenyl chloroacetate[169] (and of the positively charged acetylimidazolium ion[167]). The hydroxide ion has usually large negative deviations[166, 168] in log k in contrast to general base catalysis but in accordance with other nucleophilic catalyses, possibly because of its solvation in the ground state[170]. The Brønsted coefficient β has relatively high values, e.g. about 0·8 in the hydrolysis of p-nitrophenyl acetate[167, 168].

2. The leaving group

Although the initial addition of N depends on its nucleophilicity, the partitioning of the intermediate 7 (equation 24) to form reactants or products is related to the relative stabilities of N and R^2O^- as stated by Wiberg[171]. In order to make a nucleophilic reaction possible, the basicity of R^2O^- should be relatively weak. Thus Oakenful, Riley, and Gold[160] observed that the hydrolysis of aryl acetates catalysed by the acetate ion begins to take place by nucleophilic catalysis when the pK_a of the corresponding phenol is < 8 and becomes predominant when pK_a < 5. Bruice, Bruno and Chou[170] found that the reaction rates of p-nitrophenyl acetate (pK_a of p-nitrophenol is 7·14) and δ-thiolvalerolactone (pK_a of δ-thiovaleramide is 10·0) with several bases are equal, k_1 (equation 24) determining the rate, whereas some other nucleophiles react more rapidly with p-nitrophenyl acetate, probably because the partitioning of the tetrahedral intermediate 7 is of kinetic significance.

Kirsch and Jencks[151] explained the effect of the leaving group on the mechanism in the following way. When the leaving group is good, the attack of the nucleophile is rate-limiting. When the leaving group is made progressively worse, the second step becomes rate-limiting and the rate becomes very sensitive to the nature of the leaving group until the nucleophilic catalysis becomes slower than the general base catalysis, which is less sensitive to structural changes.

3. The nucleophile

The hydrolysis of an ester R^1COOR^2 catalysed by a carboxylate anion R^3COO^- leads to the formation of an acid anhydride intermediate 8,

$$\underset{\text{O}}{\overset{\text{O}}{\parallel}} \qquad \underset{\text{O}}{\overset{\text{O}}{\parallel}}$$

R^1C—O—CR^3. As acid anhydrides usually hydrolyse more rapidly than esters, the anhydride cannot be isolated. Its formation, however, can be demonstrated by the aid of aniline which reacts rapidly with anhydride forming acylanilide[159, 160]. Indirect evidence for the formation of an anhydride intermediate was obtained by Bender and Neveu[172] by using ^{18}O- labelled acetate as catalyst in the hydrolysis of 2,4-dinitrophenyl benzoate (9) (equation 25). The benzoic acid was found to contain about 75% of the

$$C_6H_5COOC_6H_3(NO_2)_2 + CH_3CO^*O^- \longrightarrow C_6H_5\overset{\overset{\text{O}}{\parallel}}{C}—^*O—\mid—\overset{\overset{*\text{O}}{\parallel}}{C}CH_3 \qquad (25)$$

$$\text{(9)} \qquad \qquad \overset{\text{H}_2\text{O}}{\longrightarrow} \; C_6H_5CO^*OH + CH_3CO^*OH$$

^{18}O derived from one of the ^{18}O atoms of the acetate ion.

Schonbaum and Bender[173] demonstrated that the dianion of o-mercaptobenzoic acid catalysed the hydrolysis of p-nitrophenyl acetate, and that the intermediate thioaspirin hydrolysed rapidly by intramolecular catalysis. Similarly Fuller[174] found that catechol monoanion catalysed the hydrolysis of phenyl chloroacetate. The formed intermediate, catechol monochloroacetate, hydrolysed about 800 times faster than the phenyl ester by intermolecular catalysis

The most extensively studied nucleophilic catalysts in ester hydrolysis are imidazole, pyridine, and trimethyl amine[3, 166]. In the case of the imidazole-catalysed hydrolysis of p-nitrophenyl acetate it is possible to follow spectrophotometrically the formation of both nitrophenol (at 400 mμ) and of N-acetylimidazole (at 242 mμ), and it was shown that mechanism (24) was followed[155]. Acetylimidazole has been isolated in the reaction of imidazole with p-nitrophenyl acetate[175]. The imidazole-catalysed reaction has lower activation enthalpy and entropy than alkaline and aqueous hydrolysis of p-nitrophenyl acetate[176]. Haruki, Fujii, and Imoto[177] found that amidines

$$RC\underset{NH_2}{\overset{NH}{\diagup}} \qquad (R = H, \; CH_3, \; C_6H_5, \; NH_2)$$

were especially effective catalysts hydrolysing easily even ethyl acetate and γ-butyrolactone (k being the order of 10^{-2}–10^{-3} M^{-1} sec^{-1} in 85% ethanol–water at 30°c). On the other hand, N-substituted amidines were only weak catalysts.

4. General base catalysed nucleophilic catalysis

Caplow and Jencks[178] found that the imidazole-catalysed hydrolysis of p-nitrophenyl benzoates with electron-withdrawing substituents in the acyl component had a third-order term in the rate expression. The corresponding reaction had an appreciable solvent isotope effect ($k_{H_2O}/k_{D_2O} = 1.81$) and it was concluded to be a general base catalysed nucleophilic reaction between imidazole and ester. In the catalysis by N-methylimidazole, which cannot lose a proton, no general base catalysis was detected. Bruice and Benkovic[179] observed a similar rate expression for the imidazole-catalysed hydrolysis of p-methyl- and p-methoxyphenyl acetates. This reaction had a D_2O solvent isotope effect of 2.2 and an entropy of activation of -51 E.U. which are in accordance with an imidazole-catalysed hydrolysis catalysed by another imidazole molecule via general base catalysis.

Kirsch and Jencks[180a] demonstrated that the nucleophilic reactions of imidazole with phenyl acetate, trifluoroethyl acetate, acetoxime acetate, and p-nitrophenyl toluate were catalysed by the hydroxide ion, possibly by general base catalysis. Indirect evidence for general base catalysis by water in the imidazole-catalysed hydrolysis of phenyl acetate was also presented.

D. Catalysis by Metal Ions

Metal ions often catalyse the hydrolysis of esters capable of forming metal ion complexes by an additional functional group[180b]. Thus Kroll[181] found that copper(II), cobalt(II), manganese(II), and calcium ions catalyse the hydrolysis of amino acid esters in the pH range 7.5–8.5. Bender and Turnquest[182] observed carbonyl-oxygen exchange during the copper(II) ion catalysed hydrolysis of DL-phenylalanine ethyl ester in the presence of glycine buffer. It is assumed on the basis of kinetic, spectrophotometric, and other evidence that chelated structures are responsible for the enhanced rate; in the case of cobalt(III) ion catalysed hydrolysis of glycine a complex where the ester was bound through the amino group alone was found to be ineffective[183]. Examples of the assumed chelated complexes, presented in the form given by the original authors, are in the case of nitrogen-containing esters **10** (Kroll[181]), **11** (Bender and Turnquest[182]), **12** (Ågren[184]), **13** (Alexander and Busch[183]), and **14** (Barca and Freiser[185]).

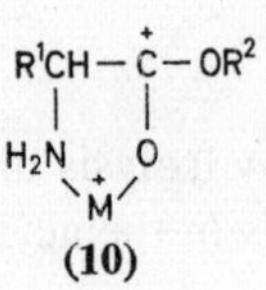

(10) 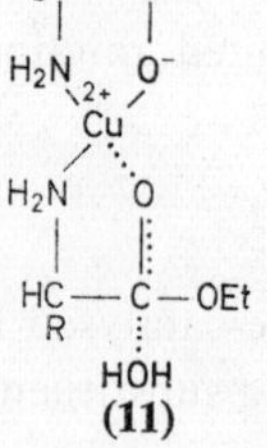   (11) 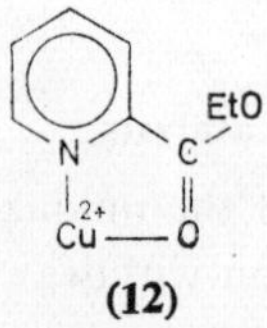   (12) 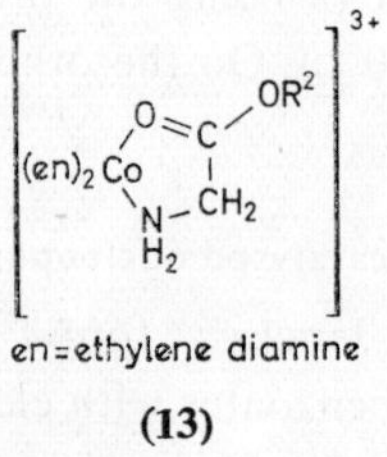   (13) 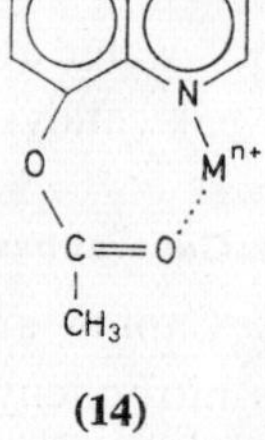   (14)

Sometimes the increased rate of the hydrolysis of a chelated ester is due to the electrostatic influence of the additional charge (Conley and Martin[186]), whereas in other cases much more pronounced rate enhancement is observed (e.g., Ågren[184] found that the rate of the hydroxide ion catalysed hydrolysis of the copper complex 12 was 10^6 times as high as that of ethyl picolinate itself).

Hydrolysis of esters with an α- or β-carboxylate ion are also found to be catalysed by metal ions through chelate formation. The oxalate ester is catalysed to a greater extent than the malonate ester[119]. For the reactions of the oxalate the structure of the transition state was postulated to be 15 by Hoppé and Prue[119], whereas Hay and Walker[187] assumed the chelate 16 to be the reactive species. Boric acid increases the rate of the alkaline hydrolysis of phenyl salicylate probably by the formation of a complex 17 (Capon and Ghosh[188]).

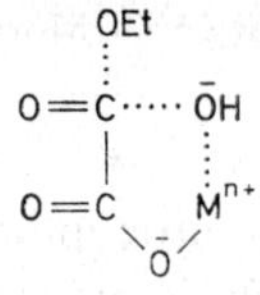

(15) 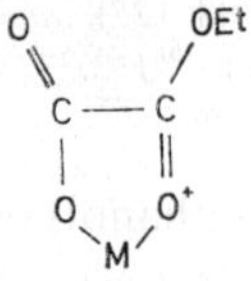   (16) 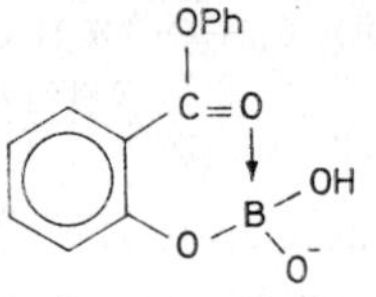   (17) 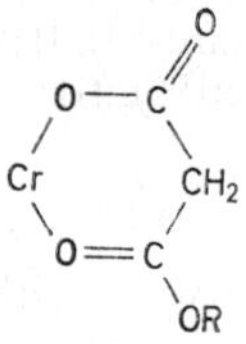   (18)

Huchital and Taube[189] observed that ester hydrolysis accompanies the reaction of the methylmalonatopentaamminecobalt(III) complex with the chromium(II) ion; a chelate, possibly 18, was assumed to be formed as an intermediate.

E. Intramolecular Catalysis

Intramolecular catalysis, also called neighbouring group participation or, if the enhanced rate is due to stabilization of the transition state, anchimeric assistance, may take place as nucleophilic or general base catalysis by basic groups, as electrophilic or general acid catalysis by acidic groups, or as bifunctional electrophilic–nucleophilic catalysis; these are all considered together in this section. Neighbouring group participation has recently been reviewed by Capon[190].

1. Carboxylate ion and carboxyl group

The rate of the hydrolysis of acetyl (Edwards[191]) and other acyl salicylic acids (19) (Garrett[192]) is independent of pH in the region from about 4 to 8 (Figure 6). As general bases do not catalyse the reaction[191] and added organ-

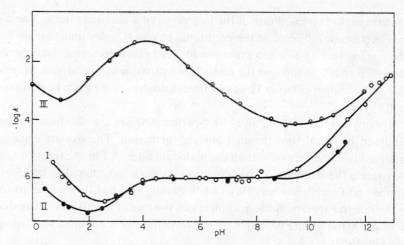

FIGURE 6. The pH profile of the logarithm of the experimental first-order rate coefficient of acyl salicylic acids. I: acetylsalicylic acid in water at 25°C (Edwards[191]), II: trimethylacetylsalicylic acid in 0·5% ethanol–water at 25°C (Garrett[192]), III: o-HOOCC$_6$H$_4$OCOCH$_2$CH$_2$COOH (**25**) in water at 25°C (Morawetz and Oreskes[193])

ic solvents either accelerate the reaction (ethanol) or have little influence (dioxane)[192], the reaction cannot be a neutral ester hydrolysis by general base catalysis. As first proposed by Chanley, Gindler, and Sobotka[194], it may be interpreted as a spontaneous reaction of the acyl salicylate anion by an intramolecular attack of the carboxylate ion on the carbonyl carbon atom producing an anhydride intermediate (**20**) (equation 26).

The hydrolysis of acetylsalicylic acid in H$_2$^{18}O yielded salicylic acid-^{18}O and acetic acid-^{18}O in a ratio which is in good agreement with that expected on the basis of mechanism (26)[195]. Fersht and Kirby[381] have recently investigated the effect of substituents in the 4 and 5 positions to the rate of hydrolysis of acetylsalicylic acid and repeated some of the above-mentioned experiments, partly with unexpected results (the acetate ion catalysed the reaction, ^{18}O did not incorporate into the produced salicylic acid). They concluded that the hydrolysis of aspirin and of most of its derivatives takes place by intramolecular general base rather than nucleophilic catalysis. On the other hand, 3,5-dinitroaspirin is considered to hydrolyse by intramolecular nucleophilic catalysis in the pH region 3–8 and by intramolecular general acid catalysis when pH is < 1[395].

Other ester hydrolyses in which an intramolecular catalysis by carboxylate ion is demonstrated to occur include the reactions of the p-nitrophenyl ester groups in acrylic acid copolymers and their aliphatic analogue, p-nitrophenyl glutarate (Morawetz and coworkers[196]), of substituted monophenyl glutarates (21), succinates (22), maleates (23), and 7-oxabicyclo[2,2,1]hept-2-ene-5,6-*exo*-dicarboxylates (24) (Gaetjens and Morawetz[197], and Bruice and coworkers[198, 199]), and of phenyl and trifluoroethyl hydrogen phthalates (Thanassi and Bruice[200]). The intramolecular hydrolysis by carboxylate anions has been found to be much more sensitive to substituent effects than is the intermolecular catalysis; the substituent effects are mainly due to differences in activation entropy[197]. The effect of the configuration is seen from the relative rates 1, 230, 10^4, and 5×10^4 for the esters 21, 22, 23, and 24, respectively[198]. Alkyl and aryl substituents

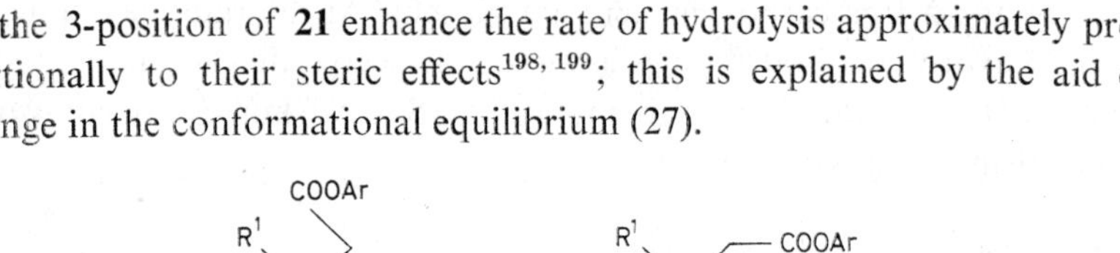

(21)   (22)   (23)   (24)

in the 3-position of 21 enhance the rate of hydrolysis approximately proportionally to their steric effects[198, 199]; this is explained by the aid of change in the conformational equilibrium (27).

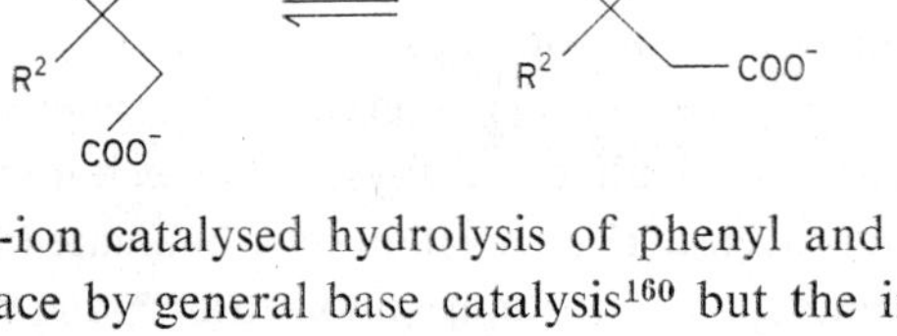

(27)

As the acetate-ion catalysed hydrolysis of phenyl and p-chlorophenyl acetates takes place by general base catalysis[160] but the intramolecularly catalysed hydrolysis of monophenyl maleates takes place by nucleophilic catalysis[198], there is a change of mechanism from general base in the intermolecular reaction to nucleophilic catalysis in the intramolecular reaction[201].

Whereas phenyl and trifluoroethyl hydrogen phthalates hydrolyse by intramolecular nucleophilic catalysis (equation 28)[200], ethyl hydrogen phthalate (Ågren, Hedsten, and Jonsson[47]), methyl hydrogen 3,6-dimethylphtha-

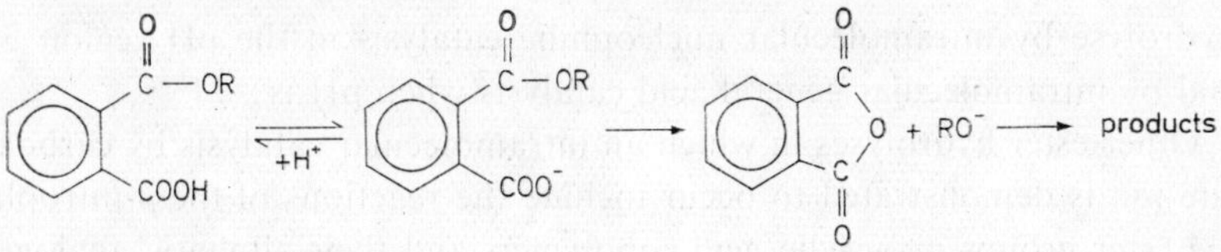

$$\text{(28)}$$

late (Eberson[202]), and methyl and 2'-chloroethyl hydrogen phthalates (Thanassi and Bruice[200]) react by electrophilic neighbouring carboxyl group participation. The change in mechanism takes place when the pK_a of the alcohol is about 13·5; in the case of the monophthalate of 2-propyn-1-ol ($pK_a = 13·55$) the rate of reaction is not dependent on the fraction of ester in the acidic and anionic forms[200]. Also the hydrolysis of ethyl hydrogen maleate and citraconate, which has been found to be independent of pH in aqueous solution[73], obviously takes place by carboxyl participation[202]. The mechanism of the carboxyl group promoted hydrolysis cannot be given with certainty because the formation of an anhydride intermediate on the reaction path has not been demonstrated and several mechanisms are possible[200].

A bifunctional electrophilic–nucleophilic catalysis was observed by Morawetz and Oreskes[193] in the hydrolysis of the ester 25. The pH–rate profile is bell-shaped (Figure 6) with a maximum at pH 3·8 where both a carboxylate ion and a free carboxylic acid are present. At the rate maximum the ester 25 hydrolyses 24,000 times as fast as the acetylsalicylate anion and 66 times as fast as the corresponding diester. Several kinetically equivalent mechanisms are possible (equation 29).

2. Hydroxyl group

A neighbouring hydroxyl group has been found to catalyse the hydrolysis of carboxylic esters (for earlier references, see references 89 and 203); recent examples are the alkaline hydrolyses of catechol monoacetate (Hansen[204]), monochloroacetate (Fuller[174]), monocinnamate (Shalitin and Bernhard[205]), and monobenzoate (Capon and Ghosh[188]) the rates of which are

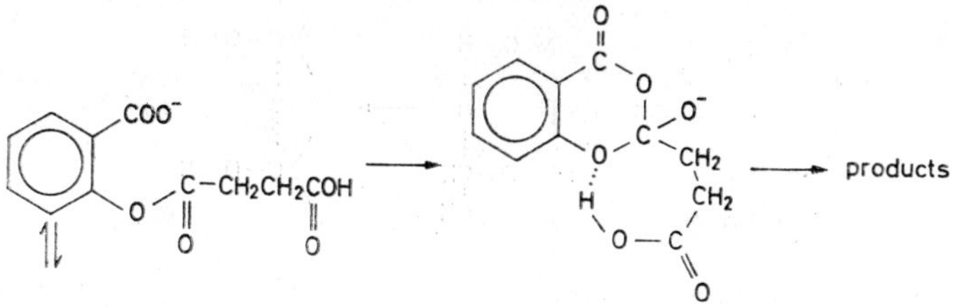

(25)

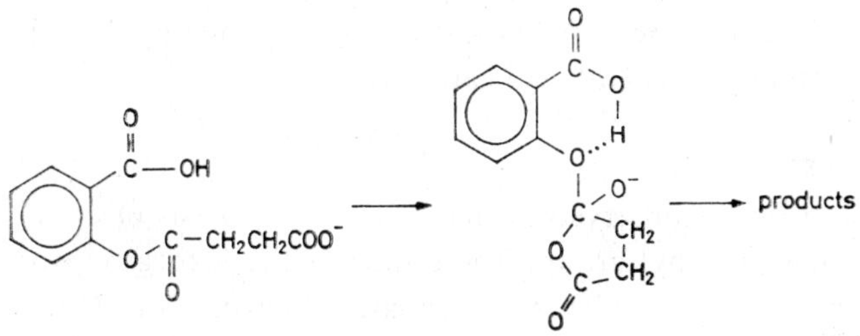

(29)

from 200 to 800 times higher than those of the corresponding phenyl or
o-methoxyphenyl esters. Bender, Kézdy, and Zerner[203] performed an extensive kinetic investigation of the hydrolysis of p-nitrophenyl 5-nitrosalicylate
and related esters without the o-hydroxyl group or with an o-methoxyl
group. The 5-nitrosalicylate differed from the other esters by having two
regions where the rate was pH-independent. The pH-independent reaction in the alkaline region (pH = 7–10) was concluded to be a reaction between water and the ionized ester rather than a kinetically equivalent reaction between the hydroxide ion and the unionized ester. This was based on
the fact that several other nucleophiles react with salicylate esters at the
same rate as with corresponding benzoate esters, in contrast to the hypothetical hydroxide ion reaction. The intramolecular general base catalysis (30)
was proposed as the most probable mechanism for the reaction.

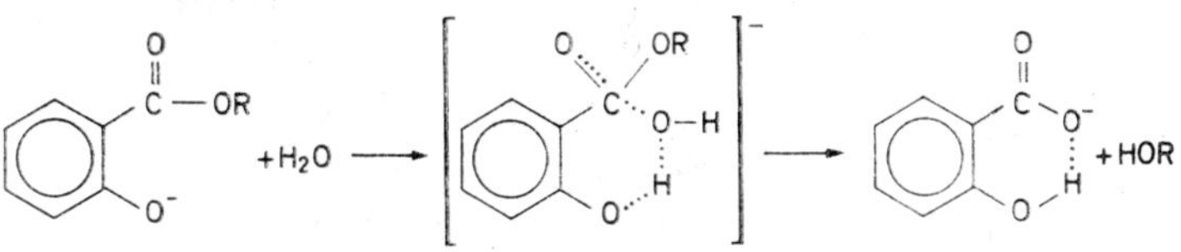

(30)

In the case of esters with less acidic neighbouring alcoholic hydroxyl
groups, such as cis-2- and cis-3-hydroxycyclopentyl acetates, the weakly
assisted hydrolysis at the pH range 8–10 is probably due to an internal solvation of the transition state for attack of the hydroxide ion at the carbonyl
group (Bruice and Fife[89]) the mechanism being given by equation (31)[190].

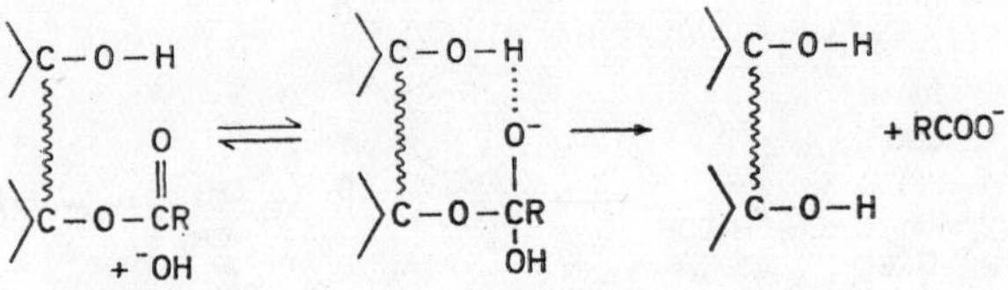

$$\begin{array}{c}
\text{(31)}
\end{array}$$

Kupchan, Eriksen and Friedman[206] have stated that the methanolysis of 1,3-diaxial hydroxyacetates, such as coprostane-3β, 5β-diol 3-monoacetate, which is subject to general base catalysis, is facilitated by the neighbouring hydroxyl group via general acid catalysis.

3. Carbonyl group

Enhanced rate is observed in the alkaline hydrolysis of esters having a neighbouring carbonyl group. These include the keto groups in the acyl component of methyl o-benzoylbenzoates (Newman and Hishida[207]) and methyl 1-benzyl-2-acetyl-6-oxo-10-hydroxy-cis-decahydroisochinolin-9-carboxylate (Becker, Schneider and Steinleitner[208]) and in the alkyl component of 2-oxo-1-methylpropyl acetate (Hansen[209]), substituted 2-oxopropyl benzoates (Schätzle and coworkers[210]), and the cinnamoyl ester of o-hydroxyacetophenone (Shalitin and Bernhard[211]); the rate enhancement caused by the keto group varies from 10 to 10^4. The aldehyde groups in methyl o-formylbenzoate (Bender and coworkers[212]) and o-formylphenyl cinnamate[211] have a still stronger effect. The introduction of the remote 19-aldehyde group of strophanthidin 3-acetate [206] has a smaller effect.

The mechanism (32) is proposed for the hydrolyses of methyl o-benzoyl

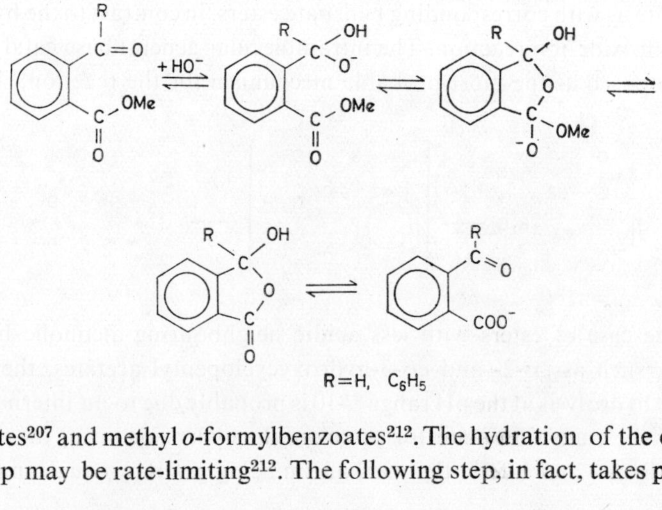

$$\text{(32)}$$

$$R = H, \quad C_6H_5$$

benzoates[207] and methyl o-formylbenzoates[212]. The hydration of the carbonyl group may be rate-limiting[212]. The following step, in fact, takes place by hydroxyl group catalysis. Evidence for mechanism (32) is obtained from steric considerations[207] and from the spectrophotometric observation and isolation of 3-morpholinophthalide in the case of methyl *o*-formylbenzoate when the catalyst is morpholine rather than the hydroxide ion[212]. The magnitude and sign of Hammett's reaction constant found for the hydrolysis of 3'- and 4'-substituted methyl 2-benzoylbenzoates ($\varrho = 2\cdot07$ in 70 vol.% dioxane-water at 30°C) are in accordance with the intramolecular mechanism (Bowden and Taylor[382]). A similar mechanism may be operative in the other cases [206], even if the possibility of electrophilic catalysis cannot be excluded when the carbonyl group is in the alkyl component of the ester[210, 211].

4. Nitrogen-containing groups

Bruice and coworkers found that imidazole groups catalysed intramolecularly the hydrolysis reactions of esters both in the acyl component as in aryl 4-4'-imidazolylbutyrates[213] (**26**; for the *p*-nitrophenyl ester the rate was 30,000 times that of *p*-nitrophenyl acetate) and in the phenyl component as in 2-(4'-imidazolyl)phenyl acetate [214] (a thousandfold rate) and also in an aliphatic alkyl component as in 4-(2'-acetoxyethyl)imidazole[213] (a tenfold rate) and 1-methyl-5-hydroxymethylimidazolyl acetate[215] (k_{OH} 40 to 220 times that for the alkaline hydrolysis of *p*- and *m*-nitrobenzyl acetates, respectively). On the other hand, methyl 4-4'-imidazolylbutyrate does not undergo hydrolysis with imidazole participation[213].

From the pH dependence of the rate, the observed small kinetic solvent isotope effects, the spectrophotometric observation of a lactam (**27**), and other evidence, it was concluded that mechanism (33), in which the imida-

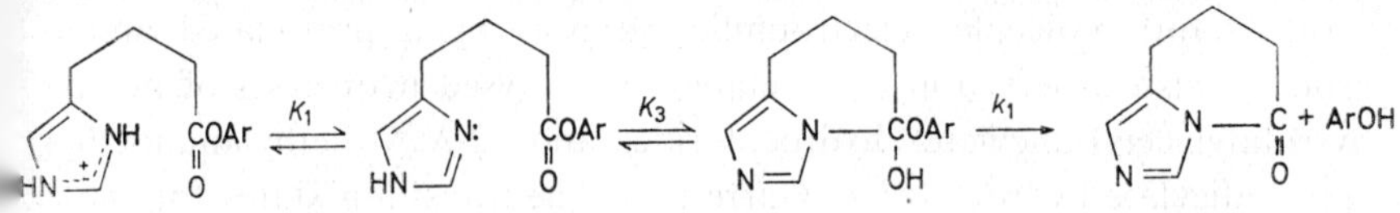

(26) (27) (33)

zole group acts as a nucleophile, is valid for phenyl (and thiol) esters around neutrality In the case of 2-(4'-imidazolyl)phenyl acetate and aliphatic esters with the imidazolyl group in the alkyl component no intermediate could be observed and the rate-limiting step may be the nucleophilic attack of the nitrogen on the acyl group. The mechanism of the reaction is thus analogous to that postulated for the pH-independent hydrolysis of 8-acetoxyquinoline (**28**) by Freiser and coworkers (equation 34)[185, 216].

35 C.C.A.E.

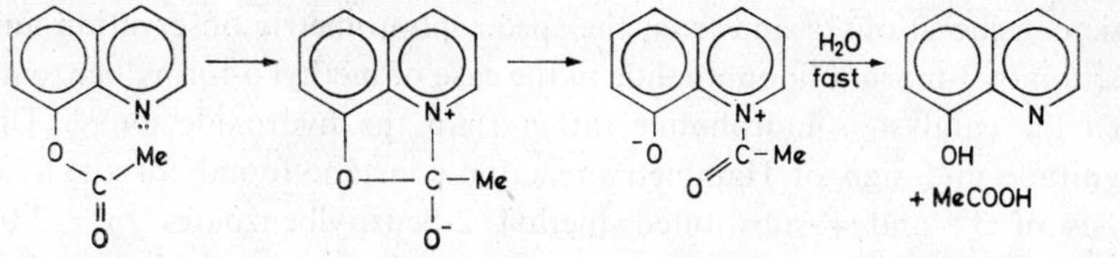

(34)

(28)

Bruice and Benkovic[201] compared the intermolecular catalysis by trimethyl amine of aryl acetates to the intramolecular reactions of the corresponding aryl 4-(N,N-dimethylamino) butyrates and 5-(N,N-dimethylamino) valerates. The activation enthalpies were found to be invariable for all the esters, therefore the activation entropy alone determined the relative rates. The formation of a five- or six-membered cyclic intermediate was found to be favoured by 16 and 13 E.U., respectively, as compared with the intermolecular reaction.

The protonated amino group acts as an intramolecular acid catalyst in the hydrolyses of 2-(N,N-dimethylamino)ethyl benzoate[47] and 2- and 3-(N,N-dialkylamino)alkyl acetates[217, 218], which hydrolyse 10–30 times as fast as the corresponding quaternary ammonium compounds, and methyl 3- and 4-(N,N-dialkylamino)carboxylates[218], when the rate increases less. On the contrary, the rates of the alkaline hydrolysis of protonated dimethylglycine esters are 2–4 times lower than those of the corresponding quaternary esters of trimethylglycine. Whether the ammonium group of the ester is tertiary or primary, is less important. The rate coefficient for the protonated ester is 20–40 times higher than that of the unprotonated form[383]. Intramolecular electrophilic catalysis by a protonated amino group is also observed in the acetate-ion catalysed hydrolyses of methyl pyrrolidylacetylsalicylate hydrochloride and 2-N,N-diethylaminoethyl acetylsalicylate hydrochloride (Garrett[219]). The transition states for these bifunctional catalyses may be written[3] as 29 and 30.

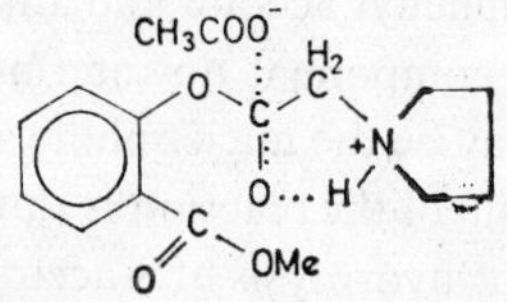

(29) 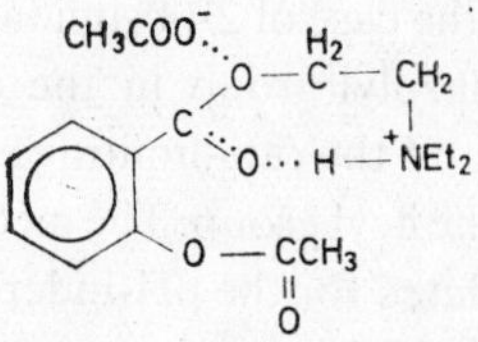　　　　(30)

The β-benzyl ester of N-benzyloxycarbonyl-L-aspartyl-L-seryl amide has in dioxane–water a rate of alkaline hydrolysis 10^7 times higher than that of benzyl propionate, and k_{OH} for the ester 31 is 5000 M^{-1} sec^{-1} (Bernhard

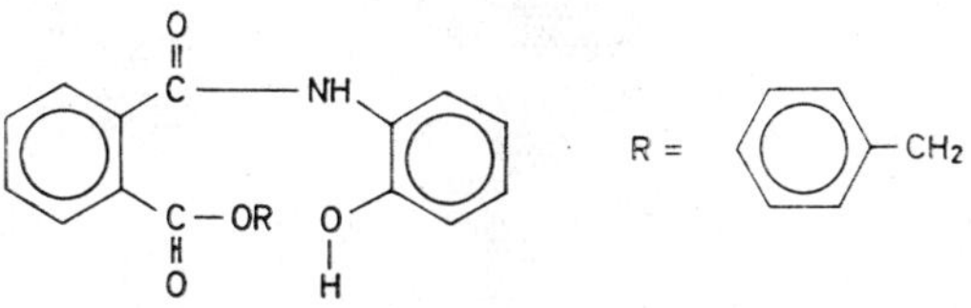

(31)

and coworkers[220, 205]). An imide intermediate (32) is formed which hydrolyses at a rate of one tenth of that for the ester. According to the proposed mechanism (35)

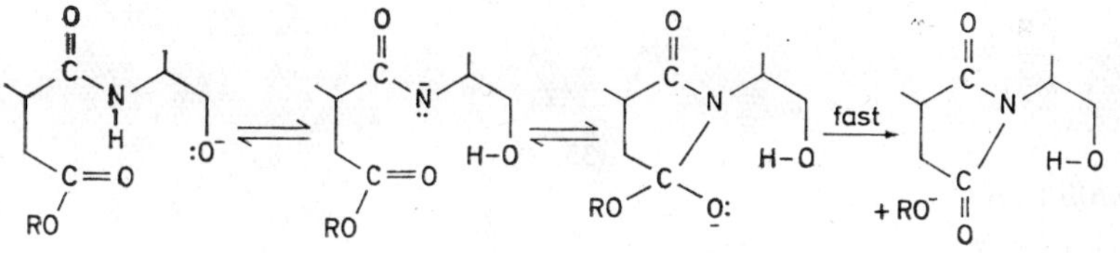

(32)

the ionized alcoholic hydroxyl group of the seryl component or the phenolic hydroxyl of 31 abstracts a proton from the amido nitrogen which then makes a nucleophilic attack on the ester carbonyl group.

Menger and Johnson[221] found that p-nitrophenyl o-methanesulphonami-dobenzoate (33) hydrolyses in slightly alkaline solutions more rapidly than the *para* compound.

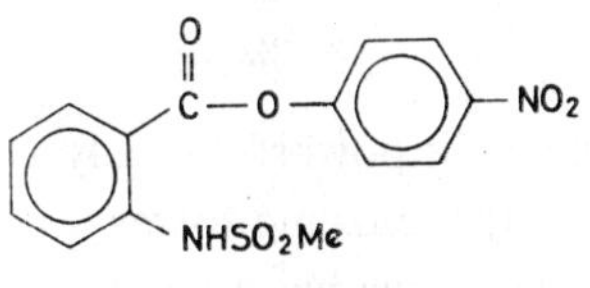

(33)

As the solvent isotope effect k_{H_2O}/k_{D_2O} was 2·02, the reaction was concluded to take place by general base catalysis or general acid–specific hydroxide ion catalysis.

5. Other groups

The early observed rapidity[222] of the hydrolysis of the phenyl ester of salicyl phosphate (34) was interpreted by Chanley, Gindler and Sobotka[194]

35*

to be due to nucleophilic catalysis by the neighbouring phosphate ion (equation 36).

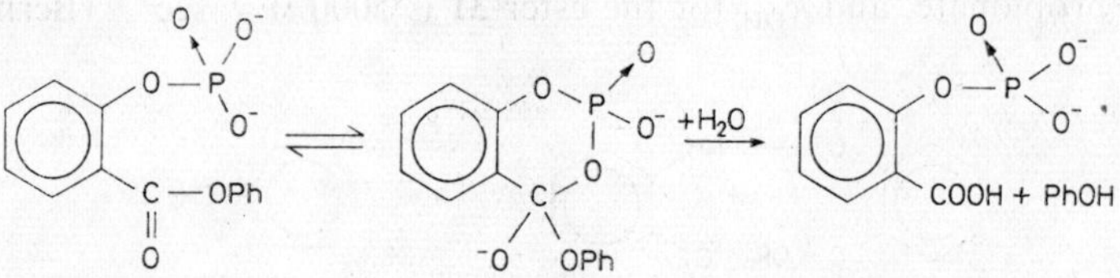

(36)

(34)

The slightly higher rates of alkaline hydrolysis of $R_3MCH_2CH_2COOEt$ (R = Me, Et) when M is Sn, Si, or Ge rather than C was stated by Drenth[223] to be caused by a stabilization of the transition state by coordination as in 35.

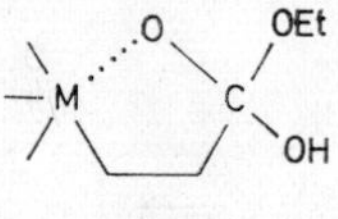

(35)

F. Inhibition

Inhibition of a reaction may be considered as negative catalysis. If a nucleophile reacts with an ester forming an unreactive compound, the hydrolysis of the ester is inhibited. Also the formation of a complex between an ester and a reagent may inhibit hydrolysis. Thus Connors and Mollica demonstrated an inhibition of the hydrolysis of methyl *trans*-cinnamate by imidazole[224] and other heterocyclic compounds[225], and Menger and Bender[226] of the hydrolysis of p-nitrophenyl 3-indoleacrylate and p-nitrophenyl 3-indoleacetate by the 3,5-dinitrobenzoate ion.

G. Enzymic Catalysis

Enzymic catalysis is the most powerful catalysis of carboxylic acid derivatives. It is believed that the enzyme forms a complex with the reacting species, the substrate. The enzyme has an active site, probably with several functional groups, which thus come in the neighbourhood of the reacting centre. The following, usually rate-limiting, stage of the reaction thus resembles an intramolecular reaction. The functional groups of the enzyme are in the side-chains of proteins and may be carboxylate ions (aspartate and glutamate), alcoholic hydroxyls (serine and threonine), phenolic hydroxyls (tyrosine), thiols (cysteine), amines (lysine), guanidines (arginine), and imidazole groups (histidine). Also the —CO—NH— grouping of the peptide bond may itself be effective. All of these groups are known to act as catalysts in ester hydrolysis approaching in effectivity in some cases to that observed in enzymic catalysis. Polymers and micelles containing several functional groups are found to be good models for enzymic reactions[384]. Most of the above-mentioned investigations of catalysis in ester hydrolysis are performed as models for enzymic catalysis. This large and controversial field will, however, not be considered in this connexion.

IV. HYDROGEN-ION CATALYSED HYDROLYSIS AND ESTERIFICATION BY ACYL-OXYGEN FISSION (A_{AC})

The hydrogen-ion catalysed or acid hydrolysis of esters and the esterification of carboxylic acids are the reverse of each other and have identical mechanisms in their respective directions, in accordance with the principle of microscopic reversibility. They are here considered side by side, the more common bimolecular mechanism ($A_{AC}2$) first and the unimolecular mechanism ($A_{AC}1$) in the last section only. Many experimental methods and theories are analogous to those in alkaline hydrolysis and are discussed only shortly and without special reference to section II.

A. Reaction Mechanism

1. Evidence for acyl-oxygen fission and the addition-elimination mechanism

Holmberg[19] showed that acetoxysuccinic acid retained the configuration of the asymmetric group during acid hydrolysis. E. H. and C. K. Ingold[20] found that no isomerization took place during acid hydrolysis and esterification in the case of 1- and 3-methylallyl acetates. The case of acid hydrolysis to which the ^{18}O method has been applied include methyl hydrogen succinate[16], diphenylmethyl formate[227], γ-butyrolactone[24], the trifluoroacetates of methanol[28], phenol[29], and (with concurrent alkyl-oxygen fission) diphenylmethanol[29], and bornyl acetate[30]. In esterification it was employed for the reaction between benzoic acid and methanol[228].

The main evidence for the addition–elimination mechanism is obtained from the observed concurrent hydrolysis and carbonyl-oxygen exchange, which also in the case of acid hydrolysis takes place approximately to the same extent as in alkaline hydrolysis in spite of the much higher rate of the latter[32].

2. Hydrogen ion catalysis

As the reaction is catalysed by the hydrogen ion, the ester in hydrolysis and the acid in esterification must become protonated. It is generally believed that there exists a rapid pre-equilibrium before the rate-limiting addition of water or alcohol to the carbonyl group. Carboxylic acids and esters are only feebly basic; the pK_a of the protonated forms of substituted benzoic acids[229] is between $-6\cdot6$ and $-8\cdot0$, and ethyl benzoate[230] has almost the same basicity ($pK_a = -7\cdot36$) as benzoic acid[229] ($pK_a = -7\cdot26$). Protonation thus takes place only to a small extent even in moderately concentrated acid solutions, and the reaction rate is approximately proportional to hydrogen ion concentration (see section IV.C.3). The observed rates are shown by Martin[385] to be reasonable in spite of the very low concentration of the protonated ester.

Divers opinions have been expressed as to where the proton is attached. From the effect of substitution on the pK_a of substituted benzoic acids, and from the Roman spectrum of benzoic acid in sulphuric acid it is concluded that the carbonyl oxygen becomes protonated owing to resonance

$$RC\begin{smallmatrix}OH\\+\\OH\end{smallmatrix}$$

stabilization (Stewart and Yates[229] and Hosoya and Nagakura[231]). Nuclear magnetic resonance spectra of methyl formate dissolved in very strong acids indicate that the protonation of esters also takes place chiefly to the carbonyl-oxygen atom (Fraenkel[232]). This may be due to the resonance form

$$\begin{smallmatrix}{}^-O\\ \\R^1\end{smallmatrix}\!\!\diagdown C{=}O^+\diagup{}^{R^2}$$

which increases the relative basicity of the carbonyl oxygen. The reacting species is not necessarily the more protonated one[227], but it is now generally believed to be the one protonated on the carbonyl-oxygen atom. Also in the non-aqueous strongly-acid media $HF\text{-}BF_3$[386] and $HSO_3F\text{-}SbF_5$[387] the protonation of several esters is found by n.m.r. to take place exclusively on the carbonyl oxygen.

When the hydrogen-ion activity in the solution becomes sufficiently high, an appreciable fraction of the ester becomes protonated. When the ester or acid is fully-protonated the rate should be independent of hydrogen-ion concentration. Jaques[233] and Lane[234] have shown that the rate of hydrolysis of ethyl acetate has a maximum in 50–60% sulphuric acid and decreases rapidly at higher concentrations. It was calculated that ethyl acetate was half-protonated in 77% sulphuric acid[234] (thus pK_a is $-6\cdot93$). The observed decrease in rate after the maximum is caused mainly by decreasing water activity. A plot of log $(k_\psi/[\text{EH}^+])$ (where EH^+ is the protonated ester and k_ψ is the first-order rate coefficient at constant $[H^+]$) versus log a_{H_2O}

was found to be a straight line with a slope of approximately two. Carbonyl-oxygen exchange with water was observed also in the region of the rate maximum[234]. Yates and McClelland[388] have found similar rate maxima for the hydrolysis of acetates of primary and secondary alcohols and of phenols at about the same sulphuric acid concentration. They concluded that two water molecules are needed in the formation of the transition state.

3. Reaction mechanism and kinetics

The mechanism for the bimolecular acid-catalysed ester hydrolysis and esterification ($A_{AC}2$) may be written[3] as equation (37). Figure 7 represents a free energy diagram corresponding to the mechanism (37) with reasonable

$$
\begin{array}{ccc}
\underset{E}{\overset{O}{\underset{\|}{R^1C}}\!-\!OR^2+H^+} & \underset{\text{fast}}{\overset{K_1}{\rightleftharpoons}} & \underset{EH^+}{R^1C\!\!\overset{OH}{\underset{OR^2}{\diagdown}}} \\
\end{array}
$$

$$
\overset{+H_2O}{\underset{\text{fast}}{\overset{k_1}{\rightleftharpoons}}} \underset{(36)}{R^1C\underset{OH_2^+}{\overset{OH}{\underset{|}{-}}OR^2}} \overset{\text{fast}}{\underset{\text{fast}}{\rightleftharpoons}} \underset{(37)}{R^1C\underset{OH}{\overset{OH}{\underset{|}{-}}OR^2}+H^+} \tag{37}
$$

$$
\overset{\text{fast}}{\underset{\text{fast}}{\rightleftharpoons}} \underset{(38)}{R^1C\underset{OH}{\overset{OH}{\underset{|}{-}}\overset{+}{O}HR^2}} \overset{\text{fast}}{\underset{+R^2OH}{\underset{k_2}{\rightleftharpoons}}} \underset{AH^+}{R^1C\overset{OH}{\underset{OH}{\diagdown}}} \overset{\text{fast}}{\underset{K_2}{\rightleftharpoons}} \underset{A}{R^1COOH+H^+}
$$

values for an ordinary ester like ethyl benzoate. The transition state must be close to **36** because the carbonyl-oxygen exchange, which must take place through the symmetrical intermediate **37**, is slower than the hydro-

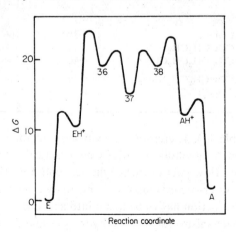

FIGURE 7. Free energy diagram for the acid hydrolysis of an ordinary ester in dilute aqueous solution (for the notations, see equation 37)

lysis[31]. It is to be remembered that the free energy difference between the ground state and the transition state determines the rate of reaction independently of the actual mechanism.

The mechanism (37) leads to the rate equation (38) for the hydrolysis (and to an analogous expression for the esterification).

$$-\frac{d[E]}{dt} = \frac{k_1 K_1 [E][H_2O][H^+]}{1+\alpha} - \frac{k_2 K_2 [A][R^2OH][H^+]}{1+1/\alpha}$$
$$= k_H [E][H^+] - k_E [A][H^+] \tag{38}$$

Here $k_H = k_1 K_1 [H_2O]/(1+\alpha)$ and $k_E = k_2 K_2 [R^2OH]/(1+1/\alpha)$ are the experimental second-order rate coefficients for the hydrolysis and esterification, respectively, and α depends on those rate coefficients in (37) whose values are assumed to be high. The equilibrium constant K (equation 39)

$$K = \frac{k_E [H_2O]}{k_H [R^2OH]} = \frac{[E][H_2O]}{[A][R^2OH]} \tag{39}$$

for the esterification reaction does not differ much from unity for ordi-

TABLE 11. Data for the equilibria of the esterification of acetic acid at 25°C (Jencks and Gilchrist[235]). For the equilibrium constant K (equation 39) all concentrations are given in mole/l. but for the free energy difference ΔG^0 the activity of pure water is taken as unity

Ester	K	ΔG^0 (cal/mole)
CH_3COOEt	3·38	1660
$CH_3COOCH_2CH_2OMe$	1·42	2180
$CH_3COOCH_2CH_2Cl$	0·46	2840
$CH_3COOCH_2CH_2\overset{+}{N}Me_3$	0·394	2940
$CH_3COOCH_2CF_3$	0·013	4970

nary esters (Table 11). Therefore the hydrolysis in aqueous solutions and the esterification in alcoholic solutions go practically to completion in kinetic conditions (low ester or carboxylic acid concentration). In reactions taking place in water–alcohol mixtures and in synthetic conditions, however, the reverse reaction has to be taken into account.

Alternative mechanisms for the acid hydrolysis and the esterification are presented as differing from (37) mainly in respect to the number of participating water molecules and possible intermediates. The apparent order of reaction in respect to water, which in moderately concentrated sulphuric acid was observed to be two (section IV.A.2), was explained by Lane[234] with the aid of the cyclic transition state **39**. Other proposed cyclic transition states[3] include **40** (Syrkin and Moiseev[236]) and **41** (Palm and coworkers[237]).

(39)   (40)   (41)

The Arrhenius equation is found to be obeyed when no changes in reaction mechanism take place. The values of activation enthalpy and entropy are in general low, e.g., the following kinetic data are obtained[238] for ethyl acetate in water at 25°C: $k_H = 1 \cdot 07 \times 10^{-4}\,M^{-1}\,sec^{-1}$, $E = 16 \cdot 76$ kcal/mole, $\log A = 8 \cdot 32$, $\Delta G^{\ddagger} = 22 \cdot 87$ kcal/mole, $\Delta H^{\ddagger} = 16 \cdot 17$ kcal/mole, and $\Delta S^{\ddagger} = -22 \cdot 5$ E.U. The negative value of the activation entropy clearly points to a bimolecular reaction.

The effect of pressure on the acid hydrolysis of alkyl acetates was studied as early as 1896 by Rothmund[239] and in 1897 by Bogojawlensky and Tammann[240]. The more recent investigation by Osborn and Whalley[241] gave the values of $-9 \cdot 1 \pm 0 \cdot 7$ and $-9 \cdot 3 \pm 0 \cdot 7$ cm³/mole for the activation volume $\Delta V^{\ddagger}$ of the acid hydrolyses of methyl and ethyl acetates, respectively, in water at 35°C. They are in accordance with a bimolecular addition mechanism if the proton adds to the carbonyl-oxygen atom[242]. On the basis of a study of the pressure effect on the hydrolysis of methyl acetate in acetone-water mixtures, Baliga and coworkers[243] stated that it could be more advantageous to employ activation parameters at constant volume, $\Delta A_v^{\ddagger}$, $\Delta U_v^{\ddagger}$, $\Delta S_v^{\ddagger}$, and $\Delta C_v^{\ddagger}$, instead of those at constant pressure, $\Delta G_p^{\ddagger}$, $\Delta H_p^{\ddagger}$, $\Delta S_p^{\ddagger}$, and $\Delta C_p^{\ddagger}$. Ultrasonic sound was found by Chen and Kalback[389] to increase the rate of the acid hydrolysis of methyl acetate by increasing the frequency factor.

4. Catalysis

Usually only the hydrogen ion is found to catalyse ester hydrolysis and esterification in acid solutions. Intramolecular electrophilic catalysis by different groups and intermolecular catalysis by Lewis acids, however, can take place in ester hydrolysis (sections III.D and E), whereas the reported cases for general acid catalysis in ester hydrolysis and esterification have often been explained as due to salt effects[3].

The rate expression for the esterification of acetic acid in methanol was found by Rolfe and Hinshelwood[244] to include a term of second order with respect to acetic acid, indicating that the reaction between acetic acid and methanol was catalysed by another acetic acid molecule. A similar dependence on acid concentration was found by Gordon and Scantlebury[245] in the 'uncatalysed' polyesterification of adipic acid with pentaerythritol. In the hydrolysis of n-pentyl formate, Olsen[246] observed a clear rate enhancement in phosphoric acid–hydrogen phosphate buffer solutions with increasing buffer concentration. As the buffer ratio was not varied, no distinction could be made between the proposed general acid catalysis by phosphoric acid and the general base catalysis by the hydrogen phosphate ion, for which, however, the observed rate coefficient seems to be too high.

Cationic exchange resins are found to catalyse the hydrolysis of esters and the esterification of acids, but usually no marked difference between the resin catalysis and the hydrogen ion catalysis has been found[3]. Affrossman and Murray[247] observed that while the catalytic effect of a sulphonic acid resin on the hydrolysis of propyl and allyl acetates in general only depends on the fraction of the resin in the acidic form, the effect of a resin which is partially exchanged with a silver ion is more effective with an unsaturated ester because of the increased concentration of the ester within the resin due to complexing of the silver ion and the unsaturated ester. Yoshikawa and Kim[248] observed that water-soluble polymeric sulphonic acids were in acetone–water poorer, but in water better, catalysts than hydrochloric acid and that the efficiency was the better the more hydrophobic the ester was.

B. Structural Effects

Only weak polar effects are to be expected in acid hydrolysis and esterification because those polar effects which facilitate the addition of a proton retard the nucleophilic attack of water in hydrolysis or of alcohol in esterification. On the other hand, strong steric retardation is to be expected because the reaction is bimolecular.

1. Polar effects: meta- and para-substituted benzene derivatives

The small influence of polarity is seen from the following relative rate coefficients for the acid hydrolysis of substituted phenyl acetates $CH_3COOC_6H_4X$ in 60 vol. % acetone–water at 25°C (Tommila and Sippola[249]): m-Me $= 0.98$, p-Me $= 1.07$, p-Cl $= 0.87$, H $= 1$. Differences in activation energy ($E = 17.2 \pm 0.05$) and in frequency factor (log A

$= 8.0 \pm 0.08$) are small and irregular. The corresponding data for p-nitrophenyl acetate are: relative rate $= 0.74$, $E = 16.74$ kcal/mole, log A $= 7.56$ (Martinmaa and Tommila[390]).

The Hammett equation (14) may be applied to data for acid hydrolysis and esterification although the fit is sometimes not good. The following small values[55, 56] of the reaction constant ϱ illustrate the unimportance of the polar factors: -0.58 for $ArCOOH + MeOH + H^+$ in methanol at $25°C$, $+0.56$ for $ArCOOH + cyclo-C_6H_{11}OH + H^+$ in cyclohexanol at $55°C$, $+0.11$ for $ArCOOEt + H^+$ in 60% acetone–water at $100°C$, $+0.05$ for $ArCH = CHCOCOOMe + H^+$ in 50% dioxane–water at $25°C$, and -0.20 for $CH_3COOAr + H^+$ in 60% acetone–water at $25°C$.

2. Resonance effects

Resonance effects should be similar in alkaline and acid hydrolyses and in esterification because the resonance stabilization of the ground state is disturbed in the transition state in each case. The data in Table 12 indicate

TABLE 12. Kinetic data for the acid hydrolysis of ethyl esters RCOOEt in water at 40°C

R	$10^5 k_H$ ($M^{-1} sec^{-1}$)	E (kcal/mole)	log A	Reference
CH_3CH_2	41·3	15·87	7·69	238
C_6H_5	0·23 (Me ester)	—	—	250
$CH_2{=}CH$	.1·68	19·13	8·59	73
$CH{\equiv}C$	1·16	16·22	4·39	251
cis-EtOCOCH$=$CH	1·17^a	19·58^a	8·73^a	73
$trans$-EtOCOCH$=$CH	4·08^a	18·57^a	8·57^a	73
$trans$-HOCOCH$=$CH	1·96	18·64	8·30	73

^{a} Without statistical corrections.

that this expected retardation by conjugation with the carbonyl group is observed.

3. Steric effects

Table 13 contains values of the rate coefficiens for the acid hydrolysis of acyl- and alkyl-substituted esters and for the esterification of acids. The substituents are arranged in order of increasing electronegativity. It can be seen that there is no correlation with the polarity of the substituent group but that steric effects are decisive.

TABLE 13. Rate coefficients for the acid hydrolysis of ethyl carboxylates RCOOEt and alkyl acetates CH_3COOR in water at 25°C and for the esterification of carboxylic acids RCOOH with ethanol in ethanol at 25°C

R	σ^*	E_s	RCOOEt	
			$10^4 k_H$ (M^{-1} sec^{-1})	Reference
i-Pr	−0·19	−0·47	0·573	252
n-Bu	−0·13	−0·39	—	—
i-Bu	−0·12	−0·93	0·240	253
n-Pr	−0·11	−0·36	0·687	238
Et	−0·10	−0·07	1·12	238
Me	0	0	1·07	238
$ClCH_2CH_2$	0·38	−0·90	0·163	254
H	0·49	+1·24	30·7	238
$MeOCH_2$	0·52	−0·19	0·655	252
$MeCOCH_2$	0·60	—	0·165	255
$ClCH_2$	1·05	−0·24	0·808	84
MeCO	1·65	—	1·20	252
Cl_2CH	1·94	−1·54	1·20	256

R	CH_3COOR		RCOOH	
	$10^4 k_H$ (M^{-1} sec^{-1})	Reference	$10^2 k_E$ (M^{-1} sec^{-1})	Reference
i-Pr	0·617	252	0·289	258
n-Bu	1·10	252	0·496	258
i-Bu	0·92	257	0·0959	258
n-Pr	1·10	238	0·500	258
Et	1·07	238	1·02	258
Me	1·09	238	1·47	258
$ClCH_2CH_2$	0·817	84	—	—
H	—	—	40·9	259
$MeOCH_2$	—	—	—	—
$MeCOCH_2$	0·335	252	—	—
$ClCH_2$	0·354	150	0·959	260

The concept of steric hindrance was established by Meyer[261] who found that the esterification of *ortho*-disubstituted benzoic acids is very slow or even prevented irrespectively of the nature of the substituents. On the basis of structural effects in the acid hydrolysis of aliphatic esters, Palomaa[252] proposed that cyclic species of 'higher order' are formed, an idea further developed by Smith and McReynolds[262]. Newman[263] paid attention to the especially strong steric effect of atoms in position number 6 from the carbonyl oxygen (Newman's 'Rule of Six'). The transition state theory was applied to steric hindrance by Hughes[264], and Becker[265] treated steric effects theoretically in ester hydrolysis and esterification.

The Taft equation (17) may be applied to the acid hydrolysis of esters and to the esterification of acids in the case of aliphatic and *ortho*-substituted aromatic compounds but with the same limitations as in alkaline hydrolysis (see section II.B.6).

4. Separation of polar and steric effects

Ingold[266] stated in 1930 that steric factors have the same effect in both acid- and base-catalysed hydrolyses of an ester, and the same applies for resonance effects. On this basis a value is obtained for the polar substituent constant σ^* from the Taft–Ingold equation (40)[267],

$$\sigma^* \equiv \frac{1}{2\cdot48} [\log (k/k^0)_B - \log (k/k^0)_A] \tag{40}$$

where the subscripts A and B, respectively, refer to otherwise identical acid and alkaline hydrolyses of esters. The factor 2·48 is a reaction constant introduced to get the σ^* values on the same scale as the Hammett σ values[65].

Taft[267] further concluded that the relative rates of esters in acid hydrolysis are independent of polar factors and defined the steric substituent constant E_s by equation (41). For reactions in which both polar and steric effects

$$E_s \equiv \log (k/k^0)_A \tag{41}$$

vary with the structure the equation (42) may be valid. ϱ^* and δ are reaction constants that measure the susceptibility of the reaction series

$$\log (k/k^0) = \varrho^*\sigma^* + \delta E_s \tag{42}$$

to the polar and steric effects, respectively[65]. Values for σ^* and E_s are given in Table 13.

A treatment of kinetic data for several series of alkyl carboxylates indicated that equation (41) did not correlate well in the case of the acid hydrolysis of alkyl formates, in contrast with acetates, propionates, and butyrates[268]. This was attributed to differences in steric effects of the alkyl groups in different reaction series and to pronounced polar effects in the hydrolysis of formic esters. On the other hand, it has been concluded that, although the assumption made in deriving the Taft–Ingold equation (40) may be valid for aliphatic systems, for *ortho*-substituted benzoates (Chapman, Shorter, and Utley[269]) and heterocyclic compounds (ten Thije and Janssen[270]) the transition states for acid and alkaline hydrolyses are not as similar as suggested, probably because the effect of solvation is dissimilar and depends on the polar nature of the substituent[269, 271]. The reaction constant 2·48 in equation (40) does vary depending on the reaction in question (Bowden[271]). Wells[272] proposed that a common scaling of the Hammett and Taft–Ingold equations requires multiplication of the σ^* values by a factor of 0·74.

Hancock and coworkers extended the Taft–Ingold equation for hyperconjugation by correcting the steric substituent constant E_s in (41) for R^1 in R^1COOR^2. In the obtained equation (43) E_s^c is the corrected steric substi-

$$E_s \equiv E_s^c + h(n-3) \tag{43}$$

tuent constant, h the reaction constant for hyperconjugation ($= 0·306$), and n the number of α hydrogens[273]. When R^2 is varied, a significant improvement was achieved when the change in the Newman's six number (i.e., in the number of atoms in the six position from the carbonyl-oxygen atom) was taken into account[274]. The equation (44)

$$\log (k/k^0) = \varrho^* \sigma^* + \delta E_s^c + \nu(\Delta 6) \tag{44}$$

proposed when the structure in R^2 is varied thus contains ($\Delta 6$) which is the six number of a substituent in the acyl component minus the six number of the same substituent in the alkyl component. Talvik and Palm[275] further extended the Taft–Ingold equation for R^1 by also taking into account carbon–carbon hyperconjugation (equation 45), where n_H and n_C are the numbers of α-C—H and α-C—C bonds, respectively.

$$\log (k/k^0) = \varrho^* \sigma^* + \delta E_s^c + h \left(n_H - 3 + \frac{n_C}{2·5} \right) \tag{45}$$

C. Medium Effects

1. Solvent isotope effects

The acid ester hydrolysis takes place in deuterium oxide at a rate about 50% higher than in ordinary water (for early references, see Wiberg[276]). Because D_2O is a weaker base than H_2O the ester is relatively more protonated in D_2O than in H_2O. If the kinetic isotope effects in the subsequent stages are weaker than in the preequilibrium step, a higher rate of hydrolysis in D_2O is observed. Salomaa, Schaleger, and Long[153] examined the acid hydrolysis of ethyl formate and methyl acetate in H_2O–D_2O solvent mixtures at 25°C. The solvent isotope effects k_{D_2O}/k_{H_2O} for these esters were found to be 1·57 and 1·68, respectively. The data for the H_2O–D_2O mixtures were in accordance with a transition state with three or more exchangeable protons, affirming the view that the acid hydrolysis of esters is not a simple A2 reaction.

2. Salt effects

Neutral salts like alkali halides usually slightly increase the rate of the acid hydrolysis of esters such as ethyl acetate (Robinson[277]), methyl chloroacetate (McTigue and Watkins[278]), and γ-butyrolactone (Long, Dunkle, and McDevit[279]). On the other hand, salts with large anions like sodium perchlorate sometimes decrease the rate[278, 279]. The activity coefficient f_E of the ester is usually found to change in the same direction but even more than the rate of hydrolysis[278, 279]. According to the Brønsted equation (18), $k/k^0 = f_E f_{H^+}/f_{\ddagger}$, salts mainly cause a change in f_E whereas the ratio $f_{H^+}/f_{\ddagger}$ varies to a smaller degree and in the opposite direction; probably $f_{\ddagger}$ changes in the same direction as f_E[279]. McTigue[280] has developed a theory for kinetic salt effects, which can be used for determining the transition state hydration number from salt effects. Values varying from 2 (γ-butyrolactone) to 5 (ethyl acetate) were calculated but stated to be uncertain[278].

3. Reaction in moderately concentrated acids

The rate equation (38) for acid ester hydrolysis would require proportionality between the rate and the hydrogen ion concentration. According to the Zucker–Hammett hypothesis[281], however, a difference is to be expected in moderately concentrated acid solutions depending on whether the transition state contains only a proton or a proton and a water molecule in addition to the reacting species[282]. In the first-mentioned case log k_ψ (k_ψ is the observed first-order rate coefficient at constant [H^+]) should be directly proportional to Hammett's acidity function H_0 (46). If, on the other hand, the reaction is bimolecular, the rate should be more nearly proportional to the concentration of the acid than to h_0.

$$H_0 = -\log h_0 = -\log \frac{a_{H^+} f_B}{f_{BH^+}} \tag{46}$$

The rates of acid hydrolysis of ordinary esters in moderately concentrated acids (up to 10 M HCl, HClO$_4$, or H$_2$SO$_4$) have been found to be more nearly proportional to the hydrogen ion concentration than to h_0, although the second-order rate coefficients k_H in general increase with increasing acid concentration (cf. the effects of salts described in the preceding section). Such esters as have been studied are, e.g., methyl formate[283], n-pentyl formate[246], methyl acetate[284], ethyl acetate[9, 283], isopropyl acetate[285], methyl benzoate[246, 286], monoglyceryl esters of benzoic and anisic acids[286], ethyl

acetoacetate[287], γ-butyrolactone[279], methylene and ethylidene diace-
tates[288, 289], and vinyl acetate[290]*.

It is no longer believed that the Zucker–Hammett hypothesis is generally valid, but the above-mentioned criteria can be applied in special
cases to distinguish between A1 and A2 mechanisms. A plot of log k_ψ
against log $[H_3O^+]$ or $-H_0$, however, is seldom linear with slope 1. Bunnett[293] proposed therefore that a plot of (log $k_\psi + H_0$) against log a_{H_2O}
might be a better criterion for the mechanism. The obtained slopes w of
the experimental straight lines for $A_{AC}2$ ester hydrolyses lay between 4
and 7. They are higher than the values of 2–3 characteristic for a nucleophilic attack by water but smaller than the values for general acid catalysis of leaving group ejection from a covalent hydrate intermediate[293].
This may depend on the complicated nature of the experimental rate coefficient (equation 38), which includes the rate coefficients both for the nucleophilic attack of water (k_1) and for the proton transfer reactions (α and
K_1); cf. also the discussion by Martin[294].

More recently Bunnett and Olsen[295] proposed the use of a plot of
(log $k_\psi + H_0$) for reactions of weakly basic substrates, or of log k_ψ for
reactions of strongly basic substrates, against $(H_0 + \log [H^+])$. The slope
ϕ of the obtained straight line (equation 47) characterizes the response

$$\log k_\psi + H_0 = \phi(H_0 + \log[H^+]) + \log k_H \qquad (47)$$

of the reaction rate to changing acid concentration. It was stated that
mechanistic interpretation of ϕ values for reactions of weakly basic substrates is fraught with uncertainty. The values for $A_{AC}2$ hydrolyses range
from 0·74 to 1.

4. Solvent effects

The rate of the acid hydrolysis of an ester like ethyl acetate depends
only slightly on the concentration of an added organic solvent when the
water content is relatively high, but the Arrhenius parameters vary with
the solvent composition in quite a complicated manner (Figure 8) resem-

---

* Kiprianova and Rekasheva[291] stated on the basis of ^{18}O experiments that vinyl
acetate hydrolyses in acid solutions by alkyl-oxygen fission following a mechanism
similar to that of vinyl ether hydrolysis. Landgrebe[292] also proposed an A_{AL} mechanism with an initial hydration of the carbon–carbon double bond. The above-mentioned and other results (structural effects, solvent and structural kinetic isotope effects,
solvent effects, thermodynamic data of activation) obtained by Yrjänä[290], however,
are similar for vinyl acetate and other simple esters and give strong evidence for the
$A_{AC}2$ mechanism also in the case of vinyl acetate. This was recently confirmed by the
^{18}O method (Euranto and Hautoniemi, unpublished results).

bling that found in alkaline hydrolysis (section II.C.3). Harned and Ross[296] determined both the rate of hydrolysis and the activity coefficient of methyl acetate in dioxane–water mixtures and found that the rate coefficient and the activity coefficient of the ester diminish with increasing dioxane content, whereas the activity coefficients of hydrogen chloride, water, and the transition state increase, the last-mentioned least. The situation is thus similar to that found in salt solutions (section IV.C.2) and the change in the activity coefficient of the ester is the most important factor.

Attempts have been made to determine the order of reaction with respect to water by measuring the rate in solutions with low water content.

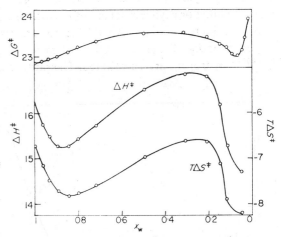

FIGURE 8. Variation of the free energy $\Delta G^{\ddagger}$, enthalpy $\Delta H^{\ddagger}$, and entropy $\Delta S^{\ddagger}$ of activation for the acid hydrolysis of ethyl acetate in acetone–water mixtures at 25°C as a function of the mole fraction x_w of water (Tommila and Hella[238])

Friedman and Elmore[297] found that the hydrolysis of methyl acetate in acetone solutions with similar ester, water, and sulphuric acid concentrations was first order both in water and ester. Koskikallio[298] observed that the rate of hydrolysis of ethyl acetate in dioxane–water mixtures was almost independent of water concentration when it was below 1 M and perchloric acid was the catalyst but decreased when sulphuric acid or hydrogen chloride were catalysts. In the first-mentioned case it was concluded that there was a rapid pre-equilibrium proportional to $[H_2O]^{-2}$ and that the reaction was of second order in water. Hydrogen chloride is a weak acid in these conditions, and the protonation step of the ester was concluded to be rate-limiting. The equilibrium constant for esterifications was found to be constant over the whole range of solvent compositions studied. It is to be remembered that specific solvation effects may be important in these low water concentrations.

In dimethyl sulphoxide–water mixtures the acid hydrolysis of ethyl acetate differs from that in acetone–water mixtures in having a rate maximum at about $x_w = 0.9$ after which the rate decreases continuously (Tommila and Murto[299]). Activation enthalpy and entropy have minima at about $x_w = 0.8$ (as in alkaline hydrolysis) and maxima at about $x_w = 0.4$ (in contrast with alkaline hydrolysis).

D. Esters with Electronegative Substituents

The rates of the acid hydrolysis of carboxylic esters decrease if one α hydrogen atom in the alkyl or acyl component of the ester is substituted by a halogen atom, but increase when two (Table 13, *cf.* also the value 1.3×10^{-4} M^{-1} sec^{-1} for k_H of $CH_2ClCOOCH_2Cl$ in water at $25°C$[13, 300]) or three[28, 29, 275, 301] α hydrogens are substituted. The rates of hydrolysis of

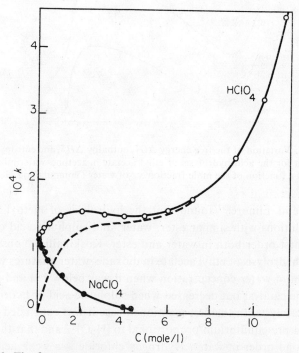

FIGURE 9. The first-order rate coefficients for the hydrolysis of chloromethyl chloroacetate in aqueous perchloric acid and sodium perchlorate solutions at 25°C. The dotted curve is that calculated for the hydrogen-ion catalysed reaction (Euranto and Cleve[300])

the last-mentioned and some other esters with highly negative substituents such as ethyl nitroacetate[301] are higher than expected[11, 275, 301] on the basis of Taft equations (42) and (45). The hydrolysis of these esters has also been studied in high acid concentrations where a rate maximum was found for methyl[28] and phenyl[29] trifluoroacetate, chloromethyl chloroacetate[300] (Figure 9), and mono-o-nitrophenyl oxalate[302]. Both the high rate (Talvik[303]) and the rate maximum (Bunton and coworkers[28, 304]) have been explained to be due to salt effects. As the exceptional acid hydrolysis is always accompanied by a relatively rapid neutral hydrolysis, the salt effect of the added acid on the neutral hydrolysis certainly affects the measured total rate of reaction. In the case of chloromethyl chloroacetate the rate was measured in presence of several salts which all had similar negative electrolyte effects[13]. If the salt effects of sodium perchlorate and perchloric acid are assumed to be identical, the true rate coefficient for the hydrogen-ion catalysed hydrolysis is obtained by substrating the rate coefficient for a sodium perchlorate solution from that for a perchloric acid solution of equal concentration (Figure 9). The resulting curve has no maximum but it is still of unusual shape, and especially the rapid increase in rate at high acid concentrations seems to be difficult to explain on the basis of salt effects. Therefore it was assumed[300] that the mechanism of the acid hydrolysis in dilute aqueous acid solutions is not $A_{AC}2$, which may, however, dominate in concentrated solutions.

The change in mechanism is also suggested by other exceptional kinetic properties found in the acid hydrolysis of the esters in question[305]: the rate decreases with increasing acetone content more rapidly than in the case of other simple esters, neutral salts considerably diminish the rate (cf. section IV.C.2.), the kinetic solvent isotope effect is the opposite to that for other esters ($k_{D_2O}/k_{H_2O} = 0.6$), and the activation enthalpy and entropy are smaller than usual. All these properties resemble those for neutral ester hydrolysis. Bunton showed in the case of methyl[28] and phenyl[29] trifluoroacetates that acyl-oxygen fission takes place. On the basis of the above criteria it was proposed[305] that the rate-limiting step is the same as in neutral ester hydrolysis (k_1 in equation 23), but that the added acid catalyses the expulsion of the leaving group. In order that this could be observed, however, the partitioning of the intermediate must affect the observed rate coefficient. Some observed structural effects in neutral ester hydrolysis[13] and isotope effects in the accompanying carboxyl–oxygen exchange[40b] seem to support this.

E. Unimolecular Hydrolysis and Esterification ($A_{AC}1$)

The unimolecular $A_{AC}1$ mechanism (equation 48) differs from the

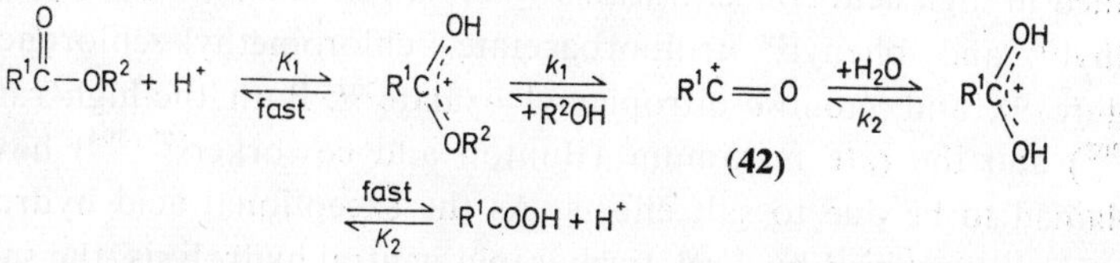

$$
\underset{\text{fast}}{\overset{K_1}{\rightleftharpoons}}
$$

(48)

$$
\underset{K_2}{\overset{\text{fast}}{\rightleftharpoons}} R^1COOH + H^+
$$

bimolecular reaction (37) in that the protonated ester or acid decomposes into an acylium ion (42). In contrast to the $A_{AC}2$ reaction, rates of the $A_{AC}1$ reactions should be greatly increased by electropositive substituents in the acyl component R^1 and retarded by electropositive substituents in the alkyl component R^2; they should be insensitive to steric retardation and highly sensitive to the solvent. The reaction should be especially favoured by sulphuric acid and other solvents where acylium ions are known to be stable. The rate coefficient should be proportional to h_0 rather than to $[H^+]^6$.

The first evidence of the occurrence of the $A_{AC}1$ mechanism was obtained by Treffers and Hammett[306], who observed that 2,4,6-trimethylbenzoic acid (mesitoic acid) dissolved in sulphuric acid to form four particles (equation 49). Methyl mesitoate, in contrast to methyl benzoate, may be

$$
RCOOH + 2\,H_2SO_4 \longrightarrow RCO^+ + H_3O^+ + 2\,HSO_4^-
$$

(49)

hydrolysed by dissolving it in sulphuric acid and pouring the mixture into water and methyl mesitoate may be prepared by pouring a sulphuric acid solution of mesitoic acid into methanol (Newman[307]). From the ultraviolet spectrum of mesitoic acid in concentrated sulphuric acid it is concluded that mesitoyl cation $Me_3C_6H_2CO^+$ is present[231].

The kinetics of the hydrolysis of methyl mesitoate in aqueous perchloric and sulphuric acid solutions was investigated by Chmiel and Long[286] who found that a plot of $\log k_\psi$ versus H_0 gave a straight line with slope 1·2. Bender, Ladenheim, and Chen[308] found that the values of the activation enthalpy and entropy in 11·5 M sulphuric acid were $21·5 \pm 1$ kcal/mole and $+6 \pm 5$ E.U., respectively. The positive activation entropy is in accordance with a unimolecular process. No carbonyl-oxygen exchange was found to take place. In the hydrolysis of methyl 4-substituted-2,6-dimethylbenzoates in 9·70 M sulphuric acid, electron-donating 4-substituents were found to accelerate the reaction[309].

Leisten[310] found that the rates of the hydrolyses of methyl and ethyl benzoates in nearly anhydrous sulphuric acid are independent of water concentration. Structural effects for substituted alkyl benzoates[311] in 99·9% sulphuric acid indicate that all methyl esters hydrolyse by the $A_{AC}1$ mechanism, while ethyl benzoates with electronegative nitro groups and isopropyl benzoate[310] hydrolyse by the $A_{AL}1$ mechanism. The hydrolysis of ethyl acetate in sulphuric acid solutions containing more than 85% acid has been found[233] to be almost independent of the activity of water, indicating that the mechanism is of A1 type. Whether it is the proposed $A_{AL}1$ or the $A_{AC}1$ mechanism, as in the case of ethyl benzoate, cannot be decided from the given data. However, Yates and McClelland[388] more recently found that the dependence of rate on water activity is different for different esters and concluded that in the case of primary alkyl and phenyl esters a change takes place from $A_{AC}2$ to $A_{AC}1$ mechanism.

Nuclear magnetic resonance measurements have shown that in $HF-BF_3$ (Hogeveen[386]) and in HSO_3F-SbF_5 (Olah, O'Brien, and White[387]) methyl esters RCOOMe of different acids cleave with acyl-oxygen fission when acylium ions are formed. The rate was found to vary with R in the following way: $H \ll Me < Et \approx i-Pr > t-Bu$[386,387]. The observed maximum in rate was caused by the opposite effects of activation enthalpy and entropy which both decreased in the above-mentioned series[387].

The hydrolysis of β-butyrolactone in strongly acid solutions was shown by Olson and coworkers with the aid of the optically active $(+)-\beta$-butyrolactone[312] and of the ^{18}O method[313] to take place by acyl-oxygen fission. Long and Purchase[314] found that its rate in concentrated perchloric and sulphuric acid solutions was proportional to h_0 indicating that the mechanism was $A_{AC}1$.

Bunnett's w values[293], $-1·10$ and $-2·47$ for methyl mesitoate and $+0·39$ and $-1·18$ for β-butyrolactone in sulphuric and perchloric acid solutions, respectively, as well as the corresponding ϕ values[295] of $-0·25$, $-0·42$, $+0·10$, and $-0·22$ differ clearly from those found for $A_{AC}2$ reactions (section IV.C.3) and fall in the region characteristic for reactions in which water is not involved in the transition state.

V. HYDROLYSIS AND ESTERIFICATION BY ALKYL-OXYGEN FISSION

A. Acid-Catalysed Reactions ($A_{AL}1$)

1. Mechanism and evidence for alkyl-oxygen fission

Unimolecular reaction by alkyl-oxygen fission can take place in the hydrolysis of esters and in the esterification of acids (equation 50) if the carbonium ion (43) is sufficiently stable. It can be stabilized by strongly

$$R^1C\overset{\displaystyle O}{\overset{\|}{—}}O—R^2+H^+ \underset{\text{fast}}{\overset{K_1}{\rightleftharpoons}} R^1C\overset{\displaystyle +OH}{\overset{\|}{—}}O—R^2 \underset{+R^1COOH}{\overset{k_1}{\rightleftharpoons}} R^2\overset{+}{} \overset{+H_2O}{\underset{k_2}{\rightleftharpoons}}$$

$$R^2OH_2^+ \underset{K_2}{\overset{\text{fast}}{\rightleftharpoons}} R^2OH+H^+ \qquad (43) \qquad\qquad (50)$$

electropositive groups in R^2 or by resonance.

Evidence for alkyl-oxygen fission in acid hydrolysis and esterification is obtained from the following facts. Racemization takes place when the α carbon atom of R^2 is asymmetric (esterification of optically active 2-octanol with acetic acid[315]; hydrolysis of optically active methylethylisohexylmethyl acetate[316]). Ether rather than ester is formed in the alcoholysis of esters (methanolysis of t-butyl benzoates[317]). Formation of alkenes (hydrolysis of t-butyl benzoate[318]) or isotopic hydrogen exchange of the hydrogens of the alkyl component (hydrolysis of s- and t-alkyl trifluoroacetates[319]) accompanies the hydrolysis. The ^{18}O method has been used in the case of t-butyl acetate[320, 321], diphenylmethyl trifluoroacetate[29], p-methoxydiphenylmethyl acetate[322], and triphenylmethyl acetate[323]. More indirect methods for distinguishing the $A_{AL}1$ and $A_{AC}2$ mechanisms are considered in the following in connexion with different types of esters.

A review of alkyl-oxygen heterolysis in carboxylic esters was given by Davies and Kenyon[324].

2. Reactions with intermediate carbonium ions

The hydrolysis of t-butyl acetate in water is a typical $A_{AL}1$ reaction yielding the intermediate Me_3C^+ carbonium ion. It differs in many respects from the $A_{AC}2$ reactions of ordinary esters: k_H (1.2×10^{-4} M^{-1} sec^{-1} in water at 25°C when 85% of the reaction takes place by the $A_{AL}1$ mechanism)[321] is higher than expected on the basis of structural effects in $A_{AC}2$ mechanism (cf. Table 13), $\Delta H^\ddagger$ is considerably higher (26·9 kcal/mole)[321], $\Delta S^\ddagger$ is positive (+13 E.U.)[321], $\Delta V^\ddagger$ is almost zero (0·0±1·0 cm³/mole at 60°C)[241], electropositive substituents in the alkyl component increase the rate (k_H for t-amyl acetate in water at 25°C is $4.4 \times 10^{-4} M^{-1} sec^{-1}$) [325], added salts regardless of their nature increase the rate and the influence is more pronounced (McTigue and Watkins[278]; they also calculated that the transition state hydration number was nearly zero), the logarithm of the rate coefficient in moderately concentrated acids is a linear function of H_0 (log $k_\psi = -1.13H_0 - 3.85)^{285}$, Bunnett's w $(-1.17)^{293}$ and ϕ $(-0.21)^{295}$ are negative indicating that water is not involved in the rate-limiting step, and organic solvent components diminish the rate more strongly[321]. Also, added dimethyl sulphoxide decreases the rate[391].

The esters of tertiary alcohols in general hydrolyse by the $A_{AL}1$ mechanism. Variation in temperature, solvent composition, or the structure of the ester, however, often changes the mechanism. Thus the percentage of the $A_{AL}1$ mechanism for t-butyl acetate is 85 at 25°C in water but 99 at 85°C in water[321] and 36 at 25°C in 70% dioxane–water[326]. While t-butyl acetate reacts in water at 25°C predominantly by the $A_{AL}1$ mechanism, the corresponding hydrolysis of t-butyl formate was found by Salomaa[327] to take place by the $A_{AC}2$ mechanism as indicated by the low values of activation energy (14.8 kcal/mole) and entropy (-21 E.U.). The change in mechanism is due to the fact that steric effects, which slow down the $A_{AC}2$ reaction in the case of acetates, are of much lesser importance for the formates. Yrjänä[328] studied the acid hydrolysis of several aliphatic alkoxy-substituted tertiary alkyl esters and found that diethyl-β-methoxyethyl-methyl acetate hydrolysed in 56% acetone–water by the $A_{AL}1$ mechanism ($\Delta H^{\ddagger} = 27.0$ kcal/mole, $\Delta S^{\ddagger} = +7.9$ E.U.) whereas methyl-bis(methoxy-methyl)methyl acetate reacted in the same solvent by the $A_{AC}2$ mechanism ($\Delta H^{\ddagger} = 17.0$ kcal/mole, $\Delta S^{\ddagger} = -26.8$ E.U.). In the case of dimethyl-methoxymethylmethyl acetate (log k_ψ was neither a linear function of log [H^+] nor H_0 in aqueous hydrogen chloride solutions at 25°C) and diethylmethoxymethylmethyl acetate ($\Delta H^{\ddagger} = 20.4$ kcal/mole, $\Delta S^{\ddagger} = -16$ E.U., the Arrhenius equation is not accurately followed in 56% acetone–water) the reaction took place concurrently by both mechanisms. From the data for the hydrolysis of dimethylmethoxymethylmethyl acetate over an extensive temperature range, the relative rates for both mechanisms could be calculated[329].

The tertiary β-isovalerolactone also hydrolyses by the $A_{AL}1$ mechanism in contrast to β-butyrolactone. This is indicated by the linear dependence of log k_ψ on H_0 (Bunnett's w^{293} is $+0.06$ and ϕ^{295} is $+0.01$) and its thousand times higher rate of hydrolysis (Liang and Bartlett[330]).

If the alkyl component of an ester contains a group which can by resonance stabilize the carbonium ion, the $A_{AL}1$ hydrolysis is preferred as compared with corresponding saturated compounds. Thus Harvey and Stimson found that α-methylallyl 2,4,6-trimethylbenzoate and benzoate, but not acetate, hydrolysed in 60% acetone–water by the $A_{AL}1$ mechanism owing to the resonance $\left\{ MeC\overset{+}{-}CH{=}CH_2 \longleftrightarrow MeC{=}CH\overset{+}{-}CH_2 \right\}$ in the formed intermediate carbonium ion[331]. Similarly diphenylmethyl 2,4,6-trimethylbenzoate and benzoate, but not formate, react in acetone–water mixtures by the $A_{AL}1$ mechanism[332]. Substituents in one of the phenyl groups of diphenylmethyl p-nitrobenzoates have the expected influences (Silver[333]).

Recent studies in concentrated sulphuric acid solutions have shown that the rate of acid hydrolysis of esters has in general a maximum followed by a minimum at still higher concentrations, and that a change in mechanism takes place (cf. section IV.E). In the case of secondary alkyl and benzyl acetates the reaction taking place after the minimum occurs by alkyl-oxygen fission[388]. It was concluded that in this $A_{AL}1$ hydrolysis water is actually released in the rate-limiting step. In the non-aqueous HSO_3F-SbF_5 system esters of ethanol and secondary and tertiary alcohols were found to cleave by alkyl-oxygen fission, the rate increasing in the mentioned order[387].

3. Reactions with intermediate alkoxymethyl cations

Alkoxymethyl esters (44) are esters as well as acetals. On the basis of the higher values for the temperature coefficients in their acid hydrolyses as compared with ordinary esters, Salmi[256, 334] concluded that the reaction took place as an acetal hydrolysis, which in this case is an $A_{AL}1$ hydrolysis (equation 51) as suggested by Salomaa[335]. The formed alkoxymethyl cation

$$\underset{\textbf{(44)}}{R^1\overset{O}{\overset{\|}{C}}O{-}CH_2OR^2} + H^+ \underset{\text{fast}}{\overset{\text{fast}}{\rightleftharpoons}} R^1\overset{+OH}{\overset{\|}{C}}O{-}CH_2OR^2 \xrightarrow[-R^1COOH]{\text{slow}} \qquad (51)$$

$$\underset{\textbf{(45)}}{\left\{ CH_2{=}\overset{+}{O}R^2 \longleftrightarrow \overset{+}{C}H_2{-}OR^2 \right\}} \xrightarrow[+H_2O]{\text{fast}} H^+ + HOCH_2OR^2 \xrightarrow{\text{fast}} H^+ + HCHO + R^2OH$$

(45) is greatly stabilized by its mesomeric structure[336]. Salomaa[335, 337, 338] presented the following evidence for this mechanism in the case of methoxymethyl and ethoxymethyl acetates: the rate of hydrolysis is more closely related to h_0 than to $[H^+]$ (Bunnett's w values[293] vary from -0.67 to -3.57 and ϕ values[295] from -0.06 to -0.44), the reactions in methanol–water or ethanol–water mixtures yield acetals in addition to formaldehyde, and the values of activation energy (ca. 22·5 kcal/mole) and frequency

factor ($\log A = 14$) are high. In the case of methoxymethyl formate the $A_{AC}2$ mechanism takes place concurrently ($w = +1.24$, $\phi = +0.15$, the Arrhenius equation is not followed accurately). Similarly Salomaa and Linnantie[339] found that the hydrolysis of β-chloroethoxymethyl formate took place predominantly by the $A_{AC}2$ mechanism (accompanied by 30% $A_{AL}1$ in water at 25°C), but that of the acetic ester by the $A_{AL}1$ mechanism. Salomaa[340] observed further that the effect of R in the $A_{AL}1$ hydrolysis of alkoxymethyl acetates CH_3COOCH_2OR correlated well with its effect both in the acid-catalysed hydrolysis of dialkoxymethanes $CH_2(OR)_2$ and in the ethanolysis of alkoxymethyl chlorides $ROCH_2Cl$, which is to be expected because the same alkoxymethyl ion $CH_2\overset{+}{=}OR$ is formed in all these cases. The solvent isotope effect[341] for the acid hydrolysis of methoxymethyl acetate in water at 25°C ($k_{D_2O}/k_{H_2O} = 2.23$) is higher than that for an $A_{AC}2$ hydrolysis but lower than for the hydrolysis of acetals (2.8).

In the case of cyclic acetal esters, 1,3-dioxolone-4

$$H_2C_{(2)}\begin{array}{c} \diagup O-C=O \\ | \\ \diagdown O-CH_2 \end{array}$$

and its methyl-substituted derivatives[342], the $A_{AL}1$ hydrolysis is so much retarded on conformational grounds that dioxolones that have an unsubstituted $C_{(2)}$ atom hydrolyse by the $A_{AC}2$ mechanism ($\Delta H^\ddagger = 16$ kcal/mole, $\Delta S^\ddagger = -21$ to -26 E.U., $w = +4.8$). On the other hand, 2, methyl-substituted dioxolones have values of $\Delta H^\ddagger$ (17–20 kcal/mole)-$\Delta S^\ddagger$ (-4 to -9 E.U.), w (about 1), and k_{D_2O}/k_{H_2O} (1.68) which are between those for the $A_{AC}2$ and $A_{AL}1$ hydrolyses. However, as they hydrolyse much slower than the corresponding dioxolanes in contrast to the corresponding open-chain compounds, these mechanisms cannot take place concurrently. It was concluded that the reaction involves a water–carbonyl addition equilibrium followed by a rate-limiting unimolecular decomposition of the intermediate formed or that the rate-limiting stage is the addition of water[342]. In the case of γ-ethoxy-γ-butyrolactone the reaction takes place by the $A_{AL}1$ mechanism (Fife[343]) because in this case there are no stereochemical reasons which would retard the unimolecular reaction.

B. Unimolecular Solvolysis ($B_{AL}1$)

The unimolecular hydrolysis of neutral ester molecules by alkyl-oxygen fission ($B_{AL}1$) is in fact a unimolecular nucleophilic substitution reaction

(S_N1) at the saturated alkyl carbon atom. The mechanism can be written as equation (52).

$$R^1COOR^2 \underset{+R^1COO^-,\ \text{fast}}{\overset{\text{slow}}{\rightleftharpoons}} {}^+R^2 \underset{\text{slow}}{\overset{\text{fast}}{\underset{}{\overset{+H_2O}{\rightleftharpoons}}}} R^2\overset{+}{O}H_2 \xrightarrow[\text{fast}]{+R^1COO^-} R^1COOH + R^2OH \quad (52)$$

Electropositive groups in R^2 and electronegative groups in R^1 should accelerate the reaction, which should not be sensitive to steric retardation. The reaction rate should decrease rapidly with diminishing polarity of the solvent[6].

The first evidence for alkyl-oxygen fission in ester hydrolysis was obtained by Kenyon and coworkers[344], who found that the hydrolysis of optically active hydrogen phthalates of 1,3-dimethylallyl, 1-methyl-3-phenylallyl, and 1-phenyl-3-methylallyl alcohols gave in concentrated sodium hydroxide solutions optically pure alcohols, but in aqueous sodium carbonate solutions inactive alcohols. In the case of the phenylmethylallyl esters, in addition to their parent alcohols, the isomers formed by allylic rearrangement were also obtained by using low hydroxide ion concentrations. The explanation is (equation 53) that when $[OH^-]$ is high the me-

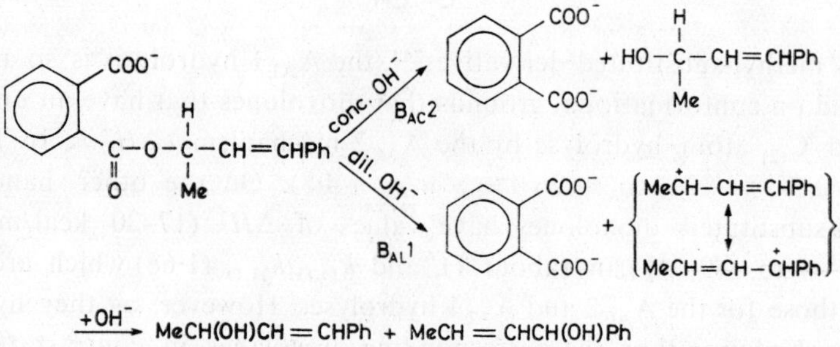

COO⁻
H
C—O—C—CH=CHPh
O
Me
conc. OH⁻
B_AC2
COO⁻
+ HO—C—CH=CHPh
COO⁻
Me
dil. OH⁻
B_AL1
COO⁻
COO⁻
MeĊH—CH=CHPh
MeCH=CH—ĊHPh
+OH⁻
MeCH(OH)CH=CHPh + MeCH=CHCH(OH)Ph

(53)

(46)

chanism is $B_{AC}2$ but when it is low the mesomeric cation (**46**) is produced by the $B_{AL}1$ mechanism[324].

The reaction products in the hydrolysis of an ester by the $B_{AL}1$ and $B_{AC}2$ mechanisms are the same but in alcoholysis the carbonium ion ${}^+R^2$ formed in the $B_{AL}1$ solvolysis reacts with the alcohol R^3OH producing the ether R^2OR^3 rather than the ester R^1COOR^3 which would have formed if acyl-oxygen fission had taken place. Thus triphenylmethyl benzoate was found by Hammond and Rudesill[345] to react with ethanol yielding ethyl triphenylmethyl ether in the reaction (54), whose rate was independent of

$$PhCOOCPh_3 \xrightarrow{-PhCOO^-} Ph_3C^+ \xrightarrow{+EtOH} Ph_3COEt + H^+ \quad (54)$$

sodium ethoxide apart from a salt effect. For further references to early, mainly non-kinetic studies, see the review of Davies and Kenyon[324].

In Table 14 examples of esters which have been found to undergo a $B_{AL}1$ hydrolysis are listed together with kinetic data. Structural and solvent effects are seen to be in accordance with the above-mentioned expectations. The rate of reaction is the higher the more stabilized is the the carbonium ion formed. Especially strong stabilization is caused by cyclopropyl groups, e.g. tricyclopropylmethyl benzoate is hydrolysed much more rapidly than triphenylmethyl benzoate, the rate being 10^7 times higher than that of triisopropylmethyl benzoate[352]. The strain release obtained in the $B_{AL}1$ solvolyses of esters with bicyclic fused cyclopropyl rings in the alkyl component makes them still much more reactive. Thus bicyclo[2.1.0]-pentane-1-methyl p-nitrobenzoate was found by Dauben and Wiseman[392] to hydrolyse in 60% acetone–water at 50°c to yield 3-methylenecyclopentanol with a rate 400,000 times higher than that of cyclopropylmethyl p-nitrobenzoate. Electronegative substituents in the acyl components of p-methoxybenzhydryl benzoates increase the rate considerably; the reaction gives a good correlation with Hammett's σ constants with the value $+1·8$ for ϱ in 89% acetone–water at 90 and 70°c[354].

The hydrolysis has in some cases been shown not to be catalysed by sodium hydroxide or other bases (triphenylmethyl acetate[323], t-butyl 2,4,6-triphenylbenzoate[26]). Formation of unsaturated compounds by elimination accompanying the hydrolysis is demonstrated in the case of t-alkyl esters of 2,4,6-triphenylbenzoic acid[26] and trifluoroacetic acid[346]. Activation energy for the hydrolysis is usually high and the activation entropy nearly zero (Table 14). In one case (the hydrolysis of benzhydryl p-nitrobenzoate in 70% acetone–water by Fox and Kohnstam[349]) accurate measurements were performed at several temperatures; it was noticed that the Arrhenius equation was not followed and the heat capacity of activation $\Delta C_p^{\ddagger}$ was found to be negative, which is in accordance with other reactions when a polar transition state is formed from a neutral molecule. Solvent polarity influences the rate strongly as in S_N1 reactions in general. Thus the, in the high water content region, approximately linear part of the plot of $\log k_0$ versus $\log [H_2O]$ for the hydrolysis of t-butyl hydrogen o-phthalate in acetone–water mixtures was found to have a slope of $5·8$[347]. The solvent isotope effect is small; the ratio k_{H_2O}/k_{D_2O} was found by Martin and Scott[393] to be $1·21$ for the hydrolysis of t-butyl trifluoroacetate in water at 10°c.

TABLE 14. Kinetic data for the hydrolysis of esters by the $B_{AL}1$ mechanism at 25°C (w = water, a = acetone, d = dioxane)

Ester	Solvent	$10^6 k_0$ (sec⁻¹)	$\Delta H^‡$ (kcal/mole)	$\Delta S^‡$ (e.u.)	$\Delta C_p^‡$	Reference
CF₃COOBu-t	70% a–w	0·266	23·8	−4·2	—	346
2,4,6-Ph₃C₆H₂COOBu-t	70% d–w	—	31·8	−3·3	—	26
o-HOOCC₆H₄COOBu-t	w	5·06[a]	26·9	+2·5	—	347
p-ClC₆H₄OCH₂COOCHPhCH=CH₂	20% EtOH–w	7·8	—	—	—	348
p-NO₂C₆H₄COOCHPh₂	70% a–w	0·0032[b]	28·2[b]	−5·9[b]	−26[b]	349
CH₃COOCPh₃	80% d–w	—	23·0	−3·7	—	323
	80% a–w	1·45	—	—	—	350
	50% a–w	72	—	—	—	350
PhCOOCPh₃	50% a–w	120	—	—	—	351
PhCOOC(cycloPr)₃	95% d–w	123	—	—	—	352
CH₃COOCH₂OMe	w	1·75[c]	21·2	−11·4	—	341
CH₃COOCHMeOMe	w	269[d]	—	—	—	341
CH₃ClCOOCH₂OMe	w	16·4	20·0	−8·6	—	341
cis-2,5-dimethyldioxolone	w	58·0	18·2	−12·4	—	353
γ-EtO-γ-butyrolactone	w	34[e]	16·9[e]	−18·5[e]	—	343

[a] At 60°C.
[b] At 50°C.
[c] At 35°C.
[d] At 5°C.
[e] At 30°C.

C. Bimolecular Hydrolysis ($B_{AL}2$)

The bimolecular hydrolysis of esters by alkyl-oxygen fission ($B_{AL}2$), which is actually a bimolecular nucleophilic substitution (S_N2) at the saturated alkyl carbon atom, is an unusual reaction because the carbonyl carbon atom is more susceptible to nucleophilic attack, and hydrolysis by the $B_{AC}2$ mechanism is thus more rapid and masks the $B_{AL}2$ mechanism (55).

$$R^1\overset{\overset{O}{\|}}{C}O\!-\!OR^2 + H_2O \underset{}{\overset{slow}{\rightleftharpoons}} R^1\overset{\overset{O}{\|}}{C}O^- + R^2OH_2^+ \overset{fast}{\longrightarrow} R^1COOH + R^2OH \qquad (55)$$

The first system in which the $B_{AL}2$ mechanism was observed was the hydrolysis of β-lactones. Optically active β-malolactonic acid was observed by Cowdrey and coworkers[355] and β-butyrolactone by Olson and Miller[312] to undergo hydrolysis in neutral and slightly acid solutions with inversion of configuration. Alkyl-oxygen fission was confirmed in the last-mentioned case by Olson and Hyde[313] by using the ^{18}O method. Similar kinetic behaviour in the case of β-propiolactone indicated the same mechanism (Long and Purchase[314]). It was confirmed by the observation of Bartlett and Rylander[356] that the reaction product from β-propiolactone and methanol in initially neutral solution was methoxypropionic acid rather than methyl β-hydroxypropionate. The exceptional behaviour of β-lactones must be due to greater release of the ring strain by the S_N2 attack on the alkyl carbon atom than by the hydration of the carbonyl group in the $B_{AC}2$ mechanism[5]. In the hydrolysis of β-butyrolactone a general base catalysis has been observed in both the $B_{AL}2$ and $B_{AC}2$ mechanisms by carbonate, borate, and phosphate ions[312, 357]. The activation energy and entropy of the $B_{AL}2$ hydrolysis of β-propiolactone[314] are relatively high ($E = 19.5$ kcal/mole, $\Delta S^{\ddagger} = -15$ E.U.). Deuterium oxide solvent isotope effect is small ($k_{H_2O}/k_{D_2O} = 1.15$, Butler and Gold[358]).

If the rate of the $B_{AC}2$ reaction of an ester is sufficiently diminished by steric hindrance, the existence of a $B_{AL}2$ hydrolysis might become observable. Although the alkaline hydrolysis of methyl 2,4,6-trimethylbenzoate was shown to take place by the $B_{AC}2$ mechanism (Bender and Dewey[359]), the still more hindered esters, methyl 2,4,6-tri-t-butylbenzoate and methyl 2-methyl-4,6-di-t-butylbenzoate, were found by Barclay, Hall and Cooke[360] to follow second-order kinetics and to hydrolyse with alkyl-oxygen fission. The reaction of methyl 2,4,6-tri-t-butylbenzoate in 90% methanol–water was slow ($k_{OH} = 1.2 \times 10^{-4}$ M^{-1} sec^{-1} at 95°C) and had high values of activation energy (25.1 kcal/mole) and frequency factor (log $A = 10.93$).

A special case of the $B_{AL}2$ mechanism, which is not an S_N2 reaction, is the hydrolysis of 4(5)-hydroxymethylimidazolyl acetate (47) studied by Bruice and coworkers[215, 361]. The proposed mechanism (56), which leads to second-order kinetics, is mechanistically an elimination (*E*lcB) reaction

$$
\underset{(47)}{
\begin{array}{c}
\text{CH}_2-\text{OCCH}_3 \\
\end{array}
}
\quad
\underset{+\text{H}^+}{\overset{K}{\rightleftarrows}}
\quad
\begin{array}{c}
\text{CH}_2-\text{OCCH}_3 \\
\end{array}
\quad
\overset{k}{\longrightarrow}
\tag{56}
$$

$$
\begin{array}{c}
\text{CH}_2 \\
\end{array}
+ \text{CH}_3\text{COO}^-
\quad \xrightarrow{+\text{H}_2\text{O}} \quad
\begin{array}{c}
\text{CH}_2\text{OH} \\
\end{array}
+ \text{CH}_3\text{COOH}
$$

followed by addition of water or, if only the rate-limiting step is considered, a $B_{AL}1$ reaction of the anionic form of the ester.

VI. HYDROLYSES INVOLVING NOT ONLY THE CARBOXYL GROUP

This section contains a discussion of some hydrolytic reactions of esters in which neither the acyl-oxygen nor the alkyl-oxygen bond is the first one to rupture. These reactions do not therefore belong to any of the mechanistic groups of Ingold's classification.

A. Dialkyl Dicarbonates

Dialkyl dicarbonates (48) are esters as well as acid anhydrides. Their hydrolysis was recently studied by Kivinen[362, 363], who also reviewed the earlier literature. The rate of reaction is only slightly influenced by added acids. The kinetic data obtained for the neutral hydrolysis of diethyl dicarbonate ($k_0 = 4 \cdot 2 \times 10^{-4}$ sec^{-1}, $\Delta H^{\ddagger} = 12 \cdot 9$ kcal/mole, $\Delta S^{\ddagger} = -39$ E.U., $\Delta C_p^{\ddagger} = -45$ cal mole^{-1} deg^{-1}, $k_{\text{H}_2\text{O}}/k_{\text{D}_2\text{O}} = 2 \cdot 50$, all in water at 25°C) and solvent effects resemble those for neutral hydrolysis of ordinary carboxylic esters (section III. A) and of open-chain acid anhydrides. It was concluded that the reaction leads to a transition state like (49), which is

$$
\underset{(48)}{
\begin{array}{c}
\text{O} \quad\; \text{O} \\
\| \quad\;\; \| \\
\text{R}^1\text{O}-\text{C}-\text{O}-\text{C}-\text{OR}^2 \\
\end{array}
}
\underset{\substack{+\text{O}-\text{H} \\ | \\ \text{H}+\text{H}_2\text{O}}}{\rightleftharpoons}
\left[
\begin{array}{c}
\text{O} \quad\; \text{O} \\
\vdots \quad\;\; \| \\
\text{R}^1\text{O}-\text{C}-\text{OCOR}^2 \\
\vdots \\
\text{O}\cdots\text{H}\cdots\text{OH}_2 \\
| \\
\text{H}
\end{array}
\right]
\underset{(49)}{}
\rightleftharpoons
\underset{\substack{\text{OH}+\text{H}_3\text{O}^+}}{
\begin{array}{c}
\text{O}^- \quad \text{O} \\
| \quad\;\; \| \\
\text{R}^1\text{O}-\text{C}-\text{OCOR}^2 \\
\end{array}
}
\tag{57}
$$

$$
\longrightarrow \text{R}^1\text{OH} + \text{R}^2\text{OH} + 2\,\text{CO}_2 + \text{H}_2\text{O}
$$

similar to that proposed for neutral ester hydrolysis (equation 23). It cannot be decided on the basis of kinetic data or from reaction products whether the ester or anhydride bond is ruptured in the product-forming reaction but the last-mentioned possibility seems to be more likely.

Also the hydrolysis of ordinary acid anhydrides could be considered as a special case of ester hydrolysis because acid anhydrides, $R^1COOCOR^2$, may be regarded as esters with the carbonyl group R^2CO as alkyl component. A discussion of anhydride hydrolysis is, however, outside the scope of this chapter.

B. Alkyl Haloformates

The chloroformates ClCOOR are esters as well as acyl chlorides. Their solvolytic reactions are much slower than those of ordinary acyl chlorides (e.g. the rate coefficients for the alcoholysis of acetyl chloride[364] and ethyl chloroformate[365] in ethanol at 25°C are 0·147 and $2·11 \times 10^{-5}$ sec^{-1}, respectively) because the ground state is stabilized by resonance (58).

$$R-O-\overset{\overset{\displaystyle O}{\|}}{C}-Cl \longleftrightarrow R-\overset{+}{O}=\overset{\overset{\displaystyle O^-}{|}}{C}-Cl \tag{58}$$

The rates for the alcoholysis of alkyl chloroformates in methanol were found by Leimu[366] to vary with R in the order $CH_2ClCH_2 \gg Me > Et \approx$ n-Pr $>$ i-Pr. Ethyl fluoroformate was found by Hudson and Green[367] to hydrolyse in 85% acetone–water about 30 times faster than the chloroformate. Hydroxide ion catalyses the reaction strongly (k_{OH}/k_0 for the hydrolysis of ethyl chloroformate at 0°C is 3×10^7 in 82% acetone–water and 3×10^6 in 15% acetone–water[368]). Mercuric ion has no influence on the rate of hydrolysis of n-butyl chloroformate in 50 vol.% dioxane–water indicating that the S_N1 mechanism is of no importance under these conditions (Hall and Lueck[369]). Nucleophilic catalysis in the hydrolysis of ethyl chloroformate in 85% acetone–water was observed by Hudson and Green[367], the order in rates being: acetoxime $>$ OH$^-$ $>$ PhO$^-$ $>$ NO$_2^-$ $>$ N$_3^-$ $>$ F$^-$ $>$ H$_2$O $>$ Br$^-$, I$^-$, CNS$^-$. Catalysis by para-substituted phenols was found[370] to follow Brønsted's law (22) with the value of 0·78 for β.

The above-mentioned observations are most easily understood if the reaction of primary chloroformates, at least in solvents of low ionizing power, takes place by a bimolecular addition–elimination mechanism (59).

$$RO-\overset{\overset{\displaystyle O}{\|}}{C}-Cl+H_2O \underset{fast}{\overset{slow}{\rightleftharpoons}} RO-\overset{\overset{\displaystyle O^-}{|}}{\underset{\underset{\displaystyle +OH_2}{|}}{C}}-Cl \overset{fast}{\longrightarrow} RO-\overset{\overset{\displaystyle O}{\|}}{\underset{\underset{\displaystyle OH}{|}}{C}}+Cl^-+H^+ \tag{59}$$

If the addition of water is general base catalysed, the slow step is identical with that in neutral ester hydrolysis (23). One can estimate that the observed rate for ethyl chloroformate ($k_0 = 3 \cdot 89 \times 10^{-4}$ sec^{-1} in water at $25°C^{365}$) is of the same order of magnitude that a $B_{AC}2$ reaction would have (the corresponding value is $7 \cdot 95 \times 10^{-4}$ sec^{-1} — see Table 10 — for methyl trichloroacetate[145] with an acyl group of presumably nearly equal electronegativity). As chlorine is a better leaving group than an alkoxyl group, the reaction of alkyl chloroformates takes place by acyl halide rather than by ester hydrolysis.

Other kinetic data for the hydrolysis of ethyl chloroformate in water ($E = 17 \cdot 9$ kcal/mole, $\Delta S^{\ddagger} = -24$ E.U., $k_{H_2O}/k_{D_2O} = 1 \cdot 95$, $\partial(\log k_0)/\partial(\log [H_2O]) = 2 \cdot 3 - 2 \cdot 4$ for different organic solvent–water mixtures, according to Kivinen[363, 365]) differ considerably from those for neutral ester hydrolysis (Table 10). The small influence of solvent composition on rate resembles that in an S_N2 reaction, but the kinetic data and structural effects are difficult to explain on this basis. The most likely explanation seems to be that in solvents of low polarity the addition–elimination mechanism (59) predominates (in 80 vol.% acetone–water the kinetic data for ethyl chloroformate[365], $\Delta H^{\ddagger} = 14 \cdot 3$ kcal/mole and $\Delta S^{\ddagger} = -38$ E.U., resemble those for neutral ester hydrolysis). When the solvent polarity increases, the relative importance of bond-breaking increases and the free energy difference between transition and ground states diminishes less than in the case of neutral ester hydrolysis. In solvents of high polarity (*e.g.* in 99% formic acid–water mixture the rate order is ClCOOMe < ClCOOEt < ClCOO-Pr-i and E for ethyl chloroformate is 25 kcal/mole[368]) the mechanism changes to a reaction with rate–limiting ionization. This mechanism already prevails in 65% acetone–water in the case of chloroformates of secondary alcohols according to Green and Hudson[371], who prefer mechanism (60) for the reaction. Minato[372] denies, however, the possibility of acylium ion

$$\underset{\underset{\text{ROC—Cl}}{\overset{\text{O}}{\|}}}{} \rightleftharpoons \text{ROC}{=}\text{O} + \text{Cl}^- \xrightarrow{+H_2O} \begin{cases} H_3O^+ + \text{olefin} + CO_2 + Cl^- \\ R\overset{+}{O}H_2 + CO_2 + Cl^- \\ HCl + ROCOOH \longrightarrow ROH + CO_2 \end{cases} \quad (60)$$

formation on the ground of the relatively slow rate of hydrolysis of alkyl chloroformates as compared with acetyl and benzoyl chloride.

Queen[394] has recently determined the thermodynamic parameters of activation for the hydrolysis of several alkyl chloroformates in water. The observed small values for $\Delta H^{\ddagger}$, $\Delta S^{\ddagger}$, $\Delta C_p^{\ddagger}$, and k_{D_2O}/k_{H_2O} for phenyl and methyl chloroformates are in accordance with the addition–elimination mechanism whereas the observed increase of the relatively large values of $\Delta C_p^{\ddagger}$ with temperature in the case of ethyl and propyl chloroformates is probably an indication of a change in mechanism toward one with higher activation enthalpy.

C. α-Haloalkyl Esters

Drushel and Bancroft[373] stated that α-chloroethyl acetate and propionate undergo hydrolysis at such a high rate that a kinetic study is impossible. The hydrolysis of α-haloalkyl esters $R^1COOCXR^2R^3$ (X is Cl, Br, or I, R's are alkyl or aryl groups or hydrogens) was not thereafter studied kinetically until after 1956 when the present author performed a series of investigations. Böhme and coworkers[374] determined the rate of hydrolysis of chloromethyl acetate and benzoate. The reaction can take place either as an ester hydrolysis or as a nucleophilic substitution of halogen (equation 61), which in both cases lead to the same products. The alkaline

$$
R^1COOCR^2R^3 + H_2O \begin{array}{c} \nearrow^{slow} \\ \searrow_{slow} \end{array}
\begin{array}{c} R^1COOH + HO-CR^2R^3 \\ | \\ X \\ R^1COOCR^2R^3 + HX \\ | \\ OH \end{array}
\begin{array}{c} \searrow^{fast} \\ \nearrow_{fast} \end{array}
R^1COOH + HX + R^2R^3CO \qquad (61)
$$

hydrolysis takes place by the $B_{AC}2$ mechanism[14, 375] (cf. Table 4) and the acid hydrolysis either by the normal $A_{AC}2$ mechanism[141, 150, 268, 376] (cf. Table 13) or by the exceptional reaction of esters that have negative substituents[300, 305] (section IV. D). These reactions are accompanied by a relatively fast uncatalysed hydrolysis (Figure 1) which can be either neutral ester hydrolysis[13, 141, 150] (section III. A, Table 10), if the ester has several electronegative substituents, or displacement of halogen. These mechanisms can easily be distinguished experimentally because neutral ester hydrolysis has especially low activation enthalpy and entropy (Table 10) as compared with those for nucleophilic substitution reactions. When both reactions occur concurrently, the activation energy is found to increase with increasing temperature[377].

Nucleophilic displacement of halogen can take place either as a bimolecular S_N2 or a unimolecular S_N1 reaction. The hydrolysis of halomethyl esters has been found to have more or less S_N2 character[150] as can be seen, e.g., from the relatively slight influence of solvent polarity on rate (Figure 10). The mechanism can thus be written in the form of equation (62).

$$
RCOOCH_2X + H_2O \xrightarrow{slow} RCOOCH_2OH + HX \xrightarrow{fast} RCOOH + CH_2O + HX \qquad (62)
$$

37 C.C.A.E

Typical kinetic data for this S_N2 or borderline solvolysis are exemplified by the following values for the hydrolysis of chloromethyl acetate[13, 150]
in water at 25°C: $k_0 = 1.53 \times 10^{-5}$ sec^{-1}, $\Delta H^{\ddagger} = 22.8$ kcal/mole,
$\Delta S^{\ddagger} = -4$ E.U., $\Delta C_p^{\ddagger} = -62$ cal mole^{-1} deg^{-1}, and $\partial(\log k_0)/\partial(\log$
$[H_2O]) = 3.3$. The structure of the acyl component R of the esters

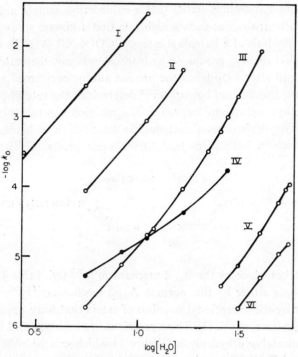

FIGURE 10. The logarithmic dependence of the rate coefficient on molar water
content in the solvolysis of α-haloalkyl esters in acetone–water mixtures at 25°C
(Euranto and coworkers[13, 141, 150, 378]). I: α-chloroisopropyl acetate (S_N1), II:
α-chlorocyclohexyl acetate (S_N1), III: α-chloroethyl acetate (S_N1), IV: α-chloroisopropyl trichloroacetate ($B_{AC}2$), V: chloromethyl chloroacetate ($B_{AC}2$), VI:
chloromethyl acetate (S_N2, 1)

RCOOCH$_2$Cl has a relatively slight influence on the rate, the observed
order in rates being[150, 379]: Ph $\ll$ Me $<$ Et $\approx$ n-Pr. Bromomethyl acetate
has been found to hydrolyse 40–70 times faster than chloromethyl acetate.
Salt effects are usually small but the rate of hydrolysis of chloromethyl
acetate is considerably increased by added bromide or iodide ions. This
effect can be interpreted as nucleophilic catalysis because rapidly reacting
bromo- or iodomethyl acetate is formed from chloromethyl acetate by a

nucleophilic halogen exchange that accompanies the hydrolysis. Corresponding nucleophilic inhibition by chloride ion is observed in the hydrolysis of bromomethyl acetate and by perchlorate, nitrate, and sulphate ions in the hydrolysis of chloromethyl acetate[150].

Substitution of one of the α hydrogen atoms of the halomethyl group of halomethyl esters by an alkyl group leads to a change to the S_N1 mechanism (63) as indicated by the over thousandfold increase in rate and

$$R^1COOCR^2R^3X \xrightarrow{slow} \left\{ R^1\overset{O}{\overset{\|}{C}}\!-\!O\!-\!\overset{+}{C}R^2R^3 \longleftrightarrow R^1\overset{O}{\overset{\|}{C}}\!-\!\overset{+}{O}\!=\!CR^2R^3 \right\} + X^-$$

$$(63)$$

$$\xrightarrow[\text{fast}]{+H_2O} R^1COOC(OH)R^2R^3 + HX \xrightarrow{fast} R^1COOH + HX + R^2R^3CO$$

by more pronounced solvent effects in the hydrolysis[150] (Figure 10). The kinetic data for the hydrolysis of α-chloroethyl acetate in 40 wt.% acetone–water at 25°C are typical for this reaction[150]: $k_0 = 1\cdot95\times10^{-3}$ sec^{-1}, $\Delta H^{\ddagger} = 18\cdot0$ kcal/mole, $\Delta S^{\ddagger} = -10\cdot6$ E.U., and $\partial(\log k_0)/\partial(\log [H_2O]) = 5\cdot9$. Added salts in general increase the rate but in the case of common-ion salts a small mass-law effect is observed[150, 376, 379].

The most pronounced structural effect is the above-mentioned large increase in rate when α hydrogens are substituted by methyl groups; the relative rates (not measured under the same conditions, see Figure 10) of CH_3COOCH_2Cl, $CH_3COOCHMeCl$, and $CH_3COOCMe_2Cl$ are about 1, 10^3, and 10^6, respectively[150, 378]. Variation in the acyl component R^1 of esters $R^1COOCHMeCl$[376, 379] and $R^1COOCMe_2Cl$[378] causes the expected variation in the rate of hydrolysis: $CH_2Cl < H < Ph < Me < Et$. On the other hand, variation of R^2 in the alkyl component of esters $CH_3COO\text{-}CMeR^2Cl$ and $CH_3COOCEtR^2Cl$ gives the rate order[378] Me > n-Pr > Et > i-Pr, which is almost the opposite to that expected on the basis of inductive effects, so that the Baker–Nathan order is followed and the hyperconjugation effect determines the rate sequence. Also in the case of esters $CH_3COOCR^2R^3Cl$ with R^2 and R^3 being H, CH_2Cl, or CH_3, the almost equal rates of hydrolysis for R^3 being H or CH_2Cl can be understood only when taking into account the number of β hydrogens that are able to contribute to the Baker–Nathan effect[141]. The influence of the halogen on rate is similar as in the S_N2 reaction: the bromoesters hydrolyse 25–75 times as fast as the corresponding chloroesters[376, 378].

VII. REFERENCES

1. M. Berthelot and L. Péan de Saint-Gilles, *Ann. Chim. Phys.*, [3] **65**, 385 (1862); [3] **66**, 5 (1862); [3] **68**, 225 (1863).
2. C. M. Guldberg and P. Waage, *Avhandl. Norske Videnskaps-Akad. Oslo, I. Mat. Naturv. Kl.*, 35 (1864).
3. M. L. Bender, *Chem. Rev.*, **60**, 53 (1960).
4. L. P. Hammett, *Physical Organic Chemistry*, McGraw-Hill, New York, 1940, Chap. 4, 6, 7 and 9.
5. J. Hine, *Physical Organic Chemistry*, 2nd ed., McGraw-Hill, New York, 1962, Chap. 12.
6. C. K. Ingold, *Structure and Mechanism in Organic Chemistry*, Cornell University Press, Ithaca, New York, 1953, Chap. 14. 47, pp. 751–782.
7. R. Wegscheider, *Z. Physik. Chem. (Leipzig)*, **41**, 52 (1902).
8. A. Skrabal and A. Zahorka, *Monatsh. Chem.*, **53–54**, 562 (1929).
9. P. Salomaa, *Suomen Kemistilehti*, **32 B**, 81 (1959).
10. H. M. Dawson and W. Lowson, *J. Chem. Soc.*, 2146 (1928).
11. E. K. Euranto and A.-L. Moisio, *Suomen Kemistilehti*, **37 B**, 92 (1964).
12. N. J. Cleve and E. K. Euranto, *Suomen Kemistilehti*, **37 B**, 126 (1964).
13. E. K. Euranto and N. J. Cleve, *Acta Chem. Scand.*, **17**, 1584 (1963).
14. E. K. Euranto and R. A. Euranto, *Suomen Kemistilehti*, **35 B**, 96 (1962).
15. A. Skrabal, *Z. Elektrochem.*, **33**, 322 (1927).
16. S. C. Datta, J. N. E. Day and C. K. Ingold, *J. Chem. Soc.*, 838 (1939).
17. E. D. Hughes, C. K. Ingold and S. Masterman, *J. Chem. Soc.*, 840 (1939).
18. J. N. E. Day and C. K. Ingold, *Trans. Faraday Soc.*, **37**, 686 (1941).
19. B. Holmberg, *Chem. Ber.*, **45**, 2997 (1912).
20. E. H. Ingold and C. K. Ingold, *J. Chem. Soc.*, 756 (1932).
21. C. Prévost, *Ann. Chim. (Paris)*, **10**, 147 (1928).
22. O. R. Quayle and H. M. Norton, *J. Am. Chem. Soc.*, **62**, 1170 (1940).
23. M. Polanyi and A. L. Szabo, *Trans. Faraday Soc.*, **30**, 508 (1934).
24. F. A. Long and L. Friedman, *J. Am. Chem. Soc.*, **72**, 3692 (1950).
25. D. N. Kursanov and R. V. Kudryavtsev, *Zh. Obshch. Khim.*, **26**, 1040 (1956).
26. C. A. Bunton, A. E. Comyns, J. Graham and J. R. Quayle, *J. Chem. Soc.*, 3817 (1955).
27. C. A. Bunton, G. Israel, M. M. Mhala and D. L. H. Williams, *J. Chem. Soc.*, 3718 (1958).
28. C. A. Bunton and T. Hadwick, *J. Chem. Soc.*, 3248 (1958).
29. C. A. Bunton and T. Hadwick, *J. Chem. Soc.*, 943 (1961).
30. C. A. Bunton, K. Khaleeluddin and D. Whittaker, *J. Chem. Soc.*, 3290 (1965).
31. M. L. Bender, *J. Am. Chem. Soc.*, **75**, 5986 (1953).
32. M. L. Bender, *J. Am. Chem. Soc.*, **73**, 1626 (1951).
33. M. L. Bender, R. D. Ginger and J. P. Unik, *J. Am. Chem. Soc.*, **80**, 1044 (1958).
34. M. L. Bender and R. D. Ginger, *Suomen Kemistilehti*, **33 B**, 25 (1960).
35. M. L. Bender and R. J. Thomas, *J. Am. Chem. Soc.*, **83**, 4189 (1961).
36. M. L. Bender, H. Matsui, R. J. Thomas and S. W. Tobey, *J. Am. Chem. Soc.*, **83**, 4193 (1961).
37. C. A. Bunton and D. N. Spatcher, *J. Chem. Soc.*, 1079 (1956).
38. S. L. Johnson, *Tetrahedron Letters*, 1481 (1964); *J. Am. Chem. Soc.*, **86**, 3819 (1964).
39. R. L. Burwell Jr. and R. G. Pearson, *J. Phys. Chem.*, **70**, 300 (1966).
40a. L. R. Fedor and T. C. Bruice, *J. Am. Chem. Soc.*, **86**, 5697 (1964); **87**, 4138 (1965).

40b. M. L. Bender and H. d'A. Heck, *J. Am. Chem. Soc.*, **89**, 1211 (1967).
41. S. G. Smith and M. O'Leary, *J. Org. Chem.*, **28**, 2825 (1963).
42. T. M. Lowry, *J. Chem. Soc.*, **127**, 1371 (1925).
43. R. B. Warder, *Chem. Ber.*, **14**, 1361 (1881).
44. H. Tsujikawa and H. Inoue, *Bull. Chem. Soc. Japan*, **39**, 1837 (1966).
45. A. H. Saadi and W. H. Lee, *J. Chem. Soc. (B)*, 1 (1966).
46. E. Sacher and K. J. Laidler, *Can. J. Chem.*, **42**, 2404 (1964).
47. A. Ågren, U. Hedsten, and B. Jonsson, *Acta Chem. Scand.*, **15**, 1532 (1961).
48. E. Tommila, A. Koivisto, J. P. Lyyra, K. Antell and S. Heimo, *Ann. Acad. Sci. Fennicae, Ser. A II*, No. 47, (1952).
49. E. Cohen and H. F. G. Kaiser, *Z. Physik. Chem. (Leipzig)*, **89**, 338 (1915).
50. K. J. Laidler and D. Chen, *Trans. Faraday Soc.*, **54**, 1026 (1958).
51. M. L. Bender and R. B. Homer, *J. Org. Chem.*, **30**, 3975 (1965).
52. L. W. Dittert, *Dissertation Abstr.*, **22**, 1837 (1961).
53. I. Christenson, *Acta Chem. Scand.*, **18**, 904 (1964).
54. E. Tommila, L. Brehmer and H. Elo, *Ann. Acad. Sci. Fennicae, Ser. A LIX*, No. 9, (1942).
55. H. H. Jaffé, *Chem. Rev.*, **53**, 191 (1953).
56. P. R. Wells, *Chem. Rev.*, **63**, 171 (1963).
57. J. E. Leffler and E. Grunwald, *Rates and Equilibria of Organic Reactions*, John Wiley and Sons, New York and London, 1963, Chap. 7.
58. S. Ehrenson in *Progress in Physical Organic Chemistry* (Ed. S. G. Cohen, A. Streitwieser, Jr. and R. W. Taft), Vol. 2, Interscience Publishers, New York, 1964, p. 195.
59. C. D. Ritchie and W. F. Sager in *Progress in Physical Organic Chemistry* (Ed. S. G. Cohen, A. Streitwieser, Jr. and R. W. Taft), Vol. 2, Interscience Publishers, New York, 1964, p. 323.
60. J. J. Ryan and A. A. Humffray, *J. Chem. Soc. (B)*, 842 (1966).
61. B. Jones and J. Robinson, *J. Chem. Soc.*, 3845 (1955).
62. P. R. Wells and W. Adcock, *Australian J. Chem.*, **19**, 221 (1966).
63. E. Tommila and C. N. Hinshelwood, *J. Chem. Soc.*, 1801 (1938).
64. F. H. Westheimer and R. P. Metcalf, *J. Am. Chem. Soc.*, **63**, 1339 (1941).
65. R. W. Taft, Jr., in *Steric Effects in Organic Chemistry* (Ed. M. S. Newman), John Wiley and Sons, London, 1956, Chap. 13, p. 556.
66. H. van Bekkum, P. E. Verkade and B. M. Wepster, *Rec. Trav. Chim.*, **78**, 815 (1959).
67. R. W. Taft, Jr. and I. C. Lewis, *Tetrahedron*, **5**, 210 (1959); R. W. Taft, Jr., S. Ehrenson, I. C. Lewis and R. E. Glick, *J. Am. Chem. Soc.*, **81**, 5352 (1959); R. W. Taft, Jr., *J. Phys. Chem.*, **64**, 1805 (1960); R. W. Taft, Jr. and I. C. Lewis, *J. Am. Chem. Soc.*, **81**, 5343 (1959).
68. R. O. C. Norman, G. K. Radda, D. A. Brimacombe, P. D. Ralph and E. M. Smith, *J. Chem. Soc.*, 3247 (1961).
69. Y. Yukawa, Y. Tsuno and M. Sawada, *Bull. Chem. Soc. Japan*, **39**, 2274 (1966).
70. E. Tommila and S. Hietala, *Acta Chem. Scand.*, **8**, 257 (1954).
71. A. Findlay and W. E. S. Turner, *J. Chem. Soc.*, **87**, 747 (1905).
72. E. Tommila, *Suomen Kemistilehti*, **25 B**, 37 (1952).
73. L. Pekkarinen, *Ann. Acad. Sci. Fennicae, Ser. A II*, No. 62, (1954).
74. E. A. Halonen, *Acta Chem. Scand.*, **9**, 1492 (1955).
75. W. J. Svirbely and I. L. Mador, *J. Am. Chem. Soc.*, **72**, 5699 (1950).
76. R. P. Bell and B. A. W. Coller, *Trans. Faraday Soc.*, **61**, 1445 (1965).
77. J. Barthel and G. Bäder, *Z. Physik. Chem. (Frankfurt)*, **48**, 109 (1966).
78. E. Tommila and H. Sternberg, *Suomen Kemistilehti*, **19 B**, 19 (1946).
79. W. P. Jencks and J. Carriuolo, *J. Am. Chem. Soc.*, **83**, 1743 (1961).

80. J. D. H. Homan, *Rec. Trav. Chim.*, **63**, 181 (1944).
81. A. Koivisto, *Ann. Acad. Sci. Fennicae, Ser. A II*, No. 73, (1956).
82. E. Tommila and S. Maltamo, *Suomen Kemistilehti*, **28 B**, 118 (1955).
83. A. Skrabal and A. M. Hugetz, *Monatsh. Chem.*, **47**, 17 (1926).
84. E. Tommila, S. Hietala and J. Nyrkiö, *Suomen Kemistilehti*, **28 B**, 143 (1955).
85. A. Skrabal and M. Belavić, *Z. Physik. Chem. (Leipzig)*, **103**, 451 (1923).
86. C. H. DePuy and L. R. Mahoney, *J. Am. Chem. Soc.*, **86**, 2653 (1964).
87. R. P. Bell and B. A. W. Coller, *Trans. Faraday Soc.*, **60**, 1087 (1964).
88. J. E. Leffler, *J. Org. Chem.*, **20**, 1202 (1955).
89. T. C. Bruice and T. H. Fife, *J. Am. Chem. Soc.*, **84**, 1973 (1962).
90. H. M. Humphreys and L. P. Hammett, *J. Am. Chem. Soc.*, **78**, 521 (1956).
91. R. Leimu, R. Korte, E. Laaksonen and U. Lehmuskoski, *Suomen Kemistilehti*, **19 B**, 93 (1946).
92. J. Barthel and G. Bäder, *Z. Physik. Chem. (Frankfurt)*, **48**, 114 (1966).
93. R. A. Fairclough and C. N. Hinshelwood, *J. Chem. Soc.*, 538 (1937).
94. E. Tommila and S. Maltamo, *Suomen Kemistilehti*, **28 B**, 73 (1955).
95. H. A. Smith and H. S. Levenson, *J. Am. Chem. Soc.*, **61**, 1172 (1939).
96. L. L. Schaleger and F. A. Long, in *Advances in Physical Organic Chemistry* (Ed. V. Gold), Vol. 1, 1963, p. 1.
97. R. W. A. Jones and J. D. R. Thomas, *J. Chem. Soc.*, *(B)*, 661 (1966).
98. C. M. Stevens and D. S. Tarbell, *J. Org. Chem.*, **19**, 1996 (1954).
99. K. Kindler, *Ann. Chem.*, **464**, 278 (1928).
100. E. Tommila, A. Nurro, R. Murén, S. Merenheimo and E. Vuorinen, *Suomen Kemistilehti*, **32 B**, 115 (1959); E. Tommila, *Ann. Acad. Sci. Fennicae, Ser. A 57*, No. 13, (1941); E. Tommila, J. Paasivirta, and K. Setälä, *Suomen Kemistilehti*, **33 B**, 187 (1960).
101. R. Huisgen and H. Ott, *Tetrahedron*, **6**, 253 (1959).
102. E. L. Eliel, *Stereochemistry of Carbon Compounds*, McGraw-Hill, New York, 1962.
103. N. B. Chapman, R. E. Parker and P. J. A. Smith, *J. Chem. Soc.*, 3634 (1960); E. A. S. Cavell, N. B. Chapman and M. D. Johnson, *J. Chem. Soc.*, 1413 (1960).
104. R. Gelin, S. Gelin and C. Boutin, *Compt. Rend.*, **262**, 1084 (1966).
105. R. Fuchs and J. A. Caputo, *J. Org. Chem.*, **31**, 1524 (1966).
106. R. W. Taft, Jr., *J. Am. Chem. Soc.*, **74**, 2729 (1952).
107. G. A. Ropp and V. F. Raaen, *J. Chem. Phys.*, **20**, 1823 (1952); **22**, 1223 (1954).
108. M. L. Bender and M. S. Feng, *J. Am. Chem. Soc.*, **82**, 6318 (1960).
109. E. A. Halevi and Z. Margolin, *Proc. Chem. Soc.*, 174 (1964).
110. W. F. K. Wynne–Jones, *Chem. Rev.*, **17**, 115 (1935).
111. T. C. Bruice, T. H. Fife, J. J. Bruno and P. Benkovic, *J. Am. Chem. Soc.*, **84**, 3012 (1962).
112. S. Arrhenius, *Z. Physik. Chem. (Leipzig)*, **1**, 110 (1887).
113. J. Spohr, *Z. Physik. Chem. (Leipzig)*, **2**, 194 (1888); S. D. Wilson and E. M. Terry, *J. Am. Chem. Soc.*, **50**, 1250 (1928); J. E. Quinlan and E. S. Amis, *J. Am. Chem. Soc.*, **77**, 4187 (1955).
114. F. A. Long and W. F. McDevit, *Chem. Rev.*, **51**, 119 (1952).
115. K. A. Connors and M. L. Bender, *J. Org. Chem.*, **26**, 2498 (1961).
116. E. F. J. Duynstee and E. Grunwald, *Tetrahedron*, **21**, 2401 (1965).
117. Fr. Becker and H. Hoffmann, *Z. Physik. Chem. (Frankfurt)*, **50**, 162 (1966).
118. L. Smith, *Z. Physik. Chem. (Leipzig)*, **177 A**, 131 (1936).
119. J. I. Hoppé and J. E. Prue, *J. Chem. Soc.*, 1775 (1957).
120. A. Indelli, G. Nolan, Jr. and E. S. Amis, *J. Am. Chem. Soc.*, **82**, 3237 (1960).
121. A. Indelli, *Trans. Faraday Soc.*, **59**, 1827 (1963).
122. A. Indelli, V. Bartocci, F. Ferranti and M. G. Lucarelli, *J. Chem. Phys.*, **44**, 2069 (1966).

123. R. F. Nielsen, *J. Am. Chem. Soc.*, **58**, 206 (1936).
124. R. P. Bell and F. J. Lindars, *J. Chem. Soc.*, 4601 (1954).
125. G. Aksnes and J. E. Prue, *J. Chem. Soc.*, 103 (1959).
126. R. F. Hudson, *J. Chem. Soc.*, *B*, 761 (1966).
127. E. S. Amis, *Solvent Effects on Reaction Rates and Mechanisms*, Academic Press, New York and London, 1966, Chap. 2 and 3.
128. S. Villermaux, J. Villermaux and R. Gibert, *J. Chim. Phys.*, **63**, 1356 (1966).
129. M. L. Bender and W. A. Glasson, *J. Am. Chem. Soc.*, **81**, 1590 (1959).
130. E. M. Arnett, W. G. Bentrude, J. J. Burke and P. McC. Duggleby, *J. Am. Chem. Soc.*, **87**, 1541 (1965).
131. E. Tommila, L. Takanen and K. Salonen, *Suomen Kemistilehti*, **31 B**, 37 (1958).
132. E. Tommila and I. Palenius, *Acta Chem. Scand.*, **17**, 1980 (1963); E. Tommila, *Ann. Acad. Sci. Fennicae Ser. A II*, No. 139, (1967); E. Tommila and M.–L. Savolainen, *Suomen Kemistilehti*, **40 B**, 212 (1967); E. Tommila and J. Martinmaa, *Suomen Kemistilehti*, **40 B**, 216 (1967).
133. A. J. Parker, *Quart. Rev. (London)*, **16**, 163 (1962).
134. E. Tommila and M.–L. Murto, *Acta Chem. Scand.*, **17**, 1947 (1963).
135. D. D. Roberts, *J. Org. Chem.*, **30**, 3516 (1965).
136. D. D. Roberts, *J. Org. Chem.*, **29**, 2039 (1964); **31**, 4037 (1966).
137. E. Tommila, *Suomen Kemistilehti*, **37 B**, 117 (1964).
138. W. Roberts and M. C. Whiting, *J. Chem. Soc.*, 1290 (1965).
139. T. C. Bruice and S. J. Benkovic, *Bioorganic Mechanisms*, Vol. I, W. A. Benjamin, New York, 1966, Chap. 1; S. L. Johnson in *Advances in Physical Organic Chemistry*, (Ed. V. Gold), Vol. 5, Academic Press, London, 1967, p. 237.
140. A. Skrabal and A. Zahorka, *Monatsh. Chem.*, **48**, 459 (1927).
141. E. Euranto, *Suomen Kemistilehti*, **35 B**, 18 (1962).
142. P. Salomaa and S. Laiho, *Acta Chem. Scand.*, **17**, 103 (1963).
143. A. Skrabal and M. Rückert, *Monatsh. Chem.*, **50**, 369 (1928).
144. E. J. Salmi and T. Suonpää, *Chem. Ber.*, **73**, 1126 (1940).
145. M. H. Palomaa, E. J. Salmi and R. Korte, *Chem. Ber.*, **72**, 790 (1939).
146. R. L. Heppolette and R. E. Robertson, *Can. J. Chem.*, **44**, 677 (1966).
147. R. E. Robertson, A. Stein and S. E. Sugamori, *Can. J. Chem.*, **44**, 685 (1966).
148. C. A. Bunton, N. A. Fuller, S. G. Perry and V. J. Shiner, *J. Chem. Soc.*, 2918 (1963).
149. A. Moffat and H. Hunt, *J. Am. Chem. Soc.*, **81**, 2082 (1959); **79**, 54 (1957).
150. E. Euranto, *Ann. Univ. Turku.*, *Ser. A I*, No. 31 (1959).
151. J. F. Kirsch and W. P. Jencks, *J. Am. Chem. Soc.*, **86**, 837 (1964).
152. E. K. Euranto, *Suomen Kemistilehti*, **38 A**, 25 (1965).
153. P. Salomaa, L. L. Schaleger and F. A. Long, *J. Am. Chem. Soc.*, **86**, 1 (1964).
154. B. M. Anderson, E. H. Cordes and W. P. Jencks, *J. Biol. Chem.*, **236**, 455 (1961).
155. D. M. Brouwer, M. J. v. d. Vlugt and E. Havinga, *Koninkl. Ned. Akad. Wetenschap.*, *Proc.*, *Ser.* **B 60**, No. 4 (1957).
156. W. P. Jencks and J. Carriuolo, *J. Biol. Chem.*, **234**, 1280 (1959).
157. J. M. Holland and J. G. Miller, *J. Phys. Chem.*, **65**, 463 (1961).
158. E. K. Euranto, *Nord. Kemistmötet, 11th Meeting 1962*, Turku, 1963, p. 239.
159. A. R. Butler and V. Gold, *J. Chem. Soc*, 1334 (1962).
160. D. G. Oakenfull, T. Riley and V. Gold, *Chem. Comm.*, 385 (1966).
161. J. M. Briody and D. P. N. Satchell, *J. Chem. Soc.*, 3778 (1965).
162. M. L. Bender, E. J. Pollock and M. C. Neveu, *J. Am. Chem. Soc.*, **84**, 595 (1962).
163. F. Covitz and F. H. Westheimer, *J. Am. Chem. Soc.*, **85**, 1773 (1963).
164. J. N. Brønsted, *Chem. Rev.*, **5**, 231 (1928).
165. S. L. Johnson, *J. Am. Chem. Soc.*, **84**, 1729 (1962).
166. M. L. Bender and B. W. Turnquest, *J. Am. Chem. Soc.*, **79**, 1656 (1957).

167. W. P. Jencks and J. Carriuolo, *J. Am. Chem. Soc.*, **82**, 1778 (1960).
168. T. C. Bruice and R. Lapinski, *J. Am. Chem. Soc.*, **80**, 2265 (1958).
169. K. Koehler, R. Skora and E. H. Cordes, *J. Am. Chem. Soc.*, **88**, 3577 (1966).
170. T. C. Bruice, J. J. Bruno and W.–S. Chou, *J. Am. Chem. Soc.*, **85**, 1659 (1963).
171. K. B. Wiberg, *J. Am. Chem. Soc.*, **77**, 2519 (1955).
172. M. L. Bender and M. C. Neveu, *J. Am. Chem. Soc.*, **80**, 5388 (1958).
173. G. R. Schonbaum and M. L. Bender, *J. Am. Chem. Soc.*, **82**, 1900 (1960).
174. E. J. Fuller, *J. Am. Chem. Soc.*, **85**, 1777 (1963).
175. W. Langenbeck and R. Mahrwald, *Chem. Ber.*, **90**, 2423 (1957).
176. T. C. Bruice and G. L. Schmir, *J. Am. Chem. Soc.*, **79**, 1663 (1957).
177. E. Haruki, T. Fujii and E. Imoto, *Bull. Chem. Soc. Japan*, **39**, 852 (1966).
178. M. Caplow and W. P. Jencks, *Biochem.*, **1**, 883 (1962).
179. T. C. Bruice and S. J. Benkovic, *J. Am. Chem. Soc.*, **86**, 418 (1964).
180a. J. F. Kirsch and W. P. Jencks, *J. Am. Chem. Soc.*, **86**, 833 (1964).
180b. D. H. Busch, *Advan. Chem. Ser.*, **37**, 1 (1963); M. L. Bender, *Advan. Chem. Ser.*, **37**, 19 (1963); M. M. Jones, *Advan. Chem. Ser.*, **49**, 153 (1965).
181. H. Kroll, *J. Am. Chem. Soc.*, **74**, 2036 (1952).
182. M. L. Bender and B. W. Turnquest, *J. Am. Chem. Soc.*, **79**, 1889 (1957).
183. M. D. Alexander and D. H. Busch, *J. Am. Chem. Soc.*, **88**, 1130 (1966).
184. A. Ågren, *Acta Pharm. Suecica*, **2**, 87 (1965).
185. R. H. Barca and H. Freiser, *J. Am. Chem. Soc.*, **88**, 3744 (1966).
186. H. L. Conley, Jr. and R. B. Martin, *J. Phys. Chem.*, **69**, 2914, 2923 (1965).
187. R. W. Hay and N. J. Walker, *Nature*, **204**, 1189 (1964).
188. B. Capon and B. C. Ghosh, *J. Chem. Soc. (B)*, 472 (1966).
189. D. H. Huchital and H. Taube, *J. Am. Chem. Soc.*, **87**, 5371 (1965).
190. B. Capon, *Quart. Rev. (London)*, **18**, 45 (1964).
191. L. J. Edwards, *Trans. Faraday Soc.*, **48**, 696 (1952); **46**, 723 (1950).
192. E. R. Garrett, *J. Am. Chem. Soc.*, **79**, 3401 (1957).
193. H. Morawetz and I. Oreskes, *J. Am. Chem. Soc.*, **80**, 2591 (1958).
194. J. D. Chanley, E. M. Gindler and H. Sobotka, *J. Am. Chem. Soc.*, **74**, 4347 (1952).
195. M. L. Bender, F. Chloupek and M. C. Neveu, *J. Am. Chem. Soc.*, **80**, 5384 (1958).
196. P. E. Zimmering, E. W. Westhead, Jr. and H. Morawetz, *Biochim. Biophys. Acta*, **25**, 376 (1957); H. Morawetz and P. E. Zimmering, *J. Phys. Chem.*, **58**, 753 (1954).
197. E. Gaetjens and H. Morawetz, *J. Am. Chem. Soc.*, **82**, 5328 (1960).
198. T. C. Bruice and U. K. Pandit, *J. Am. Chem. Soc.*, **82**, 5858 (1960).
199. T. C. Bruice and W. C. Bradbury, *J. Am. Chem. Soc.*, **87**, 4846 (1965).
200. J. W. Thanassi and T. C. Bruice, *J. Am. Chem. Soc.*, **88**, 747 (1966).
201. T. C. Bruice and S. J. Benkovic, *J. Am. Chem. Soc.*, **85**, 1 (1963).
202. L. Eberson, *Acta Chem. Scand.*, **18**, 2015 (1964).
203. M. L. Bender, F. J. Kézdy and B. Zerner, *J. Am. Chem. Soc.*, **85**, 3017 (1963).
204. B. Hansen, *Acta Chem. Scand*, **17**, 1375 (1963).
205. Y. Shalitin and S. A. Bernhard, *J. Am. Chem. Soc.*, **86**, 2291 (1964).
206. S. M. Kupchan, S. P. Eriksen and M. Friedman, *J. Am. Chem. Soc.*, **84**, 4159 (1962); **88**, 343 (1966).
207. M. S. Newman and S. Hishida, *J. Am. Chem. Soc.*, **84**, 3582 (1962).
208. H. G. O. Becker, J. Schneider and H.–D. Steinleitner, *Tetrahedron Letters*, 3761 (1965).
209. B. Hansen, *Svensk Kem. Tidskr.*, **75**, 421 (1963).
210. E. Schätzle, H. Urheim, M. Thürkauf and M. Rottenberg, *Helv. Chim. Acta*, **46**, 2418 (1963).

211. Y. Shalitin and S. A. Bernhard, *J. Am. Chem. Soc.*, **86**, 2292 (1964).
212. M. L. Bender and M. S. Silver, *J. Am. Chem. Soc.*, **84**, 4589 (1962); M. L. Bender, J. A. Reinstein, M. S. Silver and R. Mikulak, *J. Am. Chem. Soc.*, **87**, 4545 (1965).
213. T. C. Bruice and J. M. Sturtevant, *J. Am. Chem. Soc.*, **81**, 2860 (1959).
214. G. L. Schmir and T. C. Bruice, *J. Am. Chem. Soc.*, **80**, 1173 (1958).
215. T. C. Bruice and J. L. Herz, *J. Am. Chem. Soc.*, **86**, 4109 (1964).
216. C. R. Wasmuth and H. Freiser, *Talanta*, **9**, 1059 (1962).
217. B. Hansen, *Acta Chem. Scand.*, **16**, 1927 (1962).
218. G. Aksnes and P. Frøyen, *Acta Chem. Scand.*, **20**, 1451 (1966).
219. E. R. Garrett, *J. Am. Chem. Soc.*, **79**, 5206 (1957); **80**, 4049 (1958).
220. S. A. Bernhard, A. Berger, J. H. Carter, E. Katchalski, M. Sela and Y. Shalitni, *J. Am. Chem. Soc.*, **84**, 2421 (1962).
221. F. M. Menger and C. L. Johnson, *Tetrahedron*, **23**, 19 (1967).
222. A. Michaelis and W. Kerkhof, *Chem. Ber.*, **31**, 2172 (1898).
223. W. Drenth, *Rec. Trav. Chim.*, **85**, 455 (1966).
224. K. A. Connors and J. A. Mollica, Jr., *J. Am. Chem. Soc.*, **87**, 123 (1965).
225. J. A. Mollica, Jr. and K. A. Connors, *J. Am. Chem. Soc.*, **89**, 308 (1967).
226. F. M. Menger and M. L. Bender, *J. Am. Chem. Soc.*, **88**, 131 (1966).
227. C. A. Bunton, J. N. E. Day, R. H. Flowers, P. Sheel and J. L. Wood, *J. Chem. Soc.*, 963 (1957).
228. I. Roberts and H. C. Urey, *J. Am. Chem. Soc.*, **60**, 2391 (1938).
229. R. Stewart and K. Yates, *J. Am. Chem. Soc.*, **82**, 4059 (1960).
230. J. Hine and R. P. Bayer, *J. Am. Chem. Soc.*, **84**, 1989 (1962).
231. H. Hosoya and S. Nagakura, *Spectrochim. Acta*, **17**, 324 (1961).
232. G. Fraenkel, *J. Chem. Phys.*, **34**, 1466 (1961).
233. D. Jaques, *J. Chem. Soc.*, 3874 (1965).
234. C. A. Lane, *J. Am. Chem. Soc.*, **86**, 2521 (1964).
235. W. P. Jencks and M. Gilchrist, *J. Am. Chem. Soc.*, **86**, 4651 (1964).
236. Ya. K. Syrkin and I. I. Moiseev, *Usp. Khim.*, **27**, 717 (1958).
237. V. A. Palm, U. L. Haldna, A. I. Talvik and A. E. Mei, *Zh. Fiz. Khim.*, **36**, 2499 (1962).
238. E. Tommila and A. Hella, *Ann. Acad. Sci. Fennicae, Ser. A II*, No. 53, (1954).
239. V. Rothmund, *Z. Physik. Chem. (Leipzig)*, **20**, 168 (1896).
240. A. Bogojawlensky and G. Tammann, *Z. Physik. Chem. (Leipzig)*, **23**, 13 (1897).
241. A. R. Osborn and E. Whalley, *Can. J. Chem.*, **39**, 1094 (1961).
242. E. Whalley, *Trans. Faraday Soc.*, **55**, 798 (1959).
243. B. T. Baliga, R. J. Withey, D. Poulton and E. Whalley, *Trans. Faraday Soc.*, **61**, 517 (1965).
244. A. C. Rolfe and C. N. Hinshelwood, *Trans. Faraday Soc.*, **30**, 935 (1934).
245. M. Gordon and G. R. Scantlebury, *J. Chem. Soc. (B)*, 1 (1967).
246. F. P. Olsen, *Ph.D. Thesis*, Brown University; *Dissertation Abstr.*, **25**, 4414 (1965).
247. S. Affrossman and J. P. Murray, *J. Chem. Soc. (B)*, 1015 (1966).
248. S. Yoshikawa and O.-K. Kim, *Bull. Chem. Soc. Japan*, **39**, 1515 (1966).
249. E. Tommila and M. Sippola, *Suomen Kemistilehti*, **29 B**, 64 (1956).
250. R. Löwenherz, *Z. Physik. Chem. (Leipzig)*, **15**, 389 (1894).
251. E. A. Halonen, *Acta Chem. Scand.*, **10**, 1355 (1956).
252. M. H. Palomaa, *Ann. Acad. Sci. Fennicae, Ser. A IV*, No. 2 (1913).
253. A. de Hemptinne, *Z. Physik. Chem. (Leipzig)*, **13**, 561 (1894).
254. W. A. Drushel, *Am. J. Sci.*, [4] **34**, 69 (1912).
255. A. Skrabal and A. Zahorka, *Monatsh. Chem.*, **46**, 559 (1925).
256. E. J. Salmi, *Chem. Ber.*, **72**, 1767 (1939).
257. B. van Dijken, *Rec. Trav. Chim.*, **14**, 106 (1895).
258. B. V. Bhide and J. J. Sudborough, *J. Indian Inst. Sci.*, Sect. A8, 89 (1925).

259. H. Goldschmidt, H. Haaland and R. S. Melbye, *Z. Physik. Chem. (Leipzig)*, **143**, 278 (1929).
260. H. Goldschmidt and O. Udby, *Z. Physik. Chem. (Leipzig)*, **60**, 728 (1907).
261. V. Meyer, *Chem. Ber.*, **27**, 510 (1894); **28**, 1254 (1895).
262. H. A. Smith and J. P. McReynolds, *J. Am. Chem. Soc.*, **61**, 1963 (1939).
263. M. S. Newman, *J. Am. Chem. Soc.*, **72**, 4783 (1950); K. L. Loening, A. B. Garrett and M. S. Newman, *J. Am. Chem. Soc.*, **74**, 3929 (1952).
264. E. D. Hughes, *Quart. Rev. (London)*, **2**, 107 (1948).
265. Fr. Becker, *Z. Naturforsch.*, **14a**, 547 (1959); **15b**, 251 (1960); **16b**, 236 (1961).
266. C. K. Ingold, *J. Chem. Soc.*, 1032 (1930).
267. R. W. Taft, Jr., *J. Am. Chem. Soc.*, **74**, 3120 (1952); **75**, 4231 (1953).
268. E. Euranto, *Ann. Univ. Turku.*, *Ser. A I*, No. 42 (1960).
269. N. B. Chapman, J. Shorter and J. H. P. Utley, *J. Chem. Soc.*, 1291 (1963).
270. P. A. ten Thije and M. J. Janssen, *Rec. Trav. Chim.*, **84**, 1169 (1965).
271. K. Bowden, *Can. J. Chem.*, **44**, 661 (1966).
272. P. R. Wells, *J. Phys. Chem.*, **69**, 1787 (1965).
273. C. K. Hancock, E. A. Meyers and B. J. Yager, *J. Am. Chem. Soc.*, **83**, 4211 (1961).
274. C. K. Hancock, B. J. Yager, C. P. Falls and J. O. Schreck, *J. Am. Chem. Soc.*, **85**, 1297 (1963).
275. I. V. Talvik and V. A. Palm, *Reaktsionnaya Sposobnost Organ. Soedin.*, *Tartusk Gos. Univ.*, **1**, 108 (1964); I. V. Talvik, *Reaktsionnaya Sposobnost Organ. Soedin.*, *Tartusk Gos. Univ.*, **1 (2)**, 241 (1964).
276. K. B. Wiberg, *Chem. Rev.*, **55**, 713 (1955).
277. R. A. Robinson, *Trans. Faraday Soc.*, **26**, 217 (1930).
278. P. T. McTigue and A. R. Watkins, *Australian J. Chem.*, **18**, 1943 (1965).
279. F. A. Long, F. B. Dunkle and W. F. McDevit, *J. Phys. Chem.*, **55**, 829 (1951).
280. P. T. McTigue, *Trans. Faraday Soc.*, **60**, 127 (1964).
281. L. Zucker and L. P. Hammett, *J. Am. Chem. Soc.*, **61**, 2791 (1939).
282. F. A. Long and M. A. Paul, *Chem. Rev.*, **57**, 935 (1957).
283. R. P. Bell, A. L. Dowding and J. A. Noble, *J. Chem. Soc.*, 3106 (1955).
284. M. Duboux and A. de Sousa, *Helv. Chim. Acta*, **23**, 1381 (1940).
285. P. Salomaa, *Suomen Kemistilehti*, **32 B**, 145 (1959).
286. C. T. Chmiel and F. A. Long, *J. Am. Chem. Soc.*, **78**, 3326 (1956).
287. K. J. Pedersen, *Acta Chem. Scand.*, **15**, 1718 (1961).
288. R. P. Bell and B. Lukianenko, *J. Chem. Soc.*, 1686 (1957).
289. P. Salomaa, *Acta Chem. Scand.*, **11**, 247 (1957).
290. T. Yrjänä, *Suomen Kemistilehti*, **39 B**, 81 (1966).
291. L. A. Kiprianova and A. F. Rekasheva, *Dokl. Akad. Nauk SSSR*, **144**, 386 (1962).
292. J. A. Landgrebe, *J. Org. Chem.*, **30**, 2997 (1965).
293. J. F. Bunnett, *J. Am. Chem. Soc.*, **83**, 4956, 4968, 4973, 4978 (1961).
294. R. B. Martin, *J. Am. Chem. Soc.*, **84**, 4130 (1962); *J. Phys. Chem.*, **68**, 1369 (1964).
295. J. F. Bunnett and F. P. Olsen, *Can. J. Chem.*, **44**, 1899, 1917 (1966).
296. H. S. Harned and A. M. Ross, Jr., *J. Am. Chem. Soc.*, **63**, 1993 (1941).
297. H. B. Friedman and G. V. Elmore, *J. Am. Chem. Soc.*, **63**, 864 (1941).
298. J. Koskikallio, *Suomen Kemistilehti*, **35 B**, 62 (1962).
299. E. Tommila and M.-L. Murto, *Acta Chem. Scand.*, **17**, 1957 (1963).
300. E. K. Euranto and N. J. Cleve, *Reaktsionnaya Sposobnost Organ. Soedin.*, *Tartusk Gos. Univ.*, **2 (1)**, 183 (1965).
301. I. V. Talvik and V. A. Palm., *Reaktsionnaya Sposobnost Organ. Soedin.*, *Tartusk Gos. Univ.*, **2 (1)**, 110 (1965).
302. M. L. Bender and Y.-L. Chow, *J. Am. Chem. Soc.*, **81**, 3929 (1959).
303. I. V. Talvik, *Reaktsionnaya Sposobnost Organ. Soedin.*, *Tartusk Gos. Univ.*, **2 (4(6))**, 229 (1965).

304. P. W. C. Barnard, C. A. Bunton, D. Kellerman, M. M. Mhala, B. Silver, C. A. Vernon and V. A. Welch, *J. Chem. Soc., B*, 227 (1966).
305. E. K. Euranto, *Tidsskr. Kjemi, Bergvesen Met.*, **25**, 214 (1965).
306. H. P. Treffers and L. P. Hammett, *J. Am. Chem. Soc.*, **59**, 1708 (1937).
307. M. S. Newman, *J. Am. Chem. Soc.*, **63**, 2431 (1941).
308. M. L. Bender, H. Ladenheim and M. C. Chen, *J. Am. Chem. Soc.*, **83**, 123 (1961).
309. M. L. Bender and M. C. Chen, *J. Am. Chem. Soc.*, **85**, 37 (1963).
310. J. A. Leisten, *J. Chem. Soc.*, 1572 (1956).
311. D. N. Kershaw and J. A. Leisten, *Proc. Chem. Soc.*, 84 (1960).
312. A. R. Olson and R. J. Miller, *J. Am. Chem. Soc.*, **60**, 2687 (1938).
313. A. R. Olson and J. L. Hyde, *J. Am. Chem. Soc.*, **63**, 2459 (1941).
314. F. A. Long and M. Purchase, *J. Am. Chem. Soc.*, **72**, 3267 (1950).
315. E. D. Hughes, C. K. Ingold and S. Masterman, *J. Chem. Soc.*, 840 (1939).
316. C. A. Bunton, E. D. Hughes, C. K. Ingold and D. F. Meigh, *Nature*, **166**, 680 (1950).
317. S. G. Cohen and A. Schneider, *J. Am. Chem. Soc.*, **63**, 3382 (1941).
318. J. G. Hawke and V. R. Stimson, *J. Chem. Soc.*, 4676 (1956).
319. N. V. Fok, D. N. Kursanov, V. N. Setkina, E. V. Bikova and R. V. Koudriavtsev, *Reaktsionnaya Sposobnost Organ. Soedin., Tartusk Gos. Univ.*, **3** (3(9)), 178 (1966).
320. C. A. Bunton, A. E. Comyns and J. L. Wood, *Research (London)*, **4**, 383 (1951).
321. K. R. Adam, I. Lauder and V. R. Stimson, *Australian J. Chem.*, **15**, 467 (1962).
322. C. A. Bunton and T. Hadwick, *J. Chem. Soc.*, 3043 (1957).
323. C. A. Bunton and A. Konasiewicz, *J. Chem. Soc.*, 1354 (1955).
324. A. G. Davies and J. Kenyon, *Quart. Rev. (London)*, **9**, 203 (1955).
325. M. H. Palomaa, E. J. Salmi, J. I. Jansson and T. Salo, *Chem. Ber.*, **68**, 303 (1935).
326. C. A. Bunton and J. L. Wood, *J. Chem. Soc.*, 1522 (1955).
327. P. Salomaa, *Acta Chem. Scand.*, **14**, 577 (1960).
328. T. Yrjänä, *Ann. Univ. Turku.*, *Ser. AI*, No. 54 (1962).
329. T. Yrjänä, *Suomen Kemistilehti*, **37 B**, 108 (1964).
330. H. T. Liang and P. D. Bartlett, *J. Am. Chem. Soc.*, **80**, 3585 (1958).
331. G. J. Harvey and V. R. Stimson, *Australian J. Chem.*, **15**, 757 (1962).
332. G. J. Harvey and V. R. Stimson, *J. Chem. Soc.*, 3629 (1956).
333. M. S. Silver, *J. Am. Chem. Soc.*, **83**, 404 (1961).
334. E. J. Salmi, *Ann. Univ. Turku.*, *Ser. A III*, No. 3 (1932).
335. P. Salomaa, *Acta Chem. Scand.*, **11**, 132 (1957).
336. B. G. Ramsey and R. W. Taft, *J. Am. Chem. Soc.*, **88**, 3058 (1966).
337. P. Salomaa, *Acta Chem. Scand.*, **11**, 141, 235 (1957).
338. P. Salomaa, *Acta Chem. Scand.*, **11**, 239 (1957).
339. P. Salomaa and R. Linnantie, *Acta Chem. Scand.*, **14**, 586 (1960).
340. P. Salomaa, *Suomen Kemistilehti*, **33 B**, 11 (1960).
341. P. Salomaa, *Acta Chem. Scand.*, **19**, 1263 (1965).
342. P. Salomaa, *Acta Chem. Scand.*, **20**, 1263 (1966).
343. T. H. Fife, *J. Am. Chem. Soc.*, **87**, 271 (1965).
344. J. Kenyon, S. M. Partridge and H. Phillips, *J. Chem. Soc.*, 85 (1936); 207 (1937); H. W. J. Hills, J. Kenyon and H. Phillips, *J. Chem. Soc.*, 576 (1936).
345. G. S. Hammond and J. T. Rudesill, *J. Am. Chem. Soc.*, **72**, 2769 (1950).
346. A. Moffat and H. Hunt, *J. Am. Chem. Soc.*, **80**, 2985 (1958).
347. L. Pekkarinen, *Ann. Acad. Sci. Fennicae*, *Ser. A II*, No. 85 (1957).
348. G. Meyer, P. Viout and P. Rumpf, *Compt. Rend.*, **262**, C 1099 (1966).
349. J. R. Fox and G. Kohnstam, *J. Chem. Soc.*, 1593 (1963).
350. C. G. Swain, T. E. C. Knee and A. MacLachlan, *J. Am. Chem. Soc.*, **82**, 6101 (1960).
351. C. G. Swain, C. B. Scott and K. H. Lohmann, *J. Am. Chem. Soc.*, **75**, 136 (1953).

352. H. Hart and P. A. Law, *J. Am. Chem. Soc.*, **86,** 1957 (1964).
353. P. Salomaa and K. Sallinen, *Acta Chem. Scand.*, **19,** 1054 (1965).
354. M. S. Silver and G. C. Whitney, *J. Org. Chem.*, **28,** 2479 (1963).
355. W. A. Cowdrey, E. D. Hughes, C. K. Ingold, S. Masterman and A. D. Scott, *J. Chem. Soc.*, 1252 (1937).
356. P. D. Bartlett and P. N. Rylander, *J. Am. Chem. Soc.*, **73,** 4273 (1951).
357. A. R. Olson and P. V. Youle, *J. Am. Chem. Soc.*, **73,** 2468 (1951).
358. A. R. Butler and V. Gold, *J. Chem. Soc.*, 2212 (1962).
359. M. L. Bender and R. S. Dewey, *J. Am. Chem. Soc.*, **78,** 317 (1956).
360. L. R. C. Barclay, N. D. Hall and G. A. Cooke, *Can. J. Chem.*, **40,** 1981 (1962).
361. T. C. Bruice and T. H. Fife, *J. Am. Chem. Soc.*, **83,** 1124 (1961).
362. A. Kivinen, *Suomen Kemistilehti*, **38 B,** 106, 143, 159, 207 (1965).
363. A. Kivinen, *Suomen Kemistilehti*, **38 B,** 205 (1965).
364. E. K. Euranto and R. S. Leimu, *Acta Chem. Scand.*, **20,** 2028 (1966).
365. A. Kivinen, *Acta Chem. Scand.*, **19,** 845 (1965).
366. R. Leimu, *Ann. Univ. Turku.*, *Ser.* A IV, No. 3 (1935); *Chem. Ber.*, **70,** 1040 (1937).
367. R. F. Hudson and M. Green, *J. Chem. Soc.*, 1055 (1962).
368. E. W. Crunden and R. F. Hudson, *J. Chem. Soc.*, 3748 (1961).
369. H. K. Hall, Jr. and C. H. Lueck, *J. Org. Chem.*, **28,** 2818 (1963).
370. R. F. Hudson and G. Loveday, *J. Chem. Soc.*, 1068 (1962).
371. M. Green and R. F. Hudson, *J. Chem. Soc.*, 1076 (1962).
372. H. Minato, *Bull. Chem. Soc. Japan.*, **37,** 316 (1964).
373. W. A. Drushel and G. R. Bancroft, *Am. J. Sci.*, [4] **44,** 371 (1917).
374. H. Böhme, H. Bezzenberger, M. Clement, A. Dick, E. Nürnberg and W. Schlephack, *Ann. Chem.*, **623,** 92 (1959).
375. E. Euranto, *Suomen Kemistilehti*, **35 B,** 25 (1962).
376. E. Euranto, *Suomen Kemistilehti*, **33 B,** 41 (1960).
377. N. J. Cleve, *Suomen Kemistilehti*, **39 A,** 111 (1966).
378. E. K. Euranto, *Suomen Kemistilehti*, **39 A,** 110 (1966); *Acta Chem. Scand.*, **21,** 721 (1967).
379. E. K. Euranto and T. Yrjänä, *Suomen Kemistilehti*, **38 B,** 215 (1965).
380. P. Papoff and P. G. Zambonin, *Talanta*, **14,** 581 (1967).
381. A. R. Fersht and A. J. Kirby, *J. Am. Chem. Soc.*, **89,** 4853, 4857, 5960, 5961 (1967).
382. K. Bowden and G. R. Taylor, *Chem. Comm.*, 1112 (1967).
383. G. Aksnes and P. Frøyen, *Acta Chem. Scand.*, **21,** 1507 (1967); R. W. Hay and L. J. Porter, *J. Chem. Soc. (B)*, 1261 (1967); M. R. Wright, *J. Chem. Soc. (B)*, 1265 (1967).
384. R. L. VanEtten, J. F. Sebastian, G. A. Clowes and M. L. Bender, *J. Am. Chem. Soc.*, **89,** 3242, 3253 (1967); C. G. Overberger, J. C. Salamone and S. Yaroslavsky, *J. Am. Chem. Soc.*, **89,** 6231 (1967); A. Ochoa-Solano, G. Romero and C. Gitler, *Science*, **156,** 1243 (1967).
385. R. B. Martin *J. Am. Chem. Soc.*, **89,** 2501 (1967).
386. H. Hogeveen, *Rec. Trav. Chim.*, **86,** 816 (1967).
387. G. A. Olah, D. H. O'Brien and A. M. White, *J. Am. Chem. Soc.*, **89,** 5694 (1967).
388. K. Yates and R. A. McClelland, *J. Am. Chem. Soc.*, **89,** 2686 (1967).
389. J. W. Chen and W. M. Kalback, *Ind. Eng. Chem. Fundamentals*, **6,** 175 (1967).
390. J. Martinmaa and E. Tommila, *Suomen Kemistilehti*, **40 B,** 222 (1967).
391. B. G. Cox and P. T. McTigue, *Australian J. Chem.*, **20,** 1815 (1967).
392. W. G. Dauben and J. R. Wiseman, *J. Am. Chem. Soc.*, **89,** 3545 (1967).
393. J. G. Martin and J. M. W. Scott, *Chem. Ind. (London)*, 655 (1967).
394. A. Queen, *Can. J. Chem.*, **45,** 1619 (1967).
395. T. St. Pierre and W. P. Jencks, *J. Am. Chem. Soc.* **90,** 3817 (1965).

CHAPTER **12**

The decarboxylation reaction

Louis W. Clark

Western Carolina University, Cullowhee, North Carolina, U. S. A

I. SCOPE OF THE DECARBOXYLATION REACTION

A. The Carbon Dioxide-Oxygen Cycle in Nature

The carbon dioxide–oxygen cycle is one of the most interesting phenomena in nature. In this process green plants change the energy of sunlight into chemical energy, and animals make use of the chemical energy thus stored in the plant foods to carry on their various functions. In the first stage of the cycle, the green plant, with the aid of sunlight and chlorophyll, uses mainly carbon dioxide and water to manufacture starches and other substances. In this stage energy is absorbed and oxygen is released into the atmosphere. In the second stage the animal organism, with the aid of enzymes and vitamins, converts the plant foods ultimately into carbon dioxide and water. In this stage energy is released and oxygen is extracted from the atmosphere. The long range global operation of this cycle preserves the various components of the atmosphere at nearly constant proportions (carbon dioxide, 0·035% by volume, oxygen, 20·99% by volume).

Studies of the first stage of the carbon dioxide–oxygen cycle (carboxylation) by Calvin, Bassham and Benson[1] using carbon dioxide labeled with ^{14}C have shown that one of the first substances formed during photosynthesis is phosphoglyceric acid, $CH_2(OPO_3H_2)CHOHCOOH$. The radioactive carbon is found to be more abundant in the carboxyl group than in the other two carbon atoms of the molecule.

In the second stage of the cycle (decarboxylation), the carbon dioxide is produced during animal catabolism either by spontaneous decarboxylation of β-keto acids, or by oxidative decarboxylation of α-keto acids[2].

Laboratory studies on the decarboxylation reaction have very often been prompted by a desire to gain a better understanding of the complex factors operating in the second stage of the carbon dioxide–oxygen cycle.

B. Synthesis of Organic Compounds

The decarboxylation reaction plays an important role in the synthesis of a large variety of organic compounds. Space permits the mention of only a few examples.

Some alkanes may be produced by heating a mixture of the sodium salt of a monocarboxylic acid with sodium hydroxide. The following equation illustrates this method of preparation:

$$CH_3COONa + NaOH \longrightarrow CH_4 + Na_2CO_3$$

Although this type of reaction has been frequently cited as a preparative method for preparing alkanes, March[3] has pointed out that simple aliphatic acids (except acetic acid) give relatively poor yields of the corresponding alkanes by this procedure.

Alkanes, as well as other compounds, may be prepared by the Kolbe synthesis. For example, 10-methyloctadecanoic acid has been prepared by the electrolysis of a mixture of the sodium salts of 3-methylundecanoic acid and monomethyl suberic acid in methanol[4].

Unsymmetrical ketones have been prepared by the decarboxylation of acids containing no α-hydrogen atoms by heating with aerogel thoria catalyst at 490°[5].

A series of α-naphthylalkylacetic acids have been prepared by hydrolysis and decarboxylation of the corresponding α-naphthyl alkyl malonic esters[6]. The basic esters of these acids were found to be effective antispasmodics.

Alkyl or aryl halides may be prepared by the decarboxylation in the presence of chlorine or bromine, of the silver salts of carboxylic acids in inert solvents[7−9].

A series of aromatic and heterocyclic hydrazines have been prepared by application of the malonic acid decarboxylation reactions[10].

The decarboxylation of α-nitrophenylacetic acid has been used to prepare α-nitrotoluene[11].

Heating adipic acid with barium hydroxide gives cyclopentanone[12].

The usefullness of this method of synthesis depends upon two factors: (1) the introduction of a carboxyl group at a desired site in the molecule by oxidation or by other methods, and (2) the possibility of the removal of a carboxyl group under favorable circumstances. There appears to be no limit to the variety of compounds which can be produced by the proper combination of these two operations.

II. MECHANISMS OF THE DECARBOXYLATION REACTION

A. Heterolytic Fission

1. Keto acids and their anions

a. Keto acids. In 1908 Bredig and Balcom[13] studied the decarboxylation of camphor-3-carboxylic acid into carbon dioxide and camphor in water, as well as in a variety of non-aqueous solvents (aniline, ethanol, benzene, phenetole, ether.)

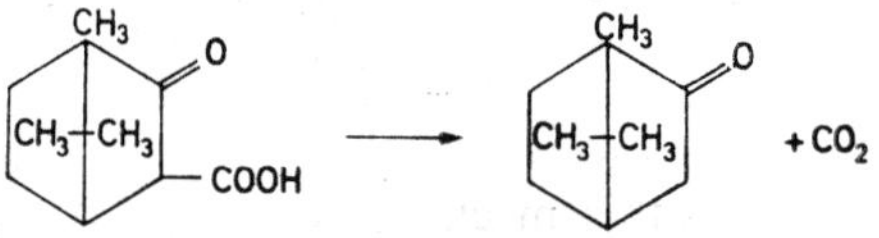

They found that the unionized acid at 98° in water decomposed about 34 times faster than did the anion, and that the rate of reaction of the unionized acid in the different solvents at this temperature increased with increasing basicity of the solvent. These results indicate that the reaction is bimolecular and involves mainly the undissociated acid.

In 1932 Pedersen[14] proposed the hypothesis that a dipolar ion is the intermediate in the decarboxylation of dimethylacetoacetic acid in solution. However, in 1941, Westheimer and Jones[15] pointed out that, if Pedersen's proposal was correct, the reaction should proceed more rapidly the higher the dielectric constant of the medium. In studying the decarboxylation of this compound in water and in water–alcohol and water–dioxane mixtures they found the rate to be essentially independent of the solvent. These findings invalidated the dipolar ion hypothesis and lead them to propose a chelate structure as the intermediate:

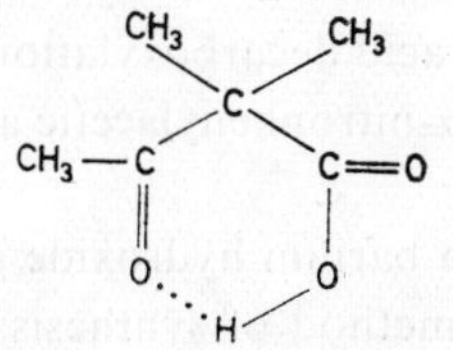

In 1947 King[16] proposed a similar mechanism for the decarboxylation of malonic acid and its derivatives:

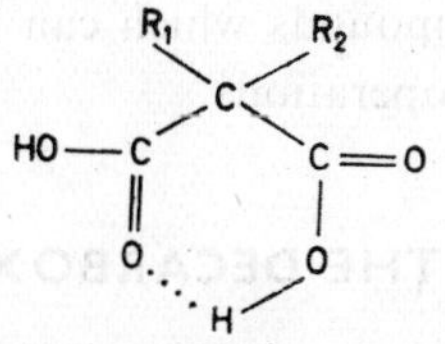

In 1948 Schenkel and Schenkel-Rudin[17] advanced the hypothesis that the transition state in the decarboxylation of β-keto acids was the enol form of the acid:

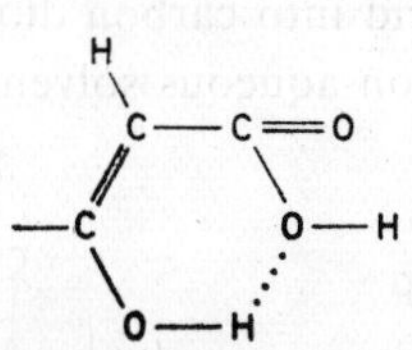

They cited as evidence for this mechanism the fact that camphenonic acid, which is incapable of enolizing, cannot be decarboxylated[18], This suggestion cannot be accepted, however, since many β-keto acids which cannot enolize (e.g., dimethylacetoacetic acid[14, 15] and diethylmalonic acid[19, 20]) readily undergo decarboxylation.

Further light was thrown on the probable mechanism of the decarboxylation of β-keto acids by the studies of Fraenkel, Belford and Yankwich in 1954 on the decarboxylation of malonic acid in dioxane and in quinoline–dioxane mixtures in which they found that the rate of reaction was proportional to the concentration of the quinoline[21]. Near 100°, extrapolation of

TABLE 1. Pseudo first-order rate constants for the decarboxylation of malonic acid in dioxane–quinoline mixtures at 99·6°

Medium	Concentration of quinoline (mole/1)	$k \times 10^5$ (sec^{-1}) with dioxane present	corrected for dioxane	Reference
(1) Dioxane	0	0·66	0	21
(2) Dioxane–quinoline	0·27	2·8	2·14	21
(3) Dioxane–quinoline	0·53	4·3	3·64	21
(4) Dioxane–quinoline	1·59	11·8	11·14	21
(5) Dioxane–quinoline	4·24	29·5	28·84	21
(6) Quinoline	7·75	—	52·5	22

the data to 100% quinoline yielded a value corresponding very closely with that obtained by measuring the reaction in pure quinoline. Details of these findings are shown in Table 1 and in Figure 1.

It will be seen in Table 1 that at the temperature of the experiments (99·6°) malonic acid undergoes decarboxylation at a slow but measureable rate in pure dioxane (1). When a small quantity of quinoline is added to the dioxane (the temperature remaining constant) the rate of the reaction increases (2). As more and more quinoline is added the rate continues to increase. The bottom row of the table shows the calculated rate of reaction at 99·6° in pure quinoline based upon data subsequently obtained by

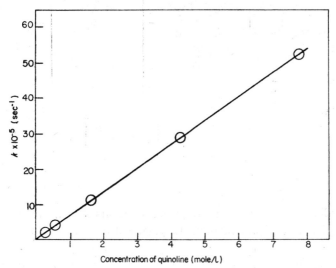

FIGURE 1. Rate of decarboxylation of malonic acid at 99·6°C in dioxane–quinoline mixtures (corrected for dioxane), based on data in Table 1.

Clark[22], since this gives somewhat better agreement than that furnished by Fraenkel, Belford and Yankwich[21].

Dioxane itself appears to act as a moderate catalyst for the decarboxylation of malonic acid as shown by the results of line (1) of Table 1. Therefore, when malonic acid is added to a mixture of dioxane and quinoline, two reactions evidently take place simultaneously — the slow decarboxylation involving dioxane, and the faster decarboxylation involving quinoline. Subtraction of the effect of dioxane from the total effect yields the contribution of quinoline. The corrected rate constants thus calculated are shown in the last column of Table 1.

Figure 1 is a plot of the corrected rate constants for the decarboxylation of malonic acid in the dioxane quinoline mixtures at 99·6° versus the molar concentration of quinoline, based upon the data in Table 1. The fact that the plot of the data yields a straight line shows that, in an ideal inert solvent, at constant temperature, the rate constant for the decarboxylation of malonic acid is a linear function of the concentration of quinoline*. These results demonstrate unequivocally that the reaction is bimolecular and lead Fraenkel, Belford and Yankwich to propose the following mechanism for the decarboxylation of malonic acid in the presence of quinoline and related solvents:

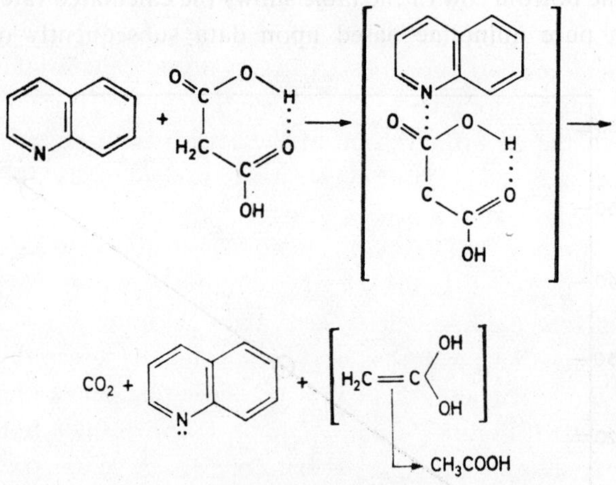

According to their proposal, the reaction occurs through the formation of an activated complex involving a polarized carbonyl carbon atom of

---

* The equation for the line in Figure 1 is $k_{-----} = 6\cdot78 \times 10^{-5}M$, where M is the concentration of the quinoline in mole/l. This equation enables one to calculate the rate constant at this temperature for any concentration of quinoline.

malonic acid (present in the ketone chelate structure as proposed by West-heimer and Jones[15], as well as by King[16]) and the nucleophilic nitrogen atom in the quinoline nucleus. The subsequent fast cleavage of malonic acid in the activated complex is favored by two important circumstances: (1) the partial neutralization of the effective positive charge on the polarized car-bonyl carbon atom of malonic acid due to the propinquity of the unshared pair of electrons on the nitrogen atom in quinoline, and (2) the hydrogen bridge in the six-membered ring of the chelate structure facilitating the transfer of the hydrogen atom from one oxygen atom to the other. In the kinetics of the reaction the rate-determining step is the formation of the activated complex, via a nucleophilic attack on the π-bonded carbon atom This step is essentially the addition of one molecule of the amine (the nucleophilic agent) to one molecule of malonic acid (the electrophilic agent, or substrate) forming a nucleophile–electrophile pair, and is thus a nucleophilic addition reaction[23]. After the formation of the activated complex the decarboxylation reaction itself evidently ensues rapidly through heterolytic fission yielding a proton and a carbanion[24]. The latter quickly rearranges to acetic acid. The decarboxylation of malonic acid is thus quite similar to a bimolecular nucleophilic substitution reaction (S_N2)[25].

The decarboxylation of other β-keto acids in nucleophilic solvents prob-ably takes place by a mechanism similar to that of the decarboxylation of malonic acid in quinoline. Evidence has also been obtained which indicates that α-keto acids probably undergo decarboxylation in nucleophilic solvents by a mechanism quite similar to that of β-keto acids[26]. The transition state in the case of the decarboxylation of the α-keto acids would necessarily show a five-membered chelate ring structure instead of the six-membered ring in case of the β-keto acids.

b. *Keto acid anions.* It has been mentioned previously (see section II. A. 1. a) that, in the case of the decarboxylation of camphor-3-carboxylic acid in water at 98° the unionized acid reacts about 34 times as fast as does the anion[13]. The fact that the anions of β-keto acids are susceptible to decar-boxylation, but less readily than the corresponding acids, appears to be entirely general. Pedersen[27] found that at 18° dimethylacetoacetic acid reacts in water solution 180 times as fast as its anion. Widmark[28] observed that acetoacetic acid in water at 37° decomposes 53 times as fast as its anion. Hall[29] observed that malonic acid in water at 98° decomposes 10 times as fast as its anion.

The behavior of α-keto acids and their anions parallel that of β-keto acids and their anions. Dinglinger and Schröer[30] found that the mono-anion of oxalic acid does not lose carbon dioxide as easily as the acid itself.

38*

Wiig[31] observed that the rate of decarboxylation of oxalic acid decreased on the addition of water to 100% sulfuric acid. Clark observed that at 100° the decarboxylation of oxalic acid takes place 21 times as fast in *m*-cresol (where little ionization would be expected) as it does in the basic, ionizing solvent quinoline[32, 33]. He found also that the decarboxylation of oxamic acid at 100° proceeds 15 times as fast in octanoic acid as it does in quinoline[34, 35]. Furthermore, the enthalpies of activation for these decarboxylation reactions are higher in the basic medium than in the acid medium, as shown in Table 2.

TABLE 2. Kinetic data for the decarboxylation of oxalic acid and oxamic acid in acid and in basic media

Solvent	Oxalic acid [32, 33]			Oxamic acid [34, 35]		
	$\Delta H^{\ddagger}$ (kcal/mole)	$\Delta S^{\ddagger}$ (E.U./mole)	$k_{100°} \times 10^7$ (sec^{-1})	$\Delta H^{\ddagger}$ (kcal/mole)	$\Delta S^{\ddagger}$ (E.U./mole)	$k_{100°} \times 10^7$ (sec^{-1})
Quinoline	38·9	15·75	2·24	47·0	37·5	1·66
m-Cresol	26·9	−11·0	47·8	—	—	—
Octanoic acid	—	—	—	30·5	−1·25	24·6

The results shown in Table 2 suggest that in quinoline ionization of these two acids occurs and the anions undergo decarboxylation, whereas in the acid media, ionization is repressed and the undissociated acids decompose.

The activated complexes involved in the decarboxylation of the anions of α- and β-keto acids probably closely resemble those of the unionized acids:

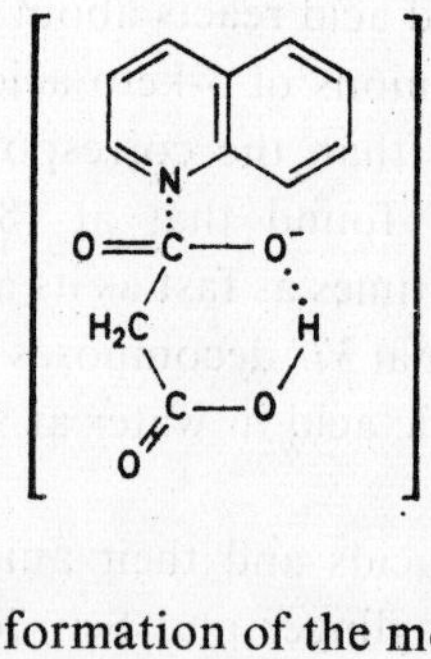

The loss of a proton in the formation of the monoanion increases the electron density on the oxygen of the carboxylate residue thus decreasing the

effective positive charge on the polarized carboxyl carbon atom. This leads to a weakening of the attraction between the nucleophile and the substrate and accounts for the increase in the enthalpy of activation[36].

2. Hydroxy acids and their anions

Results of studies by Franke and Brathun[37], and by Locke[38], on the decarboxylation of dihydroxymaleic acid indicate that the monoanion at 40° in water suffers decarboxylation 40 times faster than the undissociated acid. Brown, Hammick and Scholefield[39] studied the decarboxylation of several hydroxybenzoic acids in resorcinol and proposed a bimolecular mechanism for the reaction. Clark has studied the decarboxylation of 2,4-dihydroxybenzoic acid in amines, glycols[40], phenols, and acids[41]. His results show that at 100° the rate constant for the decarboxylation of 2,4-dihydroxybenzoic acid is 65 times greater in quinoline than it is in octanoic acid, as shown in Table 3.

TABLE 3. Kinetic data for the decarboxylation of 2,4-dihydroxybenzoic acid in acid and basic media

Solvent	$\Delta H^{\ddagger}$ (kcal/mole)	$\Delta S^{\ddagger}$ (E.U./mole)	$k_{100°} \times 10^7$ (sec^{-1})	Reference
Quinoline	34·4	5·95	9·55	40
Octanoic acid	32·8	−6·9	0·15	41

These results suggest that 2,4-dihydroxybenzoic acid ionizes in quinoline and the anion suffers decarboxylation, whereas in octanoic acid, ionization is repressed and the undissociated molecule decomposes. The supposition that the anion is involved in the decarboxylation of 2,4-dihydroxybenzoic acid in quinoline is supported by the fact that the enthalpy of activation of the reaction in quinoline is higher than it is in the acid solvent, as was noted in the parallel cases of oxalic acid and oxamic acid (see section II. A. 1. b). The fact that the anions of the β-hydroxyacids appear to be more unstable than the undissociated acids, in contrast to the behavior of α- and β-keto acids (section II. A. 1. b) may be a result of the greater hydrogen-bonding potential of the hydroxyl group as opposed to the keto group.

The mechanism of the decarboxylation of α- and β-hydroxy acids and their anions is probably quite similar to that of the decarboxylation of α- and β-keto acids and their anions. The rate-determining step in the decarboxylation of the hydroxy acids is no doubt the formation of an activated complex as illustrated by the following example:

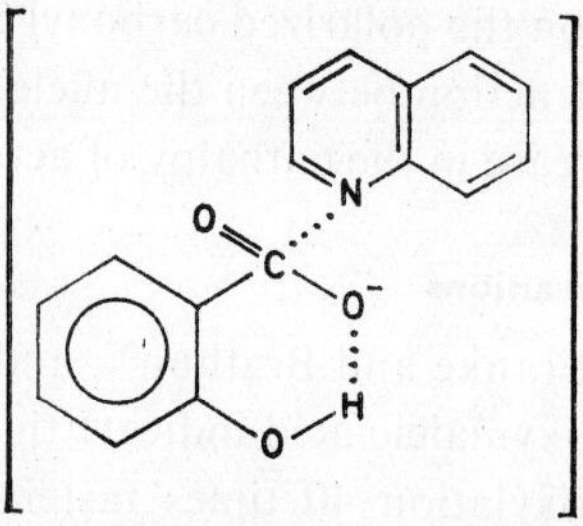

$$
\left[
\begin{array}{c}
\text{(structure of quinoline-salicylate anion complex)}
\end{array}
\right]
$$

3. The anions of trichloroacetic acid and related compounds

In 1884 Silberstein[42] observed that trichloroacetic acid is split quantitatively into chloroform and carbon dioxide when heated in organic bases such as aniline, dimethylaniline, quinoline and pyridine:

$$CCl_3COOH \longrightarrow CHCl_3 + CO_2$$

In 1906 Goldschmidt and Bräuer[43] studied the kinetics of the reaction in aniline solution and found it to be a first-order reaction. The same reaction was investigated by Salmi and Korte in dioxane–water mixtures[44]. Groote studied the decarboxylation of tribromoacetic acid in water[45], and Auerback, Verhoeck and Henne, the decarboxylation of the sodium salt of trifluoracetic acid in ethylene glycol[46]. Kappanna and coworkers studied the decarboxylation of trichloroacetic acid in water[47] and in pure aniline as well as in aniline–benzene and aniline–toluene mixtures[48]. Verhoeck and coworkers studied trichloroacetic acid and its salts in water, water–ethanol, formamide–water and aniline[49-52]. They suggested that the reaction is a unimolecular decomposition of the anion. Clark has studied the same reactions in glycerol[53], aromatic amines[54], and in fatty acids[55-57]. These investigations have established the fact that the unionized trihaloacetic acid is stable, decarboxylation involving the trihaloacetate anion. Studies of the reaction in various polar solvents indicate that the reaction is probably bimolecular and not unimolecular as postulated by Verhoeck and coworkers. The data shown in Table 4 will make this point clear.

As has been pointed out (section II. A. 1. a) malonic acid apparently does not ionize in aromatic amines and certainly not in acid solvents. Instead, the undissociated diacid combines with the nucleophile to form the activated complex. In bimolecular, nucleophilic addition reactions, any increase in the nucleophilicity of the reagent increases the attraction between it and the substrate, thus lowering the enthalpy of activation[36]. The profound differences in the nucleophilicities of acids as compared with aromatic amines is clearly demonstrated by the data in Table 4. Here it will be seen that the

TABLE 4. A comparison of the kinetic data for the decarboxylation of malonic acid and the trichloroacetate ion in various solvents

	Malonic acid[22] [58] [59]			Trichloroacetate ion[55] [56]		
	$\Delta H^{\ddagger}$ (kcal/mole)	$\Delta S^{\ddagger}$ (E.U./mole)	$k_{100°} \times 10^4$ (sec^{-1})	$\Delta H^{\ddagger}$ (kcal/mole)	$\Delta S^{\ddagger}$ (E.U./mole)	$k_{100°} \times 10^4$ (sec^{-1})
(1) Aniline	26·9	−4·46	1·3	24·5	−2·57	78
(2) o-Toluidine	25·7	−7·05	2·0	23·8	−6·82	29
(3) Quinoline	26·74	−2·37	5·0	23·98	−2·41	214
(4) 8-Methylquinoline	24·4	−10·47	2·1	22·31	−8·43	100
(5) Hexanoic acid	32·5	3·2	0·037	39·3	22·1	0·065

enthalpies of activation for the decarboxylation of malonic acid are considerably lower in the amines (lines 1, 2, 3 and 4) than in the acid (line 5).

The amines also differ among themselves as regards nucleophilicity. The methyl group in o-toluidine exerts both an inductive and a steric effect. The inductive effect increases the effective negative charge on the nitrogen atom, in other words it increases the nucleophilicity of the reagent. This is reflected by the fact that the enthalpy of activation for the decarboxylation of malonic acid is lower in o-toluidine than it is in aniline (compare lines 1 and 2 of Table 4).

The steric effect of the methyl group in the *ortho* position is an example of the *ortho* effect[60]. The approach of the electrophile to the nitrogen is hindered, thus decreasing the probability of the formation of the activated complex. The operation of this effect is demonstrated by the entropy of activation. The decrease in the entropy of activation in the decarboxylation of malonic acid on going from aniline to o-toluidine (see lines 1 and 2 of Table 4) is evidence for the steric effect in this particular reaction. The inductive and steric effects are also clearly demonstrated by the data in Table 4 for the decarboxylation of malonic acid in quinoline and in 8-methylquinoline (lines 3 and 4 of Table 4). The fact that the enthalpy of activation for the decarboxylation of malonic acid is lower in 8-methylquinoline than in quinoline is evidence for the inductive effect of the methyl group.

The data in Table 4 for the decarboxylation of the trichloroacetate anion in the various solvents exactly parallel those for malonic acid. Again the enthalpies of activation are lower in the basic solvents than in the acid, and the enthalpy of activation is lower in o-toluidine than in aniline, and is lower in 8-methylquinoline than in quinoline, pointing to the inductive effect of the methyl group. The steric effect of the methyl group is similarly demonstrated by the corresponding decreases in the entropies of activation.

These results indicate that the rate-determining step in the decarboxylation of the trichloroacetate anion in polar solvents must be, like that of unionized malonic acid, a bimolecular nucleophilic addition reaction. By analogy with malonic acid the activated complex may be formulated as follows:

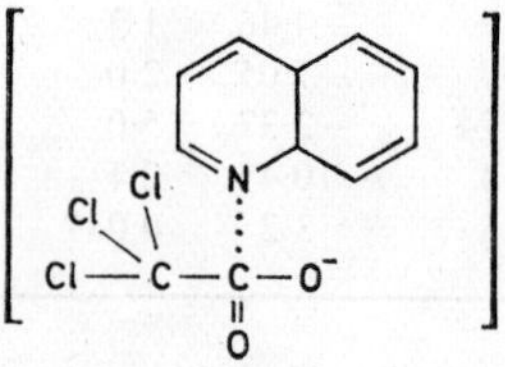

Dibromomalonic acid ($HOOCCBr_2COOH$) is somewhat analogous in structure to the trihaloacetic acids, inasmuch as all three α-hydrogens in acetic acid have been replaced by electron-attracting groups. Muus[61] has observed that the undissociated dibromomalonic acid is stable, and that the monoanion takes part in the decarboxylation reaction as in the case of the trihaloacetic acids. Like malonic acid and oxalic acid the dianion of dibromomalonic acid does not undergo decarboxylation. The activated complex for the decarboxylation of dibromomalonic acid monoanion is probably entirely analogous to that of the trihaloacetic acid.

Another group of compounds which bears a certain resemblance to malonic acid is the aliphatic α-nitro acids, (e.g. O_2NCH_2COOH). Pedersen[62-64] has observed that the monoanion of α-nitroacetic acid and related acids is involved in the decarboxylation reaction. The undissociated acid, like that of trichloroacetic acid, is stable. The decarboxylation of these compounds probably follows the same course as that of the trichloroacetate anion.

Moelwyn-Hughes and Hinshelwood[65], and Verhoeck[66] have investigated the decarboxylation of trinitrobenzoic acid and its salts in various solvents, and have found that this compound resembles the trihaloacetic acids in that the unionized molecule is stable, the anion participating in the decarboxylation reaction. It is reasonable to suppose that the decarboxylation of compounds of this class resembles that of the trichloroacetate anion.

4. Heterocyclic acids

Hammick and coworkers have studied the decarboxylation of quinaldinic acid (1)[67], isoquinaldinic acid (2)[67], picolinic acid (3)[68], and cinchoninic acid (4)[69]

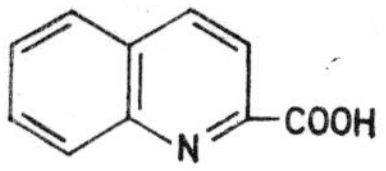

(1) 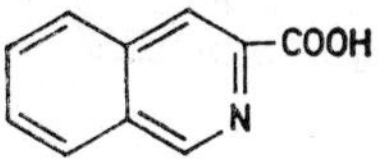 (2) 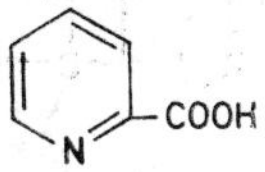 (3) 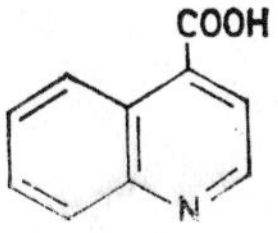 (4)

in the presence of aldehydes, ketones, esters, quinoline, and aromatic nitro compounds. Carbinols were found as by-products of the reaction in aldehydes and ketones, as shown by the following example:

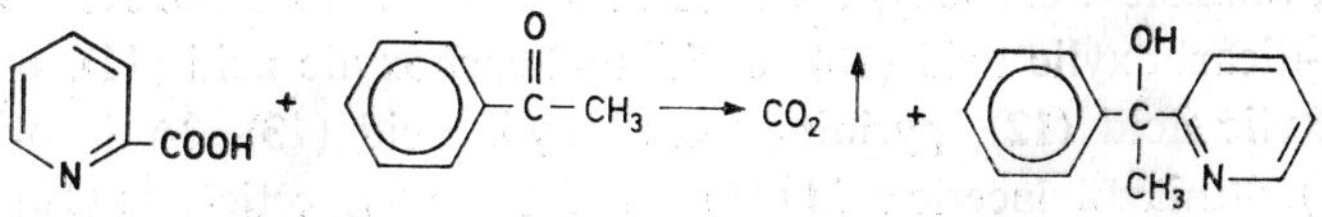

They found also[70] that quinaldinic acid suffers decarboxylation in basic solvents more rapidly than in acidic solvents. They explained these results by postulating a zwitterion structure for the transition state. By loss of carbon dioxide this would yield an intermediate carbanion which would be capable of interacting with an electrophilic agent. Their proposed mechanism is shown in the following example:

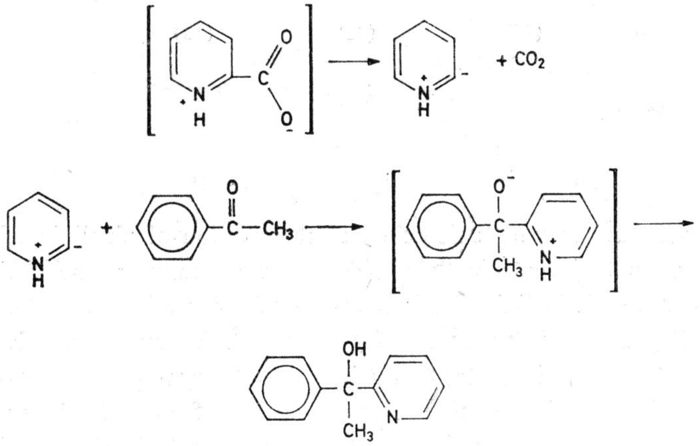

Doering and Pasternak[71] have shown that α-methyl-α-2-pyridylbutyric acid (5) and 4-pyridylacetic acid (6) probably exist as zwitterions in the transition state.

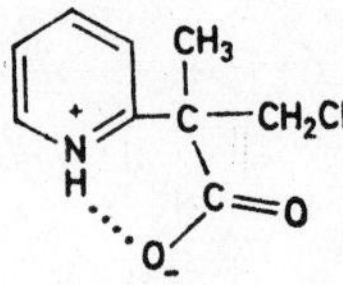

(5) 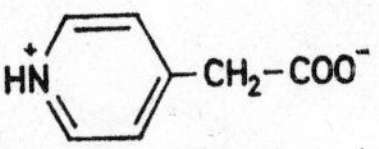   (6)

Schenkel and coworkers[72-75] studied the decarboxylation of a number of α- and β-heterocyclic acids in a variety of solvents and found that the reaction took place more rapidly in basic media than in acid solvents. The compounds which they investigated were as follows: thiazole-2-carboxylic acid (7), thiazole-4-carboxylic acid (8), thiazole-5-carboxylic acid (9), thiazole-4,5-dicarboxylic acid (10), pyridine-2-carboxylic acid (11), pyridine-3-carboxylic acid (12), pyridine-4-carboxylic acid (13), 2-thiazolylacetic acid (14), 4-thiazolylacetic acid (15), and 5-thiazolylacetic acid (16).

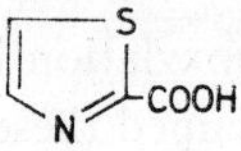

(7) 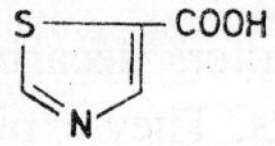   (8) 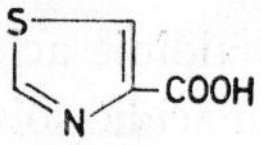   (9) 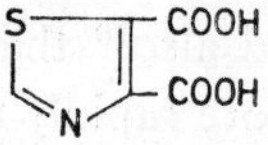   (10)

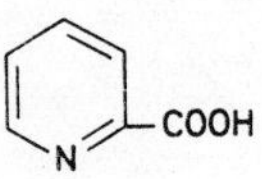

(11) 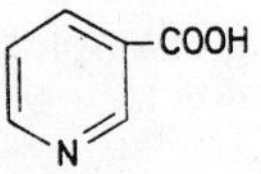   (12) 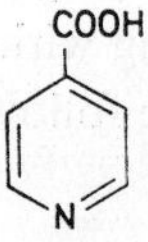   (13)

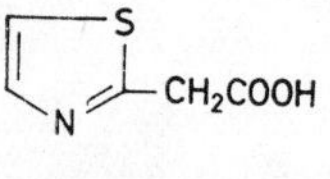

(14) 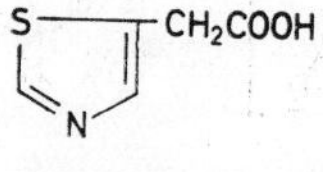   (15) 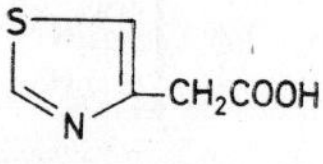   (16)

Cantwell and Brown studied the decarboxylation of picolinic acid (11) and methyl-substituted picolinic acids in p-dimethoxybenzene[76]. The decarboxylation of picolinic acid in a variety of polar solvents has been studied by Cantwell and Brown[77] and by Clark[78, 79]. A careful analysis of the experimental results indicates that it is not necessary to postulate a zwitterion as an intermediate in the decarboxylation of α- and β-heterocyclic acids. Instead the results indicate that there appears to be no essential difference in the mechanism of the decarboxylation of heterocyclic acids and that of the keto acids, hydroxy acids and trihaloacetate anions previously discussed. Support for these conclusions in provided by the data in Tables 5 and 6.

TABLE 5. Activation parameters for the decarboxylation of picolinic acid and malonic acid in several ethers

Solvent	Picolinic acid		Malonic acid		Reference
	$\Delta H^{\ddagger}$ (kcal/mole)	$\Delta S^{\ddagger}$ (E.U./mole)	$\Delta H^{\ddagger}$ (kcal/mole)	$\Delta S^{\ddagger}$ (E.U./mole)	
Phenetole	35·9	3·4	29·0	−6·0	78
β-Chlorophenetole	33·1	−2·0	27·8	−7·9	78
p-Bromoanisole	32·7	−2·5	—	—	77
p-Dimethoxybenzene	32·0	−4·6	27·1	−10·8	78

TABLE 6. Activation parameters for the decarboxylation of picolinic acid and of malonic acid in acid and basic solvents

Solvent	Picolinic acid[79]		Malonic acid[58, 59]	
	$\Delta H^{\ddagger}$ (kcal/mole)	$\Delta S^{\ddagger}$ (E.U./mole)	$\Delta H^{\ddagger}$ (kcal/mole)	$\Delta S^{\ddagger}$ (E.U./mole)
Octanoic acid	34·3	0·5	34·8	8·9
N-Ethylaniline	38·1	9·5	26·0	−6·6
N,N-Diethylaniline	36·8	6·7	25·4	−8·24

Table 5 shows the activation parameters for the decarboxylation of picolinic acid and of malonic acid in several mixed ethers. The solvents are arranged in the order of decreasing enthalpies of activation for the picolinic acid reaction. This is also the order of increasing nucleophilicity of these four solvents. The enthalpy of activation for the decarboxylation of picolinic acid decreases from 35·9 kcal/mole in phenetole to 32·0 kcal/mole in p-dimethoxybenzene. For the reaction in this series of solvents the entropy of activation decreases as the complexity of the reagent increases. Exactly the same trend is exhibited by the data for malonic acid. The enthalpy of activation for this reaction decreases from 29·0 kcal/mole in phenetole to 27·1 kcal/mole in p-dimethoxybenzene. Also the entropy of activation for this reaction decreases as the complexity of the solvent increases.

The rate-determining step in the decarboxylation of malonic acid in polar solvents has been shown to be the formation of an activated complex by way of a bimolecular nucleophilic addition reaction i.e. a nucleophilic attack on a π-bonded carbon atom (see section II. A. 1. a). The differences in the enthalpies and entropies of activation for the decarboxylation of malonic acid in the various ethers shown in Table 5 are consistent with the proposed mechanism. Since the enthalpy and entropy changes for the decarboxylation of picolinic acid in these solvents parallel those of malonic acid the conclusion is inescapable that the mechanism of the decarboxylation of picolinic acid in ethers is similar to that of malonic acid, and it appears highly probable that the unionized, five-membered, chelate structure of picolinic acid, and not the zwitterion, is involved in the decarboxylation of this acid in neutral solvents.

Further insight into the details of the mechanism of the decarboxylation of picolinic acid in polar solvents may be gained from Table 6 which presents data for the decarboxylation of picolinic acid and malonic acid in acid and basic solvents.

It will be seen in Table 6 that the enthalpy of activation for the decarboxylation of malonic acid is lower in the basic solvents, N-ethylaniline and N,N-diethylaniline, than it is in octanoic acid. Here there is no question of ionization of malonic acid in the basic media. The unionized, neutral molecule in the six-membered, chelate structure is involved in the formation of the activated complex in both the acid and basic solvents, as demonstrated previously (see section II. A. 1. a). The decrease in the enthalpy of activation for the malonic acid reaction on passing from the acid to the basic solvents is evidently a result of the greater nucleophilicity of the amines as compared with the acid.

In contrast to the bahavior of malonic acid, the enthalpies of activation for the decarboxylation of picolinic acid are *higher* in the basic solvents than in the acid. In this respect, picolinic acid resembles oxalic acid and oxamic acid (see section II. A. 1. b). The interpretation of the results obtained in the decarboxylation of oxalic and of oxamic acid was that in the basic solvents, the anion is involved in the rate-determining step. The higher enthalpy of activation for the anion reaction arises from the fact that, with the loss of the proton, the negative charge on the oxygen tends to partially neutralize the effective positive charge on the polarized carbonyl carbon atom.

The supposition that the mechanism of the reaction involving the anion of picolinic acid is quite similar to that involving the neutral molecule is supported by the fact that the enthalpy of activation and the entropy of activation for the picolinic acid reaction are both lower in N,N-diethylaniline than in N-ethylaniline (see Table 6). The behavior of picolinic acid in these two solvents parallels that of the trichloroacetate anion reaction in similar solvents (see section II. A. 3.). The differences in the enthalpies and entropies of activation for the reaction in these two amines are consistent with a bimolecular, nucleophilic addition reaction involving the picolinic acid anion.

In the decarboxylation of keto acids, hydroxy acids, trihaloacetic acids, and heterocyclic acids, and the corresponding monoanions, the loss of carbon dioxide by heterolytic fission always yields a carbanion as an intermediate. The production of carbinols in the case of the decarboxylation of picolinic acid and other imino acids dissolved in carbonyl-type compounds[67-69] is readily explicable by this circumstance without recourse to a zwitterion hypothesis.

TABLE 7. Pseudo first-order rate constants for the decarboxylation of picolinic acid in acid and basic media[79]

Solvent	$k_{100°} \times 10^3$ (sec^{-1})
Octanoic acid	8·05
N-ethylaniline	4·41
N,N-diethylaniline	6·17

As shown in Table 7, the rate constants at 100° for the decarboxylation of picolinic acid in the basic, ionizing solvents are lower than the one for the reaction in octanoic acid. If we accept the hypothesis that picolinic acid is ionized in the two amines, but not in the acid, we can see that the unionized picolinic acid at this temperature is less stable (with respect to decarboxylation) than the anion. In this respect picolinic acid resembles the keto acids (see section II. A. 2, and Table 2). (At temperatures below the isokinetic temperature the reaction having the lowest enthalpy of activation has the highest rate[80]. This accounts for the fact that the decarboxylation of picolinic acid at 100° takes place faster in N,N-diethylaniline than it does in N-ethylaniline as shown in Table 7.)

5. Miscellaneous examples of decarboxylation not covered in the preceding sections

The various acids thus far discussed which are capable of losing carbon dioxide by heterolytic fission fall into two distinct classes:

(1) Those whose anions are less stable (with respect to decarboxylation) than the unionized acids;

(2) those whose anions are more stable (with respect to decarboxylation) than the unionized acids.

In class (1) belong some acids which are very resistant to decarboxylation but whose anions readily lose carbon dioxide (e.g. trichloroacetic acid, dibromomalonic acid, α-nitroacetic acid, trinitrobenzoic acid, amino acids, and related compounds; see section II. A. 3), as well as some acids which undergo cleavage although not as easily as their corresponding mono-anions (*e.g.*, dihydroxymaleic acid and dihydroxybenzoic acid; see section II. A. 2). The keto acids (section II. A. 1) and the heterocyclic acids (section II. A. 4.) constitute class (2).

Structurally, members of class (2) are distinguished by a double-bonded oxygen or nitrogen atom in the α or β position. Members of class (1) have a carboxyl group attached to an aromatic nucleus, or to a carbon atom containing one or more electron-attracting groups other than keto or imino. It thus appears that the behavior of a particular acid (whether according to class 1 or class 2) depends to some extent on the strength of the acid and to some extent on its structure. A preponderance of strongly electron-attracting groups in the α position will tend to increase the strength of the acid promoting ionization. The stronger the acid the less stable (with respect to decarboxylation) will be the resulting anion. The presence of relatively weak electron-attracting groups in the α or β position (*e.g.* keto groups or imino groups) will favor the formation of unionized 5- or 6-membered chelate rings by intramolecular hydrogen bonding. These acids will not be as strong as the members of class (1) and hence will not ionize as readily. The unionized acid will be less stable (with respect to decarboxylation) than the anion. In representatives of both class (1) and class (2) all degrees of stability of both acid and anion are possible from an acid which is very stable (very little tendency to lose carbon dioxide) and whose anion is correspondingly very unstable (*e.g.* trichloroacetic acid[49]) to an acid which is extremely unstable (strong tendency to lose carbon dioxide) and whose anion is almost completely inert (*e.g.* dimethylacetoacetic acid[27]). In the decarboxylation of the unionized as well as the ionized form of both types of acids the mechanism is evidently the same in every case — the formation of an activated complex by way of a nucleophilic bimolecular addition reaction, followed by loss of carbon dioxide by heterolytic cleavage. The rate-determining step in every case appears to be the formation of the activated complex via a nucleophilic attack on a π-bonded carbon atom.

Electron-attracting substituents in the α or β position inducing decarboxylation are not limited to the types already discussed. Several other types of compounds which illustrate the principles adduced above will be briefly mentioned. Cram and Wingrove[81] found that the anion of (1)-2-benzenesulfonyl-2-methyloctanoic acid is readily decarboxylated although the unionized acid is stable: $CH_3(CH_2)_5C(CH_3)(SO_2C_6H_5)COOH$.

This compound is analogous to α-nitroacetic acid (see section II. A. 3), since the electron-withdrawing α-benzenesulfonyl group promotes ionization and facilitates decarboxylation of the anion. This compound thus fits into class (1) of the foregoing classification.

Gleason and Dougherty observed that unionized o-benzoylbenzoic acid (17) undergoes decarboxylation more readily than the anion[82]:

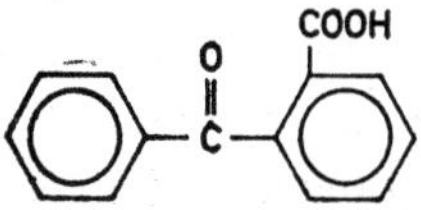

(17)

This compound is evidently vinylogous with the α-keto acids and is thus a member of class (2).

It has been suggested that the mechanism of the decarboxylation of α,β- and β,γ-unsaturated acids is similar to that of β-keto acids[83]. This group of compounds therefore would belong to class (2) of the foregoing classification.

Tommila and Kivinen[85] found that acetylene dicarboxylic acid in water at 70° decomposes about 30% faster than its anion. This places the compound in class (2), but it is obviously close to the borderline where there would be little difference in the relative stabilities of the unionized or ionized forms.

The decarboxylation of anthracene-9-carboxylic acid (18) has been stu-

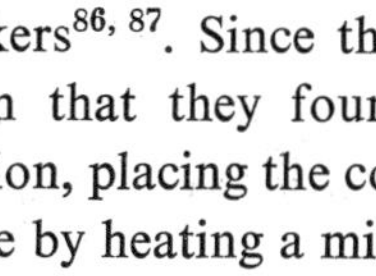

(18)

died by Schenkel and coworkers[86, 87]. Since this is a relatively weak acid, it is not surprising to learn that they found the unionized acid to be more easily split than the anion, placing the compound in class (2).

The production of methane by heating a mixture of sodium acetate and soda lime (see section I.B) no doubt takes place by heterolytic splitting. This reaction is analogous to the ketone decarboxylation reaction illustrated by the production of acetone by heating calcium acetate:

$$CH_3COO\text{---}Ca\text{---}OCOCH_3 \longrightarrow CaCO_3 + CH_3COCH_3$$

The latter reaction also probably takes place by an ionic mechanism.

B. Homolytic Fission

It is as well to consider a few decarboxylation reactions which involve homolytic rather than heterolytic fission. A free radical mechanism has been invoked to explain the Hunsdieker reaction[88], *i.e.*, the conversion of silver salts of carboxylic acids to the corresponding halides by treatment with elemental bromine or iodine in an inert solvent. Homolytic fission is also involved in the Kolbe synthesis[89]. For example, in the electrolysis of sodium acetate, the acetate anion moves to the anode where it is changed to a neutral acetate radical by the loss of an electron. It is believed that the acetate radical then loses carbon dioxide by homolytic fission forming a methyl radical. Ethane then results from the combination of two methyl radicals. Finally, photochemical decomposition of carboxylic acids may be mentioned as another example of a decarboxylation reaction taking place by homolytic fission[90, 91].

III. DISCUSSION OF SEVERAL SPECIFIC EXAMPLES

A. Decarboxylation

1. Malonic acid

The activation parameters for the decarboxylation of malonic acid in the molten state and in various aliphatic and aromatic acids are shown in Table 8. In view of the amphiprotic character of the hydroxyl group, the solvent molecule in these reactions probably functions as the nucleophilic agent, while malonic acid acts as the electrophile in the formation of the activated complex. It was shown in section II. A. 4 (Table 6) that an increase in the nucleophilicity of the solvent causes a decrease in the enthalpy of activation for the malonic acid reaction. This same principle

TABLE 8. Activation parameters for the decarboxylation of malonic acid in the molten state and in several acid solvents

Solvent	$\Delta H^{\ddagger}$ (kcal/mole)	$\Delta S^{\ddagger}$ (E.U./mole)	$\Delta F^{\ddagger}_{135°}$ (kcal/mole)	Reference
Melt	35·8	11·9	31·1	92
Propionic acid	33·6	6·1	31·1	93
Hexanoic acid	32·5	3·2	31·2	94
Benzoic acid	30·4	−1·8	31·1	59
Heptanoic acid	29·7	−3·4	31·1	59
Decanoic acid	26·6	−11·0	31·1	94
d,1-2-Methylpentanoic acid	26·45	−11·1	31·0	59

is illustrated in Table 8. It will be seen that in general the enthalpy of activation for the decarboxylation of malonic acid decreases with increasing chain length of the alkyl residue on the nucleophile. Furthermore, the entropy of activation decreases as the complexity of the solvent increases.

According to the absolute reaction rate equation (1) the rate constant of a reaction depends upon the free energy of activation[95]. The free energy

$$k = \frac{\varkappa T}{h} e^{-\Delta F^{\ddagger}/RT} \tag{1}$$

of activation at any temperature may be calculated by means of the equation (2):

$$\Delta F^{\ddagger} = \Delta H^{\ddagger} - T \Delta S^{\ddagger} \tag{2}$$

Values of the free energy of activation for the decarboxylation of malonic acid in each of the solvent at 135° (the melting point of malonic acid), calculated by means of equation (2), are shown in the third column of Table 8.

It is interesting to note in Table 8 that the free energy of activation at the melting point of malonic acid is approximately the same in the various solvents (about 31·1 kcal/mole). If this value of the free energy of activation at 135° (31·1 kcal/mole) is substituted in the absolute reaction rate equation (1), the rate constant at this temperature may be calculated. The result of such a calculation yields $k_{135°} = 0.00019$ sec^{-1}, corresponding to a half-life of 61 min. Since the rate constant at 135° is the same for the decarboxylation of malonic acid in the molten state as it is for the reaction in the acid media, these results indicate that, at the melting point of malonic acid, the rate of reaction is not affected by the presence of acidic solvents.

The temperature at which a given reaction proceeds at the same rate in different solvents is known as the isokinetic temperature of the reaction series[80]. When a reaction is carried out at some temperature *below* the isokinetic temperature, the rate constant is greater the *lower* the enthalpy of activation. When it is carried out at some temperature *above* the isokinetic temperature, the rate constant is greater the *higher* the enthalpy of activation[80].

It is often found that a change in type of solvent produces a new reaction series having the same isokinetic temperature as the first but a different rate constant[96]. An illustration of this effect is afforded by the data for the decarboxylation of malonic acid in alkanols shown in Table 9.

TABLE 9. Activation parameters for the decarboxylation of malonic
acid in alkanols[97, 98]

Solvent	$\Delta H^{\ddagger}$ (kcal/mole)	$\Delta S^{\ddagger}$ (E.U./mole)	$\Delta F^{\ddagger}_{135°}$ (kcal/mole)
n-Butyl alcohol	27·2	−4·4	29·0
n-Hexyl alcohol	26·0	−7·6	29·1
2-Ethylhexanol-1	24·8	−10·4	29·05
Di-isobutylcarbinol	24·8	−10·7	29·13
Cyclohexanol	23·0	−15·0	29·12

The changes in the enthalpy and entropy of activation for the reaction
in the various alcohols shown in Table 9 reflect the differences in the
inductive and steric effects of the different groups attached to the hydroxyl moiety. The enthalpy as well as the entropy of activation for the reaction is higher in the simplest solvent, n-butyl alcohol, than in any of the
others. Both parameters then decrease regularly with increasing complexity
of the solvents. The data in Table 9 serve as cogent arguments for the
bimolecularity of the reaction.

The free energies of activation at 135° for the decarboxylation of malonic acid in the various alcohols, shown in the third column of the table,
are approximately equal (about 29·1 kcal/mole). The rate constant at
135° calculated in the usual manner, is 0·00216 sec⁻¹, corresponding to
a half-life of about 5·5 min. We find therefore that the decarboxylation
of malonic acid in alcohols has the same isokinetic temperature as that
for the reaction in acids but that the reaction at 135° in alcohols takes place
about 11 times as fast as it does in the acid solvents.

A third series again having an isokinetic temperature of 135° is found in
aromatic amines as shown in Table 10. The average free energy of activation at 135° is about 28·75 kcal/mole. This corresponds to a value of

TABLE 10. Activation parameters for the decarboxylation of malonic
acid in amines[58]

Solvent	$\Delta H^{\ddagger}$ (kcal/mole)	$\Delta S^{\ddagger}$ (E.U./mole)	$\Delta F^{\ddagger}_{135°}$ (kcal/mole)
m-Toluidine	28·4	−0·8	28·72
Aniline	26·9	−4·45	28·70
N-Methylaniline	26·6	−5·33	28·77
N,N-Diethylaniline	25·4	−8·24	28·76
2,6-Dimethylaniline	23·1	−14·0	28·80

$k_{135°}$ of 0·0034 sec^{-1}, representing a half-life of 3·3 min. Thus, the decarboxylation of malonic acid at 135° takes place 18 times as fast in amines as it does in the acid solvents at this temperature. Thus the decarboxylation of malonic acid forms a separate reaction series in each group of solvents, all having the same isokinetic temperature, while the rate constant at the isokinetic temperature increases with increasing nucleophilicity of the solvents.

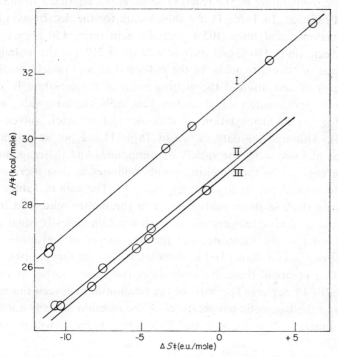

FIGURE 2. Plot of enthalpy versus entropy of activation for the decarboxylation of malonic acid in acids (line I), in alkanols (line II) and in amines (line III). Slope of the lines: 408°K (135°C).

A plot of enthalpy of activation versus entropy of activation for the decarboxylation of malonic acid in the three types of solvent (based on the data in Tables 8, 9, and 10) forms three parallel lines, the slope of the lines (which corresponds to the isokinetic temperature) being equal to the melting point of malonic acid (see Figure 2). The intercept of the isokinetic temperature line on the zero entropy of activation axis yields $\Delta F°$, the free energy of activation at the isokinetic temperature for all the reactions conforming to the line[96]. The smaller the value of $\Delta F°$ the faster is

39*

the reaction at the isokinetic temperature. In other words, the lower the line in Figure 2 the faster is the reaction at the isokinetic temperature. It will be noted that the three lines are arranged in the order of increasing nucleophilicity of the homologous series.

2. Other acids

The relationship between the melting point of the substrate and the isokinetic temperature of the reaction series is not restricted to the malonic acid reaction. In Table 11 are shown data for the decarboxylation of n-hexylmalonic acid (m.p. 105°), oxanilic acid (m.p. 150°), cyclohexylmalonic acid (m.p. 180°), and oxamic acid (m.p. 210°) in the molten state and in one or two fatty acids. In the molten state and in acid media, the free energy of activation at the melting point of the substrate is a constant and approximately equal to 31·1 kcal/mole, the same value as that found for the decarboxylation of malonic acid in acidic solvents (see Table 8). Although the data shown in Table 11 are not as complete as those for malonic, acid, the variety of compounds and the range of temperature covered by their melting points militates against ascribing the results to coincidence or to experimental error. The data in Tables 8 and 11 indicate that, in those acids studied in the molten state and in acid media, the isokinetic temperature of each reaction series is equal to the melting point of the substrate, and the free energy of activation at the melting point is a constant (31·1 kcal/mole). It is interesting to observe that the melting points of these five acids differ from one another by integral multiples of 15 degrees. The basis of the relationship between the melting points and the isokinetic temperatures of the reaction series is not immediately apparent. It is to be hoped that future research will furnish an explanation for it.

Many apparently unrelated reaction series involving decarboxylation exhibit identical isokinetic temperatures. The decarboxylation of benzylmalonic acid and its derivatives in aromatic amines forms a reaction series with an isokinetic temperature of 105° (see Table 12). The same isokinetic temperature is shown by the decarboxylation of malonanilic acid in the cresols (Table 13). Three different reaction series each having an isokinetic temperature of 150° are formed by the decarboxylation of 2,4-dihydroxybenzoic acid, benzylmalonic acid, and malonanilic acid in quinoline and its derivatives, as shown in Table 14. These isokinetic temperatures, like the melting points of the acids mentioned previously, differ from one another by integral multiples of 15 degrees.

In Tables 8, 9 and 10 we observed that several different reaction series

TABLE 11. Activation parameters for the decarboxylation of several acids in the molten state and in acid media[a]

Solvent	n-Hexylmalonic acid[35, 100] (m.p. 105°)			Oxanilic acid[59, 99] (m.p. 150°)			Cyclohexylmalonic acid[35] (m.p. 180°)			Oxamic acid[35] (m.p. 210°)		
	$\Delta H^\ddagger$	$\Delta S^\ddagger$	$\Delta F^\ddagger_{105^\circ}$	$\Delta H^\ddagger$	$\Delta S^\ddagger$	$\Delta F^\ddagger_{150^\circ}$	$\Delta H^\ddagger$	$\Delta S^\ddagger$	$\Delta F^\ddagger_{180^\circ}$	$\Delta H^\ddagger$	$\Delta S^\ddagger$	$\Delta F^\ddagger_{210^\circ}$
Melt	32·2	2·8	31·1	40·1	21·4	31·05	—	—	—	—	—	—
Hexanoic acid	31·24	0·54	31·1	—	—	—	31·82	1·94	31·09	30·5	−1·25	31·1
Octanoic acid	30·75	−1·0	31·1	36·73	13·3	31·1	33·7	5·61	31·16	30·5	−1·25	31·1

[a] Units: $\Delta H^\ddagger$ (kcal/mole); $\Delta S^\ddagger$ (eu/mole); $\Delta F^\ddagger$ (kcal/mole).

Louis W. Clark

TABLE 12. Activation parameters for the decarboxylation of benzylmalonic acid in amines[102]

Solvent	$\Delta H^{\ddagger}$ (kcal/mole)	$\Delta S^{\ddagger}$ (E.U./mole)	$\Delta F^{\ddagger}_{105°}$ (kcal/mole)
Aniline	19·8	−21·64	28·0
N-Ethylaniline	21·9	−15·8	27·9
N-s-Butylaniline	26·56	−3·6	27·92
o-Toluidine	29·9	5·0	28·0
N,N-Diethylaniline	38·4	27·4	28·05

TABLE 13. Activation parameters for the decarboxylation of malonanilic acid in cresols[103]

Solvent	$\Delta H^{\ddagger}$ (kcal/mole)	$\Delta S^{\ddagger}$ (E.U./mole)	$\Delta F^{\ddagger}_{105°}$ (kcal/mole)
m-Cresol	33·2	9·4	29·64
p-Cresol	34·0	11·54	29·63
o-Cresol	35·5	15·5	29·64

having the same isokinetic temperature were formed using the same substrate (malonic acid) and varying the type of solvent. As the nucleophilicity of the solvent increased the free energy of the reaction at the isokinetic temperature decreased. In Table 14 we observe three different reaction series having the same isokinetic temperature (150°) formed by using the same type of solvent but varying the substrate. We may deduce in this case that the free energies of activation at the isokinetic temperature are lower the greater the attraction between the substrate and the solvent.

Not only can a change in type of solvent (or substrate) produce a new reaction series having the same isokinetic temperature as the first, but, under suitable conditions, it may result in the formation of a new reaction series having a different isokinetic temperature. Table 11 illustrated this occurrence as a result of change in the substrate. The other case is illustrated by the data in Tables 13 and 14. The decarboxylation of malonanilic acid in the cresols forms a reaction series with an isokinetic temperature of 105° (see Table 13) whereas the reaction in quinoline and its derivatives (Table 14) constitutes a reaction series with an isokinetic temperature of 150°. It is probable that the decarboxylation of malonanilic acid as well as that of benzylmalonic acid differs in some points from that of

TABLE 14. Activation parameters for the decarboxylation of several acids in quinoline and its derivatives

Solvents	2,4-dihydroxybenzoic acid[101]			Malonanilic acid[102]			Benzylmalonic acid[103]		
	$\Delta H^\ddagger$	$\Delta S^\ddagger$	$\Delta F^\ddagger_{150°}$	$\Delta H^\ddagger$	$\Delta S^\ddagger$	$\Delta F^\ddagger_{150°}$	$\Delta H^\ddagger$	$\Delta S^\ddagger$	$\Delta F^\ddagger_{150°}$
Quinoline	34·5	5·95	32·0	21·0	−17·5	28·4	19·9	−19·94	28·35
8-Methylquolinine	22·9	−21·8	32·1	28·5	0·4	28·33	26·4	−4·6	28·35

^{a}Units: $\Delta H^\ddagger$ (kcal/mole), $\Delta S^\ddagger$ (e.u./mole), $\Delta F^\ddagger$ (kcal/mole).

malonic acid[103]. The first two acids appear to function as nucleophiles rather than electrophiles*.

We have seen that malonic acid and its derivatives are related to β-keto acids, whereas oxanilic acid and oxamic acid are related to α-keto acids. It is therefore logical to suppose that the results obtained in studies of the decarboxylation of malonic acid, oxanilic acid, *etc.*, will apply also to the behavior of α- and β-keto acids in general. Furthermore, in view of the fact that the decarboxylation of heterocyclic acids, trihaloacetic acids, hydroxy acids, and other types of acids, evidently takes place by a mechanism similar to that of the keto acids, it is not unlikely that similar relationships may apply to all decarboxylations involving bimolecular nucleophilic addition and heterolytic fission.

The mechanism of the decarboxylation reaction is not unlike the general S_N2 mechanism by which many other organic reactions takes place[23]. All these reactions are alike with respect to the rate-determining step, which consists in the formation of an activated complex involving an electrophile–nucleophile pair. In view of these circumstances it is reasonable to surmise that the principles adduced above for the decarboxylation reaction may apply to other types of bimolecular heterolytic reactions. Evidence for the plausibility of this assumption is presented in the following sections.

B. Relation Between Decarboxylation and Other Bimolecular Heterolytic Reactions

Meloche and Laidler[104] studied the acid- and base-catalyzed hydrolysis of benzamide and its derivatives. Rate constants for the various reactions were measured over about a 50° range of temperature in 0·025 M benzenesulfonic acid and in 0·025 M sodium hydroxide solutions using an equal concentration of reactant in a 60–40 by volume alcohol–water mixture. The activation parameters and isokinetic temperatures for the various reactions are shown in Table 15. (The isokinetic temperatures shown in Table 15 were not reported by Meloche and Laidler, but were calculated from their data by the present writer.)

It will be seen in Table 15 that a different reaction series is formed by the hydrolysis of benzamide and each of its derivatives in solutions of varying pH. Three of these series (lines 3, 4 and 5) have the same isokinetic

---

* Nucleophilic addition and electrophilic addition are essentially the same reaction— it is merely a matter of convention to distinguish between them. In either case the rate-determining step is the same, the formation of a nucleophile–electrophile pair in the transition state.

TABLE 15. Activation parameters[104] and isokinetic temperatures for the acid- and base-catalyzed hydrolysis of benzamide and its derivatives

Substrate	m.p.	Acid hydrolysis		Alkaline hydrolysis		Isokinetic temperature
		$\Delta H^{\ddagger}$ (kcal/mole)	$\Delta S^{\ddagger}$ (E.U./mole)	$\Delta H^{\ddagger}$ (kcal/mole)	$\Delta S^{\ddagger}$ (E.U./mole)	(°C)
(1) p-Nitrobenzamide	200°	23·9	−12·5	15·4	−30·0	210°
(2) p-Chlorobenzamide	180°	23·0	−14·1	16·9	−29·0	135°
(3) o-Methylbenzamide	—	26·4	−10·8	22·4	−21·3	105°
(4) p-Methylbenzamide	—	21·3	−18·9	18·2	−27·2	105°
(5) Benzamide	130°	22·6	−14·7	18·0	−26·8	105°

temperature (105°), while in the case of the other two series (lines 1 and 2) each has a different isokinetic temperature (135°, 210°). With the three series having the same isokinetic temperature it will be seen in Figure 3 that the free energy of activation at the isokinetic temperature decreases as the electrophilicity of the substrate increases. The introduction of strongly electron-attracting groups on the aromatic nucleus of the substrate (chloro-, nitro-) increases the effective positive charge on the carbonyl carbon atom, raising the isokinetic temperature. For each reaction series shown in Table 15 the rate of hydrolysis at the isokinetic temperature is independent of pH.

It is interesting to note that the isokinetic temperatures of the various reaction series involving *hydrolysis* shown in Table 15 are identical with those of the reaction series formed by the *decarboxylation* of n-hexylmalonic acid, malonic acid, and oxamic acid in acid media (see section III. A. 1). Furthermore, the increase in the isokinetic temperature of the different reaction series shown in Table 15 parallels the increase in the melting point of the substrate; similar behavior was observed with acids. These results may show that some principles deduced from studies of the decarboxylation reaction apply to other bimolecular heterolytic reactions, too.

In the S_N2 alcoholysis of acyl halides, the alcohol is the nucleophilic agent and the acyl halide is the electrophile. The rate-determining step is probably analogous to that for the decarboxylation and hydrolysis reactions, namely, the formation of an activated complex via a nucleophilic attack on a π-bonded carbon atom:

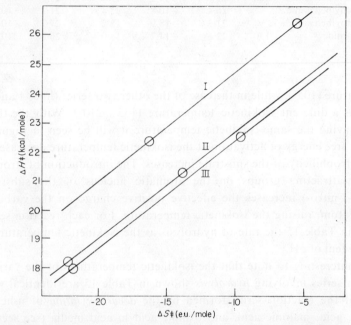

Kivinen[105] has made an extended study of the kinetics of the ethanolysis of substituted benzoyl chlorides in various mixed solvents. (Diluting the reaction mixture with a non-polar solvent, such as benzene, would be expected to affect a reaction in two ways: (1) by lowering the overall

FIGURE 3. Enthalpy versus entropy of activation plot for the acid and base-catalyzed hydrolysis of benzamide and its derivatives, based upon the data in Table 15 (lines 3,4 and 5). Line I: o-Methylbenzamide; line II: p-Methylbenzamide; line III: Benzamide. (Slope of the lines: 378°K or 105°C).

dielectric constant of the medium the electrostatic attraction between reactants would increase, thus lowering the enthalpy of activation; (2) it would tend to decrease the probability of effective collisions between reactant molecules, thus lowering the entropy of activation.) Kivinen's results strongly suggest that solvolysis reactions obey the same laws as decarboxylation and hydrolysis. For example, he found that the isokinetic temperature for the ethanolysis of m-nitrobenzoylchloride in various

TABLE 16. Free energies of activation at the isokinetic temperature ($-15°$) for the ethanolysis of substituted benzoyl chlorides in benzene solutions[105]

Electrophile	$\Delta F^{\ddagger}_{15°}$ (kcal/mole)
p-Methoxybenzoyl chloride	22·65
m-Methoxybenzoyl chloride	22·15
m-Chlorobenzoyl chloride	21·4
m-Nitrobenzoyl chloride	20·38

ethanol–benzene mixtures is $-15°$, and that for the ethanolysis of the same substrate in ethanol–ether mixtures is $-75°$.

Table 16 shows that the free energy of activation at the isokinetic temperature for four aroyl halides decreases as the *electrophilicity of the substrate* increases. It is interesting to compare these results with the data for the decarboxylation of malonic acid in several different homologous series (see Tables 8, 9 and 10) in which it was observed that the free energy of activation at the isokinetic temperature decreases as the *nucleophilicity of the solvent* increases.

In three ethanolysis reactions of aroyl halides in ethanol/ether mixtures having different isokinetic temperatures (m-nitrobenzoyl, $-75°$, m-bromobenzoyl, $-90°$, and m-methylbenzoyl, $-135°$), the isokinetic temperatures increase with increasing electrophilicity of the substrate, and parallel the increase in the melting points of the substrates. For each of these three reaction series the free energy of activation at the isokinetic temperature is constant, (18·4 kcal/mole). It will be recalled that analogous results were observed in the case of the decarboxylation of substituted malonic acids in acid media (see Table 11).

It is interesting to note that the isokinetic temperatures for the ethanolysis, hydrolysis and decarboxylation reactions discussed above differ from each other by integral multiples of 15. Although the real significance of these numbers is not readily apparent, they possibly indicate a fundamental connection between the various types of heterolytic reactions.

C. Conclusion

The results outlined in the preceding sections indicate that an increase in the mutual attraction between the electrophile–nucleophile pair taking part in a bimolecular heterolytic reaction (caused either by an increase in the effective positive charge on the electrophile or by an increase in the effective negative charge on the nucleophile) may either (1) increase the rate constant at the isokinetic temperature (without changing the isokinetic temperature), or (2) raise the isokinetic temperature. Low isokinetic temperatures are associated with reaction series involving electrophile–nucleophile pairs having relatively weak attractions for one another. Such reactions will ordinarily take place at temperatures *above* the isokinetic temperature, hence their speed will be governed by the entropy of activation, not the enthalpy of activation[95]. High isokinetic temperatures are associated with reaction series involving electrophile–nucleophile pairs having strong attractions for one another. Such reactions will ordinarily take place at temperatures *below* the isokinetic temperature, hence their speed will be governed by the enthalpy of activation, not the entropy of activation[95].

The above principles not only provide a deeper insight into the intricacies of the decarboxylation reaction but also furnish a basis for a broader understanding of other bimolecular heterolytic reactions. Their future application should aid in simplifying, organizing, correlating and systematizing the complex field of organic reactivity.

IV. REFERENCES

1. M. Calvin, J. A. Bassham and A. A. Benson, *Federation Proc.* **9**, 524 (1950).
2. P. Karlson, *Introduction to Modern Biochemistry*, 2nd ed., Academic Press, New York, 1965, p. 205.
3. J.. March, *J. Chem. Ed.*, **40**, 212 (1963).
4. R P. Linstead and B. C. L. Weedon, *J. Chem. Soc.*, 1538 (1953).
5. A. L. Miller, N. C. Cook and F. C. Whitmore, *J. Am. Chem. Soc.*, **72**, 2733 (1950).
6. F. F. Blicke and R. F. Feldkampf, *J. Am. Chem. Soc.*, **66**, 1087 (1944).
7. R. T. Arnold and P. Morgan, *J. Am. Chem. Soc.*, **70**, 4248 (1948).
8. P. Wilder, Jr. and A. Winston, *J. Am. Chem. Soc.*, **75**, 5370 (1953).
9. A. C. Cope and M. E. Synerholm, *J. Am. Chem. Soc.*, **72**, 5228 (1950).
10. A. Weissberger and H. D. Porter, *J. Am. Chem. Soc.*, **66**, 1849 (1944).
11. A. H. Blatt, (Ed.), *Organic Syntheses*, Coll. Vol. 2, John Wiley and Sons, New York, 1943, p. 512.
12. Ref. 11, Volume 1, p. 192.
13. G. Bredig and R. W. Balcom, *Chem. Ber.* **41**, 740 (1908).
14. K. J. Pedersen, *J. Phys. Chem.*, **38**, 559 (1932).
15. F. H. Westheimer and W. A. Jones, *J. Am. Chem. Soc.*, **63**, 3283 (1941).
16. J. A. King, *J. Am. Chem. Soc.*, **69**, 2738 (1947).
17. H. Schenkel and M. Schenkel–Rudin, *Helv. Chim. Acta*, **31**, 514 (1948).
18. J. Bredt, *Ann. Acad. Sci. Fennicae Ser. A* **29**, No. 2 (1927).
19. A. L. Bernoulli and W. Wege, *Helv. Chim. Acta*, **2**, 511 (1919).
20. A. L. Bernoulli and H. Jakubowics, *Helv. Chim. Acta*, **4**, 1018 (1921).
21. G. Fraenkel, R. L. Belford and P. E. Yankwich, *J. Am. Chem. Soc.*, **76**, 15 (1954).

22. L. W. Clark, *J. Phys. Chem.*, **62**, 500 (1958).
23. C. K. Ingold, *Structure and Mechanism in Organic Chemistry*, Cornell University Press, Ithaca, New York, 1953, pp. 212–214.
24. B. R. Brown, *Quart. Rev. (London)*, **5**, 131 (1951).
25. C. K. Ingold, *Structure and Mechanism in Organic Chemistry*, Cornell University Press, Ithaca, New York, 1953, p. 206.
26. L. W. Clark, *J. Phys. Chem.*, **65**, 2271 (1961).
27. K. J. Pedersen, *J. Am. Chem. Soc.*, **51**, 2098 (1929).
28. A. Widmark, *Acta Med. Scand.* **53**, 393 (1920).
29. G. A. Hall, Jr., *J. Am. Chem. Soc.*, **71**, 2691 (1949).
30. A. Dinglinger and E. Schröer, *Z. Physik. Chem. (Leipzig)*, **179**, 401 (1937); *A* **181**, 375 (1938).
31. E. O. Wiig, *J. Am. Chem. Soc.*, **52**, 4737 (1930).
32. L. W. Clark, *J. Phys. Chem.*, **61**, 699 (1957).
33. L. W. Clark, *J. Phys. Chem.*, **62**, 633 (1958).
34. L. W. Clark, *J. Phys. Chem.*, **65**, 659 (1961).
35. L. W. Clark, *J. Phys. Chem.*, **71**, 302 (1967).
36. K. J. Laidler, *Chemical Kinetics*, 2nd ed., McGraw-Hill, New York, 1965, p. 242.
37. S. Franke and F. Brathun, *Ann. Chem.* **487**, 1 (1931).
38. A. Locke, *J. Am. Chem. Soc.*, **46**, 1246 (1924).
39. B. R. Brown, D. L. Hammick and A. J. B. Scholefield, *J. Chem. Soc.*, 778 (1950).
40. L. W. Clark, *J. Phys. Chem.*, **67**, 2831 (1963).
41. L. W. Clark, *J. Phys. Chem.*, **69**, 3565 (1965).
42. H. Silberstein, *Chem Ber.* **17**, 2664 (1884).
43. H. Goldschmidt and R. Bräuer, *Chem. Ber.*, **39**, 109 (1906).
44. O. Salmi and R. Korte, *Ann. Acad. Sci. Fennicae, Ser. A* **54**, No. 10 (1940).
45. O. de Groote, *Bull. Soc. Chim. Belges.* **37**, 225 (1928).
46. I. Auerback, F. H. Verhoeck and J. Henne, *J. Am. Chem. Soc.*, **72**, 299 (1950).
47. A. N. Kappanna, *Z. Physik. Chem. (Leipzig) A* **158**, 355 (1932).
48. H. W. Patwardhan and A. N. Kappanna, *Z. Physik. Chem. (Leipzig)*, **A 166**, 51 (1933).
49. F. H. Verhoeck, *J. Am. Chem. Soc.*, **56**, 571 (1934).
50. F. H. Verhoeck, *J. Am. Chem. Soc.*, **67**, 1062 (1945).
51. G. A. Hall, Jr., and F. H. Verhoeck, *J. Am. Chem. Soc.*, **69**, 613 (1947).
52. C. N. Cochran and F. H. Verhoeck, *J. Am. Chem. Soc.*, **69**, 2987 (1947).
53. L. W. Clark, *J. Am. Chem. Soc.*, **77**, 3130 (1955).
54. L. W. Clark, *J. Phys. Chem.*, **63**, 99 (1959).
55. L. W. Clark, *J. Phys. Chem.*, **63**, 1760 (1959).
56. L. W. Clark, *J. Phys. Chem.* **64**, 1758 (1960).
57. L. W. Clark, *J. Phys. Chem.*, **64**, 917 (1960).
58. L. W. Clark, *J. Phys. Chem.*, **62**, 79 (1958).
59. L. W. Clark, *J. Phys. Chem.*, **68**, 3048 (1964).
60. L. P. Hammett, *Physical Organic Chemistry*, McGraw-Hill, New York, 1940, p. 204.
61. L. Muus, *J. Phys. Chem.*, **39**, 343 (1935).
62. K. J. Pedersen, *J. Phys. Chem.*, **38**, 559 (1934).
63. K. J. Pedersen, *Trans. Faraday Soc.*, **23**, 316 (1927).
64. K. J. Pedersen, *Acta Chem. Scand.*, **1**, 437 (1947).
65. E. A. Moelwyn-Hughes and C. N. Hinshelwood, *J. Chem. Soc.*, **128**, 186 (1931).
66. F. H. Verhoeck, *J. Am. Chem. Soc.*, **61**, 186 (1939).
67. P. Dyson and D. L. Hammick, *J. Chem. Soc.*, 1727 (1937).
68. M. R. F. Ashworth, R. P. Daffern and D. L. Hammick, *J. Chem. Soc.*, 809 (1939).
69. B. R. Brown, *J. Chem. Soc.*, 2577 (1949).

70. B. R. Brown and D. L. Hammick, *J. Chem. Soc.*, 659 (1949).
71. W. E. Doering and V. Z. Pasternak, *J. Am. Chem. Soc.*, **72**, 143 (1950).
72. H. Schenkel and M. Schenkel–Klein, *Helv. Chim. Acta*, **31**, 924 (1948).
73. H. Schenkel and A. Klein, *Helv. Chim. Acta*, **28**, 1219 (1945).
74. H. Schenkel and A. Klein, *Helv. Chim. Acta*, **29**, 476 (1946).
75. H. Schenkel and E. Mory, *Helv. Chim. Acta*, **33**, 16 (1950).
76. N. H. Cantwell and E. V. Brown, *J. Am. Chem. Soc.*, **74**, 5967 (1952).
77. N. H. Cantwell and E. V. Brown, *J. Am. Chem. Soc.*, **76**, 15 (1954).
78. L. W. Clark, *J. Phys. Chem.*, **66**, 125 (1962).
79. L. W. Clark, *J. Phys. Chem.*, **69**, 2277 (1965).
80. S. L. Friess, E. S. Lewis, A. Weissberger, (Ed.), *Technique of Organic Chemistry*, Vol. 8, Pt. 1, *Investigations of Rates and Mechanisms of Reactions*, 2nd ed., Interscience Publishers, New York, 1961, p. 207.
81. D. J. Cram and A. S. Wingrove, *J. Am. Chem. Soc.*, **85**, 1100 (1963).
82. H. Gleason and J. Dougherty, *J. Am. Chem. Soc.*, **51**, 310 (1929).
83. R. T. Arnold, O. C. Elmer and R. M. Dodson, *J. Am. Chem. Soc.*, **72**, 4359 (1950).
84. R. B. Woodward and E. C. Kornfeld, *J. Am. Chem. Soc.*, **70**, 2508 (1948).
85. E. Tommila and A. Kivinen, *Suomen Kemistilehti* **24B**, 46 (1951).
86. H. Schenkel, *Helv. Chim. Acta*, **29**, 436 (1946).
87. H. Schenkel and A. Klein, *Helv. Chim. Acta*, **28**, 1211 (1945).
88. R. G. Johnson and R. K. Ingham, *Chem. Rev.*, **56**, 250 (1956).
89. L. F. Fieser and M. Fieser, *Advanced Organic Chemistry*, Reinhold Publishing Company, New York, 1961, p. 110.
90. M. Burton, *J. Am. Chem. Soc.*, **58**, 692 (1936).
91. M. Burton, *J. Am. Chem. Soc.*, **60**, 831 (1938).
92. L. W. Clark, *J. Phys. Chem.*, **67**, 138 (1963).
93. L. W. Clark, *J. Phys. Chem.*, **64**, 41 (1960).
94. L. W. Clark, *J. Phys. Chem.*, **64**, 692 (1960).
95. S. Glasstone, K. J. Laidler and H. Eyring, *The Theory of Rate Processes*, McGraw-Hill, New York, 1941, p. 22.
96. J. E. Leffler, *J. Org. Chem.*, **20**, 1202 (1955).
97. L. W. Clark, *J. Phys. Chem.*, **64**, 508 (1960).
98. L. W. Clark, *J. Phys. Chem.*, **64**, 677 (1960).
99. L. W. Clark, *J. Phys. Chem.*, **66**, 1543 (1962).
100. L. W. Clark, *J. Phys. Chem.*, **67**, 2602 (1963).
101. L. W. Clark, *J. Phys. Chem.*, **67**, 2831 (1963).
102. L. W. Clark, *J. Phys. Chem.*, **70**, 627 (1966).
103. L. W. Clark, *J. Phys. Chem.*, **68**, 2150 (1964).
104. I. Meloche and K. J. Laidler, *J. Am. Chem. Soc.*, **73**, 1712 (1952).
105. A. Kivinen, *Ann. Acad. Sci. Fennicae Ser. A. II*, No. 108 (1961).

CHAPTER 13

Ortho esters

E. H. CORDES

Indiana University, Bloomington, Indiana, U.S.A.

I. INTRODUCTION

The chemistry of ortho esters has not been systematically explored Virtually no physicochemical studies have been undertaken concerning the ortho ester function: thus bond-lengths, bond energies, dipole moments, and light absorption properties relevant to this function are largely unknown. An abundance of such information is available for most of the other functional groups of organic chemistry. Although a large number of organic reactions with a variety of ortho esters have been probed experimentally, most of these studies have concentrated on the synthesis of a certain compound or class of compounds and have, consequently, not dealt with these processes either in terms of overall scope or in terms of mechanism. The only reaction which is an exception to this statement is that of ortho ester hydrolysis, a reaction which has been examined in some detail for the past four decades. Failure to examine the reactions of ortho esters in depth has undoubtedly shrouded much interesting chemistry. Reactions of these species must, for the most part at least, proceed via carbonium ion intermediates and, in consequence, ortho esters must share a great deal of the fascinating chemistry typical of other substrates capable of generating reasonably stable carbonium ions.

We begin this review with a consideration of synthetic methods leading to ortho esters. Then, prior to considering reactions with nucleophilic reagents, attention is directed to two aspects of ortho ester chemistry which are basic to further considerations: their protonation and their conversion to carbonium ions. The next reaction explored is that of hydrolysis, to which a major share of attention is directed as a result of our rather satisfactory understanding of this process. Subsequently, we consider reactions of amines, carbon acids, and other reagents. Throughout, we will be concerned with the mechanistic aspects of these reactions as well as with their actual or potential use for the synthesis of novel organic compounds.

Ortho acids, the hydrates of carboxylic acids and the parent compounds of ortho esters, are thermodynamically less stable than the hydrates of carbonyl substrates and, unlike the latter species, do not exist in appreciable concentrations at equilibrium in solution. Furthermore, ortho acids have not been isolated. They are, in consequence, of little direct interest and will not be considered further.

Few previous reviews of this subject have been published. Of particular note is the exhaustive monograph of the chemistry of aliphatic ortho esters by Post which develops the subject matter up to 1943[1]. A review concerning mechanisms and catalysis for ortho ester hydrolysis has been published recently[2].

II. SYNTHESIS OF ORTHO ESTERS

There exist two generally useful methods for the synthesis of ortho esters: alcoholysis of imido ester hydrochlorides and the reaction of 1,1,1-trihalides with alkoxides. Of these, the former is of greater overall utility; we consider it first. In addition to these methods, there are several reactions which have proved to be of value in isolated instances. These, together with the more common methods, have been previously discussed[1, 3].

A. From Imido Ester Hydrochlorides

Preparation of ortho esters through the alcoholysis of imido ester hydrochlorides was introduced by Pinner in 1883[4, 5] and is generally known as the Pinner synthesis. This method has several advantages. In the first place, the imido ester hydrochlorides are readily available in excellent yield from partial alcoholysis of the parent nitriles according to equation

$$R^1CN + R^2OH + HCl \longrightarrow R^1—C(OR^2)=NH . HCl \qquad (1)$$

(1). These reactions are usually conducted by the addition of dry hydrogen chloride to an equimolar mixture of nitrile and alcohol in ether or other inert solvent[6−9] or by the addition of dry nitrile to a solution of hydrogen chloride in the alcohol[10, 11]. Anhydrous conditions are important for maximum yields. In the second place, the alcoholysis of the imido ester hydrochlorides can usually be readily effected in satisfactory yield (equa-

$$R^1—C(OR^2)=NH . HCl + 2 R^3OH \longrightarrow R^1—C(OR^2)(OR^3)_2 + NH_4Cl \qquad (2)$$

tion 2)[12−14]. The success of this reaction requires that the imido ester hydrochloride be free of excess acid and be anhydrous. Equation (2) ordinarily requires a few days if conducted at room temperature but may be accomplished in a few hours by refluxing in ether[15]. If the latter alternative is chosen, care must be taken to avoid temperatures above 40° since these lead to decomposition of the hydrochloride[16, 17]. Finally, both mixed ($R^2 \neq R^3$; equation 2) and normal ortho esters may be prepared by this method. Many ortho esters obtained by the Pinner synthesis have been collected, together with their physical properties, by Post[1].

B. From Polyhalides

The earliest ortho ester synthesis is that of Williamson and Kay who, in 1854, reported the preparation of orthoformates from the action of sodium methoxide on chloroform[18]. The historical development of this method has been outlined by Post[1]. The overall course of these reactions is indicated in equation (3). The reactions are generally carried out using the tri-

$$R^1—CX_3 + 3 NaOR^2 \longrightarrow R^1—C(OR^2)_3 + 3 NaX \qquad (3)$$

chlorides and are most successful if chloroform, benzotrichloride or other substrates not possessing a hydrogen atom on the carbon adjacent to that carrying the three halogens are used. Thus, a variety of alkyl orthoformates have been prepared from chloroform and sodium alkoxides[1, 3, 19] and benzotrichloride is easily converted to ortho esters by these reagents[20]. In addition, several orthoacetates have been derived from 1,1,1-trichloroethane[1].

Although these reactions are ordinarily conducted in the absence of catalysts, Hill and coworkers have found ferric chloride to be an effective promoter for the synthesis of ortho esters derived from fluoroalcohols and either chloroform or benzotrichloride[20]. Numerous early attempts to prepare orthocarbonates from sodium alkoxides and carbon tetrachloride failed uniformly although orthoformates were obtained in several cases[1]. The use of ferric chloride as a catalyst has recently resulted in the production of orthocarbonates from carbon tetrachloride and fluoroalcohols[20]. This reaction is of limited utility for the synthesis of orthocarbonates since it fails for ordinary alcohols which are decomposed by the catalyst. The usual method of obtaining orthocarbonates is through a related reaction, that between sodium alkoxides and chloropicrin (equation 4)[21]. These products have also been obtained from sodium alkoxides and thiocarbonyl perchloride (Cl_3CSCl)[22].

$$O_2N—C(Cl)_3 + 4\ NaOR \longrightarrow C(OR)_4 + 3\ NaCl + NaNO_2 \tag{4}$$

Synthesis of unusual ortho esters in which this function is adjacent to an amide group has been reported by Russian workers who used 1-alkoxy-1,1-dichloro compounds as starting materials (equation 5)[23].

$$2\ ArONa + R^1OC(Cl)_2\overset{\overset{\displaystyle O}{\|}}{C}—NR_2^2 \longrightarrow R^1—O—\overset{\overset{\displaystyle OAr}{|}}{\underset{\underset{\displaystyle OAr}{|}}{C}}—\overset{\overset{\displaystyle O}{\|}}{C}—NR_2^2 + 2\ NaCl \tag{5}$$

C. Miscellaneous Syntheses

1. From ketene acetals

McElvain and coworkers have clearly demonstrated that addition of primary and secondary alcohols to ketene acetals in the presence of acid leads readily to ortho esters (equation 6)[24-27]. This method is well adapted

$$\underset{R^2}{\overset{R^1}{>}}\!\!=\!\!\underset{OR^4}{\overset{OR^4}{<}} + R^3OH \xrightarrow{H^+} R^1R^2HC—C(OR^4)_2(OR^3) \tag{6}$$

to the production of mixed ortho esters. It has not found wide applicability for general synthesis since the ketene acetals are not usually readily available. Indeed, at least two methods of synthesis of these ketals require ortho esters as starting material. However, in isolated instances, the use of ketene acetals can lead to novel and interesting products. For example, McElvain and McKay have prepared the cyclic ortho esters **1, 2,** and **3** from the cyclic ketene acetal, 2-methoxy-5,6-dihydropyran[28].

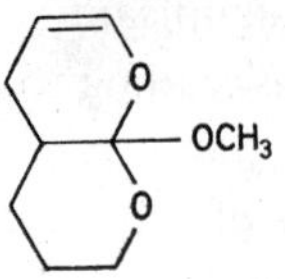

(1) 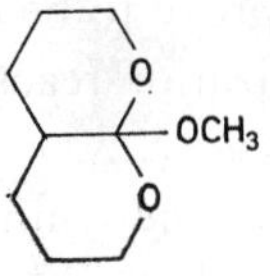 (2) 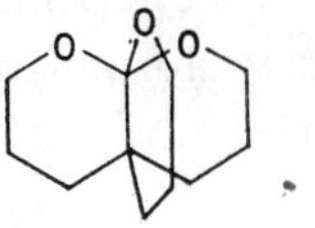 (3)

In a related reaction, Kuryla and Leis have reported the synthesis of either the ketene acetal or the ortho ester, depending on reaction conditions and nature of the alcohol used, from 1,1-dichloroethylene and β-alkoxy alcohols[29].

2. Form orthothioformates

Ethyl orthothioformate, easily prepared from ethyl formate and ethyl mercaptan, provides a convenient starting point for the synthesis of orthoformates via exchange reactions with alcohols in the presence of Lewis acid catalysts (equation 7)[30].

$$HC(SEt)_3 + 3\ ROH \xrightarrow{ZnCl_2} HC(OR)_3 + 3\ HSEt \qquad (7)$$

3. By ortho ester exchange

Treatment of ortho esters with alcohols generally leads to radical interchange[31]. Such reactions are promoted by the addition of acids. If one permits the reaction to proceed to equilibrium, a mixture of two normal ortho esters, all possible mixed ortho esters, and the two alcohols is obtained[31]. However, the reactions may be driven to completion by continuous removal through distillation of the lower boiling alcohol product.

Alcohol interchange reactions between ethyl orthoformate and certain triols has led to a series of bicyclic ortho esters[32]. That between ethyl ortho esters and t-butyl hydroperoxide produces peroxy ortho esters (equation 8)[33]. Finally, such reactions may be used to synthesize ortho esters of nuc-

$$R-C(OEt)_3 + HOO-t\text{-Bu} \longrightarrow R-\underset{\underset{OEt}{|}}{\overset{\overset{OEt}{|}}{C}}-O-O-t\text{-Bu} + EtOH \qquad (8)$$

leic acid derivatives[34]. A typical example is provided in equation (9)

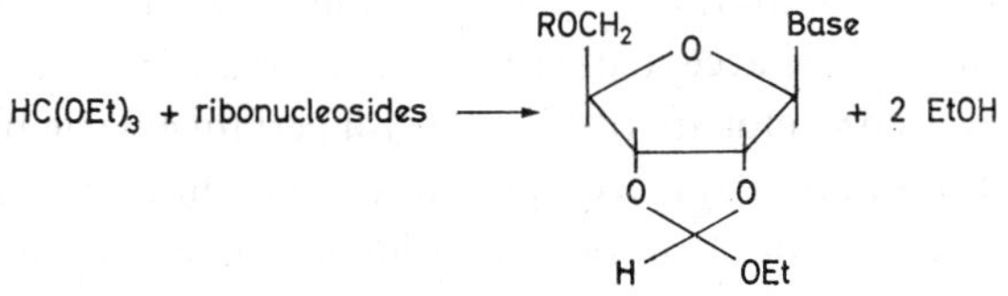

$$HC(OEt)_3 + ribonucleosides \longrightarrow \qquad + 2\ EtOH \qquad (9)$$

40*

These cyclic ortho esters may prove to be of significant value as suitably-blocked substrates for polymerization reactions leading to nucleic acids or related substances.

Several additional methods for the synthesis of ortho esters by methods of little interest at this time have been discussed by Post[1].

III. BASICITY OF ORTHO ESTERS

Ortho esters are weak oxygen bases. A quantitative understanding of their basic properties would be highly desirable from the standpoint of fuller understanding of their reactivities since their reactions usually involve acid catalysis. The acid lability of ortho esters precludes direct measurement of their extent of protonation in aqueous solution as a function of medium acidity, the most straightforward manner of evaluating basicity for weak acids. As an alternative for cases of this type, it is possible to employ positions of the O—H or O—D stretching frequencies in the infrared for suitable hydrogen donors as a measure of basicity of weak oxygen bases to which they hydrogen bond[35, 36]. Not only is the magnitude of the frequency shift for each of the compounds of interest relative to a suitable non-basic standard a measure of relative basicity but approximate values of pK_a for the conjugate acids may also be obtained by use of a linear relationship between $\Delta\mu$ and pK_a derived from a compilation of such studies[35].

The first such measurements using ortho esters and related substrates were those of West, Whatley, and Lake who employed the O—H stretching frequency of phenol as a measure of basicity[37]. These results clearly indicate the following order of basicity: ethers > ketals > ortho esters > orthocarbonates. That is, increasing substitution of alkoxy for alkyl functions lowers basicity as would be expected on the basis of greater inductive electron withdrawal by the alkoxy functions. In an effort to place these conclusions on a more quantitative basis, similar measurements were made on the compounds studied by West and on others using CH_3OD as the hydrogen donor[38]. It is for this compound that the most reliable correlation between $\Delta\mu$ and pK_a has been established[35]. In Table 1, the positions of the O—D stretching frequencies together with the derived values of pK_a for the conjugate acids of the indicated substrates are compiled. These data are in qualitative agreement with those of West and his coworkers. The absolute values of pK_a should be accepted with some reservations since deviations from the $\Delta\mu$ versus pK_a plot are rather large for certain oxygen bases. However, the variations in pK_a from compound to compound are certainly real and the magnitudes of the changes are considered to be quantitatively reliable due to the degree of structural similarity in the compounds involved. Furthermore, the values of pK_a are remarkably similar to those estimated on the basis of the assumption that the effects of polar substituents on the basicity of oxygen bases are similar to those for nitrogen bases (Table 1)[39]. The basicity of methyl-t-butyl ether is similar to those obtained for related ethers by other means[35]. Thus, the indicated values of pK_a are probably not greatly in error. We shall return to the question of ortho ester basicity somewhat later in this chapter in connection with structure–reactivity correlations for hydrolysis of these substrates.

IV. REACTIONS

A. Carbonium Ion Formation

Alkoxymethyl carbonium ions have been prepared in two ways. The initial syntheses are due to Meerwein and coworkers who obtained fluoroborates of dialkoxy- and trialkoxymethyl carbonium ions by the abstraction of hydride ion from ketals and ortho esters[40, 41]. The hydride abstractors used were generally salts of stable carbonium ions; a typical example is indicated in equation (10). Products of these reactions may be obtained as stable crystalline preparations. Subsequently, Ramsey and Taft have de-

$$\begin{array}{c} R^1 \\ \diagdown \\ \diagup \quad C \diagdown \\ H \qquad OR^2 \end{array} + (C_6H_5)_3C^+BF_4^- \longrightarrow \begin{array}{c} R_1 \\ \diagdown \\ + \diagup \quad C \diagdown \\ OR^2 \end{array} BF_4^- + (C_6H_5)_3C\!-\!H \qquad (10)$$

monstrated that alkoxymethyl cations may be generated directly from ketals, ortho esters, and orthocarbonates in strongly acidic solutions[42].

TABLE 1. Basicities of a series of weak oxygen bases

Compound	Position of O–D stretching band of deuteriomethanol (μ)	$\Delta\mu$	$pK_a{}^a$ (measured)	$pK_a{}^b$ (estimated)
Benzene	3·73	–	–	–
Carbon tetrachloride	3·73	–	–	–
Methyl t-butyl ether	3·87	0·14	−3·7	–
2,2-Dimethoxypropane	3·83	0·12	−5·2	−4·6
Methyl orthoacetate	3·82	0·09	−6·5	−6·4
Methyl orthocarbonate	3·80	0·07	−8·4	−8·5
Benzaldehyde diethyl acetal	3·84	0·11	−5·7	−5·8

^{a} Obtained from the linear relationship between $\Delta\mu$ and pK_a of reference 35.
^{b} Estimated on the basis of polar substituent effects as outlined by Bunton and DeWolfe in reference 39.

Thus, trialkoxy- and *p*-fluorophenyldialkoxymethyl carbonium ions are generated through the addition of the appropriate orthocarbonate (equ-

$$C(OR)_4 + 3 H_2SO_4 \longrightarrow (RO_3)C^+ + ROSO_3H + H_3O^+ + 2 HSO_4^- \qquad (11)$$

ation 11) or ortho ester to sulfuric acid. Orthocarbonates are also converted to the carbonium ions in trifluoroacetic acid. Formation of dialkoxymethyl and alkoxydimethyl carbonium ions requires rather more drastic conditions: 30% $SO_3.H_2SO_4$. Evidence for the stoichiometry indicated in equation (11) is compelling.

(i) Ethyl orthocarbonate in sulfuric acid gives an 'i' factor of $5 \cdot 1 \pm 0 \cdot 1$.

(ii) Proton magnetic resonance spectra of solutions of orthocarbonates have integrated absorption intensities identical to those predicted on the basis of equation (11). Furthermore, positions of signals for these solutions are similar to a composite of those for $(RO)_3C^+BF_4^-$ plus ROH in sulfuric acid.

(iii) The infrared spectra of sulfuric acid solutions of ethyl orthocarbonate are similar to those recorded for the salts prepared by Meerwein and coworkers.

In addition to direct demonstrations of the existence of the postulated carbonium ions, Taft and Ramsey have observed that dilute solutions of methyl orthocarbonate in anhydrous methanol exhibit only a single methyl proton resonance absorption in the presence of 0·5 to 1·0 M HCl[43]. The coalescence of the methyl peaks of the ortho ester and methanol is most reasonably interpreted as reflecting the rapid and reversible formation of the trimethoxymethyl carbonium ion. The life-time of this carbonium ion in anhydrous methanol solutions containing unit concentrations of ortho ester and acid is estimated to be approximately 0·01 sec.

The original investigations of Meerwein and coworkers established that dialkoxy- and trialkoxymethyl carbonium ions possess large energies of stabilization. The simple fact that salts of these carboniums ions can be readily isolated bears strong testimony to this effect. Furthermore, these workers have established that the reaction between equimolar methyl orthocarbonate and the triphenylmethyl carbonium ion proceeds to completion with transfer of a methoxy function as indicated in equation (12)[40, 41]. This observation, made in a heterogeneous system, has been confirmed by

$$(C_6H_5)_3C^+ + C(OCH_3)_4 \longrightarrow {}^+C(OCH_3)_3 + (C_6H_5)_3COCH_3 \qquad (12)$$

Ramsey and Taft under homogeneous conditions[42]. These findings indicate that the stabilization of the trimethoxymethyl carbonium ion is greater than that for the triphenylmethyl cation.

TABLE 2. Stabilization energy
relative to CH_3^+ of polysubstituted
methoxy cations[44]

Ion	Stabilization energy (± 3 kcal/mole)
$CH_3OCH_2^+$	66
$(CH_3O)_2C^+H$	85
$(CH_3O)_3C^+$	90

The stabilization energies of alkoxymethyl carbonium ions have been probed by Taft and his associates. From appearance potential measurements on a series of substituted methoxymethanes, these workers have derived the stabilization energies for such cations summarized in Table 2[44, 45]. Two points are important here. In the first place, it is clear that, in the gas phase at least, substantial stabilization energies relative to the methyl cation are observed. Put another way, the $CH_3OCH_2{}^+$ species has 33 kcal/mole more stabilization energy than the $CH_3CH_2{}^+$ cation; similarly the $(CH_3O)_2CH^+$ and $(CH_3O)_3C^+$ cations have about 18 kcal/mole more stabilization energy than $(CH_3)_3C^{+}$[44, 45]. In the second place, there is a marked saturation of the stabilization observed in the cation as hydrogens are successively replaced by methoxy functions. Introduction of the first such function involves a change in stabilization energy of 66 kcal/mole, of the second 19 kcal/mole, and of the third only about 5 kcal/mole. Similar effects have been noted for stabilization energies of triphenylmethyl anions[46]. We shall return to this point later in considerations of the relationship between structure and reactivity for hydrolysis of these substrates.

Relative stabilization energies of alkoxymethyl carbonium ions in the gaseous phase derived from appearance potential measurements have been confirmed for the liquid phase as well using nuclear magnetic resonance spectroscopy. The fluorine n.m.r. shielding for the p-$FC_6H_4C(C_6H_5)_2{}^+$ ion and for those ions derived from this one by successive replacement of phenyl groups by methoxy functions has been determined by Ramsey and Taft; the results are collected in Table 3[42]. The fluorine n.m.r. shielding clearly increases with increasing methoxy group substitution corresponding to increasing electron density at fluorine and its bonded carbon atom[47]. For the case of triphenylmethyl cations, a linear relationship has been found between fluorine n.m.r. shielding and stabilization energy[48]. Thus, the results in the present case indicate increasing stabilization energy with increasing

TABLE 3. Fluorine n.m.r. shielding in p-fluorosubstituted phenylmethoxymethyl cations in sulfuric acid[42]

Ion	δ (ppm)[a]
p-FC$_6$H$_4$C(C$_6$H$_5$)$_2{}^+$	$-30 \cdot 0$
p-FC$_6$H$_4$C(C$_6$H$_5$)(OCH$_3$)$^+$	$-27 \cdot 2$
p-FC$_6$H$_4$C(OCH$_3$)$_2{}^+$	$-20 \cdot 2$

[a]Relative to external tetrachlorotetrafluoro-cyclobutane, 40% (wt) in carbon tetrachloride.

methoxy group substitution in accord with the results pertinent to the gas phase.

Extended investigations of the nuclear magnetic resonance, infrared, and ultraviolet spectra of alkoxymethyl carbonium ions have led to the following additional conclusions[42]:

(i) the (CH$_3$O)$_3$C$^+$, (CH$_3$O)$_2$CCH$_3{}^+$, and (CH$_3$O)$_2$CH$^+$ ions are essentially coplanar;

(ii) the carbonium carbon–methoxy oxygen bond has appreciable double bond character (0·2 to 0·3 π bond order) and the energy of activation for rotation about this bond in (CH$_3$O)$_2$CCH$_3^+$ is roughly 11 kcal/mole;

(iii) the conversion of ortho esters and ketals to methoxymethyl carbonium ions causes the proton n.m.r. of the OCH$_3$ groups to shift to lower fields by about 1·8 ppm in accordance with considerations based on delocalization of positive charge to oxygen.

B. Hydrolysis

Studies concerned with mechanisms and catalysis for the hydrolysis of acetals, ketals, and ortho esters have been productive in the development of a general understanding of these topics for reactions in aqueous solution. Indeed, pioneering studies on general acid–base catalysis, solvent deuterium isotope effects, reaction kinetics in strongly acidic media, and structure–reactivity correlations have used these substances as substrates. Such early studies together with significant recent developments have clearly established the principal mechanistic and catalytic features of these hydrolytic processes. These are summarized below. A more comprehensive discussion of this topic has been recently published[2].

In acidic aqueous solutions, ortho esters hydrolyze according to the overall stoichiometry indicated in equation (13). These reactions occur with

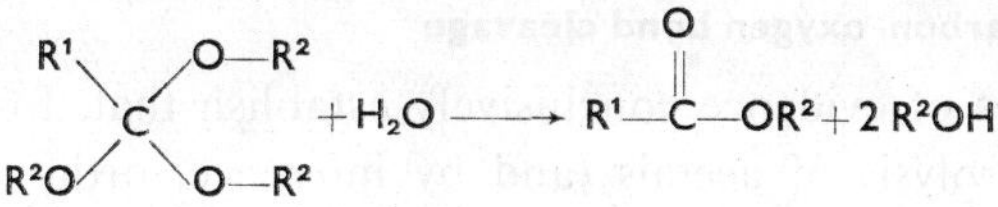

$$
\begin{array}{c}
R^1 \\
\diagdown \\
C \\
\diagup \diagdown \\
R^2O O{-}R^2
\end{array}
\raisebox{1ex}{$\overset{\displaystyle O{-}R^2}{}$}
+ H_2O \longrightarrow R^1{-}\overset{\displaystyle O}{\underset{\displaystyle \|}{C}}{-}OR^2 + 2\,R^2OH \qquad (13)
$$

the rupture of two covalent bonds to carbon and involve at least two proton transfer reactions as well and, hence, must be multi-step.

The first step in ortho ester hydrolysis in which the making or breaking of covalent bonds to carbon is involved may be visualized as occuring via one of the four transition states indicated below. Each of these transition states is pictured, for the sake of clarity, as having arisen from the conjugate acid of the substrate. Kinetic studies indicate the presence of a proton or its kinetic equivalent in the transition state but leave uncertain the question of the timing of proton transfer relative to cleavage of the C—O bond. We return to this point below. Transition states **4** and **5** picture these hydrolyses

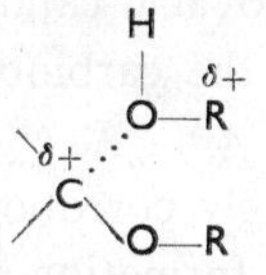

(4) 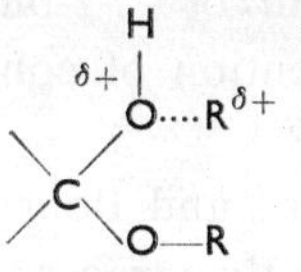   **(5)** 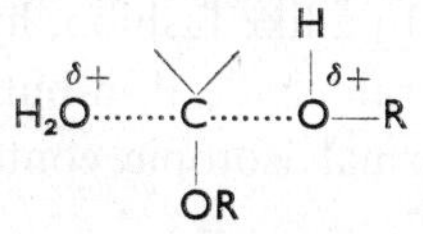   **(6)** 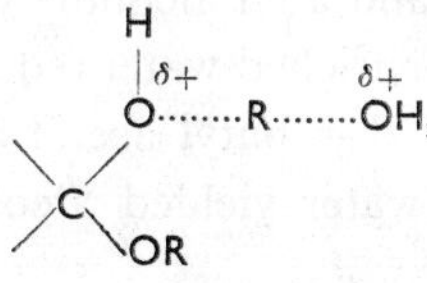   **(7)**

as occurring via unimolecular decomposition of the conjugate acids of the substrates with cleavage of the carbonyl carbon–oxygen and alcohol carbon–oxygen bonds, respectively (A1 reactions). The corresponding carbonium ions are the immediate products. Transition states **6** and **7** include the participation of water as nucleophilic reagent with carbon–oxygen bond cleavage at the sites indicated (A2 reactions). The immediate products are identical to those formed from addition of one molecule of water to the carbonium ions generated from transition states **4** and **5**. Distinction between these transition states involves (1) localization of the site of C—O bond cleavage and (2) identification of the immediate product of C—O cleavage as a carbonium ion or its hydrate. We shall consider these topics in sequence.

1. The site of carbon–oxygen bond cleavage

Several lines of evidence conclusively establish that, for most cases at least, the hydrolysis of acetals (and by inference, ortho esters as well) proceeds with cleavage of the carbonyl carbon–oxygen bond. The earliest convincing evidence for this point of view is the important work of Lucas and his associates on the hydrolysis of acetals derived from optically active alcohols. For example, hydrolysis of the D(+)-2-octanol acetal of acetaldehyde in dilute aqueous phosphoric acid yields 2-octanol having the same optical rotation as the original alcohol from which the acetal was synthesized[49]. This finding excludes formation of the alkyl carbonium ion (transition state 5), in which case substantial or complete racemization of the alcohol would be expected, and also an A2 reaction involving nucleophilic attack of solvent on the alcohol (transition state 7), in which case optical inversion of the alcohol would be expected. Similarly, the formal, acetal, and carbonate derived from D(−)-2,3-butanediol and the acetal derived from D(+)-2-butanol undergo acid-catalyzed hydrolysis with complete retention of configuration at the carbinol carbon of the alcohol[50, 51].

Stasiuk, Sheppard and Bourns have strongly corroborated the above conclusion in an isotope tracer study of acetal formation and hydrolysis[52]. The condensation of benzaldehyde and n-butyraldehyde, enriched in ^{18}O, with n-butyl and allyl alcohols yielded acetals of normal isotopic abundance and ^{18}O-enriched water (equation 14). In a like fashion, hydrolysis of benzaldehyde di-n-butyl acetal and n-butyraldehyde di-n-butyl acetal in ^{18}O-enriched water yielded alcohols of normal isotopic content (the re-

$$R^2-C\overset{\overset{\displaystyle ^{18}O}{\big\|}}{\underset{H}{\diagdown}} + 2\,R^1OH \longrightarrow R^2-C\overset{\overset{\displaystyle OR^1}{\diagup}}{\underset{OR^1}{\diagdown}}H + H_2{}^{18}O \qquad (14)$$

verse of equation 14). Thus, these reactions clearly proceed with carbonyl carbon–oxygen bond cleavage (or formation).

Less experimental work on the site of carbon–oxygen bond cleavage has been reported for the cases of ketal and ortho ester hydrolysis. One would expect these substrates to behave in a fashion similar to that of acetals. Some very early work on the hydrolysis of ketals tends to bear out this supposition[2]. Recently, Taft has studied the hydrolysis of methyl orthocarbonate in $H_2{}^{18}O$. While most of the ^{18}O does appear in the carbonyl group of dimethyl carbonate as expected, there is appreciable formation of $CH_3{}^{18}OH$ and CH_3-O-CH_3, i.e. methylation of the nucleophiles water and methanol either by the orthocarbonate or the corresponding carbonium ion[53]. These products almost certainly arise in bimolecular reactions involving alcohol carbon–oxygen bond cleavage (transition state **7** or a variant thereof).

In summary, the data indicated above and reasonable extrapolations thereof strongly suggest that, in the preponderant majority of cases, acid-catalyzed hydrolysis of acetals, ketals, and ortho esters occurs with cleavage of the carbonyl carbon–oxygen bond. We now turn to a consideration of the distinction between the two transition states, **4** and **6,** which involve bond cleavage of this type.

2. Molecularity

Several independent lines of evidence strongly suggest that the acid-catalyzed hydrolysis of ortho esters proceeds by a reaction pathway *not* involving solvent as nucleophilic reagent, *i.e.* that **4** describes the transition state for the initial reaction in which covalent bonds to carbon are broken (equation 15). These lines of evidence derive from studies on (i) the reaction kinetics, (ii) structure–reactivity correlations, (iii) entropies of activa-

$$
\begin{array}{c}
\underset{R^2O}{\overset{R^1}{\diagdown}}C\underset{\diagdown OR^2}{\overset{\diagup OR^2}{}} + H^+ \underset{+R^2OH}{\overset{-R^2OH}{\rightleftharpoons}} \underset{R^2O}{\overset{R^1}{\diagdown}}C\underset{}{\overset{\diagup OR^2}{}} + \underset{-H_2O}{\overset{+H_2O}{\rightleftharpoons}} \underset{R^2O}{\overset{R^1}{\diagdown}}C\underset{\diagdown OH}{\overset{\diagup OR^2}{}} + H^+ \\[3mm]
\longrightarrow R^1{-}\overset{\overset{\textstyle O}{\|}}{C}{-}OR^2 + R^2OH + H^+ \qquad (15)
\end{array}
$$

tion, (iv) volumes of activation, (v) isotope effects, and (vi) rate and product studies in the presence of added nucleophilic reagents. We shall consider the results of these studies sequentially.

The hydrolysis of the substrates in question is almost invariably dependent upon acid catalysis. That is, the rate laws for reactions in dilute aqueous solution have the form

$$
k_{\text{obs}} = k_{\text{H}^+}(\text{H}^+) + \sum_i k_{\text{HA}_i}(\text{HA})_i \qquad (16)
$$

in which the terms in the summation on the right-hand side of equation (16) are frequently negligible. The complete dependence of these reactions on acid catalysis suggests that water does not participate as a nucleophilic reagent. If water were able to expel alcohol from the protonated substrates in nucleophilic reactions then one might expect that hydroxide ion (or other nucleophilic reagent) would expel alkoxide ion from the corresponding free bases. Since the latter reactions are not observed, one suspects

TABLE 4. A summary of linear free energy correlations for acetal, ketal and ortho ester hydrolysis

	Substrates	Number of substituents	Solvent	Temperature	Correlation obeyed	"ϱ"	Reference
1.	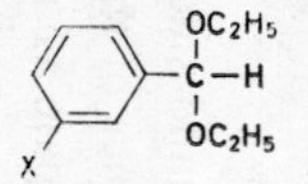	4	50% Aqueous dioxane	30°	$\log \dfrac{k}{k_0} = \sigma\varrho$	$\varrho = -3{\cdot}35$	54
2.	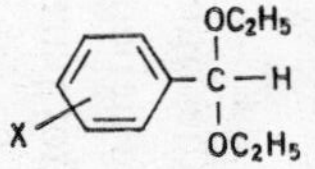	8	50% Aqueous dioxane	30°	$\log \dfrac{k}{k_0} = \varrho[\sigma + r(\sigma^+ - \sigma)]$	$\varrho = -3{\cdot}35$ $r = 0{\cdot}5$	**54** 55
3.	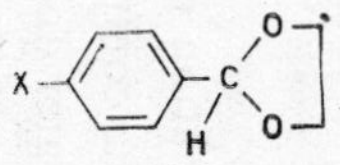	5	50% Aqueous dioxane	30°	$\log \dfrac{k}{k_0} = \varrho[\sigma + r(\sigma^+ - \sigma)]$	$\varrho = -3{\cdot}25$ $r = 0{\cdot}5$	54 55
4.	$R_1R_2C(OC_2H_5)_2{}^{a}$	23	49·6% Aqueous dioxane	25°	$\log \dfrac{k}{k_0} = (\sum \sigma^*)\,\varrho^* + 0{\cdot}54(\Delta n)$	$\varrho^* = -3{\cdot}60$	56
5.	$H_2C(OR)_2$	7	Water	25°	$\log \dfrac{k}{k_0} = \sigma^*\varrho^*$	$\varrho^* = -8{\cdot}3$	57 58
6.	$H_2C(OR)_2$	6	Water	25°	$\log \dfrac{k}{k_0} = E_a\varrho$	$\varrho = -3{\cdot}35$	57 59
7.	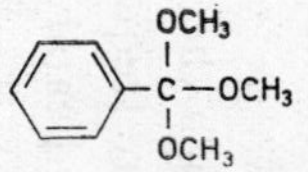	5	70% Aqueous methanol	30°	$\log \dfrac{k}{k_0} = \sigma\varrho$	$\varrho = -2{\cdot}0$	60

^{a} R_1R_2 chosen so as to preclude direct π conjugation with the reaction center.

that the former reactions do not occur either. This is, of course, a naïve argument and provides only weak evidence against nucleophilic participation by water.

At this point, attention is directed to structure–reactivity correlations which exist within individual reaction series (*i.e.* relative hydrolysis rates for methyl acetals of alkyl aldehydes). Inter-series comparisons are deferred for the moment (section IV. B. 4). We include structure–reactivity correlations for hydrolysis of acetals and ketals, closely related reactions, as well.

In several instances, second-order rate constants for reactions of interest here are correlated by one or more linear free energy relationships. These cases are collected in Table 4. Second-order rate constants for acetal and ketal hydrolysis are very sensitive to structural alterations in both the aldehyde and alcohol moieties. Such rate constants for hydrolysis of a series of *m*-substituted diethyl acetals of benzaldehyde are correlated by the Hammett σ constants and a ϱ value of $-3 \cdot 35$[54]. This ϱ value is consistent with and support for rate-determining carbonium ion formation since, in this case, electron donation from a polar substituent will both favor pre-equilibrium substrate protonation and stabilize the carbonium ion developing in the transition state. For compounds substituted in the *para* position with groups capable of electron donation by resonance, second-order rate constants fall somewhat above the line established by the *m*-substituted compounds when plotted against the σ constants and somewhat below a corresponding line when plotted against the σ^+ constants. Data of this type may be treated according to the considerations of Yukawa and Tsuno who have suggested a linear free energy correlation of the form shown in equation (17)[55].

$$\log \frac{k}{k_0} = \varrho[\sigma + r(\sigma^+ - \sigma)]. \tag{17}$$

The second-order rate constants for the hydrolysis of *p*- and *m*-substituted benzaldehyde diethyl acetals are well correlated by equation (17) and values of ϱ and r of $-3 \cdot 35$ and $0 \cdot 5$, respectively, as illustrated in Figure 1. A very similar situation exists for the hydrolysis of 2-(*p*-substituted phenyl-)-1,3-dioxolanes (Table 1 and Figure 1)[54]. The fact that these reaction rates are correlated by a set of substituent constants intermediate between σ and σ^+ is fully consistent with rate-determining carbonium ion formation.

Kreevoy and Taft have found that second-order rate constants for the hydrolysis of 24 diethyl acetals and ketals of non-conjugated aldehydes and ketones are well-correlated, with one exception, by the linear free energy

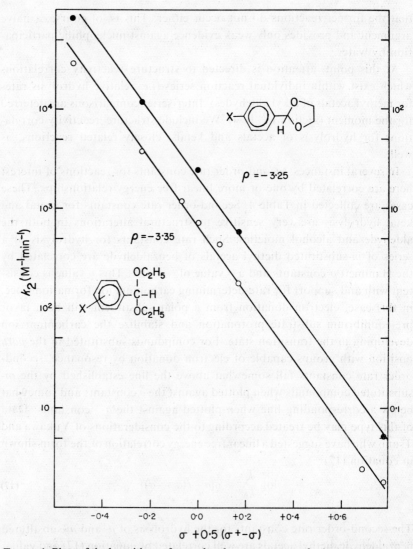

FIGURE 1. Plots of the logarithms of second-order rate constants for hydrolysis of substituted benzaldehyde diethyl acetals (○) and 2-(substituted phenyl)-1,3-dioxolanes (●) against $\sigma + 0.5(\sigma^+ - \sigma)$. The values on the left ordinate refer to the benzaldehyde acetals and those on the right to the dioxolanes. Constructed from the data of Fife and Jao[54].

relationship

$$\log \frac{k}{k_0} = \left(\sum \sigma^* \right) \varrho^* + (\Delta n)\, h \qquad (18)$$

in which σ^* is the sum of the appropriate polar substituent constants[58], Δn is the difference in the total number of α-hydrogen atoms in the carbo-

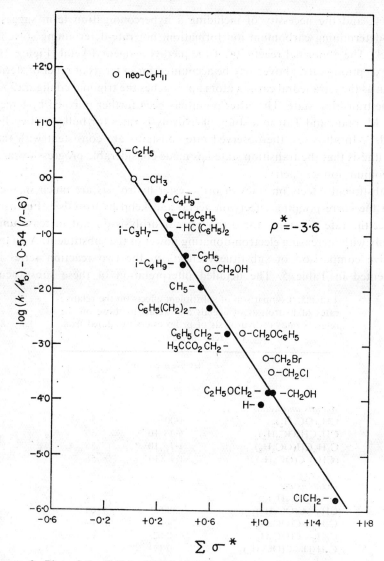

FIGURE 2. Plot of $[\log(k/k_0) - 0.54(\Delta n)]$ against $\Sigma\sigma^*$. ●: Acetals $RCH(OC_2H_5)_2$; ○: Ketals $R(CH_3)C(OC_2H_5)_2$. R is given with each point; n = number of α-hydrogen atoms[56]. [Reprinted by permission from M. M. Kreevoy and R. W. Taft, Jr., *J. Am. Chem. Soc.*, **7**, 5590 (1955)]

nyl moiety and the six in the standard of comparison, diethyl acetonal, and h is an empirical constant measuring the facilitating effect of a single hydrogen on the rate[56]. This structure–reactivity correlation is illustrated n Figure 2 for the case $h = 0.54$. Both the magnitude of the value of ϱ^*,

-3.60, and the necessity of including a hyperconjugation term suggest rate-determining carbonium ion formation, not rate-determining solvent attack. The abnormal reactivity of the methyl neopentyl ketal (Figure 2), the exception noted above, may be accounted for in terms of relief of steric strain as the tetrahedral carbon atom approaches the trigonal configuration in the transition state. The latter point has been further pursued by Kreevoy, Morgan, and Taft in a study of hydrolysis rates for bulky and cyclic ketals[61]. In all cases, the observed rate constants are consistent with the hypothesis that the transition state has made considerable progress toward carbonium ion geometry.

Substituent effects on rates of ortho ester hydrolysis are much smaller than the corresponding effects on acetal and ketal hydrolysis[39]. Furthermore, the rate constants for ortho ester hydrolysis do not increase uniformly with increasing electron-donating power of the substituent. A quantitative comparison of substituent effects in the two reaction series is presented in Table 5. The detailed interpretation of these substituent

TABLE 5. Comparison of substituent effects on the relative rates of hydrolysis of acetals and ketals to those on the relative rates of hydrolysis of ortho esters (modified from reference 39)

Compound	Relative hydrolysis rate	Reference
Acetals and ketals[a]		
$CH_2(OC_2H_5)_2$	1.00[b]	56
$CH_3CH(OC_2H_5)_2$	6.0×10^3	56
$C_6H_5CH(OC_2H_5)_2$	1.7×10^5	56
$(CH_3)_2C(OC_2H_5)_2$	1.8×10^7	56
Ortho esters[c]		
$H{-}C(OC_2H_5)_3$	1.00[d]	64
$CH_3{-}C(OC_2H_5)_3$	38.5	64
$C_2H_5{-}C(OC_2H_5)_3$	24.3	64
$C_6H_5{-}C(OC_2H_5)_3$	0.62	62
$C_2H_5O{-}C(OC_2H_5)_3$	0.17	64

[a] Reactions in 49.6% aqueous dioxane at 25°.
[b] $k_2 = 4.13 \times 10^{-5}$ M^{-1} sec^{-1}.
[c] Reactions in water at 25°.
[d] $k_2 = 5.38 \times 10^2$ M^{-1} sec^{-1}.

effects is deferred to a later section. Suffice it to say at this point that these effects are consistent with the intermediacy of carbonium ions in ortho ester hydrolysis. The systematic study of substituent effects in benzoic acid ortho ester hydrolysis[60] (entry 7, Table 4) is badly clouded by an unfortunate choice of solvent and the value of ϱ obtained cannot be firmly relied upon[62].

The use of entropies of activation as a criterion of mechanism for acid-catalyzed reactions in aqueous solution has been reviewed by Schaleger and Long[63]. Briefly stated, experience indicates that reactions proceeding with unimolecular decomposition of the protonated substrate (A1) usually exhibit entropies of activation near zero or somewhat positive while, in contrast, those proceeding with nucleophilic attack of solvent on the protonated substrate (A2) usually exhibit corresponding values which are large and negative. That bimolecular reactions should exhibit more negative entropies of activation than unimolecular reactions is reasonable in view of the loss of rotational and translational freedom of the water molecule in the transition state. However, variability in the ΔS accompanying the protonation reaction may cloud the picture and differences in entropies of activation are not always large enough to permit unambiguous conclusions. A compilation of data relevant to the substrates under consideration is presented in Table 6. These data, in light of the above generalization,

TABLE 6. Entropies of activation for hydrolysis of acetals, ketals, and ortho esters

Compound	$\Delta S^{\ddagger}$	Reference
Acetals and ketals		
Dimethoxymethane	$+6\cdot8$	65
Diethoxymethane	$+7\cdot0$	65, 66
Dimethoxyethane	$+13$	65
1,3-Dioxolane	$-0\cdot6$	67
2,2-Dimethyl-1,3-dioxolane	$+7\cdot9$	67
2,4,4,5,5-Pentamethyl-1,3-dioxolane	$-3\cdot8$	67
Substituted benzaldehyde diethyl acetals	$+0\cdot7--+2\cdot0$	54
2-(Substituted phenyl)-1,3-dioxolanes	$-6\cdot9--9\cdot6$	54
Ortho esters		
Ethyl orthoformate	$+6--+8$	65, 68
Methyl orthobenzoate	$+8\cdot4$	69
Ethyl orthobenzoate	$-0\cdot3$	62
Ethyl orthoacetate	$+5\cdot5$ (40% dioxane)	62, 70

lend additional support to the concept of carbonium ion intermediates for acetal, ketal, and ortho ester hydrolysis.

Volumes of activation, like entropies of activation, may be employed as an empirical criterion of reaction molecularity. Typical values of $\Delta V^{\ddagger}$ for

acid-catalyzed reactions considered to be unimolecular are in the range
-2 to $+6$ cm^3/mole while those for reactions considered to be bimolecular
(with nucleophilic participation of solvent) are in the range -6 to
-10 cm^3/mole[71]. This result is intuitively reasonable since, in the
unimolecular case, some loosening of a covalent bond will have occurred
in the transition state with an attendant overall increase in volume of the
reacting species, while, in the bimolecular case, the partial formation of
a covalent bond between the substrate and water in the transition state may
result in an overall decrease in volume of the reacting species. Volumes
of activation are probably a more reliable guide to mechanism than the
corresponding entropies in that the volume changes accompanying the
pre-equilibrium protonation seem less susceptible to variation than do the
entropy changes. In Table 7, volumes of activation for several reactions of

TABLE 7. Volumes of activation for certain acid-catalyzed
hydrolytic reactions

Substrate	T (°C)	$\Delta V^{\ddagger}$ (cm^3/mole)	Reference
Dimethoxymethane	25	$-0\cdot5$	65
Diethoxymethane	25	$0\cdot0$	65
Dimethoxyethane	0	$+1\cdot5$	65
	15	$+1\cdot8$	65
Ethyl orthoformate	0	$+2\cdot4$	65

interest are recorded. In each case, the value falls into the range typical
of reactions involving unimolecular decomposition of the protonated species.

Solvent deuterium isotope effects on the rate of hydrolysis of certain
acetals and ortho esters are collected in Table 8. Most of these values fall
in the range $k_{D_3O^+}/k_{H_3O^+} = 2\text{–}3$. Such solvent deuterium isotope
effects probably primarily reflect the isotope effect on the pre-equilibrium
protonation reaction. Rate increases of 2 to 3 fold in D$_2$O compared to
H$_2$O are typical of acid-catalyzed reactions considered to be unimolecular
(A1) and are similar to those predicted theoretically. For example, Bunton
and Shiner have calculated a deuterium solvent isotope effect for acetal
hydrolysis of $2\cdot5$[72]. Of particular note are the isotope effects on the hydrolysis of ethyl orthocarbonate. With the hydrated proton as catalyst, the
isotope effect is small and with acetic acid as catalyst, it is actually less than
unity. These results suggest the involvement of proton transfer in the
transition state (*cf.* discussion below).

TABLE 8. Kinetic solvent deuterium isotope effects for acid-catalyzed acetal, ketal, and ortho ester hydrolysis[a]

Substrate	Solvent	T (°C)	k_{D^+}/k_{H^+}	Reference
1,1-Dimethoxyethane	Water	25°	2·70	73
1,1-Diethoxyethane	Water	25°	2·66	74
1,1-Diethoxyethane	50% Dioxane	25°	3·1	75
1,1-Diethoxyethane	Water	15°	2·61	76
2-Methyl-1,3-dioxolane	Water	25°	2·79	73
2-Phenyl-1,3-dioxane	10% Aceto-nitrile	25°	3·1	77
Ethyl orthoformate	Water	25°	2·05	74
Ethyl orthoformate	Water	25°	2·35	78
Ethyl orthoformate	Water	15°	2·70	68
Ethyl orthoformate	Water	35°	2·31	68
Ethyl orthobenzoate	Water	25°	2·3	62
Methyl orthobenzoate	Water	25°	2·2	69
Ethyl orthocarbonate	Water		1·4	79
Ethyl orthocarbonate	Water		0·7	79

[a] Catalysts compared with substrates 1 to 13 were H_3O^+ and D_3O^+; with substrate 14 CH_3COOH and CH_3COOD.

Shiner and Cross have measured the *secondary* deuterium isotope effect resulting from α-deuteration in the carbonyl component on the rates of hydrolysis of the diethyl ketals of acetone, methyl ethyl ketone, methyl isopropyl ketone and phenoxyacetone[80]. For the fully α-deuterated substrates, a value of k_H/k_D of 1·1 to 1·25 was obtained in each case. These results may be attributed to either the greater relative inductive electron-donating power of H compared to D or the greater relative hyperconjugative electron donating power of H compared to D or to both[81]. Regardless of the precise explanation, these results substantiate the earlier conclusion that electron donation accelerates ketal hydrolysis, as expected in terms of rate-determining carbonium ion formation.

Each of the criteria indicated above provides support for the thesis that the initial step involving carbon–oxygen bond cleavage for the hydrolysis of acetals, ketals, and ortho esters involves unimolecular decomposition of the protonated substrates rather than a bimolecular reaction involving solvent as nucleophilic reagent. Taken together, these criteria constitute a strong case for this conclusion. Considerable further support is provided by studies of the hydrolysis of methyl orthobenzoate and ethyl orthocarbonate in the presence of added nucleophilic reagents.

The first-order rate constants for the decomposition of methyl orthobenzoate in slightly acidic aqueous solution are independent of the concentration of added hydroxylamine and semicarbazide under conditions in which an appreciable fraction of the ortho ester yields amine addition products rather than methyl benzoate[69]. This result is illustrated in Figure 3. For example, in the presence of 0·9 M hydroxylamine at pH 5·45, approximately 85% of the methyl orthobenzoate yields a hydroxylamine addition product, probably N-hydroxymethyl benzimidate, yet the first-order rate constant (0·0228 min^{-1}) is not appreciably different from that

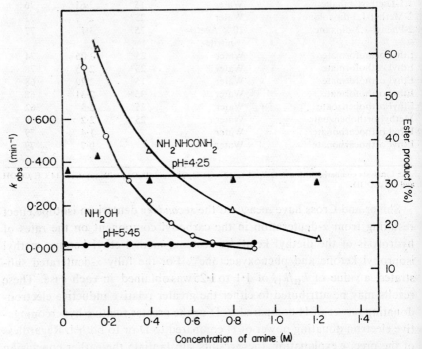

FIGURE 3. First-order rate constants (closed points, left ordinate) for the acid-catalyzed decomposition of methyl orthobenzoate at 25° and ionic strength 0·50 plotted against the concentration of hydroxylamine (circles) and semicarbazide (triangles). In addition, the fraction of methyl orthobenzoate yielding ester product (open points, right ordinate) is plotted against the concentration of these amines.

measured in the absence of hydroxylamine (0·0210 min^{-1}). Methyl benzoate does not react with hydroxylamine under the conditions of these reactions at an appreciable rate. Furthermore, the fraction of methyl orthobenzoate yielding methyl benzoate as product may be accurately calculated assuming that the conjugate acid of the ortho ester undergoes a unimolecular decomposition yielding an intermediate carbonium ion which is then rapidly partitioned between water, yielding methyl benzoate, and amine, yielding amine addition product. On this basis, the free base of hydroxylamine is calculated to be 2000 fold, and the free base form of semicarbazide 275 fold, more reactive toward the carbonium ion than water. These results strongly suggest that the solvent does *not* participate as a nucleophilic reagent in the rate-determining step of acid-catalyzed methyl orthobenzoate hydrolysis. The above experiments are closely related to the rate-product criterion originally employed by Ingold and his coworkers for the identification of unimolecular solvolysis reactions[82].

Related experiments have been performed by Kresge and Preto for the case of ethyl orthocarbonate hydrolysis, a reaction subject to modest general acid catalysis[83]. These workers observed that the rate of hydrolysis of this substrate is the same in the presence of sodium iodide as in the presence of sodium perchlorate. Using the Swain–Scott equation, a reasonable calculation reveals that, were the hydrolysis in fact bimolecular, the concentration of iodide ion employed should have more than doubled the rate through a direct nucleophilic displacement reaction. Since no rate increase was observed, one concludes that the hydrolysis reaction is, in fact, not bimolecular. Thus, this study fully corroborates that performed employing methyl orthobenzoate as substrate.

The lines of evidence cited thus far provide a convincing case for two preliminary and closely related conclusions. First, the transition state for the initial reaction involving carbon–oxygen bond cleavage for the hydrolysis of acetals, ketals, and ortho esters is characterized by carbonyl carbon–oxygen bond cleavage and is unimolecular. That is, this transition state is closely related to **4** (or a variant thereof) and, second, that equation (15) is the minimum reaction path required to account for these reactions. A closer examination of these points is presented below.

3. Rate-determining step

Several times in the above discussion it has been pointed out that certain findings (*e.g.* entropies and volumes of activation) are evidence not only for a reaction path involving the formation of a carbonium ion but for *rate-determining* unimolecular formation of a carbonium ion. However, it does *not* follow from this conclusion that formation of the carbonium ion, the first reaction in equation (15), is necessarily rate-determining. This is true since the decomposition of the tetrahedral intermediate, the terminal step in equation (15), is also almost certainly unimolecular and yields a species possessing carbonium ion character. This decomposition may be considered to yield the conjugate acid of a carbonyl compound (equation 19a) or the carbonyl compound itself (equation 19b). Thus there exist two closely related species both formed in unimolecular processes on the reac

$$
\begin{array}{l}
\underset{\text{OR}}{\overset{\text{OH}}{\diagdown}}\!\!\!\!\!\!\!\!C
\quad\underset{-\text{H}^+}{\overset{+\text{H}^+}{\rightleftharpoons}}\quad
\underset{\overset{\text{OR}}{\underset{\text{H}}{+}}}{\overset{\text{OH}}{\diagdown}}\!\!\!\!\!\!\!\!C
\quad\xrightarrow{-\text{ROH}}\quad
\overset{\diagdown}{\underset{\diagup}{C}}\!\!-\overset{+}{\text{OH}}
\quad\longleftrightarrow\quad
\overset{\diagdown}{\underset{\diagup}{C}}\!\!=\!\overset{+}{\text{OH}}
\qquad\qquad (19\text{a})
\end{array}
$$

$$
\begin{array}{l}
\underset{\text{OR}}{\overset{\text{OH}}{\diagdown}}\!\!\!\!\!\!\!\!C
\quad\underset{-\text{H}^+}{\overset{+\text{H}^+}{\rightleftharpoons}}\quad
\underset{\overset{+}{\underset{\text{H}}{\text{OR}}}}{\overset{\text{OH}\ :\text{B}}{\diagdown}}\!\!\!\!\!\!\!\!C
\quad\xrightarrow[-\text{BH}]{-\text{ROH}}\quad
\overset{\diagdown}{\underset{\diagup}{C}}\!\!-\text{O}^-
\quad\longleftrightarrow\quad
\overset{\diagdown}{\underset{\diagup}{C}}\!\!=\!\text{O}
\qquad\qquad (19\text{b})
\end{array}
$$

tion pathway and the formation of either may be rate-determining. Mechanisms involving rate-determining formation of either the carbonium ion or the product (or its conjugate acid) are consistent with substantially all of the experimental information detailed above. In view of these considerations, it is somewhat surprising that the initial step in equation (15) has been nearly unanimously agreed on in the literature as the rate-determining step.

The conclusion that formation of the carbonium ion is rate-determining is in accordance with expectations based on chemical arguments. In solutions containing little alcohol, rate-determining reaction of this carbonium ion with solvent is extremely unlikely since this requires that alcohol reacts with the carbonium ion, regenerating starting material, more rapidly than water reacts with the carbonium ion, yielding products. Since the *rate constants* for reaction of alcohol and water are almost certainly about the same (see below), the *rate* for the latter reaction must be greater than that for the former. A similar argument suggests that tetrahedral intermediate decomposition, by either route (19a) or (19b), is not rate-determining. Since the overall equilibrium constant for interconversion of, for example, methyl orthobenzoate and dimethyl orthobenzoate should be about unity, the latter would be present in much greater concentration than the former were equilibrium established, due to the high concentration of water relative to methanol. Since the *rate constants* for decomposition of these species should be about equal (or that for dimethyl orthobenzoate may actually be appreciably the larger if reaction occurs by route (19b) as a similar route is unavailable to methyl orthobenzoate), the *rate* for dimethyl orthobenzoate decomposition should be greater than the corresponding quantity for methyl orthobenzoate. These conclusions are fully corroborated in kinetic studies of ketal and ortho ester hydrolysis conducted in the presence of deuterated alcohols which are described immediately below.

A study of the kinetics and product composition for the hydrolysis of methyl ketals and methyl ortho esters in methanol-d_4-deuterium oxide mixtures (equation 20), using proton magnetic resonance spectroscopy, has provided a simple and straightforward experimental distinction between several of the possible rate-determining steps for these reactions[84, 85].

$$\text{>}C\text{<}^{OCH_3}_{OCH_3} + D^+ \rightleftharpoons \text{>}C^+\text{—}OCH_3 + CH_3OD \qquad (20a)$$

$$\text{>}C\text{<}^{OCH_3}_{OCD_3} + D^+ \rightleftharpoons \text{>}C^+\text{—}OCH_3 + CD_3OD \qquad (20b)$$

$$\text{>}C\text{<}^{OCH_3}_{OCD_3} + D^+ \rightleftharpoons \text{>}C^+\text{—}OCD_3 + CH_3OD \qquad (20c)$$

$$\text{>}C\text{<}^{OCD_3}_{OCD_3} + D^+ \rightleftharpoons \text{>}C^+\text{—}OCD_3 + CD_3OD \qquad (20d)$$

$$\text{>}C^+\text{—}OCH_3 + D_2O \longrightarrow \text{>}C{=}O + CH_3OD + D^+ \qquad (20e)$$

$$\text{>}C^+\text{—}OCD_3 + D_2O \longrightarrow \text{>}C{=}O + CD_3OD + D^+ \qquad (20f)$$

The experimental quantities determined in the study of, for example, methyl orthobenzoate hydrolysis by this method include the first-order rate constants for the disappearance of the methoxy protons of the ortho ester, $k_{\text{ortho ester}}$, for the appearance of the methyl protons of methanol, k_{MeOH}, and for the appearance of the methoxy protons of the carboxylic ester, k_{ester}. Since the proton resonance singlets for each of these groups are well separated, the rate constants can be determined simultaneously. In addition, the ratio of integrated proton intensities at infinite time of the products to some internal, time-independent, standard provides a quantitative measure of product composition for many substrates.

Both the product composition and the relative magnitudes of the various rate constants are functions of the nature of the rate-determining step. If carbonium ion formation were rapid and reversible (*i.e.* carbonium ion formation not rate-determining), the methoxy groups of the starting material would be rapidly exchanged for deuteriomethoxy groups through reaction of the carbonium ion with solvent deuteriomethanol. The ortho ester would be converted more slowly to carboxylic ester product. Thus, $k_{\text{ortho ester}}$ and k_{MeOH} would be considerably larger than k_{ester}. Furthermore, little or no carboxylic ester product containing methoxy protons would be formed since virtually all of the ortho ester would have been converted into the corresponding deuterated material in the pre-equilibrium exchange reactions. In contrast, if carbonium ion formation were rate-determining, methanol would not be exchanged for deuteriomethanol in a pre-equilib-rium reaction; hence, $k_{\text{ortho ester}}$, k_{MeOH}, and k_{ester} would be nearly identical. In addition, only protio carboxylic ester would be produced as reaction product.

Studies of this type have been carried out using 2,2-dimethoxypropane, 6,6,6-trimethoxyhexanonitrile, methyl orthobenzoate, and methyl orthocar-bonate as substrates[85]. The course of the hydrolysis of methyl orthobenzo-ate as a function of time is indicated in Figure 4. As may be judged qualita-

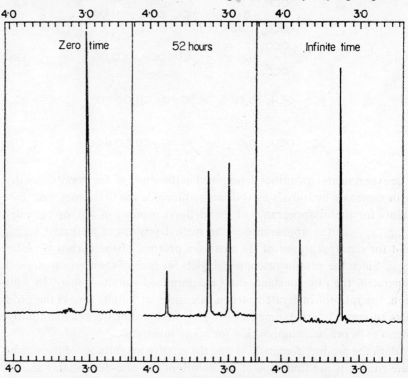

FIGURE 4. Initial, intermediate, and final proton magnetic resonance spectra for the hydrolysis of methyl orthobenzoate in an equimolar mixture of deuterium oxide and methanol-d_4[85]. Methoxy protons of ortho ester appear at 3·0 ppm, those of methyl benzoate at 3·8 ppm, and those of methanol at 3·25 ppm.

tively from this figure, the rate of disappearance of the methoxy protons of the ortho ester is comparable to the rate of appearance of the corresponding protons of methyl benzoate. Furthermore, a rather substantial amount of methyl benzoate, as opposed to deuteriomethyl benzoate, is formed in the reaction as judged from the intensity of the appropriate signal in the infinite time spectrum. These results are just those predicted on the basis of rate-

TABLE 9. First-order rate constants for the hydrolysis of ketals and ortho esters in deuteriomethanol–deuterium oxide solutions at 25°[a]

Substrate	$\dfrac{(CD_3OD)}{(D_2O)}$	Buffer	$k_{\text{ortho ester}}$ or k_{ketal} (min^{-1})	k_{MeOH} (min^{-1})	k_{ester} or k_{ketone} (min^{-1})
2,2-Dimethoxypropane	1	0·006 M acetate, 50% base	$1·4 \times 10^{-3}$	$1·3 \times 10^{-3}$	$8·1 \times 10^{-4}$
6,6,6-Trimethoxyhexanonitrile	1	0·006 M acetate, 80% base	$4·2 \times 10^{-3}$	$4·1 \times 10^{-3}$	$4·2 \times 10^{-3}$
Methyl orthobenzoate	1	0·006 M acetate, 50% base	$2·2 \times 10^{-4}$	$2·9 \times 10^{-4}$	$2·6 \times 10^{-4}$
Methyl orthobenzoate	2	0·014 M formate, 50% base	$1·1 \times 10^{-3}$	$1·4 \times 10^{-3}$	$1·3 \times 10^{-3}$
Methyl orthocarbonate	1	0·006 M acetate, 50% base	$8·3 \times 10^{-4}$	$7·4 \times 10^{-4}$	$8·0 \times 10^{-4}$
Methyl orthocarbonate	4	0·014 M formate, 50% base	$1·3 \times 10^{-3}$	$1·1 \times 10^{-3}$	$1·1 \times 10^{-3}$

[a] $k_{\text{ortho ester}}$ is the rate constant for the disappearance of the methoxy protons of ortho ester; k_{ketal} is the corresponding constant for the ketal; k_{MeOH}, k_{ester}, and k_{ketone} are rate constants for the appearance of methoxy protons of the indicated species.

determining carbonium ion formation. The quantitative results of this experiment, and of similar experiments using the other substrates noted above, are collected in Table 9. In each case, the characteristic rate constants are similar. A quantitative treatment of the kinetics of reactions of this type reveals that the relative magnitudes of the various rate constants and the product composition patterns are those expected for rate-determining carbonium ion formation provided that deuterium oxide and deuteriomethanol are about equally reactive toward the carbonium ion[85]. The latter conclusion is in accordance with the relative reactivities of water and methanol toward, for example, the *t*-butyl and benzhydryl carbonium ions[86-88].

We consider that the above results clearly establish that the first step indicated in equation (15) is rate-determining for at least those substrates explicitly studied and provide strong evidence for similar behavior in the cases of related substrates. It is important to recognize, however, that the first step in equation (15) is, in fact, composed of two steps: protonation and decomposition. The data cited above do *not* distinguish between rate-determining protonation and pre-equilibrium protonation. It is also possible, of course, that protonation and carbon–oxygen bond cleavage may be concerted processes.

4. Structure–reactivity correlations

The notion of rate-determining decomposition of the protonated substrates for acetal and ketal hydrolysis, A1 reaction paths, is fully consistent with all of the known facts concerning these reactions including the strong accelerating effect of electron-donating polar substituents (Table 4). Thus, those substrates yielding the most stable carbonium ions react most rapidly. The situation is not quite so straightforward for the hydrolysis of ortho esters in the respect that rates of hydrolysis do not parallel the expected stabilities of the corresponding carbonium ions. As noted in Table 5, ethyl orthocarbonate is less reactive than ethyl orthobenzoate which is, in turn, less reactive than ethyl orthoacetate. Thus, for this limited set of substrates, the rates are actually inversely related to carbonium ion stabilities. Neglecting the now excluded possibility that these reactions are A2, there remain several possible explanations for the failure of substituents capable of electron donation by resonance to accelerate the rate of ortho ester hydrolysis.

Firstly, DeWolfe and Jensen have noted that rates for ortho ester hydrolysis follow the *inductive* effect of the substituents and, on this basis, have suggested that the transition state for these reactions is reached so early

along the reaction coordinate that the central carbon is essentially tetrahedral therein[63]. If this were the case, electron donation by resonance would
be of little importance and the rates would be expected to follow the inductive effects alone. We do not find this hypothesis appealing particularly in
view of the large Brønsted α value for general acid catalysis of orthocarbonate hydrolysis (see below). This point of view seems to have been abandoned by at least one of the above authors[39].

A second possibility is that a saturation effect occurs. That is, it is possible
that the transition state for a dialkoxymethyl carbonium ion is so stabilized
by these functions that the remaining aryl, alkyl, or alkoxy substituent does
not lend appreciable further stabilization. The appearance potential and
fluorine n.m.r. measurements of Taft and coworkers discussed above (section IV. A) lend some support to this point. However, some additional stabilization does seem to occur on introduction of, for example, the third
alkoxy function. At any event, saturation effects, by themselves, cannot
account for the observed rate decrease on increasing alkoxy substitution.
Furthermore, saturation effects have not previously been noted in carbonium ion reactions in solution as evidenced by the fact[89] that the relative
rates of solvolysis of $C_6H_5CH_2Cl$, $(C_6H_5)_2CHCl$, and $(C_6H_5)_3CCl$ are
approximately $1 : 10^5 : 10^9$. Nevertheless, it does seem likely that a saturation effect accounts, in part, for the observed order of reactivities in the
present case.

Thirdly, Hine has suggested that the relative rates of hydrolysis of ethers,
ketals, ortho esters, and orthocarbonates can be accounted for on the basis
of the summation of substituent effects upon starting material and transition state employing the concept of (double bond)–(no bond) resonance[90].
Consider, for example, the cases of ethyl orthocarbonate and ethyl orthoacetate. With the former substrate, a total of 12 (double bond)–(no bond)
resonance structures may be written while only six will contribute to the
structure of the latter substrate.

Since stabilization by (double bond)–(no bond) resonance will be of lesser
importance in the transition state (and may completely disappear in the

carbonium ions themselves), this factor will tend to decrease the reactivity of the orthocarbonate relative to the orthoacetate. This argument may be extended to account for the relative reactivities of ethyl orthobenzoate and ethyl orthoacetate as well since substituents which donate electrons by resonance will themselves be able to participate in (double bond)–(no bond) resonance. Thus, one would expect structure **8** to make a greater contribution to the stability of methyl orthobenzoate than structure **9** to the stability of ethyl orthoacetate. The force of these arguments is weakened some-

(8)   (9)

what by the lack of thoroughly reliable information as to the quantitative importance of (double bond)–(no bond) resonance for systems involving oxygen–oxygen interactions. Furthermore, the relative basicities of ketals, ortho esters, and orthocarbonates, (section III) suggest that (double bond)–(no bond) resonance may be of minor importance in stabilization of these species. This follows since such resonance should be much more important in the unprotonated than in the protonated species; thus (double bond)–(no bond) resonance should markedly affect the observed basicities. The fact that these basicities are nicely accounted for in terms of polar effects alone (Table 1) therefore provides suggestive evidence against significant ground state stabilization by such resonance. The final assessment of its importance in accounting for the reactivities in question must await reliable thermochemical data for these substrates.

Fourthly, the most basic substrates are observed to react most rapidly: to what extent do the basicity differences account for the observed order of reactivity? In the first place, the basicities differ, as one goes from ketal to orthocarbonate, by only three orders of magnitude. Since 2,2-dimethoxypropane is about 10^3 times as reactive as methyl orthocarbonate[56, 64], the differences in basicity are barely adequate to account for the magnitude of the rate difference and certainly cannot explain why the opposite order of reactivity is not, in fact, observed. In the second place, the differences in basicity will exert their full effect on observed rates only if the reactions proceed with pre-equilibrium proton transfer (or if the Brønsted α value for general acid catalysis is unity). It is likely that, for some of the substrates at least, the protonation reaction is not a pre-equilibrium process[39]. This brings us directly to our last point.

Finally, one may argue that ortho ester hydrolysis does not proceed via an A1[91] reaction path but involves proton transfer in the rate-determining step[39]. Such reactions may be regarded as electrophilic displacements on oxygen and are designated S_E2 reactions. The postulation of an S_E2 reaction mechanism does not, in itself, account for the observed structure–reactivity relationships. It does introduce novel considerations into the evaluation of such effects and provides a new framework into which the factors enumerated above may be placed. We shall consider the arguments suggesting that these reactions are, in fact, S_E2 in character in some detail below.

5. Catalysis

The hydrolysis of acetals, ketals, and ortho esters is subject to catalysis by the hydrated proton, as noted above. Certain of these substrates are subject to catalysis by other species as well. These include acids in general and certain detergents.

Ortho ester hydrolysis was the first reaction demonstrated to be subject to the phenomenon of general acid catalysis. In 1929 Brønsted and Wynne-Jones found that the first-order rate constants for hydrolysis of ethyl orthocarbonate, ethyl orthoacetate, and ethyl orthopropionate in aqueous solution increased with increasing buffer concentration at constant pH[62]. The hydrolyses of ethyl orthoformate and methyl orthobenzoate are, in addition, subject to general acid catalysis in aqueous dioxane and aqueous methanol solutions respectively[92, 60]. The Brønsted α value for general acid-catalyzed ethyl orthocarbonate hydrolysis is near 0·70[65, 83] as is that for methyl orthobenzoate hydrolysis[60]. Although rather few systematic studies have been carried out, it seems likely that the hydrolysis of all ortho esters will be characterized by large values of α. In contrast to the case for ortho ester hydrolysis, there seem to be no reports of general acid catalysis for acetal and ketal hydrolysis.

The finding of general acid catalysis for ortho ester hydrolysis immediately raises the question of the mechanism of the hydronium ion catalyzed reaction. The following possibilities exist. Firstly, substrate protonation is a pre-equilibrium reaction and the subsequent decomposition of the conjugate acid of the substrate is rate-determining (transition state 10). Secondly, substrate protonation is the rate-determining step and the subsequent decomposition is rapid (transition state 11). Thirdly, substrate protonation and decomposition are independent reactions and occur with approximately the same rate (transition states 10 and 11 both important). Fourthly, proton transfer to the substrate and substrate decomposition are concerted processes (transition state **12**). The last possibility may itself be refined into subcategories which are considered below. It is to be noted that each of the above transition states is related to transition state **4** originally proposed and whose basic features are correct for acetal and ketal hydrolysis at least. Interesting calculations recently reported by Burton and DeWolfe strongly

$$
\begin{array}{ccc}
\text{OR} \ \text{H}\delta+ & \text{OR} & \text{OR} \\
\ |\ _{\delta+}\ | & \ |\ _{\delta+} & \ |\ _{\delta+} \\
-\text{C}\cdots\text{OR} & -\text{C}\cdots\text{O}\cdots\text{H}\cdots\text{B} & -\text{C}\cdots\text{O}\cdots\text{H}\cdots\text{B} \\
\ | & \ |\ \ \ | & \ |\ \ \ | \\
\text{OR} & \text{OR} \ \ \text{R} & \text{OR} \ \ \text{R} \\
(10) & (11) & (12)
\end{array}
$$

suggest that the first alternative is incorrect[39]. These workers have estimated the basicities of ortho esters by employing the reasonable assumption that the structural effects on the basicity of oxygen bases parallel those on the basicity of nitrogen bases. Since the basicity of aliphatic amines is correlated by the aliphatic substituent constants of Taft, σ^*, with a ϱ^* value of approximately 3[93], the basicities of ortho esters may be obtained by use of the appropriate substituent constants, the above assumption, and the values of pK_a of $-3\cdot6$ for diethyl ether and $-3\cdot8$ for dimethyl ether[35]. The estimated values of pK_a are $-5\cdot4$ for 1,1-dimethoxyethane, -7 for ethyl orthoformate, $-7\cdot6$ for ethyl orthobenzoate, and $-8\cdot5$ for ethyl orthocarbonate (see also Table 1). If we formulate ortho ester, ketal, and acetal hydrolysis as

$$S + H_3O^+ \underset{k_{-1}}{\overset{k_{-1}}{\rightleftharpoons}} SH^+ + H_2O$$

$$SH^+ \xrightarrow{k_{-2}} \text{products} \tag{21}$$

the steady state assumption yields the rate law

$$k_H = \frac{k_1 k_2}{k_{-1} + k_2} \tag{22}$$

in which k_H is the second-order rate constant for the hydronium ion catalyzed reaction. k_{-1} must have a value near 10^{10} M^{-1} sec^{-1}, the diffusion-controlled upper limit, since it involves proton transfer from a very strong acid to water[94]. Using this value and the knowledge that $K_a = k_{-1}/k_1$, we can estimate k_2 from equation (22). For example, consider the case of 1,1-dimethoxyethane hydrolysis. Using the value of pK_a of $-5\cdot6$, k_1 is calculated to be $2\cdot5 \times 10^4$ M^{-1} sec^{-1}. Since k_H for this substrate is 3×10^{-5} M^{-1} sec^{-1}, k_2 is calculated to be about 10 sec^{-1}. This value is clearly much less than that for k_{-1} consistent with the formulation of acetal hydrolysis as an A1 reaction. However, a similar calculation for ortho ester hydrolysis, using any of the substrates indicated above, yields values of k_2 near 10^{10} M^{-1} sec^{-1}. Hence, for these reactions k_{-1} is not large with respect to k_2 and, hence, protonation cannot be a pre-equilibrium process.

In addition, it seems unlikely that proton transfer itself is largely rate-determining. This would require that the rate of cleavage of a covalent bond to carbon, k_2, be much larger than that for the diffusion-controlled loss of a proton from the protonated substrate, k_{-1}. The factor setting the upper limit for processes of the former type is not the time constant for the vibration of a single bond but the rate at which the products diffuse from each other following covalent bond cleavage. This rate constant is not much larger than that for diffusion-controlled bimolecular reactions and hence, k_2 cannot be appreciably larger than k_{-1}. We are left with the latter two alternatives indicated above.

There exists at this point no definitive basis for a choice between a reaction pathway involving the successive formation of **10** and **11** and one occurring via **12** or a variant thereof. Inasmuch as ortho ester hydrolysis is known to be subject to general acid catalysis, we are inclined toward the view that the hydronium ion catalyzed reaction involves general acid catalysis as well. These points are pursued more deeply in reference 2.

The hydrolysis of methyl orthobenzoate is subject to marked catalysis by dilute aqueous solutions of sodium lauryl sulfate and other anionic detergents as illustrated in Figure 5[95, 96]. In contrast, this reaction is slightly inhibited by cationic detergents such as cetyl trimethylammonium bromide (Figure 5). Such catalysis is considered to involve a pre-equilibrium association of ortho ester and detergent as a result of hydrophobic bond formation between the species involved[97]. Consistent with this hypothesis is the observation of saturation of substrate with catalyst and saturation of catalyst with substrate. Thus, at low substrate concentrations, the first-order rate constants are independent of this variable but decrease thereafter with increasing substrate concentration and eventually approach the value for this reaction in the absence of detergents. These kinetics are entirely comparable to the Michaelis–Menten kinetics typically observed in enzyme-catalyzed reactions. The role of hydrophobic interactions is further emphasized by the finding that the hydrolysis of ethyl orthovalerate and ethyl orthopropionate, but not that of ethyl orthoformate, is subject to modest catalysis by sodium lauryl sulfate[96].

Following complex formation between detergent and ortho ester, catalysis of bond-breaking reactions may result either from (i) electrostatic stabilization of the developing carbonium ion in the transition state by the negative charges of the micelle; (ii) a locally increased concentration of hydronium ions in the immediate vicinity of the substrate–detergent complex; or (iii) general acid catalysis by sodium lauryl sulfuric acid.

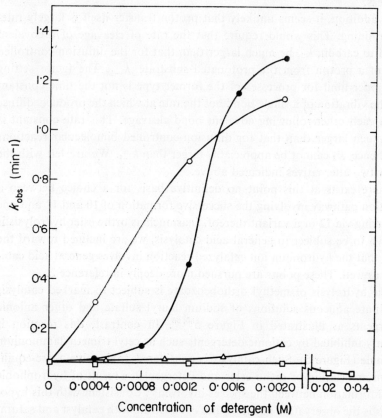

FIGURE 5. First-order rate constants for the hydrolysis of methyl orthobenzoate in aqueous solution at 25° and pH 4·76 plotted as a function of the concentration of sodium lauryl sulfate (O) sodium oleyl sulfate (●), sodium heptadecyl sulfate (Δ), and cetyltrimethylammonium bromide (□)[96].

C. With Amines

The reaction of ortho esters with amines seems to be a general process. A variety of amines have been demonstrated to react with a variety of ortho esters although the nature of the products and the extent of reactions can be sensitive functions of substrate structure and reaction conditions, particularly the presence or absence of acidic catalysts. Little in the way of mechanistic study has been devoted to these reactions. Only the finding that the rate of reaction of methyl orthobenzoate in aqueous solutions of weakly basic amines is independent of the concentration of amine (section IV. B. 2)[69] suggests that these reactions proceed via the carbonium ions derived from the ortho ester substrates. In the usual solvents used for synthetic reactions between ortho esters and amines, which have modest ion-solvating powers, the assumption of a carbonium ion intermediate may be too much of a simplification. Such reactions are probably better regarded as nucleophilic attack of the amine on loose ion pairs. Regardless of the extent of formation of a free carbonium ion, it is probably safe to conclude, that for most cases at least, departure of the alcohol precedes attack of amine. This follows from the well-known resistance of these substrates to direct nucleophilic attack, from the results on the hydrolysis reactions discussed in section IV. B, and from the kinetic observation indicated above.

We initiate discussion of reactions of amines with ortho esters by considering some rather straightforward examples and subsequently, turning attention to some rather more complicated cases.

Perhaps the best studied reaction is that of ethyl orthoformate with primary aromatic amines. Study of these reactions has been pursued since the pioneering observation of Walther and Claisen in 1895 until rather recently[98-103]. Most of the modern work is the product of the efforts of Roberts and coworkers[104-109]. The principal conclusions of these investigations may be summarized as follows. In the first place, the reaction yields principally imidates in the presence of acid and principally symmetrical amidines in the absence of acid. The synthesis of these products almost certainly follows the two-step reaction pathway outlined in equations (23) and (24). Thus the imidate is the initial product and its subsequent aminolysis yields the amidine. The early suggestion by Claisen that the formami-

$$R^1C(OEt)_3 + R^2NH_2 \underset{r}{\overset{f}{\rightleftharpoons}} R^1C(=NR^2)OEt + 2\,EtOH \qquad (23)$$

$$R^1C(=NR^2)OEt + R^2NH_2 \underset{r}{\overset{f}{\rightleftharpoons}} R^1C(=NR^2)NHR^2 + EtOH \qquad (24)$$

dine is an intermediate in the formation of the formimidate in the reaction between aniline and ethyl orthoformate[98] has been carefully excluded by Roberts and DeWolfe, who established that the amidine does not react with the ortho ester under conditions in which the imidate is obtained in high yields in the aminolysis reaction[107]. It is worth noting that amidines do react with ethyl orthoformate in the presence of acid catalysts to yield imidates although the products are very likely derived from alcoholysis of the amidine involving alcohol produced in decomposition of the ortho ester[104]. In the second place, reaction (24f) must be more rapid than (23f) since, in the absence of acid, only the amidine is obtained from the reaction of aniline even with a large excess of ortho ester[107]. The explanation provided in the literature to explain the production of imidate in the presence of acid, *i.e.* that the amidine forms rapidly in reactions (23f) and (24f) and then gradually reverts to imidate via reaction (24r) (considered to be highly acid dependent)[107, 110], must be wrong since it violates the principle of microscopic reversibility. The observation can be simply accounted for in terms of the effect of increasing acidity on the equilibrium position of equation (24). One expects this to shift to the left with increasing acidity since there are two species on the left side and only one on the right which have marked basic character. It is worth noting that the *N*-substituted phenyl alkyl forminidate products are converted, by thermal rearrangement, to *N*-alkyl formanilides[111]. Hydrolysis of these products yields *N*-alkyl anilines (equation 25); these reactions appear to be a suitable route for the preparation of both the formanilides and anilines.

$$
HC(\!=\!NAr)OR \xrightarrow{\Delta} \overset{\overset{\displaystyle O}{\|}}{HC}\!-\!\underset{\underset{\displaystyle R}{|}}{N}Ar \xrightarrow{H_2O} HCO_2H + ArNHR \tag{25}
$$

The reactions of aromatic amines with ortho esters have been extended to ethyl orthoacetate by DeWolfe[112] and by Taylor and Ehrhart[110]. The former worker has described synthesis of a variety of *N,N'*-diarylacetamidines from heating two moles of aromatic amine with one mole of ethyl orthoacetate in the presence of a small amount of *p*-toluenesulfonic acid[112]. The results of the latter group suggest that either amidines or imidates can be prepared in these reactions: the amidines from reaction of two moles of amine with one of ortho ester in the presence of one mole of acetic acid, and the imidates from an equimolar mixture of amine and ortho ester either in the absence of acid or in the presence of a trace of acid[110]. Typical results are indicated in Table 10. In contrasting these results with those obtained using orthoformates, one concludes that reaction (24f), rapid and not sensitive to acid in the latter case, must be slow and highly dependent on acid in that of orthoacetates.

The above reactions may, in part, be extended to aliphatic amines as well. Thus, treatment of one mole of ethyl orthoformate or ethyl orthoacetate with two moles of an aliphatic amine in the presence of one mole of acetic acid yields the appropriate formamidines and acetamidines[110]. These reactions do not permit isolation of the imidates, however. Reaction of aliphatic diamines with ortho esters yields the cyclic amidine products.

TABLE 10. Reaction of aromatic amines with ethyl orthoacetate[110]

	Mole ratio of amine to ortho ester	Mole ratio of acetic acid to ortho ester	Reaction time (h)	Amidine (%)	Imidate (%)
$C_6H_5NH_2$	2	1/12	2	76	—
	1	1/24	1·5	5·2	88
	2	0	1·5	trace	69
	1	0	1·5	trace	59
$o\text{-}CH_3OC_6H_4NH_2$	2	1	1·3	72	—
	2	1/12	1·75	50	—
	1	1/12	1·5	1·7	87
	0·66	0	5	—	90
	2	0	1·5	—	59
$o\text{-}CH_3C_6H_4NH_2$	2	1	1·5	57	—
$o\text{-}ClC_6H_4NH_2$	2	2/3	1·5	30	—

Thus, that between ethylenediamine and ethyl ortho esters produces imidazoline[110, 113] and that between trimethylenediamine and ortho esters yields tetrahydropyrimidines (equation 26)[113].

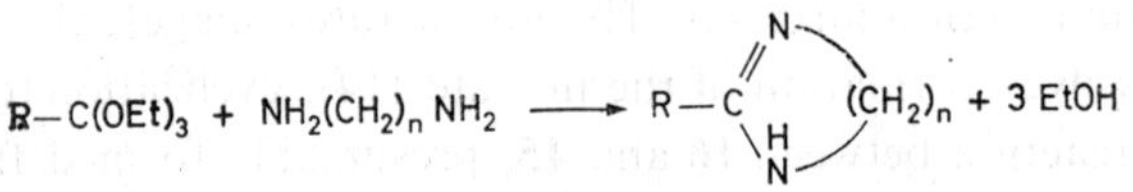

$$R-C(OEt)_3 \ + \ NH_2(CH_2)_n\,NH_2 \longrightarrow R-C\underset{N}{\overset{N}{\diagdown}}(CH_2)_n \ + \ 3\ EtOH \qquad (26)$$

A related reaction has been used for a synthesis of purines (equation 27)[114].

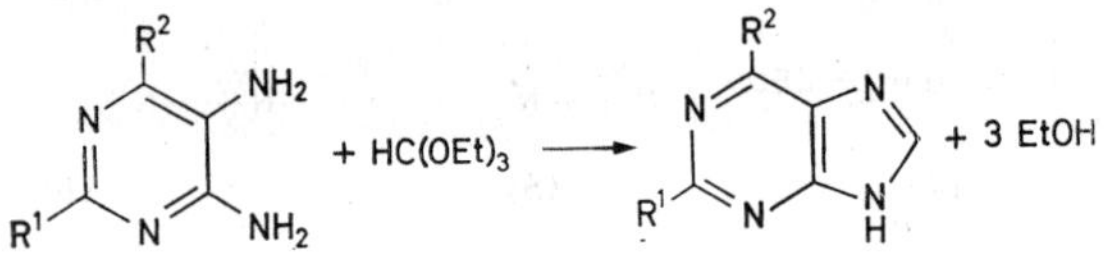

$$+ \ HC(OEt)_3 \longrightarrow \qquad + \ 3\ EtOH \qquad (27)$$

These simple reactions involving ordinary aromatic or aliphatic amines have been extended to less simple amines with the production of less simple products. Typical examples include reactions with hydrazines and ureas.

Ainsworth has demonstrated the synthesis of 2-phenyl-1,3,4-oxadiazoles from acyl hydrazines and ethyl orthoformate (equation 28)[115]. Similarly, if thioacyl hydrazines are used substrates, thiadiazoles are obtained.

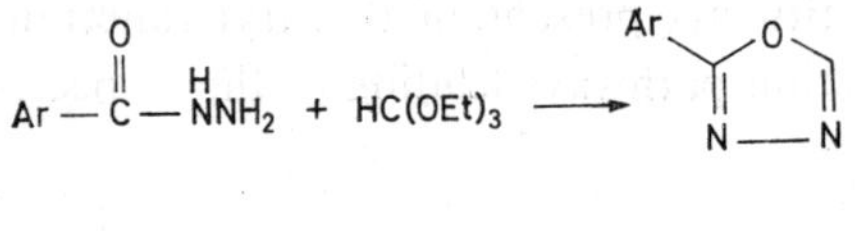

$$Ar-\overset{O}{\overset{\|}{C}}-\overset{H}{N}NH_2 \ + \ HC(OEt)_3 \longrightarrow \qquad (28)$$

It has proved possible to isolate 1-acyl-2-ethoxymethylene hydrazines (13) from reaction mixtures and to convert these materials to oxadiazoles thermally. Thus, species of the type 13 are likely intermediates in the overall reaction processes. Synthesis of these materials is, of course, exactly analogous to those for imidates discussed above and their conversion to oxadiazoles involves a simple cyclization.

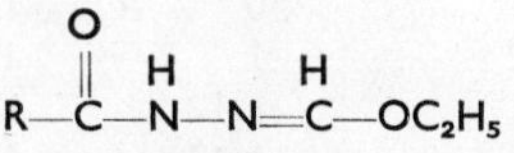

$$
\begin{array}{c}
\text{O}\\
\|\quad \text{H}\quad\ \text{H}\\
\text{R—C—N—N}{=}\text{C—OC}_2\text{H}_5
\end{array}
$$

(13)

In a related process, thiosemicarbazide has been converted to N,N'-bis-(1,3,4-thiadiazole-2)-formamidine in a reaction with ethyl orthoformate (equation 29)[116]. Three possible intermediates, 14, 15, and 16, have been

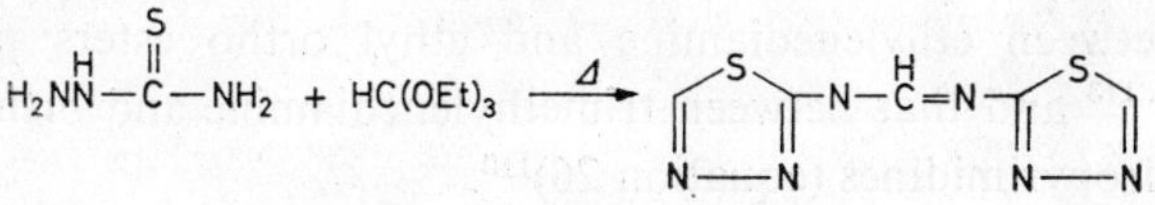

$$
\begin{array}{c}
\text{S}\\
\|\quad\ \text{H}\\
\text{H}_2\text{NN—C—NH}_2
\end{array}
+\ \text{HC(OEt)}_3 \xrightarrow{\ \Delta\ }
$$

(29)

isolated from reaction mixtures. These structures suggest that the overall reaction involves formation of the imidate (14), cyclization to 16, and an aminolysis reaction between 16 and 15, presumably formed from decomposition of 16, yielding the amidine.

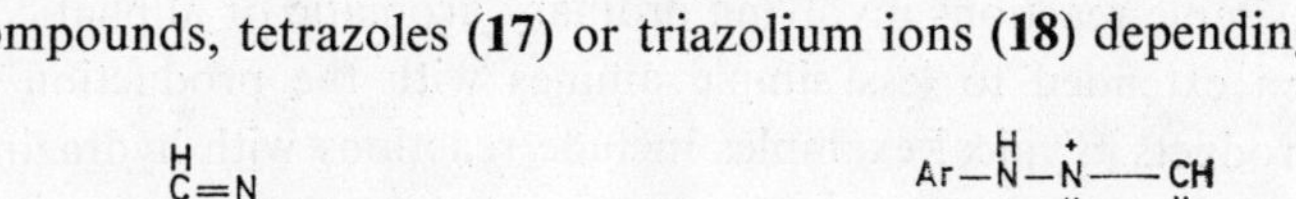

$$
\begin{array}{c}
\text{S}\quad\text{H}\quad\ \text{H}\\
\|\quad\ |\quad\ |\\
\text{H}_2\text{N—C—N—N}{=}\text{C—OEt}
\end{array}
$$

(14)   (15)   (16)

The reaction of ortho esters with aryl hydrazines yields simple imidate-type compounds, tetrazoles (17) or triazolium ions (18) depending on the

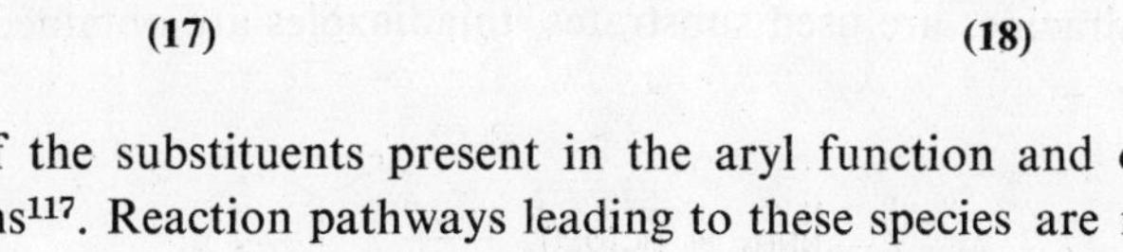

(17)   (18)

nature of the substituents present in the aryl function and on reaction conditions[117]. Reaction pathways leading to these species are not entirely clear.

Reaction of ortho esters with ureas follows the general course outlined above. Thus, treatment of one mole of ethyl orthoformate with two moles of substituted urea yields N,N'-dicarbamylformamidines (equation 30)[118].

$$\text{HC(OEt)}_3 + 2\,\text{RN}\overset{\underset{\textstyle H}{|}}{-}\overset{\overset{\textstyle O}{\|}}{C}\text{--NH}_2 \longrightarrow \text{RN}\overset{\overset{\textstyle O}{\|}}{-}\overset{\underset{\textstyle H}{|}}{C}\text{--N}=\overset{\underset{\textstyle H}{|}}{C}\text{--N}\overset{\overset{\textstyle O}{\|}}{-}\overset{\underset{\textstyle H}{|}}{C}\text{--NR} \qquad (30)$$

These compounds are intermediates for the synthesis of pyrimidines[118]. Of special interest are those reactions with ureas in the presence of excess acetic anhydride from which dialkoxymethylureas may be isolated (equ-

$$\text{R}^1\text{N}\overset{\underset{\textstyle H}{|}}{-}\overset{\overset{\textstyle O}{\|}}{C}\text{--NH}_2 + \text{R}^2\text{C(OEt)}_3 \xrightarrow{\text{(CH}_3\text{C)}_2\text{O}} \text{R}^1\text{N}\overset{\overset{\textstyle O}{\|}}{-}\overset{\underset{\textstyle H}{|}}{C}\text{--N}\overset{\underset{\textstyle H}{|}}{-}\overset{\underset{\textstyle R^2}{|}}{C}\text{(OEt)}_2 + \text{EtOH} \qquad (31)$$

ation 31)[119]. Compounds of this type are presumably intermediates in virtually all of the aminolysis reactions indicated thus far, though they have not been generally isolated. Furthermore, they bear a distinct resemblance to the tetrahedral intermediates considered to be involved in the aminolysis of esters; thus studies of their modes of decomposition may shed light on this interesting class of reactions.

Treatment of ethyl orthoformate with isocyanates in the presence of Lewis acid catalysts yields compounds structurally related to the dialkoxymethyl ureas as indicated in equation (32)[120]. The dialkoxymethyl carba-

$$\text{HC(OEt)}_3 + \text{R}\text{--N}{=}\text{C}{=}\text{O} \xrightarrow{\text{BF}_3} \text{H}\overset{\overset{\textstyle OEt}{|}}{\underset{\underset{\textstyle OEt}{|}}{C}}\overset{}{\underset{\underset{\textstyle R}{|}}{N}}\text{--CO}_2\text{Et} \qquad (32)$$

mates themselves undergo interesting reactions including condensation with carbon acids[120]: an example is indicated in equation (33). We shall

$$\text{H}\overset{\overset{\textstyle OEt}{|}}{\underset{\underset{\textstyle OEt}{|}}{C}}\overset{\overset{\textstyle R}{|}}{N}\text{--CO}_2\text{Et} + \text{NC}\text{--CH}_2\text{--CO}_2\text{Et} \longrightarrow \overset{\text{EtO}_2\text{C}}{\underset{\text{NC}}{}}\!\!\!\!C{=}\overset{\overset{\textstyle H}{|}}{\underset{\underset{\textstyle }{}}{C}}\overset{\overset{\textstyle R}{|}}{N}\text{--CO}_2\text{Et} + 2\,\text{EtOH} \qquad (33)$$

consider some related reactions in section IV. D.

The reaction of carboxylic acid amides and ethyl orthoformate in the presence of acidic catalysts proceeds with replacement of each of the ethoxy functions with an amide function as indicated in equation (34)[121]. This product may arise from addition of a third mole of amide to an amidine-type structure formed by a reaction pathway analogous to those previously described.

$$R-\underset{\underset{O}{\|}}{C}-NH_2 + HC(OEt)_3 \xrightarrow{H^+} HC\left(NH-\underset{\underset{O}{\|}}{C}-R\right)_3 + 3\,EtOH \qquad (34)$$

Little work has been done concerning the addition of secondary amines to ortho esters. The reaction of piperidine with ethyl orthoacetate in the presence of p-toluenesulfonyl chloride yields a nitrogen analog of a ketene acetal (equation 35)[122]. McElvain and Tate have reported the synthesis of

$$MeC(OEt)_3 + 2\,HN\overbrace{} \xrightarrow{p-MeC_6H_4SO_2Cl} CH_2{=}C\left[N\overbrace{}\right]_2 + 3\,EtOH \qquad (35)$$

N,N-disubstituted amides from secondary amines and ethyl orthoacetate under different conditions[123]. The pathway for this reaction is not clear.

D. With Carbon Nucleophilic Reagent

Ortho esters react quite generally with carbon nucleophilic reagents. A particularly simple example is provided by the reaction of ortho esters with hydrogen cyanide in the presence of zinc chloride or other acidic catalyst to yield 1-cyano-1,1-dialkoxyalkanes (equation 36)[124]. Although no

$$R^1C(OR^2)_3 + HCN \rightleftharpoons R^1C(OR^2)_2CN + R^2OH \qquad (36)$$

mechanistic studies have been carried out on this or related reactions it seems reasonable to conclude that they proceed via acid-promoted formation of the carbonium ions derived from the ortho esters followed by attack of the carbon nucleophile. Such a reaction may well be the first step in somewhat more complicated processes which resemble Knoevenagel reactions. For example, malononitrile reacts with ethyl orthoacetate at room temperature and in the absence of catalyst as indicated in equa-

$$CH_3C(OEt)_3 + CH_2(CN)_2 \longrightarrow CH_3-\underset{\underset{\|}{OEt}}{C}{=}\underset{\underset{CN}{\diagdown}}{C}{\nearrow}^{CN} + 2\,EtOH \qquad (37)$$

tion (37)[125]. Numerous examples of reactions of this type are known[126-129]. In certain cases in which the intermediate product cannot break down to yield the 1-alkoxy olefin compounds, reactions may proceed with participation of the ortho ester as alkylating agent rather than acylating agent[130-133]. A typical example is indicated in equation (38).

$$\phi-\underset{\underset{H}{|}}{\overset{\overset{CN}{|}}{C}}-\underset{\overset{\|}{O}}{C}-R + HC(OEt)_3 \longrightarrow \phi-\overset{\overset{CN}{|}}{C}{=}\overset{\overset{OEt}{|}}{C}-R + EtOH + HCO_2Et \qquad (38)$$

Acylation of carbon nucleophiles is, in certain cases, promoted by acetic anhydride. Jones has suggested that the initial product in these reactions is an dialkoxymethyl acetate indicated in equation (39)[134]. Such compounds

$$HC(OEt)_3 + (CH_3CO)_2O \rightleftharpoons CH_3 \overset{\overset{\displaystyle O}{\|}}{C}-O-\overset{\displaystyle H}{\underset{}{C}}(OEt)_2 + CH_3CO_2Et \qquad (39)$$

were prepared many years ago and demonstrated to be active alkylating agents[135]. They have subsequently been demonstrated to possess powers of acylation as well[136].

Among reactions of this type are a series of processes which involve condensation of ortho esters with diazoesters involving rearrangement according to the general reaction of equation (40)[137-141]. These reactions are promoted by Lewis acid catalysts. They almost certainly involve reactions of the diazoester carbanions and not of the carbenes derived by loss of nitrogen.

$$HC(OMe)_3 + N_2CH-CO_2Et \xrightarrow{BF_3 \cdot Et_2O} (MeO)_2CH-\overset{\overset{\displaystyle OMe}{|}}{CH}-CO_2Et \qquad (40)$$

In the presence of Lewis acid catalysts, ortho esters react readily with unsaturated ethers[142-145]. Most of these reactions can be understood in terms of the following reaction pathway:

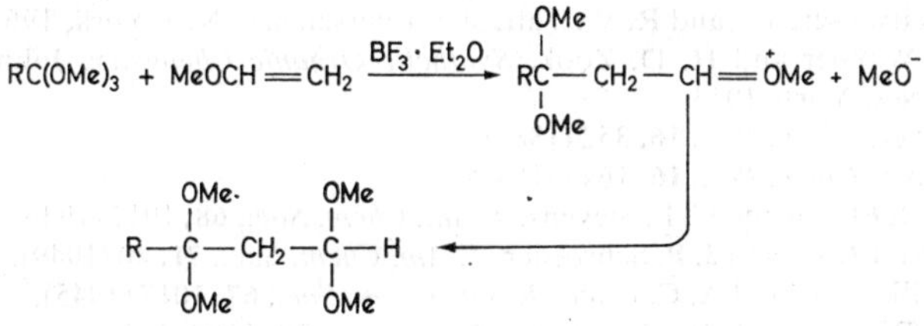

$$RC(OMe)_3 + MeOCH = CH_2 \xrightarrow{BF_3 \cdot Et_2O} RC \overset{\overset{\displaystyle OMe}{|}}{\underset{\underset{\displaystyle OMe}{|}}{}}-CH_2-CH = \overset{+}{O}Me + MeO^-$$

$$R-\overset{\overset{\displaystyle OMe}{|}}{\underset{\underset{\displaystyle OMe}{|}}{C}}-CH_2-\overset{\overset{\displaystyle OMe}{|}}{\underset{\underset{\displaystyle OMe}{|}}{C}}-H \longleftarrow$$

$$(41)$$

E. Miscellaneous Reactions

Ortho esters undergo a number of reactions which have been explored in just one or two instances. We shall summarize these at this point.

Treatment of ortho esters with lithium aluminum hydride[146] or diisobutyl aluminum hydride[147] results in their reduction to acetals.

Under strenuous conditions, it is possible to cause ortho esters to undergo elimination reactions with formation of ketene acetals[148]. Thus treatment of methyl orthophenylacetic acid with ethyl sodium or aluminum methoxide yields the methyl ketal of phenyl ketene (equation 42).

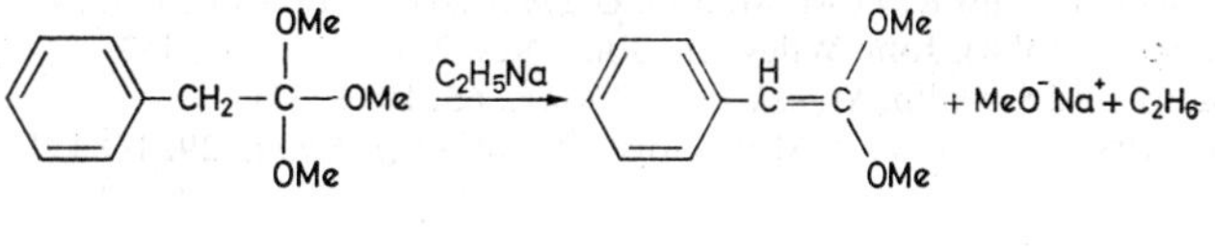

$$\langle \rangle -CH_2-\overset{\overset{\displaystyle OMe}{|}}{\underset{\underset{\displaystyle OMe}{|}}{C}}-OMe \xrightarrow{C_2H_5Na} \langle \rangle -\overset{\displaystyle H}{C}=C\overset{\displaystyle OMe}{\underset{\displaystyle OMe}{\diagdown}} + MeO^-Na^+ + C_2H_6 \qquad (42)$$

Ortho esters react with active acylating agents in the interesting fashion indicated in equation (43)[149]. The mechanism for these reactions is obscure.

$$HC(OAr)_3 + RC\overset{O}{\overset{\|}{-}}Cl \longrightarrow R\overset{O}{\overset{\|}{-}}C\overset{}{-}OAr + (ArO)_2CH\overset{}{-}Cl$$

$$HC(OR)_3 + CH_3C\overset{O}{\overset{\|}{-}}CN \longrightarrow CH_3C\overset{}{-}OR + (RO)_2C\overset{H}{\overset{}{-}}CN \quad (43)$$

Reaction of orthoformates with phosphorus pentoxide yields a mixture of triethyl phosphate and tetraethyl pyrophosphate[150]. Related reactions occur with phosphorus pentasulfide.

Hypophosphite esters are obtained through mixing ortho esters with hypophosphorus acid at room temperature[151]. This reaction is the only known convenient synthesis of such esters.

The free radical reactions which typically occur between acetals and di-t-butyl peroxide also occur with ortho esters[152].

V. REFERENCES

1. H. W. Post, *The Chemistry of the Aliphatic Ortho Esters*, Reinhold Publishing Co., New York, 1943.
2. E. H. Cordes, *Progress in Physical Organic Chemistry*, Vol. 4 (Eds. S. G. Cohen, A. Streitwieser, Jr., and R. W. Taft, Jr.), Interscience, New York, 1967, p.1.
3. R. B. Wagner and H. D. Zook, *Synthetic Organic Chemistry*, John Wiley and Sons, New York, 1953.
4. A. Pinner, *Chem. Ber.*, **16**, 352 (1883).
5. A. Pinner, *Chem. Ber.*, **16**, 1643 (1883).
6. S. M. McElvain and C. L. Stevens, *J. Am. Chem. Soc.*, **68**, 1917 (1946).
7. S. M. McElvain and J. P. Schroeder, *J. Am. Chem. Soc.*, **71**, 40 (1949).
8. S. A. Glickman and A. C. Cope, *J. Am. Chem. Soc.*, **67**, 1017 (1945).
9. M. M. Rising and T. W. Zee, *J. Am. Chem. Soc.*, **50**, 1208 (1928).
10. A. W. Dox, *Organic Syntheses*, Coll. Vol. I. 2nd ed. (Ed. A. H. Blatt), John Wiley and Sons, New York, 1944, p. 5.
11. M. J. Hunter and M. L. Ludwig, *J. Am. Chem. Soc.*, **84**, 3491 (1962).
12. S. M. McElvain, *Chem. Rev.*, **45**, 453 (1949).
13. P. P. T. Sah, *J. Am. Chem. Soc.*, **50**, 516 (1928).
14. L. G. S. Brooker and F. L. White, *J. Am. Chem. Soc.*, **57**, 2480 (1935).
15. S. M. McElvain and J. W. Nelson, *J. Am. Chem. Soc.*, **64**, 1825 (1942).
16. S. M. McElvain and B. Fajrado-Pinzon, *J. Am. Chem. Soc.*, **67**, 690 (1945).
17. S. M. McElvain and C. L. Stevens, *J. Am. Chem. Soc.*, **69**, 2663 (1947).
18. Williamson and Kay, *Ann. Chem.*, **92**, 346 (1854).
19. P. P. T. Sah and T. S. Ma, *J. Am. Chem. Soc.*, **54**, 2964 (1932).
20. M. E. Hill, D. T. Carty, D. Tegg, J. C. Butler and A. F. Stang, *J. Org. Chem.*, **30**, 411 (1965).
21. J. D. Roberts and R. E. McMahon, *Organic Syntheses*, Coll. Vol. IV, 2nd Edition, (Ed. N. Rabjohn), John Wiley and Sons, New York, 1963, p. 457.
22. B. Smith and S. Delin, *Svensk Kem. Tidskr.*, **65**, 10 (1953).
23. A. V. Kirsanov and V. P. Molosnova, *Zh. Obshch. Khim.*, **29**, 1684 (1959).

24. F. Beyerstedt and S. M. McElvain, *J. Am. Chem. Soc.*, **58**, 529 (1936).
25. F. Beyerstedt and S. M. McElvain, *J. Am. Chem. Soc.*, **59**, 1273 (1937).
26. S. M. McElvain and D. Kundiger, *J. Am. Chem. Soc.*, **64**, 254 (1942).
27. S. M. McElvain, H. I. Anthers and S. H. Shapiro, *J. Am. Chem. Soc.*, **64**, 2525 (1942).
28. S. M. McElvain and G. R. McKay, Jr., *J. Am. Chem. Soc.*, **77**, 5601 (1955).
29. W. C. Kuryla and D. G. Leis, *J. Org. Chem.*, **29**, 2773 (1964).
30. W. E. Mochel, C. L. Agre and W. E. Hanford, *J. Am. Chem. Soc.*, **70**, 2268 (1948).
31. H. Post and E. Erickson, *J. Am. Chem. Soc.*, **55**, 3851 (1933).
32. G. Crank and F. W. Eastwood, *Australian J. Chem.*, **17**, 1385 (1964).
33. A. Rieche, E. Schmitz, and E. Beyer, *Chem. Ber.*, **91**, 1942 (1958).
34. C. B. Reese and J. E. Sulston, *Proc. Chem. Soc.*, 214 (1964).
35. E. M. Arnett, *Progress in Physical Organic Chemistry*, Vol. I, (Eds. S. G. Cohen, A. Streitwieser, Jr. and R. W. Taft, Jr.) Interscience, New York 1963.
36. L. P. Hammett, *J. Chem. Phys.*, **8**, 644 (1940).
37. R. West, L. S. Whatley and K. J. Lake, *J. Am. Chem. Soc.*, **83**, 761 (1961).
38. T. Pletcher and E. H. Cordes, unpublished results.
39. C. A. Bunton and R. H. DeWolfe, *J. Org. Chem.*, **30**, 1371 (1965).
40. H. Meerwein, K. Bodenbenner, P. Borner, F. Kunert and K. Wunderlich, *Ann. Chem.*, **632**, 38 (1960).
41. H. Meerwein, V. Hederich, H. Morschel and K. Wunderlich, *Ann. Chem.*, **635**, 1 (1960).
42. B. G. Ramsey and R. W. Taft, *J. Am. Chem. Soc.*, **88**, 3058 (1966).
43. R. W. Taft and B. G. Ramsey, personal communication.
44. R. H. Martin, F. W. Lampe and R. W. Taft, *J. Am. Chem. Soc.*, **88**, 1353 (1966).
45. R. W. Taft, R. H. Martin and F. W. Lampe, *J. Am. Chem. Soc.*, **87**, 2490 (1965).
46. L. D. McKeever and R. W. Taft, *J. Am. Chem. Soc.*, **88**, 4544 (1966).
47. R. W. Taft, E. Price, I. R. Fox, I. C. Lewis, K. K. Anderson and G. T. Davis, *J. Am. Chem. Soc.*, **85**, 3146 (1963).
48. R. W. Taft and L. D. McKeever, *J. Am. Chem. Soc.*, **87**, 2489 (1965); see also pertinent references cited therein.
49. J. M. O'Gorman and H. J. Lucas, *J. Am. Chem. Soc.*, **72**, 5489 (1950).
50. H. K. Garner and H. J. Lucas, *J. Am. Chem. Soc.*, **72**, 5497 (1950).
51. E. R. Alexander, H. M. Busch and G. L. Webster, *J. Am. Chem. Soc.*, **74**, 3173 (1952).
52. F. Stasiuk, W. A. Sheppard and A. N. Bourns, *Can. J. Chem.*, **34**, 123 (1956).
53. R. W. Taft, Jr., personal communication.
54. T. H. Fife and L. K. Jao, *J. Org. Chem.*, **30**, 1492 (1965).
55. Y. Yukawa and Y. Tsuno, *Bull. Chem. Soc. Japan*, **32**, 971 (1959).
56. M. M. Kreevoy and R. W. Taft, Jr., *J. Am. Chem. Soc.*, **77**, 5590 (1955).
57. A. Skrabal and H. H. Eger, *Z. Physik. Chem. (Leipzig)*, **122**, 239 (1926).
58. R. W. Taft, Jr., *J. Am. Chem. Soc.*, **75**, 4231 (1953).
59. R. W. Taft, Jr., *J. Am Chem. Soc.*, **74**, 3120 (1952).
60. H. Kwart and M. B. Price, *J. Am. Chem. Soc.*, **82**, 5123 (1960).
61. M. M. Kreevoy, C. R. Morgan and R. W. Taft, Jr., *J. Am. Chem. Soc.*, **82**, 3064 (1960).
62. J. N. Brønsted and W. F. K. Wynne-Jones, *Trans. Faraday Soc.*, **25**, 59 (1929).
63. R. H. DeWolfe and J. L. Jensen, *J. Am. Chem. Soc.*, **85**, 3264 (1963).
64. L. L. Schaleger and F. A. Long, *Advances in Physical Organic Chemistry*, Vol. 1, (Ed. V. Gold), Academic Press, New York, 1963, p. 1.
65. J. Koskikallio and E. Whalley, *Trans. Faraday Soc.*, **55**, 809 (1959).
66. P. M. Leininger and M. Kilpatrick, *J. Am. Chem. Soc.*, **61**, 2510 (1939).
67. P. Salomaa and A. Kankaanperä, *Acta. Chem. Scand.*, **15**, 871 (1961).

68. F. Brescia and V. K. LaMer, *J. Am. Chem. Soc.*, **62**, 612 (1940).
69. J. G. Fullington and E. H. Cordes, *J. Org. Chem.*, **29**, 970 (1964).
70. M. Kilpatrick, Jr. and M. L. Kilpatrick, *J. Am. Chem. Soc.*, **53**, 3698 (1931).
71. E. Whalley, *Advances in Physical Organic Chemistry*, Vol. 2. (Ed. V. Gold), Academic Press, New York, 1964, p. 93.
72. C. A. Bunton and V. J. Shiner, Jr., *J. Am. Chem. Soc.*, **83**, 3207 (1961).
73. M. Kilpatrick, *J. Am. Chem. Soc.*, **85**, 1036 (1963).
74. J. C. Hornel and J. A. V. Butler, *J. Chem. Soc.*, 1361 (1936).
75. M. M. Kreevoy and R. W. Taft, Jr., *J. Am. Chem. Soc.*, **77**, 3146 (1955).
76. W. J. C. Orr and J. A. V. Butler, *J. Chem. Soc.*, 330 (1937).
77. M. L. Bender and M. S. Silver, *J. Am. Chem. Soc.*, **85**, 3006 (1963).
78. F. Brescia and V. K. LaMer, *J. Am. Chem. Soc.*, **60**, 1962 (1938).
79. W. F. K. Wynne-Jones, *Trans. Faraday Soc.*, **34**, 245 (1938).
80. V. J. Shiner, Jr. and S. Cross, *J. Am. Chem. Soc.*, **79**, 3599 (1957).
81. V. J. Shiner, Jr., *Tetrahedron*, **5**, 243 (1959).
82. For a discussion see: J. Hine, *Physical Organic Chemistry*, 2nd Ed., McGraw-Hill Book Co., New York, 1962.
83. A. J. Kresge and R. J. Preto, *J. Am. Chem. Soc.*, **87**, 4593 (1965).
84. A. M. Wenthe and E. H. Cordes, *Tetrahedron Letters*, 3163 (1964).
85. A. M. Wenthe and E. H. Cordes, *J. Am. Chem. Soc.*, **87**, 3173 (1965).
86. L. C. Bateman, E. D. Hughes and C. K. Ingold, *J. Chem. Soc.*, 881 (1938).
87. F. Spieth, W. C. Ruebsamen and A. R. Olson, *J. Am. Chem. Soc.*, **76**, 6253 (1954).
88. N. T. Farinacci and L. P. Hammett, *J. Am. Chem. Soc.*, **59**, 2542 (1937).
89. A. Streitwieser, Jr., *Chem. Rev.*, **56**, 571 (1956).
90. J. Hine, *J. Am. Chem. Soc.*, **85**, 3239 (1963).
91. H. Kwart and L. B. Weisfeld, *J. Am. Chem. Soc.*, **80**, 4670 (1958).
92. R. H. DeWolfe and R. M. Roberts, *J. Am. Chem. Soc.*, **76**, 4379 (1954).
93. H. K. Hall, Jr., *J. Am. Chem. Soc.*, **79**, 5441 (1957).
94. M. Eigen, *Angew. Chem. Intern. Ed. Engl.*, **3**, 1 (1964).
95. J. G. Fullington and E. H. Cordes, *Proc. Chem. Soc.*, 224 (1964).
96. M. T. A. Behme, J. G. Fullington, R. Noel and E. H. Cordes, *J. Am. Chem. Soc.*, **87**, 266 (1965).
97. W. Kauzmann, *Advan. Protein Chem.*, **14**, 1 (1959).
98. L. Claisen, *Ann. Chem.*, **287**, 360 (1895).
99. R. Walther, *J. Prakt. Chem.*, **52**, 429 (1895); **53**, 472 (1896).
100. C. Goldschmidt, *J. Chem. Soc.*, **82**, 785 (1902).
101. F. B. Dains and E. W. Brown, *J. Am. Chem. Soc.*, **31**, 1148 (1909).
102. F. B. Dains, O. O. Malleis, and J. T. Meyers, *J. Am. Chem. Soc.*, **35**, 970 (1913).
103. Y. Mizuno and M. Nishimura, *J. Pharm. Soc. Japan*, **68**, 58 (1948).
104. R. M. Roberts, *J. Am. Chem. Soc.*, **71**, 3848 (1949).
105. R. M. Roberts, *J. Am. Chem. Soc.*, **72**, 3603 (1950).
106. R. M. Roberts, R. H. DeWolfe and J. H. Ross, *J. Am. Chem. Soc.*, **73**, 2277 (1951).
107. R. M. Roberts and R. H. DeWolfe, *J. Am. Chem. Soc.*, **76**, 2411 (1954).
108. E. C. Taylor and W. A. Ehrhart, *J. Am. Chem. Soc.*, **82**, 3138 (1960).
109. R. M. Roberts, T. D. Higgins, Jr. and P. R. Noyes, *J. Am. Chem. Soc.*, **77**, 3801 (1955).
110. E. C. Taylor and W. E. Ehrhart, *J. Org. Chem.*, **28**, 1108 (1963).
111. R. M. Roberts and P. J. Vogt, *J. Am. Chem. Soc.*, **78**, 4778 (1956).
112. R. H. DeWolfe, *J. Org. Chem.*, **27**, 490 (1962).
113. H. Baganz and L. Domaschke, *Chem. Ber.*, **95**, 1840 (1962).
114. J. A. Montgomery and C. Temple, Jr., *J. Org. Chem.*, **25**, 395 (1960).
115. C. Ainsworth, *J. Am. Chem. Soc.*, **77**, 1148 (1955).

116. C. Ainsworth, *J. Am. Chem. Soc.*, **78,** 1973 (1956).
117. C. Runti and C. Nisi, *J. Med. Chem.*, **7,** 814 (1964).
118. C. W. Whitehead, *J. Am. Chem. Soc.*, **75,** 671 (1953).
119. C. W. Whitehead and J. J. Traverso, *J. Am. Chem. Soc.*, **77,** 5872 (1955).
120. (a) H. V. Brachel and R. Merton, *Angew. Chem.*, **74,** 872 (1962); (b) C. W. White-head and J. Traverso, *J. Am. Chem. Soc.*, **50,** 962 (1958).
121. H. Bredereck, F. Effenberger and H. J. Treiber, *Chem. Ber.*, **96,** 1505 (1963).
122. H. Baganz and L. Domaschke, *Chem. Ber.*, **95,** 2095 (1962).
123. S. M. McElvain and B. E. Tate, *J. Am. Chem. Soc.*, **67,** 202 (1945).
124. J. G. Erickson, *J. Am. Chem. Soc.*, **73,** 1338 (1951).
125. V. J. Pascual and M. Ballester, *Annales Real Soc. Espan. Fis. Quim.*, **44B,** 1293 (1948).
126. V. J. Pascual and F. Servatosa, *Annales Real Soc. Espan. Fis. Quim.*, **50B,** 471 (1954).
127. E. B. Knott, *J. Chem. Soc.*, 1482 (1954).
128. R. G. Jones, *J. Am. Chem. Soc.*, **74,** 4889 (1952).
129. J. Pascual and F. Serratosa, *Chem. Ber.*, **85,** 686 (1952).
130. P. B. Russell and N. Whittaker, *J. Am. Chem. Soc.*, **74,** 1310 (1952).
131. L. Claisen, *Chem. Ber.*, **26,** 2729 (1893); **29,** 1005 (1896); **31,** 1019 (1898).
132. A. Michael, *J. Am. Chem. Soc.*, **57,** 159 (1935).
133. F. Arndt, L. Loewe and M. Ozansoy, *Chem. Ber.*, **73,** 779 (1940).
134. R. G. Jones, *J. Am. Chem. Soc.*, **73,** 3684 (1951).
135. H. W. Post and E. R. Erickson, *J. Org. Chem.*, **2,** 260 (1937).
136. R. C. Fuson, W. E. Parham and L. J. Reed, *J. Org. Chem.*, **11,** 194 (1946).
137. A. Schönberg and K. Praefcke, *Tetrahedron Letters*, 2043 (1964).
138. A. Schönberg and K. Praefcke, *Chem. Ber.*, **99,** 2371 (1966).
139. A. Schönberg, K. Praefcke and J. Kohtz, *Chem. Ber.*, **99,** 3076 (1966).
140. A. Schönberg, K. Praefcke and J. Kohtz, *Chem. Ber.*, **99,** 2433 (1966).
141. A. Schönberg and K. Praefcke, *Chem. Ber.*, **99,** 196 (1966).
142. J. W. Copenhaver, *U.S. Pat.* 2,500,486 (*Chem. Abstr.* **44,** P5379 (1950)).
143. J. W. Copenhaver, *U.S. Pat.* 2,527,533 (*Chem. Abstr.* **45,** P1623 (1951)).
144. L. A. Yanovskaya, V. F. Kucherov and B. A. Rudenko, *Izv. Akad. Nauk SSSR, Otd. Khim. Nauk,* 2182 (1962). (*Chem. Abstr.*, **58,** 12410 (1963)).
145. I. N. Nazarov, S. M. Makin and B. K. Kruptsov, *Zh. Obshch. Khim.*, **29,** 3683 (1959).
146. C. J. Claus and J. L. Morgenthau, Jr., *J. Am. Chem. Soc.*, **73,** 5005 (1951).
147. L. I. Zakharin and I. M. Khorlina, *Izv. Akad. Nauk SSSR, Otd. Khim. Nauk,* 2255 (1959); (*Chem. Abstr.*, **54,** 10837 (1959)).
148. S. M. McElvain and J. T. Venerable, *J. Am. Chem. Soc.*, **72,** 1661 (1950).
149. H. Böhme and R. Neidein, *Chem. Ber.*, **95,** 1859 (1962).
150. K. C. Brannock, *J. Am. Chem. Soc.*, **73,** 4953 (1951).
151. S. J. Fitch, *J. Am. Chem. Soc.*, **86,** 61 (1964).
152. E. S. Huyser and D. T. Wang, *J. Org. Chem.*, **27,** 4696 (1962).

CHAPTER 14

Peracids and peresters

SVEN-OLOV LAWESSON and GUSTAV SCHROLL

University of Aarhus, Denmark

I. GENERAL COMMENTS

During recent years peracids have found an increased use as oxidizing agents and peresters have also become a useful synthetic tool for the organic chemist. In this review only peroxycarboxylic acids and esters are treated and efforts have been made to cover the most important and relevant data in this field. The reader is referred to earlier reviews[1-6] for more detailed accounts, and for other types of related peroxy compounds attention is called to a recent review by Sosnovsky and Brown[7].

II. PREPARATION OF PERACIDS

A. From Carboxylic Acids or Acid Anhydrides and Hydrogen Peroxide

The most often-used method for the preparation of aliphatic peracids is the reaction of hydrogen peroxide with a carboxylic acid or its anhydride. In the absence of a strong acid catalyst the attainment of equilibrium is impractically slow. A catalyst is thus necessary and sulphuric acid is the

$$\underset{\displaystyle RC-OH}{\overset{\displaystyle O}{\|}} + H_2O_2 \underset{}{\overset{H^+}{\rightleftharpoons}} \underset{\displaystyle RC-OOH}{\overset{\displaystyle O}{\|}} + H_2O$$

most common one. Other catalysts used are methanesulphonic acid[8], ethanesulphonic acid or other alkanesulphonic acids, p-toluenesulphonic acid hydrate[9], boron trifluoride or boron trifluoride monohydrate[10, 11] and pyridine oxide[12]. A difficulty encountered in preparing aliphatic peracids containing six or more carbon atoms is that the parent carboxylic acid is not soluble enough for a smooth and fast reaction. Parker and coworkers[13] have circumvented these difficulties by using 95% sulphuric acid as catalyst, cosolvent and reaction medium. Higher peracids can also be synthesized from the corresponding methyl esters which are more soluble than the corresponding acids[14]. Lower peracids are also conveniently prepared from carboxylic anhydride or boric carboxylic anhydride[15]. Monoperphthalic acid, prepared from phthalic anhydride and hydrogen peroxide, has been described by Böhme[16] and three modifications of the method have been recently published[17-19].

B. From Acid Chlorides and Sodium Peroxide

Peracids can also be obtained by treating an acid chloride with sodium peroxide in aqueous alcohol[20] or in aqueous tetrahydrofuran[21].

$$HOO^- + R\overset{O}{\underset{||}{C}}Cl \longrightarrow R\overset{O}{\underset{||}{C}}OOH + Cl^-$$

In general the reactions between acid chlorides and hydrogen peroxide are more suitable for the preparation of diacyl peroxides than of peracids.

C. From Organic Peroxides and Alcoholates

Diaroyl peroxides are in general easily available solids, which are relatively stable and easy to purify. By their basic hydrolysis, followed by acidification, the corresponding peracids are obtained in high yields. Perbenzoic acid which has been studied extensively, is prepared by treating benzoyl peroxide with sodium methoxide[22]. Substituted perbenzoic acids can be obtained by the same procedure.

$$C_6H_5\overset{O}{\underset{||}{C}}OO\overset{O}{\underset{||}{C}}C_6H_5 \xrightarrow{CH_3ONa} C_6H_5\overset{O}{\underset{||}{C}}OONa \xrightarrow{H^+} C_6H_5\overset{O}{\underset{||}{C}}OOH$$

D. By Oxidation of Aldehydes

Since 1900 it has been known that oxidation of an aldehyde produces the peracid[23] as an intermediate. Bäckström suggested that the peracid is formed by a reaction[24], the first step of which, the activation of the aldehyde, is initiated by light[25] or catalysed by heavy metals[26]. The kinetics of the autoxidation of benzaldehyde[27] account for the postulate that benzaldehyde and perbenzoic acid rapidly give the adduct which then decomposes to give benzoic acid.

$$C_6H_5\overset{O}{\underset{||}{C}}OOH + C_6H_5CHO \longrightarrow C_6H_5-\overset{O}{\underset{||}{C}}OO\overset{OH}{\underset{|}{C}}HC_6H_5 \longrightarrow 2\ C_6H_5COOH$$

The chain reaction of Bäckström[24] is formulated as follows:

$$C_6H_5\overset{.}{\underset{\underset{O}{||}}{C}} \xrightarrow{O_2} C_6H_5\overset{}{\underset{\underset{O}{||}}{C}}OO\cdot \xrightarrow{C_6H_5CHO} C_6H_5COOH + C_6H_5\overset{.}{\underset{\underset{O}{||}}{C}}$$

Autoxidation reactions of aldehydes for the laboratory preparation of peracids are not often used, but quite a few patents are known for industrial processes.

III. PHYSICAL PROPERTIES AND STRUCTURE OF PERACIDS

Generally, peracids are used in oxidation processes without being isolated. However, several peracids have been isolated in analytically pure form. Peracetic, perpropionic and perbutyric[1] acids are distillable. A great variety of solid peracids derived from aliphatic (mono- and dibasic), aromatic and heterocyclic acids are now known. m-Chloroperbenzoic acid is commercially available and is frequently used in syntheses. The lowest peracids can explode in pure form but the higher homologues are relatively stable and can be stored at a low temperature for a long time without decomposition[28].

It has been shown by infrared, x-ray diffraction and molecular weight studies[29] that long-chain aliphatic peracids exist in solution as intramolecularly chelated monomers in a five-membered ring but that in the solid state they are dimeric.

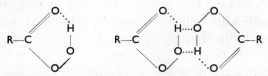

A study of the dipole moments of peracids[30] supports the suggested structures of peracids. However, certain modifications of the original theories have been necessary[31]. Several other infrared investigations of peracids have also been done[32-36], the ultraviolet absorption spectrum of peracetic acid[37] has been published and thermochemical studies of autoxidation of benzaldehyde have been reported[38, 39].

In further support of the cyclic structure ascribed to peracids are the dissociation constants of some organic peracids[40] which were shown to lie between those of the corresponding acids and hydroperoxides. A consequence of a chelated structure is that the peracids should be considerable weaker than the corresponding carboxylic acids, ($pK_A = 7\cdot1$–$8\cdot2$, the value for the corresponding carboxylic acids is $pK_A = 2\cdot6$–$4\cdot9$).

IV. REACTIONS

A. Epoxidation of Unsaturated Compounds

By reaction of peracids with olefins, oxiranes (epoxides) are formed[41]. The epoxidation (and hydroxylation) of unsaturated compounds was reviewed by Gunstone[42] and Malinovsky[43]. The reaction conditions are mild, both polar and non-polar solvents are used, and a great variety of peracids have been investigated as epoxidizing agents. This *cis* addition is facilitated by electron-releasing groups and decreased (or completely inhibited) by electron-attracting groups. Concerning the latter group, it is stated[44] that with ethyl crotonate trifluoroperacetic acid is the only known peracid that is effective. However, other methods are known for the epoxidizing of such electrophilic olefins[45].

It has been established[46] that the rate of epoxidation of olefins with peracids in general is first-order each with respect to olefin and to peracid, and that the reaction is not acid-catalysed. The favoured mechanism for the nonacid-catalysed peracid oxidation is due to Bartlett[47] and others[48-52]:

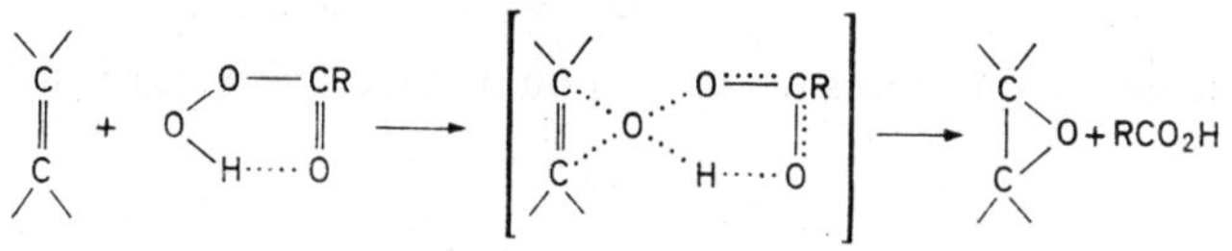

Alternative mechanisms have been discussed by Swern[53] and Huisgen, Grashey and Sauer[54].

Cycloalkenes may react with peracids in a different manner. 1-Alkyl-cyclohexenes and perbenzoic acid yield a mixture of the epoxide and a 2-alkylcyclohexanone (formed by rearrangement of the epoxide[55]), whereas 1-ethoxycyclohexene and perbenzoic acid yield 2-benzoyloxycyclo-hexanone[56].

β-Piperidinostyrene reacts with perbenzoic acid leading to an epoxide which by rearrangement forms piperidinoacetophenone[57].

B. Hydroxylation of Unsaturated Compounds

When epoxidizing unsaturated compounds with a peracid, the corresponding carboxylic acid is always produced:

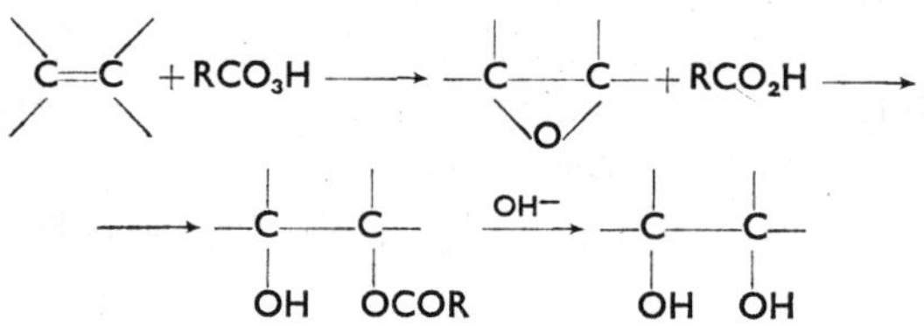

If the conditions are more vigorous (long reaction time, higher temperature) and a strong acid catalyst is present[58], the O-monoacylglycol is isolated, and is readily and in high yields converted to the α-glycol by saponification. The hydroxylation reaction is stereospecific and yields the *trans*-glycols. Different peracids have been used, either preformed or prepared and utilized *in situ*. It has been reported that trifluoroperacetic acid is very reactive in the hydroxylation of olefins[59]. Electrophilic olefins also undergo smooth hydroxylation with trifluoroperacetic acid[60].

C. Cleavage Reaction

1. Oxidation of carbonyl compounds

The Baeyer–Villiger oxidation of aldehydes and ketones has been reviewed as late as 1963[61]. In this reaction an acyclic ketone is oxidized with hydrogen peroxide or a peracid under mild conditions to give an ester in reasonable yields.

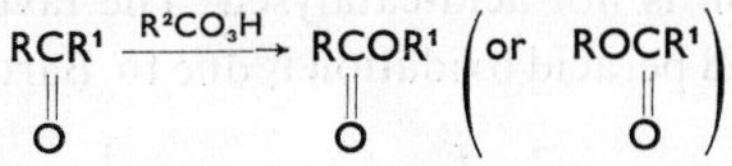

$$RCR^1 \xrightarrow{R^2CO_3H} RCOR^1 \; \left(or \; ROCR^1 \right)$$
$$\overset{\|}{O} \qquad\qquad \overset{\|}{O} \qquad\quad \overset{\|}{O}$$

A cyclic ketone produces the corresponding lactone as for instance:

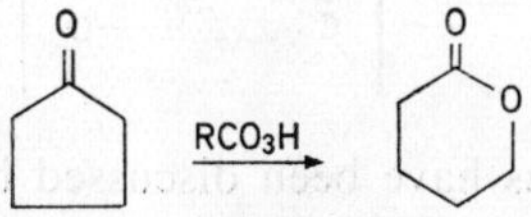

This method has proved useful in analytical[62] and synthetic, as well as degradative, studies and has been applied successfully to different carbonyl compounds such as aldehydes, aliphatic, alicyclic, aromatic and heterocyclic ketones and α, β-unsaturated ketones as well as to related compounds such as α-diketones, α-keto acids and α-keto carboxylic esters. A great variety of oxidation agents have been tried and a few comments may be pertinent. Simple ketones produce unstable and dangerous peroxides (also polymeric peroxides) when oxidized with common peracids. Trifluoroperacetic acid, however, converts most ketones to esters in high yields[61, 63]. Simple aliphatic aldehydes react in the same way. Acetals, acylals and dioxolanes of α-,β-unsaturated aldehydes have been oxidized with peracetic acid[64], and in most cases the corresponding epoxides are produced. When non-cyclic acetals[64] are used, the oxidation products are mainly the corresponding unsaturated esters as for instance:

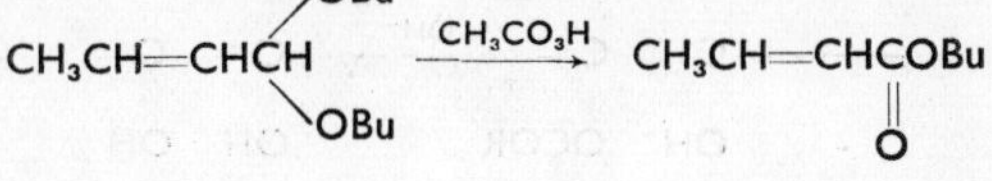

$$CH_3CH=CHCH\overset{OBu}{\underset{OBu}{\diagdown}} \xrightarrow{CH_3CO_3H} CH_3CH=CHCOBu$$
$$\overset{\|}{O}$$

Acetals of saturated aldehydes are also converted to the corresponding esters by treatment with peracetic acid[64].

The favoured mechanism for the Baeyer–Villiger reaction[61] is indicated below:

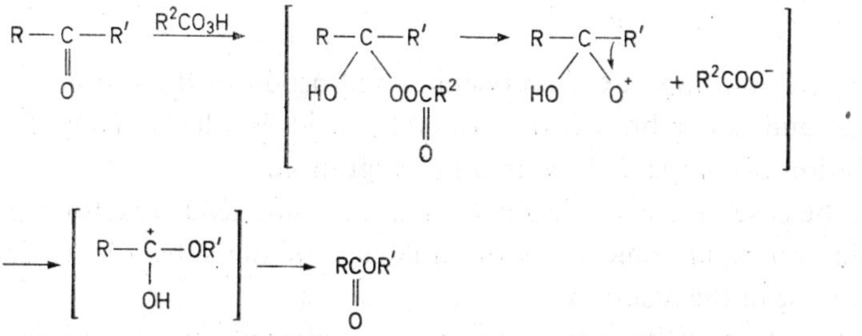

$$R-\overset{\overset{\displaystyle \|}{O}}{C}-R' \xrightarrow{R^2CO_3H} \left[R-\underset{\underset{HO}{}}{C}\overset{\overset{R'}{}}{\underset{\underset{OOCR^2}{\overset{\|}{O}}}{}} \longrightarrow R-\underset{\underset{HO}{}}{C}\overset{\overset{R'}{}}{\underset{\underset{O^+}{}}{}} + R^2COO^- \right]$$

$$\longrightarrow \left[R-\underset{\underset{OH}{}}{\overset{+}{C}}-OR' \right] \longrightarrow \underset{\underset{O}{\|}}{R}COR'$$

The peracid is first added to the carbon–oxygen double bond; then the oxygen–oxygen bond is heterolytically cleaved after which a 1,2 shift from carbon to oxygen occurs.

Investigations on the mechanism and/or kinetics of the oxidation of ketones[65-68] and aldehydes[69, 70] have recently been carried out. The stereo-chemistry[71-73] of the oxidation of ketones with perbenzoic acid has also been studied. Thus in the reaction of optically active methyl α-phenylethyl ketone, the α-phenylethyl group migrates with complete retention of its stereochemical configuration[72].

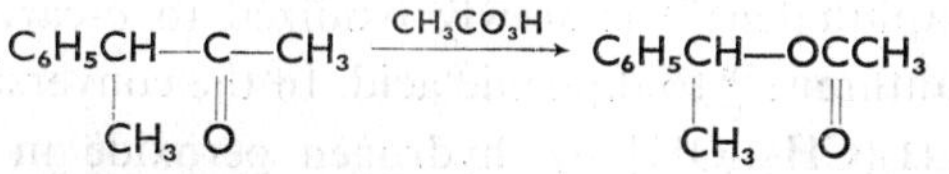

$$C_6H_5CH-\underset{\underset{CH_3}{}}{\overset{\overset{}{}}{C}}-\underset{\underset{O}{\|}}{C}CH_3 \xrightarrow{CH_3CO_3H} C_6H_5CH-\underset{\underset{CH_3}{}}{\overset{\overset{}{}}{}}O\underset{\underset{O}{\|}}{C}CH_3$$

2. Oxidation of unsaturated compounds

It is known that peracetic acid attacks only the double bond in 4-phe-nylethynyl-3-heptene[74], leaving the acetylenic bond intact:

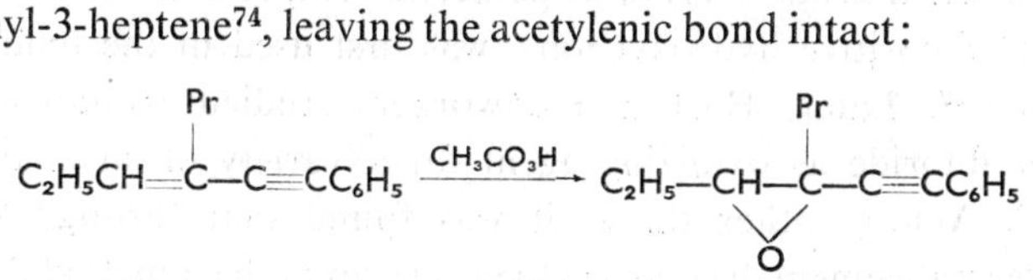

$$C_2H_5CH=\overset{\overset{Pr}{|}}{C}-C\equiv CC_6H_5 \xrightarrow{CH_3CO_3H} C_2H_5-CH-\overset{\overset{Pr}{|}}{C}-C\equiv CC_6H_5$$

Other similar cases are also known[75, 76]. On the other hand, a great variety of acetylenic compounds[77-83] have been oxidized with peracids. The general scheme of these reactions has not yet been outlined. Contrary to olefins no epoxides have been isolated from the reaction between acety-lenes and peracids[82], although that had earlier been claimed[78, 79].

Acetylenic hydrocarbons react in two ways:

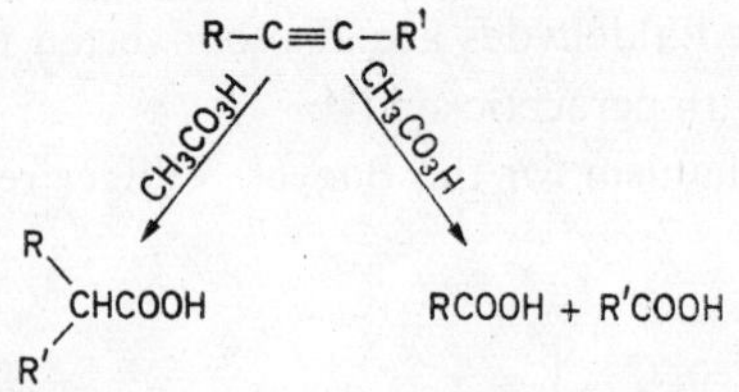

$$R-C\equiv C-R'$$

$$
\underset{\substack{R' \\ R}}{\overset{R}{\diagdown}}CHCOOH \qquad\qquad RCOOH + R'COOH
$$

Oxidative splitting at the acetylenic bond produces the normal carboxylic acids, and an α-branched carboxylic acid is alternatively formed by oxidation accompanied by an alkyl migration.

In the case of acetylenic alcohols also simultaneous reactions have been found, where in some cases the influence of the substituents determines the course of the reaction.

Attempts have been made to epoxidize allenes[84-86]. In a recent report[86] tetramethylallene was treated with peracetic acid and it is proposed that the allene oxide as well as the dioxide are molecules of finite existence.

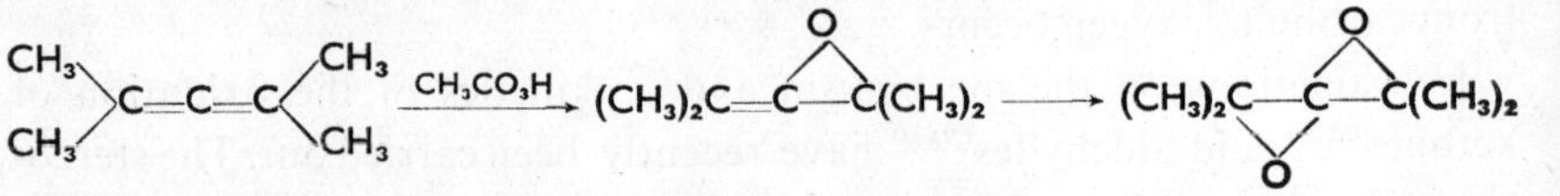

However, none of these intermediates were isolated.

3. Oxidation of aromatic compounds

Peracids react with aromatic rings[87]. Benzene in general is quite stable to peracids, naphthalene[88] is slowly oxidized to o-carboxyallocinnamic acid and phenanthrene[89] to diphenic acid. In the conversion of mesitylene to mesitol [$C_6H_2(CH_3)_3OH$] by hydrogen peroxide in acetic–sulphuric acid the hydroxyl cation was suggested to be the attacking species[90]. In related oxidations a Lewis acid (boron fluoride etherate) has been used instead of mineral acids[91]. Trifluoroperacetic acid, realized to be an excellent source of positive hydroxyl ions, was first used in the oxidation of alkylbenzenes[92]. Later, Hart and coworkers studied trifluoroperacetic acid–boron fluoride as oxidizing agent for a variety of aromatic compounds[93-98]. Among other things it was found that through Wagner–Merwein rearrangement hexamethylbenzene gives hexamethyl-2,4-cyclohexadienone in 90% yield.

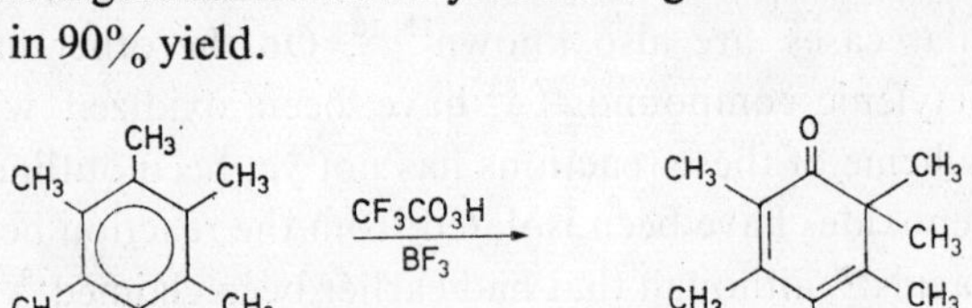

From less-substituted benzenes phenols are produced by electrophilic attack at an unsubstituted ring position:

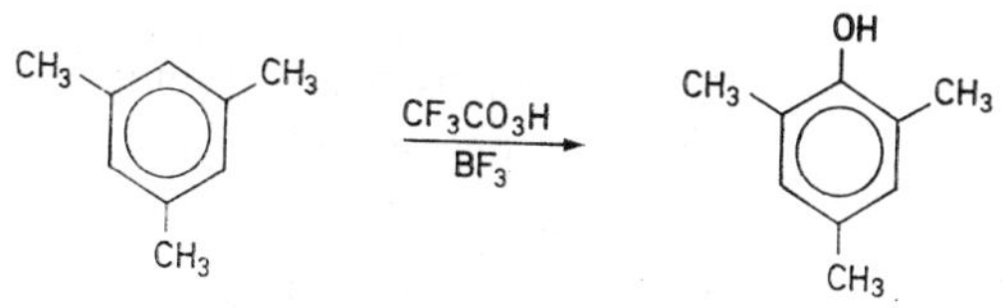

The peracid oxidation of aryl ethers up to 1951 has been reviewed[99], and since then similar studies of aryl ethers have also been performed[100–102].

D. Oxidation of Organic Sulphur Compounds

In organic synthesis mercaptocompounds are generally not oxidized by peracids, as mild oxidizing agents are necessary to avoid further undesired reactions.

Sulphides are smoothly and in high yields converted to sulphoxides or sulphones, depending upon the amount of peracid used.

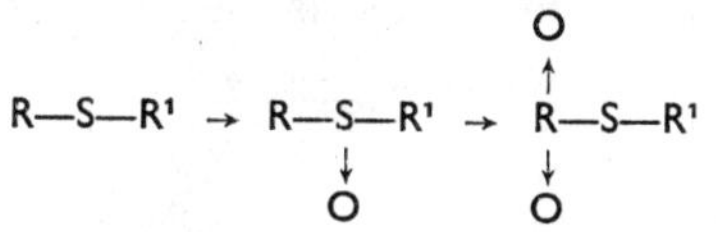

$$R\text{—}S\text{—}R^1 \rightarrow \underset{\underset{O}{\downarrow}}{R\text{—}S\text{—}R^1} \rightarrow \underset{\underset{O}{\downarrow}}{\overset{\overset{O}{\uparrow}}{R\text{—}S\text{—}R^1}}$$

The kinetics and mechanisms of such oxidations have been extensively studied[103–110]. The kinetics[103] of the oxidation of p, p'-dichlorodiphenylsulphide by various *para*-substituted perbenzoic acids display a second-order rate. The absence of salt effects and the preference for a solvent with low dielectric constant indicates a mechanism involving a nucleophilic attack of the hydrogen-bonded form of the peracid on the sulphide. The Hammett ϱ value was found to be $+1\cdot05$ for an isopropanolic solution of the reagents.

Similarly, the oxidation of phenyl sulphoxide by perbenzoic acid is considered to be a nucleophilic attack of the perbenzoic acid on the sulphur atom of the sulphoxide[104]. The ϱ values for $ArCO_3H$ oxidation of $R^1C_6H_4SOR$ in dioxane–water solution was found to be $+0\cdot64$ ($R^1 = CH_3$) and $+0\cdot54$ ($R^1 = C_6H_5$). The rate of oxidation of sulphoxides is rather dependent on the pH[109, 110], and it is found that sulphoxides are oxidized at a higher rate in alkaline solution. From the kinetic data[110] it seems likely that two mechanisms take place, one involving the hydrogen-bonded peracid, and one involving the peracid anion.

The preparation of optically active sulphoxides by oxidation of sulphides with optically active percamphoric acid has been reported[111]. The same general principle[112, 113] has been used for the preparation of other optically active sulphoxides.

Disulphides, when treated with a peracid at a temperature below 55°, are smoothly converted to thiosulphinates[114, 115].

$$R\text{---}S\text{---}S\text{---}R \longrightarrow R\text{---}\overset{\overset{\displaystyle O}{\uparrow}}{S}\text{---}S\text{---}R$$

Tri-, and tetrasulphides have also been investigated[115] and they are all oxidized to the corresponding sulphoxides.

Thiosulphonic esters are converted by peracids[116] to α-disulphones.

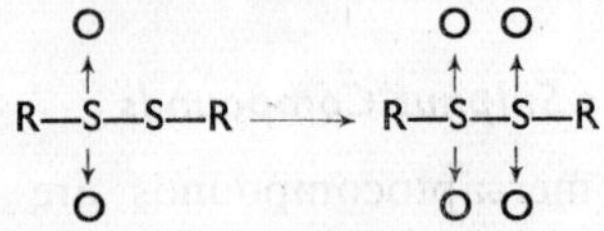

$$R\text{---}\overset{\overset{\displaystyle O}{\uparrow}}{\underset{\underset{\displaystyle O}{\downarrow}}{S}}\text{---}S\text{---}R \longrightarrow R\text{---}\overset{\overset{\displaystyle O}{\uparrow}}{\underset{\underset{\displaystyle O}{\downarrow}}{S}}\text{---}\overset{\overset{\displaystyle O}{\uparrow}}{\underset{\underset{\displaystyle O}{\downarrow}}{S}}\text{---}R$$

E. Oxidation of Organic Nitrogen Compounds

1. Amines to nitroso or nitro compounds

It has long been recognized that aromatic primary amines are oxidized by peracids to the nitroso compounds via the hydroxylamines:

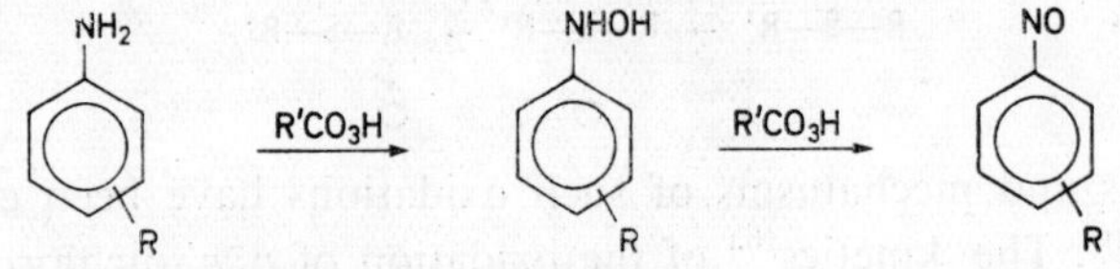

In a recent paper[117] certain mechanisms are reviewed and a new mechanism based on extended investigations is proposed.

Aromatic primary amines are oxidized to nitroaromatics[118] by peracetic acid when the reaction temperature is higher than when preparing the corresponding nitroso compounds. Trifluoroperacetic acid is an excellent oxidation agent in most cases[119, 120]. Permaleic acid is also a convenient oxidation agent for certain amines[121].

Very few investigations on reactions of aliphatic amines with peracids are known. From the meagre data available it would appear that both nitroso[122] and nitro[118] compounds are formed.

2. Tertiary amines to amine oxides

The oxidation of tertiary amines to amine oxides by peracids is a general reaction with a broad scope. Aliphatic amines are easiest to oxidize, the aromatic ones more difficult and the aromatic nitrogen heterocycles still more difficult to convert to amine oxides. A kinetic investigation on the perbenzoic acid oxidation of tertiary amines has been made[129] and a review[123] is available covering different aspects of this reaction.

Problems concerning peracid oxidation of compounds containing both sulphur and nitrogen are discussed in the literature[114, 124, 125]. In most cases the oxidation of organic sulphur compounds to sulphoxides is faster than the oxidation of nitrogen compounds to N-oxides.

3. Nitroso compounds to nitro compounds

The oxidation of nitroso compounds with peracetic acid has recently been studied mechanistically[126], and it is suggested that oxidation of nitrosobenzene by peracids follows a path similar to that for the oxidation of amines[117]. Chloroperacetic acid oxidizes nitrosobenzene more rapidly than does peracetic acid[117]. Peracetic acid does not react with o-dinitrosobenzene but trifluoroperacetic acid smoothly oxidizes it to o-dinitrobenzene[127]. By using anhydrous trifluoroperacetic acid, N-nitrosoamines have been converted to the nitramines in high yields[119, 128].

4. Oximes to nitro compounds

A quite general procedure for the oxidation of primary and secondary aliphatic, alicyclic and aromatic oximes to the corresponding nitro compounds has been described[60]. The oxidizing agent is trifluoroperacetic acid, and the solvent is acetonitrile, which is presumed to function as a base:

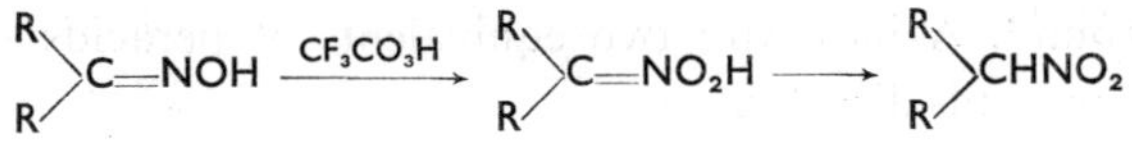

$$\begin{array}{ccc} R \\ \diagdown \\ \diagup C{=}NOH \xrightarrow{\ CF_3CO_3H\ } \\ R \end{array} \begin{array}{c} R \\ \diagdown \\ \diagup C{=}NO_2H \longrightarrow \\ R \end{array} \begin{array}{c} R \\ \diagdown \\ \diagup CHNO_2 \\ R \end{array}$$

5. Aldehyde hydrazones to azoxy compounds

The reaction of aromatic peracids with aldehyde hydrazones has been shown to produce azoxy compounds[60, 130–133]. The phenyl hydrazone is first converted to the enolic azo compound, which is then oxidized. The same type of oxidation with peracetic acid has also been accomplished[134].

6. Azo compounds to azoxy compounds

Azo compounds are oxidized smoothly to azoxy derivatives in high yields. In recent studies of peracid oxidation of substituted azobenzenes[135], good procedures for the preparation of azoxybenzenes are described. Problems encountered here are that isomeric compounds are possible.

m-Nitroazobenzene for instance gives two isomeric azoxy compounds[136]. 2-Phenylazopyridine gives an amine oxide azoxy compound[137].

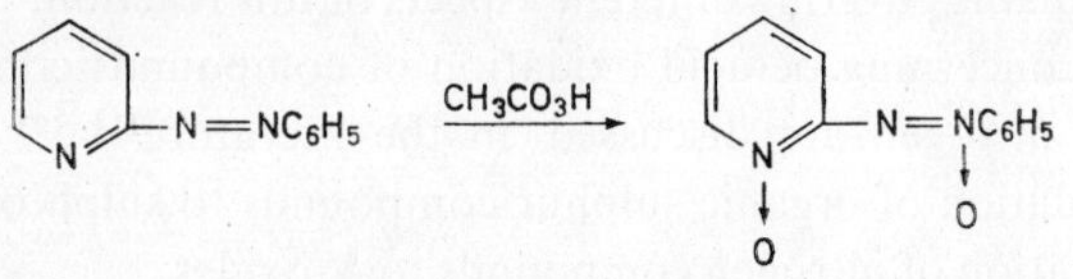

7. Oxidation of Schiff's bases and related compounds

Quite independently three different groups of investigators have found that by treatment of imines with peracids oxaziranes are formed[119, 138-140].

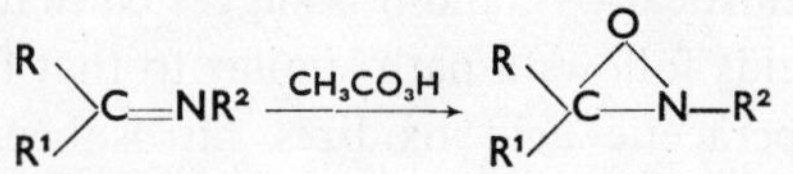

Oxaziranes react smoothly with peracids to give nitrosoalkane dimers[122], and Emmons further showed that the same dimers are produced by reaction of the imines with two equivalents of peracids[122].

Azines have also been oxidized by peracids and give the azine monoxide[141].

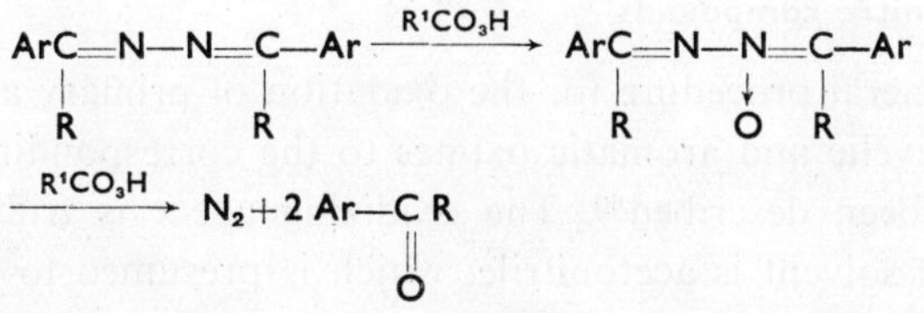

Further treatment of the amine oxide with peracid regenerates the carbonyl compound. Azines with two equivalents of peracids produce the carbonyl compounds directly.

The reaction of ketenimines with peracids yields a mixture of ketone, isonitrile and α-acyloxyamide[142]. It is suggested that epoxidation of the carbon–carbon double bond leads to an epoxyimine which may decompose to an isonitrile or a carbonyl compound or may react with carboxylic acid (or its anion):

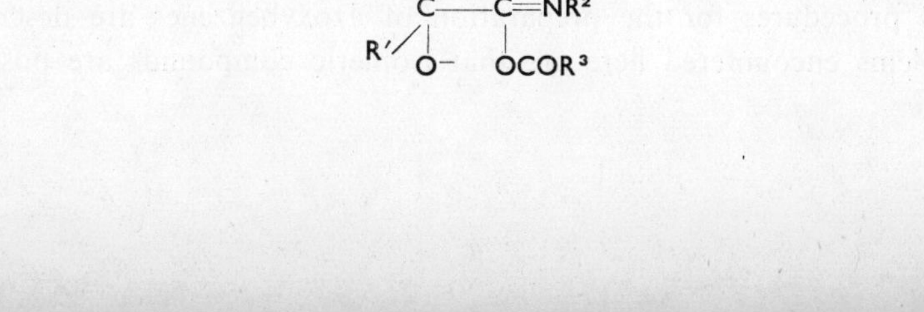

F. Oxidation of Organic Iodine Compounds

Different peracids have been found to oxidize aromatic and aliphatic iodine compounds[143, 144] to the iodoso compounds, which then usually form the diester or are oxidized to the iodoxy compounds. For example, iodobenzene and peracetic acid usually form the diacetate, whereas iodobenzene and perbenzoic acid form the iodoxybenzene:

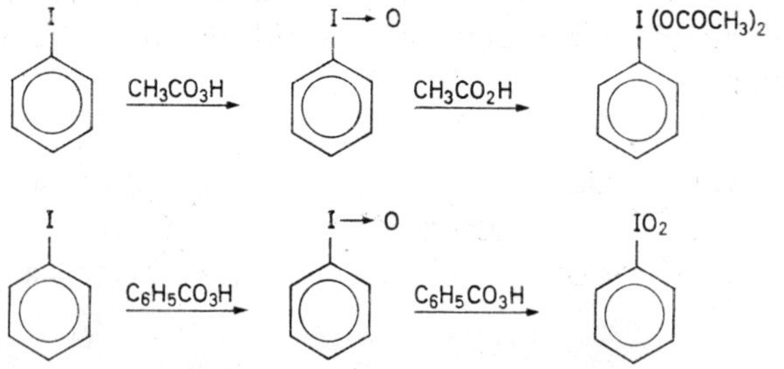

The kinetics of the perbenzoic acid oxidation of aromatic iodine compounds has been investigated by Böeseken and Wicherlink[145].

V. PREPARATION OF PERESTERS

Peresters are generally prepared by acylation of an alkylhydroperoxide, and acid chloride, ketene, diketene, or anhydride are used as acylation agents. Baeyer and Villiger[146] prepared the first perester by treating a barium salt of a hydroperoxide with an acid chloride under neutral conditions. As peresters of primary and secondary alkyl hydroperoxides are sensitive to base, the above method is very convenient in these cases. Also alkyl hydroperoxides, which are sensitive to acid or are difficult to acylate according to the common methods, are most conveniently acylated via their salts[147]. Peresters of tertiary alkylhydroperoxides, which are insensitive to base, are best prepared according to the Schotten–Baumann procedure in mild alkaline solution[148]. Pyridine[149] has also been used as the base when acylating hydroperoxides with acid chlorides. The formed hydrogen chloride can also be removed in vacuum when an alkyl hydroperoxide is reacted with an acid chloride[150] without any base present.

A very elegant method for the preparation of peresters has been worked out by Rüchardt[151] (see also reference 152), in which the carboxylic acid is used as follows:

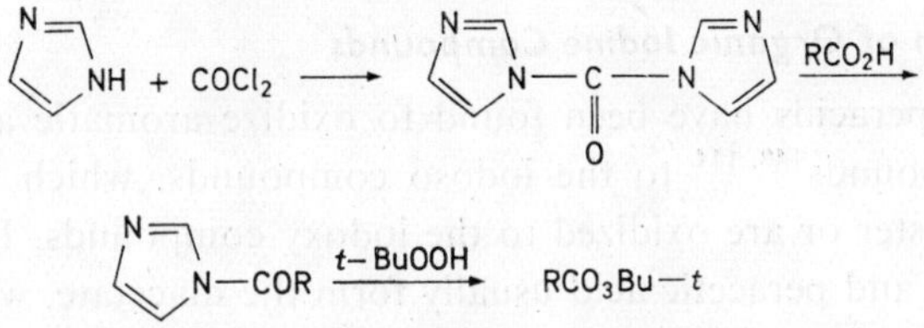

It is thus not necessary to prepare the acid chloride. The preparation of alkylidene peresters[153, 154] by treating the acid and alkylhydroperoxide with arylsulphonyl chloride at 0°C also avoids the use of the acid chloride or anhydride:

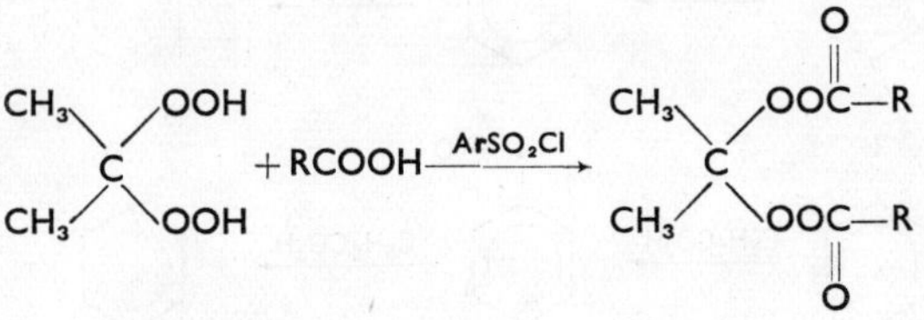

t-Butyl performiate has also been prepared directly from formic acid and t-butyl hydroperoxide, in which procedure the water formed is removed by azeotropic distillation[155]. The same perester is also prepared by treating t-butyl hydroperoxide with the mixed anhydride from formic and acetic acids[156]. Under mild conditions ketene yields t-butyl peracetate[157]:

$$CH_2{=}C{=}O + t\text{-}BuOOH \longrightarrow CH_3CO_3Bu\text{-}t$$

and diketene produces the corresponding perester of acetoacetic acid[157].

Peresters of carbonic acids are known. Phosgene and ethyl chloroformiate react with t-butyl hydroperoxide and sodium peroxide as follows:[158, 159]

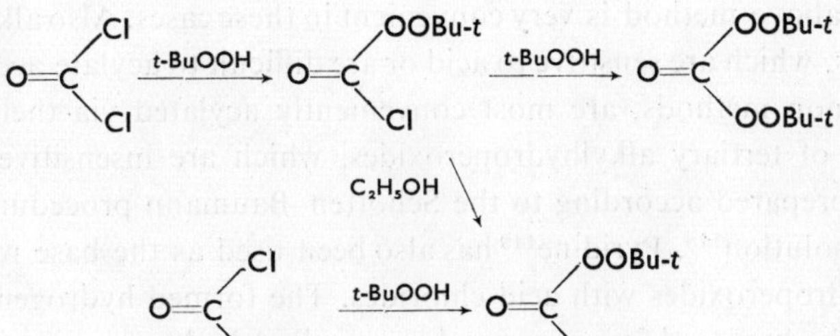

$$2\,RO_2C{-}Cl + Na_2O_2 \longrightarrow RO_2C{-}OO{-}COOR + 2\,NaCl$$

Similarly, by reacting alkyl or aryl isocyanates with hydroperoxides different types of peresters are formed[158, 160–165]:

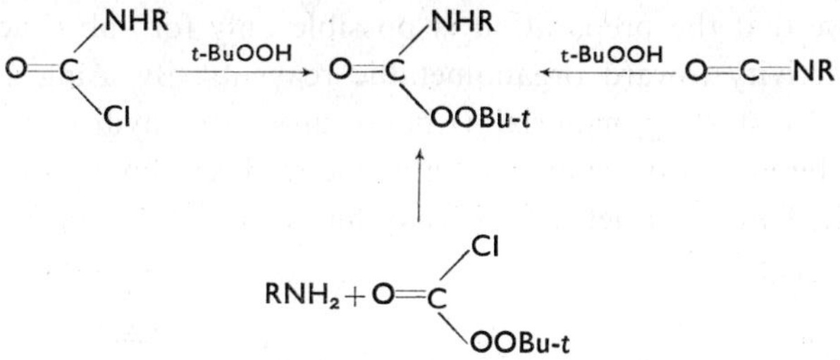

$$RNH_2 + O{=}C\Big\langle{}^{Cl}_{OOBu\text{-}t}$$

Amines and peresters of chloroformic acid also give the same type of compounds[166-168].

VI. CHEMICAL PROPERTIES OF PERESTERS

A. Reaction with Grignard Reagents

Until recently very little work had been done on reactions between Grignard reagents and peroxy compounds and only isolated cases of such reactions were known[169-174]. t-Butyl perbenzoate reacts with sodium methoxide in methanol to give methyl benzoate and the sodium salt of t-butyl hydroperoxide[175]. Contrary to the reaction of esters with Grignard reagents, t-butyl perbenzoate reacts with a variety of Grignard reagents to give good yields of t-butyl ethers and benzoic acid[176-179]:

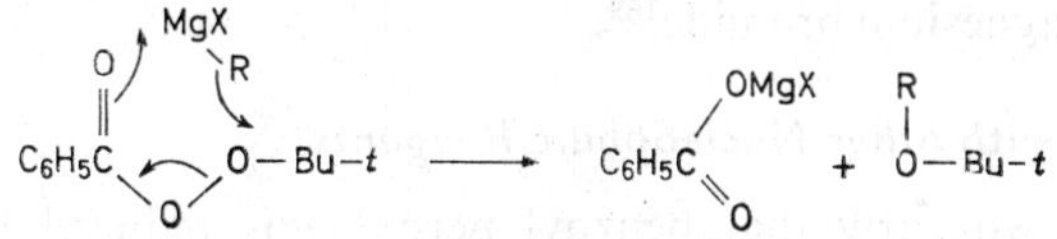

t-Butyl peracetate[180] can be used instead of t-butyl perbenzoate. This simple method of introducing a t-butoxy group into an aromatic compound and the subsequent smooth elimination of isobutylene gives a route to hydroxyaromatics[180] from the corresponding haloaromatic compounds. In heterocyclic chemistry, potential hydroxythiophenes have been prepared as follows[180, 181]:

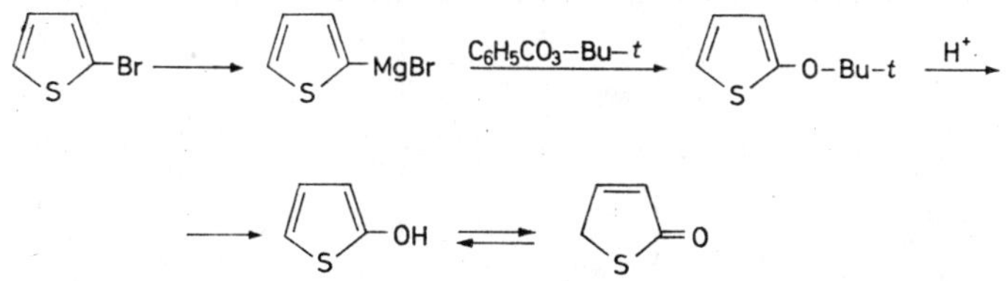

Hydroxythiophenes can otherwise be prepared by hydrogen peroxide oxidation of thiophene boronic acids[182-185]. However, this method is restricted in the sense that the preparation is possible only for substituents showing no reactivity toward organometallic reagents. By using a *t*-butoxythiophene as starting material (*t*-butyl group as hydroxyl protecting group), different substituents can be introduced and finally isobutylene is eliminated. Thus, a variety of hydroxythiophenes[186–188] can be prepared, as for instance:

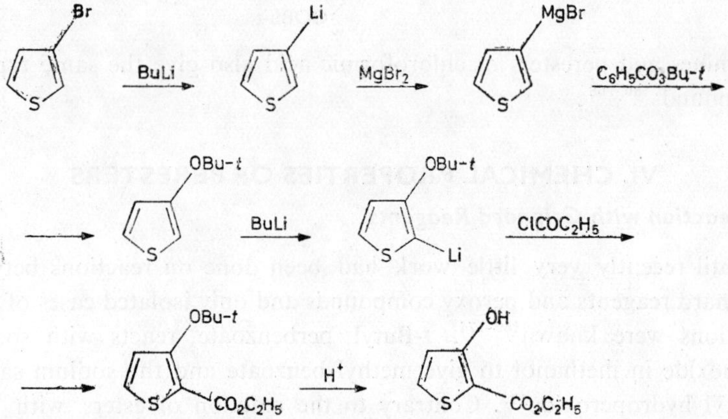

As organolithium compounds attack peresters at the carbonyl group, they have to be converted to the Grignard reagents by treatment with anhydrous magnesium bromide[189].

B. Reactions with other Nucleophilic Reagents

It was known early that benzoyl peroxide is reduced by triphenylphosphine[190] to benzoic anhydride. Horner and Jurgeleit[191] extended this investigation to a variety of peroxides, and also *t*-butyl perbenzoate:

$$C_6H_5\underset{\underset{O}{\|}}{C}{-}O{-}O{-}Bu\text{-}t + (C_6H_5)_3P \longrightarrow C_6H_5\underset{\underset{O}{\|}}{C}{-}O{-}Bu\text{-}t + (C_6H_5)_3\,PO$$

The reaction is considered to proceed by nucleophilic attack on oxygen to form an ion pair which then reacts to give the final product[191]:

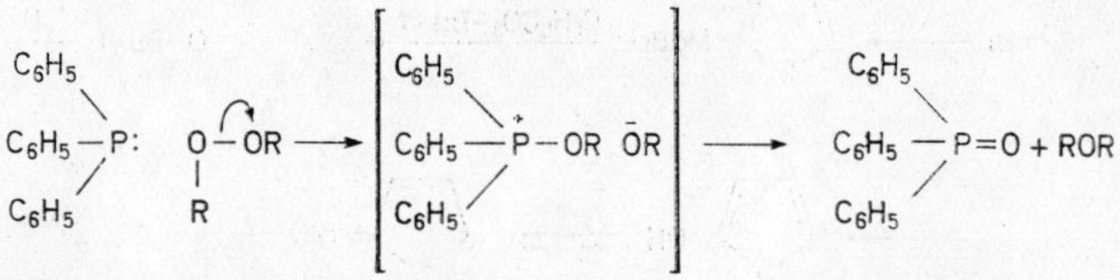

This mechanism has also been confirmed by Greenbaum, Denney and Hoffmann[192].

The reaction of the sodium derivatives of malonic esters with benzoyl peroxide in non-hydroxylic solvents yields *O*-benzoyltartronates[193–195]. Similarly, the benzoyloxy group has been introduced into the methylene group of other active methylene compounds[196–200]. Instead of benzoyl peroxide *t*-butyl peracetate can be used, whereby the *t*-butoxy group is introduced[193, 197]:

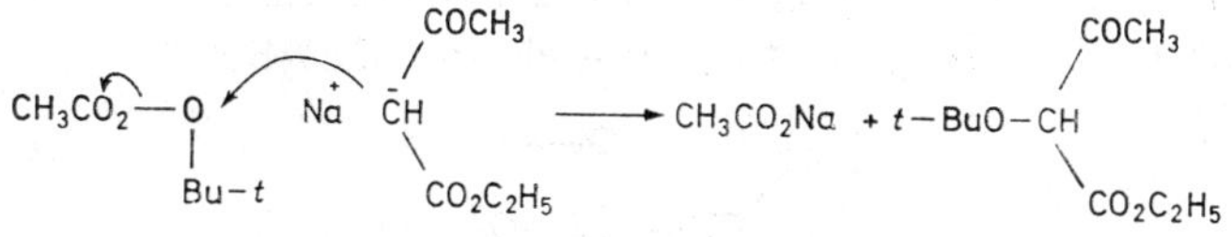

C. Friedel–Crafts Oxygenation Reactions

The decomposition of aroyl peroxides in the presence of aluminium chloride has been investigated[201–204] but no similar studies of peresters seem to have been done. However, quite recently peroxydicarbonates have attracted considerable attention as oxygenating agents in work by Kovacic and coworkers. They report oxygenation of toluene to cresols in 50% yield with diisopropyl peroxydicarbonate and aluminium chloride[205]. Other substrates (anisole[206], alkylbenzenes[206], aromatic ethers[206], halobenzenes[207], biphenyl[207], naphthalene[207]) have been shown to undergo the same reaction. As no reaction occurs in the absence of catalyst, the peroxide–catalyst complex may play a crucial role. Further, since no products arising from free-radical reactions are present, as the reaction occurs at low temperature (0–5°C) under which condition the peroxide itself is stable, and also based on other reasons (isomer distribution and kinetics), an electrophilic reaction mechanism is suggested:

$$RO{-}\overset{\overset{\displaystyle O}{\|}}{C}{-}OO{-}\overset{\overset{\displaystyle O\cdots AlCl_3}{\|}}{C}OR \xrightarrow{ArH} RO\overset{\overset{\displaystyle O}{\|}}{C}OO\overset{\overset{\displaystyle O}{\|}}{C}OAlCl_2 + ArR + HCl$$

$$RO\overset{\overset{\displaystyle O}{\|}}{C}OO\overset{\overset{\displaystyle O\cdots AlCl_3}{\|}}{C}OAlCl_2 \xrightarrow{ArH} ArOCO_2R + (Cl_2Al)_2\,CO_3 + HCl$$

$$ArO\overset{\overset{\displaystyle O\cdots AlCl_3}{\|}}{C}OR \xrightarrow{ArH} ArOCO_2AlCl_2 + ArR + HCl$$

$$ArOCO_2AlCl_2 \xrightarrow[HCl]{H_2O} ArOH + CO_2 + AlCl_3$$

Alternatively, aryl-oxygen bond formation may be the first step:

$$RO\overset{\overset{\displaystyle O}{\|}}{C}{-}OO{-}\overset{\overset{\displaystyle O\cdots AlCl_3}{\|}}{C}OR \xrightarrow{ArH} ArOCO_2R + ROCO_2AlCl_2 + HCl$$

In most cases there is strong evidence for primacy of heterolytic oxygen–oxygen cleavage rather than alkylation. Depending on the nature of the substrates, different oxygenated materials are obtained which can easily be converted to the phenol stage by hydrolysis.

Another type of perester, t-butylperoxy isopropyl carbonate, has also been reacted with aromatic compounds[205] under Friedel–Crafts conditions to produce phenols. It is suggested that aluminium chloride initially coordinates at the carbonyl group to give the electrophile t-BuO$^+$.

$$\begin{array}{c} \text{CH}_3 \\ \phantom{\text{CH}_3}\diagdown \\ \phantom{\text{CH}_3\diagdown}\text{CH}\!-\!\text{OCO}\!-\!\text{O}\!-\!\text{Bu-}t \\ \text{CH}_3\diagup \qquad\quad \delta^-\ \ \delta^+ \end{array}$$

An aryl t-butyl ether is postulated as the intermediate which then is dealkylated in the presence of the catalyst. Evidence is presented for heterolytic fission of the peroxide linkage and an electrophilic reaction mechanism.

D. Uncatalysed Perester Decomposition

Peresters derived from t-alkyl hydroperoxides can undergo different fragmentation and rearrangement reactions:

$$\text{R}\!-\!\overset{\text{O}}{\overset{\|}{\text{C}}}\!-\!\text{OOCR}'_3 \longrightarrow \begin{cases} \text{R}\overset{\text{O}}{\overset{\|}{\text{C}}}\!-\!\text{O}\cdot + \cdot\text{OR}'_3 \\ \text{R}\cdot + \text{CO}_2 + \cdot\text{OCR}'_3 \\ \overset{\text{R}''\cdot}{\longrightarrow} \cdot\text{OCR}'_3 + \text{other products} \\ \text{R}\!-\!\overset{\text{O}}{\overset{\|}{\text{C}}}\!-\!\overset{\text{R}'}{\underset{\text{OR}'}{\text{C}}}\!-\!\text{R}' \end{cases}$$

Among the peresters t-butyl peracetate (R = CH$_3$) and t-butyl perbenzoate belong to a class in which neither group R· forms a particularly stable radical. The decomposition rate varies little with the nature of R[209–211]. If R possesses appreciable stability as a free radical, carbon dioxide and the radicals R· and R$'_3$C—O· are formed. The rate of decomposition depends on the resonance stability of the radial R[212]. A similar concerted fission of at least two bonds occurs in the decomposition of di-t-butylperoxalate to give only carbon dioxide and products of the t-butoxy radicals[213, 214].

The Criegee decomposition of peresters can occur under extremely mild conditions. Just by standing a perester is converted to an acylal[215, 216]:

The reaction rate depends strongly on the dielectric constant of the solvent and must therefore involve charged species as intermediates. Also the rate of rearrangement increases with the strength of the acid RCO_3H. A synchronous mechanism, as outlined below, suggests that the fission of the oxygen–oxygen and carbon–carbon bonds is concurrent with the formation of the carbon-oxygen bond.

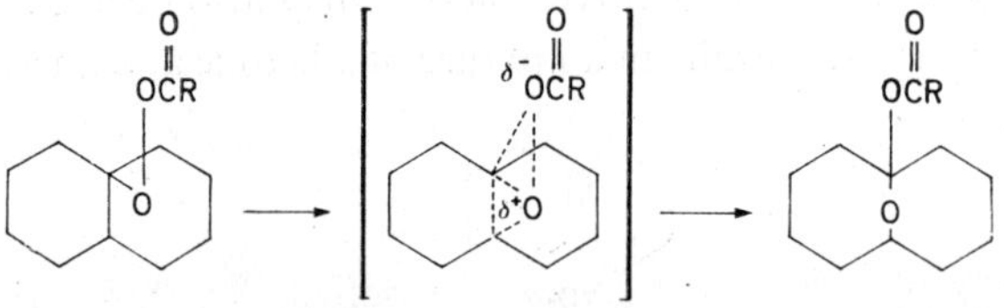

If the reaction is carried out in the presence of foreign anions, none of them is incorporated into the resulting ester[217]. Denney and coworkerst by labelling the carbonyl oxygen of the perester with ^{18}O, showed tha, virtually all of the ^{18}O in the final product is found in the carbonyl group[218, 219]. These results imply that the synchronous mechanism is correct or alternatively the reaction proceeds through an ion pair, in which the identity of the acyl oxygen atoms is preserved[220].

Simple peresters derived from t-alkyl hydroperoxides generally do not undergo the Criegee reaction. However, when a phenyl group is present at the α position of the hydroperoxide, a rearrangement is observed[147]:

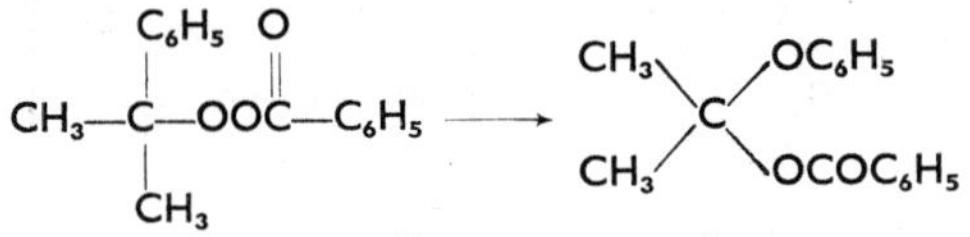

The perbenzoate of triphenylmethyl hydroperoxide is so labile that only the rearrangement product is isolated[221]:

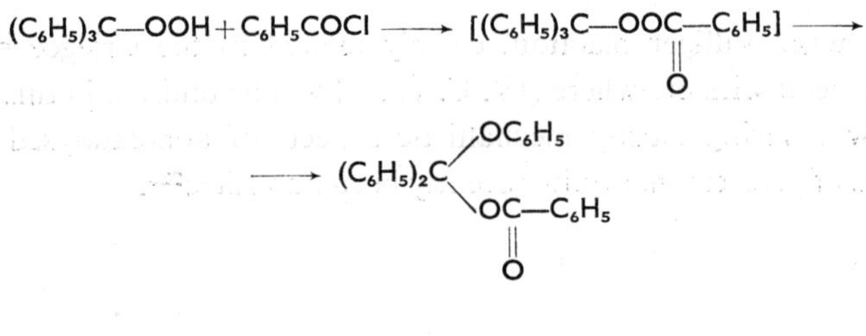

Similarly certain silylperbenzoates[222] rearrange:

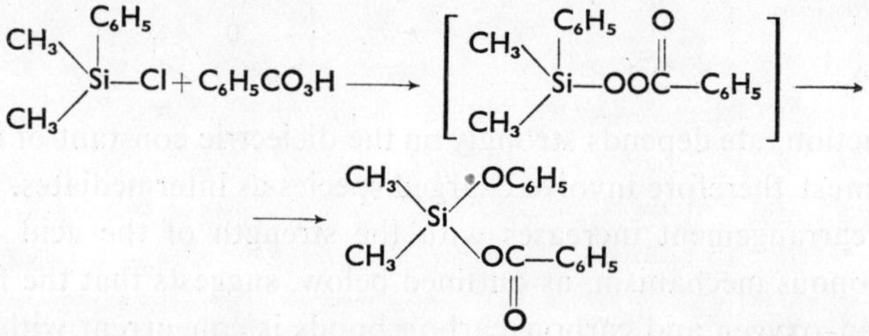

$$
\begin{array}{c}
CH_3\diagdown \; \overset{C_6H_5}{|} \\
\quad Si-Cl + C_6H_5CO_3H \longrightarrow \\
CH_3\diagup
\end{array}
\left[
\begin{array}{c}
CH_3\diagdown \; \overset{C_6H_5}{|} \quad \overset{O}{\parallel} \\
\quad Si-OOC-C_6H_5 \\
CH_3\diagup
\end{array}
\right] \longrightarrow
$$

$$
\longrightarrow
\begin{array}{c}
CH_3\diagdown \quad \diagup OC_6H_5 \\
\qquad Si \\
CH_3\diagup \quad \diagdown OC-C_6H_5 \\
\qquad\qquad\qquad \overset{\parallel}{O}
\end{array}
$$

Esters of primary and secondary alkyl hydroperoxides may undergo the Criegee reaction but a competitive reaction may also occur. Autoxidation of tetralin for instance results in a mixture of a ketone and an acylal[223]:

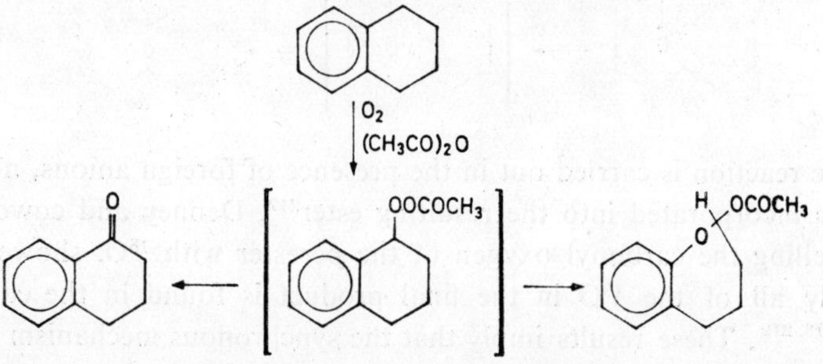

In the case of s-butyl peracetate, which gives acetic acid and butanone-2 as sole products a cyclic concerted process is postulated[224]:

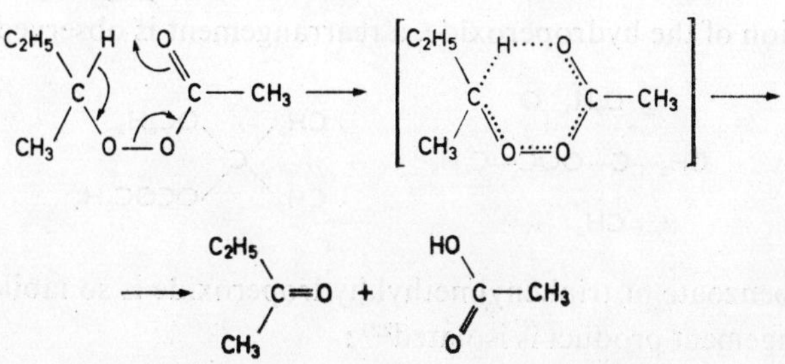

The Baeyer–Villiger reaction, closely related to the Criegee reaction, has been dealt with elsewhere (IV. C. 1) and will be omitted in this section. A review covering mostly mechanistic aspects of non-catalysed decompositions of peresters has quite recently been published[225].

E. The Perester Reaction

1. Introduction

t-Butyl perbenzoate is a commercially available perester. It is less prone to the Criegee reaction than are the corresponding bicyclic compounds. Only when warmed in the presence of a trace of perchloric acid[209] does it decompose, giving acetone and benzoic acid. In the absence of a catalyst the thermal decomposition of t-butyl perbenzoate results in a complex mixture of products[210, 226]. t-Butyl perbenzoate is also decomposed in boiling benzene in the presence of cuprous bromide[226], and only methyl benzoate and acetone are isolated. The following chain-reaction is postulated:

$$C_6H_5COOOBu\text{-}t + Cu^+ \longrightarrow C_6H_5COOCu^+(II) + t\text{-BuO}\cdot$$

$$t\text{-BuO}\cdot \longrightarrow CH_3\cdot + \underset{\underset{CH_3}{|}}{CH_3CO}$$

$$CH_3\cdot + C_6H_5CO_3t\text{-Bu} \longrightarrow C_6H_5COOCH_3 + t\text{-BuO}\cdot$$

Different copper salts have been used and found to be efficient catalysts[226-228]; cobalt salts are less effective[226], and salts such as $ZnCl_2$ and $MgBr_2$ are ineffective[226]. The copper salt catalysed decomposition of t-butyl perbenzoate in different substrates revealed substitution in the latter by the benzoyloxy group. This is called the perester reaction:

$$\underset{|}{\overset{|}{R}}\!-\!\overset{O}{\overset{\|}{C}}\!-\!H + C_6H_5\overset{O}{\overset{\|}{C}}\!-\!OOBu\text{-}t \xrightarrow{Cu^+/Cu^{2+}} \underset{|}{\overset{|}{R}}\!-\!\overset{O}{\overset{|}{C}}\!-\!O\overset{O}{\overset{\|}{C}}\!-\!C_6H_5 + t\text{-BuOH}$$

t-Butyl peracetate is another commercially available perester, which only has been studied in isolated cases.

2. Reactions of peresters with various classes of compounds

a. Olefins. As indicated earlier, peroxides and olefins give complex mixtures of products[210]. However, when a catalytic amount of cuprous bromide is present, t-butyl perbenzoate and cyclohexene predominantly yield 3-benzoyloxycyclohexene:

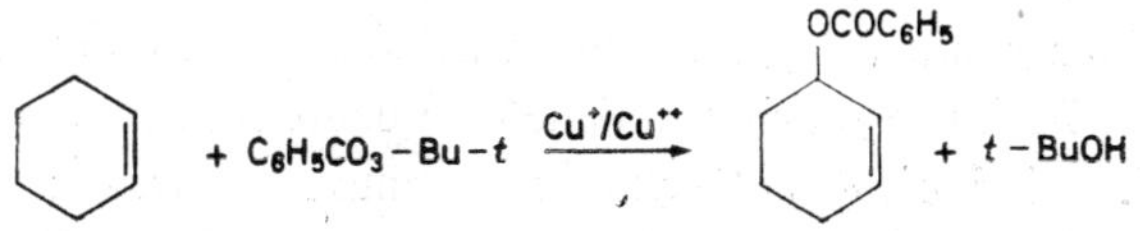

The benzoyloxy group is thus introduced into the allylic position[226, 229, 230]. As only t-butyl perbenzoate and t-butyl peracetate are commercially available, only two types of esters are easily prepared. However, if excess of a carboxylic acid is present in the reaction mixture, the allylic ester of this acid is obtained[226]:

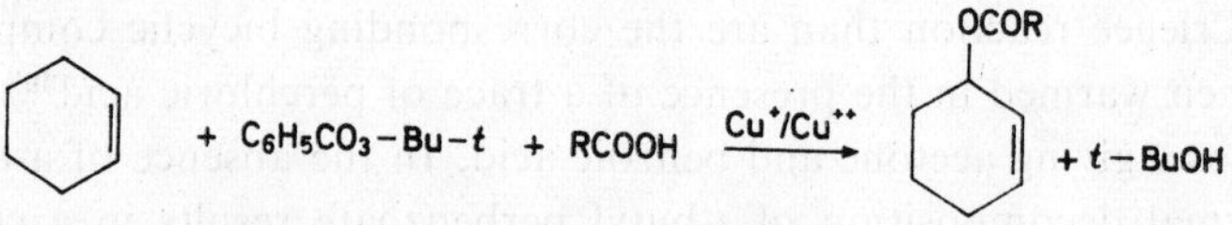

It was first thought that the perester reaction was very selective; this has later been only partly confirmed[231, 232]. For instance, when solvent is present, allylbenzene produces small amounts of cinnamyl benzoate in addition to the main product, the 3-acyloxyderivative[231].

Kochi[233] always found small amounts of the 1-acyloxyderivative when reacting a terminal olefin with *t*-butyl perbenzoate or peracetate. 2-Butene[233] containing an internal double bond, gives 84–94% 3-acyloxy-1-butene, showing in this case that almost complete isomerization has occurred.

$$CH_3{-}CH{=}CH{-}CH_3 + C_6H_5CO_3Bu\text{-}t \xrightarrow{Cu^+/Cu^{2+}}$$

$$\underset{\underset{84\text{–}94\%}{OCOC_6H_5}}{CH_3{-}\overset{|}{CH}{-}CH{=}CH_2} + \underset{\underset{6\text{–}15\%}{OCOC_6H_5}}{CH_3{-}CH{=}CH{-}\overset{|}{CH_2}}$$

Denney, Appelbaum and Denney[228] have also found the same rearrangement with internal olefins.

A quite different rearrangement was observed by Story[234, 235] when reacting norbornadiene with *t*-butyl perbenzoate:

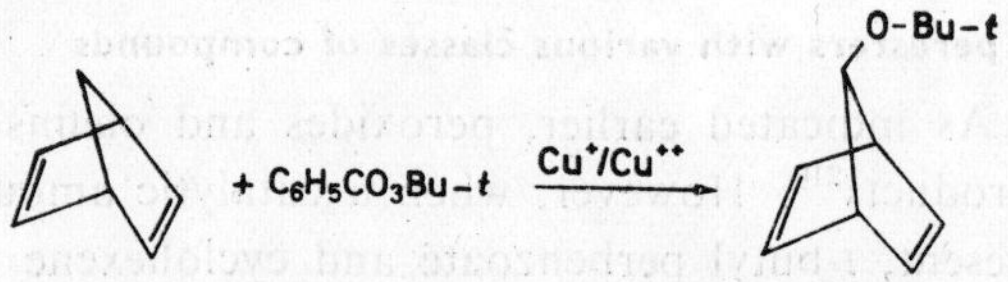

Instead of a benzoate a *t*-butyl ether is obtained. In an attempt to explain this anomalous result (2D)-norbornadiene was reacted with *t*-butyl perbenzoate and 7-*t*-butoxynorbornadiene with deuterium in all positions was produced[236]. This indicates attack of the *t*-butoxy radical on all four olefinic carbons followed by rearrangement of the norbornadiene skeleton.

b. Non-olefinic hydrocarbons. Kharasch and Fono[237] showed that cumene reacts with *t*-butyl perbenzoate in the following way:

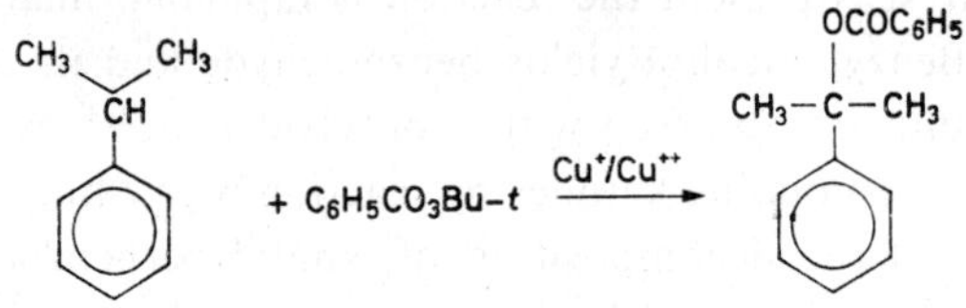

$$CH_3-CH(CH_3) \text{ [benzene ring]} + C_6H_5CO_3Bu\text{-}t \xrightarrow{Cu^+/Cu^{++}} CH_3-\underset{\underset{\text{[benzene ring]}}{OCOC_6H_5}}{\overset{OCOC_6H_5}{C}}-CH_3$$

The corresponding reaction with t-butyl peracetate has also been achieved[238]. Only hydrocarbons with benzylic hydrogen have been investigated but the experimental material is insufficient for wide generalizations. Triphenylmethane[226] does not give well-defined products, diphenylmethane[227] yields only the dimer, 1,1,2,2-tetraphenylethane, and tetralin[230] the expected ester product. Optically active 2-phenylbutane yields inactive 2-phenyl-2-benzoyloxybutane[239].

c. *Alcohols and mercaptans.* The reaction between alcohols and benzoyl peroxide was first studied by Gelissen and Hermans[203]. Bartlett and Nozaki[240] explained the formation of carbonyl compounds in the following way, and this mechanism was also supported by Urry and coworkers[241]:

$$R_2CHOH + C_6H_5COO \cdot \longrightarrow R_2\dot{C}OH + C_6H_5COOH$$

$$R_2\dot{C}OH + (C_6H_5COO)_2 \longrightarrow R_2\underset{OCOC_6H_5}{COH} + C_6H_5COO \cdot$$

$$R_2\underset{OCOC_6H_5}{COH} \longrightarrow R_2CO + C_6H_5COOH$$

The uncatalysed reaction of t-butoxy radicals with alcohols with available α-hydrogen atoms yields 1,2-dihydroxy derivatives by dimerization [241-243]:

$$RCH_2OH \xrightarrow{t\text{-}BuO \cdot} \begin{array}{c} R-CH-OH \\ | \\ R-CH-OH \end{array}$$

t-Butyl perbenzoate decomposes slowly in alcohols and with benzyl alcohol benzaldehyde and benzoic acid are formed[230, 244]. A disproportionation of the benzyl alcohol radical:

$$2 C_6H_5\dot{C}HOH \longrightarrow C_6H_5CHO + C_6H_5CH_2OH$$

or a fragmentation of the hemiacylal:

$$\begin{array}{c} C_6H_5CHOH \\ | \\ OCOC_6H_5 \end{array} \longrightarrow C_6H_5CHO + C_6H_5COOH$$

would account for these results.

44*

When a catalyst is present the reaction is rapid in primary and secondary alcohols. Benzyl alcohol yields benzaldehyde and its dibenzylacetal and similar results are observed with n-butylaldehyde[244]. Secondary alcohols yield the corresponding ketones by reaction with t-butyl perbenzoate. In t-butyl alcohol the decomposition of t-butyl perbenzoate is slow[244].

The reactions of t-butyl perbenzoate with thiols[243] and thiophenols[230] are not modified by copper salt catalysts, since the only products are the corresponding disulphides.

d. Benzylic ethers. When benzyl ethers are heated with di-t-butyl peroxide, dimerization or fragmentation or both occur[245]:

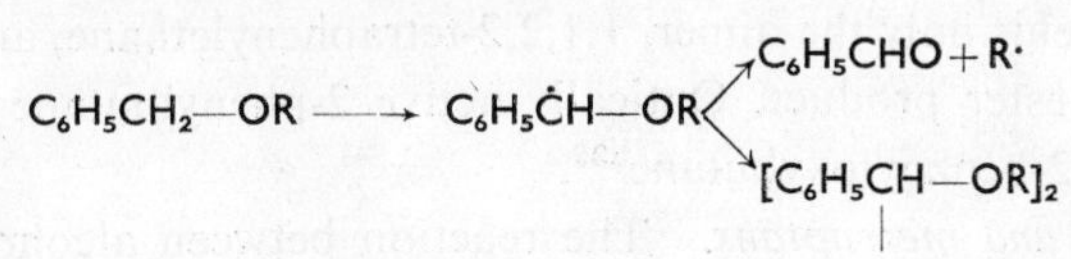

$$C_6H_5CH_2-OR \longrightarrow C_6H_5\overset{.}{C}H-OR \begin{cases} C_6H_5CHO + R\cdot \\ [C_6H_5CH-OR]_2 \end{cases}$$

(R = methyl, ethyl, isopropyl, t-butyl, phenyl, benzyl or diphenylmethyl). Irrespective of the group R, the attack of the free t-butoxy radical occurs on the methylene group between the heteroatom and the phenyl group. The cleavage of monosubstituted dibenzyl ethers by free t-butoxy radicals has later been performed by Huang and Yeo[246]. By reacting a series of benzyl ethers[178] with t-butyl perbenzoate in the presence of catalytic amounts of cuprous chloride the benzoyloxy group is introduced, giving the acylal:

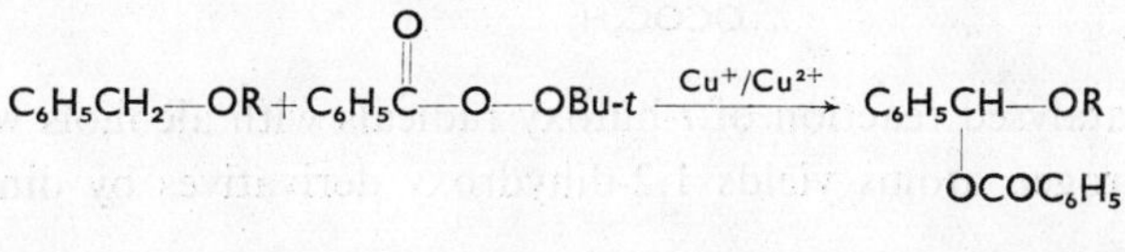

$$C_6H_5CH_2-OR + C_6H_5\overset{\overset{O}{\|}}{C}-O-OBu\text{-}t \xrightarrow{Cu^+/Cu^{2+}} \underset{\underset{OCOC_6H_5}{|}}{C_6H_5CH-OR}$$

(R = methyl, ethyl, isopropyl, t-butyl and phenyl). With dibenzyl ether no acylal is formed but instead benzaldehyde and benzaldehyde dibenzylacetal are isolated. It is proposed that the α-benzoyloxyderivative first formed undergoes an oxygenalkyl heterolysis whereby the unsymmetrical acetal is formed, which thereafter gives the dibenzylacetal:

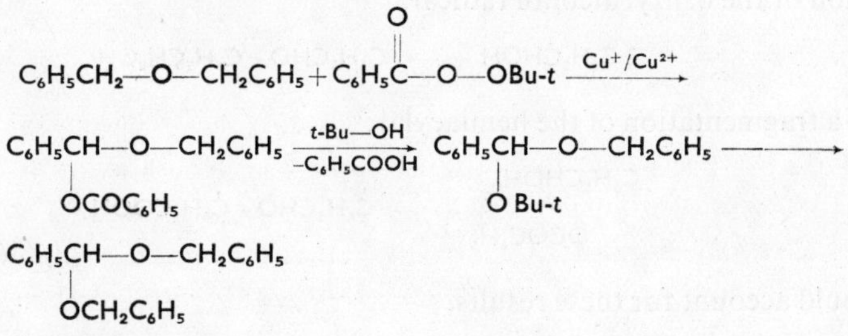

$$C_6H_5CH_2-O-CH_2C_6H_5 + C_6H_5\overset{\overset{O}{\|}}{C}-O-OBu\text{-}t \xrightarrow{Cu^+/Cu^{2+}}$$

$$\underset{\underset{OCOC_6H_5}{|}}{C_6H_5CH-O-CH_2C_6H_5} \xrightarrow[-C_6H_5COOH]{t\text{-Bu}-OH} \underset{\underset{O\,Bu\text{-}t}{|}}{C_6H_5CH-O-CH_2C_6H_5} \longrightarrow$$

$$\underset{\underset{OCH_2C_6H_5}{|}}{C_6H_5CH-O-CH_2C_6H_5}$$

The benzaldehyde isolated may have been formed from decomposition of an acetal or by its reaction with radicals[247, 248].

e. *Dialkyl and aryl alkyl ethers.* α-Benzoyloxy ethers have been formed by decomposing benzoyl peroxide in ethers[249-251], the mechanism of which has been elucidated by Denney and Feig[252]. In these investigations no fragmentation of the intermediate ether radical has been observed, which would indicate that there is a difference between benzyl ethers and simple aliphatic ethers in this respect. This difference was further accentuated when *t*-butyl peroxide was reacted with n-butyl ether, giving the dimer as the main product[253].

A series of dialkyl and aryl alkyl ethers have been reacted with *t*-butyl perbenzoate in the presence of cuprous chloride as catalyst, and as with benzyl ethers, the benzoyloxy group is introduced into the α position of the heteroatom[179, 253-255]. Free-radical reactions with aryl alkyl ether[256] have hitherto only given ill-defined products.

If excess alcohol is present during the reaction, an acetal is obtained, thus providing a convenient way of transforming ethers into aldehydes[253]:

$$RCH_2-O-CHR \xrightarrow{R'-OH} RCH_2-O-CHR \xrightarrow{R'OH}$$
$$\quad\quad\;\; | \quad\quad\quad\quad\quad\quad\quad\quad\quad | $$
$$\quad\quad OCOC_6H_5 \quad\quad\quad\quad\quad\quad OR'$$

$$R-CH\begin{array}{l} OR' \\ OR' \end{array}$$

f. *Cyclic ethers.* Anomalous results are obtained when tetrahydrofuran or tetrahydropyran react with *t*-butyl perbenzoate with a catalyst present. No benzoyloxy derivatives are isolated but instead 2-*t*-butoxytetrahydrofuran[255, 257] and 2-*t*-butoxytetrahydropyran are obtained[255, 257]. It is suggested that the benzoyloxy group is first introduced into the α position of the ether. If excess of an alcohol ROH other than *t*-butanol is present in the reaction mixture, the RO group is introduced into the cyclic ether[257]:

$$\overset{\frown}{\underset{O}{\bigcirc}} + C_6H_5CO_3Bu-t + ROH \xrightarrow{Cu^+/Cu^{++}} \overset{\frown}{\underset{O}{\bigcirc}}-OR + C_6H_5COOH + t-BuOH$$

Other cyclic ethers (1,3-dioxane[227], 1,4-dioxane[227, 255] and isochroman[227]) have also been reacted with peresters but generally only the acyloxy derivatives are produced.

g. *Sulphides.* Straight-chain aliphatic and benzylic sulphides, as well as aryl alkyl sulphides, react smoothly with *t*-butyl perbenzoate in the presence of a copper salt to give α-benzoyloxy sulphides[179, 243, 258] without oxidation of the sulphide to sulphoxide or sulphone:

$$RCH_2\text{—}S\text{—}CH_2\text{—}R + C_6H_5CO_3 Bu\text{-}t \xrightarrow{Cu^+/Cu^{2+}} RCH_2\text{—}S\text{—}\underset{\underset{OCOC_6H_5}{|}}{CH}\text{—}R$$

The benzyl sulphides do not show any tendency to fragment as the benzyl ethers do, neither do the benzoyloxy sulphides react with alcohols to give alkoxy sulphides; mercaptals[259] are isolated instead. Pyrolysis produces vinyl sulphides[179, 260].

Of the cyclic sulphides investigated, tetrahydrothiophen[227, 258] and 1,4-thioxane[227] react with t-butyl perbenzoate with a copper salt present, to give the corresponding benzoates.

h. Aldehydes, ketones and esters. Decomposition of di-t-butyl peroxide in benzaldehyde produces 1,2-dibenzoyloxy-1,2-diphenylethane according to the following reaction[261]:

$$C_6H_5CHO + \cdot OBu\text{-}t \longrightarrow t\text{-}BuOH + C_6H_5\overset{\cdot}{C}{=}O$$

$$C_6H_5\overset{\cdot}{C}{=}O + C_6H_5CHO \longrightarrow C_6H_5COO\text{—}\overset{\cdot}{CH}\text{—}C_6H_5 \longrightarrow$$

$$\begin{array}{c} C_6H_5COO\text{—}CH\text{—}C_6H_5 \\ | \\ C_6H_5COO\text{—}CH\text{—}C_6H_5 \end{array}$$

The same compound is obtained from benzaldehyde and t-butyl perbenzoate when no catalyst is used[230]. With a catalyst present t-butyl perbenzoate yields benzoic anhydride[230]:

$$C_6H_5CHO + C_6H_5CO_3Bu\text{-}t \xrightarrow{Cu^+/Cu^{2+}} (C_6H_5CO)_2O + t\text{-}BuOH$$

Aliphatic aldehydes as well as ketones give no well-defined products on reaction with t-butyl perbenzoate with or without a catalyst present.

Benzyl acetate reacts with t-butyl perbenzoate in the presence of a copper salt to give the corresponding acylal[230] (α-acetoxybenzyl benzoate):

$$C_6H_5CH_2\text{—}O\overset{\overset{O}{\|}}{C}\text{—}CH_3 + C_6H_5CO_3Bu\text{-}t \xrightarrow{Cu^+/Cu^{2+}} C_6H_5CH\overset{OCOCH_3}{\underset{OCOC_6H_5}{<}} + t\text{-}BuOH$$

The same acylal is also produced from benzyl benzoate with t-butyl peracetate under the same conditions. Without a catalyst no acylals are formed.

Simple aliphatic esters give no well-defined products. Active methylene compounds such as diethyl malonate and ethyl acetatoacetate give mainly the corresponding *t*-butoxy derivatives[227] (*cf*. section V. B).

i. Organic nitrogen compounds. The reaction between different amines and *t*-butyl perbenzoate leads only to rapid decomposition of the perester. However, dimethylamine and *t*-butyl perbenzoate yield an anomalous product, bis(*p*-dimethylaminophenyl)methane[230], irrespective of whether a catalyst is present or not. *N,N*-dimethylformamide, on the other hand, produces the normal product[227].

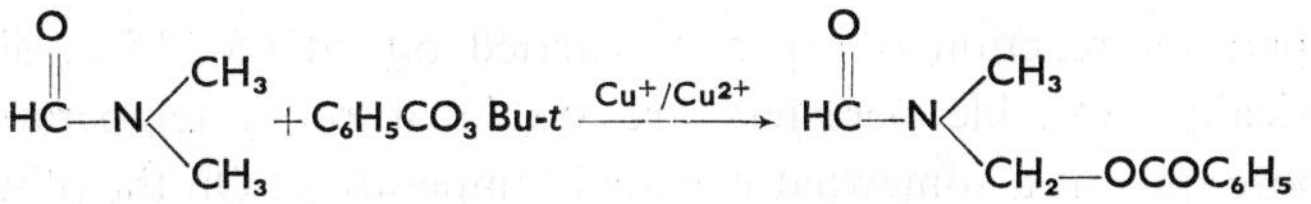

$$\underset{HC-N}{\overset{O}{\overset{\|}{}}}\overset{CH_3}{\underset{CH_3}{<}} + C_6H_5CO_3\,Bu\text{-}t \xrightarrow{Cu^+/Cu^{2+}} \underset{HC-N}{\overset{O}{\overset{\|}{}}}\overset{CH_3}{\underset{CH_2-OCOC_6H_5}{<}}$$

j. Organic silicon compounds. Only a very few results are available[262], as, for instance:

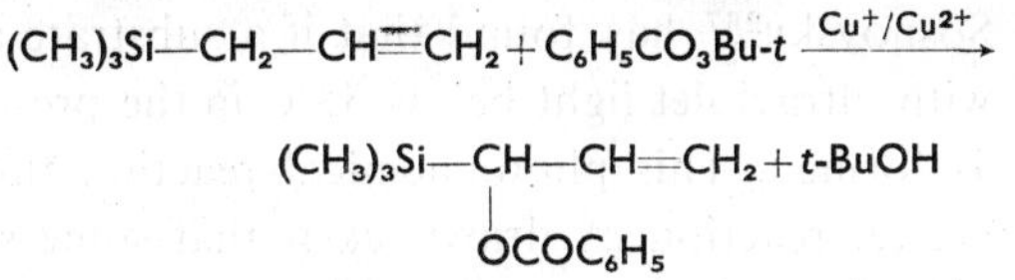

$$(CH_3)_3Si-CH_2-CH{=}CH_2 + C_6H_5CO_3Bu\text{-}t \xrightarrow{Cu^+/Cu^{2+}}$$

$$(CH_3)_3Si-\underset{OCOC_6H_5}{\underset{|}{CH}}-CH{=}CH_2 + t\text{-}BuOH$$

k. Mechanistic considerations. The copper salt catalyst has a profound influence on the decomposition of *t*-butyl perbenzoate or peracetate in the presence of different substrates. Certainly no single mechanism can account for all the results of the perester reaction. Depending upon the substrates and experimental conditions various mechanisms may be operative. Since reviews are available[263, 264] only a short outline will be given here.

Kharasch, Sosnovsky and Yang[226] suggested a chain mechanism involving concerted steps:

$$C_6H_5\overset{O}{\overset{\|}{C}}OOBu\text{-}t + Cu^+ \longrightarrow C_6H_5COOCu^+(II) + t\text{-}BuO\cdot$$

$$RH + t\text{-}BuO\cdot \longrightarrow R\cdot + t\text{-}BuOH$$

$$R\cdot + C_6H_5COOCu^+(II) \longrightarrow C_6H_5COOR + Cu^+$$

As no isomerization of olefins was observed these authors[226] believed that the radicals were not free but formed a complex with the metal ion catalyst.

Kochi[233, 265], who found isomerization of olefins in their reaction with *t*-butyl perbenzoate in the presence of copper salt, believes that the radicals formed are free and suggests a ligand transfer process. Similar conclusions have been made by two research groups[228, 239] who investigated optically active substrates; the final products are optically inactive which indicates that conventional free-radical intermediates are at hand. Walling and Zavitsas[238] have confirmed Kochi's results but prefer a 'free' carbonium ion rather than a ligand transfer process. The results of Goering and Mayer[266] are also consistent with the Kochi mechanism. Another modified mechanism has been proposed by Beckwith and Evans[231].

3. Photochemical reactions

The perester reaction is generally carried out at 65–115°C, since the commerically available peresters are stable at room temperature and decompose only at a somewhat elevated temperature. On the other hand, some of the acyloxy derivatives formed are so thermally sensitive that a relatively low reaction temperature is preferred to avoid side-reactions. With substrates of low boiling points the perester reaction has to be run in an autoclave. Sosnovsky[267] has found that if a substrate and a perester are irradiated with ultraviolet light below 35°C in the presence of copper ions, an ester is formed. This photochemical reaction thus extends the scope of the perester reaction. A drawback is that some substrates (sulphides, olefins) deactivate the catalyst.

F. Aromatic Oxygenation with Dialkyl Peroxydicarbonate–Metal Salt

Razuvaev, Kartashova and Boguslavskaya[268] reacted toluene with dicyclohexyl peroxydicarbonate in the presence of ferric chloride and isolated tolyl cyclohexyl carbonate. A heterolytic mechanism was postulated. These experiments have been repeated[269] and extended to studies of related systems. Although few experimental data are available, it is concluded that alkoxycarboxy radicals are formed in the reaction. In another note[270] the preliminary results from aromatic oxygenation with diisopropyl peroxydicarbonate–cupric chloride are described. This seems to be the first case of oxidation of an oxy radical to a species exhibiting oxonium ion character, which then effects aromatic substitution. It should be added that peroxydicarbonates[159, 271–274] and related compounds[158] have only been investigated in isolated cases (see also section VI. D and reference 225). The reason may be that this class of peroxides is hazardous to handle.

VII. ACKNOWLEDGEMENT

The authors gratefully acknowledge support from Korn og Foderstof-kompagniet A/S, Viby J, Denmark.

VIII. REFERENCES

1. D. Swern, *Chem. Rev.*, **45**, 1 (1948).
2. M. Hudlický, *Chem. Listy*, **46**, 567 (1952).
3. V. N. Belov and L. A. Kheifits, *Usp. Khim.*, **25**, 969 (1956).
4. S. Havel, *Chem. Listy*, **53**, 811 (1959).
5. S. Havel, *Chem. Listy*, **53**, 928 (1959).
6. V. J. Karnojitzki, *Chim. Ind. (Paris)*, **92**, 381 (1964).
7. G. Sosnovsky and J. H. Brown, *Chem. Rev.*, **66**, 529 (1966).
8. L. S. Silbert, E. Siegel and D. Swern, *J. Org. Chem.* **27**, 1336 (1962).
9. F. P. Greenspan, R. J. Gall and D. G. MacKellar, *J. Org. Chem.*, **20**, 215 (1955).
10. A. Gross, *U.S. Pat.* 2, 806, 045 (1957); *Chem. Abstr.*, **52**, 3373 (1958).
11. *Brit. Pat.* 905, 877 (1962); *Chem. Abstr.*, **59**, 8599 (1963).
12. *Brit. Pat.*, 891, 211 (1962); *Chem. Abstr.*, **57**, 4601 (1962).
13. W. E. Parker, C. Riccuti, C. L. Ogg and D. Swern, *J. Am. Chem. Soc.*, **77**, 4037 (1955).
14. D. Swern and W. E. Parker, *U.S. Pat.* 2, 813, 885 (1957); *Chem. Abstr.*, **52**, 3372 (1958).
15. J. d'Ans and W. Frey, *Chem. Ber.*, **45**, 1845 (1912).
16. H. Böhme, *Org. Syn., Coll. Vol.*, **3**, 619 (1955).
17. E. E. Royals and L. L. Harrell, Jr., *J. Am. Chem. Soc.*, **77**, 3405 (1955).
18. S. Linholter and P. Sørensen, *Acta Chem. Scand.*, **12**, 1331 (1958).
19. G. B. Payne, *J. Org. Chem.*, **24**, 1354 (1959).
20. A. Kergomard and J. Philibert–Bigou, *Bull. Soc. Chim. France*, 334 (1958).
21. M. Vilkas, *Bull. Soc. Chim. France*, 1401 (1959).
22. L. S. Silbert, E. Siegel and D. Swern, *Org. Syn.*, **43**, 93 (1963).
23. A. V. Bayer and V. Villiger, *Chem. Ber.*, **23**, 1569 (1900).
24. H. L. J. Bäckström, *Z. Physik. Chem. (Leipzig)*, **25B**, 99 (1934).
25. H. L. Bäckström, *J. Am. Chem. Soc.*, **49**, 1460 (1927).
26. A. H. Cook, *J. Chem. Soc.*, 1768 (1938).
27. E. Briner and A. Lardon, *Helv. Chim. Acta*, **19**, 1062 (1936).
28. D. C. Noller and D. J. Bolton, *Anal. Chem.*, **35**, 887 (1963).
29. D. Swern, L. P. Witnauer, C. R. Eddy and W. E. Parker, *J. Am. Chem. Soc.*, **77**, 5537 (1955).
30. J. R. Rittenhouse, W. Lobunez, D. Swern and J. G. Miller, *J. Am. Chem. Soc.*, **80**, 4850 (1958).
31. D. Swern and L. S. Silbert, *Anal. Chem.*, **35**, 880 (1963).
32. W. H. T. Davison, *J. Chem. Soc.*, 2456 (1951).
33. P. A Giguere and A. W. Olmos, *Can. J. Chem.*, **30**, 821 (1952).
34. G. J. Minkoff, *Proc. Roy. Soc. (London)* **A224**, 176 (1954).
35. E. R. Stephens, P. L. Hanst and R. C. Doerr, *Anal. Chem.*, **29**, 776 (1957).
36. A. A. Shubin, *Tr. Fiz. Inst. Akad. Nauk SSSR*, **9**, 125 (1958); *Chem. Abstr.*, **53**, 9812 (1959).
37. P. A. Giguere and A. W. Olmos, *Can. J. Chem.*, **34**, 689 (1956).
38. E. Briner and P. de Chastmay, *Helv. Chim. Acta*, **37**, 626 (1954).
39. E. Briner and P. de Chastmay, *Helv. Chim. Acta*, **37**, 1904 (1954).

40. A. J. Everett and G. J. Minkoff, *Trans. Faraday Soc.*, **49**, 410 (1953).
41. N. Prileschajew, *Chem. Ber.*, **42**, 4811 (1909).
42. F. D. Gunstone, *Adv. Org. Chem.*, **1**, 103 (1960).
43. M. Malinovsky, *Usp. Khim.*, **27**, 622 (1958).
44. W. D. Emmons and A. S. Pagano, *J. Am. Chem. Soc.*, **77**, 89 (1955).
45. F. C. Frostick, Jr., B. Phillips and P. S. Starcher, *J. Am. Chem. Soc.*, **81**, 3350 (1959).
46. A. Rosowsky, in *Heterocyclic Compounds with Three- and Four-membered Rings* (Ed. A. Weissberger), Part I, Interscience Publishers, New York, 1964, p. 1.
47. P. D. Bartlett, *Record Chem. Progr. (Kresge-Hooke Sci. Lib.)*, **11**, 47 (1950).
48. B. M. Lynch and K. H. Pausacker, *J. Chem. Soc.*, 1525 (1955).
49. D. R. Campbell, J. O. Edwards, J. MacLachlan and K. Polgar, *J. Am. Chem. Soc.*, **80**, 5308 (1958).
50. H. B. Henbest and R. A. L. Wilson, *J. Chem. Soc.*, 1958 (1957).
51. R. Albrecht and C. Tamm, *Helv. Chim. Acta*, **40**, 2216 (1957).
52. M. L. Sassiver and J. English, *J. Am. Chem. Soc.*, **82**, 4891 (1960).
53. D. Swern, *J. Am. Chem. Soc.*, **69**, 1692 (1947).
54. R. Huisgen, R. Grashey and J. Sauer in *The Chemistry of Alkenes* (Ed. S. Patai), Interscience Publishers, London, 1964, p. 739.
55. R. Filler, *J. Am. Chem. Soc.*, **81**, 659 (1959).
56. C. L. Stevens and J. Tazuma, *J. Am. Chem. Soc.*, **76**, 715 (1954).
57. A. Kirrmann, P. Duhamel and R. Nouri–Bimarghi, *Ann. Chem.*, **691**, 33 (1966).
58. D. Swern, *Org. Reactions*, **7**, 378 (1953).
59. W. D. Emmons, A. S. Pagano and J. P. Freeman, *J. Am. Chem. Soc.*, **76**, 3472 (1954).
60. W. D. Emmons and A. S. Pagano, *J. Am. Chem. Soc.*, **77**, 4557 (1955).
61. P. A. S. Smith, in *Molecular Rearrangements* (Ed. P. de Mayo), Part I, Interscience Publishers, New York, 1963, p. 457.
62. M. F. Hawthorne, *Anal. Chem.*, **28**, 540 (1956).
63. W. D. Emmons and G. B. Lucas, *J. Am. Chem. Soc.*, **77**, 2287 (1955).
64. D. L. Heywood and B. Phillips, *J. Org. Chem.*, **25**, 1699 (1960).
65. E. Hedaya and S. Winstein, *Tetrahedron Letters*, **13**, 563 (1962).
66. R. R. Sauers and G. P. Ahearn, *J. Am. Chem. Soc.*, **83**, 2759 (1961).
67. J. Meinwald and E. Frauenglass, *J. Am. Chem. Soc.*, **82**, 5235 (1960).
68. M. F. Hawthorne and W. D. Emmons, *J. Am. Chem. Soc.*, **80**, 6398 (1958).
69. Y. Ogata and I. Tabushi, *Bull. Chem. Soc. Japan*, **32**, 108 (1959); *Chem. Abstr.*, **54**, 2910 (1960).
70. Y. Ogata, I. Tabushi and H. Akimoto, *J. Org. Chem.*, **26**, 4803 (1961).
71. R. B. Turner, *J. Am. Chem. Soc.*, **72**, 878 (1950).
72. K. Mislow and J. Brenner, *J. Am. Chem. Soc.*, **75**, 2318 (1953).
73. J. A. Berson and S. Suzuki, *J. Am. Chem. Soc.*, **81**, 4088 (1959).
74. N. M. Malenok and I. V. Sologub, *J. Gen. Chem. USSR (Eng. Transl.)* **23**, 1181 (1953); *Chem. Abstr.*, **47**, 12210 (1953).
75. I. G. Tishcheinko and M. G. Gurevich, *Zhidko aznoe Okislenie Nepredel'nykh Organ. Soedin., Sb.*, No. 1, 85 (1961); *Chem. Abstr.* **58**, 3307 (1963).
76. R. A. Raphael, *J. Chem. Soc.*, 44 (1949).
77. J. Böeseken and G. Sloof, *Rec. Trav. Chim.*, **49**, 95 (1930).
78. H. H. Schlubach and W. Richau, *Ann. Chem.*, **588**, 195 (1954).
79. H. H. Schlubach and V. Franzen, *Ann. Chem.*, **577**, 60 (1952).
80. H. H. Schlubach and V. Franzen, *Ann. Chem.*, **578**, 220 (1952).
81. V. Franzen, *Ann. Chem.*, **587**, 130 (1954).
82. V. Franzen, *Chem. Ber.*, **87**, 1219, 1478 (1954).
83. V. Franzen, *Chem. Ber.*, **88**, 717 (1955).

84. J. Böeseken, *Rec. Trav. Chim.*, **54,** 657 (1935).
85. V. I. Pansevich–Kolyada and Z. B. Idelchik, *J. Gen. Chem. USSR (Eng. Transl.)*, **24,** 1601 (1954); *Chem. Abstr.*, **49,** 12428 (1955).
86. J. K. Crandall and W. H. Machleder, *Tetrahedron Letters*, **48,** 6037 (1966).
87. I. M. Roitt and W. A. Waters, *J. Chem. Soc.*, 3060 (1949).
88. J. Böeseken and G. Sloof, *Rec. Trav. Chim.*, **49,** 100 (1930).
89. R. E. Dean, E. N. White and D. McNeill, *J. Appl. Chem. (London)*, **3,** 469 (1953).
90. D. H. Derbyshire and W. A. Waters, *Nature*, **165,** 401 (1950).
91. J. D. McClure and P. H. Williams, *J. Org. Chem.*, **27,** 24 (1962).
92. R. D. Chambers, P. Goggin and W. K. R. Musgrave, *J. Chem. Soc.*, 1804 (1959).
93. C. A. Buehler and H. Hart, *J. Am. Chem. Soc.*, **85,** 2177 (1963).
94. H. Hart and C. A. Buehler, *J. Org. Chem.*, **29,** 2397 (1964).
95. A. J. Waring and H. Hart, *J. Am. Chem. Soc.*, **86,** 1454 (1964).
96. H. Hart, C. A. Buehler, A. J. Waring and S. Meyerson, *J. Org. Chem.*, **30,** 331 (1965).
97. H. Hart, P. M. Collins and A. J. Waring, *J. Am. Chem. Soc.*, **88,** 1005 (1966).
98. H. Hart and R. M. Lange, *J. Org. Chem.*, **31,** 3776 (1966).
99. H. Fernholz, *Chem. Ber.*, **84,** 110 (1951).
100. H. Fernholz and G. Piazolo, *Chem. Ber.*, **87,** 578 (1954).
101. S. L. Friess, A. H. Soloway, B. K. Morse and W. C. Ingersoll, *J. Am. Chem. Soc.*, **74,** 1305 (1952).
102. H. Davidge, A. G. Davies, J. Kenyon and R. F. Mason, *J. Chem. Soc.*, 4569 (1958).
103. C. O. Overberger and R. W. Cummins, *J. Am. Chem. Soc.*, **75,** 4250 (1953).
104. H. H. Szmant, H. F. Harnsberger and F. Krahe, *J. Am. Chem. Soc.*, **76,** 2185 (1954).
105. A. Cerniani and G. Modena, *Gazz. Chim. Ital.*, **89,** 843 (1959).
106. G. Modena and P. E. Todesco, *J. Chem. Soc.*, 4920 (1962).
107. A. Cerniani, G. Modena and P. E. Todesco, *Gazz. Chim. Ital.*, **90,** 3 (1960).
108. G. M. Gasperini, G. Modena and P. E. Todesco, *Gazz. Chim. Ital.*, **90,** 12 (1960).
109. R. Curci and G. Modena, *Tetrahedron Letters*, **25,** 1749 (1963).
110. R. Curci and G. Modena, *Tetrahedron Letters*, **14,** 863 (1965).
111. A. Mayr, F. Montanari and M. Tramontini, *Ricerca Sci., Suppl.*, **30,** 746 (1960); *Chem. Abstr.*, **55,** 1499 (1961).
112. K. Balenović, N. Bregant and D. Francetić, *Tetrahedron Letters*, **6,** 20 (1960).
113. K. Balenović, I. Bregovec, D. Francetić, I. Monković and V. Tomasić, *Chem. Ind. (London)*, 469 (1961).
114. H. Bretschneider and W. Klötzer, *Monatsh. Chem.*, **81,** 589 (1950).
115. Y. Minoura, *Nippon Gomu Kyokaishi*, **32,** 177 (1959); *Chem. Abstr.* **54,** 8694 (1960).
116. H. Gilman, L. E. Smith and H. H. Parker, *J. Am. Chem. Soc.*, **47,** 851 (1925).
117. K. M. Ibne–Rasa and J. O. Edwards, *J. Am. Chem. Soc.*, **84,** 763 (1962).
118. W. D. Emmons, *J. Am. Chem. Soc.*, **79,** 5528 (1957).
119. W. D. Emmons and A. F. Ferris, *J. Am. Chem. Soc.*, **75,** 4623 (1953).
120. W. D. Emmons, *J. Am. Chem. Soc.*, **76,** 3470 (1954).
121. R. White and W. D. Emmons, *Tetrahedron Letters*, **17,** 31 (1962).
122. W. D. Emmons, *J. Am. Chem. Soc.*, **79,** 6522 (1957).
123. C. C. J. Culvenor, *Rev. Pure Appl. Chem.*, **3,** 83 (1953).
124. A. Risaliti, *Ann. Chim. (Rome)*, **46,** 199 (1956).
125. G. Kobayshi, Y. Kuwayama and S. Okamura, *Yakugaku Zasshi*, **83,** 234 (1963); *Chem. Abstr.*, **59,** 5144 (1963).
126. K. M. Ibne–Rasa, C. G. Lauro and J. O. Edwards, *J. Am. Chem. Soc.*, **85,** 1165 (1963).

127. J. H. Boyer and S. E. Ellzey, *J. Org. Chem.*, **24**, 2038 (1959).
128. W. D. Emmons, *J. Am. Chem. Soc.*, **76**, 3468 (1954).
129. J. Foucart, J. Nasielski and E. Vander Danckt, *Bull. Soc. Chim. Belges*, **75**, 17 (1966).
130. B. Witkop and H. M. Kissman, *J. Am. Chem. Soc.*, **75**, 1975 (1953).
131. B. M. Lynch and K. H. Pausacker, *J. Chem. Soc.*, 2517 (1953).
132. B. M. Lynch and K. H. Pausacker, *J. Chem. Soc.*, 1131 (1954).
133. B. M. Lynch and K. H. Pausacker, *J. Chem. Soc.*, 3340 (1954).
134. B. T. Gillis and K. F. Schimmel, *J. Org. Chem.*, **27**, 413 (1962).
135. B. T. Newbold, *J. Org. Chem.*, **27**, 3919 (1962).
136. G. Leandri and A. Risaliti, *Ann. Chim. (Rome)*, **44**, 1036 (1954).
137. M. Colonna and A. Risaliti, *Gazz. Chim. Ital.*, **85**, 1148 (1955).
138. H. Krimm, *Chem. Ber.*, **91**, 1057 (1958).
139. W. D. Emmons, *J. Am. Chem. Soc.*, **79**, 5739 (1957).
140. L. Horner and E. Jürgens, *Chem. Ber.*, **90**, 2184 (1957).
141. L. Horner, W. Kirmse and H. Fernekess, *Chem. Ber.*, **94**, 279 (1961).
142. H. Kagen and I. Lillien, *J. Org. Chem.*, **31**, 3728 (1966).
143. B. A. Arbusow, *J. Prakt. Chem.*, **131**, 357 (1931).
144. J. Böeseken and *G.C.C.C.* Schneider, *Proc. Acad. Sci. Amsterdam*, **33**, 827 (1930); *Chem. Abstr.*, **25**, 923 (1931).
145. J. Böeseken and E. Wicherlink, *Rec. Trav. Chim.*, **55**, 936 (1936).
146. A. v. Baeyer and V. Villiger, *Chem. Ber.* **34**, 738 (1901).
147. H. Hock and H. Kropf, *Chem. Ber.*, **88**, 1544 (1955).
148. N. An Milas and D. M. Surgenor, *J. Am. Chem. Soc.*, **68**, 643 (1946).
149. R. Criegee, *Chem. Ber.*, **77**, 22 (1944).
150. N. A. Milas, D. G. Orphanos and R. J. Klein, *J. Org. Chem.*, **29**, 3099 (1964).
151. R. Hecht and C. Rüchardt, *Chem. Ber.*, **96**, 1281 (1963).
152. H. A. Staab, W. Rohr and F. Graf, *Chem. Ber.*, **98**, 1122 (1965).
153. N. A. Milas and A. Golubović, *J. Am. Chem. Soc.*, **81**, 3361, 5824, 6461 (1959).
154. N. A. Milas and A. Golubović, *J. Org. Chem.*, **27**, 4319 (1962).
155. C. Rückhardt and R. Hecht, *Chem. Ber.*, **97**, 2716 (1964).
156. R. E. Pincock, *J. Am. Chem. Soc.*, **86**, 1820 (1964).
157. D. Harman, *U.S. Pat.* 2,608,570 (1952); *Chem. Abstr.*, **48**, 3387 (1954).
158. A. G. Davies and K. J. Hunter, *J. Chem. Soc.*, 1808 (1953).
159. F. Strain, W. E. Bissinger, W. R. Dial, H. Rudoff, B. J. de Witt, H. C. Stevens and J. H. Langston, *J. Am. Chem. Soc.*, **72**, 1254 (1950).
160. H. Esser, K. Rastädter and G. Reuter, *Chem. Ber.*, **89**, 685 (1956).
161. N. M. Lapshin, B. N. Moryganov, G. A. Razuaev, A. V. Ryabov and M. L. Khidekel, *Vysokomolekul. Soedin.*, **3**, 1794 (1961); *Chem. Abstr.*, **56**, 13074 (1962).
162. M. Lederer and O. Fuch, *Ger. Pat.* 1,029,818 (1958); *Chem. Abstr.*, **54**, 18353 (1960).
163. E. L. O'Brien, F. M. Beringer and R. B. Mesrobian, *J. Am. Chem. Soc.*, **79**, 6238 (1957).
164. E. L. O'Brien, F. M. Beringer and R. B. Mesrobian, *J. Am. Chem. Soc.*, **81**, 1506 (1959).
165. C. J. Pedersen, *J. Org. Chem.*, **23**, 252 (1958).
166. E. Hedaya, R. L. Hinman, L. M. Kibler and S. Theodoropulos, *J. Am. Chem. Soc.*, **86**, 2727 (1964).
167. T. Koenig and W. Brewer, *J. Am. Chem. Soc.*, **86**, 2728 (1964).
168. T. Koenig and W. Brewer, *J. Am. Chem. Soc.*, **86**, 4072 (1964).
169. H. Gilman and C. E. Adams, *J. Am. Chem. Soc.*, **47**, 2816 (1925).
170. E. Müller and T. Töpel, *Chem. Ber.*, **72**, 273 (1939).

171. T. W. Campbell, W. Burney and T. L. Jacobs, *J. Am. Chem. Soc.*, **72**, 2735 (1950).
172. C. D. Hurd and H. J. Anderson, *J. Am. Chem. Soc.*, **75**, 5124 (1953).
173. C. Walling and S. A. Buckler, *J. Am. Chem. Soc.*, **77**, 6032 (1955).
174. G. A. Baramki, H. S. Chang and J. T. Edward, *Can. J. Chem.*, **40**, 441 (1962).
175. N. A. Milas and D. M. Surgenor, *J. Am. Chem. Soc.*, **68**, 642 (1946).
176. S.–O. Lawesson and N. C. Yang, *J. Am. Chem. Soc.*, **81**, 4230 (1959).
177. C. Frisell and S.–O. Lawesson, *Org. Syn.*, **41**, 91 (1961).
178. S.–O. Lawesson and C. Berglund, *Arkiv Kemi*, **16**, 287 (1960).
179. S.–O. Lawesson, C. Berglund and S. Grönwall, *Acta Chem. Scand.*, **15**, 249 (1961).
180. S.–O. Lawesson and C. Frisell, *Arkiv Kemi*, **17**, 393 (1961).
181. C. Frisell and S.–O. Lawesson, *Org. Syn.*, **43**, 55 (1963).
182. A.–B. Hörnfeldt and S. Gronowitz, *Acta Chem. Scand.*, **16**, 789 (1962).
183. A.–B. Hörnfeldt and S. Gronowitz, *Arkiv Kemi*, **21**, 239 (1963).
184. A.–B. Hörnfeldt, *Arkiv Kemi*, **22**, 211 (1964).
185. A.–B. Hörnfeldt, *Acta Chem. Scand.*, **19**, 1249 (1965).
186. H. J. Jakobsen, E. H. Larsen and S.–O. Lawesson, *Tetrahedron*, **19**, 1867 (1963).
187. H. J. Jakobsen and S.–O. Lawesson, *Tetrahedron*, **21**, 3331 (1965).
188. H. J. Jakobsen and S.–O. Lawesson, *Tetrahedron*, **22**, 871 (1967).
189. S. Gronowitz, *Arkiv Kemi*, **16**, 363 (1960).
190. F. Challenger and V. K. Wilson, *J. Chem. Soc.*, 209 (1927).
191. L. Horner and W. Jurgeleit, *Ann. Chem.*, **591**, 138 (1955).
192. M. A. Greenbaum, D. B. Denney and A. K. Hoffmann, *J. Am. Chem. Soc.*, **78**, 2563 (1956).
193. S.–O. Lawesson, T. Busch and C. Berglund, *Acta Chem. Scand.*, **15**, 260 (1961).
194. E. H. Larsen and S.–O. Lawesson, *Org. Syn.*, **45**, 37 (1965).
195. S.–O. Lawesson, C. Frisell, D. Z. Denney and D. B. Denney, *Tetrahedron*, **19**, 1229 (1963).
196. S.–O. Lawesson and C. Frisell, *Arkiv Kemi*, **17**, 409 (1961).
197. S.–O. Lawesson, M. Andersson and C. Berglund, *Arkiv Kemi*, **17**, 429 (1961).
198. S.–O. Lawesson, P. G. Jönsson and J. Taipale, *Arkiv Kemi*, **17**, 441 (1961).
199. S.–O. Lawesson, M. Dahlén and C. Frisell, *Acta Chem. Scand.*, **16**, 1191 (1962).
200. G. Näslund, A. Senning and S. O. Lawesson, *Acta Chem. Scand.*, **16**, 1324 (1962).
201. A. F. A. Reynhart, *Rec. Trav. Chim.*, **46**, 54 (1927).
202. A. F. A. Reynhart, *Rec. Trav. Chim.*, **46**, 62 (1927).
203. H. Gelissen and P. H. Hermans, *Chem. Ber.*, **58**, 479 (1925).
204. J. T. Edwards, H. S. Chang and S. A. Samad, *Can. J. Chem.*, **40**, 804 (1962).
205. P. Kovacic and S. T. Morneweck, *J. Am. Chem. Soc.*, **87**, 1566 (1965).
206. P. Kovacic and M. E. Kurz, *J. Am. Chem. Soc.*, **87**, 4811 (1965).
207. P. Kovacic and M. E. Kurz, *J. Org. Chem.*, **31**, 2011 (1966).
208. P. Kovacic and M. E. Kurz, *J. Org. Chem.*, **31**, 2459 (1966).
209. A. T. Blomquist and A. F. Ferris, *J. Am. Chem. Soc.*, **73**, 3408 (1951).
210. P. D. Bartlett and R. R. Hiatt, *J. Am. Chem. Soc.*, **80**, 1398 (1958).
211. A. T. Blomquist and I. A. Berstein, *J. Am. Chem. Soc.*, **73**, 5546 (1951).
212. P. D. Bartlett and D. M. Simons, *J. Am. Chem. Soc.*, **82**, 1753 (1960).
213. P. D. Bartlett, E. P. Benzing and R. E. Pincock, *J. Am. Chem. Soc.*, **82**, 1762 (1960).
214. P. D. Bartlett and R. E. Pincock, *J. Am. Chem. Soc.*, **82**, 1769 (1960).
215. R. Criegee, *Chem. Ber.*, **77**, 722 (1944).
216. R. Criegee and R. Caspar, *Ann. Chem.*, **560**, 127 (1948).
217. H. L. Goering and A. C. Olson, *J. Am. Chem. Soc.*, **75**, 5853 (1953).
218. D. B. Denney, *J. Am. Chem. Soc.*, **77**, 1706 (1955).
219. D. B. Denney and D. Z. Denney, *J. Am. Chem. Soc.*, **79**, 4806 (1957).
220. S. Winstein and G. C. Robinson, *J. Am. Chem. Soc.*, **80**, 169 (1958).

221. H. Wieland and J. Maier, *Chem. Ber.*, **64**, 1205 (1931).
222. E. Buncel and A. G. Davies, *J. Chem. Soc.*, 1550 (1958).
223. R. H. Snyder, H. J. Shine, K. A. Leibrand and P. O. Tawney, *J. Am. Chem. Soc.*, **81**, 4299 (1959).
224. L. J. Durham, L. Glover and H. S. Mosher, *J. Am. Chem. Soc.*, **82**, 1508 (1960).
225. C. Rückhardt, *Fortschr. Chem. Forsch.*, **6**, 251 (1966).
226. M. S. Kharasch, G. Sosnovsky and N. C. Yang, *J. Am. Chem. Soc.*, **81**, 5819 (1959).
227. C. Berglund and S.–O. Lawesson, *Arkiv Kemi*, **20**, 225 (1963).
228. D. Z. Denney, A. Appelbaum and D. B. Denney, *J. Am. Chem. Soc.*, **84**, 4969 (1962).
229. M. S. Kharasch and G. Sosnovsky, *J. Am. Chem. Soc.*, **80**, 756 (1958).
230. G. Sosnovsky and N. C. Yang, *J. Org. Chem.*, **25**, 899 (1960).
231. A. L. J. Beckwith and G. W. Evans, *Proc. Chem. Soc. (London)*, 63 (1962).
232. D. B. Denney, D. Z. Denney and G. Feig, *Tetrahedron Letters*, **15**, 19 (1959).
233. J. K. Kochi, *J. Am. Chem. Soc.*, **84**, 774 (1962).
234. P. R. Story, *J. Am. Chem. Soc.*, **82**, 2085 (1960).
235. P. R. Story, *J. Org. Chem.* **26**, 287 (1961).
236. P. R. Story, *Tetrahedron Letters*, **9**, 401 (1962).
237. M. S. Kharasch and A. Fono, *J. Org. Chem.*, **23**, 324 (1958).
238. C. Walling and A. Zavitsas, *J. Am. Chem. Soc.*, **85**, 2084 (1963).
239. P. A. Hallgarten, T. I. Wang and N. C. Yang, unpublished results.
240. P. D. Bartlett and K. Nozaki, *J. Am. Chem. Soc.*, **69**, 2299 (1947).
241. W. H. Urry, F. W. Stacey, E. S. Huyser and O. O. Juveland, *J. Am. Chem. Soc.*, **76**, 450 (1954).
242. K. Schwetlick, W. Geyer and H. Hartmann, *Angew. Chem.*, **72**, 779 (1960).
243. S.–O. Lawesson and C. Berglund, *Acta Chem. Scand.*, **15**, 36 (1961).
244. S.–O. Lawesson and C. Berglund, *Arkiv Kemi*, **17**, 485 (1961).
245. R. L. Huang and S. S. Si-Hoe, in *Vistas in Free-radical Chemistry* (Ed. W. A. Waters), Pergamon Press, London, 1959, p. 242.
246. R. L. Huang and O. K. Yeo, *J. Chem. Soc.*, 3190 (1959).
247. E. S. Huyser, *J. Am. Chem. Soc.*, **25**, 1820 (1960).
248. S.–O. Lawesson and T. Busch, *Arkiv Kemi*, **17**, 421 (1961).
249. W. E. Cass, *J. Am. Chem. Soc.*, **68**, 1976 (1946).
250. W. E. Cass, *J. Am. Chem. Soc.*, **69**, 500 (1947).
251. K. Nozaki and P. D. Bartlett, *J. Am. Chem. Soc.*, **68**, 1686 (1946).
252. D. B. Denney and G. Feig, *J. Am. Chem. Soc.*, **81**, 5322 (1959).
253. S.–O. Lawesson and C. Berglund, *Arkiv Kemi*, **17**, 465 (1961).
254. S.–O. Lawesson and C. Berglund, *Angew. Chem.*, **73**, 65 (1961).
255. G. Sosnovsky, *Tetrahedron*, **13**, 241 (1961).
256. K. M. Johnston and G. H. Williams, *J. Chem. Soc.*, 1168, (1960).
257. S.–O. Lawesson and C. Berglund, *Arkiv Kemi*, **17**, 475 (1961).
258. G. Sosnovsky, *Tetrahedron*, **18**, 15 (1962).
259. G. Sosnovsky, *Tetrahedron*, **18**, 903 (1962).
260. G. Sosnovsky and H. J. O'Neill, *J. Org. Chem.*, **27**, 3469 (1962).
261. F. F. Rust, F. N. Seubold and N. E. Vaugham, *J. Am. Chem. Soc.*, **70**, 3258 (1948).
262. G. Sosnovsky and H. J. O'Neill, *Compt. Rend.*, **254**, 704 (1962).
263. G. Sosnovsky and S.–O. Lawesson, *Angew. Chem.*, **76**, 218 (1964).
264. S.–O. Lawesson and G. Sosnovsky, *Svensk Kem. Tidskr.*, **75**, 568 (1963).
265. J. K. Kochi, *Tetrahedron*, **18**, 483 (1962).
266. H. L. Goering and U. Mayer, *J. Am. Chem. Soc.*, **86**, 3753 (1964).
267. G. Sosnovsky, *Tetrahedron*, **21**, 871 (1965).

268. G. A. Razuvaev, N. A. Kartashova and L. S. Boguslavskaya, *J. Gen. Chem. USSR*, **34**, 2108 (1964).
269. P. Kovacic and M. E. Kurz, *Chem. Comm.*, 431 (1966).
270. P. Kovacic and M. E. Kurz, *J. Am. Chem. Soc.*, **88**, 2068 (1966).
271. S. G. Cohen and D. B. Sparrow, *J. Am. Chem. Soc.*, **72**, 611 (1950).
272. H. C. McBay, O. Tucker and A. Milligan, *J. Org. Chem.*, **19**, 1003 (1954).
273. H. C. McBay and O. Tucker, *J. Org. Chem.*, **19**, 869 (1954).
274. J. R. Crano, *J. Org. Chem.*, **31**, 3615 (1966).

CHAPTER **15**

Thiolo, thiono and dithio acids and esters

MATTHYS J. JANSSEN

The University, Groningen, Netherlands

I. STRUCTURE AND GENERAL PROPERTIES

A. Introduction

This chapter discusses thio acids, dithio acids and their esters. Although the purpose is a general review of the available information, emphasis is placed on the chemistry of thio compounds in comparison to the oxygen analogues. Much of the detailed knowledge, especially from earlier studies, will not be included. In the choice of literature citations preference is given to later papers so that much pioneer research will not be referred to. Those active in this field are constantly made aware of the fact that in sulphur chemistry much is due to work done before 1900. However, this work can be easily traced via more recent papers and in particular through Reid's review[1], which contains fairly complete lists of sulphur compounds known before 1958. For details of preparation and reactions, Volume 9 of Houben Weyl's series[2] is unsurpassed.

The study of thio- and dithiocarboxylic acids is still a comparatively unexplored field in organic chemistry. Although many of these acids have an unpleasant smell, this cannot be the sole cause of this lack of interest. It is more likely that it is one of the fields which have been passed by accidentally. More is known about the derivatives which have found applications, such as the xanthates and dithiocarbamates. These compounds are included in this review in so far as they illustrate the properties of dithio acids.

Since thio and dithio acids are derivatives of thiols and thiones, short discussions of these simpler molecules are given in the following sections.

B. Structure and Physical Properties of Thiones

Whereas thiols and alcohols have at least enough in common to be recognized as related molecules, thio ketones seem very much different from ketones. They are characterized by their very high reactivity[3]. The thiocarbonyl group was once pictured as a diradical, because of spontaneous trimerization[4].

Since Campaigne's review[3] remarkable progress has been made especially in Mayer's laboratory[5] enabling aliphatic thio ketones to be more thoroughly studied. They appear to resemble ketones in more respects than was previously realized[6].

Fortunately, aromatic thio ketones and thiocarbonyl compounds in which the thione group is attached to heteroatoms are much more stable. The most stable representatives of the thiones are undoubtedly the thio ureas. A quantum-mechanical model for thione compounds[7, 8] was proposed, based on two fairly obvious principles: (i) the sulphur atom is less inclined than oxygen to form double bonds, (ii) the electronegativity of sulphur is about midway between that of carbon and oxygen.

Using these principles, spectral properties, resonance energies and charge distributions of a variety of thiocarbonyl compounds were calculated, giving reasonable agreement with experiment[7-10], with the aid of an extended Hückel method. The introduction of special postulates, distinguishing thiocarbonyl compounds from carbonyl compounds (such as 3d-orbital participation[11]) proved unnecessary. Some calculated resonance energies are given in Table 1. The fourth column gives the resonance energies in kcal/mole by calibrating the calculated value for a thio amide against the experimental value for the rotation barrier found by Walter for N-methyl-N-benzyl formamide[12]. In a recent study Sandström found a substantial relation between heights of rotation barriers and loss of π-electron energies for a number of amides and thio amides[13]. It is highly probable that the stabilizing factor against oligomerization of heterosubstituted thiones is the resonance energy. The carbon–sulphur double bond has just a slightly smaller binding energy than two single carbon–sulphur bonds[14] (115 as compared with $2 \times 61 \cdot 5$ kcal/mole), so that even a comparatively small energy contribution would render the monomeric form more stable.

In 1964, the same method was used by Janssen and Sandström[15] to compare thiones with carbonyl compounds. Somewhat surprisingly it appeared that with the first compounds of Table 1 the resonance energy of the carbonyl compound is larger but with the last ones the thiones are

45*

TABLE 1. Calculated resonance energies (R. E.) of
thiones, $XYC = S$[7, 8]

X	Y	R.E. (β unit)	R. E. (kcal/mole)
R^a	R^a	(0)	
Cl	Cl	0.286^b	6.5
RS	R	0.380	8.6
RO	R	0.766	17.4
R_2N	R	1.106	$(25.1)^c$
RS	RS	0.690	15.6
RO	RS	1.054	23.9
R_2N	RS	1.350	30.6

^{a} R means group incapable of resonance (alkyl, H).
^{b} J. Sandström, unpublished calculations.
^{c} Experimental value[12].

more stabilized. The point of intersection lies somewhere near the dithio esters (X, Y = RS, R). It depends, of course, on the exact choice of parameters but is not very sensitive to changes as long as the two principles given above are followed. The general trend in the dipole moments strongly supports this result.

The results of Janssen and Sandström[15] may be translated into terms of higher polarizability of the thione group; a substituent capable of mesomeric electron release will be more effective towards a thiocarbonyl than towards a carbonyl group.

Using a slightly different calculation method but with the same parameters, Fabian and Mehlhorn[16-20] were able to relate a number of additional physical properties (half-wave potentials for polarographic reduction and some hundred spectra) with theoretical quantities. Thus the treatment of thiocarbonyl compounds as analogous to carbonyl compounds seems soundly based.

In Table 2 a more complete comparison is made between carboxylic and (di)thiocarboxylic esters[15, 21]. The calculated resonance energies are again calibrated against Walter's results[12]. Two conclusions arise which may seem surprising at first sight: (i) thiono esters are more stabilized by resonance than oxo esters, and (ii) the thioacyl sulphur bears a higher negative charge than the acyl oxygen if the alkoxy or the thiol part is the same.

The first point may seem unlikely in view of the generally assumed high reactivity of the thiono esters. However, recent work has established that

TABLE 2. Resonance energies (R.E.) and charges of esters and thio esters[15, 21]

$$RC\overset{\diagup X}{\diagdown YR}$$

Structure	R.E. (β unit)	R.E. (kcal/mole)	Charge on X	Charge on Y
RCOOR	0·596	13·5	−0·231	+0·136
$RC\overset{\diagup O}{\diagdown SR}$	0·243	5·5	−0·250	+0·088
$RC\overset{\diagup S}{\diagdown SR}$	0·766	17·4	−0·366	+0·171
RCSSR	0·380	8·6	−0·270	+0·149

thiono acids are hydrolysed slower than esters[22] at least in acid-catalysed hydrolysis.

Many observations on thione compounds can be explained by the theory given. Although outside the scope of this review, a few may be mentioned: the greater tendency for enolization of thio ketones[6] and dithio esters[23]; the strong nucleophilicity and basicity of sulphur in thio amides and thio ureas (exemplified by the tendency to form hydrogen bonds); and the dimerization of thiophosgene (similar to the trimerization of thio ketones) but not of the subsequent structures of Table 1 (but reappearing in trifluoromethyl trifluorodithioacetate)[24].

C. General Properties of Thio Acids and Dithio Acids

1. Physical properties

Three basic structural units have to be considered in this section: the thiolocarboxylic acids (1), the thionocarboxylic acids (2) and the dithio acids (3).

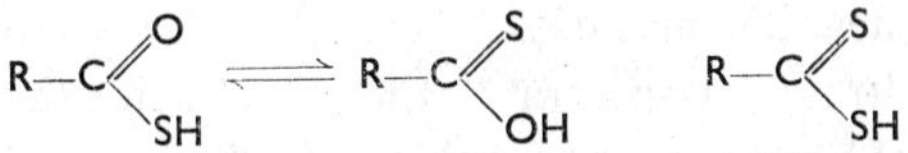

$$R-C\overset{\diagup O}{\diagdown SH} \rightleftharpoons R-C\overset{\diagup S}{\diagdown OH} \qquad R-C\overset{\diagup S}{\diagdown SH}$$

$$\quad (1) \qquad\qquad (2) \qquad\qquad (3)$$

Obviously 1 and 2 exist in a fast tautomeric equilibrium. Even at an early date, most authors agreed that the thiolo form (1) best described the properties of the acids and this opinion has been strengthened by advanced spectroscopical studies. Gordy[25] found only the thiolo acid by electron diffraction; infrared spectra show carbonyl and thiol frequencies and no hydroxyl bonds[26-28], and the ultraviolet spectrum resembles that of a thiolo ester and not that of a thiono ester[29]. Reports that favour a sizeable amount of the thiono acid tautomer from infrared[28] or ultraviolet[30] spectral data are not based on an unambiguously assigned band and rest only on one type of observation. The most definite bands for a thiono acid would be the OH stretching band in infrared spectra or the $n \to \pi^*$ absorption band in the near ultraviolet, expected around 370 mμ by analogy with thiono esters[31]. Neither of these bands has ever been noticed in the vapour phase or in a variety of solvents.

A theoretical analysis by adding bond energies[32] favours **1** by about 10 kcal/mole but this difference is offset by the larger resonance energy of **2** (Table 2). Since both quantities are highly uncertain in their precise values, a prediction on theoretical grounds cannot be made.

$$
\begin{array}{c}
\text{S} \\
/\!/ \\
\text{RC}\!:\ (-) \\
\backslash\!\backslash \\
\text{O}
\end{array}
$$

(4)

The salts derived from both thio acids contain the mesomeric anion **4**. Since **4** exhibits an i.r. band in the carbonyl region[33] it is generally assumed that most of the negative charge resides on the sulphur atom. This is in agreement with the postulate of the tendency of the carbon–sulphur bond to remain a single bond although the higher electronegativity of oxygen suggests that the other structure should dominate. Apparently the former effect prevails.

Ultraviolet studies on thio acids have been made by Hantzsch and Scharf[34], Koch[35], and Hirabayashi and Mazume[36]. They exhibit a strong band near 225 mμ and a shoulder ($n \to \pi^*$ band) around 270 mμ. The anion absorbs strongly near 250 mμ.

Whereas the ultraviolet spectra of the thiolo acids (and esters) are not very interesting, the red colour of dithio acids has invited spectrochemical investigations for many years. Dithioacetic acid has a low intensity band at 460 mμ[37] which shifts upon ionization to 450 mμ[7]. This band is assigned to an $n \to \pi^*$ absorption of the thione group on the basis of its intensity and its solvent effects. Due to more extensive conjugation the bands shift to longer wavelengths in dithiobenzoic acid[38] (538 mμ). As expected the spectra of the acids resemble those of the esters. At shorter wavelengths more intense bands due to $\pi \rightarrow \pi^*$ transitions exist.

2. Nucleophilic properties

Although individual classes will be treated in subsequent sections, a short generalized treatment of properties typical of all thio acids will be presented here, limited to two topics: nucleophilic character of the anions and radical reactions.

The comparison of the nucleophilic character of oxygen and sulphur compounds is a popular field of study but few quantitative studies have been made. Bunnett found that reaction (1) proceeded much faster with sulphur (C_6H_5SNa) than with oxygen (CH_3ONa) nucleophiles[39, 40] and concluded that a thiolate ion was about 1000 times stronger as a nucleophile than an alkoxide ion (the exact figure depending on the nature of X).

$$2,4\text{-}(NO_2)_2C_6H_3X + NaSC_6H_5 \longrightarrow 2,4\text{-}(NO_2)_2C_6H_3SC_6H_5 + NaX \qquad (1)$$
$$X = F, Cl, Br, I$$

This result, valid for hydroxylic solvents, may be profoundly influenced by a change in solvent. Kooyman[41] reviewed evidence that the free enthalpy of solvation in aqueous solvents is higher by about 8 kcal/mole for alkoxide ions compared with thiolate ions. If we would go so far as to assume that the solvation of the ions is of decisive importance for the solvent's effects upon reactions such as (1) then in less polar solvents the ratio of the rate constants of equation (1) (k_S/k_O) would diminish from 1000:1 to the extreme value of 1:100 in which case the alkoxide ion would be more reactive than the thiolate ion. No quantitative data are available in non-polar solvents.

In general, nucleophilicity increases with basicity but another factor is involved as well. This has been described in terms of polarizability[42, 43] or of oxidation potential[44-48]. The latter approach seems successful for a quantitative description and a number of parameters for sulphur nucleophiles have been computed[48]. The high oxidizability of thiolate anions will offset the lower basicity and make sulphur compounds strong nucleophiles. However, oxidation potentials are again determined in aqueous solvents and extension to non-polar systems has not been made.

The best-known description based on polarizability is Pearson's concept of hard and soft acids and bases[42]. It distinguishes two classes of bases: those which have a strong tendency to bind with protons are called hard and those which preferentially combine with heavy metal ions are called soft. Typical examples of hard bases are water or the fluoride anion, whereas hydrogen sulphide or the iodide ion are soft. The relation with polarizability is clear. Likewise, protons and boron trifluoride are hard acids and heavy metal ions or the heavier halogens (bromine, iodine) are soft acids. Evidence exists that a hard base has a strong tendency to bind with hard acids and vice versa.

Although Pearson's concept has not been treated in quantitative terms as yet, it is an extremely useful tool to rationalize the differences in reactivity of oxygen and sulphur compounds. An oxygen nucleophile, being in general a hard base, will react preferentially with comparatively hard electrophilic sites whereas sulphur nucleophiles show enhanced reactivity when the electrophilic site is more polarizable (or softer). Thus the relative nucleophilicity of oxygen and sulphur compounds depends on the type of substrate and a general statement as to which type of compounds is the better reagent is not possible.

It should be mentioned that the same conclusions follow from the Edwards–Davis treatment[44-48], since the Edwards equation contains reaction parameters which represent the response of a given substrate (in a given reaction) towards basicity and the oxidation potential of the reagent.

In the following sections we will frequently encounter evidence against the general statement that sulphur compounds are better nucleophiles than the corresponding oxygen derivatives. In many cases the apparent discrepancy is logically removed if the above-mentioned dual concept is applied.

The analysis in terms of oxidation potentials predicts differences in relative nucleophilicity depending on the extent of charge transfer in the transition state. Thus for substitutions at a saturated carbon atom, reactivity of the nucleophile is mainly governed by the oxidation (polarizability) term whereas for reactions at a carbonyl group the basicity is most important[45]. In terms of soft acids the saturated carbon centre is softer than the carbonyl group.

The relative activities of oxygen and sulphur nucleophiles are less ambiguous in the case of the anions of dithio acids. The difference in basicity between the dithiocarboxylate and carboxylate anions is much smaller than in the system thiolate/alkoxide. The complex formation with heavy metals (because of softness of the base) is enormous in particular with the dithio acids (or put alternatively, they become easily oxidized) so that dithiocarboxylate ions will in most cases be very strongly nucleophilic when compared to the carboxylates. However, one must be prepared for exceptions even here.

3. Radical reactions

Reactions of organosulphur radicals are well known with thiols and thio acids. Dithio acids have been studied less extensively but a few examples of radical-initiated additions exist. With alcohols or oxygen acids they are rare.

The importance of organosulphur radical reactions in synthetic chemistry is due to the high chain-transfer activity: instead of starting a polymerization chain they just add to double bonds.

$$RSH \longrightarrow RS\cdot + H\cdot$$
$$RS\cdot + CR_2{=}CR_2 \longrightarrow RSCR_2CR_2^{\cdot} \tag{2}$$
$$RSCR_2CR_2^{\cdot} + RSH \longrightarrow RSCR_2CR_2H + RS\cdot$$

Stacey and Harris recently reviewed this field[49]. Cunneen[50] gave the order of increasing reactivity of organosulphur radicals as

$$\text{i-AmSH} < \text{PhSH} < \text{HO}_2\text{CCH}_2\text{SH} < \text{CH}_3\text{COSH} < \text{ClCH}_2\text{COSH} < \text{Cl}_2\text{CHCOSH} < \text{Cl}_3\text{CCOSH} \tag{3}$$

Apparently, increasing acidity increases reactivity. From the fact that thiols isomerize olefins it follows that the additions are reversible[27, 28].

D. General Properties of Thiolo, Thiono and Dithio Esters

1. Physical properties

In the spectra of thiolo esters u.v. absorption occurs around 230 mμ[35, 53], with a shoulder at longer wavelengths; the i.r. spectra show absorption at expected frequencies, the carbonyl band being shifted to lower wavenumbers in comparison with esters. This shift has been interpreted[11] as involving resonance with the structure $^-S{=}C{-}O^+$ but evidence is overwhelming for the overall polarity $^+S{=}C{-}O^-$ so that probably the presence of the heavier sulphur atom is responsible for the displacement.

Spectra of the thiono and dithio esters have been studied extensively[7, 14, 31, 54], in particular the $n \rightarrow \pi^*$ bands at long wavelengths are characteristic of these compounds, although they can be (and have been) easily missed if spectra are taken at the dilutions usual for u.v. spectroscopy. Some absorption maxima are collected in Table 3.

The u.v. absorption of simple thiones can be described in good approximation by simple quantum–mechanical methods (c.f. section I. B). As expected, the $n \rightarrow \pi^*$ absorption shifts to lower wavelengths when electron-donating heteroatoms are attached to the thiono group but to higher wavelengths upon conjugation with a phenyl group.

TABLE 3. Absorption maxima of thiono esters and dithio esters (in cyclohexane).

Compound	λ_{max}	log ε	λ_{max}	log ε	λ_{max}	log ε	Reference
$CH_3CSOC_2H_5$	377	1·29	241	3·92	—	—	7, 31
$C_6H_5CSOCH_3$	418	2·04	287	2·02	—	—	17
$C_2H_5OCSOC_2H_5$	310ᵃ	—	228	3·55	—	—	7, 31
$CH_3CSSC_2H_5$	460	1·26	306	4·09	221	3·94	7, 31
$C_2H_5OCSSC_2H_5$	357	1·72	278	4·12	246	3·92	7, 31
$(CH_3)_2NCSSCH_3$	343	1·71	277	4·02	224	4·12	7
$CH_3SCSSCH_3$	429	1·45	303	4·21	238	3·55	7, 31
$C_6H_5CSSCH_3$	504	2·11	329ᵃ	3·8	—	—	17

ᵃShoulder.

Infrared studies have centred around the assignment of the $C{=}S$ stretching frequency; apparently investigators tacitly assumed that it should be strong and characteristic by analogy with carbonyl bands. The free $C{=}S$ group is supposed to have its band around $1100\ cm^{-1}$ [55] which unfortunately is the region where many single bonds have their stretching frequencies. In a number of thiono compounds strong bands are found in the region between 1000 and 1200 cm^{-1} which have been assigned to $C{=}S$ vibrations[56-60], but the shifts with substitution by heteroatoms cannot always be correlated with the expected effects on force constants. It becomes increasingly clear that the carbon–sulphur stretching frequency seldom exists as such in the majority of compounds and that combined vibrations take its place.

2. Chemical properties

Compared to the extensive studies on carboxylic esters the number of investigations on thiolo esters is extremely small. It is no wonder that up till now only vague ideas have existed about the details of thiolo ester reactions. Although thiolo esters are less stable thermodynamically than oxygen esters, they are never hydrolysed appreciably faster than carboxylic esters in alkaline media (and even slower by acids) (cf. section IV). Thiono esters give the same result[22] (section V). The rate of hydrolysis of dithio esters has not been investigated quantitatively but dithiomalonic ester derivatives have been shown to hydrolyse preferentially at the carbonyl group[23, 60]:

$$C_6H_5CH{=}C\underset{CO_2C_2H_5}{\overset{CS_2C_2H_5}{\big\langle}} \xrightarrow{KOH} C_6H_5CH{=}CHCS_2C_2H_5 + CO_2 + C_2H_5OH \qquad (4)$$
$$(5)$$

Product **5** is a red oil which decomposes on attempted distillation. Ethanethiol was only found in traces.

Apparently the dithio ester group is very stable even against alkaline hydrolysis.

II. THIO ACIDS

A. Preparation

The normal way of preparing thio acids consists of the action of hydrogen sulphide on derivatives of carboxylic acids such as acid halides, anhydrides, diacyl sulphides and esters. Most reactions are equilibrium reactions (equations 5 and 6). High temperatures tend to drive off hydrogen sulphide so that working under pressure may be necessary.

$$RCOOR + H_2S \rightleftharpoons RCOSH + ROH \qquad (5)$$
$$RCOOCOR + H_2S \rightleftharpoons RCOSH + RCOOH \qquad (6)$$

Fortunately many reactions run smoothly under mild conditions[2] either uncatalysed (*e.g.* with acid halides) or catalysed by bases (*e.g.* with anhydrides).

A variety of solvents, such as pyridine, methylene chloride *etc.* are used. Acid catalysis has been applied for the preparation of aliphatic thio acids by the action of hydrogen sulphide on anhydrides under pressure[61–63]. Judging from the reaction conditions, it seems that hydrogen sulphide reacts faster than water in these reactions.

An interesting case is the reaction between esters and NaSH. Hirabayashi and coworkers[64], extending older findings[65], found that NaSH in absolute ethanol sulphydrolyses aromatic esters of aliphatic carboxylic acids to thio acids in 70% yield but with aliphatic esters 50–70% of the carboxylic acid and ethyl esters were formed by hydrolysis and ethanolysis.

$$C_{17}H_{35}COOCH_3 + NaSH \xrightarrow{C_2H_5OH} C_{17}H_{35}COOH + C_{17}H_{35}COOC_2H_5 \qquad (7)$$

$$C_{17}H_{35}COOC_6H_5 + NaSH \xrightarrow{C_2H_5OH} C_{17}H_{35}COSH + (C_{17}H_{35}COS)_2 \ (\sim 70\%)$$
$$+ C_{17}H_{35}COOH + C_{17}H_{35}COOC_2H_5 \ (30\%) \qquad (8)$$

No appreciable amounts of thiols have been found in reactions (7) and (8) so that sulphydrolysis with attack on the alcoholic part of the ester is not involved. The authors consider the relative basicities of the leaving groups ($C_6H_5O^-$, CH_3O^- and SH^-) to be the main reason for the course of the reaction.

Under different conditions, in particular in DMF as solvent thiolate ions and NaHS attack esters with preferential attack at the alkyl carbon atom[66].

A similar selectivity is found[1, 2] when acid chlorides react with Na_2S and NaHS in ethanol. Aromatic acyl chlorides yield thio acids smoothly but aliphatic acyl chlorides are solvolysed faster by ethanol (however, several aliphatic thio acids although contaminated by acids have been prepared by this method under careful control of reaction conditions[67]).

Judging from the results with NaHS and Na_2S it appears that a fine balance exists between the reactivities of the sulphur and oxygen nucleophiles (probably both negatively charged) towards esters and acyl halides. Possibly, the aromatic substrates, being 'softer', are more reactive towards the sulphide (*cf.* section I. C. 2). A careful analysis which takes the ionization equilibria into account could throw more light on this fundamental problem. At the same time kinetic and thermodynamic factors should be distinguished carefully.

Other methods for the preparation of thio acids are of much less importance. The oldest is reported by Kekulé[68] and consists of the sulphurization of carboxylic acids by P_2S_5. The reaction is only of historical interest for the preparation of thio acids although it is still used for obtaining many thiono compounds.

A method with general applicability is the reaction of Grignard reagents with COS (equation 9)[69]; thioacetic acid is obtained by the action of hydrogen sulphide on ketene[70] (equation 10).

$$RMgCl + COS \longrightarrow RCOSMgCl \xrightarrow{H_2O} RCOSH \tag{9}$$

$$CH_2{=}C{=}O + H_2S \longrightarrow CH_3COSH \tag{10}$$

Thio acids cannot be prepared by hydrolysis of thiolo esters; the only products are a carboxylic acid and a thiol but sulphydrolysis of thiolo esters leads of course to thio acids. The reaction is a variant of equation (6), and is sometimes useful for the preparation of unstable acids *in situ*. Wieland and coworkers[71, 72] prepared aminothio acids by this method.

Thiono derivatives may be expected to give thio acids on hydrolysis (equation 11). However, thiono compounds are difficult to make (and

$$RC\overset{\displaystyle S}{\underset{\displaystyle Z}{\diagdown}} + H_2O \longrightarrow RCOSH + HZ \tag{11}$$

unpleasant to handle) with the exception of thio amides. When conditions are well chosen, thio acids may be obtained, but structural limitations exist. Seydel[73] reports the successful preparation of 2-ethylisothionicotinic acid by alkaline hydrolysis of the thio amide. Acid hydrolysis yielded the amide (equation 12). Several aromatic thio amides failed to give the thio

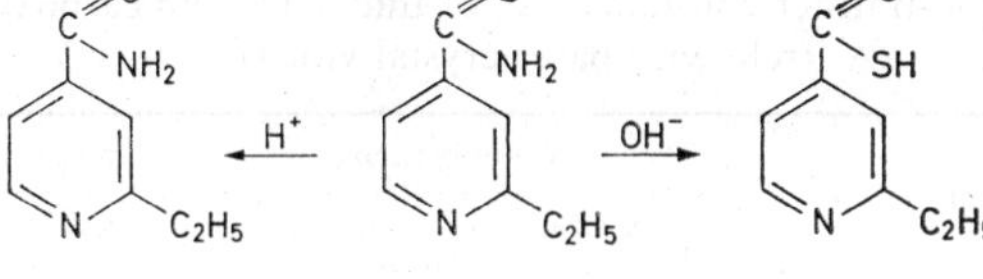

$$\tag{12}$$

acid.

B. Acidity

The acid dissociation constants in water of only two thio acids have been measured[74-77]. The results are compared with the corresponding carboxylic acids[74, 78, 79] in Table 4. As expected thio acids are the stronger acids, although the difference is much smaller ($\sim$ 1·5 units) than that between thiols and alcohols ($\sim$ 6 units). The smaller difference is largely caused by

TABLE 4. Acid dissociation constants of thio acids and carboxylic acids
(in water at 25°).

Compound	pK_a (X = S)	Reference	pK_a (X = O)	Reference
CH_3COXH	3·33, 3·41	74–77	4·76	74
C_6H_5COXH	2·48	77	4·20	78, 79

the decreased resonance stabilization of the RCOS⁻ anion, but may partly
be due to differences in solvation.

Ioffe, Sheinker and Kabachnik[80] studied the acidity of thio acids and
carboxylic acids in non-polar media by determining the proton transfer
between the acids and an indicator (crystal violet) in benzene and chlorobenzene. Protolysis constants (K_p) refer to the formation of ion pairs which
are not dissociated appreciably. The K_p values show the expected trend
upon substitution in the acids but the comparison between carboxylic
acids and thio acids gives no straightforward results. On the whole, thio
acids seem to be weaker in benzene and stronger in chlorobenzene, but
in both cases the apolar solvent makes the relative acid strength of thio
acids (*i.e.* compared with carboxylic acids) smaller. This is the opposite of
what would be expected: the lesser solvation of thiocarboxylate ions would
tend to make the thio acids relatively weak in water and the dimerization
of carboxylic acids in apolar solvents would work in the same direction,
strengthening thio acids relatively in apolar solvents. It is doubtful, therefore, whether the proton-transfer results can be interpreted in terms of
acid strength alone. The relevant data are given in Table 5.

TABLE 5. Proton-transfer constants pK_p of thio acids and carboxylic acids[80]
(reference base: crystal violet).

Compound	pK_p (in benzene)		pK_p (in chlorobenzene)	
	X = S	X = O	X = S	X = O
CH_3COXH	−0·60	−1·24	−2·89	−1·96
C_6H_5COXH	−0·86	−1·79	−3·23	−2·51
p-BrC_6H_4COXH	−1·24	−2·50	−3·62	−3·13
p-$CF_3C_6H_4COXH$	−1·36	−3·22	−3·84	−3·85

Hydrogen-bonded dimers of thio acids in non-polar solutions do occur,
although dimer formation is less than with carboxylic acids. A recent
infrared study by Ginzburg and Loginova[81] shows both hydrogen-bonded

and free SH and CO groups when thioacetic acid and thiobenzoic acid are dissolved in carbon tetrachloride. With thioacetic acid, the dimer bands disappear below a concentration of 1·4 mole/l, whereas 10^{-4} M solutions of acetic acid in carbon tetrachloride still contain an appreciable amount of dimer[82]. This large difference could in part be due to the smaller negative charge on the acyl oxygen in the case of thio acids (Table 2), but this cannot be the sole cause, since the hydrogen-bonding ability of thioacids is small[83] towards ethers also.

It must be concluded that the SH group is less able to form hydrogen bridges with a reference base (at least towards oxygen bases) than the OH group, although it is a stronger acid.

C. Substitution Reactions

The thiocarbonyl group, as can be expected, is liable to attack by nucleophilic agents. With water, thio acids are slowly hydrolysed to carboxylic acids. This reaction is the basis for the well-known use of thioacetic acid (which is more convenient than working with hydrogen sulphide) in the precipitation of metals as sulphides. The still more popular reagent thioacetamide acts by being hydrolysed to thioacetic acid in water with subsequent liberation of hydrogen sulphide[84].

Other metals are precipitated as their thioacetates which are often unstable and decompose to metal sulphides. Since metal ions catalyse the decomposition of thio acids it is hard to obtain pure thio acetates of most heavy metals.

The hydrolysis of thio acids can be either spontaneous or catalysed by acids or bases. The spontaneous hydrolysis of thioacetic acid proceeds about 20 times faster than oxygen exchange in acetic acid[77], but the acid-catalysed reaction goes slower with the thio acid[77]. The base-catalysed reaction of thioacetic acid is faster again[84] (see section IV. B). The mechanistic aspects of this and similar reactions will be covered in section IV.

Alcohols and amines drive off hydrogen sulphide likewise leading to esters and amides; the reactions go faster than with carboxylic acids[85]. In particular with amines, reaction is so rapid that thio acids resemble acid halides in this respect. Intramolecular substitution was found to be so fast for amino thio acids $H_3N^+(CH_2)_nCOS^-$, when $n = 3$ or 4[86], that sulphydrolysis of the phenyl esters with triethyl amine and hydrogen sulphide gave no acid at all, pyrrolidone and piperidone being the sole products. With longer or shorter chains sulphydrolysis yielded the aminothio acid in good yield.

Thiols react in the same way to give thiolo esters[87] (equation 13) but

$$R'COSH + R''SH \rightleftharpoons R'COSR'' + H_2S \qquad (13)$$

the esters are more easily obtained from the reaction of thio acids with alkyl halides. Dithiosuccinic acid is reported to lose hydrogen sulphide spontaneously[88] (equation 14) to give 2,5-dioxotetrahydrothiophene.

$$HSOCCH_2CH_2COSH \longrightarrow \begin{array}{c} CH_2-CO \\ | \qquad\qquad S + H_2S \\ CH_2-CO \end{array} \qquad (14)$$

A different type of substitution reaction consists of the reaction of thio acids with a great variety of substrates that are liable to nucleophilic substitutions. If the reaction centre in the substrate is a carbon atom the product is a thiolo ester (see section III. A. 1). An interesting reaction is illustrated in equation (15) where the product is probably formed via a double substitution reaction[89].

$$(C_6H_5)_2CCl_2 + 2\,HSOCCH_3 \longrightarrow (C_6H_5)_2C(SOCCH_3)_2 \longrightarrow (C_6H_5)_2CS + (CH_3CO)_2S \qquad (15)$$

Substitution may also occur on acid chlorides[90] (equation 16), esters or acids. The latter reactions are equilibria. They may be of synthetic use when the reaction can be forced into the desired direction by removing the product (equation 17)[91].

$$RCOSH + R'COCl \longrightarrow RCOSCOR' + HCl \qquad (16)$$

$$\underset{\text{solvent}}{C_6H_5COSH} + CH_3COOH \rightleftharpoons C_6H_5COOH + \underset{\text{distilled off}}{CH_3COSH} \qquad (17)$$

D. Addition Reactions

Addition of thio acids to unsaturated hydrocarbons results in good yields of thiolo esters (equation 18). The reaction may be used for the synthesis of

$$R'COSH + R''CH=CH_2 \longrightarrow R'COSCH_2CH_2R'' \qquad (18)$$

thiols, polythiols or unsaturated thiols which are in general more easily prepared by thio acid addition and subsequent hydrolysis than through direct addition of hydrogen sulphide, which gives unwanted side-reactions. An interesting example is the preparation of 2,3-dithiosuccinic[93] acid:

$$HOOCC{\equiv}CCOOH \longrightarrow HOOCCH(SAc)CH(SAc)COOH \longrightarrow$$
$$\longrightarrow HOOCCH(SH)CH(SH)COOH \qquad (19)$$

With carbonyl compounds dithioacylals are obtained easily[94], with nitriles thioamides[95] are formed:

$$C_6H_5CHO + 2\,CH_3COSH \longrightarrow C_6H_5CH(SOCCH_3)_2 \qquad (20)$$

$$C_6H_5CN + CH_3COSH \xrightarrow{H^+} C_6H_5C{\begin{array}{c} NH \\ \diagdown \\ SCOCH_3 \end{array}} \longrightarrow C_6H_5CSNH_2 + CH_3COOH \qquad (21)$$

Reaction with a ketene gives a diacylsulphide (equation 22)[70]. A few more additions to cumulated systems have been described and others should probably succeed. The method can be used to distinguish between thiocyanates and isothiocyanates[97, 98].

$$CH_2=C=O + CH_3COSH \longrightarrow CH_3COSCOCH_3 \qquad (22)$$

$$RSCN + CH_3COSH \longrightarrow RSC(NH)SCOCH_3 \longrightarrow RS_2CNHCOR$$
$$[\longrightarrow CH_3COSR + HCNS] \qquad (23)$$

$$RNCS + CH_3COSH \longrightarrow RNHCS_2COCH_3 \longrightarrow CH_3CONHR + CS_2 \qquad (24)$$

Because of the strong nucleophilic action and easy radical formation of thio acids two mechanisms are possible. Nucleophilic addition may be expected with activated (*i.e.* negatively substituted) double or triple bonds by analogy with additions of thiols which occur both with basic catalysts (if activated olefins are used) and under influence of radical initiators[99], heat or light. Since both give anti-Markovnikov orientation, the products give no clue to the mechanism. However, reaction conditions and stereochemistry do. It is certain that both kinds of reaction occur, but the interesting borderline cases have only been sparingly studied so that many mechanistic pronouncements rest on unsure ground. Electrophilic addition of thio acids has not been recorded.

With non-activated olefins, reactions are strongly affected by light, heat or radical initiators (or inhibitors) leaving no doubt as to the free-radical course. The reactions lead to terminal attachment of the thiocarboxylic group irrespective of the polarity of the substituents.

$$(CH_3)_2C=CH_2 + RCOSH \longrightarrow (CH_3)_2CHCH_2SCOR \qquad (25)$$

$$NCCH=CH_2 + RCOSH \longrightarrow NCCH_2CH_2SCOR \qquad (26)$$

Addition to triple bonds (equation 27) proceeds analogously and radi-

$$CH_3C\equiv CH + RCOSH \longrightarrow CH_3CH=CHSCOR \qquad (27)$$

cal initiators are active again. However, a parallel and perhaps important nucleophilic addition has never been disproved and stereochemical arguments may favour an ionic course (see below). An activated acetylene has been shown by Alkema and Arens to be liable to both nucleophilic and radical addition of ethanethiol[100] (equations 28 and 29). Nucleophilic reactions of thiols with acetylenes have been reviewed by Truce[101]. The addition of thiolacetic acid to ethoxyacetylene gives 1-ethoxy-1-(acetylthio)-ethene[102] even without added base.

$$HC \equiv COC_2H_5 + C_2H_5SH \xrightarrow{\text{base}} H_2C = C \Big\langle \begin{matrix} OC_2H_5 \\ SC_2H_5 \end{matrix} \tag{28}$$

$$HC \equiv COC_2H_5 + C_2H_5SH \xrightarrow{\text{no catalyst}} C_2H_5SCH = CHOC_2H_5 \tag{29}$$

Adding two equivalents of thio acid to a terminal acetylene yields 1,2-bis-acylthioalkanes[103, 104] together with the monoaddition product; with disubstituted acetylenes the reaction stops with the mono-adduct unless activating groups are present[105]. Thus dimethylacetylene dicarboxylate yields the bis-adduct[93] (cf. equation 19).

Additions to highly electrophilic unsaturated systems such as ketenes[96] (equation 22), isocyanates etc. are certainly of ionic nature.

Radical reactions of thiols and thio acids have been found to be only partially (or not at all) stereospecific, thio acids being even less stereospecific than thiols[49, 106, 107]. Thus both cis- and trans-2-chloro-2-butene give the same mixture of threo- (90%) and erythro-2-(acetylthio)-3-chlorobutane (10%). The reaction with thioacetic acid[53] and 2-chlorocyclohexene gives both cis- and trans-2-chlorocyclohexylthiolacetate; since the cis form amounts to 70%, some preference for trans addition is shown[1, 53, 106]. 80% trans addition was also found for 4-t-butyl-1-methyl-cyclohexene[107]. The issue is further complicated by the reversibility of the addition: Neureiter and Bordwell[52] found isomerization of cis-2-chloro-2-butene and cis-4-methyl-2-pentene when the olefins were irradiated in the presence of catalytic amounts of thio acids.

Reactions of thiols with acetylenes have been found to proceed mostly via trans addition[49] especially at low temperatures. Truce[101] has pointed out that with nucleophiles trans addition occurs, whereas the stereoselectivity of radical addition is much lower. A recent review by Prilezhayeva[108] confirms this suggestion. The temperature dependence might be due to the decreasing importance of free-radical reactions in the cold. Adducts are reported to isomerize on standing[100] but careful study could clarify the presence or absence of competing mechanisms. The addition of thioacetic acid to methyl propiolate[93] gave both cis- and trans-methyl β-acetylthioacrylate, but no detailed studies have been reported.

Thiolacetic acid was found to give the diadduct with allene by a terminal attack under influence of u.v. light at room temperature[109].

Selective addition of thio acids to the allyl groups of diallyl maleate and fumarate has been found[110] (equation 30a). The reaction proceeds without isomerization of the central double bond so that even a reversible addition to the acid part is excluded. This reaction illustrates the electrophilic character of the radical. With basic catalysts, Michael reaction takes place at the central double bond (together with isomerization to diallyl fumarate, see equation 30b).

$$2 \text{ CH}_3\text{COSH} + \begin{array}{c} \text{HC} \overset{\textstyle \diagup}{} \text{COOCH}_2\text{CH} = \text{CH}_2 \\ \| \\ \text{HC} \overset{\textstyle \diagdown}{} \text{COOCH}_2\text{CH} = \text{CH}_2 \end{array} \longrightarrow \begin{array}{c} \text{HC} \overset{\textstyle \diagup}{} \text{COO(CH}_2)_3\text{SCOCH}_3 \\ \| \\ \text{HC} \overset{\textstyle \diagdown}{} \text{COO(CH}_2)_3\text{SCOCH}_3 \end{array} \qquad (30a)$$

$$\text{CH}_3\text{COSH} + \begin{array}{c} \text{HC} \overset{\textstyle \diagup}{} \text{COOCH}_2\text{CH} = \text{CH}_2 \\ \| \\ \text{HC} \overset{\textstyle \diagdown}{} \text{COOCH}_2\text{CH} = \text{CH}_2 \end{array} \xrightarrow{\text{OH}^-} \begin{array}{c} \text{CH}_3\text{COS} - \text{CH} \overset{\textstyle \diagup}{} \text{COOCH}_2\text{CH} = \text{CH}_2 \\ | \\ \text{H}_2\text{C} \overset{\textstyle \diagdown}{} \text{COOCH}_2\text{CH} = \text{CH}_2 \end{array} \qquad (30b)$$

E. Miscellaneous

Thio acids are easily oxidized to disulphides (equation 31) and the aromatic acids especially must be protected against oxygen.

$$2 \text{ RCOSH} \longrightarrow (\text{RCOS})_2 \qquad (31)$$

Many oxidizing agents do the same, such as iodine, hydrogen peroxide, metal ions *etc.* Thio acids can be titrated with iodine. When oxidized strongly the sulphur is removed altogether and carboxylic acids result. This reaction presumably proceeds via unstable acylsulphonyl intermediates. Reduction with hydrogen and Raney nickel gives the alcohol or aldehyde, but conditions for selectivity have not been studied in detail. With lithium aluminium hydride a thiol is obtained.

Mikolajczyk[112] reports the desulphurization of thioacetic acid when it is kept with an equivalent amount of dimethylsulphoxide at 20° for one week.

$$\text{CH}_3\text{COSH} + (\text{CH}_3)_2\text{SO} \xrightarrow{70\%} \text{CH}_3\text{COOH} + (\text{CH}_3)_2\text{S} + \text{S} \qquad (32)$$

Condensation reactions of thioacids are not well known although Claisen condensations of the esters were studied in some detail (section III. B. 2). Behringer and Grimm[113] describe the formation of a trithiapentalene (6) in 9% yield from acetylacetone and thioacetic acid under the influence of sodium acetate. Sulphur is found in the reaction mixture.

$$\begin{array}{c} \text{O} \quad \text{O} \\ \| \quad \| \\ \text{CH}_3\text{CCH}_2\text{CCH}_3 \end{array} + \text{CH}_3\text{COSH} \xrightarrow[100°]{\text{NaOAc}} \begin{array}{c} \text{S} - \text{S} - \text{S} \\ \diagup \quad \| \quad \diagdown \\ \text{H}_3\text{C} \qquad \qquad \text{CH}_3 \end{array} \qquad (33)$$

$$(6)$$

A condensation in which the methyl group of thioacetic acid is involved is also described by Behringer (equation 34). At 20° 7 is the final product,

46*

arising from a normal (nucleophilic) addition of thioacetic acid to the triple bond.

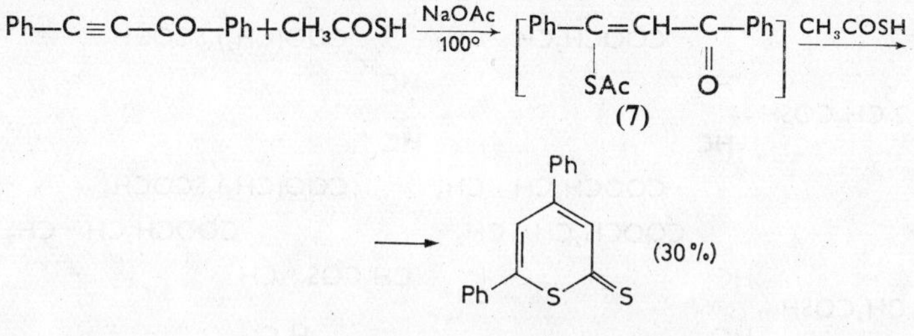

$$Ph-C\equiv C-CO-Ph+CH_3COSH \xrightarrow[100°]{NaOAc} \left[\begin{array}{c} Ph-C=CH-C-Ph \\ | \quad\quad\; || \\ SAc \quad\;\; O \end{array} \right] \xrightarrow{CH_3COSH}$$

(7)

(30%)

(34)

(8)

At higher temperature **7** reacts with thioacetic acid with formation of **8**, so that reaction (34) may be assumed to proceed via formation of **7**. The interesting feature of the ring closure is that the sulphur atom (which finally becomes the thione sulphur) is not eliminated instead of the oxygen.

That the carbonyl group of thio acids has some ketocarbonyl character (due to low resonance in the COS group) can been seen by the formation of dithioacylals[114]. However, the reaction does not proceed via intermediate benzaldehyde when thiobenzoic acid is used in equation (35).

$$3\; RCOSH + base \longrightarrow RCH(SCOR)_2 + H_2O + S$$

(35)

The easy formation of orthothio esters of formic acid is another example (equation 36)[115].

$$HCOOH + RSH \longrightarrow [HCOSR] \longrightarrow HC(SR)_3$$

(36)

III. THIOLO ESTERS

A. Preparation

1. Substitution reactions

Reaction of thiols with a number of carboxylic acid derivatives yields thiolo esters smoothly (equation 37). Acid chlorides, anhydrides, esters and even the acids themselves can be used (but not amides[2]).

These reactions, reach an equilibrium, and therefore yields are often

$$RSH + R'C\begin{array}{c} O \\ \diagdown \\ Z \end{array} \rightleftharpoons R'C\begin{array}{c} O \\ \diagdown \\ SR \end{array} + HZ$$

(37)

dependent on the possibility of shifting the equilibrium in the desired direction. In the case of amides the equilibrium lies so much towards the left that preparation from these compounds is not practical. With carboxylic acids the equilibrium is also rather unfavourable.

Reactions with acid halides can often be carried out without a (basic) catalyst, but higher yields result with the use of tertiary amines or pyridine. With anhydrides a basic catalyst is generally used. Details for preparing unsaturated esters have been described[6].

Instead of thiols several metal thiolates can be used such as RSNa, RSMgI[117], (RS)$_2$Pb[118], t-BuHgSR[119], and Me$_3$SiSR[120]. An example is given in equation (38); the COOR′ group is probably not essential.

$$RCOCl + (CH_3)_3SiSCH_2COOR' \longrightarrow RCOSCH_2COOR' + (CH_3)_3SiCl \qquad (38)$$

From this example it is clear that the thiol need not be present as an ion (the Si—S bond being highly covalent) so that acylation may be expected to take place also on sulphur compounds other than thiols. Indeed acyl halides and anhydrides yield β-substituted thiolo esters with ethylene sulphides[121, 122] (equation 39) and the reaction of acetyl bromide with thio ethers is reported to give thiolo esters[123]. Gustus and Stevens[124] report that acetyl iodide reacts in the same way (equation 40) with thio ethers; the same reaction with ethers was found to be possible and even faster.

$$CH_3COCl + CH_2\!\!-\!\!CH_2 \longrightarrow CH_3COSCH_2CH_2Cl \qquad (39)$$
$$\underset{S}{\diagdown\diagup}$$

$$CH_3COI + C_2H_5SC_2H_5 \longrightarrow CH_3COSC_2H_5 + C_2H_5I \qquad (40)$$

A substitution similar to (39) and (40) leads to thiolo esters by the reaction between acetyl chloride and a dithioacetal[125].

$$CH_2(SCH_3)_2 + CH_3COCl \longrightarrow CH_3COSCH_3 + ClCH_2SCH_3 \qquad (41)$$

With the unsymmetrical substrate ethoxymethyl ethyl sulphide the sulphur atom and not the oxygen atom is attacked by acetyl chloride:

$$C_2H_5OCH_2SC_2H_5 + CH_3COCl \longrightarrow C_2H_5OCH_2Cl + CH_3COSC_2H_5 \text{ (70\%)} \qquad (42)$$

It seems that here the sulphur atom is the better nucleophile; on the other hand the observations of Gustus and Stevens[124] lead to the reverse conclusion. Systematic investigations on the relative nucleophilicity of ethers and thio ethers towards acyl halides have not been made.

The acetylenic esters, until recently unknown, have been prepared by the action of alkynyllithium, sulphur and an acyl bromide. The reaction proceeds via a lithium alkynylthiolate[126].

$$RC\!\!\equiv\!\!CLi + S_8 \longrightarrow RC\!\!\equiv\!\!CSLi + R'COBr \longrightarrow R'COSC\!\!\equiv\!\!CR \qquad (43)$$

The second general method for synthesizing thiolo esters is of a type which has only rare examples in the carboxylic field. It consists of the nucleophilic attack of thiocarboxylate anions upon a great variety of substrates[1, 2].

Under alkaline conditions thio acids react with alkyl halides (equation 44), with ethylene oxides (giving β-hydroxy thiolo esters), ethylene sulphides, ethylene imines, alkyl tosylates and even with activated alcohols[127] (equation 45). The latter reaction can be considered as a Mannich-type

$$RCOS^- + R'Cl \longrightarrow RCOSR' + Cl^- \qquad (44)$$

$$RCOSH + R'_2NCH_2OH \longrightarrow CH_3COSCH_2NR'_2 \qquad (45)$$

reaction since it takes place when sodium thiocarboxylate is added to a mixture of a secondary amine and formaldehyde. Non-activated alcohols cannot be used, since they alcoholyse the thiocarboxylate group (section II. C). In the carboxylic field only the tosylate group is a good enough leaving group to be substituted. These reactions leave no doubt about the powerful nucleophilic action of the thiocarboxylate group. Interesting studies could be made of nucleophilicities and leaving-group tendencies with thio acids as model compounds, especially in borderline cases as equation (45).

2. Addition reactions

Since Holmberg's studies[92] the addition of thio acids to unsaturated systems has shown increasing promise as a synthetic route to thiolo esters.

As was described in section II. D, additions to ethylenes and acetylenes are mostly carried out under conditions favourable for radical additions and the orientation is in an anti-Markovnikov sense. For preparative details the reader should consult the literature[1, 2, 92, 102–110].

Addition of a thiol to ketene yields a thiolo ester[128] (equation 46):

$$H_2C{=}C{=}O + RSH \longrightarrow CH_3COSR \qquad (46)$$

When a thiol is added to a nitrile, an imidothiolo ester is formed which after hydrolysis gives a thiolo ester[129]. With hydrocyanic acid a thioloformate is obtained[130]. The addition is acid-catalysed.

$$RSH + R'CN \xrightarrow{H^+} R'C{\bigg\langle}{\stackrel{\displaystyle NH}{\displaystyle SR}} \xrightarrow{H_2O} R'COSR + NH_3 \qquad (47)$$

(R' = H, alkyl, aryl)

B. Properties

1. Thiolo esters as acylating agents

By far the most important property of thiolo esters is their tendency to act as acylating agents. In this respect thiolo esters resemble acyl halides, in accordance with the low degree of resonance in the —COS— group.

Since the discovery of the role of acetyl coenzyme A in biological acylation, the reactions of model substances have been extensively studied, in particular in their reactions with alcohols and amines.

In general acyl groups tend to migrate from sulphur to oxygen[131] or nitrogen[132, 133] as is borne out by reactions (48) and (49). Although these reactions go to completion, equilibria are in fact involved, the constants of

$$HOCH_2CH_2SCOCH_3 \longrightarrow CH_3COOCH_2CH_2SH \qquad (48)$$

$$CH_3COSC_2H_5 + H_2NC_6H_5 \longrightarrow CH_3CONHC_6H_5 + C_2H_5SH \qquad (49)$$

which have been measured in a number of cases. Thus the alcoholysis (50) is estimated to have an equilibrium constant of 134 which is approached

$$C_6H_5COSC_2H_5 + C_2H_5OH \rightleftharpoons C_6H_5COOC_2H_6 + C_2H_5SH \qquad (50)$$

experimentally from both sides[134]. Unfortunately the temperature at which the equilibrium is reached is uncertain since the reaction mixtures were heated to over 200°.

The equilibrium constant for reaction (51) was measured by Jencks and coworkers[135] in water at 39° and found to be 56.

$$HO(CH_2)_3SCOCH_3 \rightleftharpoons CH_3COO(CH_2)_3SH \qquad (51)$$

Transacylation to water (*i.e.* hydrolysis) has been studied by Reid who measured the equilibrium compositions of equimolar quantities of benzoic acid and alkylthiols and also of alkylthiobenzoate and water. The equilibrium mixture contained about 15% of the thiolo ester in both cases[136] (for alkyl = ethyl, and n-propyl) which lead to equilibrium con-

$$R'COSR'' + H_2O \rightleftharpoons R'COOH + R''SH \qquad (52)$$

stants for reaction (52) of around 30 (for carboxylic esters the same constant is 0·25). For the methyl esters K was found to be higher (~ 80)[136]. The temperature of equilibration was around 200°.

Esterifications have also been studied by Chablay[137] at 175°. The amount of thiolo ester formed ranged from 12% for primary thiols and 8% for secondary thiols to 3–4% for *t*-butylthiol.

The transacylation constants (or the reciprocal of the esterification constants) for the reactions of both alkyl thioloacetates and alkyl thiolobenzoates are: K (primary thiol) $\cong 60$, $K_{sec} \cong 150$, $K_{tert} \cong 10^4$ [137]. The same constants for alkyl acetates or benzoates are: $K_{prim} \approx K_{sec} \cong 0·5$, $K_{tert} \cong 200$[138]. The ratio of the constants for hydrolysis of thiolo esters and carboxylic esters is consistent with the equilibrium constant for alcoholysis at high temperatures.

The equilibrium for the aminolysis of thiolo esters lies much further to the side of the acetylamino derivatives[139] but in acid solutions the reaction is driven backwards because of protonation of the free amino group. In

$$H_2NCH_2CH_2SCOCH_3 \rightleftharpoons CH_3CONHCH_2CH_2SH \quad (53b: S = O) \quad (53a)$$

equilibrium (53a) a kinetic study by Martin and Parcell[140] led to an equilibrium constant of 5×10^7 (assuming pK_a of protonated β-aminoethyl thioacetate to be $9\cdot1$).

Unfortunately a similar analysis[141] of the O → N transacylation in a system (53b) (equal to (53a) but with oxygen instead of sulphur) gave an equilibrium constant of $1\cdot5\times10^3$ whereas from reactions (51) and (53a) $K_{O\rightarrow N}$ should be around 10^6. This discrepancy is not yet resolved satisfactorily, although in a later publication Martin[142] recommends abandoning his value for the O → N constant because this system according to Hansen[143] contains an irreversible step.

Hansen's criticism may have far-reaching consequences for the detailed analyses of the systems (53a) and (53b), (see section IV. D).

Apart from this unsolved question, evidence is overwhelming that compounds are successively more stable in the sequence S-acyl, O-acyl, N-acyl. This order is in agreement with the increasing resonance energy in the systems. If it is assumed that differences in solvation energies are of secondary importance and mean values for the equilibrium constants are taken as $K_{S\rightarrow O} \approx 60$ and $K_{S\rightarrow N} \approx 5\times10^7$, differences in resonance energies between the thiocarboxyl group and the carboxyl group of ca. $2\cdot5$ kcal/mole and between the thiocarboxyl and the amide group of ca. 11 kcal/mole can be estimated.

The kinetics of the transacylation will be covered in section IV.

2. Other reactions

Thiolo esters bearing α-hydrogen atoms may act as the anionic reagent in Claisen and related condensations (equation 54). Baker and Reid[144] used sodium but yields were much better (up to 70%) when isopropylmagnesium bromide was employed[145]. When ethyl acetate is subjected to the latter reaction conditions the normal Grignard reaction takes place with formation of the tertiary alcohol. Sterically hindered esters (t-butyl acetate) undergo Claisen condensation with isopropylmagnesium bromide[146] in 42% yield. With t-butyl thioacetate Claisen condensation takes place more slowly, which was attributed by Cronyn, Chang and Wall[145] to the greater stability of the organometallic compound 9 due to the greater acidity of the thiolo ester. This is also borne out by the acidities of aceto-

$$CH_3COSC_2H_5 + {}^-CH_2COSC_2H_5 \longrightarrow CH_3COCH_2COSC_2H_5 + C_2H_5S^- \quad (54)$$

$$i\text{-}PrMgBr + CH_3COSC(CH_3)_3 \longrightarrow BrMgCH_2COSC(CH_3)_3 + C_3H_3 \quad (55)$$

$$(9)$$

thioacetates compared with acetoacetates. The available data[147] are:

$$CH_3COCH_2CO_2CH_3 \quad pK_a = 10\cdot5$$
$$CH_3COCH_2COSCH_2CH_2NHCOCH_3 \quad pK_a = 8\cdot5 \qquad (56)$$

The higher acidity of α-hydrogen atoms in thiolo esters finds a logical explanation in the lesser resonance within the —COS— group and the inductive effect (often underestimated) of the sulphur atom. That thiolo esters react faster in ester condensations than esters has been explained by the greater acidity of the former but at the same time evidence exists that it also is a more reactive substrate for the attacking anion. Thus with acetone, ethyl acetate condenses to acetylacetone whereas ethyl thioacetate prefers condensation with itself to ethyl acetothioloacetate[117]. The important condensation of acetate groups in acetyl-CoA has been reviewed by Bruice[148].

With Grignard reagents the thiolo ester function is attacked giving tertiary alcohols and thiol[117, 149] unless the esters are sterically hindered and metallation occurs as described above (equation 55). The analogous zinc reagent can also apparently be formed from thiolo esters, since Reformatsky reactions can be carried out[150] with zinc as well as with the Grignard reagent. The latter with cyclohexanone gives a 60% yield of 10:

$$BrMgCH_2COSC(CH_3)_3 \; + \; \langle \text{cyclohexanone} \rangle{=}O \; \longrightarrow \; \langle \text{cyclohexane} \rangle \begin{matrix} OH \\ CH_2CCSC(CH_3)_3 \end{matrix} \qquad (57)$$

(10)

Oxidation of thiolo esters can give several products but is of minor synthetic use. Dependent upon the oxidizing agent and on the structure, either carboxylic acids or sulphonic acids result. Halogens give a multitude of products, partly by oxidation (disulphides, thiolsulphonic esters[151]) and partly by halogenation (with Cl_2: sulphenylchlorides RSCl, sulphonyl chlorides RSO_2Cl, alkylsulphur trichlorides $RSCl_3$)[152]. The presence of water has a profound influence upon the products formed.

Reduction leads mostly to alcohols or aldehydes. The action of Raney nickel has been extensively studied with the purpose of finding the conditions favouring smooth reduction to alcohols or selective reduction to the aldehyde stage. The latter is achieved with a partly deactivated catalyst (deactivated by treatment with boiling acetone[153] or ethanol[154]). However, alcohols may be obtained in excellent yields[154-157]. When the Raney nickel is completely freed of adsorbed hydrogen disulphides and hydrocarbons result[158, 159]. With lithium aluminium hydride the main products are the alcohol[160-162] and thiol, with lithium hydrides aldehydes result[163].

A new class of thiolo esters, β-thioketothiolo esters, has recently been prepared and studied by Duus, Jakobsen and Lawesson[96].

IV. KINETICS AND MECHANISM OF SUBSTITUTION REACTIONS AT THE THIOCARBOXYL FUNCTION

A. General

The discovery of the role of coenzyme A as a natural acylating agent[164] and the following application of thiolo esters as reactive starting materials for the synthesis of oligopeptides[133, 165] have led to the description of the thiolo ester group as energy rich. As seen previously the free energy change of transacylation from sulphur to oxygen or nitrogen is always negative and this should be in fact the meaning of the term 'energy rich'[135]. It follows that in reaction (58) the rate of alcoholysis of a thiolo ester (k_1) should be higher than the rate of thiolysis of an ester (k_{-1}).

$$R^1COSR^2 + R^3OH \underset{k_{-1}}{\overset{k_1}{\rightleftharpoons}} R^1COOR^3 + R^2SH \qquad (58)$$

One would expect thiocarboxyl groups to be more reactive towards nucleophiles than the carboxyl group and in a qualitative way this seems to be borne out by the very much faster aminolysis of thio acids. Thiobenzoic acid reacts with aniline spontaneously but benzoic acid reacts only on prolonged heating[85]. Likewise hydroxylamine needs heating with oxygen esters whereas thioloesters react spontaneously at room temperature. Unfortunately comparable kinetic measurements have not been performed with N-nucleophiles, but the qualitative evidence seems quite compelling.

The hydrolysis of thio acids and thiolo esters has been studied more extensively with direct reference to the comparable reaction with the oxygen analogues. A discussion is therefore best related to this reaction.

In principle esters can be hydrolysed by attack on the carbonyl group (acyl fission) or on the alkyl component (alkyl fission). Isotope experiments have shown that the first path dominates, except in cases where a specially stabilized carbonium ion is formed as with t-butyl, benzyl or triphenylmethyl esters[166]. With thiolo esters product analysis gives the type of fission immediately. It has been found that t-butyl thioloacetate yields t-butylthiol and acetic acid (i.e. acyl fission occurs) both in acid and alkaline hydrolysis[167, 168]. Triphenylmethyl thiolobenzoate gives alkyl fission upon acid but not upon alkaline hydrolysis[168, 169] (equations 59 and 60).

$$CH_3COSCR_3 + H_2O \xrightarrow{H^+} CH_3COSH + HOCR_3 \quad (R = C_6H_5) \Big\}$$

$$CH_3COSCR_3 + H_2O \xrightarrow{H^+} CH_3COOH + HSCR_3 \quad (R = CH_3) \Big\} \tag{59}$$

$$CH_3COSCR_3 + H_2O \xrightarrow{OH^-} CH_3COOH + HSCR_3 \quad (R = CH_3, C_6H_5) \tag{60}$$

Apparently the preference for acyl fission is even greater with thiolo esters than with oxygen esters.

The mechanism of carboxylic ester hydrolysis has been reviewed by Bender[170]. The reaction can be either spontaneous or catalysed by acid or base. If Bender's scheme is translated for thiolo esters (or the acids) equations (61) and (62) result.

$$R{-}COSR' \underset{A}{\overset{H_3O+}{\rightleftharpoons}} \left[R{-}C\overset{OH}{\underset{SR'}{\diagup}} \right]^+ + H_2O \underset{k_1}{\overset{k_{-1}}{\rightleftharpoons}} \underset{B \atop +OH_2}{R{-}\overset{OH}{\underset{|}{C}}{-}SR'} \overset{H_2O}{\underset{H_3O+}{\rightleftharpoons}} \underset{C}{R{-}\overset{OH}{\underset{OH}{\overset{|}{C}}}{-}SR'}$$

$$H_3O+ \Big\Updownarrow H_2O \quad (61)$$
$$D$$

$$RCOOH + HSR' \underset{F}{\overset{H_3O+}{\rightleftharpoons}} \left[R{-}C\overset{OH}{\underset{OH}{\diagup}} \right]^+ + HSR' \underset{k_2}{\rightleftharpoons} \underset{E \atop +}{R{-}\overset{OH}{\underset{OH}{\overset{|}{C}}}{-}SHR'}$$

$$RCOSR' + OH^- \underset{k_{-1}}{\overset{k_1}{\rightleftharpoons}} \underset{B \atop OH}{R{-}\overset{O^-}{\underset{|}{C}}{-}SR'} \overset{k_2}{\rightleftharpoons} \underset{E}{RCOOH + SR'^-} \underset{F}{\overset{OH^-}{\rightleftharpoons}} \tag{62}$$

$$RCOO^- + SR'^- + H_2O$$

The evidence for the tetrahedral intermediates with carboxylic esters is well founded[170]. A similar intermediate in the hydrolysis of ethyl trifluoro-thioloacetate is postulated on kinetic evidence[171]. Alcoholysis, amino-lysis and hydroxylaminolysis of certain thiolo esters support this view (see below). Thus it may be assumed that schemes (61) and (62) are indeed valid for reactions at the thiocarboxyl group.

Obviously the experimental evidence for the tetrahedral addition com-plex is obtained from studies where its stability is above a certain limit. It is not at all certain that the description via an intermediate complex is of general validity. Thus ester hydrolysis—and in fact all substitutions on (thio)carboxylic acid derivatives—may be either a two-step reaction or proceed directly as an S_N2 reaction, dependent on the life-time of the complex. In the first case and if protonation equilibria are assumed to be

rapid the kinetics of the forward reaction in (61) and (62) can be given by
a steady-state equation (63).

$$k = \frac{k_1 k_2}{k_{-1} + k_2} \tag{63}$$

If $k_{-1} \ll k_2$ only the formation of the complex is kinetically important
(at least as long as steady-state conditions are fulfilled) but when k_{-1} becomes relatively large the decomposition of the intermediate makes an
essential contribution to the overall reaction as well. Thus a discussion of
the influence of structure upon rate becomes complicated because not
only are direct influences involved but also a change in mechanism may
occur with variation of structure. In the comparison of thiolocarboxylic
and carboxylic groups another factor must be also considered in acid-
catalysed reactions. Although not of direct kinetic importance, the fast
equilibria determine the concentrations of the reacting species. Thus the
first protonation pre-equilibrium A is reflected in the rate constant k_1.
Since thiolo esters (and acids) are less basic than oxygen esters (and acids)
acid-catalysed reactions of the former are slowed down relative to the
latter. In the author's view substitution reactions on thiocarboxylate
groups present an example where simple analysis in terms of nucleophilicity of the reagent and leaving-group tendencies fail.

B. Hydrolysis of Thio Acids

Thio acids are hydrolysed either spontaneously[77] or under the influence
of acidic[77] and basic[84] catalysts. Comparison with carboxylic acids can be
obtained by studying [18]O exchange of the latter[172, 173]. Data exist only for
thioacetic acid and thiobenzoic acid. They are given together with those of
acetic acid and benzoic acid in Table 6.

TABLE 6. Hydrolysis rates of thio acids and [18]O exchange rates of carboxylic acids.

	Acid-catalysed (70°) $k_2 \times 10^3$ (l/mole sec)	Spontaneous (70°), $k_1 \times 10^5$ (sec^{-1})	Base-catalysed, $k_2 \times 10^4$ (l/mole sec)
CH_3COSH	4·25[77]	5·4[77]	1·6[84] (90°)
CH_3COOH	38·5[172]	< 0·25[172]	< 0·2[172] (100°)
C_6H_5COSH	0·2[77]	3·3[77]	—
C_6H_5COOH	0·25[173]	—	—

The spontaneous hydrolysis of thioacetic acid is faster than that of
acetic acid, but the acid-catalysed reaction is appreciably slower. With

thiobenzoic acid only the acid-catalysed rates can be compared, where the difference is negligible.

The base-catalysed hydrolysis of thioacetic acid cannot be compared with a similar reaction of acetic acid since the former is of the expression: rate $= k_2(CH_3COS^-)(OH^-)$ whereas with acetic acid a reaction of first order both in acetate ions and hydroxylic ions is not observed. However an upper limit for such a reaction [*i.e.* of the rate $= k_2'(CH_3COO^-)(OH^-)$] can be derived from the data given by Llewellyn and O'Connor[172]. At the highest pH measured it is not yet detected although the concentration of the acetate ions is 2×10^6 that of acetic acid. Since a contribution of 10% would not have remained undetected the upper limit for the bimolecular rate constant of the reaction between acetate or hydroxyl ions would be of the order of 2×10^{-5} 1/mole sec. Thus the base-catalysed reaction (as the spontaneous) is appreciably faster with thioacetic acid.

The differences between acetic acid and thioacetic acid have been rationalized by Hipkins and Satchell[77] in terms of bond breaking of C—O versus C—S bonds. The SH$^-$ group is a better leaving group, and will be removed easier in spontaneous of base-catalysed reactions. Likewise the lower basicity (or nucleophilicity) of the leaving group would make it less susceptible to acid catalysis. In acid, salts were found to exert a negative salt effect on the oxygen exchange of acetic acid but a positive one on the hydrolysis of thioacetic acid[77]. Complications as to the real predominance of the bond-breaking step (E in equation 61) and the existing ionization equilibria are not all understood.

Hydrolysis of thiolacetic anhydride and thiolbenzoic anhydride has also been studied by Hipkin and Satchell[174]. Here comparison with oxygen derivatives points to the importance of bond formation in determining the relative rates. The quantities k_S/k_O are given in Table 7.

TABLE 7. Relative reactivities towards hydrolysis
(60% v/v dioxane/water at 25°).

Anhydride	k_S/k_O	
	Spontaneous	Acid-catalysed
CH_3COSCH_3	0·5	small
$C_6H_5COSCOC_6H_5$	3	~ 0·4

C. Hydrolysis of Thiolo Esters

The acid-catalysed hydrolysis of thiolo esters is slower by a factor between 20 and 30 than that of the oxygen esters, but only a slight difference exists in alkaline hydrolysis, except when steric effects decrease the rate of hydrolysis of carboxylic esters. The results obtained by direct comparison of thiolo esters and esters in 62% acetone are compiled in Tables 8 and 9. Only one temperature has been taken, although in the original papers a temperature range was studied[175, 167, 168]. Although in acid hydrolysis the pseudo-monomolecular rate constants have been tabulated in the papers of Tarbell and coworkers[167,168], they have apparently been divided by the concentration of the acid, so that they are in fact the bimolecular constants. This seems to be indicated by the agreement of the value for ethyl acetate with that obtained earlier. It is concluded[167, 168] that on the whole the entropies of activation in acid hydrolysis do not show much variance but that the activation energies differ (Table 9), with t-butyl acetate being an exception. In alkaline hydrolysis, the entropies of activation for the thiolo esters are consistently less negative and the activition energies are higher.

In aqueous solutions a number of rate constants have been determined[176]. Comparison with hydrolysis of esters[177, 178] gives the same overall picture: acid hydrolysis of thiolo esters is about 10 times slower, alkaline hydrolysis is of about equal speed. There is a marked dependence on the nature of the groups present. A study in aqueous dioxane again gives the same overall result[179]. The rate in acid is found not to be proportional to the concentration of the acid; it seems from the data presented[179] that non-linearity is only important in acid concentrations above 0·1–0·2 mole/l in 70% (v/v) aqueous dioxane where apparently some kind of acidity function must be used. Böhme studied the effect of aromatic and aliphatic groups both in the acyl and alkyl part. This is illustrated in Table 10.

The slower acid hydrolysis of thiolo esters is mainly important in aliphatic esters; the difference seems to decrease when first one and then two aromatic groups are introduced. The decreasing value of k_O/k_S parallels the results with thio acids and anhydrides (section IV. B).

Alkaline hydrolysis rates of thiolo esters are again found to be very similar to those of the oxygen esters[179]. Ethyl thiolobenzoate was hydrolysed in 50% dioxane at 25° with $k_2 = 0·26$ l/mole min, ethyl benzoate with $k_2 = 0·29$ l/mole min. The second-order rate constant of phenyl thioacetate under the same conditions was $k_2 = 56$ l/mole min, again of the same order as phenyl acetate (exact value not given).

TABLE 8. Alkaline hydrolysis of thio esters and esters (CH_3COXR) in 62% (w%) acetone

R	k_2(l/mole min) X = S	k_2(l/mole min) X = O	Temperature (°c)	Reference	$\Delta H^‡$ (kcal/mole) X = S	$\Delta H^‡$ (kcal/mole) X = O	$\Delta S^‡$ (e.u.) X = S	$\Delta S^‡$ (e.u.) X = O
CH_3	2·41	3·96	20	167	13·1	12·2	−22	−24
C_2H_5	1·54	1·75	20	175	14·4	12·0	−19	−27
		2·13						
i-C_3H_7	0·818	0·289	20	167	17·6	12·2	− 9	−29
i-C_4H_9	0·717	0·676	20	167	18·5	12·4	− 6	−27
i-C_4H_9	0·222	0·0280	30	167	17·0	14·3	−14	−29
$CH_2CH{=}CH_2$	0·89	0·54	0	168	17·9	9·9	− 3	−33
$CH_2C_6H_5$	0·62	0·59	0	168	16	14·1	−12	−18
$C(C_6H_5)_3$	0·13	instantly	0	168	18	—	− 5	—

TABLE 9. Acid hydrolysis of thiolo esters and esters (CH_3COXR, 62% acetone).

R	$k_2 \times 10^4$ (l/mole min) X = S	$k_2 \times 10^4$ (l/mole min) X = O	Temperature (°c)	Reference	$\Delta H^‡$ (kcal/mole) X = S	$\Delta H^‡$ (kcal/mole) X = O	$\Delta S^‡$ (e.u.) X = S	$\Delta S^‡$ (e.u.) X = O
CH_3	1·94	52	30	167	17·1	15·7	−29	−27
C_2H_5	1·34	42·7	30	175	18·1	16·0	−27	−27
		42·6		167				
i-C_3H_7	1·096	20·0	30	167	19·7	16·3	−22	−27
i-C_4H_9	0·996	30·9	30	167	20·5	16·1	−19	−27
t-C_4H_9	0·823	8·00	30	167	20·7	23·2	−19	− 6
$CH_2CH{=}CH_2$	6·3	69	42	168	16·3	17·3	−32	−24
$CH_2C_6H_5$	2·6	49	40	168	19·8	17·1	−22	−24·5
$C(C_6H_5)_3$	76	—	40	168	29·7	—	−16	—

TABLE 10. Pseudo first-order constants of acid hydrolysis of thiolo esters and esters in 70% (v/v) aqueous dioxane[179].

Ester	$k_1 \times 10^4$ (min^{-1})		k_1 (X = O)/k_1 (X = S)	Temperature (°C)
	(X = S)	(X = O)		
$CH_3COXC_2H_5$	1·1	31	28	40
$CH_3COXC_6H_5$	4·5	36	8	45
$C_6H_5COXC_2H_5$	1·7	13	8	100
$C_6H_5COXC_6H_5$	3·2	5·7	1·8	100

D. Aminolysis of Thiolo Acids and Esters

Kinetic studies of the reaction of aniline and thiobenzoic acid have been made by Hawkins, Tarbell and Noble[85]. The reaction in chlorobenzene was found to be first order in each aniline and thiobenzoic acid. Catalysis by benzoic acid was found in this solvent but the rate increases less than proportionally with the concentration of benzoic acid. The authors assume a hydrogen-bonded complex between thiobenzoic acid and benzoic acid to be the reactive substrate in the catalysed reaction. The activation energy was found to be 7·4 kcal/mole. The difference in reactivity between thiobenzoic acid and benzoic acid is illustrated by the fact that the reaction of thiobenzoic acid in chlorobenzene can be conveniently followed at temperatures between 50° and 70°, whereas benzoic acid shows no appreciable reaction after a number of hours on a steam bath. The same reaction in aqueous acetic acid also gave second-order kinetics up to around 70% completion of the reaction if the water content was lower than 30% (v/v). Acetanilide was not formed. In aniline as the solvent the reaction was second order in thiobenzoic acid with a rate constant of 0·5 l/mole min at 59·5°. Additions of anilinium hydrochloride did not effect the rate of the reaction. The following mechanism was proposed[85].

$$2\ C_6H_5COSH \xrightarrow{\text{slow}} C_6H_5COSH_2^+ + C_6H_5COS^-$$
$$C_6H_5COSH_2^+ + C_6H_5NH_2 \xrightarrow{\text{fast}} C_6H_5CONHC_6H_5 + H_2S + H^+$$
(64)

This mechanism seems rather dubious for those accustomed to think of rapid proton transfer. An alternative possibility such as the rapid formation of a complex between aniline and thiobenzoic acid and the rate-determining removal of hydrogen sulphide under catalysis by thiobenzoic acid should be considered. Relevant in this respect might be the observation of Litvinenko and Oleinik[180] that benzoylation of aniline by benzoyl chloride in benzene is catalysed by thioacetic acid and thiobenzoic acid but not by dithioacetic acid. Base catalysis is indicated by this study.

Reaction of aliphatic carboxylic acids with aniline at 100° in aniline as a solvent proceeds much slower. This reaction is also second order in acid $k_2 = 2 \times 10^{-4}$ l/mole min[181].

Reactions of other amines with special reference to amino acids have been studied[182-185] but no comparison with oxygen analogues has been made.

A comparative study of the aminolysis of the thiocarboxyl and carboxyl-group in aqueous solutions was made by Connors and Bender[186] with ethyl-p-nitro(thio)benzoate. Concurrent alkaline hydrolysis was studied and the bimolecular rate constants for hydrolysis of the thiolo ester and the ester were evaluated as 0·50 and 0·61 l/mole sec respectively. The rate equation for aminolysis by n-butylamine was described in terms of a simple bimolecular reaction plus acid- and base-catalysed reactions. Acid catalysis in the aminolysis of ethyl p-nitrothiobenzoate was found to be unimportant but general base catalysis terms (OH$^-$ and butylamine) were kinetically the most important. With ethyl p-nitrobenzoate no detectable aminolysis was found under the conditions of the experiments (25·6°C and ionic strength 0·5) so that again no comparison of the rate constants can be made, but evidently thiolo esters react faster than oxygen esters by several orders of magnitude. To explain the much greater speed of aminolysis of thiolo esters compared with the approximately equal rate of hydrolysis Connors and Bender suggested that in aminolysis the return reaction of the tetrahedral intermediate, i.e. k_{-1} of equation (61), dominates in the cases of carboxylic ester aminolysis.

Studies on ω-hydroxy and amino alkylthiolacetates (11) give pertinent evidence for the existence of a tetrahedral complex in acyl transfer reactions and also on the relative rates of bond breaking.

11 readily gives acyl transfer according to equation (65) if $n = 2$ or 3 but not with higher values of n[135,187,188]. This observation suggests anchimeric assistance by formation of a five- or six-membered ring.

$$CH_3COS(CH_2)_nXH \longrightarrow CH_3COX(CH_2)_nSH \qquad (65)$$
$$(11, X = O, NH)$$

In reaction (65) (X = NH) 2-methylthiazoline (12) was found to be involved as an intermediate[128,189]. A complete scheme for this was given and studied kinetically by Martin and coworkers[140,190]. Their scheme, apart from protonation equilibria, is given in equation (66). The tetrahedral intermediate 14 was postulated on kinetic grounds and all reactions were assumed to be equilibria. All equilibrium constants were determined (see

$$
\begin{array}{ccc}
\textbf{(11)} & \textbf{(14)} & \textbf{(13)} \\
\end{array}
$$

$$
H_2N(CH_2)_2SCOCH_3 \underset{k_6}{\overset{k_5}{\rightleftarrows}} \underset{HO}{\overset{S}{\underset{C}{\diagdown}}}\underset{CH_3}{\overset{NH}{\diagup}} \underset{k_4}{\overset{k_3}{\rightleftarrows}} HS(CH_2)_2NHCOCH_3
$$

$$k_2 \updownarrow k_1$$

$$\underset{CH_3}{\overset{S}{\diagdown}}\overset{\overset{+}{N}H}{\diagup}$$

$$\textbf{(12)}$$

$$\tag{66}$$

section III. B. 1) and many rate constants, as well. For full details the original papers should be consulted, but for our discussion it is important to note that the ratio k_5/k_3 is found to be 1·2. Thus the rate of splitting of C—S and C—N bonds is nearly the same. In the oxazoline system[141] (equation 66 with O instead of S) the same ratio is found to be $6\cdot4\times10^6$, giving evidence for a tremendous preference for C—N cleavage. These results are certainly surprising and might be taken for evidence of preferred return to starting products with esters as discussed by Connors and Bender[186]. A further analysis of the ratios k_5/k_2 is tempting but the reaction k_2 involves hydrogen ions and a comparison is probably not justifiable. It must be kept in mind that an incompatibility in the equilibrium constant still exists so that the results given should be applied with caution (*cf.* discussion in section III. B. 1).

Kinetic evidence for a tetrahedral non-cyclic intermediate was found by Bruice and Fedor[191] in the hydroxylaminolysis of thiolo lactones and thiolo esters (equation 67). The authors deduce from their experiments that the intermediate **15** is more stable than that formed from oxygen esters which

$$
\begin{array}{ccc}
\underset{\underset{\diagdown_{CH_2}\diagup}{CH_2 \quad S}}{CH_2-C=O} + H_2NOH \longrightarrow \underset{\underset{\diagdown_{CH_2}\diagup}{CH_2 \quad S}}{\overset{NHOH}{CH_2-C\diagdown_{OH}}} \longrightarrow \underset{\underset{\diagdown_{CH_2}\diagup}{CH_2 \quad SH}}{\overset{NHOH}{CH_2-C\diagdown_{O}}}
\end{array}
\tag{67}
$$

$$\textbf{(15)}$$

is consistent again with a great leaving tendency of amines in the tetrahedral intermediates of oxygen esters. Both the transition states for nucleophilic attack leading to **15** and for departure of the leaving group are found to be kinetically important.

Evidence for general acid/base catalysis together with a spontaneous reaction is found by the same authors but this interpretation has been criticized by Martin and Henkle[192] who interpreted these results in terms of a general base-catalysed reaction. Of course, when protonation equilibria take part in the reaction (and they are abundant in reaction 67) acid and base catalysis cannot be distinguished kinetically and recourse must be taken to evidence of an indirect nature. The mechanism postulated for hydroxylaminolysis has also been applied to hydrazinolysis[193].

Thiocarboxylic anhydrides have recently been aminolysed together with carboxylic anhydrides by Hipkin and Satchell[174]. They found first-order reaction in amine and anhydride (proving absence of base catalysis) in 90% (v/v) dioxane–water with the rate constants given in Table 11.

TABLE 11. Relative reactivity of thiolo anhydrides and anhydrides towards aminolysis in 90% (v/v) aqueous dioxane at 25°C[174]

Amine R in $RC_6H_4NH_2$	Benzoic anhydrides $(C_6H_5CO)_2X$			Acetic anhydrides $(CH_3CO)_2X$		
	10^2k_2 (l/mole sec)		$k_2(S)/k_2(O)$	10^2k_2 (l/mole sec)		$k_2(S)/k_2(O)$
	X = S	X = O		X = S	X = O	
m-Cl	4·0	0·055	73	—	—	—
p-Cl	—	—	—	11·0	0·87	13
H	8·4	0·131	65	30·0	3·75	8
p-CH$_3$	35	0·55	64	—	—	—
p-CH$_3$O	132	2·6	51	—	—	—

The ratio of the rate constants k_S/k_O is much smaller here than with esters. Its variation was rationalized in terms of relative importance of bond formation and bond breaking in reaction (68). The authors have not analysed their results with respect to a passible tetrahedral intermediate, but assume a single-step reaction.

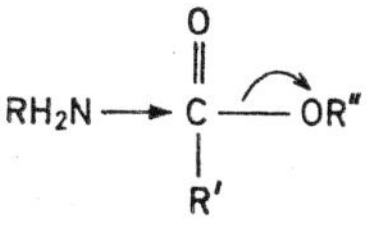

$$RH_2N \longrightarrow \overset{\overset{\displaystyle O}{\displaystyle \|}}{\underset{\underset{\displaystyle R'}{\displaystyle |}}{C}} \overset{\frown}{\longrightarrow} OR'' \tag{68}$$

E. Summary and Conclusions

Although a number of data have been collected, only a preliminary attempt can be made to describe the mechanism of nucleophilic attack at the thiocarboxyl group.

47*

In terms of a simple S_N2 reaction only two parameters are involved: bond formation and bond breaking. It is not easy to see how the difference in rate of attack of nitrogen and oxygen nucleophiles could be described in these terms. With the assumption of an intermediate as in equations (61) and (62), many more parameters become adjustable so that rationalizations can always be found when so few experimental data are available as in the present case. However, even erroneous interpretations may have their use in animating more detailed studies.

In terms of the relative rate constants (*i.e.* the rate constant for a reaction at the thiolocarboxylate group divided by the rate constant for the same reaction at a carboxylate group) the main observations are: k (rel.) for base-catalysed hydrolysis is of the order of 1; k (rel.) for acid-catalysed hydrolysis is around 0·04 for aliphatic esters but higher, up to 0·5, when aromatic groups are present (Table 10); k (rel.) for aminolysis is very large.

The most satisfactory explanation for the last result seems to be the suggestion of Connors and Bender[186] that in carboxylic acid derivatives the return reaction is important in aminolysis. Probably the lower basicity of thiolo esters makes them less susceptible to acid-catalysed hydrolysis. Nevertheless it remains remarkable that a thiolo ester (which is less stabilized by resonance) is not hydrolysed much faster than an ester. Possibly the softer character of a thiolo ester makes it less susceptible to the action of the hard base (water or hydroxyl ion). This might explain the observation that the presence of phenyl groups causes a decrease in the relative rate constant.

The rate constants for (thiolo) ester hydrolysis combined with equilibrium constants permit the evaluation of esterification rates. The equilib-

$$R^1SH + R^2COOH \underset{k_h}{\overset{k_e}{\rightleftharpoons}} R^2COSR^1 + H_2O \qquad K_{SH} = k_e/k_h \qquad (69)$$

$$R^1OH + R^2COOH \underset{k'_h}{\overset{k'_e}{\rightleftharpoons}} R^2COOR^1 + H_2O \qquad K_{OH} = k'_e/k'_h \qquad (70)$$

rium constants of reactions (69) and (70) have only been determined at high temperatures (*cf.* section III. B. 1) but the relative constant K(rel.) = K_{SH}/K_{OH} is equal to the O → S transacylation constant, which was measured in aqueous solution[135] at 39° to be $1·8 \times 10^{-2}$ (section III. B. 1).

Since k_e (rel.) $\equiv k_e/k'_e = k_h$(rel.)K(rel.), k_e(rel.) for the (hypothetical) base-catalysed esterification becomes 0·02 and for the acid-catalysed esterification 8×10^{-4}. Thus a thiol is remarkably slow in its reaction with a carboxylic acid in spite of its assumed high nucleophilicity. The acid-catalysed esterification is described by equation (61) (when read backwards) and the enormous difference points to the importance of step E as the rate-determining step since the others can hardly be expected to be so sensitive to the presence of a sulphur atom.

Apparently the attack of a protonated acid (which will be a very hard acid) on a thiol is very slow. Several observations mentioned in the previous sections have already shown that reactions of thiols whith acyl compounds are slower than expected from the generalization that the sulphur atom is a better nucleophile than the oxygen atom. The conclusion of section I. C. 2 is strengthened that this generalization is not allowed, but that the nature of the substrate can change the relative reactivities of thiols and alcohols. For recent paper with relevant discussion see R. K. Chaturdevi, A. E. MacMahon and G. L. Schmir, *J. Am. Chem. Soc.*, **89**, 6984 (1967).

V. THIONO ESTERS

This class is one of the least studied groups of compounds. They can be prepared by sulphydrolysis of imino esters[2, 58, 194-196] in ether or in a cold aqueous solution when ether is present to take up the thiono ester immediately. The product contains thioamide and yields are general not higher than about 20% (equation 71).

$$CH_3C(NH)OR + H_2S \rightleftharpoons CH_3CSOR + NH_3 \qquad (71)$$

Thiono esters react in cold ethereal and benzene solutions with ammonia or primary amines[197] and this reaction probably proceeds via the imino ester[198]. The amidine is obtained in the presence of ammonium salt; with

$$CH_3CSOC_2H_5 + H_2NR + H_3NRCl \xrightarrow{\text{ethanol}} \left[CH_3C \begin{array}{c} \nearrow NR \\ \searrow OC_2H_5 \end{array} \right]$$

$$\longrightarrow CH_3C \begin{array}{c} \nearrow NR \\ \searrow NHR \end{array} \cdot HCl + C_2H_5OH + H_2S \qquad (72)$$

an amine only, the thioamide is the main product.

In a systematic investigation of reaction (71) Reynaud and Moreau[199] found the reaction to be strongly catalysed by amines. With ammonia the only product formed is the thioamide (90% yield), with tertiary amines (pyridine, triethylamine) up to 55% of aliphatic thiono esters may be obtained, provided that the reaction is not prolonged unnecessarily, because the ammonium hydrosulphide formed reacts with the product with formation of the thioamide.

The reaction scheme proposed by Reynaud and Moreau is given in equation (73).

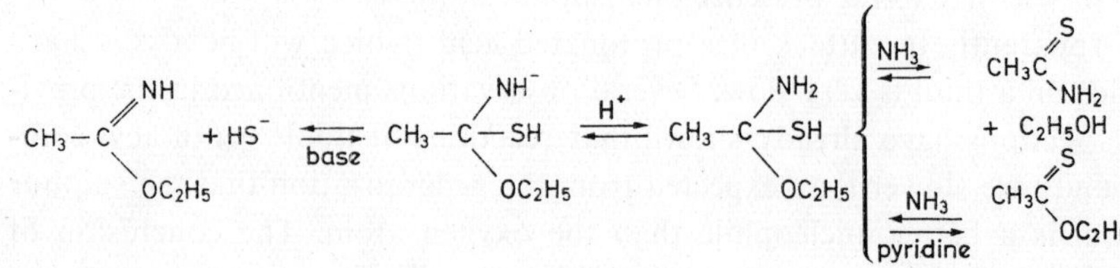

$$\tag{73}$$

A second method of preparing thiono esters consists of the reaction of Grignard reagents on chlorothionoformates[200-202] (which are obtained from alcohols with an excess of thiophosgene[203]) (equation 74). Yields are low.

$$RMgI + ClC\underset{OR'}{\overset{S}{\diagdown}} \longrightarrow RC\underset{OR'}{\overset{S}{\diagdown}} + MgICl \tag{74}$$

Aromatic thionoacyl chlorides may be alcoholysed to thionoesters[204-206] but aliphatic thiono chlorides are too unstable to be used with success[58].

$$C_6H_5C\underset{Cl}{\overset{S}{\diagdown}} + ROH \longrightarrow C_6H_5CSOR + HCl \tag{75}$$

Only recently the synthesis of a number of aromatic thioacid chlorides has been described in detail (with yields up to 80%) and the kinetics of their methanolysis have been studied[206]. 0·1 molar solutions in methanol react with pseudo-unimolecular kinetics so that apparently hydrochloric acid does not influence the reaction. The results are compared with the methanolysis and ethanolysis of benzoyl chlorides[207, 208]. Some relevant data are collected in Table 12.

Apart from o-methylthiobenzoyl chloride which reacts extremely fast, methanolysis is slower with the thionoacid chlorides than with the acid chlorides. This is certainly an unexpected result in view of the difficulties in synthesizing the compounds! Scheithauer and Mayer[206] concluded from the entropy of the reaction that at least some thiono chlorides were alcoholysed in a unimolecular reaction. The extreme reactivity of the ortho methyl derivative should be caused by non-planarity of the conjugated system. As discussed in section I. B thiocarbonyl compounds derive their

TABLE 12. Alcoholysis of thiobenzoyl chlorides and benzoyl chlorides[206]

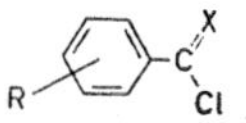

R	Methanolysis						Ethanolysis		$(X = O)$
	$k \times 10^3$ at 0° (min^{-1})		E_a (kcal/mole)		$\Delta S^{\ddagger}$ (20°) (e.u.)		$k \times 10^3$ at 0° (min^{-1})	$\Delta H^{\ddagger}$ (kcal/mole)	$\Delta S^{\ddagger}$ (e.u.)
	$X = S$	$X = O$	$X = S$	$X = O$	$X = S$	$X = O$			
H	2·2	26·5	16·9	14·9	−17·5	−19·9	4·4	15·4	−21·6
o-CH$_3$	>10^3	137	—	—	—	—	16·1[a]	16·7	−14·3
m-CH$_3$	3·3	29·5	17·4	—	−14·9	—	3·75	16·5	−17·9
p-CH$_3$	4·0	17·8	21·9	—	+2·1	—	3·35	15·0	−23·6
p-Cl	1·35	43·1	14·8	—	−26·1	—	8·4	14·0	−25·3

[a] Ethanolysis of thioanalogue $(X = S)$ approximately 67·8

stability from conjugation. The instability of o-methylthiobenzoyl chloride and the aliphatic thioacyl chlorides support this model.

A remarkable synthesis of thiono esters has been described by Banks and Cohen[209]. When a suitable thiolo ester is aminolysed an 1-alkoxy-enethiol is split off which of course isomerizes to the thiono ester (equation 76).

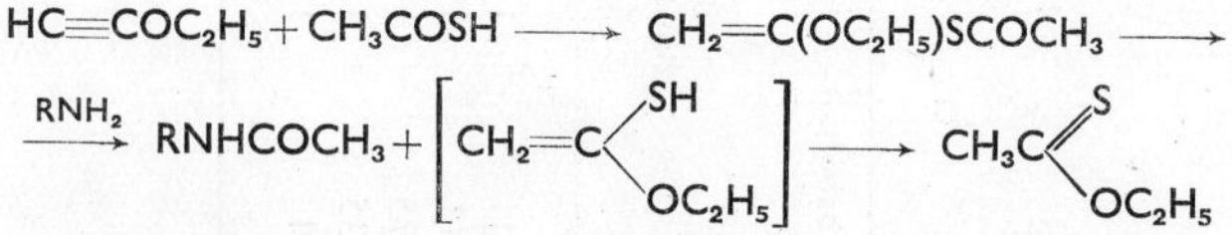

$$HC{\equiv}COC_2H_5 + CH_3COSH \longrightarrow CH_2{=}C(OC_2H_5)SCOCH_3 \longrightarrow$$

$$\xrightarrow{RNH_2} RNHCOCH_3 + \left[CH_2{=}C{\diagdown}^{SH}_{OC_2H_5} \right] \longrightarrow CH_3C{\diagdown}^{S}_{OC_2H_5} \qquad (76)$$

Thiono esters are immediately desulphurized by silver and mercury ions[2, 194]. Hydrolysis is also known to give desulphurized products but thio acids may be obtained as well.

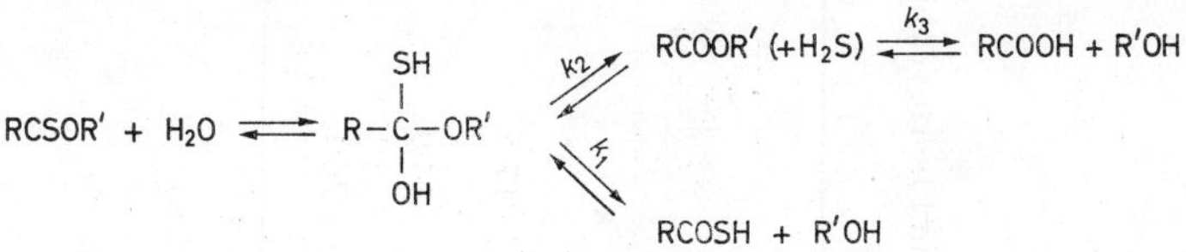

$$RCSOR' + H_2O \rightleftharpoons R{-}\underset{\underset{OH}{|}}{\overset{\overset{SH}{|}}{C}}{-}OR' \quad \xrightarrow{k_2} \quad RCOOR' \ (+H_2S) \underset{}{\overset{k_3}{\rightleftharpoons}} RCOOH + R'OH \qquad (77)$$
$$\xrightarrow{k_1} RCOSH + R'OH$$

$$(16)$$

A systematic investigation of reaction (77) has been made by Smith and O'Leary[22]. Its particular interest lies in the transition complex **16** which gives different products dependent on the leaving group. Results are given in Table 13, where k_1, k_2 and k_3 have the same meanings as they have in equation (77).

TABLE 13. Rate and product distribution of the hydrolysis of ethyl thiono benzoate in 40% benzoate (k in l/mole sec)

	NaOH (25°)	HCl (125°)
10^4k_1	57 ± 4	$1{\cdot}0\pm0{\cdot}5$
10^4k_2	16 ± 1	$5{\cdot}3\pm0{\cdot}3$
10^4k_3	55 ± 1	$5{\cdot}9\pm0{\cdot}9$
% Ethylbenzoate	22	85
% Thiobenzoic acid	78	15

Ethyl thionobenzoate is hydrolysed slightly faster than ethyl benzoate both in acid- and base-catalysed hydrolysis but the mode of decomposition of the intermediate **16** is greatly dependent on the reaction conditions.

It is of interest to compare reaction (77) with the hydrolysis of thiolo esters since the intermediate complex is greatly similar (*cf.* equations 61 and 62). The reaction with thiolo esters involves breakage of the C—S bond, *i.e.* a reaction given in the case of thionobenzoate by k_2.

In the thiolo ester/ester case this reaction proceeded about 8 times slower with the sulphur-containing esters in acid-catalysed, and about equally fast in base-catalysed, reactions. With thiono esters the C—S cleavage is about equally fast in acid and three times slower in alkaline medium. Apparently, notwithstanding the tetrahedral complexes being quite similar, they behave totally differently. It seems as if the relation is much better when the relative rate constant (K_S/K_O) for the hydrolysis of thiolo ester is compared with k_1/k_3 of Table 13. Unless this is accidental it would mean that not the heteroatom in the leaving group is important but rather the fact as to whether it is an XR or XH substituent.

In a study of ionization and ion-pair return Smith[210] subjected benzhydryl thionobenzoate to ethanolysis at 100°. 17% of the ester was solvolysed to thiobenzoic acid and benzhydryl ethyl ether so that apparently alkyloxygen fission occurs. 83% was found as benzhydryl thiolobenzoate originating from recombination of the benzhydryl and thiobenzoate ions (reaction 78).

$$(C_6H_5)_2CHOC\underset{C_6H_5}{\overset{S}{<}} \xrightarrow{C_2H_5OH}$$

$$(C_6H_5)_2CHSC\underset{C_6H_5}{\overset{O}{<}} \quad 83\%$$

$$\xrightarrow{C_2H_5OH} [(C_6H_5)_2CH]^+ [C_6H_5CSO]^-$$

$$(C_6H_5)_2CHOC_2H_5 \quad 17\%$$
$$+ C_6H_5COSH$$

(78)

Similar rearrangements take place in a number of solvents with allyl and substituted allyl thionobenzoates. Kinetics have been measured[211] at 100° and correlation with solvent ionizing power[212] is reasonable. The mechanism is described in terms of allylic rearrangements. When heated sufficiently long at temperatures well above 200°, most thiono esters rearrange to thiolo esters, although ethyl thionobenzoate can be distilled at 240° without rearrangement. The reaction is accelerated when the ester contains a tertiary amino function[194, 213]. Anchimeric assistance has been proposed to be responsible for this increase in reactivity.

Claisen-type condensations have been reported for thiono esters. Ethyl thionoacetate yields thioacetothionoacetate which, like all thiocarbonyl compounds has a strong tendency to enolize[214].

$$2 \ CH_3CSOC_2H_5 \longrightarrow CH_3C\underset{CH_2CSOC_2H_5}{\overset{S}{\diagdown}} \rightleftharpoons CH_3C(SH)\!\!=\!\!CHCSOC_2H_5 \quad (79)$$

Uhlemann and Müller[215] obtained monothiodibenzoylmethane from acetophenone and methyl thionobenzoate by the action of sodium amide.

$$C_6H_5COCH_3 + C_6H_5CSOCH_3 \xrightarrow{\ NaNH_2\ } C_6H_5CSCH_2COC_6H_5 \quad (80)$$

VI. DITHIO ACIDS

A. General

Dithio acids are not very well known. They are often extremely unpleasant to work with and their limited stability is another barrier to prospective students. Somewhat more is known from the esters but most of the knowledge in this field is derived from xanthic ($ROCS_2H$) and dithiocarbamic (R_2NCS_2H) acid derivatives which have important applications in several fields. The xanthates are used in flotation, and in viscose industry, and dithiocarbamates as rubber vulcanizers and as pesticides. Both classes have been applied in analysis of metals because of their strong complexing ability. The dithiocarbamates have been extensively reviewed by Thorn and Ludwig[216]. Salts and esters of xanthic and dithiocarbamic acids may be expected to be more stable than those of dithiocarboxylic acids themselves because of extensive conjugation in the OCS and NCS parts of the molecules.

$$\left.\begin{array}{ccc} R\overset{+}{-}O\!\!=\!\!C\overset{S^-}{\underset{SR'}{\diagdown}} \longleftrightarrow R\!-\!O\!-\!C\overset{S}{\underset{SR'}{\diagdown}} \longleftrightarrow ROC\overset{S^-}{\underset{SR}{\diagdown}} \\ \textbf{(17)} \qquad\qquad\qquad\qquad \textbf{(18)} \ {}^+ \\[2mm] R_2\overset{+}{N}\!\!=\!\!C\overset{S^-}{\underset{SR}{\diagdown}} \longleftrightarrow R_2N\!-\!C\overset{S}{\underset{SR}{\diagdown}} \longleftrightarrow R_2N\!-\!C\overset{S^-}{\underset{SR}{\diagdown}} \\ \textbf{(19)} \qquad\qquad\qquad\qquad \textbf{(20)} \ {}^+ \end{array}\right\} \quad (81)$$

Resonance structures **17** and **19** are expected to be more important than **18** and **20** (*cf.* section I. B). Experimentally this is verified in esters and salts.

Free acids decompose in hydroxylic solvents as can be expected for carbonic and carbamic acid analogues (reactions 82 and 83). Nevertheless free H_2NCS_2H was recently prepared by Gatlow and Hahnkamm[217]

$$ROCS_2H \longrightarrow ROH + CS_2 \quad (82)$$

$$R_2NCS_2H \longrightarrow R_2NH + CS_2 \quad (83)$$

The carbodithio acids are red liquids or solids, whilst the xanthic acids are of course yellow.

B. Preparation

By far the most important method of preparation of dithio acids consists of the reaction between carbon disulphide and an appropriate nucleophilic reagent. For dithiocarboxylic acids Grignard[29, 218-220] or other organometallic reagents[221] may be used. Alternatively, if active hydrogen is present a strong base is able to create the nucleophilic carbanion[222]. With nitromethane, for example, potassium hydroxide and carbon disulphide give potassium nitrodithioacetate[223]. Sodium cyanide with carbon disulphide yields cyanodithioformates[224]. The yields are in general poor for aliphatic dithio acids.

$$RMgBr + CS_2 \longrightarrow RCS_2MgBr \qquad (84)$$

$$CH_3COCHNaCO_2C_2H_5 + CS_2 \longrightarrow CH_3COCH(CO_2C_2H_5)CS_2Na \qquad (85)$$

$$CH_3NO_2 + KOH + CS_2 \longrightarrow O_2NCH_2CS_2K \qquad (86)$$

To obtain xanthates[225, 226] and dithiocarbamates[216] alcohols and amines are used in alkaline media.

$$C_2H_5OH + CS_2 + OH^- \longrightarrow C_2H_5OCS^- + H_2O \qquad (87)$$

$$2\,(C_2H_5)_2NH + CS_2 \longrightarrow (C_2H_5)_2NCS_2^- \cdot (C_2H_5)_2\overset{+}{N}H_2 \qquad (88)$$

The reaction of carbon disulphide with sodium phenoxide is reported to yield sodium p-hydroxyphenyl dithioacetate in analogy with the Kolbe synthesis[227, 228]. In the presence of aluminium chloride carbon disulphide combines with less reactive aromatic (or heteroaromatic) systems with formation of dithioates[229]. An alternative method used for preparing dithio acids consists of the sulphurization of aldehydes[218] (by H_2S_2) or trichloro compounds[230, 231] (by K_2S or KSH). Some special methods have been described but they were not tested for wider applications. They are reviewed in Reid's book[1].

Recently the preparation of dithioacetic acid in 45% yield from acetic acid and P_2S_5 has been reported[232]. With the newer methods it is possible to obtain 50–70% of dithioacids in a reasonably pure state.

C. Acidity

All dithio acids are fairly strong acids, stronger than the corresponding carboxylic acids or thio acids (Table 14). Although free xanthic and dithiocarbamic acids are unstable they have been obtained in the free state and

TABLE 14. Acid dissociation constants of dithio acids (in water).

	Temperature (°C)	pK_a	Reference
CH_3CS_2H	25	2·55	37
$C_2H_5OCS_2H$	23	1·52	233
	23	1·62	234
	23	1·70	235
H_2NCS_2H	0	~2·95	37
	20	2·95	217
$C_2H_5SCS_2H$ (21)	probably 0	1·55	37
$HSCS_2H$		~2·7	236, 237

acid dissociations have been evaluated by extrapolating to zero time. Dithiocarbamic acid decomposes faster than the xanthic acids so that the uncertainty of its dissociation constant is larger. The decomposition of ethyl trithiocarbonic acid (21) is rather slow. The variation of the acid dissociation constant is consistent with inductive and mesomeric effects.

By the crystal-violet technique (section II. B) protonation constants in benzene and chlorobenzene were derived by Ioffe, Sheinker and Kabachnik for dithio acids as well as for thio acids[80] (Table 15). The discussion given earlier is also pertinent to these data.

TABLE 15. Proton-transfer constants pK_p of dithio acids and carboxylic acids[80] (reference base: crystal violet)

Compound	pK_p (in benzene)		pK_p (in chlorobenzene)	
	X = S	X = O	X = S	X = O
CH_3CX_2H	−1·15	−1·24	−3·24	−1·96
$C_2H_5CX_2H$	−0·94	−0·67	−3·18	−1·27
$p\text{-}BrC_6H_4CX_2H$	−2·38	−2·50	−4·63	−3·13

Hydrogen bonding seems to be virtually absent in dithio acids. Although direct studies are not reported, the limited solubility of dithioacetic acid in water is significant.

D. Salt Formation

Dithioates of heavy metals are extremely useful in analysis. They are generally coloured and insoluble in water but may be extracted by a variety of organic solvents. Thus metal ions can be detected and determined quantitatively by a simple colorimetric analysis after extraction. For trace analysis of copper, cobalt, iron, nickel and many more metals dithio-carbamates have been frequently used[216, 238–241]. By adjusting pH or add-ing sequestering agents selectivity can be achieved.

Because of the easy and quantitative formation of xanthates and dithio-carbamates, analysis of alcohols, amines or carbon disulphide can also be based on the formation of heavy metal xanthates or dithiocarbamates and subsequent colorimetric determination.

Contrary to thio acids where heavy metal salts decompose easily to sul-phides, salts of dithio acids are obtained as such. Apparently the extra stabilization by chelate formation (22) is sufficient to prevent metal sul-phide formation. Even the silver and mercury dithiocarbamates are stable compounds (although not monomeric). X-ray studies have proved[242–245] that the carbon–sulphur distances in compounds like 22 are equal.

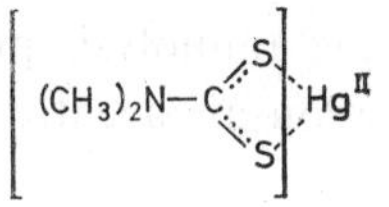

$$\left[(CH_3)_2N-C \underset{\diagdown S}{\overset{\diagup S}{\cdots}} Hg^{II} \right]$$

(22)

Sometimes the salt obtained has the metal in a changed valency state. Thus from cobalt(II) ions and sodium dialkyl dithiocarbamates, cobalt(III) dialkyl dithiocarbamates are invariably obtained and copper(II) is preci-pitated as copper(I) xanthates with simultaneous oxidation of another xanthate ion to dixanthogen (equation 89)[246]. Copper (II) dithiocarba-mates can be obtained unless electron-attracting substituents are attached to the nitrogen atoms[247].

$$Cu^{2+} + 2\,ROCS_2^- \longrightarrow ROCS_2Cu^{I} + \tfrac{1}{2}(ROCS_2)_2 \qquad (89)$$

E. Organic Chemistry of Dithio Acids

Many reactions resemble those of thio acids. With alkyl halides or other agents liable to nucleophilic substitution they give esters smoothly[230, 231, 248] like all thio acids.

With ammonia carbodithio acids are aminolysed to give thio amides (equation 90)[219, 220, 249–251]. Hydrazines[252] and substituted hydrazines

$$RCS_2H + NH_3 \longrightarrow RCSNH_2 + H_2S \qquad (90)$$

(alkyl and aryl hydrazines[253, 254] and semicarbazides[255]) react in the same way. At higher temperatures reaction sometimes goes further with re-moval of both sulphur atoms[254, 255] and formation of the corresponding nitrogen derivative of the aldehyde (equation 91). Hydroxylamine yields

$$RCS_2H + H_2NNHCONH_2 \longrightarrow RCH{=}NNHCONH_2 + H_2S + S \qquad (91)$$

the thiol oxime[254, 256] but here also reaction can proceed to formation of

$$RCS_2H + H_2NOH \longrightarrow RC(NOH)SH \longrightarrow RCH{=}NOH \longrightarrow RCN \qquad (92)$$

aldoximes and nitriles[255].

Xanthates may be aminolysed[257] but are more resistant than carbodithio acids. Not the C—S but the C—O bond is split (equation 92)[258].

$$C_2H_5OCS_2K + C_6H_5NH_2 \longrightarrow C_6H_5NHCS_2K + C_2H_5OH \qquad (93)$$

This may be caused by hydrolysis of the xanthate so that free carbon disulphide reacts with the amine. With xanthic esters preferential C—S cleavage occurs (section VII. B).

Aminolysis of dithiocarbamates seems to be most difficult of all and ammonium salts or salts of amines can be recrystallized in fairly high boiling solvents without decomposition. Only dithiocarbamates derived from primary amines yield the aminolysis product (thiourea) on heating with an amine, but this reaction apparently proceeds via isothiocyanates (equation 94)[259, 260].

$$RNHCS_2Na \longrightarrow RNCS + HSNa$$
$$RNCS + HNR_2^1 \longrightarrow RNHCSNR_2^1 \qquad (94)$$

Oxidation to disulphides is smooth and in particular the carbodithio acids must be guarded against oxidation by air[261]. The easiest way to obtain the disulphides is by oxidation with iodine or chlorine. In contrast with thio acids, dithio acids are not easily titrated iodometrically since the oxidation may proceed further so that end-points are unsharp. Recently it has been found that dithiobenzoic acid is susceptible to Clemmensen reduction with formation of benzylthiol[262].

Kinetics of reactions of carbodithio acids have not been reported, although qualitatively it is known that they are liable to slow hydrolysis.

Hydrolysis of xanthates and dithiocarbamates in acid solution yields carbon disulphide. The reactions have frequently been studied kinetically and all authors agree that it is a simple monomolecular reaction if corrections are made for the dissociation into ions. However, the mechanism has not been studied in detail and it is still unknown whether it is a decomposition of the free acid (equation 95) or the attack of a hydronium ion on the anion (equation 96), both reactions obeying the same kinetics.

$$ROCS_2^- + H^+ \overset{K_a}{\rightleftharpoons} ROCS_2H \xrightarrow{slow} ROH + CS_2 \qquad (95)$$

$$ROCS_2H \overset{K_a}{\rightleftharpoons} ROCS_2^- + H^+ \longrightarrow \underset{R}{\overset{H}{>}}O{-}CS_2^- \xrightarrow{slow} ROH + CS_2 \qquad (96)$$

For xanthic acids Iwasaki and Cooke[235, 263] favour the first scheme whereas Klein, Bosarge and Norman[264] together with earlier investigators[265, 266] favour the latter. All authors assume that the decrease in rate at acidities above 0·5 M hydrochloric acid is due to a protonation of the xanthic acid[263-266] but this seems highly improbable in view of the very

$$ROCS_2H + H^+ \xrightleftharpoons[K_{a_2}]{} \underset{H}{\overset{R}{>}}\overset{+}{O}CS_2H \tag{97}$$

low basicity of xanthic esters[267]. From the kinetics of the decomposition pK_{a2} was found to be around $-0·8$. For xanthic esters, however, protonation does not occur in acid solutions weaker than 6 M[267].

For dithiocarbamic acids (equations 95 and 96 with $RO = R_2N$) Zahradník and Zuman[268, 269] favour the second mechanism in a series of equilibria.

$$R_2NCS_2H \xrightleftharpoons[H^+]{K_S} R_2N—CS_2^- \xrightleftharpoons[H^+]{K_N} \underset{H}{R_2N—CS_2^-} \xrightleftharpoons{slow} R_2NH + CS_2 \xrightleftharpoons[H^+]{} R_2\overset{+}{N}H_2 + CS_2 \tag{98}$$

Scheme (98) has the advantage that it offers a simple explanation for the formation of dithiocarbamates in alkaline solution where the ionization equilibrium of the amine is shifted to the left and that of the dithiocarbamate towards the anion.

Soni and Trivedi[270, 271] reported two dissociation constants for diethyldithiocarbamic acid as shown in equation (99). Their latest values[271] for the dissociation constants are $pK_1 = 7·8$ and $pK_2 = 8·0$. However, in

$$(C_2H_5)_2\overset{+}{N}H—CS_2H \xrightarrow{-2H^+} C_2H_5NCS_2^- \tag{99}$$

their studies no attention was paid to decomposition and the two protons taken up by dithiocarbamate ions must refer to the uptake of protons according to reaction (98). An estimate of the equilibrium constant K_N based on the loss of resonance energy in the dithiocarbamate ion would lead to an acid dissociation constant: $pK_N \approx -2$. That (99) is improbable is also confirmed by the fact that dithiocarbamic esters are only protonated in 6 M sulphuric acid[272].

In apolar solvents dithiocarbamic acids are reasonably stable. Bode and Neumann[273] report half-lives of 25 minutes in carbon tetrachloride and 5 hours in chloroform when the solutions are shaken with 1 N hydrochloric acid.

Alkaline decomposition of xanthates has been reported frequently but the course of the reaction was obscure until Ingram and Toms studied alcoholysis and hydrolysis systematically[274].

In alcoholysis the primary product is diethyl thionocarbonate (23) which points to the existence of the tetrahedral intermediate 24.

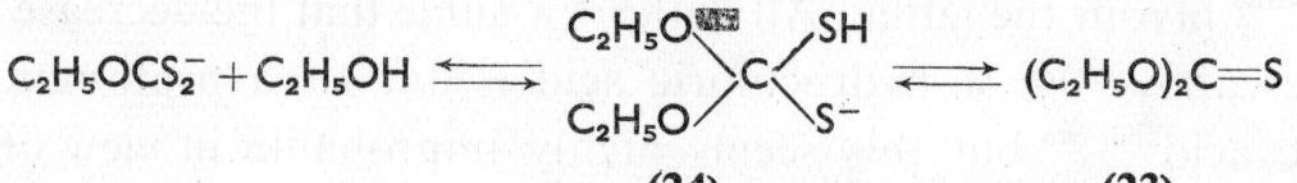

$$C_2H_5OCS_2^- + C_2H_5OH \rightleftharpoons \underset{(24)}{C_2H_5O}\overset{C_2H_5O}{\underset{C_2H_5O}{>}}C\overset{SH}{\underset{S^-}{<}} \rightleftharpoons \underset{(23)}{(C_2H_5O)_2C{=}S} \quad (100)$$

In water a similar intermediate 25 is postulated. In the presence of two equivalents of silver nitrate the salt $Ag_2S_2C(OH)(OC_2H_5)$ is obtained. With hydrogen ions the intermediate is supposed to take up a proton to form 26 which is either converted to ethylxanthic acid or to ethanol and carbon disulphide (equation 101). Thus (101) seems a more elaborate version of (95). It is important to realize that the kinetics are only obeyed if the rate-determining step is the breakdown of 26 without participation of another hydronium ion. Since the removal of an ethoxide ion is not very probable, proton transfer from sulphur to the alkoxide oxygen must be a preequilibrium. This of course brings the mechanism towards scheme 96). The real difference between (95) and (96) is the question of whether he water molecule participates in the transition complex. No experiments have been performed to solve this question.

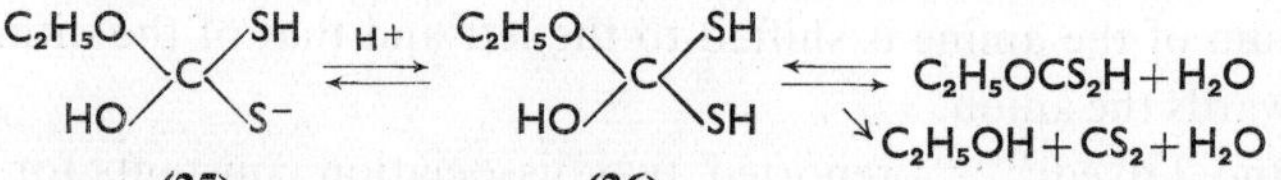

$$\underset{(25)}{\underset{HO}{\overset{C_2H_5O}{>}}C\overset{SH}{\underset{S^-}{<}}} \overset{H^+}{\rightleftharpoons} \underset{(26)}{\underset{HO}{\overset{C_2H_5O}{>}}C\overset{SH}{\underset{SH}{<}}} \rightleftharpoons \begin{array}{l} C_2H_5OCS_2H + H_2O \\ C_2H_5OH + CS_2 + H_2O \end{array} \quad (101)$$

VII. DITHIOCARBOXYLIC ESTERS

A. Preparation

The easiest and therefore almost exclusively applied method for the synthesis of dithio esters is the reaction of a dithio acid salt with an alkyl halide.

$$CH_3MgX + CS_2 \longrightarrow CH_3CS_2MgX + RBr \longrightarrow CH_3CS_2R + MgXBr$$

$$(102)^{219, 230, 262, 275-277}$$

$$C_2H_5OH + NaOH + CS_2 \longrightarrow C_2H_5OCS_2Na + RCl \longrightarrow C_2H_5OCS_2R + NaCl \quad (103)^{278}$$

$$(CH_3)_2NH + CS_2 + NaOH \longrightarrow (CH_3)_2NCS_2Na + RCl \longrightarrow (CH_3)_2NCS_2R + NaCl$$

$$(104)^{216}$$

Variations are possible—dimethyl sulphate may be used[279] and reactions with other substrates instead of alkyl halides have been applied in isolated cases. A synthesis via a Friedel–Crafts dithiocarboxylation has been described giving 43% dithioester[280]:

$$C_6H_6 + CS_2 + ClCH_2Si(CH_3)_3 \xrightarrow{AlCl_3} C_6H_5CS_2CH_2Si(CH_3)_3 \quad (105)$$

A second general, although less used, synthesis proceeds via acid chlorides; for carbodithioic acids the difficulty of preparing the acid chlorides makes it unattractive but for xanthates[203] and dithiocarbamates[281-284] thiophosgene, which is readily accessible, may be used:

$$C_2H_5OH + Cl_2CS \longrightarrow C_2H_5OCSCl + C_2H_5SH \longrightarrow C_2H_5OCS_2C_2H_5 \quad (106)$$

$$C_2H_5SH + Cl_2CS \longrightarrow C_2H_5SCSCl + C_2H_5OH \longrightarrow C_2H_5OCS_2C_2H_5 \quad (107)$$

$$C_2H_5SH + Cl_2CS \longrightarrow C_2H_5SCSCl + R_2NH \longrightarrow R_2NCS_2C_2H_5 \quad (108)$$

$$R_2NH + Cl_2CS \longrightarrow R_2NCSCl + C_2H_5SH \longrightarrow R_2NCS_2C_2H_5 \quad (109)$$

Reactions with thiophosgene are extremely versatile since the groups are attached stepwise. The first step occurs in general when the reagents are added together, the second needs the presence of alkali[118, 282]. The method can be used for the synthesis of any compound $AC\overset{\displaystyle S}{\underset{\displaystyle B}{\Big\langle}}$ in which A and or B are thiolo, amino (or alkylated amino) or alkoxy groups. Reactions can be carried out by adding the reagents in any sequence (see equations 106–109) so that expensive chemicals can be used in the final stage of the reaction. With ammonia and primary amines route (109) gives thiocyanic acid or isothiocyanates in the first step which give the final product by addition of the second reagent (equation 110).

$$RNH_2 + Cl_2CS \longrightarrow RN\!=\!C\!=\!S + 2HCl + C_2H_5SH \longrightarrow RHNCS_2C_2H_5 \quad (110)$$
R = H, alkyl

Parallel to the synthesis of thiono esters (section V) dithio esters may be obtained by reaction of iminothio ethers[58, 194, 262, 276, 277, 285, 286] and hydrogen sulphide or from thioamides and a thiol[287]. Probably the same reaction scheme is involved.

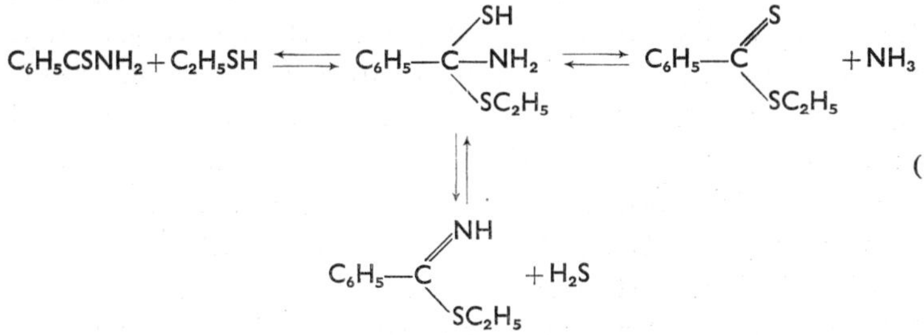

$$(111)$$

As the aminolysis of the dithio ester is a rapid reaction, acid must be added to bind the ammonia in order to obtain the dithio ester from a thio amide (the iminothio acid forms a hydrochloride as well, so when starting from this compound acid does not help). An extremely versatile reagent has been found by Eilingsfeld, Seefelder and Weidinger[288] in immonium chlorides (27). The reactions are similar to those in scheme (111) but they have a wider scope and are better adapted for synthetic purposes. The routes to relevant compounds are shown in equation (112).

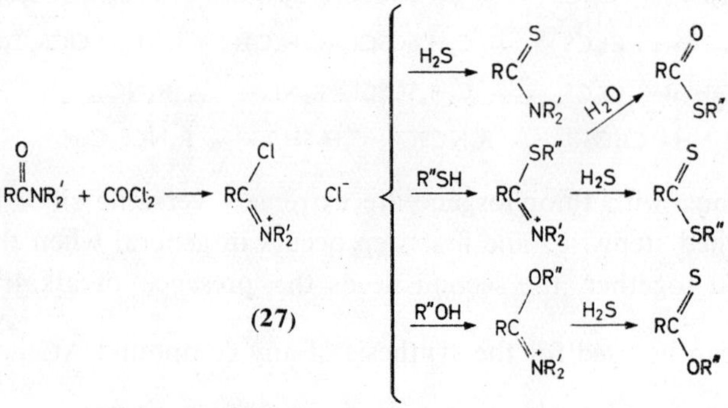

$$(112)$$

Parallel to the preparation of thiolo esters dithio esters can be made by addition of the corresponding acids to unsaturated compounds but this method has seldom been put into practice[289, 290].

$$R_2NCS_2Na + CH_2{=}CHCN \longrightarrow R_2NCS_2CH_2CH_2CN \qquad (113)$$

B. Properties

The influence of the dithiocarboxyl group as a substituent has been determined by a study of the acid dissociation constants of acetic acids of the type RCS_2CH_2COOH[291] (R = alkyl, CH_3O, C_2H_5S, $(CH_3)_2N$). The data are given in Table 16. Replacement of the group $R = CH_3$ makes the acids weaker in the sequence $R = C_2H_5S$, CH_3O, $(CH_3)_2N$. Since the three last groups are all more electronegative (which would mean acid-strengthening) than the methyl group, the dissociation constants are clearly governed by the strong mesomeric electron release of these substituents R. The sequence of decreasing acid strength (S, O, N) is the same as that of increasing resonance. Compared with thioglycolic acid ($pK_a = 3{\cdot}68$) the acids of Table 16 are all stronger.

Dithiocarboxylic esters are hydrolysed in acid and basic solutions. Kinetics seem not to have been studied, but in competitive experiments it has been shown[23, 60] that an ester group is hydrolysed faster in alkaline solution than a dithiocarboxylic ester group (section I. D. 2). According to old investigations on xanthic esters the thiol group is split off when the

TABLE 16. Acid dissociation constants at 25°
of some acids RCS_2CH_2COOH[291].

Compound[a]	pK_a
$CH_3CS_2CH_2COOH$	2·90
$CH_3OCS_2CH_2COOH$	3·09
$C_2H_5SCS_2CH_2COOH$	2·91
$(CH_3)_2NCS_2CH_2COOH$	3·32

[a] Compare CH_3COOH, $pK_a = 4·76$.

hydrolysis is stopped at the half-ester[292, 293]. This is in accordance with

$$C_2H_5OCS_2C_2H_5 \xrightarrow{OH^-} C_2H_5OCOS^- + C_2H_5SH \qquad (114)$$

the behaviour on aminolysis (see below). Dithiocarbamic esters are extremely resistant towards acid hydrolysis[289].

$$R_2NCS_2CH_2CH_2CN \xrightarrow{H^+} R_2NCS_2CH_2CH_2COOH$$
$$\xrightarrow[H_2SO_4]{conc.} R_2NCS_2CH_2CH_2CONH_2 \qquad (115)$$

Aminolysis of carbodithio esters runs (as can be expected) smoothly, yielding thio amides[230]. The reaction has been used to thioacylate amino acids and esters[294].

$$C_6H_5CS_2CH_2COONa + H_2NCH_2CO_2CH_3 \longrightarrow C_6H_5CSNHCH_2CO_2CH_3 \qquad (116)$$

With hydrazine, esters of dithio acids react rapidly at 0° provided the lower esters are used[277]. The reactions of mono- and dialkylhydrazines have been studied as well[295].

Xanthates are aminolysed with splitting of the C—S bond (equation 117). The method is useful for the preparation of thionocarbamic esters[296]. Dithiocarbamic esters are very difficult to aminolyse (equation 118)[297].

$$C_2H_5OCS_2C_2H_5 + NH_3 \longrightarrow C_2H_5OCSNH_2 + C_2H_5SH \qquad (117)$$

$$(CH_3)_2NCS_2CH_2CO_2CH_3 \xrightarrow[100°]{NH_3} (CH_3)_2NCS_2CH_2CONH_2 \ (67\%) \qquad (118)$$
$$(28)$$

With hydrazine at room temperature the carboxylic ester group in **28** is attacked and only in refluxing alcohol is the dithiocarbamate group slowly hydrazinolysed[297, 298].

Thiono and dithio esters as well as their derivatives like thioureas have nucleophilic thiono sulphur atoms. Because of increase of resonance in the series SCS < OCS < NCS, the thioureas exhibit this property most pronouncedly. Thus with methyl iodide thiourea gives methyl isothiourea (equation 119).

$$CH_3I + (NH_2)_2C{=}S \longrightarrow [(NH_2)_2CSCH_3]^+I^- \qquad (119)$$

Analogous reactions have been described for xanthates (equation 120)[299] and thionocarbamates (equation 121)[300].

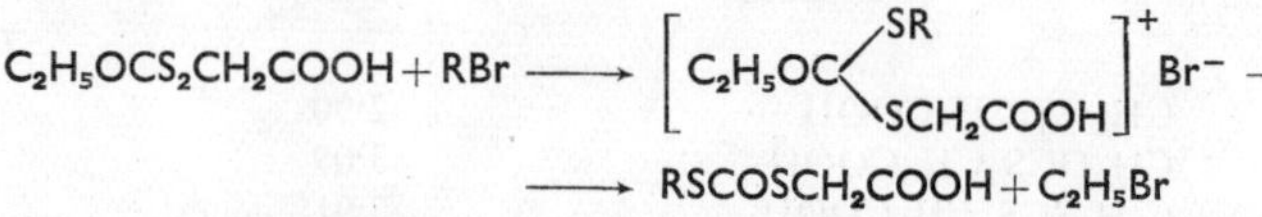

$$C_2H_5OCS_2CH_2COOH + RBr \longrightarrow \left[C_2H_5OC \begin{smallmatrix} SR \\ \\ SCH_2COOH \end{smallmatrix} \right]^+ Br^- \longrightarrow \tag{120}$$

$$\longrightarrow RSCOSCH_2COOH + C_2H_5Br$$

$$C_2H_5OCSNH_2 + CH_3I \longrightarrow CH_3SCONH_2 + C_2H_5I \tag{121}$$

A number of reactions in the dithio ester field belong to the same class. With starting materials of suitable structure, heterocyclic compounds are formed[301, 302].

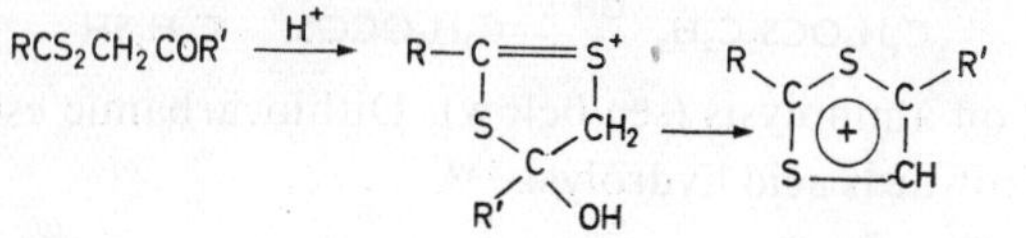

$$RCS_2CH_2COR' \xrightarrow{H^+} R-C{=\!\!=}S^+ \quad \underset{S-\!\!-CH}{\overset{R\diagdown}{C}} \underset{|}{\overset{S}{\diagup}} \underset{|}{\overset{\diagup R'}{C}} \tag{122}$$

$$C_6H_5CS_2CH_2CN \longrightarrow C_6H_5C{=\!\!=}S^+ \quad \underset{S-\!\!-CH}{\overset{C_6H_5\diagdown}{C}} \underset{|}{\overset{S}{\diagup}} \underset{|}{\overset{\diagup C-NH_2}{C}} \tag{123}$$

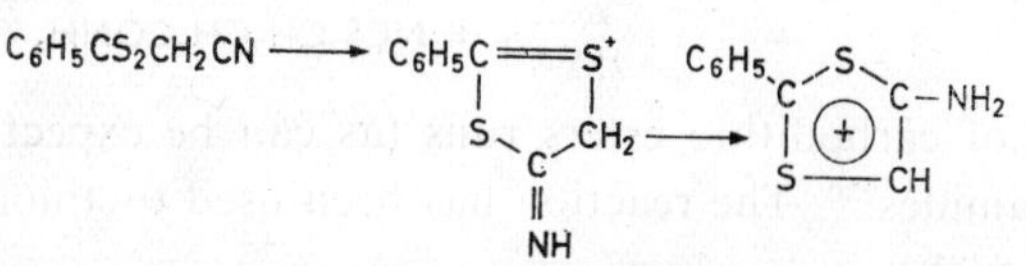

The Schönberg rearrangement[303] is supposed to follow the same mechanism[304] involving the nucleophilic attack of sulphur on the alkyl group (equation 124). It takes place when thionocarbonates are heated to 275–300°.

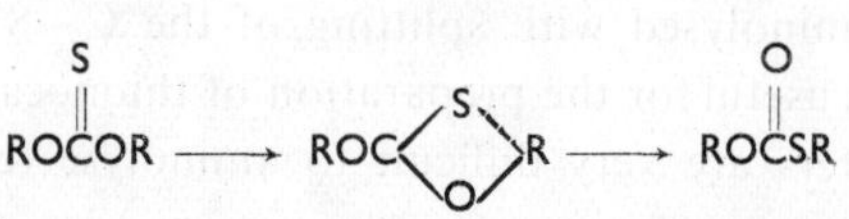

$$\underset{ROCOR}{\overset{S}{\|}} \longrightarrow ROC \underset{O}{\overset{S}{\diagdown\diagup}} R \longrightarrow \underset{ROCSR}{\overset{O}{\|}} \tag{124}$$

By carrying out the reactions in the vapour phase Kwart and Evans obtained very high yields of thiolo esters[305]. With a thionocarbamate over 80% of the corresponding thiolo esters were formed at a temperature of 400°! The reaction has been used for the synthesis of thiophenold and other aromatic thiols from phenols[306].

$$\underset{(C_2H_5)_2NCOR}{\overset{S}{\|}} \longrightarrow \underset{(C_2H_5)_2NCSR}{\overset{O}{\|}} \tag{125}$$

The well-known Chugaev reaction of xanthates may be considered a 1,6-cyclization related to the 1,4-cyclization in the Schönberg rearrangement.

The nucleophilic action of the thiono sulphur is particularly illustrated in the reaction of dithio esters under circumstances where carbanions can be formed in the α position to the ester group. Alkylation of the carbanion takes place at the sulphur atom[307]. The analogous oxygen esters are normally alkylated at the carbon atom:

$$RCH_2C\overset{\displaystyle S}{\underset{\displaystyle SR'}{\big<}} \xrightarrow{\ NH_2^-\ } R\bar{C}HC\overset{\displaystyle S}{\underset{\displaystyle SR'}{\big<}} + R''Br \longrightarrow RCH{=}C\overset{\displaystyle SR''}{\underset{\displaystyle SR'}{\big<}} \qquad (126)$$

$$RCH_2C\overset{\displaystyle O}{\underset{\displaystyle OR'}{\big<}} \xrightarrow{\ NH_2^-\ } R\bar{C}HC\overset{\displaystyle O}{\underset{\displaystyle OR'}{\big<}} + R''Br \longrightarrow RR''CHC\overset{\displaystyle O}{\underset{\displaystyle OR'}{\big<}} \qquad (127)$$

The interesting implications of the Clemmensen reduction of dithio esters have been discussed by Mayer, Scheithauer and Kunz[262]. By reduction of the thiocarbonyl group, thio ethers are formed in yields varying from 30–60%. Thiono esters are hydrolysed under Clemmensen conditions, thiolo esters and carboxylic esters are resistant against Clemmensen reduction. A fair relation was found between the amount of thio ether formed and the polarographic half-wave potential[262].

Oxidation of dithio carboxylic esters by peracetic acid leads to the thiono-S-oxide[308]; more than one equivalent gives further oxidation at the thiolo sulphur atom[309] (equation 128).

$$C_6H_5CS_2C_6H_5 \longrightarrow C_6H_5{-}C\overset{\displaystyle SO}{\underset{\displaystyle SC_6H_5}{\big<}} \longrightarrow$$

$$C_6H_5{-}C\overset{\displaystyle SO}{\underset{\displaystyle \underset{\displaystyle O}{\overset{\displaystyle \|}{S}}{-}C_6H_5}{\big<}} \longrightarrow C_6H_5{-}C\overset{\displaystyle SO}{\underset{\displaystyle SO_2C_6H_5}{\big<}} \qquad (128)$$

VIII. ACKNOWLEDGEMENTS

The author is indebted to Professor R. Mayer and Drs. J. Sandström, D. P. N. Satchell and S. Scheithauer for permission to include unpublished material, and to Professor J. Strating for many comments on the manuscript.

IX. REFERENCES

1. E. E. Reid, *Organic Chemistry of Bivalent Sulfur Compounds*, Vol. 4., Chemical Publishing Company, New York, 1962.
2. *Houbel–Weyl's Methoden der Organischen Chemie* (Ed. Eugen Müller), 4th ed., Vol. 9, Georg Thieme Verlag, Stuttgart, 1955. In particular A. Schöberl and A. Wagner, Chap. 23.

3. E. Campaigne, in *The Chemistry of the Carbonyl Group* (Ed. S. Patai), Interscience, London, 1966, Chap. 17.
4. A. Schönberg, *Chem. Ber.*, **62**, 195 (1929).
5. S. Bleisch and R. Mayer, *Chem. Ber.*, **99**, 1771 (1966).
6. R. Mayer, in *Organosulfur Chemistry* (Ed. M. J. Janssen), Interscience, New York, 1967, Chap. 13.
7. M. J. Janssen, 'The electronic structure of organic thione compounds', *Ph.D. Thesis*, Utrecht, 1959.
8. M. J. Janssen, *Rec. Trav. Chim.*, **79**, 1066 (1960).
9. J. Sandström, *Acta Chem. Scand.*, **16**, 1616 (1962); **17**, 678 (1963).
10. J. Sandström and B. Uppström, *Acta Chem. Scand.*, **19**, 2432 (1965).
11. A. W. Baker and G. H. Harris, *J. Am. Chem. Soc.*, **82**, 1923 (1960).
12. W. Walter, G. Maerten and H. Rose, *Ann. Chem.*, **691**, 25 (1966).
13. J. Sandström, *J. Phys. Chem.*, **71**, 2318 (1967).
14. S. Sunner, *Acta Chem. Scand.*, **9**, 837 (1955).
15. M. J. Janssen and J. Sandström, *Tetrahedron*, **20**, 2339 (1964).
16. A. Mehlhorn and J. Fabian, *Z. Chem.*, **5**, 420 (1965).
17. J. Fabian, *Ph.D. Thesis*, Technical University, Dresden, 1965.
18. A. Mehlhorn, J. Fabian and R. Mayer, *Z. Chem.* **5**, 21 (1965).
19. J. Fabian, A. Mehlhorn and R. Mayer, *Z. Chem.*, **5**, 22 (1965).
20. J. Fabian, H. Viola and R. Mayer, *Tetrahedron*, **23**, 4323 (1967).
21. J. Sandström and M. J. Janssen, unpublished calcuatlions.
22. S. G. Smith and M. O'Leary, *J. Org. Chem.*, **28**, 2825 (1963).
23. P. V. Laakso, *Suomen Kemistilehti*, **17B**, 1 (1944).
24. W. J. Middleton, E. G. Howard and W. H. Sharkey, *J. Org. Chem.*, **30**, 1375 (1965).
25. W. Gordy, *J. Chem. Phys.*, **14**, 560 (1946).
26. N. Sheppard, *Trans. Faraday Soc.*, **45**, 693 (1949).
27. W. W. Crouch, *J. Am. Chem. Soc.*, **74**, 2926 (1952).
28. R. Mecke and H. Spiesecke, *Chem. Ber.*, **89**, 1110 (1956).
29. F. Bloch, *Compt. Rend.*, **206**, 679 (1938).
30. Yu N. Sheinker, S. T. Ioffe and M. I. Kabachnik, *Izv. Akad. Nauk SSSR, Otd. Khim. Nauk*, 1571 (1960); English translation p. 1463.
31. M. J. Janssen, *Rec. Trav. Chim.*, **79**, 464 (1960).
32. T. L. Cottrell, *The Strength of Chemical Bonds*, Butterworths, London, 1954.
33. R. A. Nyquist and W. J. Potts, *Spectrochim. Acta*, **15**, 514 (1959).
34. A. Hantzsch and E. Scharf, *Chem. Ber.*, **45**, 3570 (1913).
35. H. P. Koch, *J. Chem. Soc.*, 387 (1949).
36. Y. Hirabayashi and T. Mazume, *Bull. Chem. Soc. Japan*, **38**, 171 (1965).
37. A. Hantzsch and W. Bucerius, *Chem. Ber.*, **59**, 793 (1926).
38. K. Issleib and W. Gründler, *Z. Chem.*, **6**, 318 (1966).
39. J. F. Bunnett and W. D. Merritt, *J. Am. Chem. Soc.*, **79**, 5967 (1957).
40. J. F. Bunnett, C. F. Hauser and K. V. Nahabedian, *Proc. Chem. Soc.*, 305 (1961).
41. E. C. Kooyman, in *Organosulfur Chemistry* (Ed. M. J. Janssen), Interscience, New York, 1967, Chap. 1.
42. R. G. Pearson, *J. Am. Chem. Soc.*, **85**, 3533 (1963).
43. J. F. Bunnett, *J. Am. Chem. Soc.*, **79**, 5969 (1957).
44. J. O. Edwards, *J. Am. Chem. Soc.*, **76**, 1540 (1954); **78**, 1819 (1956).
45. R. F. Hudson, *Structure and Mechanism in Organo-Phosphorus Chemistry*, Academic Press, London, 1965, Chap. 4.
46. J. O. Edwards and R. G. Pearson, *J. Am. Chem. Soc.*, **84**, 16 (1962).
47. R. E. Davis, *Survey of Progress in Chemistry* (Ed. A. Scott), Vol. 2, Academic Press, New York, 1964, p. 189.

48. R. E. Davis, H. Nakshbendi and A. Ohno, *J. Org. Chem.*, **31**, 2702 (1966).
49. F. W. Stacey and J. F. Harris, *Organic Reactions* (Ed. A. C. Cope), Vol. 13, John Wiley and Sons, New York, 1963, Chap. 4.
50. J. I. Cunneen, *J. Chem. Soc.*, 36, 134 (1947).
51. C. Walling and W. Helmreich, *J. Am. Chem. Soc.*, **81**, 1144 (1959).
52. N. P. Neureiter and F. G. Bordwell, *J. Am. Chem. Soc.*, **82**, 5354 (1960).
53. G. Cilento, *J. Am. Chem. Soc.*, **75**, 3748 (1953).
54. R. Mayer and S. Scheithauer, *Chem. Ber.*, **98**, 829 (1965).
55. E. Spinner, *Spectrochim. Acta*, **15**, 95 (1959).
56. B. Bak, L. Hansen-Nygaard and C. Pedersen, *Acta Chem. Scand.*, **12**, 1451 (1958).
57. L. J. Bellamy and P. E. Rogasch, *J. Chem. Soc.*, 2218 (1960).
58. M. Renson and J. Bidaine, *Bull. Soc. Chim. Belg.*, **70**, 519 (1961).
59. M. L. Shankaranarayana and C. C. Patel, *Spectrochim. Acta*, **21**, 95 (1965).
60. S. Scheithauer, *Ph.D. Thesis*, Technical University, Dresden, 1967.
61. J. C. McCool, *U.S. Pat.* 2,568,020 (1951); *Chem. Abstr.*, **46**, 3557h (1952); *U.S. Pat.* 2,587,580 (1952); *Chem. Abstr.*, **46**, 10192d (1952).
62. P. Noble, Jr., and D. S. Tarbell, *Org. Syntheses*, Coll. Vol. 4 (Ed. N. Rabjohn), John Wiley and Sons, New York, 1962, p. 924.
63. E. K. Ellingboe, *Organic Syntheses* Coll. Vol. 4. (Ed. N. Rabjohn), John Wiley and Sons, New York, 1962, p. 928.
64. Y. Hirabayashi, M. Mizuta and T. Mazume, *Bull. Chem. Soc. Japan*, **38**, 320 (1965).
65. V. Auger and M. Billy, *Compt. Rend.*, **136**, 555 (1903).
66. W. R. Vaughan and J. B. Baumann, *J. Org. Chem.*, **27**, 739 (1962).
67. B. Tchoubar, *Bull. Soc. Chim. France*, 792 (1947).
68. F. A. Kekulé, *Ann. Chem.*, **90**, 311 (1854).
69. F. Weigert, *Chem. Ber.*, **36**, 1007 (1903).
70. W. W. Crouch, *U.S. Pat.* 2,639,293 (1953); *Chem. Abstr.*, **48**, 3387f (1954).
71. T. Wieland and D. Sieber, *Nature*, 242 (1953).
72. T. Wieland, D. Sieber and W. Bartmann, *Chem. Ber.*, **87**, 1093 (1954).
73. J. Seydel, *Tetrahedron Letters*, 1145 (1966).
74. W. Ostwald, *Z. Physik. Chem. (Leipzig)*, **3**, 170 (1889).
75. J. Juillard, *Bull. Chem. Soc. France*, 3069 (1964).
76. M. M. Kreevoy, B. E. Eichinger, F. E. Stary, E. A. Katz and J. H. Sellstedt, *J. Org. Chem.*, **29**, 1641 (1964).
77. J. Hipkin and D. P. N. Satchell, *Tetrahedron*, **21**, 835 (1965).
78. B. Saxton and H. F. Meier, *J. Am. Chem. Soc.*, **56**, 1918 (1934).
79. F. G. Brockman and M. Kilpatrick, *J. Am. Chem. Soc.*, **56**, 1483 (1934).
80. S. T. Ioffe, Yu. N. Sheinker and M. I. Kabachnik, *Izv. Akad. Nauk SSSR, Otd. Khim. Nauk*, 1561 (1960); English translation, p. 1454.
81. I. M. Ginzburg and L. A. Loginova, *Optika i Spektroskopiya*, **20**, 241 (1966); *Chem. Abstr.*, **65**, 1594b (1966).
82. G. M. Barrow and E. A. Yerger, *J. Am. Chem. Soc.*, **76**, 5248 (1954).
83. K. B. Sandell, *Naturwissenshaften*, **53**, 330 (1966).
84. M. Cefola, S. Peter, P. S. Gentile and A. V. Celiano, *Talanta*, **9**, 537 (1962).
85. P. J. Hawkins, D. S. Tarbell and P. Noble, *J. Am. Chem. Soc.*, **75**, 4462 (1953).
86. T. Wieland and K. Freter, *Chem. Ber.*, **87**, 1099 (1954).
87. L. N. Owen and J. H. Chapman, *J. Chem. Soc.*, 579 (1950); L. N. Owen and P. Bladon, *J. Chem. Soc.*, 585 (1950).
88. P. Weselsky, *Chem. Ber.*, **2**, 518 (1869).
89. A. Schönberg, O. Schütz and S. Nickel, *Chem. Ber.*, **61**, 1375 (1928).
90. Q. Mingoia, *Gazz. Chim. Ital.*, **55**, 713 (1925).
91. I. A. M. Ford and S. A. M. Thompson, *Nature*, **185**, 96 (1960).

92. B. Holmberg and E. Schjånberg, *Arkiv Kemi*, **14A**, No. 7 (1940).
93. L. N. Owen and M. U. S. Sultanbawa, *J. Chem. Soc.*, 3109 (1949).
94. H. Behringer and G. F. Grunwald, *Ann. Chem.*, **600**, 23 (1956).
95. S. Ishikawa, *Sci. Papers Inst. Phys. Chem. Res. (Tokyo)*, **7**, 293 (1928); *Chem. Abstr.*, **22**, 1343 (1928).
96. F. Duus, P. Jakobsen and S.–O. Lawesson, *Tetrahedron*, **24**, 5323 (1968).
97. H. L. Wheeler and H. F. Merriam, *J. Am. Chem. Soc.*, **23**, 283 (1901); **24**, 439 (1902).
98. H. L. Wheeler and T. B. Johnson, *J. Am. Chem. Soc.*, **24**, 680 (1902).
99. W. A. Pryor, *Mechanism of Sulfur Reactions*, McGraw-Hill Book Company, New York, 1962, p. 71.
100. H. J. Alkema and J. F. Arens, *Rec. Trav. Chim.*, **79**, 1257 (1960).
101. W. E. Truce, *Organic Sulfur Compounds*, (Ed. N. Kharasch), Pergamon Press, Oxford, 1961, Chap. 12.
102. G. R. Banks and D. Cohen, *Proc. Chem. Soc.*, 83 (1963).
103. H. Bader, L. C. Cross, I. Heilbron and E. R. H. Jones, *J. Chem. Soc.*, 619 (1949).
104. H. Behringer, *Ann. Chem.*, **564**, 219 (1949).
105. P. S. Fitt and L. N. Owen, *J. Chem. Soc.*, 2240 (1957).
106. H. L. Goering, D. I. Relyea and D. W. Larsen, *J. Am. Chem. Soc.*, **78**, 348 (1956).
107. F. G. Bordwell, Ph. S. Landis and G. S. Whitney, *J. Org. Chem.*, **30**, 3764 (1965); F. G. Bordwell and W. A. Hewett, *J. Am. Chem. Soc.*, **49**, 3493 (1957).
108. E. N. Prilezhayeva, in *Organo Sulfur Chemistry* (Ed. M. J. Janssen), Interscience, New York, 1967, Chap. 4.
109. A. A. Oswald, K. Griesbaum, D. N. Hall and W. Naegele, *Am. Chem. Soc., Div. Petrol. Chem. Preprints*, **9**, 29 (1964); *Chem. Abstr.*, **64**, 12474h (1966).
110. A. A. Oswald and W. Naegele, *J. Org. Chem.*, **31**, 830 (1966).
111. M. W. Cronyn and J. Jiu, *J. Am. Chem. Soc.*, **74**, 4726 (1952).
112. M. Mikolajczyk, *Angew. Chem.*, **78**, 393 (1966); *Intern. Ed.* **5**, 419 (1966).
113. H. Behringer and A. Grimm, *Ann. Chem.*, **682**, 188 (1965).
114. H. Behringer and G. F. Grunwald, *Ann. Chem.*, **600**, 23 (1956).
115. B. Holmberg, *Ann. Chem.*, **353**, 131 (1907).
116. A. A. Schleppnik and F. B. Zienty, *J. Org. Chem.*, **29**, 1910 (1964).
117. H. Hepworth and H. W. Clapham, *J. Chem. Soc.*, **119**, 1188 (1921).
118. H. Rivier and P. Richard, *Helv. Chim. Acta*, **8**, 490 (1925).
119. H. Rheinboldt, F. Mott and E. Motzkus, *J. Prakt. Chem.*, [2] **134**, 257 (1932).
120. M. Rimpler, *Chem. Ber.*, **99**, 1528 (1966).
121. W. E. Davies and W. E. Savige, *J. Chem. Soc.*, 317 (1950).
122. V. V. Alderman, M. M. Brubaker and Wm. E. Hanford, *U.S. Pat.* 2,212,141 (1940); *Chem. Abstr.*, **35**, 4632 (1941).
123. A. Cahours, *Bull. Soc. Chim. France*, [2] **25**, 563 (1876); *Compt. Rend.* **81**, 1163 (1876).
124. E. L. Gustus and P. G. Stevens, *J. Am. Chem. Soc.*, **55**, 378 (1933).
125. H. Boehme and J. Roehr, *Ann. Chem.*, **648**, 21 (1961).
126. H. E. Wijers, P. P. Montijn, L. Brandsma and J. F. Arens, *Rec. Trav. Chim.*, **84**, 1284 (1965).
127. S. Searles Jr., S. Nukina and E. R. Magnuson, *J. Org. Chem.*, **30**, 1920 (1965).
128. R. Kuhn and G. Quadbeck, *Chem. Ber.*, **84**, 844 (1951).
129. C. M. Himel, *U.S. Pat.* 2,458,075 (1949); *Chem. Abstr.*, **43**, 3444i (1949).
130. J. Houben and R. Zivadinovitsch, *Chem. Ber.*, **69**, 2352 (1936).
131. L. W. C. Miles and L. N. Owen, *J. Chem. Soc.*, 817 (1952).
132. C. E. Dalgliesch and F. G. Mann, *J. Chem. Soc.*, 559 (1947).
133. R. Schwyzer, *Helv. Chim. Acta*, **36**, 414 (1953).
134. E. E. Reid, *Am. Chem. J.*, **43**, 489 (1910).

135. M. P. Jencks, S. Cordes and J. Carriuolo, *J. Biol. Chem.*, **235**, 3608 (1960).
136. L. S. Pratt and E. E. Reid, *J. Am. Chem. Soc.*, **37**, 1934 (1915).
137. A. Chablay, *Compt. Rend.*, **263**, 157 (1966).
138. A. Chablay, *Compt. Rend.*, **239**, 172 (1954); **240**, 2528 (1955).
139. T. Wieland and E. Bokelmann, *Ann. Chem.*, **576**, 20 (1952).
140. R. B. Martin and A. Parcell, *J. Am. Chem. Soc.*, **83**, 4830 (1961).
141. R. B. Martin and A. Parcell, *J. Am. Chem. Soc.*, **83**, 4835 (1961).
142. R. B. Martin, R. I. Hedrick and A. Parcell, *J. Org. Chem.* **29**, 3197 (1964).
143. B. Hansen, *Acta Chem. Scand.*, **17**, 1307 (1963).
144. R. B. Baker and E. E. Reid, *J. Am. Chem. Soc.*, **51**, 1567 (1929).
145. M. W. Cronyn, M. P. Chang and R. A. Wall, *J. Am. Chem. Soc.*, **77**, 3031 (1955).
146. J. C. Shivers, B. E. Hudson and C. R. Hauser, *J. Am. Chem. Soc.*, **65**, 2051 (1943).
147. L. Wessely and F. Lynen, *Federation Proc.*, **12**, 685 (1953).
148. T. C. Bruice, *Organic Sulfur Compounds* (Ed. N. Kharasch), Pergamon Press, Oxford, 1961, Chap. 35.
149. H. Gilman, J. Robinson and N. J. Beaber, *J. Am. Chem. Soc.*, **48**, 2715 (1926).
150. N. V. Organon, *Dutch Pat.* 65,696 (1950); *Chem. Abstr.*, **44**, 7350g (1950).
151. I. B. Douglass and T. B. Johnson, *J. Am. Chem. Soc.*, **60**, 1486 (1938).
152. I. B. Douglass and C. E. Osborne, *J. Am. Chem. Soc.*, **75**, 4582 (1953).
153. G. B. Spero. A. V. McIntosh Jr. and R. H. Levin, *J. Am. Chem. Soc.*, **70**, 1907 (1948).
154. M. L. Wolfrom and J. V. Karabinos, *J. Am. Chem. Soc.*, **68**, 724, 1455 (1946).
155. A. V. McIntosh Jr., E. M. Meinzer and R. H. Levin, *J. Am. Chem. Soc.*, **70**, 2955 (1948).
156. V. Prelog, J. Norymberski and O. Jeger, *Helv. Chim. Acta*, **29**, 360 (1946).
157. O. Jeger, J. Norymberski, S. Szpilfogel and V. Prelog, *Helv. Chim. Acta*, **29**, 684 (1946).
158. H. Hauptmann and B. Wladislaw, *J. Am. Chem. Soc.*, **72**, 707, 710 (1950).
159. H. Hauptmann, B. Wladislaw, L. L. Nazario, and W. F. Walter, *Ann. Chem.*, **576**, 45 (1952).
160. M. S. Newman, M. W. Renoll and I. Auerbach, *J. Am. Chem. Soc.*, **70**, 1023 (1948).
161. H. Hauptmann and P. A. Bobbio, *Chem. Ber.*, **93**, 280 (1960).
162. P. A. Bobbio, *J. Org. Chem.*, **26**, 3023 (1961).
163. P. Brandt, *Acta Chem. Scand.*, **3**, 1050 (1949).
164. F. Lynen, E. Reichert and L. Rueff, *Ann. Chem.*, **574**, 1 (1951).
165. T. Wieland and W. Schäfer, *Angew. Chem.*, **63**, 146 (1951).
166. C. K. Ingold, *Structure and Mechanism in Organic Chemistry*, Cornell University Press, Ithaca, 1953, Chap. 14.
167. P. N. Rylander and D. S. Tarbell, *J. Am. Chem. Soc.*, **72**, 3021 (1950).
168. B. K. Morse and D. S. Tarbell, *J. Am. Chem. Soc.*, **74**, 416 (1952).
169. Y. Iskander, *Nature*, **155**, 141 (1945).
170. M. L. Bender, *Chem. Rev.*, **60**, 53 (1960).
171. L. R. Fedor and T. C. Bruice, *J. Am. Chem. Soc.*, **87**, 4138 (1965).
172. D. R. Llewellyn and Ch. O'Connor, *J. Chem. Soc.*, 545 (1964).
173. C. A. Bunton, D. H. James and J. B. Senior, *J. Chem. Soc.*, 3364 (1960).
174. J. Hipkin and D. P. N. Satchell, *J. Chem. Soc.*, 1057 (1965); 345 (1966).
175. J. R. Schaefgen, *J. Am. Chem. Soc.*, **70**, 1308 (1948).
176. L. H. Noda, S. A. Kuby and H. A. Lardy, *J. Am. Chem. Soc.*, **75**, 913 (1953).
177. A. Skrabal and A. M. Hugetz, *Monatsh. Chem.*, **47**, 17 (1926).
178. L. Smith and H. Olsson, *Z. Physik. Chem. (Leipzig)*, **118**, 99 (1925).
179. H. Böhme and H. Schran, *Chem. Ber.*, **82**, 453 (1949).

180. L. M. Litvinenko and N. M. Oleinik, *Kataliz i Katalizatory, Akad. Nauk, Ukr. SSR Resp. Mezhved. sb.* 120 (1965); *Chem. Abstr.*, **64**, 4890a (1966).
181. H. Goldschmidt and R. Bräuer, *Chem. Ber.*, **39**, 97 (1906).
182. V. V. Koningsberger and J. T. G. Overbeek, *Koninkl. Ned. Akad. Wetensch., Proc.*, **B58**, 49 (1955).
183. J. T. G. Overbeek and V. V. Koningsberger, *Koninkl. Ned. Akad. Wetensch., Proc.*, **B58**, 266 (1955).
184. P. J. Hawkins and D. S. Tarbell, *J. Am. Chem. Soc.*, **75**, 2982 (1953).
185. D. S. Tarbell and D. P. Cameron, *J. Am. Chem. Soc.*, **78**, 2731 (1956).
186. K. A. Connors and M. L. Bender, *J. Org. Chem.*, **26**, 2498 (1961).
187. T. Wieland and H. Hornig, *Ann. Chem.*, **600**, 12 (1956).
188. J. S. Harding and L. N. Owen, *J. Chem. Soc.*, 1528, 1536 (1954).
189. R. Kuhn, G. Quadbeck and E. Röhm, *Chem. Ber.*, **86**, 468 (1953).
190. R. B. Martin and R. I. Hedrick, *J. Am. Chem. Soc.*, **84**, 106 (1962).
191. T. C. Bruice and L. R. Fedor, *J. Am. Chem. Soc.*, **86**, 4886 (1964).
192. R. B. Martin and L. P. Henkle, *J. Phys. Chem.*, **68**, 3438 (1964).
193. L. R. Fedor and T. C. Bruice, *J. Am. Chem. Soc.*, **86**, 4117 (1964).
194. S. A. Karjala and S. M. McElvain, *J. Am. Chem. Soc.*, **55**, 2966 (1933).
195. J. P. Jepson, A. Lawson and V. D. Lawton, *J. Chem. Soc.*, 1791 (1955).
196. R. Roger and D. G. Neilson, *Chem. Rev.*, **61**, 179 (1961).
197. P. Reynaud, R. C. Moreau and J. P. Samama, *Bull. Soc. Chim. France*, 3623 (1965).
198. P. Reynaud, R. C. Moreau and J. C. Tétard, *Compt. Rend., Ser. C.*, **262**, 665 (1966).
199. P. Reynaud and R. C. Moreau, *Bull. Soc. Chim. France*, 2999 (1964).
200. M. Delépine, *Compt. Rend.*, **150**, 1608 (1910); **153**, 281 (1911).
201. M. Delépine, *Bull. Soc. Chim. France*, **7**, 722 (1910); **9**, 904 (1911).
202. M. Delépine, *Ann. Chim. Phys.* **25**, 556 (1912).
203. M. Rivier, *Bull. Soc. Chim. France*, **35**, 837 (1906).
204. H. Staudinger and J. Siegwart, *Helv. Chim. Acta*, **3**, 824 (1920).
205. W. Meiser, *Ger. Pat.* 725,883 (1942); *Chem. Abstr.*, **37**, 5985[3] (1943).
206. S. Scheithauer and R. Mayer, *Chem. Ber.*, **98**, 838 (1965).
207. J. F. Norris, E. V. Fasce and C. J. Staud, *J. Am. Chem. Soc.*, **57**, 1415 (1935).
208. J. F. Norris and H. H. Young Jr., *J. Am. Chem. Soc.*, **57**, 1420 (1935).
209. G. R. Banks and D. Cohen, *Proc. Chem. Soc.*, 83 (1963).
210. S. G. Smith, *Tetrahedron Letters*, 979 (1962).
211. S. G. Smith, *J. Am. Chem. Soc.*, **83**, 4285 (1961).
212. S. G. Smith, A. H. Fainberg and S. Winstein, *J. Am. Chem. Soc.*, **83**, 618 (1961).
213. T. Taguchi, Y. Kawazoe, K. Yoshihira, H. Kanayama, M. Mori, K. Tabata and K. Harano, *Tetrahedron Letters*, 2717 (1965).
214. J. V. Koštíř and V. Král, *Chem. Listy*, **41**, 92 (1947); *Chem. Abstr.*, **47**, 3240f (1953).
215. E. Uhlemann and H. Müller, *Angew. Chem.*, **77**, 172 (1965).
216. G. D. Thorn and R. A. Ludwig, *The Dithiocarbamates and Related Compounds*, Elsevier Publishing Co., Amsterdam, 1962.
217. G. Gatlow and V. Hahnkamm, *Angew. Chem.*, **78**, 334 (1966).
218. R. W. Bost and O. L. Shealy, *J. Am. Chem. Soc.*, **73**, 24, 25 (1951).
219. A. Kjaer, *Acta Chem. Scand.*, **4**, 1347 (1950); **6**, 327 (1952).
220. G. Alliger, G. E. P. Smith, Jr., E. L. Carr and H. P. Stevens, *J. Org. Chem.*, **14**, 962 (1949).
221. O. Emmerling, *Chem. Ber.*, **28**, 2882 (1895).
222. F. Scheffer and R. Kickuth, *Ger. Pat.* 1,136,697 (1962); *Chem. Abstr.*, **58**, 9036a (1963).

223. E. Freund, *Chem. Ber.*, **52**, 542 (1919).
224. G. Bähr and G. Schleitzer, *Chem. Ber.*, **88**, 1771 (1955).
225. E. Treiber, *Monatsh. Chem.* **82**, 53 (1951).
226. M. Bögemann, S. Petersen, O. E. Schultz and H. Söll, *Houben Weyl's Methoden der Organischen Chemie*, 4th ed., Vol. 9, Georg Thieme Verlag, Stuttgart, 1955, p. 811.
227. E. Lippmann, *Monatsh. Chem.*, **10**, 617 (1889).
228. C. Schall, *J. Prakt. Chem.*, **54**, 415 (1896).
229. H. Jörg, *Chem. Ber.*, **60**, 1466 (1927).
230. B. Holmberg, *Arkiv Kemi*, **A17**, No. 23 (1944).
231. F. Kurzer and A. Lawson, *Org. Syn.*, **42**, 100 (1962).
232. H. Kitagawa, A. Kaji and T. Iwaki, *Japan Pat.* 13,965 (1962); *Chem. Abstr.*, **59**, 9913c (1963).
233. H. von Halban and W. Hecht, *Z. Elektrochem.*, **24**, 65 (1918).
234. I. Iwasaki and S. R. B. Cooke, *J. Phys. Chem.*, **63**, 1321 (1959).
235. I. Iwasaki and S. R. B. Cooke, *J. Am. Chem. Soc.*, **80**, 285 (1958).
236. G. Gattow and B. Krebs, *Angew. Chem.*, **74**, 29 (1962).
237. G. Gattow and B. Krebs, *Z. Anorg. Allgem. Chem.*, **325**, 15 (1963).
238. G. Eckert, *Z. Anal. Chem.*, **155**, 23 (1957).
239. K. Gleu and R. Schwab, *Angew. Chem.*, **62**, 320 (1950).
240. M. Delépine, *Bull. Soc. Chim. France*, 1 (1958).
241. V. Sedivec and J. Flek, *Chem. Listy*, **52**, 545 (1958).
242. E. A. Shuzam and V. M. Levine, *Kristallografiya*, **5**, 257 (1960).
243. A. Pignedoli and G. Peyronel, *Gazz. Chim. Ital.*, **92**, 745 (1962).
244. M. Franzini, *Z. Krist.*, **118**, 393 (1963).
245. R. Bally, *Compt. Rend.*, **257**, 425 (1963).
246. H. Debus, *Ann. Chem.*, **72**, 1 (1849); **75**, 121 (1850).
247. M. J. Janssen, *Rec. Trav. Chim.*, **76**, 827 (1957).
248. R. W. Bost and W. J. Mattox, *J. Am. Chem. Soc.*, **52**, 332 (1930).
249. A. R. Todd, F. Bergel, Karimullah and R. Keller, *J. Chem. Soc.*, 361 (1934).
250. H. Wuyts and L. C. Kuang, *Bull. Soc. Chim. Belg.*, **42**, 153 (1933).
251. H. B. König, W. Siefken and H. A. Offe, *Chem. Ber.*, **87**, 825 (1954).
252. K. A. Jensen and C. L. Jensen, *Acta Chem. Scand.*, **6**, 957 (1952).
253. H. Wuyts, *Bull. Soc. Chim. Belg.*, **38**, 195 (1929).
254. H. Wuyts, *Bull. Soc. Chim. Belg.*, **39**, 58 (1930).
255. L. I. Smith and J. Nichols, *J. Org. Chem.*, **6**, 489 (1941).
256. H. Wuyts and H. Koeck, *Bull. Soc. Chim. Belg.*, **41**, 196 (1932).
257. E. Mameli, F. K. Richter and F. D'Angeli, *Atti, Ist. Veneto Sci, hettere Arti, Classe Sci. Mat. Nat.*, **110**, 99, 101, 103 (1952); *Chem. Abstr.*, **49**, 186i, 187a (1955).
258. S. J. C. Snedker, *J. Soc. Chem. Ind.*, **44**, 74 (1925).
259. L. C. Raiford and G. M. McNulty, *J. Am. Chem. Soc.*, **56**, 680 (1934).
260. D. C. Schroeder, *Chem. Rev.*, **55**, 181 (1955).
261. J. Houben and H. Pohl, *Chem. Ber.*, **40**, 1303, 1725 (1907).
262. R. Mayer, S. Scheithauer and D. Kunz, *Chem. Ber.*, **99**, 1393 (1966).
263. I. Iwaski and S. R. B. Cooke, *J. Phys. Chem.*, **68**, 2031 (1964).
264. E. Klein, J. K. Bosarge and I. Norman, *J. Phys. Chem.*, **64**, 1666 (1960).
265. G. M. Lewis, *Dissertation*, New York University, 1947, cited. in ref. 264.
266. D. G. H. Ballard, C. H. Bamford, K. L. Gray and E. D. Totman, cited in ref. 264.
267. M. J. Janssen, unpublished observations.
268. R. Zahradník and P. Zuman, *Chem. Listy*, **52**, 231 (1958).
269. P. Zuman and R. Zahradník, *Z. Physik. Chem. (Leipzig)*, **208**, 135 (1957).

270. K. P. Soni and A. M. Trivedi, *J. Indian Chem. Soc.*, 37, 349 (1960).
271. K. P. Soni, A. M. Trivedi and I. M. Bhatt, *J. Indian Chem. Soc.*, 43, 85 (1966).
272. M. J. Janssen, *Rec. Trav. Chim.*, 81, 650 (1962).
273. H. Bode and F. Neumann, *Z. Anal. Chem.*, 169, 410 (1959).
274. G. Ingram and B. A. Toms, *J. Chem. Soc.*, 117 (1961).
275. J. Houben and K. M. L. Schultze, *Chem. Ber.*, 44, 3226 (1911).
276. H. Jörg, *Chem. Ber.*, 60, 1466 (1927).
277. K. A. Jensen and Ch. Pedersen, *Acta Chem. Scand.*, 15, 1087 (1961).
278. M. Bögemann, S. Petersen, D. E. Schultz and H. Söll, in *Houben Weyl's Methoden der Organischen Chemie* (Ed. E. Müller), 4th ed., Vol. 9, Georg Thieme Verlag, Stuttgart, 1955, Chap. 24.
279. G. Dupont, J. Lévy and M. Marot, *Bull. Soc. Chim. France*, 53, 393 (1933).
280. P. D. George, *J. Org. Chem.*, 26, 4235 (1961).
281. J. J. D'Amico and M. W. Harman, *J. Am. Chem. Soc.*, 77, 476 (1955).
282. H. Böhme, *Chem. Ber.*, 74, 248 (1941).
283. W. Ried, H. Hillenbrand and G. Oertel, *Ann. Chem.*, 590, 123 (1954).
284. R. H. Goshorn, W. W. Lewis Jr., E. Jaul and E. J. Ritter, *Organic Syntheses*, Coll. Vol. IV (Ed. N. Rabjohn) John Wiley and Sons, New York, 1963, p. 307.
285. D. A. Peak and F. Stansfield, *J. Chem. Soc.*, 4067 (1952).
286. C. S. Marvel, P. de Radzitzky and J. J. Brader, *J. Am. Chem. Soc.*, 77, 5997 (1955).
287. E. E. Reid, *Int. Congr. Appl. Chem.*, 25, 423 (1911).
288. H. Eilingsfeld, M. Seefelder and H. Weidinger, *Chem. Ber.*, 96, 2671 (1963).
289. R. Delaby, R. Damiens and R. Seyden-Penne, *Compt. Rend.*, 238, 121 (1954).
290. C. M. Buess, *J. Am. Chem. Soc.*, 77, 6613 (1955).
291. M. J. Janssen, *Rec. Trav. Chim.*, 82, 931 (1963).
292. R. Schmitt and L. Glutz, *Chem. Ber.*, 1, 166 (1868).
293. F. Salomon, *J. Prakt. Chem.*, 6, 433 (1873); 8, 114 (1874).
294. F. Kurzer, *Chem. Ind.*, 1333 (1961).
295. K. A. Jensen, H. R. Baccaro, D. Buchardt, G. E. Olsen, Ch. Pedersen and J. Taft, *Acta Chem. Scand.*, 15, 1109 (1961).
296. B. Holmberg, *J. Prakt. Chem.*, 84, 634 (1911); *Chem. Ber.*, 59, 1558 (1926).
297. G. Nachmias, *Ann. Chim. (Paris)*, 7, 584 (1952).
298. K. A. Jensen, *J. Prakt. Chem.*, 159, 189 (1941).
299. E. Bülmann and J. Bjerrum, *Chem. Ber.*, 50, 503 (1917).
300. H. L. Wheeler and B. Barnes, *Am. Chem. J.*, 22, 141 (1899); 24, 60 (1900).
301. E. Campaigne and N. W. Jacobsen, *J. Org. Chem.*, 29, 1703 (1964).
302. M. Ohta and M. Sugiyama, *Bull. Chem. Soc. Japan*, 36, 1437 (1963).
303. A. Schönberg and L. Vargha, *Chem. Ber.*, 63, 178 (1930).
304. D. H. Powers and D. S. Tarbell, *J. Am. Chem. Soc.*, 78, 70 (1956).
305. H. Kwart and E. R. Evans, *J. Org. Chem.*, 31, 410 (1966).
306. M. S. Newman and H. A. Karnes, *J. Org. Chem.*, 31, 3980 (1966).
307. P. J. W. Schuijl, L. Brandsma and J. F. Arens, *Rec. Trav. Chim.*, 85, 1263 (1966).
308. B. Zwanenburg, L. Thys and J. Strating, *Tetrahedron Letters*, 3453 (1967).
309. B. Zwanenburg, L. Thys and J. Strating, in press.

CHAPTER 16

Directive and activating effects of CO₂H and CO₂R groups in aromatic and aliphatic reactions

G. KOHNSTAM and D. L. H. WILLIAMS

University of Durham, England

I. INTRODUCTION

This chapter considers the effect of the groups CO_2H, CO_2R and CO_2^- on rates of chemical reactions, equilibrium constants, and on the composition of the products; the inclusion of CO_2^- groups in the discussion was thought desirable as many reactions of CO_2H compounds have been examined in basic media. These substituents have received less attention than some other electron-attracting groups (*e.g.* NO_2) but there are sufficient results to provide valuable information about structure–reactivity relations and about reaction mechanisms. In fact, too many studies have been reported to allow a comprehensive discussion of all the work but it is hoped that the examples chosen will indicate the general behaviour of these substituents.

Differences between the behaviour of the different carboalkoxy groups (CO_2R) are always very small and often undetectable. These groups generally act in a similar manner to CO_2H and the effect of one can usually be predicted fairly reliably from the known influence of the other, but the order of their kinetic effects in electrophilic reactions is usually the reverse of that expected from polar considerations.

The present discussion follows lines similar to those of an earlier article in this series on the activating and orienting effects of alkoxy groups[1]. Details of many of the considerations which form the basis of some of the present conclusions have therefore been omitted from this chapter but they can be obtained by reference to the earlier work[1] and to the original sources there cited.

A. The Polar Effects of Carboxyl Groups

The polar character of a substituent is usually an important factor in determining its effect on the stabilities (standard free energies) of the initial, transition, and final states of chemical reactions and therefore often largely controls its effect on rate coefficients and equilibrium constants. All the substituents of interest in this chapter represent various forms of the carboxyl group (**1**, $X = OH$, OR, O^-) where the electron displacements which occur within the group govern its overall polar properties. It is

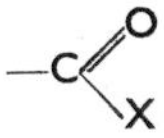

$$-C{\Large\diagup}^{\displaystyle O}_{\displaystyle X}$$

(1)

therefore convenient to consider these internal effects before discussing the different carboxyl groups as a whole (*cf.* reference 2).

1. Internal effects

All the relevant groups X in **1** release electrons by the conjugative effect ($+T$) and attract them by the inductive effect ($-I$) in the orders:

$$+T \text{ effect} \quad O^- > OH > OR \text{ (OEt } > OMe) > H$$
$$-I \text{ effect} \quad OH > OR \text{ (OMe } > OEt) > H > O^-$$

The I and T effects of H are conventionally taken as zero so that the sequence for the $-I$ effect shows the electron repulsion of O^- by this mechanism. Differences between the various alkoxy groups are small and probably arise from the different capacities of Me, Et, ... for inductive and hyperconjugative electron repulsion which will affect the $+T$ donation and $-I$ attraction by the ethereal oxygen atom in different directions[3]. Similar considerations explain the difference between the $-I$ effects of OH and OR but do not account for the greater $+T$ effect of OH which has been ascribed to hyperconjugation[4] in H—O— and to differences between the solvation stabilization[5] of H—O$^+=$ and R—O$^+=$.

The carbonyl oxygen atom attracts electrons powerfully by both the conjugative and the inductive effect so that the electron distribution in **1** will be directly affected by conjugative displacements originating at X (as in **2**) and, to a smaller extent, by the inductive and conjugative relay[3] of any inductive effect exerted by this group. Work on other systems has shown

$$-C \overset{O}{\underset{X}{\bigg\lVert}}$$

(2)

that OH and OR always act as overall electron donors in the order of their $+T$ effects when conjugated to an electron-demanding centre[3], and electron release by X should therefore increase in the order

$$H < OR \text{ (OMe } < OEt) < OH < O^- \tag{1}$$

The same order will apply to the electron displacements shown in **2** and, therefore, to the importance of valence bond form **3** in the final structure.

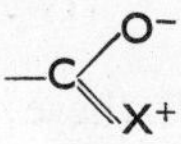

$$-C \overset{O^-}{\underset{X^+}{\diagup}}$$

(3)

These considerations are supported by the exaltation of the optical refraction associated with the various forms of the carboxyl group[6], and by the frequencies and intensities of the $C{=}O$ stretching vibration which are available for a large number of carbonyl derivatives[7-16]. It is generally agreed that the frequency is reduced by electron release from X and vice versa, that the frequency also depends on the masses of groups attached to the central carbon atom, and that the intensities (or extinction coefficients) reflect the magnitude of the conjugative electron displacements to the carbonyl oxygen atom (as shown in 2); however, different authors attach different weight to the importance of the I and T effects of X in determining the frequency. Results for some relevant methyl and phenyl compounds are shown in Table 1. It can be seen that the intensity of the vibra-

TABLE 1. The carbonyl stretching vibration of $YC{\overset{O^{7,9-14}}{\underset{X}{\diagup}}}$

X	Y = CH₃			Y = Ph		
	ν (cm⁻¹)	$10^{-4} A$ (l mole⁻¹ cm⁻²)	ε (l mole⁻¹ cm⁻¹)	ν (cm⁻¹)	$10^{-4} A$ (l mole⁻¹ cm⁻²)	ε (l mole⁻¹ cm⁻¹)
H	1729	2·1	—	1706	2·6	—
OMe	1750	3·6	—	1727	3·5	—
OEt	1742	3·5	—	1720	3·6	—
OH	1717	4·1	411	1695	4·9	523
O⁻	1560	—	745	—	—	—

tion (or the extinction coefficient) increases in the order (1) expected for resonance within the carboxyl group. The frequencies suggest the same sequence for overall electron release by X except for the inversion of the positions occupied by aldehydes and esters, presumably because the effect of $+T$ donation by OR is not sufficient to outweigh the enhancement of the frequency by its inductive electron attraction. The lower frequencies observed for the acids underline the different $+T$ effects of OH and OR, and the small difference between the electron donation by OMe and OEt is clearly apparent; esters containing larger alkyl groups however show virtually the same frequencies as ethyl esters[8]. A particularly striking example of the $+I+T$ effect of O⁻ is provided by the relatively small frequency of the carboxylate ions.

Comparison of the results for the corresponding methyl and phenyl derivatives shows clearly that the properties associated with the carbonyl stretching vibration do not depend solely on the nature of the group X.

C.C.A.E.

Additional electron displacements towards the carbonyl oxygen atom can be initiated by inductive (4) and hyperconjugative (5) donations in the aliphatic series, and by $+T$ release (6) in aromatic compounds and other

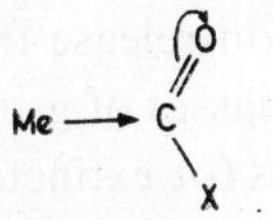

(4) 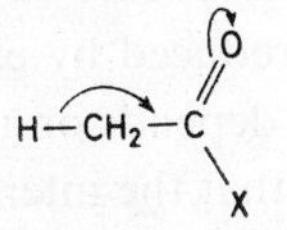   (5) 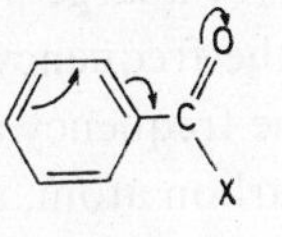   (6)

conjugated systems. If these additional effects contribute substantially to the electron displacements, the relatively large difference between the polarizabilities of methyl and phenyl with respect to electron demand should result in considerable differences between the two series, contrary to the observations (see Table 1). Moreover, both series follow a similar sequence as X is altered and it therefore seems likely that the figures given in Table 1 give a good indication of the effect of X on the electron displacements within the various carboxyl groups.

Other evidence supports this conclusion. Thus, Hückel calculations have shown the existence of a good correlation between the bond order and the carbonyl stretching frequency of various carboxyl groups[17], studies of the carbonyl $n \rightarrow \pi^*$ transition have provided a further demonstration for substantial electron donation when X=OH[18], and transitions in both the infrared and ultraviolet regions correlate well with the substituent constants σ_I and σ_R (see section I. D.) which indicate the magnitude of inductive and conjugative displacements initiated by X[19].

2. Overall polar effects

The electron displacements shown in 5 and 6 represent conjugative attraction $(-T)$ by the carboxyl group as a whole. Similarly, the displacements in 4 correspond to inductive attraction, conveniently termed a $-I_T$ effect since other factors also contribute to the inductive effect exerted by the group as a whole (see below).

Both the $-I_T$ and the $-T$ effect of carboxyl groups arise from the same cause, conjugative electron attraction by the carbonyl oxygen atom. As the group X competes with the rest of the molecule in supplying electrons to this site, the $-I_T$ and $-T$ effects should follow the converse of the sequence (1) for resonance within these groups; i.e. sequence (2). These

$$CHO > CO_2R \; (CO_2Me > CO_2Et) > CO_2H > CO_2^- \qquad (2$$

considerations also suggest that neither of these two effects will be large

The same conclusions regarding the conjugative effect can be obtained by the alternative approach which regards the compound as the resonance hybrid of valence bond forms like **3**, **7** and **8**, significant contributions by **7** and **8** corresponding to $-T$ attraction by the carboxyl group. Chang-

(7)   (8)

es in the nature of X which stabilize **3** will automatically decrease the importance of **7** and **8** in the final structure, thus leading to sequence (2) for the $-T$ effect, and the greater stability of **3** relative to **7** or **8** when $X =$ $= OR$, OH, O⁻ also suggests that the $-T$ attraction of these carboxyl groups is unlikely to be very large. Similar considerations have already been advanced to explain the relatively small sensitivity of the strength of benzoic acid to the introduction of substituents in the aromatic ring[3].

In addition to the $-I_T$ effect, a contribution to the inductive effect of carboxyl groups will arise from the $-I$ attraction of the carbonyl oxygen atom and the inductive attraction or repulsion of the substituents X. This effect (I_I) is not expected to be large and should result in attraction, decreasing in the order (3), with electron donation by CO_2^-.

$$CO_2H > CO_2R \, (CO_2Me > CO_2Et) > CHO > CO_2^- \tag{3}$$

It can therefore be concluded that, as a first appoximation, the various carboxyl groups should act as overall electron attractors ($-I-T$) following the sequence (2) for the $-I_T$, $-T$ effects when they are attached to a system with a good capacity for electron release towards the substituent. Under the converse conditions, CO_2H should attract more powerfully than CO_2R while CO_2^- repels electrons. These predictions are amply supported by experiment; examples can be found throughout this chapter. All the available evidence suggests that the electron requirements of the carbonyl oxygen atom are largely satisfied by displacements within the carboxyl group so that the group as a whole is not very strongly polar. Thus, the effect of a CO_2H substituent usually differs only a little from the effect of CO_2R, and CO_2^- invariably acts as a weak electron donor or attractor. The influence of CO_2R is usually highly insensitive to the nature of the alkyl group, R, although the considerations advanced in the preceding paragraphs suggest that CO_2Me should attract electrons more strongly than CO_2Et. However, other evidence[3] suggests only small differences between the polar properties of OMe and OEt, and it must there fore be concluded that these differences are too small to affect the behaviour of the carboxyl derivatives.

B. Physical Properties

Dipole moments of some relevant compounds are given in Table 2 where the electron-attracting properties of CO_2H are clearly demonstrated by the values for phenol (9) and 4-hydroxybenzoic acid (10). The same conclusion can be reached by noting that acetic acid has a greater dipole moment than formic acid in the gaseous phase[20], though it must be pointed out that this situation is reversed in solution[21]. Comparison of the dipole moment of 10 and 4-nitrophenol (11), and nitrobenzene (12) and benzoic acid (13), shows that CO_2H attracts electrons considerably less efficiently than NO_2 but differences between the various carboxyl groups cannot usually be recognized by these measurements, for example, 13 and its ethyl ester (14) have the same dipole moment.

TABLE 2. Dipole moments of relevant compounds[21]

Compound	Dipole moment
HOC_6H_5 (9)	1·86
4-$HOC_6H_4CO_2H$ (10)	2·76
4-$HOC_6H_4NO_2$ (11)	5·43
$C_6H_5NO_2$ (12)	4·00
$C_6H_5CO_2H$ (13)	1·80
$C_6H_5CO_2Et$ (14)	1·80

A more precise indication of the sequence of electron attraction by a series of substituents is given by their effect on the maximum for the principal absorption (^1L_A) of benzene at 203·5 mμ; increased resonance in the phenyl group displaces the maximum to greater wavelengths and increases the extinction coefficient[22]. The examples given in Table 3 again show that NO_2 attracts electrons much more powerfully than the carboxyl groups and, also, that the changes of λ_{max} decrease in the order expected from the sequence (2) predicted in section I.A.2 for decreasing electron withdrawal by the various carboxyl groups when they are attached to an electron releasing system; the small difference between benzoic acid (13) and its ethyl ester (14) is particularly noteworthy. Other transitions in substituted aromatic systems can be similarly interpreted[23].

TABLE 3. The displacement of the principal absorption maximum of benzene (203·5 mμ) by substituents[a]

Compound	$\Delta\lambda_{max}$ (mμ)
$C_6H_5NO_2$ (12)	65·0
C_6H_5CHO (15)	46·0
$C_6H_5CO_2Et$ (14)	27·0
$C_6H_5CO_2H$ (13)	26·5
$C_6H_5CO_2^-$ (16)	20·5
4-MeOC$_6$H$_4$CO$_2$H (17)	53·0
4-MeOC$_6$H$_4$CO$_2^-$ (18)	43·5
3-MeOC$_6$H$_4$CO$_2$H (19)	26·5

[a] Data for methoxy compounds from reference 24, others from reference 23.

Although the various carboxyl groups do not attract electrons particularly strongly, they are nevertheless significantly polarizable with respect to electron donation by the rest of the molecule. Thus, the introduction of 4-MeO greatly enhances $\Delta\lambda_{max}$ for benzoic acid and its conjugate base (see Table 3)..The atomic refraction constant of the carbonyl oxygen atom in a phenyl compound is considerably greater than in the corresponding alkyl compound[25] while the carbonyl stretching frequency is less in the aromatic system (see Table 1), and the value of this frequency for a large number of substituted benzoic and toluic acids is linearly related to the pK[26] and to the free energy of activation for esterification[27].

The good correlation between acidity and the carbonyl stretching frequency of the acid shows that the polar effect of substituents on the strength of benzoic acid arises essentially from their effect on the stability of the initial state in the ionization and that they change the stability of the benzoate ion by only a small amount. This conclusion greatly facilitates the discussion of the substituted acids (see section II.B.4); other arguments in favour of this view have been given elsewhere[3]. The strengths of substituted benzoic acids form the basis for assigning values to the Hammett σ constants (see section I.D.) of substituents. No precise physical significance can be attached to these constants but it is generally agreed that they reflect the overall electron donation or attraction of the substituent in benzoic acid. Alternatively, σ could equally well be regarded as a measure of the overall electron attraction exerted by the carboxyl group on the relevant substituted phenyl residue, thus demonstrating the variation of this attraction (the polarizability) over a wide range of electron donation towards the central carbon atom. It must however be stressed again that the polarizability of CO_2H (and other carboxyl groups) is not large and very much smaller than that of centres carrying a partial positive charge, such as the incipient carbonium ion in the transition state of S_N1 solvolysis. For example, the introduction of a 4-methoxy group alters the free energy of benzoic acid by 0·3 kcal at 25° [28] and the free energy of the transition state in the hydrolysis of 4'-nitrodiphenylmethyl chloride by 7·7 kcal[29].

It must be stressed that the spectral properties of molecules do not always provide a clear indication of the polar character of substituent groups. For example, the introduction of 3-OMe in benzoic acid lowers the dissociation constant by about 30%[28]. This result is generally discussed in terms of the destabilization of the acid by the $-I$ methoxy group in a system where the conjugative electron release of this group cannot be relayed to the reaction centre except by a highly inefficient second order effect[3], but it can equally validly be taken to show the reduction in the electron attraction by CO_2H when attached to an electron-demanding residue. On the other hand, the maximum wavelength (λ_{max}) for the principal absorption (^1L_A) is hardly altered by this substitution in benzoic acid (see Table 3) or in nitrobenzene[30]. Similarly, a 4-OMe substituent substantially increases λ_{max} of the benzoate ion (see Table 3), contrary to the conclusion that the polar properties of CO_2^- are relatively insensitive to the nature of other substituents present in the molecule (see previous paragraph). As this conclusion was based on properties directly associated with the (arboxyl group, it seems that the extent of resonance in the phenyl groups cas indicated by λ_{max}) does not entirely reflect the electron attraction of a carboxyl substituent.

C. Proximity Effects

The properties of the various carboxyl groups are often modified, sometimes substantially, by their immediate surroundings. Studies of the dissociation constants of carboxylic acids have provided much relevant information about CO_2H and CO_2^-, and are therefore discussed below; CO_2R can be regarded as analogous to CO_2H, as a first approximation.

1. Steric inhibition of resonance

So far the discussion has concentrated essentially on the overall electron-attracting properties of the various carboxyl groups without distinguishing between the separate contributions made by their $-I$ and $-T$ effects. The importance of the conjugative attraction ($-T$) is however demonstrated by systems where steric factors inhibit resonance so that the capacity of these groups for overall electron attraction is decreased, though some authors have doubted whether the magnitude of the steric effects is always as large as has been claimed (see section I.C.3).

The strength of benzoic acid is almost invariably increased by the introduction of *ortho* substituents, irrespective of their polar character, although the acidity is reduced by electron-donating *para* substituents. This is illustrated in Table 4 for the methyl compounds. It is generally considered that the carboxy group is twisted out of the plane of the aromatic ring by adjacent substituents and that the resulting loss of coplanarity reduces its resonance interactions with the phenyl system by increasing the energy of structures like **8**[31, 32]. The substituted acid should therefore have a smaller stability than the parent compound where steric inhibition of resonance does not occur. Similar considerations will, of course, apply to

TABLE 4. The effect of methyl
substituents on the dissociation
constant of benzoic acid in water

Substituents	K_X/K_H	
— (13)[28]	1·00	
2-Me (20)[28]	1·96	
3-Me (21)[28]	0·85	
4-Me (22)[28]	0·67	
2,6-Me$_2$ (23)[33]	9·09	$K(23)/K(20) = 4·63$
2,4-Me$_2$ (24)[33]	1·05	$K(24)/K(20) = 0·53$
2,4,6-Me$_3$ (25)[33]	5·86	$K(25)/K(23) = 0·64$

the benzoate ions but the relatively small $-T$ effect of the carboxylate ion group ensures that the destabilizing steric effect of *ortho* substitution is much less for the anion than for the undissociated acid. Acid strengthening is therefore to be expected, and the opposing effect of the relay of inductive and hyperconjugative repulsion by *o*-methyl groups in the acid (see **26**) is apparently not large enough to counterbalance the consequences of

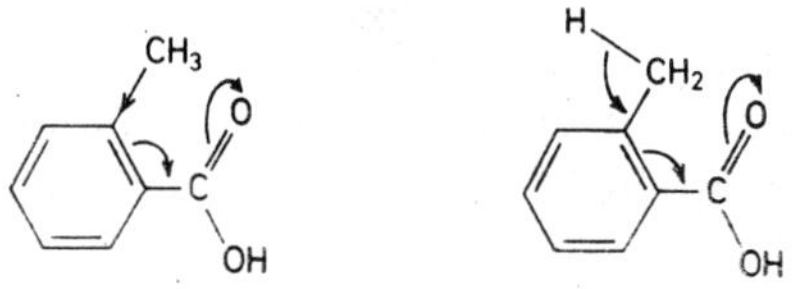

(26)

the loss of resonance.

Analogous arguments account for the fact that *cis* unsaturated carboxylic acids are usually stronger than their *trans* stereoisomers[31, 32], and steric inhibition of resonance in various *ortho*-substituted benzoic acids has also been deduced from considerations of their ultraviolet[24, 34, 35] and infrared[35] spectra. Similarly, the relatively large increase in the acidity of *o*-toluic acid on the introduction of an additional *o*-methyl group (Table 4, 20, 23) appears to be consistent with the view that resonance interactions generally decrease rapidly when the angle between the two planes of the relevant groups is increased to substantial values[36].

These considerations demand a marked decrease in the overall electron attraction of the carboxy group (or its esters) when other substituents occupy adjacent sites in the molecule. As a result the strengths of *ortho*-substituted benzoic acids should be less sensitive to the introduction of other substituents than the parent compound*, but this requirement is not always obeyed; for example, 4-methylation decreases the dissociation constants of benzoic acids in the order 2-Me > 2,6-Me$_2$ > H (see Table 4). It is therefore necessary to consider the other modifications in the polar properties of carboxyl groups which may be caused by their immediate environment.

2. Hydrogen bonding

The carbonyl oxygen atom of carboxyl groups can be considered to be associated with a small partial negative charge and should therefore form hydrogen bonds with suitable acceptors in its vicinity, as in **27**; the dimerization of carboxylic acids in solution is an example of this type of interaction. In carboxylate ions, internal resonance distributes the unit of

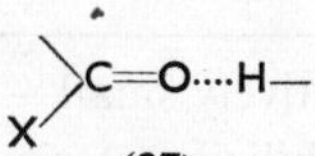

$$\underset{X}{\diagup}C=O\cdots H-$$

(27)

negative charge equally between the two oxygen atoms and each of these centres can therefore form a hydrogen bond (**28**) which will be much stronger

$$\underset{\frac{1}{2}-O}{\diagup}C\cdots O^{\frac{1}{2}-}\cdots H-$$

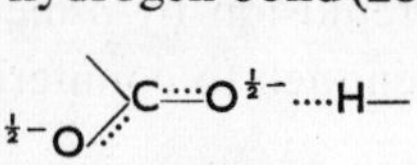

(28)

er than that in **27** under otherwise identical conditions. The formation of a single hydrogen bond, as in **28**, may change the charge distribution of

---

* Alternatively this conclusion can be expressed by noting that the reaction parameter ϱ (see section I.D) should be less for an *ortho*-substituted series than for benzoic acid[37].

the ionic group slightly but this will not alter the basic fact that **28** represents a much more stable structure than **27**.

The properties of carboxyl groups may therefore be modified by intramolecular hydrogen bonding when a suitable group occupies an adjacent site in the same molecule. Thus, x-ray diffraction measurements on solid maleic acid suggest[38] that one of the hydrogen atoms ($H_{(1)}$) is linked to an oxygen atom of the neighbouring carboxy group while the other ($H_{(2)}$) forms a similar bond with a neighbouring molecule, as in **29**. Much stronger internal hydrogen bonding occurs in the hydrogen maleate ion (**30**)[39].

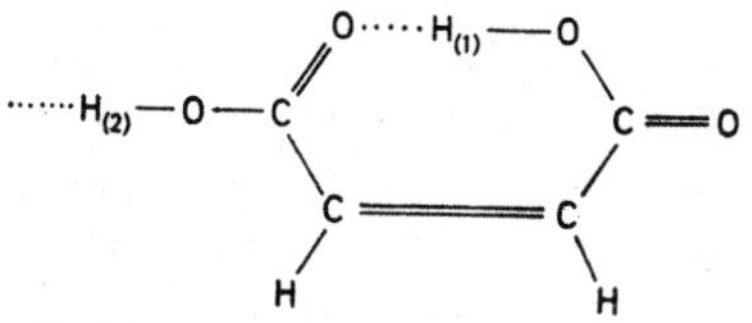

(29)*

It has been strongly argued that the greater stability of **30**, relative to

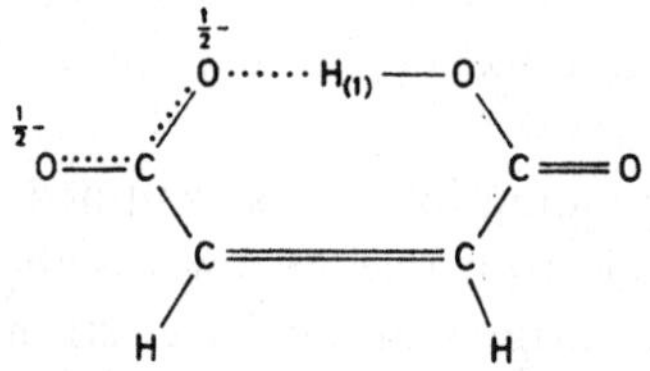

(30)*

29, is largely responsible for the substantial decrease in the first acid dissociation constant when maleic acid is replaced by its *trans* stereoisomer, fumaric acid[32]. On this interpretation, steric inhibition of resonance has only a small effect on the stability of *cis* olefinic dicarboxylic acids but it is convenient to defer a more detailed examination of this problem until the general discussion of the effect of carboxyl substituents on the ionization of carboxylic acids (section II.B).

Intramolecular hydrogen bonding in *o*-hydroxybenzoic acids was first postulated over 30 years ago[40, 41]. The original arguments were based on kinetic considerations which have been criticized[32], but the objections vanish if the stabilities of the various structures are considered instead, as

---

* The hydrogen-bonded complexes (**29** and **30**) will be three-dimensional structures. The planar representation now adopted for the sake of clarity does not, therefore, indicate the relative bond lengths.

now. Relative to benzoic acid, the dissociation constant of the 2-hydroxy derivative* is considerably greater than that of compounds containing much bulkier 2-substituents, arguing strongly against steric inhibition of resonance as the principal acid-strengthening factor in salicylic acid[43]. On the other hand, internal hydrogen bonding would stabilize the anion (32) much more than the undissociated acid (31), and the large increase in acidity on the introduction of a second o-OH is consistent with the greater stability of 33, relative to 32[44]. For similar reasons analogous structures involving C—H hydrogen bonding were assumed to account for the acid-

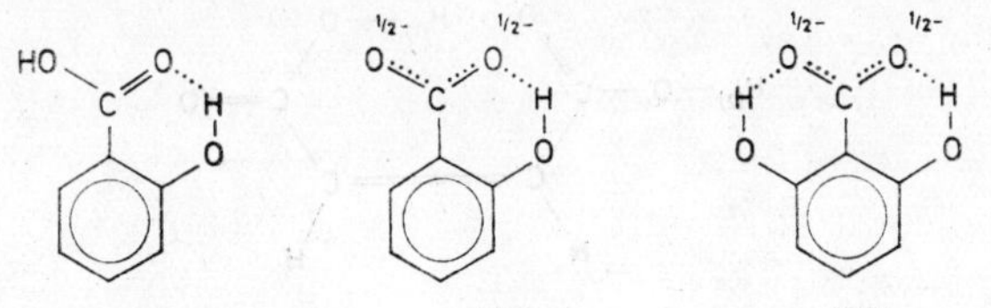

(31)   (32)   (33)

strengthening effect of o-Me[45] but this suggestion was later abandoned[33, 46], and it is noteworthy that no evidence for this type of interaction in o-toluate ions could be found in a recent study of the proton magnetic resonance spectrum[47].

Internal hydrogen bonding in *ortho*-substituted benzoic acids (*e.g.* 31) will be facilitated by electron release towards the central atom but the overall capacity of the carboxy group for electron attraction will be virtually unaffected if, as seems likely, this type of interaction does not greatly alter the stability of the acid. Consistent with this view, the introduction of 4-OH reduces the dissociation constants of benzoic and salicylic acids by almost the same amounts.

3. Solvation effects

There is much evidence that the effect of *meta*- and *para*-substituents on the strength of benzoic acid usually increases as the water content of an aqueous organic solvent is reduced[48], resulting in a larger value of the reaction parameter ϱ (see section I.D) in the poorer ionizing medium. Solvation of the carboxylate ion group (34) will be aided by electron accession

(34)

* Where no specific reference is given, values of acid dissociation constants can be found in the compilation by Kortüm, Vogel and Andrussow[42].

to the oxygen atoms so that the overall capacity of this group for attracting electrons from the rest of the molecule can be expected to decrease as the water content of the medium is reduced. As a result, an electron-releasing substituent will destabilize the substituted benzoate ion (relative to the parent ion) more in an aqueous organic solvent than in water, thus tending to give a lower relative acidity in the less aqueous solvent; the converse applies to an electron-attracting substituent. The effect of solvent changes on the stability of the carboxy compound can be neglected in this context as the neutral acid will be much less heavily solvated than its negatively charged conjugate base.

These considerations have also been applied to the discussion of the importance of steric inhibition of resonance by o-Me, a topic which has received much attention in the past since many authors regard the acid strengthening effect of this group as too large to arise solely from steric factors. The introduction of 4-substituents alters the acidity of 2-methyl- and 2,6-dimethylbenzoic acids a little more than the acidity of benzoic acid, in water and in 50% ethanol (see Table 4 and reference 37); the converse behaviour should have been observed if resonance between CO$_2$H and the aromatic ring were substantially less in the o-methyl derivatives. It was therefore concluded[37] that o-Me causes some steric hindrance to solvation of the anions but that steric inhibition of resonance in the acid is only significant when two such groups are present, in agreement with other evidence from quite different studies that small deviations from coplanarity have little effect on resonance[3, 36]. Similarly, a reduction in the electron attraction of CO$_2^-$ (or an increase in electron repulsion) can be expected to result from steric hindrance to the solvation when other groups occupy an adjacent site in the molecule, though this point has not been discussed explicitly in the literature.

The acid-strengthening effect shown by nearly all ortho-substituents also depends markedly on the nature of the solvent[46]. It usually decreases as the water content is reduced, in contrast to the behaviour of electron-attracting meta and para substituents, and even acid-weakening has been reported for some systems; e.g. for ortho methylation (benzene as solvent)[49] and ortho methoxylation (aqueous acetone)[46]. These observations have been interpreted in terms of the reduction in the effective size of the solvated carboxy group caused by the decreased solvation in the less aqueous solvent[46]; the resulting smaller deviations from coplanarity in the ortho compounds are considered to permit enhanced resonance between CO$_2$H and the aromatic ring. Alternatively the findings could be explained by assuming that the ortho substituent causes more steric hindrance to the solvation of the carboxylic acid than to the solvation of its conjugate base but this seems unlikely. In any case, neither suggestion can account for the effect of 4-substituents on the strengths of *ortho* substituted acids. Each of the two sets of observations can be rationalized separately, but no single hypothesis which embraces both has so far been produced.

In the present chapter, interest in this problem is mainly centred on the effect of adjacent groups on the polar properties of the various carboxy substituents, and the experimental evidence already mentioned suggests strongly that this effect is small, though electron attraction by CO_2^- is probably reduced a little under these conditions. Other results, discussed in section II.B, and the many examples of the validity of the additivity principle[3] in predicting the strengths of 2,4-substituted acids support this conclusion.

4. General considerations

Any discussion of the effect of substituents on reactivity in terms of their polar properties can of course only be qualitative, particularly in the present section where relatively small changes have been considered. It is noteworthy that generally, but not invariably, *meta* and *para* substituents affect the strength of benzoic acid mainly by altering the standard enthalpy change if they behave as electron donors and by altering the entropy change if they attract electrons, while the converse applies to *ortho* substituents[50]. Any rationalization of these observations in terms of polar effects is difficult but the practical success and the simplicity of basing qualitative discussions of reactivity on the polar properties of substituents seems to be sufficient to outweigh the disadvantages of such a procedure. Completely analogous considerations apply to the use of the Hammett and other substituent constants (see section I.D).

Other proximity effects (*e.g.* field effects, neighbouring group participation) which can be considered to modify the behaviour of the various carboxylate groups are considered in those sections which deal with the systems where these effects play an important role.

D. Substituent Constants

The effects of substituents (X) on the rate coefficients or equilibrium constants (k) of a wide variety of reactions are correlated remarkably well by the Hammett equation[51] (equation 4) which has stimulated an

$$\log(k_X/k_H) = \sigma\varrho \qquad (4)$$

immense amount of work in this general field. The reaction parameter, ϱ, reflects the general sensitivity of the velocity or equilibrium to the introduction of substituents, is independent of the nature of X, and has often been taken as a combined measure of the polar requirements at the reaction centre and of the facility of the system for meeting these requirements by the relay of polar effects initiated by substituents. This interpretation must be regarded as an oversimplification as it is strongly opposed by some observations[3]. On this simple view, the substituent constant σ reflects the overall polar character of the substituent and should therefore be independent of the nature of the process under consideration. It was, however, soon realized that no single set of σ's could account for all reactions and several sets have now been proposed, each particularly suitable to a given type of reaction.

The Hammett constants, σ, are based on the ionization of benzoic acids[48a, 51], although some of the original values have now been revised. It is generally agreed that these constants represent a combined measure of the I and T effects of *para* substituents (and mainly the I effect of *meta* substituents) since the carboxy group does not attract electrons very strongly. A significant capacity for $-T$ attraction by a substituent should however be enhanced, relative to its inductive effect, when the group is conjugated to a strongly electron-releasing centre, and a different scale (σ^-) was therefore proposed for such systems[48a, 51]. Similarly, a third scale (σ^+) is considered to apply when the substituent can conjugate with a powerfully attracting site[52]. Values which are believed to include no resonance contributions (σ^{n}[53], σ^0[54]) have been derived from studies in which the substituent is insulated from the reaction centre, although it has been claimed that yet another set (σ_G) provides a more valid reflection of these conditions[55]. In addition, substituent constants which contain only contributions arising from the inductive effect (σ_I) have been obtained from chemical properties of alicyclic[56], aliphatic[57], and aromatic compounds[54, 58], and from nuclear magnetic resonance studies, but it has not been possible to establish a single set (σ_R) indicating the magnitude of resonance effects. A more precise scale for these effects (σ_R^0) could however be obtained for systems in which direct resonance with the reaction centre is prohibited and can be considered to indicate the ability of the substituents to conjugate with a benzene ring[54a, 59].

The discussion of substituent effects in terms of linear free energy relations like equation (4) and its extensions has been reviewed several times recently[60–63], and the present authors' interpretation has also been summarized[1]; details can be obtained from these sources. It is recognized that equation (4) does not usually predict k_X/k_H within the limits of the experimental error and cannot therefore be regarded as completely quantitative, as expected for a simple expression. Substantial deviations can arise when highly polarizable substituents are employed but this does not apply to the various carboxyl groups now being considered. However, these substituents are relatively heavily solvated in hydroxylic media and may therefore show apparently abnormal effects when the solvent differs from that in which the substituent constants were determined.

Substituent constants for a large number of groups have been compiled by Ritchie and Sager[63]; those relevant to the present discussion are given in Table 5. Other authors have reported slightly different mean values but this does not affect the qualitative conclusions which can be drawn as the figures are usually only reliable to ± 0.05–0.1 units[61]. A very weakly polar

TABLE 5. Substituent constants for carboxyl groups

	CO_2R		CO_2H		CO_2^-	
	meta	*para*	*meta*	*para*	*meta*	*para*
σ_I (al)	0·30	—	—	—	—	—
σ_I (ar)	0·34	—	0·34	—	−0·14	—
σ_I (n.m.r.)	0·21	—	—	—	0·05	—
σ	0·37	0·45	0·37	0·45	−0·10	0·00
σ^0	0·36	0·46	—	—	—	—
σ^n	—	0·38	—	0·41	—	—
σ_G	—	0·47	—	—	—	—
σ_R^0	0·08	—	—	—	—	—
σ^+	0·37	0·48	0·32	0·42	−0·03	−0·03
σ^-	—	0·68	—	—	—	—

character for CO_2^- is indicated by the fact that the substituent constants differ only slightly from zero, while CO_2H and CO_2R seem to attract electrons to nearly the same extent and less powerfully than $NO_2[\sigma(p\text{-}CO_2H)$ $=0.45$, $\sigma(p\text{-}NO_2)=0.82]^{63}$; differences between CO_2Me and CO_2Et (not shown) are usually negligibly small. The small differences between the constants for *meta* and *para* substituents, the small value of σ_R^0, and the similarity of the figures associated with the various scales (except σ^- and σ_R^0) all suggest a relatively small $-T$ effect for these groups. As expected, CO_2R and CO_2H are only significantly polarized when they are conjugated to an electron-demanding centre ($\sigma^- > \sigma$) but even under these conditions the presence of OR, or OH, reduces the overall electron attraction below that found for other proups containing $C{=}O$ ($p\text{-}CO_2R$, σ^- $= 0.68$; $p\text{-}CHO$, $\sigma^- = 1.13^{63}$). However, σ^- increases in the order CO_2Me $< CO_2Et < CO_2H^{48a}$ *i.e.* the converse of the sequence (2) predicted for the variation of the $-T$ effects (section I.A.2). On the other hand, the figures do not represent recent data and errors of 0.05 units in σ^- for CO_2H and CO_2Me would make all the values virtually identical.

Rate coefficients and equilibrium constants calculated for the various carboxyl derivatives from equation (4) are usually in good agreement with those observed but some exceptions have been noted[53]. Moreover, proximity effects often complicate the chemical behaviour of the present substituents and it is therefore convenient not to base the subsequent discussion solely on the Hammett relation.

II. EQUILIBRIA

A. General Considerations

Apart from the polar properties of the various carboxyl groups and the proximity effects which have already been mentioned (sections I.A. and I.C.), two additional factors can also modify the influence of the substituents on chemical processes. These factors are often of particular importance in determining the equilibrium state and are therefore discussed below, before the available results are considered.

1. Statistical factors

Consider the ionization of the carboxylic acid (35, dissociation constant K_H) and its monocarboxy derivative (36, dissociation constants K_1, K_2). Two equivalent acidic centres are located in 36 but its conjugate base (37) contains only one site for proton acceptance. The effect of CO_2H

$H(CH_2)_n\, CO_2H$	(35)	$CO_2H(CH_2)_n\, CO_2H$	(36)
$CO_2^-(CH_2)_n\, CO_2H$	(37)	$CO_2^-(CH_2)_n\, CO_2^-$	(38)

on the strength of 35 therefore involves the statistically corrected first dissociation constant of 36, and is given by equation (5). Similarly, the effect of CO_2^- on the strength of 35 is obtained from the dissociation constant of

$$K^*_{CO_2H}/K_H = K_1/2K_H \qquad (5)$$

37 (K_2 of 36) by equation (6) which allows for the fact that the acid (37) contains only one dissociable proton while its ionization product (38) has

$$K^*_{CO_2^-}/K_H = 2K_2/K_H \qquad (6)$$

two equivalent basic centres. A more rigorous discussion in terms of the symmetries of 36–38 leads to the same conclusion.

In general, the statistically corrected equilibrium constant for the transfer of one proton to the solvent (K^*) is related to the observed acidity (K_1) by equation (7) when the compound contains p equivalent acidic

$$K^* = q/pK_1 \qquad (7)$$

centres and its conjugate base q equivalent sites where a proton can be attached. The difficulties which arise when the equivalent sites are associated with the same atom (e.g. a carbon acid, XCH_3) have been discussed by Bell[64b] and the present chapter therefore follows the common procedure of regarding p and q as the number of equivalent sites associated with different atoms in the molecule.

Statistically corrected equilibrium constants, K^*, are quoted throughout this section, where relevant. The necessity of these parameters has long been appreciated and, for example, the σ constants of CO_2H and CO_2^- (which are derived from the ionization constants of the benzoic acids) have been obtained on this basis. It must however be stressed that while the use of K^* facilitates the interpretation of results, care must be exercised in applying these interpretations to the prediction of experimental observations. For example, the very weakly polar character of CO_2^- should ensure that the introduction of this group at some distance from the reaction centre will have only a small effect on the acidity, but statistical factors lead to the observation that the second dissociation constant (K_2) of terephthalic acid has only half the value of the dissociation constant of benzoic acid.

2. Direct field effects

A polar group can affect the stability of the system in which it is situated by direct electrostatic interaction through space with another polar centre, i.e. independently of inductive or conjugative relay through the intervening bonds. This was first recognized by Bjerrum[65] who employed a simple electrostatic model to calculate the equilibrium constants for the double ionization (8) of aliphatic dicarboxylic acids on the assumption that the magnitude of this direct field effect in the anion is entirely respon-

$$H_2A \rightleftharpoons 2H^+ + A^{2-} \qquad (8)$$

sible for the difference between its standard free energy and that of the neutral acid. Several modifications of the original expression have been proposed (for summaries see references 44, 64a and 66) but the importance of direct electrostatic interactions in these systems is generally accepted.

In the present discussion it is sufficient to note that the direct field effect decreases with increasing distance between the polar centres, stabilizing a zwitterion (*e.g.* $NH_3^+CH_2CO_2^-$) and destabilizing a doubly-charged species (*e.g.* $CO_2^-CH_2CO_2^-$). The smaller effects arising from direct charge–dipole interactions (*e.g.* in $CO_2HCH_2CO_2^-$) also decrease more rapidly with increasing separation[44, 66], and the assumption of a spatial relay of inductive effects[56] represents either the same phenomenon or the even smaller direct dipole–dipole interactions. It must however be recognized that the close proximity of polar groups, leading to large direct field effects, also causes steric hindrance to solvation which will be greater for ionic than for dipolar groups and will be particularly marked when other bulky substituents are present (see section I.C.3, references 50, 67 and sources there cited).

B. The Ionization of Carboxylic Acids

It has already been pointed out that the polar effects of substituents should alter the strengths of carboxylic acids mainly by changing the stability of the acid, rather than its conjugate base, but recent determinations of the changes of the standard thermodynamic functions for the ionization of non-aromatic compounds have led to the conclusion that the acidity is often largely controlled by direct field effects and the solvation changes accompanying dissociation[50, 67]. The various features which can alter reactivities on the introduction of CO_2H, CO_2R and CO_2^- groups are particularly well illustrated by the ionization of carboxylic acids, and results for a number of different acid types are therefore discussed. In all cases dissociation constants refer to the statistically corrected values, K^*, obtained from the observed values via equation (7). Much of the information has been quoted from two compilations[42, 44]. Neither is entirely free from errors and all the figures employed were therefore taken from the original sources cited in these summaries.

In some instances the dissociation constants of a dicarboxylic acid (**39**, $X = CO_2H$) are known but no figures are available for the compound from which it has been derived (**39**, $X = H$). If the dibasic acid is symmet-

$$X \ldots \ldots \ldots CO_2H$$

$$(39)$$

rical (*e.g.* compound **36**), the ratio $K^*_{CO_2H}/K^*_{CO_2^-}$ represents the effect of CO_2H relative to CO_2^- on the strength of the parent compound, and is therefore useful in the present discussion. However, the first dissociation constant (K_1) of the unsymmetrical acid (**40**, $R = Alk$, $X = Y = CO_2H$)

can only give the effect of $X = CO_2H$ on the acidity of **40** ($X = H$, $Y = CO_2H$)

$$X\ CRH \ldots\ldots CH_2Y$$

(40)

while the second dissociation constant (K_2) shows the change in the strength of **40** ($X = CO_2H$, $Y = H$) on the substitution, $Y = CO_2^-$. No information about substituent effects can then be obtained from a consideration of K_1/K_2, although a knowledge of this ratio is useful in discussing any stabilizing contribution arising from internal hydrogen bonding in **40** ($X = CO_2H$, $Y = CO_2^-$).

1. Intramolecular hydrogen bonding

Hydrogen bonding between carboxy and carboxylate ion groups attached to the same molecule, as in the hydrogen maleate ion (**30**), will obviously stabilize the acid anion and may therefore be important in determining the influence of substituents ($X = CO_2H$, CO_2^-) on the strength of carboxylic acids (**39**, $X = H$). The general methods which have been employed for the recognition of these interactions are outlined below, together with some examples. Others are considered in connection with the relevant parent acids in later subsections, but the renewed interest shown during the last few years makes it impossible to discuss all the results in detail. Further information can be obtained from recent sources[50, 67, 68] and the references there cited.

Direct evidence (i.r. and n.m.r.) for internal hydrogen bonding of some acid anions in aqueous solution is available[69], though it has been pointed out that the magnitude of the chemical shift may not always reflect the strength of this linkage[67]. In most cases, however, the existence of these interactions is inferred from the effect of substituents ($X = CO_2H$, CO_2Et, CO_2^-) on the strength of the parent acid (**39**, $X = H$), and some authors have concentrated solely on the substituted derivatives.

Arguments have been based on the changes in thermodynamic properties arising from the loss of a proton. Internal hydrogen bonding in an acid anion should decrease the entropy of the system less than the formation of similar linkages between the substrate and the solvent, thus increasing ΔS^0 for the first ionization of **40** ($X = CO_2H$)[70a], and abnormal values of ΔH^0, ΔS^0 and ΔC_p^0 have been interpreted in terms of these interactions. Similar conclusions have been based on deviations from the linear relation between ΔG^0 and ΔS^0 which is shown by many carboxylic acids[50, 67], but there seems to be no agreement about the size of the deviation which can be considered significant. For example, internal hydrogen bonding is not envisaged for the hydrogen malonate ion by some[67], while others[50, 70] take the opposite view in spite of the fact that spectral measurements[71] do not support this conclusion. Similarly, it has been argued[67] that these interactions occur to a smaller extent in the hydrogen maleate ion than is generally believed (*cf.* reference 50).

Comparisons of ionization constants have provided most of the evidence. The acid anion of a dicarboxylic acid can be regarded as the conjugate base of **39** (X = CO₂H) or as the acid (**39**, X = CO₂⁻), so that any unusual stabilization of this species should result in unexpectedly large acid strengthening or weakening on the introduction of CO_2H or CO_2^- in the parent compound (**39**, X = H). Data for both substituents must, however, be available before any valid conclusions can be reached. The formation of internal hydrogen bonds requires that the two carboxyl groups be near each other, conditions which may lead to abnormal solvation changes on ionization and to large direct field effects in the dicarboxylate anion. Both factors will decrease $K_{CO_2^-}^*/K_H^*$, often substantially (see section II.A.2).

TABLE 6. The effect of substituents on the acidity of acrylic acid $(K_X^*/K_H^*)^a$

	Substituent	*cis*	*trans*	Reference
Water at 25°	CO₂H	108·5	7·10	72
	CO₂Et	14·8	7·10	72
	CO₂⁻	0·08164	0·883	72
50% EtOH at 25°	CO₂H	1000	7·24	73
	CO₂⁻	0·001	0·437	73

^{a} K_H in 50% EtOH was calculated from the data in water [42] by assuming that the effect of the solvent change on K was the same as for other similar compounds.

The application of these considerations is illustrated in Table 6 for derivatives of acrylic acid (**41**, X = H). Internal hydrogen bonding in the hydr-

$$X\ CH{=}CHCO_2H$$

(**41**)

ogen maleate ion (**41**, X = *cis* CO₂⁻) is indicated by other evidence which has already been discussed, but is not envisaged for the hydrogen fumarate ion (**41**, X = *trans* CO₂⁻). This conslusion is greatly strengthened by the fact that the effect of *cis* CO₂H and *cis* CO₂⁻ on the acidity is very much greater than when these groups are introduced in the *trans* position, in water and in 50% ethanol. A more detailed discussion of the acrylic system is deferred till section II.B.5, but it can be stated now that the *trans*

50*

substituents show a 'normal' effect and that only a small part of the acid strengthening by *cis* CO_2H can arise from steric inhibition of resonance. The ratio $K^*_{CO_2H}/K^*_{CO_2^-}$ which represents the statistically corrected equilibrium constant for process (9) should also be large if intramolecular

$$H_2A + A^{2-} \rightleftharpoons 2 HA^-$$ (9)

hydrogen bonding occurs in the acid anion. This approach is inevitable when the dissociation constant of the parent acid is not known, but it has almost invariably been employed in discussions of the problem[†] in spite of the fact that a substantial destabilization of the dicarboxylate anion by direct field or solvation effects will also increase the ratio. Helpful further information can usually be obtained from the equilibrium constant for double ionization (8), $K^*_{CO_2H}K^*_{CO_2^-} = K_1K_2$, which does not depend on any special features of the acid anion[32], but this additional criterion has only rarely been employed.

Table 7 illustrates the application of this approach to the introduction of CO_2H and CO_2^- in cyclopropane 1-carboxylic acids. The substantial changes in acidity on 1-substitution, and the accompanying large $K^*_{CO_2H}/K^*_{CO_2^-}$ and 'normal' $K^*_{CO_2H}K^*_{CO_2^-}$ all argue strongly that the two substituents alter the dissociation constant mainly by stabilizing the acid anion (**42**, X = CO_2^-) through the formation of an internal hydrogen bond. In the 2-derivatives (**43–48**) $K^*_{CO_2H}K^*_{CO_2^-}$ (or K_1K_2 for the unsymmetrical compounds) is 10–50 times greater for the *cis* systems than for its *trans* isomer, probably because the direct field effect and steric hindrance to solvation (in some of the compounds) are more pronounced for the dicarboxylate anion involving the smaller distance between the charged centres. A substantially larger factor should therefore separate $K^*_{CO_2H}/K^*_{CO_2^-}$ (or K_1/K_2) for the two isomers if internal hydrogen bonding is significant in the *cis* acid anion. On this view, such interactions appears to be negligibly small in **43, 44, 46** and **47** (X = *cis* CO_2^-), and also in the acid anions of other 1- and 1,2 derivatives[74] of **43** (X = CO_2H) which are not shown in Table 7. The same conclusion is demonstrated more simply by compound **43** (X = H) where the effects of *cis* and *trans* substituents on the acidity differ by a much greater factor when X = CO_2^- than when X = CO_2H. On the other hand, a strong stabilization of the acid anions **45** and **48** (X = *cis* CO_2^-) is clearly indicated (see also reference 75).

These results have been interpreted[74] as indicating that the facility for internal hydrogen bonding in the acid anions of dicarboxylic acids is at a

† Most authors consider K_1/K_2, the ratio of the first dissociation constant of the dicarboxylic acid to the second. For a symmetrical acid like **36**, $K_1/K_2 = 4K^*_{CO_2H}/K^*_{CO_2^-}$.

TABLE 7. The effect of substituents (X) on the strength of cyclopropane 1-carboxylic acids in water at $25°$ [42,74]. (Compounds **43–48** are shown with the 2-substituent (X or CO_2H) *cis* to the 1-carboxy group. Data for the *trans* isomers refer to the position of these two groups).

(a) Symmetrical derivatives

			K_X^*/K_H^*		$K_{CO_2H}^*/K_{CO_2^-}^*$	$1/K_{CO_2H}^* K_{CO_2^-}^*$
			CO_2H	CO_2^-		
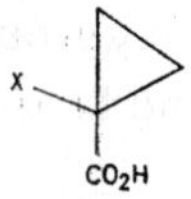 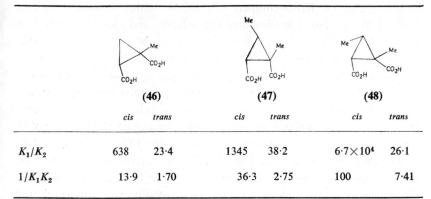	(42)		503	0·005	10^5	$1·82×10^9$
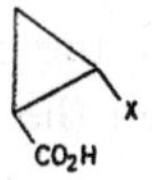	(43)	*cis*	9·26	0·030	308	16·2
		trans	5·03	1·11	4·8	0·76
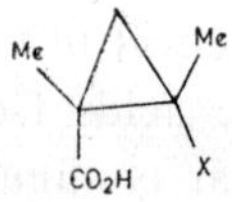	(44)	*cis*	—	—	78	41·7
		trans	—	—	8·6	0·73
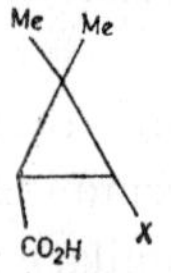	(45)	*cis*	—	—	$2·3×10^5$	45·4
		trans	—	—	8·0	1·31

(b) Unsymmetrical derivatives

	(46)		(47)		(48)	
	cis	*trans*	*cis*	*trans*	*cis*	*trans*
K_1/K_2	638	23·4	1345	38·2	$6·7×10^4$	26·1
$1/K_1 K_2$	13·9	1·70	36·3	2·75	100	7·41

maximum for some critical separation (d_{max}) between the two interacting groups and is progressively reduced as the actual distance (d) differs increasingly from d_{max}. It was assumed that d was close to d_{max} in compound **42**. An examination of molecular models showed that this distance was considerably smaller in the *cis* 1,2-diacid (**43**), and smaller still in the presence of methyl groups at the 1- and 2-positions (compounds **44**, **46**); a *trans* 3-methyl group (compound **47**) had little effect on *d*. The experimental evidence suggests that there is little if any internal hydrogen bonding in the acid anions of **43**, **44**, **46** or **47**, consistent with the view that steric crowding prevents these interactions. On the other hand, a *cis* 3-methyl group (compounds **45**, **48**) appears to force the two carboxy groups apart so that *d* now has a value close to that found for **42** and favours the intramolecular stabilization of the acid anion, as observed.

Other information supports the existence of an optimum geometry for these interactions. Hydrogen bonding becomes significant in the acid anion of malonic and succinic acids when the distance between the two carboxyl groups is reduced by the introduction of substituents (see Tables 9 and 10) but a reduction of the separation has the converse effect in the first ionization product of unsaturated *cis* 1,2-dicarboxylic acids (see section II.B.5), and it has recently been suggested[68] that a linear configuration O ... H ... O of length 2·45 Å will lead to the strongest internal hydrogen bonds.

The considerations advanced in the previous paragraph show that the geometry of an acid can play a very important role in determining the influence of CO_2H and CO_2^- substituents on its strength. Similarly, the solvent in which the system is examined can govern the magnitude of the present substituent effects to an unusually large extent. Internal hydrogen bonding must inevitably compete with the formation of similar linkages between the carboxyl groups and the solvent, and should therefore be enhanced or reduced by changing the facility of the medium for interaction with the polar groups. Thus, internal hydrogen bonding of hydrogen maleate ions can be detected in water, but not in dioxan[69a], and appears to be particularly strong in 50% aqueous ethanol where the effects of *cis* CO_2H and *cis* CO_2^- on the acidity of acrylic acid are 10–15 times greater than in water, while acid strengthening by *trans* CO_2H is nearly the same in both solvents and acid weakening by *trans* CO_2^- only twice as large in the ethanolic medium (see Table 6). Similar results for some other systems[73, 76] also demonstrate the superiority of the less aqueous solvent for promoting these interactions. Again, care is necessary when interpretations are based solely on $K^*_{CO_2H}/K^*_{CO_2^-}$ since the change to 50% ethanol will also increase this ratio by the loss of solvation stabilization which increases in the order (6) and is greatest for the dicarboxylate anion. The effect of the solvent change on $K^*_{CO_2^-}/K^*_H$ for *trans* substitution in acrylic acid (Table

6) probably arises from this cause, and other more striking examples are known (see section I.C.3).

The similar polar properties of CO_2H and CO_2R suggest that the two groups should have very similar effects on the strength of an acid, *i.e.* $K^*_{CO_2H}/K^*_{CO_2R} \simeq 1$. While this is usually observed[44], a larger value of this ratio can be expected when the acid anion of the dicarboxylic acid is stabilized by internal hydrogen bonding, since such interactions cannot occur in the ionization product of the monoester. This criterion has been successfully employed on a number of occasions and fully supports the other evidence when it is applied to the effects of substitution in acrylic acid (Table 6); $K^*_{CO_2H}/K^*_{CO_2Et} = 15\cdot3$ (*cis*), $1\cdot0$ (*trans*). On the other hand, values significantly greater than unity (up to 4) have been reported when other observations argue strongly against intramolecular hydrogen bonding in the acid anion. Here, the different effects of CO_2H and CO_2R on the acidity probably arise from the further steric hindrance to the solvation of the ionization product, caused by the presence of an alkyl group in the vicinity of a sterically crowded reaction centre (see section II.B.3).

2. Aliphatic acids

The effect of ω substituents $(X = CO_2H, CO_2R, CO_2^-)$ on the strength of the aliphatic acids (**49**, $X = H$) is shown in Table 8. Resonance between

$$X(CH_2)_n \, CO_2H$$
$$(49)$$

X and the reaction centre cannot occur in these systems, and the decrease of $K^*_{CO_2H}/K^*_H$ with increasing chain length is qualitatively consistent with inductive attraction by CO_2H as the principal factor responsible for acid strengthening. In agreement with this interpretation, an iodo group which has a similar $-I$ effect[†] also alters the dissociation constant[44] of **49** ($X = H$, $n = 1, 2, 3$) by nearly the same amount as CO_2H. On the other hand, a closer examination of the figures shows that the consequences expected for inductive relay are not always met. Thus, $K^*_{CO_2H}/K^*_H$ decreases by a smaller factor on passing from formic to acetic acid than when the acetic and propionic systems are compared, and the ratio is greater than unity even when the substituent is far removed from the reaction centre. A small stabilizing direct field effect in the acid anion could be responsible for this observation but a much larger contribution should then occur in the lower members of the series. Recent determinations of the changes in

[†] The substituent constant, σ_I, is almost the same for iodo and carboxy groups[63].

TABLE 8. K_X^*/K_H for $X(CH_2)_nCO_2H$ in water
at 25° [42, 44].

n	$X = CO_2H$	CO_2R	CO_2^-
0	151·2	—	0·584
1	39·8	25·7[a]	0·230
2	2·32	2·25[b]	0·346
3	1·51	—	0·513
4	1·29	—	0·538
5	1·19	—	0·586
7	1·17	—	0·632

[a] R = Et.
[b] R = Me.

thermodynamic properties resulting from ionization have, in fact, led to the conclusion that the strength of these acids is usually controlled mainly by these direct electrostatic interactions[50, 67] though the behaviour of malonic acid remains anomalous[50, 70a].

The acid-strengthening effect of ω-CO_2R in propionic acid (**49**, $n = 2$) is similar to that found for ω-CO_2H, as expected. A larger difference is found when these groups are introduced into acetic acid, but this observation could be fortuitous since the results were obtained in the last century and may be subject to error.

Acid weakening invariably results from the introduction of CO_2^- in **49** (X=H). This group can exert a small $+I$ effect, but the destabilization of the dicarboxylate anion by the direct field effect appears to be the predominant factor controlling the acidity of **49** (X=CO_2^-). The values of $K_{CO_2^-}^*/K_H^*$ generally increase with increasing chain length, in agreement with the expected diminution of direct charge–charge interactions as the separation is increased[64a, 66]. The unexpectedly small acid-weakening effect of CO_2^- substitution in formic acid probably arises from the geometry of the oxalate ion. Direct electrostatic interaction between the charged centres should only lead to a moderate destabilization since the two polar groups point in opposite directions, and the ion also appears to be more favourable to solvation by water than the malonate ion.

3. Alkyl-substituted aliphatic acids

The presence of alkyl groups (R) in an aliphatic acid can have a profound influence on the changes in its strength caused by the introduction of substituents (X=CO_2H, CO_2R, CO_2^-) when the steric requirements of R force X close to the reaction centre. Abnormal effects of X on the acidity become more marked as the compression of the two carboxyl groups in the substituted derivative is progressively enhanced by increasing the size of R, or the number of such groups. The first signs are a decrease in the stability of the dicarboxylate anion which results from increases in the direct field and solvation effects, and is characterized by an unusually small $K_{CO_2^-}^*/K_H^*$ with 'normal' values for $K_{CO_2H}^*/K_H^*$ and $K_{CO_2R}^*/K_H^*$. At a later stage, the two carboxyl groups are sufficiently close together to permit internal hydrogen bonding in the acid anion (increased $K_{CO_2H}^*/K_H^*$, further reductions in $K_{CO_2^-}^*/K_H^*$, normal $K_{CO_2R}^*/K_H^*$), and finally additional steric hindrance to the solvation of the ionization product of the monoester becomes noticeable (reductions in $K_{CO_2R}^*/K_H^*$).

TABLE 9. K_X^*/K_H for $XCR^1R^2CO_2H$ in water at $25°$ [42, 44].

R^1	R^2	$X = CO_2H$	$X = CO_2Et$	$X = CO_2^-$
H	H	39·8	25·7	0·230
Me	H	31·6	28·7	0·225
Et	H	35·1	21·8	0·194
n-Pr	H	35·8	—	0·200
i-Pr	H	34·8	—	0·160
t-Bu	H	34·7[76]	—	0·011[76]
Me	Me	25·4	20·3	0·127
Et	Me	46·7	—	0·049
Et	Et	178	11·1	0·0055
n-Pr	Et	192	—	0·0038
n-Pr	n-Pr	209	—	0·0030

Table 9 shows that the presence of a single alkyl group in acetic acid (50, $R^2 = H$) has little effect on K_X^*/K_H^* until the t-butyl derivative when subtantial acid weakening by CO_2^- is found. Similarly, the difference between the steric requirements of the dimethyl and methylethyl acetic acids is merely reflected in a decrease of $K_{CO_2^-}^*/K_H^*$, but the acids containing two

$$XCR^1R^2CO_2H$$
$$(50)$$

larger alkyl groups also show the enhanced acid strengthening by CO_2H and the relatively large $K_{CO_2H}^*/K_{CO_2}^*$ indicative of internal hydrogen bonding in the acid anion. It is also noteworthy that $K_{CO_2Et}^*/K_H^*$ has only half its normal value in the diethyl compound and that no additional considerations are required to account for the results observed[77] when other bulky alkyl groups are present.

Similarly, one β-alkyl group in butyric acid (**51**, $R^2 = H$) has little influence on the acid-strengthening and -weakening effects of γ-CO_2H

$$XCH_2CR^1R^2CH_2CO_2H$$

(**51**)

and γ-CO_2^-,[†] but the results[42, 44] for the dialkyl compounds suggest strongly that the substituent X is now sufficiently close to the reaction centre to allow internal hydrogen bonding when $X = CO_2^-$. On the other hand, the bulky t-butyl groups in **52** do not appear to cause similar compressions since normal values of K_X^*/K_H^* are observed[73].

$$XCHBu^tCH_2CHBu^tCO_2H$$

(**52**)

TABLE 10. K_X^*/K_H for XCHRCHRCO$_2$H in 50% aqueous ethanol at 25°[73]

	R	$X = CO_2H$	$X = CO_2Et$	$X = CO_2^-$
H		1·74	—	0·085
Me	racemic	3·80	0·87	0·016
	meso	4·47	0·97	0·056
Et	racemic	6·16	0·69	$3·5 \times 10^{-3}$
	meso	1·51	—	0·053
i-Pr	racemic	347	0·75	$2·3 \times 10^{-5}$
	meso	1·66	1·28	0·020
t-Bu	racemic	661	0·21	$7·5 \times 10^{-7}$
	meso	1·18	0·94	0·020
Me$_2$[a]		93·3	2·05	$1·9 \times 10^{-3}$
Et$_2$[a]		105	0·57	$9·5 \times 10^{-5}$

[a] For XCR$_2$CR$_2$CO$_2$H.

The β-carboxyl derivatives of α,β-dialkylpropionic acids can exist in two conformations, **53a** and **53b**. Steric factors can be envisaged to influ-

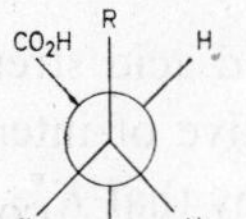

(**53a**) 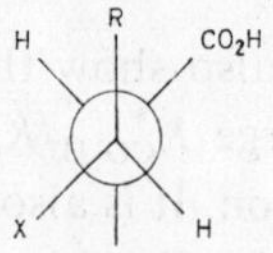   (**53b**)

[†] From the corrected dissociation constants[78], quoted in reference 42.

ence the effect of substituents (X) on the acidity when the substituted derivative is in the racemic form (53a) but much less marked abnormalities should be observed for the *meso* acids (53b). The results in Table 10 show that K_X^*/K_H^* is, in fact, virtually independent of the nature of the alkyl group in the *meso* compounds†. However, when the substituted derivative corresponds to the racemic form (53a), $K_{CO_2^-}^*/K_H^*$ is already smaller than usual when R = Me, while increases in the size of the alkyl group progressively enhance $K_{CO_2H}^*/K_H^*$ (and $K_{CO_2H}^*/K_{CO_2Et}^*$) and decrease $K_{CO_2^-}^*/K_H^*$. The very large acid-strengthening and -weakening effects of CO₂H and CO_2^- in 53a (R = *t*-Bu) are particularly striking as is the five-fold decrease in acidity on introducing CO₂Et. Tetraalkyl propionic acids (54) also

$$XCR_2CR_2CO_2H$$
(54)

show the abnormal values of K_X^*/K_H^* expected when the substituent lies close to the reaction centre, and the anticipated changes in these ratios when the size of the alkyl group is increased (see Table 10).

Similar, though less extensive, results are available for the effects of the substituents on the ionization of alkylpropionic acids in water[42, 44], and it is worth noting that all the observations on these compounds are fully consistent with the considerations advanced at the beginning of this subsection.

4. Alicyclic acids

The influence of 1-CO₂H and 1-CO_2^- on the strengths of some alicyclic carboxylic acids is shown in Table 11. It has already been pointed out in section II.B.1 that the very large substituent effects in the ionization of cyclopropane carboxylic acid probably arise from internal hydrogen bonding in the acid anion (42, X = CO_2^-), but it is also interesting to find that the three-membered ring system apparently leads to a similar separation between the two carboxyl groups in 42 as the presence of bulky alkyl groups in its aliphatic analogue; 42 and 58[77] show nearly the same value

---

† Dissociation constants were not available for all the parent compounds (53a or 53b X = H) in 50% ethanol, and the values employed were therefore calculated from the published figures for water[42] and 50% methanol[32] on the basis of reasonable assumptions. The error should certainly be less than two-fold but it might account for some of the variations observed in the *meso* series, and also for the fact that CO₂Et almost invariably appears to weaken the propionic acid (Table 10). It must however be stressed that comparisons between the *meso* and racemic derivatives of any one acid are not affected by any errors in the value assumed for K_H^*.

TABLE 11. K_X^*/K_H^* for alicyclic acids in water at 25° [42]

Acid		$X = CO_2H$	$X = CO_2^-$
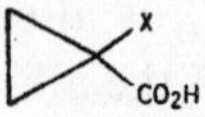	(42)	503	0·005
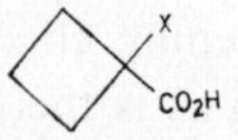	(55)	22·7	0·161
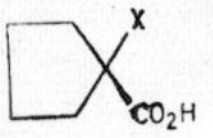	(56)	28·6	0·161
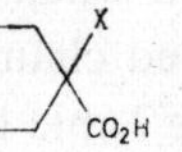	(57)	14·0	0·124

of $K_{CO_2H}^*/K_{CO_2^-}^*$. The steric requirements of alicyclic rings decrease rapidly

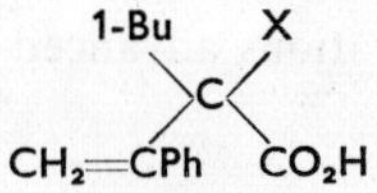

with increasing size, and it is therefore not surprising to find that K_X^*/K_H^* for 55–57 is similar to the values for the substituted derivatives of dimethylacetic acid (see Table 9).

Other substituted cyclopropane carboxylic acids (see Table 7) have already been discussed in section II.B.1. Data are also available for the effects of *cis* and *trans* substitution by CO_2H and CO_2^- in the various ring positions of 55–57 (X = H), and a full discussion of the results is available[32†]. No startling new features emerge. The acid-strengthening effect of *cis* 2-CO_2H is appreciable greater in cyclopropane carboxylic acid than in the compounds containing larger alicyclic rings but it is unlikely that this observation arises from internal hydrogen bonding in the acid anion (43, X = CO_2^-). Direct electrostatic interaction between CO_2H and CO_2^- sometimes has a significant effect on the acidity, and acid-weakening by CO_2^-

† Values originally reported for some of the derivatives of cyclobutane carboxylic acid, however, probably refer to quite different compounds[79].

in **55, 56** (X = H) is always greater for *cis* than for *trans* substitution since the charged centres of the dicarboxylate ion are nearer to each other in the *cis* isomer. The contrary observation in the cyclohexane system has been interpreted in terms of the conformations of the various species involved[32].

5. Unsaturated acids

The effects of *cis* and *trans* substitution on the dissociation constant of acrylic acid (**41**, X = H) were shown in Table 6 to demonstrate the consequences of internal hydrogen bonding in the hydrogen maleate ion (**41**, X = *cis* CO$_2^-$), but the results also provide other information.

Acid-strengthening by *trans* CO$_2$H is greater than when the same substituent is introduced in propionic acid (see Table 8) where the two carboxyl groups are also separated by two carbon atoms, possibly because the substituent can now undergo attractive conjugative interaction with the π electrons of the olefinic linkage. As a result, these electrons will be less free to conjugate with the electron-demanding reaction centre, and the presence of the $-T$ carboxy group in fumaric acid should therefore destabilize this acid more than its conjugate base (the hydrogen fumarate ion), since CO$_2$H attracts electrons more powerfully than CO$_2^-$. Alternatively, the additional acid-strengthening effect of *trans* CO$_2$H (relative to the effect of carboxy substitution in propionic acid) could be held to arise from a smaller stabilization of the acid anion by direct electrostatic interaction, since a greater distance now separates the two polar groups. The identical effect of *trans* CO$_2$Et on the acidity can be explained in similar terms, but it is a little surprising that even the small difference between the polar properties of CO$_2$H and CO$_2$Et is not reflected in the dissociation constants of a system containing the substituents close to the reaction centre (*cf.* substitution in benzoic acid, Table 13).

Analogous considerations suggest that the introduction of *trans* CO$_2^-$ should result in a small acid-strengthening contribution, arising either from the $-T$ effect of this group or from stabilization of the acid (**41**, X = CO$_2^-$) by direct electrostatic interactions, but this will be opposed by the acid-weakening consequences of the $+I$ effect and the interaction between the charged centres in the fumarate ion. As $K_{CO_2^-}^* / K_H^* = 0.88$ for *trans* substitution (Table 6), the direct field effect is clearly not large in the dicarboxylate ion.

It is also noteworthy that *trans* 2-CO$_2$H and 2-CO$_2^-$ alter the dissociation constant of cyclopropane carboxylic acid by amounts very similar to those found for *trans* substitution in acrylic acid (see Tables 6 and 7). No resonance interactions can occur in the cyclopropane system and it would therefore seem that such interactions are also of little importance in determining the acidities of fumaric acid and its conjugate base. However, the agreement between the behaviour of the two systems could be fortuitous.

Arguments have already been advanced in section II.B.1 to show that the effects of cis CO_2H, CO_2^-, and CO_2Et on the acidity of acrylic acid, strongly suggest the formation of stabilizing internal hydrogen bonds in the hydrogen maleate ion, as originally suggested by Hunter[80]. The twofold difference between $K^*_{CO_2Et}/K^*_H$ for cis and trans substitution is of the same order as that found for K^*_{Me}/K^*_H and K^*_{Ph}/K^*_H[32], and probably arises from steric inhibition of resonance in the cis derivatives. This effect can therefore only account for a small fraction of the 100-fold acid strengthening by cis CO_2H, a conclusion which is supported by other considerations. Dicarboxylic acids occupy a special position with respect to the consequences of the suppression of resonance which require the destabilisation of both the acid and its first ionization product. An improbably large effect of this type would therefore have to be assumed in order to account for the observed $K^*_{CO_2H}/K^*_H$. Other arguments have been given by Hammond[32].

The effect of 1-substituents on the strengths of some 1,2-unsaturated acids is shown in Table 12; benzoic acid (60) has been included to facilitate the comparisons. Unfortunately the ionization constants, K^*_H, for 61–63 are not known but it is thought very likely that the estimated value which was employed ($K^*_H = 3\cdot16\times10^{-5}$) is correct to within a factor of two. Errors of this magnitude would not affect the general conclusion that the acid-strengthening and -weakening effects of CO_2H and CO_2^- are at a maximum when the O...H...O length lies between 2·1 and 2·6Å. The values of $K^*_{CO_2R}/K^*_H$ confirm that the abnormally large influence of CO_2H and CO_2^- on the acidity of 61–63 (X = H) arises from internal hydrogen bonding in the acid anions. Normal substituent effects are found for the ionization of the cyclohexene and benzoic acids (59, 60) where X is closer to the reaction centre than in the other compounds, strongly suggesting that stabilizing internal interactions between CO_2^- and CO_2H will only occur if the separation between these groups exceeds a minimum value[68, 74].

The O...H...O distance in the hydrogen maleate ion (41, X = CO_2^-) is not greatly in excess of the minimum required for internal hydrogen bonding, and any reduction of this distance caused by the introduction of alkyl groups should therefore reduce acid strengthening and weakening by CO_2H and CO_2^- because the conditions for internal interactions have be-

TABLE 12. K_X^*/K_H^* for 1-substitution in 1,2 unsaturated acids in water at $25°$[44, 68]

Acid		d^a	$X=CO_2H$	$X=CO_2R$	$X=CO_2^-$
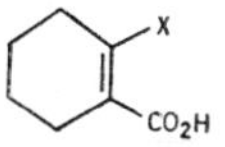	(59)	1·44	11·75	—	0·059
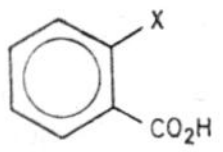	(60)	1·60	8·92	10·5	0·124
cis XCH=CHCO₂H	(41)	1·68	108·5	14·8	0·016
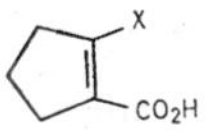	(61)^b	1·92	362·4	35·5	0·0034
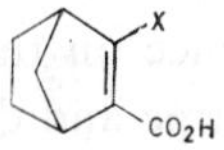	(62)^b	2·11	758·6	20·4	0·0006
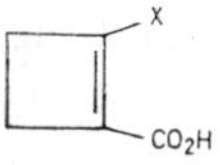	(63)^b	2·63	1200	15·1	0·0015

^{a} O...H...O length in the acid anion, in Å units.
^{b} Estimated values of K_H^* employed for these compounds.

come less favourable. Combination of the ionization constants of sub-stituted acrylic[42] and maleic[81] acids confirms these predictions: $K_{CO_2H}^*/K_H^*$ is decreased by the introduction of one methyl group, and further diminish-ed in the dimethyl compound, while the converse applies to $K_{CO_2^-}^*/K_H^*$.

6. Aromatic acids

The values of K_X^*/K_H^* for substitution in benzoic acid (64, $X=H$) are given in Table 13 and show that the acid-strengthening effect of CO₂R

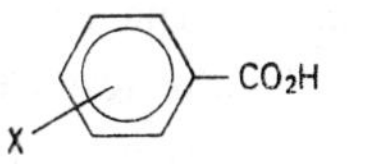

(64)

is a little greater than that of CO₂H at all three sites, consistent with the slightly better facility of the extra group for overall electron attraction.

Too much significance should, however, not be attached to this observation (or to any other small differences between values of K_X^*/K_H^*) since different sets of workers have not always reported precisely the same dissociation constants for the same acids.

TABLE 13. K_X^*/K_H^* for substituted benzoic acids in water at $25°$[a]

X	o	m	p
CO_2H	8·92[83]	1·91[84]	2·30[84]
CO_2R	10·49[80]	2·24[b]	2·69[b]
CO_2^-	0·124[83]	0·794[84]	1·11[84]

[a] Dissociation constant of benzoic acid from reference 82.
[b] From values for benzoic acid[85] and its derivatives[86] in 50% ethanol, taking $\varrho = 1·570$[53].

As expected, *ortho* substituents have the greatest influence on the dissociation constants, but the effects are not exceptionally large and do not suggest any stabilization of the acid anion (**64**, $X = o\text{-}CO_2^-$) by internal hydrogen bonding. The separation between the substituent and the reaction centre is a little less in the aromatic ester (**64**, $X = o\text{-}CO_2R$) than in the corresponding *cis* derivative of acrylic acid, and greater steric hindrance to the solvation of the ionization product of the phenyl compound probably accounts for the smaller value of $K_{CO_2R}^*/K_H^*$ (see Tables 6 and 13). This observation also supports the view (see section II.B.5) that steric inhibition of resonance, though significant, is not the major factor responsible for the change in the strength of a carboxylic acid when another carboxyl group is introduced in the immediate vicinity of the reaction centre. The eight-fold acid-weakening effect of $o\text{-}CO_2^-$ is fully consistent with the operation of a destabilizing direct field effect in the phthalate ion.

The acid-strengthening effect of *para* substituents is a little greater than when the additional carboxyl group is introduced in the *meta* position (see Table 13). If these differences are genuine, they could be considered to arise from the more efficient relay of the $-T$ effect of a substituent conjugated to the reaction centre, but the reduced importance of the direct field effect in the *para* dicarboxylate ion will also contribute to the difference between the effects of m- and $p\text{-}CO_2^-$. It must however be stressed that the direct electrostatic interactions clearly do not greatly affect the stabilities of the *meta* and *para* derivatives.

The ionization sequence of benzene polycarboxylic acids can be obtained from the observed[87] dissociation constants ($K_1, K_2, ...$) by noting that

K_1, K_2, ... represent proton exchange by the CO_2H groups in the order of their acidity in the species under consideration[†].

Corrected equilibrium constants then allow an estimation of the effect of a single substituent ($X = CO_2H$, CO_2^-) on the ionization of the 1-CO_2H group in an aromatic acid (68) which also contains other carboxyl sub-

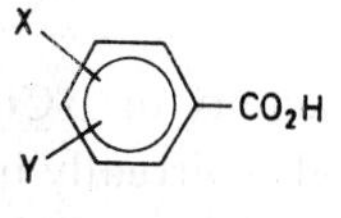

(68)

stituents (Y). Table 14 summarizes the findings of this analysis when $X = Y = CO_2H$.

TABLE 14. The effects of 2-, 3-, and 4-CO_2H on the ionization of the 1-carboxy group in benzene polycarboxylic acids in water at 25°[a]

2-CO_2H		3-CO_2H		4-CO_2H	
Other groups	$K^*_{CO_2H}/K^*_H$	Other groups	$K^*_{CO_2H}/K^*_H$	Other groups	$K^*_{CO_2H}/K^*_H$
—	8·92	—	1·91	—	2·30
4-CO_2H	10·35	5-CO_2H	2·09	2-CO_2H	2·67
5-CO_2H	12·47	6-CO_2H	2·67	2,6-$(CO_2H)_2$	2·63
		2,6-$(CO_2H)_2$	2·75	2,5-$(CO_2H)_2$	2·00
6-CO_2H	2·85	4,6-$(CO_2H)_2$	2·00	2,3,6-$(CO_2H)_3$	1·21
3,6-$(CO_2H)_2$	2·80	2,4,5-$(CO_2H)_3$	1·21	2,3,5,6-$(CO_2H)_4$	1·30
4,6-$(CO_2H)_2$	2·94	2,4,5,6-$(CO_2H)_4$	1·30		

[a] From the dissociation constants[87] (dicarboxylic acids[83, 84]) via equation (7), the 1,2,4- and 1,2,3,4,5- acids were assumed to contain two and three equivalent dissociable protons, respectively.

The effect of 2-CO_2H on the acidity is virtually independent of the number and positions of other carboxy substituents, except that the presence of 6-CO_2H invariably decreases acid-strengthening by a factor of 3.[‡] Apparently

[†] For example, K_1 reflects the proton exchange by 1- or 2-CO_2H in 65, and by 1-CO_2H in 66, since *ortho* substituents strengthen the acid more than *meta* or *para* groups. Similarly the destabilization of dicarboxylate ions by direct field effect ensures that K_2 for 66 represents the ionization of 4-CO_2H in 67.

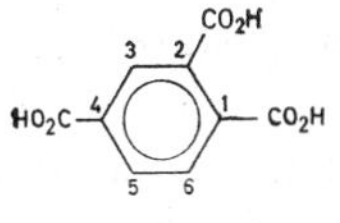

(65) 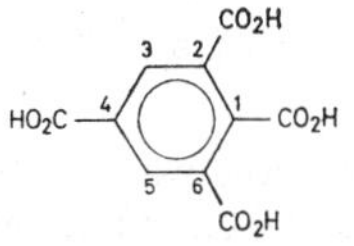   (66) 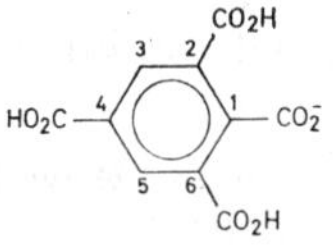   (67)

[‡] The converse of this behaviour is found[44] for 2- and 2,6-methylation.

51 C.C.A.E.

any steric suppression of resonance is already complete in the *ortho* dicarboxylic acid so that $K^*_{2\text{-}CO_2H}/K^*_H$ for a system containing 6-CO_2H represents the influence of the polar and solvation effects invoked by 2-substitution. The approximately ten-fold increase in the acidity of the 6-H compounds on the introduction of 2-CO_2H therefore suggests that, at best, steric inhibition of resonance can only cause three-fold acid-strengthening in this system.

In general, changes in the acidity of 1-CO_2H resulting from the introduction of 3- or 4-CO_2H are also virtually independent of the presence of other carboxy substituents[†]. As $K^*_{4\text{-}CO_2H}/K^*_H$ is not affected in any specific manner by the steric situation at the reaction centre it seems very likely that the $-T$ effect of 4-CO_2H plays only a minor part in determining the acidity at the 1-position.

The effect of a CO_2^- substituent on the strength of **68** can be obtained from the second dissociation constants, and reveals no unexpected features. Approximately three-fold acid weakening is observed on the introduction of 2-CO_2^-[‡], probably because the substituent causes steric inhibition of resonance, while 3- and 4-CO_2^- hardly alter the acidity at the reaction centre. Higher dissociation constants provide information about the effect of CO_2H or CO_2^- on the ionization of 1-CO_2H in molecules containing other carboxy and carboxylate ion substituents, but the interpretation is now complicated by the consequences of direct field effects involving several charged centres, and is therefore not attempted.

C. Nitrogen and Oxygen Acids

1. Alkylammonium ions

The effect of substituents ($X = CO_2^-$, CO_2R) on the equilibrium constant for the ionization (10) of some alkylammonium ions (**69**) is illustrated

$$X(CH_2)_n\,NH_3^+ \rightleftharpoons X(CH_2)_n\,NH_2 + H^+ \qquad (10)$$
$$\textbf{(69)} \qquad\qquad \textbf{(70)}$$

in Table 15. No figures can be presented for $X = CO_2H$ since the first ionization of **69** ($X = CO_2H$) produces the zwitterion (**69**, $X = CO_2^-$).

Acid strengthening by CO_2R decreases with increasing chain length, suggesting that the introduction of a $-I$ substituent stabilizes the conju-

---

† The anomalous behaviour observed when the substituted derivative contains five or six bulky groups probably arises from additional destabilization by steric factors when the substituent (X) is introduced into a highly crowded molecule.

‡ This applies to the 6-H compound; $K^*_{2\text{-}CO_2^-}/K^*_H$ for systems containing 6-CO_2H cannot be obtained from the observed dissociation constants.

TABLE 15. K_X/K_H for $X(CH_2)_n NH_3^+$
in water at 25°[42, 44, 88]

n	$X=CO_2^-$	$X=CO_2Me$
1	5·75	912
1[a]	4·47	—
2	1·86	31·6[b]
3	0·93	—
4	0·68	—

[a] For $XCHMeNH_3^+$.
[b] For CO_2Et.

gate base (70) relative to the acid (69) by invoking interactions like those in structure 71 with the polarizable NH$_2$ group. On this view, the relatively

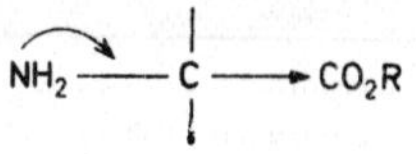

$$NH_2 \longrightarrow C \longrightarrow CO_2R$$

(71)

large values of K_{CO_2R}/K_H show that the stability of the amino compound (70) is fairly sensitive to the presence of electron-attracting substituents (see also Table 16) but the possibility of substantial contributions from direct field effects and steric hindrance to the solvation of the acid (69) is not entirely excluded by the results in Table 15.

The variation of $K_{CO_2^-}^*/K_H^*$ with changing n is surprising. No resonance can occur between the substituent and the reaction centre so that the acid (69, $X=CO_2^-$) should be stabilized relative to the parent compound by the relay of the $+I$ effect of the substituent and by direct electrostatic interaction between the two oppositely charged centres[†]. Both effects should result in acid-weakening which decreases with increasing chain length, contrary to the figures in Table 15, and it can only be assumed that stabilizing direct interactions between CO$_2^-$ and the polarizable NH$_2$ group in the ionization product (70) are sufficiently strong to counterbalance the factors which stabilize the acid (69, $X=CO_2^-$).

[†] This interaction is usually considered to explain the relatively large equilibrium constant for process (11) when $n = 1, 2$.

$$CO_2H(CH_2)_n NH_3^+ \rightleftharpoons CO_2^-(CH_2)_n NH_3^+ + H^+ \qquad (11)$$

51*

2. Anilinium ions

TABLE 16. K_X/K_H for $XC_6H_4NH_3^+$, $XC_5H_4NH^+$, and XC_6H_4OH in water at $25°$[42, 44, 88]

Substrate		$X=CO_2H$	$X=CO_2Me$	$X=CO_2^-$
$XC_6H_4NH^+$	2-	295	240[a]	—
(72)	3-	28·8	8·71	—
	4-	148	159[a]	—
	2-	15,000	—	—
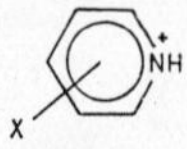	3-	912	—	—
(73)	4-	4,600	—	—
XC_6H_4OH	2-	—	—	(0·001)
(74)	3-	—	—	1·02
	4-	—	30·2[b]	3·63

[a] Similar values have been reported for CO_2Et[44].
[b] Similar values have been reported for CO_2Et, CO_2Pr-n, and CO_2CH_2Ph[44].

The relatively large and similar effects of CO_2H and CO_2R[‡] on the acidity of the anilinium ion (Table 16) probably arise from a substantial stabilization of the ionization product (75) by conjugative displacements. Thus, structures like 76 and 77 ($Y=NH_2$) can make a considerable contri-

bution to 75 ($X=4-CO_2H$, $4-CO_2R$), but the juxtaposition of two positive charges in the anilinium ion ($Y=NH_3^+$) will make 76 highly unstable while 77 ($Y=NH_3^+$) cannot exist. The smallest acid strengthening is shown by the 3-substituents, consistent with stabilization of the aniline compound (75) by structures like 78 and by second-order conjugative interaction[3]

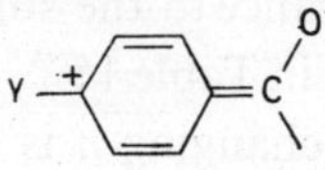

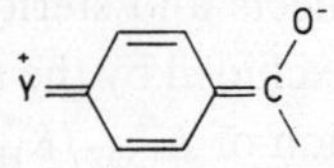

‡ The three-fold difference between the effects of $3-CO_2H$ and $3-CO_2Me$ could arise from errors in K_{CO_2Me}; this constant was determined in 1907.

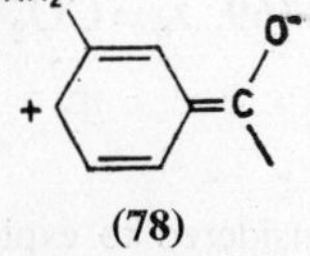

between the NH_2 and CO_2H (or CO_2R) groups. No additional effects need to be invoked to account for the consequences of 2-substitution. In particular, internal hydrogen bonding in 75 (X=2-CO_2H) is ruled out by the observation that $K_{CO_2H} \sim K_{CO_2Me}$ but the dissociation constants of the corresponding N-alkylanilinium ions show that such interactions become increasingly more important as the number of alkyl groups is increased.

The assumption, made in the previous paragraph, that 4-aminobenzoic acid is subject to considerable resonance stabilization appears, at first sight, to be contradicted by the relatively small decrease in the dissociation constant of benzoic acid on the introduction of 4-NH_2. It must however be remembered that now the parent compound is also stabilized by a contribution from structure 76 (Y=H) so that, as a first approximation, only 77 (Y=NH_2) is responsible for the difference between the stabilities of the parent compound and its substituted derivative. This conclusion can be expressed in an alternative manner by noting that the resonance between CO_2H and the aromatic ring in benzoic acid will be greatly hindered by the presence of the positive charge in the 4-anilinium ion.

3. Pyridinium ions

Table 16 shows that K_{CO_2H}/K_H for pyridinium ions (73) follows the same pattern as for the anilinium system (72) but the values are now considerably larger, presumably because structures like 79 and 80 are of greater importance in the conjugate base of 73 than 76 and 77 (Y=NH_2) in the

(79)   (80)

aniline compounds. There is little other evidence to suggest that pyridine is particularly efficient at responding to the requirements of a suitably situated electron-demanding centre, though the large values of K_F/K_H and K_{Cl}/K_H[44] certainly point to the same conclusion.

4. Phenols

The approximately 30-fold acid-strengthening of phenol (74, X=H) by the introduction of 4-CO_2R (see Table 16) is probably a direct consequence of the different capacities of O^- and HO for conjugative electron release. As a result the stabilization of the phenate ion (81) by structure 82 will be greater than the stabilization of the phenol derivative (74) by 83. Simi-

$$CO_2RC_6H_4O^-$$

(81)   (82) 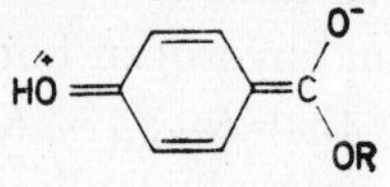   (83)

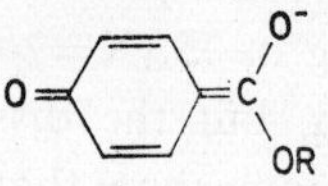

larly, the substantial electron-donating power of O^- allows a significant increase in the dissociation constant by $4\text{-}CO_2^-$ and represents one of the few unambiguous examples of the operation of the $-T$ effect of this group. Internal hydrogen bonding in the salicylate ion (see structure 32, section I.C.2) and a destabilizing direct field effect in its conjugate base account for the very small acidity of the $2\text{-}CO_2^-$ derivative, while the absence of any significant effect on the introduction of $3\text{-}CO_2^-$ has also been observed in other acids (*cf.* Table 13) and requires no further comment.

D. Carbon Acids

Proton transfer from a carbon atom (12) is greatly facilitated by the

$$X-\overset{|}{\underset{|}{C}}-H \underset{}{\overset{k_1}{\rightleftharpoons}} X-\overset{|}{\underset{|}{C}}{}^- + H^+ \qquad (12)$$

presence of carboxyl groups (X) since structure 84 is much more stable than the carbanion (85). The structure of the ionization product will

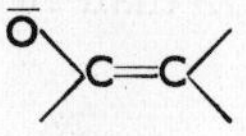

(84) 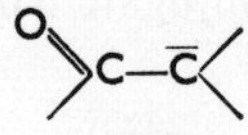   (85)

therefore always contain a large contribution from 84 and will be identical with it when no other groups are attached to the central carbon atom. Substituents (X) will generally not affect the free energy of the carbon acid to any significant extent so that the equilibrium constant for the ionization (12) should reflect the ability of X to stabilize the anion by conjugative electron attraction. The acid-strengthening effect of X should therefore follow the same sequence as the $-T$ effect, *i.e.* $MeCO > CO_2Et > CO_2Me > CO_2H$ (see section I.A.2). Two carboxyl substituents can be expected to increase the acidity more than one such group since the anion will be further stabilized by contributions from several structures like 84 and also by the spreading of the negative charge[64a].

Results relevant to the present discussion are summarized in Table 17 (from the compilation by Bell[64b]). Statistically corrected rate coefficients

TABLE 17. The effect of substituents (X) on the rate and equilibrium constants for the ionization of carbon acids in water at 25°

Acid	$K^*_{CO_2Et}$	K^*_{MeCO}	$K^*_{CO_2Et}/K^*_H$	k^*_1 (sec^{-1})		
				CO$_2$Et	CO$_2$H	MeCO
CH$_3$X (86)	3×10^{-25}	5×10^{-21}	$\sim10^{23}$	—	3×10^{-13}	$2\cdot3\times10^{-10}$
CO$_2$EtCEtHX (87)	1×10^{-15}	2×10^{-13}	—	3×10^{-7}	—	$7\cdot5\times10^{-6}$
CO$_2$EtCH$_2$X (88)	5×10^{-14}	$2\cdot1\times10^{-11}$	$1\cdot6\times10^{11}$	$2\cdot5\times10^{-5}$	—	$1\cdot2\times10^{-3}$
CO$_2$EtCBrHX (89)	—	—	—	3×10^{-4}	—	6×10^{-3}
MeCOCEtHX (90)	2×10^{-13}	1×10^{-11a}	—	$7\cdot5\times10^{-6}$	—	8×10^{-5a}
MeCOCH$_2$X (91)	$2\cdot1\times10^{-11b}$	1×10^{-9b}	$4\cdot2\times10^{9}$	$1\cdot2\times10^{-3b}$	$1\cdot2\times10^{-1}$	$1\cdot7\times10^{-2}$
MeCOCBrHX (92)	—	1×10^{-7}	—	6×10^{-3}	—	$2\cdot3\times10^{-2}$
NO$_2$CH$_2$X (93)	$1\cdot5\times10^{-6}$	8×10^{-6}	$2\cdot5\times10^{4}$	$6\cdot3\times10^{-3}$	—	$3\cdot7\times10^{-2}$

[a] For (MeCO)$_2$CHMe.   [b] For X=CO$_2$Me: $K^*_X=1\times10^{-10}$, $k_1 = 3\cdot3\times10^{-2}$.

for the ionization (k_1^*) are also shown as the transition state for this process is stabilized by the same factors as the anion. Findings for acetyl substituents have been included for comparison.

The figures in Table 17 confirm almost all the predictions made earlier in this subsection and represent one of the clearest examples of the $-T$ effect of carboxyl groups. Thus, K_X^* and k_1^* for a given system decrease in the expected order as X is varied[†], two carboxyl substituents lead to a stronger acid than one (*cf.* **86** and **88** or **91**), and the acid-strengthening and rate enhancement shown by bromo groups (*cf.* **88** and **89, 91** and **92**) are consistent with an additional stabilization of the anion or transition state by inductive electron attraction. The acid-weakening and retarding effects of alkyl groups (*cf.* **88** and **87, 90** and **91**) could arise from steric hindrance to the solvation of the anion or, possibly from C—C hyperconjugation (as in **94**) with the carboxyl group in the acid and in its anion; the resulting acid-weakening contribution arising from the stabilization of the acid would be further enhanced in the anion where the existence of

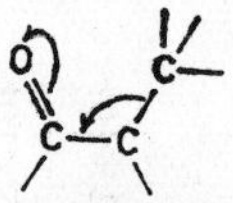

(94)

similar interactions would reduce its stabilization by structures like **84**. The present systems would be particularly favourable to the operation of C—C hyperconjugation, an effect which is by no means universally accepted.

Further evidence for the importance of the $-T$ effect of carboxyl substituents (X) is provided by results for keto–enol equilibria. An enol (**95**) will be stabilized by conjugation with X in the order MeCO > CO_2Et, but no such interactions occur in its keto isomer (**96**). In agreement with this

(95)   (96)

view, the enolization constant of acetylacetone (**91**, X = MeCO) exceeds that of ethyl acetoacetate (**91**, X = CO_2Et) by a factor of 10 in the vapour phase[89] and by a factor of 40 in water[90]. The larger difference between the

---

[†] However, k_1 for the ionization of **91** (X = CO_2H) is anomalous.

ionization constants of these two compounds (see Table 17) can be considered to result from the different capacities of O^- and HO for conjugative electron release.

III. NUCLEOPHILIC SUBSTITUTION

A. Aliphatic Compounds

1. General considerations

Nucleophilic substitution at a saturated carbon atom (process 13) can occur by one of two general mechanisms. These reactions have been discussed on innumerable occasions and only the special features which

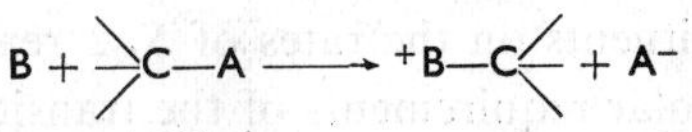

$$\text{B} + \;\underset{}{\overset{}{>}}\text{C--A} \longrightarrow {}^+\text{B--C}\!\!< + \text{A}^- \tag{13}$$

$$(97)$$

are particularly relevant to the present problem will therefore be considered in this section. Further details can be obtained from standard works and from the sources cited in references 2 and 91.

It is convenient to adopt Ingold's approach[92] which distinguishes between a bimolecular process (S_N2) when covalent participation by the reagent (B) is an essential feature of the rate-determining step, and a unimolecular process (S_N1) where this step only requires covalency change by the substrate (97). This simple classification includes reactions via carbonium ion, ion pair, and cyclic intermediates which may undergo further rapid reactions.

In reaction by mechanism S_N1, the rate-determining heterolysis of the C—A linkage should obviously be facilitated by the presence of substituents (X) which can release electrons towards the reaction centre, as in **98**; electron attraction by X can be expected to show the converse effect. Thus,

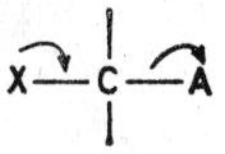

$$\text{X} \overset{\frown}{} \overset{|}{\underset{|}{\text{C}}} \overset{\frown}{} \text{A}$$

$$(98)$$

the introduction of CO_2H and CO_2R should retard reaction by this mechanism while CO_2^- should give a small acceleration or retardation, depending on whether the $+I$ or $-T$ effect predominates in the system under consideration. Some suitably situated substituents may also act as internal nucleophiles and thus facilitate C—A fission (*i.e.* S_N1 reaction*)

---

* Alternatively, the reaction can be regarded as involving an internal bimolecular process (see reference 92).

by direct bonding with the central carbon atom (as in **99**), resulting in the formation of a cyclic intermediate (**100**). This effect, which is termed neigh-

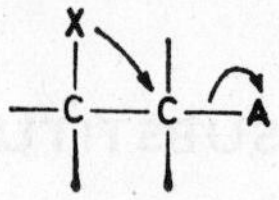

(99) 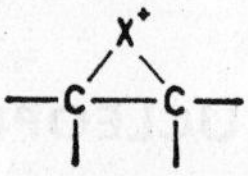   (100)

bouring-group participation or anchimeric assistance, is usually responsible for the relatively large acceleration of some reactions by the introduction of CO_2^-. Many other examples are known and a general review is available[93].

The effect of substituents on the rates of S_N2 reactions cannot usually be predicted as the polar requirements of the transition state are ambiguous[92]. Electron release towards the reaction centre will hinder the approach of the nucleophile (B) but will also facilitate C—A fission (see **101**) while electron withdrawal reverses this situation. As a result, a reaction can be

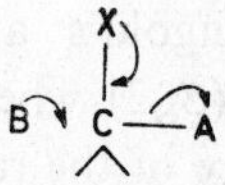

(101)

accelerated by both electron-attracting and electron-donating substituents if the relative importance of the bond-making and bond-breaking processes in the transition state depends markedly on the polar properties of the substituent employed; *e.g.* in the reaction between benzyl bromides and bromide ions[94]. Similarly, it might be expected that S_N2 reaction with an anion would be facilitated more by CO_2H and CO_2R than reaction with a neutral molecule since bond formation will be favoured when the nucleophile is negatively charged.

2. Alkyl halides

The effect of CO_2R, CO_2H and CO_2^- on the rates of S_N2 reactions has only rarely been studied, and several of the reported results are not reliable. Some of the more accurate data are summarized in Table 18 but the figures are subject to some error since each parent compound and its substituted derivative were studied by different workers under different conditions, so that k_X/k_H had to be obtained by interpolation or extrapolation of experimental results. However, the figures show clearly that CO_2R accelerates the reaction of the halides with anions while the equally power-

TABLE 18. Relative rates (k_X/k_H) for S_N2 reactions of alkyl halides. Data for reaction in water at 25°, except where indicated. [The second reference for each system refers to the parent compound $(X=H)$]

Reaction	$X=CO_2R$	$X=CO_2H$	$X=CO_2^-$	Reference
$XCH_2Br+S_2O_3^{2-}$	3·87	—	0·29	95, 96
XCH_2Cl+I^{-a}	8·83	—	—	97, 98
XCH_2Br+OH^-	—	—	0·48	99, 100
XCH_2Cl+OH^{-b}	—	—	0·15	101, 100
XCH_2Br+H_2O	—	0·10	0·15	102, 103
$XCHMeBr+H_2O$	—	0·18	$(1·83)^c$	104, 103

^{a} In acetone.
^{b} At 60°.
^{c} Mechanism S_N1 for the derivative.

ful electron attraction by CO_2H retards the reaction with water, as expected (see section III.A.1). All the S_N2 reactions listed are retarded by CO_2^- but it is noteworthy that the retardation is less than that shown by CO_2H when water is the nucleophile, consistent with the view that the bond-breaking process is the major factor controlling the rate of reaction of the substituted derivative with water.

The introduction of a CO_2^- group into alkyl halides other than the methyl compounds almost invariably accelerates their solvolysis, and there is much evidence that the reaction of the substituted derivative then occurs unimolecularly[93, 105]. It was originally proposed that the rate-determining step in the hydrolysis of these carboxylate ions involves the formation of a carbonium ion (102) which is stabilized by direct electrostatic interaction between the oppositely charged centres[105], but anchimeric assistance by the substituent leading to an intermediate lactone (103) has also been pro-

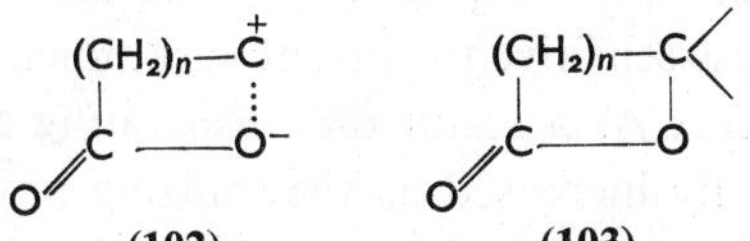

(102)   (103)

posed[106]. Although β-, γ-, and δ-lactones (103, $n = 1, 2, 3$) have been obtained from the corresponding halogenocarboxylate ions, these lactones are rapidly hydrolysed under most conditions by nucleophilic attack on the carbonyl carbon atom. No α-lactone (103, $n = 0$) has yet been prepared.

Both explanations account for the retention of optical configuration in the hydrolysis product of the α-bromopropionate ion (104)[105, 106] and for the fact[105, 107, 108] that the accelerating effect of the introduction of

$$BrCHMeCO_2^-$$
(104)

CO_2^- in alkyl halides arises from an increase in the entropy of activation, $\Delta S^{\ddagger}$. In many cases, the parent compound undergoes S_N2 solvolysis and an algebraically greater $\Delta S^{\ddagger}$ might therefore be expected for the substituted derivative which reacts by a unimolecular mechanism (*cf.* references 103 and 109). However, the increase in the rate of S_N1 solvolysis of neopentyl bromide (**105**, X = H) on making X = CO_2^- also arises from an increase in $\Delta S^{\ddagger}$ [107]. This has been explained in terms of reaction via the car-

$$Me_3CCHXBr$$
(105)

bonium ion (**102**) by assuming that some of the solvent molecules required for the solvation of the cationic centre in the transition state are already attached to the substrate in the initial state as part of the solvation shell of the CO_2^- group[107]. Alternatively, the halogenocarboxylate ion may form the lactone (**103**) in the rate-determining step. The different views about the factors controllings $\Delta S^{\ddagger}$ in solvolysis[103, 109] then both predict that the formation of the transition state (**106**) from a negatively charged substrate

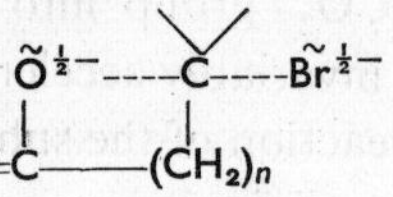

$$\tilde{O}^{\frac{1}{2}-}----\overset{\diagup}{\underset{|}{C}}----\tilde{Br}^{\frac{1}{2}-}$$
$$O=C\overset{|}{-\!\!-\!\!-\!\!-}(CH_2)_n$$
(106)

will show a larger $\Delta S^{\ddagger}$ than the S_N1 solvolysis of a structurally similar neutral halide.

Similarly, the reduction in the solvation requirements of the transition state on the introduction of CO_2^- should ensure that the rate of hydrolysis of the derivative is less sensitive to solvent variations than the parent compound[106b], as observed. As a result the accelerating effect of a CO_2^- substituent is substantially increased as the ionizing power of the solvent is reduced. This is illustrated below for α substitution in isopropyl bromide (for other examples, see Table 20); rate coefficients for the parent com-

---

* The accompanying change in the activation energy often tends to retard the hydrolysis of the substituted derivative ($E_{CO_2^-} > E_H$). As a result $k_{CO_2^-}/k_H$ increases as the temperature is raised. Thus, for the hydrolysis of **104** in water (data for ethyl bromide from reference 103):

Temperature (°c)	25	50	64
$k_{CO_2^-}/k_H$	1·09[108]	1·83[104]	2·24[105b, 106b]

pound were obtained from the data for reaction in water[103] and the appropriate parameters in the Swain equation for the effect of solvent changes on the rate[110]:

Solvent	H$_2$O, 25°	60% aq. EtOH, 25°	MeOH, 44°
$k_{CO_2^-}/k_H$	31·6[103]	329[107]	1100[106b]

Values of k_X/k_H (X = CO$_2^-$) for the hydrolysis of the derivatives of primary alkyl bromides **107**, (X = H) are summarized in Table 19. Some care

$$X(CH_2)_n CH_2 Br$$

(107)

is necessary when comparing the results since the parent bromides do not react by the same mechanism as the carboxylate ions, but it is reasonable to assume that the (hypothetical) $S_N 1$ reactivities of the ethyl and higher halides (**107**; X = H, $n \geqslant 1$) do not differ greatly from each other so that the values of $k_{CO_2^-}/k_H$ can be taken as a measure of the facilitation of C—Br heterolysis by a ω-CO$_2^-$ substituent.

TABLE 19. The effect of ω-CO$_2^-$ on the rates of hydrolysis of n-alkyl bromides X(CH$_2$)$_n$CH$_2$Br in water at 25°. (Carboxylate ion from reference 108, except where indicated. Data for the parent halides from reference 103)

n	$k_{CO_2^-}/k_H$
0	0·139[105]
1	9·14
2	272·1[a]
3	226·5[b]
4	6·23
5	1·62

[a] For the chloride at 37·5°[93].
[b] For the chloride at 50°[93].

The values of $k_{CO_2^-}/k_H$ attain a large maximum (*ca.* 270) when $n = 2$ and the sequence of the figures argues strongly in favour of anchimeric assistance by the CO$_2^-$ group to the hydrolysis of the carboxylate ion when $n = 1, 2, 3, 4$; *i.e.* reaction via the lactones (**103**). As in the reactions of other compounds which are facilitated by the formation of a ring system containing an oxygen atom (for sources see reference 93), neighbouring group participation in the ω-halogenocarboxylate ions appears to be most efficient when it leads to five-membered ring but the formation of a six-membered ring appears to be only slightly less favourable in the present systems, unlike others (cf. reference 93).

TABLE 20. Relative rates of solvolysis (k_X/k_H, $X = \omega - CO_2^-$) for $X(CH_2)_n CR^1R^2Br$. (Data for carboxylate ions at 25° are from reference 108, except where indicated. Data for alkyl bromides are from reference 103 for reaction in water, corrected for other solvents via the appropriate parameters in Swain's equation[110])

n	Solvent	R^1	R^2	k_X/k_H	Reference
0	H_2O	H	H	0·139	105
		Me	H	1·09	—
		n-Bu	H	1·92	—
		CO_2^-	H	0·64	—
		$CH_2CO_2^-$	H	13·2	106b
		Me	Me	31·6	103
		CO_2^-	Me	30·4 (64°)	105
	50% EtOH	t-Bu	H	1370	107
	MeOH	Me	H	233 (64°)	106b
	50% EtOH	t-Bu	CO_2^-	2300	107
	60% EtOH	Me	Me	329	107
	MeOH	Me	Me	1100 (44·3°)	106b
1	H_2O	H	H	9·14	—
		Me	H	7·64 (45°)	106b
		n-Pr	H	9·21	—
		CO_2^-	H	111	—
2		H	H	272 (37·5°)[a]	—
		Me	H	93·4	—
		Et	H	125·4	—

[a] For the chloride at 37·5°[93].

The acceleration of solvolysis caused by the change $X = H$ to $X = CO_2^-$ in 108 often depends markedly on the nature of the other α groups (R^1,

$$X(CH_2)_n CR^1R^2Br$$

(108)

R^2). Table 20 shows that k_X/k_H for α substitution by CO_2^- follows the sequence (14) which also represents the order of increasing electron release

$$R^1, R^2 = H < R^1 = Alk, \qquad R^2 = H < R^1, R^2 = Alk \qquad (14)$$

towards the reaction centre. This observation has been considered to show the enhancement of anchimeric assistance by CO_2^- as the driving force for heterolysis is increased[106b], a view which implies that the unimolecular solvolysis of α-halogenocarboxylate ions (**108**, $n = 0$)* occurs via the α-lactone (**103**, $n = 0$). On the other hand, it can also be seen in Table 20 that α-alkylation does not increase k_X/k_H for the introduction of β- and γ- CO_2^- groups (**108**; $n = 1, 2$) although there is strong evidence for believing that the carboxylate ion derivatives form the corresponding lactones in the rate-determining step. Similar solvolytic behaviour has been found for systems involving the analogous methoxy-5 participation[111] but the proposed explanation[1] requires that $k_{CO_2^-}/k_H$ for **108** ($n = 0$, $R^1 = R^2 = Me$) should not be larger than for **108** ($n = 0$, $R^1 = Me$, $R^2 = H$), contrary to the observations (Table 20). No α-lactone has yet been isolated and it is noteworthy that the observed variations of $k_{CO_2^-}/k_H$ with increasing electron donation towards the reaction centre are consistent with the rate-determining formation of a carbonium ion (**102**, $n = 0$) in the unimolecular reactions of the α-halogenocarboxylate ions while the solvolysis of the β and γ derivatives proceeds via the lactone. The alternative view requires the assumption that the formation of α-lactones (but not the formation of β- and γ- lactones) is facilitated by the presence of electron-releasing α groups (R^1, R^2).

The abnormally large values of k_X/k_H for **108** ($n = 0$, $R^1 = CH_2CO_2^-$) and **108** ($n = 1$, $R^1 = CO_2^-$) probably arise from the fact that both α- and β-CO_2^- groups are available to facilitate the rate-determining heterolysis of the substituted derivatives ($X = CO_2^-$), while the two parent compounds undergo reaction by mechanisms which differ from those operating for the other unsubstituted members ($X = H$) of the present series.

3. Arylalkyl compounds

Changes in the S_N1 reactivity of phenylmethyl halides (**109**) caused by

(**109**)

the introduction of *m*- or *p*-CO_2R, CO_2H, and CO_2^- can be expected to reflect the polar properties of the substituent. This is illustrated in Table 21 for the hydrolysis of α,α-dimethylbenzyl chloride (**109**, $R^1 = R^2 = Me$). The small accelerating effect of CO_2^- shows that the $-T$ attraction is not quite sufficient to counterbalance the relay of the $+I$ effect and any direct electrostatic interaction (see section II.A.2) between the substituent and

* These ions react unimolecularly if $X = CO_2^-$; $R^1 = Alk$, CO_2^-; $R^2 = H$, Alk, CO_2^-.

the strongly electron-demanding reaction centre in the transition state. Retardation of hydrolysis is greater for CO_2R than for CO_2H, as expected from the greater capacity of CO_2R for $-T$ and $-I-T$ electron attrac-

TABLE 21. Relative rates (k_X/k_H) for the S_N1 solvolysis of $XC_6H_4CMe_2Cl$ in 90% acetone at $25°$[112]

X	meta	para
CO_2Me	0·0212	0·0061
CO_2Et	0·0217	0·0065
CO_2H	0·0345	0·0124
CO_2^- [a]	1·35	1·28

[a] In EtOH; the solvent change has virtually no effect on k_{CO_2H}/k_H[113].

tion (see section I.A.2), and the observation that both substituents show more deceleration in the *para* than in the *meta* position merely reflects the well-known difference in the efficiency of conjugative relay from these two sites. The extremely small difference between the kinetic effects of CO_2Me and CO_2Et fully confirms the prediction of almost identical polar properties for the two groups.

In the absence of complicating features, the present substituents should affect the rates of the S_N1 reactions of **109** in much the same way if they are introduced in the *ortho* or *para* positions. Steric inhibition of resonance between the substituent and the phenyl ring could however occur in the *ortho* derivatives, diminishing the relay of the $-T$ effect to the reaction centre and therefore enhancing the rate relative to the *para* compound. In addition, the present *ortho* substituents may well provide anchimeric assistance to the heterolysis of C—Hal in compounds like **109**. It has already been pointed out (section III.A.2) that neighbouring group participation by CO_2^- is an energetically favourable process when it leads to a five-membered ring (γ-lactone), and it is therefore possible, that the S_N1 reaction of **109** (X = o-CO_2^-) may proceed via the phthalide (**110**). This

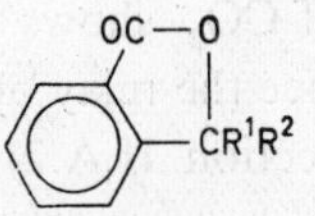

(110)

view is supported by the experimental evidence and it is noteworthy that o-CO$_2$H and o-CO$_2$R can also facilitate nucleophilic substitution in **109** by participating in the formation of a five-membered ring (see below).

Unfortunately there is very little information about the effect of the present *ortho* substituents on the S_N1 reactions of phenylmethyl compounds (**109**). The introduction of o-CO$_2$Ph hardly affects the rate of solvolysis[114a] of diphenylmethyl bromide ($k_X/k_H = 0.88$), but the substantial retardation for the same substitution in the *para* position ($k_X/k_H = 0.01$) is of the magnitude found in other S_N1 reactions. Moreover, the *ortho* derivative does not form the expected alcohol but 3-phenylphthalide (**110**, R^1 = Ph, R^2 = H), probably via the rate-determining production of **111**[114a].

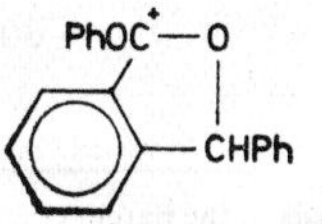

(111)

The suggested mechanism requires participation of the carbonyl oxygen atom of CO$_2$Ph in the formation of the ring system (CO-5 participation) and there is evidence that a keto group can facilitate nucleophilic substitution[115, 116] and also other reactions (see section VI.B.2) by similar interactions. The same explanation could account for the fact that a maximum is observed[116] in the rates of hydrolysis of the carboethoxyalkyl chlorides (**112**) when $n = 2$ but anchimeric assistance by the ethereal oxygen

$$CO_2Et(CH_2)_nCH_2Cl$$

(112)

atom (see structure **114**) is not excluded by the available information.

Substituent effects in the solvolysis of benzyl and diphenylmethyl bromides are shown in Table 22. Benzyl bromide reacts by mechanism S_N2 under these conditions[109, 117] though bond breaking is probably of greater importance in the transition state than bond formation with the nucleophile. The rate can therefore be expected to be much less sensitive to electron displacements caused by the introduction of substituents than when mechanism S_N1 is operating (see section III.A.1), and it is not surprising to find that p-CO$_2$R and p-CO$_2$H retard solvolysis by a much smaller factor than when the same substitution is carried out in diphenylmethyl bromide.

Introduction of the *ortho* substituents accelerates the solvolysis of benzyl bromide in the order (15) and it is noteworthy that the hydrolysis

$$H < CO_2R < CO_2H < CO_2^-$$ (15)

TABLE 22. Relative rates (k_X/k_H) for the reactions of diphenylmethyl, benzyl, and methylbenzyl bromides with aqueous organic solvents. (The figures below were obtained from the available results[114a,c] by making reasonable assumptions about the effect of solvent changes on k for those derivatives which were studied in different media from the parent compound. The experimental values for CO_2^- derivatives refer to alkaline solutions and the figures now quoted were obtained by assuming $k_{CO_2^-}/k_H = 1$ for p-CO_2^- and the same rate constant for the bimolecular reactions of the *ortho* and *para* compounds with OH^-.)

X	$XC_6H_4CHPhBr$ $(S_N1)^a$	$XC_6H_4CH_2Br$ $(S_N2)^a$	$XCH_2C_6H_4CH_2Br$ $(S_N2)^{a,\,b}$
o-CO_2Me	0.878^c	1.42	0.322
p-CO_2Me	0.010^c	0.393	0.394
o-CO_2H	—	36.2	2.24
p-CO_2H	—	0.414	0.722
o-CO_2^-	—	>300	~15
p-CO_2^-	—	~1	~1

^{a} The mechanism refers to the hydrolysis of the parent compound.
^{b} For these compounds, k_X/k_H indicates the effect of X on the reactivity of o- or p-methylbenzyl bromide.
^{c} For CO_2Ph.

of the o-CO_2H derivative produces phthalide (**110**, $R^1 = R^2 = H$)[114c], thus strongly suggesting some form of anchimeric assistance by the substituent. Neighbouring group participation by the carbonyl or ethereal oxygen atom would involve reaction via structures **113** or **114** (R = H, Alk) for the o-CO_2R and o-CO_2H compounds, and reaction via **110** for the o-carboxyl-

(113)   (114)

ate ion. Arguments against this type of anchimeric assistance in the hydrolysis of benzyl bromides have been based on the fact that the ratio k_o/k_p for CO_2Me is much smaller than this ratio for CO_2Ph in the S_N1 hydrolysis of the diphenylmethyl bromides[114a]. However, the difference between these ratios arises entirely from the large difference between the retarding effects of p-CO_2R (see Table 22) which is by no means unexpected when benzyl bromide and its p-CO_2Me derivative undergo S_N2 hydrolysis, and it is therefore doubtful whether a comparison of k_o/k_p provides any useful information about the mechanism operating in the hydrolysis of the o-CO_2Me compound. The substantial increase in $\Delta S^{\ddagger}$ for hydrolysis (*ca.* 5–6 cal/deg.) resulting from the introduction of o-CO_2R and o-CO_2H

in benzyl bromide[114a] (but not from the corresponding *para* substitutions) suggests, at least, a greater tendency for unimolecular reaction in the *ortho* derivatives (*cf.* references 109 and 118). Similar observations for the effect of CO_2^- on the hydrolysis of alkyl bromides have already been interpreted in terms of anchimeric assistance by the substituent (section III.A.2), and it is therefore considered likely that the *ortho* substituted benzyl bromides undergo solvolysis via transition states which are stabilized by contributions from structures like **110**, **113**, or **114**. This view is supported by the fact that the order of reactivities (15) is also the order expected for increasing neighbouring group participation by the substituents.

Table 22 also shows the effect of substitution in the methyl group of methylbenzyl bromides (**115**) on the rate of solvolysis. In the *para* compounds, the introduction of CO_2R retards the reaction more than the introduction of CO_2H, probably because the greater electron attraction by CO_2R has the greater effect on the overall electron release by the substituted methyl groups towards the reaction centre. The sequence (16) of

$$CO_2R < H < CO_2H < CO_2^- \qquad (16$$

reactivities for the *ortho* compounds is similar to that found for *ortho* substitution in benzyl bromide (sequence 15) apart from the inversion of the positions of CO_2R and H. It is therefore tempting to suggest that CO_2H and CO_2^- substituents facilitate the hydrolysis of the *ortho* compounds (**115**) by participating in the formation of a six-membered ring system, analogous to **113** or **114**. This type of anchimeric assistance has already been recognized in other reactions (see section III.A.2) and generally occurs less readily than the formation of the five-membered rings. In agreement with these observations, the results in Table 22 show that the relative rates (k_X/k_H) for *ortho* substitution in benzyl bromide are larger than those for substitution in the *ortho* compounds (**115**). On the other hand, the hydrolysis of the CO_2H derivative is known to produce the 'normal' hydroxy compound[114c] but it may well be that the cyclic intermediate undergoes ring opening under the prevailing reaction conditions.

The introduction of 1-substituents in 9-bromofluorene (**116**, X = H)

(116)

is generally considered as analogous to *ortho* substitution in diphenyl-methyl halides. However, the 4-derivatives undergo solvolysis about three or four times more rapidly than the 1-substituted compounds when X $= CO_2Me, CO_2H$ and about 50% more rapidly when $X = CO_2^{-114b, c}$. These observations argue strongly against any anchimeric assistance even by $1 - CO_2^-$; unfavourable steric factors are probably responsible[114b, c]. The introduction of 1-and 4-CO_2Me retards solvolysis by a much smaller factor than expected for S_N1 solvolysis ($k_X/k_H = 0.08$ and 0.5, respectively) and is accompanied by a decrease in E and a proportionally even larger decrease in $\Delta S^{\ddagger}$. It therefore seems very likely that the two carbomethoxy derivatives undergo S_N2 solvolysis, though the parent compound probably reacts unimolecularly (*cf.* reference 109 and 118).

B. Nucleophilic Aromatic Substitution

Most of the nucleophilic aromatic substitutions which have been studied involve the bimolecular attack of a base (B) on a carbon–halogen bond. There is ample evidence that bond breaking is of little importance in the rate-determining step, and structures like **117** can therefore be expected to contribute significantly to the transition state. However, the reactions rarely proceed at measurable rates unless at least one electron-attracting substituent (Y) is present, presumably because the substantial contributions from structures like **118** and **119** are essential to reduce the free energy of the activated complex to accessible values. The rates are highly sensitive to the presence of further electron-attracting groups which

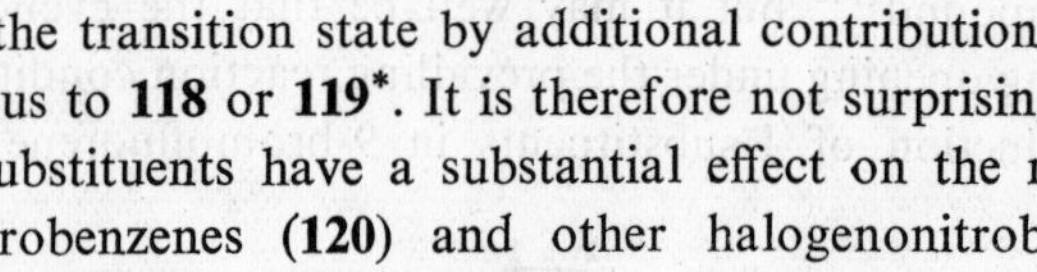

(117)   (118)   (119)

can stabilize the transition state by additional contributions from structures analogous to **118** or **119***. It is therefore not surprising to find that the present substituents have a substantial effect on the reactivities of 1-chloro-2-nitrobenzenes (**120**) and other halogenonitrobenzenes (see

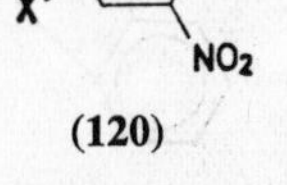

(120)

---

* For example, $k_{4-NO_2}/k_H > 10^6$ for the reaction of **120** with PhS^- in methanol[119].

Table 23), and that even the weak conjugative attraction by 4-CO_2^- is sufficient to stabilize the strongly electron-repelling reaction centre in the transition state. The substituents usually alter the rates mainly by changing the activation energy[†] and it must therefore be recognized that the values of k_X/k_H may depend markedly on the temperature.

TABLE 23. Relative rates (k_X/k_H) for reactions of 1-Cl-2-$NC_2O_6H_3X$ with bases

Reactions: I, with MeO^- in MeOH at 50°[120]
II, with PhS^- in MeOH at 35°[119]
III, with $C_5H_{11}N$ in C_6H_6 at 45°[121]

Substituent	I	II	III
4-CO_2^-	7·12, 4·45[a], 15·4[b]	—	2·56[c]
4-CO_2H	—	—	14·3
4-CO_2Me	1990, 345[a]	3600	923
5-CO_2H	—	—	0·54
5-CO_2Me	—	65	5·26
6-CO_2^-	0·373[d], 0·026[e]	—	—
6-CO_2H	—	—	15·1[f]
6-CO_2Me	174[d], 5·4[e]	83	55·3[f]

[a] For 1-Cl-2,6-$(NO_2)_2C_6H_2X$.   [b] For 1-F-2-$NO_2C_6H_3X$ at 25°[122].
[c] For 1-Br-2-$NO_2C_6H_3X$ at 35°[123].   [d] For 1-Cl-4-$NO_2C_6H_3X$.
[e] For 1-Cl-2,4-$(NO_2)_2C_6H_2X$.   [f] At 60°[124].

The nature of the alkyl group (R) has virtually no effect on $k_{4\text{-}CO_2R}/k_H$[119], but the presence of two nitro groups in the substrate leads to smaller accelerations by 4-CO_2Me and 4-CO_2^- than the same substitutions in the mononitro compounds, **120** (see Table 23). Other systems also show a decreasing sensitivity to the introduction of accelerating substituents as the reactivity of the parent compound is increased[3] and the present observations are fully consistent with the explanation which has ben proposed[1].

Although polar considerations suggest that CO_2H should accelerate reaction less than CO_2Me, the observed differences between the effects of these two substituents are larger than would have been expected. The retarding action of 5-CO_2H is particularly striking, and it has been suggested that a substantial fraction of each CO_2H compound is present in the form of its conjugate base in the reaction mixture[121]. Direct electrostatic interaction between the 5-CO_2^- group and the nucleophile (piperidine) would hinder reaction but such an interpretation would require an

[†] Exceptions are found for some 6-substituted compounds.

even greater retardation by 6-CO_2H, contrary to the observations (see Table 23). On the other hand, hydrogen bonding between 6-CO_2^- and the amine may be responsible (*cf.* reference 125).

The introduction of 6-CO_2Me or 6-CO_2^- facilitates the reaction of **121** with methoxide ions more than the same substitutions in **122** (see Table 23), presumably because of the greater steric hindrance to the approach of

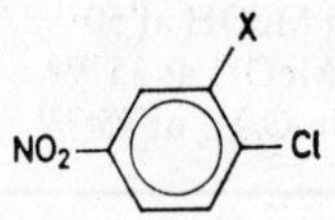

(121) 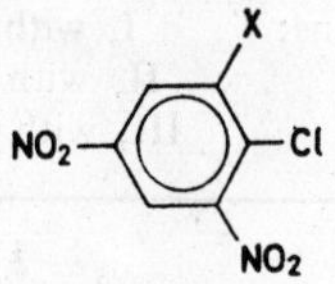   (122)

the nucleophile to that reaction centre in **122**. Retardation by 6-CO_2^- has been discussed in terms of electrostatic repulsion between the substituent and the methoxide ion[120], but this explanation suffers from the difficulty that the observed values of k_X/k_H arise from large increases in both energy and entropy of activation*. Normally such observations would be taken to show a change of mechanism on substitution and it is attractive to consider that **121** (X = 6-CO_2^-) reacts via a cyclic intermediate like **123**. How-

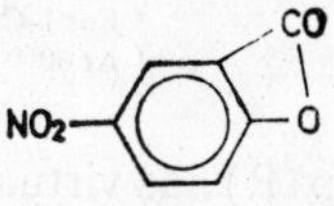

(123)

ever, the literature does not mention that the required first-order kinetics are observed and it is also noteworthy that the reaction product is that expected from bimolecular attack by methoxide ions[120]. Further work on this system is clearly necessary before the difficulty can be resolved.

All the results in Table 23 show that reaction with bases is accelerated much more by 4-CO_2Me than by 6-CO_2Me substituents. Steric inhibition of resonance between the 6-substituent and the phenyl ring has been suggested to account for these observations but other factors may also contribute since the changes in the activation parameters show no regular features (see Table 24). 6-Substitution may decrease $\Delta S^{\ddagger}$ substantially (**121** + MeO$^-$, **120** + PhS$^-$), leave its value unaltered (**120** + $C_5H_{11}N$), or even cause an increase (**122** + MeO$^-$). Any explanation of these observations

---

* For **121**, X = 6-CO_2^-, $E_X - E_H = 4\cdot7$ kcal and $\Delta S_X^{\ddagger} - \Delta S_H^{\ddagger} = 12\cdot8$ cal/deg. As a result, acceleration by the substituent is observed at temperatures greater than 100°c.

must inevitably be highly speculative and is therefore beyond the scope of this article.

TABLE 24. The effect of 6-CO_2Me on the activation parameters in nucleophilic aromatic substitution (X = 6-CO_2Me)

Substrate		Nucleophile	$\dfrac{E_X - E_H}{2\cdot3\,RT}$	$\dfrac{\Delta S_X^{\ddagger} - \Delta S_H^{\ddagger}}{2\cdot3\,R}$	Reference
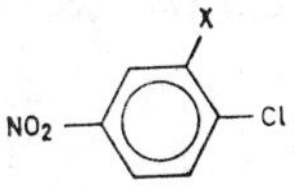 NO_2—⟨ring: X, Cl⟩	(121)	MeO⁻	4·73	−2·32	120
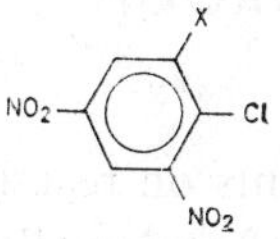 NO_2—⟨ring: X, Cl, NO_2⟩	(122)	MeO⁻	0·09	0·62	120
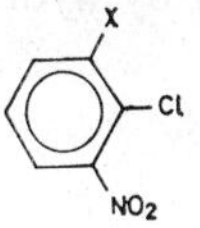 ⟨ring: X, Cl, NO_2⟩	(120)	PhS⁻	3·05	−1·13	119
		$C_5H_{11}N$	1·77	−0·03	124

The products of nucleophilic aromatic substitution in the pentafluorobenzoate ion (124, B = F)[125] differ a little from those which would have

$$C_6F_4BCO_2^-$$
(124)

been expected on the basis of the effect of CO_2^- on the reactivities of nitrohalogenobenzenes. Reaction with methoxide ions gives only the *para* product (124, B = 4-MeO), probably because electrostatic repulsion hinders attack at the *ortho* position[125]. On the other hand, attack by amines gives *ortho* and *para* products in roughly similar amounts, though *para* substitution appears to require the smaller activation energy. Hydrogen bonding between the base and the CO_2^- group has been postulated to account for the formation of the *ortho* product[125].

The decomposition of aryldiazonium ions (125) represents an example of aromatic S_N1 reaction. The process requires the electron displacement shown in 125 in the rate-determining step and should therefore be retar-

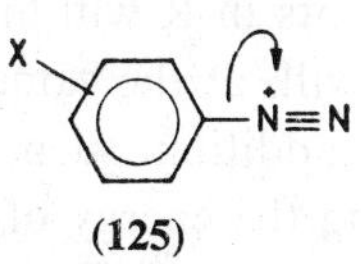

X—⟨benzene ring⟩—$\overset{+}{N} \equiv N$

(125)

ded by electron-attracting substituents (X) in the order $k_H > k_{m-X} > k_{p-X} \approx k_{o-X}$, as observed[126] for reaction in water when $X = CO_2H$:

X	$o\text{-}CO_2H$	$m\text{-}CO_2H$	$p\text{-}CO_2H$
k_X/k_H	0·12	0·55	0·19

It is noteworthy that the values of k_X/k_H are rather larger than might have been expected in a strongly electron-demanding system (*cf.* section III. A. 2).

IV. REACTIONS AT THE CARBONYL GROUP

A. General Considerations

Almost all studies of the effect of the present substituents on reactivity at carbonyl carbon atoms have involved the solvolysis of carboxylic es- ters (126). The reactions are usually catalysed by bases and by hydrogen

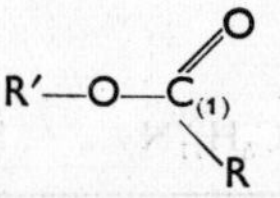

$$R'-O-C_{(1)} \overset{\displaystyle O}{\underset{\displaystyle R}{<}}$$

(126)

ions, and may either involve rupture of the R′–O linkage (alkyl-oxygen fission, AL) or fission of the $C_{(1)}$–O bond (acyl-oxygen fission, AC). Most of the examples relevant to this section involve the bimolecular mecha- nism associated with acyl-oxygen fission; $B_{AC}2$ for base-catalysed and neutral hydrolysis, $A_{AC}2$ for the acid-catalysed reaction.

The rate-determining step for $B_{AC}2$ solvolysis is illustrated below (equa- tion 17), and it can be seen that the activation process has ambiguous polar

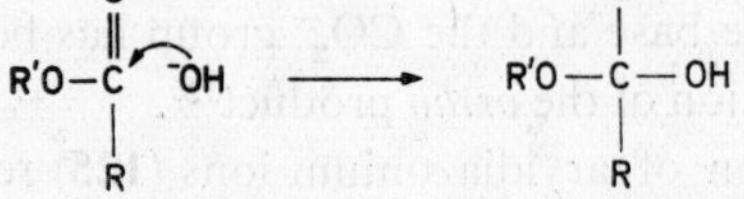

$$R'O-C \overset{\displaystyle O}{\underset{\displaystyle R}{|}} \ \ddot{O}H \longrightarrow R'O-C \overset{\displaystyle O^-}{\underset{\displaystyle R}{\underset{\displaystyle |}{|}}} OH \tag{17}$$

requirements, as in the S_N2 reactions of halides (see section III.A.2). Electron-attracting substituents in R will hinder electron accession to the carbonyl oxygen atom but will, at the same time, tend to facilitate bond formation with the base. In addition, such substituents in R will destabi- lize the ester (126) by raising the energy of the valence bond form (127), thus reducing its contribution to the structure of **126**. It can therefore be

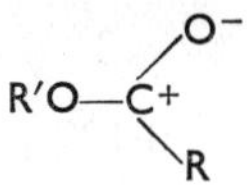

$$R'O-\overset{\displaystyle\nearrow O^-}{\underset{\displaystyle\searrow R}{C^+}}$$

(127)

concluded that CO_2H and CO_2R substituents in R will accelerate $B_{AC}2$ hydrolysis, but their effect will probably be relatively small. However, a CO_2^- group should retard reaction with a negatively charged base by direct electrostatic interaction through space.

In the acid-catalysed process ($A_{AC}2$) the protonated ester (**128**) is in rapid equilibrium with its conjugate base and reacts with the solvent in the rate-determining step, as shown in equation (18) for hydrolysis. The polar requirements are again ambiguous, and the proportion of the ester

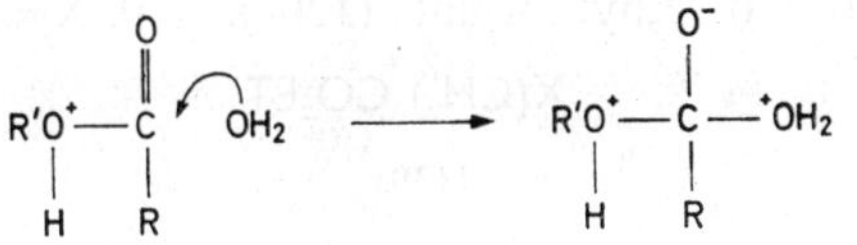

$$R'\overset{+}{O}-\overset{\displaystyle\overset{O}{\|}}{\underset{\displaystyle\underset{R}{|}}{\underset{H}{C}}}\,\,OH_2 \longrightarrow R'\overset{+}{O}-\overset{\displaystyle\overset{O^-}{|}}{\underset{\displaystyle\underset{R}{|}}{\underset{H}{C}}}-\overset{+}{O}H_2 \qquad (18)$$

(128)

present in the protonated form (**128**) appears to be the most important factor controlling the rates of structurally similar esters. Electron-attracting substituents in R should therefore retard $A_{AC}2$ solvolysis but their effect will clearly be small unless they are situated close to the ethereal oxygen atom.

Further details about the 'normal' mechanisms of ester hydrolysis can be obtained from standard works[127], but CO_2H and CO_2^- groups in R may also facilitate neutral solvolysis by intramolecular catalysis, which results in the formation of the acid anhydride[128]. This aspect is considered further in later pages of this section but it must be pointed out now that, in general, the acid- and base-catalysed processes appear to be energetically more favourable than intramolecular catalysis.

B. Bimolecular Hydrolysis

The effects of the present substituents on the rates of $A_{AC}2$ and $B_{AC}2$ hydrolysis are in complete agrement with those expected from the polar considerations outlined in the preceding paragraphs. Thus the introduciton of ω-CO_2Me in the primary aliphatic esters (**129**, X = H) retards $A_{AC}2$ hydrolysis[129], the magnitude of the effect diminishing with increasing

$$X(CH_2)_nCO_2Me \qquad (129)$$

distance of the substituent from the reaction centre:

n	0	2	3	4
k_{CO_2Me}/k_H	0·066	0·197	0·820	0·911

The relatively large effect of the substituent on the reactivity of methyl propionate (**129**, $n = 2$) is a little surprising in view of the poor relay of polar effects through a saturated carbon chain. However, analogous behaviour has also been found in the effect of ω-CO_2H on the strengths of primary carboxylic acids (see section II.B.2) and it may well be that the substituents exert their effect at least partly by direct electrostatic interaction through space. It is also noteworthy that **129** ($X = CO_2H$) reacts at almost the same rate as **129** ($X = CO_2Me$) when $n \geqslant 2$[129], although the two substituents do not attract electrons to the same extent. Differences between $X = CO_2Me$ and $X = CO_2Et$ are also negligibly small[129].

The $B_{AC}2$ reaction of ethyl formate (**130**, $n = 0$, $X = H$) with hydroxide

$$X(CH_2)_nCO_2ET$$

(**130**)

ions is accelerated about 150 times by the introduction of CO_2Et[130], as expected for an electron-attracting substituent near the reaction centre. Similarly, the approximately eight-fold retarding effect[131] of a CO_2^- group in ethyl acetate (**130**, $n = 1$, $X = H$) is fully consistent with electrostatic repulsion between the substituent and the reagent in the transition state of the rate-determining step (equation 17).

The effect of these two groups on the rate of alkaline hydrolysis of ethyl

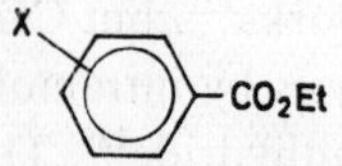

(**131**)

benzoate (**131**, $X = H$) is shown below for reaction at $25°$[132]; the experimental rate constants for the dicarboethoxy compounds (**131**, $X = CO_2Et$) have been statistically corrected:

	ortho	meta	para
k_{CO_2Et}/k_H	0·23	1·61	3·55
$k_{CO_2^-}/k_H$	0·11	0·94	1·0

The *meta* and *para* compounds show the value of k_X/k_H which would have been expected for $B_{AC}2$ reaction from the polar properties of the substituents, and the retarding effect of $o\text{-}CO_2^-$ arises mainly from an increase in the activation energy, consistent with electrostatic interaction between the negative reagent and a negative charge located near the reaction centre. At first sight, the decelerating effect of $o\text{-}CO_2Et$ might be held to reflect steric inhibition of resonance between the substituent and the aromatic ring, but the smaller reactivity of this compound arises entirely from a decreased entropy of activation. It is therefore more likely that k_X/k_H reflects the smaller probability for the approach of the reagent to the sterically more crowded reaction centre.

C. Intramolecular Catalysis

There is now much evidence that the hydrolysis of carboxyl derivatives can be greatly accelerated by the presence of suitably situated CO_2^- and CO_2H groups in the reacting molecule. The substituents act as intramolecular catalysts for the formation of the acid anhydride (see reactions 19 and 20)* which usually undergoes rapid hydrolysis to the dicarboxylic acid or its conjugate base. In some cases, however, the production of the anhydride occurs rapidly, and it is the hydrolysis of the intermediate which determines the rate; anhydrides which are stable under hydrolytic conditions are also known.

$$\tag{19}$$

$$\tag{20}$$

Evidence for abnormal acceleration by CO_2^- or CO_2H substituents is usually obtained from the pH-rate profile for hydrolysis over that range of pH where the acid- and base-catalysed reactions make no significant

---

* Alternatively, the reactions can be regarded as examples of anchimeric assistance by substituents.

contribution to the rate; it is for this reason that the term 'neutral' hydrolysis has been adopted in the present discussion. The available results show that the rate of anhydride formation depends very markedly on the steric and conformational properties of the reactant, and it is therefore not surprising to find that intramolecular catalysis by both CO_2^- and CO_2H has never been observed in the same system. In addition, anhydride formation with CO_2^- can be expected to be favoured if the departing anion (RO^- in reaction 19) is stabilized by resonance (e.g. $R = 4\text{-}BrC_6H_4$) and it is interesting to find that most of the examples of internal CO_2^- catalysis require the separation of relatively stable anions (see Table 26). Other compounds where this process involves the less stable alkoxide or amide ions only show intramolecular catalysis by CO_2H (see Table 25) but no systematic studies on the effect of varying the departing group under otherwise constant conditions have yet been reported.

TABLE 25. Relative rates (k_{CO_2H}/k_H) for the effect of CO_2H groups on the neutral hydrolysis of esters and amides

Substrate		k_{CO_2H}/k_H	Reference
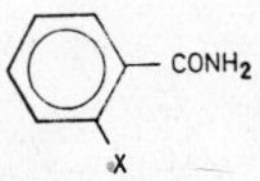 (structure) —CONH₂, X	(132)	10^5	133
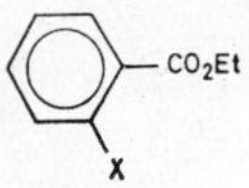 (structure) —CO₂Et, X	(133)	~ 200	134
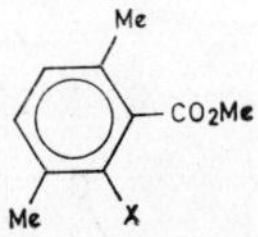 (structure) Me, —CO₂Me, Me, X	(134)	$> 2 \times 10^4$	135

^a This factor expresses the rate of the derivative, relative to methyl benzoate; a larger value can therefore be expected for k_{CO_2H}/k_H.

Table 25 gives some examples of intramolecular catalysis by CO_2H; in all cases the rates are very much larger when the substituent (X) is CO_2H than when it is CO_2^-. The large acceleration of the hydrolysis of benzamide (132, X = H) on the introduction of o-CO_2H is replaced by retardation when the same substituent is in the *para* position, and the reaction of **132** ($X = CO_2H$) containing ^{13}C with $H_2^{18}O$ provides indirect evidence for hydrolysis via the anhydride[133]. Similarly, succinanilic acid (**135**, $X = CO_2H$) reacts very much more rapidly than the parent compound (**135**, $X = H$)[136].

$$XCH_2CH_2CONHPh$$

(**135**)

The large difference between the accelerating effect of the CO_2H group in an alkyl benzoate (**133**, $X = H$) and its 3,6-dimethyl derivative (**134**, $X = H$) underlines the great sensitivity of the free energy of activation for cyclization to small changes in the separation between the CO_2R and CO_2H groups, though a small part of this difference may result from the use of different alkyl groups in the two esters. An earlier study had reported intramolecular catalysis by CO_2^-, and not by CO_2H, in the hydrolysis of methyl hydrogen-phthalate[137], but the relatively small acceleration (*ca.* ten-fold) on the introduction of the substituent and the later results (see Table 25) have led to the suggestion[135] that catalysis by a component of the buffer employed could be responsible for the earlier observation.

The carboxylate ions listed in Table 26 all undergo hydrolysis much more rapidly than their conjugate acids, the *p*-bromophenyl glutarate ion (**136**) being the least reactive. It is therefore very likely that compounds **137–143** all react via the anhydride by scheme (19). This has been particularly convincingly demonstrated for the *exo*-3,6,-*end*oxotetrahydrophthalate system (**143**) where the rate of hydrolysis of the anhydride actually controls the overall rate of solvolysis.[138] The results for the glutarates (**136–139**) and succinates (**140 and 141**) have been discussed in terms of the effective distance between the two reactive groups (CO_2R and CO_2^-)[138], but the much greater reactivity of the succinates relative to the glutarates could also arise partly from different stabilities of the five- and six-membered cyclic anhydrides. Any such difference would inevitably be reflected in the free energy of the transition state of the intramolecular process.

Eberson has pointed out[139] that the sequence of increasing values of k_{rel} (Table 26) also reflects the order of the ratios of the first to the second dissociation constants of the corresponding dicarboxylic acids (K_1/K_2). As the abnormally large values of K_1/K_2 of these compounds are considered to indicate internal hydrogen bonding in the first ionization product (see section II.B.1), it would appear that similar factors favour this process and the rather different intramolecular catalysis of ester hydrolysis by CO_2^-. This slightly surprising conclusion derives support from the fact that some of the acids now considered are partly in the form of their an-

TABLE 26. Relative rates (k_{rel}) for ester hydrolysis subject to intramolecular catalysis by CO_2^- [138]. (R = 4-BrC_6H_4, except when indicated; $k_{rel} = 1$ for the derivative of glutaric acid (136))

Substrate		k_{rel}
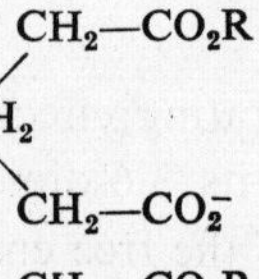 $CH_2\!-\!CO_2R$ CH_2 $CH_2\!-\!CO_2^-$	(136)	1
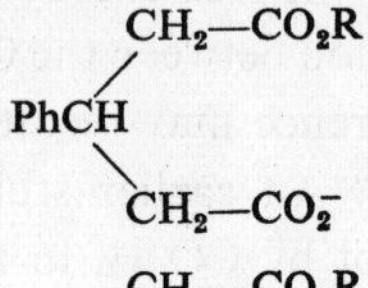 $CH_2\!-\!CO_2R$ $PhCH$ $CH_2\!-\!CO_2^-$	(137)	3·38
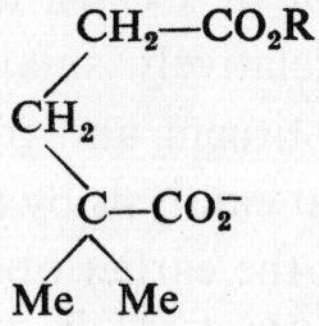 $CH_2\!-\!CO_2R$ CH_2 $C\!-\!CO_2^-$ Me Me	(138)	3·64
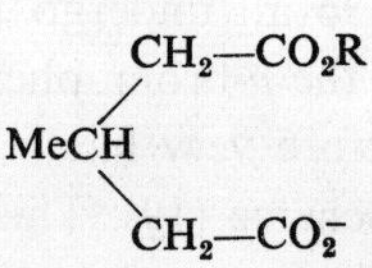 $CH_2\!-\!CO_2R$ $MeCH$ $CH_2\!-\!CO_2^-$	(139)	4·41
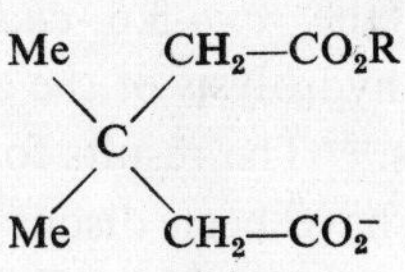 Me $\quad CH_2\!-\!CO_2R$ C Me $\quad CH_2\!-\!CO_2^-$	(140)	19·3
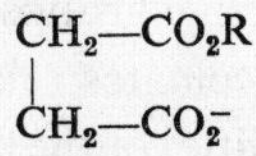 $CH_2\!-\!CO_2R$ $CH_2\!-\!CO_2^-$	(141a)	232
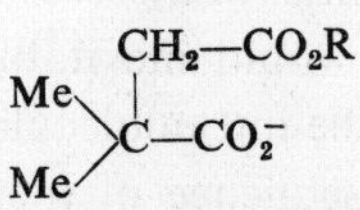 $CH_2\!-\!CO_2R$ Me $\quad$ C$-\!CO_2^-$ Me	(141b)	728
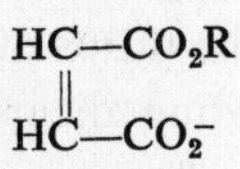 $HC\!-\!CO_2R$ $\parallel$ $HC\!-\!CO_2^-$	(142)[a]	10^4
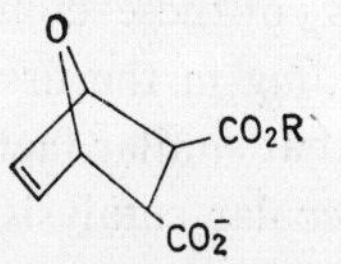 (bicyclic structure with O, CO_2R, CO_2^-)	(143)[a]	5×10^5

[a] R = 4-$MeOC_6H_4$; k_{rel} calculated from the rate of the corresponding derivative of succinic acid.

hydrides in aqueous solution[81]. Further support arises from the behaviour of the derivatives of phthalic acid; there is no evidence for internal hydrogen bonding in the first ionization product (see section II.B.5), and the hydrolysis of the half-ester is subject to intramolecular catalysis by CO_2H but not by CO_2^-[134].

The methyl ester, **144**, forms the substituted succinic anhydride relati-

$$X(CHBu-t)_2CO_2Me$$

(144)

vely easily in aqueous solution when $X = CO_2H$ but not when $X = CO_2^-$[139]. This complete reversal of the behaviour of other 4-bromophenyl succinates (see Table 26, compounds **141a** and **141b**) could be connected with the greater stability of $4\text{-}BrC_6H_4O^-$, relative to MeO^-, but the different steric situation resulting from the presence of the two bulky t-butyl groups in **144** could equally well be responsible.

V. ELECTROPHILIC AROMATIC SUBSTITUTION

A. General Considerations

Electrophilic substitution in aromatic systems has been widely studied. Early work in this field concentrated on the nitration of hydrocarbons and their derivatives but other reactions such as halogenation, acylation and mercuration have now been investigated, and the isotopic replacement of hydrogen as well as the rupture of carbon–metalloid and carbon–metal bonds have attracted much attention. The reactions have been particularly popular for the assessment of the effects of polar groups on the rates, since the influence of a particular substituent varies widely from one system to another. However, the present substituents have received relatively little attention, possibly because they do not show extreme polar properties. Moreover, most of the reactions involve the use of acid solutions so that virtually no information is available about the behaviour of CO_2^-; indeed, there must be some doubt about the extent to which the results observed for CO_2H and CO_2R substituents in strongly acid media reflect the behaviour of compounds which are partly present in the form of their conjugate acids. Only some of the features of electrophilic aromatic substitution which are relevant to the present discussion are outlined below; details can be obtained from several of the excellent reviews which are available[62, 140].

1. Mechanism

It seems very likely that the replacement of a group A attached to an aromatic system by an electrophilic reagent E involves the rate-determining formation of an intermediate (145) which decomposes rapidly to give the products (*cf.* references 21 and 141). Structures like 146 and 147,

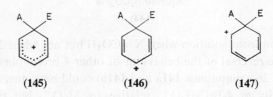

(145)   (146)   (147)

which can therefore be expected to contribute substantially to the transition state, will be destabilized by the presence of electron-attracting substituents, so that the introduction of CO_2H or CO_2R groups in the phenyl ring should retard the reaction in the order *ortho ~ para > meta*, while the converse should apply to the proportion of substitution products obtained when A = H. This simple conclusion is based solely on polar considerations and often requires minor modification since steric and other factors may result in different reactivities at sites *ortho* and *para* with respect to the substituent[3]. It is also noteworthy that the electron-attracting powers of CO_2H and CO_2R suggest a reduced reactivity and a greater proportion of *meta* substitution products for the CO_2R derivative, but this is not usually observed in electrophilic substitution (see section V. B.).

Not all electrophilic aromatic substitutions necessarily occur in the manner outlined above. Thus, the large rate of rupture of C—Sn and C—Pb bonds has led to arguments for a rate-determining nucleophilic attack on an intermediate[142], and the effect of substituents on the rate could then well be different from that observed in reaction by the more common mechanism. However, the introduction of CO_2H also retards these processes (see Table 28).

2. Partial rate factors

The effect of substituents (X) on the rate of electrophilic attack on benzene is conveniently discussed in terms of the partial rate factors (f^X) which compare the reactivities of the various positions in the substituted derivative with that of one of the six possible reaction sites in benzene. The partial rate factors are obtained from the overall rate relative to benzene (k_X/k_H) and the isomer proportions via equations (21)–(23). Once

$$f_o^X = \frac{6k_X}{2k_H} \frac{\% \, ortho}{100} \tag{21}$$

$$f_m^X = \frac{6k_X}{2k_H} \frac{\% \, meta}{100} \tag{22}$$

$$f_p^X = \frac{6k_X}{k_H} \frac{\% \, para}{100} \tag{23}$$

the partial rate factors are known, the reactivity of sites in polysubstituted compounds can be calculated from the additivity principle, which assumes that the free energy of activation is the sum of the free energies of activation at the corresponding sites in the monosubstituted compounds. For example, the partial rate factor for the replacement of one of the four equivalent hydrogen atoms in **148** is $f_o^X f_m^X$. It most however be pointed out that the additivity principle does not always apply with complete accuracy.

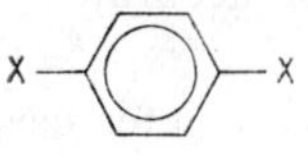

(148)

When a group other than hydrogen is replaced, the partial rate factor for the substituent (f_p^X for the reaction of p-XC₆H₄A) is identical with the observed relative rate, k_X/k_H. The results are however often reported as partial rate factors for such compounds in order to facilitate comparison with the effect of the same substituent on the rate of replacement of hydrogen in other reactions.

3. Selectivity

The effect of substituents on the rate of electrophilic substitution reactions is often correlated satisfactorily by the Hammett relation (equation 4) if the substituent constants σ^+ are employed (see section I.D). Although highly polar substituents have given anomalous results in some systems[3], the intermediate reactivities of CO₂R and CO₂H derivatives generally lie close to the best straight line log k_X against σ_X^+ when more extreme groups have also been studied.

An equivalent approach[62, 143] discusses the reactions of monosubstituted benzenes in terms of the selectivity of the electrophilic reagent which is defined by the selectivity factor, S_f (equation 24). Comparison of equations (4) and (24) shows that S_f is a direct measure of the reaction

$$S_f = \log \left(f_p^{Me}/f_m^{Me} \right) \tag{24}$$

parameter, ϱ, so that an increase in selectivity should increase the retarding effect of CO_2H and CO_2R groups, and also the proportion of *meta* substitution in benzene derivatives. Partial rate factors can be calculated to a reasonable degree of approximation from the selectivity relation (S_f) or from the extended selectivity relation (ϱ).

B. Monosubstituted Benzenes

Isomer proportions for electrophilic substitution in carboxy- and carboethoxyphenyl compounds are given in Table 27. Figures for the displacement of groups other than hydrogen (reactions 2 and 6) were calculated from the published rates and have been included in order to provide a wider range for comparison; the values are of course hypothetical. It must also be pointed out that only the *meta* and *para* compounds were studied in reactions (2) and (6) so that an assumed rate for the *ortho* compound had to be used in the calculations. The *meta:para* ratios for these compounds are, however, accurate.

TABLE 27. Isomer proportions (%) for substitution in carboxy derivatives

Reaction	ortho	meta	para
(a) Chlorination (HO_2CPh, Cl_2)[144]	0	100	0
(b) Deboronation ($EtO_2CC_6H_4B(OH)_2$, Br_2)[145] a	38·0	55·7	6·3
(c) Nitration: (i) (HO_2CPh, HNO_3)[146, 147]	18·5	80·2	1·3
(ii) (EtO_2CPh, HNO_3)[146]	28·3	68·4	3·3
(d) Bromination (HO_2CPh, Br^+)[148]	0	100	0
(e) Iodination (HO_2CPh, I^+)[149]	0	100	0
(f) Degermylation ($HO_2CC_6H_4GeEt_3$, H^+)[150] b	33·7	57·7	8·6

a Calculated from rates for *meta* and *para* derivatives, assuming $k_o/k_p = 6$.
b Calculated from rates for *meta* and *para* derivatives, assuming $k_o/k_p = 4$.

The reactions in Table 27 have been arranged in the order of their decreasing selectivity[62] but the differences between reactions (b)–(e) are extremely small. It is therefore a little surprising to find that the proportion of *meta* products does not decrease with decreasing selectivity, as expected (see section IV.A.3), and that reactions of similar selectivity can show either entirely *meta* substitution (reactions d and e) or about 20% *ortho* and *para* substitution (reaction ci). This observation illustrates yet again the limitations in the application of linear free energy relations to the comparison of highly similar reactions. In the present example, the fact that selectivity factors deduced from results with accelerating substituents (usually Me) have been applied to the findings for retarding groups may also be partly responsible.

In nitration (reaction c) ethyl benzoate gives a smaller proportion of *meta* product than benzoic acid, and the *meta* : *para* ratio for deboronation (reaction b) of a CO_2Et derivative is less than for CO_2H compounds in other reactions of similar selectivity. The required reactivity sequence ($CO_2Et > CO_2H$) is the converse of that expected from polar considerations (see section I.A.2) but has often been observed in electrophilic substitution (see Table 28). No explanation has yet been offered.

The proportion of *meta* substitution on the nitration of ethyl benzoate is increased over that reported in Table 27 when a mixture of nitric and sulphuric acids is employed, and reduced when the reagent is acetyl nitrate[151], contrary to expectation since work with electron-donating groups suggests that acetyl nitrate is the more selective reagent[3]. However, all the nitration studies involving the present substituents (and other electron-attracting groups) show that reactivity is greater at the site adjacent to the substituent than in the *para* position, the converse of the situation found when electron-releasing groups are employed[3]. It is widely believed that this 'ortho effect' is primarily steric in origin,[152] possibly because the stability of the *p*-phenonium ion (149) is more sensitive to the polar character of a substituent (X) than the *o*-phenonium ion (150).

(149)   (150)

The CO_2^- group does not appear to favour substitution at any one particular position. The statistical distribution of products (*ortho* : *meta* : *para* : = 40 : 40 : 20) found for the chlorination of benzoate ions by hypochlorous acid[153] may reflect an even spread of the activation arising from the negative charge over the three available positions but a virtual cancellation of $-I$ and $+T$ effects may equally well be responsible.

Partial rate factors have only rarely been determined for systems of interest in the present discussion; results are shown in Table 28 where the compounds have been arranged in the order of decreasing selectivity. It can be seen that the rates tend to increase as S_F decreases but several exceptions can be found; the anomalous behaviour of ethyl benzoate on nitration (reaction c) is particularly striking. Some uncertainty arises in

TABLE 28. Partial rate factors for aromatic substitution

Reaction[a]	CO_2H		CO_2Et		
	f_m	f_p	f_o	f_m	f_p
(a) Chlorination $(XPh, Cl_2)^{154}$	—	0·0008	—	—	0·0015[b]
(b) Deboronation $(XC_6H_4B(OH)_2, Br_2)^{145}$	—	—	—	0·044	0·010
(c) Nitration $(XPh, AcNO_3)^{155}$	—	—	0·0026	0·0079	0·0009
(d) Bromination $(XPh, Br^+)^{156}$	0·0225	—	—	—	—
(g) Desilylation $(XC_6H_4SiMe_3, H^+)^{157}$	—	0·0015	—	—	—
(f) Degermylation $(XC_6H_4GeEt_3, H^+)^{150}$	0·0177	0·0052	—	—	—
(h) Destannylation $(XC_6H_4Sn(C_6H_{11})_3, H^+)^{158}$	—	0·030	—	—	—

[a] Reactions having the same number as in Table 27 refer to the same process.

[b] Estimated from other observations via the additivity principle.

the values for reaction (a) which were calculated from other results via the additivity principle, since other evidence suggests that this principle may not be valid for electrophilic substitution in the presence of deactivating groups[156]. All the other features of Table 28 have, in effect, already been discussed in this subsection.

C. Disubstituted Benzenes

The electrophilic substitution reactions of carboxyphenyl compounds containing additional groups argue strongly against the validity of the additivity principle for these systems. This has been strikingly demonstrated for the bromination of phthalic acid (151) by Br^+, when $k/k_{benzene} = 4·9 \times 10^{-5}$. Comparison with the rate of reaction of benzoic acid under the same conditions shows that this observation is only consistent with the additivity principle if benzoic acid gives 30% ortho and para substitution[156]; quantitative meta bromination is however observed (see Table 27, and reference 156). Similarly, the large difference in the reactivity of

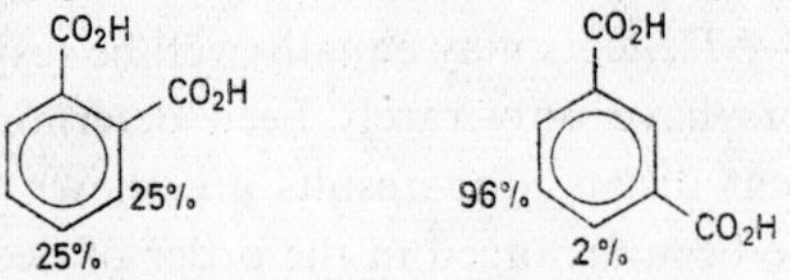

(151)　　　　　　　　(152)

benzoic acid with respect to ortho and para nitration (see Table 27) would not lead to the observed isomer proportions which are shown in **151** and **152** for nitration[159] if the additivity principle applied.

Positive bromination of the substituted anisoles (**153**) occurs essentially in the position adjacent to MeO, and the relative rate is decreased by a

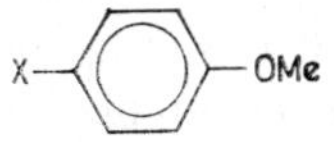

X—⟨◯⟩—OMe

(153)

factor of six when the substituent X is altered from Br to CO$_2$Et[160]. However the partial rate factors for *m*-Br and *m*-CO$_2$H are almost the same in degermylation[62], and the large difference between the rates in a reaction of larger selectivity remains surprising especially when it is remembered that CO$_2$Et is usually more reactive in electrophilic substitution than CO$_2$H. On the other hand, the isomer proportions for the nitration[159] of chlorobenzoic acids (shown in **154** and **155**) are roughly consistent with the require-

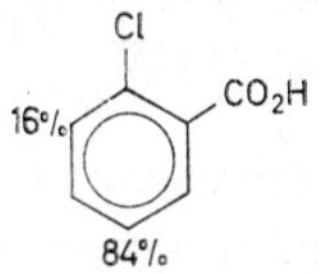

(154) 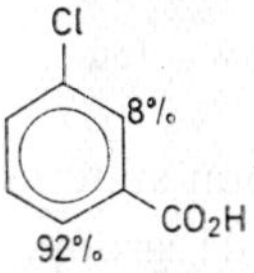   **(155)**

ments of the additivity principle in the light of the results observed in the reactions of chlorobenzene and benzoic acid.

D. Side-chain Substituents

The introduction of CO$_2$Et in the side-chain of toluene reduces the rate of nitration, as expected for an electron-attracting substituent, but the reaction still occurs more rapidly than when benzene is the substrate. The partial rate factors (see below) are virtually the same for reaction with nitric acid[161, 162] and acetyl nitrate[163], and provide a clear example of the effect of changing polar properties in the substituent; the small change in the reactivity at the *meta* position is particularly striking:

	f_o	f_m	f_p
PhCH$_3$[161]	42	2·5	58
PhCH$_2$CO$_2$Et[162]	4·6	1·2	10·4

Similarly, the phenylacetic ester gives a larger proportion of *meta* product than toluene on chloromethylation, though the reaction remains predominantly *ortho*- and *para*- directing[164]. For nitration, the successive replacement of the alkyl hydrogen in toluene by CO_2Et increases the fraction of *meta* substitution[146, 165]. However, the *meta* compound only becomes the major product when three CO_2Et groups are present and it can therefore be concluded that only at this stage does the group attached to the aromatic system behave as an overall electron attractor relative to hydrogen.

The nitration of cinnamic acid (156) occurs almost entirely at the *ortho* and *para* positions[166], although the rate is only one-tenth the rate of nitration of benzene[167]. This suggests that the group $CH=CHCO_2H$ behaves as a $-I+T$ substituent[168], like the halogens, and it is noteworthy that structures like 157 apparently make a significant contribution to the struc-

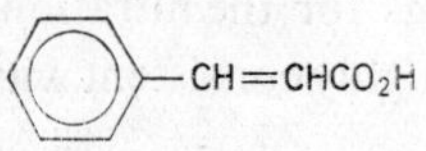

(156) 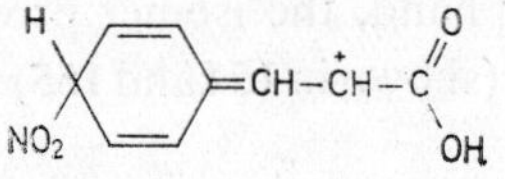   (157)

ture of the transition state in spite of the presence of the electron-attracting $(-I-T)$ CO_2H group. Similarly, the nitration of phenylpropiolic acid and its ethyl ester is mainly *ortho*, *para*-directing[169]; here too the isomer proportions suggest that CO_2Et attracts electrons less strongly than CO_2H:

	ortho	meta	para
PhC⋮CCO₂H	27	8	65
PhC⋮CCO₂Et	36	6	58

VI. ADDITION TO UNSATURATED SYSTEMS

A. General Considerations

Addition to olefins has been examined on innumerable occasions; but acetylenes have been less frequently studied. The initial attacking agent may be a radical, nucleophile or electrophile but there is very little information about the effect of the present substituents on radical addition. Similarly, though many nucleophilic additions are known for compounds containing carboxy or carboalkoxy groups (*e.g.* the Michael reaction, acid-catalysed halogenation), the presence of these groups or other powerful electron attractors is often an essential prerequisite for reaction, so that the vast majority of the available results cannot be discussed in terms of the activating effects of CO_2H and CO_2R. This section therefore concentrates mainly on electrophilic addition, but even here most results of interest in the present discussion refer to a single process (halogenation), and the amount of relevant information is rather limited.

The detailed nature of the processes involved in electrophilic addition has been the subject of much discussion and disagreement. The nature of the attacking reagent, the reaction conditions, and the structure of the substrate can all exert a profound influence on the reaction path, but common basic principles apply to all the additions. These are briefly outlined below; details can be obtained from standard works[170].

For brevity's sake, the term electrophilic addition has been omitted from section headings but the consideration of nucleophilic addition is always indicated by a specific statement.

1. Mechanism

The initial and (generally) rate-determining step in electrophilic addition can be regarded as involving the attack of a cation (E^+) or the positive site of a dipolar molecule (E—Y) to form a carbonium ion (see schemes 25 and 26); halogen molecules act as though they were dipolar

$$\tag{25}$$

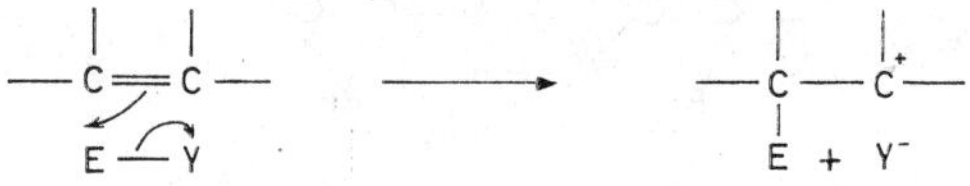

$$\tag{26}$$

species. It must be stressed that schemes (25) and (26) represent oversimplifications. Thus, there is good reason to believe that bond fission of a molecular electrophile (E—Y) sometimes occurs after the rate-determining step and that the carbonium ion is often more accurately described as a resonance hybrid of the classical structures (158–160). In general, these three

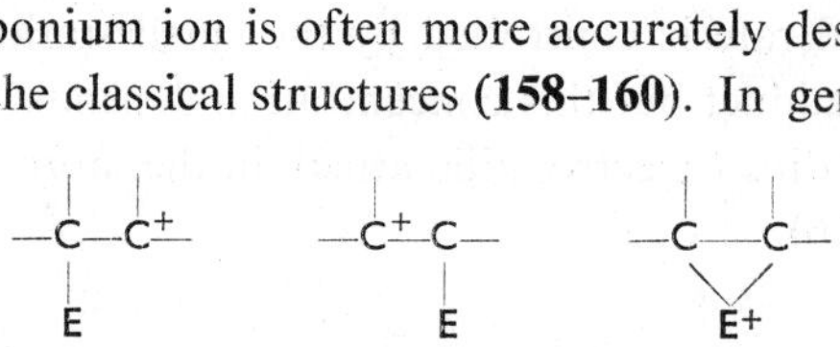

(158)   (159)   (160)

structures do not make equal contributions and the available evidence suggests that **160** frequently represents the intermediate to a good approximation. A similar conclusion arises from the assumption that the rate-determining step requires the formation of a π complex between the substrate and the electrophile, but the merits or demerits of the rival views which have been expressed are beyond the scope of this discussion. For the present purposes, the simplified schemes (25) and (26) form a sufficient basis for interpretation, subject to the various relatively minor modifications which will be indicated as they arise.

2. Reaction order

Reaction by schemes (25) and (26) requires second-order kinetics, as usually observed. A third-order rate law (second-order with respect to the electrophile) is however found on occasions, the principal examples of present interest being the reaction with hydrochloric acid in aprotic solvents and molecular halogenation in acetic acid (under some conditions) if the halogen contains at least one bromine or iodine atom. However, the effects of structural changes in the olefin on the rate of the third-order process are strikingly similar to their effects on similar second-order reactions* and it therefore seems likely that the higher order does not indicate a fundamental change of mechanism. It is believed that a halogen adds reversibly to the olefin by octet expansion of one of its atoms to form an adduct (*e.g.* scheme 27) which only ejects the second halogen atom when another molecule arrives to complete the addition. A similar adduct with

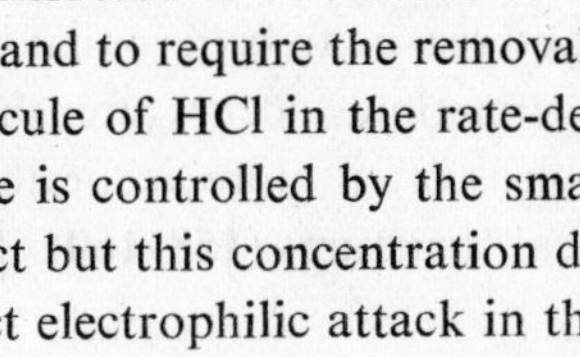

$$\underset{\substack{| \\ Br \\ | \\ Br}}{C=C} \; \rightleftharpoons \; \underset{\substack{| \\ Br \\ | \\ Br}}{-C-C-} \tag{27}$$

a molecule of HCl is considered to be formed reversibly in hydrochlorination in aprotic solvents and to require the removal of the chlorine atom as HCl_2^- by another molecule of HCl in the rate-determining step. In both cases, the observed rate is controlled by the small stationary concentration of the initial adduct but this concentration depends on the same factors as the rate of direct electrophilic attack in the more normal mode of reaction (scheme 26).

* This applies particularly to third-order bromination and second-order chlorination.

3. Kinetic substituent effects

Electrophilic attack will obviously be aided by electron accession to the reaction centre in the substrate, and substituents containing carbonyl groups should therefore retard electrophilic addition in the order of their increasing capacity for electron attraction (sequence 28). In practice, the decelerating effects of CO_2H and CO_2R are usually substantial and an abnormal reactivity of the CHO derivative (CHO$>$$CO_2H$) is generally

$$CO_2^- < CO_2R < CO_2H < CHO \qquad (28)$$

accepted as a strong indication of nucleophilic addition, as opposed to the electrophilic process.

On the other hand, the reactivities of the CO_2H and CO_2R derivatives almost invariably contradict the predictions of the sequence (28), as in electrophilic aromatic substitution. It is, of course, recognized that a simple polar approach to problems of reaction rates only represents an approximation and will therefore not necessarily allow the quantitative prediction of results for similar substituents. Other examples of behaviour not expected from polar considerations are frequently found in electrophilic addition. Thus, the introduction of COPh retards addition less than the introduction of CO_2H (though the converse would have been predicted from the electron-releasing properties of Ph and OH) and other electron-attracting groups often have a smaller effect on the rate than would have been suggested by a comparison of their polar properties (or the substituent constants σ^+) with those of CO_2H.

The kinetic effect of electron-withdrawing substituents is illustrated below for the molecular chlorination of PhCH$=$CHX in acetic acid[171]; k_2 in l mole^{-1} min^{-1}:

X	H	COPh	CO_2Et	CO_2H	CHO	NO_2
k_2	$\approx 10^7$	61	10	4·9	1·8	0·02

4. Products

The subsequent fate of the carbonium ion formed in the rate-determining step (schemes 25 and 26) depends on its structure and environment. In the present discussion, the principal relevant processes are the direct reactions of the carbonium ion with the nucleophiles present and cyclization leading to lactone formation, though there is evidence that the formation of lactones does not proceed via this intermediate but involves anchimeric assistance by CO_2^- (or CO_2R) to the rate-determining addition.

Details of these processes and of the orientation of the products are more conveniently considered in connection with specific examples in later subsections.

The stereochemical configuration of the products normally corresponds to *trans* addition. This could arise from steric hindrance to *cis* attack by the nucleophile on carbonium ions with structures like **158**, while a significant contribution from structure **160** would completely exclude this type of addition. It has also been considered possible that bonding with the nucleophile follows the attainment of the transition state for addition of the electrophile so rapidly that the site for *cis* attack by the nucleophile is sterically shielded to a greater extent than in the carbonium ion **158**[172].

On the other hand, *cis* addition is certainly not unknown. Rotation in the carbonium ion **158** about the axis of the former double bond, followed by *trans* addition, would lead to a *cis* product and has been proposed for special cases. Cyclic and even four-centre transition states like **161** have been considered[172] but would appear to be rare. Several sets of wor-

$$
\begin{array}{c}
\text{E}\cdots\text{Y} \\
\vdots \quad \vdots \\
-\text{C}\cdots\text{C}- \\
| \quad | \\
\textbf{(161)}
\end{array}
$$

kers have suggested that the rate-determining step can give a classical carbonium ion in the form of an ion pair **(162)** where the nucleophile (Y^-) is held on the same side of the original double bond as the entering electrophile[172, 173]. This ion pair can collapse to give the *cis* adduct or rearrange to the isomeric ion pair **(163)** which gives the *trans* adduct. On

$$
\begin{array}{ccc}
\text{E} \quad \text{Y}^- & & \text{E} \\
| \quad | & & | \\
-\text{C}-\text{C}^+ & \qquad & -\text{C}-\text{C}^+ \\
| \quad | & & | \\
& & \text{Y}^- \\
\textbf{(162)} & & \textbf{(163)}
\end{array}
$$

this view, *cis* addition should normally occur in addition through a classical carbonium ion, and the more usual predominance of *trans* addition can be ascribed to steric hindrance to the collapse of the original ion pair **(162)**[173a].

5. Nucleophilic addition

Experimental evidence is now available in support of the prediction[174] that even electrophilic reagents like molecular halogen can allow the realization of nucleophilic addition if the olefinic bond is sufficiently polar-

TABLE 29. Second-order rate coefficients for the reaction of *trans* RCH=CHX with bromine in acetic acid at 24°[175]

R	X	Catalyst		
		None[a]	0·0125 M H_2SO_4[b]	0·1 M HBr
Ph	COPh	0·33	32	$>10^3$
	CO₂Et	0·033	$<0·15$	—
	CHO	0·006	27	$>10^3$
	NO₂	7×10^{-5}	0·005	1·0
Me	CO₂H	0·0021	0·009	0·051
	CHO	0·0014	$>10^3$	$>10^4$

[a] For several compounds, figures were obtained from the rates of chlorination and the approximately 300-fold difference between the rates of chlorination and bromination[175].
[b] Corrected for the uncatalysed reaction.

ized by an electron-demanding group. Acid-catalysed addition represents the most relevant example in the present discussion, and is illustrated in

$$RCH=CHX$$
$$(164)$$

Table 29 for the 'molecular' bromination of substituted ethylenes (164). It can be seen that the catalysed reaction occurs very much more rapidly for 164 (X = CHO) than for 164 (X = CO₂Et, CO₂H), in complete contrast to the behaviour when no strong acid is present. The polar properties of the substituent X are, however, not the only factor controlling the rate of the catalysed process (the rate is relatively small when X = NO₂) and it is usually believed that the slow step involves nucleophilic addition to one of the mesomeric forms (165) of the protonated substrate (166) which

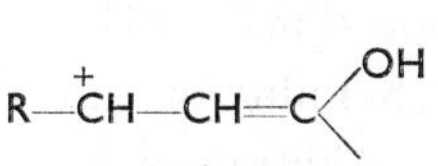

(165) 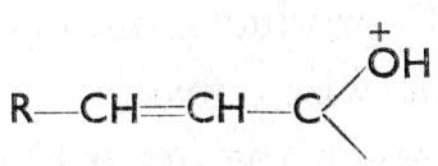   (166)

is in rapid equilibrium with its conjugate base $\left(164, X=C\begin{array}{c}O\\\\\end{array}\right)$. This view is supported by the fact that HBr and HCl are more efficient catalysts than stronger acids like H_2SO_4 (see Table 29), consistent with the formation of Br_3^- or Br_2Cl^- in the presence of halogen acids; these species can be expected to be stronger nucleophiles than Br_2.

This difference between the catalytic efficiency of HBr and H_2SO_4 forms the principal basis for believing that the slow acid-catalysed reactions of the carboxy and carboethoxy compounds also involve nucleophilic addition. Comparisons between the reactivities of **164** (X = H) and **164** (X = CO$_2$H, CO$_2$Et) by this mechanism are obviously not possible.

B. Reaction Paths and Products

1. 1,2-Addition

It has already been pointed out (section VI.A.1) that electrophilic addition to a substituted olefin (**167**) gives an adduct which can be regarded as the hybrid of three classical structures (**168–170**). Electron-attracting

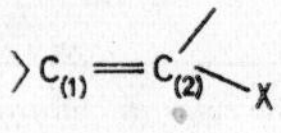

(167) 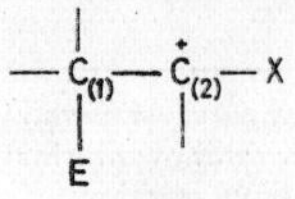   (168)

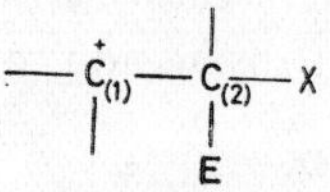

(169) 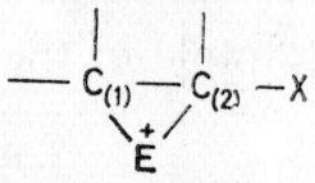   (170)

substituents (X) can be expected to destabilize **168** more than **169** while electron donation by X will stabilize both structures, the effect being greater in **168**. If **170** makes a negligibly small contribution to the structure of the intermediate carbonium ion, the initial adduct should therefore correspond to electrophilic attack at C$_{(2)}$ if X is a carboxyl derivative and at C$_{(1)}$ if an electron-repelling substituent like Me is employed. An overwhelming contribution by **170** will give the products expected from equal rates of electrophilic attack at C$_{(1)}$ and C$_{(2)}$, and intermediate cases will also give both products, but in different proportions.

Under otherwise constant conditions, the stability of the cyclic structure (**170**) increases in the order (29) so that the addition of hydrogen halides

$$E = H \ll Cl < Br < I \tag{29}$$

(HY) should give a single product, determined by the polarization of the olefin. This prediction is usually obeyed in practice. Thus, propenoic acid (**171**) gives the 3-halogenoaliphatic acid (**172**) while isopropyl halides are

$$CH_2{=}CHCO_2H$$
(171)

$$YCH_2CH_2CO_2H$$
(172)

formed in the addition to propene[176]. Similarly, the reaction of **171** with hypochlorous acid (where the electrophile can be regarded as Cl$^+$) gives a mixture of the two possible products[177], consistent with structure **170** as an important component of the initial adduct. On the other hand, the iodo-chlorination of 2-butenoic acid (**173**) gives only the 2-iodo compound (**174**)[178], corresponding to purely polar control of the structure of the car-

$$CH_3CH{=}CHCO_2H \qquad CH_3CHClCHICO_2H$$
$$(173) \qquad\qquad (174)$$

bonium ion. Presumably the extra stabilization of **169** by the methyl group is sufficient to ensure that the adduct is mainly in this form, irrespective of any stabilization of **170** when E = I.

A carboxy substituent insulated from the olefinic centre (**167**, X = CH$_2$CO$_2$H) behaves as an electron-donating group. Thus the reaction of 3-butenoic acid (**175a**) with HY gives only 3-halogenobutyric acid (**175b**) and similar results have been reported for 4-pentenoic and 5-hexenoic

$$CH_2{=}CHCH_2CO_2H \qquad CH_3CHYCH_2CO_2H$$
$$(175a) \qquad\qquad (175b)$$

acid[179]. These observations confirm the findings in electrophilic aromatic substitution (see section V.D) that the introduction of CO$_2$H or CO$_2$Et into CH$_3$ does not reverse the overall polar properties of the alkyl group. Another factor may, however, contribute in addition reactions. The two classical carbonium ions **168** and **169** will be stabilized by hyperconjugation involving the C$_{(2)}$—H and C$_{(1)}$—H bonds, respectively, and the poor relay of the polar properties of CO$_2$H through a saturated carbon chain (as in **175**) may well have a much smaller effect on the stability of **168** than the hyperconjugation interaction with four suitably placed C—H bonds so that the energy of **168** could easily be less than that of the alternative classical structure (**169**) where such interactions can only occur with one C—H linkage.

When CH$_3$ and (CH$_2$)$_n$CO$_2$H groups are located on opposite sides of an olefinic linkage, as in acids like **176a**, each of the two possible classical carbonium ions which may result from electrophilic addition can be stabilized by hyperconjugation with three C—H bonds. The greater electron repulsion by CH$_3$, relative to (CH$_2$)$_n$CO$_2$H, then controls the orientation of the initial addition which forms the ion **176b** (for examples see section VI.B.2).

$$CH_3CH{=}CH(CH_2)_nCO_2H \qquad CH_3\overset{+}{CH}{-}CHE(CH_2)_nCO_2H$$
$$(176a) \qquad\qquad (176b)$$

It has already been pointed out that the stereochemical configuration of the products usually corresponds to *trans* addition (see section VI.A.4).

In halogenation, this observation has been ascribed to a significant contribution of the cyclic halonium ion (**170**, E = Hal) to the structure of the initial adduct[180], an explanation which is still widely accepted though others have also been proposed. In any case, even *cis* halogenation is well-known (see section VI.A.4).

The reaction of tiglic acid (**177**) with HI in chloroform gives the *trans* product (**178**) in a stereochemically pure form, but *cis* and *trans* addition appear to occur at roughly the same rate when the geometrical isomer, angelic acid (**179**), is employed since the two diastereoisomers (**178** and **180**) are formed in about equal amounts[181]. No satisfactory explanation

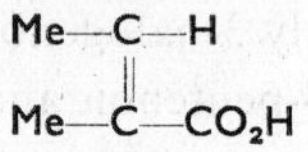

$$
\begin{array}{ccc}
\text{Me—C—H} & & \text{H—C—I} \\
\parallel & (177) & \mid \quad (178) \\
\text{Me—C—CO}_2\text{H} & & \text{Me—C—H} \\
& & \mid \\
& & \text{CO}_2\text{H}
\end{array}
$$

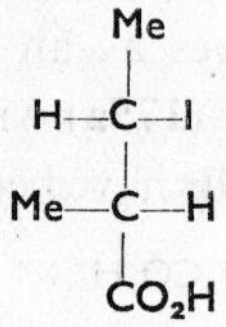

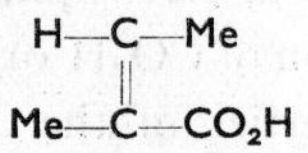

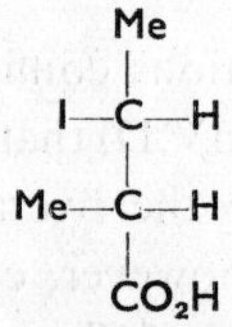

$$
\begin{array}{ccc}
\text{H—C—Me} & & \text{I—C—H} \\
\parallel & (179) & \mid \quad (180) \\
\text{Me—C—CO}_2\text{H} & & \text{Me—C—H} \\
& & \mid \\
& & \text{CO}_2\text{H}
\end{array}
$$

has yet been proposed to the exclusion of others. There is still some uncertainty about the mechanism of the reaction between unsaturated carboxylic acids and hydrogen halides (see also section VI.C.2); the rearrangement of **180** into **178** has been suggested, and the partial conversion of **179** into its more stable isomer (**177**) before addition has also been considered[182]. Similarly, the products of both *cis* and *trans* addition have recently been identified for the chlorination of methyl *trans* cinnamate, and the results have been discussed in terms of the ion pair believed to be the initial adduct[173b].

On the other hand, the addition of HCl to cyclohexene-1-carboxylic acid (**181**) has been found to give **182**, the product of 'normal' *trans* addition[188]. There must however be some doubt about the mechanism of reac-

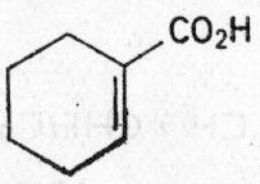

(**181**) 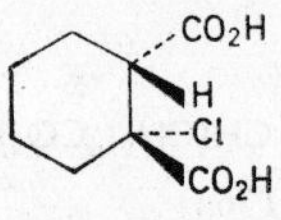   (**182**)

tions involving the addition of protons (see section VI.A.5), and the orientation of the product allows no distinction between the two possibilities; electrophilic or nucleophilic processes.

Most of the other relevant information concerning the configuration of the addition products is more conveniently discussed in the succeeding subsection.

2. Participation by neighbouring groups

It has already been pointed out during the discussion of nucleophilic substitution that a suitably situated carboxylate ion group can stabilize a carbonium ion by forming the appropriate lactone (section III.A.2). As the initial adduct of electrophilic addition can also be regarded as a carbonium ion, it is therefore not surprising to find that lactones have often been isolated from such reactions when the conditions are unfavourable to their further decomposition.

Most of the relevant observations again refer to halogenation, and several authors have considered that a carboxylate ion (**183**, say) initially forms a halonium ion which is then completely converted into the lactone (**184**) by internal nucleophilic attack of the CO$_2^-$ group (see scheme 30,

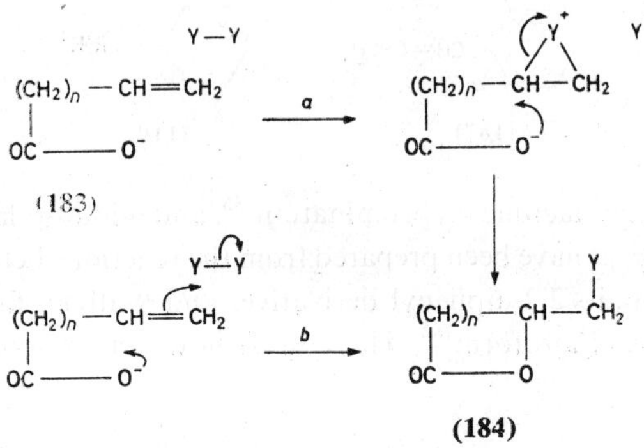

route *a*). Such an interpretation requires that the interaction of CO$_2^-$ with a carbonium ion centre is more powerful than that of a halogen, as has often been stated, but it is difficult to see why the CO$_2^-$ group should not then provide anchimeric assistance to the rate-determining addition of the electrophile at the site most favourable for this attack on polar grounds (see scheme 30, route *b*). There is much evidence in support of this view (see succeeding paragraph, and section VI.C.1) and it is therefore possible that a number of additions to unsaturated carboxylate ions may proceed via the lactone even when the intermediate has not been isolated.

For this reason systems where this possibility cannot be discounted have
been included in the present rather than the previous subsection.

The γ- and δ-lactones obtained as the products of the aqueous molecular iodination of olefinic carboxylate ions[184] show the unique orientation
which is predicted for attack of the electrophile at the site most favoured
by polar considerations. Thus, the 4-pentenoate ion (183, $n = 2$) gives
the δ-iodo-γ-lactone (184, $n = 2$), 5-hexenoate ions (183, $n = 3$) form the
ε-iodo-δ-lactone (184, $n = 3$), while the different polarization of the olefin
in the 3-pentenoate ion (176a, $n = 1$) yields the β-iodo-γ-lactone (185).
Similarly the reactions of the styrene-2-carboxylate ion (186) and its methyl derivative (187) form the phthalide (110, $R^1 = H$, $R^2 = CH_2I$ and
$CHICH_3$, respectively). Salts of cyclohexene-4-cis-1,2-dicarboxylic acid

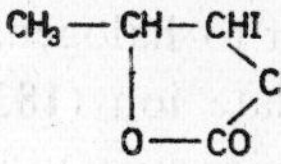

$$CH_3-CH-CHI$$
$$\underset{|}{\quad}\underset{\diagdown}{\quad}CH_2$$
$$O-CO$$

(185)

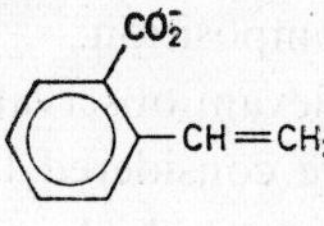

(186)

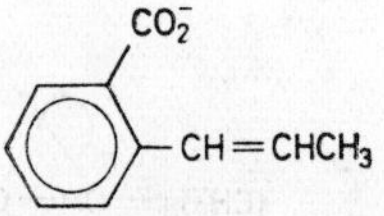

(187)

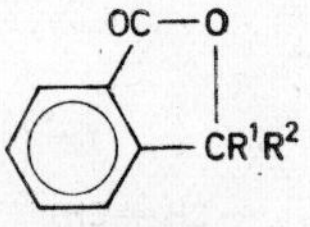

(110)

(188) also form a γ-lactone on bromination[185], and δ-iodo-γ-lactones analogous to 184 ($n = 2$) have been prepared from the reactions between ICN and
4-pentenoic acid, its 2,2-diphenyl derivative, and 9-allyl-9-fluorenecarbox-
ylic acid (189) in chloroform[186]. There must however be some uncertainty

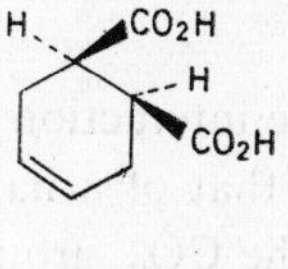

(188)

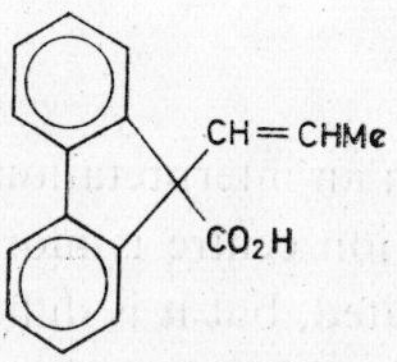

(189)

whether these reactions involve the participation by CO_2H, which has been
observed in other systems (see section III.A.3), or whether the anion is
responsible for lactone formation; any chloride ions present in the reaction
mixture could presumably act as proton acceptors.

The formation of the β-lactones (190) appears to be more difficult than

$$
\begin{array}{c}
| \quad | \\
-C-C-Hal \\
| \quad | \\
O-CO
\end{array}
$$

(190)

the production of larger ring systems in halogenation, as in other reactions. Thus, the derivatives of 190 which might have been expected from the iodination of 3- and 2-butenoate ions (see structures 175 and 173) were not in fact obtained[184], though their existence as unstable intermediates in the reaction cannot be excluded. However, the introduction of bulky substituents in acrylic acid appears to stabilize 190 as the initial adduct of electrophilic addition to the anion, and such intermediates have been isolated in the molecular halogenation of stilbene-α-carboxylate ions (191)[187], dimethylfumarate (192), and dimethylmaleate ions (193)[188].

While 192 and 193 give two different lactones, the subsequent hydro-

$$
\begin{array}{ccc}
Ph-C-CO_2^- & Me-C-CO_2^- & Me-C-CO_2^- \\
\| & \| & \| \\
H-C-Ph & ^-O_2C-C-Me & Me-C-CO_2^- \\
(191) & (192) & (193)
\end{array}
$$

lysis in neutral or slightly acid media gives the same final product, corresponding to cis addition of ClOH or BrOH to one isomer and trans addition to the other. The reactions of fumaric and maleic acids with chlorine water give mainly, but not exclusively, the products of trans addition[189], and conditions for addition to the anions can be so arranged that both isomers show stereospecific cis addition[190]. Other studies of the chlorination and bromination of fumarate and maleate ions have, however, given the same product for both isomers, trans addition to the fumarate and cis addition to the maleate[189, 191], and similar observations have recently been reported for the reactions of the diethyl esters with bromine in 1M aqueous sodium bromide[192].

No valid explanation of these observations has yet been reported (see however reference 173b). Some of the reactions of the unsubstituted dicarboxylate anions probably occur via β-lactone intermediates which may undergo ring opening by different paths under different conditions but further work is needed before the problem can be solved.

A different kind of neighbouring group participation by CO_2^- has been postulated for the hydration of fumaric acid[193]. The pH-rate profile showed that the monoanion was the most reactive species and that the transition state did not require the incorporation of an external proton. A reactive path involving the ring-chain tautomeric shift shown in scheme (31)

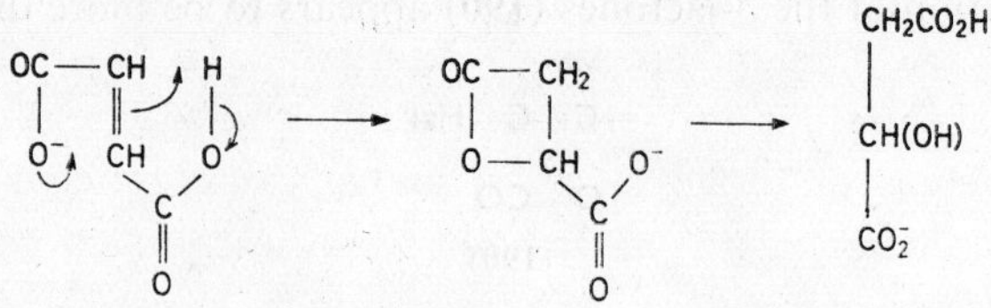

$$(31)$$

is consistent with these observations. On the other hand, maleic acid and crotonic acid undergo the more normal acid-catalysed hydration, though it is not entirely certain whether electrophilic addition is involved (see section VI.A.5).

Neighbouring group participation by CO_2R has been convincingly demonstrated for the bromination of esters of 2,2-diphenyl-4-pentenoic acid which yield the δ-bromo-γ-pentanolactone (194)[186, 194]. The product of the initial electrophilic attack probably undergoes the electron displacements shown in 195a to give an intermediate (195b) which reacts with a bromide ion to form 194 and the alkyl bromide. On the other hand, it is

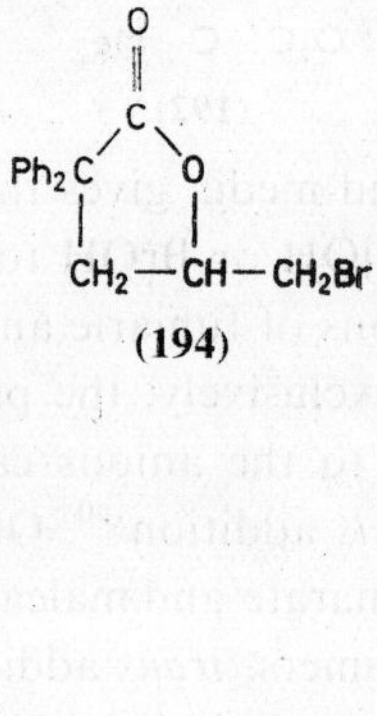

(194)

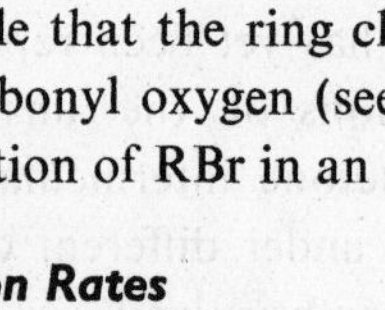

(195a)

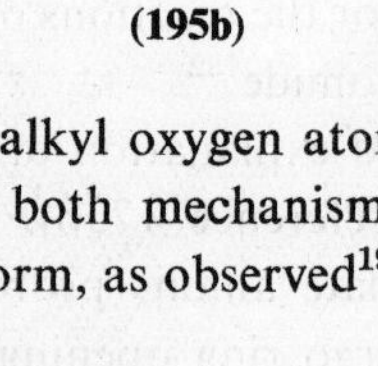

(195b)

conceivable that the ring closure involves the alkyl oxygen atom instead of the carbonyl oxygen (see section III.A.3); both mechanisms require the formation of RBr in an optically inverted form, as observed[194].

C. Reaction Rates

1. Electrophilic addition

The effect of substituents on the rate of molecular halogenation of substituted ethylenes (196) is shown in Table 30. Compounds have been

$$R^1CH{=}CR^2X$$

(196)

arranged in the order of decreasing reactivity by the parent (196, X = H) which is also the order of decreasing retardation by CO$_2$H and CO$_2$Et. This behaviour is the converse of that found in other reactions retarded by electron-attracting substituents (e.g. S_N1 solvolysis, cf. reference 112) and expected from simple polar considerations[3]. The present systems, however,

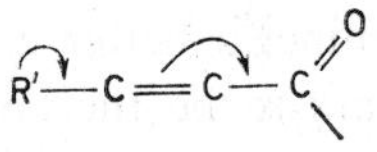

(197)

possess the special feature that the direct conjugation illustrated in 197 will progressively stabilize the substituted derivatives (196, X = CO$_2$H, CO$_2$Et) as the facility of the group R^1 for electron release is increased, resulting in a tendency to enhance the retardation of addition on the introduction of CO$_2$H or CO$_2$Et for changes like R^1 = CO$_2$H to R^1 = Me.

TABLE 30. The kinetic effect of substituents (X) on the molecular halogenation of trans R^{1}CH=CR^{2}X

R^1	R^2	$10^4 k_{CO_2H}/k_H{}^a$	$10^4 k_{CO_2Et}/k_H{}^b$
Ph	H	0·005	—
Me	H	0·010[c, d]	0·0061
H	H	0·007[c]	0·0027
Ph	Ph	0·70[c]	—
Ph	Br	2·6	—
Ph	CO$_2$H	100[c]	—
CO$_2$H	H	22	3·2[f]

[a] For chlorination or bromination in acetic acid[175].
[b] For bromination in water, k for reaction with Br$_2$[192].
[c] From bromination for X = H, chlorination for X = CO$_2$H, assuming $k(Cl_2)/k(Br_2)$ = 300.
[d] Parent compound: R^1 = n-Bu.
[e] In aqueous acetic acid[195].
[f] R^1 = CO$_2$Et.

The figures in Table 30 refer to the trans compounds when geometrica isomerism is possible. The cis carboxylic acids (196: R^1 = Ph, R^2 = H; R^1 = CO$_2$H, R^2 = H; R^1 = CO$_2$H, R^2 = Me) undergo halogenation in acetic acid three to four times as quickly as their trans isomers[195]. Steric factors may be responsible but such a conclusion is weakened by the fact that the cis and trans forms of 196 (X = R^1 = CO$_2$Et, R^2 = H) react at almost the same rate with bromine in water[192].

54*

A change from $X = CO_2H$ to $X = CO_2Et$ generally accelerates halogenation in acetic acid by 50–100%[175, 195]. Comparison of the results in Table 30 therefore shows that the rate of reaction in water is the more sensitive to the introduction of the present substituents.

If bromide ions are present, a second process which is believed to involve the initial electrophilic addition of Br_3^- can accompany the more rapid reaction with molecular bromine, a better electrophile. However, attack by the anionic reagent appears to be the energetically more favourable process in the bromination of diethyl fumarate in water, and has been interpreted[192] in terms of simultaneous electrophilic and nucleophilic attack by Br_2 and Br^-, respectively, a process which would be kinetically indistinguishable from bimolecular addition of Br_3^- to the substrate. Other results for this reaction also support the termolecular mechanism which may well be found to have a wider applicability. Whatever the nature of this additional process, the effect of the introduction of CO_2H or CO_2Et on its rate[192] is similar to that in molecular bromination, though a little smaller.

Rate coefficients for the iodination of olefinic carboxylate ions (198) in water[184b] are shown below. Unfortunately no results are available for the

$$CH_2{=}CH(CH_2)_nCO_2^-$$
(198)

n	0	1	2	3	4
k_2 (1 mole^{-1} sec^{-1})	0	0·012	78·6	10·9	0·17

reactions of the corresponding hydrocarbons under the same conditions but the variation of k_2 with chain length (n) and, particularly, the large rates found when $n = 2$, 3 are strongly reminiscent of the results observed in the hydrolysis of aliphatic ω-halogenocarboxylate ions (see section III.A.2). The present findings therefore also demand anchimeric assistance by CO_2^- in a reaction which would otherwise form a carbonium ion, when the participation by the neighbouring CO_2^- group results in a five- or six-membered cyclic structure; these intermediates are, in fact, the products of the iodination of 198, $n = 2$, 3 (see section VI.B.2). The rates for addition suggest that the formation of the γ-lactone (184, $n = 2$) is the energetically most favourable process, followed by the δ-lactone (184, $n = 3$). while assistance to the formation of other lactones is either much less efficient or does not occur at all.

Similarly, the products (see section VI.B.2) and rates of iodination of the o-alkenyl benzoate ions, 186 and 187[184b], demand neighbouring group participation by CO_2^- in the rate-determining electrophilic addition to give the phthalide (110), as in the hydrolysis of benzyl halides containing

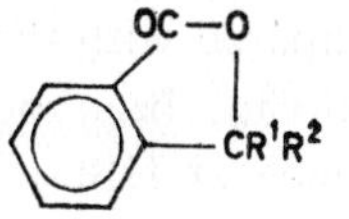

(110)

an $o\text{-}CO_2^-$ group (see section III.A.3). The rate coefficients (see below) are much larger than would have been expected without anchimeric assistance, but comparison with the formation of the analogous γ-lactone in

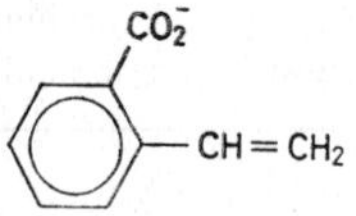

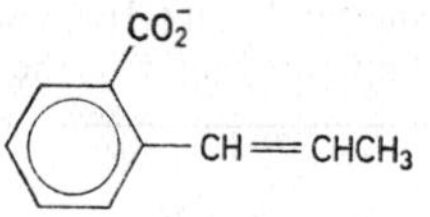

	(186)	(187)	(198, $n = 2$)
k_2:	3·12	5·3	78·6

the iodination of the 4-pentenoate ion (198, $n = 2$) shows that the production of the bicyclic compound (110) is the less favourable process. Differences between the energies of activation are responsible[184b], and probably arise from the different geometry of the two ring systems.

The introduction of CO_2H or CO_2Et into styrene (199, $X = H$) retards the formation of the epoxide (200) from reaction with perbenzoic acid

(199) 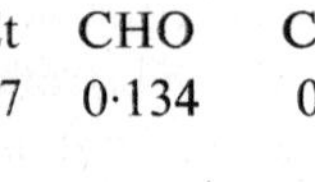   (200)

about 300-fold (see below)[196]. These reactions are generally regarded as

X	CO_2H	CO_2Et	CHO	COMe
k_X/k_H	0·0037	0·0037	0·134	0·071

electrophilic processes and the results therefore suggest that the peracid is a less efficient electrophile than bromine, since k_{CO_2H}/k_H has a much smaller value in halogenation. However, the CHO and COMe derivatives react more rapidly than the CO_2H compound, behaviour which usually indicates nucleophilic addition. It is, of course, quite possible that the CHO and CO_2H compounds form the product by different mechanisms but no conclusions can be reached on the present information.

2. Nucleophilic addition

The rates of electrophilic molecular bromination of substituted acrylic acids (**196**, $X = CO_2H$) are compared with the rates of the acid-catalysed reaction in Table 31. It has already been pointed out (section VI.A.5) that the reaction in the presence of HBr involves nucleophilic addition to the conjugate acid of the substrate in which the proton is attached to the carbonyl oxygen atom. The concentration of this species, and hence the basicity of the substrate, is clearly an imporant factor in determining the rate of the catalysed process.

TABLE 31. Comparison of the catalysed (HBr) and uncatalysed bromination of $R^1CH{=}CR^2CO_2H$[175, 195]. (The figures refer to the reaction in acetic acid at 24°).

R^1	R^2	$10^3k_{cat}{}^a$	$10^3k_{uncat}{}^a$	k_{cat}/k_{uncat}
Me_2	H^b	180	150	1·2
cis-Ph	H	70	63	1·1
trans-Me	H	51^c	2·1	24·2
Ph	CO_2H	17,500	0·63	3×10^4
H	H	440^c	0·060	7×10^3
trans-3-$NO_2C_6H_4$	H	6·0$^{c,\,d}$	0·037	160^d
trans-4-$NO_2C_6H_4$	H	3·0$^{c,\,d}$	0·016	190^d
cis-CO_2H	H	0·56^c	$3·7\times10^{-7}$	$1·5\times10^6$

^{a} Rate coefficients for catalysed (k_{cat}) and uncatalysed reaction (k_{uncat}) on the relative scale: $k_{uncat} = 1$ for allyl bromide.

^{b} $CMe_2{=}CHCO_2H$ as substrate.

^{c} From rates of chlorination; assumed 300 times greater than rates of bromination.

^{d} For the methyl ester.

The results in Table 31 have been arranged in the order of decreasing reactivity for electrophilic addition. On the whole, the same order is followed for the catalysed reaction though the rates are now less sensitive to structural changes in the substrate. Acrylic acid and its 1-carboxy-2-phenyl derivative, however, react abnormally rapidly in the presence of HBr. Electron release by the group R^1 aids electrophilic attack and also increases the basicity of the present substrates by a stabilizing contribution from structure **201** to the conjugate acid. The resulting increase in the

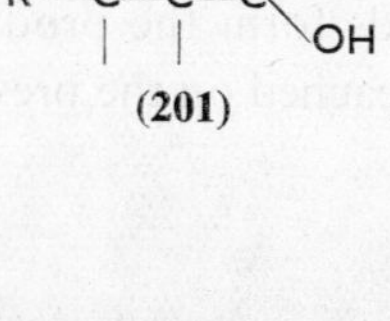

$$\overset{+}{R^1}{=}C{-}C{=}C\overset{\displaystyle OH}{\underset{\displaystyle OH}{\diagup}}$$

(201)

equilibrium concentration of the conjugate acid will tend to enhance the rate of acid-catalysed halogenation, but the increased stability of the initial state for nucleophilic attack (the conjugate acid) will have an opposing effect. In the majority of the compounds now considered, the basicity of the substrate appears to be the more important factor controlling the rate, but it must be stressed that this conclusion need not be universally applicable and may only be valid for the present, weakly basic carboxylic acids.

The abnormally large value of k_{cat} for the dicarboxylic acid ($R^1 = Ph$, $R^2 = CO_2H$) could well arise from the presence of an electron-attracting group at R^2. This would hinder electrophilic attack but would have a much smaller effect on the basicity than the same group at R^1 while, at the same time, facilitating nucleophilic addition to the conjugate acid. No similar explanation can account for the relatively rapid acid-catalysed halogenation of acrylic acid.

Considerable uncertainty exists at present about the reaction path in other acid-catalysed additions to olefinic carboxylic acids, though it seems very likely that nucleophilic addition occurs with compounds derived from acetylene (see section VI.D). The relative rates[197] for the acid-

$$CH_3CH=CHX$$
(202)

catalysed hydration of propene derivatives (**202**) at 100° (see below[*]) show a smaller retardation on the introduction of CO$_2$H than might have

X	CO$_2$H	CHO
k_X/k_H	0·08	9·1

been expected for electrophilic addition, while the acceleration caused by CHO argues strongly against a rate-determining attack by a proton on an unsaturated carbon atom. It must however be stressed that the reactions of **202** (X = CHO) and **202** (X = CO$_2$H) may not involve the same mechanism as the two processes show quite different entropies of activation.

More recent work[199] suggests that the hydration of unsaturated aldehydes and ketones involves neither electrophilic nor nucleophilic addition to the double bond (in the generally accepted sense of these terms) and that reaction occurs by a 1,4 addition process involving the carbonyl group. Highly convincing evidence in favour of electrophilic attack by a proton on olefinic carbon atoms in a carboxylic acid has been obtained[200] for the rate-determining hydration of *cis* cinnamic acid in its rearrange-

---

[*] Rates for propene (**202**, X = H) from reference 198.

ment to the *trans* isomer in sulphuric acid, but it is not yet known whether this example represents the general situation.

The lyate ion catalysed addition of hydroxylic solvents (ROH) to olefins also belongs to the class of nucleophilic additions as it seems very likely that the reactions proceed via the rate-determining addition of the base (RO⁻), followed by rapid proton transfer from the solvent to the carbanion (see scheme 32). Electron-attracting substituents (X) greatly accellerate the reaction[201a], as expected from scheme (32), but little relevant kinet-

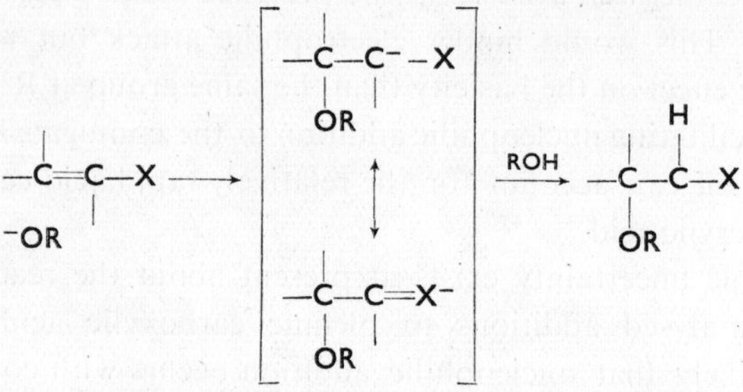

$$
\begin{bmatrix}
\overset{|}{-}C-\overset{|}{C}{}^--X \\
\quad OR \\
\\
\overset{|}{-}C-\overset{|}{C}=X^- \\
\quad OR
\end{bmatrix}
$$

(32)

ic evidence is available. The addition of methanol occurs about 15 times more rapidly when X = COMe than for X = CO₂Me[201a], and it is noteworthy that this factor is larger than that for the retardation of addition to the CHO and CO₂Et compounds (see Table 29). The base-catalysed hydration of the fumarate ion only occurs at a reasonable rate at relatively high temperatures and probably involves a similar mechanism[201b].

D. Addition to Acetylenic Acids

1. 1,2-Additions

Relative to ethylene, the π electrons in acetylene are bound more tightly to the carbon nuclei and are therefore less easily polarized, and the C—H bond is more acidic. Electrophilic addition should therefore occur more slowly with acetylenic than with olefinic compounds, while the converse can be expected to apply when addition is initiated by nucleophilic attack (*cf.* reference 202). Nevertheless, the halogenation of many acetylenic acids appears to involve the 'normal' mechanism. Thus, as expected, the molecular bromination[203] of undecynoic acid (203) occurs more rapidly than that of methyl ethylpropiolate (204) where an electron-attracting

$$HC\equiv C(CH_2)_8CO_2H \qquad EtC\equiv CCO_2Me$$

(203)   (204)

goup (CO₂Me) is directly attached to the unsaturated system.

TABLE 32. Rates of halogenation of acetylenic and olefinic acids. [The rate coefficient k_{AC} refers to the acetylenic acid (205), and k_{OC} to its olefinic analogue (206)]

R	n	k_{AC}/k_{OC}	Reference
$CH_3(CH_2)_7$	7	2×10^{-5}	203
H	8	$1 \cdot 1 \times 10^{-4}$	204
Ph	0	$0 \cdot 1$	175
CO_2H	0	41	204

The rates of halogenation of some acetylenic acids (205) are compared in Table 32 with those of the corresponding ethylene derivatives (206). Those olefins (206, $R = CH_3(CH_2)_7$, $n = 7$; $R = H$, $n = 8$) which undergo

$$RC\equiv C(CH_2)_nCO_2H \qquad RCH\equiv CH(CH_2)_nCO_2H$$
$$\text{(205)} \qquad\qquad\qquad \text{(206)}$$

rapid addition also react much more quickly than their acetylenic analogues, but the complete reversal of this behaviour in the dicarboxylic acids (205 and 206, $R = CO_2H$, $n = 0$) suggests that the two compounds no longer follow similar reaction paths. Nucleophilic addition may well represent the energetically more favourable process for the halogenation of acetylene dicarboxylic acid.

Acid-catalysed bromination has also been observed for phenyl propiolic acid (205; $R = Ph$, $n = 0$). The difference between the rate[175] and that of the ethylenic analogue (206; $R = Ph$, $n = 0$) is less than for bromination in the absence of HBr but the acetylenic derivative continues to be the less reactive compound. The converse would have been expected for direct nucleophilic addition to the substrate, but these reactions probably proceed via the conjugate acids (see section VI.C.2) and it is very likely that the difference between the electron-attracting properties of $C\equiv C$ and $C\equiv C$ linkages makes propiolic acid a weaker base than cinnamic acid. It is therefore possible that the reduced concentration of the conjugate acid of the acetylenic compound (relative to the olefin) is more than sufficient to counterbalance its increased reactivity with nucleophiles.

The hydrochlorination of α,β-acetylenic acids and esters obviously proceeds by nucleophilic attack on an unsaturated carbon atom. This is most readily demonstrated by the reactivity sequence $(33)^{205}$ but it is not yet

$$PhC\equiv CH < PhC\equiv CCO_2Me < MeO_2CC\equiv CCO_2Me \qquad (33)$$

clear whether the chloride ion reacts with the substrate or with its conjugate acid. It is also noteworthy that the chloro-olefins produced in the reactions between HCl and tetrolic (207) or acetylene dicarboxylic acids indicate entirely *trans* addition[206], while treatment of the latter compound

$$MeC\equiv CCO_2H$$

(207)

with bromine water gave 70% dibromofumaric acid (*trans* addition) and 30% of its geometrical isomer (*cis* addition)[207]. Unfortunately it is not possible to decide whether these observations result from different mechanisms for the two reactions.

2. Participation by neighbouring groups

Acetylenic compounds can also undergo reactions which involve the internal attack of CO_2H or CO_2^- on one of the unsaturated carbon atoms. Thus, the iodination of the 4-pentynoate ion (208) gives the γ-lactone (209), but the reaction occurs about 40 times more slowly than the iodolactonization of the corresponding olefin, the 4-pentenoate ion[184b]. This difference

$$CH\equiv C(CH_2)_2CO_2^- \qquad CHI=C(CH_2)_2CO$$
$$\underset{O}{\underline{\qquad\qquad}}$$

(208)   (209)

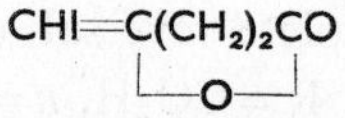

may well result from the smaller rate of electrophilic addition to acetylenes than to olefins since facility for electrophilic attack on the substrate remains an important factor, controlling the rate even when the reaction receives anchimeric assistance from neighbouring groups.

Similarly, 5-phenyl-2-penten-4-ynoic acid (210) is converted into the unsaturated δ-lactone (211) in excellent yield in strong solutions of sulphuric acid in acetic acid[208]. The reaction could involve anchimeric assist-

$$PhC\equiv CCH=CHCO_2H$$

(210)   (211)

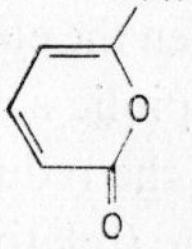

ance by CO_2H to the electrophilic attack of a proton on an acetylenic carbon atom, as indicated in 212, a mechanism which is similar to that presumably responsible for the conversion[209] of 4-pentenol (213) into 2-methyl tetrahydrofuran (214) in sulphuric acid. However, the electron

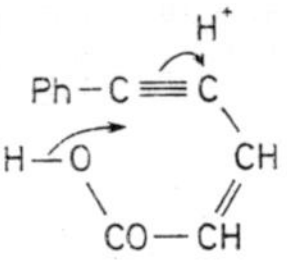

Ph—C≡C   CH$_2$=CH(CH$_2$)$_3$OH   Me

(212)   (213)   (214)

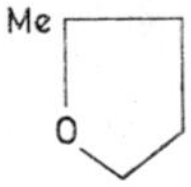

displacements shown in **212** assume preferential electrophilic addition to the acetylenic linkage in spite of the presence of C=C which usually reacts more readily with electrophiles, but anchimeric assistance by CO$_2$H (as in **212**) may well reverse the normal order of reactivity. The most obvious alternative mechanism—protonation of the carbonyl oxygen atom as the first step—also suffers from objections.

Internal nucleophilic attack by CO$_2$H on an electron-deficient centre in the same molecule has been represented as involving the 'alcoholic' oxygen atom in **212**, but some authors have assumed that the carbonyl oxygen atom is incorporated in the lactone. No distinction between the two possibilities can be made at present, as in participation by CO$_2$R (see section VI.B.2), and it is also impossible to decide whether proton loss by CO$_2$H is synchronous with cyclization (as in **212**) or whether it occurs at a later stage.

An interesting example of internal nucleophilic attack by CO$_2^-$ and CO$_2$H on the acetylenic linkage arises in the rearrangement of substituted o-carboxy-diphenylacetylenes (**215**) which form the γ-lactone (**216**) in

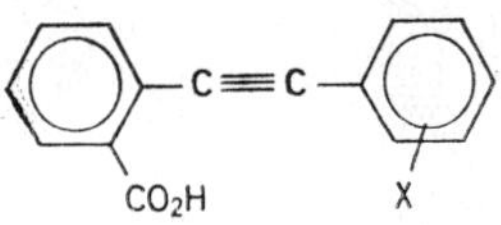

(215) 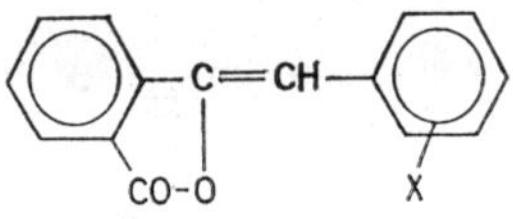   (216)

water and ethanol[210]; the kinetic results are summarized in Table 33.

For the rearrangement in water, the pH-rate profile of the parent compound (**215**, X=H) shows that the carboxylate anion is the principal reacting species at pH ⩾ 4*, and the data are fully consistent with a rate-determining cyclization involving the electron displacements shown in **217**, followed by rapid proton transfer to the resulting carbanion. The entirely different relation between the experimental rate coefficients and

---

* At higher acidities, a significant contribution to the rate arises from either the rearrangement of the carboxylic acid or an acid-catalysed process involving the carboxylate ion. The kinetics do not allow a distinction between these possibilities.

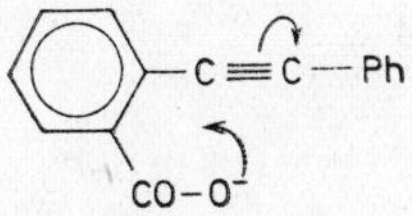

(217)

pH in lactonization of the substituted derivative (215, X = o'-CO$_2$H) demands reaction only via the monocarboxylate ion, and the approxi-

TABLE 33. Rate coefficients (10^4k in min^{-1}) for the rearrangement of o-carboxydiphenylacetylenes[210].

Solvent	Substituents		
	H	o'-CO$_2$H	p'-CO$_2$H
H$_2$O[a]	0·031	220	—
EtOH	0·38	8440	0·98
MeOH	—	600	—

[a] Rate coefficients for this solvent refer to the mono-anion.

mately 7000-fold acceleration on the introduction of the substituent is much too large to arise solely from its polar properties. The abnormally large reactivity of the o'-CO$_2$H derivative was attributed to assistance to cyclization by internal proton transfer (see 218), which is similar to the ring chain tautomerism already postulated for hydration of the hydrogen fumarate ion (see section VI.B.2).

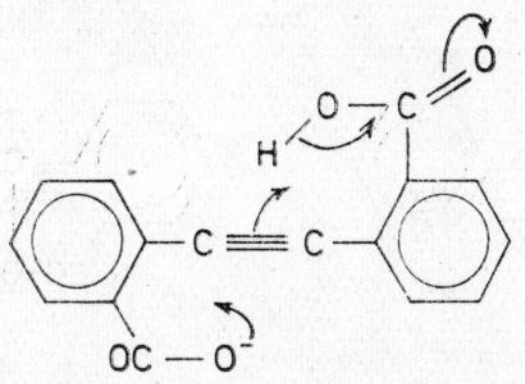

(218)

This interpretation is strongly supported by the rates of the reaction in ethanol (see Table 33) where the cyclization occurs by internal nucleophilic attack of the CO$_2$H group. Here the introduction of o'-CO$_2$H increases the rate by a factor of about 25,000 while the same substituent in the p' position only gives a three-fold acceleration, a reasonable figure for an electron-attracting centre conjugated to the site of reaction in this system. The exceptional reactivity of the o'-CO$_2$H compound is therefore clearly associated with the position of the substituent relative to the acetylenic linkage, and not with its polar properties.

Another interesting feature emerges from the results in Table 33. Several studies in aqueous solvents have shown that CO_2^- is a more efficient internal nucleophile than CO_2H (see section III.A.3), as might perhaps have been expected, but the present carboxylic acids (215) undergo more rapid reaction in ethanol than their monoanions in water. There is little reason to believe that a change from water to ethanol will drastically increase the reactivity of the monocarboxylate ions and it therefore seems that CO_2H is a better neighbouring group (relative to CO_2^-) in ethanol than in water. More extensive hydrogen bonding of CO_2H with the more polar of the two solvents now considered may be responsible for decreasing its efficiency as an internal nucleophile. The decrease in the rate of cyclization of 215 ($X = o'$-CO_2H) when the solvent is changed from ethanol to the more strongly hydrogen-bonding methanol is consistent with this view.

VII. OLEFIN ELIMINATION

Very little systematic work has been carried out on the effect of the present substituents in reactions which have not already been discussed. However a recent study[211] of the ethoxide ion catalysed transetherification of 1-substituted 2-phenoxyethanes (219) adds to the examples of the substantial change in rate which can be caused by the introduction of CO_2Et. The reaction is believed to involve the rate-determining formation of the olefin (220) which undergoes rapid nucleophilic reaction with the solvent to

$$XCH_2CH_2OPh \qquad XCH{=\!=}CH_2$$
$$\text{(219)} \qquad\qquad \text{(220)}$$

form the substituted diethyl ether.

The observation of a second-order rate law for the appearance of PhO^- is consistent with either the synchronous $E2$ elimination or with a rate-determining proton abstraction (mechanism $E1cB$)*, as illustrated in scheme (34). Electron-attracting substituents (X) can be expected to facil-

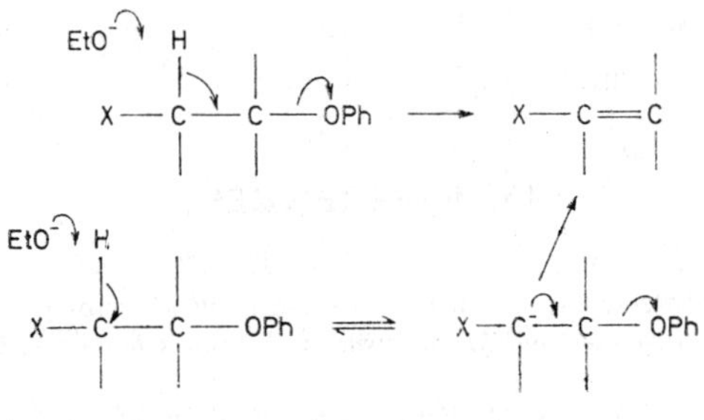

$$\tag{34}$$

* For details of olefin elimination, see reference 212.

itate proton removal from the C—H bond but will, at the same time, hinder the fission of the C—O linkage so that the effect of X on the rate is likely to be small, as in other reactions with ambiguous polar requirements (see section III.A.1). Such substituents will however greatly stabilize the carbanion (and therefore the transition state in its formation) by favouring the electron displacements shown in **221**.

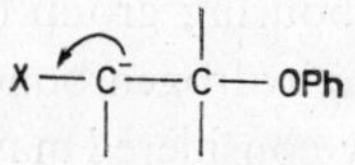

$$X \overset{\frown}{} \overset{|}{\underset{|}{C^-}} \overset{|}{\underset{|}{C}} - OPh$$

(221)

The rates found[211] for the transetherification of **219** (see below) therefore strongly suggest that the slow step occurs either by mechanism $E1cB$ or by a synchronous process in which bond fission of C—H is an over-

X	CHO	PhSO$_2$	CO$_2$Et	CO$_2^-$
k_2 (rel.)	75	1	0·16	$< 10^{-6}$

whelmingly more important factor in the transition state than the disruption of the C—O linkage, *i.e.* 'almost' $E1cB$ elimination. The approximately 500-fold acceleration caused by changing the substituent from CO$_2$Et to CHO is much larger than the retardation of addition to olefins resulting from the same alteration (see section VI.A.3), possibly because the substituent is directly attached to the reaction centre in the elimination but not in the addition reaction. On the other hand, differences between the capacities of the two groups to respond to different polar situations may be responsible.

VIII. ACKNOWLEDGEMENT

The authors' debt to Dr. C. J. M. Stirling for permission to quote unpublished results is gratefully acknowledged.

IX. REFERENCES

1. G. Kohnstam and D. L. H. Williams, in *The Chemistry of the Ether Linkage* (Ed. S. Patai), Interscience Publishers, London, 1967, Chap. 3.
2. C. K. Ingold, *Structure and Mechanism in Organic Chemistry*, G. Bell and Sons, London, 1953, p. 77.
3. G. Kohnstam and D. L. H. Williams in *The Chemistry of the Ether Linkage* (Ed. S. Patai), Interscience Publishers, and references cited therein.

4. P. W. Robertson, P. B. D. de la Mare and B. E. Swedlund, *J. Chem. Soc.*, 782 (1953); P. B. D. de la Mare, O. M. H. el Dusouqui, J. G. Tillett and M. Zeltner, *J. Chem. Soc.*, 5306 (1964).
5. C. Eaborn, *J. Chem. Soc.*, 4858 (1956).
6. C. K. Ingold, *Structure and Mechanism in Organic Chemistry*, G. Bell and Sons, London, 1953, p. 129.
7. E. J. Hartwell, R. E. Richards and H. W. Thompson, *J. Chem. Soc.*, 1436 (1948).
8. M. L. Josien, J. Lascombe and C. Vignalou, *Compt. Rend.*, **250**, 4146 (1960).
9. L. J. Bellamy, *J. Chem. Soc.*, 4221 (1955).
10. G. M. Barrow, *J. Chem. Phys.*, **21**, 2008 (1953).
11. H. W. Thompson and D. A. Jameson, *Spectrochim. Acta*, **13**, 236 (1958).
12. J. Wenograd and R. A. Spurr, *J. Am. Chem. Soc.*, **79**, 5844 (1957).
13. M. St. C. Flett, *Spectrochim. Acta*, **18**, 1537 (1962).
14. P. E. Richards and W. R. Burton, *Trans. Faraday Soc.*, **45**, 874 (1949).
15. R. N. Jones and C. Sandorfy, in *Chemical Applications of Spectroscopy* (Ed. W. West), Vol. IX of *Techniques of Organic Chemistry*, Interscience Publishers, New York, 1956, p. 470.
16. G. Berthier and J. Serre, in *The Chemistry of the Carbonyl Group* (Ed. S. Patai), Interscience Publishers, London, 1966, Chap. 1.
17. S. Forsen, *Spectrochim. Acta*, **18**, 595 (1962).
18. C. M. Moser and A. I. Kohlenberg, *J. Chem. Soc.*, 804 (1951).
19. O. Exner, M. Horák and J. Pliva, *Chem. Ind.*, 1174 (1958).
20. National Bureau of Standards, Circular 537.
21. A. L. McClellan, *Tables of Experimental Dipole Moments*, W. H. Freeman and Co., San Francisco and London, 1963.
22. L. D. Freedman and G. O. Doak, *J. Org. Chem.*, **21**, 811 (1956); H. H. Jaffé and M. Orchin, *Theory and Applications of U.V. Spectroscopy*, John Wiley and Sons, New York, 1962, p. 257.
23. W. D. Kumler and L. A. Strait, *J. Am. Chem. Soc.*, **65**, 2349 (1943).
24. W. F. Forbes, W. A. Mueller, A. S. Ralph and J. F. Templeton, *Can. J. Chem.*, **35**, 1049 (1957).
25. C. K. Ingold, *Structure and Mechanism in Organic Chemistry*, G. Bell and Sons, London, 1953, p. 128.
26. D. Peltier, A. Pichevin, P. Dizabo and M. L. Josien, *Compt. Rend.*, **248**, 1148 (1959).
27. M. St. C. Flett, *Trans. Faraday Soc.*, **44**, 767 (1948).
28. J. F. J. Dippy, *Chem. Rev.*, **25**, 151 (1939).
29. J. R. Fox and G. Kohnstam, *Proc. Chem. Soc.*, 115 (1964).
30. A. Burawoy and J. T. Chamberlain, *J. Chem. Soc.*, 2310 (1952).
31. C. K. Ingold, *Structure and Mechanism in Organic Chemistry*, G. Bell and Sons, London, 1953, p. 744.
32. G. S. Hammond, in *Steric Effects in Organic Chemistry* (Ed. M. S. Newman), John Wiley and Sons, New York, 1956, p. 425.
33. J. F. J. Dippy, S. R. C. Hughes and J. W. Laxton, *J. Chem. Soc.*, 1417 (1954).
34. E. A. Fehnel, *J. Am. Chem. Soc.*, **72**, 1404 (1950).
35. S. D. Ross, *J. Am. Chem. Soc.*, **70**, 4039 (1948).
36. See for example P. B. D. de la Mare, E. A. Johnson and J. S. Lomas, *J. Chem. Soc.*, 5317 (1964).
37. H. L. Goering, M. S. Newman and T. Rubin, *J. Am. Chem. Soc.*, **77**, 3756 (1955).
38. R. E. Rundle and M. Parasol, *J. Chem. Phys.*, **20**, 1487 (1952).
39. H. M. E. Cardwell, J. D. Dunitz and L. E. Orgel, *J. Chem. Soc.*, 3740 (1953).
40. G. E. K. Branch and D. L. Yabroff, *J. Am. Chem. Soc.*, **56**, 2568 (1934).
41. J. W. Baker, *Nature*, **137**, 236 (1936).

42. G. Kortüm, W. Vogel and K. Andrussow, *Dissociation Constants of Organic Acids in Aqueous Solution*, Butterworths, London, 1961.
43. C. K. Ingold, Structure and Mechanism in Organic Chemistry, G. Bell and Sons, London, 1953, p. 749.
44. H. C. Brown, D. H. McDaniel and O. Häfliger, *The Determination of Organic Structures by Physical Methods*, Academic Press, New York, 1955, Chap. 14.
45. J. F. J. Dippy, D. P. Evans, J. J. Gordon, R. H. Lewis and H. B. Watson, *J. Chem. Soc.*, 1421 (1937).
46. J. F. J. Dippy, S. R. C. Hughes and B. C. Kitchiner, *J. Chem. Soc.*, 1275 (1964).
47. W. J. LeNoble, J. LuValle and A. Leifer, *J. Phys. Chem.*, **66**, 1188 (1962).
48a. H. H. Jaffé, *Chem. Rev.*, **53**, 191 (1953).
48b. J. E. Leffler and E. Grunwald, *Rates and Equilibria of Organic Reactions*, John Wiley and Sons, London, 1963, Chap. 7, p. 178.
48c. Ref. 37 and 45.
49. M. M. Davis and H. B. Hetzer, *J. Res. Nat. Bur. Std.*, **60**, 569 (1958).
50. J. E. Leffler and E. Grunwald, *Rates and Equilibria of Organic Reactions*, John Wiley and Sons, London, 1963, p.p 49 and 372.
51. L. P. Hammett, *Physical Organic Chemistry*, McGraw–Hill, New York, 1940, pp. 184–193.
52. H. C. Brown and Y. Okamoto, *J. Am. Chem. Soc.*, **80**, 4979 (1958).
53. H. van Bekkum, P. E. Verkade and B. M. Wepster, *Rec. Trav. Chim.*, **78**, 815 (1959).
54a. R. W. Taft, *J. Phys. Chem.*, **64**, 1805 (1960).
54b. R. W. Taft and I. C. Lewis, *J. Am. Chem. Soc.*, **81**, 5343 (1959).
55. R. O. C. Norman, G. K. Radda, D. A. Brimacombe, P. O. Ralph and E. M. Smith, *J. Chem. Soc.*, 3247 (1961).
56. J. D. Roberts and W. T. Moreland Jr., *J. Am. Chem. Soc.*, **75**, 2167 (1953).
57. R. W. Taft, *J. Am. Chem. Soc.*, **74**, 2729, 3120 (1952); **75**, 4231 (1953).
58. R. W. Taft and I. C. Lewis, *J. Am. Chem. Soc.*, **80**, 2436 (1958); R. W. Taft, S. Ehrenson, I. C. Lewis and R. E. Glick, *J. Am. Chem. Soc.*, **81**, 5352 (1959); R. W. Taft, S. Ehrenson, I. Fox, I. C. Lewis and R. E. Glick, *J. Am. Chem. Soc.*, **82**, 756 (1960).
59. R. W. Taft, E. Price, I. R. Fox, I. C. Lewis, K. K. Anderson and G. T. Davis, *J. Am. Chem. Soc.*, **85**, 3146 (1963).
60. R. W. Taft, N. C. Deno and P. S. Skell in *Annual Review of Physical Chemistry* (Ed. H. Eyring), Vol. 9, Annual Reviews, Palo Alto, 1958, pp. 287–300; S. Ehrenson in *Progress in Physical Organic Chemistry* (Ed. S. Cohen, A. Streitweiser and R. W. Taft), Vol. 2, Interscience, New York, 1964, pp. 195–251.
61. J. E. Leffler and E. Grunwald, *Rates and Equilibria of Organic Reactions*, John Wiley and Sons, London, 1963, Chap. 7, pp. 171–262; P. R. Wells, *Chem. Rev.*, **63**, 171 (1963).
62. L. M. Stock and H. C. Brown in *Advances in Physical Organic Chemistry* (Ed. V. Gold), Vol. 1, Academic Press, London, 1963, pp. 35–154.
63. C. D. Ritchie and W. F. Sager in *Progress in Physical Organic Chemistry* (Ed. S. Cohen, A. Streitweiser and R. W. Taft), Vol. 2, Interscience, New York, 1964, pp. 323–400.
64a. R. P. Bell, *The Proton in Chemistry*, Methuen, London, 1959, Chap. 7.
64b. R. P. Bell, *The Proton in Chemistry*, Methuen, London, 1959, pp. 159 and 160.
65. N. Bjerrum, *Z. Physik. Chem. (Leipzig)*, **106**, 219 (1923).
66. C. K. Ingold, *Structure and Mechanism in Organic Chemistry*, G. Bell and Sons, London, 1953, p.p. 728–733.
67. L. Eberson and I. Wadso, *Acta Chem. Scand.*, **17**, 1552 (1963).

68. L. L. McCoy, *J. Am. Chem. Soc.*, **89**, 1673 (1967).
69a. R. E. Dodd, R. E. Miller and W. F. K. Wynne–Jones, *J. Chem. Soc.*, 2790 (1961).
69b. L. Eberson and S. Forsén, *J. Phys. Chem.*, **64**, 767 (1960); L. Eberson, *Acta Chem. Scand.*, **13**, 224 (1959).
70a. S. N. Das and D. J. G. Ives, *Proc. Chem. Soc.*, 373 (1961).
70b. H. H. Jaffé, *J. Am. Chem. Soc.*, **79**, 2373 (1957).
71. D. R. Lloyd and R. H. Prince, *Proc. Chem. Soc.*, 464 (1961).
72. G. Dahlgreen and F. A. Long, *J. Am. Chem. Soc.*, **82**, 1303 (1960).
73. L. Eberson, *Acta Chem. Scand.*, **13**, 211 (1959).
74. L. L. McCoy and G. N. Nachtigall, *J. Am. Chem. Soc.*, **85**, 1321 (1963).
75. I. Jones and F. G. Soper, *J. Chem. Soc.*, 133 (1936).
76. H. F. van Woerdin, *Rec. Trav. Chim.*, **83**, 920 (1963).
77. H. E. Merril and J. H. Wotiz, *J. Am. Chem. Soc.*, **80**, 866 (1958).
78. C. K. Ingold and H. G. G. Morhenn, *J. Chem. Soc.*, 949 (1935).
79. H. Bode, *Chem. Ber.*, **67B**, 332 (1934).
80. L. Hunter, *Chem. Ind.*, 155 (1953).
81. L. Eberson, *Acta Chem. Scand.*, **18**, 1276 (1964).
82. A. V. Jones and H. N. Parton, *Trans. Faraday Soc.*, **48**, 8 (1952).
83. W. J. Hamer, G. D. Pinching and S. F. Acree, *J. Res. Nat. Bur. Std.*, **35**, 539 (1945).
84. B. J. Thamer and A. F. Voigt, *J. Phys. Chem.*, **56**, 225 (1952); **59**, 450 (1955).
85. H. L. Goering, T. Rubin and M. S. Newman, *J. Am. Chem. Soc.*, **76**, 787 (1954).
86. J. D. Roberts and W. T. Moreland, Jr., *J. Am. Chem. Soc.*, **75**, 2267 (1953).
87. W. R. Maxwell and J. R. Partington, *Trans. Faraday Soc.*, **33**, 670 (1937).
88. E. A. Albert and E. P. Serjeant, *Ionization Constants of Acids and Bases*, Methuen, London, 1962.
89. J. B. Conant and A. F. Thompson, *J. Am. Chem. Soc.*, **54**, 4039 (1932).
90. G. Schwarzenbach and E. Felder, *Helv. Chim. Acta*, **27**, 1701 (1944).
91. C. A. Bunton, in *Studies on Chemical Structure and Reactivity* (Ed. J. H. Ridd), Methuen, London, Chap. 5 and references cited therein.
92. C. K. Ingold, *Structure and Mechanism in Organic Chemistry*, G. Bell and Sons, London, 1953, Chap. 7.
93. B. Capon, *Quart. Rev.*, **18**, 45 (1964).
94. S. Sugden and J. B. Willis, *J. Chem. Soc.*, 1360 (1951).
95. C. J. Burris and K. J. Laidler, *Trans. Faraday Soc.*, **51**, 1497 (1955).
96. E. A. Moelwyn–Hughes, *J. Chem. Soc.*, 1576 (1933).
97. J. B. Conant, R. E. Hussey and W. R. Kirner, *J. Am. Chem. Soc.*, **47**, 488 (1925).
98. Farhat–Aziz and E. A. Moelwyn–Hughes, *J. Chem. Soc.*, 1523 (1961).
99. M. W. Perrin, *Trans. Faraday Soc.*, **34**, 144 (1938).
100. I. Fells and E. A. Moelwyn–Hughes, *J. Chem. Soc.*, 398 (1959).
101. G. F. Smith, *J. Chem. Soc.*, 521 (1943).
102. E. D. Hughes, *Trans. Faraday Soc.*, **34**, 185 (1938).
103. R. E. Robertson in *Progress in Physical Organic Chemistry* (Ed. S. Cohen, A. Streitwieser, and R. W. Taft), Vol. 4, Interscience, New York, 1967.
104. P. Laurent and S. Lenoir, *Ann. Chim.*, **19**, 274 (1944).
105a. W. A. Cowdrey, E. D. Hughes and C. K. Ingold, *J. Chem. Soc.*, 1208 (1937).
105b. E. D. Hughes and N. A. Taher, *J. Chem. Soc.*, 956 (1940).
106a. S. Winstein and H. J. Lucas, *J. Am. Chem. Soc.*, **61**, 1576 (1939); S. Winstein, *J. Am. Chem. Soc.*, **61**, 1635 (1959).
106b. S. Winstein and E. Grunwald, *J. Am. Chem. Soc.*, **70**, 841 (1948).
107. J. Gripenberg, E. D. Hughes and C. K. Ingold, *Nature*, **164**, 480 (1948).
108. H. W. Heine and J. F. Lane, *J. Am. Chem. Soc.*, **73**, 1348 (1951); E. Becker, H. W. Heine and J. F. Lane, *J. Am. Chem. Soc.*, **75**, 4514 (1953).

109. G. Kohnstam in *Advances in Physical Organic Chemistry* (Ed. V. Gold), Vol. 5, Academic Press, London, 1967, p. 121.
110. J. E. Leffler and E. Grunwald, *Rates and Equilibria of Organic Reactions*, John Wiley and Sons, London, 1963, p. 302.
111. S. Winstein, E. Allred, R. Heck and R. Glick, *Tetrahedron*, **3**, 1 (1958).
112. H. C. Brown, T. Inukai and Y. Okamoto, *J. Am. Chem. Soc.*, **80**, 4969 (1958); H. C. Brown and Y. Okamoto, *J. Am. Chem. Soc.*, **80**, 4976 (1958).
113. H. C. Brown, T. Inukai and Y. Okamoto, *J. Am. Chem. Soc.*, **80**, 4972 (1958).
114. L. J. Andrews, R. M. Keefer and A. Singh, *J. Am. Chem. Soc.*, **84**, 1179 (1962); L. J. Andrews, R. M. Keefer, and R. E. Lovins, J. Am. Chem. Soc., **84**, 3959 (1962);
     L. J. Andrews, L. Chauffe and R. M. Keefer, *J. Org. Chem.*, **31**, 3758 (1966).
115. L. L. Smith and J. R. Holum, *J. Am. Chem. Soc.*, **78**, 3417 (1958); G. Baddeley, E. K. Baylis, B. G. Heaton and J. W. Rasburn, *Proc. Chem. Soc.*, 451 (1961); D. J. Parto and M. P. Serve, *J. Am. Chem. Soc.*, **87**, 1515 (1965).
116. S. Oae, *J. Am. Chem. Soc.*, **78**, 4030 (1956).
117. G. R. Cowie, H. J. M. Fitches and G. Kohnstam, *J. Chem. Soc.*, 1585 (1963).
118. B. Bensley and G. Kohnstam, *J. Chem. Soc.*, 4747 (1956).
119. L. Altieri, J. A. Brieux, A. J. Castro and A. M. Porto, *J. Chem. Soc. (B)*, 963 (1966).
120. J. Miller and V. A. Williams, *J. Am. Chem. Soc.*, **76**, 5482 (1954); *J. Chem. Soc.*, 1475 (1953); J. Miller, *J. Am. Chem. Soc.*, **76**, 448 (1954).
121. R. A. Bonelli, J. A. Brieux and W. Greizerstein, *J. Am. Chem. Soc.*, **84**, 1026 (1962).
122. H. Rouche, *Bull. Sci. Acad. Roy. Belg.*, 534 (1921); Chem. *Abstr.*, **17**, 2876 (1923).
123. E. Berliner and L. C. Monack, *J. Am. Chem. Soc.*, **74**, 1574 (1952).
124. N. E. Sharbati, *J. Org. Chem.*, **30**, 3365 (1965).
125. J. Burdon, W. B. Hollyhead and J. C. Tatlow, *J. Chem. Soc.*, 6336 (1965).
126. J. F. Bunnett and R. E. Zahler, *Chem. Rev.*, **49**, 273 (1951); M. L. Crossley, R. H. Kienle and C. H. Benbrook, *J. Am. Chem. Soc.*, **62**, 1400 (1940).
127. For example: C. K. Ingold, *Structure and Mechanism in Organic Chemistry*, G. Bell and Sons, London, 1953, pp. 752–782.; A. G. Davies and J. Kenyon, *Quart. Rev.*, **9**, 203 (1955).
128. M. L. Bender, *Chem. Rev.*, **60**, 53 (1960).
129. E. J. Salmi, *Chem. Ber.*, **72**, 1767 (1939).
130. R. Leimu, R. Korte, E. Laakonsen and U. Lekmuskoski, *Suomen Kemistilehti*, **19B**, 93 (1946); E. Tommila and H. Steinberg, *Suomen Kemistilehti*, **19B**, 19 (1946).
131. W. J. Svirbely and B. W. Lewis, *J. Phys. Chem.*, **56**, 1006 (1952); E. S. Amis and J. E. Potts, *J. Am. Chem. Soc.*, **71**, 2112 (1949).
132. E. Kivinen and E. Tommila, *Suomen Kemistilehti*, **14B**, 7 (1941); E. Tommila and S. Tommila, *Ann. Acad. Sci. Fennicae*, **A59**, No. 5, 3 (1942).
133. M. L. Bender, F. Chloupek and Y. L. Chow, *J. Am. Chem. Soc.*, **80**, 5380 (1958).
134. A. Ågren, U. Hedsten and B. Jonnson, *Acta Chem. Scand.*, **15**, 1532 (1961).
135. L. Eberson, *Acta Chem. Scand.*, **18**, 2015 (1964).
136. A. K. Herd, T. Higuchi, T. Niki and A. C. Shah, *J. Am. Chem. Soc.*, **85**, 3655 (1963).
137. M. L. Bender, F. Chloupek and M. C. Neven, *J. Am. Chem. Soc.*, **80**, 5384 (1958).
138. T. C. Bruice and U. K. Pandit, *J. Am. Chem. Soc.*, **82**, 5858 (1960); and references cited therein.
139. L. Eberson, *Acta Chem. Scand.*, **16**, 2245 (1962).
140a. C. K. Ingold, *Structure and Mechanism in Organic Chemistry*, G. Bell and Sons, London, 1953, pp. 221–305

140b.P. B. D. de la Mare and J. H. Ridd, *Aromatic Substitution*, Butterworths, London, 1959.

140c.R. O. C. Norman and R. Taylor, *Electrophilic Substitution in Benzenoid Systems*, Elsevier, London, 1965.

141. C. Eaborn, *Organosilicon Compounds*, Butterworths, London, 1960, pp. 146–157.

142. C. Eaborn and K. C. Pande, *J. Chem. Soc.*, 1566 (1960).

143. H. C. Brown and L. M. Stock, *J. Am. Chem. Soc.*, **84**, 3298 (1962); and references cited therein.

144. J. T. Bornwater and A. F. Holleman, *Rec. Trav. Chim.*, **31**, 221 (1921).

145. A. R. Hendrikson and H. G. Kuivila, *J. Am. Chem. Soc.*, **74**, 5068 (1952); L. E. Benjamin and H. G. Kuivila, *J. Am. Chem. Soc.*, **77**, 4834 (1955).

146. A. F. Holleman, *Rec. Trav. Chim.*, **18**, 267 (1899); W. J. Le Noble and G. W. Wheland, *J. Am. Chem. Soc.*, **80**, 5397 (1958).

147. B. Aliprandi, F. Cacaee and G. Ciranni, *Anal. Chem.*, **36**, 2445 (1964).

148. D. H. Derbyshire and W. A. Waters, *J. Chem. Soc.*, 564 (1950).

149. R. S. Ajemian and A. J. Boyle, *J. Org. Chem.*, **24**, 1818 (1959).

150. C. Eaborn and K. C. Pande, *J. Chem. Soc.*, 5082 (1961).

151. E. Iwata, *Nippon Kagaguku Zasshi*, **78**, 213, 348, 350 (1957); *Chem. Abstr.*, **53**, 5183 (1958).

152. P. B. D. de la Mare and J. H. Ridd, *Aromatic Substitution*, Butterworths, London, 1959, p. 82; R. O. C. Norman and G. K. Radda, *J. Chem. Soc.*, 3610 (1961).

153. S. D. Andrews and J. C. Smith, *Chem. Ind.*, 1376 (1965).

154. P. B. D. de la Mare and J. H. Ridd, *Aromatic Substitution*, Butterworths, London, 1959, p. 146.

155. C. K. Ingold and M. S. Smith, *J. Chem. Soc.*, 905 (1938).

156. P. B. D. de la Mare and I. C. Hilton, *J. Chem. Soc.*, 997 (1962).

157. F. B. Deans and C. Eaborn, *J. Chem. Soc.*, 2299 (1959).

158. C. Eaborn and J. A. Waters, *J. Chem. Soc.*, 542 (1961).

159. A. F. Holleman, quoted in reference 140(b), p. 91.

160. S. J. Branch and B. Jones, *J. Chem. Soc.*, 2921 (1955).

161. C. K. Ingold, A. Lapworth, E. Rothstein and D. Ward, *J. Chem. Soc.*, 1959 (1931).

162. C. K. Ingold and F. R. Shaw, *J. Chem. Soc.*, 575 (1949).

163. J. R. Knowles and R. O. C. Norman, *J. Chem. Soc.*, 2938 (1961).

164. J. N. Nazarov and A. V. Semenovski, *Izv. Akad. Nauk SSSR*, 100, 840, 972. (1957); S. K. Freeman, *J. Org. Chem.*, **26**, 212 (1961); and references cited therein.

165. K. E. Cooper and C. K. Ingold, *J. Chem. Soc.*, 836 (1927).

166. H. W. Underwood and E. L. Kochman, *J. Am. Chem. Soc.*, **48**, 254 (1926).

167. F. G. Bordwell and K. Rohde, *J. Am. Chem. Soc.*, **70**, 1191 (1948).

168. R. O. C. Norman and R. Taylor, *Electrophilic Substitution in Benzenoid Systems*, Elsevier, London, 1965, p. 82.

169. J. W. Baker, K. E. Cooper and C. K. Ingold, *J. Chem. Soc.*, 426 (1928).

170a.C. K. Ingold, *Structure and Mechanism in Organic Chemistry*, G. Bell and Sons, London, 1953, Chap. 12.

170b.P. B. D. de la Mare and R. Bolton, *Electrophilic Addition to Unsaturated Systems*, Elsevier, London, 1966.

171. From data collected in P. B. D. de la Mare and R. Bolton, *Electrophilic Addition to Unsaturated Systems*, Elsevier, London, 1966, p. 76.

172. P. B. D. de la Mare and R. Bolton, *Electrophilic Addition to Unsaturated Systems*, Elsevier, London, 1966, pp. 92–93.

173a.M. J. S. Dewar and R. C. Fahey, *J. Amer. Chem. Soc.*, **85**, 2245, 2248, 3645 (1963).

173b.M. C. Cabaleiro and M. D. Johnson, *J. Chem. Soc. B*, 565 (1967).

174. C. K. Ingold and E. H. Ingold, *J. Chem. Soc.*, 2354 (1931).
175. From data collected in P. B. D. de la Mare and R. Bolton, *Electrophilic Addition to Unsaturated Systems*, Elsevier, London, 1966, pp. 84, 85, and 123.
176. E. Linnemann, *Ann. Chem.*, **163**, 96 (1872); J. Wislicenus, *Ann. Chem.*, **166**, 1 (1873).
177. P. Melikov, *Chem. Ber.*, **12**, 2227 (1879); **13**, 2153 (1880).
178. C. K. Ingold and H. G. Smith, *J. Chem. Soc.*, 2742, 2752 (1931).
179. A. Michael and H. S. Mason, *J. Am. Chem. Soc.*, **65**, 683 (1943).
180. I. Roberts and G. E. Kimball, *J. Am. Chem. Soc.*, **59**, 947 (1937).
181. W. G. Young, R. T. Dillon and H. J. Lucas, *J. Am. Chem. Soc.*, **51**, 2528 (1929).
182. P. B. D. de la Mare and R. Bolton, *Electrophilic Addition to Unsaturated Systems*, Elsevier, London, 1966, p. 68. and references cited therein.
183. W. R. Vaughan, R. L. Ciavan, R. Q. Little and A. C. Schoenthaler, *J. Am. Chem. Soc.*, **77**, 1594 (1955).
184. a) E. van Tamelen and M. Shamma, *J. Amer. Chem. Soc.*, **76**, 2315 (1954); b) E. N. Rengevich, V. I. Staninets and E. A. Shilov, *Dokl. Akad. Nauk. SSSR*, **146**, 111 (1963).
185. F. V. Kucherov, A. L. Shabanov and A. S. Onishchenko, *Izv. Akad. Nauk SSSR, Otd. Khim. Nauk*, **59**, 852 (1963).
186. R. T. Arnold and K. L. Lindsay, *J. Am. Chem. Soc.*, **75**, 1048 (1953).
187. G. Besti, A. Marsili and P. L. Pacini, *Ann. Chim.*, **52**, 1070 (1962).
188. D. S. Tarbell and P. D. Barlett, *J. Am. Chem. Soc.*, **59**, 407 (1937).
189. C. K. Ingold, *Structure and Mechanism in Organic Chemistry*, G. Bell and Sons, London, 1953, pp. 662, 663.
190. P. B. D. de la Mare and R. Bolton, *Electrophilic Addition to Unsaturated Systems*, Elsevier, London, 1966, p. 99.
191. E. M. Terry and L. Eichelberger, *J. Am. Chem. Soc.*, **47**, 1067 (1925).
192. R. P. Bell and M. Pring, *J. Chem. Soc. B*, 1119 (1967), and references cited therein.
193. M. L. Bender and K. A. Connors, *J. Am. Chem. Soc.*, **84**, 1980 (1962).
194. P. N. Craig and I. H. Witt, *J. Am. Chem. Soc.*, **72**, 4925 (1950); R. T. Arnold, M. de M. Campos, and K. L. Lindsay, *J. Am. Chem. Soc.*, **75**, 1044 (1953).
195. See P. B. D. de la Mare, *Quart. Rev.*, **3**, 126 (1949).
196. D. Swern, *J. Am. Chem. Soc.*, **69**, 1692 (1947).
197. H. J. Lucas and S. Winstein, *J. Am. Chem. Soc.*, **59**, 1461 (1937); H. J. Lucas and D. Pressman, *J. Am. Chem. Soc.*, 2271 (1939).
198. V. Gold and R. S. Satchell, *J. Chem. Soc.*, 1930 (1963).
199. R. P. Bell, J. Preston and R. B. Whitney, *J. Chem. Soc.*, 1166 (1962).
200. D. S. Noyce, P. A. King, F. B. Kirby and W. L. Reed, *J. Am. Chem. Soc.*, **84**, 1632 (1962); D. S. Noyce and H. S. Avarbeck, *J. Am. Chem. Soc.*, **84**, 1644 (1962); D. S. Noyce, H. S. Avarbeck and W. L. Reed, *J. Am. Chem. Soc.*, **84**, 1647 (1962).
201. R. N. Ring, G. T. Tesoro and D. R. Moore, *J. Org. Chem.*, **32**, 1091 (1967); L. E. Erikson and R. A. Alberty, *J. Phys. Chem.*, **63**, 705 (1959).
202. P. B. D. de la Mare and R. Bolton, *Electrophilic Addition to Unsaturated Systems*, Elsevier, London, 1966, Chap. II.
203. P. B. D. de la Mare and R. Bolton, *Electrophilic Addition to Unsaturated Systems*, Elsevier, London, 1966, p. 215.
204. See P. B. D. de la Mare, *Ann. Rep. Chem. Soc.*, **47**, 126 (1950).
205. G. F. Dvorko, *Dopovidi Akad. Nauk Ukr. RSR*, 498 (1958).
206. R. Friedrich, *Ann. Chem.*, **219**, 368 (1883); A. Michael, *J. Prakt. Chem.*, **52**, 289 (1895).
207. A. Michael, *J. Prakt. Chem.*, **46**, 209 (1892).
208. A. R. Dankner, D. Dankner and T. L. Jacobs, *J. Am. Chem. Soc.*, **80**, 864 (1958).

209. R. Paul and H. Normant, *Compt. Rend.*, **216**, 689 (1943).
210. R. L. Tetsinger, E. N. Oftedall and J. R. Nazy, *J. Am. Chem. Soc.*, **87**, 742 (1965).
211. J. Crosby and C. J. M. Stirling, papers presented at a meeting of the Chemical Society, Durham, 1967.
212. For example: C. K. Ingold, *Structure and Mechanism in Organic Chemistry*, G. Bell and Sons, London, 1953, Chap. 8; D. V. Banthorpe, *Elimination Reactions*, Elsevier, London, 1963.

CHAPTER 17

Analysis of carboxylic acids and esters

T. S. MA

City University of New York, U.S.A.

I. INTRODUCTION

Methods for recognizing acids and esters in nature were known to our forefathers for millenia. In the search for food, man could easily identify acidic materials by their sour taste and was attracted by the fragrance of many naturally occurring esters. Long before the development of modern science, the techniques for isolation and purification of these substances, and their use as foodstuffs, tonic or curative agents were extensively recorded[1]. It may be of interest to mention that carboxylic acids were the first group of organic compounds investigated by the early workers in chemistry. Scheele identified tartaric acid in 1770, followed by lactic, mucic, and several vegetable acids, and had the habit of relying on smell and taste for chemical analysis.

Since carboxylic acids and esters are used in numerous industrial products and commercial materials, their analytical methods have drawn much attention and have been continuously reviewed. A large number of reports dealing with the analysis of acids and esters appear annually. Because of the limit of space and in view of the objectives of the present monograph, the analytical methods discussed in this chapter are concerned with the carboxyl (—COOH) and carboxylic ester (—COOR) functions respectively[2]. Methods recommended for the analysis of carboxylic acid and esters which are based primarily on the remaining portion of the molecule will not be included.

II. CHEMICAL METHODS FOR THE DETECTION AND CHARACTERIZATION OF CARBOXYL AND ESTER FUNCTIONS

A. Acidity Tests for the Carboxyl Function

1. Solubility test

According to the scheme for qualitative organic analysis based on solubility[3], carboxylic acids are classified into two groups. One group comprises compounds that are water-soluble and are capable of turning blue litmus red. The other group is water-insoluble but dissolves readily in 5% sodium hydroxide and also in 5% sodium bicarbonate. However, since other classes of compounds (e.g. sulfonic acids and certain phenols) may exhibit the same solubility behavior, confirmation of the carboxyl function in the sample requires more than solubility tests.

2. Liberation of carbon dioxide from sodium bicarbonate

Since the carboxyl function is a stronger acid than carbonic acid, it can liberate carbon dioxide from a solution of sodium bicarbonate. This test is best performed on a glass slide or spot plate on which one drop of 5% sodium bicarbonate is placed. When one drop of the carboxylic acid or its solution (or one crystal of the compound) is brought into contact with the reagent solution, gas bubbles are formed and can be observed conveniently by means of a simple magnifying tube[4] or under the microscope. If the sample is a water-insoluble solid, evolution of carbon dioxide bubbles from its surface indicates that it is probably a carboxylic acid and not the other classes of acidic organic compounds.

3. pH indicator test

When organic compounds are classified by the indicator method[5], carboxylic acids fall into two classes, namely strong and intermediate acidic groups. The indicator reagents for them are alizarin yellow-R mixed with bromothymol blue, and bromocresol purple mixed with thymol blue, respectively. It should be noted that the other parts of the acid molecule may influence the ionization of the carboxyl function. For instance, the pH value of trichloroacetic acid is 0·89 compared with 4·76 for acetic acid.

4. Precipitation of heavy metal salts

The lead and silver salts of carboxylic acids are only slightly soluble in water or ethanol[6]. When a solution of silver nitrate or lead nitrate is drawn into a drop of water or dilute ethanol containing the carboxylic acid (preferably in form of its sodium salt), precipitation of the salt can be easily observed by means of the magnifying tube[4]. The precipitate disappears when dilute nitric acid is added to the mixture.

B. Ferric Hydroxamate Test for Carboxyl and Ester Functions

A color reaction for the detection of the carboxyl and ester functions consists of their conversion to the corresponding hydroxamic acid. The latter is then reacted with ferric chloride to produce the highly colored ferric hydroxamate. It should be noted that, while esters can be tested directly, carboxylic acids require preliminary treatment as described in a subsequent section.

1. For esters

The compound to be tested (about 30 mg) and 0·5 ml of 1 N hydroxyl-ammonium chloride are mixed in a test tube; 2 N potassium hydroxide in methanol is added dropwise until the mixture is alkaline to litmus and then four drops more. The reaction mixture is heated just to boiling and allowed to cool. Now 2 N hydrochloric acid is introduced dropwise until the pH of the solution is approximately three. One drop of 10% ferric chloride is added. A reddish blue color indicates the formation of ferric hydroxamate from the parent carboxylic acid.

2. For carboxylic acids

In order to carry out the ferric hydroxamate test for the carboxyl function, it is necessary to prepare an ester or acid chloride first. If the diazomethane apparatus is available[7] (see section III.E), the methyl ester can be conveniently synthesized and then treated with hydroxylammonium chloride according to the directions given in the preceding paragraph.

Acid chloride is usually prepared from the carboxylic acid by refluxing with thionyl chloride for 10–20 min. The resulting acid chloride may be treated with hydroxylammonium chloride followed by ferric chloride. Merliss and Weinheimer[8] have proposed a method for converting acids in aqueous solution to acid chlorides which may then be detected by the ferric hydroxamate test. The procedure is as follows (*warning:* there is a vigorous reaction between thionyl chloride and water or potassium hydroxide). Place 2 ml of the aqueous solution (or 50–75 mg of the acid in

2 ml of water) in an 8 in. test tube and insert the tube in an ice bath. Add, in small portions, 2 ml of thionyl chloride to the tube. Occasionally shake the tube gently and keep it in the ice bath until the mixture appears homogeneous (less than 5 min, generally). While keeping the tube in an ice bath, add, in portions, 3 ml of a saturated solution of hydroxylammonium chloride in ethanol. After all of this reagent has been added, remove the tube from the ice bath and allow it to stand at room temperature for 3–5 min. Because of the large excess of thionyl chloride, a sufficient amount of the acid will be converted to the acid chloride to allow its detection by the ferric hydroxamate test. To make this test, insert a glass rod into the test tube and, with stirring, add dropwise 6 N aqueous potassium hydroxide (addition may be made more rapidly if the tube is put back in the ice bath to prevent overheating). Continue the addition of the potassium hydroxide until the pH of the mixture reaches 10 to 11 as shown by an indicator paper. As the base is added, copious quantities of potassium chloride will precipitate. This precipitated salt will not interfere with the final test and, therefore, may be neglected (it may be dissolved by adding more water). Now add 1 drop of 10% ferric chloride. The solution will turn yellow, and a yellow-to-brown precipitate may form. Add, dropwise, 12 M hydrochloric acid with stirring until the yellow color is discharged and the solution becomes essentially colorless. Finally, add 1–3 drops of 10 per cent ferric chloride. A bluish-red color is a positive test.

The reactions for the formation of ferric hydroxamate complexes from a carboxylic acid (via its acyl chloride) and an ester, respectively, may be represented by the following equations.

$$RCOCl + NH_2OH \cdot HCl \longrightarrow RCO(NHOH) + 2\,HCl \tag{1}$$

$$RCOOR' + NH_2OH \cdot HCl + KOH \longrightarrow RCO(NHOH) + R'OH + KCl + H_2O \tag{2}$$

$$RCO(NHOH) + FeCl_3 \longrightarrow Fe(RCONHO)^{2+} + Cl^- + H^+ \tag{3}$$

$$\updownarrow$$
$$Fe(RCONHO)_2^+$$
$$\updownarrow$$
$$Fe(RCONHO)_3$$

The ratio of the three colored species is dependent on the pH of the solution[9]. It should be noted that, unlike acid chlorides, esters react with hydroxylamine to form hydroxamic acids only when the reaction is carried out in an alkaline medium.

C. Characterization of the Carboxyl Function by the Preparation of Solid Derivatives

The number of reagents and methods which have been proposed for the characterization of carboxylic acids is legend. A partial list is shown in

Table 1. It will be noted that some derivatives given in the table are based on the presence of another function, beside the carboxyl, in the compound. The methods described in the following paragraphs are dependent on the reaction of the carboxyl function alone.

1. Preparation of p-toluidides

Aliphatic acids containing less than eight carbon atoms may be converted to the corresponding p-toluidides by heating the free acid with an excess of p-toluidine at 200°c for 30 min. Upon cooling, the reaction

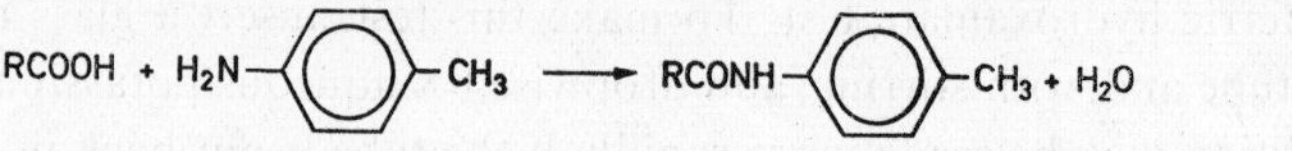

$$RCOOH + H_2N-\langle\bigcirc\rangle-CH_3 \longrightarrow RCONH-\langle\bigcirc\rangle-CH_3 + H_2O \tag{4}$$

mixture is extracted with dilute hydrochloric acid to remove the excess reagent, and with dilute sodium hydroxide to remove the unreacted acid. The p-toluidide is then recrystallized from an ethanol–water mixture.

For the higher fatty acids and aromatic carboxylic acids, it is recommended that the carboxyl function be converted to the acyl chloride by heating with an equimolar quantity of thionyl chloride at 75°c for 30 min. Benzene is then added, followed by an excess of p-toluidine, and the mixture is refluxed for another 30 min. After extraction with dilute hydrochloric acid and sodium hydroxide respectively, the benzene solution is evaporated and the residue is recrystallized.

$$RCOOH + SOCl_2 \longrightarrow RCOCl + HCl + SO_2 \tag{5}$$

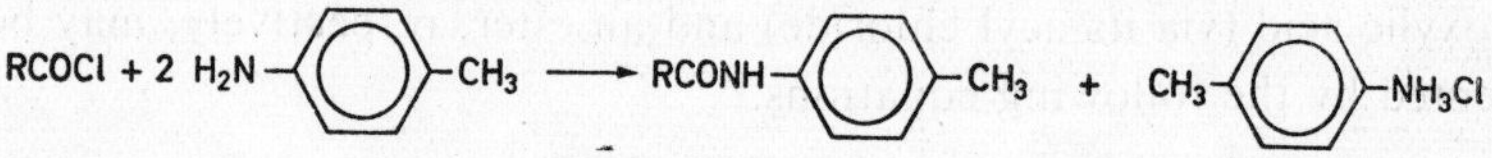

$$RCOCl + 2 H_2N-\langle\bigcirc\rangle-CH_3 \longrightarrow RCONH-\langle\bigcirc\rangle-CH_3 + CH_3-\langle\bigcirc\rangle-NH_3Cl \tag{6}$$

2. Preparation of p-bromophenacyl esters

The p-bromophenacyl esters have been used for the characterization of both liquid and solid carboxylic acids[10]. For this purpose, the carboxyl function is first converted to the sodium salt and the latter is then reacted with p-bromophenacyl bromide.

$$RCOOH + Na_2CO_3 \longrightarrow RCOONa + CO_2 + H_2O \tag{7}$$

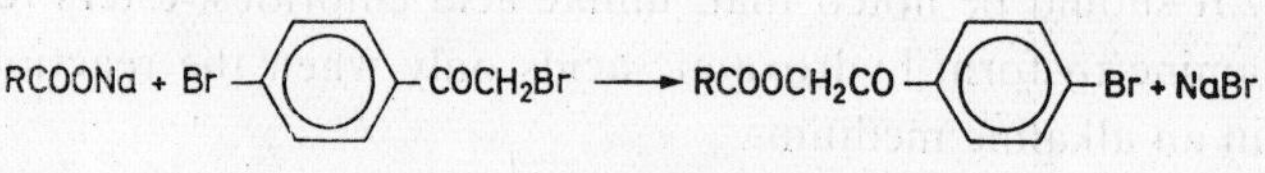

$$RCOONa + Br-\langle\bigcirc\rangle-COCH_2Br \longrightarrow RCOOCH_2CO-\langle\bigcirc\rangle-Br + NaBr \tag{8}$$

The free carboxylic acid is neutralized by means of 5% sodium carbonate (or 5% sodium hydroxide) using phenolphthalein as the indicator. The solution is then brought to the boil in order to be certain that all the acid has reacted. Dilute hydrochloric acid is now added dropwise to discharge the pink color. *p*-Bromophenacyl bromide is introduced and the mixture is heated under reflux for 2 h. After cooling, the solid is separated, washed with dilute sodium carbonate and water. The ester is recrystallized from an ethanol–water mixture.

TABLE 1. Derivatives for the identification of carboxylic acids[a]

Acid esters of dibasic acids	Hydrazides
N-Acyl-*p*-aminoazobenzenes	Hydroxamic acids
N-Acylanthranilic acids	*p*-Hydroxyanilides
N-Acylcarbazoles	Lactone derivatives
N-Acyl-2-nitro-*p*-toluidides	Methyl esters
N-Acylphenothiazines	*N*-Methylamides
N-Acylsaccharins	2-Methyl-5-isopropylanilides
N-Acyl-2-acylcarbazoles	Monoureides
p-(*N*-Acylamino)benzoic acids	Monothioureides
3-Acylaminodibenzylfurans	α-and β-Naphthylamides
2-Alkylbenzimidazole picrates	*S*-(α-Naphthylmethyl)thiuronium salts
2-Alkylbenzimidazoles	*o*-, *m*-, and *p*-Nitroanilides
Amides	*p*-Nitrobenzyl esters
Anilides	*S*-(*p*-Nitrobenzyl)thiuronium salts
N-Benzylammonium salts	Octadecylamides
S-Benzylthiuronium salts	Octadecylammonium salts
Bis(*p*-dimethylaminophenyl(ureides)	Phenacyl derivatives
p-Bromoanilides	*o*- and *p*-Phenetides
p-Bromobenzylthiuronium salts	4-Phenylazophenacyl esters
p-Bromophenacyl esters	α-Phenylethylammonium salts
α-Bromo-β-naphthylamides	Phenylhydrazides
o-Bromo-*p*-toluidides	Phenylhydrazonium salts
Carbodi-imide derivatives	*N*-Phenylimides of dibasic acids
p-Chlorobenzylthiuronium salts	*p*-Phenylphenacyl esters
p-Chlorophenacyl derivatives	Phenylmercuric salts
2,8-Diacylcarbazoles	Piperazonium salts
Diazomethane derivatives	Tetraphenylstilbonium salts
N-Diethanolamides	Thiophenylamides
Dimethylamides	*o*-Toluidides
2,4-Dinitrophenylhydrazides	*p*-Toluidides
2,4-Dinitrophenylhydrazones of the	*p*-Tolylmercuric salts
p-phenylphenacyl esters	2,4,6-Tribromoanilides
Diphenylamides	Triphenyllead salts
Dodecylamides	*p*-Xenylamides
Dodecylammonium salts	*m*-Xylidides
N-Ethanolamides	

[a] Data taken from N. D. Cheronis, J. B. Entrikin and E. M. Hodnett, *Semimicro Qualitative Organic Analysis*, John Wiley and Sons, New York, 1965.

3. Preparation of S-benzylthiuronium salts

The S-benzylthiuronium salts of more than 100 carboxylic acids have been reported[11]. The reagent (S-benzylthiuronium chloride) is easily obtained by heating benzyl chloride with thiourea in dilute alcohol[12]. The carboxyl function may be converted to its S-benzylthiuronium salt by the following procedure[13].

The carboxylic acid is neutralized with 1 N sodium hydroxide using methyl red as the indicator. The solution is then heated to 90°C and a hot aqueous solution of S-benzylthiuronium chloride is added. After mixing thoroughly, the container is immersed in an ice bath. The S-benzylthiuronium derivative of the carboxylic acid usually comes out as a crystalline solid. If an oil forms it is scratched against the walls to facilitate crystallization. Sodium chloride may be added if no precipitation occurs.

$$RCOONa + C_6H_5CH_2SC(NH_2)_2Cl \longrightarrow RCOO[(NH_2)_2CSCH_2C_6H_5] + NaCl \quad (9)$$

The S-benzylthiuronium salts should be recrystallized from anhydrous solvents, such as dioxan or alcohol. Hydrolysis of the salt takes place readily in alkaline solution and liberates the unpleasant smelling benzyl mercaptan. Besides their melting points, the infrared spectra[14] and x-ray diffraction patterns[15] of the S-benzylthiuronium salts have been suggested for the characterization of carboxylic acids.

4. Preparation of hydrazides

Carboxylic acid hydrazides may be prepared by treatment of methyl esters with hydrazine hydrate[16].

$$RCOOCH_3 + H_2NNH_2 \cdot H_2O \longrightarrow RCONHNH_2 + CH_3OH + H_2O \quad (10)$$

The methyl ester is conveniently produced by passing diazomethane into an ethereal solution of the carboxylic acid. The ether is then removed by evaporation and a solution of hydrazine hydrate in 95% ethanol is added. The reaction mixture is heated for 2 h. White crystals which come down on cooling (or on evaporation of a part of the solvent) are recrystallized from 80% ethanol. When the acid hydrazide is dissolved in water and tested with Tollens' reagent, immediate precipitation of silver occurs.

D. Characterization of the Ester Function

There is no method of preparing a derivative of an ester that can be used to characterize the whole molecule. As a rule, it is necessary to cleave the compound to yield its two components—the parent acid and parent alcohol or phenol—and then to prepare solid derivatives related to the carboxyl and hydroxyl functions respectively. If the cleavage is performed by means of a known quantity of alkali, the amount of reagent consumed may serve to characterize the ester under certain conditions, as explained below.

1. Characterization of an ester by its saponification equivalent

The cleavage of an ester in an alkaline solution may be represented by the following equation:

$$RCOOR + NaOH \longrightarrow RCOONa + R'OH \tag{11}$$

Since this reaction can be brought to completion, one equivalent of the ester will consume exactly one mole of sodium hydroxide, provided that the other parts of the ester molecule are not attacked by the alkali. Thus, if the sample is a pure compound—it may be a simple ester as depicted in equation (11), a mono ester of a dicarboxylic acid, or a mixed glyceride—measurement of the quantity of sodium hydroxide consumed by a known weight of the ester will give its equivalent weight. This is known as the saponification equivalent which is a characteristic of the compound. Understandably, the saponification equivalent cannot be used to differentiate isomeric esters. The procedure for determining saponification equivalents will be found in section III. F.

2. Conversion of esters to acid hydrazides

The preparation of acid hydrazides from methyl carboxylates has been discussed in section II. C. 4. Higher esters are converted to methyl esters by methanolysis:

$$RCOOR' + CH_3OH \longrightarrow RCOOCH_3 + R'OH \tag{12}$$

Without isolating the resulting methyl ester, the mixture may be reacted with hydrazine hydrate to form the hydrazide as follows. The ester (about 1 g) is dissolved in 5 ml of absolute methanol containing a small quantity of sodium methoxide, the latter which is conveniently prepared by adding 0·1 g of metallic sodium to the methanol prior to the introduction of the ester. The reaction mixture is heated under reflux for 30 min and then the excess methanol is distilled off. About 1 ml of 85% hydrazine hydrate is introduced and the mixture is refluxed for 15 min. If the solution becomes turbid, methanol is added to clarify the reaction mixture. After heating for 2 h, the solution is poured into an evaporating dish and the alcohol is evaporated. On cooling the residue, the crude hydrazide which precipitates is recrystallized from an alcohol–water mixture.

3. Conversion of esters to acyl *N*-(β-aminoethyl) morpholines

Carboxylates react with *N*(β-aminoethyl)morpholine as represented below:

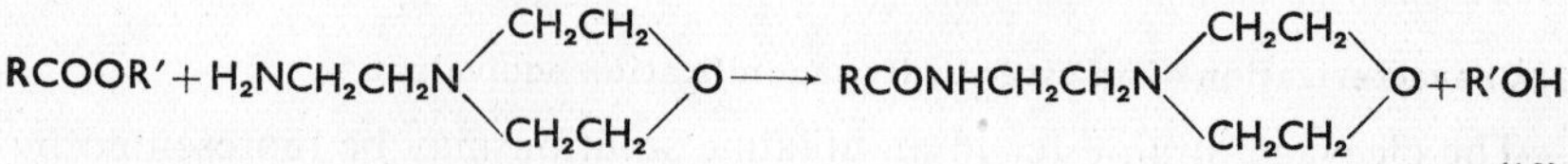

$$RCOOR' + H_2NCH_2CH_2N\begin{array}{c}CH_2CH_2\\CH_2CH_2\end{array}O \longrightarrow RCONHCH_2CH_2N\begin{array}{c}CH_2CH_2\\CH_2CH_2\end{array}O + R'OH$$

$$(13)$$

Most of the resulting acyl *N*-(β-aminoethyl)morpholines are obtained as crystalline solids[17]. If desired, they may be converted to the corresponding quaternary methyl iodides:

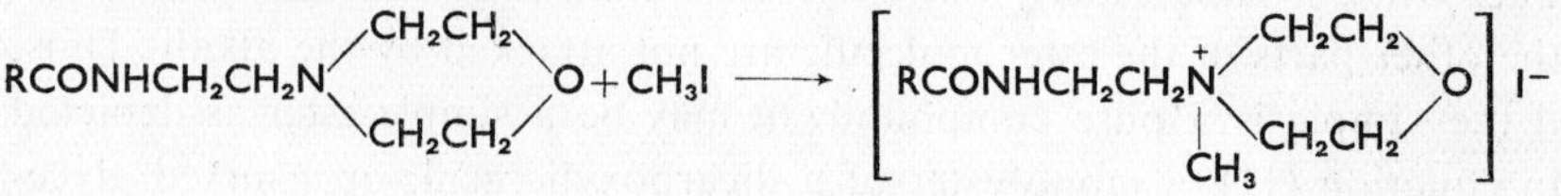

$$RCONHCH_2CH_2N\begin{array}{c}CH_2CH_2\\CH_2CH_2\end{array}O + CH_3I \longrightarrow \left[RCONHCH_2CH_2\overset{+}{\underset{CH_3}{N}}\begin{array}{c}CH_2CH_2\\CH_2CH_2\end{array}O \right] I^-$$

$$(14)$$

The acyl *N*-(β-aminoethyl)morpholine is prepared by heating about 10^{-3} equiv. of the ester with 10^{-3} mole of *N*-(β-aminoethyl)morpholine for 3 h. The reaction mixture is then cooled in an ice bath. If no precipitation occurs, 5 ml of ligroin is added and the walls of the container are scratched to induce crystallization. The crude product is recrystallized from ethanol (if the carboxyl group is aromatic or contains more than 11 carbon atoms) or from ligroin.

In order to prepare the quaternary iodide, the acyl *N*-(β-aminoethyl)-morpholine is dissolved in the minimum quantity of absolute methanol, followed by the addition of 1 ml of methyl iodide. The mixture is refluxed for 2 h. The solution is cooled in an ice bath and diethyl ether is added dropwise until precipitation is complete. The quaternary salt is recrystallized from absolute methanol.

4. Conversion of esters to 3,5-dinitrobenzoates

The alcoholic component of an ester can be converted to its 3,5-dinitrobenzoate by heating the ester with 3,5-dinitrobenzoyl chloride in the presence of pyridine for 1–2 h, as indicated in equation (15). The conversion

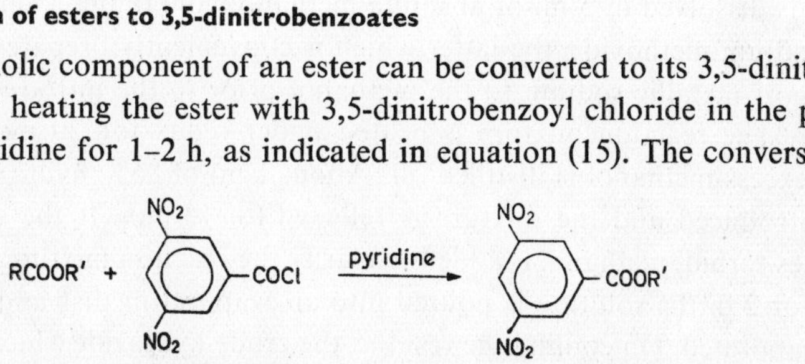

$$RCOOR' + \underset{NO_2}{\overset{NO_2}{\bigcirc}}-COCl \xrightarrow{pyridine} \underset{NO_2}{\overset{NO_2}{\bigcirc}}-COOR' \qquad (15)$$

can also be effected by refluxing the ester and 3,5-dinitrobenzoic acid with concentrated sulfuric acid as catalyst[18]. The reaction mixture is treated with water and the 3,5-dinitrobenzoate is extracted with ether. After washing with sodium carbonate solution, the ether extract is evaporated. The crude ester is purified by dissolving it in hot methanol and adding water to induce precipitation.

5. Miscellaneous methods for the characterization of the alcoholic component

When the hydrolysis of an ester is carried out in aqueous solution, if the resulting alcohol is water-insoluble it will appear as a wax or an oil. This may be separated, purified and identified. In the event that the alcohol is water-soluble and has low molecular weight, it may be separated from the reaction mixture by distillation. The distillate is tested for the $CH_3C(OH)$ structure by the iodoform reaction. If the alcoholic component can be converted to a volatile aldehyde or ketone, the distillate is treated with potassium dichromate in sulfuric acid. The carbonyl compound thus obtained is then identified by appropriate reactions[19]. For instance, a methyl ester is characterized by the formation of formaldehyde which is detected by the chromotropic acid color reaction.

III. CHEMICAL METHODS FOR THE DETERMINATION OF CARBOXYL AND ESTER FUNCTIONS

Quantitative analyses of carboxylic acids and esters are in frequent demand in the routine analytical laboratory (products control, raw material testing, *etc.*) as well as for research and development purposes (new compounds and mixtures, *etc.*) Taking advantage of the chemical reactivity of the carboxyl and ester functions, a number of 'wet methods' have been developed. Furthermore, many variations and modifications of a method are encountered in the literature[2, 20, 21]. Representative macro and micro methods are described in the following sections. The micro procedure requires about 0·1 milliequivalent (abbreviation, mequiv) of the compound for analysis; the macro procedure uses 1 mequiv or more sample in the determination.

Whereas chemical methods are based on chemical changes involving the carboxyl or ester function, the organic sample that is taken for analysis is not all destroyed. In some methods the original compound may be recovered after the determination; in other methods derivatives of the original compound can be isolated. Analytical methods which do not depend on chemical reactions and generally known as physical methods will be found in sections IV and V.

A. Determination of the Carboxyl Function by Neutralization

The carboxyl function normally exhibits a dissociation value, K_a, of the order of 10^{-5}, which is slightly larger than that of carbonic acid ($K_a = 4\times10^{-7}$). Hence, the simple method of determining carboxylic acids involves titration with a base. The standard aqueous sodium hydroxide solution may be employed, provided that an appropriate indicator is used and the slight interference of atmospheric carbon dioxide is minimized. If the carboxyl function is attached to a carbon atom with an adjacent electron-attracting substituent (*e.g.* Cl, NO_2), the acid strength of the compound is enhanced. The opposite effect is produced by an adjacent electron-donating substituent (*e.g.* NH_2). Thus, the ionization of the carboxyl function in some amino carboxylic acids may be suppressed to such an extent that no precise neutralization end-point is obtainable in aqueous media. In this case, it is necessary to perform the determination by non-aqueous titration.

1. Determination with aqueous titrant

The micro procedure for compounds that dissolve in dilute alcohol is as follows[22]. About 0·1 mequiv of the carboxylic acid is accurately weighed and transferred into a 50 ml conical Pyrex flask. 5 ml of 50% ethanol (previously neutralized) and 2 drops (0·05 ml) of 1% phenolphthalein are added. The solution is brought to the boil in order to expel carbon dioxide, and titrated whilst hot with standardized 0·01 N sodium hydroxide. When the end-point approaches, the solution is again boiled. The titration is complete when the faint pink color of the indicator persists for 30 s. If the solution has been overtitrated, 0·5–1 ml of standardized 0·01 N hydrochloric acid is accurately measured into the flask and the excess of acid is then back-titrated with the sodium hydroxide solution.

$$RCOOH + OH^- \rightleftharpoons RCOO^- + H_2O \qquad (16)$$

Dilute ethanol is used as solvent for water soluble carboxylic acids because it helps to suppress the reverse reaction shown in equation (16). For samples that do not dissolve readily in 50% ethanol, 95% ethanol, and ethanol mixed with acetone or dioxan may be employed.

When large-size samples are used for analysis, the concentration of the standard alkali solution and the volume of solvent added are increased correspondingly. Thus, 1 mequiv of the carboxylic acid (200–500 mg), is dissolved in 25 ml of 50% ethanol or other suitable solvent and the titration is performed with 0·1 N sodium hydroxide using 5 drops of 1% phenolphthalein as indicator. For the analysis of industrial and commercial materials, it is not uncommon to measure 5 g of sample into 50 ml of solvent and titrate the solution with 0·5 N alkali. Isopropyl alcohol is generally used for convenience, in place of ethanol, and the titration is carried out at ambient temperature since precision is not required.

2. Determination with non-aqueous titrant

Sodium ethoxide or methoxide and potassium methoxide solutions are recommended as titrants for the neutralization of carboxylic acids in non-aqueous media[23-26]. Quaternary ammonium hydroxides have been used[27, 28], but these titrants are less stable than the alkali alkoxides[29]. Acetone, dimethylformamide, and benzene–methanol are suitable solvents[30], while the benzene–isopropyl alcohol mixture is specified in the A.S.T.M. method[31].

Visual indicators can be used to locate the end-point. Potentiometric titration is necessary, however, when the acid strength of the compound is unknown. In this case, simultaneous determination of the titration curve and indicator color change may be carried out. Other electrometric methods such as conductometric and high-frequency titrimetry[32-34] have been proposed, but the potentiometric techhnique is in general found to be the most satisfactory[35].

It should be noted that alkali alkoxides are extremely sensitive to carbon dioxide and that the benzene–methanol mixture has a high coefficient of expansion. The apparatus shown in Figure 1 is designed to prevent the interference of atmospheric carbon dioxide during titration as well as to accomodate the volume changes of the titrant in the reservoir. The titration vessel is essentially a flat-bottomed flask which has five necks, one in line with the vertical axis of the vessel and the other four equally spaced, slightly oblique and lower. The vertical neck has a ground-glass joint that connects the tip of the Machlett buret. Two diametrically opposed side necks are for the insertion of the electrodes while the remaining side necks permit continuous flushing with nitrogen gas. If it is desired to perform the determination at a specified temperature (e.g. titration of higher fatty acids at 65°C using 0·05 N sodium ethoxide[23]), a thermometer can be fitted in the nitrogen outlet.

   a. Procedure for visual titration. The apparatus is assembled as shown in Figure 1 and the two side necks for electrodes are closed with rubber stoppers. 5 ml of dimethylformamide (or other organic solvent) is introduced by means of a pipet through the nitrogen outlet. Nitrogen gas is bubbled through the liquid while it is agitated by the magnetic stirring device. Two drops of thymol blue indicator solution are added, and any acid impurities in the solvent are neutralized by running in the standardized titrant (0·02 N sodium methoxide in benzene–methanol) until the appearance of the first permanent blue color[36]. The sample (about 0·1 mequiv) which has been accurately weighed in a weighing tube is now introduced through the nitrogen outlet and the tube is reweighed. After dissolution of the sample, the contents of the flask is titrated to a definite blue end-point.

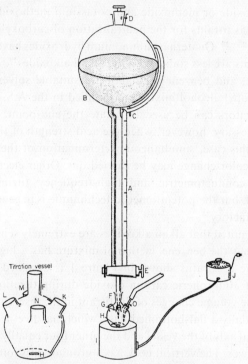

FIGURE 1. Apparatus for neutralization in non-aqueous solution A. Machlett buret; B. titrant; C. nitrogen inlet; D. excess pressure outlet; E. three-way Teflon stopcock; F. ground-glass joint; G. titration vessel; H. stirring bar; I. magnetic stirrer; J. rheostat; K, L. electrode inlets; M. nitrogen outlet; N. nitrogen inlet; O. nitrogen inlet tube. Courtesy of *Mikrochimica Acta*.

If 1 mequiv of the carboxylic acid is used, the volume of solvent delivered into the titration flask should be 25 ml and 0·1 N sodium methoxide should be the titrant. Azo violet may be used as indicator.

b. Procedure for potentiometric titration. If a 0·1 mequiv sample is employed, the volume of solvent is increased to 25 ml. Nitrogen is passed through and the solvent is agitated. The electrodes are so inserted that they are at least 2 mm below the surface of the liquid and are separated by a distance of less than 5 mm. The sample is now added through the nitrogen outlet. (For concurrent visual and potentiometric titration, the indicator solution is introduced immediately after the sample). With the potentiometer properly set, the titration is started with a large increment of standard sodium methoxide solution. The increment should be reduced when the approaching equivalence point is indicated by big jumps in voltage, and the addition of sodium methoxide should be continued until the equvivalence point is well passed. (For concurrent visual titration, the buret reading should be taken exactly at the time of color change).

3. Presentation of titration values

After the contents of the carboxyl function in the sample have been determined by neutralization, the result may be expressed in one of the following ways, depending on the purpose of the analysis. Needless to say, these expressions are interconvertible.

a. Per cent carboxyl group. This presentation is recommended when a pure sample of a new carboxylic acid is analyzed. If the analysis is carried out on the micro scale using 0.01 N alkali, the formula for calculation is as follows:

$$\% \text{ COOH} = \frac{\text{ml of } 0.01 \text{ N titrant} \times 45.02}{\text{mg of sample}}$$

b. Neutralization equivalent. This expression is generally used in the analysis of known carboxylic acids for the purpose of identification. Its value is identical to the molecular weight of the compound divided by the number of carboxyl functions in the molecule, and is calculated by the following formula:

$$\text{Neutralization equivalent} = \frac{(\text{mg of sample})}{(\text{ml of titrant}) (\text{normality of titrant})}$$

c. Per cent compound. This expression is frequently used in purity tests of carboxylic acids. It is also used for the analysis of a mixture in which the carboxylic acid is a major component. As shown by the formula for calculation given below, it is necessary to know the equivalent weight (same value as the neutralization equivalent) of the carboxylic acid.

$$\% \text{ Compound} = \frac{(\text{ml of titrant})(\text{normality of titrant})(\text{mequiv wt. in mg})(100)}{(\text{mg of sample})}$$

d. Acid value (or acid number). This term is defined as the number of milligrams of potassium hydroxide required to neutralize the free carboxylic acids present in one gram of sample:

$$\text{Acid value} = \frac{(\text{ml of titrant})(\text{normality of titrant})(56 \cdot 10)}{(\text{g of sample})}$$

It is commonly employed in expressing the acid contents of fats, oils, waxes, and other industrial materials. Whereas the molecular weight of potassium hydroxide appears in the calculation, the alkali titrant is usually a standardized sodium hydroxide solution.

B. Determination of the Carboxyl Function as Active Hydrogen

The carboxyl hydrogen reacts with methyl magnesium iodide to liberate the molar equivalent of methane as in equation (17).

$$RCOOH + CH_3MgI \longrightarrow CH_4 + RCOOMgI \tag{17}$$

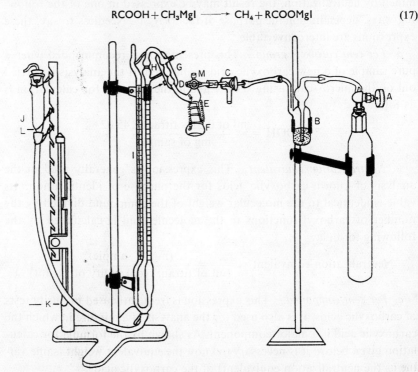

FIGURE 2. Gasometric apparatus for 0·1 mequiv samples (from reference 2, p. 152)

The gasometric apparatus shown in Figure 2 is recommended for the determination on the 0·1 mequiv range[37]. The carboxylic acid is placed in the reaction vessel F and the apparatus is filled with dry methane (or nitrogen).

The mercury level in the gasometer HI is adjusted to the zero mark. The Grignard reagent solution is then added through stopcock D. The volume of methane liberated is then measured by resetting the mercury level. Since the carboxyl function is very active, the reaction rapidly goes to completion at room temperatures. This apparatus also serves well for the analysis of the ester function (see sections H and I below) and of mixtures containing esters and free carboxylic acids.

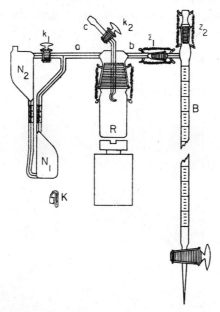

FIGURE 3. Active hydrogen apparatus of Soucek. Courtesy of *Chem. Listy*

For the analysis of 1 mequiv samples, the active hydrogen apparatus of Soucek[38] may be employed with advantage, because it eliminates the use of mercury. As shown in Figure 3, the manometer N_1N_2 and the buret B are filled with dibutyl ether. The Grignard reagent is placed in the vessel R, while the sample to be analyzed is kept in the basket K which hangs on the hook of the nitrogen inlet tube C. After the apparatus has been equilibrated, the basket is brought down by means of a magnet. The liquid in the buret is then drained out until the manometer returns to the original level and the volume of methane produced is indicated by the buret reading.

C. Determination of the Carboxyl Function by Decarboxylation

The carboxyl function in some compounds can be quantitatively decarboxylated as represented by equation (18). It is difficult, however, to obtain complete decarboxylation for most carboxylic acids.

$$RCOOH \longrightarrow RH + CO_2 \qquad (18)$$

Nevertheless, under controlled conditions, the yields of carbon dioxide are directly proportional to the amounts of carboxyl function present in the sample[39, 40]. Examples are shown in Figure 4. Since this method is specific

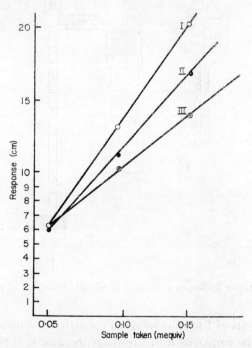

FIGURE 4. Gas chromatographic determination of the carboxyl function by decarboxylation. I. *p*-hydroxybenzoic acid; II. *o*-chlorobenzoic acid; III. Malonic acid. Courtesy of *Mikrochim. Acta.*

for the carboxyl function among all acidic substances, it is particularly useful when the analysis of carboxylic acid content in an acidic mixture is desired. The carbon dioxide liberated may be determined by titrimetry[41] or manometrically[39]. When the sample is in the 0·1 mequiv range, determination of the carbon dioxide by means of gas chromatography is recommended[40]. The apparatus for decarboxylation is shown in Figure 5. The carboxylic acid is placed in the vessel C and the reagent is added through the rubber cap D using a syringe. After the apparatus has been heated for a predeter- mined period, the carbon dioxide may be driven into a solution of barium hydroxide or into the gas chromatograph.

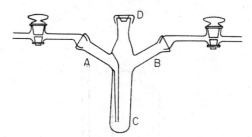

FIGURE 5. Decarboxylation apparatus. Courtesy of *Mikrochim. Acta.*

D. Determination of the Carboxyl Function using the Karl Fischer Reagent

Aliphatic carboxylic acids can be esterified with absolute methanol in the presence of boron trifluoride and the resulting water can be measured by means of the Karl Fischer reagent, as depicted below.

$$RCOOH + CH_3OH \xrightarrow{\ BF_3\ } RCOOCH_3 + H_2O \qquad (19)$$

$$H_2O + C_6H_5N \cdot I_2 + C_6H_5N \cdot SO_2 + C_6H_5N \longrightarrow 2\ C_6H_5N \cdot HI + C_6H_5NO(SO_2) \quad (20)$$

Because the esterification reaction requires heating and because atmo- spheric moisture interferes with the determination, the apparatus shown in Figure 6 has been designed for micro analysis. The reaction vessel A is con- nected by a ground-glass joint to the tip of the buret holding the Karl Fischer reagent[42]. The side arm C is parallel to the base of the flask and carries a Teflon plunger, the end of which is cut out so that the microboat B can be placed in it. To side arm D are fitted the electrodes, while the stopcock E serves as a vent to release the pressure momentarily. In the reaction vessel are placed 2 ml of absolute methanol and 1 ml of boron trifluoride dissolved in cellosolve[42]. The carboxylic acid (0·1 mequiv) is weighed in the microboat which is placed in the plunger. The plunger is then turned to drop the sample into the liquid mixture. The reaction vessel is now placed on the heating stage[43] maintained at 60°C for 0·5 h. Upon cooling, the contents of the reaction vessel are titrated with the Karl Fischer reagent after the electrodes are connected to the 'dead stop' indi- cator.

Analysis of 1 mequiv samples may be carried out in a ground-glass stop-pered flask, followed by visual titration with Karl Fischer reagent of high concentration. Considerable error will be introduced, however, in humid weather.

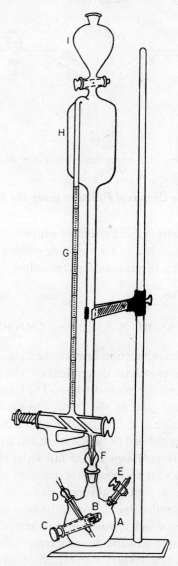

FIGURE 6. Apparatus for esterification and titration with Karl Fischer reagent. From reference 2, p. 168

E. Colorimetric Determination of Carboxyl and Ester Functions as Ferric Hydroxamates

Mention has been made in section II. B of the complexity of the ferric hydroxamate solutions resulting from carboxylic esters (see equation 3). Fortunately, the color is reproducible under controlled conditions and

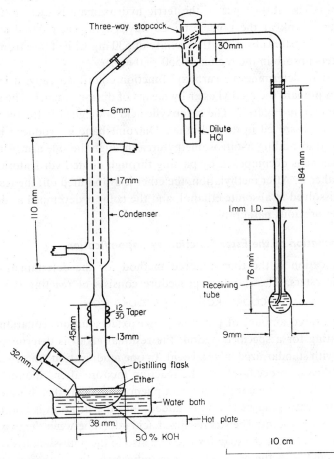

Three-way stopcock
30mm
6mm
Dilute HCl
17mm
110 mm
Condenser
184 mm
1mm I.D.
76 mm
Receiving tube
45 mm
12/30 Taper
13mm
9mm
32mm
Distilling flask
Ether
Water bath
38 mm
Hot plate
50% KOH
10 cm

FIGURE 7. Apparatus for preparation of methyl carboxylates. From reference 2, p. 191

there is a linear relationship between color intensity and the amount of the original ester. The procedure for determining esters in the 1 mequiv range is as follows[44].

The carboxylic ester is dissolved in 5 ml of absolute ethanol in a flask with a ground-glass joint. In another flask is placed 5 ml of absolute ethanol to serve as blank. Into each flask is added 3 ml of the reagent, freshly prepared by mixing equal volumes of 12·5 wt/vol.% hydroxylaminehydro-chloride in methanol and 12·5% sodium hydroxide. The mixtures are heated under reflux for 5 min and then quantitatively transferred into separate 50 ml volumetric flasks by washing with ferric perchlorate solution (0·0057 M in ethanol). After filling to the mark, the solutions are allowed to stand for 10 min. The absorbance of the ferric hydroxamate is then measured against the blank at the predetermined wavelength. The colored species from aliphatic esters absorb in the vicinity of 530 mμ while those from aromatic esters absorb in the region of 550–560 mμ.

In order to determine the carboxyl function colorimetrically, it is convenient to prepare the methyl ester by means of diazomethane[7]. The apparatus is shown in Figure 7. The carboxylic acid is weighed in the receiving tube A and dissolved in diethyl ether. Diazomethane is produced in the generator B by adding N-nitrosomethylurea through the side arm. Unused diazomethane is decomposed by passing through dilute hydrochloric acid in the beaker C. After methylation, the ether is evaporated off, the residual ester is dissolved in absolute ethanol, and the color is developed as described in the previous paragraph.

F. Determination of the Ester Function by Saponification

Saponification is the time-honored method for the determination of carboxylic esters. The standard procedure consists of reacting the ester

$$RCOOR' + KOH \longrightarrow RCOOK + R'OH \qquad (21)$$

with a measured amount of potassium hydroxide solution (equation 21). After heating for a specified period, the residual alkali is determined by titration with standardized mineral acid. In one modification, known as the double-indicator procedure[45, 46], the reaction mixture after saponification is first titrated to the neutral point with phenolphthalein; then bromophenol blue indicator and benzene are added, and the titration is continued to the green end-point. The standardized acid added between the two end-points is equivalent to the soap formed and also to the potassium hydroxide which reacted with the ester during saponification. In another modification[47], the carboxylic acid formed is liberated by ion exchange and determined by titration with a standardized base (see section III. G).

1. Macro procedure

The macro saponification procedure described below is based on the practice of an industrial analytical laboratory which handles a wide range of esters[44, 48]. Three variations of the procedure are presented, so that the reader can select the method best suited to the sample to be analyzed.

a. Saponification with aqueous potassium hydroxide. The reaction is carried out either in a 250 ml glass-stoppered conical flask or in a pressure bottle (available from B. Preiser Co., Charleston, West Virginia; spring-capped bottle from the pharmacist also suitable). The choice is dependent on the reaction conditions to be used, some examples being shown in Table 2. Into each of two flasks (or pressure bottles, for elevated temperature operation) is delivered 25·0 ml of 1 N potassium hydroxide, and isopropyl alcohol is added as specified (see Table 2). The ester (1–15 mequiv)

TABLE 2. Reaction conditions for the saponification of esters with aqueous potassium hydroxide[a]

| Compound | Reaction Conditions | | |
	2-Propanol co-solvent (ml)	Temperature (°C)	Time (min)
Butyl acetate	15	25	45
2-Butyl acetate	15	98	15
Butyl acrylate	15	25	60
Butyrolactone	–	25	15
Cyclopentenyl acetate	10	25	15
Dibutyl maleate	15	25	45
Didecyl phthalate	40	98	60
Diethyl sulfate	20	98	15
Dimethyl phthalate	20	98	30
Ethyl acetate	–	25	30
Ethyl acetoacetate	–	98	60
Ethyl formate	–	25	15
2-Ethylhexyl acetate	35	98	15
3-Heptyl acetate	35	98	90
Methyl acetate	–	25	15
Methyl methacrylate	15	98	15
4-Methyl-2-pentylacetate	30	98	30
Thiodiglycol diacetate	10	25	15

[a] Data taken from F. E. Critchfield, *Organic Functional Group Analysis*, Pergamon Press, Oxford, 1963.

is weighed into one flask, the other being a blank. If the mixture appears non-homogeneous, 5 ml of methyl alcohol is added. The flasks are then closed and kept at the prescribed temperature for 15–90 min as specified. After saponification, if a white precipitate develops, sufficient distilled water is added to dissolve it and the same volume is added to the blank. The contents of each flask is then titrated with standardized 0·5 N hydrochloric acid using phenolphthalein as indicator.

b. Saponification with potassium hydroxide in diethylene glycol. This variation is employed for the determination of chemically resistant and high molecular weight water-insoluble esters. Alkali-resistant conical flasks of 300 ml capacity with ground-glass joints are used as reaction vessels. Into each of two flasks is added 50·0 ml of 0·5 N potassium hydroxide. One flask serves as blank, while 1–15 mequiv of the ester is accurately weighed into the second flask. A few borosilicate glass beads are added to each flask which is connected to the water-cooled condenser. The solutions are heated under reflux for the predetermined period (60–120 min). After saponification, while the solutions are cooling to room temperature, the tops of the condensers are purged with nitrogen in order to prevent carbon dioxide contamination. The walls of the condensers are washed with distilled water, and the contents of the flasks are then titrated with 0·05 N hydrochloric acid to the phenolphthalein end-point.

c. Saponification in the presence of phenylhydrazine. This modification is used for the determination of vinyl esters or an ester sample which contains aldehydes. The vinyl group reacts with potassium hydroxide to produce acetaldehyde as depicted in equation (22). Incorporation of

$$RCOOCH{=}CH_2 + KOH \longrightarrow RCOOK + CH_3CHO \qquad (22)$$

phenylhydrazine in the saponification medium converts aldehydes into phenylhydrazones which are stable to alkali. Otherwise, the presence of aldehydes interferes with the determination of esters by saponification due to the fact that aldehydes are attacked by, and consume, potassium hydroxide. In each of two pressure bottles is placed 25·0 ml of 1 N potassium hydroxide, followed by 35 ml of isopropyl alcohol and 5 ml of methyl alcohol. A current of nitrogen gas is passed through the bottles for 2 min and 5 ml of phenylhydrazine (freshly distilled) is added to each bottle. One bottle being kept as blank, 1–15 mequiv of the ester is accurately weighed into the other bottle. Both bottles are allowed to stand for the time (15–60 min) and at the temperature (usually 98°C) for quantitative saponification. Upon cooling, the contents of the bottles are titrated with 0·5 N hydrochloric acid using phenolphthalein as indicator.

2. Micro procedure

Saponification in a sealed borosilicate tube[49] is recommended for micro determinations. The ester (0·1 mequiv) is weighed in the reaction tube, dissolved in 0·1–0·5 ml of a suitable solvent and 1·00 ml of 0·5 N KOH is added to it. The tube is then sealed and heated in a metal block[43] at 150°C for 1 h. Another sealed tube containing the same volume of solvent and 0·5 N potassium hydroxide is used as blank. After cooling to room temperature, the tubes are opened and their contents are titrated with standardized 0·05 N hydrochloric acid using phenolphthalein as indicator.

3. Presentation of the results

The result of quantitative analysis of an ester by saponification may be expressed in one of the following three forms. The saponification equivalent is usually reported for pure compounds, while per cent compound (or ester) indicates that the sample is a mixture. The saponification value (or saponification number) of fats, oils, or waxes is defined as the number of mg of potassium hydroxide required for the complete saponification of 1 g of sample. The respective formulae for calculation are given below, where b = ml hydrochloric acid for titration of the blank, and s = ml hydrochloric acid for titration of the sample.

$$\text{Saponification equivalent} = \frac{\text{wt. of sample in mg}}{(b-s)\,(\text{normality of HCl})}$$

$$\%\,\text{Compound} = \frac{(b-s)\,(\text{normality of HCl})(\text{mequiv wt. of ester in mg})(100)}{\text{wt. of sample in mg}}$$

$$\text{Saponification value (or number)} = \frac{(b-s)\,(\text{normality of HCl})}{\text{wt. of sample in g}}$$

G. Determination of the Carboxyl Function in an Ester

As is apparent in the above section, the precision of ester determination by saponification may suffer from several factors: (i) titration of the large excess of residual alkali, (ii) loss of alkali due to reaction with container walls and contaminants, and (iii) incomplete saponification. These difficulties can be circumvented by determining the carboxylic acid after the saponification reaction. The blank is eliminated. By running two samples at different temperatures and/or heating periods, unsatisfactory analysis due to incomplete saponification is readily detected.

The determination is conveniently performed on the 0·1 mequiv scale[47]. After saponification in the sealed tube (or in a glass-stoppered test tube if there is no danger of losing the ester by volatilization), isopropyl alcohol is added to the reaction mixture. The solution is then transferred into the ion-exchange column packed with Amberite IR-120 (or other strongly acidic resin). The amount of free carboxylic acids in the elute is determined by titration with standardized 0·01 N sodium hydroxide using the cresol red–thymol blue mixed indicator.

H. Determination of the Ester Function by Grignard Reagent

The ester function reacts with methylmagnesium iodide according to the following equation:

$$R-C\underset{OR'}{\overset{O}{\big\langle}} + 2\ CH_3MgI \longrightarrow R-\underset{CH_3}{\overset{CH_3}{\underset{|}{\overset{|}{C}}}}-O-MgI + R'OMgI \qquad (23)$$

Since the reaction products do not form gases, whereas the unreacted Grignard reagent liberates methane on treatment of the mixture with aniline, an indirect method for the determination of the ester function presents itself. The gasometric apparatus shown in Figure 2 may be used. The ester is weighed in the reaction vessel F, dissolved in an appropriate liquid and the apparatus is filled with methane (or nitrogen). A measured volume of standardized Grignard reagent solution is added by means of the syringe through the stopcock channel D. After the completion of the reaction (equation 23), the gasometer HI is adjusted to zero. Aniline is now added with a syringe and the amount of methane formed is measured.

I. Determination of the Ester Function with Lithium Aluminum Hydride

The ester function is quantitatively reduced by lithium aluminum hydride as indicated by the following equation:

$$4\ RCOOR' + 2\ LiAlH_4 \longrightarrow LiAl(OCH_2R)_4 + LiAl(OR')_4 \qquad (24)$$

Thus an ester can be determined by titration with a standardized lithium aluminum hydride solution in tetrahydrofuran[50]. The end-point is located potentiometrically, or visually with p-aminoazobenzene as indicator. Alternatively, the determination can be carried out in the gasometric apparatus shown in Figure 2, using a procedure similar to that described in section H. The excess of lithium aluminum hydride is measured by adding ethanol which liberates hydrogen gas on reacting with the residual reagent.

J. Miscellaneous Methods

Gravimetric determinations of certain carboxylic acids based on the precipitation of their metal salts have been reported[51-54]. Carboxylic acids which are stronger than hydrogen sulfide can be determined by the displacement reaction of potassium hydrosulfide[55]. Most carboxylic acids are sufficiently strong to induce the iodide–iodate reaction as shown in equation (25). The reaction rate, however, is considerably slower than that

$$6\ RCOOH + KIO_3 + 5\ KI \longrightarrow 3\ I_2 + 6\ RCOOK + 3\ H_2O \qquad (25)$$

for mineral acids. By incorporation of an excess known amount of thio-sulfate in the mixture, the reaction is brought to completion; the residual thiosulfate is then determined iodometrically[56].

Esters derived from lower alcohols may be determined by measuring their alkoxyl contents. The method is based on cleavage of the ester func-tion by hydrogen iodide, as shown in equation (26). Methyl and ethyl esters are conveniently analyzed in the alkoxyl apparatus[57, 58,].

$$RCOOR' + HI \longrightarrow RCOOH + R'I \tag{26}$$

If the parent alcohol contains three or more carbon atoms, gas chromato-graphic determination of the alkyl iodide is recommended[59].

While the carboxyl and the ester functions as such cannot be determined polarographically, attempts have been made to convert them into deriva-tives that are reducible at the dropping mercury electrode. Thus, aliphatic esters are converted to the corresponding hydroxamic acids, and the hydroxamic acid wave from the equilibrium mixture of the sample with alkaline hydroxylamine has been found to be reproducible and proportio-nal to the ester concentration[60]. This method can be adapted to the analysis of carboxylic acids through their methyl esters (see section E). Another possible route for determining the carboxyl function is to convert it to the acid chloride and then to aldehyde, the latter being easily determined in the polarograph[61].

IV. METHODS FOR THE SEPARATION OF CARBOXYLIC ACIDS AND ESTERS

Because carboxylic acids and esters encountered in natural products and industrial materials are usually mixtures, analytical separation is frequ-ently an essential step in the process of identification or determination of a sample. Prior to the development of chromatographic methods, distilla-tion was practically the only technique for separation. For instance, the Reichert–Meisel values in the analysis of fats and oils involve the distilla-tion of the fatty acids produced from the sample under specific conditions. As the gas chromatograph becomes a common tool in the analytical laboratory, it is the most popular instrument for the separation of acids and esters. On the other hand, thin-layer and paper chromatography are also used because these techniques are simple and economical. Other methods of separation are applicable to certain types of mixtures.

A. Separation by Gas Chromatography

Since James and Martin[62] reported the separation of aliphatic carboxylic acids by gas chromatography in 1952, numerous workers have published papers on this subject. Volatile acids can be injected into the gas chromatograph directly. A typical gas chromatogram is shown in Figure 8, which

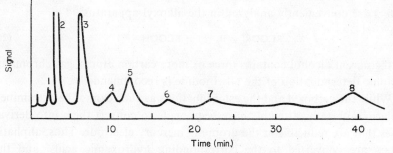

FIGURE 8. Gas chromatograph of a mixture of fatty acids; 1. formic; 2. acetic; 3. propionic; 4. isobutyric; 5. *n*-butyric; 6. an isovaleric; 7. *n*-valeric; 8. *n*-caproic. (From reference 152)

represents the separation of a free fatty acid mixture isolated from a biological sample. Non-volatile acids are first converted to a suitable ester or other volatile derivative. Column packings that have been recommended recently include Carbowax–terephthalic acid supported on Chromosorb[63], diethylene glycol adipate–phosphoric acid on Gas-chrom[64], sebacic acid on Chromosorb[65], isophthalic acid–Carbowax on glass beads[66], and dilauroyl-3,3'-thiodipropionate on stannic phosphate[67].

Methyl esters are the preferred derivatives for high-boiling and non-volatile carboxylic acids. They are conveniently prepared by the diazomethane reaction[68, 69] (see section III. E, Figure 7), and by esterification of the acids with absolute methanol in the presence of boron trifluoride[70, 71], sulfuric acid[72], or thionyl chloride[73]. On the other hand, carboxylic acids of low molecular weights have been converted to the corresponding octyl esters before gas chromatographic separation[74]. Columns containing Apiezon on Chromosorb and diethylene glycol succinate polyester on Chromosorb are recommended for the separation of alkyl esters[75, 76]. Benzenecarboxylic acids and solid aliphatic acids have been converted to their trimethylsilyl derivatives and separated on silanized Chromosorb[77, 78].

The response of a flame ionization detector to aliphatic carboxylic acids and esters has been investigated. There is good correlation between concentration and peak area for long-chain methyl esters[79]. With free fatty acids containing six or more carbon atoms, the response to the carboxyl carbon atom approaches that to a normal carbon atom on a relative weight basis[80]. Collaborative study on the quantitative analysis of volatile fatty acids has shown that the gas chromatographic method gives results as accurate as those afforded by the A.O.A.C. official method[81, 82].

B. Separation by Thin-layer Chromatography

Separation of mixtures by differential migration on a thin bed of inert material has been known for a long time. However, this simple technique did not receive much attention in the analytical laboratory until Stahl popularized it by publishing several articles about a decade ago. Its applications to the analysis of esters and acids were immediately recognized[83-85]. Various adsorbents and solvent systems are given in the monographs[86, 87].

Mono-, di-, and triglycerides of higher fatty acids have been separated on Floridin by development with heptane–dioxan or carbon tetrachloride–dioxan[88]. Saturated glycerides are best separated at 30–40°C while unsaturated esters are separated at 20°C. The separated esters can be quantitatively eluted from the thin-layer plate by means of ethanol or ethanol–chloroform mixed solvent.

While the early papers were mostly concerned with separation of esters[89-91], successful attempts at acid separation by thin-layer chromatography have been reported. Long-chain fatty acids (from lauric to stearic) are separated on Kieselguhr with 8% liquid paraffin in benzene as the stationary phase and acetic acid–liquid paraffin–water (8 : 1 : 1) as the mobile phase[92]. The acids are detected by thymolphthalein or pyrogallol red. Benzenecarboxylic acids are separated on silica gel by using various proportions of ethanol–aqueous ammonia–water as the mobile phase[93]. Separation of the isomeric di-, tri-, and tetracarboxylic acids requires an ammoniacal mobile phase (5 : 3 : 1); the penta- and hexacarboxylic acids respond to little or no ammonia and more water (3 : 1 : 7) in the mobile phase. Dicarboxylic and hydroxy carboxylic acids can be separated on cellulose with either ethyl acetate–formic acid–water (3 : 1 : 1) or butanol–formic acid–water (6 : 1 : 2), and the spots are located by methanolic solution of bromophenol blue and methyl red[94]. Several other satisfactory combinations have also been reported[95]; these include (i) polyoxyethylene glycol–Kieselguhr as adsorbent and ispropyl ether–formic acid–water (90 : 7 : 3) as developer, (ii) polyamide as adsorbent and isopropyl ether–ligroin–carbon tetrachloride–formic acid–water (50 : 20 : 20 : 8 : 1) as developer, (iii) polyamide as adsorbent and acetonitrile–ethyl acetate–formic acid (9 : 1 : 1) as developer, (iv) polyamide as adsorbent and butyl formate–ethyl acetate–formic acid (9 : 1 : 1) as developer, and (v) Kieselgel as adsorbent and isopropyl ether–formic acid–water (90 : 7 : 3) as developer.

Quantitative thin-layer chromatography of acids has been carried out by combining it with the labelled derivative technique[96]. The procedure is as follows: To the acid mixture (10 mg) dissolved in ether is added 0·2 ml of tritiated water solution. After allowing 15 min for exchange, the tritiated acids are quantitatively esterified with diazomethane in ether. The resulting mixture is applied to a layer of silica gel, and developed with benzene–ligroin (7 : 3) or 2,2,4-trimethylpentane–ether (4 : 1). The plate is then sprayed with ethanolic 'Ultraphor' (or with ethanolic 2,7-dichlorofluorescein if the silica gel is impregnated with aqueous silver nitrate). The zones are located under u.v. light and extracted for 4 h with ether–hexane (7 : 3). The extracts are evaporated. After the addition of liquid scintillator, the respective components are counted. Recoveries better than 95% are reported.

C. Separation by Paper Chromatography

It is interesting to recall how paper chromatography revolutionized the technique for separating amino acids[97]. Many reviews and thousands of publications have since appeared that are related to paper chromatographic separation of compounds containing the carboxyl and ester functions[98]. Both one- and two-dimensional methods are employed[99, 100]. Separation of fatty acids (C_1 to C_{16}) as their 4-[(4-dimethylamino) phenylazo]-phenacyl esters has been reported recently[101]. The advantages of this method are its rapidity, sensitivity, and simplicity. Because colored derivatives of the acids are used, the spots are identified without the aid of a spray reagent. The mixture of these esters is separated on paper impregnated with dimethylformamide, using various combinations of toluene and petroleum as the developer. Aliphatic dicarboxylic acids are separated on paper impregnated with methylcellulose, while their esters require prior conversion into the corresponding hydroxamic acid before analysis; the developing solvents comprise ethyl acetate, dioxan, and water[102]. When acid developing solvents (e.g. formic acid–isopentyl formate–water) are used for the separation of dicarboxylic acids, the R_f values for acids having the same number of carbon atoms increase rectilinearly with decrease in the number of carbon atoms between the carboxyl functions. A nomogram has been prepared from which this number can be derived from the empirical formula of the acid and its R_f value[103].

Quantitative paper chromatography of fatty acids has been critically evaluated[104]. A simple method for the estimation of fruit acids on the paper chromatogram consists of spraying both sides of the paper with methyl red indicator solution and measuring the absorbance of the spots[105]. Determination of dicarboxylic acids has been carried out by titrimetry as follows[106]. The sample is placed on paper strips simultaneously with a standard solution of the mixture of acids. After development with the upper layer of benzene–acetone–formic acid–water (10 : 5 : 2 : 6), the strips are dried and formic acid is removed by steaming. The standard acids are located by dipping the chromatogram in a solution of bromocresol red. From the test chromatograms, portions are cut out corresponding to the separated acids. These are extracted with water, and titrated with 0·01 N sodium hydroxide.

D. Separation by Column Chromatography

Chromatographic separation in a column, based on the principle of partition or adsorption, was much in vogue for several decades. As other methods became available, however, this analytical tool has lost some of its favor. The difficulties of using the column for separating carboxylic acids and esters are: (i) it requires special precaution in packing; (ii) equilibrium of the system is attained slowly; (iii) the separation is not readily discernible since the carboxyl and ester functions are not chromophoric groups.

The different solvent systems suitable for separation of acids and esters have been reviewed[107, 108]. In general, acids are separated by their ionization values. Thus, when the concentration of butanol in the butanol–chloroform mixture used as the mobile phase is gradually increased, the effluent contains the carboxylic acids in the order of increasing acid strength[109]. Alumina is the common packing material in the adsorption column, whereas either cellulose powder or silica gel is used for partition chromatography[110].

In the separation of carboxylic acids by partition, water or an aqueous solution is always used as the stationary liquid phase. Aliphatic acids of low molecular weights are separated by the following procedure[111]. The column is packed with wet silica gel impregnated with bromocresol green. The mixture of acids is dissolved in chloroform, transferred to the column, and the latter is eluted with chloroform–butanol. The displaced fractions are titrated with standardized alkali solution. The separation is not clean, however, and the position of zones in the column varies from sample to sample, probably dependent on the concentration of acids in the mixture.

The separation of fatty acids[112], steroid and terpenoid acids[113], and dicarboxylic acids from C_3 to C_{13} have been reported[114, 115]. Benzenecarboxylic acids are separated in a silica gel column saturated with 35–40% of water[116]. Sulfuric acid is placed on top of the packing before the sample containing the carboxylic acids is added. Using a flow rate of 3·4 ml per min, the butanol content of the butanol–chloroform mobile phase is gradually increased from 0 to 60%. The eluate is continuously monitored by conductivity measurements to indicate the emergence of the mono- to hexacarboxylic acids.

Methyl esters of linoleic and linolenic acids can be separated by adsorption on a silicic acid column[117]. The separation is improved by the use of solutes of intermediate adsorption affinities when the methyl esters of four C_{18} acids (stearic, oleic, linoleic, and linolenic) are present in the sample[118].

E. Separation by Ion-exchange Resins

It is apparent that the technique of ion exchange is applicable to the carboxyl function and not to the ester function. Carboxylic acids that are contaminated with mineral acids and/or sulfonic acids can be separated from the latter two groups by the use of a strongly acidic resin. On the other hand, a strong base–anion exchange resin can be used for the isolation of carboxylic acids from a mixture containing other classes of organic compounds[119].

Since most carboxylic acids are weak acids within a narrow range of pH values, the separation of carboxylic acid mixtures requires careful selection of the ion-exchange resin and rigid control of experimental conditions. Fortunately, a large variety of resins is available for testing, and tailor-made resins can be synthesized. The operation of the ion-exchange column can also be controlled by appropriate mechanisms. There are many examples of successful separation of carboxylic acids with ion-exchange resins[120], the most spectacular being the system for amino acids from protein hydrolyzates[121]. Figure 9 shows the sequence of the appearance of the components in a solution containing 17 amino acids.

The separation of carboxylic acids by ion-exchange resins is dependent on the type of resin, particle size, nature of the eluent, temperature, and rate of elution[122, 123]. Thus, the separation of the amino acids shown in Figure 9 requires continuous variation of pH and temperature. Nevertheless, after the ion-exchange column has been properly adjusted and the operating conditions established, the system can be depended upon to yield reproducible results. This is the basis of automatic amino acid analyzers[124].

The adsorption of acids on cationite (Pb^{2+} and Ca^{2+} forms) and by anionites (OH^- form) has been investigated under dynamic conditions using 10 carboxylic acids as test compounds[125]. The Pb^{2+} cationite adsorbs oxalic, tartaric, citric, and malonic acids and Ca^{2+} cationite absorbs oxalic and citric acids, while C_1–C_4 fatty acids, as well as maleic and succinic acids are not adsorbed. These differences form the basis of a method for separating organic acids.

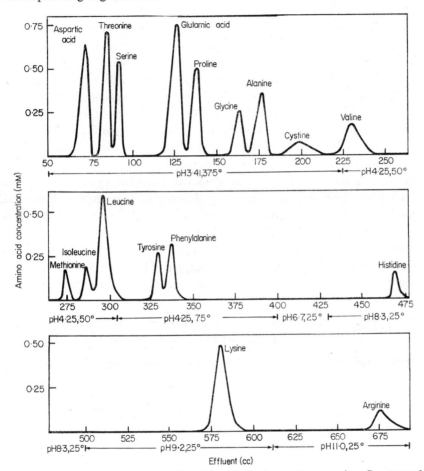

FIGURE 9. Separation of mixture of amino acids by ion-exchange resins. Courtesy of *J. Biol. Chem.*

A scheme for automatic ion exclusion partition chromatography of acids has been devised[126]. The acids are separated on a 100 cm column of sulfonated polystyrene resin (H^+ form) with water as eluent. The eluate is titrated automatically to provide a volume versus time graph from which the concentration and possible identity of each acid can be integrated.

An analysis requires about 1 h, and no pretreatment or regeneration of the column is necessary. Low concentrations of water-soluble carboxylic acids can be determined in the presence of a large excess of mineral acids by by-passing the latter from the titration cell.

F. Separation by Distillation

Distillation is occasionally used for the analysis of carboxylic acids and esters. For example, the Duclaux constant, which is based on steam distillation[127, 128], is still recommended for the separation and identification of C_1–C_6 fatty acids in a current laboratory manual[3]. The method is as follows: The carboxylic acid (10 g) is dissolved in 150 ml of distilled water. Using 0·1 N sodium hydroxide as the titrant and phenolphthalein as the indicator, the titre for 10 ml of the acid solution is recorded. Then 100 ml of the remaining solution is transferred into a 250 ml distilling flask which is closed with a rubber stopper without a thermometer. The flask is connected to a short water-cooled condenser by means of a rubber stopper in such a way that the side arm of the distilling flask extends well into the narrow portion of the condenser tube. The distillation is carried out in such a manner that drops come from the end of the condenser at a constant rate. The distillate is collected in a 10 ml graduate cylinder and three 10 ml fractions (A, B, C) are collected. The respective fractions are titrated with the 0·1 N sodium hydroxide mentioned above. The Duclaux number is then calculated by the formula:

$$\text{Duclaux number} = \frac{(\text{ml of } 0\cdot1 \text{ N NaOH for 10 ml fraction}) \times 100}{(\text{ml of } 0\cdot1 \text{ N NaOH for original 10 ml}) \times 10}$$

and the results are compared with the known values given in Table 3.

TABLE 3. Duclaux numbers[a]

Acid	Fraction		
	A	B	C
Formic	3·95	4·40	4·55
Acetic	6·8	7·1	7·4
Propionic	11·9	11·7	11·3
n-Butyric	17·9	15·9	14·6
Isobutyric	25·0	20·9	16·0
n-Valeric	24·5	20·6	17·0
Isovaleric	28·7	23·1	16·8
n-Caproic	33·0	24·0	19·0

[a] Data taken from R. L. Shriner, R. C. Fuson and D. Y. Curtin, *The Systematic Identification of Organic Compounds*, 5th ed., John Wiley and Sons, New York, 1964.

Fractional distillation can be used to separate volatile acids and esters[129]. Previously, the analytical separation of amino acids was dependent on the fractionation of their ethyl esters by distillation. Since fractional distillation is time-consuming and not precise, it is not used unless the other techniques are not applicable to the analysis of the sample in question. Analytical distillation requires relatively large amounts of the working material. With the development of the spinning band and similar precision fractionating columns, 1 g or less of liquid mixtures can be handled, but this quantity is still many times larger than that required in other methods.

G. Micellaneous Methods

Carboxylic acids are usually separated from other organic materials by extraction with aqueous sodium hydroxide or carbonate. The free acids can be extracted from the aqueous solution by means of a suitable organic solvent. Whereas the common procedures use ethyl or isopropyl ether, probably the most effective agent is tributyl phosphate[130]. Esters are separated from aqueous media in a similar manner.

Mixtures of acids and esters, respectively, can be separated by partition between two liquid phases. The optimum conditions should be ascertained before the method is used for analytical purposes. For complicated systems the countercurrent distribution technique is recommended. Thus, a mixture of 300 mg each of four higher fatty acids is quantitatively separated in the 220-tube apparatus[131].

Precipitation and fractional recrystallization may be used in special cases. One of the industrial processes for separating fatty acids involves emulsification with surface-active agents[132], but this principle has not been utilized for analytical separations.

V. SPECTROSCOPIC METHODS FOR THE ANALYSIS OF CARBOXYLIC ACIDS AND ESTERS

Great strides have been made in recent years in the application of spectroscopy to the solution of analytical problems related to the carboxyl and ester functions. With the proliferation of commercial spectroscopic instruments, the spectrometers have become readily available analytical tools. In these physical methods, the sample can be recovered after analysis, albeit with some loss due to transfer and other operations.

Spectroscopic methods are extremely useful for the detection of the carboxyl and ester functions and the identification of compounds containing these functions. Quantitative analysis, however, cannot be performed without known samples of the authentic compounds. Whereas 1 mequiv of an unknown carboxylic acid present in a mixture generates 1 mmole of methane if the determination is carried out with the chemical method using the Grignard reagent (see section III.B), there is no way to obtain this quantitative information from the spectroscopic data. Not only is it impossible to assign an absolute value to the absorbance due to 1 mequiv of certain grouping, but the wavelength of maximum absorption also varies from one compound to another. As an example, the divergence of the spectroscopic values based on the measurement of the infrared absorption of the $C=O$ stretching of a number of carboxylic esters is shown in Table 4.

TABLE 4. Effect of Structure on infrared absorption of carbonyl groups in esters[a]

Ester	Frequency (cm^{-1})	Molar extinction coefficient
Isoamyl acetate	1743	610
Propyl propionate[b]	1740	553
Ethyl butyrate[b]	1738	600
Ethyl palmitate	1738	569
Ethyl phenylacetate[b]	1740	537
Ethyl α-bromobutyrate	1744	504
Ethyl trichloroacetate	1770	720
Methyl methacrylate	1727	672
Methyl acrylate	1735	584
2-Ethylhexyl acrylate[b]	1728	634
Diallyl fumarate	1730	576[c]
Diallyl maleate	1738	433[c]
Dibutyl oxalate[b]	1746	459[c]
Diethyl adipate[b]	1739	586[c]
Dibutyl phthalate	1732	523[c]
Butyl benzoate[b]	1723	767
Benzyl benzoate[b]	1725	713
Ethyl cinnamate[b]	1717	697
Methyl salicylate[b]	1684	673

[a] Data taken from R. T. Hall and W. E. Shaefer. In *Organic Analysis*, Vol. 2, Interscience, New York, 1954, p. 60.

[b] Saponification equivalent indicates 100% purity within accuracy of analysis. These samples were also found to contain negligible amounts (<0·25%) of free acid. Chemical analyses were not made on other samples.

[c] For polycarboxylic esters equivalent weight, rather than molecular weight, was used in calculating extinction coefficients.

A. Ultraviolet and Visible Spectroscopy

The carboxyl and ester functions do not absorb in the ultraviolet and visible regions of the spectrum. Therefore, if a compound is indicated by chemical tests to contain either one of these two functions and if its solution is colored or shows absorption in the ultraviolet spectrophotometer, there is clear evidence of the presence of other functional groups such as

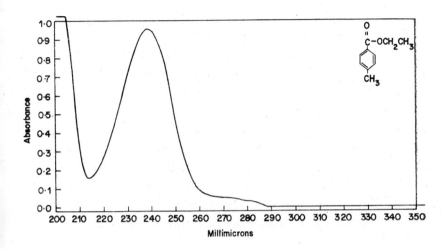

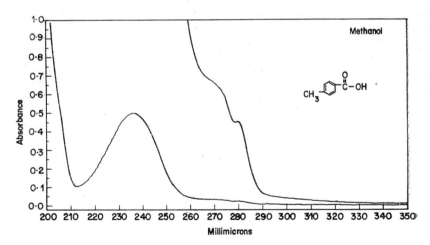

FIGURE 10. Ultraviolet spectra of *p*-toluic acid and its ethyl ester. Courtesy Sadtler Research Laboratories, Inc.

unsaturation and benzenoid structure. Figure 10 shows the ultraviolet spectra of *p*-toluic acid and ethyl *p*-toluate respectively. It is apparent that the absorption is due to the benzene nucleus and not the carboxyl or ester function.

If it is desired to analyze the carboxyl and ester functions in the visible spectrum, they can be converted to derivatives which possess strong chromophoric groups. For instance, fatty acids may be esterified to form colored products[101]; esters are converted to ferric hydroxamates (see section II.B).

B. Infrared Spectroscopy

A large number of infrared spectra have been compiled[133]. Among the 35,000 compounds covered in 1967, approximately 3500 contain the carboxyl function and 5000 possess the ester function. Identification of these two functions in the infrared spectra presents no difficulty[134, 135]. Their principal absorption bands are listed in Tables 5 and 6 respectively[136].

TABLE 5. Infrared absorption bands of the carboxyl function[a]

Atomgrouping	Frequency (cm^{-1})	Wavelength (mμ)	Intensity	Remarks
O—H				
Free OH	3550–3500	2·82– 2·86	m.	O—H stretching
Bonded OH	3300–2500	3·00– 4·00	w.	Broad band, O—H stretching
All OH	955– 890	10·47–11·24	v.	o.o.p. def.
C=O Stretching				
Saturated aliphatic acids	1725–1700	5·80–5·88	s.	
α, β-Unsaturated acids	1715–1680	5·83–5·95	s.	All acids examined as dimers in solid or liquid phase
Aryl acids	1700–1680	5·88–5·95	s.	
Intramolecular H bonded acids	1680–1650	5·95–6·06	s.	
α-Halo acids	1740–1715	5·75–5·83	s.	
Others				
Solid fatty acids	1350–1180	7·40–8·48	w.	CH$_2$ vibrations, characteristic band patterns
CO$_2$H	1440–1395	6·94–7·17	w.	Combination band of C—O stretching and OH i.p. def.
	1320–1210	7·58–8·26	s.	
Carboxylate ion CO$_2^-$	1610–1550	6·21–6·45	s.	Asymmetric stretching
	1420–1300	7·04–7·69	m.	Symmetric stretching

[a] Data taken from A. D. Cross, *Practical Infrared Spectroscopy*, 2nd ed., Butterworths, Washington, 1964.
[b] m = medium; s = strong; v = variable; w = weak

TABLE 6. Infrared absorption bands of the ester function[a]

Atomgrouping	Frequency (cm^{-1})	Wavelength (μ)	Intensity[b]	Remarks
	C=O Stretching			
Saturated aliphatic esters	1750–1735	5·71–5·76	s.	—
α, β Unsaturated and aryl esters	1730–1715	5·78–5·83	s.	—
Vinylic and phenolic esters	1800–1770	5·56–5·65	s.	—
α-Keto esters and α-diesters	1755–1740	5·70–5·75	s.	—
Enolic β-keto esters	1655–1635	6·04–6·12	s.	Chelation
o-Hydroxy (amino) benzoates, etc.	1690–1670	5·92–5·99	s.	Chelation
γ-Keto esters, non-enolic β-keto esters, and γ- (and higher) diesters	1750–1735	5·71–5·76	s.	—
	C—O Stretching			
Formates	1200–1180	8·33–8·48	s.	—
Acetates	1250–1230	8·00–8·13	s.	—
Vinylic and phenolic acetates	1220–1200	8·20–8·33	s.	—
Propionates and higher esters	1200–1170	8·33–8·55	s.	—
Esters of α, β unsaturated aliphatic acids	1310–1250	7·63–8·00	s.	—
	1180–1130	8·48–8·85	s.	—
Esters of aromatic acids	1300–1250	7·69–8·00	s.	—
	1150–1100	8·70–9·09	s.	—

[a] Data taken from A. D. Cross, *Practical Infrared Spectroscopy*, 2nd ed., Butterworths, Washington, 1964.
[b] s = strong

The carboxyl function exhibits very characteristic absorption at 3000–2500 cm^{-1} which comprises a group of small bands. The band at the highest frequency is due to OH. The C=O band is considerably stronger than that in aldehydes and ketones; it appears at 1760 cm^{-1} for the monomeric carboxyl and at 1710 for the dimer. It should be noted that most data on carboxylic acids are for the dimers in view of the strong hydrogen bonding of the carboxyl function, even in the gaseous state. Thus, the absorption at 1420 cm^{-1} and 1300–1200 cm^{-1} (both due to coupling between in-plane OH bending and CO stretching), and at 920 cm^{-1} (due to OH out-of-plane bending) are ascribed to the dimer. When an acid is converted to its salt with a metal ion or triethylamine, the carboxylate ion shows absorption bands at 1610–1550 cm^{-1} and at 1400 cm^{-1} respectively[134].

The ester function has an absorption band at 1735 cm^{-1}, the intensity of which is between those of the ketonic and carboxyl functions. It also

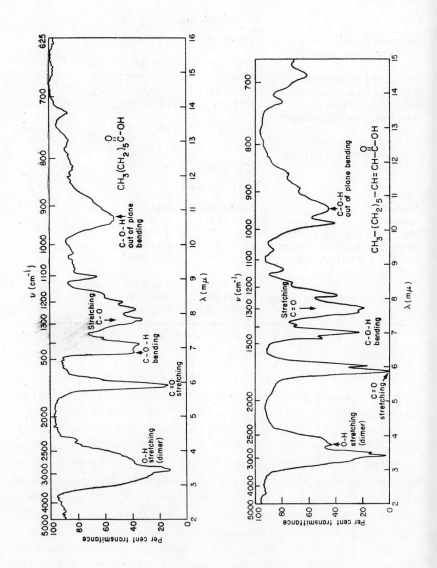

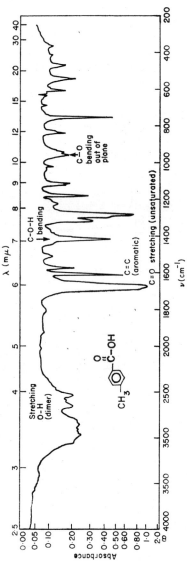

FIGURE 11. Infrared absorption spectra of carboxylic acids. Courtesy Sadtler Research Laboratories, Inc.

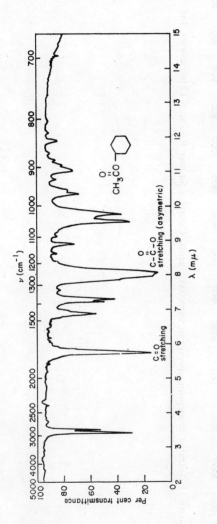

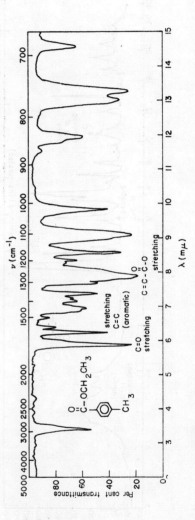

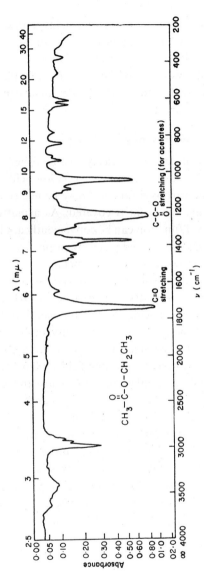

FIGURE 12. Infrared absorption spectra of esters. Courtesy Sadtler Research Laboratories, Inc.

has two bands at 1300–1050 cm^{-1}, due to asymmetrical and symmetrical stretching of the ester C—O—C. The asymmetrical stretching ('ester band') is usually stronger than the C=O band; it is fairly constant for the particular type of ester[137].

In Figure 11 are shown the spectra of three carboxylic acids representing straight-chain saturated and unsaturated aliphatic acids and benzene-carboxylic acid with an alkyl substituent. The spectra of three carboxylic esters are given in Figure 12; it shows the ethyl ester of an aliphatic acid, the cyclohexyl ester of the same acid, and an ethyl ester of a benzene-carboxylic acid with an alkyl substituent.

C. Nuclear Magnetic Resonance Spectroscopy

In the n.m.r. spectra, carboxyl hydrogen is easily distinguished from the other protons in the molecule because it appears at a very low field[138, 139]. However, for the same reason, the carboxyl function may be neglected in the n.m.r. spectra unless the whole spectrum is covered. As a matter of fact, since the proton of the carboxyl function can be conveniently analyzed by very simple procedures, its detection and determination are rarely carried out in the nuclear magnetic resonance spectrometer.

Understandably, a significant feature of the carboxyl function is its ease of proton exchange. The chemical shifts of the carboxyl OH group accompanying solvent dilution can provide useful information[140, 141]. Another aspect of the carboxyl function is the possibility of hydrogen bonding. In the case of acetic acid, the strongest hydrogen bonding occurs in the dimer, since the dimer chemical shift is to low field by 0·65 p.p.m.

The proton resonance spectra of a number of fatty acids and esters (glycerides) have been reported[142, 143]. It is shown that the signal at −3·0 p.p.m. on the n.m.r. spectra of glycerides comes from the two OCH$_2$ groups and represents four protons. Aliphatic carboxylic acids with branched alkyl groups in the α-position have been analyzed by nuclear magnetic resonance spectrometry[144]. Three types of structures are investigated, using tetramethylsilane as the internal standard. αα-Dialkyl acids (αα-dimethyl and α-methyl-α-alkyl) give no distinguishing peaks, whereas α-methyl acids do. The n.m.r. spectra of typical aliphatic and aromatic acids are shown in Figure 13.

D. Mass Spectroscopy

The fragmentation of the carboxyl function in the mass spectrometer has been throughly investigated and its identification on the mass spectra presents no special problem[145, 146]. Many mass spectra of aliphatic and aroma-

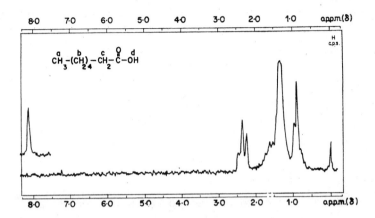

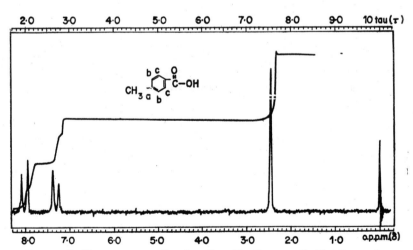

FIGURE 13. N.m.r. spectra of carboxylic acids. Courtesy Sadtler Research
Laboratories, Inc.

tic acids are known[147-149]. The molecular weight of a carboxylic acid is twice that of the parent hydrocarbon whose peaks are generally large enough to be identified. Large peaks at masses 31, 45, and 59 indicate the presence of oxygen. The peak at mass 45 is due to $(CO_2H)^+$ and always appears stronger than the peaks at masses 31 and 59. Prominent peaks occur at mass 17 and at the associated masses 16 and 18.

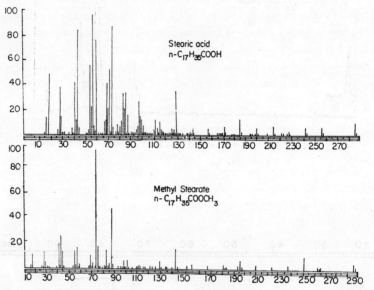

FIGURE 14. The mass spectra of stearic acid and methyl stearate. Courtesy of *Appl. Spectr.*

For aliphatic carboxylic acids, the most characteristic rearrangement ion occurs at mass 60, which is the molecular weight of acetic acid. This may form the base peak of certain fatty acids. In the spectrum of stearic acid (Figure 14), the peak at mass 60 has a height of almost 80% of the base peak. On the other hand, the presence of the benzene nucleus in aromatic carboxylic acids leads to a much increased parent peak. This peak, together with those formed from the parent ion by loss of OH and COOH, are usually the strongest in the mass spectrum of an aromatic acid.

As a rule, di- and polycarboxylic acids are thermally decomposed if they are heated to temperatures at which their vapor pressures are sufficient to plot a mass spectrum. For this reason, such compounds are generally converted to their methyl or ethyl esters before examination by mass spectrometry[150]. When two carboxyl functions are in adjacent positions in the aromatic ring, the thermal degradation product is shown to be the anhydride of the parent dicarboxylic acid[149].

Methyl carboxylates can be identified by the peak at mass 31 due to $(CH_3O)^+$, and at mass 59 due to $(CH_3OCO)^+$. Methyl esters break with the minimum of rearrangements, whereas ethyl esters sometimes lead to difficulties due to their loss of C_2H_4 which has the same mass as CO. The parent peaks from aliphatic esters are usually small[151]; aromatic esters on the other hand show clear parent peaks on their mass spectra[149]. The base peak in all cases corresponds to loss of OCH_3, and the parent ion decreases as the number of ways in which this ion can be formed becomes larger. The peak corresponding to fragmentation at the other side of the CO group to give the (p – 59) peak is of about the same intensity as the parent ion except when two ester functions are *ortho* to each other.

VI. ACKNOWLEDGEMENTS

The author thanks Dr. Robert C. Gore of the Perkin–Elmer Corporation for advice on infrared spectroscopy; Dr. F. E. Critchfield and Dr. J. B. Johnson of Union Carbide Chemical Co., for information on industrial analysis; and Sadtler Research Laboratories, Inc., for the u.v., i.r., and n.m.r. spectra.

VII. REFERENCES

1. J. Needham, *Science and Civilization in China*, Cambridge University Press, Cambridge, 1954.
2. N. D. Cheronis and T. S. Ma, *Organic Functional Group Analysis*, John Wiley and Sons, New York, 1964, p. 7.
3. R. L. Shriner, R. C. Fuson and D. Y. Curtin, *The Systematic Identification of Organic Compounds*, John Wiley and Sons New York, 1964, p. 68.
4. T. S. Ma, *J. Chinese Chem. Soc., Ser. II.*, **9**, 176 (1962).
5. N. D. Cheronis, J. B. Entrikin and E. M. Hodnett, *Semimicro Qualitative Organic Analysis*, John Wiley and Sons, New York, 1965.
6. S. Veibel, *The Identification of Organic Compounds*, 6th ed., G.E.C. Gad Publisher, Copenhagen, 1966.
7. T. S. Ma and R. Roper, *Microchem. J.*, **1**, 246 (1957).
8. F. E. Merliss and A. J. Weinheimer, unpublished paper given at the Southwest Regional Meeting of the ACS. Dallas, Texas. December, 1962; see ref. 5, p. 363.
9. G. Aksnes, *Acta Chem. Scand.*, **11**, 719 (1957).
10. M. Frankel and S. Patai, *Tables for Identification of Organic Compounds*, Chemical Rubber Publishing Co., Cleveland, 1960, p. 109.
11. A. Jart, A. J. Bigler and V. Bitsch, *Anal. Chim. Acta*, **31**, 472 (1964).
12. E. Chambers and G. W. Watt, *J. Org. Chem.*, **6**, 376 (1941).
13. A. Friediger and C. Pedersen, *Acta Chem. Scand.*, **9**, 1425 (1955).
14. A. Jart, *Acta Polytech. Scand. Chem. Met.* Ser., No. 2, 1963.
15. H. Morita and N. M. Miles, *Anal. Chem.*, **28**, 1081 (1956).

16. R. Roper and T. S. Ma, *Mikrochim. Acta*, 212 (1968).
17. R. W. Bost and L. V. Mullen, Jr., *J. Am. Chem. Soc.*, **73**, 1967 (1951).
18. W. B. Renfrow and A. Chaney, *J. Am. Chem. Soc.*, **68**, 150 (1946).
19. F. Feigl, *Spot Tests in Organic Analysis*, 6th ed., Elsevier Publishing Co., Amsterdam, 1965.
20. J. Mitchell, Jr., B. A. Montague and R. H. Kinsey, in *Organic Analysis*, Vol. 3, Interscience Publishers, New York, 1956, p. 1.
21. R. T. Hall and W. E. Shaffer, in *Organic Analysis* (Eds. J. Mitchell, I. A. Kolthoff, E. S. Proskauer, A. Weissberger), Vol. 2, Interscience Publishers, New York, 1954. p. 19.
22. N. D. Cheronis and T. S. Ma, *Organic Functional Group Analysis*, John Wiley and Sons, New York, 1964, p. 483.
23. R. B. Sandin, M. Kulka and D. W. Wooley, *Ind. Eng. Chem., Anal. Ed.*, **8**, 355 (1936).
24. J. W. Fritz, *Acid–Base Titrations in Nonaqueous Solvents*, G. F. Smith Chemical Co., Columbus, 1952, p. 28.
25. R. D. Tiwari and J. P. Sharma, *Z. Anal. Chem.*, **195**, 267 (1963).
26. C. Pries and C. J. F. Bottcher, *Anal. Chim. Acta.*, **31**, 293 (1964).
27. T. R. Williams and M. Lautenschlager, *Talenta*, **10**, 804 (1963).
28. L. P. Kulev and P. V. Kristalov, *Tr. Komis. po Analit. Khim. Akad. Nauk SSSR, Inst. Geokhim. i Analit. Khim.* **11**, 457 (1960).
29. P. Fijolka and I. Lanz, *Plaste Kautschuk*, **7**, 169 (1960).
30. S. Siggia, *Quantitative Organic Analysis Via Functional Groups*, 3rd ed., John Wiley and Sons, New York, 1963, p. 131.
31. American Society for Testing Materials, 'Petroleum Products', *A.S.T.M. Standards* Part 5. D 664-52, Philadelphia, 1952, p. 253.
32. V. A. Zarinskif and I. A. Gurev, *Zh. Analit. Khim.* **19**, 1429 (1964).
33. A. P. Kreshkov, L. M. Bykova and O. F. Kirilova, *Zh. Analit. Khim.*, **20**, 840 (1965).
34. J. E. Dubois and P. C. Lacaze, *Anal. Chim. Acta*, **33**, 602 (1965).
35. M. L. Moss, J. H. Elliott and R. T. Hall, *Anal. Chem.*, **20**, 748 (1948).
36. See ref. 2, p. 562 for detailed discussion of reagents.
37. See ref. 2, p. 565 for detailed description of the apparatus.
38. M. Soucek, *Coll. Czechoslav. Chem. Comm.*, **23**, 554 (1958).
39. M. H. Hubacher, *Anal. Chem.*, **21**, 945 (1949).
40. T. S. Ma, C. T. Shang and E. Manche, *Mikrochim. Acta*, 572 (1964).
41. L. Maros, I. Perl, M. Szakács and E. Schulek, *Acta Chim. Acad. Sci. Hung.*, **40**, 379 (1964).
42. See ref. 2, p. 613 for detailed description of the reagents.
43. T. S. Ma and R. T. E. Schenck, *Mikrochemie*, **40**, 245 (1953).
44. F. E. Critchfied, *Organic Functional Group Analysis*, Pergamon Press, Oxford, 1963.
45. W. Rieman, III, *Ind. Eng. Chem., Anal. Ed.*, **15**, 325 (1943).
46. K. Marcali and W. Rieman, III, *Ind. Eng. Chem. Anal. Ed.*, **18**, 144 (1946).
47. N. D. Cheronis and T. S. Ma, *Organic Functional Group Analysis*, John Wiley and Sons, New York, 1964, p. 513.
48. F. E. Critchfield and J. B. Johnson, private communication.
49. N. D. Cheronis and T. S. Ma, *Organic Functional Group Analysis*, John Wiley and Sons, New York, 1964, pp. 95, 511.
50. T. Higuchi, C. J. Lintner and R. H. Schlief, *Science*, **111**, 63 (1950).
51. H. Roth, in Methoden der organischen Chemie, Vol. 2 (Ed. Houben–Weyl–Müller), Thieme, Stuttgart, 1953, p. 502.
52. G. J. Goepfert, *Biochem. J.*, **34**, 1012 (1940).

53. G. A. Dalin and J. N. Haimsohn, *Anal. Chem.*, **20**, 740 (1948).
54. M. H. Swann, *Anal. Chem.*, **21**, 1448 (1949).
55. W. H. Hunter and J. D. Edwards, *J. Am. Chem. Soc.*, **35**, 452 (1913).
56. W. Ruziczka, *Chim. Anal.* **32**, 33 (1950).
57. N. D. Cheronis and T. S. Ma, *Organic Functional Group Analysis*, John Wiley and Sons, New York, 1964, p. 559.
58. A. Steyermark, *Anal. Chem.*, **20**, 368 (1948).
59. M. Schachter and T. S. Ma, *Mikrochim. Acta*, 55 (1966).
60. T. Osterud and M. Prytz, *Acta Chem. Scand.*, **15**, 1923 (1961).
61. P. Zuman, *Organic Polarographic Analysis*, Pergamon Press, Oxford, 1964.
62. A. T. James and A. J. P. Martin, *Analyst*, **77**, 915 (1952); *Biochem. J.*, **50**, 679 (1952).
63. B. Byars and G. Jordan, *J. Gas Chromatog.*, **2**, 304 (1964).
64. H. A. Swain, Jr., *J. Gas Chromatog.*, **3**, 246 (1965).
65. R. B. Jackson, *J. Chromatog.*, **16**, 306 (1964).
66. J. G. Nikelly, *Anal. Chem.*, **36**, 2244 (1964).
67. K. Konishi and Y. Kano, *Japan Analyst*, **13**, 299 (1964).
68. G. G. McKeown and S. I. Read, *Anal. Chem.*, **37**, 1780 (1965).
69. T. Briggs and S. R. Lipsky, *Biochim. Biophys. Acta*, **97**, 579 (1965).
70. S. A. Hyun, G. V. Vahouny, and C. R. Treadwell, *Analyt. Biochem.*, **10**, 193 (1965).
71. N. W. Alcock, *Anal. Biochem.*, **11**, 335 (1965).
72. M. Rogozinski, *J. Gas Chromatog.*, **2**, 328 (1964).
73. M. Gee, *Anal. Biochem.*, **37**, 926 (1965).
74. P. Dubois, *Ann. Technol. Agri.*, **14**, 179 (1965).
75. E. Bendel, B. Fell, H. Hübner, W. Meltzow and J. Tetteroo, *J. Chromatog.*, **19**, 177 (1965).
76. J. Eisner and D. Firestone, *J. Ass. Offic. Agri. Chemists*, **48**, 1191 (1965).
77. Z. I. Horii, M. Makita, I. Takeda, Y. Tamura and Y. Ohnishi, *Chem. Pharm. Bull. (Tokyo)*, **13**, 636 (1965).
78. Z. I. Harii, M. Makita and Y. Tamura, *Chem. Ind.*, 1494 (1965).
79. J. L. Moore, T. Richardson and C. H. Amundson, *J. Gas Chromatog.*, **2**, 318 (1964).
80. R. G. Ackman and J. C. Sipos, *J. Chromatog.*, **16**, 208 (1964).
81. H. Salwin, *J. Ass. Offic. Agri. Chemists.*, **48**, 628 (1965).
82. A.O.A.C., Official Methods of Analysis, 9th ed., 1960.
83. E. Stahl, *Fette, Seifen, Anstrichmittel*, **60**, 1027 (1958).
84. E. Mutschler and H. Rochelmeyer, *Arch. Pharm.*, **292**, 449 (1959).
85. H. K. Mangold and D. C. Malins, *J. Am. Oil Chemists' Soc.*, **37**, 383 (1960).
86. E. Stahl (Ed.), *Dünnschicht-Chromatographie*, Springer-Verlag, Berlin, 1962.
87. J. G. Kirchner, *Thin-layer Chromatography*, John Wiley and Sons, New York, 1967.
88. J. Pokorny and O. Herodek, *Sb. Vysoke Skoly Chem.-Technol. Praze, Oddil Fak. Potravinareske Technol.*, **8**, 87, (1964).
89. K. B. Lie and J. F. Nye, *J. Chromatog.*, **8**, 75 (1962).
90. G. B. Crump, *Nature*, **193**, 674 (1962).
91. C. B. Barrett, M. S. J. Dallas and F. B. Padley, *Chem. Ind.*, 1050 (1962).
92. J. Sliwiak and Z. Kwapniewski, *Microchem. J.*, **9**, 237 (1965).
93. J. Kolesinska, T. Urbanski and A. Wielopolski, *Chem. Anal. (Warsau)*, **10**, 1107 (1965).
94. E. Bancher and H. Scherz, *Mikrochim. Acta*, 1159 (1964).
95. E. Knappe and I. Rohdewald, *Z. Anal. Chem.*, **210**, 183 (1965).
96. G. K. Koch and G. Jurriens, *Nature*, **208**, 1312 (1965).

97. R. Condsen, A. J. P. Martin and R. M. Sygne, *Biochem. J.*, **35**, 91 (1941).
98. For example, see J. Asselineau, *Bull. Soc. Chim. France*, **19**, 884 (1952).
99. R. I. Cheftel, R. Munier and M. Macheboeuf, *Bull. Soc. Chim. Biol.*, **35**, 1085 (1953).
100. C. C. Woodward and G. S. Rabideau, *Anal. Chem.*, **26**, 248 (1954).
101. J. Churacek, J. Kopecny, M. Kulhacy and M. Jurecek, *Z. Anal. Chem.*, **208** (1965).
102. L. A. Salmin, L. A. Mirkind and A. I. Kamneva, *Zh. Analit. Khim.*, **19**, 1391 (1964).
103. B. E. Kuvaev and N. S. Imyanitov, *Zh. Analit. Khim.*, **20**, 876 (1965).
104. A. Seher, *Fette, Seifen Anstrichmittel*, **67**, 255 (1965).
105. T. S. Ma and R. Roper, *Mikrochim. Acta.*, 167 (1968).
106. R. A. Kiseleva and M. S. Dudkin, *Zavodsk. Lab.*, **31**, 1448 (1965).
107. H. Kalbe, *Z. Physiol. Chem.*, **297**, 19 (1954).
108. R. D. Hartley and G. J. Lawson, *J. Chromatog.*, **7**, 69 (1962).
109. C. S. Marvel and R. D. Rands, Jr., *J. Am. Chem. Soc.*, **72**, 2642 (1950).
110. R. Raveux and J. Bové, *Bull. Chim. France*, 369 (1957).
111. S. R. Elsden, *Biochem. J.*, **40**, 252 (1946).
112. G. Hammarberg and B. Wickberg, *Acta Chem. Scand.*, **14**, 882 (1960).
113. S. C. Pan, A. I. Laskin and P. Principe, *J. Chromatog.*, **8**, 32 (1962).
114. T. Kobayashi and S. Miyazaki, *Tokyo Shikensho Hokoku*, **55**, 71 (1960).
115. J. L. Occolowitz, *J. Chromatog.*, **5**, 373 (1961).
116. D. Salbut, J. Bimer, W. Kutkiewicz and A. Wielopolski, *Chem. Anal. (Warsaw)*, **10**, 1099 (1965).
117. R. W. Riemenschneider, S. F. Herb and P. L. Nichols, Jr., *J. Am. Oil Chemists' Soc.*, **26**, 371 (1949).
118. F. E. Kurtz, *J. Am. Chem. Soc.*, **74**, 1902 (1952).
119. F. Bryant and B. T. Overell, *Nature*, **167**, 361 (1951).
120. R. Kunin, *Elements of Ion Exchange*, Reinhold, New York, 1960.
121. S. Moore and W. H. Stein, *J. Biol. Chem.*, **192**, 663 (1951).
122. O. Samuelson, *Ion Exchangers in Analytical Chemistry*, John Wiley and Sons, New York, 1953.
123. U. B. Larsson and O. Samuelson, *J. Chromatog.*, **19**, 404 (1965).
124. Technicon, Inc., Chancey, New York.
125. I. K. Tsitovich, E. A. Konovalova and B. F. Tsarichenko, *Izv. Vysshykh Uchebn. Zavedenii Khim. i Khim. Teknol.*, **8**, 60 (1965).
126. G. A. Harlow and D. H. Morman, *Anal. Chem.*, **36**, 2438 (1964).
127. A. Duclaux, *Ann. Inst. Pasteur*, **7**, 265 (1895).
128. C. Lind, *Kemisk Maanedsbad*, **22**, 125 (1941).
129. A. W. Weitkamp and L. C. Brunstrum, *Oil Soap (Egypt)*, **18**, 47 (1941).
130. H. A. Pagel and F. W. McLafferty, *Anal. Chem.*, **20**, 272 (1948).
131. E. H. Ahrens, Jr. and L. C. Craig, *J. Biol. Chem.*, **195**, 299 (1952).
132. W. Stein and H. Hartmann, *U.S. Pat.* 2, 800, 493.
133. *Sadtler Chemical Classes Index*, The Sadtler Research Laboratories, Philadelphia, 1967.
134. K. Nakanishi, *Infrared Absorption Spectroscopy–Practical*, Holden–Day, San Francisco, 1962.
135. R. T. Conley, *Infrared Spectroscopy*, Allyn and Bacon, Boston, 1966.
136. A. D. Cross, *Practical Infrared Spectroscopy*, 2nd ed., Butterworths, Washington, 1964.
137. A. R. Katrizky, J. M. Lagowski and J. A. T. Beard, *Spectrochim. Acta*, **16**, 954 (1960).
138. J. D. Roberts, *Nuclear Magnetic Resonance*, McGraw-Hill, New York, 1959.

139. J. A. Pople, W. G. Schneider and H. J. Bernstein, *High-resolution Nuclear Magnetic Resonance*, McGraw-Hill, New York, 1959.
140. H. S. Gutowsky and A. Saika, *J. Chem. Phys.*, **21**, 1688 (1953).
141. B. N. Bhar and G. Lindström, *J. Chem. Phys.*, **23**, 1958 (1955).
142. M. J. Chisholm and C. Y. Hopkins, *Can. J. Chem.*, **35**, 358 (1957).
143. C. Y. Hopkins and H. J. Bernstein, *Can. J. Chem.*, **37**, 775 (1959).
144. W. L. Senn, Jr. and L. A. Pine, *Anal. Chim. Acta*, **31**, 441 (1964).
145. J. H. Beynon, *Mass Spectrometry and Its Application to Organic Chemistry*, Elsevier, Amsterdam, 1960.
146. A. Cornu and R. Mascot, *Analyse Organique par Spectrométrie de Masse à Haute Résolution*, Presses Universitaires de France, Paris, 1964.
147. G. P. Happ and D. W. Stewart, *J. Am. Chem. Soc.*, **74**, 4404 (1952).
148. F. W. McLafferty, *Appl. Spectr.*, **11**, 148 (1957).
149. R. S. Gohlke and F. W. McLafferty, *A.S.T.M. Committee E-14, 4th Annual Meeting, San Francisco*, 1955.
150. L. D. Quin and M. E. Hobbs, *Anal. Chem.*, **30**, 1400 (1958).
151. Mass Spectral Data Sheets, American Petroleum Institute.
152. S. D. Nogane and R. S. Juvette, Jr., *Gas Liquid Chromatography*, John Wiley and Sons, New York, 1962, p. 47.

CHAPTER **18**

Biological formation and reactions of the —COOH and —COOR groups

SHAWN DOONAN

University College, London, England

ABBREVIATIONS

ACP	:	ACYL CARRIER PROTEIN
ADP	:	ADENOSINE DIPHOSPHATE
Ala.	:	ALANINE
AMP	:	ADENOSINE MONOPHOSPHATE
Asn.	:	ASPARAGINE
Asp.	:	ASPARTIC ACID
ATP	:	ADENOSINE TRIPHOSPHATE

BChl.	:	BACTERIOCHLOROPHYLL
CDP	:	CYTIDINE DIPHOSPHATE
Chl.	:	CHLOROPHYLL
CMP	:	CYTIDINE MONOPHOSPHATE
CoA, CoASH	:	COENZYME A
CoQ_{10}	:	COENZYME Q_{10}
CTP	:	CYTIDINE TRIPHOSPHATE
Cyt.	:	CYTOCHROME
DCMU	:	3-(3,4-DICHLOROPHENYL)-1,1-DIMETHYL UREA
DOPA	:	3,4-DIHYDROXYPHENYLALANINE
DPIP	:	DICHLOROPHENOLINDOPHENOL
E.C.	:	ENZYME COMMISSION
ETF	:	ELECTRON TRANSFER FLAVOPROTEIN
ETS	:	ELECTRON TRANSPORT SYSTEM
FAD	:	FLAVINE ADENINE DINUCLEOTIDE
FMN	:	FLAVINE MONONUCLEOTIDE
FP	:	FLAVOPROTEIN
GABA	:	γ-AMINOBUTYRIC ACID
GDP	:	GUANOSINE DIPHOSPHATE
Gln.	:	GLUTAMINE
Glu.	:	GLUTAMIC ACID
Gly.	:	GLYCINE
GMP	:	GUANOSINE MONOPHOSPHATE
GSH	:	GLUTATHIONE (γ-GLUTAMYLCYSTEINYLGLYCINE)
GTP	:	GUANOSINE TRIPHOSPHATE
HMG-CoA	:	β-HYDROXY-β-METHYLGLUTARYL-CoA
IETC	:	INTERMEDIATE ELECTRON TRANSFER COMPLEX
KDC	:	α-KETOGLUTARATE DEHYDROGENASE COMPLEX
KDPG	:	2-KETO-3-DEOXY-6-PHOSPHOGLUCONATE
Leu.	:	LEUCINE
NAD^+	:	NICOTINAMIDE ADENINE DINUCLEOTIDE
NADH	:	REDUCED FORM OF NAD^+
$NADP^+$	:	NICOTINAMIDE ADENINE DINUCLEOTIDE PHOSPHATE
NADPH	:	REDUCED FORM OF $NADP^+$
NAGA	:	N-ACETYLGLUTAMIC ACID
NQNO	:	NONYL-4-HYDROXYQUINOLINE-N-OXIDE
—Ⓟ	:	—PO_3H_2 AND CHARGED FORMS
P_i	:	'INORGANIC PHOSPHATE' (H_3PO_4 AND CHARGED FORMS)

Pal.P.	:	Pyridoxal-5′-phosphate
PDC	:	Pyruvate dehydrogenase complex
PMS	:	Phenazine methosulphate
PP_i	:	'Inorganic pyrophosphate' ($H_4P_2O_7$ and charged forms)
PRPP	:	5-Phosphoribosyl pyrophosphate
PSI	:	Pigment system I
PSII	:	Pigment system II
RHP	:	Rubrum haeme protein
Ser.	:	Serine
TPP	:	Thiamine pyrophosphate
UDP	:	Uridine diphosphate
UDPG	:	Uridine diphosphoglucose
UMP	:	Uridine monophosphate
UTP	:	Uridine triphosphate

I. INTRODUCTION

Carboxylic acids and esters are ubiquitous in biological systems and it is impossible to give a complete and exhaustive coverage of these compounds in a review of this type. An attempt will be made, therefore, to deal only with those acids and esters which are involved in the main metabolic pathways of animals, plants and microorganisms, and with those which have special biological importance. Throughout, attention will be focused on the enzymes or enzyme systems catalysing the reactions under consideration and on the relevance of these reactions to the function of the organism.

The classification of enzymes has been undertaken by the Enzyme Commission of the International Union of Biochemistry[1]; where possible on first mention of a particular enzyme its classification number will be given and the recommendations of the Commission on Enzyme Nomenclature will be followed.

It should be pointed out that many biologically important carboxylic acids are polyfunctional; in particular the α-amino and α-keto acids are very widely distributed. These two functional groups have already been reviewed in this series[2, 3] and some overlap between this and the previous articles is inevitable.

It must be considered inadequate in a review designed mainly for chemists, simply to tabulate reaction types and to divorce the reactions from the metabolic pathways in which they occur. To take this course can lead to serious misconceptions. One example of this may help to emphasize the point. The enzyme malate dehydrogenase* (E.C.1.1.1.37) catalyses reaction (1).

$$
\begin{array}{ccccc}
\text{COOH} & & & \text{COOH} & \\
| & & & | & \\
\text{CHOH} & + & \text{NAD}^+ \rightleftharpoons & \text{CO} & + \quad \text{NADH} + \text{H}^+ \\
| & & & | & \\
\text{CH}_2 & & & \text{CH}_2 & \\
| & & & | & \\
\text{COOH} & & & \text{COOH} & \\
\text{L-malate} & & & \text{oxaloacetate} &
\end{array}
\qquad (1)
$$

The equilibrium constant for this reaction is about 10^{-12}, that is, the formation of L-malate from oxaloacetate is strongly favoured. However, the principal role of this enzyme in biological systems is to convert L-malate to oxaloacetate during the operation of the citric acid cycle (see later). The unfavourable equilibrium effect is overcome in this system by the continuous supply of L-malate and the continuous removal of oxalo-acetate by other components of the cycle. This intimate interrelationship of reactions is a general feature of biochemical systems which cannot, therefore, be fully understood except in the context of the integrated metabolism of the organism.

Since the overall metabolism of an organism is very complex, it is appro-priate at this stage to point out some general features and underlying principles of the process. It is useful to focus attention on a relatively small number of key compounds for the discussion of metabolic processes, and Figure 1 presents a highly simplified summary of the main pathways, which will be dealt with in this review in relation to these key intermediates.

An organism is primarily engaged in two types of metabolic processes; firstly in the breakdown (catabolism) of macromolecular compounds to provide energy for its chemical and mechanical activities and secondly in the synthesis (anabolism) of molecular species required for its structure and functions. Catabolism of the three major types of nutritive compounds (fats, carbohydrates and proteins) leads mainly to the very limited range of intermediates shown in Figure 1. Subsequently, the intermediates may be degraded further to yield carbon dioxide, water and simple nitrogenous excretion products or they may be used in the anabolic processes of the organism. It should be emphasized, however, that although the products of catabolism and the substrates of anabolism are the same, the catabolic

---

* In accordance with usual practice, carboxylic acids will be refered to as carboxyla-tes since, at physiological pH, the charged species predominates in solution. However for convenience, chemical equations will be written without reference to the charged state of the reactants.

and anabolic pathways have certain essential differences, as will be shown later.

In addition to the link between catabolic and anabolic processes provided by the central intermediates of metabolism, a further connection

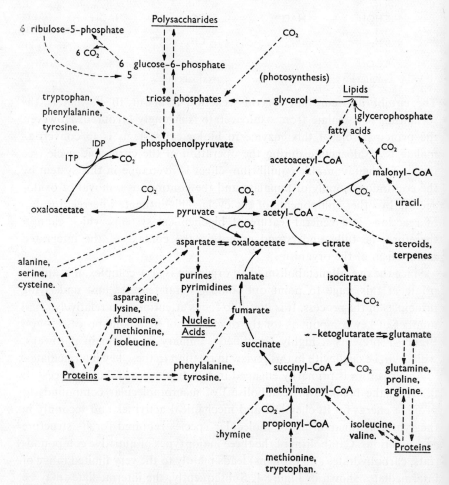

FIGURE 1. Outline of the relationships between some important metabolic processes

arises from the generation of reducing and phosphorylating agents by the former and their utilization in the latter.

Catabolism is essentially an oxidative process and is accompanied by the formation of the reduced forms of the pyridine nucleotides nicotinamide adenine dinucleotide (NAD^+) and nicotinamide adenine dinucleotide phosphate ($NADP^+$). The structures of the oxidized forms of these coenzymes are shown in Figure 2. Pyridine nucleotides are universally involved in biological redox reactions; in each case an enzyme exists which

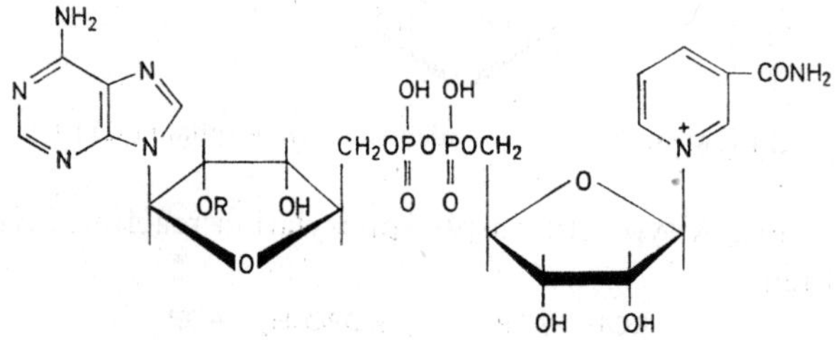

NH₂ ... OH OH ... CONH₂ ... CH₂OPO POCH₂ ... OR OH ... O O ... O ... OH OH

$$NAD^+ \quad R = -H, \qquad NADP^+ \quad R = -\overset{O}{\underset{OH}{\overset{\|}{P}}}-OH$$

FIGURE 2. Structures of the pyridine nucleotides (NAD^+ and $NADP^+$)

is specific both for the substrate and for the coenzyme. A typical example has already been presented in equation (1) above. Oxidation of the substrate is accompanied by reduction of the coenzyme as shown below.

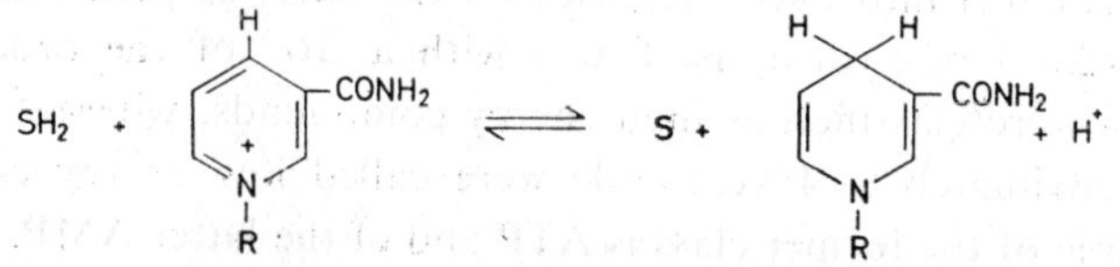

$$SH_2 \; + \; \text{(pyridinium—CONH}_2\text{)} \;\rightleftharpoons\; S \; + \; \text{(dihydropyridine—CONH}_2\text{)} \; + \; H^+$$

The reaction involves the transfer of two electrons and a hydrogen nucleus which may in theory be accomplished either by hydride ion transfer or by transfer of a hydrogen atom and an electron. Current opinion is that hydride transfer is in fact involved.

Two routes are available for reoxidation of the reduced coenzymes produced during catabolism. As a rough generalization it may be said that NADPH is employed in reductive anabolic processes whilst NADH is oxidized via the terminal respiratory system of the organism.

Probably the most important link between catabolism and anabolism is provided by the molecule adenosine triphosphate (ATP, Figure 3). Formation of this species by phosphorylation of adenosine diphosphate (ADP) during catabolism and its utilization as a phosphorylating agent during anabolism is one of the most basic themes underlying the activities of

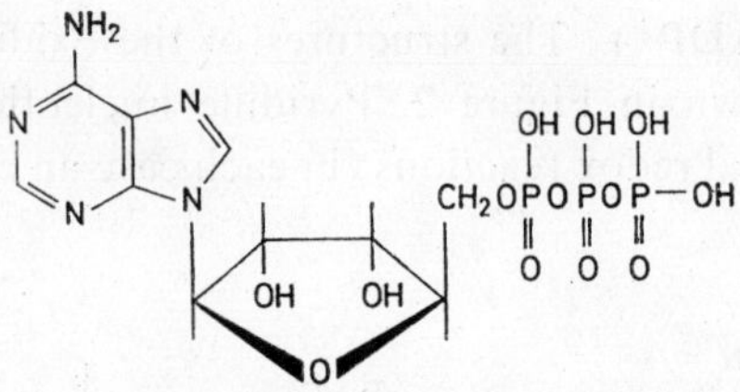

FIGURE 3. Structure of adenosine triphosphate (ATP)

living organisms. A typical transphosphorylation reaction involving ATP may be written

$$ROH + ATP \rightleftharpoons ROPO_3H_2 + ADP$$

Less frequently, ATP may function by the transfer to an acceptor of pyrophosphate, adenylate or adenosyl residues.

It may be appropriate at this point to consider certain conceptual errors which abound in the biochemical literature concerning the participation of ATP in metabolic reactions. For a definitive account of this subject, the article by Vernon[4], on which the following account is based, may be recommended.

The original source of confusion was an arbitrary division of biological phosphate esters (and certain other classes of compounds which will be mentioned later) into two categories on the basis of their standard free energies ($\Delta G°$) of hydrolysis. Esters with a $\Delta G°$ of the order of -10 kcal/mole were classified as high energy compounds, whereas those with $\Delta G°$ approximately -4 kcal/mole were called low energy compounds. An example of the former class is ATP and of the latter AMP. According to Lipmann[5], the conversion of ADP to ATP represents a mechanism for the storage of an amount of energy, equal to $\Delta G°$, which may subsequently be utilized for the promotion of energetically unfavourable reactions. The energy is visualized as being 'stored' in the terminal phosphate bond of ATP—hence the nomenclature 'high energy phosphate bond'. The concept of 'storage of free energy' in a chemical bond is incompatible with the principles of thermodynamics. The high $\Delta G°$ associated with the hydrolysis of ATP, or indeed any other pyrophosphate ester, reflects simply the formation of a complete (or a large fraction of) negative charge on the hydrolysis products with concomitant change of free energy of solvation in the system.

According to the high energy phosphate bond theory, ATP provides an 'energy link' between catabolism and anabolism; that is, catabolic reactions, characterized by their association with a decrease in free energy, are coupled to ATP synthesis whilst energetically unfavourable anabolic reactions are 'driven' by energy released during ATP hydrolysis. For example, during the anabolic degradation of 1 mole of glucose to lactic acid, two moles of ATP are produced. The stoichiometric relationship is shown in equation (2).

$$C_6H_{12}O_6 + 2\,ADP + 2\,H_3PO_4 \rightarrow 2\,CH_3CH(OH)COOH + 2\,ATP + 2\,H_2O \tag{2}$$

This reaction may be represented by the sum of equations (3) and (4).

$$C_6H_{12}O_6 \rightarrow 2\,CH_3CH(OH)COOH \tag{3}$$

$$2\,ADP + 2\,H_3PO_4 \rightarrow 2\,ATP + 2\,H_2O \tag{4}$$

The standard free energies ΔG_3°, and ΔG_4° are known to be about -50 and $+17$ kcal/mole respectively. Hence it follows that ΔG_2° is about -33 kcal/mole; consequently although the equilibrium for equation (4) lies to the left, that for equation (2) lies to the right and it must be possible to find a series of reactions, the overall result of which is the reaction represented by equation (2). The error which is frequently made is to assume that reactions (3) and (4) actually occur and that the free energy change associated with (3) is used to 'drive' reaction (4), a situation which is clearly impossible. If this view is taken, however, it is possible to calculate an efficiency for the energy transfer since the 'liberation' of 50 kcal of free energy from glucose degradation results in the 'storage' of 17 kcal of free energy in ATP, an overall efficiency of about 30%. A great deal of speculation has resulted from efficiency calculations of this sort.

The theory then goes on to describe how energy stored in ATP is utilized in anabolic reactions. Consider the formation of a compound XY, from precursors XOH and YH, which proceeds with the following stoichiometry:

$$XOH + YH + ATP \rightarrow XY + ADP + H_3PO_4 \tag{5}$$

This can clearly be split into:

$$XOH + YH \rightarrow XY + H_2O \tag{6}$$

and

$$ATP + H_2O \rightarrow ADP + H_3PO_4 \tag{7}$$

Since ΔG_5° is given by $\Delta G_6^\circ + \Delta G_7^\circ$, then reaction (5) will proceed in the direction shown provided that ΔG_6° is not more positive than about 8 kcal/mole. Proponents of the high energy phosphate bond concept would reason that the free energy of hydrolysis from reaction (7) is used to drive reaction (6), the overall result being represented by equation (5).

In fact, the probable course of reaction (5) would be as follows:

$$XOH + ATP \rightarrow XOPO_3H_2 + ADP \tag{8}$$

$$XOPO_3H_2 + YH \rightarrow X\!-\!Y + H_3PO_4 \tag{9}$$

These two reactions are linked chemically since the product of equation (8) is the reactant of equation (9). It is obvious that, given a formulation of reaction (5) in terms of reactions (8) and (9), any consideration of efficiency of ATP usage is meaningless since this is governed by the stoichiometry of the reaction, not by the energy available from ATP hydrolysis. Reaction (7) can now be considered in its true perspective; namely as a device which, together with equation (6), can be used to calculate ΔG_5° and hence the equilibrium constant of reaction (5). It must be emphasized that reaction (7) must not be allowed to proceed if the overall reaction (5) is to take place, since hydrolysis of ATP serves simply to uncouple reactions (8) and (9). Similar considerations to the above may be applied to the formation of ATP during glucose metabolism (reaction (2)). This process cannot proceed via reactions (3) and (4). The mechanism of this process will be dealt with later.

In summary, certain catabolic reactions are associated with the transfer of the phosphate radical from donor molecules to ADP to produce ATP. The ATP so formed forms a 'chemical link' with anabolic process in which it participates in a variety of reactions, some of these being trans-phosphorylation reactions such as reaction (5). The hydrolysis of ATP does not participate in anabolic processes, the only effect of this reaction being to prevent transphosphorylation reactions from occuring; consequently calculations of the efficiency of anabolic reactions based on the value of the standard free energy of hydrolysis of ATP are inappropriate.

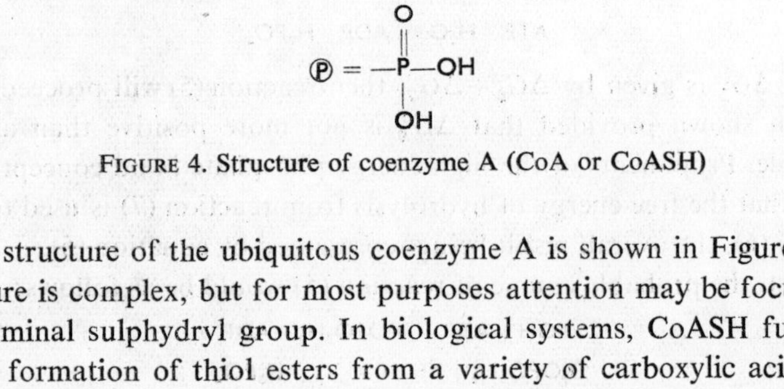

FIGURE 4. Structure of coenzyme A (CoA or CoASH)

The structure of the ubiquitous coenzyme A is shown in Figure 4; the structure is complex, but for most purposes attention may be focused on the terminal sulphydryl group. In biological systems, CoASH functions by the formation of thiol esters from a variety of carboxylic acids. For example, in the case of acetic acid, acetyl-CoA is formed:

$$CH_3COOH + CoASH \rightarrow CoASCOCH_3 + H_2O$$

The formation of acyl-CoA derivatives proceeds by a complex series of reactions involving ATP.

Thiol esters are generally more reactive then oxygen esters, and acyl-CoA derivatives undergo four main types of reactions. These are nucleophilic attack at the acyl carbon atom, addition reactions to unsaturated acyl groups, condensation reactions at the α-carbon atom and acyl transfer reactions. These reactions will be described in detail in later sections.

II. BIOLOGICAL CARBOXYLATION REACTIONS

This section will deal with the ways in which organisms incorporate carbon dioxide into their tissues. Prior to about 1940 it was generally considered that carbon dioxide fixation was limited to photosynthetic and chemosynthetic forms of life; that is, those forms which derive all their organic molecules from carbon dioxide. Speculations were made as to the possible roles of carbon dioxide fixation in heterotrophs (organisms lacking photosynthetic or chemosynthetic pathways), but manometric procedures available at that time were not sufficiently sensitive to detect a small uptake of carbon dioxide accompanied by a much greater rate of evolution of this gas.

After the isotope ^{14}C became readily available, the high sensitivity of radioactivity measurements made it a relatively simple task to detect incorporation of $^{14}CO_2$ into tissue materials. At the same time, using partition chromatography, the primary products of heterotrophic carboxylation were identified and the subsequent reactions of these compounds were examined.

The use of these techniques has shown that carboxylation reactions may be divided into two classes. In autotrophic carboxylation, carbon dioxide is incorporated into an acceptor molecule, followed by formation of other carbon compounds and ultimate regeneration of the acceptor, whereas heterotrophic carboxylation is involved as a single step in the synthesis of a specific compound. Photosynthetic and chemosynthetic organisms carry out both of these types of reaction; other organisms fix carbon dioxide only by the heterotrophic routes.

The enzymes involved in biological carboxylation reactions will be dealt with in the next section and the relevance of particular reactions to overall metabolic processes will be considered in sections II.B and II.C.

A. Classification and Mechanism of Action of the Carboxylases

Only the enzymes which are believed to be involved in autotrophic and heterotrophic carbon dioxide *fixation* will be dealt with in this section[6-9].

Four types of carboxylases may be distinguished, depending on the coenzymes involved and the form of the acceptor molecule.

1. Reductive carboxylases (E.C.1.1.1.)[6,9,10]

The enzymes in this group are dependent on NADPH and a divalent cation (Mg^{2+} or Mn^{2+}) for their action; the substrates contain a carbonyl group adjacent to the acceptor carbon atom, the carbonyl group being reduced during the reaction. The individual enzymes are:

(a) Malate dehydrogenase (decarboxylating) (E.C.1.1.1.40)[262]—originally known as the 'malic' enzyme.

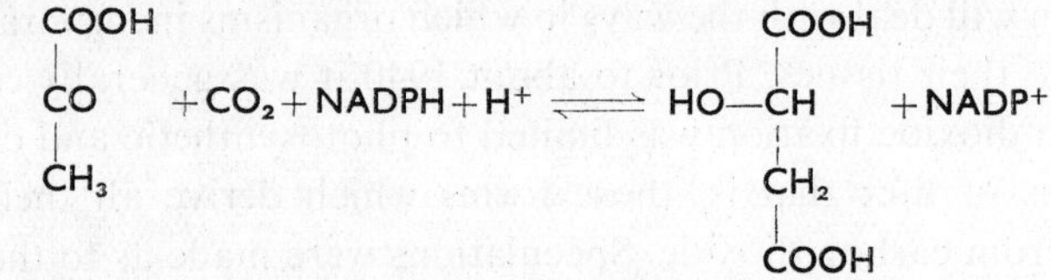

$$
\begin{array}{c}
\text{COOH} \\
| \\
\text{CO} \\
| \\
\text{CH}_3
\end{array}
\; + CO_2 + NADPH + H^+ \rightleftharpoons
\begin{array}{c}
\text{COOH} \\
| \\
\text{HO—CH} \\
| \\
\text{CH}_2 \\
| \\
\text{COOH}
\end{array}
\; + NADP^+
$$

pyruvate   L-malate

(b) Isocitrate dehydrogenase (decarboxylating) (E.C.1.1.1.42)[263]

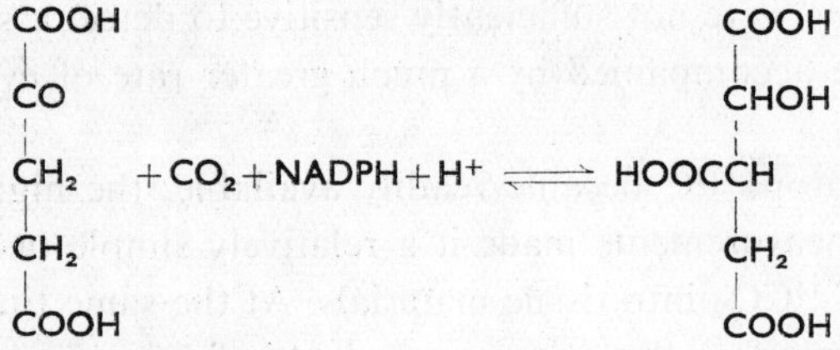

$$
\begin{array}{c}
\text{COOH} \\
| \\
\text{CO} \\
| \\
\text{CH}_2 \\
| \\
\text{CH}_2 \\
| \\
\text{COOH}
\end{array}
\; + CO_2 + NADPH + H^+ \rightleftharpoons
\begin{array}{c}
\text{COOH} \\
| \\
\text{CHOH} \\
| \\
\text{HOOCCH} \\
| \\
\text{CH}_2 \\
| \\
\text{COOH}
\end{array}
$$

α-ketoglutarate   *threo*-D$_s$-isocitrate

(Note: several systems of nomenclature are encountered for the naturally occurring form of isocitrate. The most rigorous name is αD$_s$, βL$_s$-isocitrate. The enzyme under discussion is frequently called L$_s$-isocitrate dehydrogenase, presumably since carboxylation of α-ketoglutarate results in a centre with the configuration L$_s$.)

(c) Phosphogluconate dehydrogenase (decarboxylating) (E.C.1.1.1.44)[264]

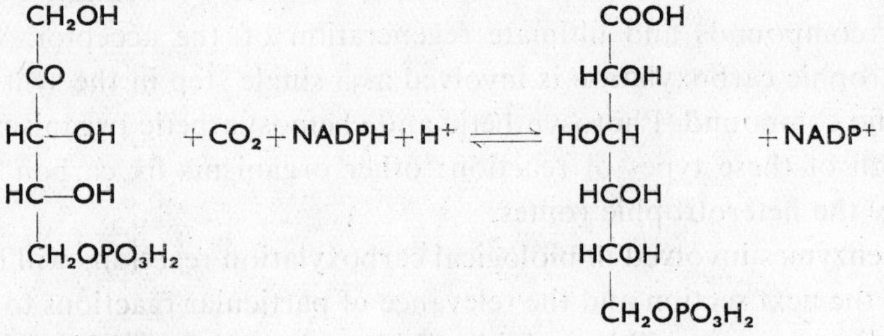

$$
\begin{array}{c}
\text{CH}_2\text{OH} \\
| \\
\text{CO} \\
| \\
\text{HC—OH} \\
| \\
\text{HC—OH} \\
| \\
\text{CH}_2\text{OPO}_3\text{H}_2
\end{array}
\; + CO_2 + NADPH + H^+ \rightleftharpoons
\begin{array}{c}
\text{COOH} \\
| \\
\text{HCOH} \\
| \\
\text{HOCH} \\
| \\
\text{HCOH} \\
| \\
\text{HCOH} \\
| \\
\text{CH}_2\text{OPO}_3\text{H}_2
\end{array}
\; + NADP^+
$$

D-ribulose-5-P   6-phospho-D-gluconate

The properties of these enzymes have been reviewed in reference 10 (Chapters 7,5 and 11 respectively). A similar set of enzymes (E.C.1.1.1.39, 41 and 43) is known for which the coenzyme is NADH. It is believed that the NADH-dependent enzymes are solely involved in biological decarboxylation reactions, whereas the NADPH enzymes may function both as carboxylases and decarboxylases. This is consistent with the known equilibrium constants of the NADPH enzymes (19·6, 1·3 and 1·9 l/mole for the carboxylation of pyruvate, α-ketoglutarate and ribulose-5-phosphate respectively at pH 7·4, 7·0 and 7·4)[6]. Whereas the malate and isocitrate dehydrogenases have been extensively studied, little is known about the phosphogluconate dehydrogenase; consequently most of what follows will apply only to the first two enzymes.

The reactions involve a dehydrogenation and decarboxylation (in the oxidative direction) and it is reasonable to assume that the corresponding β-keto acids (oxaloacetate, oxalosuccinate and 3-keto-6-phospho-D-gluconate) may be intermediates in the reactions. Consistently it has been shown that isocitrate dehydrogenase catalyses reactions (10) and (11)[265, 266],

$$\text{oxalosuccinate} \longrightarrow \alpha\text{-ketoglutarate} + CO_2 \qquad (10)$$

$$NADPH + H^+ + \text{oxalosuccinate} \longrightarrow \text{isocitrate} + NADP^+ \qquad (11)$$

whereas only reaction (12) has been demonstrated in the case of the malic enzyme[267].

$$\text{oxaloacetate} \longrightarrow \text{pyruvate} + CO_2 \qquad (12)$$

It has been clearly demonstrated, however, that free oxaloacetate and oxalosuccinate do not occur as intermediates in the overall carboxylation and decarboxylation reactions, which suggests that these species, if formed, are firmly bound to the enzyme surfaces[266, 268]. Malate and isocitrate dehydrogenases are inhibited by sulphydryl blocking reagents, protection from inhibition being afforded by prior incubation of the enzymes with the appropriate substrates. This is consistent with the suggestion that the keto acid intermediates are bound to the enzyme surfaces by covalent bonds with sulphur[10].

Enzyme-catalysed exchange of hydrogen atoms of the $C_{(3)}$-methylene of α-ketoglutarate and the terminal methyl group of pyruvate with solvent water has been observed[269], tending to implicate the enol forms of α-ketoglutarate and pyruvate as intermediates in the reactions. The requirement shown in the exchange reactions for NADPH and NADPH plus CO_2 respectively remains to be explained. Experiments have been reported using isocitrate tritiated in the β-position; failure to detect exchange of tritium with water suggests that the enol form of oxalosuccinate does not participate in the reaction.

On the basis of the facts quoted above, a tentative scheme for the mechanism of action of these enzymes may be presented:

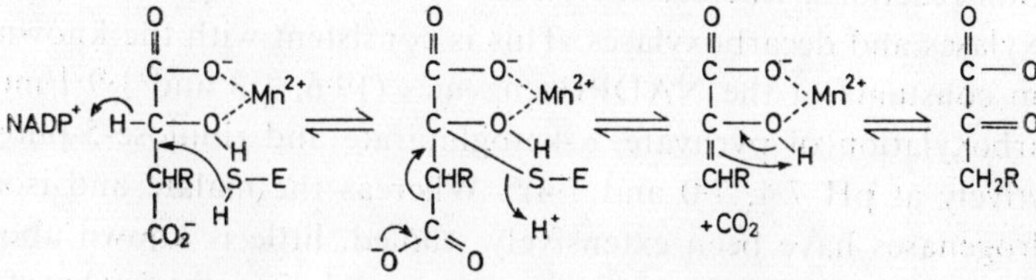

isocitrate,   oxalosuccinate,   enol forms of
malate   oxaloacetate   α-ketoglutarate,
pyruvate

It should be emphasized that this scheme is conjectural, and also that some of the facts quoted above indicate that the mechanisms of the malate and isocitrate dehydrogenases are not identical. Due to the lack of experimental data, no mechanism can at this stage be proposed for the 6-phosphogluconate dehydrogenase.

2. Enoyl carboxy-lyases (E.C.4.1.1.)[6,11]

These enzymes catalyse carboxylation of a reactive enol substrate; it is likely, however, that the reaction mechanisms of the individual enzymes are not related. The members of the group are as follows:

(a) Phosphoenolpyruvate carboxy-lyases

Four distinct enzymes have been characterized which catalyse reactions of the form:

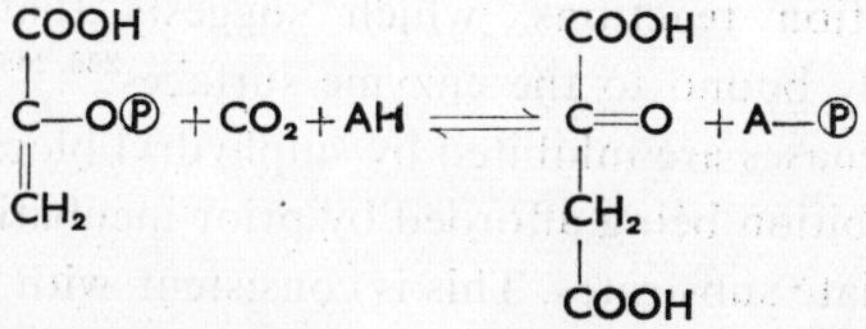

$$
\begin{array}{c}
\text{COOH} \\
|\\
\text{C}-\text{O}\textcircled{P} \\
||\\
\text{CH}_2
\end{array}
+ CO_2 + AH \rightleftharpoons
\begin{array}{c}
\text{COOH} \\
|\\
\text{C}=\text{O} \\
|\\
\text{CH}_2 \\
|\\
\text{COOH}
\end{array}
+ A-\textcircled{P}
$$

phosphophenolpyruvate   oxaloacetate

(Note: the symbol $\textcircled{P}$ denotes the phosphate radical.)

Here, AH is an acceptor for the phosphate moiety. An enzyme isolated from plants and autotrophic bacteria (E.C.4.1.1.31) transfers phosphate to water with the formation of inorganic phosphate[270, 271], whilst propionic acid bacteria contain an enzyme (phosphoenolpyruvate carboxytransphosphorylase) which employs inorganic phosphate as the acceptor and produces pyrophosphate[272]. Nucleoside diphosphates serve as acceptors for the remaining two enzymes (phosphoenolpyruvate carboxykinases). The yeast enzyme employs adenosine diphosphate (ADP)[273] and the enzyme from mammalian source (E.C.4.1.1.32) either guanosine or inosine diphosphates (GDP, IDP) as phosphate acceptors with the generation of the corresponding triphosphates[274].

These enzymes are rather poorly characterized. They show a dependence on divalent cations (Mn^{2+} or Mg^{2+}) and are inhibited by sulphydryl-blocking reagents, but these facts are insufficient to support any proposed mechanism of action. Experiments with the enzyme from plant sources have shown that ^{18}O from $NaHC^{18}O_3$ is incorporated into product oxalo-acetate and phosphate in the ratio 2:1. This is consistent with a concerted mechanism of the type shown.

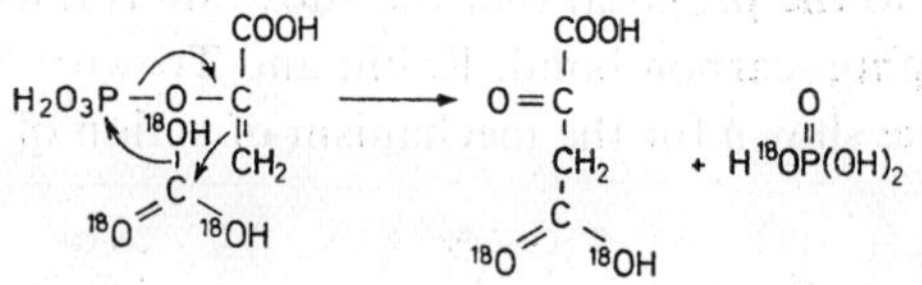

Further experimental work is required to substantiate this mechanism and to show how far, if at all, it is applicable to the other enzymes of this group.

(b) Phosphoribosyl-5-aminoimidazole carboxylase (E.C.4.1.1.20)[275]

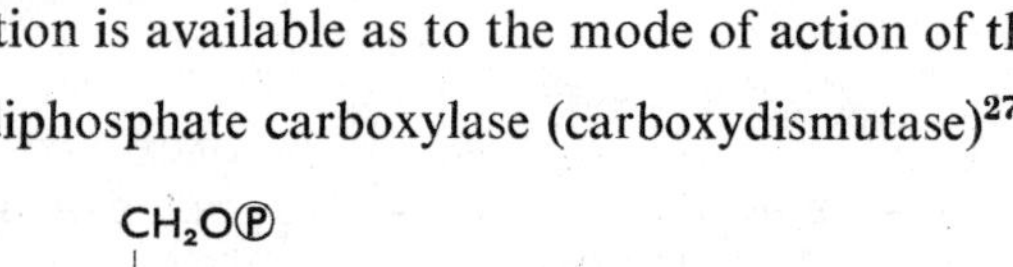

5′-phosphoribosyl-5-amino-4-imidazole

5′-phosphoribosyl-5-amino-4-imidazole carboxylate

No information is available as to the mode of action of this enzyme.

(c) Ribulose diphosphate carboxylase (carboxydismutase)[276, 277]

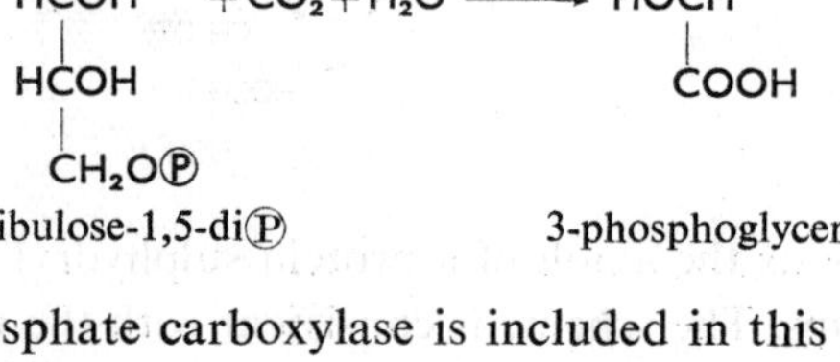

D-ribulose-1,5-di℗

3-phosphoglycerate

Ribulose diphosphate carboxylase is included in this group due to the suggestion of Bassham[12] and his colleagues that the carboxylation reaction occurs *via* the ene-diol intermediate:

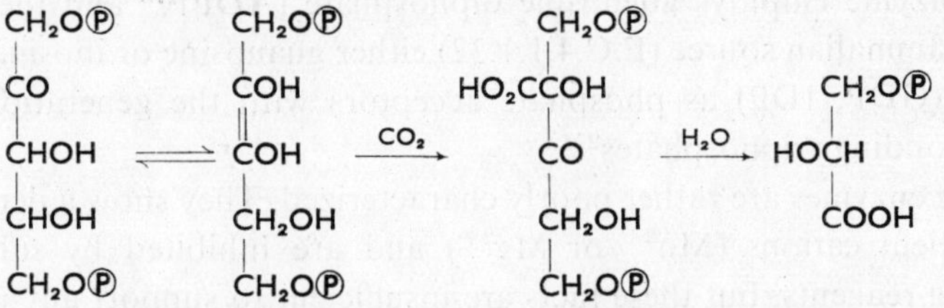

$$
\begin{array}{ccccccc}
CH_2O\text{(P)} & & CH_2O\text{(P)} & & CH_2O\text{(P)} & & CH_2O\text{(P)} \\
| & & | & & | & & | \\
CO & & COH & & HO_2CCOH & & CH_2O\text{(P)} \\
| & \rightleftharpoons & || & \xrightarrow{CO_2} & | & \xrightarrow{H_2O} & | \\
CHOH & & COH & & CO & & HOCH \\
| & & | & & | & & | \\
CHOH & & CH_2OH & & CH_2OH & & COOH \\
| & & | & & | & & \\
CH_2O\text{(P)} & & CH_2O\text{(P)} & & CH_2O\text{(P)} & &
\end{array}
$$

Attempts to demonstrate incorporation of deuterium and tritium from DOH and TOH into ribulose diphosphate have been inconclusive, and the six-carbon addition product has not been isolated. These facts, combined with the observation of inhibition of the enzyme by sulphydryl-blocking reagents has led to the proposal that the substrate is bound to the enzyme surface by a sulphur–carbon bond. Rabin and Trown[13] have recently presented the scheme shown for the mechanism of action of this enzyme.

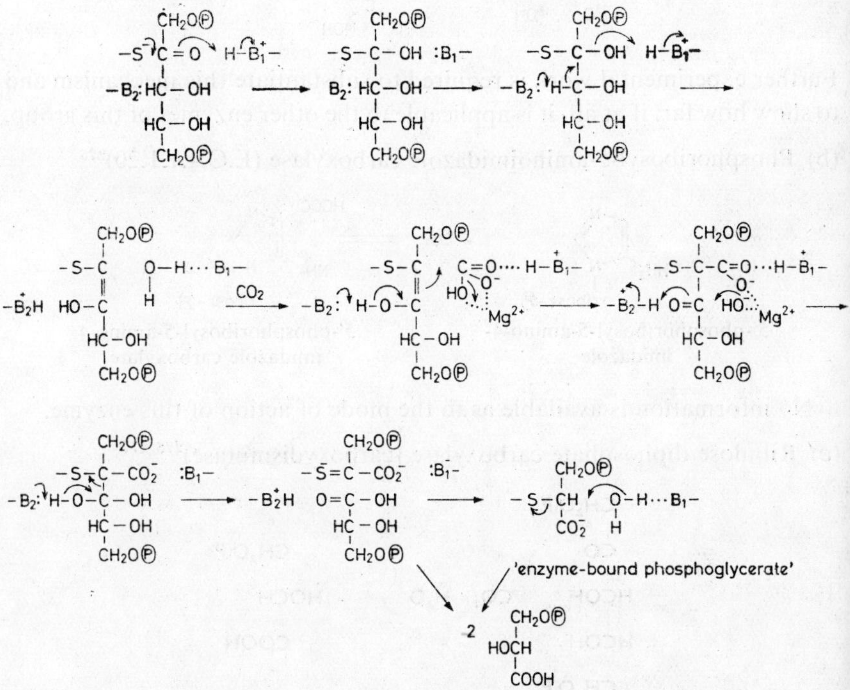

Here $-S^-$ represents the anion of a protein sulphydryl group and B_1 and B_2 are basic groups. The scheme is consistent with the requirement of the enzyme for magnesium ions and the involvement of group B_2 rules out the possibility of carboxylation of the epimeric xylulose-1,5-diphosphate. The implications of this scheme in connection with the fixation of carbon dioxide by photosynthetic organisms will be discussed later.

3. Carbon dioxide ligases (E.C.6.4.1) and carboxyltransferases (E.C.2.1.3)[8,9]

These two groups of enzymes are conveniently discussed together since they are mechanistically very similar. The first group comprises the following enzymes.

(a) Pyruvate carboxylase (E.C.6.4.1.1)

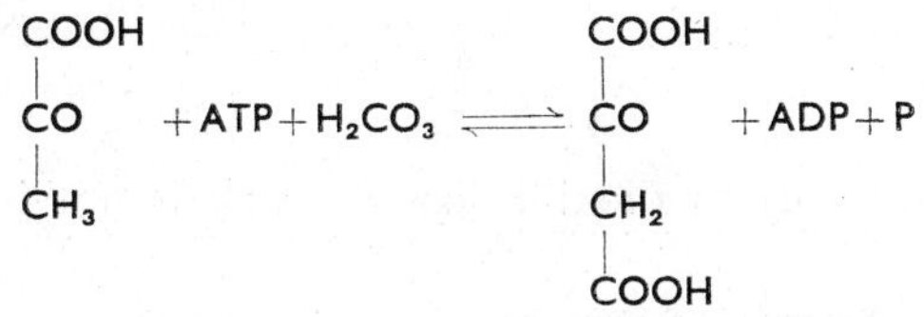

$$\begin{array}{ccc} \text{COOH} & & \text{COOH} \\ | & & | \\ \text{CO} & +\text{ATP}+\text{H}_2\text{CO}_3 \rightleftharpoons \text{CO} & +\text{ADP}+\text{P} \\ | & & | \\ \text{CH}_3 & & \text{CH}_2 \\ & & | \\ & & \text{COOH} \end{array}$$

pyruvate   oxaloacetate

[Note: the form in which carbon dioxide participates in most biological carboxylation reactions (*i.e.* as free carbon dioxide or as the bicarbonate ion) is not known, and in this review the reactive species will usually be shown for convenience as CO_2. In the present case, however, the reactive species has been identified as the bicarbonate ion (see later).]

The symbol P_i is used to denote 'inorganic phosphate'; that is, the species H_3PO_4 or any of its charged forms. Similarly, PP_i denotes 'inorganic pyrophosphate'.

(b) Acetyl-CoA carboxylase (E.C.6.4.1.2)

$$\text{CH}_3\text{COSCoA}+\text{ATP}+\text{H}_2\text{CO}_3 \rightleftharpoons \text{HOOCCH}_2\text{COSCoA}+\text{ADP}+\text{P}_i$$
acetyl-CoA   malonyl-CoA

(c) Propionyl-CoA carboxylase (E.C.6.4.1.3)

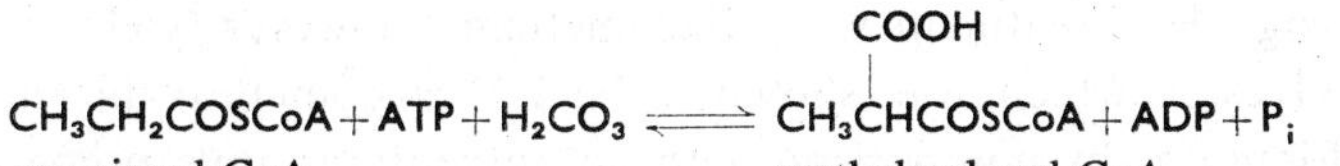

$$\begin{array}{cc} & \text{COOH} \\ & | \\ \text{CH}_3\text{CH}_2\text{COSCoA}+\text{ATP}+\text{H}_2\text{CO}_3 \rightleftharpoons \text{CH}_3\text{CHCOSCoA}+\text{ADP}+\text{P}_i \end{array}$$
propionyl-CoA   methylmalonyl-CoA

(d) Methylcrotonyl-CoA carboxylase (E.C.6.4.1.4)

$$\begin{array}{cc} \text{CH}_3 & \text{CH}_2\text{CO}_2\text{H} \\ | & | \\ \text{CH}_3\text{C}{=}\text{CHCOSCoA}+\text{ATP}+\text{H}_2\text{CO}_3 \rightleftharpoons \text{CH}_3\text{C}{=}\text{CHCOSCoA}+\text{ADP}+\text{P}_i \end{array}$$
β-methylcrotonyl-CoA β-methylglutaconyl-CoA

In each case, except the first, a CoA ester is involved in the reaction. In the case of pyruvate carboxylase, the enzyme from mammalian liver shows an absolute requirement for catalytic amounts of acetyl-CoA whereas the bacterial enzyme shows no such requirement.

Only one carboxyltransferase has so far been characterized; this is methylmalonyl-CoA transcarboxylase (E.C.2.1.3.1) which catalyses the reaction:

$$
\underset{\substack{\text{methylmalonyl-}\\\text{CoA}}}{\overset{\text{COOH}}{\underset{|}{\text{CH}_3\text{CHCOSCoA}}}} + \underset{\substack{\text{CH}_3\\\text{pyruvate}}}{\overset{\substack{\text{COOH}\\|}}{\text{CO}}} \rightleftharpoons \underset{\text{propionyl-CoA}}{\text{CH}_3\text{CH}_2\text{COSCoA}} + \underset{\substack{\text{CH}_2\\|\\\text{COOH}\\\text{oxaloacetate}}}{\overset{\substack{\text{COOH}\\|}}{\text{CO}}}
$$

In contrast to the carbon dioxide ligases, ATP is not involved in this reaction.

The distinctive feature of these enzymes is that they are dependent on biotin (Figure 5) for their action; this is shown by the complete inhibition of the enzymes by avidin, a small protein which binds biotin very firmly.

It is usually the case that the initial formulation of the mechanism of action of an enzyme which employs one of the B-group vitamins as a cofactor arises from studies of the reactions of the free cofactors. Such is the case with the biotin-dependent enzymes.

Lynen[14] and his coworkers found that in the presence of methylcrotonyl-CoA carboxylase, ATP, magnesium ions and bicarbonate, free biotin is carboxylated according to reaction (13); that is, biotin can replace the natural substrate of the enzyme.

$$\text{ATP} + \text{HCO}_3^- + \text{biotin} \xrightleftharpoons[]{\text{carboxylase}} \text{ADP} + \text{P}_i + \text{CO}_2\text{-biotin} \tag{13}$$

The CO_2–biotin adduct was found to be unstable with a half-life of about 20 minutes at pH 7 and 20°, the half-life decreasing rapidly with decreasing pH. Treatment with diazomethane, however, yielded a stable dimethyl ester which was shown to be 1'N-carbomethoxybiotin methyl ester (Figure 5) by comparison with an authentic sample prepared from biotin and methyl chloroformate. Lynen pointed out that the ureido-nitrogen of biotin has weakly acidic properties; hence carboxybiotin may be considered as an acid anhydride. If, therefore, carboxybiotin is an intermediate in the enzyme-catalysed carboxylation reactions, it could reasonably be expected to participate in an electrophilic transfer of carbon dioxide to a suitably activated acceptor. Such an acceptor is naturally provided in the case of pyruvate where the relevant carbon atom is adjacent to a carbonyl group. In the case of the other carboxylase sub-

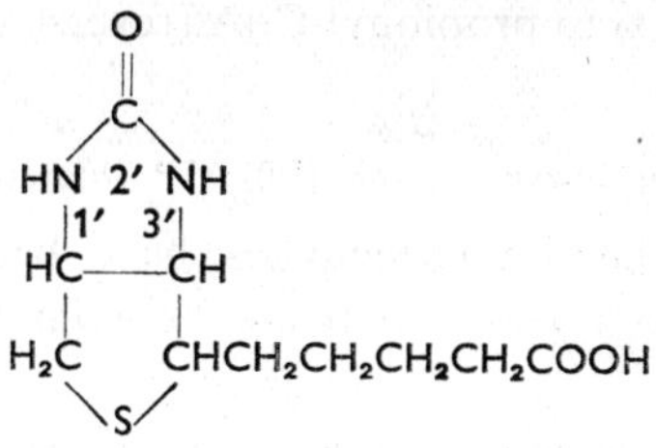

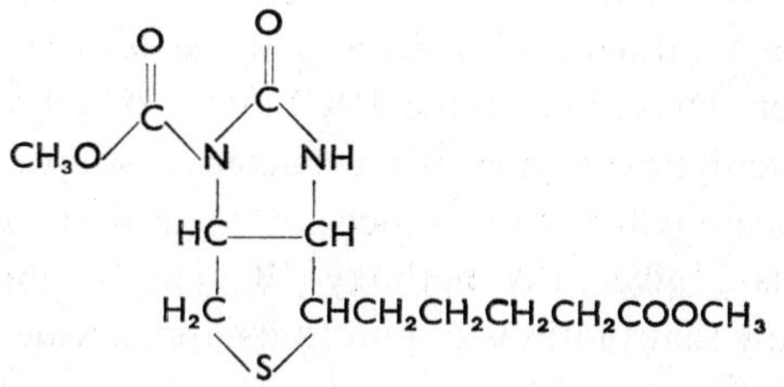

$$
\begin{array}{c}
O \\
\parallel \\
C \\
HN \quad 2' \quad NH \\
|1' \quad 3'| \\
HC———CH \\
H_2C \quad CHCH_2CH_2CH_2CH_2COOH \\
\diagdown S \diagup
\end{array}
$$

$$
\begin{array}{c}
O \quad\quad O \\
\parallel \quad\quad \parallel \\
CH_3O\diagup C \diagdown N \quad NH \\
HC———CH \\
H_2C \quad CHCH_2CH_2CH_2CH_2COOCH_3 \\
\diagdown S \diagup
\end{array}
$$

Structure of $1N'$-carbomethoxybiotin methyl ester

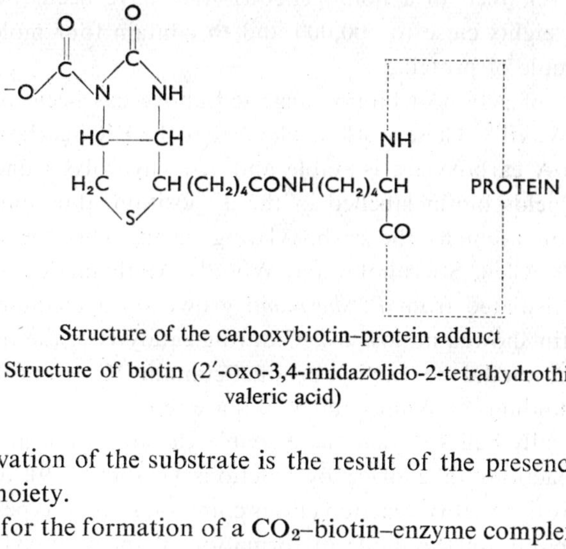

$$
\begin{array}{c}
O \quad\quad O \\
\parallel \quad\quad \parallel \\
{}^-O\diagup C \diagdown N \quad NH \\
HC———CH \\
H_2C \quad CH(CH_2)_4CONH(CH_2)_4CH \quad\quad PROTEIN \\
\diagdown S \diagup \quad\quad\quad\quad NH \\
\quad\quad\quad\quad CO
\end{array}
$$

Structure of the carboxybiotin–protein adduct

FIGURE 5. Structure of biotin (2'-oxo-3,4-imidazolido-2-tetrahydrothiophene-n-valeric acid)

strates, activation of the substrate is the result of the presence of the coenzyme A moiety.

Evidence for the formation of a CO_2–biotin–enzyme complex during the carboxylation reaction was provided by Kaziro and Ochoa[15]. Propionyl-CoA carboxylase from pig heart muscle was carboxylated by either reaction (14) or (15).

$$ATP + H^{14}CO_3^- + biotin\text{–}enzyme \underset{}{\overset{Mg^{2+}}{\rightleftharpoons}} {}^{14}CO_2\text{–}biotin\text{–}enzyme + ADP + P_i \quad (14)$$

$$
\begin{array}{c}
{}^{14}COOH \\
| \\
CH_3CHCOSCoA + biotin\text{–}enzyme \rightleftharpoons CH_3CH_2COSCoA + {}^{14}CO_2\text{–}biotin\text{–}enzyme \quad (15)
\end{array}
$$

The carboxylated enzyme was isolated and shown to contain approximately one mole of acid-labile $^{14}CO_2$ per mole of biotin. The enzyme was shown to transfer $^{14}CO_2$ to propionyl-CoA (reverse of equation 14) and to catalyse reaction (16).

$$ADP + {}^{32}P_i + {}^{14}CO_2\text{-biotin-enzyme} \underset{}{\overset{Mg^{2+}}{\rightleftharpoons}} [{}^{32}P]\,ATP + biotin\text{-enzyme} + H^{14}CO_3^- \quad (16)$$

Analogous carboxylated intermediates have been isolated from the remaining three carboxylases and from methylmalonyl-CoA transcarboxylase.

These $^{14}CO_2$-biotin-enzyme complexes are less stable than free carboxybiotin, but again treatment with diazomethane leads to a stable carbomethoxy product. Digestion of the [^{14}C]-methoxycarbonyl protein with pronase (a proteolytic enzyme from bacterial sources) liberates 1'N-methoxy-[^{14}C]-carbonylbiocytin, which on hydrolysis with the specific enzyme biotinidase gives 1'N-methoxy-[^{14}C]-carbonylbiotin and lysine. These results show that biotin is bound to its apoenzyme by a peptide link with the ε-amino group of a lysine residue (Figure 5). All the biotin enzymes so far obtained in a homogeneous state have been found to have molecular weights close to 700,000 and to contain four moles of bound biotin per mole of protein.

The mode of action of biotin suggested above has been questioned by Waite and Wakil[16]. These workers claim that the [^{14}C]-carboxylated form of acetyl-CoA carboxylase is stable and after hydrolysis under rigorous conditions yields biotin labelled in the 2' position, thus implicating the ureido-carbon atom as the carboxylating agent. This theory has been examined by Allen, Stjernholm and Wood[17]. Methylmalonyl-CoA transcarboxylase isolated from *P. shermanii* grown on a medium containing [2'-^{14}C]-biotin showed no loss of ^{14}C during catalysis of the transcarboxylase reaction, nor did the reaction product contain ^{14}C. It must be concluded that the findings of Waites and Wakil are in error.

It is generally believed that the enzymic carboxylation and transcarboxylation reactions take place by reactions (17) and (19), and (18) and (19) respectively; that is, reaction (19) is common to both types of enzymes which differ only in the mode of formation of the carboxylated enzyme intermediate.

$$H_2CO_3 + ATP + biotin\text{-enzyme} \underset{}{\overset{Mg^{2+}}{\rightleftharpoons}} CO_2\text{-biotin-enzyme} + ADP + P_i \quad (17)$$

$$methylmalonyl\text{-CoA} + biotin\text{-enzyme} \rightleftharpoons CO_2\text{-biotin-enzyme} + propionyl\text{-CoA}$$
$$(18)$$

$$CO_2\text{-biotin-enzyme} + acceptor \rightleftharpoons biotin\text{-enzyme} + carboxylated\ acceptor$$
$$(19)$$

This formulation is also supported by a variety of isotope exchange experiments. For example, in accordance with equation (17), propionyl-CoA carboxylase catalyses the exchange of $[^{32}P]ADP$ with ATP in the presence of P_i, or of $^{32}P_i$ with ATP in the presence of ADP. Both of the exchange reactions require Mg^{2+} and HCO_3^-, and are inhibited by avidin and sulphydryl reagents.

Kaziro[18] and his colleagues have studied the reactions of purified pig heart propionyl-CoA carboxylase using $H_2^{18}O$ and $NaHC^{18}O_3$ and have shown that the bicarbonate ion is the reactive species of CO_2 in reaction (17). One ^{18}O from bicarbonate was found in the liberated P_i and the remaining two atoms in the free carboxyl group of methylmalonyl-CoA. These authors propose the concerted mechanism shown for reaction (17) (cf. A2a).

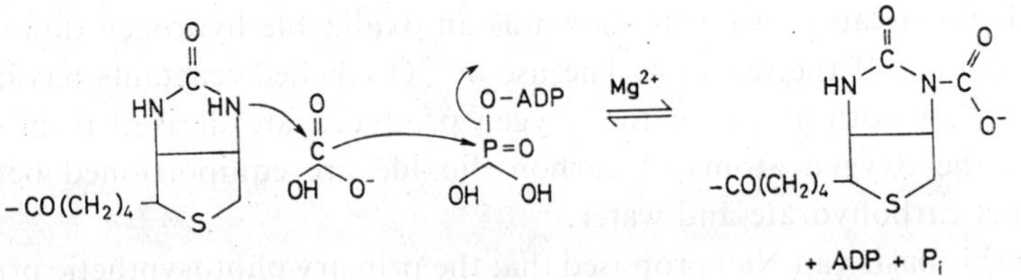

No detailed evidence is available concerning the mechanism of reactions (18) and (19), but it would seem likely, as suggested above, that reaction (19) involves a direct nucleophilic transfer of CO_2 to the activated acceptor.

B. Photosynthesis

Photosynthesis is without doubt the most important single biochemical process, since it is on this source that the rest of the biosphere (with the exception of the chemosynthetic bacteria) depends for its continued existence.

The photosynthetic apparatus is restricted to the plant family, including such primitive forms as the algae and diatoms (but not fungi) and to the photosynthetic bacteria. The latter may be divided into two main groups, the green and purple sulphur bacteria (*Thiorhodaceae*) and the purple non-sulphur bacteria (*Athiorhodaceae*). The processes in plants and bacteria are basically very similar and will be dealt with together. Many monographs and reviews of photosynthesis are available of which those given as references 19–27 may be recommended.

1. Formulation of the reaction

Basically, the photosynthetic reaction in plants may be represented by the equation

$$CO_2 + H_2O \longrightarrow [CH_2O] + O_2$$

where $[CH_2O]$ is the basic structural unit of carbohydrate material.

Non-oxygen producing photosynthesis (bacteria) has been formulated similarly:

$$CO_2 + 2H_2D \longrightarrow [CH_2O] + H_2O + 2D$$

Here H_2D is an oxidizable hydrogen donor which may be for example hydrogen sulphide in the case of sulphur bacteria, a simple organic acid in the case of non-sulphur bacteria, or hydrogen gas in some hydrogen-adapted bacteria and algae.

Van Niel[278, 279] pointed out the basic similarity between these formulations if the plant reaction is written:

$$CO_2 + 2H_2O \longrightarrow [CH_2O] + H_2O + O_2$$

In this formulation water is shown as an oxidizable hydrogen donor and as a product of the reaction. The use of ^{18}O-labelled reactants has in fact shown that both atoms of the oxygen produced are derived from water whilst the oxygen atoms of carbon dioxide are equiportioned between product carbohydrate and water.

On this basis van Niel proposed that the primary photosynthetic process, common to all phototrophs, is the photolysis of water to produce oxidizing and reducing species:

$$4H_2O + 2X + 4Y \xrightarrow{light} 4YOH + 2XH_2$$

Here, X and Y represent hypothetical oxidizing and reducing agents respectively, whose subsequent reactions may be written:

$$4YOH \longrightarrow 2H_2O + 4Y + O_2 \quad \text{(Plants)}$$

or

$$4YOH + 2H_2D \longrightarrow 4H_2O + 4Y + 2D \quad \text{(Bacteria)}$$

and

$$2XH_2 + CO_2 \longrightarrow [CH_2O] + H_2O + 2X$$

This simple representation is only partly correct. It is now known that the generation of oxidizing and reducing species in plant photosynthesis is brought about at two separate photochemical centres by an electron transport process. Photolysis of water is a consequence of this electron transport in plants and is probably not involved in bacterial photosynthesis at all.

Two very important aspects of the photosynthetic process are, however, embodied in van Niel's formulation. First, the scheme requires the generation of a stable reducing species. Secondly, this reducing species is responsible for carbon dioxide fixation by a reaction or series of reactions which are not dependent on light. This separation of photosynthesis into light and dark reactions has been demonstrated by many workers in this field

The reducing agent produced by the light reactions of photosynthesis has been identified as NADPH (Figure 2) in the case of plants, and NADH in bacteria. Of equal, or perhaps greater, importance is the generation during the electron transport process of ATP. This process, (photophosphorylation), was first demonstrated by Arnon[28] and his colleagues. The requirement for these two species (NADPH and ATP) may be seen from the overall equation for carbon dioxide fixation:

$$CO_2 + 3\ ATP + 2\ NADPH + 3\ H_2O + 2\ H^+ \longrightarrow [CH_2O] + 3\ ADP + 2\ NADP^+ + 3\ H_3PO_4 + H_2O$$

Before considering the individual reactions which contribute to overall carbon dioxide fixation, some attention must be given to the unique processes by which ATP and NADPH are produced in photosynthetic systems.

2. Composition and organisation of the photosynthetic apparatus.

Photosynthetic cells of green plants and algae (except the blue-green variety) contain protoplasmic inclusions called chloroplasts. The chloroplast as seen by electron microscopy consists of structureless stroma in which are distributed dark lamellar components called grana. Isolated lamellae have been shown to contain the photosynthetic pigments and, in the presence of suitable oxidizing agents, can catalyse oxygen evolution from water (the Hill reaction[29]). When supplemented with a preparation from the stroma of the chloroplast, reduction of $NADP^+$ and generation of ATP may be demonstrated.

Park[30] and his collaborators have shown that the lamellae of spinach chloroplasts are composed of a regular array of subunits in the form of oblate ellipsoids (dimensions 200×100 Å). These subunits have been called quantasomes and are believed to be the basic photosynthetic unit. They consist of about 50% protein, the remainder being a complex mixture of chlorophyll (a and b), carotenoids, quinones, phospholipids, galactosylglycerides and sulpholipids. Non-haeme iron, and manganese and copper ions were also found, together with two types of cytochrome (cytochrome b_6 and cytochrome f). The structure of some of these molecules[22] are shown in Figure 6; a survey of photosynthetic pigments has been given by Smith and French[31].

Less is known about the photosynthetic unit in bacteria. With a variety of bacteria, rupture of the cell wall allows the isolation of highly coloured particles (chromatophore fragments) capable of carrying out the reactions of photosynthesis. There is considerable evidence, however, that these

FIGURE 6. Structures of some photosynthetic pigments and electron transfer molecules

Chlorophyll Prosthetic Groups

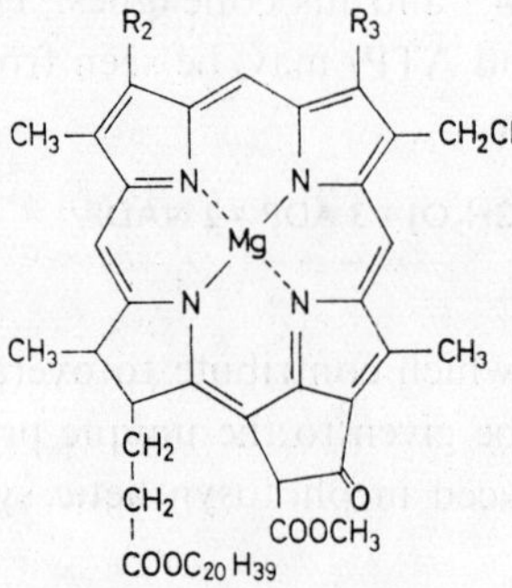

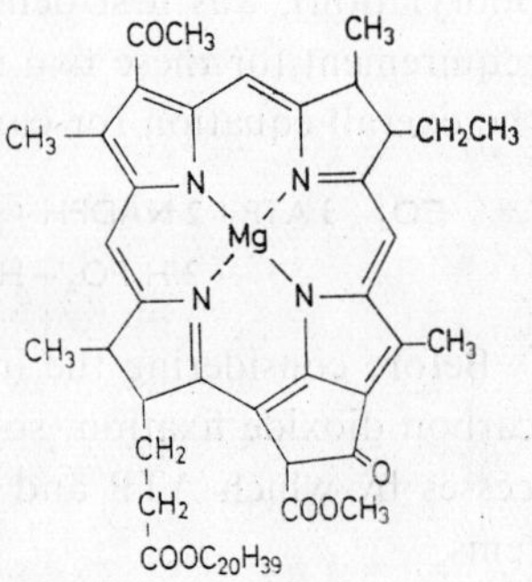

Chlorophyll a. $R_2 = -CH = CH_2$, $R_3 = -CH_3$ Bacteriochlorophyll

Chlorophyll b. $R_2 = -CH = CH_2$, $R_3 = -CHO$

Chlorophyll d. $R_2 = -CHO$, $R_3 = -CH_3$

Cytochrome Prosthetic Groups

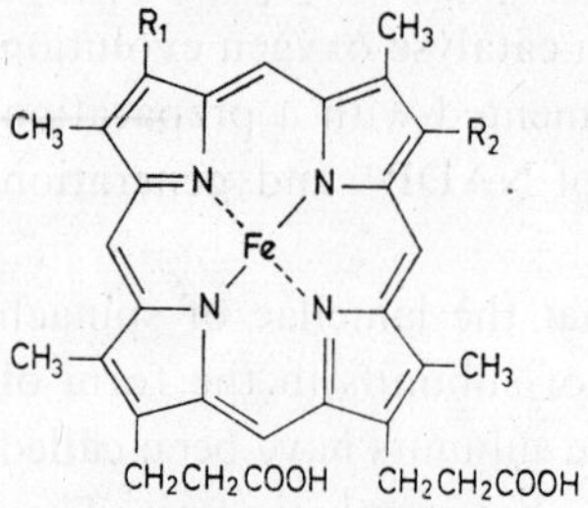

B-class (e.g. cyt. b_b)

$R_1 = R_2 = -CH = CH_2$

C-class (e.g. cyt. f)

$R_1 = R_2 = -CH\overset{CH_3}{\underset{SR}{<}}$

Biliprotein Prosthetic Groups

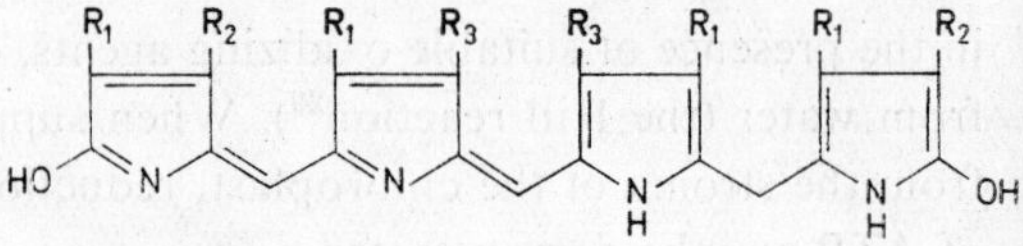

mesobiliviolin (in Phycocyanin)

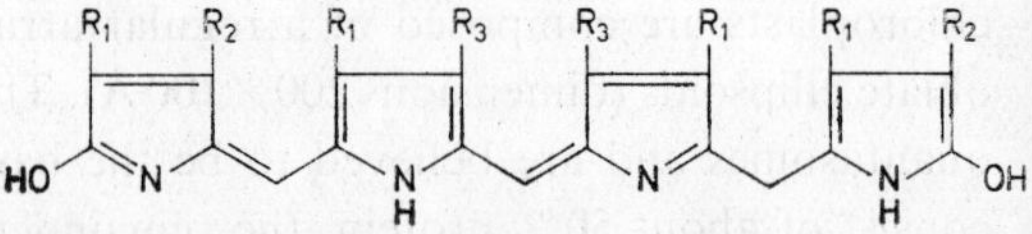

mesobilirhodin (in Phycoerythrin)

$R_1 = -CH_3$, $R_2 = -CH_2CH_3$,

$R_3 = -CH_2CH_2COOH$

β-Carotene

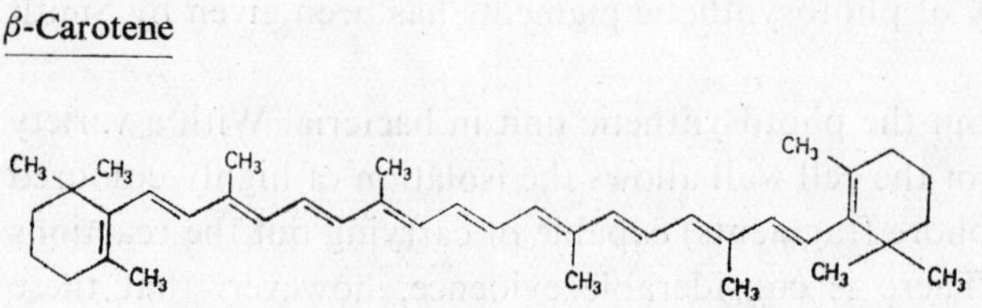

Plastoquinone

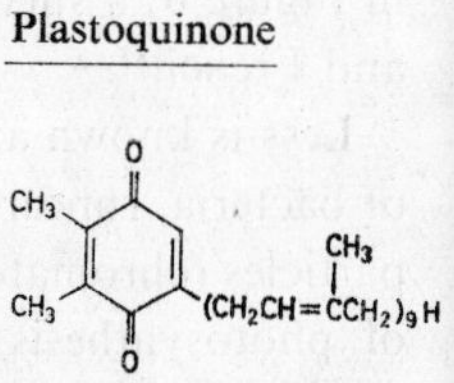

particles do not pre-exist *in vivo*, but are an artefact of the isolation procedure. Hence it is probable that, as in the case of the blue-green algae, the bacterial photosynthetic pigments are loosely associated with membrane structures in the cell.

3. The Light reactions[32-35]

It is widely held that photochemical electron transport in plants involves two distinct photochemical centres. The original evidence for this came from studies of the efficiency of photosynthesis as a function of the wavelength[33]. For example, a pronounced drop in the efficiency of photosynthesis (the red drop) was observed with *Chlorella*, *Chroococcus* and the diatom *Navicula* when light of wavelength greater than 685 mμ (the absorption maximum of chlorophyll *a in vivo*) was used. The red drop could, however, be delayed if a background of green light was supplied. This is known as an enhancement effect. Studies with *Chlorella* showed that with a background of red light absorbed by chlorophyll *a*, maximal enhancement was obtained with light of wavelength 480 and 655 mμ, corresponding to the absorption maxima of chlorophyll *b in vivo*. Similar experiments with *Navicula*, the blue-green alga *Anacystis* and the red alga *Porphyridium* showed enhancement spectra implicating fucoxanthol, phycocyanin and phycoerythrin respectively as ancillary light absorbing species in the short wavelength system.

In general, the long wavelength system (pigment system I or PSI, see Figure 7)[33] involves a long-wavelength non-fluorescent form of chlorophyll *a* absorbing at about 685 mμ and a small quantity of a special chlorophyll (P700) absorbing maximally at 700 mμ. The short wavelength system, (pigment system II, PSII)[33], contains chlorophyll *b* or chlorophyll *c* in addition to a short wavelength form of chlorophyll *a* (670 mμ), together with phycobilin pigments and carotenoids (mainly β-carotene); this system is less well characterized than PSI.

On the basis of available evidence, Vernon and Avron[33] have presented the scheme shown in Figure 7 for the light-induced electron flow in plant chloroplasts leading from the oxidation of water to the reduction of $NADP^+$. The direction of the arrows shows the direction of electron flow. Broad arrows denote photochemical processes, and dotted arrows non-physiological reactions. Absorption of light by PSII causes an electron to move against a thermochemical gradient to an acceptor, plastoquinone, ($E_0' \sim 0$ volts, pH 7) whilst the 'positive hole' causes liberation of oxygen from water ($E_0' \sim +0.8$ volts). The steps involved in the latter process are obscure, but manganese and chloride ions are known to be involved.

60*

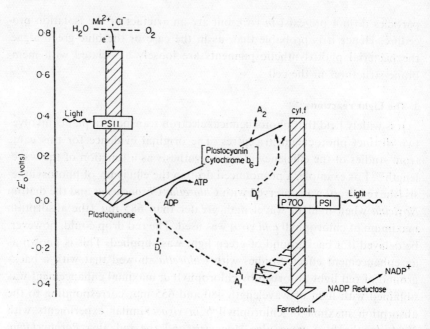

FIGURE 7. Light-induced electron flow in chloroplasts (Taken from reference 33.)

A thermochemical reaction then occurs between about 0 and 0·4 V, mediated by an incompletely characterized system called the Intermediate Electron Transfer Complex (IETC). This complex contains cytochrome b_6 and a blue copper-containing protein, plastocyanin. Vernon and Avron do not specify the arrangement of these species in the IETC, but Hill[25] has cytochrome b_6 ($E_0' = -0.03$ V) as the acceptor from plastoquinone followed by plastocyanin ($E_0' \sim +0.4$ V). In both schemes, the terminal acceptor of the IETC and donor to PSI is cytochrome f ($E_0' = +0.375$ V), or cytochrome c (algae). A light-induced electron flow through PSI with P700 as the reaction centre follows, the acceptor for this system being the small iron-containing protein ferredoxin. There is some evidence that electron transfer from P700 to ferredoxin may be *via* plastoquinone[25, 36]. The final physiological process is the transfer of electrons to $NADP^+$ ($E_0' = -0.324$ V), catalysed by the flavoprotein $NADP^+$ reductase. A comprehensive review of ferredoxin in photosynthesis is given by San Pietro and Black[37].

It must be emphasized that the precise physicochemical events following light absorption by a photosynthetic pigment are not known. It has been shown that PSI contains some 300 molecules of chlorophyll *a* for every molecule of the special chlorophyll (P700) and cytochrome f. Absorption of light by a molecule of chlorophyll a is thought to promote an electron to an excited singlet state. The excited chlorophyll molecule, either in the singlet or a triplet state, then transfers its excitation energy to the special (or *reaction centre*) chlorophyll, P700. The sequence of events may be as follows:

$$Q[n \text{ Chl.P700}]cyt.f(Fe^{2+}) \xrightarrow{h\nu} Q[(n-1) \text{ Chl.Chl*.P700}]cyt.f(Fe^{2+})$$

$$\longrightarrow Q^{-}_{red.}[n \text{ Chl.P700}^{+}]cyt.f(Fe^{2+}) \longrightarrow$$

$$Q^{-}_{red.}[n \text{ Chl.P700}]cyt.f(Fe^{3+})$$

Here, n Chl. represents the aggregate of chlorophyll a molecules. The electron donor is shown as cytochrome f and the acceptor as plastoquinone (Q). A similar situation probably pertains with PSII but the details are even less certain.

It is obvious that the photochemical process associated with either PSI or PSII cannot cause electron flow through a potential gradient greater than the energy of the quantum of absorbed light. For light of wavelength 700 mμ this is equivalent to 1780 mV for a one electron process. As can be seen from Figure 7, this value is considerably greater than the potential differences actually involved.

Compelling evidence for the scheme in Figure 7 has been educed from the effects of added electron acceptors and donors in chloroplast preparations. When PSI and PSII are operative and coupled by the IETC, oxygen evolution may be demonstrated when acceptors AI and AI' are introduced (the Hill reaction). AI type acceptors include quinones, nitrite, and ferricyanide. AI' type acceptors which, after reduction, are able to donate electrons to the IETC include phenazine methosulphate (PMS), flavine mononucleotide (FMN), vitamin K_3, and 2,6-dichlorophenolindophenol (DPIP). These acceptors can, therefore, catalyse a cyclic, light-induced electron flow through PSI. Acceptors A2 (ferricyanide, methylene blue) are reduced by PSII and IETC; in their presence, photolysis of water may be observed when PSI is inoperative. Donors DI and DI' donate electrons to the IETC and promote flow of electrons through PSI resulting in NADP$^+$ reduction without operation of PSII. The donors include the reduced forms of PMS and DPIP in the presence of ascorbate, the point of entry depending mainly on the concentration of the donor.

Available evidence suggests that only one photochemical pigment system exists in the photosynthetic bacteria, this system closely resembling PSI of plants[32]. Purple bacteria contain a variety of chlorophyll called bacteriochlorophyll (BChl., Figure 6). Minor components (BChl.b.) have been demonstrated in *Rhodospirillum rubrum* (P890) *Chromatium* (P890) and *Rhodopseudomonas spheroides* (P870), and are thought to be the reaction centre pigments in these species (equivalent to P700 in plants). The green sulphur bacteria differ from the purple bacteria in their main chlorophyll component; in this case it is chlorobium chlorophyll 660 and 650. Small amounts of a different type (P770) similar to BChl. have been isolated from *Chlorobium thiosulphatophilum* and *Chloropseudomonas ethylicum* which may represent the reaction centre molecule in these species.

Vernon[32] has given an electron transport scheme for the organism *R. rubrum* (Figure 8). In this case absorption of light promotes transfer of an electron from reduced cytochrome c_2 to ferredoxin (cf. PSI in plants)

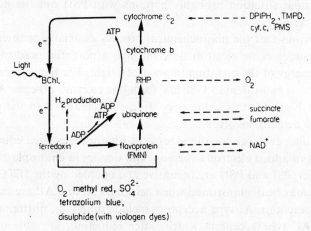

FIGURE 8. Probable electron transport sequence and reaction sites in *R. Rubrum* (Taken from reference 32.)

Reduction of NAD^+ is linked to ferredoxin oxidation by a flavoprotein, again similar to the plant system. Alternatively, a cyclic electron flow may be established from ferredoxin to cytochrome c_2 via ubiquinone, RHP (rubrum haeme protein) and cytochrome *b*. Points of entry of non-physiological acceptors and donors are shown in Figure 8. Also shown is a site for hydrogen production which arises from the presence of a hydrogenase in photosynthetic bacteria. This expression of bacterial photoreduction has been critically reviewed[35, 38]. It can be seen that an electron transport system of the type shown in Figure 8 is in some ways more flexible then the corresponding system in plants since a variety of points of entry of electrons from physiological substrates are available. Hence, although different bacterial species photometabolize various organic and inorganic substrates, it seems likely that an electron transport sequence of the type operative in *R. rubrum* could accomodate these variations.

A link may be recognized between the plant and bacterial light reaction systems in mutant forms of algae which have been adapted for photoreduction. In these species, carbon dioxide fixation is accompanied by oxidation of molecular hydrogen, and it appears that only PSI is operative. Consistently, such adopted algae do not show enhancement effects, and are similar to bacteria in this respect.

4. Photophosphorylation[25,32,33]

From the positions of the phosphorylation sites shown in Figures 7 and 8 it can be seen that photophosphorylation occurs during the dark oxidative phases of the electron transport sequences. In this it resembles the process of mitochondrial oxidative phosphorylation[39]. It seems likely that the mechanism of phosphorylation is similar in these processes, but the reactions involved are obscure. It is conventional to write the process as

$$\text{ADP} + \text{P}_i \longrightarrow \text{ATP} + \text{H}_2\text{O} \tag{20}$$

However, following the arguments given in section I, it seems highly unlikely that a direct condensation between phosphate and ADP occurs and it must be emphasized that equation (20) is a purely symbolic representation.

The locus of the phosphorylation site shown in Figure 7 has been established by studies of the three types of photophosphorylation described below.

(a) Non-cyclic photophosphorylation[28, 280].

This is the process which probably occurs *in vivo*, and requires intact coupled electron transport systems. The overall reaction may be written:

$$\text{NADP}^+ + \text{ADP} + \text{P}_i \xrightarrow{\text{light}} \text{NADPH} + \tfrac{1}{2}\text{O}_2 + \text{ATP} + \text{H}^+$$

The reaction may be observed with isolated chloroplast preparations provided that stoichiometric quantities of an acceptor (NADP$^+$, ferredoxin, or AI in Figure 7) are supplied.

Non-cyclic phosphorylation is inhibited by compounds which block the action of PSII. Examples are 3-(3,4-dichlorophenyl)-1,1-dimethyl urea (DCMU), nonyl-4-hydroxyquinoline-N-oxide (NQNO) and salicyloxime[281, 282]. Inhibition may be relieved by donors of type DI which donate electrons before the proposed site of ATP formation.

(b) Cyclic Photophosphorylation[280, 282-284].

Under strictly anaerobic conditions, a cyclic photophosphorylation reaction may be observed with isolated chloroplasts if catalytic quantities of AI′ type acceptors are added, (e.g. PMS, quinones). The overall result is

$$ADP + P_i \xrightarrow{\text{light}} ATP + H_2O$$

Phosphorylation is coupled to a cyclic electron flow with no net change in oxidation level of other components in the system. The inhibitors listed above do not affect cyclic photophosphorylation, since PSII does not contribute to the process.

(c) Pseudo-cyclic Photophosphorylation[285].

Under aerobic conditions, a variant of the cyclic process, yielding the same overall result, may be observed. Here the catalytic electron acceptor is photoreduced and subsequently reoxidized by molecular oxygen arising from the photolysis of water. FMN is a typical acceptor of this type. Inhibition by DCMU and NQNO is observed since oxygen production by PSII is required.

Under non-cyclic conditions, with acceptors such as ferricyanide, NADP$^+$ and DPIP, the formation of 1 molecule of ATP for every electron pair transferred has been observed (*i.e.* ATP/2e$^-$ = 1)[28]. However, since in the absence of photophosphorylation a basal electron flow (30–50% of the optimum) is still observed, it has been claimed that the ratio ATP/2e$^-$ for electron transport associated with phosphorylation should be considered closer to two than to unity[286]. This argument has been claimed to show that two sites of phosphorylation exist in the chloroplast electron transport sequence.

Support for this hypothesis has come from studies of phosphorylation uncoupling agents[287]; that is, compounds which inhibit photophosphorylation without diminution of electron transport. Some uncouplers are equally effective under cyclic and non-cyclic conditions, *e.g.* atebrine, chloropromazine. Others are much more effective with non-cyclic rather then cyclic photophosphorylation, *e.g.* octyl guanidine, whilst ammonium chloride, for example, uncouples non-cyclic slightly more effectively then cyclic photophosphorylation. The data tend to indicate that the loci of the cyclic and non-cyclic processes are different.

Cyclic photophosphorylation has been observed in a variety of species of bacteria[32]. The proposed sites for this process are shown in Figure 8. Again, phosphorylation occurs during the dark, oxidative phase of electron transport. It can be seen that the electron transport system of *R. rubrum* can catalyse a cyclic phosphorylation *in vivo*, with no net change in oxidation state of any other component of the system; indeed, some authors claim that this process is of far greater importance than NAD^+ reduction in bacterial systems, and others press the view that it is the only expression of the bacterial photochemical electron transport *in vivo*. Non-cyclic photophosphorylation has been observed with chromatophore fragments from *R. rubrum* in which cyclic electron flow was inhibited with 2-heptyl-4-hydroxyquinoline-*N*-oxide. Reduction of NAD^+ and associated phosphorylation could be restored by the addition of ascorbate and $DPIPH_2$.

5. The Carbon reduction cycle[20, 21, 40]

The generally accepted sequence of reactions by which carbon dioxide is incorporated into hexose by photosynthetic organisms is shown in Figure 9. The cycle is due mainly to Bassham and Calvin and was derived from studies of ^{14}C-labelling of intermediates produced by algae growing in an atmosphere of $^{14}CO_2$[21].

In terms of carbon atoms, the basic scheme may be represented as shown;

$$6\ CO_2 + 6\ C_{(5)} \longrightarrow 12\ C_{(3)} \searrow C_{(6)}$$

that is, carbon dioxide is converted to hexose by an initial reaction with a C_5 compound which is eventually regenerated by a cyclic process.

Support for the proposed reaction sequence in Figure 9 has been provided by the isolation, from a variety of photosynthetic species, of enzymes catalysing all the steps in the sequence[6]. The sequence of reactions is conveniently considered in the following stages:

(a) Phosphorylation of ribulose-5-phosphate to give the 1,5-diphosphate
(b) carboxylation of the acceptor followed by hydrolytic cleavage to 3-phosphoglycerate
(c) conversion of 3-phosphoglycerate to glyceraldehyde-3-phosphate. At this stage, the steps requiring ATP and NADPH from the electron transport sequence are concluded and are followed by
(d) regeneration of ribulose-5-phosphate and formation of product hexose by a series of steps which are not restricted to photosynthetic species. These steps are essentially the same as those which occur in the pentosephosphate cycle for glucose catabolism. Similarly, phosphoglycerate kinase, the enzyme catalysing step b, is of wide occurrence.

It is beyond the scope of this review to present the results of kinetic studies of ^{14}C-labelling which have given rise to the Calvin–Benson cycle.

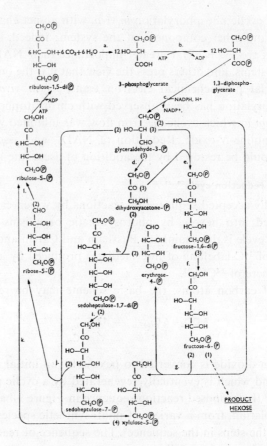

Reaction	Enzyme	E.C. number
a.	carboxydismutase (ribulose-1,5-di⑫ carboxylase)	—
b.	phosphoglycerate kinase	2.7.2.3
c.	glyceraldehydephosphate dehydrogenase	1.2.1.9
d.	triosephosphate isomerase	5.3.1.1
e.	aldolase	4.1.2.7
f.	hexosediphosphatase	3.1.3.11
g.	transketolase	2.2.1.1
h.	aldolase	4.1.2.7
i.	phosphatase (specificity uncertain)	
j.	transketolase	2.2.1.1
k.	ribulose phosphate epimerase	5.1.3.1
l.	ribosephosphate isomerase	5.3.1.6
m.	phosphoribulokinase	2.7.1.19

FIGURE 9. The Calvin–Benson cycle

These are amply dealt with in the literature[20, 21, 40]. It may be appropriate, however, to indicate some lines on which this formulation has been criticized.

Although, as has been previously stated, photosynthetic tissues contain all the enzymic activities required by the Calvin–Benson cycle, it has frequently been reported that the activities of various of these enzymes are too low to support carbon dioxide fixation at the rate observed *in vivo*. Similarly, fixation by a reconstituted enzyme system requires much higher carbon dioxide pressures and proceeds at a lower rate than *in vivo*. These objections may be countered by the assumptions that decreased enzymic activities result from the isolation procedures and that, *in vivo*, the various enzyme activities are grouped together in a single (or small number of) proteins—that is, multifunctional enzymes[34, 40]. Bassham[34] has considered ways in which such a multifunctional enzyme system, perhaps closely associated with the lipid membranes of the lamellae, could bring about the reactions of the carbon reduction cycle with high efficiency. If such an organized system did exist, then it is highly likely that the properties of individual enzymes obtained by fragmentation of the parent multifunctional system would fail to reflect the full potentialities of the system *in vivo*.

The scheme shown in Figure 9 has also been questioned as a result of the observation that the hexose produced by organisms growing on $^{14}CO_2$ is frequently asymmetrically labelled (the 'Gibbs effect'[288, 289]). Proponents[20] of the Calvin–Benson cycle have shown, however, that this effect can be explained on purely kinetic grounds. Asymmetric labelling could also arise if the two contributing triose molecules could be kept apart before condensation and hence prevented from equilibrating. In this connection, Bassham and Kirk[41] have found that carboxylation of ribulose-1,5-diphosphate in *Chlorella* produces only one molecule of free phosphoglyceric acid whereas in isolated enzyme systems and in dark periods *in vivo*, two molecules are produced. They suggest that this requirement for light indicates that the process *in vivo* is reductive, possibly involving direct participation by reduced ferredoxin.

Such a scheme would require very precise localization of the carboxylating system with respect to the electron transport sequence, and the pathway would not be observed in disrupted systems. Asymmetric labelling of hexose could arise from such a reaction sequence.

A similar discrimination between the product phosphoglycerate molecules of the carboxydismutase reaction is inherent in the mechanism proposed by Rabin and Trown[13]. This has been presented in detail in section

II A2. If this mechanism is operative, then the reaction results in the production of one free molecule of 3-phosphoglycerate and one molecule bound to the enzyme surface by a sulphur–carbon bond. The bound molecule could be liberated by hydrolysis or transferred to a neighbouring region of a multifunctional enzyme system.

After short exposure of photosynthetic organisms to $^{14}CO_2$, the radioactive label appears not only in the intermediates of the Calvin–Benson cycle but also in a variety of other compounds, principally amino acids and dicarboxylic acids. These are formed from intermediates of the cycle by known routes which will be dealt with later. It has been known for some time, however, that under certain conditions (high light intensity and oxygen level, low levels of carbon dioxide) radioactivity rapidly appears in the two-carbon fragment, glycollic acid. Similarly, ^{14}C introduced into these organisms in the form of [2-^{14}C]-glycollic acid rapidly appears in both the α and β carbon atoms of 3-phosphoglycerate. It has been claimed[42] that these processes are incompatible with the Calvin–Benson cycle and have been cited as evidence for the direct formation of a two-carbon unit from carbon dioxide as a primary process in photosynthesis.

A plausible explanation of the formation of glycollic acid as a by-product of the transketolase reaction has, however, been given by Calvin and Bassham[21]. Transketolase shows a complete dependence on the coenzyme thiamine pyrophosphate (TPP; see section IV. A.1). The coenzyme acts as an acceptor for a two-carbon fragment from the donor ketose:

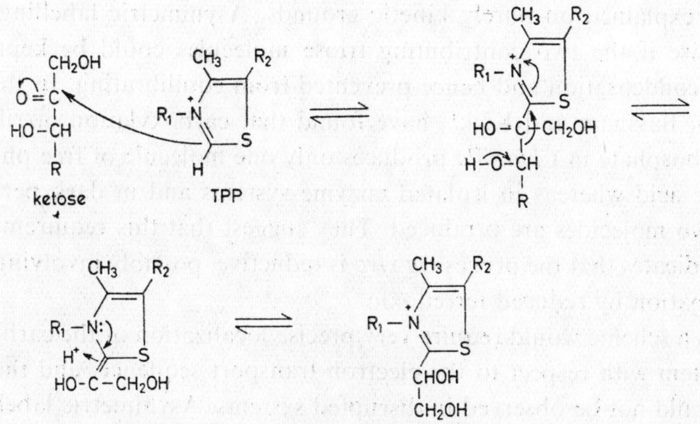

2-(1,2-dihydroxyethyl)-TPP

The transketolase reaction is completed by transfer of the two-carbon fragment to the acceptor aldose. Calvin and Bassham, however, suppose that, under certain conditions, an oxidative cleavage of the 2-(1,2-dihyd-roxyethyl)-TPP adduct may occur with the production of glycollic acid. A possible sequence of reactions is:

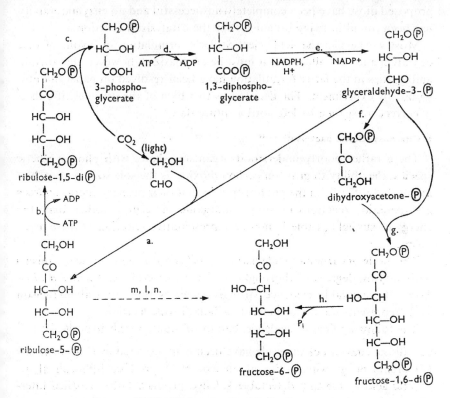

Here ES$_2$ is a hypothetical enzyme possibly associated with a lipoic acid cofactor.

Stiller[42] has claimed that the results of the studies of ^{14}C-labelling in photosynthetic organisms have been misinterpreted by Calvin and his associates. An entirely new scheme is proposed in which a diose is synthe-sized *de novo* from carbon dioxide. This diose then condenses with glycer-

FIGURE 10. Modified path of carbon in photosynthesis

aldehyde-3-phosphate to yield ribulose-5-phosphate which, after phosphorylation to the diphosphate, is carboxylated in the carboxydismutase reaction. Glyceraldehyde-3-phosphate is regenerated in the usual way. The scheme is shown in Figure 10. Product hexose is considered to arise both from the condensation of two triose phosphate molecules in the fructose diphosphate aldolase reaction (g), and from ribulose-5-phosphate by the ribulose-5-phosphate epimerase (m), transketolase (l) and transaldolase (n) reactions.

The main arguments in favour of this scheme are as follows. It does not involve those reactions of the Calvin–Benson cycle for which the enzymic activities in isolated systems are claimed to be inadequate; the scheme would inevitably give rise to asymmetrically labelled products; the formation of glycollic acid is easily explained, given a diose of the type shown in Figure 10; finally, it explains the observation by many workers that carbon dioxide is in some way involved in the light reactions of photosynthesis. On the debit side, it must be emphasized that intensive searches for the proposed diose have been completely unsuccessful and no enzymic activity has been found in living organisms for the catalysis of reaction a.

Most workers in the field reject Stiller's formulation in favour of that of Calvin and his colleagues; it must be emphasized, however, that certain ambiguities in the latter formulation have been resolved on paper but not, so far, by experiment. The decision as to which of the schemes, if either, is correct must then be left until a future date.

6. Chemosynthetic bacteria[26, 43, 44]

These rather poorly understood organisms share, with photosynthetic species, the ability to grow on carbon dioxide as the sole source of carbon atoms, but differ from the phototrophs in that their primary energy sources are oxidation reactions of simple inorganic molecules, rather than the energy of sunlight. Table 1 shows a representative selection of these organisms.

The mode of carbon dioxide fixation has been studied in these organisms with varying degrees of thoroughness. In the case of *Thiobacillus denitrificans* the available evidence strongly suggests that the Calvin–Benson cycle represents the major path of carbon dioxide fixation.

The following lines of evidence tend to substantiate this hypothesis;

(a) all the enzymes of the cycle have been demonstrated in *T. denitrificans*
(b) organisms growing in the presence of $^{14}CO_2$ produce 3-phosphoglyceric acid as the first detectable labelled product. Other labelled intermediates are formed on longer exposure to $^{14}CO_2$

(c) The pool size of ribulose-1,5-diphosphate increases when the organism is deprived of carbon dioxide, whilst the content of 3-phosphoglycerate decreases. Removal of thiosulphate ions from the medium produces the opposite effect.

TABLE 1

Organism	Oxidant	Substrate	Product
Thiobacillus thiooxidans	O_2	$S,S_2O_3^{2-}$	SO_4^{2-}
Thiobacillus denitrificans	NO_3^-	$S,S_2O_3^{2-}$	SO_4^{2-}
Thiobacillus thioparus	O_2	$S_2O_3^{2-}$	SO_4^{2-}
Nitrosomonas spp.	O_2	NH_4^+	NO_2^-
Nitrobacter spp.	O_2	NO_2^-	NO_3^-
Hydrogenomonas spp.	O_2	H_2	H_2O
Ferrobacillus ferrooxidans	O_2	Fe^{2+}	Fe^{3+}
Desulphovibrio desulphuricans*	SO_4^{2-}	H_2	S^-

* This organism may not be fully autotrophic.

Experiments with *T. thiooxidans* and *T. thioparus* have led to similar results.

The hydrogen bacteria may grow either autotrophically, when their energy requirements are met by the oxidation of hydrogen by molecular oxygen (or by nitrate ions in the case of *Micrococcus denitrificans*), or heterotrophically, by conventional mechanisms. When growing autotrophically, *Hydrogenomonas ruhlandii* contains large quantities of carboxydismutase; the amount of this enzyme is drastically reduced, however, under conditions of heterotrophic growth. Similar observations have been reported with other *Hydrogenomonas* species. Hence, although the evidence is very limited, it seems that these bacterial species are able to use the Calvin–Benson cycle for carbon dioxide fixation.

If, as suggested above, the chemosynthetic bacteria fix carbon dioxide by the Calvin–Benson cycle, then mechanisms must exist for the coupling of their substrate oxidation reactions to ATP and NADH production. The observation of cytochrome reduction associated with substrate oxidation in several of the organisms listed above strongly suggests that ATP formation occurs by a type of oxidative phosphorylation analogous to that postulated in the dark phases of photosynthesis. In support of this hypothesis, a cytochrome-containing particle has been isolated from *Nitrobacter agilis* in which ATP formation is linked to nitrite oxidation.

Except in the case of the hydrogen bacteria, the mode of formation of NADH is obscure. Indeed, it is difficult to see how in some cases the reduction of pyridine nucleotides can be directly linked to the substrate oxidation reactions. For example, in the case of the *Ferrobacilli*, the standard free energy change associated with the Fe^{2+}–Fe^{3+} reaction is about 27 kcal/mole less than that of the reduction of NAD^+. However, the mechanism of NAD^+ reduction operative in the *Hydrogenomonas* species is at least partly understood; the bacteria contain an active hydrogenase which catalyses the direct reduction of NAD^+ by molecular hydrogen:

$$NAD^+ + H_2 \xrightarrow{\text{hydrogenase}} NADH + H^+$$

C. Heterotrophic Carbon Dioxide Fixation[7-9]

As was previously pointed out, organisms which are incapable of growing on carbon dioxide as the sole source of carbon atoms, do nevertheless carry out a variety of primary carbon dioxide fixation reactions. The enzymes which may be involved in these reactions have been discussed in section II.A and it remains only to consider the relevance of those reactions to the overall metabolism. Heterotrophic carbon dioxide fixation reactions are of two main types. First, there are those reactions in which carboxylation contributes to the formation of a specific compound required for the metabolism of the organism. Secondly, reaction sequences have been described in which carbon dioxide plays a catalytic role; a carboxylation reaction is followed by a decarboxylation. Sequences of this type usually occur during the early stages of anabolic pathways where they serve as a device for the circumvention of a reaction with an unfavourable equilibrium constant.

1. Formation of dicarboxylic acids from pyruvate

It is easy to distinguish two main ways in which the formation of dicarboxylic acids (oxaloacetate or malate) by carboxylation of pyruvate plays an essential role in metabolic processes.

Reference to Figure 1 shows that oxaloacetate serves both as an essential intermediate in the cyclic series of reactions (Krebs cycle) by which acetyl-CoA is catabolized to carbon dioxide and also as a starting point for a variety of anabolic reactions. Anabolic reactions starting from oxaloacetate and other Krebs cycle intermediates cause a continuous drain on these species; the concentrations of these intermediates must be maintained at a level sufficiently high for the operation of the cycle by so-called anaplerotic sequences. The most important of these is the formation of oxaloacetate by carboxylation of pyruvate.

Carbon dioxide plays a catalytic role in the second process to be discussed. The anaerobic breakdown of glucose to pyruvate may be represented as shown in equation (21).

$$\text{glucose} \xrightleftharpoons{a} \text{3-phosphoglycerate} \xrightleftharpoons{b}$$

$$\text{phosphoenolpyruvate} \xrightarrow[\underset{\text{ADP} \quad \text{ATP}}{}]{c} \text{pyruvate} \tag{21}$$

Reaction sequences a and b in (21) are freely reversible whereas the pyruvate kinase reaction (c) lies in the direction of pyruvate formation ($\Delta G° - 6$ kcal/mole). In gluconeogenesis, (the formation of glucose from non-carbohydrate material), reactions a and b occur in reverse, but reaction c is by-passed by a process which involves carboxylation of pyruvate and a subsequent decarboxylation reaction. The first sequence of reactions proposed for this by-pass is shown in equation (22).

$$\text{pyruvate} \xrightarrow[\underset{\text{NADPH, H+} \quad \text{NADP+}}{}]{\overset{CO_2}{}} \text{malate} \xrightarrow[\underset{\text{NAD+} \quad \text{NADH, H+}}{}]{}$$

$$\text{oxaloacetate} \xrightarrow[\underset{\text{GTP} \quad \text{GDP}}{}]{\overset{CO_2}{}} \text{phosphoenolpyruvate} \tag{22}$$

The role of carboxylase is assigned to the malic enzyme, whilst phosphoenolpyruvate carboxykinase catalyses the decarboxylation step. It has been argued that although the properties of the carboxykinase and its intracellular localization (mitochondrial) are consistent with this formulation, this is not the case with the malic enzyme. The latter enzyme is localized in the cytoplasm of the cell and has a very low affinity for carbon dioxide. It is now generally believed that the sequence of events during gluconeogenesis is better described by reactions (23) and (24).

$$\text{pyruvate} + CO_2 + \text{ATP} \rightleftharpoons \text{oxaloacetate} + \text{ADP} + P_i \tag{23}$$

$$\text{oxaloacetate} + \text{GTP} \rightleftharpoons \text{phosphoenolpyruvate} + \text{GDP} + CO_2 \tag{24}$$

Here, carboxylation is catalysed by the biotin enzyme, pyruvate carboxylase. This enzyme is of mitochondrial origin, and has a high affinity for carbon dioxide. It seems likely that the phosphoenolpyruvate carboxytransphosphorylase of the propionic acid bacteria (reaction 25) is similarly involved in the formation of phosphoenolpyruvate from oxaloacetate.

$$\text{oxaloacetate} + PP_i \rightleftharpoons \text{phosphoenolpyruvate} + P_i + CO_2 \tag{25}$$

However, pyruvate carboxylase has not been identified with certainty in these species, and it is probable that reaction (23) is replaced by some other carboxylation process.

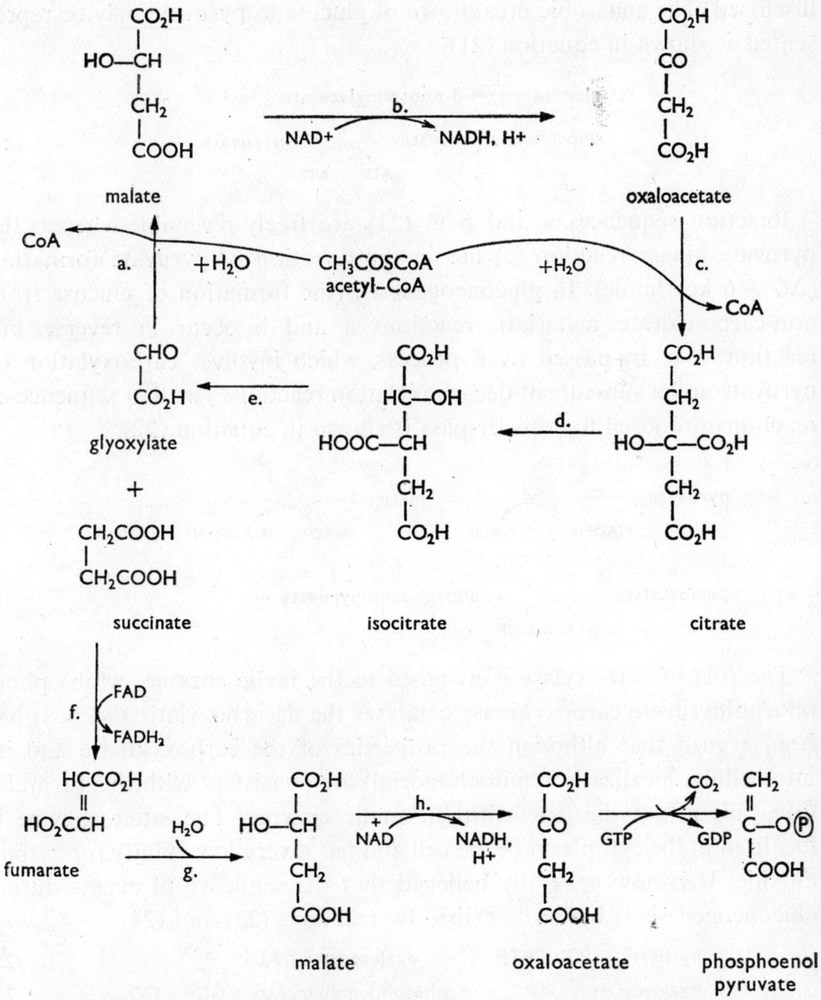

Reaction	Enzyme	E.C. Number
a.	malate synthase	4.1.3.2
b.	malate dehydrogenase	1.1.1.37
c.	citrate synthase	4.1.3.7
d.	aconitase	4.2.1.3
e.	isocitratase	4.1.3.1
f.	succinate dehydrogenase	1.3.99.1
g.	fumarase	4.2.1.2

FIGURE 11. The glyoxalate cycle

The phosphoenolpyruvate carboxylases from plants and photosynthetic microorganisms use water as the phosphate acceptor. In this case, the equilibrium favours carboxylation and the reaction is thought to provide the major route of carbon dioxide fixation when these organisms are growing in the dark.

It is interesting to compare the route for gluconeogenesis described by equations (23), (24) and (21 a, b) with that operative in organisms growing entirely on two-carbon compounds. In this case, phosphoenolpyruvate arises from succinate, which is formed from two molecules of acetyl-CoA by a series of reactions (the glyoxalate cycle) described by Kornberg and Madsen[290-292]. The reactions are shown in Figure 11, together with the names of the enzymes catalysing individual steps. Only two of the reactions (a and e) are unique to this process. Enzymes catalysing these reactions (malate synthase and isocitratase) are found only in bacteria and higher plants and are completely absent from animal tissues.

2. Propionate metabolism[7, 8, 45]

It was originally believed that the production of propionate from glucose and other metabolites by the propionic acid bacteria occurred by carboxylation of pyruvate to oxaloacetate, followed by conversion to succinate and succinyl-CoA, and finally decarboxylation to propionyl-CoA. This formulation was questioned when it was repeatedly demonstrated that carbon dioxide turnover did not take place during the process. The situation was clarified, however, by the isolation from propionic acid bacteria, of methylmalonyl-CoA transcarboxylase by Swick and Wood[46]; the demonstration of this reaction suggested the formulation shown in Figure 12 for propionate fermentation.

The scheme involves two linked cycles; one cycle is involved in the transfer of a one-carbon fragment from methylmalonyl-CoA to pyruvate and subsequent regeneration of methylmalonyl-CoA, whilst the other is concerned with CoA utilization. Oxaloacetate formed in the transcarboxylase reaction is converted to succinate by the usual route. A reaction between succinate and propionyl-CoA, catalysed by succinyl-CoA transferase, yields succinyl-CoA and propionate. Succinyl-CoA is converted to D_s-methylmalonyl-CoA by a very interesting pair of reactions. The first involves a rearrangement of the carbon skeleton, catalysed by methylmalonyl-CoA mutase; this enzyme requires a derivative of vitamin B_{12} as cofactor. The product of the mutase reaction has the configuration L_s and is converted to the D_s isomer by the enzyme methylmalonyl-CoA racemase[47].

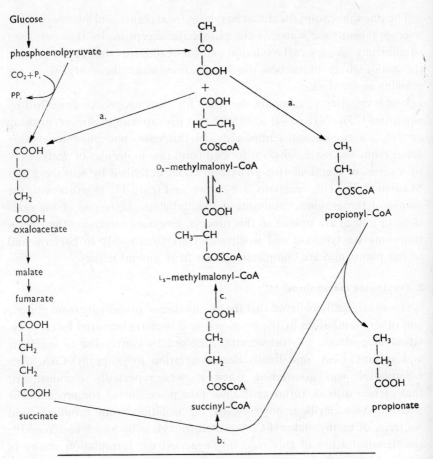

Reaction	Enzyme	E.C. number
a.	methylmalonyl-CoA transcarboxylase	2.1.3.1
b.	succinyl-CoA transferase	–
c.	methylmalonyl-CoA mutase	5.4.99.2
d.	methylmalonyl-CoA racemase	–

Overall reaction

$$\text{Pyruvate} + 2\,\text{NADH} + 2\text{H}^+ \rightleftharpoons \text{Propionate} + 2\,\text{NAD}^+ + \text{H}_2\text{O}$$

FIGURE 12. Propionate fermentation

It can be seen that free carbon dioxide is not involved in the formation of propionic acid by the pathway shown in Figure 12. Some incorporation of $^{14}\text{CO}_2$ into the product has, however, been observed. This arises from carboxylation of phosphoenolpyruvate in the carboxytransphosphorylase reaction; the oxaloacetate so formed is converted to propionate via the symmetrical succinic acid. Consistently, cell-free extracts from *P. shermanii* in which the transcarboxylase has been inhibited by avidin, metabolize pyruvate to succinate with net fixation of carbon dioxide.

Propionate is produced in plants and animals mainly as an end-product of the degradation of fatty-acids with an odd number of carbon atoms (see section VI) and from catabolism of certain of the amino acids. In these species, propionate is metabolized to pyruvate according to equation (26).

$$\text{propionate} + \text{ATP} + 2\,H_2O \longrightarrow \text{pyruvate} + \text{ADP} + P_i + 4[H] \tag{26}$$

That the interconversion occurs via a symmetric 4-carbon intermediate is shown by the equal labelling of the α and β positions of pyruvate produced from either [2-^{14}C]- or [3-^{14}C]-propionate. In contrast to the reverse process occurring in propionic acid fermentation by bacteria, carbon dioxide fixation plays an essential part in the process. Thus pyruvate produced from propionate by preparations of liver mitochondria in the presence of $^{14}CO_2$ becomes labelled in the carboxyl group. Plant and animal tissues contain no methylmalonyl-CoA carboxylase. Hence reaction (26) may be represented by the processes described by reactions (27–32).

$$\text{propionyl—CoA} + H_2O + CO_2 + \text{ATP} \rightleftharpoons \text{methylmalonyl—CoA}$$
$$+ \text{ADP} + P_i \tag{27}$$

$$D_s\text{-methylmalonyl—CoA} \rightleftharpoons L_s\text{-methylmalonyl—CoA} \tag{28}$$

$$L_s\text{-methylmalonyl—CoA} \rightleftharpoons \text{succinyl—CoA} \tag{29}$$

$$\text{succinyl—CoA} + \text{propionate} \rightleftharpoons \text{propionyl—CoA} + \text{succinate} \tag{30}$$

$$\text{succinate} + H_2O \dashrightarrow \text{oxaloacetate} + 4[H] \tag{31}$$

$$\text{oxaloacetate} \longrightarrow \text{pyruvate} + CO_2 \tag{32}$$

Alternatively, succinate may enter the Krebs cycle. It may be seen that reactions (28) to (31) are the reverse of those responsible for propionic acid fermentation. The presence of an enzyme-catalysing reaction (30) has yet to be unambiguously demonstrated in animal tissues and the possibility remains that this step is more complicated than shown here.

An alternative route for propionate metabolism exists. In this, propionyl-CoA is dehydrogenated to acrylyl-CoA. A hydration reaction yields L-lactoyl-CoA; this (or the D-isomer after racemization) undergoes hydrolysis and dehydrogenation to pyruvate. The importance of this pathway in plant metabolism appears to be greater than in animal metabolism.

3. Catabolism of leucine and isovalerate

Leucine is catabolized in animal tissue to acetyl-CoA and acetoacetate by a series of reactions which include decarboxylation and subsequent carboxylation. Lynen[48] has shown that β-methylcrotonyl-CoA is carboxylated during this process. Similarly, Knappe[49] and his coworkers have isolated β-methylcrotonyl-CoA carboxylase from a species of *Mycobacterium*

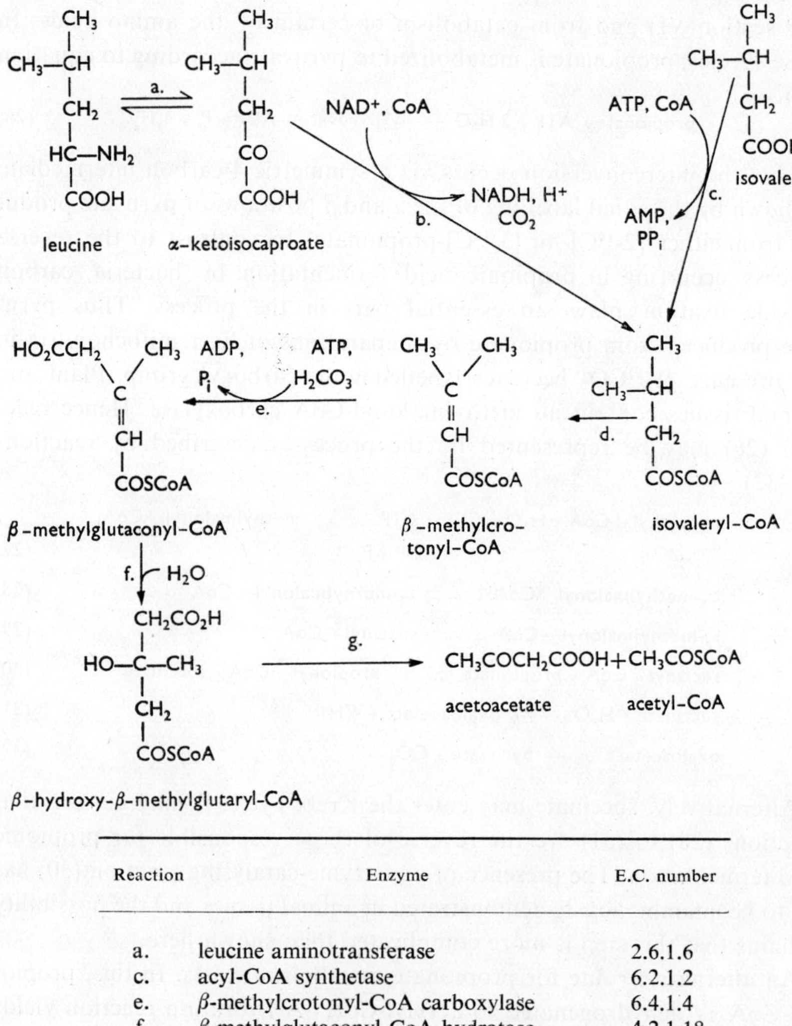

Reaction	Enzyme	E.C. number
a.	leucine aminotransferase	2.6.1.6
c.	acyl-CoA synthetase	6.2.1.2
e.	β-methylcrotonyl-CoA carboxylase	6.4.1.4
f.	β-methylglutaconyl-CoA hydratase	4.2.1.18
g.	β-hydroxymethylglutaryl-CoA lyase	4.1.3.4

FIGURE 13. Catabolism of leucine and isovalerate

grown on isovaleric acid as the sole source of carbon. The catabolism of this compound is accompanied by a net fixation of carbon dioxide. Figure 13 shows the common pathways involved.

Leucine is converted to isovaleryl-CoA by transamination, catalysed by leucine aminotransferase, followed by concerted oxidative decarboxylation and acyl-CoA formation. The enzyme responsible for the latter step has not been characterized. Isovaleryl-CoA formation from isovalerate is catalysed by a non-specific acyl-CoA synthetase. Dehydrogenation and carboxylation yield β-methylglutaconyl-CoA which is hydrated by a specific enzyme to β-hydroxy-β-methylglutaryl-CoA. Cleavage of this product by hydroxymethylglutaryl-CoA lyase produces acetyl-CoA and acetoacetate, both of which may be degraded to carbon dioxide. It should be pointed out, however, that β-hydroxy-β-methyl glutaryl-CoA is an essential intermediate in the biosynthesis of steroids and terpenes (see later). It is probable, therefore, that the degradation of leucine and isovalerate is coupled to these biosynthetic reactions in some species.

4. Biosynthesis of purine nucleotides

One feature that practically all organisms have in common is the ability to synthesize the purine and pyrimidine nucleoside phosphates which are the basic units of ribose and deoxyribose nucleic acids. The biosynthesis

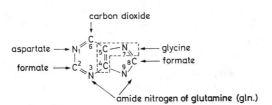

FIGURE 14. Origin of atoms of the purine skeleton

of purine nucleotides is of relevance here since one carbon atom ($C_{(6)}$) of the base skeleton is derived directly from carbon dioxide. Figure 14 shows the purine skeleton, and the origin of each of its constituent atoms (Buchanan et al.[50], Greenberg[51]).

The biosynthetic pathway from ribose-5-phosphate to inosinic acid is given in Figure 15. The sequence is complex and lack of space prohibits a detailed discussion of the individual reactions. It should be noted, however, that the two formylation reactions shown (d, j) do not involve free formate; the one-carbon fragment is transferred to the acceptor by derivatives of the coenzyme tetrahydrofolate. The carboxylation reaction (g) is catalysed by phosphoribosyl-aminoimidazole carboxylase; no details of this reaction are known.

Conversion of inosinic acid to the 6-amino derivative (adenylic acid) is achieved by a GTP-dependent reaction with aspartate and subsequent

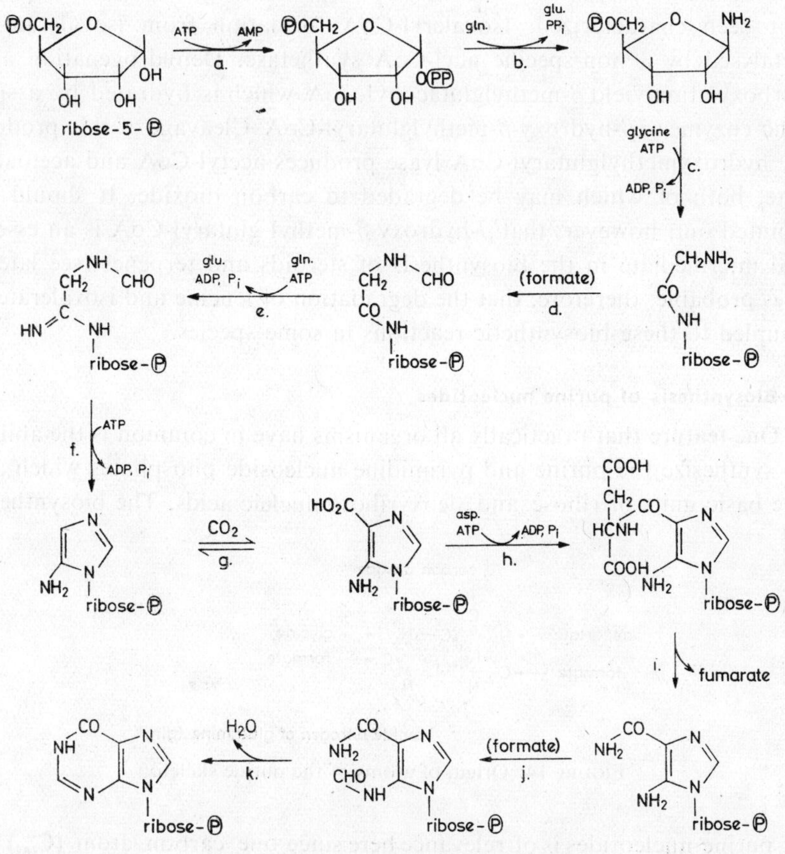

FIGURE 15. Biosynthesis of inosinic acid

elimination of fumarate precisely analogous with reactions h and i of Figure 15. The formation of guanylic acid (2-amino derivative) requires an initial oxidation to xanthylic acid (2-hydroxy derivative). Guanylic acid is then produced by an ATP-dependent amination similar to reaction c of Figure 15.

III. CARBAMIC ACID AND THE CARBAMATES

A. Biosynthesis of Carbamyl Phosphate[52]

The biosynthesis of carbamyl phosphate by fixation of carbon dioxide and ammonia is a chemical activity common to bacteria, the ureotelic vertebrates and probably the higher plants. The route by which the synthesis occurs, however, differs markedly in these groups of organisms.

Carbamyl phosphate formation was first demonstrated in the bacterium *Streptococcus faecalis* and was shown to proceed with the stoichiometry of equation (33)[293].

$$NH_3 + CO_2 + ATP \rightleftharpoons NH_2CO \cdot OPO_3H_2 + ADP \qquad (33)$$

Reaction (33) is catalysed by the enzyme carbamate kinase (E.C.2.7.2.2). Carbamic acid has been shown to be the true substrate for this enzyme and the formation of carbamylphosphate probably proceeds according to equations (34) and (35).

$$NH_3 + CO_2 \rightleftharpoons NH_2COOH \qquad (34)$$

$$NH_2COOH + ATP \rightleftharpoons NH_2CO \cdot OPO_3H_2 + ADP \qquad (35)$$

It is believed that the equilibration between carbamic acid and free carbon dioxide and ammonia (equation 34) proceeds without the intervention of an enzyme. The standard free energy change associated with reaction (34) has been estimated at $+6\,kcal/mole$; hence it can be seen that the equilibrium position of the overall reaction (33) lies well to the left[294].

Carbamate kinase has been located in mung bean mitochondria but in general the mode of formation of carbamyl phosphate in the higher plants is not known.

In distinct contrast to the situation pertaining in microorganisms, carbamyl phosphate formation in liver mitochondria proceeds according to reaction (36).

$$NH_3 + CO_2 + 2\,ATP + H_2O \longrightarrow NH_2CO \cdot OPO_3H_2 + 2\,ADP + P_i \qquad (36)$$

The reaction is catalysed by carbamyl phosphate synthetase. This enzyme has been purified from bullfrog liver (*Rana catesbiana*)[295]. The enzyme has a molecular weight of 315,000 and shows a dependence on N-acetylglutamate (NAGA) for its activity. Jones and Spector[53] have claimed that the reaction involves initial formation of an NAGA-carbon dioxide compound (equation 37); carbamyl phosphate is then formed according to equation (38).

$$ATP + CO_2 + NAGA + H_2O \rightleftharpoons ADP + P_i + NAGA—CO_2 \qquad (37)$$

$$NAGA—CO_2 + ATP + NH_3 \rightleftharpoons NAGA + ADP + NH_2CO \cdot OPO_3H_2 \qquad (38)$$

More recent work[54], however, tends to indicate that NAGA is not directly involved in the reaction sequence, but functions as an enzyme activator. It is not possible at present to decide between these proposed roles for NAGA.

It should be noted that the overall reaction (36) lies strongly in the direction of carbamyl phosphate formation, whereas the opposite is the case with reaction (33). It is unlikely that the two distinct pathways have arisen

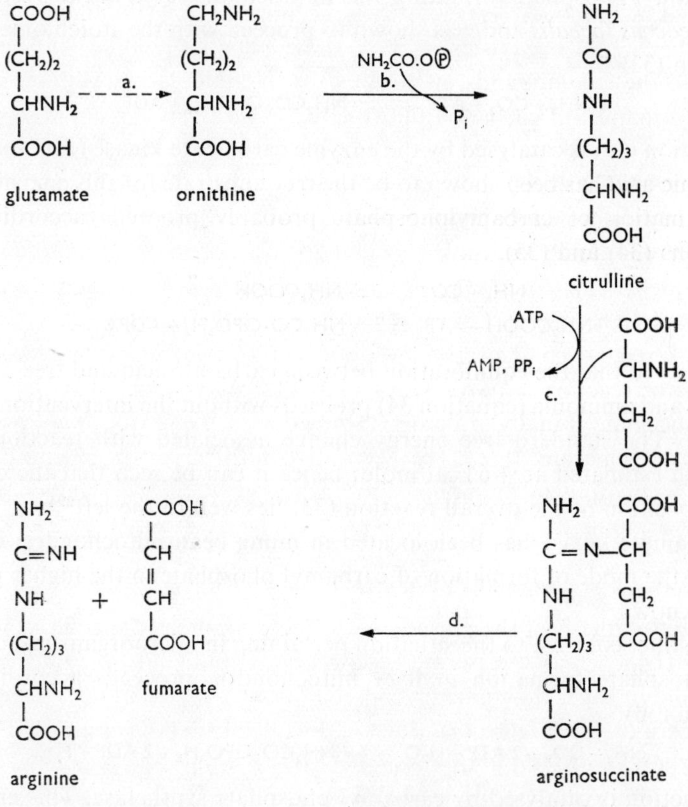

Reaction	Enzyme	E.C. Number
b.	ornithine transcarbamylase	2.1.3.3
c.	arginosuccinate synthetase	6.3.4.5
d.	arginosuccinase	4.3.2.1

FIGURE 16. Biosynthesis of arginine

as an evolutionary accident. More probably, the difference reflects the fact that ammonia is toxic to terrestial animals at very low levels whereas bacterial species can tolerate much higher concentrations of this substance. The carbamyl phosphate synthetase reaction provides an effective route for the fixation of low levels of ammonia.

B. Anabolic Functions of Carbamyl Phosphate[57]

Formation of carbamyl phosphate is the first step in two apparently unrelated anabolic sequences, namely arginine and pyrimidine nucleotide biosynthesis. All organisms so far studied employ similar biosynthetic routes to these compounds, in contrast to the differences exhibited in the initial formation of carbamyl phosphate.

1. Arginine biosynthesis[55]

The key intermediate in arginine biosynthesis is ornithine which is derived from glutamate via the γ-semialdehyde or the N-acetylglutamate-γ-semialdehyde. A condensation reaction between ornithine and carbamyl phosphate (reaction b, Figure 16) yields citrulline and inorganic phosphate. A second condensation reaction between citrulline and aspartate ensues; the reaction requires Mg^{2+} and ATP and yields AMP, PP_i and arginosuccinate. Finally arginine is formed from arginosuccinate by elimination of fumarate. It can be seen that both the nitrogen and carbon atoms originating in ammonia and carbon dioxide are retained in the product arginine.

2. Biosynthesis of pyrimidine nucleotides

The biosynthesis of the purine nucleotides has already been dealt with in section II. Whereas the purine ring system is built into a preformed N-glycosidic linkage, the immediate precursor of the pyrimidine nucleotides is free orotic acid (Figure 17). Carbon dioxide again contributes one of the ring carbon atoms, but in this case, the carbon atom is derived from carbamyl phosphate, not free carbon dioxide.

Figure 17 shows the reactions leading from carbamyl phosphate and aspartate to orotic acid and thence to uridine triphosphate (UTP) and cytidine triphosphate (CTP). Linkage of the orotic acid to the 1'-position of the ribose ring occurs by reaction with 5-phosphoribosylpyrophosphate; the enzyme responsible for this reaction is completely specific for orotate. Uridylic acid then arises from decarboxylation of orotidine-5'-phosphate and is converted to UTP by two successive phosphate transfer reactions from ATP. Cytidine triphosphate is formed by amination of the 4-position

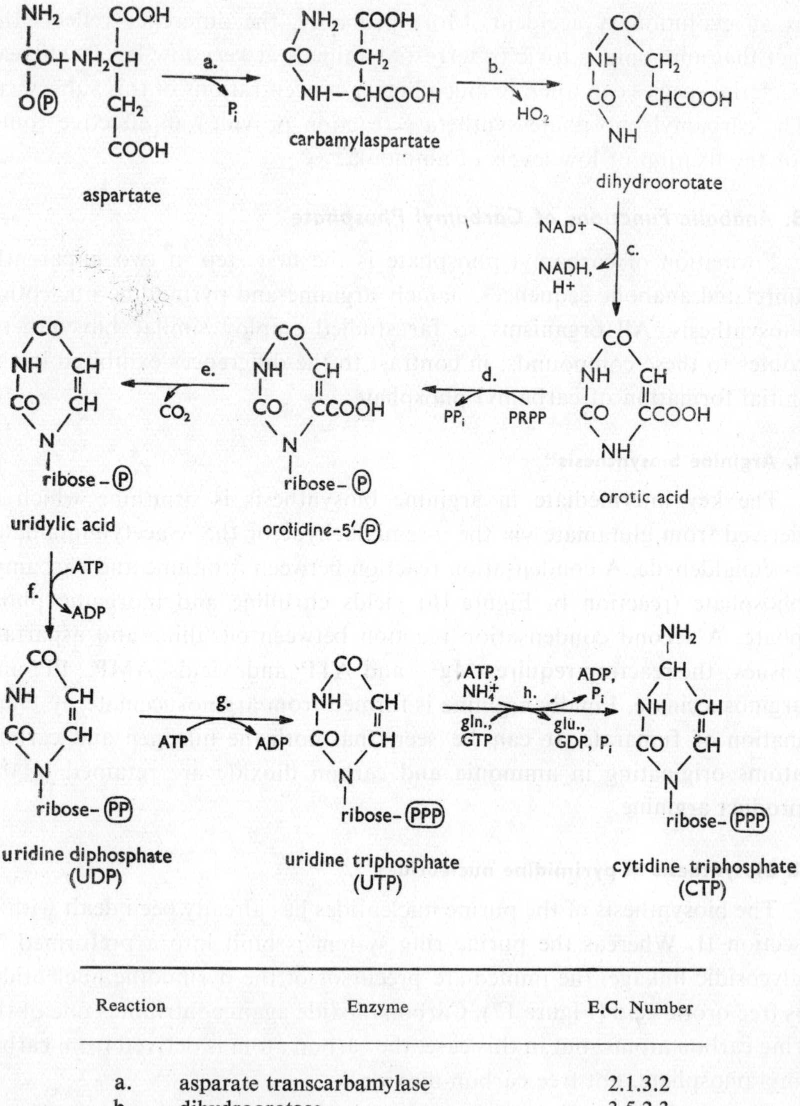

Reaction	Enzyme	E.C. Number
a.	asparate transcarbamylase	2.1.3.2
b.	dihydroorotase	3.5.2.3
c.	dihydroorotate dehydrogenase	1.3.3.1
d.	orotidine-5′-phosphate pyrophosphoryl-ase	2.4.2.10
e.	orotidine-5′-phosphate decarboxylase	4.1.1.23
f.	nucleosidemonophosphate kinase	2.7.4.4
g.	nucleosidediphosphate kinase	2.7.4.6
h.	CTP synthetase	6.3.4.2

FIGURE 17. Biosynthesis of the pyrimidine nucleotides

of the pyrimidine ring. In microorganisms, the reaction involves ammonium ions and ATP whilst glutamine and GTP are employed by the enzyme from animal sources.

Amination of UTP is the only route known for the biosynthesis of CTP. This again is in marked contrast to the situation with the purine nucleotides, where AMP and GMP are formed by independent routes from the common precursor inosinic acid.

C. The Formation of Urea from Carbamyl Phosphate

Ammonia, which is formed in relatively large amounts during the catabolism of proteins, is highly toxic to animals (the lethal blood level is 5 mg/100 ml in the rabbit). Detoxication of catabolic nitrogen[57] is achieved by conversion to either urea (ureotelic species — mammals, amphibians, fish) or uric acid (uricotelic species — birds, snakes, lizards and terrestrial gastropods); we are concerned here with the mechanism by which urea is produced in the liver of ureotelic species.

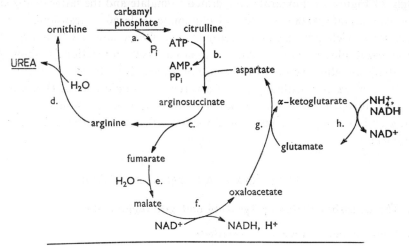

Reaction	Enzyme	E.C. Number
a–c.	see Figure 16	
d.	arginase	3.5.3.1
e.	fumarase	4.2.1.2
f.	malate dehydrogenase	1.1.1.37
g.	aspartate aminotransferase	2.6.1.1
h.	glutamate dehydrogenase	1.4.1.3

FIGURE 18. Urea formation in liver

This problem was studied by Krebs and Henseleit[58], using liver slices incubated with ammonium salts, bicarbonate ions, and lactate. Added ornithine and citrulline were found to promote urea formation, and arginine was shown to be an intermediate in the process. Moreover, the enzyme arginase (E.C. 3.5.3.1), which catalyses the hydrolysis of arginine to ornithine and urea, is found in large quantities in the liver of ureotelic species. This enzyme is either entirely absent or present in organs other than the liver (*e.g.* kidney) in non-ureotelic species. On the basis of the above evidence, urea synthesis has been formulated as shown in Figure 18.

The scheme is the same as that presented for the biosynthesis of arginine (Figure 15) except for the final hydrolytic reaction catalysed by arginase. This reaction regenerates ornithine which therefore acts catalytically in this process. Of the atoms in the product urea, the carbon and one nitrogen are derived from carbamyl phosphate, whilst the second nitrogen atom arises from aspartate. The route by which fumarate produced in the arginosuccinase reaction is converted back to aspartate is shown on the right of Figure 18. Fumarate is hydrated to malate and the latter oxidized to oxaloacetate. Oxaloacetate and glutamate undergo a transamination reaction to yield asparate and α-ketoglutarate. Glutamate may be reformed from α-ketoglutarate by a reductive ammonia fixation reaction (as shown) catalysed by glutamate dehydrogenase or alternatively by a second transamination reaction with any one of a variety of α-amino acids. Hence the second nitrogen atom of urea may arise directly from ammonia or from the α-amino group of an amino acid, the former route probably being more important.

IV. DECARBOXYLATION REACTIONS

A. The Decarboxylation of Pyruvate and α-Ketoglutarate

1. Types and mechanisms of the reactions

Decarboxylation reactions of pyruvate and α-ketoglutarate are of central importance in metabolic processes and the enzymes catalysing these reactions have been examined in some detail.

Early studies revealed a bewildering multiplicity of reaction types and products. An enzyme isolated from yeast (α-carboxylase, pyruvate decarboxylase, 2-oxo-acid carboxy-lyase 4.1.1.1), higher plants and certain microorganisms catalyses the irreversible decarboxylation of pyruvate to acetaldehyde (reaction 39)[59].

$$CH_3COCOOH \longrightarrow CH_3CHO + CO_2 \qquad (39)$$

The yeast enzyme also decarboxylates higher homologues of pyruvate up to α-ketocaproate but does not act on the analagous dicarboxylic acids. In addition to reaction (39), pyruvate decarboxylases from a variety of sources have been shown to catalyse the production of acetoin by the so-called carboligase reaction (40) and by the related reactions (41) and (42).

$$2\,CH_3CHO \longrightarrow CH_3CH(OH)COCH_3 \tag{40}$$

$$CH_3CHO + CH_3COCOOH \longrightarrow CH_3CH(OH)COCH_3 + CO_2 \tag{41}$$

$$2\,CH_3COCOOH \longrightarrow CH_3CH(OH)COCH_3 + 2\,CO_2 \tag{42}$$

Similar reactions with the higher homologues of pyruvate have been reported. Early workers assumed that reactions 40–42 were catalysed by a separate enzyme (carboligase); attempts to separate the carboligase and decarboxylase activities were unsuccessful and it is now recognized that both activities are associated with one enzyme.

Some species of bacteria (*e.g. Aerobacter aerogenes*) contain a pyruvate decarboxylase which, in the first instance, produces acetolactate by reaction (43) and then acetoin by decarboxylation reaction (44).

$$2\,CH_3COCOOH \longrightarrow \underset{\underset{\displaystyle COOH}{|}}{CH_3COC(OH)CH_3} \tag{43}$$

$$\underset{\underset{\displaystyle COOH}{|}}{CH_3COC(OH)CH_3} \longrightarrow CH_3COCH(OH)CH_3 + CO_2 \tag{44}$$

Pyruvate also undergoes a variety of oxidative decarboxylation reactions to yield acetate or, more usually, an ester of acetic acid[60]. Direct formation of acetate by an enzyme extract from *Proteus vulgaris* has been reported; a similar enzyme system is developed by an acetate-requiring auxotroph of *E. coli*. These enzyme systems are composed of a water soluble flavoprotein (FP) fraction which catalyses reaction (45), and a particulate cytochrome fraction which, in the presence of oxygen, is responsible for reoxidation of the flavoprotein.

$$CH_3COCOOH + FP + H_2O \longrightarrow CH_3COOH + FPH_2 + CO_2 \tag{45}$$

Artificial electron acceptors (*e.g.* ferricyanide) can replace the cytochrome fraction in isolated enzyme extracts.

More generally, the product of oxidative decarboxylation of pyruvate in microorganisms is either acetyl-CoA or acetyl phosphate. An extract from *Clostridium butylicum* has been shown to catalyse the formation of acetyl phosphate from pyruvate with simultaneous liberation of hydrogen (reaction 46).

$$CH_3COCOOH + P_i \longrightarrow CH_3CO\,ⓟ + CO_2 + H_2 \tag{46}$$

The reaction is thought to proceed by initial formation of acetyl-CoA (reaction 47) followed by conversion to acetylphosphate in the phosphoroclastic reaction (48). Reaction (48) is catalysed by phosphate acetyltransferase (1.3.1.8).

$$CH_3COCOOH + CoA\text{—}SH \rightleftharpoons CH_3CO\text{—}SCoA + CO_2 + H_2 \qquad (47)$$

$$CH_3CO\text{—}SCoA + P_i \rightleftharpoons CH_3CO\circledP + CoA\text{—}SH \qquad (48)$$

Acetyl phosphate so formed may equilibrate with ATP, the reaction (49) being catalysed by acetate kinase (2.7.2.1).

$$CH_3CO\circledP + ADP \rightleftharpoons CH_3COOH + ATP \qquad (49)$$

This series of reactions (47–49) provides an important route to ATP formation in bacteria growing under anaerobic conditions.

A similar enzyme system has been isolated from *E. coli*; in this case, however, the overall reaction (50) involves production of formate instead of carbon dioxide and hydrogen.

$$CH_3COCOOH + P_i \rightleftharpoons CH_3CO\circledP + HCOOH \qquad (50)$$

In both cases, the details of the redox reactions are obscure, but it is thought that ferredoxin (cf. section II. B.3) may be involved in the initial transfer of electrons.

In direct contrast to the two cases discussed above, formation of acetyl phosphate without intervention of CoA has been described in the organism *Lactobacillus delbrueckii* (equation 51).

$$RCOCOOH + P_i + O_2 \rightleftharpoons RCO\circledP + CO_2 + H_2O_2 \qquad (51)$$

The special feature of the pyruvate oxidase catalysing reaction (51) which renders participation by CoA unnecessary is not known.

In animal tissue and certain microorganisms under aerobic conditions, pyruvate is converted to acetyl-CoA by an interesting multienzyme system called the pyruvate dehydrogenase complex (PDC—see later). The reaction may be written as in reaction (52), or, including reoxidation of NADH by the terminal electron transport system, as in reaction (53).

$$CH_3COCOOH + NAD^+ + CoA\text{—}SH \xrightarrow{PDC} CH_3CO\text{—}SCoA$$
$$+ NADH + CO_2 + H^+ \qquad (52)$$

$$CH_3COCOOH + \tfrac{1}{2}O_2 + CoA\text{—}SH \xrightarrow[ETS]{PDC} CH_3CO\text{—}SCoA$$
$$+ H_2O + CO_2 \qquad (53)$$

An essentially similar enzyme system has also been characterized which catalyses the oxidative decarboxylation of α-ketoglutarate to succinyl-CoA by a reaction analagous to (52) and (53). Both the pyruvate and α-ketoglutarate dehydrogenase systems catalyse acyloin formation under anaerobic conditions. Similar enzyme systems probably exist for the oxidative decarboxylation of other keto acids (*e.g.* α-ketoisocaproate— see section II. C.3) but detailed investigations of such systems have not been carried out.

The essential feature which the enzymes discussed in this section have in common and on which a rationalization of their mode of action can be based is an absolute requirement for thiamine pyrophosphate (TPP— Figure 19).

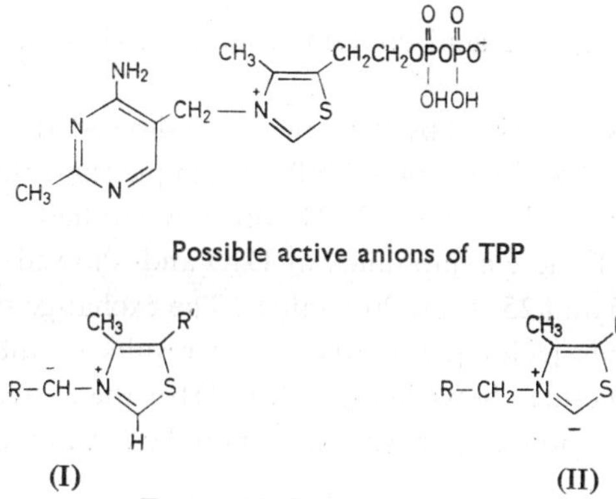

Possible active anions of TPP

(I)   (II)

FIGURE 19. Structure of TPP

Thiamine was first isolated in 1926, and the combined efforts of several groups of workers over the next decade led to the elucidation of the structure of the molecule[61]. In 1937, Loemann and Schuster[62] isolated the coenzyme of yeast pyruvate decarboxylase and identified it as TPP; subsequent workers have demonstrated the requirement for TPP in the keto acid oxidase and dehydrogenase systems and also in the transketolase and phosphoketolase reactions. In all cases, enzymic activity is destroyed by removal of TPP but may be recovered by replacement of the coenzyme in the presence of a divalent cation. The mono- and triphosphate analogues of TPP have no coenzyme activity.

Thiamine and TPP act catalytically in the decarboxylation of pyruvate and in acetoin formation in model systems (that is, in the absence of the apoenzyme). Studies of these reactions[63] led to the suggestion that the active form of thiamine is the anion formed by removal of a proton from the methylene carbon atom (Figure 19 (I)). This anion was considered to condense with the carbonyl group of the substrate to produce an intermediate which undergoes decarboxylation. Such a mechanism would require that the methylene carbon atom become deuterated during model reactions carried out in solvent D_2O. Westheimer[64] and his colleagues have examined this possibility. The thiamine was isolated and cleaved with bisulphite in D_2O (equation 54).

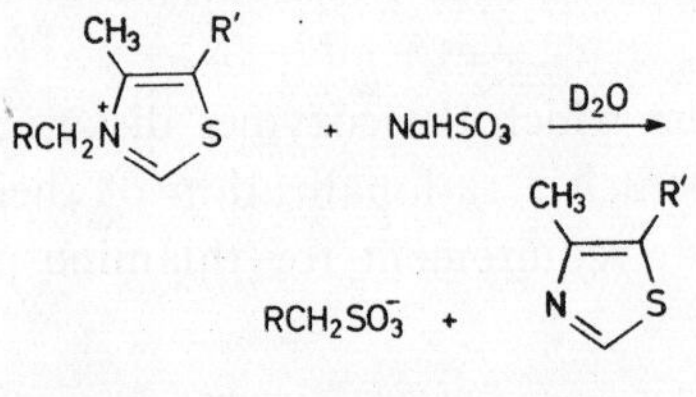

$$RCH_2\overset{+}{N} \begin{matrix} CH_3 & R' \\ & \end{matrix} S + NaHSO_3 \xrightarrow{D_2O}$$

$$RCH_2SO_3^- + N \begin{matrix} CH_3 & R' \\ & \end{matrix} S \qquad (54)$$

The resulting sulphonic acid contained no deuterium in the methylene group.

The situation was clarified by Breslow[65] who showed that the $C_{(2)}$-hydrogen atom of the thiazolium ring of TPP and simple thiazolium salts undergoes facile exchange with solvent D_2O. Breslow studied the proton magnetic resonance of these compounds in D_2O and showed the half-life of exchange at pH 5 and 25° to be 20 minutes. The exchange rate was too fast to measure by this technique at pH 7. The results of these experiments suggested the structure shown in Figure 19 (II) for the active anion of TPP.

The decarboxylation of pyruvate may now be formulated as shown in reaction (55).

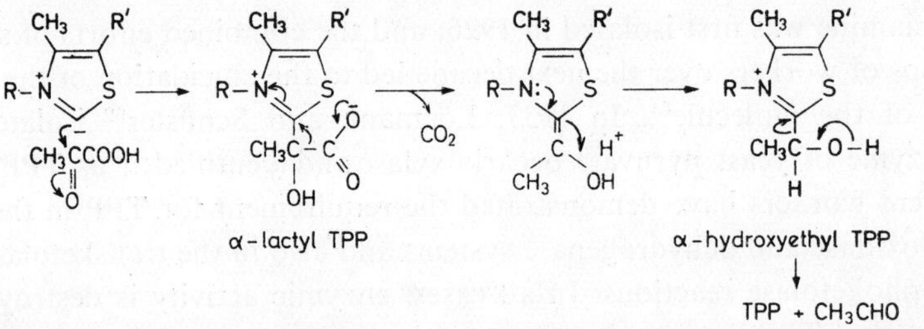

$$(55)$$

α-lactyl TPP   α-hydroxyethyl TPP

$$TPP + CH_3CHO$$

In equation (55), the reaction is shown as occurring with free TPP. The same sequence of events is believed to occur in the enzyme-catalysed reaction, but no information is available as to the role of the protein. Similarly, the nature of the binding between TPP and the apoenzyme is unknown.

Strong support for the reaction sequence shown has been provided by the isolation of α-lactyl TPP and α-hydroxyethyl TPP after incubation of pyruvate with purified yeast decarboxylase[66]. Similarly, dihydroxyethyl TPP has been identified as an intermediate in the transketolase reaction[67]. Synthetic α-hydroxyethyl TPP has been prepared by incubation of pyruvate and TPP; the derivative is cleaved to TPP and acetaldehyde by yeast decarboxylase apoprotein. The apoprotein of transketolase similarly catalyses formation of D-sedoheptulose-7-phosphate from D-ribose-5-phosphate and dihydroxyethyl TPP.

Finally, the exchange of the $C_{(2)}$-hydrogen atom of TPP required by the reaction sequence (55) has been demonstrated, and the importance of the $C_{(2)}$-hydrogen atom has been emphasized by the absence of coenzyme

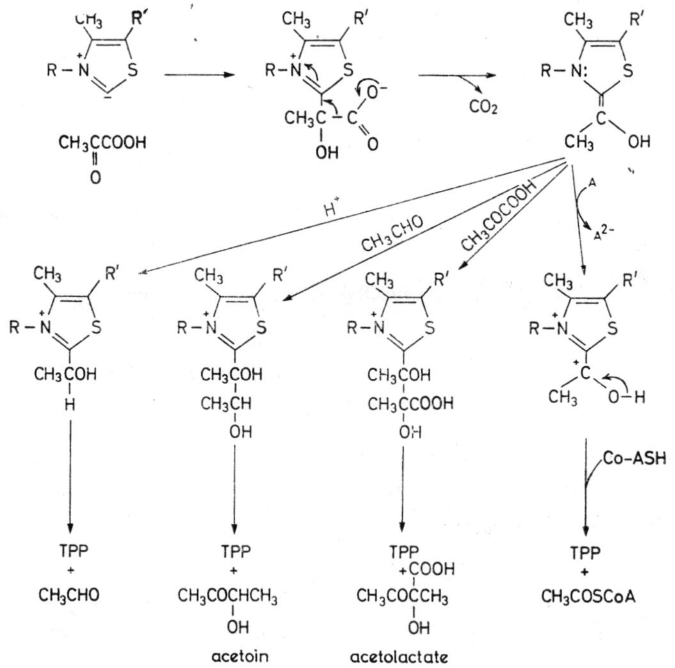

FIGURE 20. Summary of TPP-dependent reactions of pyruvate

activity in derivatives of TPP in which this atom is replaced by alkyl substituents.

A rationalization of the TPP dependent reactions of pyruvate is given in Figure 20. It must be repeated that the most important feature of the reactions, namely the role of the protein, has been ignored in this formulation.

The pyruvate and α-ketoglutarate dehydrogenase complexes (PDC and KDC), mentioned previously, are dependent on TPP for their activity, but also show an absolute requirement for lipoic acid (Figure 21). The requirement for an essential factor in addition to TPP for the oxidation of α-keto acids in microorganisms (e.g. pyruvate, α-ketobutyrate, β-methyl-α-ketobutyrate, β-methyl-α-ketoisovalerate in *Streptococcus faecalis*[68]) was recognised about 20 years ago. Crystalline lipoic acid was first isolated from the water-insoluble residue of beef liver after treatment of the material with 6 N sulphuric acid; 30 mg of lipoic acid was obtained from 10 tons of liver residue[69].

The essential feature of the lipoic acid molecule is the disulphide bond which is easily reduced to yield dihydrolipoic acid. The standard electrode potential of the lipoate–dihydrolipoate system ($-0{\cdot}3$ V at pH 7·0) is comparable with that of the NAD^+–NADH system.

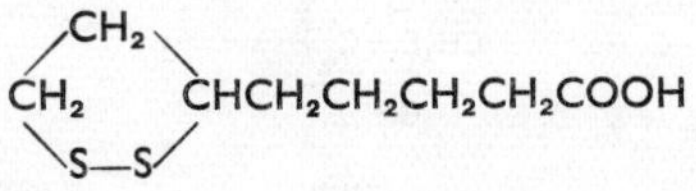
CH2
CH2
CHCH2CH2CH2CH2COOH
S—S

Nature of linkage between lipoic acid and lipoate reductase transacetylase

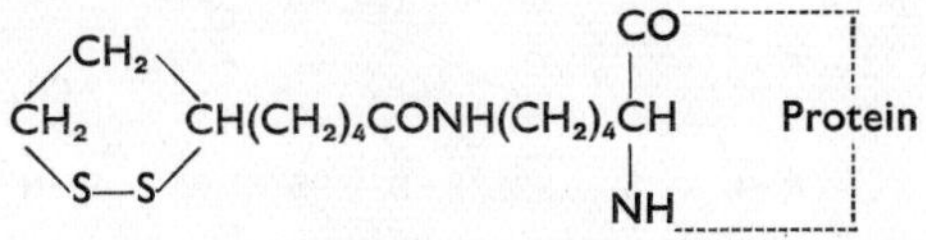
CO
CH2
CH2
CH(CH2)4CONH(CH2)4CH
Protein
S—S
NH

FIGURE 21. Structure of lipoic acid (1,2-dithiolane-3-valeric acid)

It is currently believed[60, 70] that the sequence of events in the lipoic acid-dependent oxidative decarboxylation of pyruvate and α-ketoglutarate may be described by reactions (56–60).

$$\text{R—COCOOH} + \text{E}_1\text{—TPP} \longrightarrow \text{E}_1\text{—TPP—CHOHR} + \text{CO}_2 \qquad (56)$$

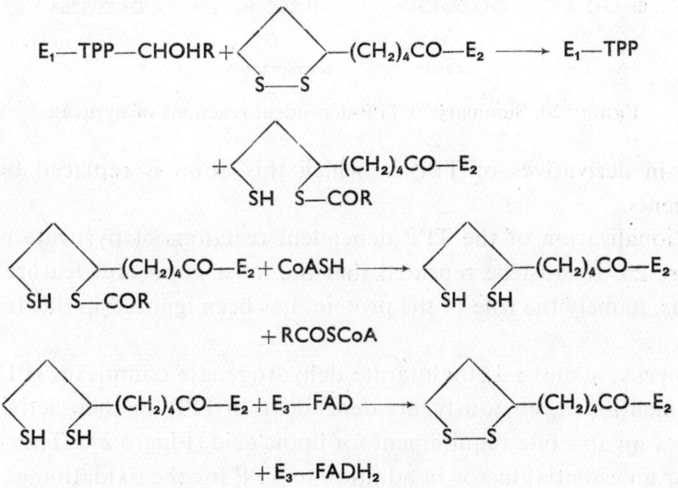
E1—TPP—CHOHR+
—(CH2)4CO—E2
E1—TPP
S—S
+
—(CH2)4CO—E2
SH S—COR
—(CH2)4CO—E2+CoASH
—(CH2)4CO—E2
SH S—COR
SH SH
+RCOSCoA
—(CH2)4CO—E2+E3—FAD
—(CH2)4CO—E2
SH SH
S—S
+E3—FADH2

(57)

(58)

(59)

$$\text{E}_3\text{—FADH}_2 + \text{NAD}^+ \longrightarrow \text{E}_3\text{—FAD} + \text{NADH} + \text{H}^+ \qquad (60)$$

The three enzymes (represented here by E_1, E_2 and E_3) catalysing reactions (56–60), exist in mammalian and bacterial cells as highly organized multicomponent units. Both the pyruvate and α-ketoglutarate dehydrogenase complexes have been isolated from *E. coli*; the 'molecular weights' of the complexes are 4.8×10^6 and 2.4×10^6 respectively. Similarly, the PDC from pigeon breast muscle has a molecular weight of 4×10^6 and the KDC from pig heart muscle has a molecular weight of 2×10^6. Isolation of the complexes from other sources is being carried out by Reed and his coworkers (unpublished work quoted in reference 70).

The PDC from *E. coli* was the first of these complexes to be resolved into its three constituent enzymes[71]. The complex is reported to contain 16 molecules of pyruvate decarboxylase (E_1, reaction 56; molecular weight 183,000; contains stoichiometric amounts of bound TPP); one aggregate of lipoate reductase-transacetylase (E_2, reactions 57 and 58; aggregate molecular weight 1.6×10^6; 64 subunits each containing one molecule of bound lipoic acid); 8 molecules of dihydrolipoate dehydrogenase (E_3, reactions 59 and 60; molecular weight 112,000; two molecules of bound FAD per protein molecule). The complex as isolated does not contain CoA or NAD^+ and these species must be added if pyruvate dehydrogenation is to proceed. Recombination experiments have been carried out with the isolated enzyme components. E_1 and E_3 do not combine but each combines independently with E_2. A mixture of all three components in the ratio 3 : 2 : 1 gives rise to a multicomponent complex very similar, both structurally and enzymically, to the native complex.

Similarly, the KDC from *E. coli* has been fractionated into three components[72]; these are α-ketoglutarate decarboxylase, lipoate reductase-transsuccinylase and dihydrolipoate dehydrogenase. Recombination experiments show that the dihydrolipoate dehydrogenase from the PDC can combine with E_1 and E_2 from the KDC and *vice versa*.

Some comments on individual reactions in the sequence (56–60) may be made. The mode of action of TPP has been dealt with above, and a direct comparison may be drawn between reaction (56) and the decarboxylation of pyruvate by yeast decarboxylase. As would be expected, both the PDC and KDC catalyse the oxidative decarboxylation of the appropriate keto acids in the presence of an added electron acceptor (*e.g.* ferricyanide, reaction 61).

$$\text{RCOCOOH} + 2\,\text{Fe(CN)}_6^{3-} + \text{H}_2\text{O} \longrightarrow \text{RCOOH} + \text{CO}_2 + 2\,\text{Fe(CN)}_6^{4-} + 2\,\text{H}^+ \quad (61)$$

The reaction requires TPP and a divalent cation, but not CoA, NAD^+ or lipoic acid. Similarly, acyloin formation is catalysed by the PDC and KDC in the presence of added aldehyde compounds.

Reaction (57) is considered to be a nucleophilic attack of the anion of the TPP–'active aldehyde' adduct, on protein-bound lipoic acid (reaction 62).

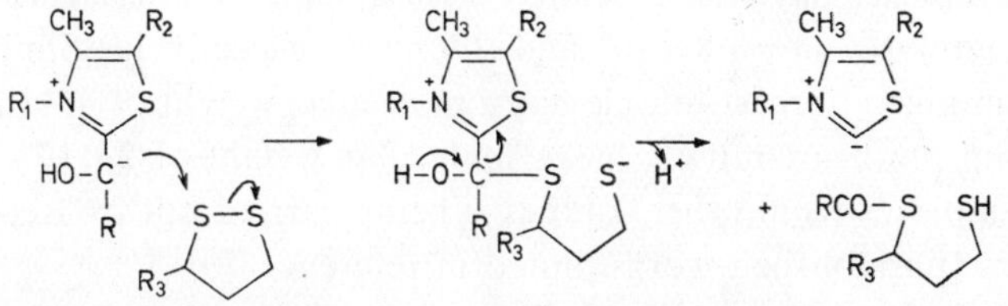

$$(62)$$

Transfer of the group RCO- to CoA (equation 58) then occurs by a nucleophilic displacement of the dihydrolipoyl enzyme from the 6-S-acyllipoate derivative. Evidence for reactions (57) and (58) based on experiments with substrate quantities of lipoic acid has been reviewed by Reed[70].

The difficulty experienced in removal of lipoic acid from the PDC or KDC suggested that the coenzyme is bound to the apoenzyme by a covalent bond; this suggestion has been substantiated by experiments with the microorganism *E. coli*. When this organism is grown aerobically in a medium containing [^{35}S]-lipoic acid, radioactivity is incorporated into both the PDC and KDC. Reed and his coworkers[73] have isolated the complexes and, after treatment of the complexes with performic acid and then hydrochloric acid (12 N), have shown the radioactivity to be associated with a 6,8-disulphonic acid derivative of ε-N-octanoyllysine. These experiments demonstrate clearly that lipoic acid is bound to its apoprotein by a peptide bond with the ε-amino group of a lysine residue (Figure 21). Reed[70] has speculated on a possible significance of the mode of binding of lipoic acid. The lipoyl moiety is required to undergo reaction at three separate sites in the complex (site of formation of the S-acyl derivative, transacylation site and site of dihydrolipoate dehydrogenation), but does not become free from the surface of its apoenzyme during these reactions. The lipoyl-lysyl linkage may provide a flexible 'arm' of length about 14 Å which allows translation of the functional part of the lipoic acid residue from one site to the next.

Reaction sequence (56–60) is concluded by reoxidation of the dihydrolipoate moiety catalysed by the flavoprotein dihydrolipoate dehydrogenase. The enzyme has been shown to be identical with Straub's diaphorase (1.6.4.3) and has been the subject of a recent review[74].

It is thought that reactions (59) and (60) are an oversimplification of the terminal redox reactions occurring in the PDC and KDC, and that reactions (63) and (64) provide a better description of the process.

$$\text{(ring)}-(CH_2)_4CO-E_2 + FAD-E_3\diagup^{S}_{\diagdown S} \longrightarrow$$

$$\text{(ring)}-(CH_2)_4CO-E_2 + \dot{H}\ FAD-E_3\diagup^{S^\bullet}_{\diagdown SH} \qquad (63)$$

$$\dot{H}\ FAD-E_3\diagup^{S^\bullet}_{\diagdown SH} + NAD^+ \longrightarrow FAD-E_3\diagup^{S}_{\diagdown S} + NADH + H^+ \qquad (64)$$

E_3 is shown as containing a catalytically active disulphide group which is reduced to a sulphydryl group and a sulphur radical; the flavine group is similarly reduced to the semiquinone radical. The involvement of a disulphide group in reactions (63) and (64) may be inferred from the facts that dihydrolipoate dehydrogenase is inhibited by arsenite, the inhibition being relieved by dithiol compounds and from the observation that reduction of the enzyme with lipoic acid or NADH results in the formation of two protein sulphydryl groups per residue of FAD. Evidence for the formation of a flavine semiquinone during the reaction has been educed from changes in the absorbtion spectrum of the enzyme in the region 500 to 600 mμ. The proposed radicals must, however, be strongly coupled since no e.s.r. signal has been detected.

2. Metabolic significance of the reactions

The oxidative degradation of organic molecules to carbon dioxide and water coupled to the generation of nucleoside triphosphates and reduced pyridine nucleotides is one of the central activities of most living organisms. It might be supposed that the reactions by which this is achieved are specific for the compound to be degraded, but this is not the case. Reference to Figure 1 shows that catabolism of the three major types of macromolecules encountered in biological systems leads to a very limited range of intermediate products. Carbohydrates are degraded to pyruvate (or triose phosphates), fats to acetyl-CoA (and some propionyl-CoA) and glycerol, and proteins to oxaloacetate, α-ketoglutarate, fumarate and succinate. It is only necessary to consider, therefore, the subsequent reactions of these metabolic intermediates.

In animal cells and microorganisms under aerobic conditions, pyruvate is oxidized by the pyruvate dehydrogenase complex to acetyl-CoA and carbon dioxide. This reaction is important from two points of view. First, the product acetyl-CoA may be degraded completely to carbon dioxide as will be described below. Secondly, acetyl-CoA is the precursor for the biosynthesis of lipids; the oxidative decarboxylation of pyruvate provides, therefore, the link between the processes of carbohydrate catabolism and lipid anabolism. In fact, very little nutritive material is stored in the form of carbohydrate, the bulk being converted to lipid and deposited in the adipose tissue.

Certain other reactions of pyruvate deserve comment. In vertebrate muscle, where production of pyruvate occurs under essentially anaerobic conditions, the terminal step of glycolysis is the reduction of pyruvate to lactic acid (equation 65).

$$CH_3COCOOH + NADH + H^+ \rightleftharpoons CH_3CH(OH)COOH + NAD^+ \qquad (65)$$

Reaction (65) is catalysed by lactate dehydrogenase (1.1.1.27). Lactate is removed from the muscle by the circulation system and subsequently reoxidized to pyruvate. Lactic acid is also the sole product of carbohydrate fermentation by the lactic acid bacteria (*Lactobacilli*). An alternative anaerobic mode of fermentation of carbohydrates occurs in yeast (*e.g. Saccharomyces cerevisiae*) and other microorganisms. Here, pyruvate is decarboxylated to give acetaldehyde (equation 66) and the product reduced to ethanol (equation 67).

$$CH_3COCOOH \longrightarrow CH_3CHO + CO_2 \qquad (66)$$
$$CH_3CHO + NADH + H^+ \longrightarrow CH_3CH_2OH + NAD^+ \qquad (67)$$

Reaction (67) is catalysed by alcohol dehydrogenase (1.1.1.1).

The complete degradation of acetate (in the form of acetyl-CoA) to carbon dioxide is carried out by the citric acid (or tricarboxylic acid) cycle. Many investigators have studied the component reactions of this cycle, but its integrated formulation shown in Figure 22 is due mainly to Krebs. Lack of space precludes a discussion of the copious evidence which has been presented in support of the reaction sequence given in Figure 22; this has been authoritatively reviewed by Krebs and Lowenstein[75]. Enzymes catalysing the reactions of the citric acid cycle have been found in all species of animals and higher plants so far examined; the enzymes are located in a highly organized fashion in sub-cellular organelles called mitochondria[76]. Many species of microorganisms are also capable of carrying out the reactions of the cycle.

The first three reactions of the cycle (a and b, Figure 22) are concerned with the formation of isocitrate from acetyl-CoA and oxaloacetate. An initial condensation reaction catalysed by citrate synthase (citrate condensing enzyme, citrogenase, oxaloacetate transacetase[77]) is followed by dehydration and hydration to produce *threo*-D_s-L_s-isocitrate; it is believed that

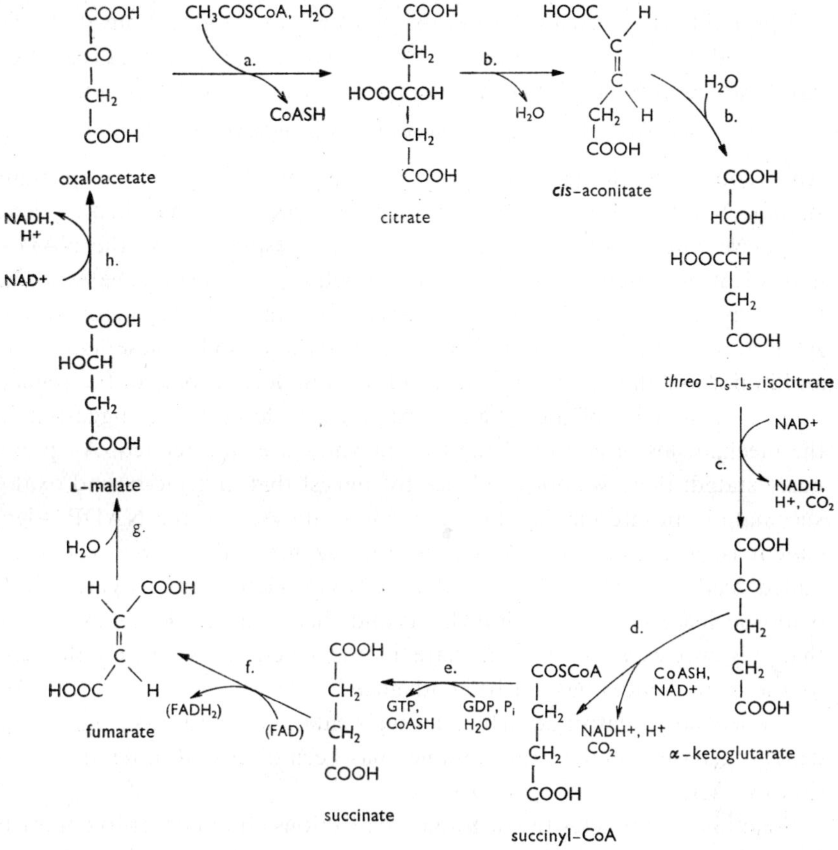

COOH   CH₃COSCoA, H₂O   COOH   HOOC   H
| | \\ /
CO   a.   CH₂   b.   C
| ──────────→ | ──────→   ‖
CH₂   CoASH   HOOCCOH   C
| | H₂O   / \
COOH   CH₂   CH₂ H
| H₂O |
oxaloacetate   COOH   COOH

citrate   cis–aconitate

NADH,   H₂O
H+   b.
h.
NAD+   COOH
                                              |
COOH   HCOH
| |
HOCH   HOOCCH
| |
CH₂   CH₂
| |
COOH   COOH

L–malate   threo –Dₛ–Lₛ–isocitrate

H₂O   g.   NAD+

H   COOH   c.
\ /
C   NADH,
‖   H+, CO₂
C
HOOC   H   COOH
                                              |
fumarate   (FADH₂)   CO
                                              |
(FAD)   COOH   CH₂
f. | e.   COSCoA   d. |
CH₂   ────── | CoASH,   CH₂
| GTP, | NAD+ |
CH₂   CoASH   CH₂   COOH
| GDP, Pᵢ |
COOH   H₂O   CH₂   NADH+, H+   α–ketoglutarate
              | CO₂
succinate   COOH

succinyl–CoA

Reaction	Enzyme	E.C. Number	$\Delta G^{o\prime}$ kcal/mole
a.	citrate synthase	4.1.3.7	−9·08
b.	aconitase	4.2.1.3	+1·59
c.	isocitrate dehydrogenase	1.1.1.41	−1·70
d.	α-ketoglutarate dehydrogenase (KDC)		−8·82
e.	succinyl-CoA synthetase	6.2.1.4	−2·12
f.	succinate dehydrogenase	1.3.99.1	∼0
g.	fumarase	4.2.1.2	−0·88
h.	malate dehydrogenase	1.1.1.37	+6·69

FIGURE 22. The citric acid cycle

both the dehydration and subsequent hydration reactions are catalysed by the same enzyme, namely aconitase[78].

Two oxidative decarboxylation reactions then take place, but it should be noticed that these are of quite distinct reaction types. The first (equation 68) is catalysed by an NAD^+-dependent isocitrate dehydrogenase.

$$\text{isocitrate} + NAD^+ \longrightarrow \alpha\text{-ketoglutarate} + NADH + H^+ + CO_2 \qquad (68)$$

An account has already been given (see section II. A.1) of the formally similar $NADP^+$-dependent enzyme; the enzyme concerned in the citric acid cycle differs from this in several important respects. First, the NAD^+-dependent enzyme is localized in the mitochondria as would be expected for an enzyme of the citric acid cycle. Secondly, the enzyme is activated by ADP and isocitrate and inhibited by NADH and ATP; these activation and inhibition effects are thought to play an important part in the regulation of the activity of the cycle. Thirdly, available evidence suggests that the mechanisms of action of the two enzymes are not related. As previously stated, there is good evidence to suggest that enzyme-bound oxalosuccinate is an intermediate in the reaction catalysed by the $NADP^+$-dependent isocitrate dehydrogenase, and the enzyme is able to decarboxylate added oxalosuccinate. In the case of the NAD^+-dependent enzyme, added oxalosuccinate is not decarboxylated and there is no evidence to suggest that this species is an intermediate in the reaction. No rationalization of these differences has yet been advanced.

The second oxidative decarboxylation is catalysed by the α-ketoglutarate dehydrogenase complex; this complex has been discussed in detail in section IV. A.1.

Reaction d completes the degradative reactions and in order to complete the cyclic process it is necessary to convert the product, succinyl-CoA, to oxaloacetate. Succinyl-CoA synthetase catalyses the formation of succinate from the CoA-thioester coupled with phosphorylation of GDP to GTP (equation 69).

$$\text{succinyl—CoA} + GDP + P_i \rightleftharpoons \text{succinate} + GTP + CoA \qquad (69)$$

This reaction is an example of substrate-level oxidative phosphorylation.

In order to replace one of the methylene groups of succinate by the keto group of oxaloacetate, the usual biochemical expedient of dehydrogenation, hydration and oxidation is employed. The dehydrogenation reaction is catalysed by the flavoprotein succinate dehydrogenase[79]. This enzyme is extremely tightly bound to the mitochondrial membrane and early attempts to solubilize it were unsuccessful. This problem has been overcome, and more detailed knowledge of the mode of action of the enzyme is to be expected. Fumarate is converted to L_s-malate in a reversible hydration reaction catalysed by fumarase[79]. Finally, oxidation of malate to oxalo-acetate is catalysed by the NAD^+-dependent malate dehydrogenase, (this enzyme is distinct from the $NADP^+$-dependent malic enzyme discussed in section II. A.1).

The overall processes represented in Figure 22 may be written as in equation (70).

$$\text{acetyl—CoA} + 3\,NAD^+ + (FAD) + GDP + P_i + 2\,H_2O \xrightarrow{\text{CAC}} 2\,CO_2 + CoA + 3\,NADH +$$
$$+ (FADH_2) + GTP + 3\,H^+ \qquad (70)$$

Equation (70) is not, however, a complete representation of the processes. For each turn of the cycle, 3 molecules of NADH and one of $FADH_2$ (*i.e.* reduced flavoprotein) are produced; if the cycle is to continue to operate these species must be reoxidized by the terminal electron transport system. A discussion of the electron transport system is beyond the scope of this article but the recent review by Green[80] may be recommended. Suffice it to say that the ETS brings about the reoxidation of NADH and $FADH_2$ by a series of electron transfers coupled to ATP synthesis (oxidative phosphorylation[81]). The overall reactions may be represented by equations (71) and (72).

$$NADH + H^+ + \tfrac{1}{2}O_2 + 3\,ADP + 3\,P_i \longrightarrow NAD^+ + 3\,ATP + 4\,H_2O \qquad (71)$$
$$FADH_2 + \tfrac{1}{2}O_2 + 2\,ADP + 2\,P_i \longrightarrow FAD + 2\,ATP + 3\,H_2O \qquad (72)$$

The yield of ATP from oxidation of $FADH_2$ is lower than that from oxidation of NADH since $FADH_2$ enters into the ETS at a later point than NADH. Combining reactions (71) and (72) with (70) gives equation (73) which describes the complete oxidation of acetyl-CoA to carbon dioxide and water.

$$CH_3COCoA + GDP + 11\,ADP + 12\,P_i + 2\,O_2 \longrightarrow 2\,CO_2 + CoA + GTP + 11\,ATP + 13\,H_2O$$
$$(73)$$

A list is given in Figure 22 of the apparent free energy changes ($\Delta G^{\circ\prime}$) associated with the reactions of the citric acid cycle. Values are those given by Johnson[82] and refer to 25°, pH 7·0, ionic strength 0·15 and 'analytical concentration' of reactants. It can be seen that the reactions fall into three classes. Reactions b, c, e, f and g are characterized by small values (either positive or negative) of $\Delta G^{\circ\prime}$ and may be considered to be freely reversible. Reaction h, however, is associated with an apparent free energy change of $+6·67$ kcal/mole; the equilibrium position of this reaction is such that the cycle tends to be reversed. This effect is overcome by the citrate synthase and α-ketoglutarate dehydrogenase reactions; both of these are characterized by large, negative values of $\Delta G^{\circ\prime}$. Summation of $\Delta G^{\circ\prime}$ values for the individual reactions gives a value of $-14 \cdot 32$ kcal/mole for the process represented by equation (70). Reactions (71) and (72) are also associated with negative free energy changes ($\Delta G_{71}^{\circ\prime} = -30$ kcal/mole, $\Delta G_{72}^{\circ\prime} = -21$ kcal/mole). Hence the oxidative degradation of acetyl-CoA coupled to electron transfer and oxidative phosphorylation (reaction 73), has an associated free energy change of -125 kcal/mole. It can be seen, therefore, that the cycle tends to operate only in the direction of carbon dioxide formation and represents a very efficient mechanism for the degradation of acetyl-CoA.

An interesting fact which does not emerge from the formulation of the citric acid cycle in Figure 22 is that on the first turn of the cycle neither of the carbon atoms derived from acetyl-CoA is liberated as carbon dioxide. The path taken by individual carbon atoms is shown in Figure 23 and it can be seen that carbon dioxide arises from the carboxyl groups of oxaloacetate. Citrate, a symmetrical molecule, behaves asymmetrically in the aconitase reaction; the enzyme discriminates between the two methylene

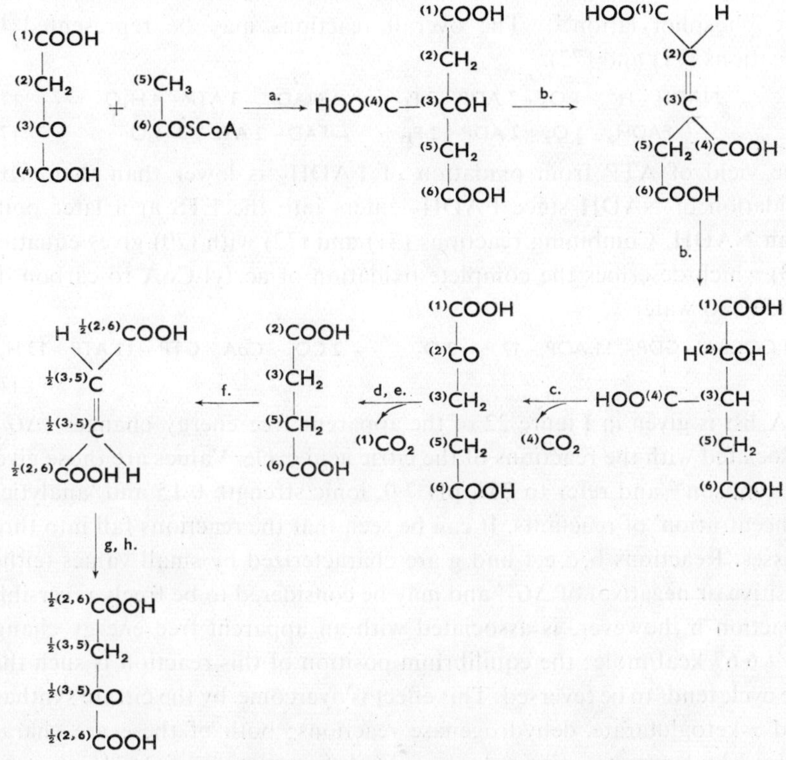

FIGURE 23. Path of carbon atoms in the citric acid cycle

groups and removes hydrogen only from the methylene derived from oxaloacetate. Hydration then yields isocitrate with the unique carbon distribution shown: it is then inevitable that the carbon atoms removed in reactions c and d are those originating at positions 4 and 1 of oxalo-acetate respectively. At the stage of fumarate formation, however, the two carboxyl carbons and the two ethylenic carbons become equivalent and randomization occurs. The course proposed for carbon atoms in Figure 23 has been amply substantiated by isotope-labelling experiments.

As was stated previously, the principal intermediate products of protein catabolism are succinate, fumarate, oxaloacetate and α-ketoglutarate; it is necessary to enquire how these substances are metabolized by the citric acid cycle. These compounds are all intermediates of the cycle and are therefore readily converted to malate or oxaloacetate. At this point a decarboxylation reaction is required. It is thought that malate and oxalo-acetate leave the mitochondria and undergo decarboxylation to pyruvate, the reaction being catalysed by the cytoplasmic malic enzyme (equation 74)

$$\text{malate} + \text{NADP}^+ \longrightarrow \text{pyruvate} + \text{CO}_2 + \text{NADPH} + \text{H}^+ \tag{74}$$

Oxaloacetate may be decarboxylated by the same enzyme. Final degradation to carbon dioxide is then carried out in the mitochondria by the pyruvate dehydrogenase complex and the citric acid cycle.

B. Decarboxylation of Amino Acids

1. Distribution and mechanism of action of the amino acid decarboxylases.

The existance of enzymes catalysing the decarboxylation of amino acids was originally implied from observations of amine formation by micro-organisms growing on media containing amino acids. The reaction may be written as equation (75).

$$\text{RCH(NH}_2)\text{COOH} \longrightarrow \text{RCH}_2\text{NH}_2 + \text{CO}_2 \tag{75}$$

Work concerning specific decarboxylases for amino acids in bacteria, in plant and in animal tissue has been reviewed by Gale[83], Blaschko[84] and Schales[85].

Present knowledge of the distribution and specificity of the amino acid decarboxylases is summarized in Table 2[86, 87]. Generally, the amino acid undergoing decarboxylation contains a third polar group (—COOH, —NH$_2$, —OH, =NH, —SO$_3$H etc.) in addition to the α-amino and carboxylic acid groups. An exception is the decarboxylase from *Proteus vulgaris* which acts on a variety of aliphatic amino acids[88]. Amino acid decarboxylases from bacterial sources are usually specific for a particular

TABLE 2. Decarboxylation of Amino Acids

Substrate[a]	Product	E.C. No.	Source[b]
1. Aminomalonate	Glycine	4.1.1.10	A
2. Aspartate (α)	β-Alanine	4.1.1.11	B
3. Aspartate (β)	α-Alanine	4.1.1.12	B
4. Leucine	Isoamylamine	4.1.1.14	B
Valine	Isobutylamine		
Isoleucine	2-Methylbutylamine		
Norvaline	Butylamine		
α-Aminobutyrate	Propylamine		
5. Glutamate	γ-Aminobutyrate	4.1.1.15	A, B, P
Cysteinesulphinate	Hypotaurine		A
Cysteate	Taurine		A
γ-Methyleneglutamate	γ-Amino-α-methylene-butyrate		P
6. 3-Hydroxyglutamate	3-Hydroxy-4-amino-butyrate	4.1.1.16	B
7. Ornithine	Putrescine	4.1.1.17	B
8. Lysine	Cadaverine	4.1.1.18	B
9. Arginine	Agmatine	4.1.1.19	B
10. Meso-2,6-diamino-pimelate	Lysine	4.1.1.20	B
11. Histidine	Histamine	4.1.1.22	A, B, P
12. (o- or p-)Aminobenzoate	Aniline	4.1.1.24	B
13. Tyrosine	Tyramine	4.1.1.25	B
3,4-Dihydroxyphenyl-alanine	3,4-Dihydroxy-phenylethylamine		
14. 3,4-Dihydroxyphenyl-alanine	3,4-Dihydroxy-phenylethylamine	4.1.1.26	A
Tyrosine	Tyramine		
5-Hydroxytryptophan	Serotonin		
Tryptophan	Tryptamine		
Phenylalanine	Phenylethylamine		
Histidine	Histamine		
15. Tryptophan	Tryptamine	4.1.1.27	B
16. 5-Hydroxytryptophan	Serotonin	4.1.1.28	B
17. Cysteinsulphinate	Hypotaurine	4.1.1.29	A
18. γ-Methyleneglutamate	γ-Amino-α-methylene-butyrate		P
19. Serine	Ethanolamine		A

[a] Substrates of a particular enzyme are listed in order of reactivity. Enzymes are named by first substrate in list. e.g. 4.1.1.26 is DOPA decarboxylase.

[b] A = animal, B = bacterial, P = plant.

substrate; enzymes from plant and animal sources seem to show a wider specificity for their substrates. In particular, an important enzyme from animal tissue (4.1.1.26) is believed to decarboxylate a broad range of aromatic amino acids. Decarboxylases from bacterial and animal sources also differ in that the pH optima of the former are generally in the range pH 2·5–6, whereas mammalian amino acid decarboxylases are most active at neutral or alkaline pH.

It should be emphasized that, with the exception of the glutamate, histidine and β-aspartate decarboxylases, none of the enzymes listed in Table 3 has been well-characterized; thus the Table may undergo future extension and modification.

Two of the reactions listed in Table 2 differ markedly from the rest. Firstly, a bacterial enzyme has been characterized which catalyses the β-decarboxylation of aspartate (equation 76).

$$\begin{array}{c} COOH \\ | \\ CHNH_2 \\ | \\ CH_2 \\ | \\ COOH \end{array} \xrightarrow{\beta\text{-decarboxylation}} \begin{array}{c} COOH \\ | \\ CHNH_2 \\ | \\ CH_3 \end{array} + CO_2 \qquad (76)$$

This reaction is related to α-decarboxylation and will be considered in detail below. Secondly, it has been shown that serine undergoes decarboxylation in animal tissues (equation 77).

$$\text{serine} \longrightarrow \text{ethanolamine} + CO_2 \qquad (77)$$

No serine decarboxylase activity has been demonstrated in such tissue, but Borkenhagen[89] and his coworkers have shown that rat liver preparations catalyse reactions (78) and (79).

$$\text{phosphatidylethanolamine} + \text{serine} \longrightarrow \text{phosphatidylserine} + \text{ethanolamine} \qquad (78)$$

$$\text{phosphatidylserine} \longrightarrow \text{phosphatidylethanolamine} + CO_2 \qquad (79)$$

It is probable, therefore, that the decarboxylation of serine arises from reactions (78) and (79).

The first indication of a cofactor requirement for enzymic decarboxylation of amino acids came from the work of Gale[90]. The production of tyrosine decarboxylase by *Streptococcus faecalis* required the presence of an unidentified factor in the medium. Bellamy and Gunsalus[91] identified the cofactor as a phosphorylated derivative of pyridoxal (Figure 24II); they showed that the tyrosine decarboxylase activity of *S. faecalis* cells grown on a medium deficient in pyridoxine (Figure 24I) was greatly increased by added pyridoxal and increased much more by a mixture of pyridoxal and ATP. This work was extended by Baddiley and Gale[92] who showed pyridoxal phosphate (Pal. P) to be the cofactor for lysine, arginine

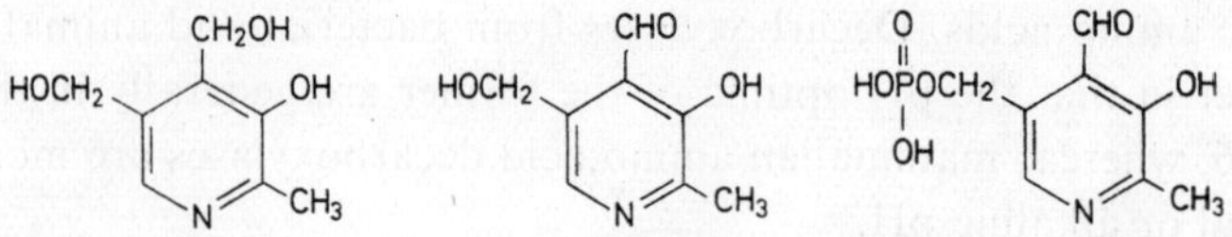

$$
\begin{array}{ccc}
\text{CH}_2\text{OH} & \text{CHO} & \overset{\displaystyle O}{\underset{\displaystyle OH}{\|}}\text{HOPOCH}_2 \quad \text{CHO}
\end{array}
$$

pyridoxine (I)   pyridoxal (II)   pyridoxal phosphate (III)

FIGURE 24.

and ornithine decarboxylases. The 5′-phosphate (Figure 24III) was synthesized in 1952[93] and shown to be the active form.

Pyridoxal phosphate was also shown to be the cofactor of glutamate aspartate transaminase which catalyses reaction (80).

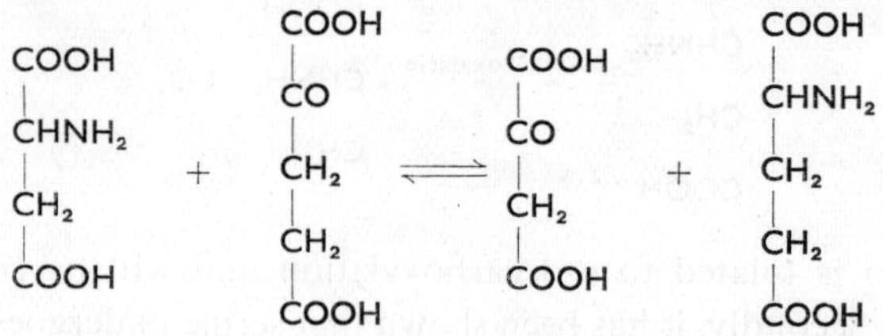

$$
\begin{array}{c}
\text{COOH} \\
|\\
\text{CHNH}_2 \\
|\\
\text{CH}_2 \\
|\\
\text{COOH}
\end{array}
\;+\;
\begin{array}{c}
\text{COOH} \\
|\\
\text{CO} \\
|\\
\text{CH}_2 \\
|\\
\text{CH}_2 \\
|\\
\text{COOH}
\end{array}
\;\rightleftharpoons\;
\begin{array}{c}
\text{COOH} \\
|\\
\text{CO} \\
|\\
\text{CH}_2 \\
|\\
\text{COOH}
\end{array}
\;+\;
\begin{array}{c}
\text{COOH} \\
|\\
\text{CHNH}_2 \\
|\\
\text{CH}_2 \\
|\\
\text{CH}_2 \\
|\\
\text{COOH}
\end{array}
\qquad (80)
$$

aspartate   α-ketoglutarate   oxaloacetate   glutamate

Subsequent investigations have shown pyridoxal phosphate to be the cofactor for a very wide range of enzyme-catalysed reactions of amino acids; this subject has been reviewed by Braunstein[94]. The participation of pyridoxal phosphate has not been unambiguously demonstrated for all reactions listed in Table 3; it seems reasonable to assume, however, that the close similarity in the reactions and substrates involved reflects a uniformity of cofactor requirement and reaction mechanism.

Free pyridoxal (or the 5′-phosphate) acts as a catalyst in many of the reactions of amino acids which are catalysed by pyridoxal phosphate dependent enzymes. In each case, the initial reaction is condensation of the amino acid with pyridoxal (phosphate) to form an aldimine intermediate[94]. The reaction scheme suggested by Westheimer[115] and by Metzler[95] and his colleagues for the decarboxylation reaction is shown in Figure 25 (reactions a–d). Formation of the aldimine (a) results in labilization of the C_α-COOH bond; elimination of carbon dioxide (b) is followed by pick-up of a proton from the solvent (c) and finally hydrolysis (d) yields the amine and free pyridoxal phosphate. A previous suggestion[96] that the reaction occurs with initial labilization of the C_α-H bond followed by loss of carbon dioxide has been refuted on the basis of the evidence presented below.

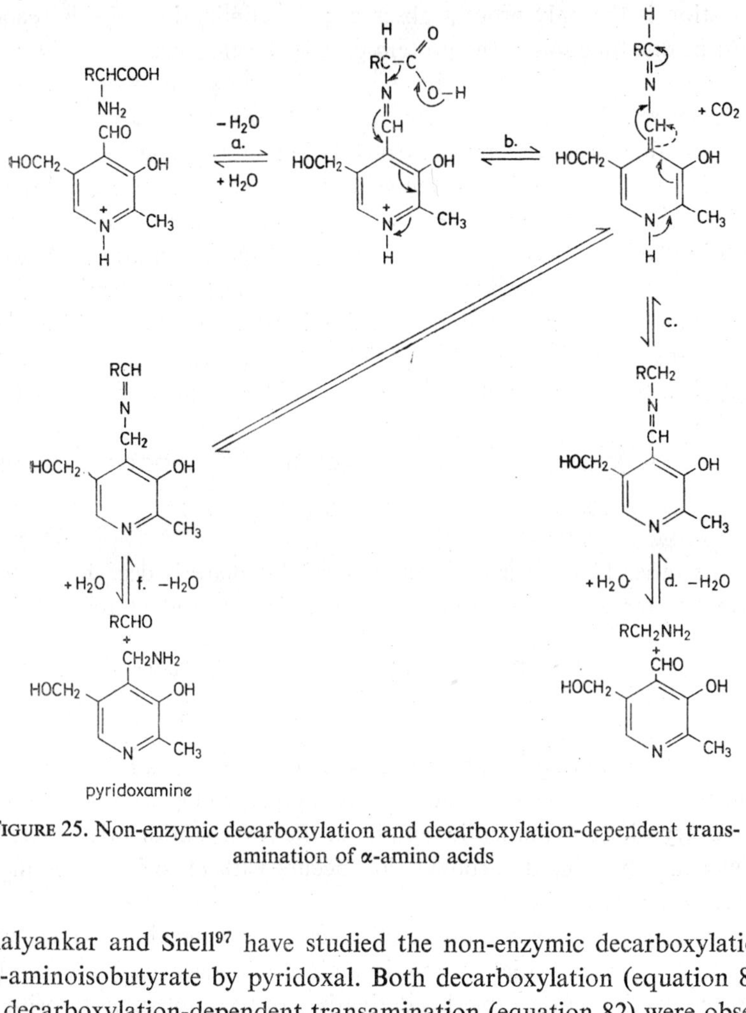

FIGURE 25. Non-enzymic decarboxylation and decarboxylation-dependent transamination of α-amino acids

Kalyankar and Snell[97] have studied the non-enzymic decarboxylation
of α-aminoisobutyrate by pyridoxal. Both decarboxylation (equation 81)
and decarboxylation-dependent transamination (equation 82) were observed.

$$\alpha\text{-aminoisobutyrate} \longrightarrow \text{isopropylamine} + CO_2 \tag{81}$$

$$\alpha\text{-aminoisobutyrate} + \text{pyridoxal} \longrightarrow \text{acetone} + CO_2 + \text{pyridoxamine} \tag{82}$$

The decarboxylation reaction is readily explained by reactions a–d,
Figure 25 (R = —CH$_3$, —H replaced by —CH$_3$). The lack of a C$_\alpha$-H
group in this case lends strong support to the suggested mechanism. Decarboxylation-dependent transamination may be considered to occur by
reactions a, b, e and f; that is, the difference lies only in the nature of the
imine undergoing hydrolysis. In the enzyme-catalysed reactions, decarb-

63   C.C.A.E.

oxylation is the only process observed; the elimination of side reactions must be one function of the protein part of the enzyme.

Support for sequence a–d in enzymic decarboxylation of amino acids has been provided by Mandeles[98] and his coworkers, who showed that the tyramine produced by decarboxylation of tyrosine in solvent D_2O contained only one atom of deuterium; any reaction sequence involving labilization of the α-hydrogen atom of tyrosine would result in the incorporation of two atoms of deuterium under these conditions. Mandeles and Hanke[99] have shown that only one of the possible enantiomeric pair of γ-monodeutero-γ-aminobutyric acids is produced by decarboxylation of glutamic acid in D_2O solvent. The product was found to exchange deuterium for hydrogen in the presence of glutamate decarboxylase (equation 83).

$$\text{HOOC(CH}_2)_2\text{CHNH}_2 \xrightarrow[\text{D}_2\text{O}]{\text{GDC}} \text{HOOC(CH}_2)_2\text{CHNH}_2 \xrightarrow[\text{H}_2\text{O}]{\text{GDC}} \text{HOOC(CH}_2)_2\text{CH}_2\text{NH}_2$$
$$\underset{\text{COOH}}{|} \qquad\qquad\qquad \underset{\text{D}}{|}$$

(83)

On the other hand, when α-monodeuteroglutamate is decarboxylated in solvent water, all the deuterium is retained in the product amine[84].

$$\text{HOOC(CH}_2)_2\text{CDNH}_2 \xrightarrow[\text{H}_2\text{O}]{\text{GDC}} \text{HOOC(CH}_2)_2\text{CDNH}_2 \qquad (84)$$
$$\underset{\text{COOH}}{|} \qquad\qquad\qquad \underset{\text{H}}{|}$$

Belleau and Burba[100], by comparison of the rate of oxidation by monoamine oxidase of the α-monodeuterotyramine produced by decarboxylation of tryosine and authentic samples of the two enantiomeric forms, have shown that tyrosine decarboxylation occurs with retention of configuration at the α-carbon atom.

Further support for the proposed reaction sequence is provided by the fact that the mammalian aromatic amino acid decarboxylase catalyses decarboxylation of the α-methyl derivatives of tryptophan, m-tyrosine and 3,4-dihydroxyphenylalanine.

The β-decarboxylation of aspartate differs from the α-decarboxylation reactions described above in several important respects. Incubation of aspartate β-decarboxylase with the substrate results in rapid formation of alanine but also in a slow inactivation of the enzyme; a similar inactivation results from incubation of the enzyme with a variety of amino acids. This inactivation has been shown to arise from conversion of the cofactor into the inactive pyridoxamine form; the modified cofactor readily dissociates from the enzyme surface. Inactivation of the enzyme is prevented or reversed by the addition of pyridoxal phosphate and may be prevented or

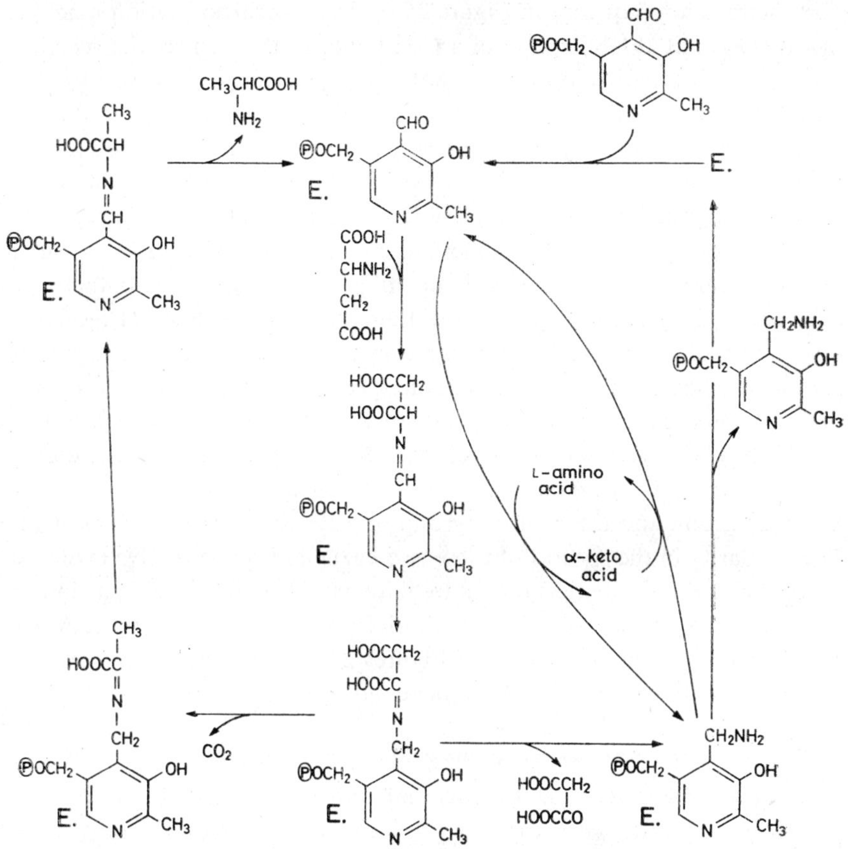

FIGURE 26. The reactions of aspartate β-decarboxylase

delayed by the addition of α-keto acids. In the latter case, small amounts of the corresponding amino acids are formed and the activating effects of keto acids may be ascribed to the reconversion of the pyridoxamine form of the holoenzyme to the pyridoxal form by a transamination reaction. Added pyridoxal phosphate exerts its activating effect by regeneration of the holoenzyme after dissociation of pyridoxamine phosphate from the enzyme surface. The reactions of aspartate β-decarboxylase are summarized in Figure 26[101, 102].

In summary, aspartate β-decarboxylase acts both as a decarboxylase and a non-specific transaminase. Although the transaminase activity is low, it is nevertheless of great importance since it controls the amount of the enzyme in the form which is active in decarboxylation. It should also be noted that the reaction sequence in Figure 26 for the β-decarboxylation reaction involves the initial labilization of a C-H bond. In this respect the reaction bears more similarity to transamination (see, for example, review in reference 86) than to α-decarboxylation.

Amino acid decarboxylases have not, in general, been obtained, either in large amounts or in homogeneous states. Some work has been done on the purified glutamate decarboxylase from *Escherichia coli*[103]. The enzyme has a molecular weight of 300,000 and contains two molecules of bound pyridoxal phosphate per molecule of apoprotein. The bound cofactor is characterized by an absorption maximum at 415 mμ below pH 5·0 (corresponding to the active form of the enzyme) which moves to 340 mμ at pH values above 5. These characteristic spectra bear a strong resemblance to those observed in the case of glutamate aspartate transaminase[86]. In this case, it has been shown conclusively that the pyridoxal form of the cofactor is bound to the enzyme by an aldimine linkage between the cofactor aldehyde group and the ε-amino group of a lysine residue in the protein. The similarity in the spectra of these two enzymes is presumptive evidence that pyridoxal phosphate is bound to glutamate decarboxylase in a similar fashion. If this is the case, then the initial reactions between the enzymes and substrates in the sequences of Figures 25 and 26 should be shown as transaldiminations rather than simple condensation reactions.

2. Biological significance of the amino acid decarboxylases

a. Bacterial enzymes   Most bacterial amino acid decarboxylases are inducible enzymes; that is, they are produced in large quantities only when the growth medium contains the appropriate substrate. Similarly, optimal production of the decarboxylases requires acid growth conditions. These observations led Gale[83] to speculate that the formation of amines from amino acids is a device used by microorganisms to counteract the effects of an acid environment. An alternative suggestion[86] is that decarboxylation of amino acids supplies at least part of the carbon dioxide required for the metabolic processes of the organism; this source of carbon dioxide might become particularly important at low pH where the solubility of the gas in the growth medium is low. These suggestions are, however, purely speculative and no clearly defined function can be ascribed to the majority of these enzymes. Specific metabolic functions can be ascribed, on the other hand, to the aspartate α-decarboxylase and the *meso*-2,6-diaminopimelate decarboxylase.

Several routes for the formation of β-alanine in microbial systems have been described, but the main route in such organisms as *E. coli*, *Azoto-*

bacter vinlandii and *Rhizobium leguminosarium*, particularly under acid growth conditions, is the α-decarboxylation of aspartate. The product, β-alanine, is mainly involved in the biosynthesis of the pantetheine side chain of coenzyme A (Figure 3). The probable sequence of reactions involves an initial condensation (reaction 85) between β-alanine and pantoic acid (derived from valine) to yield pantothenate. Pantothenate is phosphorylated by ATP and the resulting phosphopantothenate is condensed with cysteine, in a CTP-dependent reaction to produce phosphopantoth-enylcysteine (reaction 86). Formation of the side chain is completed by decarboxylation (reaction 87) to phosphopantetheine.

$$
\begin{array}{c}
\text{COOH} \\
|\\
\text{CH}_2 \\
|\\
\text{CH}_2\text{NH}_2
\end{array}
\quad + \quad
\begin{array}{c}
\text{CH}_2\text{C(CH}_3)_2\text{CHCOOH} \\
\quad |\qquad\qquad | \\
\quad \text{OH}\qquad\quad \text{OH}
\end{array}
\xrightarrow[\text{ATP AMP, PP}_i]{}
\begin{array}{c}
\text{CH}_2\text{C(CH}_3)_2\text{CHCONH(CH}_2)_2\text{COOH} \\
\quad |\qquad\qquad | \\
\quad \text{OH}\qquad\quad \text{OH}
\end{array}
$$

pantoate   pantothenate   (85)

$$
\begin{array}{c}
\text{CH}_2\text{C(CH}_3)_2\text{CHCONH(CH}_2)_2\text{COOH} \\
\quad |\qquad\qquad | \\
\quad \text{O℗}\qquad\quad \text{OH}
\end{array}
\quad + \quad
\begin{array}{c}
\text{COOH} \\
|\\
\text{CHNH}_2 \\
|\\
\text{CH}_2\text{SH}
\end{array}
\xrightarrow{\text{CTP}}
$$

phosphopantothenate   (86)

$$
\begin{array}{c}
\text{CH}_2\text{C(CH}_3)_2\text{CHCONH(CH}_2)_2\text{CONHCHCH}_2\text{SH} \\
\quad |\qquad\qquad |\qquad\qquad\qquad\qquad\qquad | \\
\quad \text{O℗}\qquad\quad \text{OH}\qquad\qquad\qquad\qquad \text{COOH}
\end{array}
$$

phosphopantothenylcysteine

$$
\begin{array}{c}
\text{CH}_2\text{C(CH}_3)_2\text{CHCONH(CH}_2)_2\text{CONHCHCH}_2\text{SH} \\
\quad |\qquad\qquad |\qquad\qquad\qquad\qquad\qquad | \\
\quad \text{O℗}\qquad\quad \text{OH}\qquad\qquad\qquad\qquad \text{COOH}
\end{array}
\xrightarrow{\text{CO}_2}
$$

                                                                                (87)

$$
\begin{array}{c}
\text{CH}_2\text{C(CH}_3)_2\text{CHCONH(CH}_2)_2\text{CONHCH}_2\text{CH}_2\text{SH} \\
\quad |\qquad\qquad | \\
\quad \text{O℗}\qquad\quad \text{OH}
\end{array}
$$

phosphopantetheine

Higher organisms lack the enzyme pantothenate synthetase; these organisms are unable to synthesize pantothenate and require a dietary source of this compound.

The enzyme *meso*-2,6-diaminopimelate decarboxylase is of particular importance in bacterial metabolism since it catalyses the terminal step in the biosynthesis of lysine (Figure 27). The sequence of reactions starting from pyruvate and aspartate-β-semialdehyde is as shown in Figure 27. It should be noted that this enzyme differs from the other bacterial amino acid decarboxylases in that the pH optimum is near neutrality; this may

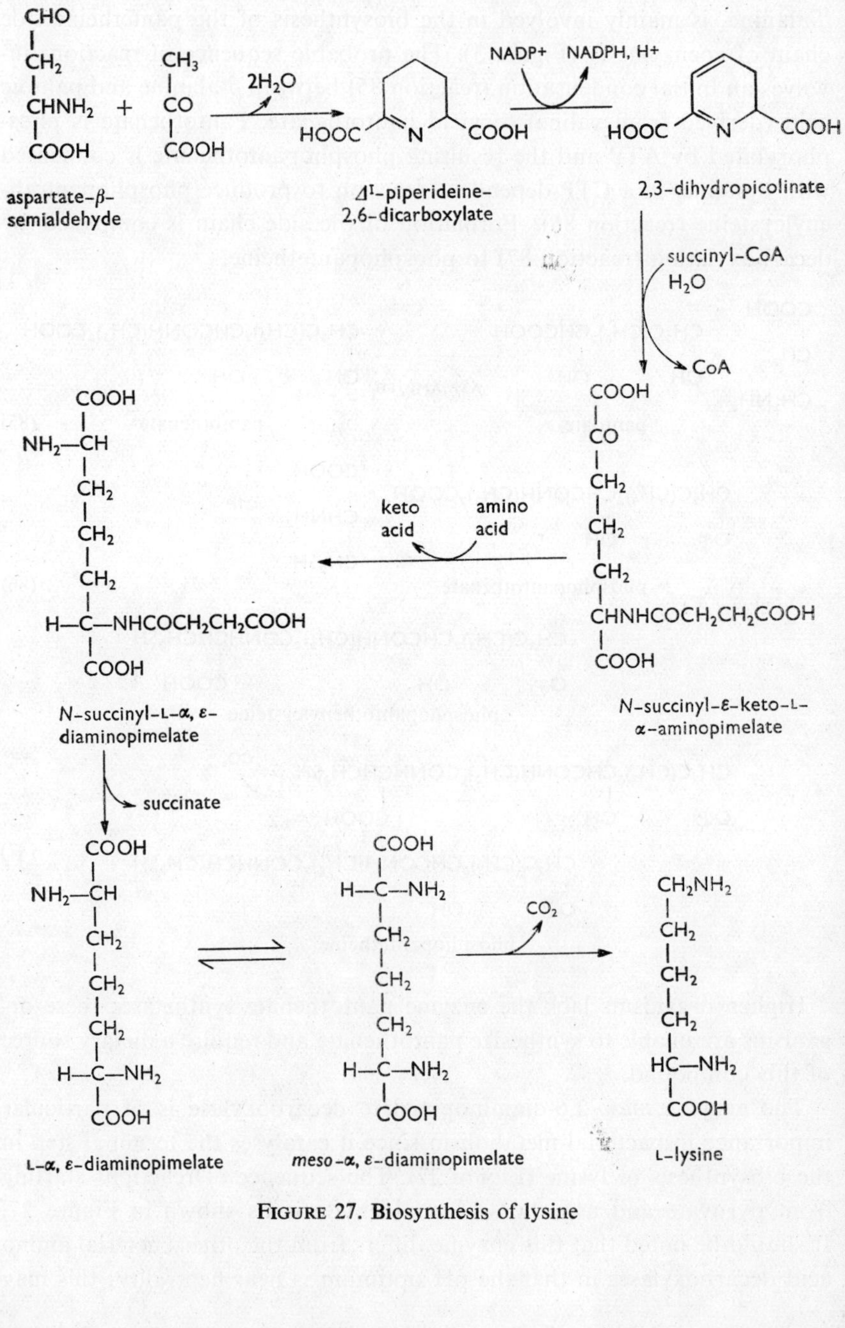

FIGURE 27. Biosynthesis of lysine

reflect the importance of the enzyme to microorganisms growing under normal conditions.

b. Mammalian enzymes   Decarboxylation of amino acids is of considerable significance in mammalian systems, since the amines produced in these reactions have profound physiological effects.

Decarboxylation of glutamate yields γ-aminobutyrate (GABA) (reaction 88). The decarboxylase is found in relatively large quantities in mammalian brain tissue and GABA is thought to play an important part in the activity of the nervous system[104, 105].

$$
\begin{array}{ccc}
\text{COOH} & & \\
| & & \text{CH}_2\text{NH}_2 \\
\text{CHNH}_2 & & | \\
| & & \text{CH}_2 \\
\text{CH}_2 & \longrightarrow & | \quad +\text{CO}_2 \\
| & & \text{CH}_2 \\
\text{CH}_2 & & | \\
| & & \text{COOH} \\
\text{COOH} & &
\end{array}
\tag{88}
$$

It seems likely that, in crustacea, GABA acts as the synaptic chemical transmitter for inhibitory nerve fibres. This role was suggested by the identification of GABA with Factor I, a substance in mammalian brain extracts which exerts an inhibitory effect on the crustacean stretch receptor neuron and whose activity parallels that of stimulation of the inhibitory nerve. A similar role for GABA in the electrical activity of the brain has been suggested but there is little evidence available to support the suggestion. Other views have been presented; for example, Grundfest[106] has postulated that GABA acts in the brain by blocking excitatory synapses and does not affect inhibitory synapses. These problems remain to be resolved.

The route by which excess GABA is removed from brain tissue is well known. A transamination reaction (89) with α-ketoglutarate yields succinic semialdehyde and glutamate; succinic semialdehyde may be oxidized to succinic acid and the latter metabolized by the citric acid cycle.

$$
\begin{array}{ccccccc}
\text{CH}_2\text{NH}_2 & & \text{COCOOH} & & \text{CHO} & & \text{CH(NH}_2)\text{COOH} \\
| & & | & & | & & | \\
\text{CH}_2 & & \text{CH}_2 & & \text{CH}_2 & & \text{CH}_2 \\
| & + & | & \rightleftharpoons & | & + & | \\
\text{CH}_2 & & \text{CH}_2 & & \text{CH}_2 & & \text{CH}_2 \\
| & & | & & | & & | \\
\text{COOH} & & \text{COOH} & & \text{COOH} & & \text{COOH}
\end{array}
\tag{89}
$$

Dopa (3,4-dihydroxyphenylalanine) decarboxylase is a key enzyme in the biosynthesis of noradrenaline and adrenaline; the biosynthetic pathway, originally suggested by Blaschko[107] is shown in Figure 28. Adrenaline is derived from noradrenaline by an *N*-methylation reaction involving *S*-adenosylmethionine.

In addition to its role as a hormone noradrenaline is believed to act as the chemical transmitter in postganglionic sympathetic nerve fibres (adrenergic fibres)[108]. These fibres are rich in dopamine and noradrenaline and

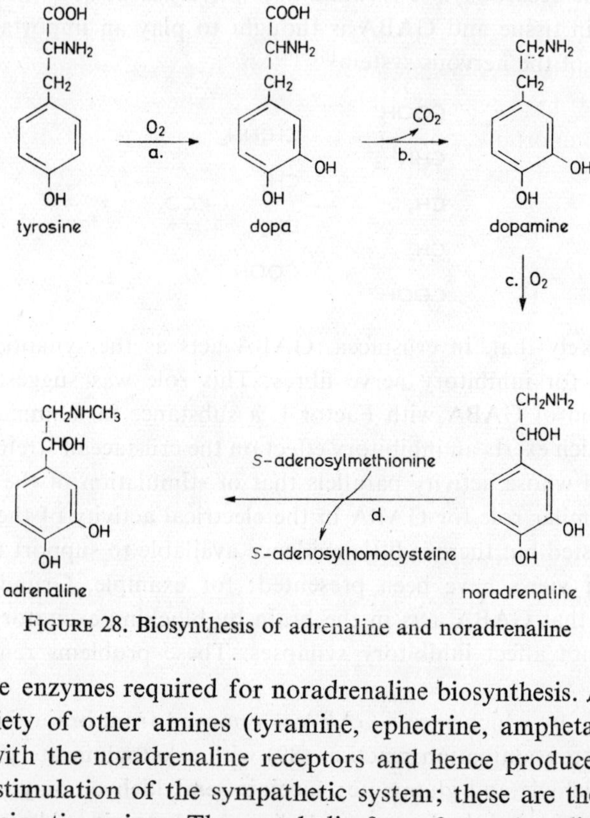

FIGURE 28. Biosynthesis of adrenaline and noradrenaline

contain the enzymes required for noradrenaline biosynthesis. Adrenaline and a variety of other amines (tyramine, ephedrine, amphetamine) can combine with the noradrenaline receptors and hence produce the same effects as stimulation of the sympathetic system; these are the so-called sympathomimetic amines. The metabolic fate of noradrenaline and the other sympathomimetic amines has been discussed by Axelrod[109].

The decarboxylation of histidine yields histamine (reaction 90).

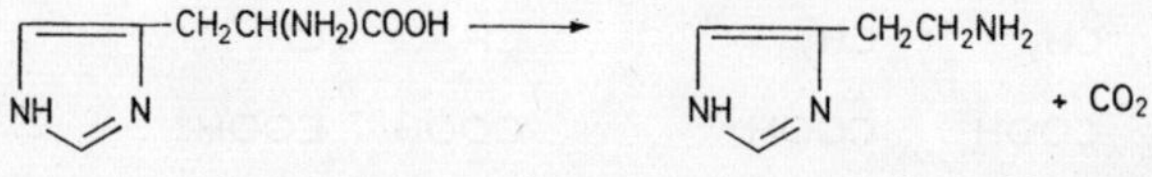

$$\tag{90}$$

Reaction (90) is catalysed both by the general aromatic amino acid decarboxylase and by a specific histidine decarboxylase. The latter enzyme is found in large quantities in mast cells of the connective tissue[110] and in rapidly developing tissue such as tumours. Histamine is stored within the mast cells in granules and may be released from these granules in large quantities in certain pathological conditions. Release of preformed histamine has been shown to be closely related with allergic reactions[111] and with experimentally induced anaphylactic shock[112].

The role of histamine in normal physiological processes is still under active investigation[111]. Administered histamine is known to stimulate acid secretion by the parietal cells of the gastric mucosa, and it is currently believed that induced histamine plays an active part in digestive processes.

Schayer[113] has reviewed studies of the catabolism of histamine *in vivo*. The most important route appears to be the conversion of histamine to methylimidazole acetate by the sequence shown in equation (91).

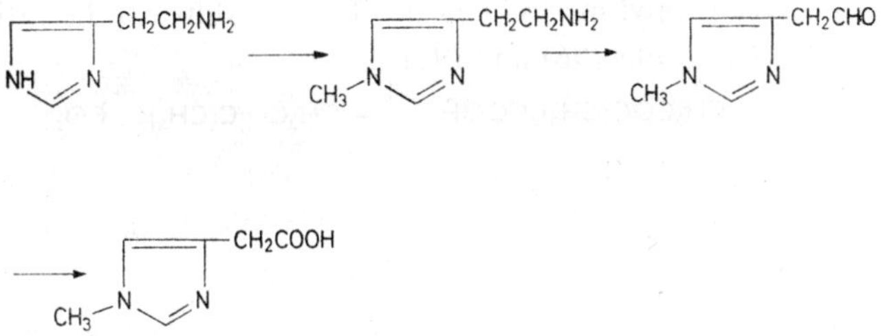

$$(91)$$

The enzymes catalysing these reactions are not well-characterized.

C. Decarboxylation of β-Keto Acids

The enzymes which catalyse β-decarboxylation reactions show no requirement for organic cofactors. Two such enzymes have been investigated in some detail, namely oxaloacetate decarboxylase (E.C. 4.1.1.3) and acetoacetate decarboxylase (E.C. 4.1.1.4); these enzymes catalyse reactions (92) and (93) respectively. Early work on these enzymes has been reviewed by Ochoa[114].

$$HOOCCH_2COCOOH \longrightarrow CH_3COCOOH + CO_2 \tag{92}$$
$$CH_3COCH_2COOH \longrightarrow CH_3COCH_3 + CO_2 \tag{93}$$

Oxaloacetate decarboxylase shows an absolute requirement for a divalent cation whereas the acetoacetate decarboxylase shows no such requirement. This difference is reflected in the nonenzymic decarboxylation of these two species; metal ions catalyse the decarboxylation of oxaloacetate but not of acetoacetate.

The decarboxylation of β-keto monocarboxylic acids may be represented by equations (94) and (95)[115]; that is, the internally hydrogen-bonded

keto acid decarboxylates to give the enol of the product which then ketonizes.

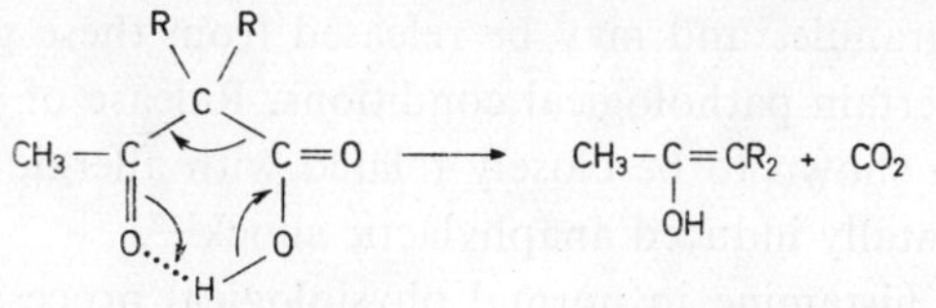

$$CH_3-C \overset{\overset{R}{\diagdown}\overset{R}{\diagup}}{\underset{\overset{\parallel}{O..}}{\underset{H}{C}}} \underset{\overset{\diagup}{O}}{C=O} \longrightarrow CH_3-C=CR_2 + CO_2 \atop \quad\quad\overset{|}{OH} \tag{94}$$

$$CH_3C(OH){=\!=}CR_2 \longrightarrow CH_3COCHR_2 \tag{95}$$

Pedersen[116] has provided good evidence for this reaction scheme in the case
of the decarboxylation of dimethylacetoacetate; he found that the addition of bromine does not alter the rate of evolution of carbon dioxide but
stoichiometric amounts of bromine are consumed during the reaction.
Under the conditions of the experiment, no significant quantity of bromine
is absorbed by isopropyl methyl ketone. These results can be rationalized
on the basis of equations (96) and (97).

$$CH_3COC(CH_3)_2COOH \longrightarrow CH_3C{=\!=}C(CH_3)_2 + CO_2 \atop \quad\quad\quad\overset{|}{OH} \tag{96}$$

$$CH_3C{=\!=}C(CH_3)_2 + Br_2 \longrightarrow CH_3COCBr(CH_3)_2 + HBr \atop \overset{|}{OH} \tag{97}$$

That the reaction involves decarboxylation of the internally hydrogen-
bonded acid rather than a form in which the carbonyl oxygen atom is
protonated is suggested by the observation that the rate of the reaction is
essentially unaffected by the polarity of the solvent[117].

Reaction (94) should obviously be facilitated by a centre of positive
charge on the β-carbon atom and Pedersen[118] suggested that the catalysis of
decarboxylation of β-ketoacids by primary amines is due to the initial
formation of a protonated Schiff's base (reaction 98). This species undergoes decarboxylation (reaction 99) and then hydrolysis to the ketone
(reaction 100).

$$\underset{\overset{\parallel}{O}\;\;+ \atop RNH_2}{CH_3CC(CH_3)_2COO^-} \underset{H_2O}{\rightsquigarrow} \underset{\overset{\parallel}{NR}}{CH_3CC(CH_3)_2COO^-} \overset{H+}{\longrightarrow} \underset{^+NHR}{CH_3CC(CH_3)_2COO^-} \tag{98}$$

$$\underset{^+NHR}{CH_3CC(CH_3)_2COO^-} \underset{CO_2}{\rightsquigarrow} \underset{NHR}{CH_3C{=\!=}C(CH_3)_2} \tag{99}$$

$$\underset{NHR}{CH_3C{=\!=}C(CH_3)_2} \overset{H+}{\longrightarrow} \underset{^+NHR}{CH_3CCH(CH_3)_2} \overset{H_2O}{\longrightarrow} CH_3COCH(CH_3)_2 + RNH_3^+ \tag{100}$$

Westheimer[119] has presented very strong evidence that the enzyme-catalysed decarboxylation of acetoacetate proceeds by a route analagous to reactions (98–100), in which the amine function is provided by a lysine side chain group in the protein. These investigations were carried out on the enzymes from *Clostridium acetobutylicum* and *C. madisonii*. The reaction scheme may be represented by equations (101) and (102) where E-NH$_2$ represents the protein-bound lysine amino group.

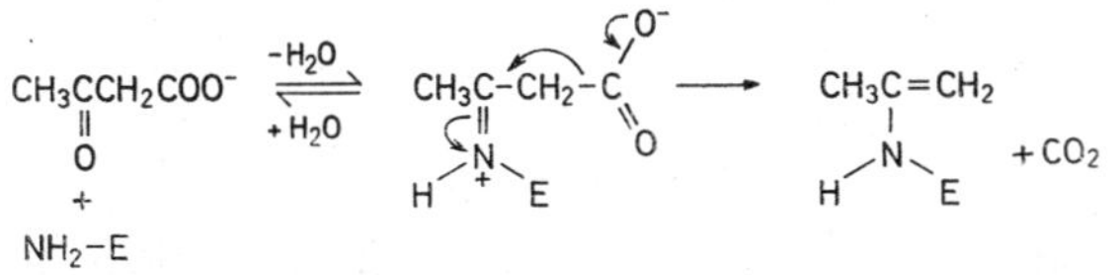

$$(101)$$

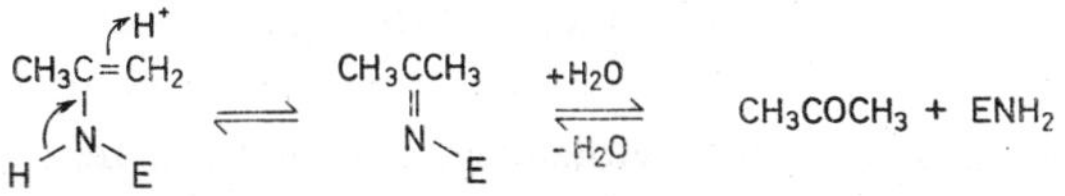

$$(102)$$

Kinetic analysis of the enzyme reaction has shown a pH dependence and an inhibitory effect by monovalent amines consistent with a cationic group at the active site.

Direct evidence for the proposed reaction sequence has been obtained. Decarboxylation of acetoacetate was shown to be accompanied by exchange of the carbonyl oxygen atom with solvent; similarly, exchange of oxygen was observed when acetone was incubated with the enzyme. Careful control experiments were carried out to allow for non-enzymic exchange reactions. Evidence for the occurence of Schiff's base intermediates was obtained by incubation of the enzyme with its substrate and subsequent addition of sodium borohydride. About 75% of the enzymic activity was destroyed compared with only 10% in control experiments in which the substrate was omitted. Similar experiments with [3-^{14}C]-acetoacetate were carried out; after treatment with borohydride, radioactivity became irreversibly bound to the protein and treatment with 6 N hydrochloric acid liberated ε-isopropyllysine which was identified chromatographically. These results can be explained by equation (103).

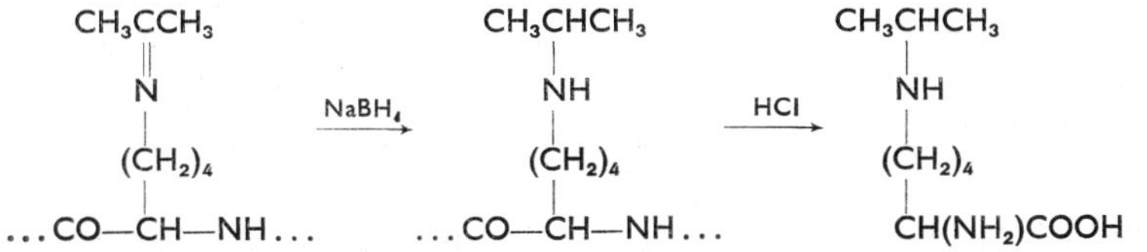

$$(103)$$

Schiff's base
intermediate   reduced form   free ε-isopropyllysine

When the enzyme-catalysed decarboxylation of acetoacetate was carried out in the presence of hydrogen cyanide, progressive inactivation of the enzyme was observed. The inactivation could be reversed by prolonged dialysis. These results are consistent with reversible addition of hydrogen cyanide to the imine bond of a Schiff's base intermediate (reaction 104). Consistently, hydrogen cyanide protected the enzyme against irreversible inactivation by borohydride.

$$\text{HCN} + \quad \begin{array}{c} \text{CH}_3\text{--C--CH}_3 \\ \| \\ \text{N} \\ \diagdown \\ \text{E} \end{array} \quad \rightleftharpoons \quad \begin{array}{c} \text{CH}_3\text{--C(CN)CH}_3 \\ | \\ \text{NHE} \end{array} \tag{104}$$

The non-enzymic decarboxylation of dimethyloxaloacetate has been studied by Steinberg and Westheimer[120]. During the course of the decarboxylation reaction at pH 3·0, a species was produced characterized by an absorption spectrum in the ultraviolet region which was markedly different from the spectra of either the starting material or the final product (α-ketoisovalerate). The spectrum showed a strong band at 260 mμ characteristic of an α, β-unsaturated carboxylic acid. By analogy with the proposed mechanism for the decarboxylation of acetoacetate, these workers postulated the scheme in equation (105).

$$\text{HOOCC} \overset{\|}{\underset{O}{}} \text{C(CH}_3)_2 \text{--C} \overset{\diagup O^-}{\underset{O}{\diagdown}} \longrightarrow \text{HOOC--C} \underset{O^-}{\overset{|}{=}} \text{C(CH}_3)_2 \overset{H^+}{\longrightarrow} \tag{105}$$

$$\underset{\substack{\| \\ O}}{\text{HOOCCCH(CH}_3)_2}$$

Polyvalent metal ions are strong catalysts for the decarboxylation of dimethyloxaloacetate. In the presence of metal ions, the reaction may be represented by equation (106); that is, the formation of the metal chelate provides a centre of positive charge which facilitates the electronic shifts leading to decarboxylation.

$$\underset{\substack{| \\ ^-O \\ \diagdown \\ M^{n+}}}{O=C} \text{---} \underset{\substack{\| \\ O \\ \diagup}}{C} \text{C(CH}_3)_2 \text{---C} \overset{\diagup O^-}{\underset{O}{\diagdown}} \xrightarrow{CO_2} \underset{\substack{| \\ ^-O \\ \diagdown \\ M^{n+}}}{O=C} \text{---} \underset{\substack{| \\ ^-O \\ \diagup}}{C} \text{=C(CH}_3)_2 \overset{H^+}{\longrightarrow} \underset{\substack{| \\ ^-O \\ \diagdown \\ M^{n+}}}{O=C} \text{---} \underset{\substack{\| \\ O}}{CCH(CH}_3)_2 \tag{106}$$

Evidence for the scheme in equation (106) is provided by the fact that decarboxylation of the monoethylester of dimethyloxaloacetate

($C_2H_5OCOCO(CH_3)_2COOH$) is not catalysed by metal ions, presumably due to its inability to form the appropriate chelate. The reaction sequence is also consistent with the colour changes associated with catalysis by ferric ions. Chelates of the two keto acids with ferric ions are yellow, whereas the ferric ion chelate of the enolate ion is blue. The ferric ion catalysed reaction is accompanied by colour changes from yellow to green to blue and finally to yellow; it is thus possible to observe the formation of the intermediate in reaction (106).

Oxaloacetate decarboxylases have been isolated from *Micrococcus lysodeikticus*[121] and *Azotobacter vinlandii*[122]; both enzymes show a dependence on metal ions, Mn^{2+} being the most effective. The requirement of the enzymes for metal ions has been taken as evidence that the main features of the enzymic and metal-catalysed reactions are similar, the principal role of the protein part of the enzyme being to bind the metal and substrate. The enzyme from *M. lysodeikticus* has been shown to catalyse the exchange of carbon dioxide with the β-carboxyl group of oxaloacetate, but not the net fixation of carbon dioxide by pyruvate. The results are consistent with a scheme such as equation (106) where the decarboxylation is reversible, but the subsequent ketonization reaction is not. It should be noted that the enzyme from *Azotobacter* does not catalyse the exchange of the β-carboxyl group of oxaloacetate to an appreciable extent; this probably reflects differences of stability of the enzyme–enolate complexes in the two cases.

Hence it would seem likely that the enzyme-catalysed decarboxylations of oxaloacetate and acetoacetate proceed by fundamentally different routes, but by routes which are closely parallel to those adopted in model systems. The difference must be dictated by the lack of a β-carboxyl group in the case of acetoacetate; this precludes the formation of a reactive metal chelate and requires that an alternative method of activation of the keto group be employed.

V. CONVERSION OF PRE-EXISTING GROUPS TO —COOH

Formation of carboxylic acids in biological systems may be achieved, *e.g.* by oxidation of an alcohol or an aldehyde. Reactions of this type are known in which either a free carboxylic acid group or acyl-CoA or acyl phosphate derivatives or lactones are formed. A brief discussion will also be given of the relatively poorly understood oxidative and hydrolytic cleavage reactions of carbon–carbon bonds which give rise to carboxyl groups.

A. Oxidation of Alcohols to Acids

In biological systems, the conversion of an alcohol to an acid is usually carried out by two separate NAD^+- (or $NADP^+$) linked enzymes (E_1 and E_2) catalysing reactions (107) and (108) (see section I).

$$RCH_2OH + NAD^+ \xrightarrow{E_1} RCHO + NADH + H^+ \tag{107}$$

$$RCHO + NAD^+ + H_2O \xrightarrow{E_2} RCOOH + NADH + H^+ \tag{108}$$

Some enzymes exist which catalyse the oxidation of alcohols to acids according to reaction (109), e.g. histidinol dehydrogenase (E.C. 1.1.1.23, reaction 110) and uridine diphosphoglucose (UDPG) dehydrogenase (E.C. 1.1.1.22, reaction 111).

$$RCH_2OH + 2 NAD^+ + H_2O \longrightarrow RCOOH + 2 NADH + 2 H^+ \tag{109}$$

$$(110)$$

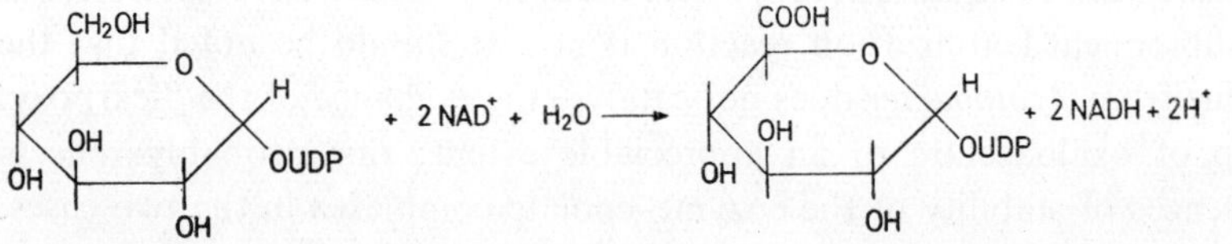

$$(111)$$

Histidinol dehydrogenase is concerned in the terminal step of histidine biosynthesis; it is interesting to note that this biosynthetic scheme is unique in the timing of formation of the carboxyl group. Adams[123] has purified the histidinol dehydrogenases from yeast, *E. coli* and *Arthrobacter histidinolvarans* and Adams and Loper[124] have obtained the enzyme from *Salmonella typhimurium* grown on histidinol.

Adams has examined the action of the enzymes on the aldehyde analogue of histidinol, namely histidinal. In the presence of NADH, histidinal is reduced to histidinol whereas histidine is formed with NAD^+. Both activities of the enzyme increased in a constant ratio during the purification procedure and could not be separated from one another. Consistently, a histidine requiring auxotroph of *E. coli* was unable to oxidize histidinol or histidinal. This evidence suggests that the histidinol dehydrogenase catalyses both the production of the aldehyde intermediate and also the oxidation of the latter to histidine. The intermediate must, however, be firmly bound to the enzyme surface since attempts to trap it with semi-carbazide were not successful.

UDP-glucose dehydrogenase was first identified in calf liver[125] and subsequently in plants and microorganisms. The enzymes from peas and calf liver have been partially purified and their properties have been reviewed[126]. The oxidation of UDP-glucose to UDP-glucuronate was clearly shown to require two equivalents of NAD$^+$. No aldehyde intermediate has been detected in the reaction and no attempts to prepare the synthetic aldehyde analogue of UDP-glucose have been reported. The enzyme is inhibited by sulphydryl group blocking-reagents and seems to be dependent on a metal ion for activity; these properties are characteristic of other better-characterized dehydrogenases.

Glucuronic acid and UDP-glucuronate play important parts in several areas of metabolism. For example, glucuronides are commonly formed in the course of detoxication and facilitate excretion of metabolites. It has been shown[127] that administration of foreign organic chemicals to the rat caused increased excretion of glucuronides and an adaptive increase in the level of UDP-glucose dehydrogenase.

Glucuronic acid is an important constituent of many structural polysaccharides. For example, the capsular polysaccharide of type 111 *Pneumococci* consists of alternating units of glucose and glucuronic acid as shown. The polysaccharide is formed by condensation of the UDP-derivatives of the monomers.

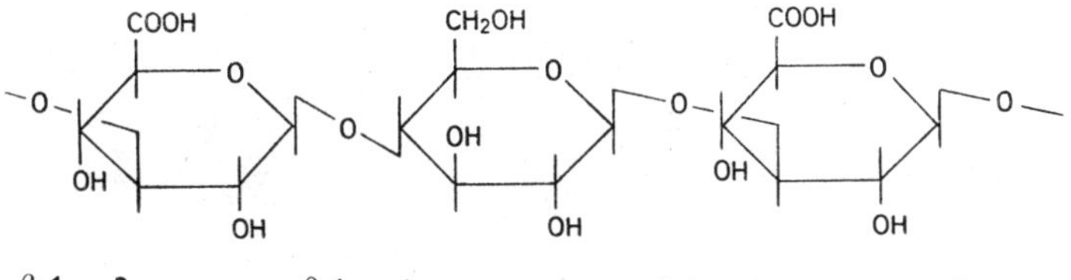

β 1 → 3 β 1 → 4 β 1 → 3 β 1 → 4

Other important glucuronate-containing polysaccharides are chondroitin (glucuronate and *N*-acetylgalactosamine) and hyaluronic acid (glucuronate and *N*-acetylglucosamine); dermatin sulphate contains *N*-acetylgalactosamine-2-sulphate and iduronate, the latter being formed from glucuronate by epimerization at $C_{(5)}$.

Attention has recently been focused on the importance of glucuronate in the catabolism of carbohydrates and as a central intermediate in the biosynthesis of ascorbate (vitamin C)[128]. The D-glucuronate-L-gulonate pathway (Figure 29) provides a route for the breakdown of glucose (via the 6-phosphate) and any other species which are convertible to intermediates in the cycle; enzymes catalysing the reactions of the cycle have been found

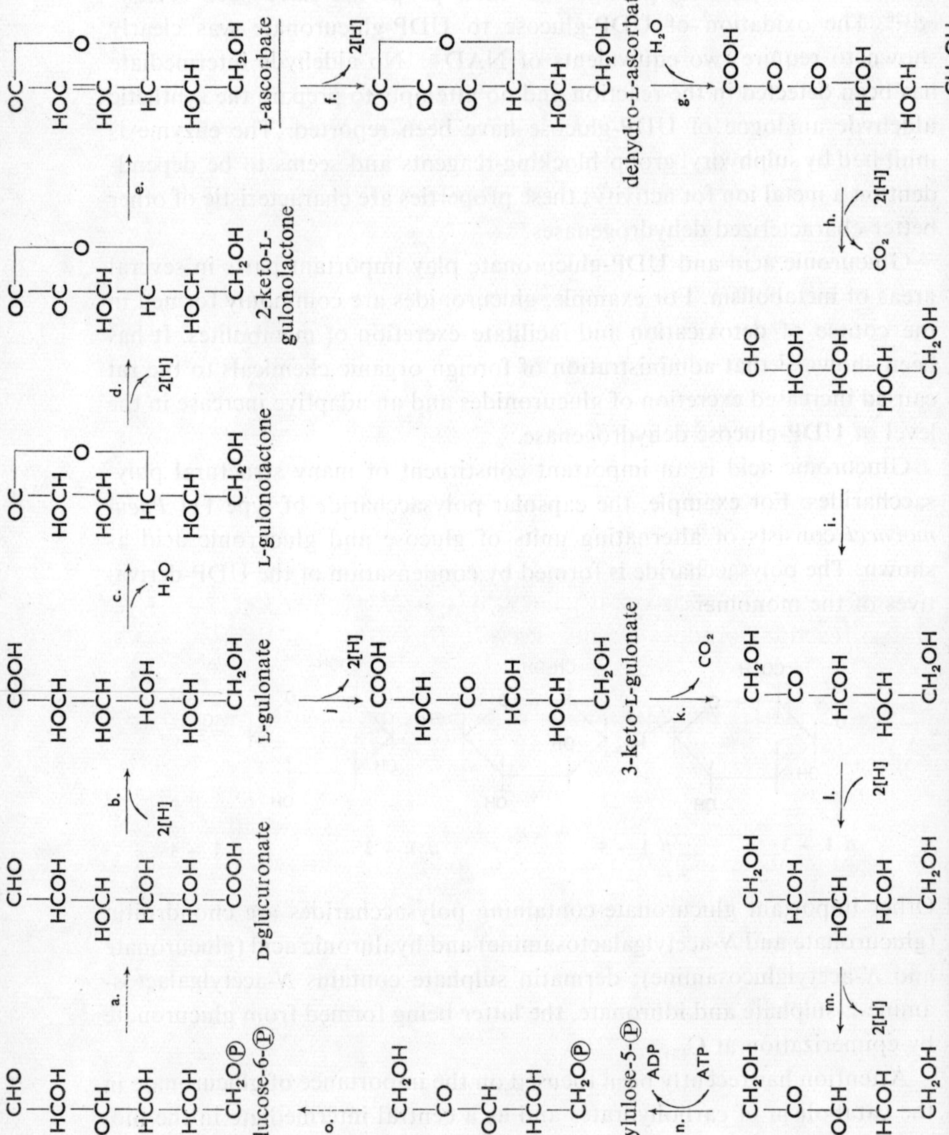

L-ascorbate

dehydro-L-ascorbate

2-keto-L-gulonolactone

L-gulonolactone

L-gulonate

3-keto-L-gulonate

D-glucuronate

D-glucose-6-P

D-xylulose-5-P

a.   b.   c. H_2O d. $2[H]$ e.   f. $2[H]$ g. H_2O

h. $2[H]$ CO_2 i.   j. $2[H]$ k. CO_2 l. $2[H]$ m. $2[H]$

n. ADP ATP   o.

in animals and the higher plants. Reaction sequence a in Figure 29 consists of the conversion of glucose-6-phosphate to the 1-phosphate, formation of UDP-glucose, oxidation to UDP-glucuronate and conversion to glucuronate by successive removal of UMP and phosphate. A reduction reaction (b) converts D-glucuronate to L-gulonate which undergoes oxidation (j) and decarboxylation (k) to L-xylulose. An interesting pair of reactions (l, m) convert L-xylulose to the D-configuration; after phosphorylation (n) six molecules of D-xylulose-5-phosphate are converted to 5 molecules of glucose-6-phosphate by the well-known pentose phosphate pathway (o).

Ascorbate arises from L-gulonate by dehydration to the lactone (c) followed by oxidation (d) and enolization (e). It is well known that a small number of species (the primates and the guinea pig) are unable to synthesize vitamin C and develop scurvy if this vitamin is absent from the diet. It has been shown that these species are able to carry out all the reactions in Figure 29 except the oxidation of L-gulonolactone to 2-keto-L-gulonolactone. Similarly, some individuals show a hereditary lack of the enzyme catalysing reaction (1), the reduction of L-xylulose; such individuals excrete large quantities of L-xylulose, a condition known as idiopathic pentosuria.

The metabolism of ascorbate is described by reactions f to i (Figure 29). These reactions produce L-xylulose which is further metabolized to glucose-6-phosphate.

B. The Formation and Metabolism of 6-Phosphogluconate

1. Formation

The conversion of glucose-6-phosphate to 6-phosphogluconate takes place in two stages, namely oxidation of the substrate to glucono-δ-lactone-6-phosphate (equation 112) and hydrolysis of the lactone to 6-phosphogluconate (equation 113).

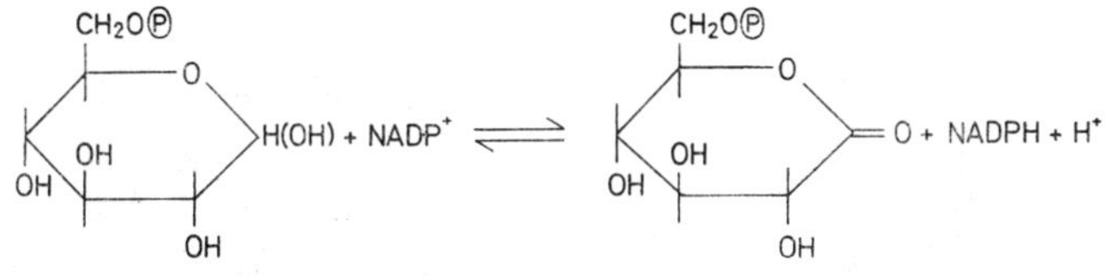

$$\text{H(OH)} + \text{NADP}^+ \rightleftharpoons \text{=O} + \text{NADPH} + \text{H}^+ \quad (112)$$

$$+ \text{H}_2\text{O} \longrightarrow \text{COOH} \quad (113)$$

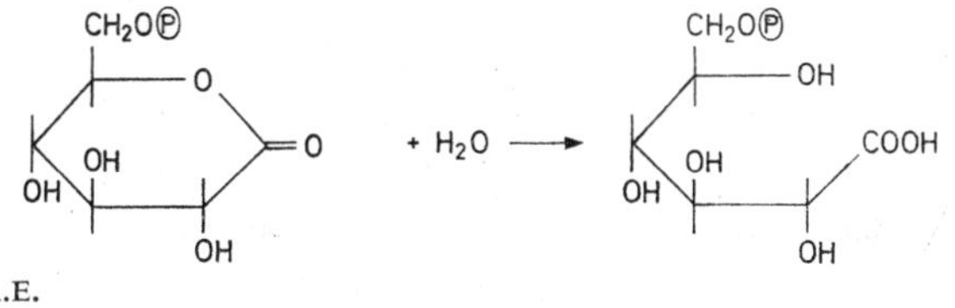

Reaction (112) is catalysed by the ubiquitous glucose-6-phosphate dehydrogenase (E.C. 1.1.1.49), an enzyme first identified in mammalian red blood cells by Warburg and Christian[129]; the properties of the enzyme have been reviewed[130]. Equation (112) is written with $NADP^+$ as the oxidizing agent; this is the case with the enzymes from yeast and mammalian tissue, whereas enzymes from some bacteria and moulds use NAD^+ and the enzyme from *Leuconostoc mesenteroides* appears to use both with equal facility.

Reaction (112) is freely reversible ($\Delta G^{\circ\prime} \sim -0.1$ kcal/mole) with the δ-lactone, while the γ-lactone is not reduced. Under normal conditions hydrolysis of the lactone is very rapid; the lactone may, however, be trapped as the hydroxamate. It is still not clear whether the enzyme acts on the α- or β-forms of the substrate or both. The enzyme from yeast has been shown to oxidize the non-phosphorylated β-D-glucopyranose which is presumptive evidence that the active form of the true substrate has the β-configuration.

The hydrolysis of gluconolactone-6-phosphate proceeds to completion ($\Delta G^{\circ\prime} \sim -5$ kcal/mole). The reaction in animals is thought to proceed without intervention of an enzyme. In plants and some microorganisms, however, reaction (113) is catalysed by gluconolactonase (E.C. 3.1.1.17) which is active both with the phosphorylated and non-phosphorylated lactones. Similar enzymes are known which catalyse the hydrolysis of L-arabono-γ-lactone, D- and L-gulono-δ-lactones and D-glucurono-γ-lactone.

In addition to the route described above, 6-phosphogluconate may be formed from free β-glucose by reactions (114–116).

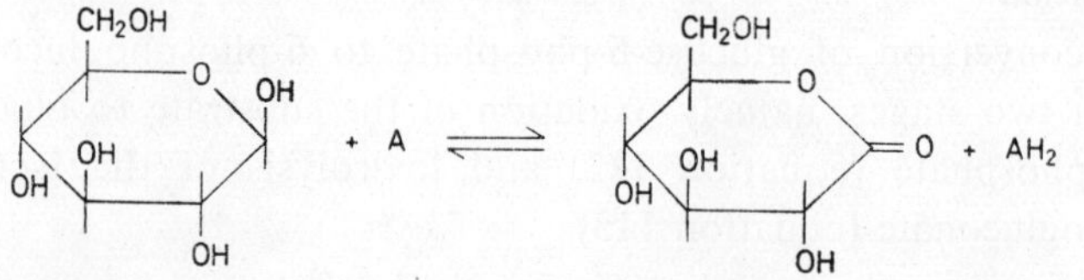

$$\text{(114)}$$

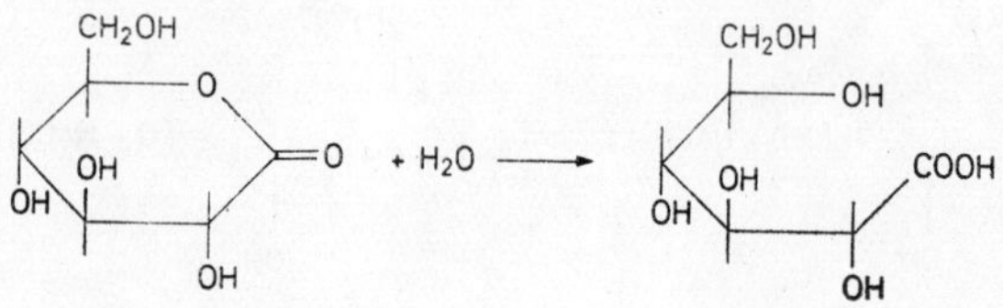

$$\text{(115)}$$

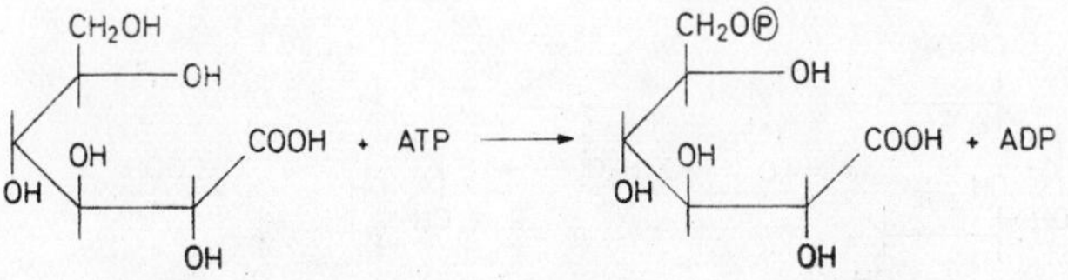

$$\text{(116)}$$

Reaction (116) is catalysed by a specific enzyme, gluconokinase (2.7.1.12). The nature of the acceptor A in equation (114) varies from species to species. An enzyme from liver employs either NAD^+ or $NADP^+$ (equation 117) as the oxidizing agents. The enzyme also oxidizes β-D-xylose and

$$\beta\text{-D-glucose} + NAD(P)^+ \rightleftharpoons \beta\text{-D-gluconolactone} \qquad (117)$$
$$+ NAD(P)H + H^+$$

6-deoxy-D-glucose. Other enzymes have been identified which oxidize L-arabinofuranose and D-galactose to their corresponding lactones.

The other important variant of reaction (114) is described by equation (118a); the intermediate hydrogen acceptor (A in equation 114) is enzyme-bound flavine adenine dinucleotide (E.FAD). Oxygen is the terminal hydrogen acceptor (reaction 118b) and the overall reaction is as in (119).

$$\beta\text{-D-glucose} + E.FAD \rightleftharpoons \beta\text{-D-gluconolactone} + E.FADH_2 \qquad (118a)$$
$$E.FADH_2 + O_2 \longrightarrow E.FAD + H_2O_2 \qquad (118b)$$
$$\beta\text{-D-glucose} + O_2 \longrightarrow \beta\text{-D-gluconolactone} + H_2O_2 \qquad (119)$$

Reactions (117) and (118) are catalysed by glucose oxidase (E.C. 1.1.3.4); the properties of the enzyme have been reviewed by Bentley[131]. Studies of the enzymology of reaction (120) were initiated by Müller[132]

$$\beta\text{-D-glucose} + H_2O + O_2 \longrightarrow \text{gluconate} + H_2O_2 \qquad (120)$$

who prepared active extracts of glucose oxidase from the moulds *Aspergillus niger* and *Penicillium glaucum*. The enzyme has subsequently been found in a variety of moulds and in *Pseudomonas*; in the latter case, the enzyme is particulate and is linked to the respiratory chain. A non-flavine enzyme has also been demonstrated in *A. niger* which does not use oxygen as the terminal acceptor; the most effective acceptor so far found is 2,6-dichlorophenolindophenol.

Bentley and Neuberger[133] have shown conclusively that glucose oxidase functions as a dehydrogenase rather than an oxygenase. Hydrogen peroxide produced during reaction (120) carried out in either $H_2^{18}O$ with $^{16}O_2$ or in $H_2^{16}O$ with $^{18}O_2$ was examined for its isotopic constitution; in the first case the product hydrogen peroxide contained no ^{18}O but in the second case it had the same ^{18}O content as the substrate oxygen. Friedberg and Kaplan[134] have shown that the $C_{(1)}$-H bond of glucose is cleaved in the rate-limiting step of reaction (120); primary isotope effects of 10 and 3·4 were observed with [1-T]-glucose and [1-D]-glucose respectively.

The enzyme has a molecular weight of 150,000 and contains two moles of FAD per mole of protein. Mason[135] has postulated that both molecules of FAD are involved in the oxidation of a single molecule of glucose with intermediate formation of flavine semiquinones (cf. the pyruvate dehydrogenase complex, section IV. A.1). A possible reaction sequence is shown in equation (121). Part of the glucose molecule around the $C_{(1)}$ carbon atom is shown and the molecule is assumed to be in the $C_{(1)}$ chair conformation. Hydrogen peroxide may be formed by reaction with the diradical form of the reduced enzyme or with the form $FAD.E.FADH_2$.

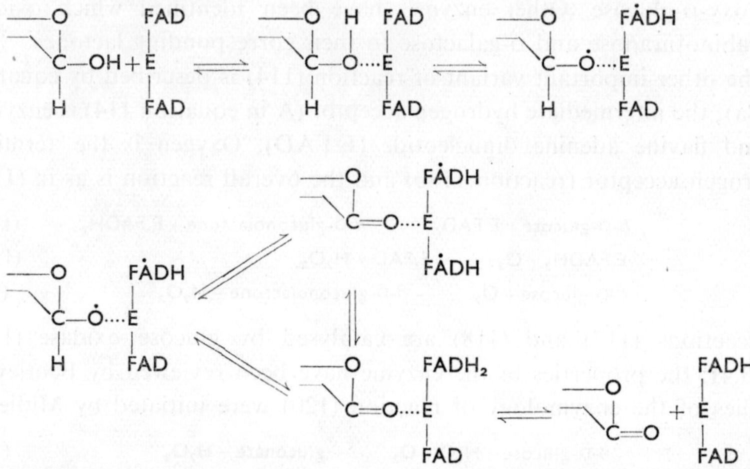

$$(121)$$

2. Metabolism

In most forms of life, two main pathways are available for the complete oxidation of hexoses to carbon dioxide The first, the Embden–Meyerhof glycolytic pathway is discussed in section V. D.1. The second, known variously as the pentose phosphate pathway, the hexose monophosphate shunt or the phosphogluconate pathway is shown in Figure 30[136].

Glucose is diverted into this pathway by the action of glucose-6-phosphate dehydrogenase (or any of the other 6-phosphogluconate-producing reactions described above). After hydrolysis of the intermediate 6-phosphogluconolactone ₍which reaction may or may not be enzyme-catalysed depending on the organism), an $NADP^+$-dependent oxidative decarboxylation reaction (reaction c) produces carbon dioxide and ribulose-5-phosphate. The properties of 6-phosphogluconate dehydrogenase have been discussed in section II. A.1.

The remainder of the reactions (d–l) are concerned with the regeneration of five molecules of glucose-6-phosphate from six molecules of ribulose-5-phosphate. Several of these reactions and enzymes are identical with those of the Calvin–Benson cycle (Figure 9).

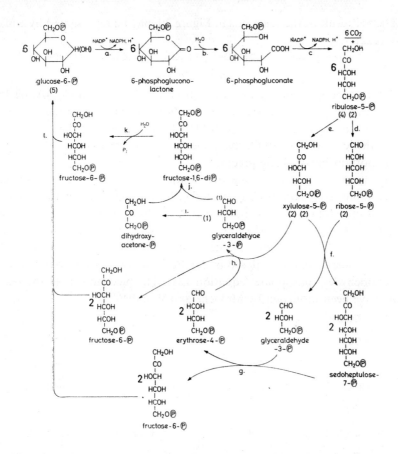

Reaction	Enzyme	E.C. Number
a.	glucose-6-phosphate dehydrogenase	1.1.1.49
b.	gluconolactonase	3.1.1.17
c.	phosphogluconate dehydrogenase	1.1.1.44
d.	ribosephosphate isomerase	5.3.1.6
e.	ribulosephosphate 3-epimerase	5.1.3.1
f, h.	transketolase	2.2.1.1
g.	transaldolase	2.2.1.2
i.	triosephosphate isomerase	5.3.1.1
j.	aldolase	4.1.2.7
k.	hexosediphosphatase	3.1.3.11
l.	glucosephosphate isomerase	5.3.1.9

FIGURE 30. The pentose phosphate pathway

The net result of the reactions in Figure 30 may be represented by equation (122).

$$glucose\text{-}6\text{-}phosphate + 12\ NADP^+ + 7\ H_2O \longrightarrow 6CO_2 + 12\ NADPH +$$
$$+ 12\ H^+ + H_3PO_4 \qquad (122)$$

NADPH produced during the reaction may be converted to NADH by the transhydrogenation reaction and the NADH oxidized by the terminal electron transport system, or the NADPH may be used in the organism's biosynthetic activities. Similarly, some of the intermediates in Figure 30 are important biosynthetic precursors.

An alternative route for the degradation of 6-phosphogluconate has been established for the genus of bacteria *Pseudomonas*, namely the Entner–Doudoroff pathway: dehydration catalysed by 6-phosphogluconate dehydrase (E.C. 4.2.1.12) followed by ketonization of the enol product yields 2-keto-3-deoxy-6-phosphogluconate, (KDPG). The latter undergoes an aldol cleavage reaction catalysed by KDPG aldolase to yield pyruvate and glyceraldehyde-3-phosphate (equation 123). The mechanisms of the reactions have been discussed by Meloche and Wood[137].

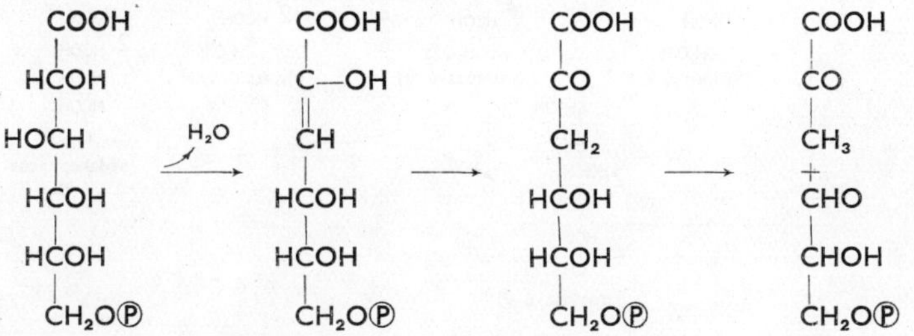

$$\begin{array}{llll}
COOH & COOH & COOH & COOH \\
HCOH & C-OH & CO & CO \\
HOCH \xrightarrow{\ H_2O\ } & CH & CH_2 \longrightarrow & CH_3 \\
HCOH & HCOH & HCOH & + \\
HCOH & HCOH & HCOH & CHO \\
CH_2O\textcircled{P} & CH_2O\textcircled{P} & CH_2O\textcircled{P} & CHOH \\
& & & CH_2O\textcircled{P}
\end{array} \qquad (123)$$

C. Oxidation of Aldehydes

1. Glyceraldehyde-3-phosphate dehydrogenase

Glyceraldehyde-3-phosphate dehydrogenase (GDPH, previously referred to as triosephosphate dehydrogenase, E.C. 1.2.1.12) catalyses an oxidative phosphorylation reaction to produce the mixed anhydride 1,3-diphosphoglycerate (equation 124).

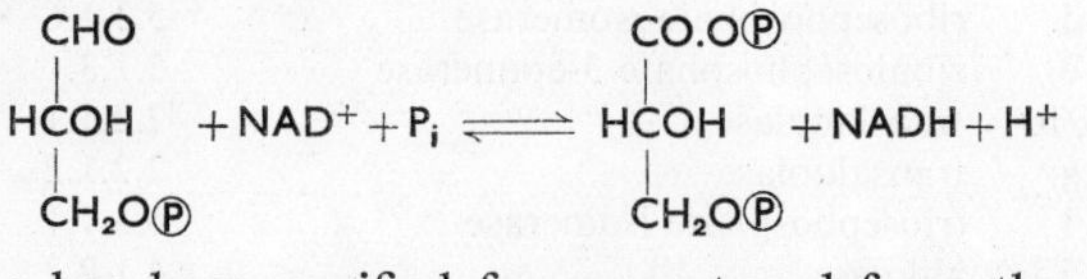

$$\begin{array}{ll}
CHO & CO.O\textcircled{P} \\
HCOH \quad + NAD^+ + P_i \rightleftharpoons & HCOH \quad + NADH + H^+ \qquad (124) \\
CH_2O\textcircled{P} & CH_2O\textcircled{P}
\end{array}$$

The enzyme has been purified from yeast and from the muscle of a variety of vertebrates. Velick and Furfine[138] have reviewed the properties of the enzyme, and the reaction mechanism has been the subject of much recent study[139-141].

GDPH has a molecular weight of 120,000–140,000 and contains 4 moles of bound NAD^+ per mole of protein. The enzyme is unique among the dehydrogenases in that the coenzyme is sufficiently firmly bound to the protein not to be removed during dialysis or crystallization. The coenzyme may, however, be removed from the protein by competitive absorption onto charcoal; complete reactivation of the enzyme then requires 4 equivalents of NAD^+. The absorption spectrum of enzyme-bound NAD^+ is of great interest. Free NAD^+ has no absorption maximum between 300 and 400 mμ whilst the reduced form, NADH, absorbs maximally at 340 mμ; enzyme-bound NAD^+ on the other hand has an absorption maximum at 365 mμ. Formation of the active complex between GDPH and NAD^+ requires free sulphydryl groups and an inactive modification of the enzyme which shows no 365 mμ absorption may be converted to the active form by incubation with sulphydryl compounds (*e.g.* glutathione). Treatment of the active enzyme with sulphydryl-blocking reagents results in the disappearance of the absorption at 365 mμ, and in some cases (*e.g.*with *p*-chloromercuribenzoate) in the dissociation of NAD^+ from the enzyme surface. Racker and Krimsky[142] have suggested that these observations may be explained by the formation of a covalent bond between a protein sulphydryl group and the $C_{(4)}$ position of the nicotinamide ring of NAD^+ (equation 125).

$$\text{E-S-H} \rightleftharpoons \text{E-S} + \text{H}^+ \qquad (125)$$

The quinoid structure would be expected to exhibit an absorption maximum in the range 300 to 400 mμ. Kosower[143] maintains, however, that the 365 mμ absorption arises from a charge-transfer interaction between the sulphydryl group and the nicotinamide ring.

There is good evidence that the enzyme-catalysed oxidation of glyceraldehyde-3-phosphate involves formation of an enzyme-*S*-acyl intermediate. Incubation of substrate levels of the enzyme with glyceraldehyde-3-phosphate and NAD^+ in the absence of inorganic phosphate results in a very rapid appearance of NADH in amounts determined by the concentration of the enzyme, accompanied by formation of the acyl-enzyme intermediate (equation 126); for convenience, only one of the four active —SH groups is shown.

$$\text{RCHO} + \text{E—SH} + NAD^+ \rightleftharpoons \text{E—S—COR} + \text{NADH} + \text{H}^+ \qquad (126)$$

A subsequent slow utilization of NAD^+ occurs as ESH is liberated from the acyl intermediate (equation 127).

$$E\text{—}S\text{—}COR + H_2O \rightleftharpoons E\text{—}SH + RCOOH \qquad (127)$$

Addition of inorganic phosphate to the reaction mixture results in a second burst of substrate oxidation by ESH liberated in the acyl transfer reaction (128)[144].

$$E\text{—}S\text{—}COR + P_i \rightleftharpoons E\text{—}SH + RCOO\circled{P} \qquad (128)$$

Catalytic quantities of arsenate can replace phosphate in reaction (128); the product arsenate anhydride undergoes rapid hydrolysis to form the free acid RCOOH and regenerate arsenate.

The acyl transfer reaction (128) has been extensively studied[142, 144, 296]. Formation of the acyl enzyme may be carried out by incubation with acetyl phosphate ($R = \text{—}CH_3$) or 1,3-diphosphoglycerate ($R = \text{—}CHOHCH_2O\circled{P}$). The enzyme catalyses a slow hydrolysis of the substrate (equation 127) or, as shown by using $^{32}P_i$, a rapid exchange of phosphate. In the presence of sulphydryl compounds (e.g. CoA) the corresponding thiol esters are formed (129). An important feature of the exchange reactions is that they

$$E\text{—}S\text{—}COR + R'\text{—}SH \rightleftharpoons ESH + RCOSR' \qquad (129)$$

require the presence of bound NAD^+ on the enzyme. The NAD^+ does not participate in oxidation-reduction reactions, and is presumably involved in catalysis of the acyl transfer reaction. This implies that regeneration of the NAD^+-form of the enzyme must take place before the acyl exchange reaction. On the basis of the evidence described above a plausible mechanism of action may be presented (equations 130–132)[139–141].

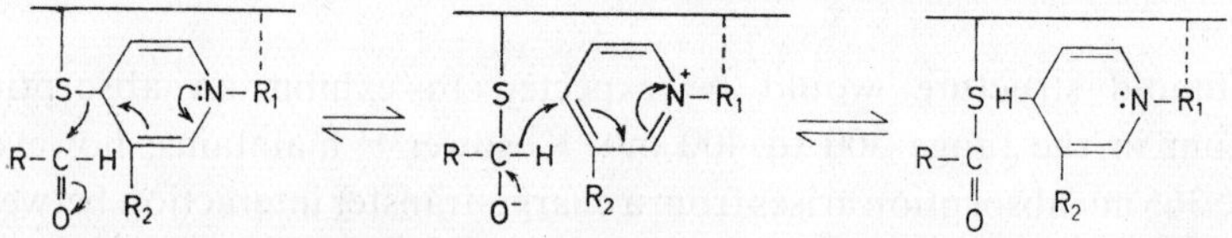

$$(130)$$

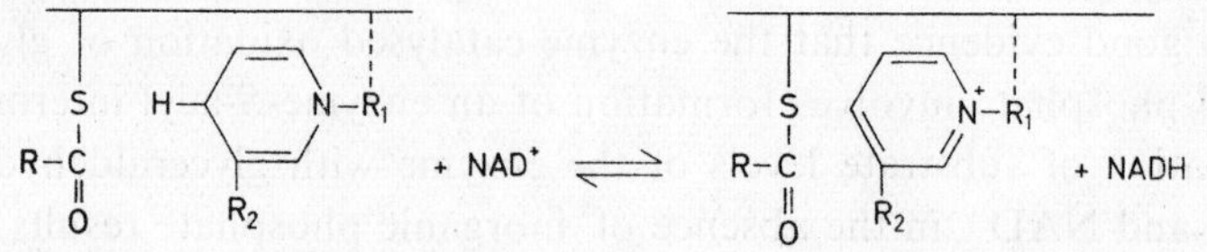

$$(131)$$

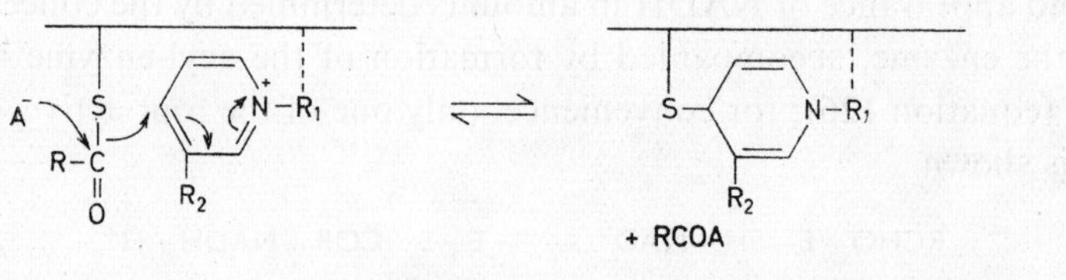

$$(132)$$

It should be noted that reaction (132) involves a nucleophilic attack on the carbonyl group by A^- which, in the physiological reaction, is the phosphate anion. This species is a poor nucleophile in aqueous solution, but its proposed role in reaction (132) has been substantiated by experiments with $[^{18}O]$-P_i. When the overall oxidative phosphorylation is carried out in the presence of $[^{18}O]$-P_i and the appropriate kinase, ^{18}O is found in the carboxyl group of the product carboxylic acid[145]. This is consistent with equation (133).

$$E-S-\underset{\underset{^{18}O-PO_3H_2}{\big|}}{\overset{\overset{O}{\|}}{C}}-R \quad \rightleftharpoons \quad ES^- + \underset{\underset{^{18}OPO_3H_2}{\big|}}{\overset{\overset{O}{\|}}{C}}-R \quad \xrightarrow{\quad ADP \quad ATP \quad} \quad RCO^{18}OH \qquad (133)$$

The conversion of glyceraldehyde-3-phosphate to 1,3-diphosphoglycerate is the essential oxidative step in the Embden–Meyerhof pathway for glucose metabolism[297-299]. The reactions leading from glucose to two molecules of pyruvate (or of lactate) are shown in Figure 31. Glucose is converted to fructose-1,6-diphosphate (reactions a–c) and the latter cleaved to two molecules of triose phosphate (d). By virtue of the fact that the triose phosphates are readily interconvertible by the isomerase reaction (e), both molecules may undergo oxidative phosphorylation to yield 1,3-diphosphoglycerate (f); transfer of phosphate to ADP (g) results in the formation of 3-phosphoglycerate. An intramolecular phosphate transfer (h) followed by dehydration (i) produces phosphoenolpyruvate. Formation of pyruvate is completed by phosphate transfer to ADP (j) and a spontaneous ketonization (k). The overall process may be written as in equation (134).

$$\text{glucose} + 2\,ADP + 2\,P_i + 2\,NAD^+ \longrightarrow 2\,\text{pyruvate}$$
$$+ 2\,ATP + 2\,NADH + 2\,H^+ + 2\,H_2O \qquad (134)$$

Pyruvate may be oxidized further by the usual routes or used in the biosynthetic processes of the organism.

Under aerobic conditions, the NADH formed in equation (134) may be reoxidized by the terminal electron transport system. Under anaerobic conditions (e.g. during muscular activity or during the growth of anaerobic bacteria), some other route for regeneration of NAD^+ is required. The most widely used reaction for this purpose is the reduction of pyruvate to lactate (equation 135). The overall process is given in equation (136).

$$\text{pyruvate} + NADH + H^+ \rightleftharpoons \text{lactate} + NAD^+ \qquad (135)$$
$$\text{glucose} + 2\,ADP + + 2\,P_i \longrightarrow 2\,\text{lactate} + 2\,ATP + + 2\,H_2O \qquad (136)$$

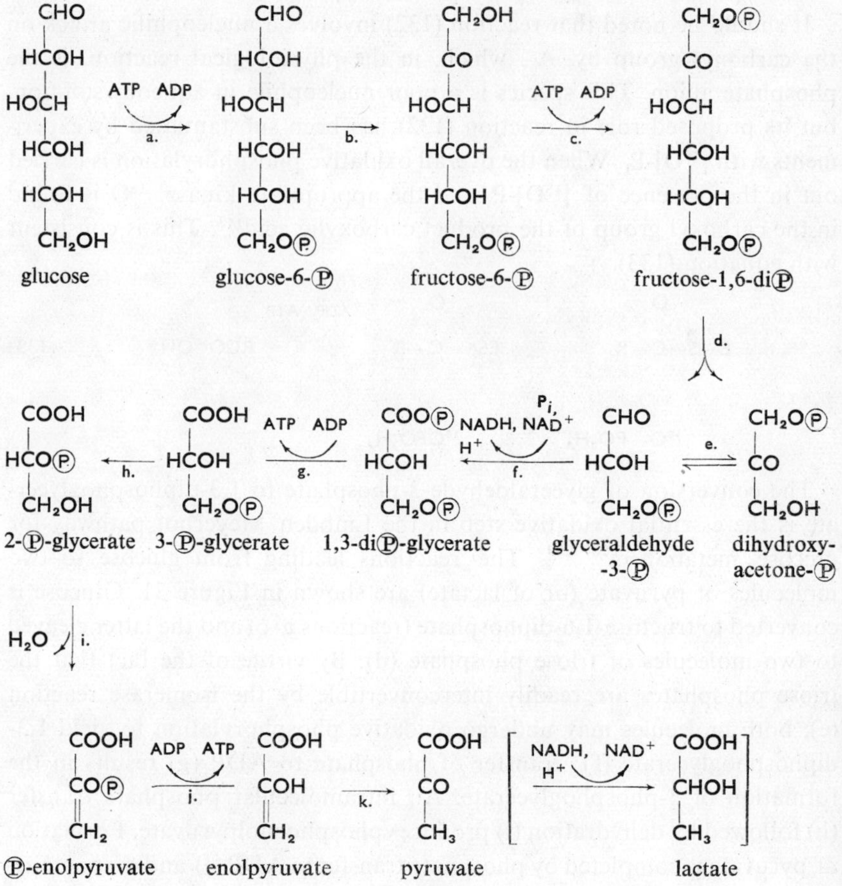

Reaction	Enzyme	E.C. Number
a.	hexokinase	2.7.1.1
b.	glucosephosphate isomerase	5.3.1.9
c.	phosphofructokinase	2.7.1.11
d.	aldolase	4.1.2.7
e.	triosephosphate isomerase	5.3.1.1
f.	glyceraldehyde-3-℗ dehydrogenase	1.2.1.12
g.	phosphoglycerate kinase	2.7.2.3
h.	phosphoglyceromutase	5.4.2.1
i.	enolase	4.2.1.11
j.	pyruvate kinase	2.7.1.40

FIGURE 31. The Embden–Meyerhof glycolytic pathway

2. Aldehyde dehydrogenases

The dehydrogenation of aldehydes may be represented in general terms by equation (137). Jakoby[146] has reviewed the enzymes catalysing reactions

$$RCHO + NAD(P)^+ + HA \rightleftharpoons RCO.A + NAD(P)H + H^+ \tag{137}$$

of the type shown, and has classified them according to the nature of HA.

The most widely distributed class of aldehyde dehydrogenases is that for which HA is water (reaction 138); these enzymes may be further sub-divided into general aldehyde dehydrogenases and specific semialdehyde dehydrogenases.

$$RCHO + NAD(P)^+ + H_2O \longrightarrow RCOOH + NAD(P)H + H^+ \tag{138}$$

General aldehyde dehydrogenases have been isolated from a variety of sources. The enzymes differ from one another in their substrate specificities, cofactor requirements and activation effects. For example, the enzyme from beef liver[147] (E.C. 1.2.1.3) oxidizes formaldehyde, acetaldehyde, glycolaldehyde, propionaldehyde, butyraldehyde, isovaleraldehyde and some aromatic aldehydes to the corresponding acids. The enzyme will not reduce $NADP^+$ and is activated by sulphydryl compounds. Rabbit liver contains two distinct aldehyde dehydrogenases of similar broad substrate specificities; one of the enzymes is sensitive to steroids (some of which activate and some inhibit) and is inhibited by magnesium ions, while the other enzyme is unaffected by these reagents[148]. Yeast contains two aliphatic aldehyde dehydrogenases. One of these enzymes (E.C. 1.2.1.5) is activated by potassium ions and requires either NAD^+ or $NADP^+$[149]; the other enzyme (E.C. 1.2.1.4) is activated by magnesium and is active only with $NADP^+$[150].

It is believed that the main biological function of these enzymes is the rapid and complete removal of toxic aldehydes from the organism. Quite low levels of aldehyde can be oxidized by these enzymes, the equilibrium for reaction (138) lying far to the right ($\Delta G^{o'} \sim -12$ kcal/mole at pH 7 and 25°).

Specific semialdehyde dehydrogenases are also widely distributed. Enzymes oxidizing succinic semialdehyde to succinic acid have been isolated from bacteria and from brain tissue. The brain enzyme is specific for NAD^+ whilst the bacterial enzymes utilize both NAD^+ and $NADP^+$. An enzyme from *Pseudomonas* oxidized γ-aminobutyraldehyde to γ-aminobutyric acid[151]. The aldehyde exists in solution in equilibrium with a cyclic Schiff's base form; present evidence suggests that the enzyme acts on the open-chain form of the substrate and is not involved in catalysing decyclization of the Schiff's base (scheme 139).

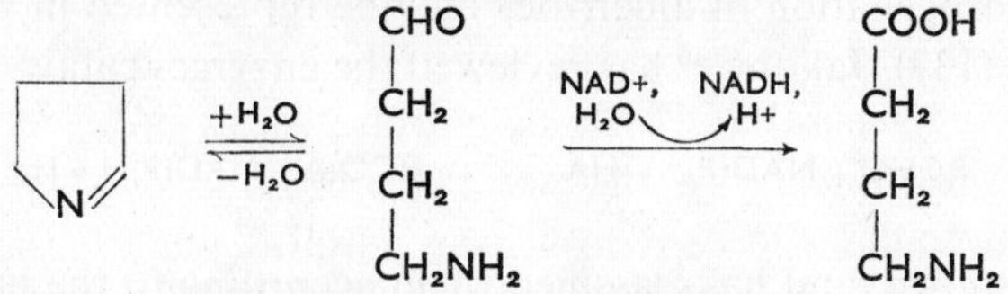

$$
\begin{array}{c}
\text{CHO} \\
|\\
\text{CH}_2 \\
|\\
\text{CH}_2 \\
|\\
\text{CH}_2\text{NH}_2
\end{array}
\quad
\xrightarrow[\text{H}_2\text{O}]{\text{NAD+, NADH, H+}}
\quad
\begin{array}{c}
\text{COOH} \\
|\\
\text{CH}_2 \\
|\\
\text{CH}_2 \\
|\\
\text{CH}_2\text{NH}_2
\end{array}
\tag{139}
$$

A similar situation exists in the case of glutamate-γ-semialdehyde which exists in solution as the cyclic form, Δ^1-pyrroline-5-carboxylate.

The second main class of aldehyde dehydrogenases includes those enzymes which catalyse the formation of thiol or phosphate esters. Acyl-CoA derivatives are formed from a variety of aliphatic aldehydes by an enzyme (E.C. 1.2.1.10) from *Clostridium kluyveri* (equation 140)[152]. This reaction is readily reversible.

$$\text{RCHO} + \text{NAD}^+ + \text{CoA} \rightleftharpoons \text{RCO.CoA} + \text{NADH} + \text{H}^+ \tag{140}$$

Also included in this group are the formaldehyde dehydrogenases of yeast[153] and bovine liver[155] (E.C. 1.2.1.1.). Although the product of the enzyme-catalysed oxidation is free formic acid, it is believed that an initial reaction with the sulphydryl group of glutathione (γ-glutamylcysteinylglycine) yields S-formylglutathione (see below).

An enzyme has been isolated from yeast (1.2.1.11) which catalyses the reversible oxidation of aspartate-β-semialdehyde to β-aspartylphosphate[154] (equation 141).

$$
\begin{array}{c}
\text{CHO} \\
|\\
\text{CH}_2 \\
|\\
\text{CHNH}_2 \\
|\\
\text{COOH}
\end{array}
+ \text{P}_i + \text{NADP}^+ \rightleftharpoons
\begin{array}{c}
\text{CO.O}\textcircled{P} \\
|\\
\text{CH}_2 \\
|\\
\text{CHNH}_2 \\
|\\
\text{COOH}
\end{array}
+ \text{NADPH} + \text{H}^+
\tag{141}
$$

Arsenate can replace phosphate, when the product is free aspartic acid. Aspartate-β-semialdehyde, formed by phosphorylation of aspartate by aspartate kinase followed by the reverse of reaction (141), is the precursor for the biosynthesis of threonine, methionine and lysine.

None of the enzymes described above have been obtained in a sufficiently pure state to allow detailed examination of reaction mechanism. The available evidence suggests, however, that all these enzymes have similar modes of action and are in many ways similar to glyceraldehyde-3-phosphate dehydrogenase.

There is tentative evidence to suggest that protein sulphydryl groups are involved in the enzyme-catalysed reaction. All aldehyde dehydrogenases are inhibited by sulphydryl-blocking reagents and some are completely inactive in the absence of added thiols. Inactivation of aldehyde dehydrogenases by sulphydryl-blocking reagents is frequently diminished or prevented by the presence of the appropriate substrate or coenzyme. For example, the bovine liver general aldehyde dehydrogenase is protected against inhibition by the presence of the natural coenzyme NAD^+ but not by $NADP^+$. Formaldehyde dehydrogenase requires glutathione (GSH) for activity and there is good evidence[155] that the initial reaction is formation of S-hydroxymethylglutathione (equation 142).

$$G{-}SH + HCHO \rightleftharpoons \overset{\displaystyle H}{\underset{\displaystyle OH}{G{-}S{-}C{-}H}} \qquad (142)$$

A subsequent oxidation reaction yields S-formylglutathione (reaction 143).

$$\overset{\displaystyle H}{\underset{\displaystyle OH}{G{-}S{-}C{-}H}} + NAD^+ \longrightarrow G{-}S{-}CHO + NADH + H^+ \qquad (143)$$

Rose[156] has demonstrated the reverse of reaction (143) with synthetic S-formylglutathione.

Aldehyde dehydrogenases are characterized by an inhibitory effect of arsenite at low concentrations, the effect only being manifested in the presence of added thiol compounds[157]. Inhibitory effects of this kind are usually interpreted as due to complex formation between arsenite and a pair of closely associated sulphydryl groups. In the present case, arsenite inhibition is competitively prevented by the appropriate substrate.

Jakoby[146] has presented a reaction scheme (reactions 144, 145) which involves intermediate formation of an enzyme-thiohemiacetal derivative followed by oxidation to the corresponding S-acyl form (reaction 144). The product is then liberated by either a hydrolytic (145a) or acyl exchange (145b) reaction depending on the enzyme involved.

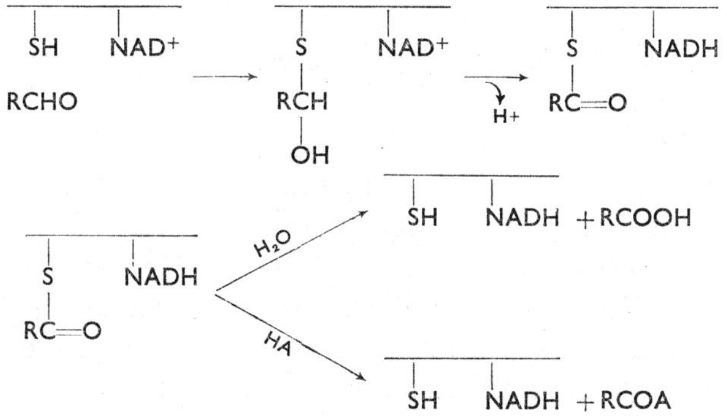

$$(144)$$

$$(145a)$$

$$(145b)$$

The scheme can be extended to accomodate formaldehyde dehydrogenase (and possibly other cases where enzymic activity is dependent on exogenous sulphydryl compounds) by substituting reactions (146) and (147) for (144).

$$RCHO + R'SH \rightleftharpoons R'SCHR$$
$$| \atop OH$$

(146)

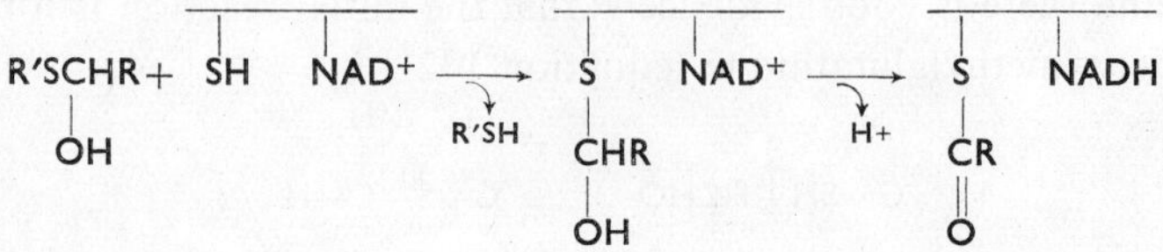

$$R'SCHR + SH \quad NAD^+ \longrightarrow S \quad NAD^+ \longrightarrow S \quad NADH$$

(147)

These schemes are by no means completely established. No role is indicated for a pair of sulphydryl groups and the significance of arsenite inhibition is not known.

3. Oxidation of aldehydes by metalloflavoproteins

The metalloflavoproteins aldehyde oxidase (E.C. 1.2.3.1) and xanthine oxidase (E.C. 1.2.3.2) catalyse reactions with the stoichiometry shown in equation (148). Aldehyde oxidase is found in mammalian liver, and

$$S + H_2O + O_2 \longrightarrow SO + H_2O_2$$

(148)

catalyses the oxidation of a wide range of purine and quinoline derivatives in addition to the oxidation of aldehydes (reaction 149).

$$RCHO + O_2 + H_2O \longrightarrow RCOOH + H_2O_2$$

(149)

Xanthine oxidase is widely distributed and has been purified from milk and from mammalian liver; the enzyme is most active in the oxidation of xanthine to urate (reaction 150) but also readily oxidizes hypoxanthine and a wide range of aldehydes, purines and pteridines. Early work on

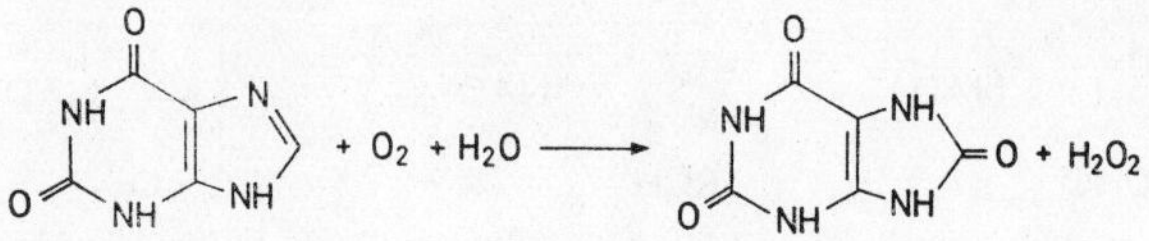

(150)

xanthine oxidase has been reviewed by Bray[158] and de Renzo[159].

In vitro, oxygen may be replaced by nitrate, ferricyanide, nitro compounds, cytochrome *c*, and a variety of dyes including methylene blue and 2,6-dichlorophenolindophenol. A similar enzyme, xanthine dehydrogenase, has been isolated from avian liver and kidney; this enzyme uses NAD^+ rather than oxygen as the electron acceptor *in vivo*.

The purified enzymes have been shown to contain molybdenum, flavine (FAD), non-haeme iron and, in the case of aldehyde oxidase, coenzyme Q_{10} (CoQ_{10})[160] (see Table 3). It appears that the non-haeme iron is associated with a stoichiometric amount of acid-labile sulphur (cf. ferredoxin, section II. B.3).

TABLE 3

Enzyme	Source	Molecular Weight	FAD	Mo	Fe	CoQ_{10}
Xanthine oxidase	Milk	300,000	2	2	8	—
	Mammalian liver		2	2	8	—
Xanthine dehydrogenase	Avian liver		2	4	16	—
Aldehyde oxidase	Mammalian liver	280,000	2	2	8	2

Formally, the reaction of oxidases involves removal of a proton and two electrons from the substrate followed by hydroxylation of the resulting carbonium ion (e.g. reaction 151). The electron transfer chain from the sub-

$$RCHO + H_2O \longrightarrow RCOOH + 2H^+ + 2e^- \tag{151}$$

strate to the acceptor has been studied using electron spin resonance spectroscopy[160–162]. It seems to be agreed that the electron transfer sequence is as shown in equation (152), with CoQ_{10} operating between FAD and

$$\text{substrate} \longrightarrow Mo^{VI} \longrightarrow FAD \longrightarrow Fe^{III}\text{-S} \longrightarrow O_2 \tag{152}$$

Fe^{III}-S in the case of aldehyde oxidase.

Bray[161] and his coworkers have shown that the initial product of reduction of Mo^{VI} is a species of Mo^V (designated $Mo^V_{\delta,\gamma}$) which isomerizes to a second species ($Mo^V_{\alpha,\beta}$) with a different e.s.r. spectrum; these forms are presumed to differ in the type or arrangement of ligands in their coordination spheres. It was also shown that at alkaline pH values, FAD is reduced to a semiquinone free radical form, FADH˙. The observations on the changes undergone by non-haeme iron were somewhat equivocal, but it seems likely that a reduction Fe^{III}-S → Fe^{II}-S takes place. Hence, in the case of xanthine oxidase, the electron transfer reactions may be tentatively represented by equation (153).

$$\begin{array}{c} S + H_2O \\ SO + 2H^+ \end{array} \Bigg) \left[\left(\begin{array}{c} Mo^V_{\delta,\gamma} \longrightarrow Mo^V_{\alpha,\beta} \\ Mo^{VI} \longleftarrow \end{array} \right) \left(\begin{array}{c} FADH˙ \\ FAD \end{array} \right) \left(\begin{array}{c} Fe^{II}\text{—S} \\ Fe^{III}\text{—S} \end{array} \right)_2 \right] \left(\begin{array}{c} H_2O_2 \\ O_2 + 2H^+ \end{array} \right. \tag{153}$$

The details of electron transfer from the iron couple to oxygen are not known.

D. Carbon–Carbon Bond Cleavage Resulting in Carboxylate Formation

1. Oxidative cleavage

The oxygenases[163, 164] (E.C. 1.99.2) are rather poorly characterized enzymes which catalyse direct incorporation of oxygen into the appropriate substrate (reaction 154).

$$S + O_2 \longrightarrow S-O_2 \qquad (154)$$

Frequently the substrates are phenolic compounds; in these cases oxidation is accompanied by ring cleavage and formation of carboxylic acid groups. Cleavage of an *ortho*-diphenol may occur either at the bond between the two hydroxyl groups leading to a dicarboxylic acid (*e.g.* oxidation of catechol to *cis, cis*-muconate catalysed by pyrocatechase[165], reaction 155), or at a position adjacent to one of the hydroxyl groups when the product is a ω-aldoacid (*e.g.* oxidation of catechol to α-hydroxymuconate semialdehyde catalysed by metapyrocatechase[166], reaction 156).

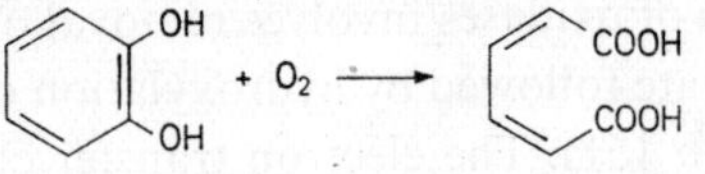

$$(155)$$

$$(156)$$

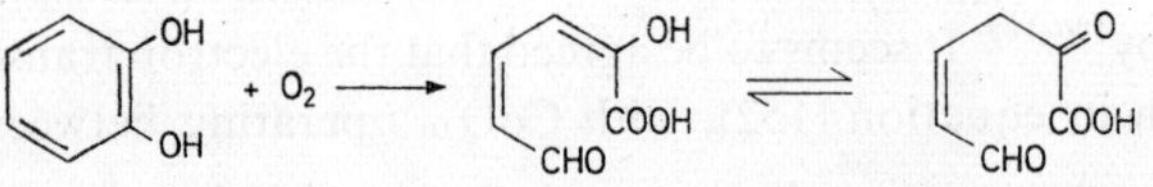

Two important and widely distributed enzymes catalyse reactions analogous to (156). Homogentisate oxygenase[167] catalyses the oxidation of homogentisate to 4-maleylacetoacetate (equation 157) and 3-hydroxyanthranilate oxygenase[165] produces 1-amino-4-formyl-1,3-butadiene-1,2-dicarboxylate (reaction 158) from 3-hydroxyanthranilic acid.

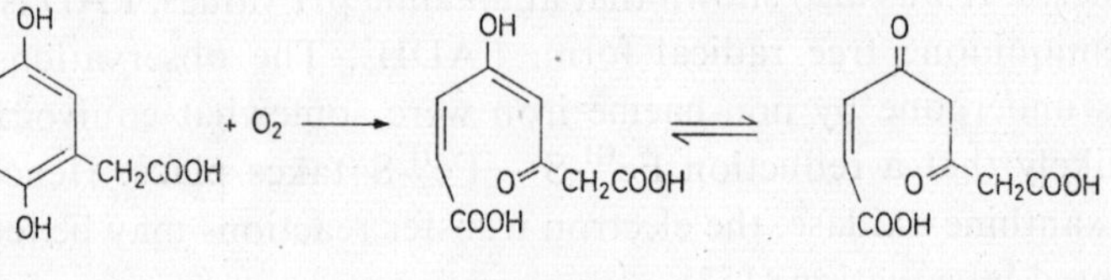

$$(157)$$

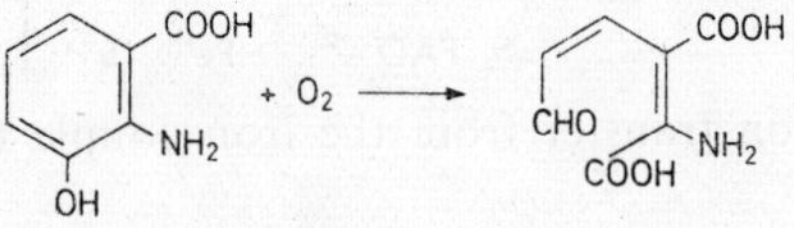

$$(158)$$

Inducible oxygenases acting on a variety of other phenolic compounds have been isolated from *Pseudomonas* species[163]. These enzymes are not well-characterized and will not be discussed here.

One case of the oxidative cleavage of an alicyclic compound has been recognized[169]. Inositol oxygenase, an enzyme in mammalian liver, catalyses the oxidation of *myo*-inositol to D-glucuronate via the open chain compound shown in equation (159). This reaction makes possible the catabol-

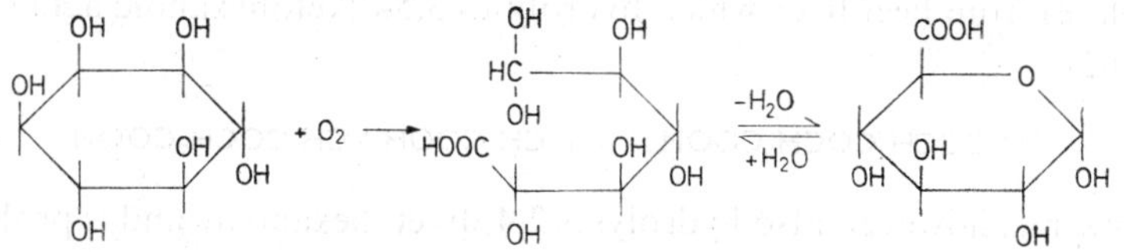

$$(159)$$

ism of inositol by the D-glucuronate-L-gulonate pathway described previously.

A similar reaction to those described above is the formation of *N*-formylkynurenine from tryptophan by tryptophan oxygenase[170] (reaction 160). In this case, however, it is the indole ring which is cleaved and the

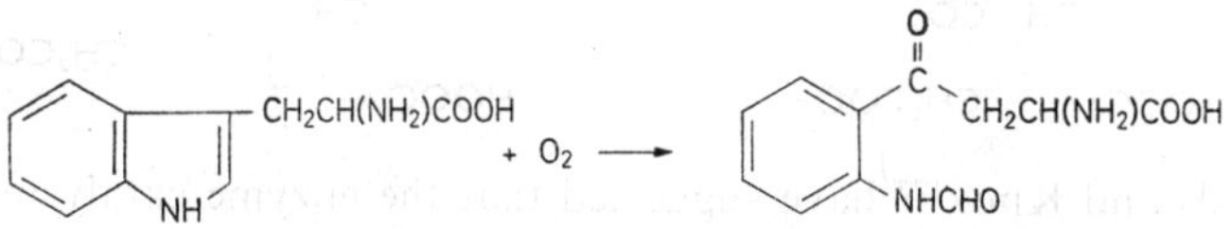

$$(160)$$

product is an *N*-formylamino ketone rather than a carboxylic acid.

Little is known concerning the mode of action of the oxygenases, and, except in the case of the pyrocatechase of *Pseudomonas fluorescens*, the enzymes have not been obtained in a state approaching purity. The latter enzyme has been shown to have a molecular weight of about 80,000 and to contain two gram ions of Fe^{II} per mole of protein. It has been shown conclusively, using $^{18}O_2$, that both atoms of oxygen in the product *cis, cis*-muconate are derived from oxygen, but no catechol-oxygen intermediate has been detected during the reaction. There is evidence that the other oxygenases contain catalytically-active Fe^{II} but the role of the metal is not clear. Tryptophan oxygenase is unique among these enzymes in that it contains a haeme prosthetic group (ferrous protoporphyrin IX).

2. Hydrolytic cleavage

The enzyme-catalysed hydrolytic cleavage of diketo acids has been known since 1948, but the specificities and mode of action of the enzymes involved are still not known. Meister and Greenstein[171] reported the presence of an enzyme in liver which cleaves 2,4-diketovalerate to acetate and pyruvate (reaction 161).

$$CH_3COCH_2COCOOH + H_2O \longrightarrow CH_3COOH + CH_3COCOOH \qquad (161)$$

The enzyme also cleaves the higher homologues of 2,4-diketovalerate to pyruvate and the corresponding fatty acid and is known as 2,4-diketo acid hydrolase. Similarly, Connors and Stotz[172] isolated an enzyme (triacetate hydrolase) from beef liver which hydrolyses 3,5-diketohexanoic acid (reaction 162).

$$CH_3COCH_2COCH_2COOH \longrightarrow CH_3COOH + CH_3COCH_2COOH \qquad (162)$$

The enzyme, however, also hydrolyses 2,4-diketohexanoate and is probably identical to 2,4-diketo acid hydrolase. Of considerable metabolic importance is an enzyme found in liver which catalyses the hydrolysis of fumaryl acetoacetate to fumarate and acetoacetate (equation 163).

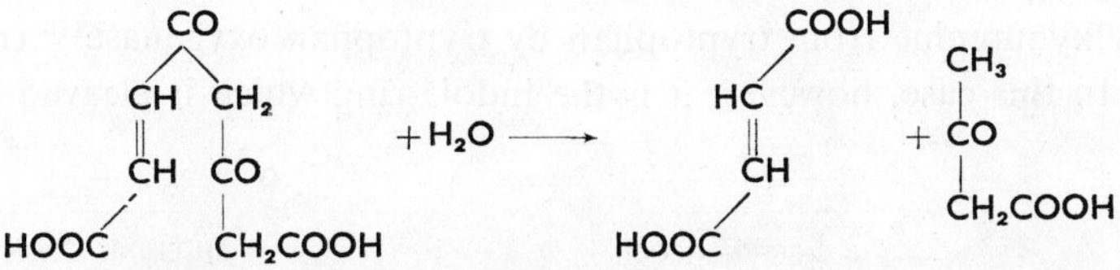

$$\begin{array}{c}
CO \\
| \\
CH \quad CH_2 \\
\| \quad \quad | \\
CH \quad CO \\
/ \quad \quad | \\
HOOC \quad CH_2COOH
\end{array}
+ H_2O \longrightarrow
\begin{array}{c}
COOH \\
/ \\
HC \\
\| \\
CH \\
/ \\
HOOC
\end{array}
\quad
\begin{array}{c}
CH_3 \\
| \\
CO \\
| \\
CH_2COOH
\end{array}
\qquad (163)$$

Edwards and Knox[173] have suggested that the enzyme catalysing reaction (163) may be the 2,4-diketo acid hydrolase.

Many moulds produce oxalate as an end product of metabolism. Hayaishi[174] and his colleagues demonstrated the presence in these species of an enzyme which catalyses the hydrolysis of oxaloacetate to oxalate and acetate (reaction 164). The enzyme is known to require Mn^{2+} for activity.

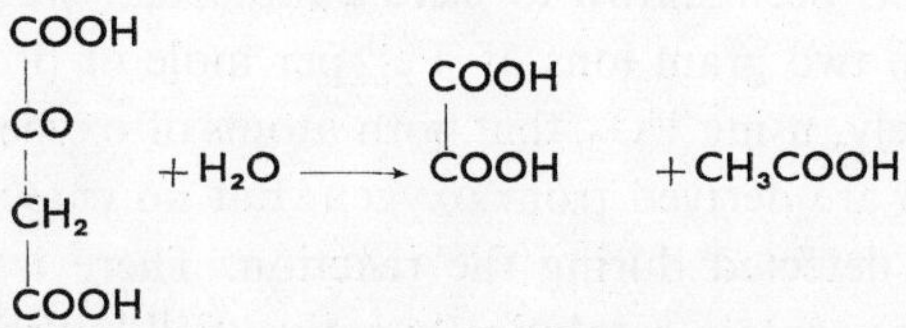

$$\begin{array}{c}
COOH \\
| \\
CO \\
| \\
CH_2 \\
| \\
COOH
\end{array}
+ H_2O \longrightarrow
\begin{array}{c}
COOH \\
| \\
COOH
\end{array}
+ CH_3COOH \qquad (164)$$

The cleavage of oxaloacetate bears a formal similarity to the hydrolytic stage of the carboxydismutase reaction (cf. section II. A.2).

An interesting enzyme, first reported by Braunstein[175] and his coworkers, catalyses the hydrolysis of kynurenine to anthranilic acid and alanine (reaction 165); the 3-hydroxy analogue of kynurenine is hydrolysed with equal facility. Kynureninase is completely dependent for its activity on pyridoxal-5′-phosphate. By analogy with other reactions for which this

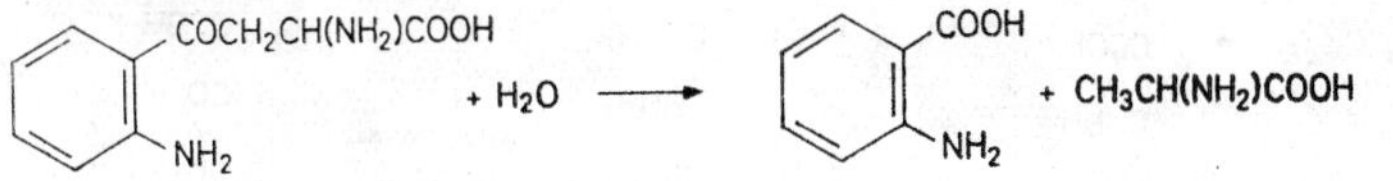

$$\text{(o-}C_6H_4(NH_2))\text{COCH}_2\text{CH(NH}_2)\text{COOH} + H_2O \longrightarrow (\text{o-}C_6H_4(NH_2))\text{COOH} + CH_3\text{CH(NH}_2)\text{COOH} \quad (165)$$

cofactor is required, Braunstein has presented a mechanism of action of the enzyme shown, in part, in equations (166) and (167). A Schiff's base is formed between the substrate and enzyme-bound pyridoxal phosphate (R—CHO) followed by conversion to the tautomeric form (equation 166).

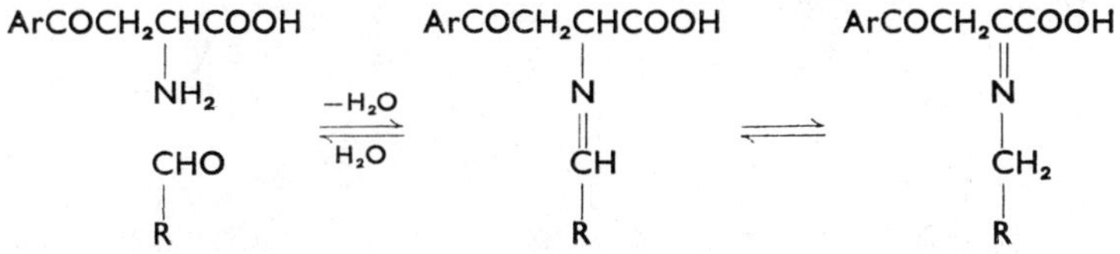

$$\underset{\underset{\underset{R}{|}}{\overset{|}{CHO}}}{\overset{|}{\underset{|}{NH_2}}}{ArCOCH_2CHCOOH} \quad \underset{H_2O}{\overset{-H_2O}{\rightleftharpoons}} \quad \underset{\underset{\underset{R}{|}}{\overset{||}{CH}}}{\overset{||}{\underset{|}{N}}}{ArCOCH_2CHCOOH} \quad \rightleftharpoons \quad \underset{\underset{\underset{R}{|}}{\overset{|}{CH_2}}}{\overset{||}{\underset{|}{N}}}{ArCOCH_2CCOOH} \quad (166)$$

$$\underset{\underset{\underset{R}{|}}{\overset{|}{CH_2}}}{\overset{||}{\underset{|}{N}}}{ArCOCH_2CCOOH} \quad \underset{-H_2O}{\overset{H_2O}{\rightleftharpoons}} \quad \overset{OH}{\underset{\underset{\underset{R}{|}}{\overset{|}{CH_2}}}{\underset{\overset{|}{N}}{\overset{|}{OH}}}}{ArC—CH_2CCOOH} \quad \rightleftharpoons \quad \overset{O}{\underset{\underset{\underset{R}{|}}{\overset{|}{CH_2}}}{\underset{\overset{|}{N}}{\overset{|}{OH}}}}{ArC} + CH_3CCOOH \quad (167)$$

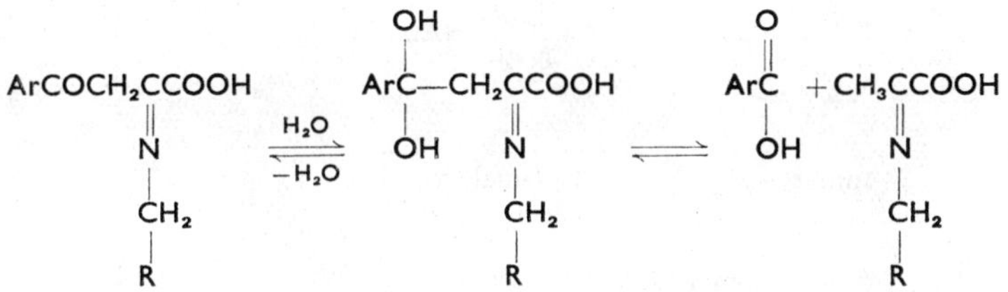

The carbonyl group is hydrated, after which cleavage between $C_{(\beta)}$ and $C_{(\gamma)}$ occurs, facilitated by the electron withdrawing effect of the Schiff's base nitrogen (scheme 167). Alanine is liberated from the Schiff's base as shown in reaction (168).

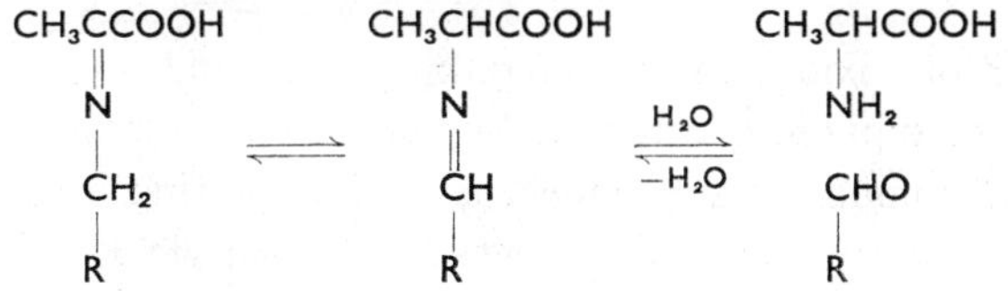

$$\underset{\underset{\underset{R}{|}}{\overset{|}{CH_2}}}{\overset{||}{\underset{|}{N}}}{CH_3CCOOH} \quad \rightleftharpoons \quad \underset{\underset{\underset{R}{|}}{\overset{||}{CH}}}{\overset{|}{\underset{|}{N}}}{CH_3CHCOOH} \quad \underset{-H_2O}{\overset{H_2O}{\rightleftharpoons}} \quad \underset{\underset{\underset{R}{|}}{\overset{|}{CHO}}}{\overset{|}{\underset{|}{NH_2}}}{CH_3CHCOOH} \quad (168)$$

3. Biological importance of the reactions

Oxidative and hydrolytic carbon–carbon bond cleavage reactions are the most important features of the pathways for the degradation of the aromatic amino acids.

Phenylalanine and tyrosine share a common degradative pathway (Figure 32). Phenylalanine is converted to tyrosine by the enzyme phenylalanine hydroxylase (reaction a; Figure 32); the reaction involves oxygen and NADPH. A transamination reaction (b) produces p-hydroxyphenylpyruvate; this intermediate undergoes an interesting series of conversions

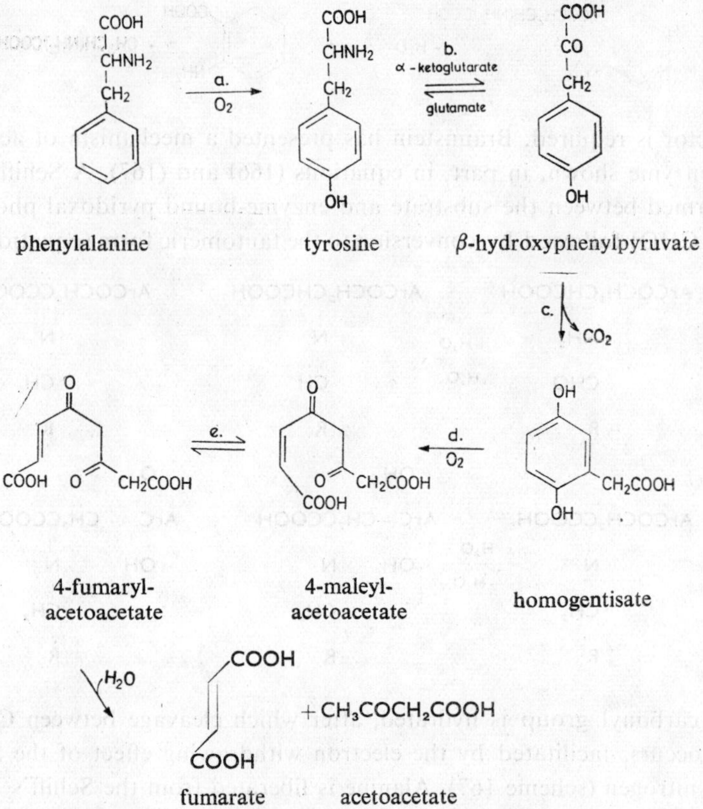

COOH
|
CHNH₂
|
CH₂

phenylalanine

a.
O₂ →

COOH
|
CHNH₂
|
CH₂

OH

tyrosine

b.
α - ketoglutarate ⇌ glutamate

COOH
|
CO
|
CH₂

OH

β-hydroxyphenylpyruvate

c. ↓ CO₂

COOH   O   CH₂COOH
O

**4-fumaryl-
acetoacetate**

e. ⇌

O   CH₂COOH
COOH

**4-maleyl-
acetoacetate**

d.
O₂ ←

OH

CH₂COOH

OH

homogentisate

↘ H₂O

COOH

COOH

fumarate

+ CH₃COCH₂COOH

acetoacetate

FIGURE 32. Metabolism of tyrosine and phenylalanine

(decarboxylation, oxidation, side chain migration and ring hydroxylation) all of which are apparently catalysed by a single copper-containing enzyme. The product, homogentisate, undergoes oxidative ring opening (d) to 4-maleylacetoacetate; a *cis-trans* isomerism (e) then yields 4-fumarylacetoacetate. Reaction e is catalysed by a specific isomerase which requires glutathione as coenzyme. Finally, 4-fumarylacetoacetate is cleaved hydrolytically to fumarate and acetoacetate (f), which may be metabolized by the usual routes.

Similar oxidative and hydrolytic carbon–carbon bond cleavage reactions are involved in some of the important reactions of tryptophan metabolism[176] (Figure 33). Tryptophan oxygenase catalyses an oxidative cleavage of the indole ring to produce N-formylkynurenine (a). Deformylation (b) catalysed by kynurenine formylase yields kynurenine which may be cleaved hydrolytically to anthranilate, the main excretion product of tryptophan in mammals, and alanine. Alternatively, kynurenine may be hydroxylated in the 3- position (d) and the product cleaved to 3-hydroxyanthranilate and alanine (e). Oxidative cleavage of 3-hydroxyanthranilate (f) produces the highly reactive intermediate 2-acroleyl-3-amino-fumarate; this may either undergo spontaneous ring closure (g) to yield quinolinate, or decarb-

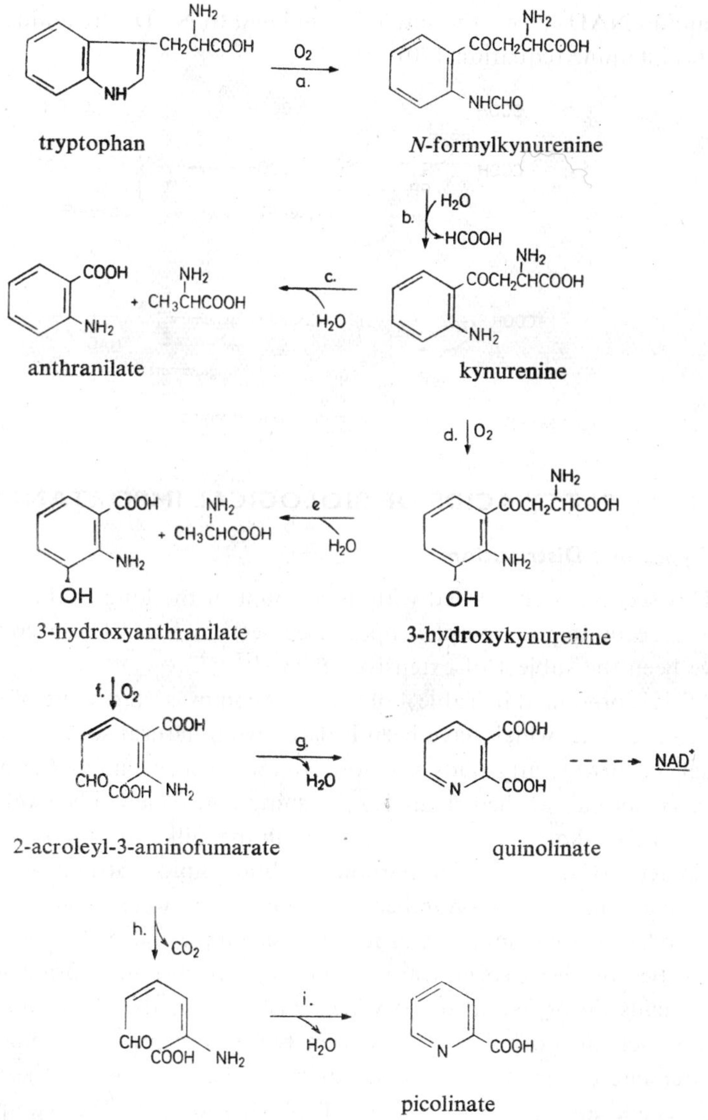

FIGURE 33. Important pathways in tryptophan degradation

oxylation at the 2- position (h), followed by ring closure (i) to yield picoli-nate. The reaction resulting in quinolinate is biologically the most impor-tant, since this species is the precursor for NAD^+ biosynthesis. In outline, quinolinate reacts with phosphoribosylpyrophosphate (PRPP) to produce quinolinate ribonucleotide which is decarboxylated to the nicotinate ana-logue (equation 169). Nicotinate ribonucleotide and ATP react to yield desamido-NAD^+, and the latter is amidated to NAD^+ by amide transfer from glutamine (equation 170).

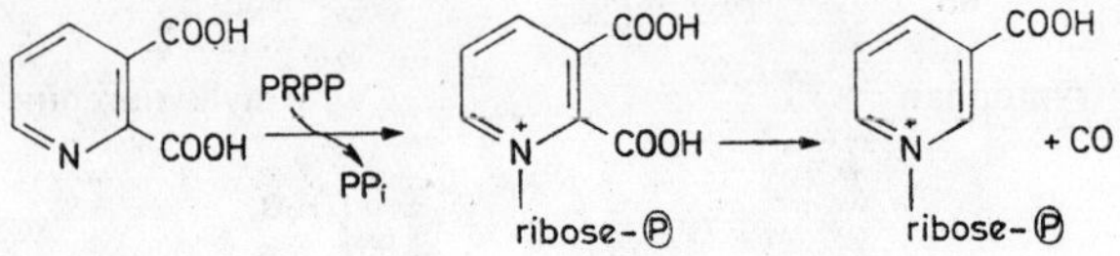

$$(169)$$

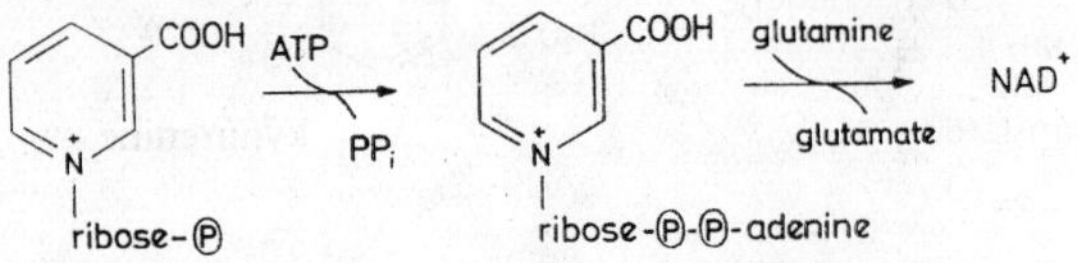

$$(170)$$

VI. FATTY ACIDS OF BIOLOGICAL IMPORTANCE

A. Types and Distribution

This section is concerned with an account of the long-chain fatty acids which occur in the complex lipids (see section VII). These compounds have been the subject of extensive reviews[177–180].

A list is presented in Table 4 of the most common, and some of the more rare, fatty acids which have been isolated from natural sources. The most widely occurring fatty acids are those containing an even number of carbon atoms in an unbranched chain, e.g., palmitic and oleic acids. Considerable quantities of short chain acids are found in the milk of ruminants and long chain acids (C_{20-27}) are of importance in brain lipids. Straight-chain fatty acids containing an odd number of carbon atoms were originally consid-ered to be very uncommon, but modern techniques have shown that small quantities of these species are of universal occurrence. Branched-chain fatty acids do not seem to be widely distributed; the occurrence of iso-valeric acid in porpoise fat has been recognized for some time, but of greater interest are the two series of branched-chain acids (the iso acids (C_{10-28}) and ante-iso acids (C_{9-31})—Table 4) shown by Weitkamp[181] to be present in wool wax.

TABLE 4. Some naturally occurring fatty acids

Fatty Acid	Structure	Occurrence
(a) Saturated		
Myristic	$CH_3(CH_2)_{12}COOH$	General
Palmitic	$CH_3(CH_2)_{14}COOH$	General
Stearic	$CH_3(CH_2)_{16}COOH$	General
'iso acids'	$CH_3CH(CH_2)_nCOOH$ $\|$ CH_3	Wool wax, microorganisms
'ante-iso acids'	$CH_3CH_2CH(CH_2)_nCOOH$ $\|$ CH_3	Wool wax, microorganisms
(b) Unsaturated		
Palmitoleic	$CH_3(CH_2)_5CH=CH(CH_2)_7COOH$	Macadamia nuts
Oleic	$CH_3(CH_2)_7CH=CH(CH_2)_7COOH$	General
Nervonic	$CH_3(CH_2)_7CH=CH(CH_2)_{13}COOH$	Brain lipid
Linoleic	$CH_3(CH_2)_3(CH_2CH=CH)_2(CH_2)_7COOH$	Seed oils
Linolenic	$CH_3(CH_2CH=CH)_3(CH_2)_7COOH$	Seed oils
Arachidonic	$CH_3(CH_2)_3(CH_2CH=CH)_4(CH_2)_3COOH$	Animal tissue
α-Eleostearic	$CH_3(CH_2)_3(CH=CH)_3(CH_2)_7COOH$	Tung oil
Mycomycin	$CH{\equiv}C-C{\equiv}CCH=C=CH(CH=CH)_2-$ CH_2COOH	*Nocardia acidophilus*
(c) Carbocyclic		
Lactobacillic	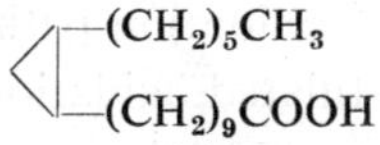 $\displaystyle\triangleleft{{-(CH_2)_5CH_3}\atop{-(CH_2)_9COOH}}$	*Lactobacilli*
Sterculic	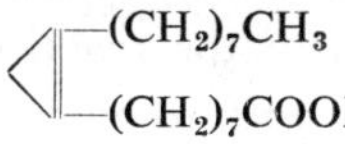 $\displaystyle\triangleleft{{-(CH_2)_7CH_3}\atop{-(CH_2)_7COOH}}$	*Sterculia*
Chaulmoogric	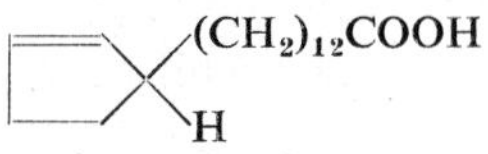 $\displaystyle\square\!\!\!\diagdown{{(CH_2)_{12}COOH}\atop{H}}$	*Flacourtiaceae*

Oleic acid is usually accompanied by small quantities of palmitoleic acid (*cis*-9-hexadecenoic acid); macadamia nuts, however, contain palmitoleic acid as a major component. Also of considerable importance is the C_{24} nervonic acid (*cis*-15-tetracosenoic acid), a major constituent of brain lipid.

Poly-unsaturated fatty acids are of some importance. Linoleic acid (*cis*, *cis*-9, 12-octadecadienoic acid) occurs widely in seed fats, as does the tri-unsaturated linolenic acid (*cis*, *cis*, *cis*-9,12,15-octadecatrienoic acid).

The former is claimed to be an essential growth factor for young mammals where it acts as a precursor for the biosynthesis of arachidonic acid (*cis, cis, cis, cis*-5,8,11,14-eicosatetraenoic acid); the importance of arachidonic acid remains to be explained. A few polyunsaturated fatty acids have been characterized in which the double bonds are conjugated; an example is α-eleostearic acid (*cis, trans, trans*-9,11,13-octadecatrienoic acid) which is a major component of tung oil. This acid is also unusual in that two of the double bonds have the *trans* configuration rather than the usual *cis*. Mention should be made of the remarkable fatty acid mycomycin (trideca-3,5,7,8-tetraene-10,12-diynoic acid) which has been isolated from cultures of *Nocardia acidophilus*; this compound contains two acetylenic linkages and is optically active by virtue of its allene structure.

A few fatty acids containing carbocyclic structures have been isolated from natural sources. Three-carbon rings are the interesting feature of lactobacillic acid (from several species of *Lactobacilli*) and sterculic acid (from seed oils of *Sterculia* species); in the former case the ring is saturated and in the latter unsaturated. Chaulmoogric acid contains the cyclopentenyl ring system; this acid is a major component of the oil from various *Flacourtiaceae* species, where it is frequently accompanied by the doubly-unsaturated gorlic acid.

B. Degradation of Fatty Acids

In all forms of life, the main route for the degradation of saturated long-chain fatty acids is the so-called β-oxidation pathway originally proposed by Knoop in 1904[182]. Knoop postulated oxidation of the fatty acid to the β-keto derivative, followed by removal of the carboxy-terminal two-carbon fragment as acetate and repetition of the reaction sequence. This theory resulted from feeding experiments with ω-phenyl-substituted fatty acids. ω-Substituted fatty acids with an even number of carbon atoms in the side chains were degraded to phenylacetate whilst those with an odd number of carbon atoms were degraded to benzoic acid. These metabolites formed condensation products with glycine (phenylaceturate and hippurate respectively, reactions 171, 172) and these products were identified in the urine. The results are consistent with the step-wise removal of two-carbon fragments from the carboxy-terminal end of the fatty acid.

$$PhCH_2COOH + NH_2CH_2COOH \longrightarrow PhCH_2CONHCH_2COOH \quad (171)$$
$$PhCOOH + NH_2CH_2COOH \longrightarrow PhCONHCH_2COOH \quad (172)$$

A critical assessment of Knoop's hypothesis could not be made until cell-free preparations were obtained which were capable of carrying out the oxidation of fatty acids. Such a preparation was obtained from guinea pig liver by Leloir and Muñoz[183] in 1939. A problem then arose, however, in that none of the expected intermediates of the oxidation process or any fatty acids of intermediate chain length could be detected. This conundrum was resolved in 1951 by Lynen and Reichert[184] who recognized that the active form of the substrate for oxidation is in fact the acyl-CoA derivative rather than the free fatty acid, and that the intermediates in the oxidation are similarly bound to CoA. This means, of course, that the intermediates of the oxidation process can never be present in quantities greater than the amount of CoA in the preparation. Once this hurdle was surmounted, progress on the confirmation of Knoop's scheme and on the isolation of enzymes catalysing the individual reactions was rapid.

The reactions involved in β-oxidation of fatty acids are shown in Figure 34. The sequence is initiated by formation of the acyl-CoA derivative (reaction a); this then undergoes α,β-elimination (b), hydration (c), oxidation (d) and finally thiolytic cleavage by CoA (e) to yield the acyl-CoA containing two carbon atoms less than the starting material and acetyl-CoA. The cycle is repeated until the four-carbon acyl derivative is obtained when thiolytic cleavage yields two molecules of acetyl-CoA. It is interesting to note the similarity between reaction sequence b–d and the sequence involved in the oxidation of succinate to oxaloacetate in the citric acid cycle (Section IV. A.2).

Lack of space precludes a detailed account of the properties of the individual enzymes catalysing the reactions in Figure 34, but some of the more interesting features will be mentioned. A general review of the enzymes has been given by Green and Wakil[185] and the properties of the individual enzymes have been described[186–190]; Bruice and Benkovic[191] have discussed the reactivity of thiolesters.

Three distinct ATP-dependent thiokinases catalysing reactions of the general form shown in equation (173) have been characterized[186].

$$RCOOH + ATP + CoASH \rightleftharpoons RCOSCoA + AMP + PP_i \qquad (173)$$

Acetate thiokinase from yeast activates acetate, propionate and acrylate but not fatty acids of greater chain length. Medium-chain fatty acid thiokinase is active with acids containing from four to twelve carbon atoms and also with the corresponding phenyl-substituted, β-hydroxy-, α,β- and β,γ-unsaturated and branched-chain acids. Finally, a long chain fatty acid thiokinase (C_8–C_{22}) has been isolated from mammalian liver. In all cases, the enzymes are specific for ATP and show a dependence on Mg^{2+} ions for activity. It is generally accepted that reaction (173) occurs

$$CH_3(CH_2CH_2)_{n-1}CH_2CH_2COOH$$

a. $\nearrow$ CoASH

$$CH_3(CH_2CH_2)_{n-1}CH_2CH_2COSCoA$$

b. $\nearrow$ EFAD
$\searrow$ EFADH$_2$

$$CH_3(CH_2CH_2)_{n-1}CH{=}CHCOSCoA$$

c. $\nearrow$ H$_2$O

$$CH_3(CH_2CH_2)_{n-1}CH(OH)CH_2COSCoA$$

d. $\nearrow$ NAD+
$\searrow$ NADH, H+

$$CH_3(CH_2CH_2)_{n-1}COCH_2COSCoA$$

e. $\nearrow$ CoASH
$\searrow$ CH$_3$COSCoA

$$CH_3(CH_2CH_2)_{n-1}COSCoA$$

f. Repeat b–e

$$CH_3(CH_2CH_2)_{n-2}COSCoA + CH_3COSCoA$$

$(n{-}3)$ cycles

$$CH_3COSCoA + CH_3COSCoA$$

Reaction	Enzyme	E.C. Number
a.	Thiokinase	6.2.1.1, 6.2.1.2, 6.2.1.3
b.	Acyl dehydrogenase	1.3.2.2
c.	Enoyl hydratase	4.2.1.17
d.	β-hydroxyacyl dehydrogenase	1.1.1.35
e.	β-ketoacyl thiolase	2.3.1.16

FIGURE 34. Oxidation of fatty acids

in two stages. In the first step (Mg^{2+}-dependent) an enzyme-bound acyl adenylate is produced (equation 174) which, in the second step, reacts with CoA to yield the acyl-CoA derivative (equation 175). Both steps are freely reversible.

$$RCOOH + ATP + E \xrightleftharpoons{Mg^{2+}} E.AMP\text{—}COR + PP_i \qquad (174)$$

$$E.AMP\text{—}COR + CoASH \rightleftharpoons RCOSCoA + AMP + E \qquad (175)$$

More recently, a GTP-specific thiokinase has been isolated from beef liver mitochondria[192] (equation 176); the specificity with respect to the acyl group is unknown.

$$RCOOH + GTP + CoASH \rightleftharpoons RCOSCoA + GDP + P_i \qquad (176)$$

Most cells capable of oxidizing fatty acids contain at least two and sometimes as many as four distinct acyldehydrogenases (reaction 177) which differ in specificity for the chain length of the acyl-CoA derivative. For example, pig liver mitochondria contain three such enzymes with specifi-

$$RCH_2CH_2COSCoA + E.FAD \rightleftharpoons RCH\text{=}CHCOSCoA + FADH_2 \qquad (177)$$

cities for C_4 to C_8, C_8 to C_{12} and C_{12} to C_{16} acyl-CoA derivatives respectively. The enzymes have, however, many important properties in common. First, the unsaturated product in each case has the *trans*-configuration; this is of importance in the next step in the reaction sequence. The acyl dehydrogenases each contain two moles of bound FAD per mole of protein; evidence is accumulating that the two molecules of FAD cooperate in the oxidation of one molecule of substrate and are reduced to the semiquinone free radical stage (cf. glucose oxidase, section V. B.1), during the reaction. In the presence of substrates, the reduced form of the enzyme will not react with any of the usual electron acceptors (oxygen, ferricyanide, *etc.*) to an appreciable extent, neither can reoxidation be brought about by direct coupling to the electron transport system of the mitochondrion. Instead, reoxidation is brought about by a specific flavoprotein (electron transfer flavoprotein[193] (ETF), reaction 178), this protein in turn being oxidized by the terminal electron transport system.

$$E.FADH_2 + ETF \rightleftharpoons E.FAD + ETFH_2 \qquad (178)$$

Enoyl hydratase[188] differs from the enzymes described above in that a single enzyme appears to hydrate all α,β-unsaturated acyl-CoA derivatives (equation 179) irrespective of chain length.

$$RCH\text{=}CHCOSCoA + H_2O \rightleftharpoons RCH(OH)CH_2COSCoA \qquad (179)$$

The hydration reaction is stereospecific and the product is the L-β-hydroxyacyl-CoA.

The next enzyme in the sequence, β-hydroxyacyl dehydrogenase (equation 180) is similarly unspecific as regards the chain length of the substrate

$$RCH(OH)CH_2COSCoA + NAD^+ \rightleftharpoons RCOCH_2COSCoA + NADH + H^+ \quad (180)$$

but is, on the other hand, completely specific for the cofactor (NAD^+) and the configuration of the substrate (L-). The equilibrium position of reaction (180) is pH-dependent, β-hydoxyacyl-CoA formation being favoured at pH 7 whilst the reverse reaction is favoured at pH values greater than 9.

The sequence of reactions in Figure 34 is concluded by thiolytic cleavage of the β-ketoacyl-CoA derivative, the thiol group of CoA displacing the terminal —$CH_2COSCoA$ moiety (reaction 181). In most tissues, a single

$$RCOCH_2COSCoA + CoASH \longrightarrow RCOSCoA + CH_3COSCoA \quad (181)$$

β-ketothiolase catalyses the cleavage of the whole range of β-ketoacyl-CoA derivatives from C_4 upwards, but some tissues (for example, heart) appear to contain in addition a thiolase specific for the cleavage of the four-carbon acetoacetyl-CoA. It should be noted that, although the reverse of reaction (181) has been demonstrated in the case of acetoacetyl-CoA production, the equilibrium position of the reaction is very strongly in favour of production of acetyl-CoA.

Reference to Figure 34 shows that three main requirements must be met to allow fatty acid oxidation to proceed *in vivo*. A supply of ATP must be available for the activation reaction, CoA must be released from the product acetyl-CoA and provision must be made for the reoxidation of $ETFH_2$ and NADH. All of these requirements may readily be met since the site of fatty acid oxidation is located in the mitochondrion. The reduced species (NADH and $ETFH_2$) are reoxidized by the terminal electron transport system and the necessary ATP is generated by concomitant oxidative phosphorylation. Acetyl-CoA is condensed with oxaloacetate to yield citrate and free CoA, the citrate then being metabolized by the citric acid cycle, whilst the CoA is available to participate further in fatty acid oxidation.

In the case of fatty acids with an odd number of carbon atoms, the β-oxidation pathway is again employed, but the final product in this case is one molecule of acetyl-CoA and one of propionyl-CoA; the metabolic fate of propionyl-CoA has already been described (section II. C.2). The mode of oxidation of unsaturated fatty acids is not yet known[194].

A final point of interest in connection with the oxidation of fatty acids is the accumulation in the liver, particularly under the abnormal conditions of diabetes, of large amounts of free acetoacetate. Lynen and his coworkers[195] argue that this compound is not formed by direct deacylation of acetoacetyl-CoA, but by the combination of reactions (182) and (183), (HMG-CoA represents β-hydroxy-β-methylglutaryl-CoA).

$$\text{acetyl-CoA} + \text{acetoacetyl-CoA} \longrightarrow \text{HMG-CoA} + \text{CoA} \qquad (182)$$

$$\text{HMG—CoA} \longrightarrow \text{acetoacetate} + \text{acetyl-CoA} \qquad (183)$$

Distinct condensing and cleaving enzymes catalyse reactions (182) and (183) respectively. Reaction (182) is of considerable importance since HMG-CoA is the precursor for the biosynthesis of mevalonate which in turn is the precursor for the biosynthesis of cholesterol and, in plants, the terpenes.

C. Biosynthesis of Fatty Acids

Early studies of fatty acid biosynthesis in microorganisms and whole animals showed that the entire carbon chain is derived from acetate. Similarly, Stadtman and Barker[196] showed that extracts of the microorganism *Clostridium kluyveri* catalysed the formation of butyric and caproic acids from ethanol or acetate; these studies implicated acetyl-CoA or acetyl phosphate as the precursor for fatty acid synthesis. The subsequent elucidation of the β-oxidation pathway of fatty acid degradation led quite naturally, therefore, to the suggestion that the biosynthesis of these species occurs by a reversal of this route. It now appears, however, that the *de novo* synthesis of fatty acids occurs by a different pathway, the reversal of β-oxidation only being of importance in the elongation of preformed fatty acids by condensation with acetyl-CoA[197]; the latter process occurs in the mitochondrion.

Wakil and his collaborators[198-200] studied the synthesis of fatty acids as catalysed by purified extracts of pigeon liver. These workers showed that the synthesizing system is of cytoplasmic rather than mitochondrial origin. The biosynthetic process showed an absolute requirement for carbon dioxide[201], and biotin was shown to be an essential component of the enzyme system[202]; these observations were rationalized by the discovery of acetyl-CoA carboxylase (reaction 184)[203] and the demonstration that malonyl-CoA rather than acetyl-CoA is the effective precursor for fatty acid bio-

$$\text{CH}_3\text{COSCoA} + \text{CO}_2 + \text{ATP} + \text{H}_2\text{O} \longrightarrow \text{COOHCH}_2\text{COSCoA} + \text{ADP} + \text{P}_i \quad (184)$$

synthesis. Several groups of workers were able to show, however, that the methyl-terminal pair of carbon atoms are derived from acetyl-CoA directly, and not via malonyl-CoA (see review by Wakil[204]).

$$CH_3COSCoA + CO_2 + ATP \xrightarrow{\text{a.}} COOHCH_2COSCoA + ADP + P_i$$

$$CH_3COSCoA + ACP\text{—}SH \xrightarrow{\text{b.}} CH_3COSACP + CoASH$$

$$HOOCCH_2COSCoA + ACP\text{—}SH \xrightarrow{\text{c.}} HOOCCH_2COSACP + CoASH$$

$$HOOCCH_2COSACP + CH_3COSACP \xrightarrow{\text{d.}} CH_3COCH_2COSACP + ACPSH + CO_2$$

$$CH_3COCH_2COSACP + NADPH + H^+ \xrightarrow{\text{e.}} CH_3CHOHCH_2COSACP + NADP^+$$

$$CH_3CHOHCH_2COSACP \xrightarrow{\text{f.}} CH_3CH\text{=}CHCOSACP + H_2O$$

$$CH_3CH\text{=}CHCOSACP + NADPH + H^+ \xrightarrow{\text{g.}} CH_3CH_2CH_2COSACP + NADP^+$$

$$CH_3(CH_2CH_2)COSACP + (n-1)HOOCCH_2COSCoA + 2(n-1)NADPH +$$

$$2(n-1)H^+ \xrightarrow[\text{c, d, e, f, g.}]{\text{Reactions}}$$

$$CH_3(CH_2CH_2)_nCOSACP + (n-1)CoASH + (n-1)CO_2 + 2(n-1)NADP^+ + (n-1)H_2O$$

Reaction	Enzyme
a.	acetyl-CoA carboxylase
b.	acetyl transacylase
c.	malonyl transacylase
d.	β-ketoacyl-ACP synthetase
e.	β-ketoacyl-ACP reductase
f.	β-hydroxylacyl-ACP dehydrase
g.	enoyl-ACP reductase

FIGURE 35. Biosynthesis of fatty acids in *E. coli*

Further progress on the elucidation of the pathway of fatty acid biosynthesis depended on the purification and characterization of the synthesizing systems. Three such systems have been studied in detail, namely those from pigeon liver[205], yeast[206] and *E. coli*[207]. The system from yeast behaves as a single multienzyme complex of molecular weight 2,300,000 which has resisted fractionation into active subunits; the system from avian liver has been fractionated into only two components. The fatty acid synthesizing system of *E. coli* has, on the other hand, been fractionated into its individual components, thus allowing a complete investigation of the reactions involved to be carried out.

The *E. coli* synthesizing system contains a small, heat-stable protein whose function is to act as a carrier of the acyl intermediates of fatty acid biosynthesis[208]. The role of the acyl carrier protein (ACP) was shown by experiments in which substrate quantities of the protein were incubated with acetyl-CoA and malonyl-CoA; the products of the reaction were acetoacetyl-ACP and carbon dioxide (equation 185), the acetoacetyl moiety being bound to ACP via a thiol ester linkage. Incubation of radio-labelled

$$ACP—SH + CH_3COSCoA + COOHCH_2COSCoA \rightleftharpoons CH_3COCH_2COS—ACP$$
$$+ CO_2 + 2\ CoASH \qquad (185)$$

acetoacetyl-ACP (prepared by reaction 185) with unlabelled malonyl-CoA, NADPH and the enzymes of the *E. coli* fatty acid synthesizing system led to the rapid production of radioactive vaccenic acid (*cis*-11,12-octadecenoic acid). Degradation of the vaccenic acid showed that the labelled acetoacetyl-ACP was incorporated intact into the methyl terminal group of the product fatty acid.

The availability of purified ACP allowed the synthesis of possible ACP-acyl intermediates of fatty acid synthesis to be undertaken; use of these intermediates then allowed the identification of the individual steps in the process (Figure 35).

Reaction a (Figure 35) is catalysed by acetyl-CoA carboxylase; the properties of this enzyme have been described in a previous section. Reactions b and c involve the transfer of acetyl and malonyl moeities respectively from CoA to ACP; it has been shown that separate enzymes exist for the catalysis of these two reactions[209]. A condensation-decarboxylation reaction (d) yields the β-ketoacyl-ACP derivative; the enzyme catalysing this reaction is specific for the acyl-ACP derivatives of acetate and malonate, and is completely inactive with the corresponding acyl-CoA compounds. It is of interest to compare the redox reaction (e) with the corresponding reaction of the β-oxidation pathway. In the present case the reductase employs NADPH as the coenzyme and the product of the reaction is the $D(-)-\beta$-hydoxyacyl-ACP derivative; the enzyme in the β-oxidation sequence is active with NADH and the β-hydroxyacyl-CoA substrate is of the opposite configuration, namely $L(+)$.

The β-ketoacyl-ACP reductase is active with the corresponding CoA esters but, in the case of acetoacetyl-CoA, for example, the rate of reduction is some sixty times less than with the true substrate. More complete substrate specificity is shown by the next enzyme in the sequence, namely β-hydroxyacyl-ACP dehydrase, which appears to be without activity with the corresponding CoA derivatives as substrates. Finally, reduction of the enoyl-ACP derivative (reaction g) is catalysed by an NADPH-specific reductase. Again it is interesting to compare this reaction with the corresponding step in the β-oxidation pathway where a flavoprotein enzyme is involved; this difference probably reflects the different subcellular localizations of the two systems, since reoxidation of reduced flavoproteins requires close contact with the mitochondrial electron transport system.

A considerable amount of information concerning the acyl binding site of ACP is available. Early work suggested that the single sulphydryl group of ACP was part of a cysteine residue, but more recent studies[210, 211] have shown that the sulphydryl-containing moiety is β-mercaptoethylamine. Hydrolysates of ACP contain, in addition to β-mercaptoethylamine, one mole each of β-alanine, pantoate and phosphate per mole of protein (9500 g). The presence of these species suggested that the prosthetic group of ACP is 4'-phosphopantetheine (cf. the side chain of CoA), a hypothesis which has been amply substantiated by Vagelos and his coworkers[208].

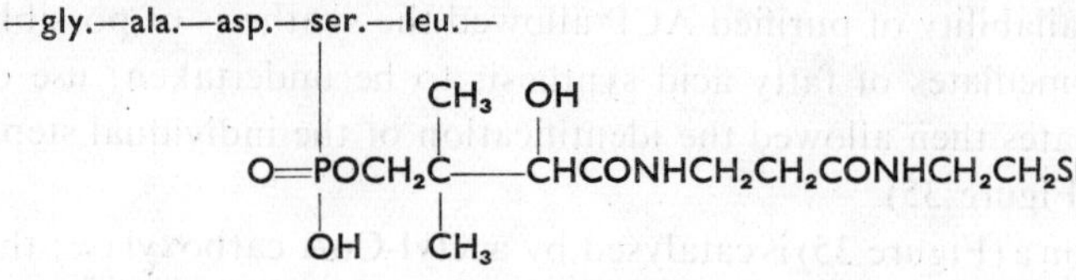

—gly.—ala.—asp.—ser.—leu.—

$$O{=}POCH_2C\text{———}CHCONHCH_2CH_2CONHCH_2CH_2SH$$

with CH_3, OH above and OH, CH_3 below.

FIGURE 36. Structure of the active site of the acyl carrier protein from *E. coli*

These authors have also shown that the 4'-phosphopantetheine residue is linked to a serine residue in the protein by a phosphate ester bond, and have established the sequence of amino acids in the tetrapeptide containing this serine residue. The structure of the acyl binding site is shown in Figure 36.

ACP has been shown to be a component of the fatty acid synthesizing systems of a variety of bacteria; in the case of the multienzyme systems from yeast and liver, however, the establishment of ACP as a functional component of the systems rests on the identification of 4'-phosphopantetheine and pantothenate as degradation products of the protein complexes. It is, however, widely held that ACP is present in the fatty acid synthesizing systems of plants, microorganisms and animals, and that the reactions in Figure 35 provide an accurate description of the biosynthetic process in these species.

Straight-chain fatty acids with an odd number of carbon atoms arise from substitution of a propionyl moiety for acetyl in the initial condensation reaction[212]. By analogy with reaction d of Figure 35 the process may be represented by reaction (186).

$$CH_3CH_2COSACP + HOOCCH_2COSACP \rightleftharpoons CH_3CH_2COCH_2COSACP + ACPSH + CO_2$$

$$(186)$$

Reduction of the β-ketovaleryl-ACP and chain elongation by successive condensations with malonyl-ACP leads to a fatty acid with an odd number of carbon atoms. Biosynthesis of the branched-chain iso and ante-iso acids may be explained in a similar fashion[213]. Substitution of isobutyryl-ACP or L-α-methylbutyryl-ACP for acetyl-ACP leads to the iso and ante-iso series respectively (reactions 187 and 188).

$$
\begin{array}{c}
\text{CH}_3\text{CHCOSACP} + \text{HOOCCH}_2\text{COSACP} \rightleftharpoons \text{CH}_3\text{CHCOCH}_2\text{COSACP} \\
\qquad | \qquad\qquad\qquad\qquad\qquad\qquad\qquad\qquad\qquad | \\
\qquad \text{CH}_3 \qquad\qquad\qquad\qquad\qquad\qquad\qquad\qquad\qquad \text{CH}_3 \qquad\qquad (187) \\
+ \text{ACPSH} + \text{CO}_2
\end{array}
$$

$$
\begin{array}{c}
\text{CH}_3\text{CH}_2\text{CHCOSACP} + \text{HOOCCH}_2\text{COSACP} \rightleftharpoons \text{CH}_3\text{CH}_2\text{CHCOCH}_2\text{COSACP} \\
\qquad\quad | \qquad\qquad\qquad\qquad\qquad\qquad\qquad\qquad\qquad\qquad | \\
\qquad\quad \text{CH}_3 \qquad\qquad\qquad\qquad\qquad\qquad\qquad\qquad\qquad\quad \text{CH}_3 \qquad (188) \\
+ \text{ACPSH} + \text{CO}_2
\end{array}
$$

Two biosynthetic routes to the unsaturated fatty acids have been recognized, one of which involves desaturation of preformed fatty acids whilst the other involves introduction of the double bond during chain elongation. The former route (reaction 189) was first demonstrated in *Saccharomyces cerevisiae* by Bloomfield and Bloch[214], and has subsequently been

$$
\text{CH}_3(\text{CH}_2)_m\text{CH}_2\text{CH}_2(\text{CH}_2)_n\text{COSCoA} \xrightarrow{\text{O}_2,\ \text{NADPH}} \text{CH}_3(\text{CH}_2)_m\text{CH}=\text{CH}(\text{CH}_2)_n\text{COSCoA}
$$
$$(189)$$

implicated in the production of unsaturated fatty acids by a variety of microorganisms, plants and animals under aerobic conditions (see reviews by Lennarz[213] and Vagelos[215]). The enzyme catalysing reaction (189) is particle-bound and has yet to be solubilized and characterized; information available at present suggests, however, that the reaction requires FAD and ferrous ions.

Evidence for a route to unsaturated fatty acids involving β,γ-dehydration of β-hydroxy acids of intermediate chain length was obtained by Bloch and his coworkers[216]. *Clostridium butyricum* grown in the presence of [1-^{14}C]-octanoic acid produced [9-^{14}C]-9,10-hexadecenoic and [11-^{14}C]-11,12-octadecenoic acids in addition to the corresponding C_{16} and C_{18} saturated fatty acids; these results were explained on the basis of the scheme shown in Figure 37. Support for this scheme was provided by Baronowsky[217] who showed that *cis*-3,4-decenoic acid was converted only to unsaturated fatty acids by *C. butyricum*.

Further evidence for the pathway shown in Figure 37 has been obtained. Crude extracts of *E. coli* catalyse the production of *trans*-α,β-decenoyl-CoA and *cis*-β,γ-decenoyl-CoA from β-hydroxydecanoyl-CoA[218]; separation of the two enzymes catalysing α,β- and β,γ-dehydration of D$(-)$-β-hydroxydecanoyl thioesters has been achieved by Bloch and his co-workers[219]. The products of α,β- and β,γ-dehydration of β-hydroxydeca-noyl-CoA do not act as precursors of long chain fatty acids; Bloch and his colleagues[220] have shown, however, that the corresponding α,β- and β,γ-unsaturated decenoyl-ACP derivatives do serve as precursors for the bio-synthesis of saturated and unsaturated fatty acids respectively. Thus the intermediates and products of the reactions shown in Figure 37 should be written as the corresponding acyl-ACP derivatives.

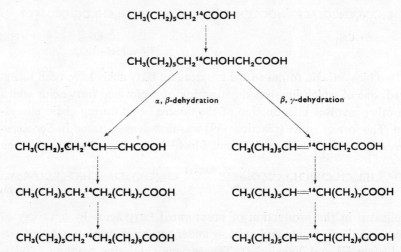

Reaction sequences represented by broken arrows are the usual reactions of chain elongation.

FIGURE 37. Biosynthesis of saturated and unsaturated fatty acids by *C. butyricum*

The biosynthesis of cyclopropane fatty acids from unsaturated straight chain species is of interest. Early work (reviewed by Kates[221]) showed that the ring methylene group is derived from the methyl residue of methionine. Zalkin[222] and his colleagues showed that extracts of *Serratia marcescens* and *C. butyricum* catalyse the incorporation of the methyl group of *S*-adenosylmethionine into cyclopropane fatty acids. In the case of extracts from *C. butyricum* an absolute requirement for a phospholipid containing an unsaturated acyl group was demonstrated, and the reaction product was shown to be a cyclopropane-containing phosphatidylethanolamine derivative. These observations led to the formulation of the biosynthetic process as shown in equation (190).

$$\begin{array}{ccc}
\text{RCOOCH}_2 & & \text{CH}_2 \quad\quad \text{RCOOCH}'_2 \\
| & & \diagup\diagdown \quad | \\
\text{CH}_3(\text{CH}_2)_m\text{CH}=\!=\text{CH}(\text{CH}_2)_n\text{COOCH} & \longrightarrow & \text{CH}_3(\text{CH}_2)_m\text{CH}-\text{CH}(\text{CH}_2)_n\text{COOCH} \\
\text{O} \quad | & & \text{O} \quad | \\
\| & & \| \\
\text{NH}_2\text{CH}_2\text{CH}_2\text{OPOCH}_2 & & \text{NH}_2\text{CH}_2\text{CH}_2\text{OPOCH}_2 \\
| & & | \\
\text{OH} & & \text{OH}
\end{array}$$

$$\begin{array}{cc}
+ \ \text{CH}_3-\overset{+}{\text{S}}\text{-adenosine} & + \ \text{S-adenosine} \\
| & | \\
\text{CH}_2 & \text{CH}_2 \quad\quad + \ \text{H}^+ \\
| & | \\
\text{CH}_2\text{CHNH}_2\text{COOH} & \text{CH}_2\text{CHNH}_2\text{COOH}
\end{array}$$

(190)

Chung and Law[223] have partially purified the cyclopropane fatty acid synthetase catalysing reaction (190). Phosphatidic acid and phosphatidylglycerol are good substrates for the enzyme, but the corresponding phosphatidylserine and lecithin derivatives are not[224]. Hildebrand and Law[225] have studied reaction (190) with positional isomers of the phospholipid substrates; generally cyclopropane fatty acyl groups are formed preferentially from unsaturated acyl groups in the β-position of phosphatidylethanolamine whilst in the case of *C. butyricum* the γ-position is favoured. These specificities are by no means complete, however, and a mixture of products is invariably formed.

VII. THE COMPLEX LIPIDS

A. Introduction: Limited Importance of Simple Esters

With a few notable exceptions, esters of short chain acids and alcohols have no recognized biological function, and hence will not be discussed in detail. Such esters are toxic to higher organisms and are removed from the system by hydrolysis (reaction 191); enzymes catalysing reaction (191)

$$R_1\text{CO.OR}_2 + H_2O \rightleftharpoons R_1\text{COOH} + R_2\text{OH} \tag{191}$$

(carboxyl esterases, E.C. 3.1.1.1) have been found in many species (animals, plants, yeasts, moulds) but have not in general been well characterized. Webb[226] has studied the substrate specificity of the carboxyl esterase from horse liver. Both aliphatic and aromatic esters are hydrolysed by this enzyme; in the case of aliphatic esters, the rate of hydrolysis increases with chain length of both the alkyl and acyl functions up to between four and six carbon atoms and then drops off rapidly. This broad specificity is consistent with the proposed role of the enzyme in detoxication processes.

One of the most widely studied enzyme-catalysed reactions is the hydrolysis of esters (*e.g.* phenol esters such as *p*-nitrophenyl acetate and the esters of *N*-substituted tryptophan and tryosine derivatives) by chymotrypsin (E.C. 3.4.4.5). Many reviews of this subject have been presented (*e.g.* Bender and Kezdy[227]) and the topic was discussed in a recent symposium[228]. This topic will not be dealt with here, since chymotrypsin is in fact a protease rather than an esterase and the true biological substrates for the enzyme are proteins and peptides.

Complex lipids containing choline will be discussed in the next section, but it is appropriate at this juncture to mention briefly acetylcholine, which is concerned with the transmission of impulses at the synaptic and effector junctions in the parasympathetic nervous system and is also involved in the passage of impulses along nerve fibres[229]. Choline is formed by three successive methylations of the base ethanolamine by *S*-adenosylmethionine (192) and the ester produced by transfer of the acetyl function from acetyl-CoA; the latter reaction is catalysed by choline acetyltrans-

$$HOCH_2CH_2NH_2 + 3\ CH_3\!\!-\!\!\overset{+}{S}\text{-adenosine} \longrightarrow HOCH_2CH_2\overset{+}{N}(CH_3)_3$$
$$\underset{R}{|} \qquad\qquad + 3\ RS\text{-adenosine} + 2\,H^+ \quad (192)$$

$$HOCH_2CH_2\overset{+}{N}(CH_3)_3 + CH_3COSCoA \longrightarrow CH_3COOCH_2CH_2\overset{+}{N}(CH_3)_3$$
$$+ CoASH \qquad\qquad\qquad (193)$$

ferase (E.C. 2.3.1.6).

Consistent with the proposed role of acetylcholine in nervous function, brain and nervous tissue contain large quantities of the enzyme cholinesterase which catalyses the hydrolysis of acetylcholine to choline and acetate. The electric organ of *Electrophorus electricus* is a particularly good source of this enzyme. Other organs of animals contain a similar enzyme, pseudocholinesterase, which catalyses the hydrolysis of analogues of acetylcholine and of simple esters such as methylbutyrate. The mechanism of action of these enzymes has been discussed by Davies and Green[230].

B. Structures, Distribution and Functions of Complex Lipids

The structures of many of the important naturally-occurring complex lipids are shown in Figure 38. The most wide spread lipids are the triglycerides. In general, depot fats consist of a complex mixture of mixed triglycerides (that is, R_1, R_2 and R_3 are different). Many attempts have been made to predict the distribution of fatty acids in triglycerides from the composition of the mixture of fatty acids obtained after hydrolysis of depot fats; the various methods have been critically reviewed by Lovern[231].

(a) Triglycerides

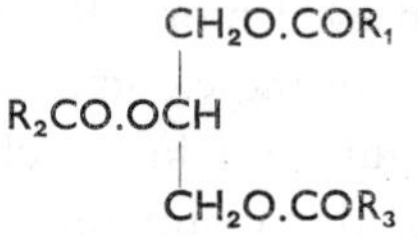

$$CH_2O.COR_1$$
$$R_2CO.OCH$$
$$CH_2O.COR_3$$

R_1CO, R_2CO, R_3CO—fatty acyl residues

(b) Phospholipids

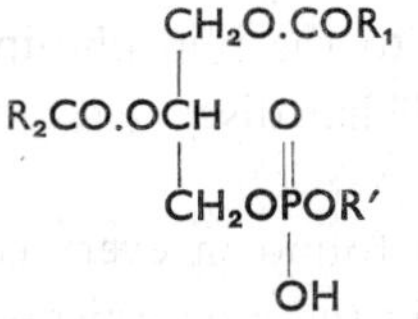

$$CH_2O.COR_1$$
$$R_2CO.OCH \quad O$$
$$CH_2OPOR'$$
$$OH$$

$R' =$ —H phosphatidic acid

$R' =$ —$CH_2CH(NH_2)COOH$ phosphatidylserine

$R' =$ —$CH_2CH_2NH_2$ phosphatidylethanolamine (cephalin)

$R' =$ —$CH_2CH_2NHCH_3$ N-methylphosphatidylethanolamine

$R' =$ —$CH_2CH_2N(CH_3)_2$ N,N-dimethylphosphatidylethanolamine

$R' =$ —$CH_2CH_2\overset{+}{N}(CH_3)_3$ phosphatidylcholine (lecithin)

$R' =$ —$CH_2CHOHCH_2OH$ phosphatidylglycerol

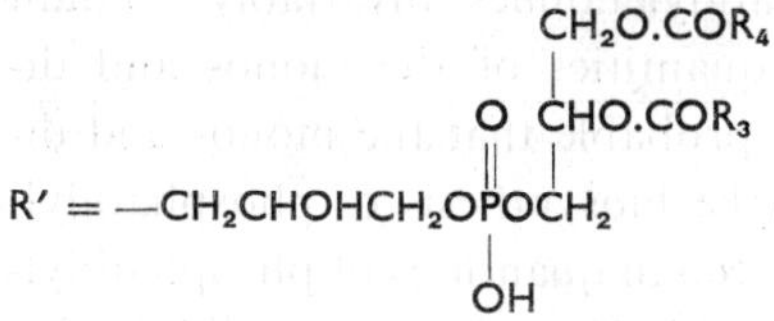

$$CH_2O.COR_4$$
$$O \quad CHO.COR_3$$
$R' =$ —$CH_2CHOHCH_2OPOCH_2$ disphosphatidylglycerol (cardiolipin)
$$OH$$

$R' =$

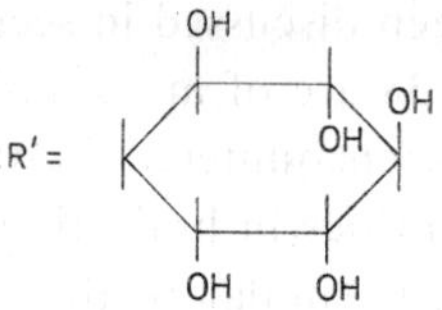

phosphatidylinositol

(c) Plasmalogens

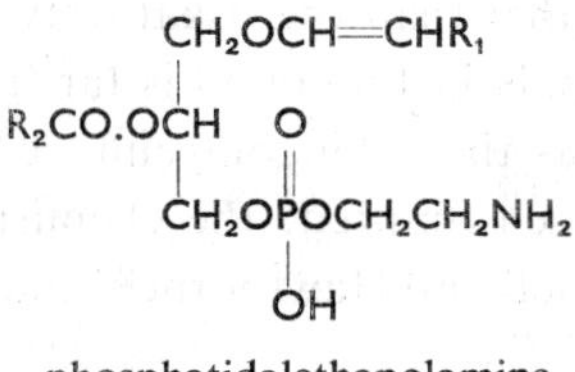

$$CH_2OCH{=}CHR_1$$
$$R_2CO.OCH \quad O$$
$$CH_2OPOCH_2CH_2NH_2$$
$$OH$$

phosphatidalethanolamine

FIGURE 38. Structures of some complex lipids

Chemical methods for the determination of the structures of triglycerides have been reviewed by the same author.

The phospholipids are an even more diverse class of compounds than the triglycerides. The parents of these compounds may be considered to be the phosphatidic acids (Figure 38) in which the β- and γ-positions of glycerol are esterified with long-chain fatty acids whilst the α-position carries a phosphate residue. Phospholipids have a centre of assymetry at the β-carbon atom and are usually of the L-configuration. A recent monograph[232] has been devoted to the phospholipids and their chemistry has been reviewed[233]; Lennarz[213] has discussed recent advances in the study of bacterial phospholipids.

Phospholipids have been found in every cellular organism so far examined, but their functions in these organisms are still far from clear. It is thought that these compounds are important structural components of cell membranes (where they may be involved in active transport phenomena) and of other functional units such as mitochondria and chloroplasts.[232]

In many tissues, the most abundant phospholipids are the phosphatidylcholines (Figure 38), that is, the diacyl derivatives of L-α-glycerylphosphorylcholine. Tissues containing phosphatidylcholines invariably contain phosphatidylethanolamines and small quantities of the mono- and dimethylphosphatidylethanolamines; it is probable that the mono- and dimethyl derivatives are intermediates in the biosynthesis of phosphatidylcholine from phosphatidylethanolamine. Small quantities of phosphatidylserine are also usually found; the relationship between phosphatidylserine and phosphatidylethanolamine has been discussed in section IV. B.1. The other phospholipids shown in Figure 38 are of more limited distribution.

A somewhat different structure is encountered in the plasmalogens (Figure 38), which occur in large quantities in brain tissue. The structure shown for ethanolamine plasmalogen is similar to that of phosphatidylethanolamine except that the substituent on the γ-position of glycerol is a long-chain vinyl ether rather than the usual fatty acyl group.

The list of complex lipids in Figure 38 is far from complete and many interesting species such as the sphingomyelins, cerebrosides, gangliosides and sulpholipids have been omitted; the chemistry of these compounds has been reviewed by Ansell and Hawthorne[232] and Hanahan and Brockerhoff[233].

C. Degradation of Complex Lipids

1. Triglycerides

In animals, the degradation of ingested triglycerides occurs mainly in the small intestine and is catalysed by the enzyme pancreatic lipase (E.C. 3.1.1.3). The positional specificity of this enzyme has been the subject of much study[234, 235]; these studies have been greatly complicated by the fact that the enzyme acts readily only on an emulsion of the triglyceride and by the readiness with which mono- and diglycerides undergo intramolecular acyl shift reactions. It is now generally agreed, however, that the enzyme

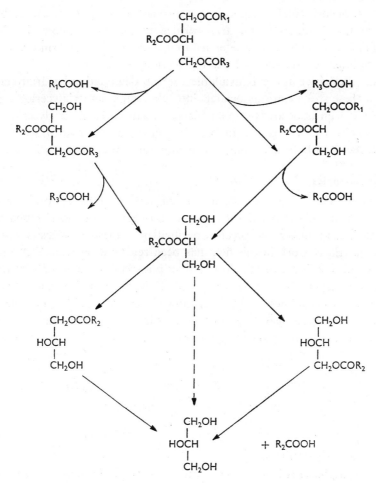

FIGURE 39. Probable route for the degradation of triglycerides by pancreatic lipase

rapidly hydrolyses the α- and α'-ester bonds in triglycerides but acts slowly, if at all, on ester linkages at the β-position. Consistently, large quantities of β-monoglycerides are found in the intestines of animals after fatty meals. The positional specificity of hydrolysis by pancreatic lipase is now so widely accepted that it is frequently made use of in studies of the structures of mixed triglycerides.

The complete degradation of triglycerides to glycerol and fatty acids by pancreatic lipase has been studied, but it is uncertain whether the enzyme catalyses the hydrolysis of ester linkages in the β-position of monoglycerides or acts on the α-esters formed by intramolecular migration of the acyl function. Benzonana and his colleagues[235] favour the latter view, but until this problem is finally settled it is appropriate to represent the pathway for complete degradation of triglycerides by the scheme shown in Figure 39. (This scheme should be extended to include possible acyl shift reactions of the intermediate α,β- and α',β-diglycerides.)

An alternative route is available for the degradation of triglycerides when these species are associated into chylomicrons (triglyceride aggregates with an associated protein moiety). In this case, the triglycerides are hydrolysed by the enzyme lipoprotein lipase, and the hydrolysis appears to proceed without accumulation of intermediate mono- or diglycerides.

2. Phospholipids

In spite of the extensive studies which have been carried out in recent years on enzymes catalysing the hydrolysis of phospholipids (*i.e.* phospholipases), the number and precise specificities of these enzymes is still uncertain. (Early work in this field has been reviewed by Kates[234].) Recent work tends to indicate that five distinct phospholipases exist[236, 237]. Four of these enzymes act on native phospholipids; the most widely accepted nomenclature for these enzymes is shown in Figure 40. The fifth enzyme, lysophospholipase, catalyses the hydrolysis of the remaining fatty acyl ester linkage of lysophospholipids (*e.g.* reaction 194). These enzymes are

$$\begin{array}{l} CH_2OCOR_1 \\ | \\ HOCH \quad O \\ | \quad \| \\ CH_2OPOR' \\ | \\ OH \end{array} + H_2O \longrightarrow \begin{array}{l} CH_2OH \\ | \\ HOCH \quad O \\ | \quad \| \\ CH_2OPOR' \\ | \\ OH \end{array} + R_1COOH \qquad (194)$$

discussed briefly below.

Phospholipase A is widely distributed, being found in animal and plant tissues and in large quantities in the venoms of snakes, wasps and bees.

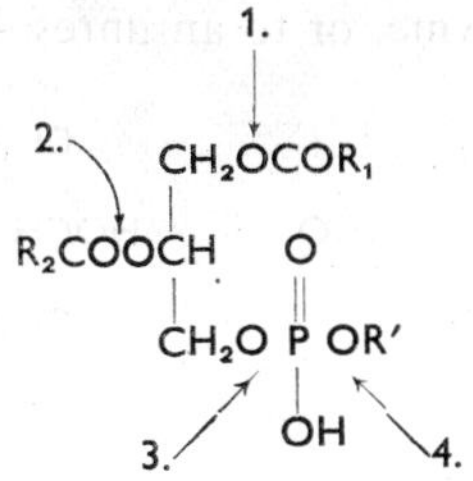

$$\begin{array}{c}
1. \\
| \\
2.\diagdown\;\;\overset{\downarrow}{CH_2OCOR_1} \\
\diagdown\;| \\
R_2COOCH\quad O \\
|\quad\quad \| \\
CH_2O\;P\;OR' \\
\diagup\;|\;\nwarrow \\
3.\diagup\;\;OH\;\;\diagdown 4.
\end{array}$$

Position of Hydrolysis	Enzyme
2.	Phospholipase A
1 and 2.	Phospholipase B
3.	Phospholipase C
4.	Phospholipase D

FIGURE 40. Specificities of the phospholipases

Few attempts have been made to obtain the enzyme in a homogeneous state and the studies reported below have mainly been carried out with impure venom extracts. The enzyme removes both saturated and unsaturated fatty acid residues from phosphatidylcholine (hence the trivial name 'lecithinase A') phosphatidylethanolamine and plasmalogens, but not from phosphatidylinositides; the hydrolysis of the latter class of compounds is not well understood. Only one fatty acid residue is removed per molecule of substrate and early work tended to suggest that the ester linkage cleaved was at the γ-position of the phosphatide. This conclusion was questioned when it became apparent that the fatty acyl residues of plasmalogens, which are cleaved by phospholipase A, occupy the β-position in the molecule (Figure 38). Exhaustive studies of the action of phospholipase A on natural and synthetic lecithins of known structure have now shown that it is, in fact, the ester linkage at the β-position which is hydrolysed[238-240].

Lysophospholipases have been found in many animal tissues[241] and in a variety of moulds such as *Penicillium notatum*[242]. These enzymes hydrolyse saturated and unsaturated fatty acyl linkages in lysophosphatidylcholines and lysophosphatidylethanolamines. None of the enzymes has been obtained in a pure state and the details of substrate specificity and the action of activators and inhibitors are somewhat uncertain.

There is still considerable doubt as to whether phospholipase B activity (the hydrolysis of both fatty ester linkages in phospholipids—reaction

195) is due to a single enzyme, or to an unresolved mixture of phospholi-

$$
\begin{array}{ccc}
\text{CH}_2\text{OCOR}_1 & & \text{CH}_2\text{OH} \\
| & & | \\
\text{R}_2\text{COOCH} \quad \text{O} & +2\,\text{H}_2\text{O} \longrightarrow & \text{HOCH} \quad \text{O} \\
| \quad\ \ \| & & | \quad\ \ \| \\
\text{CH}_2\text{OPOR}' & & \text{CH}_2\text{OPOR}' \\
| & & | \\
\text{OH} & & \text{OH}
\end{array}
\quad +\ \begin{array}{c} \text{R}_1\text{COOH} \\ \text{R}_2\text{COOH} \end{array}
\quad (195)
$$

pase A and lysophospholipase. Kates[234] has argued that the activities are the property of a single enzyme on the basis that the pH optimum of phospholipase B is much lower than that of phospholipase A and that phospholipase B is not activated by ether or by calcium ions as is phospholipase A. Hanahan[243], however, has taken the opposite view. This problem cannot be resolved until further purification of phospholipase B is carried out.

The enzyme which catalyses the hydrolysis of phosphatidylcholine to yield phosphorylcholine and a diglyceride (reaction 196) is referred to here

$$
\begin{array}{ccc}
\text{CH}_2\text{OCOR}_1 & & \text{CH}_2\text{OCOR}_1 \\
| & & | \\
\text{R}_2\text{COOCH} \quad \text{O} & +\text{H}_2\text{O} \longrightarrow & \text{R}_2\text{COOCH} \\
| \quad\ \ \| & & | \\
\text{CH}_2\text{OPOCH}_2\text{CH}_2\overset{+}{\text{N}}(\text{CH}_3)_3 & & \text{CH}_2\text{OH} \\
| & & \\
\text{OH} & &
\end{array}
$$

$$
+ \text{HO}\overset{\text{O}}{\underset{|}{\overset{\|}{\text{P}}}}\text{OCH}_2\text{CH}_2\overset{+}{\text{N}}(\text{CH}_3)_3 \quad (196)
$$
$$
\text{OH}
$$

as phospholipase C. The enzyme was first identified in the α-toxin of *Clostridium perfringens*[244] and has subsequently been found in plant and animal tissues[234]. In addition to phosphatidylcholine, the corresponding ethanolamine and serine phospholipids are hydrolysed, as in sphingo-

$$
\text{sphingomyelin} + \text{H}_2\text{O} \longrightarrow \text{N-acetylsphingosine} + \text{phosphorylcholine} \quad (197)
$$

myelin (reaction 197). The enzyme will not, however, catalyse the hydrolysis of glycerylphosphorylcholine.

The remaining enzyme, phospholipase D, is of more limited distribution having been found so far only in plant tissues. The enzyme was first identified in cabbage leaves and carrot roots by Hanahan and Chaikoff[245], and was shown to cleave choline from phosphatidylcholine (reaction 198); serine and ethanolamine phosphatides are also hydrolysed as is choline

$$
\begin{array}{c}
\text{CH}_2\text{OCOR}_1\\
|\\
\text{R}_2\text{COOCH}\quad\text{O}\\
|\qquad\parallel\\
\text{CH}_2\text{OPOCH}_2\text{CH}_2\overset{+}{\text{N}}(\text{CH}_3)_3\\
|\\
\text{OH}
\end{array}
\;+\text{H}_2\text{O}\;\longrightarrow\;
\begin{array}{c}
\text{CH}_2\text{OCOR}_1\\
|\\
\text{R}_2\text{COOCH}\quad\text{O}\\
|\qquad\parallel\\
\text{CH}_2\text{OPOH}\\
|\\
\text{OH}
\end{array}
$$

$$+\text{HOCH}_2\text{CH}_2\overset{+}{\text{N}}(\text{CH}_3)_3 \quad (198)$$

plasmalogen. Glycerylphosphorylcholine and phosphorylcholine are not hydrolysed.

In many tissues, the degradation of phospholipids leads to the production of glycerylphosphorylcholine and glycerylphosphorylethanolamine. Dawson[246] has shown that animal tissues contain a single enzyme, glycerophosphorylcholine diesterase (E.C. 3.1.4.2), which catalyses the hydrolysis of either of these compounds to glycerophosphate and the free base (*e.g.* reaction 199).

$$
\begin{array}{c}
\text{CH}_2\text{OH}\\
|\\
\text{HOCH}\quad\text{O}\\
|\qquad\parallel\\
\text{CH}_2\text{OPOCH}_2\text{CH}_2\overset{+}{\text{N}}(\text{CH}_3)_3\\
|\\
\text{OH}
\end{array}
\;+\text{H}_2\text{O}\;\longrightarrow\;
\begin{array}{c}
\text{CH}_2\text{OH}\\
|\\
\text{HOCH}\quad\text{O}\\
|\qquad\parallel\\
\text{CH}_2\text{OPOH}\\
|\\
\text{OH}
\end{array}
$$

$$+\text{HOCH}_2\text{CH}_2\overset{+}{\text{N}}(\text{CH}_3)_3 \quad (199)$$

D. Biosynthesis of Lipids

1. Phosphatidic acids

Phosphatidic acids are the key intermediates in the biosynthesis of both the triglycerides[247] and the phospholipids[243]; it is appropriate, therefore, to consider first the route by which these compounds are formed.

The precursor of the phosphatidic acids is glycerophosphate; two routes for the formation of this compound are known. Meyerhof and Kiessling[249] showed in 1933 that glycerophosphate is formed in tissue homogenates by reduction of dihydroxyacetone phosphate. Subsequent work[250] has shown that the reaction may be formulated as in equation (200); the product

$$
\begin{array}{c}
\text{CH}_2\text{OH}\\
|\\
\text{CO}\quad\text{O}\\
|\qquad\parallel\\
\text{CH}_2\text{OPOH}\\
|\\
\text{OH}
\end{array}
\;+\text{NADH}+\text{H}^+\;\rightleftharpoons\;
\begin{array}{c}
\text{CH}_2\text{OH}\\
|\\
\text{HOCH}\quad\text{O}\\
|\qquad\parallel\\
\text{CH}_2\text{OPOH}\\
|\\
\text{OH}
\end{array}
\;+\text{NAD}^+ \quad (200)
$$

of the reaction is L-α-glycerophosphate. A product with the same configuration is obtained by the direct phosphorylation of glycerol by ATP (reaction 201) catalysed by glycerol kinase (E.C. 2.7.1.30)[251]. The enzyme re-

$$
\begin{array}{ccc}
\text{CH}_2\text{OH} & & \text{CH}_2\text{OH} \\
| & & | \\
\text{CHOH} + \text{ATP} \longrightarrow & \text{HOCH} \quad \text{O} & + \text{ADP} \qquad (201)\\
| & \quad \| & \\
\text{CH}_2\text{OH} & \text{CH}_2\text{OPOH} & \\
& | & \\
& \text{OH} &
\end{array}
$$

quires Mg^{2+} ions for activity and also catalyses the phosphorylation of di-hydroxyacetone and glyceraldehyde. It is thought that the phosphorylation reaction is of importance in liver and kidney whilst the reductive route to glycerophosphate predominates in adipose tissue and the intestinal mucosa.

Conversion of glycerophosphate to a phosphatidic acid is believed to involve two acyl transfer reactions from fatty acyl-CoA derivatives to the hydroxyl groups of glycerophosphate, the reaction (202) being catalysed by glycerophosphate acyl-transferase (E.C. 2.3.1.15)[251]. The order of esterification of the hydroxyl functions is not known.

$$
\begin{array}{ccc}
\text{CH}_2\text{OH} & \quad \text{R}_1\text{COSCoA} & \text{CH}_2\text{OCOR}, \\
| & & | \\
\text{HOCH} \quad \text{O} \; + & \longrightarrow & \text{R}_2\text{COOCH} \quad \text{O} \quad + 2 \text{ CoASH} \qquad (202)\\
| \quad \| & \quad \text{R}_2\text{COSCoA} & | \quad \| \\
\text{CH}_2\text{OPOH} & & \text{CH}_2\text{OPOH} \\
| & & | \\
\text{OH} & & \text{OH}
\end{array}
$$

The origin of the acyl-CoA derivatives involved in reaction (202) is of some interest. Kornberg and Pricer[252] showed in 1953 that a soluble extract from liver catalyses the formation of acyl-CoA derivatives from fatty acids (C_5–C_{22}) in the presence of CoA and ATP (reaction 203).

$$
\text{RCOOH} + \text{CoASH} + \text{ATP} \longrightarrow \text{RCOSCoA} + \text{AMP} + \text{PP}_i \qquad (203)
$$

This route is no doubt of importance, but an alternative and so far unevaluated route may be available. It will be recalled that the product of fatty acid synthesis in *E. coli* and probably in mammalian tissues is the corresponding acyl-ACP derivative (section VI. C); it seems likely that these compounds may undergo acyl exchange reactions with CoA resulting in the production of acyl-CoA derivatives (reaction 204). The obvious advantage of reaction (204) in the biosynthesis of lipids from endogenous

$$
\text{RCOSACP} + \text{CoASH} \; \rightleftharpoons \; \text{RCOSCoA} + \text{ACPSH} \qquad (204)
$$

fatty acids is that no ATP is used during the process.

This argument may be taken a step further by assuming that acyl-ACP derivatives can replace acyl-CoA derivatives in reaction (202). Goldfine[253] has presented evidence which implicates palmitoyl-ACP in the synthesis of a lysophosphatidic acid by extracts of *E. coli*, but the general involvement of ACP derivatives in these processes has yet to be demonstrated.

Hokin and Hokin[254] have presented evidence for an alternative mode of phosphatidic acid biosynthesis. These workers have identified an enzyme in brain microsomes which catalyses the transfer of a phosphate group from ATP to the terminal hydroxyl group of diglycerides which contain unsaturated acyl groups (reaction 210). The relative contributions of this

$$
\begin{array}{l}
\text{CH}_2\text{OCOR}_1 \\
| \\
\text{R}_2\text{COOCH} \qquad + \text{ATP} \longrightarrow \\
| \\
\text{CH}_2\text{OH}
\end{array}
\qquad
\begin{array}{l}
\text{CH}_2\text{OCOR}_1 \\
| \quad\quad\ \ \text{O} \\
\text{R}_2\text{COOCH} \quad\ \ \| \qquad + \text{ADP} \\
| \quad\quad\ \ \text{CH}_2\text{OPOH} \\
\qquad\qquad\ | \\
\qquad\qquad\ \text{OH}
\end{array}
\qquad (210)
$$

and the previously described route to phosphotidic acid formation in animal tissues has not been established.

2. Triglycerides[255]

The synthesis of triglycerides from phosphatidic acid involves an initial hydrolytic cleavage of the phosphate moiety to yield a diglyceride (reaction 211); the reaction is catalysed by phosphatidate phosphatase (E.C. 3.1.3.4).

$$
\begin{array}{l}
\text{CH}_2\text{OCOR}_1 \\
| \qquad\quad\ \ \text{O} \\
\text{R}_2\text{COOCH} \quad \| \qquad + \text{H}_2\text{O} \longrightarrow \\
| \\
\text{CH}_2\text{OPOH} \\
\ \ | \\
\ \ \text{OH}
\end{array}
\qquad
\begin{array}{l}
\text{CH}_2\text{OCOR}_1 \\
| \\
\text{R}_2\text{COOCH} \qquad + \text{H}_3\text{PO}_4 \\
| \\
\text{CH}_2\text{OH}
\end{array}
\qquad (211)
$$

Subsequently, the free hydroxyl group is esterified by transfer of an acyl group from acyl-CoA (reaction 212). The biosynthetic sequence is summa-

$$
\begin{array}{l}
\text{CH}_2\text{OCOR}_1 \\
| \\
\text{R}_2\text{COOCH} \qquad + \text{R}_3\text{COSCoA} \longrightarrow \\
| \\
\text{CH}_2\text{OH}
\end{array}
\qquad
\begin{array}{l}
\text{CH}_2\text{OCOR}_1 \\
| \\
\text{R}_2\text{COOCH} \qquad + \text{CoASH} \\
| \\
\text{CH}_2\text{OCOR}_3
\end{array}
\qquad (212)
$$

rized in Figure 41.

It should be mentioned that although the principal route to the diglycerides involved in reaction (212) is via the corresponding phosphatidic acid, there is evidence[256] that monoglycerides formed in the intestine during digestion of fats may be directly acylated to yield diglycerides (reaction 213).

$$\text{monoglyceride} + \text{acyl-CoA} \longrightarrow \text{diglyceride} + \text{CoA} \qquad (213)$$

Triglyceride formation may then occur as in reaction (212).

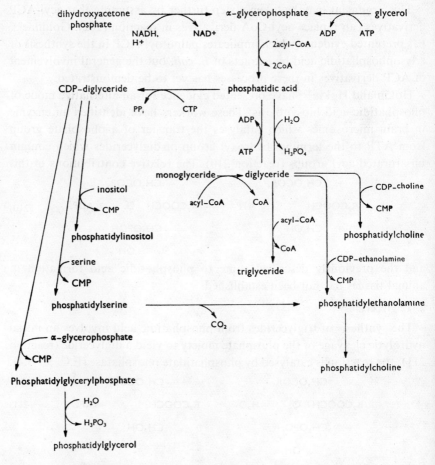

FIGURE 41. Main routes for the biosynthesis of complex lipids

3. Phospholipids

The conversion of phosphatidic acids to the various phospholipids is a subject of considerable interest, but, since it is not of direct relevance to the subject of this review, will be dealt with very briefly. A review of this topic has been given by Ansell and Hawthorne[257], and Lennarz[213] has surveyed recent studies in the biosynthesis of bacterial phospholipids.

In animal tissues, the biosynthesis of both phosphatidylcholine and phosphatidylethanolamine proceeds by transfer of the phosphorylated base from the corresponding cytidine diphosphate derivative (CDP-choline or CDP-ethanolamine) to the terminal hydroxyl group of a diglyceride; the possible sources of diglycerides have been discussed above. In the case of phosphatidylcholine, the process may be represented by equations (214–216). Choline is phosphorylated by ATP in a reaction catalysed by

$$(CH_3)_3\overset{+}{N}CH_2CH_2OH + ATP \longrightarrow (CH_3)_3\overset{+}{N}CH_2CH_2O\overset{\text{O}}{\underset{\text{OH}}{\overset{\|}{P}}}OH + ADP \qquad (214)$$

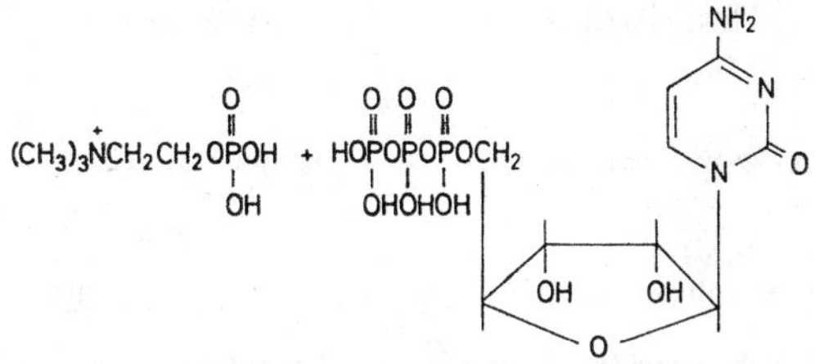

$$(215)$$

$$\rightleftharpoons (CH_3)_3\overset{+}{N}CH_2CH_2OCDP + PP_i$$

$$
\begin{array}{l}
CH_2OCOR_1 \\
| \\
R_2COOCH \qquad + (CH_3)_3\overset{+}{N}CH_2CH_2OCDP \longrightarrow CMP + \\
| \\
CH_2OH
\end{array}
$$

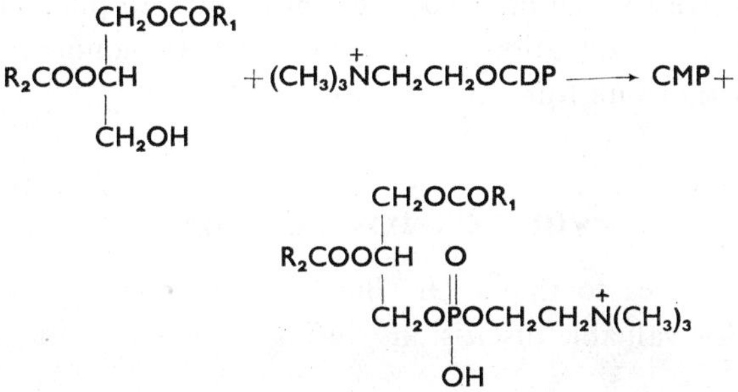

$$(216)$$

$$
\begin{array}{l}
CH_2OCOR_1 \\
| \\
R_2COOCH \quad O \\
| \qquad \| \\
CH_2O\overset{}{\underset{OH}{P}}OCH_2CH_2\overset{+}{N}(CH_3)_3 \\
\end{array}
$$

choline kinase (E.C. 2.7.1.32)[258]. The product, phosphorylcholine, reacts with CTP to produce CDP-choline with the elimination of pyrophosphate (equation 215); the reaction is catalysed by cholinephosphate cytidyltransferase (E.C. 2.7.7.15). Finally, the phosphorylcholine residue is transferred to a diglyceride by cholinephosphotransferase (E.C. 2.7.8.2) and CTP is regenerated by phosphorylation reactions involving ATP. Analogous reactions are involved in the biosynthesis of phosphatidylethanolamine.

A somewhat different reaction sequence appears to be involved in the biosynthesis of phosphatidylinositol in animal tissues. Paulus and Kennedy[259] have proposed a scheme in which the CDP-diglyceride acts as an intermediate as originally postulated by Agranoff and his coworkers[260]. CDP-diglyceride is formed by the reaction of phosphatidic acid and CTP (equation 217); transfer of the phosphatidic acid residue from CDP-diglyceride to inositol (reaction 218) results in the production of phosphat-idylinositol.

$$\text{phosphatidic acid} + \text{CTP} \rightleftharpoons \text{CDP-diglyceride} + \text{PP}_i \qquad (217)$$

$$\text{CDP-diglyceride} + \text{inositol} \longrightarrow \text{CMP} + \text{phosphatidylinositol} \qquad (218)$$

It is of interest that the biosynthetic route described by reactions (217) and (218) appears to be used in *E. coli* for the synthesis of phosphatidyl-glycerol (reactions 219, 220) phosphatidylserine (reaction 221) and hence, by the previously described decarboxylation reaction (222), phosphatidyl-ethanolamine[261].

$$\text{CDP-diglyceride} + \alpha\text{-glycerophosphate} \longrightarrow \text{phosphatidylglycerylphosphate} + \text{CMP}$$
$$(219)$$

$$\text{phosphatidylglycerylphosphate} + \text{H}_2\text{O} \longrightarrow \text{phosphatidylglycerol} + \text{H}_3\text{PO}_4 \qquad (220)$$

$$\text{CDP-diglyceride} + \text{serine} \longrightarrow \text{phosphatidylserine} + \text{CMP} \qquad (221)$$

$$\text{phosphatidylserine} \longrightarrow \text{phosphatidylethanolamine} + \text{CO}_2 \qquad (222)$$

Also of interest is the fact that the major route to phosphatidylcholine in bacteria is by methylation of phosphatidylethanolamine rather than the direct synthetic route found in animals.

VIII. ACKNOWLEDGMENT

The author wishes to thank Dr. Barbara E. C. Banks for reading this review and for valuable discussions during its preparation.

REFERENCES

1. *Report of the Commission on Enzymes of the International Union of Biochemistry*, Pergamon Press, London, 1961.
2. B. E. C. Banks in *The Chemistry of the Amino Group* (Ed. S. Patai), Interscience, London, 1968, Chap. 9.
3. F. Eisenberg Jr., in *The Chemistry of the Carbonyl Group* (Ed. S. Patai), Interscience, London, 1966, Chap. 7.
4. C. A. Vernon in *Size and Shape Changes of Contractile Polymers* (Ed. A. Wassermann), Pergamon Press, Oxford, 1960, Chap. 6.
5. F. Lipmann, *Adv. Enzymol.*, **1**, 99 (1941).
6. W. Vishniac, B. L. Horecker and S. Ochoa, *Adv. Enzymol.*, **19**, 1 (1957).
7. H. G. Wood and R. L. Stjernholm in *The Bacteria* (Eds. I. C. Gunsalus and R. Y. Stanier), Vol. 3, Academic Press, New York, 1962, Chap. 2.
8. H. G. Wood and M. F. Utter, *Essays Biochem.*, **1**, 1 (1965).
9. S. Ochoa and Y. Kaziro in *Comprehensive Biochemistry* (Eds. M. Florkin and E. H. Stotz), Vol. 16, Elsevier, Amsterdam, 1965, p. 210.

10. *The Enzymes* (Eds. P. D. Boyer, H. Lardy and K. Myrbäck), Vol. 7, Academic Press, New York, 1963.
11. M. F. Utter in *The Enzymes* (Eds. P. D. Boyer, H. Lardy and K. Myrbäck), Vol. 5, Academic Press, New York, 1961, Chap. 19.
12. J. A. Bassham, A. A. Benson, L. D. Kay, A. Z. Harris, A. T. Wilson and M. Calvin, *J. Amer. Chem. Soc.*, **76**, 1760 (1954).
13. B. R. Rabin and P. W. Trown, *Nature*, **202**, 1290 (1964).
14. F. Lynen, J. Knappe, E. Lorch, G. Jüttig and E. Ringelmann, *Angew. Chem.*, **71**, 481 (1959).
15. Y. Kaziro and S. Ochoa, *J. Biol. Chem.*, **236**, 3131 (1961).
16. M. Waite and S. J. Wakil, *J. Biol. Chem.*, **238**, 81 (1963).
17. H. G. Allen, R. Stjernholm and H. G. Wood, *J. Biol. Chem.*, **238**, 2889 (1963).
18. Y. Kaziro, L. F. Hass, P. D. Boyer and S. Ochoa, *J. Biol. Chem.*, **237**, 1460 (1962).
19. R. Hill and C. P. Whittingham, *Photosynthesis*, John Wiley and Sons, New York, 1957.
20. J. A. Bassham and M. Calvin, *The Path of Carbon in Photosynthesis*, Prentice-Hall, Englewood Cliffs, N.J., 1957.
21. M. Calvin and J. A. Bassham, *The Photosynthesis of Carbon Compounds*, Benjamin, New York, 1963.
22. M. D. Kamen, *Primary Processes in Photosynthesis*, Academic Press, New York, 1963.
23. R. K. Clayton, *Molecular Physics in Photosynthesis*, Blaisdell, New York, 1965.
24. H. Gest, A. San Pietro and L. P. Vernon (Eds.), *Bacterial Photosynthesis*, Antioch Press, Yellow Springs, Ohio, 1965.
25. R. Hill, *Essays Biochem.*, **1**, 121 (1965).
26. S. R. Elsden in *The Bacteria* (Eds. I. C. Gunsalus and R. Y. Stanier), Vol. 3, Academic Press, New York, 1962, Chap. 1.
27. T. W. Goodwin (Ed.), *Biochemistry of Chloroplasts*, Academic Press, New York, 1967.
28. D. I. Arnon, F. R. Whatley and M. B. Allen, *Science*, **127**, 1026 (1958).
29. R. Hill and R. Scarisbrick, *Nature*, **146**, 61 (1940).
30. R. B. Park and J. Biggins, *Science*, **144**, 1009 (1964).
31. J. H. Smith and C. J. French, *Ann. Rev. Plant Physiol.*, **14**, 181 (1963).
32. L. P. Vernon, *Ann. Rev. Plant Physiol.*, **15**, 73 (1964).
33. L. P. Vernon and M. Avron, *Ann. Rev. Biochem.*, **34**, 269 (1965).
34. J. A. Bassham, *Adv. Enzymol.*, **25**, 39 (1963).
35. N. I. Bishop, *Ann. Rev. Plant Physiol.*, **17**, 185 (1965).
36. R. K. Clayton, *Ann. Rev. Plant Physiol.*, **14**, 159 (1963).
37. A. San Pietro and C. C. Black, *Ann. Rev. Plant Physiol.*, **16**, 155 (1966).
38. J. G. Ormerod and H. Gest, *Bacteriol. Revs.*, **26**, 51 (1962).
39. D. E. Griffiths, *Essays Biochem.*, 1965, **1**, 91.
40. J. A. Bassham, *Ann. Rev. Plant Physiol.*, **15**, 101 (1964).
41. J. A. Bassham and M. Kirk, *Biochim. Biophys. Acta*, **43**, 447 (1960).
42. M. Stiller, *Ann. Rev. Plant Physiol.*, **13**, 151 (1962).
43. H. Lees, *Ann. Rev. Microbiol.*, **14**, 83 (1960).
44. Symposium on Autotrophy, *Bacteriol. Revs.*, **26**, 142 (1962).
45. Y. Kaziro and S. Ochoa, *Adv. Enzymol.*, **26**, 283 (1964).
46. R. W. Swick and H. G. Wood, *Proc. Natl. Acad. Sci. U.S.*, **46**, 28 (1960).
47. H. G. Wood, R. W. Kellermeyer, R. Stjernholm and S. H. G. Allen, *Ann. N.Y. Acad. Sci.*, **112**, 661 (1964).
48. F. Lynen, *J. Cellular Comp. Physiol.*, **54(SI)**, 33 (1959).
49. J. Knappe, H. G. Schliegel and F. Lynen, *Biochem. Z.*, **335**, 101 (1961).

50. J. C. Sonne, J. M. Buchanan and A. M. Delluva, *J. Biol. Chem.*, **173**, 69;81 (1948).
51. G. R. Greenberg, *J. Biol. Chem.*, **190**, 611 (1951).
52. P. P. Cohen in *The Enzymes* (Eds. P. D. Boyer, H. Lardy and K. Myrbäck), Vol. 6, Academic Press, New York, 1962, Chap. 29.
53. M. E. Jones and L. Spector, *J. Biol. Chem.*, **235**, 2897 (1960).
54. L. A. Fahien and P. P. Cohen, *J. Biol. Chem.*, **239**, 1925 (1964).
55. E. Umbarger and B. D. Davis in *The Bacteria* (Eds. I. C. Gunsalus and R. Y. Stanier), Vol. 3, Academic Press, New York, 1962, Chap. 4.
56. P. Reichard, *Adv. Enzymol.*, **21**, 263 (1959).
57. P. P. Cohen and G. W. Brown Jr. in *Comprehensive Biochemistry* (Eds. M. Florkin and E. H. Stotz), Vol. 2, Elsevier, Amsterdam, 1960, p. 161.
58. H. A. Krebs and H. Henseleit, *Z. Physiol. Chem.*, **210**, 33 (1932).
59. M. F. Utter in *The Enzymes* (Eds. P. D. Boyer, H. Lardy and K. Myrbäck), Vol. 5, Academic Press, New York, 1961, Chap. 19.
60. D. Rao Sanadi in *The Enzymes* (Eds. P. D. Boyer, H. Lardy and K. Myrbäck), Vol. 7, Academic Press, New York, 1963, Chap. 14.
61. E. P. Steyn-Parvé and C. H. Monfoort in *Comprehensive Biochemistry* (Eds. M. Florkin and E. H. Stotz), Vol. 11, Elsevier, Amsterdam, 1963, Chap. 1.
62. K. Lohmann and P. Schuster, *Biochem. Z.*, **294**, 188 (1937).
63. D. E. Metzler in *The Enzymes* (Eds. P. D. Boyer, H. Lardy and K. Myrbäck), Vol. 2, Academic Press, New York, 1960, Chap. 9.
64. K. Fry, L. L. Ingraham and F. H. Westheimer, *J. Amer. Chem. Soc.*, **79**, 5225 (1957).
65. R. Breslow, *J. Amer. Chem. Soc.*, **79**, 1762 (1957).
66. H. Holzer and K. Beaucamp, *Biochim. Biophys. Acta*, **46**, 226 (1961).
67. A. G. Datta and E. Racker, *J. Biol. Chem.*, **236**, 624 (1961).
68. D. J. O'Kane and I. C. Gunsalus, *J. Bacteriol.*, **56**, 499 (1948).
69. L. J. Reed, I. C. Gunsalus, G. H. F. Schnakenberg, Q. F. Soper, H. E. Boaz, S. F. Kern and T. V. Parker, *J. Amer. Chem. Soc.*, **75**, 1267 (1953).
70. L. J. Reed in *Comprehensive Biochemistry* (Eds. M. Florkin and E. H. Stotz), Vol. 14, Elsevier, Amsterdam, 1966, Chap. 11.
71. M. Koike, L. J. Reed and W. R. Caroll, *J. Biol. Chem.*, **238**, 30 (1963).
72. B. B. Mukherjee, J. Matthews, D. L. Horney and L. J. Reed, *J. Biol. Chem.*, **240**, 2268 (1965).
73. M. Koike, L. J. Reed and W. R. Caroll, *J. Biol. Chem.*, **235**, 1924 (1960).
74. V. Massey in *The Enzymes* (Eds. P. D. Boyer, H. Lardy and K. Myrbäck), Vol. 7, Academic Press, New York, 1963, Chap. 13.
75. H. A. Krebs and J. M. Lowenstein in *Metabolic Pathways* (Eds. D. M. Greenberg), Vol. 1, Academic Press, New York, 1960, p. 219.
76. A. Lehninger, *The Mitochondrion*, Benjamin, New York, 1964.
77. J. R. Stern in *The Enzymes* (Eds. P. D. Boyer, H. Lardy and K. Myrbäck), Vol. 5, Chap. 21, Academic Press, New York, 1961.
78. S. R. Dickman in *The Enzymes* (Eds. P. D. Boyer, H. Lardy and K. Myrbäck), Vol. 5, Academic Press, New York, 1961, Chap. 30.
79. R. A. Alberty in *The Enzymes* (Eds. P. D. Boyer, H. Lardy and K. Myrbäck), Vol. 5, Academic Press, New York, 1961, Chap. 32.
80. D. E. Green in *Comprehensive Biochemistry* (Eds. M. Florkin and E. H. Stotz), Vol. 14, Elsevier, Amsterdam, 1966, Chap. 6.
81. E. C. Slater in *Comprehensive Biochemistry* (Eds. M. Florkin and E. H. Stotz), Vol. 14, Elsevier, Amsterdam, 1966, Chap. 7.
82. M. J. Johnson in *The Enzymes* (Eds. P. D. Boyer, H. Lardy and K. Myrbäck), Vol. 3, Academic Press, New York, 1960, Chap. 21.
83. E. F. Gale, *Adv. Enzymol.*, **6**, 1 (1946).

84. H. Blaschko, *Adv. Enzymol.*, **5**, 67 (1945).
85. O. Schales in *The Enzymes* (Eds. P. D. Boyer, H. Lardy and K. Myrbäck), 1st ed., Vol. II(1), Academic Press, New York, 1951, p. 216.
86. B. M. Guirard and E. E. Snell in *Comprehensive Biochemistry* (Eds. M. Florkin and E. H. Stotz), Vol. 15, Elsevier, Amsterdam, 1964, Chap. 5.
87. A. Meister, *Biochemistry of the Amino Acids*, Vol. 1, Academic Press, New York, 1965, p. 325.
88. L. Ekladius, H. K. King and C. R. Sutton, *J. Gen. Microbiol.*, **17**, 602 (1957).
89. L. K. Borkenhagen, E. P. Kennedy and L. Fielding, *J. Biol. Chem.*, **236**, PC28 (1961).
90. E. F. Gale, *Biochem. J.*, **34**, 846 (1940).
91. I. C. Gunsalus and W. D. Bellamy, *J. Biol. Chem.*, **155**, 357 (1944).
92. J. Baddiley and E. F. Gale, *Nature*, **155**, 727 (1945).
93. J. Baddiley and A. P. Mathias, *J. Chem. Soc.*, 2583 (1952).
94. A. E. Braunstein, in *The Enzymes* (Eds. P. D. Boyer, H. Lardy and K. Myrbäck), Vol. 2, Academic Press, New York, 1960, Chap. 6.
95. D. Metzler, M. Ikawa and E. E. Snell, *J. Amer. Chem. Soc.*, **76**, 648 (1954).
96. E. Werle and W. Koch, *Biochem. Z.*, **319**, 305 (1949).
97. G. D. Kalyankar and E. E. Snell, *Biochemistry*, **1**, 594 (1962).
98. S. Mandeles, R. Koppelman and M. E. Hanke, *J. Biol. Chem.*, **209**, 327 (1954).
99. S. Mandeles and M. E. Hanke, *Abstr. 124th Meeting Amer. Chem. Soc.*, Chicago, 1953, No. 55C.
100. B. Belleau and J. Burba, *J. Amer. Chem. Soc.*, **82**, 5751 (1960).
101. A. Meister, J. S. Nishimura and A. Novogrodsky in *Chemical and Biological Aspects of Pyridoxal Catalysis* (Eds. E. E. Snell, P. M. Fasella, A. E. Braunstein and A. Rossi Fanelli), Pergamon, Oxford, 1963, p. 229.
102. A. Novogrodsky, J. S. Nishimura and A. Meister, *J. Biol. Chem.*, **238**, PC1903 (1963).
103. R. Shukuya and G. W. Schwert, *J. Biol. Chem.*, **235**, 1653 (1960).
104. K. A. C. Elliot and H. H. Jasper, *Physiol. Revs.*, **39**, 383 (1959).
105. D. R. Curtis and J. C. Watkins, *Pharmacol. Revs.*, **17**, 347 (1965).
106. H. Grundfest, *Physiol. Revs.*, **37**, 337 (1957).
107. H. Blaschko, *J. Physiol.*, **96**, 508 (1939).
108. E. Marley, *Advan. Pharmacol.*, **3**, 167 (1964).
109. J. Axelrod, *Physiol. Revs.*, **39**, 751 (1959).
110. G. B. West, *J. Pharm. Pharmacol.*, **11**, 513 (1959).
111. G. Kahlson and E. Rosengren, *Ann. Rev. Pharmacol.*, **5**, 305 (1965).
112. J. L. Mongar and H. O. Schild, *Physiol. Revs.*, **42**, 226 (1962).
113. R. W. Schayer, *Physiol. Revs.*, **39**, 116 (1959).
114. S. Ochoa, *Physiol. Revs.*, **31**, 56 (1951).
115. F. H. Westheimer in *The Enzymes* (Eds. P. D. Boyer, H. Lardy and K. Myrbäck), Vol. 1, Academic Press, New York, 1959, Chap. 6.
116. K. J. Pedersen, *J. Amer. Chem. Soc.*, **51**, 2098 (1929); **58**, 240 (1936).
117. F. H. Westheimer and W. A. Jones, *J. Amer. Chem. Soc.*, **63**, 3283 (1941).
118. K. J. Pedersen, *J. Phys. Chem.*, **38**, 559 (1934); *J. Amer. Chem. Soc.*, **60**, 595 (1938).
119. F. H. Westheimer, *Proc. Chem. Soc.*, 253 (1963).
120. R. Steinberger and F. H. Westheimer, *J. Amer. Chem. Soc.*, **71**, 4158 (1949); **73**, 429 (1951).
121. L. O. Krampitz and C. H. Werkman, *Biochem. J.*, **35**, 595 (1941).
122. G. W. E. Plaut and H. A. Lardy, *J. Biol. Chem.*, **180**, 13 (1949).
123. E. Adams, *J. Biol. Chem.*, **217**, 325 (1955).
124. J. C. Loper and E. Adams, *Fed. Proc.*, **20**, 254 (1961).

125. J. L. Strominger, H. M. Kalckar, J. Axelrod and E. S. Maxwell, *J. Amer. Chem. Soc.*, **76**, 6411 (1954).

126. J. L. Strominger, R. Okazaki and T. Okazaki in *The Enzymes* (Eds. P. D. Boyer, H. Lardy and K. Myrbäck), Vol. 7, Academic Press, New York, 1963, Chap. 8.

127. S. Hollmann and O. Touster, *Biochim. Biophys. Acta*, **26**, 338 (1962).

128. J. J. Burns in *Metabolic Pathways* (Ed. D. M. Greenberg), Vol. 1, Academic Press, New York, 1960, p. 341.

129. O. Warburg and W. Christian, *Biochem. Z.*, **242**, 206 (1931).

130. E. A. Noltmann and S. A. Kuby in *The Enzymes* (Eds. P. D. Boyer, H. Lardy and K. Myrbäck), Vol. 7, Academic Press, New York, 1963, Chap. 11.

131. H. Bentley in *The Enzymes* (Eds. P. D. Boyer, H. Lardy and K. Myrbäck), Vol. 7, Academic Press, New York, 1963, Chap. 24.

132. D. Müller, *Biochem. Z.*, **199**, 136 (1928); **205**, 111 (1929); **213**, 211 (1929); **232**, 423 (1931).

133. R. Bentley and A. Neuberger, *Biochem. J.*, **45**, 584 (1949).

134. F. Friedberg and L. Kaplan, *Abstr. Amer. Chem. Soc. 131st Meeting*, Miami, 1957.

135. H. S. Mason, *Adv. Enzymol.*, **19**, 79 (1957).

136. B. Axelrod in *Metabolic Pathways* (Ed. D. M. Greenberg), Vol. 1, Academic Press, 1960, p. 205.

137. H. P. Meloche and W. A. Wood, *J. Biol. Chem.*, **239**, 3505, 3511 (1964).

138. S. F. Vellick and C. Furfine in *The Enzymes* (Eds. P. D. Boyer, H. Lardy and K. Myrbäck), Vol. 7, Academic Press, New York, 1963, Chap. 12.

139. I. Krimsky and E. Racker, *Biochemistry*, **2**, 512 (1963).

140. A. L. Murdock and O. J. Koeppe, *J. Biol. Chem.*, **239**, 1983 (1964).

141. A. G. Hilvers, K. Van Dam and E. C. Slater, *Biochim. Biophys. Acta*, **85**, 206 (1964).

142. E. Racker and I. Krimsky, *J. Biol. Chem.*, **198**, 731 (1952).

143. E. M. Kosower, *J. Amer. Chem. Soc.*, **78**, 3497 (1956).

144. S. F. Vellick and J. E. Hayes Jr., *J. Biol. Chem.*, **203**, 545 (1953).

145. M. Cohn, *Biochim. Biophys. Acta*, **20**, 92 (1956).

146. W. B. Jakoby in *The Enzymes* (Eds. P. D. Boyer, H. Lardy and K. Myrbäck), Vol. 7, Academic Press, New York, 1963, Chap. 10.

147. E. Racker, *Physiol. Revs.*, **35**, 1 (1955).

148. E. S. Maxwell and Y. J. Topper, *J. Biol. Chem.*, **236**, 1032 (1961).

149. S. Black, *Arch. Biochem. Biophys.*, **34**, 86 (1951).

150. J. E. Seegmiller, *J. Biol. Chem.*, **201**, 629 (1953).

151. W. B. Jakoby and J. Fredericks, *J. Biol. Chem.*, **234**, 2141 (1959).

152. R. M. Burton and E. R. Stadtman, *J. Biol. Chem.*, **202**, 873 (1953).

153. A. Z. Budenstein, *The Metabolism of Formaldehyde in Yeast*, Ph.D. Dissertation, Yale University, 1955.

154. S. Black and N. G. Wright, *J. Biol. Chem.*, **213**, 39 (1955).

155. P. Strittmatter and E. G. Ball, *J. Biol. Chem.*, **213**, 445 (1951).

156. Z. B. Rose and E. Racker, *J. Biol. Chem.*, **237**, 3279 (1962).

157. W. B. Jakoby, *J. Biol. Chem.*, **232**, 89 (1958).

158. R. C. Bray in *The Enzymes* (Eds. P. D. Boyer, H. Lardy and K. Myrbäck), Vol. 7, Academic Press, New York, 1963, Chap. 22.

159. E. C. de Renzo, *Adv. Enzymol.*, **17**, 293 (1956).

160. P. Handler, K. V. Rajagopalan and V. Aleman, *Fed. Proc.*, **23**, 30 (1964).

161. G. Palmer, R. C. Bray and H. Beinert, *J. Biol. Chem.*, **239**, 2657, 2666 (1964).

162. V. Aleman, S. T. Smith, K. V. Rajagopalan and P. Handler in *Non-Haeme Iron Proteins* (Ed. A. San Pietro), Antioch Press, Yellow Springs, Ohio, 1965, p. 327.

163. O. Hayaishi in *The Enzymes* (Eds. P. D. Boyer, H. Lardy and K. Myrbäck), Vol. 8, Academic Press, New York, 1963, Chap. 13.

164. O. Hayaishi, *Ann. Rev. Biochem.*, **31**, 25 (1962).
165. O. Hayaishi, M. Katagiri and S. Rothberg, *J. Biol. Chem.*, **229**, 905 (1957).
166. S. Dagley and D. A. Stopher, *Biochem. J.*, **73**, 16 (1959).
167. D. I. Crandall, R. C. Krueger, F. Anan, Y. Yasunobu and H. S. Mason, *J. Biol. Chem.*, **235**, 3011 (1960).
168. O. Wiss, M. Simmer and H. Peter, *Z. Physiol. Chem.*, **304**, 321 (1956).
169. F. C. Charalampous, *J. Biol. Chem.*, **235**, 1286 (1960).
170. O. Hayaishi, S. Rothberg, A. H. Mehler and Y. Saito, *J. Biol. Chem.*, **229**, 889 (1957).
171. A. Meister and J. P. Greenstein, *J. Biol. Chem.*, **175**, 573 (1948).
172. W. M. Connors and E. Stotz, *J. Biol. Chem.*, **178**, 881 (1949).
173. S. W. Edwards and W. E. Knox, *J. Biol. Chem.*, **220**, 79 (1956).
174. O. Hayaishi, H. Shimazono, M. Katagiri and Y. Saito, *J. Amer. Chem. Soc.*, **78**, 5126 (1956).
175. A. E. Braunstein, E. V. Goryachenkova and T. S. Paskhina, *Biokhimiya*, **14**, 163 (1949).
176. E. Quagliariello, *Ital. J. Biochem.*, **12**, 65 (1963).
177. K. S. Markley (Ed.), *Fatty Acids*, Interscience, New York, 1960.
178. T. P. Hilditch, *The Chemical Constitution of Natural Fats*, Chapman and Hall, London, 1956.
179. J. F. Mead, D. R. Howton and J. C. Nevenzel in *Comprehensive Biochemistry* (Eds. M. Florkin and E. H. Stotz), Vol. 6, Elsevier, Amsterdam, 1965, Chap. 1.
180. J. A. Lovern in *Comprehensive Biochemistry* (Eds. M. Florkin and E. H. Stotz), Vol. 6, Elsevier, Amsterdam, 1965, Chap. 2.
181. A. W. Weitkamp, *J. Amer. Chem. Soc.*, **67**, 447 (1945).
182. F. Knoop, *Beitr. Chem. Physiol. Path.*, **6**, 150 (1904).
183. L. F. Leloir and J. M. Muñoz, *Biochem. J.*, **33**, 734 (1939).
184. F. Lynen and E. Reichert, *Angew. Chem.*, **63**, 47 (1951).
185. D. E. Green and S. J. Wakil in *Lipide Metabolism* (Ed. K. Bloch), John Wiley and Sons, New York, 1960, Chap. 1.
186. W. P. Jencks in *The Enzymes* (Eds. P. D. Boyer, H. Lardy and K. Myrbäck), Vol. 6, Academic Press, New York, 1962, Chap. 23.
187. H. Beinert in *The Enzymes* (Eds. P. D. Boyer, H. Lardy and K. Myrbäck), Vol. 7, Academic Press, New York, 1963, Chap. 17.
188. J. R. Stern in *The Enzymes* (Eds. P. D. Boyer, H. Lardy and K. Myrbäck), Vol. 5, Academic Press, New York, 1961, Chap. 31.
189. S. J. Wakil in *The Enzymes* (Eds. P. D. Boyer, H. Lardy and K. Myrbäck), Vol. 7, Academic Press, New York, 1963, Chap. 4.
190. G. Hartmann and F. Lynen in *The Enzymes* (Eds. P. D. Boyer, H. Lardy and K. Myrbäck), Vol. 5, Academic Press, New York, 1961, Chap. 22.
191. T. C. Bruice and S. J. Benkovic, *Bioorganic Mechanisms*, Benjamin, New York, 1966, Chap. 3.
192. C. R. Rossi and D. M. Gibson, *J. Biol. Chem.*, **239**, 1694 (1964).
193. F. L. Crane and H. Beinert, *J. Biol. Chem.*, **218**, 717 (1956).
194. J. F. Mead in *Lipide Metabolism* (Ed. K. Bloch), John Wiley and Sons, New York, 1960, Chap. 2.
195. F. Lynen, U. Henning, C. Bublitz, B. Sarbo and L. Kroplin-Rueff, *Biochem. Z.*, **330**, 269 (1958).
196. E. R. Stadtman and H. A. Barker, *J. Biol. Chem.*, **180**, 1085, 1095, 1169 (1949); **181**, 221 (1949); **184**, 769 (1950).
197. W. R. Harlan Jr. and S. J. Wakil, *J. Biol. Chem.*, **238**, 3216 (1963).
198. D. M. Gibson, M. I. Jacob, J. W. Porter, A. C. Tietz and S. J. Wakil, *Biochim. Biophys. Acta*, **23**, 219 (1957).

199. S. J. Wakil, J. W. Porter and D. M. Gibson, *Biochim. Biophys. Acta*, **24**, 453 (1957).
200. J. W. Porter, S. J. Wakil, A. C. Tietz, M. I. Jacob and D. M. Gibson, *Biochim. Biophys. Acta*, **25**, 35 (1957).
201. D. M. Gibson, E. B. Titchener and S. J. Wakil, *Biochim. Biophys. Acta*, **30**, 376 (1958).
202. S. J. Wakil, E. B. Titchener and D. M. Gibson, *Biochim. Biophys. Acta*, **29**, 225 (1958).
203. S. J. Wakil, *J. Amer. Chem. Soc.*, **80**, 6465 (1958).
204. S. J. Wakil, *Ann. Rev. Biochem.*, **31**, 369 (1962).
205. S. J. Wakil, *J. Lipid Res.*, **2**, 1 (1961).
206. F. Lynen, *Fed. Proc.*, **20**, 941 (1961).
207. P. Goldman, A. W. Alberts and P. R. Vagelos, *J. Biol. Chem.*, **238**, 1255 (1963).
208. P. R. Vagelos, P. W. Majerus, A. W. Alberts, A. R. Larrabee and G. P. Ailhaud, *Fed. Proc.*, **25**, 1485 (1966).
209. A. W. Alberts, P. W. Majerus, B. Talamo and P. R. Vagelos, *Biochemistry*, **3**, 1563 (1964).
210. F. Sauer, E. L. Pugh, S. J. Wakil, R. Delaney and R. Hill, *Proc. Natl. Acad. Sci. U.S.*, **52**, 1360 (1964).
211. P. W. Majerus, A. W. Alberts and P. R. Vagelos, *Proc. Natl. Acad. Sci. U.S.*, **53**, 410 (1965).
212. P. Cady, S. Abraham and I. L. Chaikoff, *Biochim. Biophys. Acta*, **70**, 118 (1963).
213. W. J. Lennarz, *Adv. Lipid Res.*, **4**, 175 (1966).
214. D. K. Bloomfield and K. Bloch, *J. Biol. Chem.*, **235**, 337 (1960).
215. P. R. Vagelos, *Ann. Rev. Biochem.*, **33**, 139 (1964).
216. G. Scheuerbrandt, H. Goldfine, P. E. Baronowsky and K. Bloch, *J. Biol. Chem.*, **236**, PC70 (1961).
217. P. E. Baronowsky, *Ph.D. Thesis*, Harvard, 1963.
218. W. J. Lennarz, R. J. Light and K. Bloch, *Proc. Natl. Acad. Sci. U.S.*, **48**, 840 (1962).
219. A. T. Norris, S. Matsumura and K. Bloch, *J. Biol. Chem.*, **239**, 3653 (1964).
220. D. J. H. Brock, L. R. Kass and K. Bloch, *Fed. Proc.*, **25**, 340 (1966).
221. M. Kates in *Advances in Lipid Research* (Eds. R. Paoletti and D. Kritchevsky), Vol. 2, Academic Press, New York, 1964, p. 17.
222. H. Zalkin, J. H. Law and H. Goldfine, *J. Biol. Chem.*, **238**, 1242 (1963).
223. A. E. Chung and J. H. Law, *Biochemistry*, **3**, 1989 (1964).
224. A. E. Chung and J. H. Law, *Biochemistry*, **3**, 967 (1964).
225. J. G. Hildebrand and J. H. Law, *Biochemistry*, **3**, 1304 (1964).
226. E. C. Webb, quoted in M. Dixon and E. C. Webb, *The Enzymes*, 2nd ed. Longmans, London, 1964, p. 219.
227. M. L. Bender and F. J. Kezdy, *J. Amer. Chem. Soc.*, **86**, 3704 (1964).
228. T. W. Goodwin, J. I. Harris and B. S. Hartley (Eds.), *Structure and Activity of Enzymes*, Academic Press, London, 1964, pp. 37, 87.
229. D. Nachmansohn in *The Harvey Lectures 1953–1954*, Academic Press, New York, 1955.
230. D. R. Davies and A. L. Green, *Adv. Enzymol.*, **20**, 283 (1958).
231. J. A. Lovern in *Comprehensive Biochemistry* (Eds. M. Florkin and E. H. Stotz), Vol. 6, Elsevier, Amsterdam, 1965, Chap. 2.
232. G. B. Ansell and J. N. Hawthorne, *Phospholipids*, Elsevier, Amsterdam, 1964.
233. D. J. Hanahan and H. Brockerhoff in *Comprehensive Biochemistry* (Eds. M. Florkin and E. H. Stotz), Vol. 6, Elsevier, Amsterdam, 1965, Chap. 3.
234. M. Kates in *Lipide Metabolism* (Ed. K. Bloch), John Wiley and Sons, New York, 1960, Chapter 5.

235. G. Benzonana, B. Entressangles, G. Marchis-Mouren, L. Paséro, L. Sarda and P. Desnuelle in *Metabolism and Physiological Significance of Lipids* (Eds. R. M. C. Dawson and D. N. Rhodes), John Wiley and Sons, New York, 1964, p. 141.
236. L. L. M. Van Deenen in *Metabolism and Physiological Significance of Lipids* (Eds. R. M. C. Dawson and D. N. Rhodes), John Wiley and Sons, New York, 1964, p. 155.
237. G. B. Ansell and J. N. Hawthorne, *Phospholipids*, Elsevier, Amsterdam, 1964, Chap. 6.
238. N. H. Tattrie, *J. Lipid. Res.*, **1**, 60 (1959).
239. D. J. Hanahan, H. Brockerhoff and E. J. Barron, *J. Biol. Chem.*, **235**, 1917 (1960).
240. G. H. de Hass and L. L. M. van Deenen, *Rec. Trav. Chim.*, **80**, 951 (1961).
241. R. M. C. Dawson, *Biochem. J.*, **64**, 192 (1956).
242. D. Faibairn, *J. Biol. Chem.*, **173**, 705 (1948).
243. D. J. Hanahan, *Prog. Chem. Fats Lipids*, **4**, 142 (1957).
244. M. G. Macfarlane and B. C. J. G. Knight, *Biochem. J.*, **35**, 884 (1941).
245. D. J. Hanahan and I. L. Chaikoff, *J. Biol. Chem.*, **168**, 233 (1947); **169**, 699 (1947); **172**, 191 (1948).
246. R. M. C. Dawson, *Biochem. J.*, **62**, 689 (1956).
247. S. B. Weiss and E. P. Kennedy, *J. Amer. Chem. Soc.*, **78**, 3550 (1956).
248. E. P. Kennedy, *Fed. Proc.*, **20**, 934 (1961).
249. O. Meyerhof and W. Kiessling, *Biochem. Z.*, **264**, 40 (1933).
250. R. W. R. Baker and G. Porcellati, *Biochem. J.*, **73**, 561 (1959).
251. A. Kornberg and W. E. Pricer, *J. Biol. Chem.*, **204**, 345 (1953).
252. A. Kornberg and W. E. Pricer, *J. Biol. Chem.*, **204**, 329 (1953).
253. H. Goldfine, *Fed. Proc.*, **25**, 405 (1966).
254. M. R. Hokin and L. E. Hokin, *J. Biol. Chem.*, **234**, 1381 (1959).
255. B. Shapiro in *Metabolism and Physiological Significance of Lipids* (Eds. R. M. C. Dawson and D. N. Rhodes), John Wiley and Sons, New York, 1964, p. 33.
256. J. R. Senior and K. J. Isselbacher, *J. Biol. Chem.*, **237**, 1454 (1962).
257. G. B. Ansell and J. N. Hawthorne, *Phospholipids*, Elsevier, Amsterdam, 1964, Chap. 5.
258. J. Wittenberg and A. Kornberg, *J. Biol. Chem.*, **202**, 431 (1953).
259. H. Paulus and E. P. Kennedy, *J. Biol. Chem.*, **225**, 1303 (1956).
260. B. W. Agranoff, R. M. Bradley and R. O. Brady, *J. Biol. Chem.*, **233**, 1077 (1958).
261. J. Kanfer and E. P. Kennedy, *J. Biol. Chem.*, **239**, 1720 (1964).
262. S. Ochoa, A. H. Mehler and A. Kornberg, *J. Biol. Chem.*, **174**, 979 (1948).
263. S. Ochoa, *J. Biol. Chem.*, **159**, 243 (1945).
264. B. L. Horecker and P. Z. Smyrniotis, *J. Biol. Chem.*, **196**, 135 (1952).
265. J. Moyle, *Biochem. J.*, **63**, 552 (1956).
266. G. Siebert, M. Carsiotis and G. W. E. Plaut, *J. Biol. Chem.*, **226**, 977 (1957).
267. W. J. Rutter and H. A. Lardy, *J. Biol. Chem.*, **233**, 374 (1958).
268. J. B. V. Salles, I. Harary, R. F. Banfi and S. Ochoa, *Nature*, **165**, 675 (1950).
269. Z. B. Rose, *J. Biol. Chem.*, **235**, 928 (1960).
270. R. S. Bandurski and C. M. Greiner, *J. Biol. Chem.*, **204**, 781 (1953).
271. T. T. Tchen and B. Vennesland, *J. Biol. Chem.*, **213**, 533 (1955).
272. P. M. L. Siu and H. G. Wood, *J. Biol. Chem.*, **237**, 3044 (1962).
273. J. B. Cannata and A. O. M. Stoppani, *J. Biol. Chem.*, **238**, 1196; 1208 (1963).
274. K. Kurahashi, R. J. Pennington and M. F. Utter, *J. Biol. Chem.*, **226**, 1059 (1957).
275. L. N. Lukens and J. M. Buchanan, *J. Biol. Chem.*, **234**, 1799 (1959).
276. A. Weissbach, B. L. Horecker and J. Hurwitz, *J. Biol. Chem.*, **218**, 795 (1956).
277. W. B. Jacoby, D. O. Brummond and S. Ochoa, *J. Biol. Chem.*, **218**, 811 (1956).
278. C. B. van Niel, *Arch. Mikrobiol.*, **3**, 1 (1931).
279. C. B. van Niel, *Adv. Enzymol.*, **1**, 263 (1941).

280. Z. Gromet-Elhanan and M. Avron, *Biochem. Biophys. Res. Commun.*, **10**, 215 (1963).
281. A. T. Jagendorf and M. M. Margulies, *Arch. Biochem. Biophys.*, **90**, 184 (1960).
282. A. T. Jagendorf and M. Avron, *Arch. Biochem. Biophys.*, **80**, 246 (1959).
283. A. Trebst and H. Eck, *Z. Naturforsch.*, **16**, 455 (1961).
284. K. Tagawa, H. Y. Tsujimoto and D. I. Arnon, *Proc. Natl. Acad. Sci. U.S.*, **50**, 544 (1963).
285. G. Forti and A. T. Jagendorf, *Biochim. Biophys. Acta*, **59**, 322 (1961).
286. D. W. Krogmann, A. T. Jagendorf and M. Avron, *Plant Physiol.*, **34**, 272 (1959).
287. M. Avron and N. Shavit, *Natl. Acad. Sci.–Natl. Res. Council Publ.* 1145, 611 (1963).
288. O. Kandler and M. Gibbs, *Plant Physiol.*, **31**, 411 (1956).
289. C. A. Fewson, P. K. Kindel and M. Gibbs, *Plant Physiol.*, **36**, Suppl. IX (1961).
290. H. L. Kornberg and H. A. Krebs, *Nature*, **179**, 988 (1957).
291. N. B. Madsen, *Biochim. Biophys. Acta*, **27**, 199 (1958).
292. H. L. Kornberg, *Symp. Soc. Gen. Microbiol.*, **15**, 8 (1965).
293. M. E. Jones, L. Spector and F. Lipmann, *Proc. 3rd Intern. Congr. Biochem.*, *Brussels*, 278 (1954).
294. M. E. Jones and F. Lipmann, *Proc. Natl. Acad. Sci. U.S.*, **46**, 1194 (1960).
295. M. Marshall, R. L. Metzenburg and P. P. Cohen, *J. Biol. Chem.*, **233**, 102 (1958).
296. J. Harting and S. F. Velick, *J. Biol. Chem.*, **207**, 857; 867 (1954).
297. F. F. Nord and S. Weiss in *The Chemistry and Biology of Yeasts* (Ed. A. H. Cook), Academic Press, New York, 1958, p. 323.
298. D. Burk, *Cold Spring Harbour Symposia Quant. Biol.*, **7**, 420 (1939).
299. B. Axelrod in *Metabolic Pathways* (Ed. D. M. Greenburg), Vol. 1, Academic Press, New York, 1960, p. 97.

Author index

This author index is designed to enable the reader to locate an author's name and work with the aid of the reference numbers appearing in the text. The page numbers are printed in normal type in ascending numerical order, followed by the reference numbers in brackets. The numbers in *italics* refer to the pages on which the references are actually listed.

Subject Index